TABLE OF ATOMIC MASSES AND ATOMIC NUMBERS

Based on the 1985 Report of the Commission on Atomic Weights of the International Union of Pure and Applied Chemistry and for the elements as they exist naturally on earth. Scaled to the relative atomic mass of carbon-12. The estimated uncertainties in values, between ±1 and ±9 units in the last digit of an atomic mass, are in parentheses after the atomic mass. (From *Pure and Applied Chemistry,* Vol. 58 (1986), pp. 1677–1692. Copyright © 1986 IUPCAC)

Element	Symbol	Atomic Number	Atomic Mass		Element	Symbol	Atomic Number	Atomic Mass	
Actinium	Ac	89	227.0278	(L)	Neodymium	Nd	60	144.24(3)	(g)
Aluminum	Al	13	26.981539(5)		Neon	Ne	10	20.1797(6)	(g, m)
Americium	Am	95	243.0614	(L)	Neptunium	Np	93	237.0482	(L)
Antimony	Sb	51	121.75(3)		Nickel	Ni	28	58.69(1)	
Argon	Ar	18	39.948(1)	(g, r)	Niobium	Nb	41	92.90638(2)	
Arsenic	As	33	74.92159(2)		Nitrogen	N	7	14.00674(7)	(g, r)
Astatine	At	85	209.9871	(L)	Nobelium	No	102	259.1009	(L)
Barium	Ba	56	137.327(7)		Osmium	Os	76	190.2(1)	(g)
Berkelium	Bk	97	247.0703	(L)	Oxygen	O	8	15.9994(3)	(g, r)
Beryllium	Be	4	9.012182(3)		Palladium	Pd	46	106.42(1)	(g)
Bismuth	Bi	83	208.98037(3)		Phosphorus	P	15	30.973762(4)	
Boron	B	5	10.811(5)	(g, m, r)	Platinum	Pt	78	195.08(3)	
Bromine	Br	35	79.904(1)		Plutonium	Pu	94	244.0642	(L)
Cadmium	Cd	48	112.411(8)	(g)	Polonium	Po	84	208.9824	(L)
Calcium	Ca	20	40.078(4)	(g)	Potassium	K	19	39.0983(1)	
Californium	Cf	98	242.0587	(L)	Praseodymium	Pr	59	140.90765(3)	
Carbon	C	6	12.011(1)	(r)	Promethium	Pm	61	144.9127	(L)
Cerium	Ce	58	140.115(4)	(g)	Protactinium	Pa	91	231.03588(2)	(Z)
Cesium	Cs	55	132.90543(5)		Radium	Ra	88	226.0254	(L)
Chlorine	Cl	17	35.4527(9)		Radon	Rn	86	222.0176	(L)
Chromium	Cr	24	51.9961(6)		Rhenium	Re	75	186.207(1)	
Cobalt	Co	27	58.93320(1)		Rhodium	Rh	45	102.90550(3)	
Copper	Cu	29	63.546(3)	(r)	Rubidium	Rb	37	85.4678(3)	(g)
Curium	Cm	96	247.0703	(L)	Ruthenium	Ru	44	101.07(2)	(g)
Dysprosium	Dy	66	162.50(3)	(g)	Samarium	Sm	62	150.36(3)	(g)
Einsteinium	Es	99	252.083	(L)	Scandium	Sc	21	44.955910(9)	
Erbium	Er	68	167.26(3)	(g)	Selenium	Se	34	78.96(3)	
Europium	Eu	63	151.965(9)	(g)	Silicon	Si	14	28.0855(3)	
Fermium	Fm	100	257.0951	(L)	Silver	Ag	47	107.8682(2)	(g)
Fluorine	F	9	18.9984032(9)		Sodium	Na	11	22.989768(6)	
Francium	Fr	87	223.0197	(L)	Strontium	Sr	38	87.62(1)	(g, r)
Gadolinium	Gd	64	157.25(3)	(g)	Sulfur	S	16	32.066(6)	(r)
Gallium	Ga	31	69.723(4)		Tantalum	Ta	73	180.9479(1)	
Germanium	Ge	32	72.61(2)		Technetium	Tc	43	98.9072	(L)
Gold	Au	79	196.96654(3)		Tellurium	Te	52	127.60(3)	(g)
Hafnium	Hf	72	178.49(2)		Terbium	Tb	65	158.92534(3)	
Helium	He	2	4.002602(2)	(g, r)	Thallium	Tl	81	204.3833(2)	
Holmium	Ho	67	164.93032(3)		Thorium	Th	90	232.0381(1)	(g, r, Z)
Hydrogen	H	1	1.00794(7)	(g, m, r)	Thulium	Tm	69	168.93421(3)	
Indium	In	49	114.82(1)		Tin	Sn	50	118.710(7)	(g)
Iodine	I	53	126.90447(3)		Titanium	Ti	22	47.88(3)	
Iridium	Ir	77	192.22(3)		Tungsten	W	74	183.85(3)	
Iron	Fe	26	55.847(3)		(Unnilhexium)	(Unh)	106	263.118	(L, n)
Krypton	Kr	36	83.80(1)	(g, m)	(Unnilpentium)	(Unp)	105	262.114	(L, n)
Lanthanum	La	57	138.9055(2)	(g)	(Unnilquadium)	(Unq)	104	266.11	(L, n)
Lawrencium	Lr	103	260.105	(L)	(Unnilseptium)	(Uns)	107	262.12	(L, n)
Lead	Pb	82	207.2(1)	(g, r)	Uranium	U	92	238.0289(1)	(g, m, Z)
Lithium	Li	3	6.941(2)	(g, m, r)	Vanadium	V	23	50.9415(1)	
Lutetium	Lu	71	174.967(1)	(g)	Xenon	Xe	54	131.29(2)	(g, m)
Magnesium	Mg	12	24.3050(6)		Ytterbium	Yb	70	173.04(3)	(g)
Manganese	Mn	25	54.93805(1)		Yttrium	Y	39	88.90585(2)	
Mendelevium	Md	101	258.10	(L)	Zinc	Zn	30	65.39(2)	
Mercury	Hg	80	200.59(3)		Zirconium	Zr	40	91.224(2)	(g)
Molybdenum	Mo	42	95.94(1)						

(g) Geologically exceptional specimens of this element are known that have different isotopic compositions. For such samples, the atomic mass given here may not apply as precisely as indicated.

(L) This atomic mass is for the relative mass of the isotope of longest half-life. The element has no stable isotopes.

(m) Modified isotopic compositions can occur in commercially available materials that have been processed in undisclosed ways, and the atomic mass given here might be quite different for such samples.

(n) Name and symbol are assigned according to systematic rules developed by the IUPAC.

(r) Ranges in isotopic compositions of normal samples obtained on earth do not permit a more precise atomic mass for this element, but the tabulated value should apply to any normal sample of the element.

(Z) Despite having no stable isotopes, the terrestrial compositions of samples of the long-lived isotopes allow a meaningful atomic mass.

GENERAL CHEMISTRY

Principles and Structure

FIFTH EDITION

GENERAL CHEMISTRY

Principles and Structure

JAMES E. BRADY

St. John's University
Jamaica, New York

WILEY
JOHN WILEY & SONS
New York · Chichester · Brisbane · Toronto · Singapore

Production Supervisor: Katharine Rubin
Design Supervisor: Sheila Granda
Editorial Supervisor: Priscilla Todd
Copy Editor: Patricia Brecht
Text Designer: Lee Goldstein
Cover Designer: Carolyn Joseph
Photo Researcher: Elyse Rieder
Illustrator: John Balbalis
Photo Research Manager: Stella Kupferberg

Library of Congress Cataloging in Publication Data:

Brady, James E.
    General chemistry: principles and structure/James E. Brady, —
5th ed.

    p. cm.
    Includes bibliographical references.
    ISBN 0-471-62131-5
    1. Chemistry.   I. Title
QD31.2.B7   1990   89-24866
540--dc20   CIP

Printed and bound by Von Hoffmann Press, Inc.

10 9 8 7 6

PREFACE

In all its editions, this book has had two principal objectives, to be interesting and easily understood by the student and to be useful to the teacher by being up-to-date with respect to trends in the teaching of chemistry. This tradition served as a guide in planning this revision as well, and keeping it in mind, I have worked toward several specific goals.

First, in view of the changing direction of the introductory chemistry course, I wished to provide an improved framework within which the instructor could emphasize the importance of descriptive chemistry without compromising the treatment of principles. Second, I wished to reduce the size of the book. As you know, chemistry textbooks have become enormous, which has added to their cost and bulk and has surely intimidated many incoming college students. And third, I wanted to improve further the explanations of concepts and to enhance the effectiveness of the text in developing problem-solving skills.

As in the past, to meet these goals I have brought together my own thoughts, input from users of previous editions, as well as suggestions of reviewers. Those familiar with the fourth edition will notice that this has led to some significant changes in the flow of topics. In spite of this, care has been taken to write chapters in a way that their order of presentation can be changed easily.

Organization

The last decade has seen a significant change in the teaching of general chemistry, with a renewed interest in stressing the facts of chemistry along with the presentation of chemical principles. The organizational structure of this edition is designed to fulfill these needs. To give the book a sharper focus and to make the rationale for the organization easier to see, it has been divided into a number of parts.

Part I, consisting of the first five chapters, presents a basic introduction to chemical concepts and the development of stoichiometric principles. It also provides a first look at some of the chemical and physical properties of the elements and their compounds. The philosophy here is that the discovery of facts precedes the development of theory. The goal of these chapters is to provide the student with a fundamental store of chemical information and an understanding of the need for theory. They also serve as a solid foundation for quantitative experiments in the laboratory.

In Part II the theme centers on the development of theoretical explanations of observed facts. It begins by introducing the student to the concept of energy, electronic structure, and chemical bonding. Then, with this as background, there is a return to a discussion of chemical facts that illustrates how theory helps explain chemical phenomena.

Part III focuses on the physical properties of substances. It deals with the states of matter and intermolecular attractions as well as the physical properties of mixtures.

In Part IV the spotlight is on factors that determine the outcome of chemical reactions. It begins with thermodynamics, followed by three chapters on equilibrium and then one on electrochemistry. Placing the chapter on kinetics following electrochemistry is a major change in this edition and was done for several reasons. First, kinetics is seen by many instructors as less important than the subjects that precede it here. Second, many users of the text have felt that the discussion of kinetics, especially reaction mechanisms, should come after the development of equilibrium concepts. And third, the presence of the kinetics chapter within the thermodynamics–equilibrium sequence seemed for many to interrupt the flow of topics.

Part V surveys the chemistry of the elements. Although there have been some changes in the sequence of topics discussed, the level and depth of coverage have changed little from the previous edition. Finally, Part VI concludes the text with a discussion of nuclear reactions and their role in chemistry.

Within this organizational structure, some substantial changes have been made in the presentation of topics. For example, thermochemistry, previously discussed in the chapter on thermodynamics, now appears in Chapter 6 at the beginning of Part II of the text. This chapter also provides an effective, class-tested treatment of the concepts of energy and temperature. The rules for assigning oxidation numbers have been condensed and made easier to remember and apply. Within the first chapter on chemical bonding, the concept of formal charge is introduced in a separate section and used as a criterion in the selection of Lewis structures. The second bonding chapter has been streamlined by moving the discussion of molecular orbital theory to an appendix. Complex ions of metals are introduced in Chapter 10 as an example of Lewis acid–base chemistry. This chapter also includes a thoroughly revised treatment of the trends in the strengths of Brønsted–Lowry acids. The chapter on gases has been condensed by combining discussions of the gas laws. In the chapter on the physical properties of solutions, the treatment of conversions among concentration units has been revamped to give students an improved structure within which to work problems. The importance of disorder as a driving force in the formation of solutions is now stressed in the discussion of the solution process. The treatment of thermodynamics has also been revised to make it more meaningful to the student. The chapter on acid–base equilibria has been totally reworked with an emphasis on the application of Brønsted–Lowry concepts. And the discussion of galvanic cells now includes the use of standard cell diagrams.

Other notable changes include the expansion of the tables of thermodynamic data within the text and the inclusion of additional tables of data in Appendix C. Although the calorie as an energy unit is described in the text and compared to the joule, energy values in tables and exercises are now entirely in joules and kilojoules.

In the discussions of solubility product and selective precipitation of ions, I have omitted sulfides and sulfide precipitations because the equilibria are apparently more complex than had been described in earlier editions. (See R. J. Myers, *J. Chem. Educ.*, **1986**, *63*, 687.) The revised treatment teaches the same concepts, but without obscuring them with unnecessary complications.

Textbook size

Those who have been teaching for some time have surely noticed that chemistry books have evolved by accretion and now contain much more material than anyone can teach in a two-semester course. Books have become so large that students resent having to carry them, are burdened by their high cost, and are frightened by the prospects of having to learn all the material in them. With this in mind, one of my principal goals was to reduce the size of the book. This was accomplished partly by streamlining discussions wherever possible (e.g., the Bohr theory and the gas laws). It also required the elimination of some topics, and in deciding what to remove from the book, I sought input from those who have used the book in the past. By consensus, topics chosen for elimination were those hardly ever covered in general chemistry.

Among the major deletions are the separate chapters dealing with organic chemistry and biochemistry. For students who need some organic chemistry at the introductory level, two sections covering elementary organic chemistry have been added to Chapter 22 following discussion of the inorganic chemistry of carbon. A section on inorganic and organic polymers has also been included in Chapter 23 following the section on the chemistry of silicon. I have also eliminated much of the historical material related to the development of the theory of atomic structure, balancing equations by the use of oxidation numbers, the valence bond treatment of the bonding in transition metal complexes, and the nomenclature of metal complexes. What remains is more than sufficient to present a rigorous, in-depth course in general chemistry.

Student orientation

In executing the changes described above, my overriding intent was to ensure that the book would continue to be an effective learning tool for students. In adding new material and revising previous discussions, I have used class-tested approaches that I have found effective in teaching my students. As in the past, I have assumed no prior student background in chemistry and have limited mathematical sophistication to simple algebra. It is assumed that students will have scientific calculators at their disposal, and the use of calculators is described in Appendix A. New terms are set in bold type when first encountered, and important equations are set in colored type to call attention to them. To help maintain student interest, an attractive design has been adopted that makes effective use of color in illustrations and photographs.

To assist students in developing problem-solving skills, the text contains many carefully worked examples. Each begins with a title that describes the nature of the problem to be solved. New to this edition is the inclusion of an *analysis* section in many of the examples. This section, which comes after the statement of the problem, explains the reasoning that goes into solving the problem. The purpose is to help students develop the ability to think through a problem before attempting to work it out. At the ends of chapters there are comprehensive problem sets organized by topic. Answers to selected exercises are given in Appendix D. I have tried to provide a sufficient number of problems of a given type so that students can work through a problem whose answer appears in Appendix D before attempting a similar one whose answer is not given.

In conclusion, it has been a source of great satisfaction that students and teachers have found this book and its supplements useful. I hope that this edition will also please you, and I invite your comments and suggestions.

JAMES E. BRADY

ACKNOWLEDGMENTS

It is with a great deal of pleasure that I thank those who have contributed in so many ways to the completion of this project. First, with affectionate appreciation, I want to thank my wife, June, and my children, Mark and Karen, who continue to put up with the inconveniences brought about by rushed deadlines. I am grateful to the staff at Wiley for their careful work, attention to detail, and cheerful spirit, especially my editor Dennis Sawicki for his guidance and encouragement; his assistant Joann Spear for keeping track of all the paperwork; the editorial supervisor Priscilla Todd; the copy editor Patricia Brecht; the picture editor Stella Kupferberg and photo researcher Elyse Rieder; the illustrator John Balbalis (best in the business); the design supervisor Sheila Granda; and the designer Lee Goldstein. Special thanks have to go to Katharine Rubin, the production supervisor, for her delightful sense of humor and ability to get me to meet those deadlines and to Professor Otis Dermer who helped with the proofreading. I particularly want to thank Professor Larry Peck of Texas A&M University for his assistance in gleaning errors from the text and problem sets that others missed. I am also grateful to my colleagues and students for their support and their many constructive suggestions, especially Drs. Ernest Birnbaum, Eugene Holleran, Neil Jespersen, Istvan Lengyel, William Pasfield, and Siao Sun. And finally, my special thanks go to the following colleagues who have helped shape this book by their thoughtful reviews and criticisms of the manuscript and their many valuable suggestions.

Jo A. Beran
Texas A & I University

Nordulf Debye
Towson State University

Roy Dowling
Louisiana Tech University

Frank Gomba
U.S. Naval Academy

Gregory Grant
University of Tennessee

Kenneth Hyde
SUNY Oswego

Delwin Johnson
St. Louis Community College at Forest Park

Frank Milio
Towson State University

Robert Orr
Delaware Valley College

M. Larry Peck
Texas A & M University

Dennis Rushforth
University of Texas at San Antonio

Joyce Shade
U.S. Naval Academy

Dean Skovlin
California State University at Northridge

Donald Titus
Temple University

Russell Trimble
Southern Illinois University

Boyd Waite
U.S. Naval Academy

J.E.B.

SUPPLEMENTS

A complete package of supplements to accompany this text is available to assist both teacher and student.

Study Guide to Accompany General Chemistry, Principles and Structure, Fifth Edition, by James E. Brady. This softcover book has been carefully structured to assist students in mastering concepts and developing problem-solving skills. It is keyed section by section to the text, and for each section there is a set of objectives, a brief review (sometimes with additional worked examples), a self-test with answers, and a glossary of new terms.

Instructor's Manual for General Chemistry, Principles and Structure, Fifth Edition, by James E. Brady. This manual, available to instructors only, contains suggestions for course scheduling and alternative topic sequences, detailed chapter objectives and chapter rationales, and suggestions for lecture demonstrations.

Laboratory Manual for General Chemistry, Principles and Structure, Fourth Edition, by Jo Beran and James E. Brady. This manual features a thorough techniques section with photographs that illustrate important apparatus and manipulations, and 46 experiments sequenced to follow the topical development of the text. For the teacher, an instructor's manual accompanies the laboratory manual.

Solutions Manual for General Chemistry, Principles and Structure, by M. Larry Peck. This softcover supplement provides detailed solutions to all the numerical problems in the text, as well as answers to all the questions.

Testbank for General Chemistry, Principles and Structure, by James E. Brady. This package of multiple-choice questions is available from Wiley at no charge for teachers who adopt this book.

Computerized Testing System for the IBM PC and Macintosh computers. This testing system with questions allows the user to prepare multiple-choice examinations. It can be obtained from Wiley without charge by instructors who adopt this book.

Transparency Acetates. Instructors who adopt this book may obtain from Wiley, without charge, a set of $8\frac{1}{2} \times 11$-inch transparencies that duplicate key illustrations in the text.

CONTENTS

PART II

THE STRUCTURE OF MATTER AND THE ORIGIN OF CHEMICAL PERIODICITY

PART III

PHYSICAL PROPERTIES AND THE STATES OF MATTER

PART IV

FACTORS THAT CONTROL THE OUTCOME
OF REACTIONS

CHAPTER 14
CHEMICAL THERMODYNAMICS   454

CHAPTER 15
CHEMICAL EQUILIBRIUM
IN GASEOUS SYSTEMS   484

CHAPTER 16
ACID–BASE EQUILIBRIA IN
AQUEOUS SOLUTIONS   512

CHAPTER 17
SOLUBILITY AND COMPLEX
ION EQUILIBRIA   558

PART VI

NUCLEAR REACTIONS AND THEIR ROLE IN CHEMISTRY

GENERAL CHEMISTRY

Principles and Structure

PART

I

THE BASICS OF CHEMISTRY AND CHEMICAL REACTIONS

CHAPTER

1

INTRODUCTION

As its title suggests, the purpose of this chapter is to introduce you to the subject of chemistry. Our goal here is to examine how science in general operates and to make you aware of the kinds of materials with which chemists work. You will learn the importance of laboratory measurements, and you will see some of the ways that chemicals are classified. Throughout all this, you will begin to be exposed to some of the jargon that chemists use. If you've had a previous course in chemistry, perhaps much of the material presented in this chapter will be familiar. Nevertheless, be sure you understand it all and have learned what is expected, because students who have mastered these fundamental concepts and learned to "speak the language" of chemistry have little difficulty later on.

1.1 WHAT IS CHEMISTRY?

Whether or not you have ever taken a course in this subject, it should come as no surprise to learn that chemistry is about chemicals. These are not abstract substances that ordinary mortals need fear. They include all the day-to-day things you touch and see and smell. Your body is composed of chemicals, as is everything that surrounds you, including this book and the air you're breathing. Chemicals are everywhere! In fact, the only place you *don't* find them is in a vacuum.

Many chemicals occur naturally in the world around us or are produced by living things, both animal and vegetable. Rocks and sand; iron, gold, silver, and copper; cotton and wool; sugar and salt: These are some examples of common materials, composed of chemicals, that humans have used for ages to build shelter, make clothing, and feed themselves.

As science progressed, it was discovered that many useful economically important natural chemicals could be manufactured from more humble starting materials, and so the chemical industry began. In this century especially, science and technology developed ways to make new chemicals that never before existed on earth. Examples are nylon and polyester plastics used to make fibers from which fabrics are woven. These new chemicals were developed because the fibers made from them have some properties that surpass those of natural fibers such as cotton and wool. Today, the number of synthetic chemicals in use in medicine, industry, and around the house is enormous, and it would be difficult to imagine what life would be like without them.

Despite the benefits that chemistry has brought, we tend to hear more about the problems. All too often we hear about toxic waste dumps or chemicals in foods and the environment that are potential cancer producers. Among the challenges faced by chemistry and technology these days, therefore, is the development of methods to control and manage the wastes that naturally accompany the production and use of the valuable new materials.

The science of chemistry

The science of chemistry is concerned with a number of aspects of chemicals. Chemists ask questions such as "What are chemicals composed of?" and "How are the characteristics or properties of chemicals determined by their compositions?"

An important part of chemistry, of course, is the study of **chemical reactions,** changes that occur when chemicals interact with each other to form new and entirely different substances. A chemical reaction is really quite an amazing thing to observe, and part of the "fun" of chemistry is watching chemical reactions take place. Since chemical reactions play such a central role in the study of chemistry, let's examine one just to see how remarkably the properties of the substances change when the reaction occurs.

Consider the substances shown in Figure 1.1. On the left is sodium. It is a metal, so soft that it can be cut with a knife. Sodium combines very rapidly with oxygen and moisture in the air, and the crust that covers the outside of the bar of sodium was formed from this reaction. If placed in liquid water, the sodium reacts violently, producing bubbles of flammable hydrogen gas and a substance called sodium hydroxide (commonly known as lye), which is extremely corrosive to skin. For this reason, contact of sodium with the skin (which is usually moist) must be avoided.

The photograph on the right in Figure 1.1 shows another substance, chlorine. It is a pale yellow-green gas that is also very corrosive to animal tissue. Severe lung damage leading to death can result if chlorine is breathed, which explains the use of this chemical as a war gas during World War I.

When chlorine and sodium are brought together, they react violently with each other, as shown in Figure 1.2 on page 4. As the reaction proceeds, the sodium and chlorine are changed into a new substance whose chemical name is sodium chloride, but which is better known to nonchemists as ordinary table salt. If you think about this reaction for a minute, you will see that it is really quite amazing. Here we begin with two chemicals, sodium and chlorine, either of which is capable of causing great bodily harm. But when sodium and chlorine are brought together, these harmful

Throughout this book, important terms are set in **bold type** when they are introduced and defined. Be sure you learn their meanings.

In chemistry, we use the word **substance** to mean the chemical material of which an object is composed. An ice cube, for example, is composed of the *substance* water, and the metal bar in Figure 1.1*a* is composed of the *substance* sodium.

The odor of liquid chlorine bleach, such as Clorox, is due to very small amounts of chlorine gas escaping from the bleach solution.

Most chemical reactions are not as violent as the reaction between sodium and chlorine.

(a)

(b)

Figure 1.1. (a) *Sodium is a metal that is so soft it can be cut with a knife. The white crust was formed by the reaction of sodium with oxygen and moisture in the air. The freshly exposed surface of the metal will tarnish quickly because of this. (b) Chlorine is a pale yellow-green gas.*

Figure 1.2. *A small piece of sodium was melted in a metal spoon before being thrust into the flask of chlorine gas. The sodium combines rapidly with the chlorine to form sodium chloride. The sodium used in this reaction had been stored in kerosene to protect it from oxygen and moisture. The reaction of chlorine with traces of kerosene still on the sodium released some black smoke, which can also be seen in the flask.*

The flame that a glassblower uses to soften glass is produced by the reaction of methane (natural gas) with pure oxygen. In the laboratory you use the same reaction to give the flame of a Bunsen burner.

properties disappear. In their place we find the properties of this new substance, common salt, a chemical that our bodies must have in order to function.

All chemical reactions display similar remarkable changes in the properties of the chemicals involved, although often the changes are not as spectacular as in the reaction between sodium and chlorine. Nevertheless, observing these changes is an important part of "doing chemistry" in the laboratory, and they are something you should watch for in your study of chemistry.

1.2 THE SCIENTIFIC METHOD

Chemistry, as you know, is said to be a science. What this means is that chemists, like other scientists, adhere to a general, overall philosophy of approach to the study of nature that is called the **scientific method.**

Actually, the scientific method is little more than a formal statement of the steps that any of us follow as we logically approach a problem. Consider, for example, an auto mechanic attempting to fix a car that won't start. First, the apparent cause of the problem is located by observing the results of one or more tests. Next, a suspect part is replaced or adjustments are made on the engine, and then an attempt is made to start the car. If the mechanic was correct in judging the cause of the problem, the job is done. If not, other tests are performed and repairs made until the car is finally running.

When we approach a problem in the sciences, we proceed in much the same way. The first step in the scientific method, therefore, can be called **observation.** This is the purpose of the experiments that we perform in the laboratory where nature can be observed under controlled conditions, so the results of our experiments are reproducible. The bits of information we obtain are called **data** and can be classified as either qualitative or quantitative. A **qualitative observation** does not have numbers associated with it. An example is the observation that *sodium reacts with chlorine to produce a white powder, sodium chloride.* This is the reaction described in the previous section. However, if we were to measure the *weight* of sodium that reacts with a given *weight* of chlorine, a numerical value for the sodium-to-chlorine weight ratio would be obtained and we would have made a **quantitative observation.** Quantitative observations are often more useful to scientists than qualitative observations because they provide much more information.

When confronted with massive amounts of information, we quite naturally begin to look for similarities and differences in an attempt to gain a view of the "big picture." So it is in the sciences, too. As data are collected, trends and generalizations are sought that help summarize and organize the information, so that it is easier to comprehend. In science, a statement that summarizes facts that come from many experiments is called a **law,** or sometimes a **natural law.**

In one sense, laws serve as a convenient means of storage for the vast amounts of data from which they are derived. But laws are really more useful than that. They also allow us to predict the outcome of as yet untried experiments. For instance, chemists have found that when methane (natural gas) is burned in pure gaseous oxygen to form carbon dioxide and water vapor, two volumes of oxygen are always needed to completely use up one volume of methane, provided the volumes are measured at the same temperature and pressure. This simple statement is a law dealing with the reaction of methane with oxygen. It allows us to predict, for example, that if we had 5 liters of methane, we would need 10 liters of oxygen gas for a complete reaction. We can make this prediction with confidence, even if nobody had ever worked with these volumes of these gases, because the law we applied is based on the results of many experiments with varying amounts of methane and oxygen.

There are times when a law is expressed in a simple verbal statement, as in the law just discussed dealing with the reaction of methane with oxygen, but very often laws are expressed in equation form. For example, it is observed that the force of attraction between two particles that carry opposite electrical charges decreases as the distance between the particles increases. This is more accurately stated in equation form by Coulomb's law,

$$F = \gamma \frac{q_1 q_2}{r^2}$$

in which F is the force of attraction, q_1 and q_2 are the charges on the particles, r is the distance between the particles, and γ is a proportionality constant.

Laws are frequently named for the people who discover them. This law is named after Charles Coulomb (1736–1806), a French scientist and inventor.

As useful as laws are at condensing large amounts of experimental data, laws do not explain *why* nature behaves as it does. Scientists, being human (despite what you may have heard to the contrary), are not satisfied with simple statements of fact and seek to explain their observations. Therefore, the second stage in the scientific method is to propose tentative explanations, or **hypotheses** that can be tested by experiment. If hypotheses are not disproven by repeated experiments, they develop into theories. Thus, a **theory** is a tested explanation of the behavior of nature. Theories always serve as guides to new experiments and are constantly being tested. If a theory is proven to be incorrect by experiment, it must either be discarded in favor of a new one or, as is often the case, modified so that all the experimental observations can be accounted for. Science develops, therefore, by a constant interplay between theory and experiment.

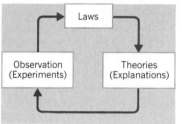

The scientific method is a cyclic process. Experiments suggest laws, which are explained by theories, which themselves suggest new experiments to be done, which produce new laws and theories, and so on.

An important point to remember about theories is that they can seldom be *proven* to be correct. Usually, the best we can do is fail to find an experiment that disproves a theory. Of course, this doesn't mean that such an experiment is not possible, so we can never be quite sure that a theory is true. There always should be a lingering doubt, and a scientist must always be careful not to confuse theory with experimental fact. Too many times in the past have incorrect theories been accepted as fact, and the progress of science slowed because of it.

The application of the scientific method isn't always as organized as it might appear from this discussion. A lot of trial and error and many false starts go on. But overall, the development of science has followed this general approach.

Although the scientific method seems like a very "cut and dried" approach, it is important to also realize that many of the most important advances in science, such as the discoveries of radioactivity by Henri Becquerel and penicillin by Alexander Fleming, came about by accident. These discoveries were really only partly accidental, however, because the people involved had learned to think "scientifically" and were aware that they had observed something new and exciting. It is also important to realize that many hours of careful work are required for scientific progress, regardless of how "lucky" you might be.

1.3 LABORATORY OBSERVATIONS: THE INTERNATIONAL SYSTEM OF UNITS

In the preceding section, you learned that the critical first step in a scientific investigation is *observation*. That's why we have laboratories. In a chemistry laboratory, some of our observations are qualitative; for example, we simply observe what happens when two chemicals such as sodium and chlorine are mixed. However, in chemistry (and all the other sciences as well), real progress requires quantitative observations, which means measurements of various kinds must be made.

A key part of any measurement is the units associated with the measured quantity. For example, to say that the length of something is "three" is meaningless unless a specific unit is associated with the number. Except in the sciences, in the United States we usually use customary "English units" of measurement. We use yards, feet, and inches to measure length; tons, pounds, and ounces to measure weight; and gallons, quarts, and pints to measure volumes of liquids. All the other industrialized nations of the world have used a metric-based system for many years, and this is also the system used in the sciences, including chemistry. The United States is in a gradual transition to these units, and you have probably noticed that metric units, as well as our customary units, appear on many consumer products.

In 1960, the General Conference of Weights and Measures, an international body, adopted and recommended for worldwide use a modified version of the older metric system. This newer system is called the **International System of Units** (abbreviated **SI** from the French *le Système International d'Unités*). Although, by now, most of the SI units have been generally accepted, some of the older non-SI metric units are still widely used by many practicing scientists. Furthermore, the existence of the older units in the scientific literature demands that we be aware of both the old and new. Most of the units used in this book are SI units, but in a few cases the more familiar older units are retained.

> The original metric system was first formulated by the French Academy of Sciences during the 1790s.

As a starting point, the SI specifies a set of seven **base units,** which are given in Table 1.1. Although some of them may not be familiar to you now, we will use most of them at one time or another during this course and we will say more about them as they are encountered.

The SI base units are very precisely defined, in almost all cases in terms of a reproducible physical phenomenon. For instance, the meter is defined now as *exactly* the distance that light travels in a vacuum in 1/299,792,458th of a second. Only the SI unit for mass, the kilogram, is defined by an object made by human hands. It consists of a platinum–iridium alloy block stored at the International Bureau of Weights and Measures in France. The photograph in Figure 1.3 shows the U.S. standard kilogram, kept at the National Bureau of Standards in Washington, D.C. As nearly as possible, it is a duplicate of the one in France and serves at least indirectly as the calibrating standard for all "weights" used for scales and balances (including those you will use in the chemistry lab) in the United States.

Figure 1.3. *The standard kilogram belonging to the National Bureau of Standards in Washington, D.C.*

> Area has the dimensions of (distance)^2.

Derived units

At first glance, the SI seems to be a very limited system of units. There are many quantities, such as area and volume, whose units do not appear in Table 1.1. In the SI, units for such quantities are obtained by appropriate combinations of the base units and are called **derived units.** The way the SI base units are combined in such cases depends on the dimensions of the measured quantities. For instance, if you wished to calculate the area of a rectangular carpet, you would multiply its length by its width. The *unit* for area is likewise obtained as the product of the *unit* for length and the *unit* for width. Since length and width are both distances, for which the SI unit is the meter (m), the SI unit for area is m^2 (meter squared or square meter).

$$\text{length} \times \text{width} = \text{area}$$
$$\text{m} \quad \times \quad \text{m} \quad = \text{m}^2$$

Similarly, speed is a ratio of distance to time. It is computed as the distance traveled divided by the elapsed time. The SI unit for speed is therefore meter/second or m/s.

Table 1.1. The seven SI base units

Physical Quantity	Name of Unit	Symbol
Mass	kilogram	kg
Length	meter[a]	m
Time	second	s
Electric current	ampere	A
Temperature	kelvin	K
Luminous intensity	candela	cd
Amount of substance	mole	mol

[a] All English-speaking nations except the United States have adopted the -re spelling for metre. Both -re and -er forms have support within the United States, although the National Bureau of Standards currently favors the -er spelling.

Working with larger and smaller units

Often, either the base units or the derived units are of a size that makes them inconvenient for ordinary measurements. The SI unit for volume, for example, is the cubic meter (m^3), which is somewhat larger than a cubic yard. For expressing volumes that are routinely measured in the laboratory, however, the cubic meter is awkward. An ordinary glass of water, for instance, is about 0.00025 m^3. The SI solves this problem by creating larger and smaller units by modifying the basic units with decimal factors and prefixes. A complete list of these factors and prefixes is given in Table 1.2. Be sure to learn those printed in color because they are the ones you will encounter most frequently in this course.

0.00025 m^3 = 250 cm^3 (That's about eight fluid ounces.)

Notice that the factors in Table 1.2 are expressed in exponential form. This type of notation (also called **scientific notation**) will be used frequently in our discussions and is reviewed in Appendix A at the back of the book. Table 1.3 illustrates how these multipliers work and how the prefixes associated with them are used to name the modified units.

If necessary, you should review Appendix A.

On the following pages are several examples that illustrate applications of the SI and conversions among SI units. These and the many other examples that you will find throughout this book are intended to teach you not only how to work out problems in chemistry, but also how to approach problem solving in general. In many cases, you will find that an example is divided into three parts. First is the statement of the problem. Then there is an analysis of the problem, where we figure out what "tools" we need and, in more complex examples, where we decide what

Table 1.2. The sixteen SI prefixes

Factor	Prefix	Symbol	Factor	Prefix	Symbol
10^{18}	exa	E	10^{-1}	deci	d
10^{15}	peta	P	10^{-2}	centi	c
10^{12}	tera	T	10^{-3}	milli	m
10^{9}	giga	G	10^{-6}	micro	μ
10^{6}	mega	M	10^{-9}	nano	n
10^{3}	kilo	k	10^{-12}	pico	p
10^{2}	hecto	h	10^{-15}	femto	f
10^{1}	deka	da	10^{-18}	atto	a

Table 1.3. Modifying the size of SI units with prefixes

Prefix	Multiplication Factor	Examples		Symbol
kilo-	1000 (10^3)	1 kilometer	= 1000 meter (10^3 m)	km
		1 kilogram	= 1000 gram (10^3 g)	kg
deci-	1/10 (10^{-1})	1 decimeter	= 0.1 meter (10^{-1} m)	dm
centi-	1/100 (10^{-2})	1 centimeter	= 0.01 meter (10^{-2} m)	cm
milli-	1/1,000 (10^{-3})	1 millimeter	= 0.001 meter (10^{-3} m)	mm
		1 millisecond	= 0.001 second (10^{-3} s)	ms
		1 milligram	= 0.001 gram (10^{-3} g)	mg
micro-	1/1,000,000 (10^{-6})	1 micrometer	= 0.000 001 meter (10^{-6} m)	μm
		1 microgram	= 0.000 001 gram (10^{-6} g)	μg
nano-	1/1,000,000,000 (10^{-9})	1 nanometer	= 0.000 000 001 meter (10^{-9} m)	nm
		1 nanogram	= 0.000 000 001 gram (10^{-9} g)	ng

must be done with them. Finally, there is the solution to the problem, where the elements given in the statement of the problem are combined with those in the analysis to arrive at an answer.

EXAMPLE 1.1. CONVERSIONS AMONG SI UNITS

PROBLEM: A desk is found to be 1437 mm wide. What is this width expressed in meters?

ANALYSIS: The problem here is to translate units of mm into meters. Of course, you first have to realize that mm means millimeters; if necessary, be sure to review the symbols for the SI base units and prefixes. Next, to solve the problem we need to know what the prefix "milli" means. You should know that milli means "$\times 10^{-3}$." Now, let's see how we combine this information to answer the question.

SOLUTION: In working a problem of this kind, we can substitute the SI multiplication factor in place of its corresponding symbol. Since m (for milli) means $\times 10^{-3}$, we substitute "$\times 10^{-3}$" in place of the m.

$$1437 \text{ mm} = 1437 \times 10^{-3} \text{ m}$$

$$\boxed{\times 10^{-3}}$$

This should be rewritten as 1.437 m. (If necessary, review how to change numbers from scientific notation to standard decimal notation in Appendix A at the end of the book.)

One of the kinds of problems that we routinely encounter is the conversion of units — for example, converting centimeters to meters, or nanometers to centimeters. In fact, many problems in chemistry can be considered to be of this type, and there is a particularly useful method that we will use in setting up unit conversion problems. It is called the **factor-label method** and is described in detail in Appendix A at the back of the book. The method is based on the notion that units undergo the same kinds of mathematical operations that numbers do. In fact, this is the basis for forming the derived units mentioned on page 6. The next two examples illustrate how the factor-label method is applied.

EXAMPLE 1.2. CONVERSIONS AMONG SI UNITS

PROBLEM: A certain person is 172 cm tall. Express this height in decimeters.

ANALYSIS: As a beginning, let's restate the problem in the form of an equation

$$172 \text{ cm} = ? \text{ dm}$$

Looking at the problem in this way, we see that the job is to change centimeters to decimeters. To solve the problem by the factor-label method, we need to have relationships that can take us from centimeters to decimeters. Reviewing Table 1.2, we can't see a way of doing this directly, but we can write the relationships

$$1 \text{ cm} = 10^{-2} \text{ m}$$

$$1 \text{ dm} = 10^{-1} \text{ m}$$

Notice that these provide a path from cm to dm. Centimeters can be converted to meters using the first relationship, and then meters can be changed to decimeters using the second.

SOLUTION: Now that we have assembled all the information needed to solve the problem and have a path from the given quantity (172 cm) to the answer, we can use the factor-label method to set up the arithmetic. In this method, we multiply the given quantity (in this case, 172 cm) by one or more conversion factors that we create using the relationships needed to solve the problem. A **conversion factor** is a fraction that we form from a valid relationship between units. For example, the first equation above allows us to construct the following two conversion factors:

$$\frac{1 \text{ cm}}{10^{-2} \text{ m}} \quad \text{and} \quad \frac{10^{-2} \text{ m}}{1 \text{ cm}}$$

When we choose a conversion factor, we look to see which units must cancel. In this case, if we multiply our given 172 cm by the second factor, the units cm will cancel and we will have converted centimeters to meters.

$$172 \cancel{\text{ cm}} \times \left(\frac{10^{-2} \text{ m}}{1 \cancel{\text{ cm}}} \right) = 1.72 \text{ m}$$

As pointed out in Appendix A, multiplying a quantity by a conversion factor is equivalent to multiplying by 1. It doesn't change the size of the quantity; it just expresses the quantity in different units.

Similarly, we use the second equation to convert meters to decimeters.

$$1.72 \cancel{\text{ m}} \times \left(\frac{1 \text{ dm}}{10^{-1} \cancel{\text{ m}}} \right) = 17.2 \text{ dm}$$

We can also "string together" conversion factors and obtain the same net result.

$$1.72 \cancel{\text{ cm}} \times \left(\frac{10^{-2} \cancel{\text{ m}}}{1 \cancel{\text{ cm}}} \right) \times \left(\frac{1 \text{ dm}}{10^{-1} \cancel{\text{ m}}} \right) = 17.2 \text{ dm}$$

to meters ↑
to decimeters

EXAMPLE 1.3. CONVERSIONS AMONG SI UNITS

PROBLEM: Calculate the number of cubic centimeters in 0.225 dm³.

ANALYSIS: Once agan, let's begin by stating the problem in the form of an equation.

$$0.225 \text{ dm}^3 = ? \text{ cm}^3$$

As before, we see that it is a unit converson problem. To solve it, we need relationships among cubic units. Examining Table 1.2, we see that such units aren't available, so we have to figure out how to create them from the units that we do have. In the previous problem we used the relationships among meters, decimeters, and centimeters.

$$1 \text{ dm} = 10^{-1} \text{ m}$$

$$1 \text{ cm} = 10^{-2} \text{ m}$$

We could also express these relationships as

$$10 \text{ dm} = 1 \text{ m}$$

$$100 \text{ cm} = 1 \text{ m}$$

The results would be the same.

A review of arithmetic operations involving numbers expressed in powers of 10 is located in Appendix A.

We always treat the prefix and unit together as a single entity when applying exponents. Thus, mm³ is *always* "($\times 10^{-3}$ m)³" and *never* "$\times 10^{-3}$ (m)³."

We can obtain cubic units by cubing each side of each equation.

$$(1 \text{ dm})^3 = (10^{-1} \text{ m})^3$$

$$1 \text{ dm}^3 = 10^{-3} \text{ m}^3$$

Notice that we've cubed both the number and the unit! Similarly,

$$(1 \text{ cm})^3 = (10^{-2} \text{ m})^3$$

$$1 \text{ cm}^3 = 10^{-6} \text{ m}^3$$

Now that we have the necessary relationships among the cubic units, we can solve the problem by letting the units serve as our guide.

SOLUTION: We multiply the given quantity 0.225 dm³ by converson factors, keeping our eye on which units have to cancel.

As explained below, 1 cubic decimeter is the same as 1 liter (1 L). The answer here tells us that 0.225 L = 225 cm³.

$$0.225 \text{ dm}^3 \times \frac{10^{-3} \text{ m}^3}{1 \text{ dm}^3} \times \frac{1 \text{ cm}^3}{10^{-6} \text{ m}^3} = 255 \text{ cm}^3$$

Thus, 0.225 dm³ = 225 cm³.

Units for laboratory measurement

In chemistry, it is necessary to routinely measure mass, length, volume, and temperature. The units that we ordinarily use to express the first three quantities are based on the **gram** (abbreviated **g**), the **meter (m),** and the **liter (L),**[1] respectively. The gram itself, which is one-thousandth of the SI base unit kilogram, is a conveniently sized unit for most laboratory measurements of mass (about which we will say more in Section 1.5). The meter, however, is slightly longer than a yard, so for most laboratory purposes we measure length in smaller units of centimeters or millimeters.

$$1 \text{ cm} = 10^{-2} \text{ m}$$

$$1 \text{ mm} = 10^{-3} \text{ m}$$

It is often easier to remember these relationships if they are expressed as

$$1 \text{ m} = 100 \text{ cm}$$

$$1 \text{ m} = 1000 \text{ mm}$$

This also means, of course, that 1 cm = 10 mm.

The liter is defined by the SI as exactly 1 cubic decimeter, which is the same as 1000 cubic centimeters (see Figure 1.4).

Sometimes you may see the abbreviation cc used for cubic centimeter instead of cm³.

$$1 \text{ L} = 1 \text{ dm}^3$$

$$1 \text{ L} = 1000 \text{ cm}^3$$

Since there are also 1000 mL in one liter, the size of the milliliter and cubic centimeter are the same—they are identical.

$$1 \text{ mL} = 1 \text{ cm}^3$$

Most laboratory glassware is graduated in milliliters.

Temperature In a formal sense, temperature is a quantity that determines the direction that heat will flow spontaneously; heat always flows from something at a higher temperature to something at a lower temperature. This is such a common

---

[1] The preferred abbreviation for the liter is "L," which is the abbreviation that we will use in this book. On some older laboratory glassware and in many other books, you will find liter abbreviated "l," as in milliliter, ml.

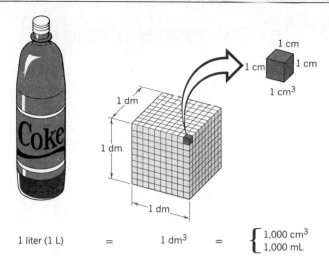

Figure 1.4. *One liter equals one cubic decimeter.* $1 L = 1 dm^3 = 1000 cm^3 = 1000 mL.$

$$1 \text{ liter (1 L)} \quad = \quad 1 \text{ dm}^3 \quad = \quad \left\{ \begin{array}{l} 1{,}000 \text{ cm}^3 \\ 1{,}000 \text{ mL} \end{array} \right.$$

Figure 1.5. *A typical laboratory thermometer.*

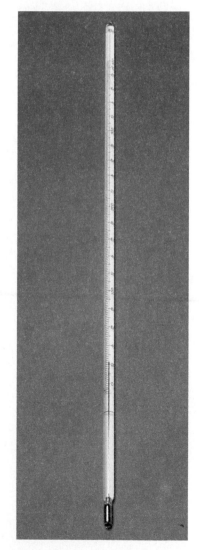

experience that the definition is hardly necessary, and you certainly recognize when something is hotter than something else.

Temperature is commonly measured with a **thermometer** (Figure 1.5), which consists of a narrow capillary tube connected at one end to a thin-walled glass bulb that's filled with some liquid (usually mercury). As the temperature of the bulb rises, the liquid expands and pushes its way up the capillary. The height of the liquid in the capillary is directly proportional to the temperature.

Two reference temperatures are chosen to make the markings on the scale of a thermometer. The height of the mercury column is marked after the thermometer is brought to each temperature and the distance between the two marks is then divided into some number of degree units, depending on the temperature scale used.

The reference temperatures used for defining the common temperature scales are the freezing point and boiling point of water. These are chosen simply for practical reasons. For any pure substance such as water, there is only one tempera-ture (under ordinary laboratory conditions) at which both liquid and solid can coexist at the *same* temperature. This temperature is called the freezing point or melting point (they are identical). If ice is placed in liquid water and stirred, some melting of the ice will occur until the remaining ice and the liquid come to the same temperature. If some heat is absorbed by the ice–liquid mixture, some of the ice will melt, but the temperature will stay the same. If some heat is removed, some liquid will freeze, but once again, the temperature will stay constant. Because the solid–liquid mixture maintains a *reproducible constant temperature,* the experimentalist has plenty of time to mark the thermometer accurately. Similarly, at any given pressure, a pure liquid will always boil at a reproducible constant temperature that is indepen-dent of how fast heat is supplied to the liquid. Thus, the boiling point also serves as a convenient reference temperature.

On the **Fahrenheit scale,** which is in common use in the United States, the freezing point of water is assigned a temperature of 32 °F and the boiling point a temperature of 212 °F. The difference between these two reference points, 180 Fahrenheit degrees, thus defines the size of the Fahrenheit degree unit. In the sciences, the temperature scale that is employed is the **Celsius scale** (at one time, called the centigrade scale). You have probably noticed the growing use of Celsius temperatures in weather reports on the radio and TV. The Celsius scale defines 0 °C as the freezing point of water and 100 °C as the boiling point of water. This means that 100 Celsius degrees are equal to 180 Fahrenheit degrees, so the Celsius degree is nearly twice as large as a degree on the Fahrenheit scale. Temperatures in degrees

Figure 1.6. *Comparison of temperature scales.*

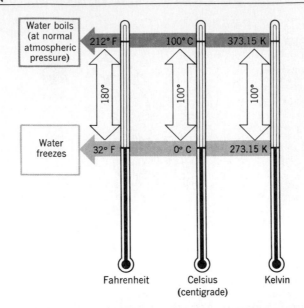

Celsius, $t_{(^\circ C)}$, are related to temperatures in degrees Fahrenheit, $t_{(^\circ F)}$, by the equation

$$t_{(^\circ C)} = \tfrac{5}{9}[t_{(^\circ F)} - 32] \tag{1.1}$$

Important equations are printed in blue. You will also find many equations are numbered for easy reference later on.

The name of the temperature scale is capitalized, but the name of the degree unit, the kelvin, is not.

The SI temperature unit is the **kelvin (K).** The size of the degree unit on the Kelvin scale is the same as on the Celsius scale. The difference between the two temperature scales is the location of the zero point. On the Kelvin scale, water freezes at 273.15 K (note that the degree symbol ° is not used in the SI unit), so the relationship between the Celsius temperature, $t_{(^\circ C)}$, and the Kelvin temperature, $t_{(K)}$, is

$$t_{(K)} = t_{(^\circ C)} + 273.15 \tag{1.2}$$

Often, we will need to express these temperatures only to the nearest whole degree, so we can use the relationship

$$t_{(K)} = t_{(^\circ C)} + 273$$

Zero kelvin is known as *absolute zero.* It is the lowest temperature that nature can allow. We will say more about it later in the book.

Thermometers are never graduated directly in the Kelvin scale, so if you need to know the Kelvin temperature, you must measure the Celsius temperature and then calculate the corresponding Kelvin temperature. Figure 1.6 illustrates the differences among these three temperature scales.

Although we will use the units just described for laboratory measurements, it is helpful to have a feeling for how large they are. Table 1.4 gives the approximate

Table 1.4. Comparison of the English and metric systems

Length	1 meter = 39.37 inches (in.)
	2.540 centimeters = 1 inch[a]
Mass	1 kilogram = 2.204 pounds (lb)
	453.6 grams = 1 pound
Volume	1 liter = 1.057 quarts (qt)
	29.57 milliliters = 1 fluid ounce (oz)
	28.32 liters = 1 cubic foot (ft^3)

A table of common English-to-English conversions is located inside the rear cover of the book.

[a] This is an exact relationship, so the numbers can be considered to have as many significant figures as you would like.

EXAMPLE 1.4. CONVERTING BETWEEN ENGLISH AND SI UNITS

PROBLEM: Calculate the number of meters in 0.200 mile.

ANALYSIS: First, let's restate the problem as an equation.

$$0.200 \text{ mi} = ? \text{ m}$$

Next, we assemble a set of relationships that will take us from miles to meters. There is more than one set that can do this, but the following is one that many would pick:

$$1 \text{ mi} = 5280 \text{ ft}$$

$$1 \text{ ft} = 12 \text{ in.}$$

$$1 \text{ m} = 39.37 \text{ in. (from Table 1.4)}$$

Note the path that they provide from miles to feet to inches to meters. Now that we've got everything we need, we can solve the problem.

SOLUTION: We use the relationships above to create conversion factors and then use them so that the units we want to get rid of will cancel.

$$0.200 \text{ mi} \times \left(\frac{5280 \text{ ft}}{1 \text{ mi}}\right) \times \left(\frac{12 \text{ in.}}{1 \text{ ft}}\right) \times \left(\frac{1 \text{ m}}{39.37 \text{ in.}}\right) = 322 \text{ m}$$

Notice that the answer has been rounded off to three digits. There is a good reason for this, as you will see in the next section.

English equivalents of some of these units and Example 1.4 illustrates how we can use them in converting between our customary U.S. units and the SI units.

1.4 MEASUREMENT AND SIGNIFICANT FIGURES

You have learned now about the importance of laboratory measurements and systems of units. In making a measurement, a scientist is actually concerned about two things. One, of course, is the size of the measured quantity, which depends on the units used to express it. The other, which is equally important, is how *reliable* the measurement is. It is this second aspect of measurement that we focus on in this section.

Significant figures

When we make a measurement, we obtain numbers by reading them from a scale of some sort on the measuring device. Because of this, there is nearly always some limitation on the number of meaningful digits that can be obtained. As an illustration, let's consider measuring the length of a block of wood with two different rulers, as shown in Figure 1.7.

Using the ruler in Figure 1.7a, we might estimate the length of wood block to be 3.2 cm. Notice that to arrive at this number, we are forced to estimate the second digit; this is, we must decide whether the length lies closer either to 3.2 or to 3.3 cm. Because we are making an estimate, some uncertainty exists in the second digit (the 2), and the third digit—the one following the 2—is completely unknown. We would find a similar situation for any other object whose length we measured with

Many instruments today have digital displays, and one might be tempted to think that there is no uncertainty associated with measurements made with these devices because it is not necessary for the experimentalist to estimate the rightmost digit when making a measurement. Actually, some uncertainties do exist; it is just that the makers of the instruments have used electronics to make the estimates for us.

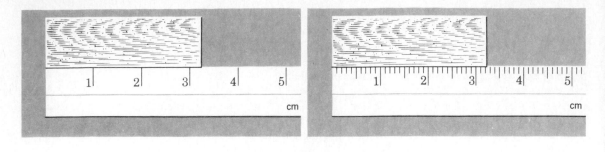

(a)   *(b)*

Figure 1.7. *Measuring the length of a block of wood with two different rulers. (a) Length = 3.2 cm. (b) Length = 3.24 cm.*

this ruler, so we can say that for measurements made with the ruler in Figure 1.7a, we are not justified in reporting numbers containing more than two figures.

When a number is written to represent the result of a measurement, it is always assumed that, unless stated otherwise, only the rightmost digit is uncertain. It is also assumed that any digits farther to the right are unknown and that all the other digits to the left are known for sure. The digits that are obtained as the result of a measurement are called **significant figures.** Thus, the measurement illustrated in Figure 1.7a yields a number with two significant figures.

In Figure 1.7b, we see the same piece of wood measured with a ruler subdivided by additional graduations. Now we see that both the 3 and the 2 are known for sure, and we can try to estimate the third digit. One estimate of the length might be 3.24 cm, although some people might judge it to be either 3.23 or 3.25 cm. This time we have two digits known for sure and one that is somewhat uncertain, so we have a measurement that contains three significant figures.

The importance of significant figures is that they indicate to us the reliability of measurements. In the determination of the length of the piece of wood previously described, we saw two different values, obtained with two different measuring devices. Our intuition tells us that we can place more confidence in the value with the greater number of significant figures. Laws and theories are derived from measured quantities, and our confidence in them is directly related to the confidence we are able to place in the data on which they are based.

EXAMPLE 1.5. THE MEANING OF SIGNIFICANT FIGURES

PROBLEM: The length of a large room was measured to be 26.0 m using a tape measure graduated in tenths of a meter.

(a) How many significant figures are there in this measurement?

(b) Have the maximum number of significant figures been obtained, considering how the tape measure is graduated?

(c) What would be wrong with reporting the length as 26 m?

SOLUTION: (a) There are three significant figures. Both digits to the left of the decimal point can be assumed to be known for sure, and the reported value suggests that only the zero to the right of the decimal point has any uncertainty. In other words, the reported value suggests that the length could lie between 25.9 and 26.1 m.

(b) The answer is no. If the markings on the tape measure are tenths of a meter apart, we should be able to estimate between the markings, so we should be able to obtain values to the nearest hundredth of a meter. In other words, if the tape measure had been used to its fullest potential, we should have obtained a value with four significant figures.

(c) If the length were reported as just 26 m, it is implied that the measurement is uncertain by at least ± 1 m. But we know the measurement is more reliable than that. If you do the work to get the third significant figure, don't waste it by reporting only two.

Precision and accuracy

In discussions of measured quantities, the words accuracy and precision are often used. The term **precision** refers to how closely measurements of the same quantity come to each other. For example, if a number of people were to repeat the measurement shown in Figure 1.7a, the values they would obtain might differ from each other by about 0.1 cm. We can say, therefore, that the lengths measured with this ruler are *uncertain* by about 0.1 cm, which means that any given measurement could be either 0.1 cm larger or 0.1 cm smaller. The way this is usually expressed is to say that the uncertainty is ±0.1 cm. On the other hand, repeated measurements using the ruler in Figure 1.7b would yield values that might differ by about 0.01 cm, so the uncertainty in these measurements would be ±0.01 cm. The measurements made with this second ruler have a *smaller amount of uncertainty* than those made with the first one, and they are therefore considered to be more precise.

In general, the more significant figures there are in a measured quantity, the greater the precision of measurement. A reported value of 3.24 cm implies a smaller amount of uncertainty, and therefore greater precision, than a reported value of 3.2 cm.

The term **accuracy** refers to how close an experimental observation lies to the true value. Generally, a more precise measurement will be a more accurate measurement. In our example above, the value 3.24 cm has a greater precision than 3.2 cm and probably is also closer to the true length (whatever that may be).

There are instances when a measurement may be precise, but not particularly accurate. The ruler in Figure 1.8, for instance, is improperly marked. Failure to notice this error, and therefore failure to compensate for it, would yield measurements that would be wrong by 1 cm, even though three significant figures can be read from the scale. To avoid problems of this type, scientists **calibrate** their instruments. This involves adjusting the instrument to give correct values while the instrument is being used to measure accurately known standards.

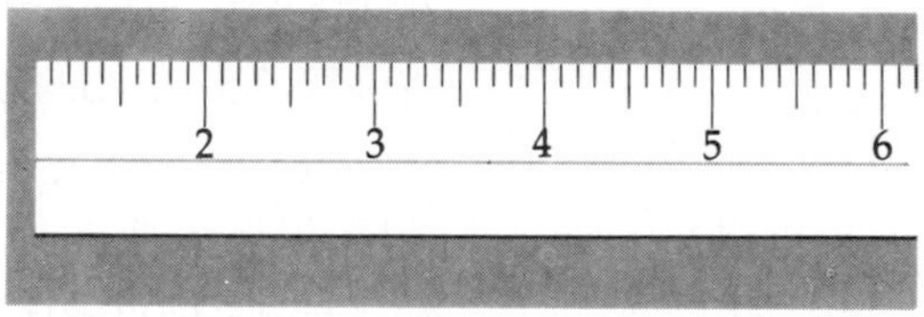

> The uncertainty in a measurement is an estimate of how much larger or smaller another measurement of the same quantity is likely to be.

> The uncertainty in a measurement is sometimes reported along with the value of the measurement itself — for example, 3.2 ± 0.1 cm or 3.24 ± 0.01 cm.

Figure 1.8. *An improperly calibrated ruler. All measurements made with this ruler will be in error by 1 cm.*

Scientific notation and significant figures

In evaluating the quality of experimental data, a routine task is counting the number of significant figures in a measurement. Normally, this amounts to little more than simply looking at the value. For instance, a measured length of 23.6 cm obviously contains three significant figures, and a measured length of 1.203 cm has four. But how about the numbers 0.00215 m and 12.30 m? Do the zeros in these measured values count as significant figures?

The general rule is that a zero *does not* count as a significant figure if its function is merely to locate the position of the decimal point. For example, consider the measured value 0.00215 m. Without the zeros to the left of the 2, we wouldn't know where the decimal point belongs and we wouldn't know how large the measurement is. Now, notice what happens to these zeros when we rewrite the number in scientific notation.

$$0.00215 \text{ m} = 2.15 \times 10^{-3} \text{ m}$$

We see that the zeros disappear; their function was just to locate the decimal point

> If you are not familiar with scientific notation and exponential arithmetic, you should be sure to study Appendix A.

and they are really not measured digits. A rule that we can apply, therefore, is that *zeros that lie to the left of the first nonzero digit are not counted as significant figures.*

How about the zero in the measured value 12.30 m? This zero *is* counted as a significant figure because it would not have been written unless the digit had been estimated to be a zero. Also note that this zero does not disappear when the number is written in scientific notation.

$$12.30 \text{ m} = 1.230 \times 10^1 \text{ m}$$

This leads to another general statement: *In a measured quantity, a zero that lies to the right of the decimal point and also to the right of the first nonzero digit always counts as a significant figure.* If we put all this together, in the measurement below only the nonzero digits plus the zeros above the asterisks count as significant figures

$$0.0054070 \text{ m}$$
$$\quad\quad\quad * *$$

and the measurement therefore contains five significant figures.

Scientific notation is especially useful for removing ambiguities that arise in counting significant figures in large numbers. For example, suppose that someone told you that the length of a field is 1200 m. How many significant figures are in this value? If the measurement had been made with a device that is reliable to the nearest meter, then the value is 1200 ± 1 m, and the measurement has four significant figures. On the other hand, if the distance had simply been estimated as 1200 m with an uncertainty of 100 m, then the value is 1200 ± 100 m, and the measurement has only two significant figures. By just looking at the number 1200, however, we cannot tell how many significant figures it contains; we can't tell whether the zeros are significant figures or whether they just locate the decimal point. This is when scientific notation comes to the rescue. If we wish to express 1200 to four significant figures, we can write it as 1.200×10^3, but if we wished to show only two significant figures, we can write 1.2×10^3. Below we see three different ways of expressing a length of 1200 m, each expressing a different number of significant figures.

$$1.200 \times 10^3 \text{ m (four significant figures)}$$
$$1.20 \times 10^3 \text{ m (three significant figures)}$$
$$1.2 \times 10^3 \text{ m (two significant figures)}$$

Significant figures and calculations

In almost all cases, the numbers that we obtain from measurement are used to calculate other quantities, and we must exercise care to report the calculated result in a way that neither overstates nor understates the amount of confidence we have in it. This means that we must be careful to report the computed value with the proper number of significant figures.

To see how problems can arise, suppose we wished to calculate the area of a rectangular carpet whose sides have been measured, using two different rulers, to be 6.2 m and 7.00 m long. We know the area is the product of these two numbers.

$$\text{area} = 6.2 \text{ m} \times 7.00 \text{ m} = 43.4 \text{ m}^2$$

The question is, "How many significant figures are justified in the answer?"

The length 6.2 m has two significant figures, which implies an uncertainty of about ± 0.1 m. Suppose an error of 0.1 m had, in fact, been made and that the scale of the ruler should have read 6.3 m. How much of an error would this have caused in the area? Let's see by recalculating the area using 6.3 m instead of 6.2 m.

$$\text{area} = 6.3 \text{ m} \times 7.00 \text{ m} = 44.1 \text{ m}^2$$

Notice that this error in the measured length would cause a change in the second

"Scientific" calculators allow you to enter numbers in scientific notation. If you have this kind of calculator, read Appendix A or the direction booklet for your calculator to be sure you can correctly enter these numbers.

digit of the answer (it changes from a 3 to a 4). An uncertainty in the second digit of the length causes an uncertainty in the second digit in the answer.

The length of the other dimension, 7.00 m, has an implied uncertainty of about ±0.01 m. If the 7.00 m were in error by this amount, instead of the other measurement, how much would this influence the answer? As before, let's recalculate the area, this time using 7.01 m.

$$\text{area} = 6.2 \text{ m} \times 7.01 \text{ m} = 43.5 \text{ m}^2$$

An error in the third digit of 7.00 m causes a change in the third digit of the answer, so if only the 7.00 m were in error, the answer would have its uncertainty in the third digit.

Because it is possible for *either* the 6.2 m or the 7.00 m to be in error, we really can't be sure of the area to anything better than the nearest ±1 m². In other words, to be on the safe side, we must assume the worst and round off the answer to two significant figures, which gives a value of 43 m². Notice that there are two significant figures in the answer, which is the same as the number of significant figures in the least precise factor in the calculation, 6.2 m.

The analysis that we have just performed leads to a general rule of thumb that we apply to calculations that involve either multiplication or division.

> *For multiplication or division, the product or quotient should not have more significant figures than are present in the least precise factor in the calculation.*

As an illustration, consider the calculation below, in which we will assume that all the numbers are measured values. A calculator gives the answer shown.

$$\frac{21.95}{3.62 \times 4.5} = 1.347452425$$

The least precise factor (i.e., the one with the fewest number of significant figures) is the value 4.5, which contains two significant figures. The answer, therefore, should contain no more than two significant figures and should be rounded[2] to 1.3.

For addition and subtraction, the procedure used to determine the number of significant figures in an answer is different. Here, the number we write as the result of a calculation is determined by the figure with the largest amount of uncertainty. For example, consider the following sum:

$$\begin{array}{r} 4.371 \text{ m} \\ +302.5 \quad \text{ m} \\ \hline 306.871 \text{ m (before rounding)} \end{array}$$

The first quantity, 4.371 m, has an uncertainty of ±0.001 m; the second an uncertainty of ±0.1 m. When we add these quantities, we expect the answer to be uncertain by at least ±0.1 m, so we must round the answer in this case to the nearest tenth; it should be reported as 306.9 m. The rule for addition and subtraction, therefore, can be stated as follows.

> *The amount of uncertainty in a sum or difference will be at least as large as the largest uncertainty in any of the terms involved in the calculation.*

For addition and subtraction, we round the answer so it has the same number of decimal places as the quantity with the fewest number of decimal places.

---

[2] When we wish to round off a number at a certain point, we simply drop the digits that follow if the first of them is less than 5. Thus, 6.2317 rounds to 6.23, if we wish only two decimal places. If the first digit after the point of round off is larger than 5, or if it is a 5 followed by other nonzero digits, then we add 1 to the preceding digit. Thus, 6.236 and 6.2351 both round to 6.24. Finally, when the digit after the point of round off is a 5 and no other digits follow the 5, then we drop the 5 if the preceding digit is even and we add 1 if it is odd. Thus, 8.165 rounds to 8.16, but 8.175 rounds to 8.18.

Exact numbers

In some calculation, we use numbers that come from definition (such as 3 ft = 1 yd or 12 in. = 1 ft) or that are the result of a direct count (such as the number of people in a room). These numbers are called **exact numbers** because they contain no uncertainty. We can think of them as having an infinite number of significant figures, and when we perform calculations involving them, we don't need to consider them in applying our rules. For example, suppose we wanted to convert the length 4.27 yd into feet. By the factor-label method, we have

$$4.27 \text{ yd} \times \left(\frac{3 \text{ ft}}{1 \text{ yd}} \right) = 12.8 \text{ ft}$$

three significant figures

Notice that the number of significant figures in the answer is determined solely by the number of significant figures in the *measured* length.

EXAMPLE 1.6. PERFORMING CALCULATIONS INVOLVING SIGNIFICANT FIGURES

PROBLEM: Perform the following calculations and report the answers to the proper number of significant figures:

(a) 9.428 g + 4.26 mL (b) 29.3 cm + 213.87 cm

(c) 144.3 cm² + (2.54 cm × 8.4 cm)

SOLUTION: (a) Using a calculator, we get

$$\frac{9.428 \text{ g}}{4.26 \text{ mL}} = 2.21314554 \text{ g/mL}$$

The number 9.428 g contains four significant figures and 4.26 mL contains three. The result of the calculation must be rounded to 2.21 g/mL because for division the answer should not contain more significant figures than are found in the factor with the fewest of them.

(b) The arithmetic is

$$
\begin{array}{r}
29.3 \text{ cm} \\
+213.87 \text{ cm} \\
\hline
243.17 \text{ cm (before rounding off)}
\end{array}
$$

The quantity 29.3 cm has an uncertainty of ±0.1 cm and 213.87 cm has an uncertainty of ±0.01 cm. The largest of these uncertainties is in the tenths place, so we must round the answer to the nearest tenth of a centimeter, which gives an answer of 243.2 cm.

(c) This part requires a little more thought. Before doing the addition, we must perform the multiplication.

$$2.54 \text{ cm} \times 8.4 \text{ cm} = 21.336 \text{ cm}^2 \text{ (before rounding)}$$

Since 8.4 contains only two significant figures, the product must be rounded to 21 cm². Now we can do the addition.

$$
\begin{array}{r}
144.3 \text{ cm}^2 \\
+\ 21 \quad \text{ cm}^2 \\
\hline
165.3 \text{ cm}^2 \text{ (before rounding)}
\end{array}
$$

The uncertainty in 144.3 is ±0.1, the uncertainty in 21 is ±1. We should therefore round the answer to the nearest ±1 cm², which gives 165 cm².

Note that if rounding off is delayed until the end of the calculation, a slightly different answer is obtained.

$$
\begin{array}{r}
144.3 \\
+\ 21.336 \\
\hline
165.636 \longrightarrow 166 \text{ (rounded)}
\end{array}
$$

This small discrepancy is sometimes called "round-off error." In working problems, if you obtain an answer that differs just slightly from the answer given in the book, don't be overly concerned. It might be just round-off error.

1.5 MATTER AND ITS PROPERTIES

In the preceding sections we've discussed the importance of observations and measurements in the development of science. With this as background, let's take a closer look now at the kinds of things that are the subjects of observations in chemistry.

Matter

All the chemicals that make up our world are examples of *matter*, whether they be found in pencils, books, hamburgers, or people. **Matter** is defined as anything that takes up space and has mass. In establishing this definition, we are very careful to specify *mass* rather than *weight*, even though we often use these two terms as if they were interchangeable.

When we use the term **mass**, we are referring to the *amount* of matter in an object. For any given object, this amount is constant and doesn't depend on where the object is. One of the ways we observe mass is as a resistance to a change in the motion of an object. An object of low mass, such as a Ping-Pong ball, is easy to move; even a light breeze will cause it to roll off a table. But an object of large mass, such as a cement truck, is difficult to get moving, and even a strong wind will not blow it away.

Weight is a measure of the force with which an object of a given mass is attracted by gravity. Unlike mass, the weight of an object is *not* a constant; it can depend quite strongly on where the object is located. For example, on the moon the gravitational attraction is only about one-sixth as great as it is on the earth, so an object on the moon has a weight that is only one-sixth as large as on earth. Even on the earth the gravitational attraction varies slightly from place to place, so the weight of an object is also slightly different from place to place. Because of this, when we want to specify the amount of matter under study, we give its mass rather than its weight.

As you learned earlier, mass is measured in units of grams (or kilograms when the mass is very large). The device that we use for such measurements is called a **balance,** and the use of the balance, oddly enough, is called *weighing*. Figure 1.9 is a drawing of a traditional two-pan balance, which actually functions by comparing weights. An object whose mass is to be measured is placed on one pan and known masses (they are usually called *weights*) are added to the other pan until the pointer comes to the center of the scale. At this point, the contents of both pans are attracted equally by gravity (i.e., they weigh the same) and because gravity is the same for

No one ever talks about "massing" an object.

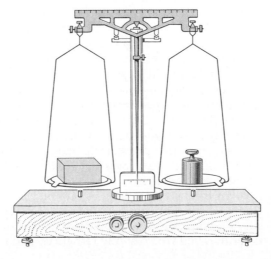

Figure 1.9. *A traditional two-pan balance. Known masses are added to the right pan until the pointer is centered. The contents of both pans have the same weight and therefore also possess the same mass.*

(a)   (b)

Figure 1.10. *Modern single-pan balances.* (a) *A typical top loading balance.* (b) *An analytical balance capable of measurements to the nearest 0.0001 g.*

both, the total of the known masses on one pan equals the mass of the object on the other pan.

Most balances used in modern labs are similar to those shown in Figure 1.10. These are called *single-pan balances* because only one pan is visible. In many single-pan balances, the known masses are located inside the case where they can't be damaged by probing fingers. The knobs move levers that add and remove weights until balance is achieved. The most modern balances operate electronically and actually don't contain weights, although they still must be calibrated against known standards.

Properties of matter

In describing a sample of matter, we specify its characteristics or properties. For example, we might note its **physical state,** which means we would specify whether it is a solid, a liquid, or a gas. We might also specify its mass, which is another measurable property of matter. We use such properties routinely to identify the many things we come across each day. For example, you might recognize your chemistry book by its color, how heavy it is, and by the printing on its cover. In chemistry, we also use the properties of substances for identification. For example, suppose you wished to know what material was used to construct the objects in Figure 1.11. Their shiny surfaces suggest that they're metallic, but the color tells you that they're not made of copper or gold. If you were informed that when a magnet is brought close, the magnet pulls at the objects, you might suspect the objects are made of iron. And if you were told that when the objects are left overnight in the rain, they begin to rust, you would be even more confident that they're made of iron.

Luster, color, magnetic attraction, and the tendency to corrode are just some of the many properties that we use to recognize and classify different samples of matter. These properties themselves can be divided into various broad categories. One way is to divide them into extensive properties and intensive properties.

An **extensive property** is one that depends on the size of the sample being studied. Examples are mass and volume; as the size of the sample increases, so does its mass and volume.

An **intensive property** is one that is independent of the size of the sample. Some examples are physical state, color, melting point, and boiling point. For example, all samples of pure copper are solid at room temperature, have a characteristic color that you would surely recognize, and melt at a temperature of 1083 °C.

Color can sometimes be deceiving, however. For example, metallic silver appears black when the silver particles are very tiny, as in a black and white photograph or negative.

Figure 1.11. *What properties could you use to determine what these objects are composed of?*

One of the interesting things about extensive properties is that their *ratio* is often independent of the sample size, so intensive properties can be created in this way. One such intensive property is called **density,** which is defined as the ratio of an object's mass to its volume.

$$\text{density} = \frac{\text{mass}}{\text{volume}} \qquad (1.3)$$

Liquid water, for example, has a density of 1.00 g/mL. This means that if we had 1.00 g of water, it would occupy 1.00 mL. If we had 20.0 g of water, it would occupy 20.0 mL, but the ratio of mass to volume would still be the same: 20.0 g/20.0 mL = 1.00 g/1.00 mL.

A sample of a liquid might be identified by its density (an intensive property) but not by either its mass or volume alone.

Normally, when a substance is heated or cooled, its volume expands or contracts. This means that the object's mass is packed into either a larger or a smaller volume, so the density actually changes with temperature. For example, at 25 °C (slightly above room temperature) the density of water is 0.9970 g/mL, while at 35 °C its density is 0.9956 g/mL. As you can see, the density doesn't change much with temperature, and it is very useful to remember that *with very little error, we can take the density of water to be 1.00 g/mL without worrying about what the temperature is.* However, you also should keep in mind that for very accurate work, the temperature does matter. Let's look now at some examples dealing with measuring and using density.

EXAMPLE 1.7. CALCULATING DENSITY

PROBLEM: An aluminum bar was weighed and found to have a mass of 14.2 g. Its volume was measured to be 5.26 mL. What is the density of aluminum?

SOLUTION: To calculate the density, you *must* know how it is defined; it's the ratio of mass to volume. Using the symbol d for density, therefore, we obtain

Learning definitions is important. If you don't know the definition of density, how could you hope to solve a problem such as this?

$$d = \frac{\text{mass}}{\text{volume}} = \frac{14.2 \text{ g}}{5.26 \text{ mL}}$$

$$= 2.70 \text{ g/mL}$$

Notice that we've rounded the answer to three significant figures to be consistent with the data.

EXAMPLE 1.8. USING DENSITY IN CALCULATIONS

PROBLEM: A certain copper coin has a mass of 3.14 g. The density of copper is 8.96 g/cm^3. What is the volume of the coin?

ANALYSIS: Let's restate the problem.

$$3.14 \text{ g copper} \Longleftrightarrow ? \text{ cm}^3$$

We have used the symbol $\Longleftrightarrow$ here instead of an equals sign because mass can't really equal volume. We will use this symbol to mean "is equivalent to," and we view the problem as asking, "How many cubic centimeters of copper are equivalent to 3.14 g of copper?"

To solve the problem, we need a relationship between mass and volume, and this is exactly what the density gives us. It tells us the mass of each cubic centimeter (or each milliliter) of the sample. In this case we see that if we had 1.00 cm^3 of copper, its mass would be 8.96 g. The density also tells us that if we had 8.96 g of copper, its volume would be 1.00 cm^3. Anytime you have a relationship like this between two quantities, you can use it to make conversion factors for use in a calculation. For the density, we can make the following two conversion factors:

> The density tells us that 8.96 g of copper is equivalent to 1.00 cm^3 of copper.

$$\frac{8.96 \text{ g copper}}{1.00 \text{ cm}^3} \quad \text{or} \quad \frac{1.00 \text{ cm}^3}{8.96 \text{ g copper}}$$

SOLUTION: Now that we've figured out how to get the answer, we must simply be careful to make sure the units cancel, and this means we must use the second of the two conversion factors above.

$$3.14 \text{ g copper} \times \left(\frac{1.00 \text{ cm}^3}{8.96 \text{ g copper}} \right) \Longleftrightarrow 0.350 \text{ cm}^3$$

The volume of the coin, therefore, is 0.350 cm^3 (or 0.350 mL).

A property that's closely related to density is **specific gravity** (often abbreviated *sp. gr.*). It is defined as the ratio of the density of a substance to the density of water.[3]

$$\text{sp. gr.} = \frac{d_{\text{substance}}}{d_{\text{water}}}$$

Because of the way specific gravity is defined, it is taken to be a unitless quantity, one that simply describes how the density of the substance compares to the density of water. Thus, a substance with a specific gravity of 2.00 is twice as dense as water.

The usefulness of specific gravity is that it allows us to compute the density of a substance in a wide variety of units by simply multiplying the specific gravity of the substance by the density of water expressed in the desired units. In this way, two tables — one containing the specific gravities of substances and the other containing

---

[3] In reporting specific gravity, the usual practice is to specify the temperatures at which the densities of the substance and the water were measured.

EXAMPLE 1.9. FINDING DENSITY FROM SPECIFIC GRAVITY

PROBLEM: Hexane, a solvent used in rubber cement, has a specific gravity of 0.668. What is the density of hexane in units of grams per milliliter and pounds per gallon, given that the density of water in these units is 1.00 g/mL and 8.34 lb/gal.

ANALYSIS: Let's begin with the definition. (Once again, you must know the definitions of quantities; otherwise, you will never know how to use them in problems!)

$$(\text{sp. gr.})_{\text{substance}} = \frac{d_{\text{substance}}}{d_{\text{water}}}$$

Since we want the density of the substance (hexane), let's solve for this quantity.

$$d_{\text{hexane}} = (\text{sp. gr.})_{\text{hexane}} \times d_{\text{water}}$$

All we have to do is multiply the specific gravity of hexane by the density of water in the desired units.

SOLUTION: For the density in units of g/mL,

$$d_{\text{hexane}} = (0.668) \times (1.00 \text{ g/mL})$$
$$= 0.668 \text{ g/mL}$$

(Notice that in units of g/mL, the numerical values of density and specific gravity are the same. If you have the specific gravity, just add the units "g/mL" to get the density.)
    For the density in units of lb/gal,

$$d_{\text{hexane}} = (0.668) \times (8.34 \text{ lb/gal})$$
$$= 5.57 \text{ lb/gal}$$

(Notice that for units other than g/mL, we must do some arithmetic.)

the density of water in a variety of units — take the place of the many tables that would otherwise be needed to express the densities of substances in all those different units.

Physical and chemical properties

This is another way of classifying properties. A **physical property** is one that can be observed without changing the chemical makeup of a substance. An example is melting point. For instance, water (ice) melts at 0 °C. This is a physical property of water. To measure the melting point, we use a thermometer to determine the temperature at which solid water changes to liquid water. This change, which is called a **physical change,** does not alter the chemical makeup of the water. Both the liquid and solid are water. Density is another example of a physical property. To determine the density of water, we measure the mass of a given volume of the liquid. Neither of these measurements changes the water into another substance; in fact, they don't change the water at all.

A **chemical property** is the tendency of a substance to undergo a particular chemical reaction. For instance, a chemical property of water is that it reacts violently with sodium, as shown in Figure 1.12, to give gaseous hydrogen and a substance called sodium hydroxide (commonly known as lye). As we observe this chemical property, the water and sodium undergo a change, a **chemical change,** that produces other substances. After we have observed the chemical property, the water and sodium are gone and these other substances have taken their place.

Figure 1.12. *Sodium reacts quite vigorously with water, liberating flammable hydrogen gas and forming sodium hydroxide, which is commonly known as lye.*

1.6 ELEMENTS, COMPOUNDS, AND MIXTURES

In the previous section, we described how the various properties of matter are categorized. Scientists have also found that matter itself falls naturally into three classifications — those that form the title to this section. Elements, compounds, and mixtures are the kinds of materials you will work with in the laboratory, so it is important that you know what they are and how to distinguish among them.

Elements are the simplest substances ever produced in chemical reactions.

Elements are substances that cannot be decomposed into simpler substances by chemical reactions. They are therefore the simplest kinds of matter that can exist under conditions that we normally find in the chemistry lab and they are the simplest forms of matter that we deal with directly. Elements serve as the building blocks for all the more complex substances that we encounter, from common table salt to extremely complex proteins. All are composed of a limited set of elements. At present, there are 108 known elements, but only a much smaller number will be of real interest to us in this course.

Elements combine with each other in chemical reactions to form compounds. A **compound** is a substance that is composed of two or more elements, and for each individual compound its elements are always present in the same proportions by mass. For instance, you probably know that the compound water is composed of two elements, hydrogen and oxygen. All samples of pure water, regardless of their source, contain these two elements in the proportion of one part by mass of hydrogen to eight parts by mass of oxygen (for example, 1.0 g hydrogen to 8.0 g oxygen). When hydrogen reacts with oxygen to form water, they always combine in this same proportion by mass. If 1.0 g of hydrogen reacts, exactly 8.0 g of oxygen reacts — no more and no less. It is also important to understand that when elements combine to form a compound, the elements lose their individual identity and the properties that are observed are those of the compound. For example, at room temperature hydrogen and oxygen are gases, but water is a liquid. Nearly all their other properties also differ.

In nature, there are very few materials that even approach being pure compounds or elements; nearly everything is a mixture.

Elements and compounds are considered to be *pure substances* because of their constancy of composition. **Mixtures,** on the other hand, can be of *variable composition.* For example, both water and sodium chloride are compounds because they have constant compositions from one sample to another. But salt can be dissolved in water in varying amounts to give many mixtures with a wide range of compositions.

Mixtures can be described as being either homogeneous or heterogeneous. A **homogeneous mixture** is called a **solution** and has uniform properties throughout. This means that if we were to examine a portion of a solution of sodium chloride in water, we would find that it has the same properties as any other portion of the same solution. We could also say that the solution consists of a single *phase.* Thus, a **phase** is defined as any part of a system that has a particular set of uniform properties and composition.

A **heterogeneous mixture** is not uniform. An example is oil and water (Figure 1.13). If we were to sample one portion of the mixture, we might find it has the properties of oil, while some other part of the mixture would have the properties of water. This mixture consists of two phases — the oil and the water. If we shake the mixture so the oil is dispersed as small droplets throughout the water (as in a salad dressing), all the oil droplets taken together still constitute just *one* phase, because the oil in one droplet has the same properties and composition as the oil in any other droplet. If we were to then add an ice cube to this "brew," we would have three phases — the ice (a solid), the water (a liquid), and the oil (another liquid). In all these examples, we can detect the presence of two or more phases because a distinct boundary exists between them.

One way to distinguish between a pure substance and a mixture of substances is to measure the melting point. The temperature of a pure substance will stay constant as it melts. Ice, as you know, melts at 0 °C and this temperature remains the same

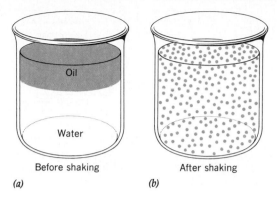

Figure 1.13. *Oil and water form a heterogeneous mixture.* (a) *Two phases are clearly evident.* (b) *After shaking, there are still two phases, the oil and the water.*

until all the ice has melted. However, if we melt something that's a mixture of substances, usually the temperature will change gradually as the solid is converted entirely to a liquid. This difference in behavior can often be used as a test to determine whether a material is pure or not; if the temperature stays constant while it melts, it is pure, but if the temperature changes while the solid melts, the sample is impure and is a mixture.

There is another significant way that mixtures differ from elements and compounds. When a mixture is prepared, the chemical properties of the components (and often their physical properties, too) do not change. When elements combine to form a compound, however, profound changes in properties occur. For example, copper and sulfur are two elements. Copper, of course, is a reddish-colored metal, a good conductor of electricity, and is relatively resistant to corrosion. Sulfur is a yellow nonmetallic solid, shown in powdered form in Figure 1.14. A mixture of sulfur and copper is easily prepared, but in the mixture (Figure 1.14*b*) we can still see evidence of the properties of copper and the properties of sulfur. The formation of the mixture has involved a physical process—a process that doesn't alter the chemical characteristics of the components.

When a mixture of copper and sulfur is heated, a chemical reaction occurs. The sulfur and copper combine to form a compound, and as we've seen before, this is accompanied by very noticeable changes in properties. After the reaction is over, we can't find anything that has the properties of either sulfur or copper. In their place is a new substance, called copper sulfide, with new properties (Figure 1.14*c*). It doesn't conduct electricity; it doesn't have the color of either copper or sulfur; it has a density that is different from that of either copper or sulfur; and its chemical properties are entirely different, too. As noted earlier, such changes are what characterize chemical reactions. The relationships among elements, compounds, and mixtures are summarized in Figure 1.15.

Figure 1.14. (a) *Alongside a crucible and its cover we see a coil of red-colored copper wire and yellow powdered sulfur.* (b) *When mixed in the crucible, the copper and sulfur retain their individual properties.* (c) *When the mixture of copper and sulfur is heated, a reaction takes place and a new substance called copper sulfide is formed. The copper sulfide has the same shape as the coiled copper wire from which is was formed, but notice that its properties differ from those of both the copper and the sulfur.*

(a)

(b)

(c)

Figure 1.15. *Classification of matter.*

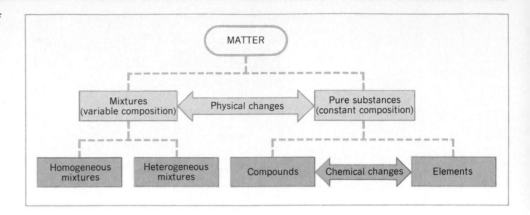

1.7 SYMBOLS, FORMULAS, AND EQUATIONS

In a certain sense, learning chemistry is like learning a foreign language such as Greek (or Russian, if you happen to be Greek). We have chemical symbols for the elements, which can be likened to an alphabet. The formulas we write using these symbols are like words, and the chemical equations are like sentences. As in learning any new language, we must begin with the alphabet.

Each of the elements has been assigned a **chemical symbol** that we can think of as a shorthand way of representing the element. The symbols consist of one or two letters that usually bear a resemblance to the English name of the element. For instance, carbon = C, chromium = Cr, chlorine = Cl, calcium = Ca, and zinc = Zn. Notice that the first letter is capitalized, but if there is a second letter, it is not. Some of the elements have symbols that do not seem to correspond at all to their English names. In almost all cases, these elements have been known since the very early days of chemistry when Latin was the universal language among scientists, and their symbols are derived from their Latin names. Some examples are potassium (L. *kalium*) = K, sodium *(natrium)* = Na, silver *(argentum)* = Ag, mercury *(hydrargyrum)* = Hg, and copper *(cuprum)* = Cu.

An alphabetical list of the elements and their chemical symbols appears on the inside front cover of the book. You will also find there a chart called the periodic table, which contains the symbols for the elements arranged in a numerical order that we will explain later. You will not need to know the symbols for all the elements, but you should begin to learn the important ones by studying the names and symbols for the first 20 elements in the periodic table.

A chemical compound is represented symbolically by its **chemical formula,** which is assembled using chemical symbols. For example, water is represented by H_2O, carbon dioxide by CO_2, methane (natural gas) by CH_4, and aspirin by $C_9H_8O_4$. Chemical formulas also show the quantitative compositions of substances. Here we take the chemical symbol to stand for the smallest particle of the element, an **atom.** (We will have a lot more to say about atoms later.) The subscripts in a formula give the relative numbers of atoms of each element in the compound (and when no subscript is written, a subscript of 1 is implied). The formula H_2O, for instance, describes a substance containing two hydrogen atoms for every oxygen atom. Similarly, the compound CH_4 contains one atom of carbon for every four atoms of hydrogen.

Often, two or more atoms are able to join tightly together so that they behave as a single particle called a **molecule.** If the atoms are of different elements, as in water (H_2O) or methane (CH_4), it is a molecule of a compound. If the atoms are of the *same* element, it is a molecule of an element. Some common and important elements that occur in nature as molecules composed of two atoms are hydrogen, H_2; oxygen, O_2,

Your teacher may add to the number of elements whose names and symbols you should memorize.

Molecules that contain just two atoms are called *diatomic* molecules.

(a)   *(b)*

Figure 1.16. (a) *Blue crystals of copper sulfate, $CuSO_4 \cdot 5H_2O$.* (b) *When heated, the blue crystals lose water. Here we see the nearly white powder of $CuSO_4$ that remains.*

nitrogen, N_2; fluorine, F_2; chlorine, Cl_2; bromine, Br_2; and iodine, I_2. Be sure you learn the formulas for these, because you will encounter them many times throughout this course.

Some chemical formulas are more complex than those described above and contain parentheses. An example is the compound ammonium sulfate (a fertilizer), which has the formula $(NH_4)_2SO_4$. The subscript "2" outside the parentheses specifies the presence of two NH_4 units — a total of two nitrogens and eight hydrogens. This means we could also write the formula $N_2H_8SO_4$, although we will see later that there are good reasons for using the parentheses.

There are certain substances that form crystals that contain water molecules when their aqueous solutions are evaporated. These crystals are called **hydrates.** For example, copper sulfate, an agricultural fungicide, forms blue crystals that have five water molecules for each copper sulfate ($CuSO_4$). Its formula is written $CuSO_4 \cdot 5H_2O$. If the blue crystals are heated, water can be driven off to leave pure $CuSO_4$, which appears almost white. (See Figure 1.16).

A **chemical equation** is written to show the chemical changes that occur during a chemical reaction. In a sense, it's a "before and after" description of the reaction. For example, the equation

$$Zn + S \longrightarrow ZnS$$

describes the reaction shown in Figure 1.17, in which zinc (Zn) reacts with sulfur (S) to produce zinc sulfide (ZnS), a substance used on the inner surfaces of TV screens. The substances on the left of the arrow are called the **reactants** and are the chemicals present before the reaction takes place. Those on the right of the arrow are called the **products** and are the substances present after the reaction is over. (In the reaction above, there is only one product.) The arrow is read as "react to yield" or simply "yield." Thus, the equation above can be read as "zinc plus sulfur react to yield zinc sulfide," or "zinc plus sulfur yield zinc sulfide," or "zinc reacts with sulfur to yield zinc sulfide."

Sometimes it is desirable, or even necessary, to indicate whether the reactants and products are solids, liquids, gases, or dissolved in a solvent such as water. This is done by placing the letters s = solid, l = liquid, g = gas, or aq = aqueous (water) solution in parentheses following the formulas of the substances in the equation. For instance, the equation

$$CaCO_3(s) + H_2O(l) + CO_2(g) \longrightarrow Ca(HCO_3)_2(aq)$$

describes a reaction between solid calcium carbonate (limestone), liquid water, and gaseous carbon dioxide to yield an aqueous solution of calcium bicarbonate. This is

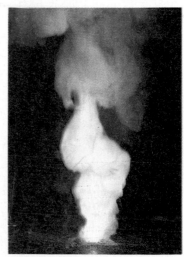

Figure 1.17. *Once ignited, a mixture of zinc and sulfur reacts rapidly to form zinc sulfide, ZnS.*

the reaction that's responsible for the dissolving of limestone by groundwater containing dissolved carbon dioxide. It is one of the causes of "hard water" and the formation of limestone caverns.

Many of the equations that we will write will contain numbers, called **coefficients,** preceding the chemical formulas. An example is the reaction of hydrogen (H_2) with oxygen (O_2) to form water.

$$2H_2 + O_2 \longrightarrow 2H_2O$$

We interpret this equation to mean that two hydrogen molecules plus one oxygen molecule (a coefficient of 1 is assumed if none is written) react to yield two molecules of water. Such an equation is said to be **balanced** because it contains the same number of atoms of each element on both sides of the arrow. You will begin to learn how to balance equations in Chapter 3. In anticipation of this, you should begin to learn now how to count atoms in formulas and equations.

EXAMPLE 1.10. COUNTING ATOMS IN A CHEMICAL EQUATION

PROBLEM: The combustion of butane, the liquid fuel that can be seen sloshing around inside a disposable cigarette lighter, follows the chemical equation

$$2C_4H_{10} + 13O_2 \longrightarrow 8CO_2 + 10H_2O$$

How many atoms of oxygen are included among the product molecules?

ANALYSIS: We have to remember two things when we count the atoms. First, we have to count how many atoms of oxygen are in each of the formulas among the products. (In doing this, we keep in mind that when no subscript is written, a subscript of 1 is assumed.) Then we have to multiply this by the number of each of these molecules as specified by the coefficients.

SOLUTION: Both the CO_2 and H_2O contain oxygen atoms. One molecule of CO_2 contains two atoms of oxygen, so 8 molecules contain a total of 16 oxygens. Similarly, one H_2O contains one oxygen, so 10 water molecules contain a total of 10 oxygens. Adding all the oxygens together gives a grand total of 26. (Notice that this is the same number of oxygens as in 13 molecules of O_2; the equation is balanced.)

REVIEW QUESTIONS AND PROBLEMS

Problems whose numbers are in blue have their answers in Appendix D at the back of the book. The more difficult problems are marked with asterisks.

General Questions

1.1 What is a chemical? Where are chemicals found in nature?

1.2 What is it about chemicals that the science of chemistry studies?

1.3 When sodium and chlorine are brought together, how do we know that a chemical reaction occurs between them?

1.4 The rusting of iron involves a slow chemical reaction in which iron is changed into a substance called iron oxide. What *observable* changes allow you to know when this reaction has occurred?

Scientific Method

1.5 What is the difference between a theory and a law?

1.6 Describe how the actions of a doctor in treating an ill patient follow the scientific method.

1.7 What distinguishes a qualitative observation from a quantitative one? Which kind is usually more useful? Why?

1.8 Define *data.*

Units and Conversions

1.9 For each of the following quantities, give the name and symbol of the SI base unit.
(a) mass   (d) length
(b) time   (e) amount of substance
(c) temperature

1.10 In English units, we specify fuel economy in units of *miles per gallon*. What would be the SI derived unit for fuel economy?

1.11 The newton (N) is the SI derived unit for force. Force is mass times acceleration and acceleration has the dimensions of (distance) ÷ (time)^2. What are the units for the newton in terms of the SI base units? (A newton is a force about equal to the weight of a large lemon.)

1.12 What power of ten is implied by the following SI prefixes? What symbol is used to stand for each of them? (Try to answer this question without referring to Table 1.2.)
(a) pico   (d) nano
(b) mega   (e) kilo
(c) centi

1.13 Give the symbol for each of the following units.
(a) milligram   (d) microsecond
(b) decimeter   (e) centigram
(c) kilosecond

1.14 Fill in the blanks with the correct number.
(a) 3.4 kg = _____ g
(b) 5.7 ns = _____ s
(c) 4.0 Mmol = _____ mol
(d) 6.4 mg = _____ g
(e) 7.2 cm = _____ m

1.15 Fill in the blanks with the correct unit.
(a) 3.2 _____ = 3.2×10^{-9} m
(b) 42 _____ = 4.2×10^{-2} m
(c) 7.3 _____ = 0.0073 g
(d) 12.5 _____ = 125 mm
(e) 3.5 _____ = 3.5×10^{-3} mL
(f) 0.84 _____ = 840 cm

1.16 Use the factor-label method to perform the following conversions.
(a) 1.40 m to cm   (f) 1885 nm to μm
(b) 2855 mm to m   (g) 0.385 kg to mg
(c) 185 mL to L   (h) 48.5 cm^2 to mm^2
(d) 0.0253 L to mL   (i) 143 mL to mm^3
(e) 195 mm to cm   (j) 345 cm^3 to L

1.17 The SI units for mass and volume are the kilogram and cubic meter. However, in the laboratory we usually use the units gram and milliliter. Why?

1.18 Use the factor-label method to perform the following conversions.
(a) 112 lb to kg   (d) 48 mi/hr to cm/s
(b) 45 m/s to mi/hr   (e) 15 yd^2 to m^2
(c) 1.00 mi^3 to m^3

1.19 After shopping for a sports car, a certain wealthy freshman decided to choose between a new two-passenger, six-cylinder Smokebelcher, which gives 21 mi/gal, and an old 1974, eight-cylinder, 10-passenger Pferdburper (with automatic transmission), which the owner guaranteed would deliver 10 km/L. On the basis of gas mileage, which car would be more economical to operate?

1.20 A student has just returned from Germany with a car purchased while on vacation. The speedometer is calibrated in kilometers per hour (km/hr). Driving away from the pier, the student notices a sign that posts the speed limit at 35 mi/hr. What is the maximum speed that can be reached, in km/hr, without having to worry about receiving a speeding ticket?

1.21 After receiving a speeding ticket, the student in the preceding question is informed by the police officer that the courthouse is located "4.3 miles straight down the road. You can't miss it." The odometer in the student's car measures kilometers (km). How far must the student travel, in kilometers, before arriving at the courthouse?

1.22 In the country of Ferdovia, the Ferds thrive on potatoes. The average Ferd earns 142 thrubs (the local currency of Ferdovia) per week and spends approximately $\frac{1}{14}$ of his yearly income on potatoes. If potatoes cost 2 thrubs per pound, how many pounds of potatoes does the average Ferd consume each year?

1.23 Why are the freezing point and boiling point of water chosen as reference temperatures for the definition of the Fahrenheit and Celsius temperature scales?

1.24 What special name is given to the temperature 0 K?

1.25 Perform the following conversions:
(a) 50 °F to °C   (d) −40 °C to °F
(b) 25 °C to °F   (e) 0 K to °F
(c) 80 K to °C

1.26 Gallium metal has one of the largest liquid ranges of any element. It melts at 30 °C and boils at 1983 °C. What are its melting and boiling points in degrees Fahrenheit?

1.27 Tungsten, used as filaments in electric light bulbs, has a melting point of 6152 °F. What is its melting point expressed in degrees Celsius and in kelvins?

1.28 Solid carbon dioxide (dry ice) has a temperature of −78 °C. What is its temperature on the Fahrenheit scale? What is its temperature on the Kelvin scale?

1.29 Normal human body temperature is 98.6 °F. In Europe, and in many hospitals in the United States, the Celsius temperature scale is used almost universally so that clinical thermometers are calibrated in °C. If you had one of these thermometers, what temperature would you expect, in °C, for a normal healthy individual? If your temperature registered 39 °C, what would your temperature be in °F?

*1.30 Naphthalene (used in mothballs) has a melting point of 80 °C and a boiling point of 218 °C. Suppose that this substance was used to define a new temperature scale on which the melting point of naphthalene was 0 °N and its boiling point was 100 °N. What would be the freezing point and boiling point of water in °N? What general equation could we use to relate temperatures in °C and °N?

Measurement and Significant Figures

1.31 Why is it important for scientists to be concerned about reporting the correct number of significant figures in their measured and calculated quantities?

1.32 Define in your own words the terms *precision* and *accuracy*.

1.33 How many significant figures are there in the following measured quantities?
(a) 1.0370 g   (d) 100.1 °C
(b) 0.000417 m   (e) 9.0010 g
(c) 0.00309 cm

1.34 Speed is defined as a ratio: distance traveled divided by the length of time required to go that distance (speed = distance/time). Suppose a car travels 346.2 mi in 6.27 hr.
(a) What is its average speed in mi/hr? (Write all the digits given by your calculator.)
(b) What is the minimum uncertainty in the distance traveled?
(c) What is the minimum uncertainty in the time?
(d) Show, by applying these uncertainties to the measured values, that the calculated speed should contain only three significant figures.

1.35 Express each of the following in scientific notation. Assume that any digits to the right of the last nonzero digit are *not* significant figures.
(a) 1250 g
(b) 13,000,000 m
(c) 60,230,000,000,000,000,000,000 atoms
(d) 214,570 mg
(e) 31.47 g

1.36 Express each of the following in scientific notation:
(a) 0.00040 °C   (d) 0.0000328 g
(b) 0.0000000003 km   (e) 0.00000000000091 m
(c) 0.002146 g

1.37 Write the following numbers in decimal notation:
(a) 3×10^{10} m   (d) 3.4×10^{-7} g
(b) 2.54×10^{-5} m (e) 0.0325×10^{6} cm
(c) 122×10^{-2} g

1.38 The length of a piece of land was measured to be 3000 m. Using scientific notation, express the measurement
(a) to two significant figures
(b) to three significant figures
(c) in cm, to two significant figures

1.39 Round the following numbers to the nearest tenth:
(a) 14.68 cm   (d) 33.45 cm
(b) 18.32 cm   (e) 8.350 cm
(c) 27.452 cm

1.40 Perform the following arithmetic and round the answers to the proper number of significant figures. Assume all the values come from measurements.
(a) 2.41 cm $\times$ 3.2 cm
(b) 4.025 m $\times$ 18.2 m
(c) 81.4 g $\div$ 104.2 cm^3
(d) 3.476 g + 0.002 g
(e) 81.4 g − 0.002 g

1.41 Perform the following arithmetic and express your answers in scientific notation rounded to the proper number of significant figures. Assume all the values come from measurements.
(a) $(2.047 \times 10^8 \text{ m}) + (14.33 \times 10^8 \text{ m})$
(b) $(12.4 \times 10^8 \text{ m}) + (92.7 \times 10^7 \text{ m})$
(c) $(42.003 \times 10^5 \text{ m}) - (3.25 \times 10^3 \text{ m})$
(d) $118.45 \text{ mol} - (0.033 \times 10^3 \text{ mol})$
(e) $1.00 \text{ °C} + (3.75 \times 10^{-8} \text{ °C})$

1.42 Perform the following arithmetic and round the answers to the proper number of significant figures. Assume all the values come from measurements.
(a) $(341.7 \text{ cm}^2 - 22 \text{ cm}^2)$
    $+ (0.00224 \text{ cm} \times 814{,}050 \text{ cm})$
(b) $(82.7 \text{ cm}^2 \times 143 \text{ cm}) + (274 \text{ cm}^3 - 0.00653 \text{ cm}^3)$
(c) $(3.53 \text{ cm}^3 \div 0.084 \text{ cm}) - (14.8 \text{ cm} \times 0.046 \text{ cm})$
(d) $(324 \text{ m} \times 0.0033 \text{ m}) + (214.2 \text{ m} \times 0.0225 \text{ m})$
(e) $(4.15 \text{ mm} + 82.3 \text{ mm}) \times (0.024 \text{ mm} + 3.000 \text{ mm})$
(f) $0.2510 \text{ m} \times (15.50 \text{ m} - 12.75 \text{ m})$

1.43 Perform the following arithmetic and express the answers with the correct units. Report the answers in scientific notation rounded to the proper number of significant figures. Assume all the values come from measurements.
(a) $(12.45 \times 10^6 \text{ cm}^2) \div (2.24 \times 10^3 \text{ cm})$
(b) $822 \text{ m} \div 0.028 \text{ hr}$
(c) $(635.4 \times 10^{-5} \text{ cm}) \div (42.7 \times 10^{-4} \text{ s})$
(d) $(31.3 \times 10^{-12} \text{ m}) \div (8.3 \times 10^{-6} \text{ m/s})$
(e) $(0.74 \times 10^{-9} \text{ mol}) \div (825.3 \times 10^{18} \text{ m}^3)$

1.44 Perform the following arithmetic and express the answers with the correct units. Report the answers in scientific notation rounded to the proper number of significant figures. Assume all the values come from measurements.
(a) $(8.3 \times 10^{-6} \text{ km}) \times (4.13 \times 10^{-7} \text{ km})$
    $\div 5.411 \times 10^{-12} \text{ km})$
(b) $[(3.125 \times 10^{-6} \text{ km/s}) + (5.127 \times 10^{-5} \text{ km/s})]$
    $\times (6.72 \times 10^8 \text{ s})$
(c) $[14.39 \text{ m}^2 + (2.43 \times 10^1 \text{ m}^2)] \div 1275 \text{ m}$
(d) $[(1.583 \times 10^{-2} \text{ km}) - (0.00255 \text{ km})]$
    $\div [142.3 \text{ s} + (0.257 \times 10^2 \text{ s})]$
(e) $(0.00425 \text{ g}) \div [(0.0008137 \text{ cm}^3)$
    $+ (2.65 \times 10^{-5} \text{ cm}^3)]$

1.45 One mile is 5280 ft (1 mi = 5280 ft). How many significant figures are in each of the numbers in this relationship?

1.46 How many inches are there in the measured length 0.2263 mi?

Properties of Matter

1.47 What is the difference between an extensive and an intensive property? Can you think of any examples not mentioned in the text?

1.48 Consider two samples of matter, one a copper penny and the other a piece of window glass. Give three properties that would be different for the two samples. Are there any properties in which they are alike?

1.49 How is mass different from weight? Why do we use mass instead of weight to specify the amount of matter in a given sample?

1.50 Define (a) *physical property* and (b) *chemical property*. How does a physical change differ from a chemical change?

1.51 One of the substances found in gasoline is called heptane. From what you know about gasoline,
(a) suggest one physical property of heptane.
(b) suggest one chemical property of heptane.

Density and Specific Gravity

1.52 What is the difference between density and specific gravity? What are their common metric units?

1.53 A block of magnesium had a mass of 14.3 g and a volume of 8.46 cm³. What is the density of magnesium in g/cm³?

1.54 Water was placed into a graduated cylinder until the volume read 25.0 mL. An irregularly shaped piece of metal weighing 50.8 g was placed in the cylinder and completely submerged. The water level rose to the 36.2-mL mark. What is the density of the metal in g/mL?

1.55 Titanium is an important structural metal used in aircraft because of its strength and light weight. A solid cylinder of titanium 2.48 cm in diameter and 4.75 cm long was weighed and found to have a mass of 104.2 g. Calculate the density of titanium in g/cm³.

1.56 Lead is a well-known "heavy" metal. It has a density of 11.35 g/cm³.
(a) What is the mass of 12.0 cm³ of lead?
(b) What is the volume in cm³ occupied by 155 g of lead?

1.57 Chloroform, $CHCl_3$, a liquid once used as an anesthetic, has a density of 1.492 g/mL.
(a) What is the volume in mL of 10.00 g of $CHCl_3$?
(b) What is the mass of 10.00 mL of $CHCl_3$?

1.58 A glass vessel that can be repeatedly filled with precisely the same volume of liquid is called a **pycnometer.** A certain pycnometer, when empty and dry, weighed 25.296 g. When filled with water at 25 °C, the pycnometer and water weighed 34.914 g. When filled with a liquid of unknown composition, the pycnometer and its contents weighed 33.485 g (also at 25 °C). At 25 °C the density of water is 0.9970 g/mL.
(a) What is the volume of the pycnometer in mL?
(b) What is the density of the unknown liquid in g/mL?

1.59 Suppose that you were going to prepare a table of densities for 100 substances in which the density of each substance is to be expressed in five sets of units. How many numerical values must appear in this table? How many numerical values would be needed to express this same information using specific gravities and the density of water in different units?

1.60 The density of water is 8.34 lb/gal. Isopropyl alcohol, sold in drugstores as rubbing alcohol, has a density of 6.56 lb/gal.
(a) What is the specific gravity of isopropyl alcohol?
(b) What is the density of isopropyl alcohol in the units g/mL?

1.61 Propylene glycol, a substance used in nontoxic antifreeze for freshwater systems of campers and boats, has a specific gravity of 1.04. Water has a density of 8.34 lb/gal. What would be the weight in lb of the contents of a 10,000-gallon tank car filled with propylene glycol?

1.62 Chemical Bond (the scientifically inclined brother of James) was faced with a chilling choice of beverage. Before him were three beakers containing clear colorless liquids. One was water, but the other two were fatally poisonous. Unaware of Chemical's scientific background, his foes thought they would give him a sporting chance in choosing which liquid to drink. He was told that the first beaker contained 275 mL of liquid, which weighed 0.275 kg. The second contained 245 mL of liquid, which weighed 389 g. The third contained 265 mL of liquid, which weighed 2.99×10^5 mg. Which liquid should our daring detective have chosen?

Elements, Compounds, and Mixtures

1.63 How does an element differ from a compound? How are elements and compounds different than mixtures?

1.64 How is a physical change different than a chemical change?

1.65 What is the definition of a solution? How many phases can be present in a solution?

1.66 Two samples of the mineral quartz obtained from different locations were analyzed. One sample was found to consist of 3.44 g of silicon and 3.91 g of oxygen. The other consisted of 6.42 g of silicon and 7.30 g of oxygen. Do these data suggest that quartz is a compound or a mixture?

1.67 There are many examples of homogeneous and heterogeneous mixtures in the world around us. How would you classify seawater, air (unpolluted), smog, smoke, club soda, black coffee, a ham sandwich?

1.68 Identify the phases that exist in a copper pan containing two iron nails, a quart of water, and four glass marbles.

1.69 A chemist was asked to determine whether a solid sample was a mixture or a pure substance. The chemist measured the melting point and found that the sample began to melt at 143 °C, but by the time the entire sample had melted, the temperature had risen to 154 °C. What conclusion could the chemist make based on these observations?

1.70 Suppose a mixture contained salt, copper powder, and iron filings. Describe how you could separate and isolate the components of this mixture.

Symbols, Formulas, and Equations

1.71 Write chemical symbols for the following:
(a) iron
(b) sodium
(c) potassium
(d) phosphorus
(e) bromine
(f) calcium
(g) nitrogen
(h) neon
(i) manganese
(j) magnesium

1.72 Give the name of each of the following elements:
(a) Ag
(b) Cu
(c) S
(d) Cl
(e) Al
(f) Au
(g) Cr
(h) W
(i) Ni
(j) Hg

1.73 How many atoms of each kind are represented in the following formulas:
(a) K_2S
(b) Na_2CO_3
(c) $K_4Fe(CN)_6$
(d) $(NH_4)_3PO_4$
(e) $Na_3Ag(S_2O_3)_2$

1.74 Plaster is composed of calcium sulfate ($CaSO_4$), which exists as a hydrate containing two water molecules for each $CaSO_4$. Write the formula for the hydrate.

1.75 Potassium alum (often just called alum) is used medically as an astringent, a substance that shrinks and drives blood from tissue. Its formula is $KAl(SO_4)_2 \cdot 12H_2O$. How many atoms of each element are represented by this formula?

1.76 Which of the following equations are not balanced?
(a) $ZnCl_2 + NaOH \rightarrow Zn(OH)_2 + NaCl$
(b) $CuCO_3 + 2HCl \rightarrow CuCl_2 + CO_2 + H_2O$
(c) $Fe_2O_3 + 2CO \rightarrow 2Fe + 2CO_2$
(d) $NH_4NO_3 \rightarrow N_2O + 2H_2O$

Additional Exercises

1.77 Perform the following arithmetic and round the answer to the correct number of significant figures: $(3.6 \times 10^4 \text{ m}) + (5.6 \times 10^7 \text{ cm})$.

1.78 Gasoline has a specific gravity of 0.684. How many pounds of gasoline fill an 18.5-gallon fuel tank?

1.79 When an object floats in water, it displaces a volume of water that has a weight equal to the object's weight. If a ship has a weight of 4255 tons, how many cubic feet of seawater will it displace? The specific gravity of seawater is 1.025. 1 ton = 2000 lb.

1.80 A sample of a metal was found to have a mass of 28.31 kg and a volume of 3.932×10^{-3} m³. Which of the following metals might the sample be?
(a) iron (density = 7.86 g/cm³)
(b) nickel (density = 8.90 g/cm³)
(c) chromium (density = 7.20 g/cm³)

1.81 In 1987 the scientific community was astounded to learn of a development of enormous practical potential when it was announced at a meeting of the American Physical Society that a material had been discovered that becomes superconducting (meaning it loses all resistance to the flow of electricity) at a temperature of about 90 K, which is about 13 K higher than the boiling point of liquid nitrogen. The only other known superconducting materials had to be cooled to near absolute zero to achieve superconducting properties. What Celsius and Fahrenheit temperatures correspond to 90 K? What is the boiling point of liquid nitrogen in °C and in kelvins?

ATOMS, MOLECULES AND MOLES

In Chapter 1 you were introduced to some important basic concepts in chemistry, such as matter and elements and compounds. You learned a little about chemical reactions and you learned about chemical formulas and chemical equations. You also learned about the importance of laboratory measurements, which was something that early scientists didn't understand. Real progress in chemistry didn't begin until scientists started to make quantitative observations of the amounts of chemicals involved in chemical reactions.

In this chapter and the next we will examine some basic chemical laws and how we came to believe in the existence of atoms. You will learn how to think in chemical terms about amounts of substances that combine in chemical reactions and you will learn how chemical formulas are determined. The ability to perform the kinds of chemical calculations introduced here and in the next chapter is an essential part of learning chemistry. Usually, when students have difficulty later in the course, their problems can be traced to gaps in their understanding of the concepts presented in these chapters. Therefore, as you study the material covered here, concentrate on *understanding* the concepts. From an understanding will follow an ability to solve problems, but if you concentrate on solving problems without understanding what you're doing, you will almost certainly find yourself in trouble later in the course.

Learning chemistry is *not* merely learning how to do chemical calculations. Chemical calculations are just one of the many tools we use to study chemical compounds and reactions.

2.1 LAWS OF CHEMICAL COMBINATION

In the very early days of chemistry, little was really known about the nature of chemical substances and chemical reactions, so perhaps it isn't surprising that many incorrect theories about the nature of matter evolved. For example, it had long been known that when a piece of wood was burned, the resulting ash weighed less than the wood itself. One theory was that something called *phlogiston* was evolved during combustion.

The phlogiston theory survived for some time until a French chemist named

Lavoisier was a victim of the French Revolution. He went to the guillotine at the age of 52 on May 8, 1794.

Antoine Lavoisier demonstrated, by experiments in which careful *measurements* of the weights of chemicals were made, that combustion involves the reaction of a substance with oxygen. He also showed, *by careful measurements,* that if the combustion reaction is carried out in a closed container, there is no net change in mass during the reaction. These observations and others made under carefully controlled conditions form the basis for the **law of conservation of mass,** which states that *in any chemical reaction, mass is neither created nor destroyed.* (When we say that something is conserved, we mean that it isn't gained or lost, but stays the same.) The law of conservation of mass is an important chemical law dealing with chemical reactions and serves as the reason that we balance chemical equations.

The definition of *compound* is a statement of the law of definite proportions.

Lavoisier's experiments prompted others to also perform careful quantitative measurements on chemical substances, and their results ultimately led to another important chemical law called the **law of definite proportions** (also known as the **law of definite composition**). This law states that *in any sample of a pure chemical substance, we always find the same elements in the same definite proportions by mass.* For example, we mentioned earlier that in any sample of pure water, regardless of its source, we always find the elements hydrogen and oxygen in the ratio of 1.00 g of H to 8.00 g of O. Thus, if a sample of water were taken that had 2.00 g of H, then it would have 16.00 g of O, but the ratio is still the same, because

$$2.00 \text{ g H}/16.00 \text{ g O} = 1.00 \text{ g H}/8.00 \text{ g O}$$

Furthermore, if we form water from hydrogen with oxygen, the elements combine in exactly this ratio, regardless of the amounts actually available. If 2.00 g of hydrogen is mixed with 8.00 g of oxygen and allowed to react, all the oxygen will be consumed, but only 1.00 g of hydrogen will react; 1.00 g of hydrogen will be left over. For water, it is impossible to vary the mass ratio of these two elements and still have water. All compounds have similar fixed ratios by mass of their elements.

2.2 DALTON'S ATOMIC THEORY AND ATOMIC MASSES

As we've noted, the laws of conservation of mass and definite proportions are based solely on experimentally measured masses of chemicals in compounds and chemical reactions. Chemists of the late eighteenth century could not help but wonder what matter must be like in order for these laws to hold, and in 1803 an English schoolteacher–scientist named John Dalton proposed an explanation, called the atomic theory, that changed the course of chemical science.

The concept of atoms was not new; Greek philosophers as early as 500 B.C. had pondered the possibility that matter might be composed of tiny indivisible particles. In fact, the word *atom* comes from the Greek *atomos* meaning indivisible. The ancient Greeks, however, had no data to explain and their proposals were little more than exercises in thought. Dalton's theory was different, though, because his had to explain the observed laws of conservation of mass and definite proportions.

The theory Dalton proposed can be expressed by the following postulates:

POSTULATES OF THE ATOMIC THEORY

1. Matter is composed of tiny indivisible particles called atoms.
2. All atoms of a given element are identical, but differ from atoms of other elements. (This means that all atoms of a given element have the same mass, but this mass differs from the masses of atoms of other elements.)
3. A chemical compound is composed of the atoms of its elements in a definite fixed numerical ratio.
4. A chemical reaction merely consists of a reshuffling of atoms from one set of combinations to another. The individual atoms themselves, however, remain intact and do not change.

The test of the theory, of course, is in its ability to explain the facts. First, let's look at the law of conservation of mass. If a chemical reaction does nothing more than take the atoms found in the reactants and redistribute them among the products, then the total number of atoms of each kind must remain the same (provided, of course, that no atoms can enter or leave the reaction vessel). Since atoms don't change mass during a reaction, the total mass of the atoms must also stay the same. In other words, the mass must stay constant during the reaction, which is exactly what the law of conservation of mass says.

The law of definite proportions is also easy to explain. To see this, let's imagine that two elements, say A and B, form a compound in which each molecule of the substance is composed of one atom of A and one of B. (A molecule, you recall, can be thought of as *a group of atoms held together tightly enough that they behave as, and can be recognized as, a single particle.*) Let's also suppose that an atom of A is twice as heavy as an atom of B, so if an atom of B were arbitrarily assigned a mass of 1 unit, then the mass of an atom of A would be 2 units. Below we see how the masses of A and B vary for various numbers of molecules.

The smallest particle of carbon dioxide is a molecule of carbon dioxide, which consists of one carbon atom and two oxygen atoms.

Number of Molecules	Number of Atoms of A	Mass of A	Number of Atoms of B	Mass of B	Mass Ratio, (mass A)/(mass B)
1	1	2 units	1	1 unit	2/1
2	2	4 units	2	2 units	4/2 = 2/1
10	10	20 units	10	10 units	20/10 = 2/1
500	500	1000 units	500	500 units	1000/500 = 2/1

Now, notice that no matter how may molecules we have, each with the same 1-to-1 ratio by atoms, *the ratio by mass is the same.* This is exactly what the law of definite proportions says: In any sample of a compound, regardless of size, the elements are always present in the same proportion (ratio) by mass.

Atomic masses

The key to the success of Dalton's atomic theory was the concept that each element has atoms with a characteristic atomic mass. It explained the laws of chemistry so well that chemists soon sought to measure atomic masses. But how could this be done? Atoms are too small to see and too small to have their masses measured individually on a laboratory balance.

The terms **atomic weight** and **atomic mass** are often used interchangeably.

In the discussion above, we examined a hypothetical compound formed from two atoms, one A and one B. We saw that the mass ratio in any sample is 2 to 1 because the masses of the atoms are in a 2-to-1 ratio. But we could just as well have reasoned backward here by saying that the only way the mass ratio could always be 2 to 1 is if the ratio of the masses of the atoms themselves is 2 to 1. We see, therefore, that by determining the mass ratio of the elements in a large sample, we can make conclusions about the mass ratio of the atoms in the compound. Let's see how this works now for real elements and a real compound.

Hydrogen and fluorine form a compound called hydrogen fluoride. Its formula is HF, so a molecule of HF contains one atom of hydrogen and one atom of fluorine. In samples of this compound, it is always found that the mass of fluorine is 19 times the mass of hydrogen. Since the atoms are present in equal numbers, we can reason backwards and conclude that each fluorine atom must be 19 times as heavy as each hydrogen atom. We have therefore established the *relative* masses of hydrogen and fluorine atoms.

If a fluorine atom is 19 times as heavy as a hydrogen atom, then the F-to-H mass ratio must be 19 : 1.

In the two examples we've just examined, we saw that by measuring the ratio of the masses of the elements, we can figure out the ratio of the masses of the atoms. But there is a very important condition here; we must *know the formula of the compound.*

A self-consistent table of atomic masses was not developed until the mid-nineteenth century.

The symbol **u** comes from the SI term **unified atomic mass unit.**

Scientists who work with substances composed of very large molecules, such as proteins, often refer to an atomic mass unit as a **dalton** (1 u = 1 dalton).

The mass in grams of one atomic mass unit has been measured to be 1.660×10^{-24} g.

The need to know formulas before the relative masses of the atoms could be determined was a major stumbling block, but a way of obtaining formulas was eventually devised and a set of relative masses of the atoms of the elements could then be obtained. However, they were just *relative* masses, telling how much heavier one particular atom was than some other atom. We want to assign numerical values to these atomic masses, and this can be done if we know the mass of an atom of one of the elements. Then the masses of the others can be calculated from the various mass ratios.

Because atoms are too small to be seen and weighed in units of grams on a balance, an atomic mass scale was devised in which mass is measured in **atomic mass units** (the SI symbol is **u**). The choice of a standard for this scale is complicated by the fact that Dalton's key concept isn't quite correct; most of the elements exist in nature as mixtures of several kinds of atoms (called *isotopes*) with slightly different masses. Fortunately, this didn't affect the net result of his theory because in any sample of an element large enough to see, the *average* mass of the enormous number of atoms present is the same, so the elements behave as if their individual atoms have masses equal to the average. However, because the relative abundance of the different isotopes of an element could conceivably change over a long period of time, it was decided to choose one isotope of one element to define how big the atomic mass unit is. This isotope is one of carbon and is called **carbon-12.** It is assigned a mass of exactly 12 u, so the atomic mass unit is defined as 1/12th of the mass of one atom of this isotope. By choosing the atomic mass unit to be this size, the atomic masses of many of the other elements come out to be approximately whole numbers.

A complete table of atomic masses is located on the inside front cover of the book and they are also given along with the symbols of the elements in the periodic table. The numbers in these tables are relative average atomic masses, which means they are the average masses, expressed in atomic mass units, of the mixtures of isotopes found in naturally occurring samples of the elements.

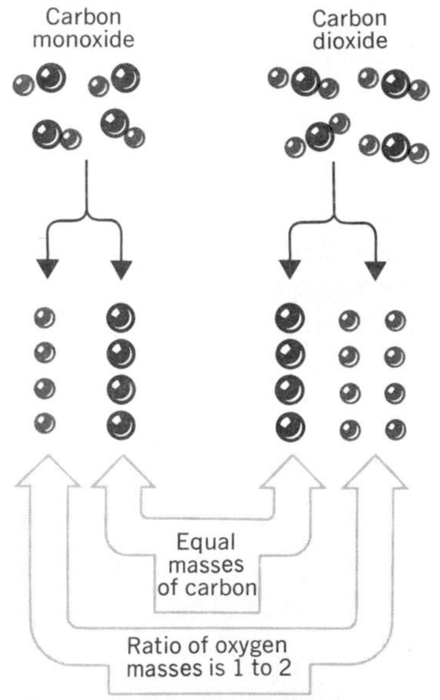

Carbon monoxide   Carbon dioxide

Equal masses of carbon

Ratio of oxygen masses is 1 to 2

Figure 2.1. *The law of multiple proportions. Here are four molecules each of carbon monoxide and carbon dioxide. Each contains the same number of carbon atoms, ☻, and thus the same mass of carbon. The masses of oxygen that are combined with the carbon are in a whole-number (1 to 2) ratio.*

The law of multiple proportions

An interesting by-product of Dalton's atomic theory was the discovery of another law of chemical combination called the **law of multiple proportions,** which can be expressed as follows: *Suppose we have samples of two compounds formed by the same two elements. If the masses of one element are the same in the two samples, then the masses of the other element are in a ratio of small whole numbers.* This law, which is mainly of historical interest, is most easily understood by looking at an example. As you probably know, carbon forms two different compounds with oxygen called carbon monoxide and carbon dioxide. In 2.33 g of carbon monoxide, we find 1.33 g of oxygen combined with 1.00 g of carbon. In 3.66 g of carbon dioxide, we find 2.66 g of oxygen combined with 1.00 g of carbon. Notice that the masses of oxygen that are combined with the same mass of carbon (1.00 g) are in a ratio of 2 to 1 (a ratio of small whole numbers).

$$\frac{2.66 \text{ g}}{1.33 \text{ g}} = \frac{2}{1}$$

This result is consistent with the atomic theory if a molecule of carbon monoxide (CO) contains 1 atom of C and 1 atom of O, and a molecule of carbon dioxide (CO_2) contains 1 atom of C and 2 atoms of O. Then, as illustrated in Figure 2.1, when we have equal numbers of molecules, we have equal numbers of carbon atoms and equal masses of carbon. But notice that there are twice as many oxygen atoms in the carbon dioxide as in the carbon monoxide, so the ratio of the oxygen masses has to be 2 to 1.

Example 2.1 below provides another illustration of the law of multiple proportions and shows how we can use the law of definite composition to our advantage in a chemical calculation.

EXAMPLE 2.1. USING THE FACTOR-LABEL METHOD IN A CHEMICAL CALCULATION

PROBLEM: Sulfur forms two compounds with fluorine. In one of them, it is found that 0.447 g S is combined with 1.06 g F, and in the other, 0.438 g S is combined with 1.56 g F. Show that these data support the law of multiple proportions.

ANALYSIS: We can't test the law immediately because we need to have the same mass of one of the elements in samples of the two compounds. What we need to do, therefore, is to imagine changing the size of one of the samples so this condition is fulfilled. There are several ways to go about this, but one is to imagine that we had a smaller sample of the second compound so that it too would have 1.06 g of fluorine. Then we could compare the masses of sulfur. Let's see if we can calculate how much sulfur would be in this smaller sample.

The law of definite composition tells us that in any sample of the second compound, the relative amounts of sulfur and fluorine will be the same. Whenever we find 0.438 g S, we will also find 1.56 g F. Thus, for this compound there is a relationship between these amounts of S and F that we call a *chemical equivalence.* Often is is convenient to express this in an equationlike form such as

$$1.56 \text{ g F} \Longleftrightarrow 0.438 \text{ g S}$$

where once again we have used the symbol $\Longleftrightarrow$ to mean "is equivalent to." (We really shouldn't use an equals sign because fluorine can't "equal" sulfur.)

The merit of a chemical equivalence is that we can use it to form conversion factors for use in calculations. From the equivalence just stated, we can make these two conversion factors

$$\frac{1.56 \text{ g F}}{0.438 \text{ g S}} \quad \text{and} \quad \frac{0.438 \text{ g S}}{1.56 \text{ g F}}$$

Let's see how we can use one of them to calculate the amount of sulfur that would be in a sample of the second compound if that sample were to contain 1.06 g of fluorine.

Some people do prefer to write this as
$$1.56 \text{ g F} = 0.438 \text{ g S}$$

The equals sign in this case is just interpreted to mean "is equivalent to."

SOLUTION: We can state our desired calculation as

$$1.06 \text{ g F} \Longleftrightarrow ? \text{ g S (in the second compound)}$$

To find the answer, we multiply 1.06 g F by the second of the conversion factors, which relates the amounts of S and F in the second compound. This allows us to cancel the units "g F."

$$1.06 \text{ g F} \times \left(\frac{0.438 \text{ g S}}{1.56 \text{ g F}} \right) \Longleftrightarrow 0.298 \text{ g S}$$

Now we know the masses of sulfur combined with the same mass (1.06 g) of fluorine in the two compounds. The next step is to examine the ratio of the masses of sulfur, which is

$$\frac{0.447 \text{ g}}{0.298 \text{ g}} = \frac{1.50}{1.00}$$

If we double both numerator and denominator, we find the ratio is 3/2, which is a ratio of small whole numbers. The data in the problem, therefore, do support the law of multiple proportions.

2.3 THE MOLE CONCEPT

In today's world, the study of chemical substances and chemical reactions requires the ability to experimentally determine the nature of the products of chemical reactions. We must be able to find out their formulas and we must be able to decide how much of the various chemicals we need when we carry out chemical reactions. In other words, we must be able to deal with elements and compounds and with chemical reactions quantitatively. **Stoichiometry** (derived from the Greek *stoicheion* = element and *metron* = measure) is the term we use in describing the quantitative aspects of chemical composition and reaction.

The atomic theory of Dalton and the development of a table of atomic masses of the elements really opened the door to stoichiometric calculations, but before this can be appreciated, we must first examine what is surely the most important concept of all in stoichiometry, the *mole*.

As you've learned, atoms react to form molecules in simple whole-number ratios. Hydrogen and oxygen atoms, for instance, combine in a 2-to-1 ratio to form water, H_2O, and carbon and oxygen atoms combine in a 1-to-1 ratio to form carbon monoxide, CO. Knowing this, suppose we wanted to make carbon monoxide from carbon atoms and oxygen atoms in such a way that there wouldn't be any atoms of either element left over. If we only wanted one molecule, we could imagine bringing together 1 atom of C and 1 atom of O. If two molecules were desired, then we would need 2 atoms of C and 2 of O, and so on for any number of molecules we wished. However, we really can't work with individual atoms because they are so tiny. Therefore, in any real-life laboratory situation we must increase the sizes of the samples to the point where we can see them and manipulate them, but we must do this in a way that maintains the proper ratio of atoms.

One way we could enlarge the amounts in a chemical reaction would be to work with dozens of atoms instead of individual atoms.

$$1 \text{ atom C} \quad + \quad 1 \text{ atom O} \quad \longrightarrow \quad 1 \text{ molecule CO}$$
$$1 \text{ dozen C atoms} + 1 \text{ dozen O atoms} \longrightarrow 1 \text{ dozen CO molecules}$$
$$\text{(12 atoms C)} \qquad \text{(12 atoms O)} \qquad \text{(12 molecules CO)}$$

Notice that the 1-to-1 ratio of *dozens* of atoms is *exactly* the same as the 1-to-1 ratio of the atoms themselves. If we were to take 2 dozen carbon atoms and 2 dozen oxygen atoms (another 1-to-1 ratio of *dozens*), we again could be sure that there would be equal numbers of atoms of carbon and oxygen (a 1-to-1 ratio of *atoms*). In fact, it doesn't matter how many dozens of each kind of atom we take; all we have to be careful about is that we have equal numbers of dozens, so that the 1-to-1 ratio by dozens and by atoms is maintained.

This is such an important concept that it's worth exploring for another case. Let's consider the substance water, H_2O. If we were dealing with individual atoms, we could write the equation

$$2 \text{ atoms of H} + 1 \text{ atom of O} \longrightarrow 1 \text{ molecule of } H_2O$$

We could then scale up the size of the reaction by working with dozens of hydrogen and oxygen atoms

$$2 \text{ dozen H atoms} + 1 \text{ dozen O atoms} \longrightarrow 1 \text{ dozen } H_2O \text{ molecules}$$

or

$$4 \text{ dozen H atoms} + 2 \text{ dozen O atoms} \longrightarrow 2 \text{ dozen } H_2O \text{ molecules}$$

or

$$6 \text{ dozen H atoms} + 3 \text{ dozen O atoms} \longrightarrow 3 \text{ dozen } H_2O \text{ molecules}$$

In each case, we maintain a 2-to-1 ratio of H to O atoms by maintaining a 2-to-1 ratio of dozens of these atoms.

It should be obvious now that if we had some way of counting atoms by the dozen, we could take dozens of them in a ratio that is exactly equal to the desired atom ratio, and in so doing we would be assured of having the proper atom ratio. Unfortunately, a dozen atoms or molecules is still much too small to work with, so we must find a still larger unit. The "chemist's dozen" is called the **mole** (abbreviated **mol**). It is composed of 6.022×10^{23} objects (we will say more about the origin of this number, called **Avogadro's number,** later).

$$1 \text{ dozen} = 12 \text{ objects}$$

$$1 \text{ mole} = 6.022 \times 10^{23} \text{ objects}$$

The dozen and the mole are similar units; each stands for a certain number of things.

The same reasoning that we can use with the dozen applies equally to the mole. The mole is simply a much larger collection.

$$1 \text{ mole of C atoms} + 1 \text{ mole of O atoms} \longrightarrow 1 \text{ mole of CO molecules}$$

or

$$\underset{(6.022 \times 10^{23} \text{ atoms C})}{1 \text{ mol C}} + \underset{(6.022 \times 10^{23} \text{ atoms O})}{1 \text{ mol O}} \longrightarrow \underset{(6.022 \times 10^{23} \text{ molecules CO})}{1 \text{ mol CO}}$$

We see that if we take one mole of carbon atoms and one mole of oxygen atoms, we have equal numbers of carbon and oxygen atoms and can construct exactly 1 mol of CO molecules with nothing left over.

The preceding discussion illustrates the most important aspect of the mole concept.

> The ratios in which individual atoms combine to form molecules are *exactly* the same as the ratios in which *moles* of these atoms combine.

For example, to form carbon tetrachloride, CCl_4, we know that

$$1 \text{ atom C} + 4 \text{ atoms Cl} \longrightarrow 1 \text{ molecule } CCl_4$$

We can immediately enlarge this to moles

$$1 \text{ mol C} + 4 \text{ mol Cl} \longrightarrow 1 \text{ mol } CCl_4$$

By taking moles of carbon and chlorine atoms in a 1-to-4 ratio, we can be sure of having one carbon atom for every four chlorine atoms. And if our goal is simply keeping the carbon-to-chlorine atom ratio equal to 1 to 4, we can work with *any* number of moles of carbon atoms, just as long as the number of moles of chlorine atoms is four times larger. Thus, if we started with 2 mol C atoms, we would need 8 mol Cl atoms

Remember, "1 mol C" means 1 mole of C atoms. What does "1 mol CCl_4" mean?

$$2 \text{ mol C} + 8 \text{ mol Cl} \longrightarrow 2 \text{ mol } CCl_4$$

or, if we started with 5 mol C atoms, we would need 20 mol Cl atoms

$$5 \text{ mol C} + 20 \text{ mol Cl} \longrightarrow 5 \text{ mol } CCl_4$$

In each case, there is a 1-to-4 ratio of carbon to chlorine both by moles and by atoms.

To summarize then, the *ratio* by which moles of substances react is the same as the *ratio* by which their atoms and molecules react. This simple idea forms the basis for *all* quantitative chemical reasoning.

Students who have difficulty learning chemistry often have not learned to think problems through in terms of moles.

EXAMPLE 2.2. DETERMINING MOLE RATIOS FROM CHEMICAL FORMULAS

PROBLEM: What mole ratio of carbon to chlorine must be chosen to prepare the substance C_2Cl_6 (hexachloroethane), an important industrial solvent?

ANALYSIS: Whenever we need to obtain a mole ratio from a chemical formula, we have to remember that the ratio by moles is *exactly* the same as the ratio by atoms. Then the problem is easy.

SOLUTION: The subscripts in the formula C_2Cl_6 give us the atom ratio, which is 2 atoms of C to 6 atoms of Cl. This means that the mole ratio must be 2 mol of C to 6 mol of Cl.

Atom ratio	Mole ratio
$\dfrac{2 \text{ atom C}}{6 \text{ atom Cl}}$	$\dfrac{2 \text{ mol C}}{6 \text{ mol Cl}}$

This is the same, of course, as a ratio of (1 mol C)/(3 mol Cl).

The atom ratio in a chemical formula such as C_2Cl_6 establishes a number of mole ratios that are useful for constructing conversion factors that can be employed in solving problems. We've already seen one of these, which can be written as either

$$\frac{2 \text{ mol C}}{6 \text{ mol Cl}} \quad \text{or} \quad \frac{6 \text{ mol Cl}}{2 \text{ mol C}}$$

There are others as well. The formula tells us that 1 molecule of C_2Cl_6 contains 2 atoms of C and that it also contains 6 atoms of Cl. We can scale this up immediately to moles: 1 mol C_2Cl_6 contains 2 mol C and 6 mol Cl. This gives two equivalencies

$$1 \text{ mol } C_2Cl_6 \Longleftrightarrow 2 \text{ mol C}$$

$$1 \text{ mol } C_2Cl_6 \Longleftrightarrow 6 \text{ mol Cl}$$

In other words, any time we have a mole of C_2Cl_6 molecules, there will be two moles of carbon atoms in it, and any time we have a mole of C_2Cl_6 molecules, there will be six moles of chlorine atoms in it. That's what we mean by an equivalence ($\Longleftrightarrow$). As described earlier, we can use these equivalencies to form conversion factors

$$\frac{1 \text{ mol } C_2Cl_6}{2 \text{ mol C}} \quad \text{or} \quad \frac{2 \text{ mol C}}{1 \text{ mol } C_2Cl_6}$$

and

$$\frac{1 \text{ mol } C_2Cl_6}{6 \text{ mol Cl}} \quad \text{or} \quad \frac{6 \text{ mol Cl}}{1 \text{ mol } C_2Cl_6}$$

The following examples show how we use them.

EXAMPLE 2.3. USING MOLE RATIOS AS CONVERSION FACTORS

PROBLEM: How many moles of carbon atoms are needed to combine with 4.87 mol Cl to form the substance C_2Cl_6?

ANALYSIS: Let's restate the problem as follows:

$$4.87 \text{ mol Cl} \Longleftrightarrow ? \text{ mol C}$$

Now we see that we need a conversion factor that will change "mol Cl" to "mol C" As noted above, this is given by the subscripts in the chemical formula.

SOLUTION: The formula allows us to construct two conversion factors relating moles of C and moles of Cl. Keeping in mind that the mole ratio is the same as the atom ratio, we have these conversion factors to choose from

$$\frac{2 \text{ mol C}}{6 \text{ mol Cl}} \qquad \frac{6 \text{ mol C1}}{2 \text{ mol C}}$$

To get rid of the units "mol Cl" we must use the one on the left.

$$4.87 \text{ mol Cl} \times \left(\frac{2 \text{ mol C}}{6 \text{ mol Cl}}\right) \Longleftrightarrow 1.62 \text{ mol C}$$

We need 1.62 mol C. Notice that we are entitled to report three significant figures here. The numbers in the mole ratio are exact numbers because atoms combine in exact whole-number ratios when they form compounds.

Of course, we would also simplify these ratios to give

$$\frac{1 \text{ mol C}}{3 \text{ mol Cl}} \quad \text{and} \quad \frac{3 \text{ mol Cl}}{1 \text{ mol C}}$$

In fact, we often do.

EXAMPLE 2.4. USING MOLE RATIOS AS CONVERSION FACTORS

PROBLEM: How many moles of carbon are in 2.65 mol C_2Cl_6?

SOLUTION: Our problem can be stated as

$$2.65 \text{ mol } C_2Cl_6 \Longleftrightarrow ? \text{ mol C}$$

From the chemical formula we get the equivalence

$$1 \text{ mol } C_2Cl_6 \Longleftrightarrow 2 \text{ mol C}$$

and then use it to construct the necessary conversion factor.

$$2.65 \text{ mol } C_2Cl_6 \times \left(\frac{2 \text{ mol C}}{1 \text{ mol } C_2Cl_6}\right) \Longleftrightarrow 5.30 \text{ mol C}$$

Thus, 2.65 mol C_2Cl_6 contains 5.30 mol C.

2.4 MEASURING MOLES OF ATOMS

In a chemical reaction, atoms or molecules combine in whole-number ratios, and we also have seen that moles of these substances react in the same whole-number ratios. In this sense, then, the mole might be called a *chemical unit*. Its size is large enough so that a mole of atoms or molecules represents an amount that is easily worked with in the laboratory, but unfortunately there is no device that allows us directly to count out atoms in multiples of Avogadro's number. We must, therefore, have a way of translating these chemical units to *laboratory units*—something that can be measured in the lab.

Earlier it was stated that a mole consists of 6.022×10^{23} objects. This rather strange number was not selected deliberately—instead, it just happens to be the number of atoms in a sample of any element that has a mass in grams that is numerically equal to that element's atomic mass.[1] For example, the atomic mass of

---

[1] Historically, the mole was first defined as the amount of an element having a mass in grams numerically equal to its atomic mass. Later, it was discovered experimentally that there are 6.022×10^{23} atoms in one mole of an element. We will see how Avogadro's number can be measured in later chapters.

Since there are equal numbers of atoms when the weights of the elements are taken in proportion to their atomic masses, we can also speak of a *pound-mole (lb-mol)* or a *ton-mole (ton-mol)*. A lb-mol of an element has a mass in pounds that is numerically equal to its atomic mass, and a ton-mol has a mass in tons that is numerically equal to its atomic mass. (Of course, since a pound and a ton are larger than a gram, the lb-mol and ton-mol have more atoms than the mole defined in terms of grams.) Thus, 12 lb of C (1 lb-mol C) contains the same number of atoms as 16 lb of O (1 lb-mol O) because the weights of C and O are taken in the same ratio as their atomic masses. Similarly, 12 tons of C (1 ton-mol C) contains the same number of atoms as 16 tons of O (1 ton-mol O). These are useful ideas when dealing with very large amounts of reactants in industrial processes.

Figure 2.2. *One mole of four different elements: sulfur, iron, mercury, and copper. Although each sample has a different mass, each contains the same number of atoms.*

carbon is 12.011, so 1 mol of carbon atoms has a mass of 12.011 g. Similarly, the atomic mass of oxygen is 15.9994, so a mole of oxygen atoms has a mass of 15.9994 g. (See Figure 2.2.)

$$1 \text{ mol C} = 12.011 \text{ g C}$$

$$1 \text{ mol O} = 15.9994 \text{ g O}$$

A table of atomic masses gives us the number of grams in 1 mol of the atoms of any element.

Thus, it is the balance that becomes our tool for measuring moles. To obtain one mole of any element, all we need to do is look up the element's atomic mass. The number we find is the number of grams of the element that we must take to have one mole of it.

Converting between grams and moles is a routine calculation that you must learn to do quickly. Some examples of this kind of calculation, along with the use of the mole in chemical calculations, are shown in the following sample problems.

EXAMPLE 2.5. CONVERTING FROM GRAMS TO MOLES

PROBLEM: How many moles of silicon, Si, are in 30.5 g of Si? Silicon is an element used to make transistors.

SOLUTION: Our problem is one of converting the units of grams of Si to moles of Si, that is, 30.5 g Si = (?) mol Si. We know from the table of atomic masses that

$$1 \text{ mol Si} = 28.1 \text{ g Si}$$

To convert from g Si to moles, we must multiply 30.5 g Si by a factor that contains the units "g Si" in the denominator, that is,

$$\frac{1 \text{ mol Si}}{28.1 \text{ g Si}}$$

Thus,

$$30.5 \text{ g Si} \times \left(\frac{1 \text{ mol Si}}{28.1 \text{ g Si}} \right) = 1.09 \text{ mol Si}$$

Therefore, 30.5 g Si = 1.09 mol Si.

EXAMPLE 2.6. CONVERTING FROM MOLES TO GRAMS

PROBLEM: How many grams of copper, Cu, are there in 2.55 mol of Cu?

SOLUTION: From the table of atomic masses,

$$1 \text{ mol Cu} = 63.5 \text{ g Cu}$$

The conversion factor must have the units mol Cu in the denominator, that is,

$$\frac{63.5 \text{ g Cu}}{1 \text{ mol Cu}}$$

Setting up the problem, we have

$$2.55 \text{ mol Cu} \times \left(\frac{63.5 \text{ g Cu}}{1 \text{ mol Cu}} \right) = 162 \text{ g Cu}$$

The mass of 2.55 mol Cu is 162 g.

EXAMPLE 2.7. USING MOLE RELATIONSHIPS

PROBLEM: How many moles of Ca are required to react with 2.50 mol of Cl to produce the compound $CaCl_2$ (calcium chloride), used to melt ice on roads in the winter.

SOLUTION: Our problem: 2.50 mol Cl ⟺ (?) mol Ca.
We know from the formula that 1 atom of Ca combines with 2 atoms of Cl. Thus,

$$1 \text{ atom Ca} \Longleftrightarrow 2 \text{ atoms Cl}$$

Because moles combine in the same ratio as atoms

$$1 \text{ mol Ca} \Longleftrightarrow 2 \text{ mol Cl}$$

We obtain the answer as follows:

$$2.50 \text{ mol Cl} \times \left(\frac{1 \text{ mol Ca}}{2 \text{ mol Cl}} \right) \Longleftrightarrow 1.25 \text{ mol Ca}$$

EXAMPLE 2.8. USING THE MOLE/MASS RELATIONSHIP

PROBLEM: How many grams of Ca must react with 41.5 g of Cl to produce $CaCl_2$?

ANALYSIS: Once again, we are asked to relate an amount of Ca to an amount of Cl, but this time we are starting with grams instead of moles. The direct relationship between Ca and Cl is obtained from the subscripts in the chemical formula, and we've seen that these give us a mole relationship:

$$1 \text{ mol Ca} \Longleftrightarrow 2 \text{ mol Cl}$$

Therefore, the first step in solving the problem is to convert grams of Cl to moles of Cl; then we can use the mole relationship to find the number of moles of Ca that are needed. To change grams of Cl to moles is done easily enough by looking up the atomic mass of Cl, which allows us to write

$$1 \text{ mol Cl} = 35.5 \text{ g Cl}$$

These two relationships now allow us to go from the given 41.5 g of Cl to moles of Ca needed. The last step is to convert "mol Ca" to "g Ca". Once again, we go to the table of atomic masses.

$$1 \text{ mol Ca} = 40.1 \text{ g Ca}$$

Now we have all the information needed to solve the problem, but before we actually go to the solution, let's diagram how it will work.

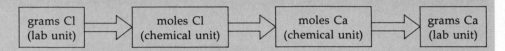

We will find that many of the stoichiometry problems that we do follow this kind of strategy. We will be given some set of lab units—grams, for instance—and we will have to calculate an amount of some other substance in lab units. However, to do this we will have to go through chemical units (moles), because the quantitative relationship between the two substances is expressed not in terms of grams, but rather as a mole ratio.

SOLUTION: We follow the outline above, being careful to set up the conversion factors so that units cancel correctly.

First we calculate how many moles of Cl are available. Then we calculate how many moles of Ca are needed. Finally, we calculate the mass of this many moles of Ca.

$$41.5 \text{ g Cl} \times \left(\frac{1 \text{ mol Cl}}{35.5 \text{ g Cl}}\right) = 1.17 \text{ mol Cl}$$

$$1.17 \text{ mol Cl} \times \left(\frac{1 \text{ mol Ca}}{2 \text{ mol Cl}}\right) \Leftrightarrow 0.585 \text{ mol Ca}$$

$$0.585 \text{ mol Ca} \times \left(\frac{40.1 \text{ g Ca}}{1 \text{ mol Ca}}\right) = 23.5 \text{ g Ca}$$

Usually, we put all these steps together as follows.

Stringing together conversion factors like this makes the arithmetic easy to perform in one step on a calculator and helps us avoid mistakes that we might otherwise make copying answers and reentering them into the calculator.

$$41.5 \text{ g Cl} \times \left(\frac{1 \text{ mol Cl}}{35.5 \text{ g Cl}}\right) \times \left(\frac{1 \text{ mol Ca}}{2 \text{ mol Cl}}\right) \times \left(\frac{40.1 \text{ g Ca}}{1 \text{ mol Ca}}\right) \Leftrightarrow 23.4 \text{ g Ca}$$

to moles of Cl

to moles of Ca

to grams of Ca

(The small discrepancy between the answers is due to our not having rounded off during the second calculation.)

In this last example we have strung together a series of conversion factors. When we set up the problem, we link these together by considering what units must be eliminated by cancellation. Thus, the first factor had to have g Cl in the denominator and took us as far as moles of Cl. The second factor had to have moles of Cl in the denominator and would give us moles of Ca if we stopped at that point. The third factor was required to have moles Ca in the denominator. At this point we stop since our units are now those of the answer, and we have only to perform the arithmetic to obtain the correct result.

Avogadro's number and the mole

We have noted that the number of particles in one mole, which is called Avogadro's number, is equal to 6.022×10^{23}. This is a tremendous number; written in ordinary decimal notation, it is 602,200,000,000,000,000,000,000. Because there are so many atoms in one mole of an element, the mass of an individual atom is extremely small. The next example illustrates how Avogadro's number can be used to calculate properties of individual atom-sized particles.

EXAMPLE 2.9. USING AVOGADRO'S NUMBER IN CALCULATIONS

PROBLEM: What is the mass of one atom of calcium?

ANALYSIS: When a calculation deals with some property of individual atoms or molecules and relates that property to things that we can measure in the laboratory, Avogadro's number is usually needed. Here we can restate the problem as

$$1 \text{ atom Ca} = (?) \text{ g Ca}$$

Notice that we are being asked to express the property of one atom (its mass) in terms of a quantity (grams) that we can measure in the laboratory. We need Avogadro's number to relate the tiny submicroscopic world of atoms to the large world of the laboratory.

SOLUTION: We have the relationships

$$6.02 \times 10^{23} \text{ atoms Ca} = 1 \text{ mol Ca (to 3 significant figures)}$$

$$1 \text{ mol Ca} = 40.1 \text{ g Ca}$$

The solution to the problem can then be set up as

$$1 \text{ atom Ca} \times \left(\frac{1 \text{ mol Ca}}{6.02 \times 10^{23} \text{ atoms Ca}} \right) \times \left(\frac{40.1 \text{ g Ca}}{1 \text{ mol Ca}} \right) = 6.66 \times 10^{-23} \text{ g Ca}$$

2.5 MEASURING MOLES OF COMPOUNDS: MOLECULAR MASS AND FORMULA MASS

As with elements, the balance can also be used indirectly to measure moles of compounds. The simplest way of obtaining the weight of one mole of a substance is merely to add up the atomic masses of all the elements present in the compound. If the substance is composed of molecules (for example, CO_2, H_2O, or NH_3), the sum of the atomic masses is called the **molecular mass** (or **molecular weight**— the terms are used interchangeably). Thus, the molecular mass of CO_2 is obtained as

$$
\begin{array}{lll}
\text{C} & 1 \times 12.0 \text{ u} = & 12.0 \text{ u} \\
2 \text{ O} & 2 \times 16.0 \text{ u} = & \underline{32.0 \text{ u}} \\
CO_2 & \text{total} & 44.0 \text{ u}
\end{array}
$$

Similarly, the molecular mass of $H_2O = 18.0$ u and that of $NH_3 = 17.0$ u. The weight of one mole of a substance is obtained simply by writing its molecular mass followed by the units, grams. Thus,

$$1 \text{ mol } CO_2 = 44.0 \text{ g}$$

$$1 \text{ mol } H_2O = 18.0 \text{ g}$$

$$1 \text{ mol } NH_3 = 17.0 \text{ g}$$

Figure 2.3 shows 1-mol samples of several different compounds.

Figure 2.3. *One mole of four different compounds. Each contains the same number of formula units, although the number of atoms is different from sample to sample. They differ because the number of atoms per formula unit is different for each compound.*

1 mol sodium chromate Na_2CrO_4

1 mol water (H_2O)

1 mol sodium chloride NaCl

1 mol copper sulfate $CuSO_4 \cdot 5H_2O$

In later chapters we will encounter many compounds that do not contain separate, distinct molecules. We will find that when certain atoms react, they often gain or lose negatively charged particles called **electrons.** Sodium and chlorine happen to react this way. When sodium chloride, NaCl, is formed from the elements, each Na atom loses one electron and each Cl atom gains one. Since Na and Cl are electrically neutral to start, these atoms acquire a charge when NaCl is formed. These are written as Na^+ (positive because Na has lost a negatively charged electron) and Cl^- (negative because Cl has gained an electron). *Atoms or groups of atoms that have acquired an electrical charge are called* **ions.** Since solid NaCl is composed of Na^+ and Cl^- ions, it is said to be an **ionic compound.**

This entire topic is explored further in later chapters. For now, it is only necessary for you to know that compounds that are ionic do not contain molecules. Their formulas simply state the ratio of the different atoms in the substance. In NaCl, the atoms are in a 1-to-1 ratio. In the ionic compound $CaCl_2$, the ratio of Ca to Cl atoms is 1 to 2 (relax—at this point, you were not expected to know that $CaCl_2$ is ionic). Rather than refer to molecules of NaCl or $CaCl_2$, we use the term **formula unit** to specify the pair of ions in NaCl (Na^+ and Cl^-) or the set of three ions in $CaCl_2$.

For ionic compounds, the sum of the atomic masses of the elements present in one formula unit is known as the **formula mass** or **formula weight.** For NaCl, this is $22.99 + 35.45 = 58.44$. One mole of NaCl (6.022×10^{23} formula units of NaCl) would contain 58.44 g NaCl. Use of the term formula mass, of course, is not restricted to ionic compounds. It can also be applied to molecular substances, in which case the terms formula mass and molecular mass mean the same thing.

Na	22.99 u
Cl	35.45 u
NaCl	58.44 u

EXAMPLE 2.10. CONVERSIONS BETWEEN MOLES AND GRAMS FOR COMPOUNDS

PROBLEM: Sodium carbonate, Na_2CO_3, is a very important industrial chemical used in making glass.

(a)   How many grams does 0.250 mol Na_2CO_3 weigh?

(b)   How many moles of Na_2CO_3 are in 132 g Na_2CO_3?

SOLUTION: To answer these questions, we need the formula mass of Na_2CO_3. We calculate this from the atomic masses of its elements

2 Na	$2 \times 23.0 =$	46.0 u
1 C	$1 \times 12.0 =$	12.0 u
3 O	$3 \times 16.0 =$	48.0 u
	Total	106.0 u

In this example, calculating the formula mass to the nearest 0.1 u gives enough significant figures to perform the calculations properly

The formula mass is 106.0 u; therefore,

$$1 \text{ mol } Na_2CO_3 = 106.0 \text{ g } Na_2CO_3$$

This can be used to make conversion factors relating grams and moles of Na_2CO_3, which we need to answer the questions.

(a)   To convert 0.250 mol Na_2CO_3 to grams, we set up the units to cancel.

$$0.250 \text{ mol } Na_2CO_3 \times \left(\frac{106.0 \text{ g } Na_2CO_3}{1 \text{ mol } Na_2CO_3}\right) = 26.5 \text{ g } Na_2CO_3$$

(b)   Again, we set the units to cancel.

$$132 \text{ g } Na_2CO_3 \times \left(\frac{1 \text{ mol } Na_2CO_3}{106.0 \text{ g } Na_2CO_3}\right) = 1.25 \text{ mol } Na_2CO_3$$

2.6 PERCENTAGE COMPOSITION

A very simple and also often very useful computation is the calculation of the **percentage composition** of a compound—that is, the percentage of the total mass (also called the percent by weight) contributed by each element. The procedure to determine the percentage composition is illustrated in Example 2.11.

EXAMPLE 2.11. CALCULATING PERCENTAGE COMPOSITION FROM A CHEMICAL FORMULA

PROBLEM: What is the percentage composition of chloroform, $CHCl_3$, a substance once used as an anesthetic?

SOLUTION: The total mass of 1 mol of $CHCl_3$ is obtained from the molecular mass

$$12.01 \text{ u} + 1.008 \text{ u} + 3 \times 35.45 \text{ u} = 119.37 \text{ u}$$

or for 1 mol of $CHCl_3$,

$$12.01 \text{ g} + 1.008 \text{ g} + 3 \times 35.45 \text{ g} = 119.37 \text{ g}$$

Then,

$$\% \text{ C} = \frac{\text{weight carbon}}{\text{weight } CHCl_3} \times 100\%$$

$$\% \text{ C} = \frac{12.01 \text{ g C}}{119.37 \text{ g } CHCl_3} \times 100 = 10.06\% \text{ C}$$

In general,

$$\% \text{ by weight} = \frac{\text{weight of part}}{\text{weight of whole}} \times 100\%$$

Similarly,

$$\% \text{ H} = \frac{1.008 \text{ g H}}{119.37 \text{ g CHCl}_3} \times 100 = 0.844\% \text{ H}$$

$$\% \text{ Cl} = \frac{106.35 \text{ g Cl}}{119.37 \text{ g CHCl}_3} \times 100 = 89.09\% \text{ Cl}$$

total % = 100.00

A similar calculation can be used to determine the mass of an element in a given sample of a compound. This is illustrated in Example 2.12.

EXAMPLE 2.12. CALCULATING THE MASS OF AN ELEMENT IN A SAMPLE OF A COMPOUND

PROBLEM: Calculate the mass of iron (Fe) in a 10.0 g sample of iron oxide, Fe_2O_3, commonly referred to as rust.

ANALYSIS: The easiest way to solve this kind of problem is to examine how we would go about calculating the mass of 1 mol of Fe_2O_3. In one mole of this compound, there are 2 mol Fe and 3 mol O. Therefore,

2 Fe: $2 \times 55.85 \text{ g} = 111.7 \text{ g Fe}$ ← this mass of Fe

3 O: $3 \times 16.00 \text{ g} = 48.0 \text{ g O}$ is in

1 mol $Fe_2O_3 = 159.7$ g Fe_2O_3 ← this mass of Fe_2O_3

Thus, the calculation of a formula mass establishes mass relationships among the elements and one mole of the compound. In this case we can write

$$111.7 \text{ g Fe} \Leftrightarrow 159.7 \text{ g Fe}_2O_3$$

and now we can use this to calculate the mass of Fe in the Fe_2O_3 sample.

SOLUTION: We use the relationship established above to make a conversion factor that changes grams of Fe_2O_3 to grams of Fe.

$$10.0 \text{ g Fe}_2O_3 \times \left(\frac{111.7 \text{ g Fe}}{159.7 \text{ g Fe}_2O_3} \right) \Leftrightarrow 6.99 \text{ g Fe}$$

Thus, in 10.0 g Fe_2O_3 there is 6.99 g Fe.

2.7 CHEMICAL FORMULAS

There are different kinds of chemical formulas, and each conveys certain kinds of information. This can include the elemental composition, the relative numbers of each kind of atom present, the actual numbers of atoms of each kind in a molecule of the substance, or the structure of a molecule of the substance. For convenience, we can classify formulas according to the amount of information that they provide.

A formula that uses the smallest set of whole-number subscripts to specify the relative number of atoms of each element present in a formula unit is called a

simplest formula. It is also called an **empirical formula** because it is normally derived from the results of some experimental analysis. The formulas NaCl, H_2O, and CH_2 are empirical formulas.

A formula that states the actual number of each kind of atom found in a molecule is called a **molecular formula.** H_2O is a molecular formula (as well as an empirical formula) since a molecule of water contains 2 atoms of H and 1 atom of O. The formula C_2H_4 is a molecular formula for a substance (ethylene) containing 2 atoms of carbon and 4 atoms of hydrogen. Note that the simplest formula of this compound is CH_2 because the carbon-to-hydrogen ratio is 1 to 2. The simplest formula CH_2 is not unique to C_2H_4, however. A substance whose empirical formula is CH_2 could have as a molecular formula CH_2, C_2H_4, C_3H_6, and so on.

A third type of formula is a **structural formula,** for example,

acetic acid (present in vinegar)

In a structural formula the dashes between the different atomic symbols represent the "chemical bonds" that bind the atoms to each other in the molecule. You will learn more about them in Chapter 8. A structural formula gives us information about the way in which the atoms in a molecule are linked together and provides information that also allows us to write the molecular and empirical formulas. Thus, for acetic acid shown above, we can also write its molecular formula ($C_2H_4O_2$) and its empirical formula (CH_2O).

The most desirable kind of formula to have, of course, is the structural formula, because it also contains all the information provided by the other two types. However, in chemistry, as in the rest of life, there is no "free lunch"—we never get something for nothing! The more information a formula conveys, the more difficult it is to arrive at experimentally. We will see how empirical and molecular formulas are derived, but most of the procedures involved in the determination of structural formulas are beyond the scope of this book.

2.8 EMPIRICAL FORMULAS AND MOLECULAR FORMULAS

The subscripts in an empirical formula give us the atom ratios in a compound. In CH_2, for example, the C-to-H atom ratio is 1 to 2, and as you've already learned, the atom ratio is exactly the same as the mole ratio. In CH_2 we find the elements in a ratio of 1 mol C to 2 mol H. This equality between atom and mole ratios is the basis for the method of determining empirical formulas; we experimentally measure the mole ratios, which then give us the atom ratios. For instance, suppose we had a sample of a compound containing only carbon and hydrogen in which these elements occurred in a ratio of 1 mol C to 3 mol H. The only way the mole ratio could be 1 to 3 is if the atom ratio is 1 to 3, so the empirical formula must be CH_3.

The next question, of course, is, "How do we obtain the mole ratios?" In *every* empirical formula calculation, we must at some point have the same basic information

> To calculate an empirical formula, we need to know the mass of each of the elements in a given mass of the compound.

Webster's defines empirical as "... founded upon experiment or experience."

It is possible for many different compounds to have the same empirical formula.

Keep this statement in mind whenever you tackle an empirical formula calculation.

Once we have the masses of the elements, we convert them to moles and then find the simplest whole-number ratio of the moles, which gives us the subscripts in the formula. The following examples illustrate how this is done.

EXAMPLE 2.13. CALCULATING AN EMPIRICAL FORMULA

PROBLEM: A sample of a brown-colored gas that is a major air pollutant is found to contain 2.34 g of N and 5.34 g of O. What is the simplest formula of the compound?

SOLUTION: In this problem we are given the mass of each element, so we can go on to calculate the number of moles of each in the sample. We know that

$$1 \text{ mol N} = 14.0 \text{ g N} \quad \text{(Why?)}$$
$$1 \text{ mol O} = 16.0 \text{ g O} \quad \text{(Why?)}$$

Therefore,

$$2.34 \text{ g N} \times \left(\frac{1 \text{ mol N}}{14.0 \text{ g N}} \right) = 0.167 \text{ mol N}$$

$$5.34 \text{ g O} \times \left(\frac{1 \text{ mol O}}{16.0 \text{ g O}} \right) = 0.334 \text{ mol O}$$

We might write our formula $N_{0.167}O_{0.334}$. It does indeed tell us the relative number of moles of N and O; however, since the formula should have meaning on a molecular level where whole numbers of atoms are combined, the subscripts must be integers. If we divide each subscript by the smallest one, we obtain

$$N_{\frac{0.167}{0.167}}O_{\frac{0.334}{0.167}} = NO_2$$

EXAMPLE 2.14. CALCULATING AN EMPIRICAL FORMULA FROM PERCENTAGE COMPOSITION

PROBLEM: What is the empirical formula of a compound composed of 43.7% P and 56.3% O by weight?

SOLUTION: It is quite common to have the results of a chemical analysis in the form of percentage composition by weight. The simplest way to proceed in such a case is to imagine having a 100-g sample of the compound. From the analysis, this sample would contain 43.7 g P and 56.3 g O (notice that the percents become grams of compound). Now that we know the masses of phosphorus and oxygen in the same sample, we convert the masses to moles and proceed as before.

$$43.7 \text{ g P} \times \left(\frac{1 \text{ mol P}}{31.0 \text{ g P}} \right) = 1.41 \text{ mol P}$$

$$56.3 \text{ g O} \times \left(\frac{1 \text{ mol O}}{16.0 \text{ g O}} \right) = 3.52 \text{ mol O}$$

The formula is

$$P_{1.41}O_{3.52} = P_{\frac{1.41}{1.41}}O_{\frac{3.52}{1.41}} = PO_{2.50}$$

We could also obtain whole-number subscripts by multiplying the values by 4, but this would give P_4O_{10}. The empirical formula, however, should have the *smallest* set of whole-number subscripts, so the proper answer is P_2O_5.

Whole numbers can be obtained by doubling each of the subscripts. Thus, the empirical formula is P_2O_5.

Usually, it is not possible to decompose a compound into its elements and then weigh them, so a different strategy must be used. As a rule, this involves causing the compound to undergo some chemical reaction that divides the elements from each other and isolates them in compounds having known chemical formulas. Such analyses require very careful work to ensure that all of a given element in the original sample is isolated in the compound of known formula. The next two examples illustrate this for typical analyses of compounds composed of the elements carbon, hydrogen, and oxygen.

EXAMPLE 2.15. DETERMINING THE EMPIRICAL FORMULA FROM A CHEMICAL ANALYSIS

PROBLEM: A 1.025-g sample of a compound that contains only carbon and hydrogen was burned in oxygen to give carbon dioxide and water vapor as products. These products were trapped separately and weighed. It was found that 3.007 g of CO_2 and 1.845 g H_2O were formed in this reaction. What is the empirical formula of the compound?

ANALYSIS: In order to calculate the empirical formula, we need to know the masses of carbon and hydrogen in a sample of the compound (in this case, in the 1.025-g sample). The combustion reaction has caused all the carbon to become part of the CO_2 and all the hydrogen to become part of the H_2O. If we can calculate the mass of carbon in the 3.007 g of CO_2, we then know how much carbon was in the original sample. Similarly, if we calculate the mass of hydrogen in the 1.845 g of H_2O, we then know how much hydrogen was in the original sample. Once we have the masses of C and H, we proceed as usual.

SOLUTION: In Section 2.6 you learned how to calculate the mass of an element in a given sample of a compound. Let's use that knowledge now to calculate the mass of carbon and hydrogen in the CO_2 and H_2O.

In 1 mol of CO_2 (44.01 g), there is 1 mol (12.01 g) of C. In the 3.007 g of CO_2, therefore, the mass of carbon is

$$3.007 \text{ g CO}_2 \times \left(\frac{12.01 \text{ g C}}{44.01 \text{ g CO}_2} \right) \Leftrightarrow 0.8206 \text{ g C}$$

Similarly, in 1 mol of H_2O (18.02 g), there are 2 mol (2.016 g) of H, so the mass of H in the 1.845 g of H_2O is

$$1.845 \text{ g H}_2O \times \left(\frac{2.016 \text{ g H}}{18.02 \text{ g H}_2O} \right) \Leftrightarrow 0.2064 \text{ g H}$$

Now we have the masses of C and H in the sample, so we can calculate the number of moles of each,

$$0.8206 \text{ g C} \times \left(\frac{1 \text{ mol C}}{12.01 \text{ g C}} \right) = 0.06833 \text{ mol C}$$

and

$$0.2064 \text{ g H} \times \left(\frac{1 \text{ mol H}}{1.008 \text{ g H}} \right) = 0.2048 \text{ mol H}$$

Now we determine the subscripts:

$$C_{0.06833}H_{0.2048}$$

Dividing by the smallest value (0.06833)

$$C_{\frac{0.06833}{0.06833}} H_{\frac{0.2048}{0.06833}}$$

gives $C_1H_{2.977}$. The empirical formula, therefore, is CH_3.

EXAMPLE 2.16. DETERMINING THE EMPIRICAL FORMULA FROM A CHEMICAL ANALYSIS

PROBLEM: A 0.1000-g sample of ethyl alcohol (grain alcohol), known to contain only carbon, hydrogen, and oxygen, was burned completely in oxygen to form the products CO_2 and H_2O. These products were trapped separately and weighed; 0.1910 g CO_2 and 0.1172 g H_2O were found. What is the empirical formula of the compound?

ANALYSIS: This is similar to the preceding problem, but with a twist. As before, we can determine the amount of C and H from the masses of CO_2 and H_2O obtained. However, the oxygen in the original sample is distributed among the CO_2 and H_2O *along with* additional oxygen required in the combustion. In other words, not all the oxygen in the products came from the original sample, so we have no direct method of obtaining the amount of oxygen in the sample. We can get it indirectly, however, because we know that the sum of the masses of C, H, and O must add up to the total mass of the sample. Therefore, after we have calculated the masses of C and H, we just add them together and subtract from the sample weight (0.1000 g) to get the mass of oxygen in the sample. Then, after we've gotten all three masses (that of C, H, and O), we follow our usual procedure, calculating the number of moles of each, and so on.

SOLUTION: First we calculate the mass of carbon and hydrogen in the CO_2 and H_2O.

$$0.1910 \text{ g CO}_2 \times \left(\frac{12.01 \text{ g C}}{44.01 \text{ g CO}_2}\right) \Leftrightarrow 0.05212 \text{ g C}$$

$$0.1172 \text{ g H}_2O \times \left(\frac{2.016 \text{ g H}}{18.02 \text{ g H}_2O}\right) \Leftrightarrow 0.01311 \text{ g H}$$

When we add the masses of C and H, we get a total of 0.06523 g, so the mass of oxygen in the sample must have been

$$\text{mass of O} = (\text{mass of sample}) - (\text{total mass of C and H})$$
$$= 0.1000 \text{ g} - 0.06523 \text{ g} = 0.0348 \text{ g O}$$

Now we convert the masses of C, H, and O to moles of each of the elements. Thus, for carbon we have

$$0.05212 \text{ g C} \times \left(\frac{1 \text{ mol C}}{12.01 \text{ g C}}\right) = 0.004340 \text{ mol C}$$

Similar calculations for the other two elements give 0.0130 mol H and 0.00218 mol O. The empirical formula is therefore

$$C_{0.00434}H_{0.0130}O_{0.00218} = C_{\frac{0.00434}{0.00218}}H_{\frac{0.0130}{0.00218}}O_{\frac{0.00218}{0.00218}}$$

Dividing, we get $C_{1.99}H_{5.96}O_1$, which rounds to C_2H_6O.

or

$$C_2H_6O$$

We approached both of the preceding examples in essentially the same way. We calculated the masses of C and H in the CO_2 and H_2O, and then later we changed these masses to moles so we could calculate the mole ratios. We might think of this as a generic approach to these kinds of problems. But perhaps you noticed that for Example 2.15 there is a shorter route to the number of moles of C and H. Because the compound in this example contained only C and H, we really didn't need to know the mass of each element. All we need to calculate is the number of moles of C in the

CO_2 and the number of moles of H in the H_2O. These are the number of moles of C and H in the original sample of the compound. Thus, for carbon

$$3.007 \text{ g } CO_2 \times \left(\frac{1 \text{ mol } CO_2}{44.01 \text{ g } CO_2} \right) \times \left(\frac{1 \text{ mol C}}{1 \text{ mol } CO_2} \right) = 0.06833 \text{ mol C}$$

and for hydrogen

$$1.845 \text{ g } H_2O \times \left(\frac{1 \text{ mol } H_2O}{18.02 \text{ g } H_2O} \right) \times \left(\frac{2 \text{ mol H}}{1 \text{ mol } H_2O} \right) = 0.2048 \text{ mol } H_2O$$

Once we have these values, we proceed as before.

In Example 2.16, we had to calculate the mass of C and H so that we could subtract their total from the mass of the sample to obtain the mass of oxygen. There was no shortcut available for this problem.

The lesson to be learned from this is that by obtaining a firm understanding of the concepts developed here, and by thinking the problem through before attempting to solve it, you can sometimes save yourself some work.

Molecular formulas

Not only does the molecular formula give us the atom ratios in a compound, it also tells us the actual number of atoms of each kind in a molecule of the substance. Now, you should recall that it is possible for more than one compound to have the same empirical formula. The molecules C_2H_4, C_3H_6, C_4H_8, and C_5H_{10} all have a 1-to-2 ratio of carbon to hydrogen atoms and the empirical formula CH_2. One important way that these substances differ, however, is that they have different molecular masses. In fact, their molecular masses are multiples of the mass of the simplest formula, CH_2, as illustrated in Table 2.1. Recognizing this, we see that if we were able in some way to learn the molecular mass of the compound, we could simply divide the molecular mass by the mass of the empirical formula unit to obtain a whole number. This number tells us how many times the empirical formula is repeated in the molecular formula. Multiplying the subscripts in the empirical formula by this integer then gives the formula for the molecule, as illustrated in Example 2.17.

Table 2.1. Molecular masses as multiples of the empirical formula mass

Formula	Molecular mass
CH_2	$14.0 = 1 \times 14.0$
C_2H_4	$28.0 = 2 \times 14.0$
C_3H_6	$42.0 = 3 \times 14.0$
C_4H_8	$56.0 = 4 \times 14.0$
C_nH_{2n}	$n \times 14.0$

Ionic compounds don't have molecular formulas because they don't contain molecules.

You will learn how molecular masses can be measured in Chapters 11 and 13.

EXAMPLE 2.17. DETERMINING THE MOLECULAR FORMULA OF A COMPOUND

PROBLEM: A colorless liquid used in rocket engines, whose empirical formula is NO_2, has a molecular mass of 92.0. What is its molecular formula?

SOLUTION: The formula mass of NO_2 is 46.0. The number of times the empirical formula, NO_2, occurs in the compound is

$$\frac{92.0}{46.0} = 2$$

The molecular formula is then $(NO_2)_2 = N_2O_4$ (dinitrogen tetroxide). N_2O_4 is the preferred answer because $(NO_2)_2$ implies a knowledge of the structure of the molecule (i.e., that two NO_2 units are somehow joined together).

REVIEW QUESTIONS AND PROBLEMS

Problems whose numbers are in blue have their answers in Appendix D at the back of the book. The more difficult problems are marked with asterisks.

Dalton's Atomic Theory and Atomic Masses

2.1 State the postulates of Dalton's atomic theory.

2.2 Which chemical laws did Dalton's theory explain? Which law did it predict?

2.3 What is the most important atomic property in Dalton's atomic theory?

2.4 Distinguish between the terms atom and molecule.

2.5 How is the currently accepted atomic mass unit defined?

2.6 If the atomic mass unit were defined in such a way that a single fluorine atom weighed 1 u, what would the atomic masses of carbon and hydrogen be?

2.7 "Elements A and B form a compound in which for every 3.00 g of A there are 16.0 g of B." Does this statement supply enough information to enable you to calculate the relative masses of atoms of A and B? Explain your answer.

* 2.8 In a certain compound, 6.92 g of X was found to be combined with 0.584 g of carbon. Given that the atomic mass of carbon is 12.0 u and the formula of the compound is CX_4, calculate the atomic mass of X.

2.9 Suppose that the element carbon has been assigned a mass of 4.00 atomic mass units. What would the atomic masses of oxygen and sulfur be?

Laws of Definite Proportions and Multiple Proportions

2.10 State, in words, the following laws:
(a) the law of conservation of mass
(b) the law of definite proportions
(c) the law of multiple proportions

2.11 Cyclopropane, a very effective anesthetic, contains the elements carbon and hydrogen combined in a ratio of 1.00 g of hydrogen to 6.00 g of carbon. If a given sample of cyclopropane was found to contain 24.0 g of hydrogen, how many grams of carbon would it contain?

2.12 Three samples of a solid substance composed of elements X and Y were prepared. The first was found to contain 4.31 g X and 7.69 g Y; the second was composed of 35.9% X and 64.1% Y; it was observed that 0.718 g X reacted with Y to form 2.00 g of the third sample. Show how these demonstrate the law of definite composition.

2.13 Two samples of Freon (a coolant used in refrigerators and air conditioners) were analyzed. In one sample, 1.00 g C was found to be combined with 6.33 g F and 11.67 g Cl. In the second sample, 2.00 g C was found to be combined with 12.66 g F and 23.34 g Cl. What are the ratios of the masses of carbon to fluorine, carbon to chlorine, and fluorine to chlorine in each of these samples. Do the data support the law of definite proportions? Explain your answer.

2.14 Copper forms two oxides. In one of them, there are 1.26 g of oxygen combined with 10.0 g of copper. In the other, there are 2.52 g of oxygen combined with 10.0 g of copper. Show that these data illustrate the law of multiple proportions.

2.15 Two compounds are formed between phosphorus and oxygen. A 1.50-g sample of one compound was found to contain 0.845 g of phosphorus, while a 2.50-g sample of the other contained 1.09 g of phosphorus. Show that these data are consistent with the law of multiple proportions.

* 2.16 In a sample of the compound MnO, 4.00 g of oxygen is combined with 13.7 g of manganese. How many grams of oxygen would be combined with 7.85 g of manganese in the compound MnO_2?

The Mole and Simple Mole Relationships

2.17 In what sense are the mole, the dozen, and the gross related to each other? How many things are in a mole of something?

2.18 What difference, if any, is there between the expressions *1 mol C* and *one mole of carbon atoms*?

2.19 Aluminum metal is resistant to corrosion because the metal at the surface combines with oxygen from the air to form a thin, tough film of aluminum oxide, Al_2O_3, that protects the metal underneath. In this compound,
(a) What is the Al to O atom ratio?
(b) What is the Al to O mole ratio?
(c) If 2 mol Al were to react to form this compound, how many moles of oxygen atoms would react?
(d) If 0.2 mol Al were to react to form this compound, how many moles of oxygen atoms would react?

2.20 In the air pollutant SO_2 (sulfur dioxide), (a) what is the sulfur-to-oxygen atom ratio, and (b) what is the mole ratio of sulfur to oxygen atoms?

2.21 For baking soda, $NaHCO_3$, give the atom ratio and the mole ratio of
(a) Na to H   (d) H to C
(b) Na to C   (e) H to O
(c) Na to O   (f) C to O

2.22 Express each mole ratio in the preceding question as a *pair* of conversion factors that could be used in calculations that employ the factor-label method.

2.23 Aluminum sulfate, $Al_2(SO_4)_3$, is a compound used in sewage treatment plants.
- (a) How many atoms of S are in one formula unit of $Al_2(SO_4)_3$?
- (b) How many moles of S are in 1 mol $Al_2(SO_4)_3$?
- (c) How many atoms of Al are in one formula unit of $Al_2(SO_4)_3$?
- (d) How many moles of Al are in 1 mol $Al_2(SO_4)_3$?
- (e) Use your answer to part b to construct a pair of conversion factors that relate moles of S to moles of $Al_2(SO_4)_3$.
- (f) Use your answer to part d to construct a pair of conversion factors that relate moles of Al to moles of $Al_2(SO_4)_3$.

2.24 Butane, the volatile liquid fuel in disposable cigarette lighters, has the formula C_4H_{10}.
- (a) What is the mole ratio of C to H in this compound?
- (b) How many moles of C are in 3.00 mol C_4H_{10}?
- (c) How many moles of H are in 0.250 mol C_4H_{10}?
- (d) How many moles of C must combine with 0.600 mol H to make this compound?
- (e) Hydrogen occurs in nature as *diatomic* molecules, H_2. How many moles of H_2 have enough H atoms to react with 0.600 mol C to give C_4H_{10}?

2.25 Ordinary sand is composed chiefly of silica, a compound in which there are two oxygen atoms for each silicon atom.
- (a) What is the formula for silica?
- (b) How may atoms of oxygen would be needed to combine with 25 atoms of silicon to form silica?
- (c) How many moles of oxygen atoms would be needed to combine with 25 moles of silicon atoms to form silica?
- (d) If you had 4.50 moles of silica, how many moles of silicon and oxygen atoms would there be in it?

2.26 How many moles of S are in 1.00 mol of As_2S_3?

2.27 How many moles of O are in 1.50 mol of Cr_2O_3?

2.28 Based only on the amount of carbon available, how many moles of CO_2 could be liberated from 1.00 mol of limestone, $CaCO_3$?

2.29 How many moles of sulfate (SO_4) are there in 1 mol of $Al_2(SO_4)_3$?

2.30 How many moles of $BaSO_4$ could be made from 1.25 mol of $Al_2(SO_4)_3$, provided that sufficient amounts of any other required chemicals are available.

Measuring Moles of Atoms and Compounds

2.31 Give the mass in grams of 1.00 mol of atoms of each of the following elements:
- (a) magnesium
- (b) carbon
- (c) iron
- (d) chlorine
- (e) sulfur
- (f) strontium

2.32 For each of the elements in the preceding question, construct a pair of conversion factors that relate *grams of the element* to *moles of atoms of the element*.

2.33 How many moles of atoms are in 50.0 g of each of the following elements?
- (a) sodium
- (b) arsenic
- (c) chromium
- (d) aluminum
- (e) potassium
- (f) silver

2.34 Why is the term formula mass preferred for some substances (e.g., NaCl) rather than molecular mass?

2.35 Calculate the formula mass of the following and round the answer to the nearest 0.1 u.
- (a) MgO
- (b) $CaCl_2$
- (c) PCl_5
- (d) S_2Cl_2
- (e) Na_3PO_4

2.36 What is the mass in grams of 1.00 mol of each of the compounds in the preceding question?

2.37 For each compound in Problem 2.35, construct a pair of conversion factors that relate *grams of compound* to *moles of compound*.

2.38 Calculate the formula mass of the following and round the answer to the nearest 0.1 u.
- (a) SiO_2 (quartz)
- (b) $Mg(OH)_2$ (in milk of magnesia)
- (c) $MgSO_4 \cdot 7H_2O$ (Epsom salts)
- (d) $Ca_2Mg_5(Si_4O_{11})_2(OH)_2$ (asbestos)
- (e) $C_6H_8O_6$ (vitamin C)
- (f) $C_{12}H_{22}O_{11}$ (sucrose – table sugar)

2.39 What is the mass in grams of 1.35 mol of caffeine, $C_8H_{10}N_4O_2$?

2.40 What is the mass in grams of 2.33 mol of penicillin, $C_{16}H_{18}O_4N_2S$?

2.41 What is the mass in grams of 6.30 mol of lead sulfate, $PbSO_4$?

2.42 What is the mass in grams of 0.144 mol of TiO_2, a pigment used in white paint?

2.43 How many moles of sodium bicarbonate, $NaHCO_3$ (baking soda), are in a 242-g sample?

2.44 How many moles of butane, C_4H_{10}, are in 1.40×10^3 g of butane?

2.45 How many moles of sulfuric acid, H_2SO_4, are in 85.3 g of H_2SO_4?

2.46 A substance with the formula $PbHAsO_4$ has been used in insecticides, and on the label it appears under the name lead arsenate. How many moles of this poison are in 25.0 g of it?

2.47 How many moles of potassium atoms are in 125 g of KCl?

2.48 How many moles of S are in 632 g of iron pyrite, FeS_2?

2.49 When coal containing iron pyrite, FeS_2, is burned, all the sulfur is converted to the air pollutant, sulfur dioxide, SO_2. How many moles of FeS_2 would have to react to produce 1.00 kg of SO_2?

Using Avogadro's Number

2.50 What is the mass, in grams, of one atom of Fe? What is the mass of one molecule of SO_2?

2.51 If a sample of ethylene, C_2H_4, contains 3.50×10^{17} atoms of carbon, how many atoms of hydrogen does it contain? How many grams of carbon and hydrogen are in the sample?

2.52 Ordinary table sugar is sucrose, $C_{12}H_{22}O_{11}$. What is the mass in grams of one molecule of sucrose? How many times heavier is a molecule of sucrose than one atom of carbon? How many molecules of sucrose are in 25.0 g of sucrose? What is the total number of *atoms* in 25.0 g of sucrose?

2.53 How many atoms of carbon are in 4.00×10^{-8} g of propane, C_3H_8?

2.54 Carbon atoms have a diameter of about 1.5×10^{-8} cm. If carbon atoms were laid in a row 3.0 cm long, what would be the total mass of carbon?

2.55 Calculate the mass of Cu required to react with 5.00×10^{20} molecules of S_8 to form Cu_2S.

Percentage Composition

2.56 Calculate the percentage composition of each of the following:
(a) $FeCl_3$
(b) Na_3PO_4
(c) $KHSO_4$
(d) $(NH_4)_2HPO_4$
(e) Hg_2Cl_2

2.57 Calculate the percentage composition of each of the following:
(a) (benzene) C_6H_6
(b) (ethyl alcohol) C_2H_5OH
(c) (potassium dichromate) $K_2Cr_2O_7$
(d) (xenon tetrafluoride) XeF_4
(e) (calcium carbonate) $CaCO_3$

2.58 For each part of Problem 2.57, calculate the mass of the first element in the formula that is present in one mole of the compound. Then for each, use this information to construct a pair of conversion factors that relate *grams of element* to *grams of the compound*.

2.59 Calculate the number of grams of (a) Fe in 15.0 g Fe_2O_3, (b) Al in 25.0 g $Al_2(SO_4)_3$, (c) Na in 16.0 g Na_2CO_3, and (d) Mg in 48.0 g $MgCl_2$.

2.60 Calculate the mass of nitrogen in 30.0 g of the amino acid glycine, $CH_2(NH_2)COOH$.

2.61 Calculate the mass of hydrogen in 12.0 g of NH_3.

2.62 A 12.5-g sample of a compound containing only phosphorus and sulfur was analyzed and found to contain 7.04 g of phosphorus and 5.46 g of sulfur. What is the percentage composition of this compound?

2.63 A 4.25-g sample of a compound that contains only carbon, hydrogen, and oxygen was burned in an atmosphere of pure O_2. This produced 9.34 g CO_2 and 5.09 g H_2O. (a) Calculate the masses of C and H that were in the original 4.25-g sample. (b) What was the mass of O in the original sample? (c) What is the percentage composition of this compound?

Chemical Formulas

2.64 State the difference between a structural formula, a molecular formula, and an empirical formula?

2.65 What is the empirical formula of each of the following?
(a) $(NH_4)_2S_2O_8$
(b) Fe_2O_3
(c) Al_2Cl_6
(d) C_6H_6
(e) $C_3H_8O_3$
(f) $C_6H_{12}O_6$
(g) Hg_2SO_4

2.66 Ethylene glycol, used as permanent antifreeze, has the structural formula

$$H-O-\underset{\underset{H}{|}}{\overset{\overset{H}{|}}{C}}-\underset{\underset{H}{|}}{\overset{\overset{H}{|}}{C}}-O-H$$

What are its molecular formula and empirical formula?

2.67 Why is the simplest formula for a substance called an empirical formula? What is the dictionary definition of empirical?

2.68 What kind of basic information must you have in order to calculate the empirical formula of a compound?

2.69 A sample of an air pollutant composed of sulfur and oxygen was found to contain 1.40 g sulfur and 2.10 g oxygen. What is the empirical formula of the compound?

2.70 The Freon propellant from an aerosol can was analyzed. A sample of it contained 0.423 g C, 2.50 g Cl, and 1.34 g F. What is the empirical formula of this substance?

2.71 What is the empirical formula of the compound described in Problem 2.62?

2.72 What is the empirical formula of the compound described in Problem 2.63?

2.73 A dry-cleaning fluid composed of carbon and chlorine was found to have the composition 14.5% C, 85.5% Cl (by mass). What is the empirical formula of this compound?

2.74 Arsenic reacts with oxygen to form a compound that is 75.7% arsenic and 24.3% oxygen, by mass. What is the empirical formula of this compound?

2.75 A 1.31-g sample of sulfur was allowed to react with an excess of chlorine to produce 4.22 g of a product that contains only sulfur and chlorine. What is the empirical formula of the compound?

2.76 A substance was found to have the following composition by mass: 60.8% sodium, 28.5% boron, and 10.5% hydrogen. What is the empirical formula of the compound?

2.77 Vanillin has the following composition by mass: carbon, 63.2%; hydrogen, 5.26%; and oxygen, 31.6%. What is the empirical formula of vanillin?

2.78 A 0.537-g sample of an organic compound containing only carbon, hydrogen, and oxygen was burned in air to produce 1.030 g CO_2 and 0.632 g H_2O. What is the empirical formula of the compound?

2.79 A 1.500-g sample of a compound composed of chromium and chlorine was dissolved in water and allowed to react with $AgNO_3$. This converted all the chlorine in the original sample into AgCl, which was collected and found to weigh 4.072 g.
  (a) How many grams of Cl are in the AgCl obtained in this analysis?
  (b) How many grams of Cr and Cl were in the original 1.500-g sample?
  (c) What is the empirical formula of the chromium–chlorine compound?

* 2.80 A 1.35-g sample of a substance containing carbon, hydrogen, nitrogen, and oxygen was burned to produce 0.810 g H_2O and 1.32 g CO_2. In a separate reaction, all the nitrogen in 0.735 g of the substance was converted to ammonia. This gave 0.284 g NH_3. Determine the empirical formula of the substance.

* 2.81 An organic compound was synthesized and a sample of it was analyzed and found to contain only C, H, N, O, and Cl. It was observed that when a 0.150-g sample of the compound was burned, it produced 0.138 g CO_2 and 0.0566 g H_2O. All the nitrogen in a different 0.200-g sample of the compound was converted to NH_3, which was found to weigh 0.0238 g. Finally, the chlorine in a 0.125-g sample of the compound was converted to Cl^- and by reacting it with $AgNO_3$, all the chlorine was recovered as AgCl. The AgCl, when dried, was found to weigh 0.251 g.
  (a) Calculate the weight percent of each element in the compound.
  (b) Determine the empirical formula for the compound.

2.82 The following are empirical formulas and formula masses (FM) for five compounds. What are their molecular formulas?
  (a) NaS_2O_3,   FM = 270.4
  (b) C_3H_2Cl,   FM = 147.0
  (c) C_2HCl,   FM = 181.4
  (d) Na_2SiO_3,   FM = 732.6
  (e) $NaPO_3$,   FM = 305.9

2.83 Citric acid, the substance that makes lemon juice sour, is composed of only carbon, hydrogen, and oxygen. When a 0.5000-g sample of citric acid was burned, it produced 0.6871 g CO_2 and 0.1874 g H_2O. The molecular mass of the compound is 192. What are the empirical and the molecular formulas for citric acid?

2.84 Polystyrene, a common plastic, is composed of many styrene units linked together as shown below.

The basic styrene unit is shown within brackets. The subscript n means that this unit is repeated many times. A particular sample of this plastic was found to have an average molecular mass of 1 million u. What is the average number of styrene units in a chain?

Additional Exercises

2.85 An atom of a certain isotope of gold is 14.9977 times as heavy as an atom of carbon-12. What is the atomic mass of this gold isotope?

2.86 How many moles of propane, C_3H_8, must be burned to give the same number of moles of CO_2 as would be produced by the combustion of 2.50 mol of butane, C_4H_{10}?

2.87 How many grams of oxygen are in 65.0 g of copper sulfate, $CuSO_4 \cdot 5H_2O$?

2.88 One of the new superconducting materials discovered in 1987 has the following composition by mass: barium, 41.23%; copper, 28.62%; yttrium, 13.35%; and oxygen, 16.81%. What is the empirical formula of this substance?

2.89 The element chromium has a density of 7.20 g/cm^3. What volume, expressed in cubic centimeters, is occupied by a single chromium atom?

* 2.90 The chromium plating on a typical automobile bumper is approximately 7.5×10^{-5} cm (about 30 millionths of an inch) thick. Given that the density of chromium is 7.20 g/cm^3, approximately how many atoms thick is the chromium plating? (For simplicity, assume cubic atoms.)

2.91 A sample of a compound of silver and sulfur weighing 0.9225 g was heated in oxygen, which changed all the sulfur to sulfur dioxide, SO_2. The SO_2 was dissolved in a solution of NaOH and then allowed to react with $KMnO_4$ and $BaCl_2$, which finally produced $BaSO_4$. The $BaSO_4$ was collected and found to weigh 0.8689 g. What is the empirical formula of the silver–sulfur compound? (This is the black compound that forms on silverware when it tarnishes.)

2.92 A compound of mercury and chlorine once used as a treatment for syphilis (before penicillin was discovered) is composed of 84.98% mercury by mass. Its molecular mass is 472. What is the molecular formula of the compound?

2.93 Iron pyrite, FeS_2, forms beautiful golden crystals that are known as "fool's gold."
  (a) How many moles of sulfur would be needed to combine with 1.00 mol Fe to form FeS_2?
  (b) How many moles of iron are needed to combine with 1.44 mol S to form FeS_2?
  (c) How many moles of sulfur are in 3.00 mol FeS_2?
  (d) How many moles of FeS_2 are needed to give 3.00 mol Fe?

3

CHEMICAL REACTIONS
AND THE MOLE CONCEPT

In the last chapter you learned how to apply the mole concept to elements and compounds. You learned how to measure moles of these substances using a balance, and you learned how, by determining the mole ratios of the elements in compounds, chemical formulas can be obtained. These are just a few aspects of stoichiometry that depend on the mole as the critical central unit in the chemical reasoning. In this chapter we will examine other applications of the mole concept as we study how to relate the amounts of chemicals involved in chemical reactions.

3.1 CHEMICAL REACTIONS AND CHEMICAL EQUATIONS

Whenever chemists begin to think of the changes that occur during a chemical reaction, they almost instinctively start with a chemical equation. As you learned earlier, such an equation presents a sort of "before and after" picture of the chemical substances involved. By examining an equation, therefore, we can get an overview of what takes place.

In order to write an equation, we must be able to write the formulas of the reactants (those substances written at the left of the arrow) and the products (those substances on the right). How a chemist might arrive at such information depends on the reason the equation is being written.

If an experiment has just been carried out, the equation might be meant to describe what has just occurred in a chemical reaction. In this case, the reactants are known because the chemist knows what chemicals were placed in the reaction vessel. The products, however, must be collected and identified (by a chemical analysis, for example) before a valid equation can be written.

Often we write chemical equations to help us plan experiments. In these cases, either we know from past experience what will occur when the reactants are mixed, or we might be making an educated guess about the outcome of the reaction. At this time, you're not expected to know or predict the outcome of reactions, so you will be told what the reactants and products are in equations. However, later in the course, you will learn how to predict certain kinds of reactions and you will be expected to learn how certain chemicals react when mixed.

One way that chemical equations are particularly useful in planning experiments is that they allow us to determine the quantitative relationships that exist among the amounts of reactants and products, and this is the topic we will explore in the pages ahead. To be helpful in this way, however, chemical equations must be balanced, which means that they must obey the law of conservation of mass by having the same number of atoms of each kind on both sides of the arrow.

Balancing chemical equations

In order to minimize errors, writing a balanced chemical equation should always be viewed as a two-step process.

Step 1. First write an unbalanced equation, being careful to write the correct formula for each substance involved. (You're not expected to know the formulas of specific compounds yet and, as noted above, you're not yet expected to predict the products of reactions. For now, these will be given to you.)

Step 2. Balance the equation by adjusting the coefficients that precede the formulas of the reactants and products so that there are the same number of atoms of each kind on both sides of the arrow.

Hydrochloric acid is added to a solution of sodium carbonate. Among the products is the gas CO_2, which is seen bubbling from the reaction mixture.

It is very important to remember that in carrying out step 2, *you must not alter the formulas of the reactants or products.* To do so changes the nature of the chemicals described by the equation, so even if you obtain a balanced equation, it can't possibly be the correct equation.

Most simple equations can be easily balanced by *inspection.* This requires examining the equation and adjusting the coefficients until equal numbers of each element are present among both the reactants and products. For example, consider the reaction illustrated in the photograph in the margin, which shows a solution of hydrochloric acid (HCl) being added to a solution of sodium carbonate (Na_2CO_3). The products of the reaction are sodium chloride ($NaCl$), gaseous carbon dioxide (CO_2), and water.

To obtain the balanced equation for the reaction, we proceed as follows.

Step 1. Write the unbalanced equation, being sure to give the correct formulas for the reactants and products.

$$Na_2CO_3 + HCl \longrightarrow NaCl + H_2O + CO_2$$

Step 2. Place coefficients in front of chemical formulas to balance the equation. This is where you will probably need some practice if you are to learn to balance an equation quickly. Although there are no set rules to tell you where to start, it is often best to seek out the most complex formula in the equation and begin there by giving it a coefficient of 1. In this case, we start with the Na_2CO_3. There are two atoms of Na in this formula, so to balance the Na we need to place a coefficient of 2 in front of NaCl on the right. This gives

$$Na_2CO_3 + HCl \longrightarrow 2NaCl + H_2O + CO_2$$

Balancing an equation by inspection proceeds partly by trial and error and becomes easier with practice.

Balance atoms other than H and O first, then balance H, and finally balance O. This is usually the best way to proceed.

Although this balances the Na, it causes the Cl to become out of balance, but we can correct this by placing a 2 in front of HCl on the left.

$$Na_2CO_3 + 2HCl \longrightarrow 2NaCl + H_2O + CO_2$$

Notice that this also brings the hydrogen into balance, and a quick count for each element reveals that the equation is now balanced.

The coefficients we obtained for the equation above aren't the only ones that will give a balanced equation. For this equation, and in fact for *any* equation, there is an infinite number of sets of coefficients that make the numbers of atoms of each kind the same on both sides of the arrow. For example, both of the following equations are also balanced.

$$2Na_2CO_3 + 4HCl \longrightarrow 4NaCl + 2H_2O + 2CO_2$$

$$5Na_2CO_3 + 10HCl \longrightarrow 10NaCl + 5H_2O + 5CO_2$$

The usual practice, however, is to consider the equation *properly balanced* only when the smallest set of whole-number coefficients is used. (However, as you will see later, there are some occasions when it is convenient to violate this rule.)

EXAMPLE 3.1. BALANCING AN EQUATION BY INSPECTION

PROBLEM: Balance the following equation for the combustion of octane, C_8H_{18}, which is a component of gasoline:

$$C_8H_{18} + O_2 \longrightarrow CO_2 + H_2O$$

SOLUTION: First, we assign C_8H_{18} (the most complex formula) a coefficient of 1. Next, we see that we need to have $8CO_2$ on the right to balance the carbon and $9H_2O$ on the right to balance the hydrogen ($9H_2O$ contains 18 H atoms because each H_2O contains 2 H atoms). This gives

$$C_8H_{18} + O_2 \longrightarrow 8CO_2 + 9H_2O$$

Now we can work on the oxygen. On the right there are 25 O atoms ($2 \times 8 + 9 = 25$). On the left the O atoms come in pairs. This means that we must have $12\frac{1}{2}$ pairs (O_2 molecules) to have 25 O atoms on the left. This gives us

$$C_8H_{18} + 12\frac{1}{2}O_2 \longrightarrow 8CO_2 + 9H_2O$$

Finally, we can eliminate the fractional coefficient by doubling each coefficient.

$$2C_8H_{18} + 25O_2 \longrightarrow 16CO_2 + 18H_2O$$

3.2 CALCULATIONS BASED ON CHEMICAL EQUATIONS

A chemical equation can be interpreted in several ways. Consider, for example, the balanced equation for the combustion of ethanol, C_2H_5OH, the alcohol that's blended with gasoline in the fuel known as *gasohol.*

$$C_2H_5OH + 3O_2 \longrightarrow 2CO_2 + 3H_2O$$

On a molecular, submicroscopic level, we can view this as a reaction between individual molecules.

1 molecule C_2H_5OH + 3 molecules $O_2 \longrightarrow$ 2 molecules CO_2 + 3 molecules H_2O

But we can just as easily scale this up to lab-sized amounts by applying the same "mole reasoning" as in the last chapter. There you learned that the ratio by which atoms of the elements are combined in a compound is exactly the same as the ratio by which moles of the atoms are combined; *the atom ratios and the mole ratios are identical.*

For a chemical reaction, we can make a similar statement. *The ratios by which the molecules react or are formed are exactly the same as the ratios by which* **moles** *of these substances react or are formed.* This means that for the combustion of ethanol, we can also write

$$1 \text{ mol } C_2H_5OH + 3 \text{ mol } O_2 \longrightarrow 2 \text{ mol } CO_2 + 3 \text{ mol } H_2O$$

Of course, we don't always have to start with 1 mol of C_2H_5OH. If we burned 2 mol of ethanol, then

$$2 \text{ mol } C_2H_5OH + 6 \text{ mol } O_2 \longrightarrow 4 \text{ mol } CO_2 + 6 \text{ mol } H_2O$$

In fact, we can use any amounts we wish, but we will always find that three times as many moles of O_2 are consumed as moles of C_2H_5OH, and for every mole of C_2H_5OH consumed, 2 mol of CO_2 and 3 mol of H_2O will be formed. We get this information from the chemical equation because

> The coefficients in a chemical equation provide the ratios by which moles of one substance react with or form moles of another.

If you understand this key statement, then you're ready to solve any stoichiometry problem dealing with a chemical reaction.

For example, the equation for the combustion of C_2H_5OH gives six chemical equivalences that we can use to form conversion factors for calculations.

$$1 \text{ mol } C_2H_5OH \Longleftrightarrow 3 \text{ mol } O_2$$

$$1 \text{ mol } C_2H_5OH \Longleftrightarrow 2 \text{ mol } CO_2$$

$$1 \text{ mol } C_2H_5OH \Longleftrightarrow 3 \text{ mol } H_2O$$

$$3 \text{ mol } O_2 \Longleftrightarrow 2 \text{ mol } CO_2$$

$$3 \text{ mol } O_2 \Longleftrightarrow 3 \text{ mol } H_2O$$

$$2 \text{ mol } CO_2 \Longleftrightarrow 3 \text{ mol } H_2O$$

Let's look at some examples that illustrate how we use this information in a chemical calculation.

EXAMPLE 3.2. USING A CHEMICAL EQUATION IN A CALCULATION INVOLVING MOLES

PROBLEM: How many moles of oxygen are needed to burn 1.80 mol C_2H_5OH according to the balanced equation

$$C_2H_5OH + 3O_2 \longrightarrow 2CO_2 + 3H_2O$$

SOLUTION: The coefficients of the equation give us the relationship

$$1 \text{ mol } C_2H_5OH \Longleftrightarrow 3 \text{ mol } O_2$$

which we can use for a conversion factor. We set up the arithmetic so the units mol C_2H_5OH cancel.

$$1.80 \text{ mol } C_2H_5OH \times \left(\frac{3 \text{ mol } O_2}{1 \text{ mol } C_2H_5OH} \right) \Longleftrightarrow 5.40 \text{ mol } O_2$$

We need 5.40 mol O_2.

EXAMPLE 3.3. USING A CHEMICAL EQUATION IN A CALCULATION INVOLVING MOLES

PROBLEM: How many moles of CO_2 will be formed when 0.274 mol C_2H_5OH burns?

SOLUTION: Now we look at the coefficients for C_2H_5OH and CO_2, which give us

$$1 \text{ mol } C_2H_5OH \Longleftrightarrow 2 \text{ mol } CO_2$$

Then we set up the arithmetic to get the proper units in the answer.

$$0.274 \text{ mol } C_2H_5OH \times \left(\frac{2 \text{ mol } CO_2}{1 \text{ mol } C_2H_5OH} \right) \Longleftrightarrow 0.548 \text{ mol } CO_2$$

EXAMPLE 3.4. USING A CHEMICAL EQUATION IN A CALCULATION INVOLVING MOLES

PROBLEM: How many moles of water will form when 3.66 mol CO_2 are produced during the combustion of C_2H_5OH?

SOLUTION: The coefficients in the equation for the combustion of C_2H_5OH tell us that

$$2 \text{ mol } CO_2 \Longleftrightarrow 3 \text{ mol } H_2O$$

Therefore,

$$3.66 \text{ mol } CO_2 \times \left(\frac{3 \text{ mol } H_2O}{2 \text{ mol } CO_2} \right) \Longleftrightarrow 5.49 \text{ mol } H_2O$$

When we actually carry out an experiment in the lab, we are normally interested in working in laboratory units such as grams. For example, we may want to know how many grams of one reactant are needed to completely use up some number of grams of another, or we may want to know how many grams of some product are formed from some given number of grams of a reactant. In working problems of this kind, we always follow essentially the same path, which can be diagrammed as shown in Figure 3.1.

Perhaps you recognize that the strategy described in Figure 3.1 is effectively the same as the one we used in Chapter 2 when working with the numbers of grams of elements combined in a compound. *We begin with lab units, change them to chemical units (moles), do our reasoning in moles, and then at the end we translate back to lab units.* Keep this sequence of operations in mind, not only for the problems that you solve in this chapter, but for others of a similar nature that you will encounter in later chapters.

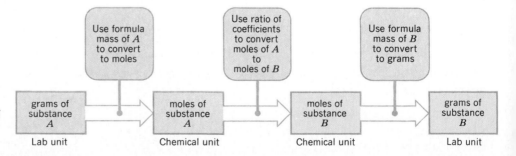

Figure 3.1. *A diagram showing in a general way how we tackle stoichiometry problems that involve a chemical reaction.*

EXAMPLE 3.5. USING A CHEMICAL EQUATION IN A CALCULATION INVOLVING GRAMS

PROBLEM: In Review Exercise 2.19 it was noted that freshly exposed surfaces of aluminum react with oxygen to form a tough film of aluminum oxide, Al_2O_3, that protects the metal underneath from further corrosion. The reaction is

$$4Al + 3O_2 \rightarrow 2Al_2O_3$$

How many grams of O_2 are required to react with 0.300 mol Al?

ANALYSIS: This problem is just slightly more difficult than the previous examples. In this case, we are seeking grams instead of moles.

$$0.300 \text{ mol Al} \Leftrightarrow ? \text{ g } O_2$$

There is no direct relationship to connect moles of Al with grams of O_2, but the balanced equation does provide a path between moles of Al and moles of O_2.

$$4 \text{ mol Al} \Leftrightarrow 3 \text{ mol } O_2$$

We can use this to calculate the number of moles of O_2 needed. Then we can use the formula mass of O_2 to change moles of O_2 to grams of O_2.

$$1 \text{ mol } O_2 \Leftrightarrow 32.0 \text{ g } O_2$$

A common error is to write the formula for oxygen as O_2 and then use 16.0 as the molecular mass. Be careful about this.

SOLUTION: First we find the number of moles of O_2 needed to react with the Al.

$$0.300 \text{ mol Al} \times \left(\frac{3 \text{ mol } O_2}{4 \text{ mol Al}} \right) \Leftrightarrow 0.225 \text{ mol } O_2$$

Next, we change moles of O_2 to grams of O_2 to get our answer.

$$0.225 \text{ mol } O_2 \times \left(\frac{32.0 \text{ g } O_2}{1 \text{ mol } O_2} \right) = 7.20 \text{ g } O_2$$

We could have combined both steps by stringing together the conversion factors as follows.

$$0.300 \text{ mol Al} \times \left(\frac{3 \text{ mol } O_2}{4 \text{ mol Al}} \right) \times \left(\frac{32.0 \text{ g } O_2}{1 \text{ mol } O_2} \right) \Leftrightarrow 7.20 \text{ g } O_2$$

Regardless of how we work the arithmetic, however, the answer is the same. We need 7.20 g O_2 to react with 0.300 mol Al.

EXAMPLE 3.6. USING A CHEMICAL EQUATION IN A CALCULATION INVOLVING GRAMS

PROBLEM: Using the chemical equation in the preceding example, calculate the number of grams of Al_2O_3 that could be formed if 12.5 g O_2 react completely with aluminum.

ANALYSIS: Once again, we can restate the problem in equation form

$$12.5 \text{ g } O_2 \Leftrightarrow ? \text{ g } Al_2O_3$$

There is no direct path from grams of O_2 to grams of Al_2O_3, but the equation does provide a path from moles of O_2 to moles of Al_2O_3. That's the central route we have to follow.

$$3 \text{ mol } O_2 \Leftrightarrow 2 \text{ mol } Al_2O_3$$

Once we realize this, then we see that we must first change 12.5 g O_2 to moles of O_2 by using the formula mass of O_2. Next, we use the relationship above to find the number of moles of

Al_2O_3, and finally, we use the formula mass of Al_2O_3 to change moles of Al_2O_3 to grams of Al_2O_3.

SOLUTION: The three relationships needed to solve the problem are

$$1 \text{ mol } O_2 = 32.0 \text{ g } O_2$$

$$3 \text{ mol } O_2 \leftrightarrow 2 \text{ mol } Al_2O_3$$

$$1 \text{ mol } Al_2O_3 = 102 \text{ g } Al_2O_3$$

We use them to construct conversion factors and arrange them so that the units cancel correctly.

$$12.5\,\cancel{g\,O_2} \times \left(\frac{1\,\cancel{mol\,O_2}}{32.0\,\cancel{g\,O_2}}\right) \times \left(\frac{2\,\cancel{mol\,Al_2O_3}}{3\,\cancel{mol\,O_2}}\right) \times \left(\frac{102\text{ g }Al_2O_3}{1\,\cancel{mol\,Al_2O_3}}\right) \leftrightarrow 26.6 \text{ g } Al_2O_3$$

This is a concrete example of the guidelines described in Figure 3.1.

Notice that in solving this problem, we made use of the mole ratio specified by the coefficients of the balanced equation. In general, there is no way to go from grams of one substance to grams of another without going through moles. The following diagram summarizes the steps involved.

3.3 LIMITING-REACTANT CALCULATIONS

When we carry out chemical reactions, we don't always take care to mix the reactants in exactly the right proportions so all the reactants are used up with nothing left over. It often happens that there will be an excess amount of one or more of the reactants, and when this happens, one reactant is completely consumed before the others are used up. For example, suppose 5 mol of H_2 and 1 mol of O_2 are mixed and allowed to react according to the equation

$$2H_2 + O_2 \longrightarrow 2H_2O$$

The coefficients in the equation tell us that the entire 1 mol of O_2 will be able to react because we have more than the necessary 2 mol of H_2 that's required. In other words, there is more than enough H_2 to react completely with all the O_2. In fact, since we began with 5 mol of H_2, we can expect that when the reaction is over, 3 mol of H_2 will be left over unreacted.

In this example, the O_2 is referred to as the **limiting reactant** because when it is gone, no further reaction can occur and no further product (H_2O) can be formed. Stated another way, in this particular mixture of 1 mol O_2 and 5 mol H_2, it is the amount of O_2 that *limits* the amount of H_2O that can be formed.

In solving "limiting-reactant" problems, we have to identify which is the limiting reactant. Then, we calculate the amount of product formed based on the amount of the limiting reactant available, as shown in the next two examples.

EXAMPLE 3.7. SOLVING LIMITING-REACTANT PROBLEMS

PROBLEM: Zinc and sulfur react to form zinc sulfide, a substance used in phosphors that coat the inner surfaces of TV picture tubes. The equation for the reaction is

$$Zn + S \longrightarrow ZnS$$

In a particular experiment, 12.0 g of Zn are mixed with 6.50 g of S and allowed to react.

(a) Which is the limiting reactant?

(b) How many grams of ZnS can be formed, based on the amount of the limiting reactant in this particular reaction mixture?

(c) How many grams of which reactant will remain unreacted in this experiment?

The reaction of zinc with sulfur is shown on page 27 in Chapter 1.

ANALYSIS: (a) To identify the limiting reactant, we choose one reactant and calculate the amount of the other that would be needed to give complete reaction. Then we compare the amount needed with the amount available to see if we really have enough. As you see below in the solution part, we find that zinc is the limiting reactant.

(b) Once we know the limiting reactant, we use it to calculate the amount of the product that can be formed. This is an ordinary stoichiometry problem of the type you learned to solve earlier.

(c) Since zinc is the limiting reactant, sulfur must be the reactant left over after the reaction is finished. Based on the amount of zinc available, we can calculate the amount of sulfur that will react. The difference between the amount of sulfur available and the amount used is the amount left over.

SOLUTION: (a) As with other stoichiometry problems, it is best to do our reasoning in moles, so we first convert the number of grams of each reactant to moles.

$$12.0 \text{ g Zn} \times \left(\frac{1 \text{ mol Zn}}{65.4 \text{ g Zn}} \right) = 0.183 \text{ mol Zn}$$

$$6.50 \text{ g S} \times \left(\frac{1 \text{ mol S}}{32.1 \text{ g S}} \right) = 0.202 \text{ mol S}$$

Now, let's pick one of these reactants and find how many moles of the other are needed for complete reaction. It doesn't really matter which one we choose, so let's work with the Zn. We have 0.183 mol Zn, and since the Zn and S react in a 1-to-1 mole ratio, we will need 0.183 mol S to use up all the Zn. We actually have 0.202 mol S in the reaction mixture, so we have more than enough sulfur; in fact, sulfur will be left over and all the Zn will react. *This means that Zn is the limiting reactant.*

We would have come to the same conclusion had we chosen the sulfur to work with. The 0.202 mol S would require 0.202 mol Zn for complete reaction, but we only have 0.183 mol Zn. Since the reaction mixture doesn't have enough zinc to use up all the sulfur, *zinc must be the limiting reactant.*

The chemical that's left over **can't** *be the limiting reactant.*

The chemical we don't have enough of is the limiting reactant.

(b) Since zinc is the limiting reactant, we use the amount of zinc available (0.183 mol) to calculate the amount of product formed. From the coefficients in the balanced equation,

$$1 \text{ mol Zn} \Longleftrightarrow 1 \text{ mol ZnS}$$

This takes us from moles of Zn to moles of ZnS. Then we use the formula mass of ZnS (97.5 g/mol) to convert to grams.

$$1 \text{ mol ZnS} = 97.5 \text{ g ZnS}$$

The solution to this part of the problem is

$$0.183 \text{ mol Zn} \times \left(\frac{1 \text{ mol ZnS}}{1 \text{ mol Zn}} \right) \times \left(\frac{97.5 \text{ g ZnS}}{1 \text{ mol ZnS}} \right) \Longleftrightarrow 17.8 \text{ g ZnS}$$

(c) Since zinc is the limiting reactant, sulfur will be left over. The amount left over is the difference between the amount originally available (0.202 mol S) and the amount that is consumed in the reaction with the Zn (0.183 mol S).

$$\text{mol S left over} = (0.202 - 0.183) \text{ mol S} = 0.019 \text{ mol S}$$

Then we change this to grams of sulfur.

$$0.019 \text{ mol S} \times \left(\frac{32.1 \text{ g S}}{1 \text{ mol S}}\right) = 0.61 \text{ g S}$$

Thus, after the reaction is finished, there will be 0.61 g S left over.

EXAMPLE 3.8. SOLVING A LIMITING-REACTANT PROBLEM

PROBLEM: Ethylene, C_2H_4, burns in air to form CO_2 and H_2O according to the equation

$$C_2H_4 + 3O_2 \longrightarrow 2CO_2 + 2H_2O$$

How many grams of CO_2 will be formed when a mixture containing 1.93 g C_2H_4 and 5.92 g O_2 is ignited?

ANALYSIS: As you've probably discovered by now, one of the most difficult tasks in solving a problem is deciding what kind of problem you're dealing with. If you are asked to solve a problem like this one, the first thing you have to do is to recognize it as a limiting-reactant problem. The main clue is that the problem gives the amounts of *both* reactants and then goes on to ask something related to the stoichiometry of the reaction (i.e., the amount of product formed).

Remember this: In a limiting-reactant problem, you are given the amounts of each reactant.

Once you realize it is a limiting-reactant problem, the next step is to change the amounts of each reactant to moles and then identify the limiting reactant. Choose one reactant and ask how many moles of the other are needed for complete reaction. Then compare the amount needed with the amount actually available and decide which is limiting.

Finally, use the number of moles of the limiting reactant actually available to calculate the amount of product formed.

SOLUTION: First we convert the given amounts of the reactants to moles.

$$1.93 \text{ g } C_2H_4 \times \left(\frac{1 \text{ mol } C_2H_4}{28.0 \text{ g } C_2H_4}\right) = 0.0689 \text{ mol } C_2H_4$$

$$5.92 \text{ g } O_2 \times \left(\frac{1 \text{ mol } O_2}{32.0 \text{ g } O_2}\right) = 0.185 \text{ mol } O_2$$

Now we pick one reactant and find how many moles of the other are needed for complete reaction. Let's choose the O_2 (although we could just as easily select the C_2H_4 to work with). How many moles of C_2H_4 are needed to use up all the O_2? The equation gives us the mole ratio between C_2H_4 and O_2.

$$1 \text{ mol } C_2H_4 \Leftrightarrow 3 \text{ mol } O_2$$

$$0.185 \text{ mol } O_2 \times \left(\frac{1 \text{ mol } C_2H_4}{3 \text{ mol } O_2}\right) \Leftrightarrow 0.0617 \text{ mol } C_2H_4$$

We could also reason that 0.0689 mol C_2H_4 requires 3 × (0.0689) mol O_2 = 0.207 mol O_2. But we only have 0.185 mol O_2, so O_2 must be the limiting reactant.

We need 0.0617 mol C_2H_4, but we have 0.0689 mol C_2H_4; we have more C_2H_4 than we need, so all the O_2 should be able to react. Therefore, O_2 must be the limiting reactant.

Finally, we use the amount of limiting reactant (0.185 mol O_2) to calculate the amount of CO_2 formed.

$$0.185 \; \cancel{\text{mol } O_2} \times \left(\frac{2 \; \cancel{\text{mol } CO_2}}{3 \; \cancel{\text{mol } O_2}} \right) \times \left(\frac{44.0 \text{ g } CO_2}{1 \; \cancel{\text{mol } CO_2}} \right) \Leftrightarrow 5.43 \text{ g } CO_2$$

The maximum amount of CO_2 that could be obtained from the reaction mixture is 5.43 g.

There is an interesting postscript to the last problem. Although not stated explicitly, our assumption was that even without sufficient oxygen to consume all the C_2H_4, any C_2H_4 that did react was converted completely to CO_2 and H_2O. If this were the case, one aspect of automotive air pollution would be removed. What actually happens when the hydrocarbon (in this case, C_2H_4) is present in excess is that some of it is converted to CO. In the internal-combustion engine, gasoline (which is composed of a mixture of hydrocarbons) is burned in a limited supply of oxygen and the incomplete combustion therefore produces a mixture of CO, CO_2, and H_2O.

3.4 THEORETICAL YIELD AND PERCENTAGE YIELD

Sometimes a given set of reactants is able to produce more than one set of products, depending on reaction conditions. In the last paragraph of the preceding section, for instance, it was pointed out that the combustion of hydrocarbons in a limited supply of oxygen produces a mixture of products. This happens because **side reactions,** reactions other than the desired reaction, produce **side products,** products other than those being sought. Usually, the formation of side products is undesirable, and three quantities that chemists are concerned with under these circumstances are the theoretical yield, the actual yield, and the percentage yield. The **theoretical yield** *of a given product is the maximum yield that could be obtained if the reactants gave only that product, with no side reactions.* The theoretical yield is a calculated quantity. It is obtained by using the balanced equation for the desired reaction to calculate the amount of product that would be formed from the amounts of reactants in a given reaction mixture. In Example 3.8, for example, we calculated the theoretical yield of CO_2, assuming that all the C_2H_4 that burned was converted entirely to CO_2 and H_2O.

The **actual yield** is the amount of product *actually* obtained in a given experiment. Normally, it is obtained by isolating the product in an experiment and weighing it. Ordinarily, there is no way of calculating this quantity; it is measured experimentally.

The theoretical yield is a computed quantity; the actual yield is obtained by measuring the amount of product actually obtained in an experiment.

The **percentage yield** is a measure of the efficiency of the reaction and is defined as

$$\text{percentage yield} = \frac{\text{actual yield}}{\text{theoretical yield}} \times 100\%$$

For example, suppose that in the experiment described in Example 3.8, only 3.48 g of CO_2 was obtained, with the remainder of the carbon as either CO, elemental carbon, or unreacted C_2H_4. The actual yield of CO_2 is the measured amount recovered from the reaction mixture, 3.48 g. The theoretical yield was calculated to be 5.43 g based on the stoichiometry of the combustion reaction, so the percentage yield is

If, from past experience, you know what the percentage yield is likely to be, then you can estimate what the actual yield will be.

$$\text{percentage yield of } CO_2 = \frac{3.48 \text{ g } CO_2}{5.43 \text{ g } CO_2} \times 100\%$$
$$= 64.1\%$$

3.5 REACTIONS IN SOLUTION

In many chemical reactions, both in the laboratory and in the world around us, one or more of the reactants are present in a solution—that is, they are dissolved in some fluid such as water. In our bodies, for instance, nutrients are dissolved in blood and are carried to our cells where they undergo the complex chain of reactions called metabolism.

Having the reactants dissolved in a solution presents some important benefits when it comes to actually carrying out a reaction. For example, if we mix crystals of solid sodium chloride, NaCl, with crystals of solid silver nitrate, $AgNO_3$, nothing much appears to happen. However, if we dissolve these compounds in water first and then mix their solutions, a rapid reaction takes place, as shown in Figure 3.2. The reason for the difference in the behavior of the solids and their solutions is not difficult to understand. When the crystals are mixed, only their outer surfaces can come in contact, which means that only a very small fraction of the reactants have any hope of reacting. However, when these substances are dissolved in water, the individual particles of each reactant are set free and become intimately mixed with molecules of the water. When the solutions are combined, they mix further and the particles of both compounds intermingle, which allows the reaction between them to occur rapidly.

The chemical equation for the reaction shown in Figure 3.2 is

$$NaCl(aq) + AgNO_3(aq) \longrightarrow AgCl(s) + NaNO_3(aq)$$

where we have used (aq) to show that the NaCl, $AgNO_3$, and $NaNO_3$ are in an aqueous solution and (s) to show that the AgCl is a solid. The heavy milky "cloud" in the reaction mixture in Figure 3.2 is caused by the appearance of the white solid AgCl. A solid that's formed in a solution as a result of a chemical reaction like this is called a **precipitate.**

A chemical reaction in solution is not always accompanied by the formation of a precipitate. In some reactions a gas is formed, as in the reaction between hydrochloric acid and sodium carbonate, which is shown in the photograph on page 59. Sometimes just a color change occurs, and sometimes it hardly seems that anything happens at all because all the reactants and products are soluble in water and are colorless.

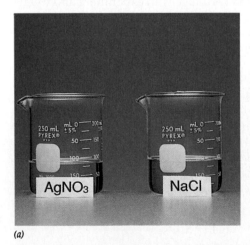

(a)

(b)

Figure 3.2.   *The reaction of sodium chloride, NaCl, with silver nitrate, $AgNO_3$, when the reactants are dissolved in aqueous solutions. (a) Separate solutions of the reactants. (b) As the solutions are combined, a rapid reaction occurs and forms insoluble white AgCl.*

(a)   (b)

Figure 3.3. *The reaction between metallic zinc, Zn, and sulfuric acid, H₂SO₄. (a) Separate beakers containing the reactants. (b) When the H₂SO₄ is added to the zinc, the metal gradually dissolves as bubbles of H₂ gas escape.*

Often it is sufficient for just one of two reactants to be in solution. In Figure 3.3, for example, we see what happens when a solution of sulfuric acid (H_2SO_4, the acid in automobile batteries) is added to zinc metal. The H_2SO_4 reacts with the zinc to give an aqueous solution of the compound zinc sulfate, $ZnSO_4$, and bubbles of hydrogen gas.

$$Zn(s) + H_2SO_4(aq) \longrightarrow ZnSO_4(aq) + H_2(g)$$

We will study reactions in solution in much greater detail in Chapter 5. Our goal in this chapter is just to examine ways of dealing with them quantitatively, and as you will see, special methods have been developed to deal with their stoichiometry in a simple way. Before discussing these methods, however, let's define some terms that we commonly use when we discuss solutions.

Terminology applied to solutions

It is quite common in talking about solutions to use the terms solute and solvent. Normally, we refer to the **solvent** as the component whose physical state doesn't change when the solution is formed. All the other components, which are dissolved in the solvent, are called **solutes.** In an aqueous salt solution, for example, the liquid water is the solvent and the solid salt that is dissolved in the liquid is the solute.

In Chapter 1 we defined a solution as a homogeneous mixture whose composition can vary. For example, the amount of salt dissolved in a given amount of water can vary from one solution to another. When we want to refer to the relative amounts of solute and solvent in a solution, we use the term **concentration.** A solution that contains a large amount of solute for a given amount of solvent is said to have a high concentration of the solute.

When we want to compare in a qualitative way the relative concentrations of solutions, we use the terms concentrated and dilute. A **concentrated solution** is one with a relatively large concentration of solute, while a **dilute solution** has a relatively small concentration. An important thing to remember, however, is that *concentrated* and *dilute* are relative terms. A solution that contains 0.01 g NaCl per liter of solution is dilute when compared to a more concentrated one that has 0.1 g NaCl per liter, but the latter solution is itself dilute compared to one that has 10 g NaCl per liter.

When water is part of a solution, it is almost always thought of as the solvent, even when the amount of water is very small. *Concentrated* H_2SO_4, for example, consists of 96% H_2SO_4 by mass, but we think of it as a lot of H_2SO_4 dissolved in a small amount of H_2O.

Molar concentration

Often, it is necessary to express the concentration of a solution quantitatively, and as you will see later in the book, there are several ways that this can be done. The one that is most useful for dealing with the stoichiometry of reactions in solution is called

Molarity is the ratio of moles of solute to liters of solution.

Molarity is an intensive quantity formed as a ratio of two extensive quantities: moles of solute and volume of solution.

molar concentration or **molarity,** symbolized by M. This is defined as the number of moles of solute in the solution divided by the volume of the solution expressed in liters.

$$\text{molarity } (M) = \frac{\text{moles of solute}}{\text{liters of solution}}$$

A solution that contains 1.00 mol NaCl in 1.00 L of solution has a molarity of 1.00 mol NaCl/(L of solution), or 1.00 M, and we say that it is a 1.00-**molar** solution. Let's look at a simple example showing how the molarity of a solution is calculated.

EXAMPLE 3.9. CALCULATING THE MOLARITY OF A SOLUTION

PROBLEM: A 2.00-g sample of sodium hydroxide, NaOH (the major ingredient in the drain cleaner Drano), was dissolved in water to give a solution with a volume of exactly 200 mL. What is the molarity of this NaOH solution?

SOLUTION: To calculate the molarity, we take the ratio of the number of moles of solute to the number of *liters* of solution. This means we have to change the amount of NaOH to moles and the volume of the solution to liters.

For working with lab-sized amounts of chemicals, it is convenient to express the formula mass in units of grams per mole.

The formula mass of NaOH is 40.0 g/mol, so

$$2.00 \text{ g NaOH} \times \left(\frac{1 \text{ mol NaOH}}{40.0 \text{ g NaOH}} \right) = 0.0500 \text{ mol NaOH}$$

When expressed in liters, 200 mL becomes 0.200 L. The molarity is therefore

The abbreviation of "solution" is soln.

$$\text{molarity} = \frac{0.0500 \text{ mol NaOH}}{0.200 \text{ L soln}}$$
$$= 0.250 \text{ mol NaOH/L}$$
$$= 0.250 \, M \text{ NaOH}$$

The reason molarity is such a useful concentration unit is because if we know the molarity of a particular solution, we can dispense a desired number of moles of the solute just by measuring out the proper volume. For example, suppose we had a large container filled with a 0.250 M solution of NaOH. Let's also suppose that in a particular reaction we needed exactly 0.250 mol of NaOH. The label on the container tells us that each liter of the solution contains 0.250 mol NaOH, so all we have to do is measure out 1.00 L of the solution and we will have the necessary 0.250 mol of NaOH. Similarly, if we needed 0.500 mol of NaOH for some other experiment, we could get it by measuring out 2.00 L of the solution, and if we needed only 0.125 mol NaOH, we could get it by measuring out 0.500 L (500 mL) of the solution. To use molarity properly, therefore, you have to learn to use it to relate moles of solute and volumes of solution, and the following examples illustrate how this is done.

EXAMPLE 3.10. CALCULATING THE VOLUME OF A SOLUTION THAT CONTAINS A SPECIFIC AMOUNT OF SOLUTE

PROBLEM: How many milliliters of 0.250 M NaOH solution are needed to provide 0.0200 mol of NaOH?

ANALYSIS: For calculation purposes, molarity provides a bridge between moles of solute and volume of solution. A label that reads 0.250 M NaOH can be translated to mean 0.250 mol NaOH/L of solution. This ratio can be used as a conversion factor, either directly as written or inverted:

$$\frac{0.250 \text{ mol NaOH}}{1.00 \text{ L soln}} \qquad \frac{1.00 \text{ L soln}}{0.250 \text{ mol NaOH}}$$

The first step is to translate the label.

We can also express the volume in milliliters and write these as

$$\frac{0.250 \text{ mol NaOH}}{1000 \text{ mL soln}} \qquad \frac{1000 \text{ mL soln}}{0.250 \text{ mol NaOH}}$$

The solution to the problem, then, involves translating the given molarity into the necessary ratio (conversion factor) so that units cancel correctly.

SOLUTION: We can restate the problem as

$$0.0200 \text{ mol NaOH} \Longleftrightarrow ? \text{ mL NaOH soln}$$

To change "mol NaOH" to milliliters of solution, we need to choose a conversion factor with "mol NaOH" in the denominator. Since we want milliliters for the answer,

$$0.0200 \text{ mol NaOH} \times \left(\frac{1000 \text{ mL soln}}{0.250 \text{ mol NaOH}}\right) \Longleftrightarrow 80.0 \text{ mL soln}$$

Thus, if we measure out 80.0 mL of this solution, it will contain the desired 0.0200 mol of NaOH.

EXAMPLE 3.11. CALCULATING THE AMOUNT OF SOLUTE IN A SOLUTION OF KNOWN MOLARITY

PROBLEM: How many grams of NaOH are in 50.0 mL of 0.400 M NaOH solution?

ANALYSIS: This time we can state the problem as

$$50.0 \text{ mL soln} \Longleftrightarrow ? \text{ g NaOH}$$

The molarity serves as a conversion factor to change "mL soln" to moles of NaOH, and then we can use the formula mass of NaOH to find the number of grams.

SOLUTION: First, we have to translate 0.400 M to the corresponding ratio of moles to volume.

$$0.400 \text{ } M \text{ means } \frac{0.400 \text{ mol NaOH}}{1000 \text{ mL soln}}$$

Now we apply the ratio as a conversion factor to cancel "mL soln"

$$50.0 \text{ mL soln} \times \left(\frac{0.400 \text{ mol NaOH}}{1000 \text{ mL soln}}\right) \Longleftrightarrow 0.0200 \text{ mol NaOH}$$

The formula mass of NaOH is 40.0 g/mol. Therefore,

$$0.0200 \text{ mol NaOH} \times \frac{40.0 \text{ g NaOH}}{1 \text{ mol NaOH}} = 0.800 \text{ g NaOH}$$

Thus, 50.0 mL of 0.400 M NaOH contains 0.800 g NaOH.

Sometimes, while you're working in the laboratory, it becomes necessary to prepare a solution that has a specific concentration. This is really not very difficult, as the next example shows.

EXAMPLE 3.12. PREPARING A SOLUTION THAT HAS A SPECIFIC MOLARITY

PROBLEM: How many grams of silver nitrate, $AgNO_3$, are needed to prepare 500 mL of a 0.300 M $AgNO_3$ solution?

ANALYSIS: What we really need to know here is how many grams of $AgNO_3$ have to be in the final solution. If we can figure this out, we can weight out this much solute and dissolve it in enough solvent to give the desired solution. This problem, therefore, is just like the preceding one.

SOLUTION: First, let's translate the molarity.

$$0.300 \ M \ AgNO_3 \text{ means } \frac{0.300 \text{ mol } AgNO_3}{1000 \text{ mL soln}}$$

In the final solution, the amount of $AgNO_3$ that must be present is

$$500 \ \cancel{\text{mL soln}} \times \left(\frac{0.300 \text{ mol } AgNO_3}{1000 \ \cancel{\text{mL soln}}} \right) = 0.150 \text{ mol } AgNO_3$$

The formula mass of $AgNO_3$ is 170 g/mol. Therefore,

$$0.150 \ \cancel{\text{mol } AgNO_3} \times \frac{170 \text{ g } AgNO_3}{1 \ \cancel{\text{mol } AgNO_3}} = 25.5 \text{ g } AgNO_3$$

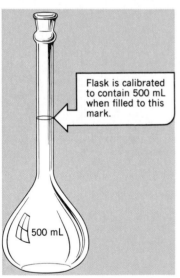

Flask is calibrated to contain 500 mL when filled to this mark.

500 mL

Figure 3.4. *A 500-mL volumetric flask.*

Here we are using dilute *as a verb.* To dilute *a solution is to make it less concentrated.*

To actually prepare the solution described in Example 3.12, we would have to dissolve the 25.5 g $AgNO_3$ in enough water to give a *final volume* of exactly 500 mL. Measuring volumes with this accuracy is accomplished using a volumetric flask (Figure 3.4). The flask contains the specified volume when it is filled to the mark etched around the neck. Figure 3.5 illustrates the sequence of steps involved in preparing the solution.

As a final point, in preparing the solution in the preceding example, we adjusted the volume of the solution to a *final volume* of 500 mL. We didn't simply add 500 mL of water to the silver nitrate, because this would give a final volume just a bit larger than 500 mL (both the solute and solvent take up space in the flask). If we had actually added 500 mL of water, the concentration would have been just a bit less than the desired 0.300 M because the solute would be spread through a slightly larger volume than expected.

3.6 PREPARING SOLUTIONS BY DILUTION

In the course of performing routine laboratory work, it is not uncommon to have to dilute solutions—to make them less concentrated by increasing the amount of solvent. For example, many laboratory chemicals are purchased in the form of concentrated aqueous solutions (Table 3.1) because that is the most economical way to buy them. Usually, however, these chemicals are too concentrated to use as purchased, and therefore they must be made more dilute.

The process of dilution involves mixing a concentrated solution with additional solvent to give some larger final volume. Throughout this process, the number of

(a) (b) (c) (d) (e)

Figure 3.5. *Preparation of a solution with a particular molarity. (a) The solute is accurately weighed into a volumetric flask. (b) Distilled water is added. (c) The mixture is swirled to dissolve the solute. (d) After more water has been added, a medicine dropper is used to add water carefully until the volume is brought exactly to the mark etched around the neck of the flask. (e) The flask is stoppered and then shaken to make the solution uniform.*

Table 3.1. Concentrated laboratory reagents

Reagent	Density (g/mL)	Percent by Mass	Molarity
Sulfuric acid (H_2SO_4)	1.84	96	18
Hydrochloric acid (HCl)	1.18	36	12
Phosphoric acid (H_3PO_4)	1.7	85	15
Nitric acid (HNO_3)	1.42	70	16
Acetic acid ($HC_2H_3O_2$)	1.05	100	17.5
Aqueous ammonia (NH_3)	0.90	28	15

Chemicals stocked in the laboratory are frequently referred to as **reagents.**

moles of the solute remains constant, and only the volume increases. This fact forms the basis for working problems dealing with dilution.

If we multiply a solution's molarity M by its volume V, we obtain the number of moles of the solute.

$$M \cdot V = \frac{\text{mol}}{\cancel{L}} \times \cancel{L} = \text{mol}$$

Since the number of moles of solute stays the same during a dilution, the product of the initial molarity and volume (M_iV_i) must be equal to the product of the final molarity and volume (M_fV_f). This gives the useful equation

$$M_iV_i = M_fV_f \qquad (3.1)$$

The next three examples illustrate how we can use this equation.

When the relationship in Equation 3.1 is used, the volumes can be in either milliliters or liters. If the milliliter is used, the product $M \cdot V$ gives *millimoles*, but during dilution the number of millimoles of solute doesn't change either.

EXAMPLE 3.13. WORKING DILUTION PROBLEMS

PROBLEM: How many milliliters of concentrated H_2SO_4(18.0 M) are required to prepare 750 mL of 3.00 M H_2SO_4 solution?

SOLUTION: Let's apply Equation 3.1.

$$M_iV_i = M_fV_f$$

$$M_i = 18.0\ M \quad M_f = 3.00\ M$$

$$V_i = ? \qquad V_f = 750\ mL$$

Solving for V_i gives

$$V_i = \frac{M_fV_f}{M_i}$$

$$V_i = \frac{(3.00\ M)(750\ mL)}{18.0\ M}$$

$$V_i = 125\ mL$$

To prepare the solution, we dilute 125 mL of the concentrated H_2SO_4 to a total final volume of 750 mL.

A very important safety note applies to Example 3.13. When concentrated chemicals are diluted, a large amount of heat is sometimes liberated. This is especially true for sulfuric acid. **To absorb this heat safely, you must** *always add the concentrated acid to the water,* **never the reverse.** If water is added to the concentrated acid, so much heat is liberated that it can cause the water to boil suddenly, spattering the acid. If you happen to be standing in the way, this could definitely ruin your whole day!

EXAMPLE 3.14. WORKING DILUTION PROBLEMS

PROBLEM: How much water must be added to 25.0 mL of 0.500 M KOH solution to produce a solution whose concentration is 0.350 M?

SOLUTION: Our equation is

$$M_iV_i = M_fV_f$$

$$M_i = 0.500\ M \quad M_f = 0.350\ M$$

$$V_i = 25.0\ mL \quad V_f = ?$$

Solving for V_f and substituting, we get

$$V_f = \frac{(0.500\ M)(25.0\ mL)}{0.350\ M}$$

$$V_f = 35.7\ mL$$

Since the initial volume was 25.0 mL, we must *add* 10.7 mL. (We are assuming that volumes are additive. For working with dilute solutions, the assumption is usually quite valid.)

EXAMPLE 3.15. WORKING DILUTION PROBLEMS

PROBLEM: Suppose that 200 mL of water were added to 300 mL of a solution labeled 0.600 M HNO_3. What will be the new concentration of the solute in the final solution?

SOLUTION: Once again,

$$M_iV_i = M_fV_f$$

$M_i = 0.600\ M \quad M_f = ?$

$V_i = 300\ mL \quad V_f = 200\ mL + 300\ mL = 500\ mL$

Solving for M_f gives

$$M_f = \frac{M_iV_i}{V_f}$$

$$= \frac{(0.600\ M)(300\ \cancel{mL})}{(500\ \cancel{mL})}$$

$$= 0.360\ M$$

The concentration of HNO_3 in the final solution is $0.360\ M$.

When we use Equation 3.1, volumes can be in any convenient unit. The only requirement is that V_i and V_f have the *same* units.

In the preceding two examples, we have made a subtle assumption—that the volumes of the solutions are additive. In other words, in Example 3.15 we assumed that 200 mL of H_2O plus 300 mL of the more concentrated solution give a final total volume of 500 mL. It turns out that this isn't quite true. For almost all purposes it is close enough, but for extremely precise work, a volumetric flask has to be used. To give a solution with a desired final molarity, we take the proper volume of the more concentrated solution and place it in the volumetric flask. Then additional solvent is gradually added until the volume is brought to the mark etched around the neck of the flask. In that way, the solute in the original concentrated solution is distributed over *exactly* the correct final volume.

For most routine laboratory work, we can assume that volumes of dilute aqueous solution are additive.

3.7 THE STOICHIOMETRY OF REACTIONS IN SOLUTION

The quantitative relationships that apply to reactions in solution are exactly the same as those for reactions that occur anywhere else: The coefficients in the balanced chemical equation provide the mole ratios needed to solve stoichiometry problems. The differences, when they exist, are in the *laboratory units* that are used to measure the amounts of the reactants.

Consider the reaction shown in Figure 3.6 in which a solution containing so-

(a)

(b)

Figure 3.6. (a) *Beakers containing clear solutions of colorless $Pb(NO_3)_2$ and yellow Na_2CrO_4. (b) The solution of Na_2CrO_4 being added to the solution of $Pb(NO_3)_2$. The reaction produces a yellow precipitate of insoluble $PbCrO_4$.*

dium chromate, Na_2CrO_4, is added to a solution of lead nitrate, $Pb(NO_3)_2$. The yellow precipitate that's formed is lead chromate, $PbCrO_4$, a compound used as a pigment in oil paints and watercolors. The equation for the reaction is

$$Pb(NO_3)_2(aq) + Na_2CrO_4(aq) \longrightarrow PbCrO_4(s) + 2NaNO_3(aq)$$

In Section 3.2, you learned to calculate the amounts of reactants needed for a given reaction by converting moles to grams. We can take the same approach to reactions in solution. In this case, the stoichiometry problems are no different than those we have done earlier. For example, if we wish to allow 0.100 mol of $Pb(NO_3)_2$ to react with 0.100 mol of Na_2CrO_4, the first step is to convert these amounts to grams using the appropriate formula masses. The formula mass of $Pb(NO_3)_2$ is 331 g/mol and the formula mass of Na_2CrO_4 is 162 g/mol, so 0.100 mol of $Pb(NO_3)_2$ weighs 33.1 g and 0.100 mol of Na_2CrO_4 weighs 16.2 g. To carry out the reaction, we would dissolve these amounts of the reactants in water and then combine the solutions to obtain the desired products.

When reactions take place in solution, however, we have another way of measuring amounts of reactants — by measuring volumes of solutions of known molarity. For example, suppose you were given 250 mL of 0.100 M $Pb(NO_3)_2$ solution. By multiplying the volume of the solution by its molarity, you obtain the number of moles of $Pb(NO_3)_2$ in the solution. The arithmetic can be done in two ways. One is to convert the 250 mL to liters (0.250 L).

$$0.250 \; \text{L soln} \times \left(\frac{0.100 \; \text{mol Pb(NO}_3)_2}{1.00 \; \text{L soln}} \right) \Longleftrightarrow 0.0250 \; \text{mol Pb(NO}_3)_2$$

The other is to express the volume in the molarity expression as 1000 mL.

$$250 \; \text{mL soln} \times \left(\frac{0.100 \; \text{mol Pb(NO}_3)_2}{1000 \; \text{mL soln}} \right) \Longleftrightarrow 0.0250 \; \text{mol Pb(NO}_3)_2$$

Both give the same answer, of course. And once you've calculated the number of moles of $Pb(NO_3)_2$, you can then go on to use the coefficients in the equation to calculate the number of moles of Na_2CrO_4 needed, or the number of moles of $PbCrO_4$ produced, or the number of moles of $NaNO_3$ produced. Let's look at some specific examples now that show how molarity and volumes are used in working stoichiometry problems for reactions in solution.

EXAMPLE 3.16. USING MOLARITY IN STOICHIOMETRY PROBLEMS

PROBLEM: Aluminum hydroxide, $Al(OH)_3$, one of the antacid ingredients in Maalox, can be prepared by the reaction of aluminum sulfate, $Al_2(SO_4)_3$, and sodium hydroxide, NaOH. The balanced chemical equation for the reaction is

$$Al_2(SO_4)_3(aq) + 6NaOH(aq) \longrightarrow 2Al(OH)_3(s) + 3Na_2SO_4(aq)$$

How many milliliters of 0.200 M NaOH solution are needed to completely react with 3.50 g $Al_2(SO_4)_3$?

ANALYSIS: The problem can be stated as follows.

$$3.50 \; \text{g Al}_2(SO_4)_3 \Longleftrightarrow ? \; \text{mL } 0.200 \; M \; \text{NaOH}$$

As in any stoichiometry problem, the amounts of reactant are related by *moles* according to the coefficients in the balanced equation. Therefore, the first step will be to change grams of

$Al_2(SO_4)_3$ to moles of $Al_2(SO_4)_3$. Then we can use the coefficients of the equation to find the number of moles of NaOH needed in the reaction, and finally we can use the molarity as a conversion factor to find the volume of the NaOH solution needed.

SOLUTION: First, we calculate the number of moles of $Al_2(SO_4)_3$ (formula mass = 342.2 g/mol).

$$3.50 \text{ g Al}_2\text{(SO}_4)_3 \times \left(\frac{1 \text{ mol Al}_2\text{(SO}_4)_3}{342.2 \text{ g Al}_2\text{(SO}_4)_3} \right) = 1.02 \times 10^{-2} \text{ mol Al}_2\text{(SO}_4)_3$$

Next, we use the coefficients to establish the necessary mole ratio.

$$1.02 \times 10^{-2} \text{ mol Al}_2\text{(SO}_4)_3 \times \left(\frac{6 \text{ mol NaOH}}{1 \text{ mol Al}_2\text{(SO}_4)_3} \right) \Longleftrightarrow 6.12 \times 10^{-2} \text{ mol NaOH}$$

The final step is to calculate the volume of the NaOH solution that contains this number of moles. Translating the molar concentration yields the two conversion factors

$$\frac{0.200 \text{ mol NaOH}}{1000 \text{ mL soln}} \quad \text{and} \quad \frac{1000 \text{ mL soln}}{0.200 \text{ mol NaOH}}$$

It is the second one that we need to make the units cancel correctly.

$$6.12 \times 10^{-2} \text{ mol NaOH} \times \left(\frac{1000 \text{ mL soln}}{0.200 \text{ mol NaOH}} \right) \Longleftrightarrow 306 \text{ mL soln}$$

To carry out the reaction, we would dissolve the 3.50 g of $Al_2(SO_4)_3$ in water and then add 306 mL of the 0.200 M NaOH solution to give complete reaction.

EXAMPLE 3.17. USING MOLARITY IN STOICHIOMETRY PROBLEMS

PROBLEM: Chalk is composed of calcium carbonate, $CaCO_3$. This water-insoluble compound is formed when a solution of calcium chloride, $CaCl_2$, is added to a solution of sodium carbonate, Na_2CO_3. The reaction is

$$CaCl_2(aq) + Na_2CO_3(aq) \longrightarrow CaCO_3(s) + 2NaCl(aq)$$

How many milliliters of 0.250 M $CaCl_2$ are needed to react completely with 50.0 mL of 0.150 M Na_2CO_3 solution?

SOLUTION: From the volume and molarity of the Na_2CO_3 solution, we can calculate the number of moles of Na_2CO_3 that are available. A concentration of 0.150 M Na_2CO_3 means

$$\frac{0.150 \text{ mol Na}_2\text{CO}_3}{1000 \text{ mL soln}}$$

Therefore,

$$50.0 \text{ mL soln} \times \left(\frac{0.150 \text{ mol Na}_2\text{CO}_3}{1000 \text{ mL soln}} \right) \Longleftrightarrow 7.50 \times 10^{-3} \text{ mol Na}_2\text{CO}_3$$

Since the coefficients of $CaCl_2$ and Na_2CO_3 are the same in the equation, the amount of $CaCl_2$ needed will also be 7.50×10^{-3} mol. The final step, therefore, is to calculate the volume of the $CaCl_2$ solution that contains this number of moles of solute. The concentration gives us these conversion factors

$$\frac{0.250 \text{ mol CaCl}_2}{1000 \text{ mL soln}} \quad \text{and} \quad \frac{1000 \text{ mL soln}}{0.250 \text{ mol CaCl}_2}$$

The second one gives us

$$7.50 \times 10^{-3} \text{ mol CaCl}_2 \times \left(\frac{1000 \text{ mL soln}}{0.250 \text{ mol CaCl}_2} \right) \Longleftrightarrow 30.0 \text{ mL soln}$$

The required volume of this $CaCl_2$ solution is 30.0 mL.

EXAMPLE 3.18. LIMITING REACTANTS IN SOLUTION REACTIONS

PROBLEM: Silver bromide, AgBr, is the principal light-sensitive chemical in photographic film. This insoluble compound forms when an aqueous solution of silver nitrate, $AgNO_3$, is mixed with an aqueous solution of calcium bromide, $CaBr_2$.

$$2AgNO_3(aq) + CaBr_2(aq) \longrightarrow 2AgBr(s) + Ca(NO_3)_2(aq)$$

How many grams of solid AgBr will be formed if 50.0 mL of 0.180 M $AgNO_3$ are mixed with 60.0 mL of 0.0850 M $CaBr_2$?

ANALYSIS: When you're given both the volume and the molarity of a solution, you're actually given the number of moles of solute. That is because the product of molarity and volume (in liters) yields moles. Since we are given the molarity and volume for *both* reactants, we effectively have the number of moles of each. In Section 3.3 you learned that in such cases it is necessary to determine the limiting reactant and then calculate the amount of product based on the amount of the limiting reactant that is available.

molarity $\times$ volume (L) = moles

$\frac{\text{moles}}{\text{liters}} \times \text{liters} = \text{moles}$

SOLUTION: Now that we know the strategy we have to use, let's solve the problem. First, we calculate the number of moles of each reactant in their respective solutions.

For $AgNO_3$

$$50.0 \text{ mL soln} \times \left(\frac{0.180 \text{ mol AgNO}_3}{1000 \text{ mL soln}} \right) \Longleftrightarrow 9.00 \times 10^{-3} \text{ mol AgNO}_3$$

For $CaBr_2$

$$60.0 \text{ mL soln} \times \left(\frac{0.0850 \text{ mol CaBr}_2}{1000 \text{ mL soln}} \right) \Longleftrightarrow 5.10 \times 10^{-3} \text{ mol CaBr}_2$$

Next, we find out which is the limiting reactant. Let's determine how many moles of $CaBr_2$ are needed to react with all the $AgNO_3$.

$$9.00 \times 10^{-3} \text{ mol AgNO}_3 \times \left(\frac{1 \text{ mol CaBr}_2}{2 \text{ mol AgNO}_3} \right) \Longleftrightarrow 4.50 \times 10^{-3} \text{ mol CaBr}_2 \text{ needed}$$

Recognizing this as a limiting-reactant problem is the first step in solving it correctly.

Notice that the amount of $CaBr_2$ available (5.10×10^{-3} mol) is larger than the amount needed. This means $CaBr_2$ will be left over, so $AgNO_3$ is the limiting reactant.

Finally, we calculate the amount of AgBr formed from the available $AgNO_3$.

$$9.00 \times 10^{-3} \text{ mol AgNO}_3 \times \left(\frac{2 \text{ mol AgBr}}{2 \text{ mol AgNO}_3} \right) \Longleftrightarrow 9.00 \times 10^{-3} \text{ mol AgBr}$$

The formula mass of AgBr is 187.8 g/mol, so

$$9.00 \times 10^{-3} \text{ mol AgBr} \times \left(\frac{187.8 \text{ g AgBr}}{1 \text{ mol AgBr}} \right) = 1.69 \text{ g AgBr}$$

The weight of AgBr formed by mixing these two solutions is 1.69 g.

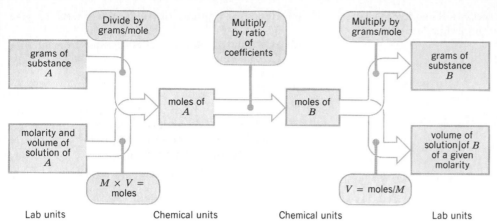

Figure 3.7. *An overview of the method for solving stoichiometry problems that concern chemical reactions. Here we see that there is more than one way to specify laboratory units for the substances involved in the reaction. The general overall procedure for solving the problem is the same, however. Only the translation between lab and chemical units differs, depending on the lab units given or asked for.*

In Figure 3.1 (page 62) we summarized an overall approach to working stoichiometry problems:

lab units $\longrightarrow$ chemical units $\longrightarrow$ chemical units $\longrightarrow$ lab units
(grams of *A*)   (mol *A*)   (mol *B*)   (grams of *B*)

Now we've seen that there is another way of specifying lab units: molarity and volume. These can now be included in the overall strategy for solving stoichiometry problems, as outlined in Figure 3.7.

REVIEW QUESTIONS AND PROBLEMS

Problems whose numbers are in blue have their answers in Appendix D at the back of the book. The more difficult problems are marked with asterisks.

Balancing Equations

3.1 What important chemical law is obeyed by a balanced chemical equation?

3.2 Balance the following equations by inspection:
(a) $ZnS + HCl \rightarrow ZnCl_2 + H_2S$
(b) $HCl + Cr \rightarrow CrCl_2 + H_2$
(c) $Al + Fe_3O_4 \rightarrow Al_2O_3 + Fe$
(d) $H_2 + Br_2 \rightarrow HBr$
(e) $Na_2S_2O_3 + I_2 \rightarrow NaI + Na_2S_4O_6$

3.3 Balance the following equations:
(a) $LaCl_3 + Na_2CO_3 \rightarrow La_2(CO_3)_3 + NaCl$
(b) $NH_4Cl + Ba(OH)_2 \rightarrow BaCl_2 + NH_3 + H_2O$
(c) $Ca(OH)_2 + H_3PO_4 \rightarrow Ca_3(PO_4)_2 + H_2O$
(d) $La_2(CO_3)_3 + H_2SO_4 \rightarrow La_2(SO_4)_3 + H_2O + CO_2$
(e) $Na_2O + (NH_4)_2SO_4 \rightarrow Na_2SO_4 + H_2O + NH_3$

3.4 Balance the following equations:
(a) $C_4H_{10} + O_2 \rightarrow CO_2 + H_2O$
(b) $C_7H_6O_2 + O_2 \rightarrow CO_2 + H_2O$
(c) $P_4O_{10} + H_2O \rightarrow H_3PO_4$
(d) $FeS_2 + O_2 \rightarrow Fe_2O_3 + SO_2$
(e) $NH_3 + O_2 \rightarrow NO + H_2O$

3.5 Balance the following equations:
(a) $Fe + HCl \rightarrow H_2 + FeCl_2$
(b) $PbO_2 + HCl \rightarrow H_2O + PbCl_2 + Cl_2$
(c) $Fe_2O_3 + H_2SO_4 \rightarrow Fe_2(SO_4)_3 + H_2O$
(d) $NO_2 + H_2O \rightarrow NO + HNO_3$
(e) $C_2H_6S + O_2 \rightarrow CO_2 + H_2O + SO_2$

Calculations Based on Chemical Equations

3.6 The reaction that takes place in an alkaline flashlight battery as the battery is discharged is

$$Zn + 2MnO_2 + 2H_2O \longrightarrow Zn(OH)_2 + 2MnO(OH)$$

According to this equation,
(a) How many moles of Zn react when 0.200 mol of MnO(OH) is formed?
(b) How many moles of H_2O react with 0.150 mol Zn?
(c) How many moles of $Zn(OH)_2$ are formed when 0.100 mol MnO_2 reacts?
(d) How many moles of H_2O react when 0.600 mol MnO_2 reacts?

3.7 Acetylene, which is used as a fuel in welding torches, is produced in a reaction between calcium carbide and water

$$CaC_2 + 2H_2O \longrightarrow Ca(OH)_2 + C_2H_2(g)$$

calcium carbide   acetylene

(a) How many moles of C_2H_2 would be produced from 2.50 mol of CaC_2?
(b) How many grams of C_2H_2 would be formed from 0.500 mol of CaC_2?
(c) How many moles of water would be consumed when 3.20 mol of C_2H_2 is formed?
(d) How many grams of $Ca(OH)_2$ are produced when 28.0 g of C_2H_2 is formed?

3.8 Consider the following balanced equation:

$$6ClO_2 + 3H_2O \longrightarrow 5HClO_3 + HCl$$

(a) How many moles of $HClO_3$ are produced from 14.3 g of ClO_2?
(b) How many grams of H_2O are needed to produce 5.74 g HCl?

3.9 White phosphorus, composed of P_4 molecules, is used in military incendiary devices because it ignites spontaneously when exposed to air. The product of reaction with oxygen is P_4O_{10}.
(a) Write a balanced chemical equation for the reaction of P_4 with air. (Remember that oxygen in the air is present as O_2 molecules.)
(b) How many moles of P_4O_{10} can be produced using 0.500 mol O_2?
(c) How many grams of P_4 are needed to produce 50.0 g P_4O_{10}?
(d) How many grams of P_4 will react with 25.0 g O_2?

3.10 Hydrazine, N_2H_4, and hydrogen peroxide, H_2O_2, have been used as rocket propellants. They react according to the equation

$$7H_2O_2 + N_2H_4 \longrightarrow 2HNO_3 + 8H_2O$$

(a) How many moles of HNO_3 are formed from 0.0250 mol N_2H_4?
(b) How many moles of H_2O_2 are required if 1.35 mol H_2O is to be produced?
(c) How many moles of H_2O are formed if 1.87 mol HNO_3 is produced?
(d) How many moles of H_2O_2 are required to react with 22.0 g N_2H_4?
(e) How many grams of H_2O_2 are needed to produce 45.8 g HNO_3?

3.11 When iron is produced from its ore, Fe_2O_3, the reaction is

$$Fe_2O_3 + 3CO \longrightarrow 2Fe + 3CO_2$$

(a) How many moles of CO are needed to produce 35.0 mol Fe?
(b) How many moles of Fe_2O_3 react if 4.50 mol CO_2 is formed?

(c) How many grams of Fe_2O_3 must react to give 0.570 mol Fe?
(d) How many moles of CO are needed to react with 48.5 g Fe_2O_3?
(e) How many grams of Fe are formed when 18.6 g CO reacts?

3.12 During the naval battles of the South Pacific in World War II, the U.S. Navy produced smoke screens by spraying titanium tetrachloride into the moist air where it reacted according to the equation

$$TiCl_4 + 2H_2O \longrightarrow TiO_2 + 4HCl$$

The dense smoke was caused by the TiO_2.
(a) How many moles of H_2O are needed to react with 6.50 mol $TiCl_4$?
(b) How many moles of HCl are formed when 8.44 mol $TiCl_4$ reacts?
(c) How many grams of TiO_2 are formed from 14.4 mol $TiCl_4$?
(d) How many grams of HCl are formed when 85.0 g $TiCl_4$ reacts?

3.13 The insecticide DDT (which ecologists now recognize as a serious environmental pollutant) is manufactured in a reaction between chlorobenzene and chloral:

$$2C_6H_5Cl + C_2HCl_3O \longrightarrow C_{14}H_9Cl_5 + H_2O$$

chlorobenzene   chloral   DDT

How many kilograms of DDT can be produced from 1000 kg of chlorobenzene?

3.14 Aspirin (which many students take after working on chemistry problems) is prepared by the reaction of salicylic acid ($C_7H_6O_3$) with acetic anhydride ($C_4H_6O_3$) according to the equation

$$C_7H_6O_3 + C_4H_6O_3 \longrightarrow C_9H_8O_4 + C_2H_4O_2$$

(aspirin)

How many grams of salicylic acid must be used to prepare two 5-grain aspirin tablets (1 g = 15.4 grains)?

3.15 Dimethylhydrazine, $(CH_3)_2NNH_2$, was used as a fuel in the Apollo lunar descent module, with liquid N_2O_4 as the oxidizer. The products of the reaction between these two in the rocket engine are H_2O, CO_2, and N_2.
(a) Write a balanced chemical equation for the reaction.
(b) Calculate the number of kilograms of N_2O_4 required to burn 50.0 kg of dimethylhydrazine.

3.16 The fermentation of sugar to produce ethyl alcohol follows the equation

$$C_6H_{12}O_6 \xrightarrow{\text{yeasts}} 2C_2H_5OH + 2CO_2$$

What is the maximum number of grams of alcohol that can be obtained from 500 g of sugar?

* 3.17 White lead, a pigment used in lead-based paints, is manufactured by the reactions

$$2Pb + 2HC_2H_3O_2 + O_2 \longrightarrow 2Pb(OH)C_2H_3O_2$$

$$6Pb(OH)C_2H_3O_2 + 2CO_2 \longrightarrow$$
$$Pb_3(OH)_2(CO_3)_2 + 2H_2O + 3Pb(C_2H_3O_2)_2$$
white lead

(a) Starting with 20.0 g of Pb, how many grams of white lead can be prepared?

(b) How many grams of CO_2 will be required if 14.0 g O_2 is consumed in the first reaction?

Assume that all the Pb in the first reaction is completely converted to the products of the second reaction.

Ton-Mole, Pound-Mole Problems

3.18 When dealing with large amounts of chemicals in industrial applications, chemical engineers can work in pound-moles or ton-moles, defined as the mass in pounds or tons, respectively, that is numerically equal to the formula mass. On this basis, what is the mass, in pounds, of 1.00 lb-mol of each of the following?

(a) CH_4 (d) NO_2

(b) Na   (e) Fe

(c) $CuSO_4$

3.19 What is the mass, in tons, of calcium in 2.40 ton-mol of $CaCl_2$?

3.20 Phosphate rock, $Ca_3(PO_4)_2$, is treated with sulfuric acid to produce phosphate fertilizer

$$Ca_3(PO_4)_2 + 2H_2SO_4 + 4H_2O \longrightarrow$$
$$Ca(H_2PO_4)_2 + 2CaSO_4 \cdot 2H_2O$$
phosphate fertilizer

How many tons of sulfuric acid are required to react with 25.0 tons of phosphate rock? Work the problem using ton-moles.

3.21 How many pounds of ammonia can be produced from 650 lb of H_2 by the reaction $N_2 + 3H_2 \rightarrow 2NH_3$. Work out the problem using lb-moles.

Limiting-Reactant Calculations

3.22 What is meant by the term *limiting reactant?* In words, describe how the limiting reactant is identified.

3.23 Consider the reaction

$$Fe(s) + 2HCl(aq) \longrightarrow FeCl_2(aq) + H_2(g)$$

In an experiment, 0.40 mol Fe and 0.75 mol HCl are combined.

(a) Which is the limiting reactant?

(b) How many moles of H_2 will be formed?

(c) How many moles of the reactant in excess will remain after the reaction has stopped?

3.24 For the reaction described in Problem 3.8, how many grams of $HClO_3$ are produced when 4.25 g of ClO_2 is allowed to react with 0.853 g of H_2O?

3.25 Aluminum (Al) reacts with sulfuric acid (H_2SO_4), which is the acid in automobile batteries, according to the equation

$$2Al + 3H_2SO_4 \longrightarrow Al_2(SO_4)_3 + 3H_2$$

If 20.0 g Al is put into a solution containing 115 g of H_2SO_4,

(a) which is the limiting reactant?

(b) how many moles of H_2 will be formed?

(c) how many grams of $Al_2(SO_4)_3$ will be formed?

(d) how many grams of the reactant in excess will be left over?

3.26 Under appropriate conditions, acetylene (C_2H_2) and HCl react to form vinyl chloride, C_2H_3Cl. This substance is used to manufacture polyvinyl chloride (PVC) plastics and has been shown to be carcinogenic. The equation for the reaction is

$$C_2H_2 + HCl \longrightarrow C_2H_3Cl$$

In a given instance, 35.0 g of C_2H_2 is mixed with 51.0 g of HCl.

(a) Which is the limiting reactant?

(b) How many grams of C_2H_3Cl are formed?

(c) How many grams of the reactant in excess remain after the reaction is completed?

3.27 Chlorofluorocarbons, commonly known as Freons, have been implicated in the gradual destruction of the earth's ozone shield. One of these, Freon-12, is a gas that is used as a refrigerant and is prepared by the reaction

$$3CCl_4 + 2SbF_3 \longrightarrow 3CCl_2F_2 + 2SbCl_3$$
Freon-12

If 150 g of CCl_4 is mixed with 100 g of SbF_3,

(a) how many grams of CCl_2F_2 can be formed?

(b) how many grams of which reactant will remain after reaction has ceased?

* 3.28 Silver tarnishes in the presence of hydrogen sulfide (rotten egg odor) and oxygen because of the reaction

$$4Ag + 2H_2S + O_2 \longrightarrow 2Ag_2S + 2H_2O$$
hydrogen   silver sulfide,
sulfide   black

How many grams Ag_2S could be obtained from a mixture of 0.950 g Ag, 0.140 g H_2S, and 0.0800 g O_2?

3.29 Phosgene, $COCl_2$, was once used as a war gas. It is poisonous because when it is inhaled, it reacts with water in the lungs to produce hydrochloric acid, HCl, which causes severe lung damage, leading ultimately to death. The chemical reaction is

$$COCl_2 + H_2O \longrightarrow CO_2 + 2HCl$$

(a) How many moles of HCl are produced by the reaction of 0.430 mol of $COCl_2$ with an excess of H_2O?

(b) How many grams of HCl are produced when 11.0 g of CO_2 is formed?

(c) How many moles of HCl will be formed if 0.200 mol of $COCl_2$ is mixed with 0.400 mol of H_2O?

*3.30 Acetylene, C_2H_2, can react with two molecules of Br_2 to form $C_2H_2Br_4$ by the series of reactions

$$C_2H_2 + Br_2 \longrightarrow C_2H_2Br_2$$

$$C_2H_2Br_2 + Br_2 \longrightarrow C_2H_2Br_4$$

If 5.00 g of C_2H_2 is mixed with 40.0 g of Br_2, what weights of $C_2H_2Br_2$ and $C_2H_2Br_4$ will be formed? Assume that all the C_2H_2 has reacted.

Theoretical Yield and Percentage Yield

3.31 What is meant by the *theoretical yield* for a given reaction mixture? What is the meaning of *percentage yield?* When you carry out a reaction, how do you determine the *actual yield?*

3.32 The reaction that's used to make the refrigerant gas Freon-12, CCl_2F_2, was described in Problem 3.27 as

$$3CCl_4 + 2SbF_3 \longrightarrow 3CCl_2F_2 + 2SbCl_3$$

In a certain experiment, 14.6 g of SbF_3 was allowed to react with an excess of CCl_4. After the reaction was finished, 8.62 g of CCl_2F_2 was isolated from the reaction mixture.
(a) What was the theoretical yield of CCl_2F_2 in this experiment expressed in grams?
(b) What was the actual yield of CCl_2F_2 in grams?
(c) What was the percentage yield of CCl_2F_2 in this experiment?

3.33 The combustion of methyl alcohol, CH_3OH, in a plentiful supply of oxygen follows the equation

$$2CH_3OH + 3O_2 \longrightarrow 2CO_2 + 4H_2O$$

When 6.40 g of CH_3OH was mixed with only a slight excess of oxygen and allowed to react, 6.12 g of CO_2 was obtained.
(a) Based on the equation above, what was the theoretical yield (in grams) of CO_2 in this experiment?
(b) What was the actual yield of CO_2 in this experiment expressed in grams?
(c) What was the percentage yield of CO_2?

3.34 In an experiment, a student allowed benzene, C_6H_6, to react with excess bromine, Br_2, in an attempt to prepare bromobenzene, C_6H_5Br. This reaction also produced, as a by-product, dibromobenzene, $C_6H_4Br_2$. On the basis of the equation

$$C_6H_6 + Br_2 \longrightarrow C_6H_5Br + HBr$$

(a) What is the maximum number of grams of C_6H_5Br that the student could have hoped to obtain from 15.0 g of benzene? (This is the theoretical yield.)
(b) In this experiment, the student obtained 2.50 g of $C_6H_4Br_2$. How many grams of C_6H_6 were *not* converted to C_6H_5Br?

(c) What was the student's actual yield of C_6H_5Br in grams?
(d) Calculate the percentage yield for this reaction.

*3.35 In a reaction between methane, CH_4, and chlorine, Cl_2, four products can be formed: CH_3Cl, CH_2Cl_2, $CHCl_3$, and CCl_4. In a particular instance, 20.8 g of CH_4 was allowed to react with excess Cl_2 and gave 5.0 g CH_3Cl, 25.5 g CH_2Cl_2, and 59.0 g $CHCl_3$. All the CH_4 reacted.
(a) How many grams of CCl_4 were formed?
(b) On the basis of available CH_4, what is the theoretical yield of CCl_4 in grams?
(c) What is the percentage yield of CCl_4?
(d) How many grams of Cl_2 reacted with the CH_4? (*Note:* The hydrogen that is displaced from the carbon also combines with Cl_2 to form HCl.)

*3.36 A chemist wishes to synthesize a certain compound that has a molecular mass of 100. The synthesis requires six consecutive steps, each giving a 50% yield (computed on a *mole* basis). If the chemist begins with 30.0 g of starting material having a molecular mass of 80.0, how many grams of final product will be obtained? How many grams of starting material will be required to produce 10.0 g of final product?

Reactions in Solution and Molar Concentration

3.37 Why are solutions used to carry out chemical reactions?

3.38 Define the following terms:
(a) precipitate   (e) concentration
(b) solution   (f) concentrated
(c) solvent   (g) dilute
(d) solute

3.39 Define molar concentration.

3.40 To make 1.00 L of a 1.00 M solution of the sugar glucose, $C_6H_{12}O_6$, requires 180 g $C_6H_{12}O_6$. Describe how you would actually go about preparing this solution.

3.41 Trisodium phosphate, Na_3PO_4, is a very powerful, but caustic, cleaning agent. If a bottle containing a solution of Na_3PO_4 were labeled 0.20 M Na_3PO_4, how could you use that information in the form of a conversion factor?

3.42 Calculate the molar concentration (molarity) of the solute in the following solutions:
(a) 0.250 mol NaCl in 0.400 L of solution
(b) 1.45 mol sucrose in 345 mL of solution
(c) 195 g H_2SO_4 in 875 mL of solution
(d) 80.0 g KOH in 0.200 L of solution

3.43 Calculate the molar concentration (molarity) of the following solutions:
(a) 1.35 mol NH_4Cl in a total volume of 2.45 L
(b) 0.422 mol $AgNO_3$ in a total volume of 742 mL
(c) 3.00×10^{-3} mol KCl in 10.0 mL of solution
(d) 4.80×10^{-2} g $NaHCO_3$ in 25.0 mL of solution

3.44 A millimole (mmol) is 10^{-3} mol. Suppose that each milliliter of a solution contained 0.250 mmol of NaCl, that is, its concentration is 0.250 mmol/mL.

(a) Calculate the concentration of this solution in units of mol/L.

(b) What is the molarity of this solution?

(c) In what way are the numerical values of concentration related when the concentration is reported in molarity, mol/L, and mmol/mL?

3.45 Suppose a solution of lithium carbonate, Li_2CO_3, a drug used to treat manic depressives, is labeled 0.150 M.

(a) How many moles of Li_2CO_3 are present in 250 mL of this solution?

(b) How many grams of Li_2CO_3 are in 630 mL of the solution?

(c) How many milliliters of the solution would be needed to supply 0.0100 mol Li_2CO_3?

(d) How many milliliters of the solution would be needed to provide 0.0800 g of Li_2CO_3?

3.46 A solution was labeled 0.375 M KOH.

(a) How many milliliters of this solution are needed to give 0.100 mol KOH?

(b) How many moles of KOH are in 45.0 mL of this solution?

(c) How many milliliters of this solution are needed to give 10.0 g of KOH?

(d) How many grams of KOH are in each milliliter of the solution?

3.47 Calcium acetate, $Ca(C_2H_3O_2)_2$, is the substance used along with methyl alcohol to make "canned heat." How many grams of calcium acetate are needed to prepare 2.00 L of a 0.250 M solution of $Ca(C_2H_3O_2)_2$?

3.48 How many grams of potassium nitrate, KNO_3, are needed to prepare 250.0 mL of a 3.000×10^{-2} M KNO_3 solution?

*3.49 The formula for Epsom salts is $MgSO_4 \cdot 7H_2O$. How many grams of Epsom salts are needed to prepare 500 mL of a solution that can be labeled 0.150 M $MgSO_4$?

Dilution of Solutions

3.50 How does the number of moles of solute in a solution change as the solution is diluted?

3.51 What is the proper procedure for diluting a solution of a concentrated reagent such as H_2SO_4 (sulfuric acid)?

3.52 How many milliliters of 18.0 M H_2SO_4 must be added to 100 mL of H_2O to give a solution of 5.00 M H_2SO_4?

3.53 How many milliliters of concentrated NH_3 must be used to prepare 250 mL of 0.500 M NH_3?

3.54 What volume (in mL) of concentrated H_2SO_4 (18.0 M H_2SO_4) must be used to prepare 400 mL of 3.00 M H_2SO_4 solution?

3.55 To what volume (in mL) must 100 mL of 0.500 M H_2SO_4 be diluted to give 0.200 M H_2SO_4?

3.56 How many milliliters of water must be added to 85.0 mL of 1.00 M H_3PO_4 to produce 0.650 M H_3PO_4?

*3.57 How many milliliters of 1.00 M HCl must be added to 50.0 mL of 0.500 M HCl to give a solution whose concentration is 0.600 M?

Stoichiometry of Reactions in Solution

3.58 In Problem 3.25, it was mentioned that aluminum reacts with sulfuric acid.

$$2Al(s) + 3H_2SO_4(aq) \longrightarrow Al_2(SO_4)_3(aq) + 3H_2(g)$$

(a) How many milliliters of 0.200 M H_2SO_4 are needed to react completely with 3.50 g Al?

(b) If a large piece of aluminum (an excess amount of Al) was placed in 400 mL of 0.200 M H_2SO_4 solution, how many moles of H_2 would be produced by this reaction?

3.59 The light-sensitive compound in photographic film is usually silver bromide, AgBr. It is made commercially by the reaction of solutions of sodium bromide, NaBr, and silver nitrate, $AgNO_3$.

$$NaBr(aq) + AgNO_3(aq) \longrightarrow AgBr(s) + NaNO_3(aq)$$

If 300 mL of 0.250 M NaBr are mixed with 200 mL of 0.400 M $AgNO_3$,

(a) which is the limiting reactant?

(b) how many grams of AgBr are formed?

3.60 Lye, which is sodium hydroxide (NaOH), can be neutralized by sulfuric acid. The reaction is

$$2NaOH(aq) + H_2SO_4(aq) \longrightarrow Na_2SO_4(aq) + 2H_2O$$

(a) How many milliliters of 0.200 M H_2SO_4 are needed to react completely with 25.0 mL of 0.400 M NaOH?

(b) How many milliliters of 0.100 M NaOH are needed to react completely with 50.0 mL of 0.270 M H_2SO_4?

(c) If 40.0 mL of 0.300 M NaOH are added to 15.0 mL of 0.350 M H_2SO_4, how many moles of Na_2SO_4 will be formed?

3.61 Magnesium hydroxide, $Mg(OH)_2$, is the white milky substance in milk of magnesia. When NaOH is added to a solution of $MgCl_2$, $Mg(OH)_2$ is formed.

$$2NaOH(aq) + MgCl_2(aq) \longrightarrow Mg(OH)_2(s) + 2NaCl(aq)$$

(a) How many milliliters of 0.300 M NaOH are needed to react with 75.0 mL of 0.200 M $MgCl_2$?

(b) How many grams of $Mg(OH)_2$ can be formed if an excess of NaOH is added to 50.0 mL of 0.600 M $MgCl_2$?

(c) How many grams of $Mg(OH)_2$ will be formed if 30.0 mL of 0.200 M $MgCl_2$ are added to 100 mL of 0.140 M NaOH solution?

*3.62 The silver nitrate ($AgNO_3$) in 20.00 mL of a certain solution was allowed to react with sodium chloride according to the equation $AgNO_3(aq) + NaCl(aq) \rightarrow AgCl(s) + NaNO_3(aq)$. The AgCl was collected, dried, and weighed to give 0.2867 g AgCl. What was the molarity of the original $AgNO_3$ solution?

* **3.63** To what volume (in mL) must 250 mL of 1.40 M H_2SO_4 be diluted to give a solution, 25.0 mL of which is able to completely react with 15.0 mL of 0.750 M NaOH? The equation for the reaction is

$$2NaOH + H_2SO_4 \longrightarrow Na_2SO_4 + 2H_2O$$

Additional Exercises

3.64 Aluminum reacts with hydrochloric acid (HCl) to give hydrogen gas (H_2) and aluminum chloride ($AlCl_3$).
(a) Write a balanced equation for this reaction.
(b) If 0.300 mol of Al reacts, how many moles of H_2 are formed?
(c) If 0.200 mol of H_2 is formed, how many moles of $AlCl_3$ are also formed?
(d) If 0.600 mol HCl reacts, how many moles of Al react?
(e) How many grams of Al can react with 9.13 g of HCl?

3.65 Suppose that 400 mL of 0.200 M NaOH were mixed with 800 mL of 0.600 M NaOH. What would be the molarity of NaOH in the final solution? Assume volumes are additive.

3.66 The reaction in Problem 3.64 can occur between solid aluminum and aqueous hydrochloric acid solution. The products are the gas H_2 and an aqueous solution of $AlCl_3$.
(a) Write the balanced equation for the reaction showing the actual physical states of the reactants and products.
(b) How many grams of aluminum can dissolve in 300 mL of 0.150 M HCl solution?
(c) After all the HCl in the solution in part b has reacted, what will be the molar concentration of $AlCl_3$ in the solution? Assume there is no change in the volume of the solution.

3.67 Suppose that 6.00 g of magnesium is placed into 350 mL of 0.800 M HCl solution where it reacts according to the equation

$$Mg(s) + 2HCl(aq) \longrightarrow MgCl_2(aq) + H_2(g)$$

(a) How many grams of H_2 will be produced in the reaction?
(b) After the reaction was finished, a student attempted to recover the $MgCl_2$ that was formed in the experiment and obtained 8.64 g of pure $MgCl_2$. What was the percentage yield of this product?

3.68 Using the reactions given in Exercise 3.17, how many kilograms of white lead will be formed from 10.0 kg of Pb if the percentage yield in the first step is 85.8%, and the percentage yield in the second step is 72.3%?

3.69 If 235 mL of 0.600 M HCl is mixed with 94.0 mL of 0.750 M Na_2CO_3, the following reaction occurs.

$$2HCl(aq) + Na_2CO_3(aq) \longrightarrow 2NaCl(aq) + CO_2(g) + H_2O$$

When the reaction is finished, what is the molar concentration of NaCl in the resultant solution? (Neglect the volume of water formed in the reaction and assume the volumes of the solutions are additive.)

THE PERIODIC TABLE AND SOME PROPERTIES OF THE ELEMENTS

Chemical calculations like the ones you studied in the last chapter are tools that chemists use to learn chemical facts — such things as the compositions and reactions of substances. These facts form the central core of chemistry and serve as the foundation on which theoretical explanations of behavior are built, so they are obviously of great importance.

The goal of this chapter is to introduce some of the properties of the elements and the reactions they undergo to form compounds. Along the way, you will see how the periodic table was developed and you will begin to discover what scientists have learned about the inner structures of atoms. And finally, you will start to learn how to name chemical compounds, a task that is essential for communication among chemists.

4.1 SOME PROPERTIES OF THE ELEMENTS

The range of properties shown by the elements is tremendous. At room temperature, some are gases, two are liquids, and the rest are solids. Some are metallic and some are not, and some have properties in between. Some are hard and some are soft; some are very dense and others have very low densities. With such variety, we have to look for ways to classify these properties so that it is possible to make some sense out of them.

One of the simplest methods of classification is to divide the elements into three categories: **metals, nonmetals,** and **metalloids**. The elements in each of these categories have certain distinctive characteristics.

Figure 4.1. *A very thin film of pure gold is transferred from a paper backing to a sculpture in Rockefeller Center in New York City. The gold provides a nontarnishing decorative finish to the work.*

Metals

Everyone has seen metals of one kind or another—an iron nail, aluminum foil, copper wire, or a "chrome-plated" bumper on a car, for example. And you are no doubt familiar with some of the properties that characterize metals, even if you haven't thought about them very much. One of these, for instance, is the distinctive appearance that metals have. They shine with a luster so characteristic that it's called a *metallic luster.*

Metals are also similar in their abilities to deform without breaking when hit with a hammer and to stretch when pulled. All metals have both of these properties to some degree. The ability to deform when hammered is called **malleability,** and some metals, such as gold, can be hammered or squeezed into extremely thin sheets. *Gold leaf* (Figure 4.1), for example, consists of gold with a small amount of silver and copper that has been beaten into sheets so thin (about 1/280,000 in.) that they are translucent; some light can actually be seen passing through them. Malleability is also a property that a blacksmith relies on when forging a horseshoe, and a silversmith uses the malleability of silver in hammering a design into a fine silver tray.

The ability of a metal to stretch when pulled from opposite directions is called **ductility.** This property is used in the manufacture of wire, which is illustrated in Figure 4.2. The metal to be made into wire, which might be steel, copper, or brass, is first formed into a rod. One end is tapered, fed through a die, and attached to a pulling device on the other side. The metal is then drawn through the die where it undergoes a reduction in size and an increase in length.

Everyone knows that metals are good conductors of electricity. They are also good conductors of heat. If you have ever touched a metal object that has been lying in the sun for a while, you know how very hot it feels. In fact, it feels much hotter than other objects alongside that are not metallic. The reason is that as your hand absorbs heat from the metal, heat travels quickly from the neighboring parts of the object to replace it. Nonmetallic objects don't feel as hot because when heat is removed by your hand, it can't be replaced rapidly, so the part of the object in contact with your hand becomes cooler.

More than 70% of the elements are metals, and although there are some similarities among them, there are many differences, too. Some metals are quite common and we encounter them nearly every day in their elemental forms, that is, uncombined with other elements. The metals mentioned previously (iron, aluminum, copper, and chromium) are only a few examples. There are other metals that are so

Figure 4.2. *Copper wire is pulled through smaller and smaller dies and becomes thinner and thinner. The ductility of copper makes this possible.*

(a)

(b)

Figure 4.3. *Two forms of the element carbon.* (a) *Graphite, a black powder.* (b) *Diamonds of gem quality.*

reactive only chemists, or chemistry students, ever have an opportunity to see them. For example, in Chapter 1 we mentioned the reaction between sodium and chlorine to form sodium chloride, common table salt. Sodium is a very reactive metal that combines rapidly not only with chlorine, but also with both oxygen and moisture.

The range of chemical reactivity of metals is very broad. Sodium is typical of one extreme, while gold is typical of the other. Jewelry is made from gold for several reasons, one of which is the fact that it doesn't tarnish when exposed to air and moisture. This very low reactivity, combined with gold's high electrical conductivity, accounts for one of this metal's most important commercial uses: the plating of electrical contacts in computers and other electronic devices.

Besides chemical reactivity, metals differ in certain physical properties such as hardness and melting point. Some metals are very hard, and some are very soft. Chromium and iron are examples of hard metals; gold and lead are examples of soft ones. Sodium is also a soft metal; in Figure 1.1 (page 3), we see it being cut with a knife. The extremes of melting point are even more impressive. Tungsten has the highest melting point of any element, 3400 °C (6150 °F). Because tungsten can be heated until it is glowing white hot without melting, it is used as the filament in electric light bulbs. Mercury has the lowest melting point of any metal, −38.9 °C, which means that it is a liquid at room temperature (about 25 °C). As you know, mercury is the fluid commonly used in thermometers.

Nonmetals

Most of the nonmetallic elements (the nonmetals) are rarely encountered in our daily activities in their elemental forms; instead, they are usually found in compounds. One nonmetal that most people have seen is carbon, which occurs in nature in two different forms (Figure 4.3). The more common variety is called graphite. This is the form that we find in charcoal briquets and the "lead" in lead pencils. The less common and more valuable form of carbon is diamond. Graphite and diamond have properties that are quite different from those that we associate with metals. Neither has the luster of a metal, and neither is malleable or ductile.

Other nonmetals that you have encountered are oxygen and nitrogen, which are the principal components of the atmosphere. Usually, you are not aware of their existence, because they are colorless gases and you can't see them. As you learned in Chapter 1, oxygen and nitrogen occur as **diatomic molecules,** molecules containing two atoms each. Other nonmetallic elements form similar molecules, and most are gases also. These are hydrogen (H_2), fluorine (F_2), and chlorine (Cl_2). Bromine (Br_2) and iodine (I_2) are also diatomic, but bromine is a liquid and iodine is a solid at room temperature (Figure 4.4).

Sulfur, another nonmetal, is the yellow powder shown in Figure 1.14 on page 25.

Figure 4.4. (a) Bromine is a dark red liquid that vaporizes easily to give orange vapors. (b) Iodine, a dark purple solid, is also easily vaporized to give purple vapors.

(a)

(b)

Just as the properties of the metals cover a broad range, so do the properties of the nonmetals. As we've seen, some are gases and one (bromine) is a liquid. There are others that are solids; carbon is just one example. Besides differing in these physical properties, nonmetals differ from each other in their chemical properties. Fluorine, for example, is extremely reactive, while helium is *inert* (totally unreactive). These differences, which are very important, will be explored in more detail later in the book.

Metalloids

Metalloids (also called *semimetals*) are elements that have properties that are intermediate between those of metals and those of nonmetals. The best-known example is the element silicon. Two others are arsenic (As) and antimony (Sb), which are shown in Figure 4.5. In terms of outward appearances, these elements have something of a metallic look about them, but their dark color gives them away. They certainly differ in appearance from typical metals such as iron or silver.

Metalloids are typically semiconductors—they conduct electricity, but not nearly as well as metals. These semiconductor properties are especially valuable in the electronics industry, because they make possible all the microelectronic devices found in hand-held calculators and microcomputers. See Figure 4.6. Except for their electrical properties, however, the metalloids are much more like nonmetals than metals.

Figure 4.5. *Elemental antimony (left) and arsenic (right).*

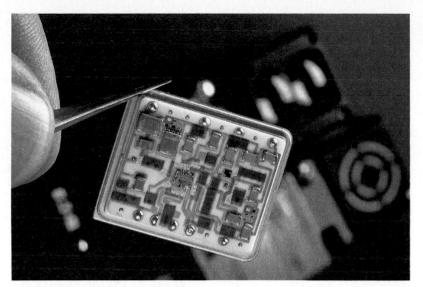

Figure 4.6. *At the center of this tiny electronic device is a microcircuit set into the surface of a tiny silicon chip. Silicon's semiconductor properties make possible microelectronic devices such as this.*

4.2 THE FIRST PERIODIC TABLE

Chemical and physical properties like those described in the last section were discovered early on in the history of chemistry. Scientists, even as early as 1800, had accumulated significant amounts of information about the elements known to them. This knowledge, however, existed for the most part as isolated and unrelated facts that needed to be correlated in some fashion before their total significance could be grasped. Early attempts at a classification of the elements met with only limited success, and it wasn't until 1869 that the forerunner of our modern periodic table was devised. This resulted from the work of two chemists, a Russian named Dmitri Mendeleev and a German named Julius Lothar Meyer. They worked independently and produced similar tables at about the same time. Mendeleev presented the results of his work to the Russian Chemical Society in the early part of 1869, but Meyer's table didn't appear until December of that same year. Mendeleev had the good fortune of presenting his first, so he is usually given credit for the periodic table.

Mendeleev was a chemistry teacher, and while he was preparing a textbook for his students, he discovered that if he arranged the elements in order of increasing atomic mass, elements with similar properties occurred at periodic intervals. For example, he could pick out the elements lithium (Li), sodium (Na), potassium (K), and rubidium (Rb). Each of these elements forms a water-soluble compound with chlorine that has the general formula MCl, where M stands for Li, Na, K, and so on. Although this is an interesting fact itself, what is especially significant is that if we examine the elements that immediately follow Li, Na, K, and Rb in the list (Be, Mg, Ca, and Sr, respectively), they too make up a group of similar elements. For example, they form the compounds $BeCl_2$, $MgCl_2$, $CaCl_2$, and $SrCl_2$. Mendeleev found this kind of phenomenon occurring repeatedly in his list of elements and he realized that he could divide the list into a series of rows that, when stacked one over the other, placed elements with similar properties in vertical columns. The result was the first periodic table, which is illustrated in Figure 4.7.

When Mendeleev constructed his table, not all the elements had yet been discovered. He realized this because in order to always have similar elements in the same column, or **group,** he was forced to leave occasional blanks in his table. It was also necessary for him to reverse the atomic-mass order of tellurium (Te) and iodine

Dmitri Mendeleev

Mendeleev was convinced that the periodic recurrence of properties was related to atomic mass. He thought that the atomic mass of Te had been measured incorrectly, so he place Te where it belonged based on its chemical properties.

	Group I	Group II	Group III	Group IV	Group V	Group VI	Group VII	Group VIII
1	H 1							
2	Li 7	Be 9.4	B 11	C 12	N 14	O 16	F 19	
3	Na 23	Mg 24	Al 27.3	Si 28	P 31	S 32	Cl 35.5	
4	K 39	Ca 40	— 44	Ti 48	V 51	Cr 52	Mn 55	Fe 56, Co 59 Ni 59, Cu 63
5	(Cu 63)	Zn 65	— 68	— 72	As 75	Se 78	Br 80	
6	Rb 85	Sr 87	?Yt 88	Zr 90	Nb 94	Mo 96	— 100	Ru 104, Rh 104 Pd 105, Ag 108
7	(Ag 108)	Cd 112	In 113	Sn 118	Sb 122	Te 128	I 127	
8	Cs 133	Ba 137	?Di 138	?Ce 140	—	—	—	— — — —
9	—	—	—	—	—	—	—	
10	—	—	?Er 178	?La 180	Ta 182	W 184	—	Os 195, Ir 197 Pt 198, Au 199
11	(Au 199)	Hg 200	Tl 204	Pb 207	Bi 208	—		
12	—	—	—	Th 231	—	U 240	—	— — — —

Figure 4.7. *The way Mendeleev arranged the elements known to him in his periodic table, which was first published in 1871. The numbers appearing with the symbols are the atomic masses.*

Eka-silicon turned out to be the element germanium, whose properties are remarkably like those predicted by Mendeleev.

(I), whose atomic masses in 1869 were thought to be 128 and 127 u, respectively. Mendeleev placed them in the table in reverse order (according to atomic masses) because their properties dictated that tellurium belongs in Group VI and iodine in Group VII. (The groups are numbered with Roman numerals for ease of identification.)

One of the benefits of Mendeleev's table was that it became possible to predict some of the properties of the missing elements. This was because elements in any particular column had to have similar properties. For example, germanium, which lies below silicon and above tin in Group IV, had not been discovered when Mendeleev assembled his table. Therefore, a space appears at this spot in the chart. On the basis of its position, Mendeleev predicted that the properties of this element, which he called "eka-silicon," should lie intermediate between those of silicon and tin.

If you look at the modern periodic table in Figure 4.8 on page 94, you will see a column of elements that did not appear in Mendeleev's table. This is the rightmost column with the heading "Noble gases." These are extremely unreactive elements that exist as colorless and odorless gases in very small amounts in the atmosphere. Because they formed no known compounds, scientists in Mendeleev's time were totally unaware of their existence. When these elements were finally discovered, it was found that the atomic mass of argon (Ar) is slightly larger than that of potassium (K). Nevertheless, potassium clearly belongs with the other elements in Group I and argon clearly belongs in the group with the other noble gases. Once again, as in the case of Te and I, it was necessary to place a pair of elements in the table in reverse atomic-mass order.

The need to switch the atomic-mass order of these two pairs of elements led scientists to realize, finally, that atomic mass was not what really determines where an element belongs in the periodic table. The true basis for the periodic table lies elsewhere, as discussed in the next section.

4.3 THE MODERN VIEW OF THE ATOM

The troubles that are encountered when the elements are arranged in atomic-mass order in Mendeleev's periodic table are removed if they are arranged in order of their *atomic numbers.* To understand atomic number, however, we first must take a brief look at the internal structures of atoms.

Dalton's vision of atoms as indivisible particles we now know to be incorrect. Experiments begun in the late nineteenth century and continued to the present day have shown that atoms are themselves composed of still simpler *subatomic* particles. Many such particles have been discovered, but the principal ones — those most important to us — are called **protons, neutrons,** and **electrons.**

Protons and electrons are electrically charged particles. They carry opposite kinds of charge, with protons having a charge that we specify as *positive* (+) and electrons having a charge that we specify as *negative* (−). One of the main features of these kinds of electrical charge is that opposite charges attract each other while charges of the same kind repel. Thus, protons attract electrons, but protons repel other protons and electrons repel other electrons. Neutrons, as their name suggests, carry no charge and therefore are electrically neutral.

In the SI, electrical charge is expressed in units of **coulombs** (symbol **C**). One coulomb is the amount of electrical charge that passes a given point in a wire when an electric current of 1 ampere flows for 1 second. In more familiar terms, if you have a 100-watt light bulb burning, it takes about 1.2 seconds for 1 coulomb of charge to pass through the bulb. This is a fairly large amount of charge, and the amount of charge carried by just one electron is much smaller; it has been measured to be 1.60×10^{-19} C. Because the charge on the electron is negative, we say that its charge is equal to -1.60×10^{-19} C. Protons carry exactly the same amount of charge as the electron; it is just of the opposite kind, so the charge carried by a proton is $+1.60 \times 10^{-19}$ C.

Whenever we deal with the electrical charge on a particle, it is always in multiples of 1.60×10^{-19} C, so it is convenient to simply define a unit of charge as equal to this amount. On this scale, an electron has one unit of negative charge (we say its charge is 1−) and a proton has one unit of positive charge (we say its charge is 1+).

These subatomic particles have another important property, their mass. Protons and neutrons are relatively heavy particles with masses of approximately one atomic mass unit (1 u). The electron, on the other hand, is a very light particle with a mass of only about 1/1836 of the mass of a proton. The properties of protons, neutrons, and electrons are summarized in Table 4.1.

The nuclear atom

The concept of an *atomic nucleus* should be familiar to anyone who has ever heard of nuclear energy. The **nucleus** is the name given to the extremely small and very dense

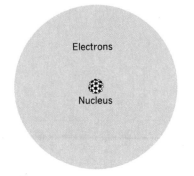

Protons and neutrons are found in the nucleus at the center of an atom. Electrons fill the space around the nucleus. In this drawing the size of the nucleus is much exaggerated. Its actual diameter is about 1/100,000th of the diameter of the atom.

● Protons
○ Neutrons

Table 4.1. Some properties of subatomic particles

	Mass		Charge	
	Grams	Atomic Mass Units (u)	Coulombs	Electronic Charge Units
Proton	1.67×10^{-24}	1.007276	$+1.602 \times 10^{-19}$	1+
Neutron	1.67×10^{-24}	1.008665	0	0
Electron	9.11×10^{-28}	0.0005486	-1.602×10^{-19}	1−

particle that experiments have shown is located at the center of every atom. Experiments have also shown that all of an atom's protons and neutrons are located in its nucleus and that the electrons are distributed in the space around the nucleus. How these electrons are arranged is very important in chemistry, and we will discuss this at some length in Chapter 7. For now, however, it is sufficient just to know that we find them outside the nucleus and that they fill most of the volume of an atom.

Matter as a whole is electrically neutral, which means that the total number of positive charges must equal the total number of negative charges.

As far as chemistry is concerned, the nucleus is of interest for two principal reasons. One is that the number of protons in the nucleus, which is referred to as an atom's **atomic number,** determines the number of electrons that the atom must have in order to be electrically neutral.

The second reason is that the mass of an atom is determined primarily by the number of protons and neutrons in its nucleus, each of which contributes approximately one atomic mass unit. These particles are so much heavier than the electrons that the mass of the nucleus is very nearly equal to the atomic mass of the atom. In addition, because the nucleus is so tiny, the density of the nuclear material is enormous, about 10^{14} g/cm^3. To give you some idea of how dense this is, if all the nuclei of all the atoms in the crude oil cargo of one of the world's largest supertankers could be crammed together until they were touching, they would occupy a volume of only about 0.004 cm^3. This is the volume of about a *tenth* of a drop of water, yet their combined mass would be over 400,000 tons!

Isotopes

As we mentioned in Chapter 2, not all atoms of the same element have identical masses as first suggested by Dalton. We referred to these different kinds of atoms as **isotopes.** The existence of isotopes is a common phenomenon, and most of the elements occur naturally as mixtures of isotopes.

As we will see later, the properties of an element are determined almost entirely by the number and distribution of its electrons around the nucleus. Therefore, it is the atomic number that serves, indirectly, to distinguish an atom of one element from an atom of another, because the number of electrons must equal the atomic number in an electrically neutral atom. In other words, an atom's atomic number identifies which element it is. If atoms of the same element differ at all in mass, it is because they have different numbers of neutrons.

The nuclear particles protons and neutrons are called **nucleons.** The mass number is the total number of nucleons.

A particular isotope of an element is identified by specifying its atomic number, given the symbol Z, and its **mass number** A. The mass number is the sum of the number of protons and the number of neutrons in the atom. The number of neutrons can therefore be obtained from the difference $A - Z$.

We represent an isotope symbolically by writing its mass number as a superscript and its atomic number as a subscript, both preceding the atomic symbol.

$$^A_Z X$$

For example, a carbon atom ($Z = 6$) that has six neutrons would be given the symbol $^{12}_6 C$. This is the carbon-12 isotope that serves as the basis of the atomic-mass scale.

^{12}C is the only atom whose atomic mass is exactly equal to its mass number.

It should be noted that, except for carbon-12, there is a difference between an isotope's mass number and its actual mass expressed in atomic-mass units. For example, the isotope ^{16}O has a mass number of 16, which means that the sum of the number of protons and neutrons is 16. However, the actual mass of an atom of ^{16}O is 15.99491 u. The reasons for this discrepancy are somewhat complicated and will be discussed in Chapter 24. The main point now is that you realize that the mass and the mass number have slightly different values.

As we've noted above, almost all the elements occur in nature as mixtures of isotopes. The element copper, for example, is found to contain two naturally occurring isotopes, $^{63}_{29}Cu$ and $^{65}_{29}Cu$, whose masses have been accurately determined to be

62.9298 and 64.9278 u, respectively. Their relative abundances are 69.09% and 30.91%. The observed average atomic mass of copper, 63.55, is obtained as an average of the isotopic masses, weighted according to the relative abundances of the isotopes, as shown in Example 4.1.

EXAMPLE 4.1. CALCULATING THE AVERAGE ATOMIC MASS FROM ISOTOPIC MASSES AND RELATIVE ABUNDANCES

PROBLEM: Using the data supplied in the paragraph above, calculate the average atomic mass of copper.

ANALYSIS: The arithmetic involved in solving this problem is very simple, once you understand what has to be done. Suppose that we had one mole of naturally occurring copper. The data above tell us that it is composed of two isotopes, ^{63}Cu and ^{65}Cu. In addition, their relative abundances tell us that 69.09% of the atoms would be ^{63}Cu and 30.91% of the atoms would be ^{65}Cu. That's the same as saying that we would have 69.09% of a mole of ^{63}Cu and 30.91% of a mole of ^{65}Cu. We can easily calculate how much these weigh, because we have the atomic masses of each of the isotopes.

SOLUTION: The mass of a mole of ^{63}Cu is 62.9298 g, so the mass of 69.09% of a mole is

$$62.9298 \text{ g} \times 0.6909 = 43.48 \text{ g}$$

The mass of a mole of ^{65}Cu is 64.9278 g, so the mass of 30.91% of a mole is

$$64.9278 \text{ g} \times 0.3091 = 20.07 \text{ g}$$

Therefore, in one mole of naturally occurring copper there is 43.48 g of ^{63}Cu and 20.07 g of ^{65}Cu, so the total mass of the mole of copper atoms is

$$43.48 \text{ g} + 20.07 \text{ g} = 63.55 \text{ g}$$

If a mole of naturally occurring copper weighs 63.55 g, then the average atomic mass of copper must be 63.55 u.

4.4 ATOMIC NUMBERS AND THE MODERN PERIODIC TABLE

When the elements are arranged in the periodic table in order of their atomic numbers, all the inconsistencies that had occurred in Mendeleev's table disappear. Tellurium and iodine, and argon and potassium fall naturally into place in their respective groups. Thus, it is really the atomic number of an element — the number of protons in the nuclei of its atoms — that determines where the element fits into the table, and since elements with similar properties are found in the same group, it must be the atomic number of an element that determines the kinds of chemical and physical properties that it has. *Why* this is so is naturally of great interest to us, but we will have to wait until Chapter 7 to explore it further. For now, let's look at how the modern periodic table is constructed so we can begin to see the way we use it to correlate chemical and physical properties.

The periodic table that is used today is shown in Figure 4.8. The numbers printed above the chemical symbols are atomic numbers, and those below are atomic masses. Like Mendeleev's table, it consists of a number of rows called **periods,** which are identified by Arabic numerals. The vertical columns are called **groups,** each containing a *family of elements.* The groups are also identified by number. The

Members of a group in the periodic table bear resemblances to each other as do the members of a family — hence, the term *family of elements.*

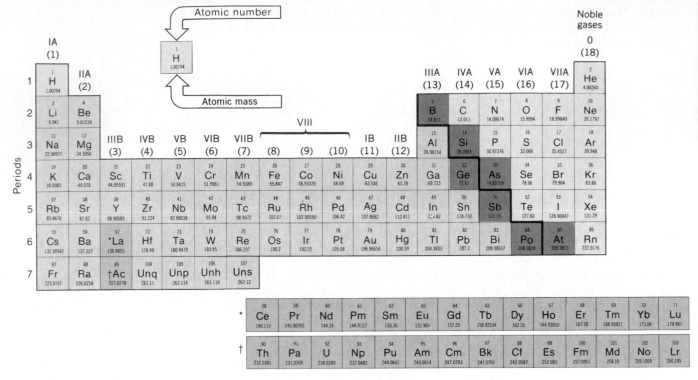

Figure 4.8. *The modern periodic table. The elements in tan are the nonmetals, those in red are the metalloids, and those in blue-gray are the metals. The two rows of elements in dark blue are the inner transition elements, which are also metals. They fit in the main body of the table after La and Ac.*

numbering system used in the United States for many years closely follows Mendeleev's and identifies each group by a Roman numeral and a letter, either A or B. These are shown at the top of each group. Recently, the International Union of Pure and Applied Chemistry (IUPAC) adopted an alternative system[1] in which the groups are numbered left to right from 1 to 18. These numbers are below the Roman numeral designations. There has been considerable controversy concerning the new system, and many chemists and chemical educators are opposed to it. Because the issue is far from settled, we will use the Roman numeral and the A and B group designations.

The groups labeled with the letter A (Groups IA through VIIA) and Group 0 are referred to collectively as the **representative elements.** Those labeled with the letter B (Groups IB through VIIB) plus Group VIII (actually composed of the three short columns in the center of the table) are called the **transition elements.** The reason for the A and B group designations is that there are some similarities between the A-group and the B-group elements, although the similarities are often weak. This will be discussed further later in the book.

Finally, there are two long rows of elements placed just below the main part of the table. These elements, known as **inner transition elements,** actually belong in the body of the table as shown in Figure 4.9. They are usually set below simply to conserve space, so the table can be printed conveniently across a page without the type being too small to read. Notice in Figure 4.9 that the first row of inner transition elements follows the element lanthanum (La) and that the second follows actinium (Ac). Because of the elements they follow, the first row (elements 58 through 71) are

The IUPAC is an international body that sets standards for the chemical sciences.

---

[1] In Europe, a different A and B group designation has been used in which IUPAC groups 1 through 7 are labeled IA through VIIA and IUPAC groups 11 through 17 are labeled IB through VIIB. In fact, one of the principal reasons that the IUPAC adopted the new system was to replace the two different systems that were in use at the same time.

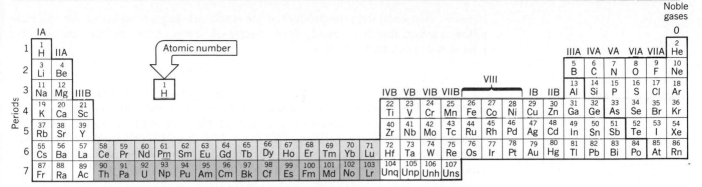

Figure 4.9. *The modern periodic table with the two rows of inner-transition elements in their proper locations.*

known as the **lanthanides,** and the second row (90 through 103) are known as the **actinides.** Often, the lanthanides are also called the **rare earth elements** because of their very low abundance in the earth's crust.

Certain families of elements are characterized by names as well as their group numbers. For example, the Group IA elements (excluding hydrogen) are known as the **alkali metals** and the Group IIA elements are known as the **alkaline earth metals.** The Group VIIA elements are the **halogens,** a name derived from the Greek meaning "salt former." Finally, the Group 0 elements are called the **noble gases** (also sometimes the *inert gases*) because of their very low degree of chemical reactivity.

Alkali comes from the Arabic alquili, meaning ashes of saltwort. Saltwort is a shrub that grows in coastal areas. Its ashes were once a major source of sodium carbonate (soda ash).

Trends in properties in the periodic table

The periodic table is probably the most valuable aid that chemists have at their disposal. It can be used to correlate all sorts of useful information, and it helps us see important variations in properties among the elements. One of the most easily observed trends is the variation in the metallic character of the elements.

In Section 4.1 you saw that we can classify an element as being a metal, a nonmetal, or a metalloid according to the properties it displays. In the periodic table, elements on the left and center are metals and those at the upper right are nonmetals (see Figure 4.8). The heavy step-shaped line drawn from boron (B) to astatine (At) roughly defines the boundary between metallic and nonmetallic behavior, and those elements adjacent to this line (except aluminum) are the metalloids. Thus, there is a gradual transition from metallic to nonmetallic properties in this region. Going from left to right across a period, the elements change from metallic to nonmetallic. In period 3, for example, aluminum is a metal, silicon a metalloid, and phosphorus a nonmetal. A similar transition occurs when going from top to bottom in a group in this region; the elements change from nonmetallic to metallic. For example, in Group IVA we find a nonmetal, carbon, at the top and a metal, lead, at the bottom. In between are two metalloids, silicon and germanium.

Most of the elements are metals.

4.5 REACTIONS OF METALS WITH NONMETALS; THE FORMATION OF IONIC COMPOUNDS

The variation in the metallic properties of the elements is just one of many trends that can be followed in the periodic table. In this and the next section we will examine another, the kinds of compounds formed by the metals and the nonmetals.

When two or more substances combine to form a single product, the reaction is sometimes called a **combination reaction.** Combination reactions often take place between elements, with the product of the reaction being a compound. An example is the reaction that has already been discussed several times before, the reaction between sodium and chlorine.

$$2Na(s) + Cl_2(g) \longrightarrow 2NaCl(s)$$

As mentioned briefly in Chapter 2, when this reaction occurs, each atom of sodium loses one electron, which is transferred to a chlorine atom. A neutral sodium atom has 11 protons inside its nucleus and 11 electrons outside, so the loss of an electron leaves the particle with one more positive charge than negative charge. As a result, the sodium ion carries a charge of $1+$ and is written as Na^+. When the chlorine atom gains an electron, it acquires an additional negative charge, so the *chloride ion* carries a charge of $1-$ and is written Cl^-. Because the compound NaCl contains these ions, it is called an ionic compound.

The reaction between sodium and chlorine is typical of many reactions between metals and nonmetals:

> Metals tend to react with nonmetals to form ionic compounds.

*A **cation**, pronounced cat'-ion, is a positive ion. An **anion**, pronounced an'-ion, is a negative ion.*

In these reactions, the metal atoms each lose one or more electrons and become positive ions, or **cations**, and the nonmetal atoms each gain one or more electrons and become negative ions, or **anions.**

The number of electrons gained or lost by an atom depends on a number of factors that we will explore later. Our goal now is to examine what these ions are and then to learn how we can use this information to write the formulas for ionic compounds.

Although hydrogen is placed in Group IA, it is not a metal like the other elements below it.

Table 4.2 lists the ions formed by most of the representative elements. Notice first that the atoms in any particular group form ions with the same charge. For example, all the ions formed by the alkali metals (Group IA) have a charge of $1+$, and all the ions formed by the halogens (Group VIIA) have a charge of $1-$. Also notice that for the metal ions, the number of positive charges is equal to the metal's group number. Thus, the Group IA metals form ions with a $1+$ charge, those in Group IIA form ions with a $2+$ charge, and aluminum in Group IIIA forms an ion with a $3+$ charge. By remembering this, you can easily use the periodic table to write the formula of the ion of one of these metals. For instance, if you needed to know the charge on the ion formed by Ba, check the periodic table. Barium is located in Group IIA, so the ion is Ba^{2+}.

We can also use the periodic table to help us remember the ions formed by the nonmetals. The number of negative charges is just the number of steps to the right that you have to go in the table to get to the noble gas column. Consider oxygen, for example. It is in Group VIA, and to get to the noble gas column requires moving two steps to the right.

For a given nonmetal, the negative charge on the anion equals 8 minus the nonmetal's group number.

Oxygen forms an ion with a negative *two* charge, O^{2-}. Similarly, to move from fluorine to the noble gas column is just one step, and fluorine forms the ion F^-.

The ions formed by the representative elements are easy to remember based on their positions in the periodic table, but unfortunately this isn't the case for the ions of the transition metals. Nor is it the case for the **post-transition** metals, the metals that are found in periods 4, 5, and 6 just to the right of the transition elements. This is because the transition and post-transition metals are able to form more than one ion, depending on the substances they react with and the reaction conditions. Some of these ions, which you should learn, are listed in Table 4.3.

REACTIONS OF METALS WITH NONMETALS; THE FORMATION OF IONIC COMPOUNDS

Table 4.2. Ions formed by the representative elements

		Group Number				
IA	IIA	IIIA	IVA	VA	VIA	VIIA
Li^+	Be^{2+}		C^{4-}	N^{3-}	O^{2-}	F^-
Na^+	Mg^{2+}	Al^{3+}	Si^{4-}	P^{3-}	S^{2-}	Cl^-
K^+	Ca^{2+}				Se^{2-}	Br^-
Rb^+	Sr^{2+}				Te^{2-}	I^-
Cs^+	Ba^{2+}					

Table 4.3. Cations formed by some transition and post-transition elements

Chromium	Cr^{2+} Cr^{3+}	Gold	Au^+ Au^{3+}
Manganese	Mn^{2+} Mn^{3+}	Zinc	Zn^{2+}
		Cadmium	Cd^{2+}
Iron	Fe^{2+} Fe^{3+}	Mercury	Hg_2^{2+} Hg^{2+}
Cobalt	Co^{2+} Co^{3+}	Tin	Sn^{2+} Sn^{4+}
Nickel	Ni^{2+}	Lead	Pb^{2+}
Copper	Cu^+ Cu^{2+}		Pb^{4+}
		Bismuth	Bi^{3+}
Silver	Ag^+		

Writing formulas for ionic compounds

The ratio in which the atoms of two elements combine to form an ionic compound is determined by the charges acquired by the cation and anion. This is because compounds are always electrically neutral, so the ions must occur in a ratio whereby the total number of positive charges is exactly equal to the total number of negative charges. This is why sodium chloride is NaCl instead of $NaCl_2$ or Na_2Cl. Only a 1-to-1 ratio of Na^+ to Cl^- ions will give an electrically neutral substance.

When we write the formula for an ionic compound, we follow several rules. Some are required by nature, as in the electrical neutrality requirement described in the preceding paragraph. Others are just customs that chemists always follow.

RULES FOR WRITING FORMULAS FOR IONIC COMPOUNDS

1. The positive ion is always written first in the formula.
2. The ratio of positive ions to negative ions must be such that the total number of positive charges equals the total number of negative charges; the formula unit must be electrically neutral.
3. The smallest set of subscripts that give electrical neutrality is always chosen. We always write empirical formulas for ionic compounds.

The reason for the third rule is that there are no molecules in an ionic compound. Each positive ion is surrounded by and attracted equally to some number of negative ions, and vice versa, so we can't say that any particular positive ion belongs to any particular negative ion. All we can do is specify the ratio in which the ions occur in the compound.

Let's look now at a few simple examples of how the rules are applied.

1. *The compound formed from calcium and oxygen.* Looking at the periodic table, we see that calcium is in Group IIA, so the calcium ion is Ca^{2+}. Oxygen is a nonmetal in Group VIA, so its ion is O^{2-}. To make an electrically neutral compound, we take the ions in a 1-to-1 ratio, and the formula is CaO.

2. *The compound formed from zinc and chlorine.* Zinc is a transition metal and it forms the ion Zn^{2+}. Chlorine is in Group VIIA, so its ion is Cl^-. To give an electrically neutral compound, we must have two Cl^- for each Zn^{2+}, so the formula is $ZnCl_2$.

Applying rule 2 for writing formulas sometimes takes a little juggling. Consider, for example, the compound formed when oxygen in the air reacts with a freshly exposed surface of aluminum to form a thin film of aluminum oxide. Aluminum forms Al^{3+} and oxygen forms O^{2-}. In the compound, we need a ratio of ions that makes the total positive and negative charges the same. This requires that we take two Al^{3+} ions for every three O^{2-} ions, because

$$\begin{array}{l} 2\ Al^{3+}\ \text{have a charge of}\ 2 \times (3+) = 6+ \\ \underline{3\ O^{2-}\ \text{have a charge of}\ 3 \times (2-) = 6-} \\ \hspace{3cm} \text{Total net charge} = 0 \end{array}$$

There is a very simple shortcut that you can use to obtain the formula of an ionic compound. It involves exchanging superscripts for subscripts.

$$Al^{3+} \qquad O^{2-}$$
$$Al_2O_3$$

This forces the total positive charge to equal the total negative charge. However, if you use this method, remember that ionic compounds always have the smallest set of whole-number subscripts in their formulas. Applying this method to writing the formula for the compound formed from Ti^{4+} and O^{2-} gives Ti_2O_4, but these subscripts are each divisible by 2. The correct formula is therefore TiO_2.

Ions that contain more than one atom

There are many substances that contain **polyatomic ions,** ions that are composed of more than one atom. An example of such an ion is the carbonate ion, CO_3^{2-}, which is present in the chalk that your teacher uses to write on the blackboard. All four of the atoms in the CO_3^{2-} ion are bound together in the same way that atoms are bound together in a molecule. Polyatomic ions are therefore similar to molecules, except that they carry electrical charges.

Some of the most common polyatomic ions are given in Table 4.4. You should learn their formulas (including their charges) and their names, because they will be encountered frequently in this and other chemistry courses that you take. As shown in Figure 4.10, some of these ions have characteristic colors that they impart to compounds containing them.

Figure 4.10. *Some polyatomic ions have characteristic colors that they impart to compounds and solutions that contain them. On the left is a solution of potassium chromate, which contains the yellow chromate ion, CrO_4^{2-}. In the center is a solution of potassium dichromate, which contains the red-orange dichromate ion, $Cr_2O_7^{2-}$. On the right is a solution of potassium permanganate, which contains the violet permanganate ion, MnO_4^-.*

Table 4.4. Some common polyatomic ions

Cations

NH_4^+	ammonium
H_3O^+	hydronium

Anions (alternative names in parentheses)

CO_3^{2-}	carbonate	ClO_2^-	chlorite
HCO_3^-	hydrogen carbonate (bicarbonate)	ClO^- (or OCl^-)	hypochlorite
$C_2O_4^{2-}$	oxalate	PO_4^{3-}	phosphate
CN^-	cyanide	HPO_4^{2-}	hydrogen phosphate
NO_3^-	nitrate	$H_2PO_4^-$	dihydrogen phosphate
NO_2^-	nitrite	CrO_4^{2-}	chromate
OH^-	hydroxide	$Cr_2O_7^{2-}$	dichromate
SO_4^{2-}	sulfate	MnO_4^-	permanganate
HSO_4^-	hydrogen sulfate (bisulfate)	$C_2H_3O_2^-$	acetate
SO_3^{2-}	sulfite		
HSO_3^-	hydrogen sulfite (bisulfite)		
ClO_4^-	perchlorate		
ClO_3^-	chlorate		

You need to know the names of the polyatomic ions so you can name compounds that contain them; you need to know their charges so you can write the formulas of compounds that contain them.

Writing the formula for a compound that contains a polyatomic ion follows the same rules that we've applied in the previous discussion. The only difference is that when two or more of the polyatomic ions occur in the formula, the formula for the ion is enclosed in parentheses. Thus, the compound composed of Ca^{2+} and PO_4^{3-} has the formula $Ca_3(PO_4)_2$:

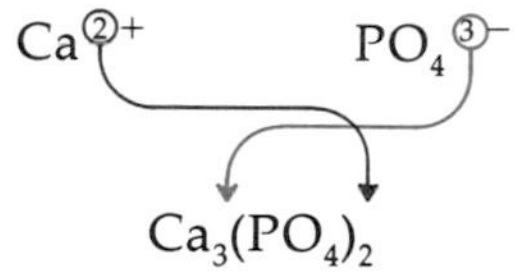

4.6 REACTIONS OF NONMETALS WITH EACH OTHER; THE FORMATION OF MOLECULAR COMPOUNDS

The nonmetals react not only with metals but also with each other. However, when two nonmetals combine to form a compound, electrically neutral molecules are formed instead of ions. An example of such a reaction occurs between oxygen and hydrogen to form water.

$$2H_2(g) + O_2(g) \longrightarrow 2H_2O(l)$$

The reaction of two nonmetals to form a compound is also a combination reaction.

The number of compounds formed among the nonmetals is enormous. There are millions of *organic compounds*, for example, whose molecules are composed chiefly of carbon and hydrogen, but that often contain other nonmetals as well (principally oxygen, nitrogen, sulfur, and the halogens). Because there are so many compounds of the nonmetals, we certainly can't discuss them all at this point. However, some observations that we can make about their makeup can be related to the positions of the nonmetals in the periodic table.

Table 4.5. Simple hydrogen compounds of the nonmetals

Group IVA	Group VA	Group VIA	Group VIIA
CH_4	NH_3	H_2O	HF
SiH_4	PH_3	H_2S	HCl
GeH_4	AsH_3	H_2Se	HBr
	SbH_3	H_2Te	HI

Simple compounds of the nonmetals with hydrogen

Hydrogen is a nonmetal that doesn't really fit well into any group based on its chemical and physical properties. It is put in Group IA for reasons that will be discussed in Chapter 7.

Except for the noble gases, all the nonmetals form compounds with hydrogen. Sometimes these compounds can be made by reacting hydrogen directly with the nonmetal, as in the reaction between hydrogen and oxygen, but often they are the products of other reactions between compounds. The specifics of such reactions will be left for later discussions. For now, let's look at just the formulas of the hydrogen compounds, which are given in Table 4.5. Here we have included compounds between hydrogen and some of the metalloids as well because in many of their chemical reactions the metalloids behave as nonmetals do.

One of the interesting things to notice in Table 4.5, which will help you remember the formulas, is that the number of hydrogens in these compounds is the same as the number of negative charges carried by the simple ions of the nonmetals. However, it is important to keep in mind that these hydrogen compounds are *not* ionic. The way that the atoms are held together within a molecule is different from the way ions are held together in an ionic compound. The attractions within molecules do not arise because of the transfer of electrons between atoms, but rather occur by the sharing of electrons between atoms—a topic that will be discussed much more fully in Chapters 8 and 9.

These compounds are known as *nonmetal hydrides.*

Some of the simple nonmetal–hydrogen compounds are very important to us. Water, of course, seems to be everywhere and is the most common of all of these chemicals. Another is methane, CH_4, which is the chief component of natural gas—a fuel used in the home for cooking and heating and used industrially for heating, the generation of electricity, and as a raw material in the manufacture of other chemicals. Ammonia, NH_3, is an important starting material in the synthesis of other nitrogen-containing compounds, a valuable fertilizer, and a chemical whose odor almost anyone would recognize because it is found in so many household cleaning products.

Simple compounds of the nonmetals with oxygen

All the nonmetals and metalloids, including some of the noble gases, form molecular compounds (called oxides) with oxygen. As with the hydrogen compounds, many oxides can be made by direct reaction of the element with oxygen, but there are also many that can only be made by reactions between compounds. For example, sulfur reacts directly with oxygen to form sulfur dioxide, SO_2. On the other hand, none of the nitrogen oxides is prepared in quantity by direct combination; instead, they are formed from other nitrogen-containing compounds.

Finding correlations between the formulas of nonmetal oxides and the positions of the nonmetals in the periodic table isn't as easy as with the hydrogen compounds because so many of these elements form more than one oxide. Nitrogen, for example, forms N_2O, NO, NO_2, N_2O_3, N_2O_4, and N_2O_5. Nevertheless, there are some similarities within groups of elements that are illustrated by the oxides listed in Table 4.6.

Table 4.6. Empirical formulas of some oxides of the nonmetals

Group IIIA	Group IVA	Group VA	Group VIA
B_2O_3	CO_2	N_2O_3	—
		N_2O_5	
	SiO_2	P_2O_3	SO_2
		P_2O_5	SO_3
	GeO_2	As_2O_3	SeO_2
		As_2O_5	SeO_3
		Sb_2O_3	TeO_2
		Sb_2O_5	TeO_3

The halogens also form compounds with oxygen, but they have not been included in this table.

Many nonmetal oxides are also very important to us. Carbon dioxide, CO_2, is a product of metabolism and we exhale it when we breathe. You are probably also aware that carbon monoxide, CO, is one of the toxic gases present in auto exhaust fumes. The sulfur oxides, SO_2 and SO_3, are produced by combustion of sulfur-containing fuels and contribute to an environmental problem called "acid rain." Some nitrogen oxides are culprits in the formation of smog and are also partly responsible for acid rain. And silicon dioxide, SiO_2, is the chemical substance in quartz and quartz sands from which glass is made.

Quartz crystals such as these are composed of silicon dioxide, SiO_2.

4.7 SOME PROPERTIES OF IONIC AND MOLECULAR COMPOUNDS

In the preceding two sections, you learned that ionic and molecular compounds are composed of quite different kinds of particles. In an ionic compound, there are discrete electrically charged particles (ions). Molecular substances, on the other hand, are composed of electrically neutral molecules. See Figure 4.11. Because of

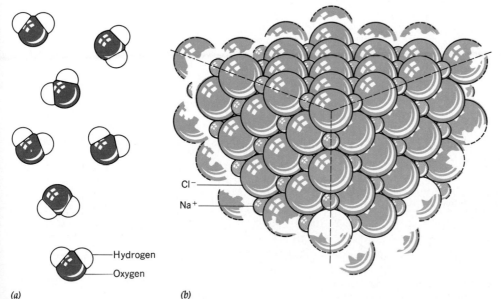

Cl⁻

Na⁺

Hydrogen
Oxygen

(a)   *(b)*

Figure 4.11. (a) *Molecules, such as those of water, consist of discrete particles, each composed of a certain fixed number of atoms.* (b) *In an ionic compound, such as NaCl, ions are packed so that each cation is surrounded by some number of anions, and each anion is surrounded by some number of cations. In NaCl, each ion is surrounded by 6 ions of opposite charge.*

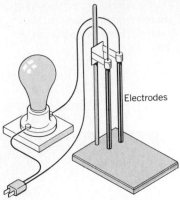

Figure 4.12. *An apparatus to test electrical conductivity. The electrodes are connected in series with the light bulb, and when electrical contact is made across the electrodes, the bulb can light.*

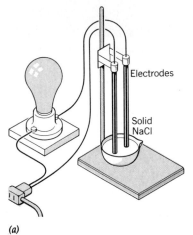

(a)

these differences, ionic and molecular substances have very different properties, which we can see if we compare representatives of each type.

Two substances that you are familiar with that belong to these two classes of compounds are sodium chloride (NaCl) and a compound called eicosane ($C_{20}H_{42}$). Sodium chloride, of course, is table salt and is an ionic compound. Eicosane is the chemical name for one of the compounds in a mixture of substances commonly called paraffin wax — the material from which candles are made, and which is used for sealing jars of homemade jams and jellies. Eicosane is a molecular compound.

If you sprinkle some salt crystals on a piece of paper and study them under a magnifying glass, you will see that sodium chloride is a white crystalline material. The crystals are brittle and shatter to a powder when you crush them. You also know that sodium chloride is soluble in water, and you may be aware that it melts at a high temperature. In fact, NaCl melts at about 800 °C and boils at over 1400 °C.

Eicosane has all the properties that we associate with wax. If you attempt to crush it, it just flattens out; wax is easily deformed. Eicosane is insoluble in water, but it does dissolve in liquids such as gasoline and paint thinner — solvents in which salt is very insoluble. In contrast to salt, eicosane melts at a low temperature of 37 °C (which just happens to be normal human body temperature), and eicosane boils at only 343 °C.

Another property that distinguishes salt from compounds like eicosane, and that reflects the kinds of particles found within it, is its ability to conduct electricity when melted. This property can easily be tested with the apparatus shown in Figure 4.12, which consists of two electrodes connected in series with the light bulb. When the apparatus is plugged into an electrical outlet and contact is made across the electrodes with something that is able to conduct electricity, the light bulb glows.

As illustrated in Figure 4.13, when the electrodes are dipped into crystals of solid salt or shavings of solid eicosane, the bulb doesn't light, which shows that neither of these solids conducts electricity. However, when both compounds are melted and the electrodes are dipped into them, the light bulb lights for the molten salt, but it remains unlit for the eicosane. Molten (liquid) NaCl conducts electricity, but molten $C_{20}H_{42}$ does not.

The differences between sodium chloride and eicosane are easy to understand when we consider the kinds of particles that they contain. In NaCl, the positive ions are surrounded by negative ions and vice versa, as illustrated in Figure 4.14. The attractions between the oppositely charged ions hold the crystal together. When a

Figure 4.13. *Neither* (a) *solid NaCl nor* (b) *solid eicosane conducts electricity.* (c) *Molten eicosane doesn't conduct either, but* (d) *molten NaCl does.*

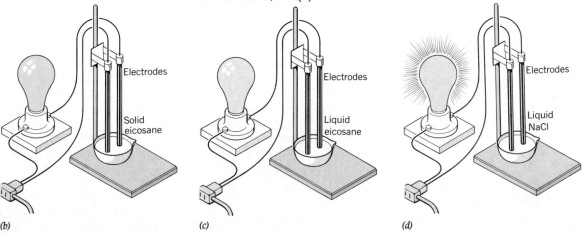

(b)   (c)   (d)

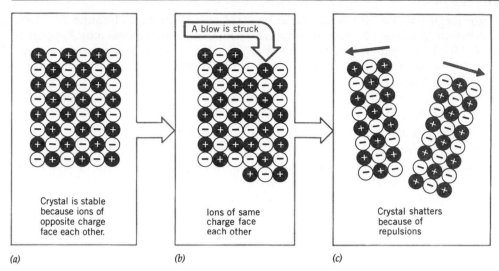

Figure 4.14. (a) *In a crystal of an ionic compound, attractions between ions of opposite charge hold the substance together.* (b) *When the crystal is struck, part of it slips past the rest. Ions of the same charge come face-to-face.* (c) *Repulsions between like-charged ions push the two parts of the crystal apart.*

crystal of salt is struck, even a small slippage of one part of the crystal past another can cause attractions to change to repulsions as ions of the same charge come face-to-face. The repulsive forces cause the crystal to break. As a result, ionic solids such as NaCl are brittle. In a crystal of a molecular substance, on the other hand, the molecules are neutral, so when this crystal is deformed, there are no sudden changes in whatever forces hold the crystal together. Therefore, the solid remains intact and just changes shape.

The electrical behavior is even easier to explain. Neither solid NaCl nor solid $C_{20}H_{42}$ conducts electricity because the ions in NaCl and the molecules in $C_{20}H_{42}$ are frozen in place. When melted, the $C_{20}H_{42}$ still is unable to conduct because the liquid contains only neutral molecules that cannot transport electrical charge. But when NaCl is melted, the ions are set free and are able to move about. These charged particles can transport charge through the liquid, so molten sodium chloride does conduct.

The differences between sodium chloride and ⁀icosane described here are typical of the differences that we observe among the properties of ionic and molecular compounds in general. Besides having learned what some of these differences are in this section, you should also have begun to see how learning about the internal structures of chemical compounds can help us understand their properties.

4.8 OXIDATION–REDUCTION REACTIONS

Oxygen reacts with most of the elements to from compounds that we call oxides, and from the time that oxygen was discovered, the term *oxidation* has been associated with that kind of reaction. Magnesium, for example, combines readily with oxygen, so a freshly exposed surface of the metal is quickly *oxidized* to give a coating of magnesium oxide, MgO. Iron is also oxidized slowly in air and forms rust, which is composed of Fe_2O_3. It has also been known since the iron age that the substance we now call iron oxide can be broken down, or *reduced,* to give the free metal. Recovery of a metal from its oxide therefore became known as *reduction.*

In modern terms, oxidation and reduction have been given broader meanings, which we can see if we analyze what happens when a metal such as iron is oxidized and its oxide is reduced. Iron oxide, Fe_2O_3, is an ionic compound composed of the ions Fe^{3+} and O^{2-}. When iron reacts with oxygen,

$$4Fe(s) + 3O_2(g) \longrightarrow 2Fe_2O_3(s)$$

When ignited, magnesium metal can burn in air, producing much light and heat and forming magnesium oxide, MgO. The magnesium oxide can be seen dangling from the burning piece of magnesium metal ribbon.

iron begins as electrically neutral atoms that lose electrons to become Fe^{3+} ions. When the oxide is reduced to give metallic iron, the reverse process must happen, so Fe^{3+} ions must gain electrons to give Fe atoms. It is this loss and gain of electrons, which occurs in many similar reactions, that is now associated with the terms oxidation and reduction.

> **Oxidation** is the loss of electrons by a substance.
> **Reduction** is the gain of electrons by a substance.

Reactions that involve oxidation and reduction are called **oxidation–reduction reactions,** or **redox reactions,** for short.

Now that we have these definitions, let's examine a reaction to see how we apply the terms. Consider the oxidation of magnesium described above.

$$2Mg(s) + O_2(g) \longrightarrow 2MgO(s)$$

The product MgO is ionic and contains the ions Mg^{2+} and O^{2-}, which are formed by the transfer of electrons from magnesium to oxygen. We can analyze this electron transfer by considering the loss and gain of electrons by the atoms separately. If we use the symbol e^- to stand for an electron, the loss of electrons by magnesium can be written

$$Mg \longrightarrow Mg^{2+} + 2e^- \quad \text{(oxidation)}$$

The change is identified as oxidation because magnesium loses electrons.

For oxygen in this reaction, we can write

$$O_2 + 4e^- \longrightarrow 2O^{2-} \quad \text{(reduction)}$$

This time the change is marked as a reduction because oxygen gains electrons. Therefore, in this reaction, magnesium is oxidized and oxygen is reduced. For this and *any* redox reaction, *both oxidation and reduction occur simultaneously.* We can never have one substance lose electrons without some other substance there to gain them. We know this because electrons are never found to be either a reactant or a product in any net chemical change. *This also requires that the total number of electrons gained be exactly equal to the total number of electrons lost.* In the reaction of Mg with O_2, two Mg atoms react for each O_2 that reacts. Two Mg atoms lose a total of $4e^-$, while one O_2 gains $4e^-$.

It is observed experimentally that oxidation and reduction always occur together, never one without the other.

Two terms that we often use in discussing redox reactions are oxidizing agent and reducing agent. *The **oxidizing agent** is the substance that takes electrons from the substance that is oxidized, thereby causing oxidation to take place.* That's what O_2 does in the reaction between Mg and O_2; it takes electrons from Mg and causes Mg to be oxidized, so O_2 is the oxidizing agent. Notice that the oxidizing agent (O_2) becomes reduced in the reaction.

*The **reducing agent** is the substance that gives electrons to the substance that is reduced, thereby causing reduction to occur.* That's what Mg does when it reacts with O_2; it gives electrons to O_2 and causes O_2 to be reduced. Notice that the reducing agent (Mg) is oxidized.

The substance reduced is the oxidizing agent; the substance oxidized is the reducing agent.

Oxidation numbers

The reactions of oxygen with magnesium and with sulfur,

$$2Mg(s) + O_2(g) \longrightarrow MgO(s)$$

$$S(s) + O_2(g) \longrightarrow SO_2(g)$$

have at least one superficial similarity. Both reactions produce oxides. However, there are significant differences between these two compounds, as discussed earlier.

Magnesium oxide is ionic, but sulfur dioxide is molecular and doesn't contain ions. Nevertheless, it seems sensible that we should be able to treat both reactions in a way that would justify the statement that oxygen oxidizes both magnesium and sulfur. In other words, we would like to treat both of these reactions as redox reactions. Oxidation numbers allow us to do this.

Oxidation numbers are numbers (either positive or negative) that we assign to atoms in a compound so we can follow the changes that take place in redox reactions. They are assigned according to a set of rules described below in which we treat every substance as if it were ionic and made up of a collection of electrically charge ions. For the purposes of assigning oxidation numbers, we do this *whether or not* the compound is ionic or molecular, which means that for molecular substances the oxidation numbers are really fictitious charges.

RULES FOR ASSIGNING OXIDATION NUMBERS

1. The oxidation number of any element in its elemental form is zero, regardless of the complexity of the molecule in which it occurs. Thus, the atoms in Ne, F_2, P_4, and S_8 all have oxidation numbers of zero.
2. The oxidation number of any **monatomic** ion (an ion composed of only one atom) is equal to the charge on the ion. The ions Na^+, Al^{3+}, and S^{2-} have oxidation numbers equal to $+1$, $+3$, and -2, respectively.
3. The sum of all the oxidation numbers of all the atoms in a compound is zero. For polyatomic ions, the sum of the oxidation numbers must equal the charge on the ion.

In addition to these basic rules, we obey the following when assigning oxidation numbers to specific atoms in a compound:

4. Fluorine has an oxidation number of -1.
5. Hydrogen has an oxidation number of $+1$.
6. Oxygen has an oxidation number of -2.

When assigning oxidation numbers, you will sometimes encounter situations in which these rules conflict with one another. When this happens, the rule higher up takes precedence.

The rules above, plus what you have learned earlier about the ions formed by various metals and nonmetals, will allow you to assign oxidation numbers to the atoms in a wide variety of substances. Let's look at a few examples.

Oxidation numbers are equal to electrical charges only for simple monatomic ions.

EXAMPLE 4.2. ASSIGNING OXIDATION NUMBERS TO ATOMS

PROBLEM: Assign oxidation numbers to each of the atoms in the following: (a) $FeCl_3$, (b) KNO_3, (c) H_2O_2, (d) $Fe_2(SO_4)_3$, (e) $Cr_2O_7^{2-}$, (f) ClO_3^-, and (g) $Na_2S_4O_6$.

SOLUTION: (a) We treat a compound formed between a metal and a nonmetal as an ionic compound. You learned that chlorine forms the negative ion Cl^-, so the oxidation number of Cl in this compound is -1. (Because oxidation numbers and charges aren't always the same, especially for molecular compounds, we want to distinguish as clearly as we can between them. Therefore, we will write the sign before the number for oxidation numbers, as in -1, but the number before the sign when expressing the actual charge on an ion, as in $1-$ for Cl^-.) The oxidation number of the Fe can then be obtained by the summation rule (rule 3).

$$
\begin{array}{rl}
\text{Cl} & 3 \times (-1) = -3 \\
\underline{\text{Fe}} & \underline{1 \times (x)\ \ = \ \ \ x} \\
& \text{sum} = \ \ \ 0
\end{array}
$$

For the sum to be zero, the oxidation number of the iron must be $+3$. Of course, we can come to this same conclusion by realizing that the iron in this compound occurs as the Fe^{3+} ion. The oxidation number must be $+3$ according to rule 2.

(b)   Potassium (Group IA) forms ions with a charge of $1+$ (K^+), so the oxidation number of K must be $+1$ (rule 2). Oxygen is assigned an oxidation number of -2 (rule 6). We get the oxidation number of N by applying rule 3.

$$
\begin{array}{lll}
\text{K} & 1 \times (+1) = +1 & \text{(rule 2)} \\
\text{O} & 3 \times (-2) = -6 & \text{(rule 6)} \\
\underline{\text{N}} & \underline{1 \times (x) \quad = \quad x} & \\
& \text{sum} = \quad 0 & \text{(rule 3)}
\end{array}
$$

For the sum to be zero, x must equal $+5$, so the oxidation number of N in this compound is $+5$.

H_2O_2 is hydrogen peroxide, a common antiseptic.

(c) H_2O_2 is a compound formed between nonmetals, so we expect it to be molecular. Since there are no ions, we can't use rule 2. Rules 5 and 6 refer to H and O, so we can use these, but there is a conflict. If we take H to be $+1$ according to rule 5, then O must be assigned an oxidation number of -1 in order for the sum of oxidation numbers to be zero. On the other hand, if we take O to be -2 according to rule 6, then H must be $+2$ for the sum to be zero. As mentioned earlier, however, when two rules conflict, we apply the one higher up on the list. Therefore, we take H to be $+1$ and the oxidation number of O in this compound is -1.

You need to know that the sulfate ion is SO_4^{2-} to figure out that the iron in $Fe_2(SO_4)_3$ is Fe^{3+}. If you don't know the charges on the various polyatomic ions, study Table 4.4 again.

(d) $Fe_2(SO_4)_3$ is an ionic compound composed of the SO_4^{2-} ion and the Fe^{3+} ion. The oxidation number of the Fe^{3+} must be $+3$. Oxygen has an oxidation number of -2 according to rule 6. Therefore,

$$
\begin{array}{lll}
\text{Fe} & 2 \times (+3) = +6 & \text{(rule 2)} \\
\text{S} & 3 \times (x) \quad = \quad 3x & \\
\underline{\text{O}} & \underline{12 \times (-2) = -24} & \text{(rule 6)} \\
& \text{sum} = \quad 0 & \\
3x + (+6) + (-24) = & 0 & \\
3x = +18 & \\
x = \quad +6 &
\end{array}
$$

The oxidation number of S must be $+6$ in this compound.

(e)   For the $Cr_2O_7^{2-}$ ion, the sum of oxidation numbers must equal the charge on the ion. Therefore,

$$
\begin{array}{lll}
\text{Cr} & 2 \times (x) \quad = \quad 2x & \\
\underline{\text{O}} & \underline{7 \times (-2) = -14} & \text{(rule 6)} \\
& \text{sum} = \quad -2 & \text{(rule 3)} \\
2x + (-14) = & -2 & \\
x = \quad +6 &
\end{array}
$$

In this ion, the oxidation number of Cr is $+6$. However, you should keep in mind that the atoms in a polyatomic ion are held together by the same kinds of attractions that hold together atoms in molecules. There are really no Cr^{6+} ions within the $Cr_2O_7^{2-}$ ion. As in molecular substances, the oxidation numbers of Cr and O here are not their actual charges.

Chlorine will only be Cl^- when it's in a compound with a metal such as NaCl or $FeCl_3$. When there is another nonmetal also present, you have to figure out the oxidation number of the Cl. The same applies to other nonmetal-containing compounds.

(f)   The ClO_3^- ion is composed of two nonmetals, which means that the ion is held together by the same kind of attractions that exist within molecules. This means that we can't just assume that the Cl exists as a Cl^- ion within the ClO_3^- ion, so we can't apply rule 2 in this case. However, rule 6 tells us that oxygen has an oxidation number of -2, so we can calculate what Cl must be in order for the sum of oxidation numbers to be equal to the charge on the ion.

$$
\begin{array}{lll}
\text{Cl} & 1 \times (x) \quad = \quad x & \\
\underline{\text{O}} & \underline{3 \times (-2) = -6} & \text{(rule 6)} \\
& \text{sum} = -1 & \text{(rule 3)}
\end{array}
$$

It's not hard to see that chlorine must have an oxidation number of $+5$.

(g) Once again, the sum of oxidation numbers must be zero. Sodium is an alkali metal and must be Na^+, so its oxidation number is $+1$. Oxygen is -2 according to rule 6. Therefore,

$$
\begin{array}{lll}
Na & 2 \times (+1) = & +2 \text{ (rule 2)} \\
S & 4 \times (x) = & 4x \\
O & 6 \times (-2) = & -12 \text{ (rule 6)} \\
\hline
 & \text{sum} = & 0 \\
(+2) + (4x) + (-12) = & 0 \\
 & 4x = & +10 \\
 & x = & +5/2
\end{array}
$$

The oxidation number of sulfur is $+5/2$. Notice that oxidation numbers don't have to be whole numbers (although they usually are).

Using oxidation numbers

In a redox reaction, there is a change in the oxidation number or **oxidation state** (we use the terms interchangeably) of two or more elements. Consider, for example, the reaction between magnesium and oxygen

$$2Mg + O_2 \longrightarrow 2MgO$$
$$0 0 +2 -2$$

where we have written the oxidation numbers of the elements below their symbols. Notice that the oxidation number of Mg changes from 0 to $+2$ and the oxidation number of O changes from 0 to -2. Thus, the oxidation of Mg is accompanied by an increase in oxidation number (an *increase* in the sense that the oxidation number of Mg is becoming *more positive*). The reduction of O_2, on the other hand, is accompanied by a decrease in the oxidation number (a *decrease* in the sense that the oxidation number of O is becoming *less positive* or more negative). This gives us another more general way of defining oxidation and reduction according to the change in oxidation number.

> **Oxidation** is an increase in oxidation number.
> **Reduction** is a decrease in oxidation number.

This is the preferred way of defining oxidation and reduction.

To be consistent with our earlier definitions, the **oxidizing agent** is the substance that is reduced and the **reducing agent** is the substance oxidized.

By using these new definitions, we can now see that the reaction between sulfur and oxygen is also a redox reaction. Let's rewrite the equation with the oxidation numbers below the symbols.

$$S + O_2 \longrightarrow SO_2$$
$$0 0 +4 -2$$

Notice that the oxidation number of sulfur increases from 0 to $+4$, so the sulfur is oxidized. The oxidation number of O decreases from 0 to -2, so the substance O_2 is reduced. This means that O_2 is the oxidizing agent and S the reducing agent.

EXAMPLE 4.3. IDENTIFYING OXIDATION AND REDUCTION IN A REACTION

PROBLEM: In the following reaction, which substance is oxidized and which is reduced? Which substance is the oxidizing agent and which is the reducing agent?

$$14HCl + K_2Cr_2O_7 \longrightarrow 2KCl + 2CrCl_3 + 3Cl_2 + 7H_2O$$

SOLUTION: Let's begin by writing the oxidation numbers under each of the chemical symbols.

$$14HCl + K_2Cr_2O_7 \longrightarrow 2KCl + 2CrCl_3 + 3Cl_2 + 7H_2O$$
$$\underset{+1\ -1}{} \quad \underset{+1\ +6\ -2}{} \quad \underset{+1\ -1}{} \quad \underset{+3\ -1}{} \quad \underset{0}{} \quad \underset{+1\ -2}{}$$

Notice that some Cl doesn't change its oxidation number. We are interested in the Cl that does, however, and this Cl changes from -1 to 0.

Now we look for changes in oxidation numbers. We see that the oxidation number of Cl changes from -1 to 0. This is an increase in oxidation number (becoming less negative is the same as becoming more positive), so the substance oxidized is HCl. We also see that Cr changes from $+6$ to $+3$. This is a decrease in oxidation number, so the $K_2Cr_2O_7$ is reduced.

The oxidizing agent is the substance that's reduced; it is $K_2Cr_2O_7$. The reducing agent is the substance that's oxidized; it is HCl.

In the preceding example, notice that we have used oxidation numbers assigned to the various atoms to identify the changes corresponding to oxidation and reduction, but for identifying the substances oxidized and reduced, the entire formulas for the substances were specified. This is because oxidation numbers are just a bookkeeping tool we use to follow redox reactions. The actual chemical changes generally involve entire molecules or polyatomic ions, not just individual atoms within these particles, and we indicate this by specifying the entire formulas for the substances oxidized and reduced.

Oxidation numbers make it possible for us to identify many different reactions as redox reactions. As you will discover, this is one of the more important classes of reactions, and they will be encountered frequently throughout the remainder of the book.

4.9 NAMING CHEMICAL COMPOUNDS

Communication among scientists is essential. Without it there really is no sense in doing scientific research. For chemists, a key part of this communication is describing the chemicals used in experiments, and for that we need a way of naming compounds. You've already seen some chemical names in the preceding sections in this and earlier chapters. Now that you have learned how to write the formulas for a variety of compounds, you are ready to study how these names come about.

If you leaf through a chemical reference book, such as the *Handbook of Chemistry and Physics*, you will find an enormous number of compounds listed. Yet these represent just a small fraction of all the compounds that have been discovered, and each year the list grows longer. Naming these compounds presents a real challenge because it is desired that each unique substance have its own unique name. In addition, names can't be chosen in a haphazard way; otherwise, there would be no possible way for anyone to remember them all. Because of this, a systematic procedure for naming chemical compounds has been developed.

A system of names used in a particular branch of knowledge, like chemistry, is called *nomenclature*.

In this section you will see, in a somewhat abbreviated way, how **inorganic compounds** are named. These are compounds whose structures are not determined primarily by the linking together of carbon atoms. Such carbon compounds are called **organic compounds** and are discussed in Chapter 22.

You will discover during your study of chemistry that not every compound is named according to the system. Some very common compounds such as water and ammonia, NH_3, were known long before the systematic nomenclature was developed and are best recognized by their common names. Common names are also used for extremely complex compounds whose names derived on a systematic basis are very long and cumbersome.

The rules of chemical nomenclature are established by the IUPAC.

Binary compounds of a metal and a nonmetal

A **binary compound** is composed of atoms of only two different elements. When one is a metal and the other a nonmetal, the metal is named first, followed by the nonmetal. For metals that only occur in one oxidation state, the English name of the metal is always used. The name of the second element is obtained by adding the suffix *-ide* to its stem, as shown in Table 4.7. Some typical examples are

$NaCl$	sodium chloride
SrO	strontium oxide
Al_2S_3	aluminum sulfide
Mg_3P_2	magnesium phosphide

Many metals, especially those of the transition elements, are commonly found to exist in more than one positive oxidation state. There are two methods that have been used to name compounds of these elements. In the older method, the suffixes *-ic* and *-ous* are used to specify the higher and lower oxidation states. Thus, the +2 and +3 oxidation states of chromium are specified as

Cr^{3+}	chromic ion	$CrCl_3$	chromic chloride
Cr^{2+}	chromous ion	$CrCl_2$	chromous chloride

When the metal has a symbol derived from the Latin name for the element, its Latin stem is generaly used. For example, iron has two common oxidation states, +2 and +3, corresponding to Fe^{2+} (ferrous ion) and Fe^{3+} (ferric ion). Other examples are given in Table 4.8. Notice that this method only differentiates between the higher and lower oxidation states; it doesn't specify what the oxidation states actually are, which is one of its principal weaknesses.

The preferred method of naming ions like those in Table 4.8 is called the **Stock system,** named after the German chemist Alfred Stock who developed it. In the Stock system, a Roman numeral equal to the metal's oxidation state is placed in

Table 4.7. Names of monatomic anions derived from nonmetals

Group IVA	Group VA	Group VIA	Group VIIA
C^{4-} *carbide*[a]	N^{3-} *nitride*	O^{2-} *oxide*	F^- *fluoride*
Si^{4-} *silicide*	P^{3-} *phosphide*	S^{2-} *sulfide*	Cl^- *chloride*
	As^{3-} *arsenide*	Se^{2-} *selenide*	Br^- *bromide*
		Te^{2-} *telluride*	I^- *iodide*

[a] Carbon also forms a number of complex carbides, for example, C_2^{2-} in CaC_2.

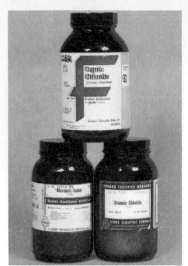

Bottles of chemicals labeled according to the older system of nomenclature.

Notice that there isn't a space between the name of the metal and the Roman numeral in parentheses.

Table 4.8. Metals that commonly form two ions

Chromium	Manganese	Iron	Cobalt	Lead
Cr^{2+} chromous	Mn^{2+} manganous	Fe^{2+} ferrous	Co^{2+} cobaltous	Pb^{2+} plumbous
Cr^{3+} chromic	Mn^{3+} manganic	Fe^{3+} ferric	Co^{3+} cobaltic	Pb^{4+} plumbic

Copper	Tin	Mercury		
Cu^+ cuprous	Sn^{2+} stannous	Hg_2^{2+} mercurous (note that there are two Hg atoms)		
Cu^{2+} cupric	Sn^{4+} stannic	Hg^{2+} mercuric		

parentheses following the English name of the element. Thus, Fe^{2+} is named iron(II) ion and Fe^{3+} iron(III) ion. The alternate names for $FeCl_2$ and $FeCl_3$ are therefore

$$FeCl_2 \quad \text{ferrous chloride} \quad \text{or} \quad \text{iron(II) chloride}$$
$$FeCl_3 \quad \text{ferric chloride} \quad \text{or} \quad \text{iron(III) chloride}$$

When the Stock system is used, it is important to remember that the Roman numeral is the oxidation state of the metal (the charge on the ion) and that it is not necessarily the subscript of the metal ion. For example,

$$Cu_2O \quad \text{copper(I) oxide}$$
$$CuO \quad \text{copper(II) oxide}$$

Although the Stock system is preferred today, it is still necessary to know the older system as well. For example, if an experiment calls for iron(III) chloride, it is quite likely that you will find it in a bottle labeled ferric chloride.

Binary compound of nonmetals

For naming a binary compound formed by two nonmetals, a third system of nomenclature is followed. This system uses Greek prefixes to indicate the numbers of atoms of each kind in one molecule of the substance. These prefixes, along with their meanings, are

di-	two	penta-	five	octa-	eight
tri-	three	hexa-	six	nona-	nine
tetra-	four	hepta-	seven	deca-	ten

In the name of the compound, the first element in the formula is given its ordinary English name. The second element is specified by adding the suffix *-ide* to the root of the element's name. The name for the compound P_4O_{10} illustrates how the system works.

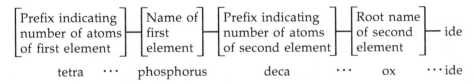

Thus, P_4O_{10} is tetraphosphorus decaoxide. Some other examples are

NO_2	nitrogen dioxide
N_2O_4	dinitrogen tetroxide
	(the *a* in *tetra* is dropped for ease of pronunciation)
N_2O_5	dinitrogen pentoxide

PCl_3 phosphorus trichloride
PCl_5 phosphorus pentachloride
S_2Cl_2 disulfur dichloride

In some instances, the prefix *mono-,* meaning *one,* is also used when it is desired to avoid ambiguity.

CO_2 carbon dioxide
CO carbon monoxide

Compounds containing polyatomic ions

In Section 4.5 we saw that many ions contain more than one atom and are therefore referred to in general as polyatomic ions. These species enter into ionic compounds as discrete units and generally stay intact in most chemical reactions. A list of these is given in Table 4.4. As with binary compounds, substances that contain these ions are always named with the positive ion first. Some examples are

Na_2CO_3 sodium carbonate $Ba(OH)_2$ barium hydroxide
$Ca(C_2H_3O_2)_2$ calcium acetate $(NH_4)_2SO_4$ ammonium sulfate

The Stock system is also preferred when the metal can form more than one positive ion.

Usually, the *-ide* ending specifies a monatomic anion, such as Cl^- (chlor*ide*). Hydroxide ion (OH^-) and cyanide ion (CN^-) are two exceptions.

	Stock System	Old System
$MnSO_4$	manganese(II) sulfate	manganous sulfate
$Fe_2(C_2O_4)_3$	iron(III) oxalate	ferric oxalate

Binary acids

Among the important classes of compounds that we will discuss later in the book are substances called acids. Perhaps you have already encountered some common ones in the laboratory—hydrochloric acid, for example—and you surely have experienced the sour taste of citric acid in lemon juice and acetic acid in vinegar. As we will see, acids are substances that release H^+ ions when they are dissolved in water.

An important kind of acid is formed when a binary compound of hydrogen and a nonmetal (for example, hydrogen chloride, HCl, or hydrogen sulfide, H_2S) is dissolved in water. Water solutions of these compounds are called binary acids (or sometimes *hydro acids*). They are named as *hydro . . . ic acid,* where the stem of the name of the nonmetal is inserted in place of the dotted line. Examples are

$HF(aq)$ hydro*fluor*ic acid
$HCl(aq)$ hydro*chlor*ic acid
$HBr(aq)$ hydro*brom*ic acid
$HI(aq)$ hydro*iod*ic acid
$H_2S(aq)$ hydro*sulfur*ic acid

The gas HF is *hydrogen fluoride;* its aqueous solution is called *hydrofluoric acid.* Similar statements apply to the other nonmetal hydrides in this list.

When an acid is allowed to react with hydroxide ion (a reaction called **neutralization**), an ionic compound is formed. For example,

$$NaOH + HCl \longrightarrow H_2O + NaCl$$

$$2KOH + H_2S \longrightarrow 2H_2O + K_2S$$

Ionic compounds such as NaCl and K_2S are called salts. The word salt is not reserved just for sodium chloride, although this is the common household name for NaCl. In chemistry, the term **salt** is used for *any* ionic compound not containing oxide ion or

Anions such as Cl^- can be formed directly from the elements, as when Cl_2 reacts with Na to form NaCl, or they can be formed by neutralization of an acid with a base.

hydroxide ion. Notice that the salts formed from *hydro . . . ic acids* contain monatomic anions that end in the suffix *-ide.*

Oxoacids

Oxoacids are acids that contain hydrogen, oxygen, and at least one other element (usually a nonmetal). Sulfuric acid, H_2SO_4, is an example. Notice that the hydrogen is specified first in the formula (we will say more about this in the next chapter). Often, an element is able to form more than one oxoacid. Sulfur, for example, forms the two acids H_2SO_4 and H_2SO_3, which differ in the oxidation state of the sulfur ($+6$ in H_2SO_4 and $+4$ in H_2SO_3) as well as in the number of oxygens. In naming these acids, we give the one having the element in the higher oxidation state the suffix *-ic* and the acid having the element in the lower oxidation state the suffix *-ous.* Thus, we have

$$H_2SO_4 \quad \text{sulfuric acid}$$
$$H_2SO_3 \quad \text{sulfurous acid}$$

Notice that the prefix *hydro* is not used in naming these oxoacids.

Compounds produced by the neutralization of oxoacids contain the polyatomic ions given in Table 4.4. Notice that the anion derived from the "ic" acid has a name that ends in *-ate*, whereas the anion that comes from the "ous" acid ends in *-ite.*

H_2SO_4	sulfuric acid	$SO_4{}^{2-}$	sulfate
H_2SO_3	sulfurous acid	$SO_3{}^{2-}$	sulfite
HNO_3	nitric acid	$NO_3{}^-$	nitrate
HNO_2	nitrous acid	$NO_2{}^-$	nitrite
$HClO_3$	chloric acid	$ClO_3{}^-$	chlorate
$HClO_2$	chlorous acid	$ClO_2{}^-$	chlorite

Some nonmetals (especially the halogens) form more than two oxoacids. This is the case with chlorine, bromine, and iodine. For these acids, we use the prefix *hypo-* for the acid that has fewer oxygens than the "ous" acid, and we use the prefix *per-* for the acid that has more oxygens than the "ic" acid. For example,

hypochlorous acid	$HClO$	ClO^-	hypochlorite
chlorous acid	$HClO_2$	$ClO_2{}^-$	chlorite
chloric acid	$HClO_3$	$ClO_3{}^-$	chlorate
perchloric acid	$HClO_4$	$ClO_4{}^-$	perchlorate

Acid salts

Partial neutralization of an acid that is capable of furnishing more than one H^+ per acid molecule gives salts that are called **acid salts.** Some examples are shown in the table at the left. When only one acid salt is formed (as with H_2SO_4 or H_2CO_3), the salt can be named by adding the prefix *bi-* to the name of the anion of the acid.

$$NaHSO_4 \quad \text{sodium } \textit{bi}\text{sulfate}$$
$$NaHCO_3 \quad \text{sodium } \textit{bi}\text{carbonate}$$

The salt can also be named by specifying the presence of the H by writing "hydrogen."

$NaHSO_4$	sodium hydrogen sulfate
NaH_2PO_4	sodium dihydrogen phosphate
Na_2HPO_4	sodium hydrogen phosphate
	(disodium hydrogen phosphate)

Note the use of the prefix *di-* to indicate the number of hydrogen atoms (as well as to remove ambiguity as to the number of Na in the last formula).

Although polyatomic ions are formed by neutralization reactions of oxoacids, they can also be made by other reactions we haven't discussed yet.

Parent Acid	Typical Acid Salts
H_2SO_4	$NaHSO_4$
H_2CO_3	$NaHCO_3$
H_3PO_4	NaH_2PO_4
	Na_2HPO_4

This common household product uses sodium hydrogen sulfate (sodium bisulfate) as its active ingredient. The manufacturer calls it "sodium acid sulfate."

REVIEW QUESTIONS AND PROBLEMS

Problems whose numbers are in blue have their answers in Appendix D at the back of the book. The more difficult problems are marked with asterisks.

Metals, Nonmetals, and Metalloids

4.1 Define *malleability*. Why is this an important property to a blacksmith?

4.2 Define *ductility*. What commercial process makes use of this property?

4.3 Besides malleability and ductility, give three other properties of metals.

4.4 Compare gold, iron, and sodium with respect to their reactivity toward air and moisture. Why isn't iron used to make jewelry?

4.5 The reaction of sodium with water produces gaseous hydrogen and sodium hydroxide, NaOH. Write a balanced chemical equation for this reaction.

4.6 Only two metals are colored. Which are they? (You should be able to answer this based on firsthand experience.)

4.7 Why do metals feel so hot compared to nonmetallic objects, when left in the summer sun?

4.8 Give an industrial application of gold that relies on this metal's very low degree of chemical reactivity.

4.9 Which metal has the highest melting point? What is one of its uses? Which metal has the lowest melting point? What is one of its uses?

4.10 Which nonmetals occur as diatomic molecules? Write their formulas. Which of these are gases, which is a liquid, and which is a solid?

4.11 What are the names of the two elemental forms of carbon? How are they alike, and how are they different?

4.12 Referring to Figure 1.14 on page 25, compare those properties of copper and sulfur that are evident from the photograph. Is it evident which is a metal and which is a nonmetal?

4.13 Which two nonmetals are the principal constituents of the atmosphere?

4.14 Which is the most chemically reactive nonmetal?

4.15 In what ways are the metalloids similar to metals, and how do they differ from metals?

The First Periodic Table

4.16 What was the basis on which Mendeleev constructed his periodic table?

4.17 Why were there blanks left in Mendeleev's periodic table?

4.18 Why was Mendeleev able to predict the properties of the elements that belonged in the blanks in his table?

4.19 Besides Tc and I, Ar and K, find *three* other places in the modern periodic table where elements occur in reverse atomic-mass order.

4.20 Which group of elements appears in the modern periodic table, but did not appear in Mendeleev's table? Why?

The Modern View of the Atom

4.21 What is the definition of a coulomb? From the measured electrical charge on an electron, calculate the number of coulombs of charge carried by 1.00 mol of electrons.

4.22 Use the data in Table 4.1 to calculate the mass in grams of 1.00 mol of electrons? Hydrogen atoms each contain one electron. From the atomic mass of hydrogen, calculate the percentage of the mass of an "average" hydrogen atom that is contributed by an electron.

4.23 In terms of atomic structure, what is an atom's atomic number?

4.24 What distinguishes an atom of nitrogen from an atom of oxygen?

4.25 Aluminum atoms form ions having a charge equal to $+4.8 \times 10^{-19}$ C. How should we write the symbol for this ion?

4.26 The element selenium forms an ion with a charge of -3.2×10^{-19} C. How should we write the symbol for this ion?

4.27 The proton (the nucleus of a hydrogen atom) has a mass of 1.67×10^{-24} g. Suppose that its diameter is 1.00×10^{-13} cm. Calculate the density of the nucleus, assuming it is spherical in shape. *Note:* A sphere's volume V can be calculated from its radius r by the equation

$$V = \tfrac{4}{3} \pi r^3$$

4.28 If the diameter of a nucleus is approximately 10^{-13} cm and the diameter of an atom is approximately 2×10^{-8} cm, calculate approximately the percentage of the volume of an atom that is occupied by its nucleus.

4.29 The earth has a mass of 6.59×10^{21} tons and a diameter of approximately 8000 mi. What would the diameter of the earth (in miles) be if it had the same mass but was composed entirely of nuclear material? (Use the density of nuclear material calculated in Problem 4.27.)

4.30 Where are neutrons found in an atom? Are there any atoms that have no neutrons?

4.31 For each of the following, give the approximate mass in u and the charge in units of 1.60×10^{-19} C:
(a) proton   (b) neutron   (c) electron

Isotopes

4.32 Explain why the atomic masses of some elements (such as Cl or Cu, for instance) are so far from whole numbers.

4.33 What is the difference between an atom's mass number and its actual atomic mass?

4.34 What are the numbers of protons, neutrons, and electrons in each of the following: $^{132}_{55}Cs$, $^{115}_{48}Cd^{2+}$, $^{194}_{81}Tl$, $^{105}_{47}Ag^{+}$, $^{78}_{34}Se^{2-}$?

4.35 What are the numbers of protons, neutrons, and electrons in each of the following: $^{131}_{56}Ba$, $^{109}_{48}Cd^{2+}$, $^{36}_{17}Cl^{-}$, $^{63}_{28}Ni$, $^{170}_{69}Tm$?

4.36 Write the appropriate symbol for each of the following isotopes:
(a) $Z = 26$, $A = 55$ (d) $Z = 71$, $A = 170$
(b) $Z = 37$, $A = 86$ (e) $Z = 70$, $A = 169$
(c) $Z = 81$, $A = 204$

4.37 How many neutrons are in each of the atoms in Question 4.36?

4.38 The element Eu occurs naturally as a mixture of 47.82% $^{151}_{63}Eu$, whose mass is 150.9 u, and 52.18% $^{153}_{63}Eu$, whose mass is 152.9 u. Calculate the average atomic mass of Eu.

4.39 Naturally occurring boron consists of two isotopes, ^{10}B with a mass of 10.01294 u and ^{11}B with a mass of 11.00931 u. The abundance of ^{10}B is 19.6% and that of ^{11}B is 80.4%. Calculate the average atomic mass of B.

4.40 Naturally occurring lead is composed of four isotopes. Their abundances and masses are given below. Calculate the average atomic mass of lead.

Isotope	Mass (u)	Abundance (%)
^{204}Pb	203.973	1.48
^{206}Pb	205.9745	23.6
^{207}Pb	206.9759	22.6
^{208}Pb	207.9766	52.3

* 4.41 Naturally occurring chlorine is composed of ^{35}Cl with atomic mass of 34.96885 u and ^{37}Cl with atomic mass of 36.96590 u. The average atomic mass of Cl is 35.453. What are the percentages of each isotope in naturally occurring chlorine?

* 4.42 Ordinary silver is a mixture of two isotopes, ^{107}Ag with a mass of 106.9041 u and ^{109}Ag with a mass of 108.9047 u. The average relative atomic mass of silver is 107.868 u. What are the relative abundances, expressed as percents, of these two silver isotopes?

The Modern Periodic Table

4.43 Make a sketch of the modern periodic table in which the lanthanides and actinides are placed in the body of the table in their proper locations.

4.44 What do we call a column of elements in the periodic table? What do we call a row of elements?

4.45 Which of the following are representative elements: Mg, Ti, Fe, Se, Ni, Br?

4.46 Which of the following are transition elements: Sr, Ru, As, W, Ag, Al?

4.47 Which elements constitute the inner transition elements?

4.48 Which of the following elements is a halogen: Na, Ca, Fe, F, As?

4.49 Write the formulas of the molecules formed by the halogens in their elemental states.

4.50 Which of the following elements is an alkali metal: Br, K, O, S, N?

4.51 Which of the following elements is an alkaline earth metal: Cu, B, Ba, Ne, Se?

4.52 Which of the following are metals: Ta, Nd, Se, F, Cs?

4.53 Write the symbols for the metalloids.

4.54 How does the metallic character of the elements vary from left to right across a period? How does it vary from top to bottom to Group VA?

4.55 Identify the metals, nonmetals, and metalloids in period 4. Identify them in Group VA.

4.56 From what you've learned in this and earlier chapters, which one of the following metals would be especially reactive toward oxygen and moisture: Fe, Pt, K, Mn, or Sn? On what information do you base your answer?

4.57 Radium, a highly radioactive element, forms a water-soluble compound with chlorine. From what you've learned in this chapter, what would you expect its formula to be?

4.58 You encounter certain metals quite often on a day-to-day basis. Identify those in the following list that are transition metals:
(a) aluminum   (e) tin
(b) iron   (f) silver
(c) chromium   (g) gold
(d) copper   (h) lead

Reactions of Metals with Nonmetals

4.59 What is a *combination reaction*?

4.60 Why do we write only empirical formulas for compounds such as CaO, NaCl, and BaF_2?

4.61 After studying Section 4.5, use a periodic table to help you write the symbols for the ions formed by (a) strontium, (b) sodium, (c) sulfur, (d) aluminum, and (e) bromine.

4.62 Use a periodic table to help you write the symbols for the ions formed by (a) potassium, (b) nitrogen, (c) magnesium, (d) oxygen, and (e) calcium.

4.63 What ions are commonly formed by (a) iron, (b) copper, (c) tin, (d) zinc, and (e) chromium?

4.64 Write the formulas of all the ionic compounds formed by the following metals with chloride ion, Cl^-, and with sulfide ion, S^{2-}:
(a) chromium
(b) manganese
(c) iron
(d) cobalt
(e) nickel
(f) copper
(g) silver

4.65 Write the formulas of all the ionic compounds formed by the following metals with bromide ion, Br^-, and oxide ion, O^{2-}:
(a) gold
(b) zinc
(c) cadmium
(d) silver
(e) tin
(f) lead
(g) bismuth

4.66 Give the correct name for each of the following polyatomic ions: (a) NH_4^+, (b) CO_3^{2-}, (c) CrO_4^{2-}, (d) SO_3^{2-}, (e) $C_2H_3O_2^-$.

4.67 Write the correct formula, including the charge, for each of the following polyatomic ions:
(a) cyanide ion
(b) perchlorate ion
(c) permanganate ion
(d) nitrate ion
(e) phosphate ion
(f) hydroxide ion
(g) oxalate ion
(h) dichromate ion
(i) sulfate ion
(j) bicarbonate ion
(k) sulfite ion
(l) nitrite ion

4.68 Write formulas for compounds composed of the following pairs of ions: (a) Na^+, CO_3^{2-}; (b) Ca^{2+}, ClO_3^-; (c) Sr^{2+}, S^{2-}; (d) Cr^{3+}, Cl^-; (e) Ti^{4+}, ClO_4^-.

4.69 Write the formulas for the ionic compounds formed by each of the metals in Table 4.3 with the anions specified in Question 4.66. (This is a lot of work, but well worth the effort.)

4.70 Write formulas for compounds composed of the following pairs of ions:
(a) Fe^{3+}, HPO_4^{2-}
(b) K^+, N^{3-}
(c) Ni^{2+}, NO_3^-
(d) Cu^{2+}, $C_2H_3O_2^-$
(e) Ba^{2+}, SO_3^{2-}

Reactions Between Nonmetals

4.71 How do compounds formed between two nonmetals differ from those formed between a metal such as barium and a nonmetal such as fluorine?

4.72 Use the periodic table and what you have learned in Section 4.6 to write the formulas for the simplest compounds of the following elements with hydrogen: (a) phosphorus, (b) sulfur, (c) bromine, (d) silicon, (e) antimony.

4.73 Lead forms two compounds with oxygen. Based on the position of lead in the periodic table, suggest what their formulas might be.

4.74 Arsenic forms two compounds with chlorine, $AsCl_3$ and $AsCl_5$. What chlorine compounds are expected for phosphorus?

4.75 In the simplest compounds of the nonmetals with fluorine, the number of fluorine atoms is the same as the number of hydrogen atoms in the simplest nonmetal hydrides. What are the formulas for the simplest fluorine compounds of the following elements: (a) carbon, (b) nitrogen, (c) oxygen, (d) arsenic, (e) chlorine?

Properties of Ionic and Molecular Compounds

4.76 In general, how do the melting points of ionic and molecular compounds compare?

4.77 Why are ionic compounds more brittle than molecular compounds?

4.78 Which of the following will likely conduct electricity when melted: (a) PCl_3, (b) AlF_3, (c) OF_2, (d) CaF_2, (e) SeO_2?

4.79 For each of the following pairs of substances, choose the chemical that has the higher melting point: (a) OF_2 or CaF_2, (b) PCl_3 or $AlCl_3$, (c) TiO_2 or CO_2, (d) NaH or HCl.

Oxidation–Reduction, Oxidation Numbers

4.80 Give two modern definitions of the terms *oxidation* and *reduction*. Which of the definitions is preferred?

4.81 How to the terms *oxidation number* and *oxidation state* differ?

4.82 Explan why the reaction between calcium and oxygen to form CaO is a redox reaction.

4.83 In your own words, write the six rules for assigning oxidation numbers to atoms in a chemical formula. (If you can, avoid peeking at the rules on page 105.)

4.84 What happens to the oxidizing agent in a redox reaction? What happens to the reducing agent?

4.85 Assign oxidation numbers to each atom in $KClO_2$, $BaMnO_4$, Fe_3O_4, O_2F_2, IF_5, $HOCl$, $CaSO_4$, $Cr_2(SO_4)_3$, O_3, and Hg_2Cl_2.

4.86 Assign oxidation numbers to each atom in H_2SO_4, CBr_4, OF_2, KO_2, $CrCl_3$, Mn_2O_7, $KMnO_4$, $H_2C_2O_4$, $KClO_3$, and $LiNO_3$.

4.87 Identify the following changes as either oxidation or reduction:
(a) MnO_2 to MnO_4^-
(b) BiO_3^- to Bi^{3+}
(c) SO_2 to SO_3
(d) OCl^- to ClO_3^-
(e) N_2O_4 to N_2O

4.88 For each of the following reactions, identify (1) the substance oxidized, (2) the substance reduced, (3) the oxidizing agent, and (4) the reducing agent.
(a) $2HNO_3 + 3H_3AsO_3 \rightarrow 2NO + 3H_3AsO_4 + H_2O$
(b) $NaI + 3HOCl \rightarrow NaIO_3 + 3HCl$
(c) $2KMnO_4 + 5H_2C_2O_4 + 3H_2SO_4 \rightarrow 10CO_2 + K_2SO_4 + 2MnSO_4 + 8H_2O$

(d) $6H_2SO_4 + 2Al \rightarrow Al_2(SO_4)_3 + 3SO_2 + 6H_2O$

(e) $K_2Cr_2O_7 + 14HCl \rightarrow 2KCl + 2CrCl_3 +$
     $3Cl_2 + 7H_2O$

4.89 For each of the following reactions, identify (1) the substance oxidized, (2) the substance reduced, (3) the oxidizing agent, and (4) the reducing agent.

(a) $NaIO_3 + 5NaI + 6HCl \rightarrow 6NaCl + 3I_2 + 3H_2O$

(b) $3Cu + 8HNO_3 \rightarrow 3Cu(NO_3)_2 + 2NO + 4H_2O$

(c) $Cu + 4HNO_3 \rightarrow Cu(NO_3)_2 + 2NO_2 + 2H_2O$

(d) $Cu + 2H_2SO_4 \rightarrow CuSO_4 + SO_2 + 2H_2O$

(e) $3SO_2 + 2HNO_3 + 2H_2O \rightarrow 3H_2SO_4 + 2NO$

Naming Compounds

4.90 Name the following (when appropriate, use *both* the Stock system and the old system):

(a) NaBr   (f) P_4O_6
(b) CaO   (g) $AsCl_5$
(c) $FeCl_3$ (h) $Mn(HCO_3)_2$
(d) $CuCO_3$ (i) $NaMnO_4$
(e) CBr_4 (j) O_2F_2

4.91 Write chemical formulas for the following compounds:

(a) aluminum nitrate
(b) iron(II) sulfate
(c) ammonium dihydrogen phosphate
(d) iodine pentafluoride
(e) phosphorus trichloride
(f) dinitrogen tetroxide
(g) potassum permanganate
(h) magnesium hydroxide
(i) hydrogen selenide
(j) sodium hydride

4.92 Name the following, using the Stock system when appropriate:

(a) Cr_2O_3 (f) $AlPO_4$
(b) $Mg(H_2PO_4)_2$ (g) Mg_3N_2
(c) $Cu(NO_3)_2$ (h) PbC_2O_4
(d) $CaSO_4$ (i) $(NH_4)_2CO_3$
(e) $Ba(OH)_2$ (j) $K_2Cr_2O_7$

4.93 Write formulas for the following:

(a) titanium(IV) oxide
(b) silicon tetrachloride
(c) calcium selenide
(d) potassium nitrate
(e) aluminum sulfate
(f) nickel(II) bicarbonate
(g) sodium hydrogen sulfate
(h) ammonium dichromate
(i) calcium acetate
(j) strontium hydroxide

4.94 Stannous fluoride is used in some brands of toothpaste to help fight tooth decay. What is the formula of this compound?

4.95 Name the following, using the Stock system when appropriate:

(a) $SrCl_2$ (f) $Zn(C_2H_3O_2)_2$
(b) $Ca(NO_3)_2$ (g) $HBrO_3$
(c) CuS (h) $HgBr_2$
(d) $Sn_3(PO_4)_2$ (i) $CoSO_4$
(e) $Ni(ClO_3)_2$ (j) KH_2AsO_4

4.96 Name the following, using the Stock system when appropriate:

(a) ClF_3 (f) H_2Te
(b) N_2O_5 (g) $H_2Te(aq)$
(c) $(NH_4)_2HPO_4$ (h) Mg_3P_2
(d) $SeCl_4$ (i) SnS_2
(e) S_4N_4 (j) OF_2

4.97 Write formulas for the following:

(a) lead(IV) acetate
(b) sodium selenide
(c) barium phosphate
(d) hydrogen iodide
(e) hydriodic acid
(f) phosphorus tribromide
(g) calcium hypochlorite
(h) silver oxalate
(i) chromic acid (*hint:* What's the name of CrO_4^{2-}?)
(j) silicon tetrafluoride

4.98 Write formulas for the following:

(a) stannous fluoride
(b) aluminum phosphide
(c) bromic acid
(d) strontium dihydrogen phosphate
(e) vanadium(III) oxide
(f) antimony pentachloride
(g) nitrogen monoxide
(h) cuprous bromide
(i) cupric bromide
(j) ammonium bisulfite

4.99 Write formulas for the following:

(a) trisulfur nonaoxide   (f) periodic acid
(b) iodine pentachloride   (g) lithium hypoiodite
(c) chromium(III) sulfate   (h) mercuric nitrate
(d) ferric sulfate   (i) auric sulfate
(e) plumbous sulfide   (j) bismuth(V) oxide

4.100 Write formulas for the following:

(a) tetraarsenic decaoxide
(b) dinitrogen
(c) disulfur dinitride
(d) pentasulfur hexanitride
(e) manganese(III) cyanide
(f) hydrocyanic acid

4.101 Write formulas for the following:

(a) ferric sulfate   (e) stannic chloride
(b) ferrous chloride   (f) cobaltous hydroxide
(c) mercurous nitrate   (g) auric chloride
(d) cuprous chloride   (h) chromic acetate

4.102 Write the names for each of the compounds in Exercise 4.69. When appropriate, follow the Stock system of nomenclature.

CHEMICAL REACTIONS
IN AQUEOUS SOLUTION

In Chapter 3 you learned that many reactions are carried out in solution simply for practical reasons. When the reactants are dissolved in a solvent, their particles become uniformly distributed throughout the solution and are able to intermingle freely. This permits the reaction to occur as rapidly as possible.

One of the most important solvents for chemical reactions is water. It certainly is a common substance, and it is also a good solvent for many different kinds of chemicals, both ionic and molecular. In fact, this ability to dissolve so many different chemicals, to one degree or another, is one of the major concerns of modern society, as it struggles with increasing problems of water pollution in heavily populated areas. Much attention has been paid to reactions in aqueous solutions, partly because of the general availability of water as a solvent in the lab, partly because of potential reactions of water pollutants in the environment, and also because water is the medium in which biochemical reactions take place.

5.1 SOLUTION TERMINOLOGY

Many of the terms that we use in discussing solutions were introduced to you in Chapter 3. These include the terms *solvent* and *solute.* The solvent is generally taken to be the substance present in the solution in largest proportion, and all the other substances are considered to be solutes. In solutions that contain water, however, water is almost always thought of as the solvent even if it is present in relatively small amounts. For example, a mixture composed of 96% H_2SO_4 and 4% H_2O by mass is called "concentrated sulfuric acid," which implies that a large amount of sulfuric acid is *dissolved in* a small amount of water — water is taken to be the solvent and H_2SO_4 the solute.

Another set of terms mentioned in Chapter 3 is *concentrated* and *dilute.* A concentrated solution has a *relatively large* proportion of solute to solvent, and a dilute solution has a *relatively smaller* proportion of solute to solvent. The emphasis here is that these are relative terms: A particular solution is considered concentrated

Figure 5.1. *When a small seed crystal of sodium acetate is added to a supersaturated solution of the salt, the excess solute crystallizes rapidly until the solution is just saturated. The crystallization shown in this sequence of photographs took less than 10 seconds.*

when compared to some other solution that has a lower proportion of solute to solvent.

In most cases, there is a limit to the amount of a solute that can dissolve in a given amount of solvent at a particular temperature. For example, if we add sodium chloride to 100 mL of water at 0 °C, only 35.7 g of the salt will dissolve, regardless of how much we place into the water. Any excess NaCl will simply settle to the bottom of the container. A solution that contains as much dissolved solute as it can hold while in contact with excess solute is said to be a **saturated solution,** and the amount of solute needed to give a saturated solution with a given amount of solvent is called the **solubility** of that particular solute. Thus, the solubility of sodium chloride in water at 0 °C is 35.7 g of NaCl per 100 mL of water. Usually a solute's solubility changes with temperature. For example, at 100 °C the solubility of NaCl is 39.1 g/100 mL of H_2O. This means that we should always specify the temperature when stating the solubility.

If a particular solution contains less solute than is needed for saturation, it is said to be an **unsaturated solution.** An example would be a solution of 20 g of NaCl in 100 mL of H_2O at 0 °C. An unsaturated solution is capable of dissolving more solute—in this case, an additional 15.7 g of NaCl could be dissolved in each 100 mL.

It is important to remember that the terms saturated and unsaturated are not related in any direct way to the terms concentrated and dilute. For example, a saturated solution of silver chloride at room temperature contains only 0.000089 g AgCl/100 mL of water and would certainly be considered dilute. On the other hand, it would take about 500 g of lithium chlorate, $LiClO_3$, to form a saturated solution in 100 mL of water at this same temperature. A solution containing 400 g of $LiClO_3$ generally would be considered concentrated, even though it is unsaturated.

Finally, there are some substances that frequently form **supersaturated solutions,** solutions that contain more solute than ordinarily required for saturation. Sodium acetate, $NaC_2H_3O_2$, is an example. At 0 °C, this compound is soluble in water to the extent of 119 g/100 mL, but its solubility increases greatly with increasing temperature. If a hot unsaturated solution that contains more than 119 g of $NaC_2H_3O_2$ per 100 mL is cooled to 0 °C, the excess solute should separate from the solution and settle to the bottom, but usually it does not. The excess solute remains in the solution, and the solution is supersaturated. Supersaturated solutions are unstable; if a tiny crystal of $NaC_2H_3O_2$ is added, additional solute crystallizes on this "seed" crystal until the concentration of the solution drops to the point of saturation (see Figure 5.1).

If a particular supersaturated solution contains 150 g $NaC_2H_3O_2$ per 100 mL, 31 g of the solute will crystallize from each 100 mL when a tiny seed crystal is added.

5.2 ELECTROLYTES

Water is generally a good solvent for ionic compounds, and water solutions (aqueous solutions) that contain these substances have some unusual properties, one of which is that they conduct electricity. This can be demonstrated with a conductivity appa-

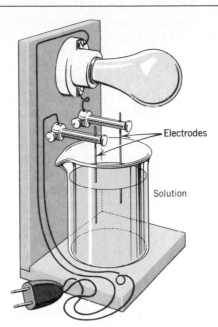

Figure 5.2. *The conductivity apparatus described in Chapter 4 can also be used to test the electrical conductivity of solutions, as illustrated here.*

ratus like that shown in Figure 5.2. If the electrodes are dipped into pure water, the bulb doesn't light because pure water is a very poor conductor of electricity. However, if a soluble ionic compound, such as NaCl, is added to the water, the bulb begins to glow brightly as soon as the solid begins to dissolve. A compound such as NaCl that gives an electrically conducting solution is said to be an **electrolyte.**

How do we explain the electrical conductivity of aqueous solutions of ionic compounds? When these substances dissolve in water, the ions that are held rigidly in place in the solid become separated and are able to roam about in the solution, essentially free of each other. We say that the compound **dissociates,** or breaks apart to yield the ions, and it is the presence of these freely moving ions that permits the solution to conduct electricity.

This explanation of electrolytes was first proposed by Svante Arrhenius, one of Sweden's most famous chemists. It's interesting to note that he was nearly denied his doctoral degree in 1884 at the university in Upsala, Sweden, for making such a revolutionary suggestion. Nevertheless, his theory has withstood the test of time because it explains so successfully the unusual properties of salt solutions.

Arrhenius received the 1903 Nobel prize in chemistry for his theory of electrolytes.

When an ionic compound dissociates in water, its ions are not really entirely free. Instead, they become surrounded by water molecules and are said to be **hydrated.** We indicate this by writing (aq) after the formulas of the ions. For example, the dissociation of sodium chloride that occurs when the solid is dissolved in water can be represented by the equation

$$NaCl(s) \longrightarrow Na^+(aq) + Cl^-(aq)$$

(aq) = aqueous solution
(s) = solid

Often, for simplicity, we will leave off the labels (s) and (aq) if doing so creates no confusion.

The production of ions in solutions is not limited to ionic compounds. There are many molecular substances that react with water to produce ions and therefore qualify as electrolytes. Hydrogen chloride is a typical example. When HCl gas is dissolved in water, the following reaction takes place:

$$HCl(g) + H_2O \longrightarrow H_3O^+(aq) + Cl^-(aq)$$

This reaction is properly called an **ionization reaction** because it produces ions

The name proton is frequently used to mean hydrogen ion. The removal of hydrogen's single electron gives just a bare proton.

where there was none before. (However, it is often referred to as dissociation, simply to avoid having to use different terms for ionic and molecular electrolytes.) The reaction occurs by the transfer of a proton, or hydrogen ion (H^+), from the HCl molecule to the water molecule to produce the **hydronium ion,** H_3O^+, and a chloride ion, Cl^-. Thus, even though hydrogen chloride exists as discrete electrically neutral molecules when pure (liquid HCl doesn't conduct electricity), when it is dissolved in water, it produces ions by a chemical reaction and thereby becomes an electrolyte.

H^+ is the active ingredient in H_3O^+.

As we will see, the hydronium ion is one of the most important species to consider in discussions of chemical reactions in aqueous solutions. It is useful to think of it as an H^+ ion, or a proton, that has associated itself with a water molecule. We can do this because when the hydronium ion reacts, it does so by giving up the proton, which leaves the water molecule as one of the products. In a sense, the H_2O of the hydronium ion just serves as a carrier for the H^+ ion. For this reason, the hydronium ion is often written simply as H^+, and we often speak of the H_3O^+ ion as the hydrogen ion. By leaving out the H_2O of the H_3O^+ ion, we can write the dissociation of HCl as

$$HCl(aq) \longrightarrow H^+(aq) + Cl^-(aq)$$

Even though we write H^+, *always* remember that there is at least one, and probably several more, H_2O molecules associated with the proton in aqueous solutions.[1]

EXAMPLE 5.1. WRITING EQUATIONS FOR IONIZATION REACTIONS

PROBLEM: Hydrogen bromide, HBr, is an electrolyte in water and gives hydrogen ion and bromide ion in solution. Write an equation for the ionization reaction.

SOLUTION: The reaction is between HBr molecules and water molecules.

$$HBr(aq) + H_2O \longrightarrow H_3O^+(aq) + Br^-(aq)$$

If we write the hydronium ion in its abbreviated form as H^+, the equation can be written

$$HBr(aq) \longrightarrow H^+(aq) + Br^-(aq)$$

Strong and weak electrolytes—chemical equilibrium

In general, ionic compounds are strong electrolytes. They are essentially 100% dissociated.

The two examples of electrolytes discussed above, NaCl and HCl, are essentially completely dissociated in aqueous solution; 1 mol of NaCl gives 1 mol of Na^+ ions and 1 mol of Cl^- ions, and 1 mol of HCl gives 1 mol of H_3O^+ ions and 1 mol of Cl^- ions. Substances such as NaCl and HCl, which are completely dissociated in solution, are called **strong electrolytes.**

There are also many molecular substances that have no tendency at all to undergo ionization when dissolved in water. Sugar and ethyl alcohol are two relatively common examples. When these compounds are dissolved in water, their molecules just mix with the water molecules to give a uniform solution, but the solution doesn't contain any ions because the solute doesn't react with water. Since these solutes do not produce ions in a solution, the solution doesn't conduct electricity and these solutes are therefore called **nonelectrolytes.**

Between the extremes of strong electrolytes and nonelectrolytes exists a large collection of compounds called **weak electrolytes.** These compounds produce

[1] There is, in fact, evidence that suggests that the H^+ ion may exist as $H_9O_4^+$, that is, $H_3O(H_2O)_3^+$, in aqueous solution.

aqueous solutions that conduct electricity, but only very weakly. An example is acetic acid, $HC_2H_3O_2$, which is the substance that gives vinegar its sour taste. If the electrodes of the conductivity apparatus are dipped into a solution of this solute, the bulb glows dimly.

In a solution of acetic acid, only a small fraction of all the acetic acid molecules are present as ions, produced by the reaction

$$HC_2H_3O_2(aq) + H_2O \longrightarrow H_3O^+(aq) + C_2H_3O_2^-(aq)$$

For example, in a 1.0 M solution of $HC_2H_3O_2$, only about 0.42% of the solute has undergone this reaction. The rest of the acetic acid exists as uncharged molecules.

The reason for the limited degree of dissociation of weak electrolytes is worth discussing at this point because it illustrates one of the most important concepts in chemistry, one to which we will devote three chapters at a later time (Chapters 15, 16, and 17).

In a solution of acetic acid, molecules of $HC_2H_3O_2$ are constantly colliding with molecules of water and, in each encounter, there is a certain probability that a proton will be transferred from an $HC_2H_3O_2$ molecule to a water molecule to yield H_3O^+ and $C_2H_3O_2^-$ ions. There are also encounters, in this solution, between acetate ions and hydronium ions. When these ions meet, there happens to be a high probability that an H_3O^+ ion will lose a proton to a $C_2H_3O_2^-$ ion to reform $HC_2H_3O_2$ and H_2O molecules. Thus, in this solution there are two reactions occurring simultaneously,

$$HC_2H_3O_2 + H_2O \longrightarrow H_3O^+ + C_2H_3O_2^- \tag{1}$$

and

$$H_3O^+ + C_2H_3O_2^- \longrightarrow HC_2H_3O_2 + H_2O \tag{2}$$

When the rate at which the ions are formed by reaction 1 is equal to the rate at which they disappear by reaction 2, their concentrations in the solution will no longer change with time. In fact, the concentrations of all the species wil remain constant from this point on, even though if we followed any particular $C_2H_3O_2$ unit in the solution, it would sometimes exist as a $C_2H_3O_2^-$ ion and at other times as a $HC_2H_3O_2$ molecule. Such a state of affairs is called **equilibrium.** It is said to be a **dynamic equilibrium** because things are continually happening in the solution—two reactions are taking place: ions reacting to yield molecules and molecules reacting to produce ions.

Dynamic implies activity. The opposing reactions don't cease at equilibrium; they just occur at equal rates (equal speeds).

To indicate chemical equilibrium in a reacting system, we use a set of double arrows, $\rightleftharpoons$, in the chemical equation. Thus, the equilibrium that we have been discussing is expressed as

$$HC_2H_3O_2(aq) + H_2O \rightleftharpoons H_3O^+(aq) + C_2H_3O_2^-(aq)$$

The use of this notation implies that the *forward reaction* (the reaction going from left to right) is occurring at the same rate as the *reverse reaction* (the reaction going from right to left). In a solution of acetic acid, these rates just happen to become equal when only a small fraction of all the acetic acid molecules has undergone ionization. We say that the extent of ionization is small and that the **position of equilibrium,** the relative proportions of the reactants and products, *lies to the left,* in the direction of the molecular form of the solute. In other words, almost everything is present in nonionized form.

The specific chemical properties of a substance determine the relative tendencies to occur of the forward and reverse reactions.

Some additional examples of weak electrolytes are found in Table 5.1 on the next page. Notice that water itself is included in this table because it is actually a very weak electrolyte by virtue of the reaction

$$H_2O + H_2O \rightleftharpoons H_3O^+(aq) + OH^-(aq)$$

This slight ionization of water plays a very important role in many chemical reactions in which water is a solvent. Special attention is given to this topic in Chapter 16.

Table 5.1. Some weak electrolytes

Substance	Dissociation Reaction	Percent Dissociation of the Solute in a 1.00 M Solution
Water	$H_2O + H_2O \rightarrow H_3O^+ + OH^-$	1.8×10^{-7} (55.5 moles of H_2O per liter)
Acetic acid	$HC_2H_3O_2 + H_2O \rightarrow H_3O^+ + C_2H_3O_2^-$	0.42
Ammonia	$NH_3 + H_2O \rightarrow NH_4^+ + OH^-$	0.42
Hydrogen cyanide	$HCN + H_2O \rightarrow H_3O^+ + CN^-$	2.0×10^{-3}

For strong electrolytes, such as HCl, the reaction of the ions to form neutral molecules has nearly no tendency to occur. When an H_3O^+ ion encounters a Cl^- ion in the solution, nothing happens to them. Therefore, when HCl is dissolved in water, only the forward reaction takes place, and in just an instant all the HCl is converted to ions. The solute has become 100% ionized. When we write the equation for the reaction of a strong electrolyte with water, we omit the reverse arrow because the reverse reaction doesn't occur. Therefore, for HCl we write

$$HCl(aq) + H_2O \longrightarrow H_3O^+(aq) + Cl^-(aq)$$

The concept of a dynamic equilibrium is very important. All processes, both chemical and physical, tend to move toward a state of equilibrium, and we will use this concept many times in later chapters to analyze physical changes as well as chemical reactions.

EXAMPLE 5.2. WRITING EQUATIONS FOR THE IONIZATION OF WEAK ELECTROLYTES

PROBLEM: Formic acid, $HCHO_2$, is a chemical that fire ants inject into their victims when they bite. It is a weak electrolyte that gives H_3O^+ and CHO_2^- ions in water. Write an equation for its ionization.

SOLUTION: Since one of the ions formed in the solution is H_3O^+, the $HCHO_2$ is reacting just like acetic acid does. In addition, since $HCHO_2$ is a weak electrolyte, the ionization reaction is an equilibrium. Therefore, the proper equation is

$$HCHO_2(aq) + H_2O \rightleftharpoons H_3O^+(aq) + CHO_2^-(aq)$$

We can also write the equation as follows if we use the abbreviated form of the hydronium ion:

$$HCHO_2(aq) \rightleftharpoons H^+(aq) + CHO_2^-(aq)$$

5.3 REACTIONS BETWEEN IONS

Many of the chemical reactions that you carry out in the laboratory portion of your chemistry course involve electrolytes dissolved in water. Usually, these reactions take place between the ions present in the solution and can therefore be called **ionic reactions.** A typical example is the reaction that occurs when solutions of sodium chloride and silver nitrate are mixed, which is shown in Figure 5.3. As one solution is added to the other, a white solid, silver chloride, is formed. If the sodium chloride solution contains 1 mol of NaCl and the silver nitrate solution contains 1 mol of

$AgNO_3$, then 1 mol of AgCl is formed and the solution contains 1 mol of dissolved $NaNO_3$. If desired, we can separate the AgCl from the solution by filtering the mixture as shown in Figure 5.4. When the **filtrate,** the clear solution that passes through the filter, is evaporated, crystals of sodium nitrate are left behind.

The chemical equation for the change that has occurred in this reaction is

$$AgNO_3(aq) + NaCl(aq) \longrightarrow AgCl(s) + NaNO_3(aq)$$

This kind of reaction, in which cations and anions have changed partners, is called **metathesis,** or **double replacement** (Cl^- has replaced NO_3^- and NO_3^- has replaced Cl^-).

The equation above is called a **molecular equation,** because all the reactants and products are written as if they were molecules. (Of course, you know that ionic substances don't exist as molecules either in the solid state or in solution. We just call this a molecular equation because we haven't indicated the presence of ions.)

A more accurate representation of the reaction as it actually occurs is obtained if we take into account what happens to the solutes when they are dissolved in water. As noted earlier, any soluble ionic compound exists in solution, not as molecules, but instead as ions dispersed throughout the solvent. These compounds are 100% dissociated. Therefore, in water the NaCl exists as Na^+ and Cl^- ions. Similarly, in its solution, the $AgNO_3$ exists as Ag^+ and NO_3^- ions. When the two solutions are mixed, solid AgCl is formed by a combination of the Ag^+ and Cl^- ions. We call such a solid, formed in a solution as the result of a chemical reaction, a **precipitate.** The solution that exists after the formation of AgCl contains just Na^+ and NO_3^- ions, and is therefore a solution of sodium nitrate. To show those substances that are completely dissociated in the reaction, we can rewrite the equation as

$$Ag^+(aq) + NO_3^-(aq) + Na^+(aq) + Cl^-(aq) \longrightarrow AgCl(s) + Na^+(aq) + NO_3^-(aq)$$

This is called the **ionic equation** and is obtained by writing the formulas of any soluble strong electrolytes in "dissociated form" and the formulas of any insoluble substances in "molecular form."

If you examine the ionic equation for this reaction, you see that Na^+ and NO_3^-

Figure 5.3. *The white solid AgCl is formed when a solution of NaCl is poured into a solution of $AgNO_3$.*

NaCl exists as ions in the solid and liquid states and in aqueous solutions.

The formula for the precipitate is given in "molecular form" because the ions are no longer free of each other but are trapped together in the solid.

(a)   (b)

Figure 5.4. *A precipitate can be separated from a solution by filtration. (a) The mixture containing the precipitate is guided onto filter paper held in a glass funnel by pouring it down a glass stirring rod. The beaker below catches the filtrate that passes through the filter. (b) The last traces of the precipitate are rinsed from the beaker by a stream of distilled water from a wash bottle.*

do not actually undergo any change. The same Na^+ and NO_3^- ions are present after the reaction as before and they have, in a sense, "just gone along for the ride." For this reason, ions that do not change during a reaction are called **spectator ions.** Sometimes, when we write ionic equations, we omit the spectator ions so we can concentrate just on those ions that are involved in the reaction. This gives the **net ionic equation.** For the reaction between NaCl and $AgNO_3$, the net ionic equation is obtained by removing the spectator ions Na^+ and NO_3^- as follows,

$$Ag^+(aq) + \cancel{NO_3^-(aq)} + \cancel{Na^+(aq)} + Cl^-(aq) \longrightarrow AgCl(s) + \cancel{Na^+(aq)} + \cancel{NO_3^-(aq)}$$

which gives

$$Ag^+(aq) + Cl^-(aq) \longrightarrow AgCl(s) \qquad \text{(net ionic equation)}$$

The net ionic equation is useful because it focuses our attention on the changes that actually occur during a reaction. Furthermore, it also allows us to generalize what we have learned, so we can predict the outcome of other similar reactions. For example, the net ionic equation here tells us that any substance that produces Ag^+ ion in solution will react with any substance that produces Cl^- ion in solution, and that the product will be a precipitate of AgCl. Therefore, if we mix a solution of potassium chloride with a solution of silver fluoride (AgF is soluble in water, even though AgCl is not[2]), we expect to obtain a precipitate of AgCl. This is, in fact, precisely what happens. The molecular, ionic, and net ionic equations for the reaction are as follows:

$$AgF(aq) + KCl(aq) \longrightarrow AgCl(s) + KF(aq) \qquad \text{(molecular)}$$
$$Ag^+(aq) + F^-(aq) + K^+(aq) + Cl^-(aq) \longrightarrow AgCl(s) + K^+(aq) + F^-(aq) \qquad \text{(ionic)}$$
$$Ag^+(aq) + Cl^-(aq) \longrightarrow AgCl(s) \qquad \text{(net ionic)}$$

Each of these kinds of equations is useful in its own way; none is the best way of representing the reaction. The form that we use in any particular instance depends on what aspect of the reaction we wish to focus our attention. If we want to think about the net chemical change that occurs, the net ionic equation is best. However, if we want to work with these chemicals in carefully measured amounts, it is the stoichiometry of the molecular equation that is important.

EXAMPLE 5.3. WRITING IONIC AND NET IONIC EQUATIONS FOR METATHESIS REACTIONS

PROBLEM: If a solution of sodium hydroxide is added to a solution of iron(III) chloride, a precipitate of iron(III) hydroxide is formed. After the reaction, the solution contains dissolved sodium chloride. Write a molecular equation for the metathesis reaction and then write the appropriate ionic and net ionic equations for the reaction.

SOLUTION: The first step is to write the correct formulas of the reactants and products. (You should be able to do this based on what you learned in the last chapter.) Sodium hydroxide is NaOH; iron(III) chloride is $FeCl_3$; iron(III) hydroxide is $Fe(OH)_3$; and sodium chloride is NaCl. The balanced molecular equation is

$$3NaOH(aq) + FeCl_3(aq) \longrightarrow Fe(OH)_3(s) + 3NaCl(aq)$$

---

[2] The solubility of AgCl is very low, 0.000089 g/100 mL, and AgCl can be considered, for most purposes, to be insoluble; that is, the amount of AgCl in solution can generally be considered negligible. For comparison, the solubility of AgF in water is approximately 185 g/100 mL at room temperature.

To write the ionic equation, we "dissociate" all the soluble compounds and write the formula of the insoluble $Fe(OH)_3$ in molecular form. This gives

$$3Na^+ + 3OH^- + Fe^{3+} + 3Cl^- \longrightarrow Fe(OH)_3(s) + 3Na^+ + 3Cl^-$$

[For simplicity, we've omitted the (aq) after the formulas of the ions in solution. However, we've kept (s) after the formula for the iron(III) hydroxide to call attention to the fact that it is a precipitate.]

To obtain the net ionic equation, we drop any spectator ions from the ionic equation. Notice that there are $3Na^+$ on both sides and also $3Cl^-$ on both sides. If we omit these, we get the net ionic equation.

$$3OH^- + Fe^{3+} \longrightarrow Fe(OH)_3(s)$$

Usually, we prefer to write the cation first. This equation would normally be written

$$Fe^{3+} + 3OH^- \longrightarrow Fe(OH)_3(s)$$

5.4 ACID–BASE REACTIONS

Among the kinds of substances you learned to name in the last chapter are compounds that we call acids and bases. Arrhenius also recognized these substances as electrolytes, and they include some of our most important and common chemicals (see Figure 5.5). For example, the sour taste of vinegar and lemon juice comes from the acids they contain, acetic acid and citric acid, respectively. Ascorbic acid (vitamin C) is an essential part of our diets. And sulfuric acid, the acid in automobile batteries, ranks well above any other industrial chemical in annual production.

Among important bases are ammonia, found in cleaning products around the home, and sodium hydroxide, available on supermarket shelves under the name *lye* and present in oven cleaners and drain cleaners such as Drano. Even milk of magnesia, which is taken to soothe an upset (acid) stomach, is a base.

Acids and bases have certain properties that help us identify them. For example, a solution of an acid has a sour taste. That's why lemon juice and vinegar are sour. On the other hand, bases such as the magnesium hydroxide in milk of magnesia have a bitter taste. (*Caution:* Even though these are general properties of acids and bases, you should *never* test for acids and bases by tasting a chemical in the laboratory. It might not be healthy!) Another property of acids and bases is their effect on

Figure 5.5. *Common substances that contain acids and bases.*

Acids

Bases

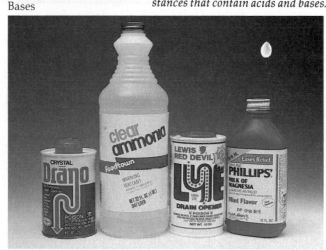

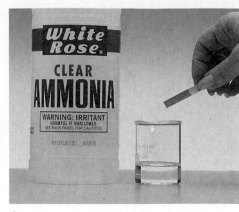

Figure 5.6. *Litmus is an indicator that turns red when made acidic and blue when made basic.* (a) *Lemon juice is acidic.* (b) *Ammonia is basic.*

(a)   (b)

indicators, chemicals whose colors depend on the acidity or basicity of their solutions. A typical example is the dye called litmus. Litmus is a chemical that has a blue color in a basic solution and a pink color in an acidic solution. It is common to have vials of litmus paper — thin strips of paper impregnated with litmus — available in the lab to test the acidity or basicity of solutions. If pink litmus paper is dipped into a solution that is basic, it turns blue, and if blue litmus paper is dipped into a solution that is acidic, it turns pink, as shown in Figure 5.6.

Stated in modern terms, the Arrhenius definitions of acids and bases are as follows: *An **acid** is a substance that increases the concentration of hydronium ion, H_3O^+, in an aqueous solution, and a **base** is a substance that increases the concentration of hydroxide ion, OH^-.* Let's examine more closely some examples of each of these kinds of chemicals.

Acids

In general, acids are molecular substances that produce hydronium ion by reaction with water. For example, hydrogen chloride is an acid, because when it is dissolved in water, it reacts with the solvent to produce H_3O^+

$$HCl(aq) + H_2O \longrightarrow H_3O^+(aq) + Cl^-(aq)$$

If we use H^+ as an abbreviation for the hydronium ion and leave out the molecule of water that carries the H^+, we can also express this reaction as

$$HCl(aq) \longrightarrow H^+(aq) + Cl^-(aq)$$

As you learned in Section 5.2, HCl is a strong electrolyte, meaning that it is essentially 100% dissociated in solution. Therefore, in any reasonably concentrated solution of HCl there is a high concentration of H_3O^+, and because of this, HCl is called a **strong acid.**

There are also many acids that are weak electrolytes. Acetic acid, $HC_2H_3O_2$, is an example. Recall that this acid reacts with water according to the equation

$$HC_2H_3O_2(aq) + H_2O \rightleftharpoons H_3O^+(aq) + C_2H_3O_2^-(aq)$$

or more simply

$$HC_2H_3O_2(aq) \rightleftharpoons H^+(aq) + C_2H_3O_2^-(aq)$$

This is an equilibrium, and in a solution of $HC_2H_3O_2$ only a small fraction of the solute is dissociated into ions. This means that the concentration of H_3O^+ in the solution is low. As a result, acetic acid and other acids that are weak electrolytes are called **weak acids.**

Notice that in the formula for acetic acid, $HC_2H_3O_2$, hydrogen appears twice — once at the beginning of the formula and once in the middle. In writing the formula for an acid, it is common practice to write the *acidic hydrogens,* those that are able to transfer to water molecules to give H_3O^+ ions, first in the formula. Other hydrogens that are not "acidic" are written later. Even though the acetic acid molecule contains four hydrogen atoms, only one of them is able to react with a water molecule to form H_3O^+.

Both HCl and $HC_2H_3O_2$ are able to furnish only one hydrogen ion (or proton) per molecule of the acid. Such acids are said to be **monoprotic acids.** There are also many acids that are able to furnish more than one proton per molecule of the acid. As a class, they are referred to as **polyprotic acids.** Two examples are sulfuric acid, H_2SO_4 and phosphoric acid, H_3PO_4.

Sulfuric acid is also called a **diprotic acid** because each molecule of it is able to give up two protons. This happens in two distinct steps

$$H_2SO_4(aq) \longrightarrow H^+(aq) + HSO_4^-(aq)$$

$$HSO_4^-(aq) \rightleftharpoons H^+(aq) + SO_4^{2-}(aq)$$

Similarly, phosphoric acid, which is an example of a **triprotic acid,** dissociates in three separate steps.

$$H_3PO_4(aq) \rightleftharpoons H^+(aq) + H_2PO_4^-(aq)$$

$$H_2PO_4^-(aq) \rightleftharpoons H^+(aq) + HPO_4^{2-}(aq)$$

$$HPO_4^{2-}(aq) \rightleftharpoons H^+(aq) + PO_4^{3-}(aq)$$

Notice that the second step in the dissociation of H_2SO_4 is an equilibrium (only about 10% of the HSO_4^- is actually dissociated). Despite this, sulfuric acid is considered a strong acid because the first dissociation step is complete. Phosphoric acid, which is used in manufacturing fertilizers and detergents and various soft drinks such as colas, is a weak acid because all three of its dissociation steps are equilibria that do not proceed very far toward completion.

All the acids that we have examined so far contain hydrogen in their formulas. However, there are other substances that do not contain hydrogen, yet still produce acidic solutions when dissolved in water. A common example is carbon dioxide. When dissolved in water (in a carbonated beverage, for example), it reacts as follows:

$$CO_2(g) + H_2O \rightleftharpoons H_2CO_3(aq)$$

The compound H_2CO_3 is called carbonic acid; it is a weak diprotic acid that dissociated in two steps, the first of which is

$$H_2CO_3(aq) \rightleftharpoons H^+(aq) + HCO_3^-(aq)$$

This particular acid, incidently, is responsible for the large limestone caverns that exist in various parts of the world. Groundwater, made acidic by CO_2 dissolved in it from the atmosphere, trickles through the limestone ($CaCO_3$) with which it reacts.

$$CaCO_3(s) + H_2CO_3(aq) \longrightarrow Ca^{2+}(aq) + 2HCO_3^-(aq)$$

The product of the reaction is *soluble* calcium bicarbonate that flows away in the groundwater, leaving huge caverns behind, like that shown in Figure 5.7.

The reaction of CO_2 with water is typical of many of the nonmetal oxides; they react with water to form oxoacids. Another example is sulfur dioxide, an air pollutant released by burning sulfur-containing fuels. It reacts with water as follows:

Not all nonmetal oxides are able to react with water. Carbon monoxide and nitrogen monoxide are two that do not.

$$SO_2(g) + H_2O \longrightarrow H_2SO_3(aq)$$

This is one of the reactions that is responsible for *acid rain,* an environmental problem that is particularly severe in the northeastern United States, southern Can-

Figure 5.7. *Groundwater made acidic by dissolved carbon dioxide gradually dissolves limestone from underground deposits, creating immense caverns like the one shown in this photograph.*

Anhydride comes from the Greek *anhydros* meaning *without water.*

ada, and heavily industrialized areas of Europe. Rainwater, passing through the atmosphere polluted by sulfur dioxide and other oxides, such as SO_3 and NO_2, becomes acidic by reactions of this type.

A nonmetal oxide that is able to react with water to form an oxoacid is called an **acidic anhydride,** which means "acid without water." Table 5.2 contains a list of common acids and acidic anhydrides, and it also shows their reactions with water. Those that are marked with an asterisk are strong acids; you should learn these. It happens that there are very few strong acids, so the list is short. Then, if you come across an acid that isn't on the list of strong ones, you can be fairly sure it is weak.

Bases

There are two principal kinds of bases: ionic hydroxides and molecular substances that react with water to produce OH^-. Sodium hydroxide and calcium hydroxide are typical ionic hydroxides. In the solid state they consist of a metal ion and hydroxide ion, and when they dissolve in water, they dissociate.

$$NaOH(s) \longrightarrow Na^+(aq) + OH^-(aq)$$

$$Ca(OH)_2(s) \longrightarrow Ca^{2+}(aq) + 2OH^-(aq)$$

As is typical of ionic compounds when they dissolve in water, this dissociation is complete, so ionic metal hydroxides are **strong bases.**

The most common molecular substance that is a base is ammonia, NH_3. It reacts with water in an equilibrium.

$$NH_3(aq) + H_2O \rightleftharpoons NH_4^+(aq) + OH^-(aq)$$

In this case, a proton is transferred from a water molecule to an ammonia molecule. After the H_2O loses an H^+, the particle left behind is a hydroxide ion, OH^-.

The reaction of NH_3 with water is an equilibrium, and only a small fraction of the NH_3 placed into the solution is present as NH_4^+ and OH^- ions. Ammonia is a weak electrolyte, and because its solutions have relatively few OH^- ions, it is also said to be a **weak base.** In general, molecular bases are weak bases.

Aqueous solutions of ammonia are sometimes called ammonium hydroxide, which suggests the formula NH_4OH. The species NH_4OH does not exist, however.

Table 5.2 Some acids and bases

Acids (Those followed by an asterisk are strong electrolytes and are completely ionized in aqueous solution.)

Monoprotic acids	HF	hydrofluoric acid	$HClO_3$	chloric acid*	
$HX \rightarrow H^+ + X^-$	HCl	hydrochloric acid*	$HClO_4$	perchloric acid*	
	HBr	hydrobromic acid*	HIO_4	periodic acid*	
	HI	hydroiodic acid*	HNO_3	nitric acid*	
	HOCl	hypochlorous acid	HNO_2	nitrous acid	
	$HClO_2$	chlorous acid	$HC_2H_3O_2$	acetic acid	
Diprotic acids	H_2SO_4	[a]sulfuric acid*	H_2S	hydrosulfuric acid	
$H_2X \rightarrow H^+ + HX^-$	H_2SO_3	sulfurous acid	H_3PO_3	phosphorous acid (only two	
$HX^- \rightarrow H^+ + X^{2-}$	H_2CO_3	carbonic acid		hydrogens can be removed	
	$H_2C_2O_4$	oxalic acid		as protons)	
Triprotic acids	H_3PO_4	phosphoric acid (orthophosphoric acid)			
$H_3X \rightarrow H^+ + H_2X^-$					
$H_2X^- \rightarrow H^+ + HX^{2-}$					
$HX^{2-} \rightarrow H^+ + X^{3-}$					
Typical acidic oxides	SO_2	$SO_2 + H_2O \rightarrow H_2SO_3$			
(nonmetal oxides)	SO_3	$SO_3 + H_2O \rightarrow H_2SO_4$			
	N_2O_3	$N_2O_3 + H_2O \rightarrow 2HNO_2$			
	N_2O_5	$N_2O_5 + H_2O \rightarrow 2HNO_3$			
	P_4O_6	$P_4O_6 + 6H_2O \rightarrow 4H_3PO_3$			
	P_4O_{10}	$P_4O_{10} + 6H_2O \rightarrow 4H_3PO_4$			

Bases

Molecular bases	NH_3	[b]ammonia	$(NH_3 + H_2O \rightleftarrows NH_4^+ + OH^-)$
(weak bases)	N_2H_4	hydrazine	$(N_2H_4 + H_2O \rightleftarrows N_2H_5^+ + OH^-)$
	NH_2OH	hydroxylamine	$(NH_2OH + H_2O \rightleftarrows NH_3OH^+ + OH^-)$
Ionic bases	metal hydroxides		$M(OH)_n \rightarrow M^{n+} + nOH^-$
(strong bases)	NaOH		
	$Ca(OH)_2$		
Typical basic oxides	Na_2O	} $M_2O + H_2O \rightarrow 2MOH$	
(metal oxides)	K_2O		
	CaO	} $MO + H_2O \rightarrow M(OH)_2$	
	SrO		
	BaO		

[a] The second dissociation of H_2SO_4 is that of a weak acid.

[b] Aqueous solution of ammonia is sometimes referred to as ammonium hydroxide. Commercially prepared solutions of NH_3, for example, are labeled in this way.

Earlier in this section you learned that nonmetal oxides are acidic anhydrides. Metal oxides have just the opposite property—they are **basic anhydrides.** An example is sodium oxide, Na_2O. This is actually an ionic compound containing Na^+ and O^{2-} ions, and when it is dissolved in water, the oxide ion reacts to form hydroxide ion.

$$O^{2-} + H_2O \longrightarrow 2OH^-$$

This reaction is complete, so any soluble metal oxide immediately reacts with water to give a solution of its hydroxide.

$$Na_2O(s) + H_2O \longrightarrow 2Na^+(aq) + 2OH^-(aq)$$

Table 5.2 also contains a list of bases and basic anhydrides, and it shows their reactions with water.

Neutralization

The most important reaction that acids and bases undergo is their reaction with each other, a reaction called **neutralization.** In aqueous solutions the neutralization reaction between a strong acid and a strong base takes the form of the net ionic equation

$$H_3O^+(aq) + OH^-(aq) \longrightarrow 2H_2O$$

or if we use H^+ as an abbreviation for the hydronium ion,

$$H^+(aq) + OH^-(aq) \longrightarrow H_2O$$

If we examine the molecular equation for a typical acid–base reaction, the reaction between sodium hydroxide and hydrochloric acid,

$$NaOH(aq) + HCl(aq) \longrightarrow NaCl(aq) + H_2O$$

The driving force behind this reaction is strong enough to cause insoluble bases to dissolve in acids and insoluble acids to dissolve in bases.

we can arrive at another useful generalization. *The products of a neutralization reaction in aqueous solution are a salt and water.*

The formation of water in a neutralization reaction has such a strong tendency to occur that acids also will react with both insoluble metal hydroxides and oxides. For example, the neutralization of stomach acid, which is HCl, by insoluble magnesium hydroxide in milk of magnesia follows the net ionic equation

$$Mg(OH)_2(s) + 2H^+(aq) \longrightarrow Mg^{2+}(aq) + 2H_2O$$

In a somewhat similar reaction, acid is used to remove rust (Fe_2O_3) from the surface of steel before it is given a protective coating of tin or zinc. Even though the iron oxide is insoluble, it reacts with the acid.

Insoluble metal oxides will even dissolve in weak acids.

$$Fe_2O_3(s) + 6H^+(aq) \longrightarrow 2Fe^{3+}(aq) + 3H_2O$$

Acid salts

You also learned to name acid salts in Chapter 4.

In the reaction of NaOH with HCl, there is only one possible salt that can be formed. When a base reacts with a polyprotic acid, however, two or more salts may be possible, corresponding to different extents of neutralization of the acid. Consider, for example, the reaction of NaOH with H_2SO_4. If 2 mol of NaOH are added to a solution that contains 1 mol of H_2SO_4, complete neutralization of the acid takes place and the salt Na_2SO_4 is formed.

$$2NaOH(aq) + H_2SO_4(aq) \longrightarrow Na_2SO_4(aq) + 2H_2O$$

On the other hand, if only 1 mol of NaOH is added to 1 mol of H_2SO_4, the products are 1 mol of water and 1 mol of $NaHSO_4$.

$$NaOH(aq) + H_2SO_4(aq) \longrightarrow NaHSO_4(aq) + H_2O$$

Similar reactions can occur with triprotic acids such as H_3PO_4, and salts such as the following can be obtained:

Na_3PO_4	trisodium phosphate (sodium phosphate)
Na_2HPO_4	disodium hydrogen phosphate
NaH_2PO_4	sodium dihydrogen phosphate

Recall that salts that are the result of the partial neutralization of a polyprotic acid are called **acid salts.** Acid salts such as $NaHSO_4$ are acidic and can undergo further neutralization. Thus, $NaHSO_4$ is able to react with NaOH.

$$NaHSO_4(aq) + NaOH(aq) \longrightarrow Na_2SO_4(aq) + H_2O$$

5.5 WHY METATHESIS REACTIONS OCCUR

Recall that a metathesis reaction (also called a double replacement reaction) involves two compounds in solution and an exchange of cations between the two anions. An example is the reaction between silver nitrate and sodium chloride, for which we can write the following molecular, ionic, and net ionic equations:

Molecular equation

$$AgNO_3(aq) + NaCl(aq) \longrightarrow AgCl(s) + NaNO_3(aq)$$

Ionic equation

$$Ag^+(aq) + NO_3^-(aq) + Na^+(aq) + Cl^-(aq) \longrightarrow AgCl(s) + Na^+(aq) + NO_3^-(aq)$$

Net ionic equation

$$Ag^+(aq) + Cl^-(aq) \longrightarrow AgCl(s)$$

The reason this reaction occurs is because a precipitate of silver chloride is formed. In a sense, it is the formation of the precipitate that is the driving force for the reaction, and we can tell that a net reaction takes place because when we drop the spectator ions from the ionic equation, a net ionic equation exists. We can contrast this to the situation that exists when solutions of KCl and $NaNO_3$ are mixed. First, let's attempt to write a molecular equation for a metathesis reaction. We do this by exchanging cations between the chloride ion and the nitrate ion.

The spectator ions are Na^+ and NO_3^-.

$$KCl + NaNO_3 \longrightarrow KNO_3 + NaCl$$

Both of the reactants in this equation are soluble in water, as are both of the products. Let's make a habit of indicating this by writing (*aq*) after the formulas of the soluble substances.

$$KCl(aq) + NaNO_3(aq) \longrightarrow KNO_3(aq) + NaCl(aq)$$

To write metathesis equations correctly, you must be able to recognize polyatomic ions and you must know their electrical charges. If necessary, review Table 4.4 on page 99.

Now, let's construct the ionic equation. To do this, we write all soluble strong electrolytes in dissociated form. Because all the compounds in the equation are salts, we know they are all strong electrolytes, so the ionic equation is

$$K^+(aq) + Cl^-(aq) + Na^+(aq) + NO_3^-(aq) \longrightarrow K^+(aq) + NO_3^-(aq) + Na^+(aq) + Cl^-(aq)$$

When we compare the left and right sides of this equation, we see that they are the same except for the sequence in which we have written the ions. If we cross out the spectator ions, there is nothing left, which tells us that *there is no net chemical change in this mixture.* Therefore, we would say that KCl does not react with $NaNO_3$ when their solutions are mixed. All we obtain is a mixture of the ions.

The formation of a precipitate is just one of *three* general driving forces in metathesis reactions. A net ionic equation also will exist if one of the products of the reaction is a weak electrolyte, or if one of the products is a gas. Let's examine all three of these in a bit more detail.

If a net ionic equation exists after dropping spectator ions, you know that an ionic reaction will occur between the reactants.

Precipitation reactions

In the examples given above, we see that the formation of a precipitate in the reaction between $AgNO_3$ and NaCl prevents us from canceling all the ions as spectator ions. As a result, a net ionic equation exists. On the other hand, if all the reactants and products are soluble, all the ions cancel and there is no reaction. In principle, therefore, if we knew the solubilities of all the compounds that could be formed between pairs of cations and anions, we could *predict,* based on the formation of a precipitate, when chemical reactions would occur.

Predicting metathesis reactions based on solubilities is not as simple as it may seem at first glance. This is because there is no sharp distinction between soluble and insoluble compounds. Certainly, substances such as sodium chloride would be considered soluble, and substances such as silver chloride insoluble. But between these extremes there are compounds like $PbCl_2$ and $AgC_2H_3O_2$, which are of intermediate solubility, and therefore are referred to as *partially soluble*, or *slightly soluble*.

Whether or not a precipitate of any particular salt will form when solutions of reactants are mixed depends on the concentrations of the ions that make up the salt. If these ion concentrations are large enough to make the reaction mixture supersaturated with respect to the salt, then a precipitate will form. If the reacture mixture isn't supersaturated, then no precipitate will form. Generally, a salt is said to be insoluble if a precipitate will form even if the concentrations of its ions are very small. However, if the potential product of the reaction is of moderate solubility, the concentrations of the ions have to be large to achieve supersaturation of the salt, so a reaction will only be observed if the solutions of the reactants are fairly concentrated.

This entire discussion has been on a qualitative level. A quantitative treatment of solubilities of ionic solids is reserved until Chapter 17. For now, we will use the following **solubility rules** as a rough guide in predicting the course of metathesis reactions. In constructing these rules, we have divided all salts into only two categories: soluble and insoluble. In doing so, we've included the slightly soluble substances among the insoluble compounds, the assumption being that our predictions will apply to reaction mixtures in which the ion concentrations are approximately 0.01 *M* or larger.

> If the concentrations of the ions are 0.01 *M* or larger, even slightly soluble salts will precipitate and a net reaction will be observed.

Solubility rules

SALTS THAT ARE SOLUBLE

1. All salts of the alkali metals are soluble.
2. All salts of the ammonium ion, NH_4^+, are soluble.
3. All salts of the following anions are soluble: nitrate ion (NO_3^-), chlorate ion (ClO_3^-), perchlorate ion (ClO_4^-), and acetate ion ($C_2H_3O_2^-$).

SALTS THAT ARE GENERALLY SOLUBLE, WITH SOME EXCEPTIONS

4. All chlorides, bromides, and iodides are soluble except those of Ag^+, Pb^{2+}, and Hg_2^{2+}. (Note that mercury in the +1 oxidation state exists as the diatomic ion Hg_2^{2+}.)
5. All sulfates (SO_4^{2-}) are soluble except those of Ca^{2+}, Sr^{2+}, Ba^{2+}, and Pb^{2+}.

SALTS THAT ARE GENERALLY INSOLUBLE, WITH SOME EXCEPTIONS

6. All metal oxides are insoluble, except for those of the alkali metals, Ca^{2+}, Sr^{2+}, and Ba^{2+}. Recall that soluble metal oxides are basic anhydrides and react with water to give hydroxide ion.

$$O^{2-} + H_2O \longrightarrow 2OH^-$$

Therefore, soluble metal oxides react with water to give their hydroxides in solution. For example,

$$CaO(s) + H_2O \longrightarrow Ca^{2+}(aq) + 2OH^-(aq)$$

> A solution of a soluble metal oxide is really a solution of that metal's hydroxide. For example, a solution of Na_2O is really a solution of NaOH because of the reaction of O^{2-} with water.

7. All hydroxides are insoluble, except for those of the alkali metals, Ca^{2+}, Sr^{2+}, and Ba^{2+}.
8. All carbonates (CO_3^{2-}), phosphates (PO_4^{3-}), sulfides (S^{2-}), and sulfites (SO_3^{2-}) are insoluble, except for those of NH_4^+ and the alkali metals. (Of course, since you know that all ammonium salts and all salts of the alkali metals are soluble, you already know these exceptions.)

EXAMPLE 5.4. USING THE SOLUBILITY RULES

PROBLEM: Are the following salts soluble or insoluble, based on the solubility rules given above: (a) Na_2S, (b) $AgBr$, (c) $Cu(C_2H_3O_2)_2$?

SOLUTION: (a) Na_2S contains the ions Na^+ and S^{2-}. The cation is that of an alkali metal, so we need go no further. All salts of the alkali metals are soluble. Therefore, we conclude that Na_2S is soluble.

(b) $AgBr$ contains the ions Ag^+ and Br^-. This time we recognize the anion as one listed among the "soluble with exceptions." All salts of Br^- are soluble *except* those of Ag^+, Pb^{2+}, and Hg_2^{2+}. Since the cation is among the exceptions, this salt is insoluble.

(c) $Cu(C_2H_3O_2)_2$ contains the ions Cu^{2+} and $C_2H_3O_2^-$. Salts of the acetate ion, $C_2H_3O_2^-$, are included in the soluble category. Therefore, we know that $Cu(C_2H_3O_2)_2$ is soluble.

You should practice applying the solubility rules by working the exercises at the end of the chapter. Once you know the rules, you will be ready to tackle problems like those illustrated in the next two examples.

EXAMPLE 5.5. APPLYING THE SOLUBILITY RULES TO PREDICTING METATHESIS REACTIONS

PROBLEM: Will a chemical reaction occur if solutions of $FeCl_3$ and KOH are mixed? If so, what insoluble product or products are formed?

ANALYSIS: You've seen that a net ionic equation exists when a metathesis reaction occurs, and that none exists when there is no reaction. Therefore, to answer the question, we must attempt to write a net ionic equation. We begin with a molecular equation, change it to an ionic equation, and then drop spectator ions. If anything is left, a chemical reaction takes place, and we can easily identify the products.

SOLUTION: We first must write an appropriate molecular equation starting with the two reactants given in the question. To construct this equation properly, we must be very careful to write correct formulas for the products. This requires that we determine which ions are in each reactant and then assemble the correct formulas of the products by exchanging anions.

$FeCl_3$ is composed of Fe^{3+} and Cl^- ions, and KOH is composed of K^+ and OH^- ions. When we exchange anions between the cations, the correct formulas for the products are $Fe(OH)_3$ and KCl. Now we can construct the balanced molecular equation.

$$FeCl_3 + 3KOH \longrightarrow 3KCl + Fe(OH)_3$$

On the basis of the solubility rules we conclude that KCl is soluble, but $Fe(OH)_3$ is not. Therefore, we expect a precipitate of $Fe(OH)_3$ to form in the reaction mixture while the KCl remains in solution in dissociated form. The ionic equation for the reaction is

$$Fe^{3+}(aq) + 3Cl^-(aq) + 3K^+(aq) + 3OH^-(aq) \longrightarrow Fe(OH)_3(s) + 3K^+(aq) + 3Cl^-(aq)$$

Dropping spectator ions (K^+ and Cl^-) gives the net ionic equation.

$$Fe^{3+}(aq) + 3OH^-(aq) \longrightarrow Fe(OH)_3(s)$$

Since there is a net ionic equation, a reaction does occur and the equation tells us that one insoluble product, $Fe(OH)_3$, is formed.

EXAMPLE 5.6. APPLYING THE SOLUBILITY RULES IN PREDICTING METATHESIS REACTIONS

PROBLEM: Is a chemical reaction expected to occur if solutions of NH_4NO_3 and $Pb(C_2H_3O_2)_2$ are mixed?

SOLUTION: To answer the question, we must again determine whether there is a net ionic equation. First, we begin with a molecular equation, exchanging NO_3^- for $C_2H_3O_2^-$, and vice versa when we write the products.

$$2NH_4NO_3 + Pb(C_2H_3O_2)_2 \longrightarrow Pb(NO_3)_2 + 2NH_4C_2H_3O_2$$

Solubility rules 2 and 3 tell us that all the reactants *and* the products are soluble. The ionic equation is therefore

$$2NH_4^+(aq) + 2NO_3^-(aq) + Pb^{2+}(aq) + 2C_2H_3O_2^- \longrightarrow$$
$$Pb^{2+}(aq) + 2NO_3^-(aq) + 2NH_4^+(aq) + 2C_2H_3O_2^-(aq)$$

When we cancel spectator ions, there is nothing left. There is no net ionic equation, so we can conclude that there is no metathesis reaction between these two reactants.

Reactions in which a weak electrolyte is formed

In any solution of a weak electrolyte, it is observed that only a small fraction of the solute is dissociated (ionized). Most of it exists in the form of molecules, rather than as ions. In a 1 M solution of acetic acid, for example, only about 0.42% of the acid exists as H^+ and $C_2H_3O_2^-$ ions, which means that the other 99.58% is present in the solution as molecules of $HC_2H_3O_2$. The reason for this, as you may recall, is that there is little tendency for the molecules of $HC_2H_3O_2$ to react with the solvent to form ions, but there is a strong tendency for the ions to react with each other to form molecules. As a result, the rates of the opposing reactions in the equilibrium

$$HC_2H_3O_2(aq) \rightleftharpoons H^+(aq) + C_2H_3O_2^-(aq)$$

can only become equal if most of the $C_2H_3O_2$ units are present in molecules.

If H^+ and $C_2H_3O_2^-$ ions are brought together in large numbers in a solution, an unstable situation exists. There is essentially no $HC_2H_3O_2$, so the ions will be combining at a much faster rate than they are being produced. Therefore, in just a brief instant nearly all the ions will disappear as they are replaced in the solution by molecules of the acid.

This is precisely what happens when solutions of the strong electrolytes HCl and $NaC_2H_3O_2$ are mixed. The ionic equation for the reaction is

$$H^+(aq) + Cl^-(aq) + Na^+(aq) + C_2H_3O_2^-(aq) \longrightarrow Na^+(aq) + Cl^-(aq) + HC_2H_3O_2(aq)$$

Notice that we've written the formula for acetic acid in nonionized (molecular) form. This is because most of the $HC_2H_3O_2$ exists this way. If we now cancel spectator ions, the net ionic equation is

$$H^+(aq) + C_2H_3O_2^-(aq) \longrightarrow HC_2H_3O_2(aq)$$

The driving force in this reaction — the reason a reaction occurs — is the decrease in the number of ions that takes place when the two fully dissociated reactants form the partially dissociated product.

One of the most important classes of reactions in aqueous solution that takes place by the formation of a weak electrolyte is acid–base neutralization. This reaction, you recall, involves the formation of water, a *very weak* electrolyte. If solutions of HCl and NaOH are mixed, the net ionic equation for the neutralization is

$$H^+(aq) + OH^-(aq) \longrightarrow H_2O$$

Water is such a weak electrolyte that its formation can cause insoluble oxides to dissolve in acids and weak acids to react with bases. For example, you learned earlier that iron(III) oxide dissolves in a strong acid such as HCl as follows:

Molecular equation

$$Fe_2O_3(s) + 6HCl(aq) \longrightarrow 2FeCl_3(aq) + 3H_2O$$

Ionic equation

$$Fe_2O_3(s) + 6H^+(aq) + 6Cl^-(aq) \longrightarrow 2Fe^{3+}(aq) + 6Cl^-(aq) + 3H_2O$$

Net ionic equation

$$Fe_2O_3(s) + 6H^+(aq) \longrightarrow 2Fe^{3+}(aq) + 3H_2O$$

The reaction of acetic acid with sodium hydroxide follows these equations.

Molecular equation

$$HC_2H_3O_2(aq) + NaOH(aq) \longrightarrow NaC_2H_3O_2(aq) + H_2O$$

Ionic equation

$$HC_2H_3O_2(aq) + Na^+(aq) + OH^-(aq) \longrightarrow Na^+(aq) + C_2H_3O_2^-(aq) + H_2O$$

Net ionic equation

$$HC_2H_3O_2(aq) + OH^-(aq) \longrightarrow C_2H_3O_2^-(aq) + H_2O$$

EXAMPLE 5.7. WRITING EQUATIONS FOR REACTIONS THAT INVOLVE THE FORMATION OF A WEAK ELECTROLYTE

PROBLEM: Copper(II) oxide dissolves slowly in acetic acid to give a solution of copper(II) acetate. Write molecular, ionic, and net ionic equations for the reaction.

SOLUTION: The reactants are CuO and $HC_2H_3O_2$. We also know that an acid reacts with a metal oxide to give a salt and water. Therefore, the molecular equation is

$$CuO + 2HC_2H_3O_2 \longrightarrow Cu(C_2H_3O_2)_2 + H_2O$$

Now let's work on the ionic equation. To do this, we have to write all soluble strong electrolytes in dissociated form, and all weak electrolytes and insoluble compounds in molecular form. From the solubility rules, we know that CuO is insoluble in water (rule 6), and that $Cu(C_2H_3O_2)_2$ is soluble (rule 3). The ionic equation therefore becomes

$$CuO(s) + 2HC_2H_3O_2(aq) \longrightarrow Cu^{2+}(aq) + 2C_2H_3O_2^-(aq) + H_2O$$

This is also the net ionic equation because there are no spectator ions to cancel.

Reactions in which a gas is formed

In some cases, the molecular species formed in a metathesis reaction is either insoluble in water and bubbles out as a gas, or it decomposes into a substance that evolves as a gas. For example, when HCl is added to a solution of Na_2S, one of the products is the weak electrolyte H_2S. However, H_2S is a gas that has a low solubility in water, so it bubbles out of the reaction mixture. The molecular equation for the reaction is

$$2HCl(aq) + Na_2S(aq) \longrightarrow H_2S(g) + 2NaCl(aq)$$

Figure 5.8. *Carbon dioxide is evolved when aqueous HCl is added to a solution of Na₂CO₃.*

for which the net ionic equation is

$$2H^+(aq) + S^{2-}(aq) \longrightarrow H_2S(g)$$

Another example is the reaction between HCl and Na_2CO_3, which produces the weak electrolyte H_2CO_3.

$$2HCl(aq) + Na_2CO_3(aq) \longrightarrow H_2CO_3(aq) + 2NaCl(aq)$$

The ionic equation for the reaction is

$$2H^+(aq) + 2Cl^-(aq) + 2Na^+(aq) + CO_3^{2-}(aq) \longrightarrow H_2CO_3(aq) + 2Na^+(aq) + 2Cl^-(aq)$$

and the net ionic equation is

$$2H^+(aq) + CO_3^{2-}(aq) \longrightarrow H_2CO_3(aq)$$

Carbonic acid, H_2CO_3, is unstable in large concentrations and decomposes readily to give CO_2 and H_2O. Carbon dioxide is not very soluble in water, and it escapes as a gas as shown in Figure 5.8. The equation for the decomposition of H_2CO_3 is

$$H_2CO_3(aq) \longrightarrow H_2O + CO_2(g)$$

Therefore, the net overall ionic equation for the reaction between hydrogen ion and carbonate ion is

$$2H^+(aq) + CO_3^{2-}(aq) \longrightarrow H_2O + CO_2(g)$$

A similar reaction occurs when bicarbonates are treated with acids.

$$H^+(aq) + HCO_3^-(aq) \longrightarrow H_2CO_3(aq) \longrightarrow H_2O + CO_2(g)$$

Both soluble and insoluble carbonates react with acids to liberate carbon dioxide.

The formation of CO_2 in the reaction of acids with carbonates is a sufficiently strong driving force for reaction to cause insoluble carbonates to react. For example, insoluble calcium carbonate, found in limestone, reacts with acids such as HCl to release CO_2.

$$CaCO_3(s) + 2HCl(aq) \longrightarrow CaCl_2(aq) + CO_2(g) + H_2O$$

The ionic and net ionic equations are

$$CaCO_3(s) + 2H^+(aq) + 2Cl^-(aq) \longrightarrow Ca^{2+}(aq) + 2Cl^-(aq) + CO_2(g) + H_2O$$

and

$$CaCO_3(s) + 2H^+(aq) \longrightarrow Ca^{2+}(aq) + CO_2(g) + H_2O$$

The reactions of acids with carbonates and bicarbonates are quite common. For instance, some people take sodium bicarbonate (baking soda) to soothe an upset stomach. The bicarbonate ion reacts with stomach acid (HCl) to give carbon dioxide, which makes the person burp. Another example is the product Alka Seltzer, which contains (among other things) sodium bicarbonate and a weak acid, citric acid. When the tablet is placed in water and begins to dissolve, the citric acid reacts with the sodium bicarbonate to give the fizz.

A list of some other gases that are evolved in metathesis reactions is given in Table 5.3. Among these, ammonia deserves special mention. Ammonia has quite a large solubility in water, so if it is formed in only small amounts, very little actually leaves the solution. Its presence can still be detected, however, because enough ammonia escapes from the solution to give a perceptible odor.

Acid spills in the lab can be safely neutralized by sprinkling solid $NaHCO_3$ on them, because the hydrogen ion combines with the bicarbonate ion to form harmless CO_2 and H_2O.

When an Alka-Seltzer tablet dissolves in water, sodium bicarbonate reacts with citric acid to give bubbles of CO_2.

Table 5.3. Gases that are formed in metathesis reactions

Gas	Typical Reaction in Which It Is Produced
CO_2	$Na_2CO_3 + 2HCl \rightarrow H_2CO_3 + 2NaCl$ $H_2CO_3 \rightarrow H_2O + CO_2(g)$ *Net equation:* $CO_3^{2-} + 2H^+ \rightarrow CO_2(g) + H_2O$
SO_2	$Na_2SO_3 + 2HCl \rightarrow H_2SO_3 + 2NaCl$ $H_2SO_3 \rightarrow H_2O + SO_2(g)$ *Net equation:* $SO_3^{2-} + 2H^+ \rightarrow H_2O + SO_2(g)$
NH_3	$NH_4Cl + NaOH \rightarrow NH_3(g) + H_2O + NaCl$ *Net equation:* $NH_4^+ + OH^- \rightarrow NH_3(g) + H_2O$
H_2S	$Na_2S + 2HCl \rightarrow H_2S(g) + 2NaCl$ *Net equation:* $S^{2-} + 2H^+ \rightarrow H_2S(g)$
NO NO_2	$NaNO_2 + HCl \rightarrow HNO_2 + NaCl$ $2HNO_2 \rightarrow H_2O + NO_2(g) + NO(g)$ *Net equation:* $2NO_2^- + 2H^+ \rightarrow H_2O + NO_2(g) + NO(g)$

EXAMPLE 5.8. WRITING EQUATIONS FOR REACTIONS THAT INVOLVE THE FORMATION OF A GAS

PROBLEM: Write a net ionic equation for the reaction between NH_4NO_3 and $Ba(OH)_2$

SOLUTION: We begin as usual by writing a double replacement reaction.

$$2NH_4NO_3 + Ba(OH)_2 \longrightarrow 2NH_4OH + Ba(NO_3)_2$$

The molecular species NH_4OH (ammonium hydroxide) does not really exist; it is fictitious. It actually is $NH_3 + H_2O$ (note that there are one N, five H, and one O in both NH_4OH and $NH_3 + H_2O$). We should therefore rewrite the equation as

$$2NH_4NO_3 + Ba(OH)_2 \longrightarrow 2NH_3 + 2H_2O + Ba(NO_3)_2$$

The ionic equation is obtained by applying the solubility rules and also noting that NH_3 is a gas (Table 5.3).

$$2NH_4^+(aq) + 2NO_3^-(aq) + Ba^{2+}(aq) + 2OH^-(aq) \longrightarrow 2NH_3(g) + 2H_2O + Ba^{2+}(aq) + 2NO_3^-(aq)$$

Finally, canceling spectator ions gives

$$2NH_4^+(aq) + 2OH^-(aq) \longrightarrow 2NH_3(g) + 2H_2O$$

which becomes

$$NH_4^+(aq) + OH^-(aq) \longrightarrow NH_3(g) + H_2O$$

when the coefficients are reduced to the simplest set of whole numbers.

In this section we have developed ways for you to predict the outcome of a large number of ionic reactions. *To apply these methods, you must know the solubility rules on page 132, you must be able to recognize weak electrolytes as taught in Sections 5.2 and 5.3, and you must know the contents of Table 5.3.* Learning these things will require a lot of work and practice on your part, but it is worthwhile because the knowledge gained is a valuable part of what you should take with you from this course.

5.6 REDOX REACTIONS IN SOLUTION; BALANCING EQUATIONS BY THE ION–ELECTRON METHOD

Metathesis reactions are not the only kind that occur between ions in aqueous solutions. Many redox reactions also involve ionic reactants, products, or both. As you will see, redox reactions are often more complex than metathesis reactions and their equations are difficult to balance by inspection. One of the goals of this section, therefore, is to teach you a method for writing and balancing net ionic equations for such reactions. We will also look very briefly at some common laboratory oxidizing and reducing agents.

Although our focus here will be on laboratory chemicals, it is important to realize that ionic oxidizing agents and reducing agents find many practical uses. An example is liquid laundry bleach, such as Clorox, which is a dilute solution of sodium hypochlorite, NaOCl. In solution, this compound dissociates to give Na^+ and OCl^- ions. The hypochlorite ion (OCl^-) is a powerful oxidizing agent that is able to oxidize many colored compounds to give colorless products. The hypochlorite ion is also the "active ingredient" in many products used to fight mildew and in many of the chemicals added to swimming pools to "chlorinate" the water. The OCl^- kills molds and other microorganisms (germs) by oxidizing them. Batteries of all sorts also employ redox reactions to generate electricity, and the reactions that take place within them occur in an aqueous environment that contains ionic reactants. The reactions that happen in batteries will be discussed in Chapter 18.

The ion–electron method

You've seen that net ionic equations are useful for focusing our attention on the key changes that occur in metathesis reactions. As you might imagine, they are also helpful in analyzing redox reactions between ions. A procedure that is particularly well suited for balancing net ionic equations for redox reactions is called the **ion–electron** method. It is based on the principle of divide and conquer: The equation is split into two simpler parts, called **half-reactions,** which are balanced separately and then recombined to give the final balanced net ionic equation.

As an example, consider the reaction between solutions of $SnCl_2$ and $HgCl_2$, which gives insoluble Hg_2Cl_2 as one product and Sn^{4+} in solution as the other. In applying the ion–electron method, we begin by writing a *skeleton equation* that shows only those substances that are actually involved in the reaction; we leave out any spectator ions. For the reaction at hand, the reactants are Sn^{2+}, Hg^{2+}, and Cl^-. The products are Hg_2Cl_2 and Sn^{4+}. The skeleton equation is therefore

$$Sn^{2+} + Hg^{2+} + Cl^- \longrightarrow Sn^{4+} + Hg_2Cl_2$$

Next, we divide the equation into two half-reactions. These are

$$Sn^{2+} \longrightarrow Sn^{4+}$$

$$Hg^{2+} + Cl^- \longrightarrow Hg_2Cl_2$$

Dividing the skeleton equation into the half-reactions is the most difficult part of using the ion–electron method, so let's review how it is accomplished. Notice that in the skeleton equation there are *two* products. We choose one of them as a product in one of the half-reactions, and the other as the product in the second half-reaction. Now look at the first half-reaction. Since there is tin on the right, we almost must have tin on the left — the same elements must appear on opposite sides of the arrow; otherwise, it would be impossible to balance the half-reaction. For the second half-reaction, the product contains both mercury and chlorine, so we require both mercury (Hg^{2+}) and chlorine (Cl^-) on the left.

The next step is to balance the half-reactions, and there are two requirements to be met here. *In order for any equation involving ions to be balanced,*

1. There must be the same number of atoms of each kind on both sides of the equation.
2. The net charge must be the same on both sides of the equation.

The first requirement is met by placing coefficients into the half-reactions to balance the atoms. The first equation already contains one tin on each side, so nothing needs be done to it. The second equation is balanced according to atoms by placing coefficients of 2 in front of both Hg^{2+} and Cl^-. This gives

$$Sn^{2+} \longrightarrow Sn^{4+}$$

$$2Hg^{2+} + 2Cl^- \longrightarrow Hg_2Cl_2$$

The second requirement is met by balancing the half-reactions according to charge. This is accomplished by adding as many electrons as necessary to the more positive side (or sometimes the less negative side). For the first half-reaction, we add two electrons to the right to give a net charge of $2+$ on both sides.

$$Sn^{2+} \longrightarrow Sn^{4+} + 2e^-$$

Charge: $2+$ $(4+)+(2-)=2+$
                              (net)

For the second half-reaction, we must add two electrons to the left to give a net charge of zero on both sides.

$$2e^- + 2Hg^{2+} + 2Cl^- \longrightarrow Hg_2Cl_2$$

Charge $\underbrace{(2-)+(4+)+(2-)}_{Net = 0}$ 0

This gives the two balanced half-reactions

$$Sn^{2+} \longrightarrow Sn^{4+} + 2e^-$$

$$2e^- + 2Hg^{2+} + 2Cl^- \longrightarrow Hg_2Cl_2$$

In recombining the half-reactions, we make use of one of the basic facts about redox reactions: *The total number of electrons gained must **always** be the same as the total number lost.* In the two half-reactions we have here, this condition is met (we will see later what to do if the condition isn't met), so we can just add the two half-reactions together as follows:

$$Sn^{2+} \longrightarrow Sn^{4+} + 2e^-$$
$$\underline{2e^- + 2Hg^{2+} + 2Cl^- \longrightarrow Hg_2Cl_2}$$
$$2e^- + Sn^{2+} + 2Hg^{2+} + 2Cl^- \longrightarrow Sn^{4+} + Hg_2Cl_2 + 2e^-$$

Finally, we cancel anything that's the same on both sides of the equation. In this and

Removing the electrons from the overall equation is just like removing spectator ions from an ionic equation, which you learned to do earlier.

any other equation that you balance by this method, the electrons *must* cancel; if they don't, you've made a mistake. The finished equation is

$$Sn^{2+} + 2Hg^{2+} + 2Cl^- \longrightarrow Sn^{4+} + Hg_2Cl_2$$

Notice that both the atoms and the charge are in balance.

Reactions that involve H$^+$ or OH$^-$

In many redox reactions in aqueous solution, H$^+$ or OH$^-$ are either consumed or produced. These reactions also usually involve water as either a product or reactant. For example, if concentrated hydrochloric acid is poured over potassium permanganate ($KMnO_4$), the chloride ion from the acid is oxidized to elemental chlorine by the permanganate ion, which is reduced to give Mn^{2+} in the solution. During this reaction, hydrogen ion is consumed as it combines with the oxygen atoms of the permanganate to give molecules of water. During the reaction, therefore, the amount of hydrogen ion in the solution decreases along with the amounts of chloride and permanganate ions.

Not only are H$^+$ and OH$^-$ reactants or products in many reactions, but their presence or absence can also affect the other products of the reactions. For example, if permanganate ion is used as an oxidizing agent in an acidic solution, the reduction product is generally Mn^{2+}. But if the solution is basic, the permanganate ion is reduced to insoluble MnO_2. Therefore, when we carry out a redox reaction, it is important to know something about the acidity or basicity of the solution.

Of course, the hydrogen ion is more accurately represented as H_3O^+, but for simplicity we use the hydronium ion's abbreviation, H$^+$.

When we use the ion–electron method for balancing equations, it is not necessary to know whether H$^+$ or OH$^-$ is a reactant or a product, or whether water is consumed or produced. All we need to know is whether the reaction is taking place in an acidic or a basic solution. The act of balancing the equation then tells us which of these species, if any, are involved.

Reactions that occur in an acidic solution

In any acidic aqueous solution, two of the major species are H_2O and H$^+$. These can be used in the ion–electron method to help us balance hydrogen and oxygen atoms in half-reactions. The general approach is essentially the same as that used on balancing the equation for the reaction between Sn^{2+}, Hg^{2+}, and Cl^-. It is very systematic and can be broken down into the following steps:

If you follow these rules, you will *always* obtain a properly balanced equation.

STEPS IN THE ION–ELECTRON METHOD FOR ACIDIC SOLUTIONS

Step 1. Divide the skeleton equation into half-reactions.

Step 2. Balance atoms other than oxygen and hydrogen.

Step 3. Next, balance the oxygens in each half-reaction by adding water molecules to the side that needs oxygen atoms. Add one H_2O for each oxygen needed.

It is important to follow these steps in the sequence given here.

Step 4. Balance the hydrogen atoms in each half-reaction by adding H$^+$ to the side that needs hydrogen. Add one H$^+$ for each hydrogen needed.

Step 5. Balance the charge in each half-reaction by adding electrons to the appropriate side.

Step 6. Multiply each half-reaction by appropriate factors to make the number of electrons gained equal the number lost.

Step 7. Add the two half-reactions together.

Step 8. Cancel anything that is the same on both sides of the equation.

(a)

(b)

(a) Dark purple crystals of KMnO₄ in the flask on the right will be treated with concentrated hydrochloric acid. (b) The oxidation of Cl⁻ by MnO₄⁻ in the strongly acidic solution yields yellow-green chlorine gas.

To see how these rules apply, let's balance the equation for the reaction of HCl with KMnO₄, which we mentioned above. In this reaction Cl⁻ is oxidized to Cl₂ and MnO₄⁻ is reduced to Mn²⁺. The skeleton equation, which does not contain either H⁺ or H₂O, is

$$Cl^- + MnO_4^- \longrightarrow Cl_2 + Mn^{2+} \quad \text{(acidic solution)}$$

You will always be told whether the solution is acidic or basic.

Step 1. *Divide the equation into half-reactions.*

$$Cl^- \longrightarrow Cl_2$$
$$MnO_4^- \longrightarrow Mn^{2+}$$

Except for hydrogen and oxygen, each side of a given half-reaction must have the same elements, otherwise, it can't be balanced.

Step 2. *Balance atoms other than H and O.* In the first half-reaction we place a coefficient of 2 in front of Cl⁻. We don't have to do anything to the second half-reaction.

$$2Cl^- \longrightarrow Cl_2$$
$$MnO_4^- \longrightarrow Mn^{2+}$$

Step 3. *Balance oxygens by adding H₂O to that side that needs O.* In the second half-reaction, there are 4 oxygens on the left and none on the right. We therefore have to add 4H₂O to the right side. Then the oxygens will balance.

$$2Cl^- \longrightarrow Cl_2$$
$$MnO_4^- \longrightarrow Mn^{2+} + 4H_2O$$

Step 4. *Balance hydrogens by adding H⁺ to the side that needs H.* The right side of the second half-reaction has a total of 8 hydrogens; there is none on the left. Therefore, we add 8H⁺ to the left side. (Be sure you remember to write the + sign on the hydrogen ion.)

$$2Cl^- \longrightarrow Cl_2$$
$$8H^+ + MnO_4^- \longrightarrow Mn^{2+} + 4H_2O$$

Note that all the atoms now balance.

Step 5. *Balance the charge by adding electrons.* In this step we make the net charge the same on both sides by following the same procedure as in the reaction between Sn²⁺, Hg²⁺, and Cl⁻. In the first half-reaction we have to add 2 electrons to the right side. In the second half-reaction we must add 5 electrons to the left side. (Before we add electrons, the net charge on the left is 7+ and the net charge on the right is 2+. To make them the same, we add 5e⁻ to the left.)

$$2Cl^- \longrightarrow Cl_2 + 2e^-$$
$$5e^- + 8H^+ + MnO_4^- \longrightarrow Mn^{2+} + 4H_2O$$

Step 6. *Make the number of electrons gained equal the number lost.* This can be accomplished by multiplying the first half-reaction through by 5 and the second through by 2. There will then be $10e^-$ lost and gained.

$5 \times (2e^-) = 10e^-$
$2 \times (5e^-) = 10e^-$

$$5(2Cl^- \longrightarrow Cl_2 + 2e^-)$$

$$2(5e^- + 8H^+ + MnO_4^- \longrightarrow Mn^{2+} + 4H_2O)$$

Step 7. *Add the two half-reactions.* When we do this, we will not bother to bring down the electrons, because we know they will cancel. They have to — we went to the trouble to make them the same in each half-reaction.

$$5(2Cl^- \longrightarrow Cl_2 + 2e^-)$$
$$\underline{2(5e^- + 8H^+ + MnO_4^- \longrightarrow Mn^{2+} + 4H_2O)}$$
$$10Cl^- + 16H^+ + 2MnO_4^- \longrightarrow 5Cl_2 + 2Mn^{2+} + 8H_2O$$

Step 8. *Cancel anything that is the same on both sides.* In this case there is nothing to be canceled, so we are finished. The final balanced equation is

$$10Cl^- + 16H^+ + 2MnO_4^- \longrightarrow 5Cl_2 + 2Mn^{2+} + 8H_2O$$

The following is another example.

EXAMPLE 5.9. BALANCING AN EQUATION FOR A REACTION THAT TAKES PLACE IN AN ACIDIC SOLUTION

PROBLEM: Balance the following equation by the ion–electron method.

$$Cr_2O_7^{2-} + H_2S \longrightarrow Cr^{3+} + S \quad \text{(acidic solution)}$$

SOLUTION: We simply follow the steps outlined above.

Step 1. Divide the equation into half-reactions.

$$Cr_2O_7^{2-} \longrightarrow Cr^{3+}$$

$$H_2S \longrightarrow S$$

Step 2. Balance the Cr in the first equation.

$$Cr_2O_7^{2-} \longrightarrow 2Cr^{3+}$$

$$H_2S \longrightarrow S$$

Step 3. Balance oxygens with H_2O.

$$Cr_2O_7^{2-} \longrightarrow 2Cr^{3+} + 7H_2O$$

$$H_2S \longrightarrow S$$

Step 4. Balance hydrogens with H^+.

$$14H^+ + Cr_2O_7^{2-} \longrightarrow 2Cr^{3+} + 7H_2O$$

$$H_2S \longrightarrow S + 2H^+$$

Step 5. Balance the charge with electrons.

$$6e^- + 14H^+ + Cr_2O_7^{2-} \longrightarrow 2Cr^{3+} + 7H_2O$$

$$H_2S \longrightarrow S + 2H^+ + 2e^-$$

Steps 6 and 7. Multiply the second equation through by 3 to make the electrons gained equal to the electrons lost. Then add the half-reactions together.

$$6e^- + 14H^+ + Cr_2O_7{}^{2-} \longrightarrow 2Cr^{3+} + 7H_2O$$
$$\underline{3 \times (H_2S \longrightarrow S + 2H^+ + 2e^-)}$$
$$14H^+ + Cr_2O_7{}^{2-} + 3H_2S \longrightarrow 2Cr^{3+} + 7H_2O + 3S + 6H^+$$

Step 8. Cancel anything that's the same on both sides. Six H^+ can be canceled (removed) from each side, which leaves $8H^+$ on the left. The final equation is therefore

$$8H^+ + Cr_2O_7{}^{2-} + 3H_2S \longrightarrow 2Cr^{3+} + 7H_2O + 3S$$

Redox reactions in basic solutions

We have just seen that H_2O and H^+ are used to balance half-reactions that occur in acidic solution. In a basic solution the dominant species are H_2O and OH^-, so these are the species that should be used to achieve material balance. Although you can use H_2O and OH^- directly,[3] the simplest technique is to first balance the reaction as if it occurred in acidic solution and then perform the conversion described below to adjust it to conform to conditions in basic solution.

Suppose we wished to balance the following half-reaction taking place in a basic solution:

$$Pb \longrightarrow PbO$$

First we balance it as though it occurred in an acidic solution.

$$H_2O + Pb \longrightarrow PbO + 2H^+ + 2e^-$$

The conversion to basic solution follows these three steps.

Step 1. For each H^+ that must be eliminated from the equation, add an OH^- to both sides of the equation. In this example, we have to eliminate $2H^+$, so we add $2OH^-$ to each side.

$$H_2O + Pb + 2OH^- \longrightarrow PbO + 2H^+ + 2OH^- + 2e^-$$

Step 2. Combine H^+ and OH^- to form H_2O. We have $2H^+$ and $2OH^-$ on the right, which give $2H_2O$.

$$H_2O + Pb + 2OH^- \longrightarrow PbO + \underbrace{2H_2O}_{2H^+ + 2OH^-} + 2e^-$$

Step 3. Cancel any H_2O that are the same on both sides. We can cancel one H_2O from each side. The final balanced half-reaction in basic solution is

$$Pb + 2OH^- \longrightarrow PbO + H_2O + 2e^-$$

---

[3] To balance half-reactions in **basic solution,** the following rules can be used:
1. *To balance a hydrogen atom, we add* **one** H_2O *molecule to the side of the half-reaction deficient in hydrogen, and to the other side we add* **one** *hydroxide ion.*
2. *To balance one oxygen atom, we add* **two** *hydroxide ions to the side deficient in oxygen and* **one** H_2O *molecule to the other side.*

EXAMPLE 5.10. BALANCING AN EQUATION FOR BASIC SOLUTION

PROBLEM: Balance the following reaction for the oxidation of plumbite ion, $Pb(OH)_3^-$, to lead dioxide by hypochlorite ion in basic solution:

$$Pb(OH)_3^- + OCl^- \longrightarrow PbO_2 + Cl^-$$

SOLUTION: First, the equation is balanced as though it occurred in acidic solution. We begin by dividing it into half-reactions.

$$Pb(OH)_3^- \longrightarrow PbO_2$$

$$OCl^- \longrightarrow Cl^-$$

Now we balance them according to atoms.

$$Pb(OH)_3^- \longrightarrow PbO_2 + H_2O + H^+$$

$$2H^+ + OCl^- \longrightarrow Cl^- + H_2O$$

Then we balance the charge by adding electrons.

$$Pb(OH)_3^- \longrightarrow PbO_2 + H_2O + H^+ + 2e^-$$

$$2e^- + 2H^+ + OCl^- \longrightarrow Cl^- + H_2O$$

Since the number of electrons lost in the first half-reaction is already equal to the number gained in the second, we can add them.

$$2H^+ + OCl^- + Pb(OH)_3^- \longrightarrow Cl^- + 2H_2O + PbO_2 + H^+$$

Next, we cancel substances that are the same on both sides. The equation is now balanced for acidic solution.

$$H^+ + OCl^- + Pb(OH)_3^- \longrightarrow Cl^- + 2H_2O + PbO_2$$

Next, we perform the three-step conversion to basic solution. First we add to *each side* the same number of OH^- as there are H^+ in the equation.

$$\underbrace{H^+ + OH^-}_{\text{forms } H_2O} + OCl^- + Pb(OH)_3^- \longrightarrow Cl^- + 2H_2O + PbO_2 + OH^-$$

On the left, H^+ and OH^- become H_2O.

$$H_2O + OCl^- + Pb(OH)_3^- \longrightarrow Cl^- + 2H_2O + PbO_2 + OH^-$$

Then we cancel one H_2O from each side to get the final balanced equation.

$$OCl^- + Pb(OH)_3^- \longrightarrow Cl^- + H_2O + PbO_2 + OH^-$$

Some common oxidizing and reducing agents

In the lab, it is often necessary to cause an oxidation or a reduction of some chemical that's being used in an experiment. There are many oxidizing and reducing agents that could conceivably be used, but in making the choice, we would probably want to choose one that is easy to use. This tends to exclude certain substances most of the time. For example, chlorine (Cl_2) is quite a powerful oxidizing agent, but it can't be used in an open laboratory because it is a poisonous gas; special precautions must be taken to use chlorine safely. Therefore, in most applications we would avoid chlorine as an oxidizing agent of choice.

The three most common oxidizing agents in the laboratory are the permanga-

Figure 5.9. *A solution of purple KMnO$_4$ is poured into a stirred acidified solution that contains Fe^{2+} ion. The reduction product of the MnO$_4^-$ ion is the nearly colorless Mn^{2+} ion.*

nate ion, MnO_4^-, the chromate ion, CrO_4^{2-}, and the dichromate ion, $Cr_2O_7^{2-}$. These are all very powerful oxidizing agents and salts that contain them should be handled with care. They are quite effective in supplying oxygen and should not be allowed to contact organic material because of the potential for fire.

Permanganate Ion The permanganate ion itself is purple, and that is the color exhibited by solutions containing this ion (Figure 5.9). Generally, the permanganate ion is available as the purplish-black potassium salt, $KMnO_4$.

When permanganate ion functions as an oxidizing agent, the manganese is reduced from the $+7$ oxidation state. However, the oxidation state of Mn in the product depends on the acidity of the solution. If the reduction takes place in a strongly acidic solution, the manganese is reduced to the nearly colorless Mn^{2+} ion according to the half-reaction

$$8H^+(aq) + MnO_4^-(aq) + 5e^- \longrightarrow Mn^{2+}(aq) + 4H_2O$$

This can lead to a very dramatic color change if the other reactants and products in the redox reaction have little or no color themselves, as shown in Figure 5.9. Here we see a solution of $KMnO_4$ being poured into a solution that contains Fe^{2+} and a large concentration of sulfuric acid. The purple of the MnO_4^- ion disappears almost immediately as the solutions mix and the following reaction occurs:

$$5Fe^{2+}(aq) + 8H^+(aq) + MnO_4^-(aq) \longrightarrow Mn^{2+}(aq) + 4H_2O + 5Fe^{3+}(aq)$$

If the permanganate ion is reduced in a neutral or slightly basic solution, the product of the reduction is generally insoluble manganese dioxide, MnO_2. The half-reaction for this reduction is

$$2H_2O + MnO_4^-(aq) + 3e^- \longrightarrow MnO_2(s) + 4OH^-(aq)$$

Chromate Ion and Dichromate Ion The sodium and potassium salts of these ions are commonly found on laboratory shelves. Both chromate and dichromate ions contain chromium in the $+6$ oxidation state, and they can be converted from one to the other by adjusting the acidity of the solution. If a solution containing the yellow chromate ion is made acidic, the CrO_4^{2-} is converted to the red-orange $Cr_2O_7^{2-}$.

$$\underset{\text{chromate ion}}{2CrO_4^{2-}(aq) + 2H^+(aq)} \longrightarrow \underset{\text{dichromate ion}}{Cr_2O_7^{2-}(aq) + H_2O}$$

Solutions of Na$_2$CrO$_4$ (left) and Na$_2$Cr$_2$O$_7$ (right).

On the other hand, if the solution contains dichromate ion and is made basic, the $Cr_2O_7{}^{2-}$ is converted to $CrO_4{}^{2-}$.

$$Cr_2O_7{}^{2-}(aq) + 2OH^-(aq) \longrightarrow 2CrO_4{}^{2-}(aq) + H_2O$$

Because of these reactions, when used in an acidic solution, the active oxidizing agent is $Cr_2O_7{}^{2-}$. If the solution is basic, however, the oxidizing agent is $CrO_4{}^{2-}$.

Regardless of the acidity of the solution, when these ions act as oxidizing agents, the chromium is reduced to the $+3$ oxidation state. However, the formula of the product that contains the chromium does depend on how acidic or basic the solution is. In acidic solutions, the chromium is reduced to Cr^{3+} ion. In slightly basic solutions, the reduction product is insoluble $Cr(OH)_3$. And in a very basic solution, the chromate ion is reduced to the ion $CrO_2{}^-$, which is called the chromite ion.

Acidic solution

$$6e^- + 14H^+(aq) + Cr_2O_7{}^{2-}(aq) \longrightarrow 2Cr^{3+}(aq) + 7H_2O$$

Slightly basic solution

$$3e^- + 4H_2O + CrO_4{}^{2-}(aq) \longrightarrow Cr(OH)_3(s) + 5OH^-(aq)$$

Very basic solution

$$3e^- + 2H_2O + CrO_4{}^{2-}(aq) \longrightarrow CrO_2{}^-(aq) + 4OH^-(aq)$$

Laboratory reducing agents can be chosen from a wide variety of substances. If a strong reducing agent is needed, one of the more reactive metals can be used — magnesium or zinc, for example. When these metals react, they form positive ions and thereby supply electrons to other substances. For example, magnesium is oxidized as follows (and thereby serves as a reducing agent):

$$Mg \longrightarrow Mg^{2+} + 2e^-$$

There are drawbacks to using a metal such as this, however, because the reaction must take place at the surface of the metal. This makes it difficult to control the reaction. Usually, when redox reactions are carried out in solution, reducing agents are chosen that are soluble in water. In this way, they can react with the oxidizing agent in a homogeneous environment. Among the most common reducing agents are the sulfite ion, $SO_3{}^{2-}$, the bisulfite ion, $HSO_3{}^-$, and the thiosulfate ion, $S_2O_3{}^{2-}$.

Sulfites and Bisulfites Salts that contain the sulfite ion, $SO_3{}^{2-}$, or the bisulfite ion, $HSO_3{}^-$, are often used as convenient reducing agents. These anions come from the neutralization (complete or partial) of sulfurous acid, H_2SO_3.

When sulfite or bisulfite ions are oxidized, the product is sulfate ion. If the solution is basic, the reactant is $SO_3{}^{2-}$, regardless of whether the original solute contained sulfite or bisulfite ion. This is because the bisulfite ion is itself slightly acidic, and in the presence of base it is neutralized to $SO_3{}^{2-}$. On the other hand, if the solution is acidic, the form of the reactant is $HSO_3{}^-$ or even H_2SO_3. This is because protons are forced onto the $SO_3{}^{2-}$ ion in the presence of acid, with the number of H^+ added depending on how acidic the reaction mixture is.

The oxidation of bisulfite ion in an acidic solution follows the half-reaction

$$HSO_3{}^-(aq) + H_2O \longrightarrow SO_4{}^{2-}(aq) + 3H^+(aq) + 2e^-$$

The oxidation of sulfite ion in a basic solution occurs more easily than the oxidation of bisulfite in an acidic solution, so in basic solution it is a better reducing agent. The reduction takes place according to this half-reaction.

$$SO_3{}^{2-}(aq) + 2OH^-(aq) \longrightarrow SO_4{}^{2-}(aq) + H_2O + 2e^-$$

Thiosulfate Ion Another sulfur-containing reducing agent that has laboratory applications is the thiosulfate ion, $S_2O_3^{2-}$. If attacked by a strong oxidizing agent, the $S_2O_3^{2-}$ ion is oxidized to sulfate ion. That is what happens, for example, if chlorine gas is bubbled into a solution of $Na_2S_2O_3$. The reaction is

$$4Cl_2(aq) + S_2O_3^{2-}(aq) + 5H_2O \longrightarrow 8Cl^-(aq) + 2SO_4^{2-}(aq) + 10H^+(aq)$$

This reaction makes $S_2O_3^{2-}$ especially useful for trapping Cl_2 gas that might otherwise be released into the atmosphere.

Another important reaction of thiosulfate ion is with iodine, I_2, which is a less powerful oxidizing agent than Cl_2. In this case the oxidation product of the $S_2O_3^{2-}$ is the $S_4O_6^{2-}$ ion (tetrathionate ion).

$$I_2(aq) + 2S_2O_3^{2-}(aq) \longrightarrow 2I^-(aq) + S_4O_6^{2-}(aq)$$

5.7 THE STOICHIOMETRY OF IONIC REACTIONS

Toward the end of Chapter 3 we discussed some of the applications of stoichiometry to chemical reactions in solution, and if you look back at page 75, you will see that these included metathesis reactions. At that time, however, we could not treat them as ionic reactions because you had not yet been introduced to the concepts of ions and electrolytes. Now that we've overcome that restriction, we can examine how stoichiometric principles can be applied to ionic substances and reactions in solution.

Concentrations

In working quantitatively with solutes in a solution, we need to know the solute's concentration. There are many ways of expressing concentration, and each has its advantages for specific applications. One way of expressing concentration, for example, is percent composition by mass. Traditionally, this has been called *weight/ weight percent* and is indicated by writing the symbol (w/w) following the percent sign. For example, concentrated sulfuric acid is composed of 96% H_2SO_4 by mass, which is written as 96% (w/w) H_2SO_4. This way of specifying concentration gives the composition of the solution in parts per hundred by mass. In other words, it tells us how many grams of the solute are present in 100 g of solution.

EXAMPLE 5.11. WORKING WITH WEIGHT PERCENT AS A CONCENTRATION UNIT

PROBLEM: How would you prepare a solution that is 5.00% (w/w) NaCl in water?

SOLUTION: The concentration unit tells us that there should be 5.00 g NaCl in 100 g of the solution. To prepare the solution, we add 95.0 g of water to 5.00 g of NaCl. Since the density of water is very close to 1.00 g/mL, we can use 95.0 mL of water and save the trouble of weighing the water.

A somewhat similar unit that is frequently used to express very small concentrations (e.g., the concentrations of impurities or pollutants in air or water) is **parts per million, ppm,** by volume or by mass. For instance, a typical carbon monoxide level in heavy smog is approximately 40 ppm by volume, whereas a typical nitrogen oxide concentration is about 0.2 ppm by volume.

Analytical methods have become so sensitive that some substances can be detected at concentration levels of parts per billion (ppb).

EXAMPLE 5.12. CALCULATING CONCENTRATION IN PARTS PER MILLION

PROBLEM: A 500-L sample of air with a density of 1.20 g/L was found to contain 2.40×10^{-3} g of the pollutant SO_2. What is the concentration of the SO_2 in the air, expressed in percent by mass and in parts per million (ppm) by mass.

SOLUTION: To calculate either of these quantities, we need to know the total mass of the air sample, which we can calculate from the volume and the density.

$$\text{mass of air} = 500 \cancel{L} \times \frac{1.20 \text{ g}}{1.00 \cancel{L}} = 600 \text{ g air}$$

The percentage by mass is calculated as follows:

$$\% \text{ by mass} = \frac{\text{mass of } SO_2}{\text{mass of air}} \times 100\%$$

$$= \frac{2.40 \times 10^{-3} \cancel{g}}{600 \cancel{g}} \times 100\%$$

$$= 4.00 \times 10^{-4} \%$$

The concentration in parts per million is calculated as

$$\text{ppm by mass} = \frac{\text{mass of } SO_2}{\text{mass of air}} \times 10^6 \text{ ppm}$$

$$= \frac{2.40 \times 10^{-3} \cancel{g}}{600 \cancel{g}} \times 10^6 \text{ ppm}$$

$$= 4.00 \text{ ppm}$$

Notice that the concentration of the SO_2 expressed as a percent is very small, which makes the value difficult to visualize and compare with other values. The same concentration expressed in parts per million, however, is a more easily comprehended number, which is why parts per million (or even parts per billion) is used for very small concentrations.

One of the most useful ways of expressing concentration is in moles per liter or molarity. This was treated rather extensively in Section 3.5, and if necessary, you should review this section to freshen your memory. Molarity is particularly well suited for dealing with the stoichiometry of reactions in solution, and some of the aspects of this were examined in Chapter 3.

Molarity is useful because it expresses the amount of solute in chemical units: moles.

When you work with ionic compounds and reactions in solution, one of the kinds of calculations that you should be able to perform routinely is determining the molarity of a particular ion in a solution of a strong electrolyte. For example, suppose you were given a solution labeled "1.00 M $CaCl_2$." What are the concentrations of Ca^{2+} ion and Cl^- ion in this solution? To answer this question, you have to realize that the given concentration refers to the number of moles of the salt per liter, and you also have to remember that such salts are completely dissociated in an aqueous solution. In this case, there is 1.00 mol of $CaCl_2$ per liter, and $CaCl_2$ dissociates in solution as follows:

$$CaCl_2 \longrightarrow Ca^{2+}(aq) + 2Cl^-(aq)$$

One mole of $CaCl_2$ dissociates to give *one* mole of Ca^{2+} and *two* moles of Cl^-, so there is one mole of Ca^{2+} per liter and two moles of Cl^- per liter. Therefore, we also could label this solution "1.00 M Ca^{2+}" and "2.00 M Cl^-."

EXAMPLE 5.13. CALCULATING THE CONCENTRATION OF AN ION IN A SOLUTION

PROBLEM: What are the aluminum ion and sulfate ion concentrations in a 0.240 M solution of $Al_2(SO_4)_3$?

SOLUTION: Each formula unit of $Al_2(SO_4)_3$ dissociates into two Al^{3+} ions and three SO_4^{2-} ions. Therefore, the number of moles of Al^{3+} is two times the number of moles of $Al_2(SO_4)_3$ specified. Similarly, the number of moles of SO_4^{2-} is three times the number of moles of $Al_2(SO_4)_3$ that are given. This means that

$$Al^{3+} \text{ concentration} = 2 \times (0.240 \; M) = 0.480 \; M$$

$$SO_4^{2-} \text{ concentration} = 3 \times (0.240 \; M) = 0.720 \; M$$

Stoichiometry

Once you have learned to calculate the concentrations of ions in a solution of a salt, you can easily use a net ionic equation to solve stoichiometry problems. This is illustrated in Example 5.14.

EXAMPLE 5.14. USING A NET IONIC EQUATION IN STOICHIOMETRY PROBLEMS

PROBLEM: Consider the net ionic equation for the reaction of chloride ion with permanganate ion in an acidic solution.

$$2MnO_4^-(aq) + 10Cl^-(aq) + 16H^+(aq) \longrightarrow 2Mn^{2+}(aq) + 5Cl_2(g) + 8H_2O$$

How many milliliters of 0.350 M $CaCl_2$ solution are needed to give 1.25 g of Cl_2 gas?

SOLUTION: First we calculate the number of moles of Cl_2. From the atomic mass table, the formula mass of Cl_2 is 70.9. Therefore,

$$1.25 \; g\;Cl_2 \times \left(\frac{1 \text{ mol } Cl_2}{70.9 \; g\;Cl_2} \right) = 1.76 \times 10^{-2} \text{ mol } Cl_2$$

From the coefficients in the equation, it is clear that the formation of this amount of Cl_2 would require the oxidation of $2(1.76 \times 10^{-2}) = 3.52 \times 10^{-2}$ mol Cl^-.

$(1.76 \times 10^{-2} \text{ mol } Cl_2)$

$\times \left(\dfrac{10 \text{ mol } Cl^-}{5 \text{ mol } Cl_2} \right)$

$= 3.52 \times 10^{-2} \text{ mol } Cl^-$

Now we have to calculate now many milliliters of the $CaCl_2$ solution are needed. A simple way of doing this is first to calculate the chloride ion concentration in the $CaCl_2$ solution. Since one $CaCl_2$ gives two Cl^- ions, the chloride ion concentration is two times the given concentration of the salt.

$$Cl^- \text{ concentration} = 2 \times (0.350 \; M) = 0.700 \; M \; Cl^-$$

Recall that molarity gives us a conversion factor that can be used in either of two ways. in this case we can write

$$\frac{0.700 \text{ mol } Cl^-}{1000 \text{ mL soln}} \quad \text{or} \quad \frac{1000 \text{ mL soln}}{0.700 \text{ mol } Cl^-}$$

We use the second factor to obtain the answer.

$$3.52 \times 10^{-2} \; mol\;Cl^- \times \left(\frac{1000 \text{ mL soln}}{0.700 \; mol\;Cl^-} \right) \Longleftrightarrow 50.3 \text{ mL soln}$$

The volume of $CaCl_2$ solution needed is 50.3 mL.

The same principles that we applied in solving the problem in Example 5.14 can be used in working limiting-reactant problems involving ionic reactions. This is illustrated by the following example.

EXAMPLE 5.15. WORKING LIMITING-REACTANT PROBLEMS FOR IONIC REACTIONS

PROBLEM: Suppose that 20.0 mL of 0.150 M $Al_2(SO_4)_3$ are added to 30.0 mL of 0.200 M $BaCl_2$ solution, yielding a precipitate of $BaSO_4$.

$$Ba^{2+}(aq) + SO_4^{2-}(aq) \longrightarrow BaSO_4(s)$$

(a)   How many grams of $BaSO_4$ will be formed in the reaction?

(b)   What will be the concentrations of any ions remaining in the reaction mixture after the reaction is complete?

ANALYSIS: Notice that in the statement of the problem we've been given the molarity and volume of each of the reactants. It's important that you remember that the product of molarity and volume gives moles of solute, because then you can realize that we've actually been given the number of *moles* of each reactant, and this is what tells us that it is a limiting-reactant problem.

To solve a limiting-reactant problem, we need to determine which is the limiting reactant and then base our calculation of the amount of product formed on the amount of the limiting reactant available. We can also figure out which reactant is left over.

Because the chemical equation given is a net ionic equation, the best approach to solving this problem is to calculate the number of moles of each ion in the solution and then proceed from there, as shown below.

SOLUTION: The number of moles of aluminum sulfate in its solution is

$$20.0 \text{ mL soln} \times \left(\frac{0.150 \text{ mol } Al_2(SO_4)_3}{1000 \text{ mL soln}} \right) \Longleftrightarrow 3.00 \times 10^{-3} \text{ mol } Al_2(SO_4)_3$$

and the number of moles of barium chloride in its solution is

$$30.0 \text{ mL soln} \times \left(\frac{0.200 \text{ mol } BaCl_2}{1000 \text{ mL soln}} \right) \Longleftrightarrow 6.00 \times 10^{-3} \text{ mol } BaCl_2$$

Therefore, on the basis of the formulas of the salts, we have

Al^{3+}	6.00×10^{-3} mol
SO_4^{2-}	9.00×10^{-3} mol
Ba^{2+}	6.00×10^{-3} mol
Cl^-	1.20×10^{-2} mol

Now we are ready to solve both parts of the problem.

(a)   The reaction involves Ba^{2+} and SO_4^{2-}, which react in a 1-to-1 mole ratio. Examining the numbers of moles of each, we can see that there is more sulfate ion than needed to react with all the barium ion. Therefore, Ba^{2+} is the limiting reactant, and we calculate the amount of product formed based on it.

$$6.00 \times 10^{-3} \text{ mol } Ba^{2+} \times \left(\frac{1 \text{ mol } BaSO_4}{1 \text{ mol } Ba^{2+}} \right) \times \left(\frac{233.4 \text{ g } BaSO_4}{1 \text{ mol } BaSO_4} \right) \Longleftrightarrow 1.40 \text{ g } BaSO_4$$

(b)   To calculate the concentrations of the ions in the final solution, we have to take into account the fact that one solution dilutes the other when the two of them are mixed.

$\left(\dfrac{\text{mol}}{\text{L}} \right) \times \text{L} = \text{mol}$

6.00×10^{-3} mol Ba^{2+} requires 6.00×10^{-3} mol SO_4^{2-}.

Therefore, we need the final volume. This is the sum of the volumes of the two solutions that were mixed, and it equals 50.0 mL, or 0.0500 L.

The concentrations of the spectator ions, Al^{3+} and Cl^-, are

$$\frac{6.00 \times 10^{-3} \text{ mol } Al^{3+}}{0.0500 \text{ L soln}} = 0.120 \ M \ Al^{3+}$$

$$\frac{1.20 \times 10^{-2} \text{ mol } Cl^-}{0.0500 \text{ L soln}} = 0.240 \ M \ Cl^-$$

$$\text{Molarity} = \frac{\text{moles of ion}}{\text{total volume of solution}}$$

In the final solution there is also some leftover sulfate ion that wasn't able to react with Ba^{2+}. The number of moles of this ion in the final solution is equal to the initial number of moles minus the number of moles that reacted.

$$\text{moles } SO_4{}^{2-} \text{ remaining} = (9.00 \times 10^{-3} \text{ mol}) - (6.00 \times 10^{-3} \text{ mol})$$
$$= 3.00 \times 10^{-3} \text{ mol } SO_4{}^{2-}$$

The sulfate ion concentration is therefore

$$\frac{3.00 \times 10^{-3} \text{ mol } SO_4{}^{2-}}{0.0500 \text{ L soln}} = 0.0600 \ M \ SO_4{}^{2-}$$

Because Ba^{2+} was the limiting reactant, it was used up completely, so its concentration is practically zero. (Actually, a very small amount of $BaSO_4$ still remains in solution. In Chapter 17, this will be treated quantitatively. For now, however, you can take the concentration of the limiting reactant to be essentially zero after the reaction is over.)

5.8 CHEMICAL ANALYSIS AND TITRATIONS

One of the problems frequently faced in chemistry is the determination of the composition of a substance, or of a mixture of substances. For example, in Chapter 2 you learned that the calculation of an empirical formula requires information about the amounts of each of the elements combined in a sample of the compound. When a substance appears as the product of a chemical reaction, it doesn't simply stand up and declare its composition—this information must be obtained experimentally. The name that we give to the experimental determination of chemical composition is **chemical analysis.**

Chemical analysis is not just something of interest to a research chemist who makes new compounds. Chemical, pharmaceutical, and cosmetic companies employ chemists whose sole job it is to analyze samples of the products being manufactured, so that quality can be maintained. Similarly, mining companies rely on analytical chemists to determine the compositions of the raw materials taken from the earth, so that they can be properly processed to obtain the desired end products. And environmental chemists use chemical analysis to determine whether pollutants are present in the air and water, and if so, what their concentrations are. In fact, one very important facet of chemistry is the investigation of new and better ways to achieve chemical analyses, particularly of compounds present in mixtures where difficulties in separating the components present a major obstacle to success.

Ionic reactions in solution can frequently be used to advantage in performing chemical analyses, where the kind of reaction used depends on the specific nature of the material being analyzed. Sometimes a precipitation reaction is convenient, as in Example 5.16 below. In this type of analysis, a reaction is chosen that allows us to separate the target of the analysis from the rest of the sample being studied. In Example 5.16, for instance, the goal is to analyze for sodium sulfate in the presence of sodium chloride. The reaction that is chosen is one that allows the separation of the sulfate from the chloride, so the amount of sulfate can be determined. That is why Ba^{2+} is added to a solution of the sample—$BaSO_4$ is insoluble but $BaCl_2$ is soluble.

EXAMPLE 5.16. USING A PRECIPITATION REACTION IN A CHEMICAL ANALYSIS

PROBLEM: A white powder was known to be a mixture of NaCl and Na_2SO_4. A sample of the powder weighing 1.244 g was dissolved in water and a solution of $Ba(NO_3)_2$ was added until the precipitation of $BaSO_4$ was complete. The reaction mixture was filtered carefully to be sure that none of the precipitate was lost, and the $BaSO_4$ was then dried and found to weigh 0.851 g. What was the percentage by mass of Na_2SO_4 in the original sample?

ANALYSIS: This is really a very simple problem. Let's examine how we will solve it. First, we can calculate the number of moles of $BaSO_4$ that were obtained. From this, we can calculate the number of moles of Na_2SO_4 in the sample. We can do this because all the sulfate ion in the Na_2SO_4 was recovered in the $BaSO_4$ precipitate. Once we know the number of moles of Na_2SO_4 in the sample, we can calculate its mass in grams and then the percentage by mass.

SOLUTION: Now that we know how to proceed, the calculations are simple. The number of moles of $BaSO_4$ (formula mass 233.4) is

$$0.851 \text{ g } \cancel{BaSO_4} \times \left(\frac{1 \text{ mol } BaSO_4}{233.4 \text{ g } \cancel{BaSO_4}} \right) = 3.65 \times 10^{-3} \text{ mol } BaSO_4$$

$BaSO_4$ is formed in the reaction

$$Ba^{2+}(aq) + SO_4{}^{2-}(aq) \longrightarrow BaSO_4(s)$$

3.65×10^{-3} mol $BaSO_4$

$\times \left(\dfrac{1 \text{ mol } SO_4{}^{2-}}{1 \text{ mol } BaSO_4} \right)$

$\times \left(\dfrac{1 \text{ mol } Na_2SO_4}{1 \text{ mol } SO_4{}^{2-}} \right)$

$= 3.65 \times 10^{-3}$ mol Na_2SO_4

so each mole of $BaSO_4$ is formed from one mole of sulfate ion. The number of moles of $SO_4{}^{2-}$ in the sample, therefore, is also 3.65×10^{-3} mol. From the formula Na_2SO_4, we see that each mole of this compound yields one mole of sulfate, so the number of moles of Na_2SO_4 is 3.65×10^{-3}, too. The formula mass of Na_2SO_4 is 142.0, so the number of grams of Na_2SO_4 in the original sample was

$$3.65 \times 10^{-3} \text{ } \cancel{\text{mol } Na_2SO_4} \times \left(\frac{142.0 \text{ g } Na_2SO_4}{1 \text{ } \cancel{\text{mol } Na_2SO_4}} \right) = 0.518 \text{ g } Na_2SO_4$$

The percentage by mass of Na_2SO_4 in the sample is obtained as

$$\text{percentage } Na_2SO_4 = \frac{\text{mass of } Na_2SO_4}{\text{mass of sample}} \times 100\%$$

Substituting values gives

$$\text{percentage } Na_2SO_4 = \frac{0.518 \text{ g } Na_2SO_4}{1.244 \text{ g sample}} \times 100\%$$

$$= 41.6\% \text{ } Na_2SO_4$$

The sample is 41.6% Na_2SO_4.

Titrations

Titration is an analytical procedure that allows us to measure the amount of one solution needed to react exactly with the contents of another solution. Such analyses, which involve the measurements of volumes of solutions of reactants, are called **volumetric analyses.** In a titration, one of the solutions that contains a reactant is

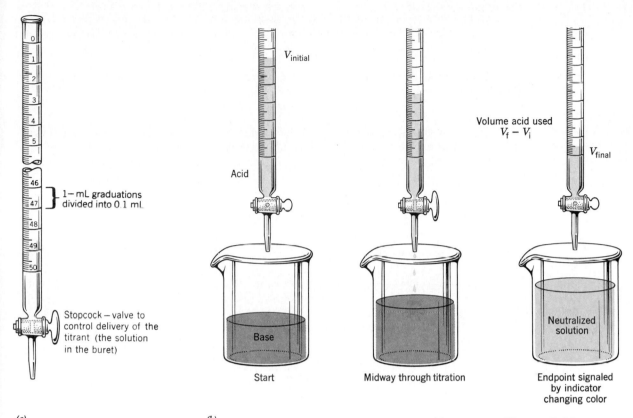

(a) *(b)*

Figure 5.10. (a) *A buret.* (b) *The titration of an acid by a base in which an acid-base indicator is used to signal the end of the reaction.*

placed in a **buret,** a long tube fitted at one end with a valve (called a **stopcock**) and precisely graduated in milliliters and tenths of milliliters (see Figure 5.10). The solution in the buret is called the **titrant,** and during a titration this solution is delivered slowly through the stopcock into another vessel that contains a solution of the other reactant. The titrant is added until complete reaction is signaled by an **indicator,** a substance that generally is added to the solution in the receiving vessel and which undergoes some sort of color change when the reaction is over. The color change of the indicator marks the **endpoint** in the titration, so named because it is at this point that the delivery of the titrant is stopped and the volume of the titrant used in the reaction recorded.

One kind of reaction that is frequently used in titrations is acid-base neutralization. Normally, the solution of the base is the titrant and the solution of the acid is placed in the receiving flask or beaker. The indicator is a substance that has one color in an acidic solution and another color in a basic solution. Litmus is such a substance — you may recall that in the presence of an acid, litmus is pink, and in the presence of a base, litmus is blue. Another indicator you may use in the laboratory is phenolphthalein (the second "ph" is silent when you pronounce it). Phenolphthalein is colorless in an acidic solution and pink in a basic solution.

EXAMPLE 5.17. CALCULATIONS INVOLVING TITRATIONS

PROBLEM: A solution of sodium hydroxide was prepared having a concentration of approximately 0.1 *M*. It was desired to measure the concentration of this solution accurately, so a 20.00 mL portion of a 0.1000 *M* solution of HCl was carefully measured into a beaker and a few drops of phenolphthalein were added to it. A buret was filled with the NaOH solution and used to titrate the HCl solution. The titration required 18.47 mL of the base in order to reach the endpoint. What is the molarity of the NaOH solution?

SOLUTION: Since the problem deals with a chemical reaction, let's begin with a balanced chemical equation.

$$NaOH(aq) + HCl(aq) \longrightarrow NaCl(aq) + H_2O$$

Determining the exact concentration of a solute in a solution is called *standardizing the solution*. A standard solution is one with a precisely known concentration of solute.

To solve the problem, we need to know the exact ratio of moles of NaOH to liters of solution. In the titration, we used 18.47 mL of the NaOH solution. If we can calculate the amount of NaOH in this volume, we will then have all the information needed to calculate the molarity of the solution. We can get this information from the amounts of chemicals used in the reaction.

First, we calculate the number of moles of HCl in the solution in the beaker *before* the titration was begun. This is obtained from the acid's initial volume and molarity.

$$20.00 \; \text{mL HCl soln} \times \left(\frac{0.1000 \; \text{mol HCl}}{1000 \; \text{mL HCl soln}} \right) \Longleftrightarrow 2.000 \times 10^{-3} \; \text{mol HCl}$$

2.000 × 10⁻³ mol NaOH was in 18.47 mL of titrant.

The coefficients of the equation tell us that the HCl and NaOH combine in a 1-to-1 mole ratio. Therefore, the number of moles of NaOH that reacted with the HCl in order to reach the endpoint was also 2.000×10^{-3} mol. Finally, we take the ratio of the number of moles of NaOH to the number of liters of solution that contained it. Since 18.47 mL equals 0.01847 L, the molarity is

$$\frac{2.000 \times 10^{-3} \; \text{mol NaOH}}{0.01847 \; \text{L soln}} = 0.1083 \; M \; \text{NaOH}$$

Acid–base titrations find many uses in chemical analyses. This is illustrated in the next example.

EXAMPLE 5.18. USING ACID–BASE TITRATIONS IN A CHEMICAL ANALYSIS

PROBLEM: A sample of an analgesic drug was analyzed for aspirin, a monoprotic acid, $HC_9H_7O_4$, by titration with a base. In a titration, a 0.500-g sample of the drug required 21.50 mL of 0.100 M NaOH for complete neutralization. What percentage by mass of the drug was aspirin?

SOLUTION: Let's begin with a chemical equation for the reaction.

$$HC_9H_7O_4 + NaOH \longrightarrow NaC_9H_7O_4 + H_2O$$

From the volume of the base and its concentration, we can calculate the number of moles of base consumed.

$$21.50 \; \text{mL soln} \times \left(\frac{0.100 \; \text{mol NaOH}}{1000 \; \text{mL soln}} \right) \Longleftrightarrow 0.00215 \; \text{mol NaOH}$$

0.00215 mol NaOH
$\times \left(\dfrac{1 \; \text{mol HC}_9\text{H}_7\text{O}_4}{1 \; \text{mol NaOH}} \right)$
= 0.00215 mol HC₉H₇O₄

The coefficients of $HC_9H_7O_4$ and NaOH are the same in the chemical equation, so the number of moles of each that react are the same. Therefore, the number of moles of $HC_9H_7O_4$ in the sample was 0.00215 mol $HC_9H_7O_4$. The formula mass of the aspirin is 180.2, so the number of grams of aspirin in the sample was

$$0.00215 \; \text{mol HC}_9\text{H}_7\text{O}_4 \times \left(\frac{180.2 \; \text{g HC}_9\text{H}_7\text{O}_4}{1 \; \text{mol HC}_9\text{H}_7\text{O}_4} \right) = 0.387 \; \text{g HC}_9\text{H}_7\text{O}_4$$

The percentage of aspirin is found as

$$\text{percentage aspirin} = \frac{\text{mass aspirin}}{\text{mass sample}} \times 100\%$$

Substituting values gives

$$\text{percentage aspirin} = \frac{0.387 \text{ g } HC_9H_7O_4}{0.500 \text{ g sample}} \times 100\%$$
$$= 77.4\%$$

5.9 EQUIVALENT WEIGHTS AND NORMALITY

In certain types of reactions there is always the same clearly defined and predictable ratio in which the reactants combine. For example, in any acid–base neutralization, an amount of acid that furnishes 1 mol of H^+ *always* reacts completely with an amount of base that furnishes 1 mol of OH^-. This is assured by the stoichiometry of the neutralization reaction

$$H^+ + OH^- \longrightarrow H_2O$$

Similarly, in a redox reaction the amount of the substance that loses 1 mol of electrons will *always* react exactly with the amount of another substance that gains 1 mol of electrons. This is because the numbers of electrons lost and gained *must* be exactly the same.

The recognition of such similarities in the stoichiometry of acid–base reactions and of redox reactions leads us to the concept of an **equivalent** (we will use the abbreviation **eq**). If we have a reaction between two substances—call them A and B—the equivalent is always defined so that 1 equivalent of A (1 eq A) reacts with exactly 1 equivalent of B (1 eq B). **Equivalents** *always* **react in a 1-to-1 ratio.**

$$1 \text{ eq } A \Longleftrightarrow 1 \text{ eq } B$$

The equivalent is not an SI unit, and many practicing chemists have gotten away from the concept. Nevertheless, the idea of equivalents is still widely used in other sciences, so it is necessary for you to learn about it.

Remember, equivalents *always* react one for one.

Acid–base reactions

An equivalent of an acid is defined as the amount of acid that furnishes 1 mol of H^+. An equivalent of a base furnishes 1 mol of OH^-. Notice that this definition assures that 1 eq acid will react with exactly 1 eq base. Let's see how this applies to some typical acids and bases.

An equivalent of base can also be defined as the amount of base that reacts with 1 mol of H^+.

Consider the reaction of H_2SO_4 with NaOH.

$$H_2SO_4 + 2NaOH \longrightarrow Na_2SO_4 + 2H_2O$$

In this reaction, 1 mol H_2SO_4 furnishes 2 mol H^+, so 1 mol H_2SO_4 must be the same as 2 eq H_2SO_4. For the base, 1 mol NaOH furnishes 1 mol OH^-, so 1 mol NaOH = 1 eq NaOH. For acids and bases, the number of equivalents per mole is the same as the number of H^+ supplied per molecule of acid or the number of OH^- supplied per formula unit of base. Another way of stating this is

$$\text{number of eq} = \text{number of mol} \times n \tag{5.1}$$

where n is a whole number; it is the number of H^+ provided by one molecule of acid or the number of OH^- provided by one formula unit of base in the particular reaction being studied.

The number of equivalents is never less than the number of moles.

EXAMPLE 5.19. DETERMINING THE NUMBER OF EQUIVALENTS OF AN ACID

PROBLEM: How many equivalents are there in 0.400 mol H_3PO_4 if this acid is completely neutralized to give PO_4^{3-}? How many equivalents are there if it is converted to HPO_4^{2-}?

For a polyprotic acid, you must know the extent to which the acid is neutralized in order to calculate the number of equivalents per mole.

SOLUTION: One molecule of H_3PO_4 must supply three H^+ if it is to become PO_4^{3-}; therefore, $n = 3$. Substituting into Equation 5.1 gives

$$\text{number of eq} = 0.400 \times 3$$
$$= 1.20$$

Thus, 0.400 mol H_3PO_4 is 1.20 eq when it reacts to give PO_4^{3-}.

If the original H_3PO_4 was only converted to HPO_4^{2-}, each molecule of acid would give $2H^+$, so $n = 2$. Therefore, 0.400 mol H_3PO_4 is 0.800 eq when it is converted to HPO_4^{2-}.

For simplicity, we will choose reactions in which the base is always completely neutralized.

In Example 5.19 we see that the number of equivalents per mole of acid depends on the extent to which the acid is neutralized in a particular reaction. Although this is a concern with acids, you can assume that the number of equivalents per mole for a base is the same as the number of OH^- in one formula unit of the base. Thus, 1 mol $Ba(OH)_2$ would be 2 eq $Ba(OH)_2$, and 1 mol $Al(OH)_3$ would be 3 eq $Al(OH)_3$.

A quantity used in many stoichiometric calculations involving equivalents is the **equivalent weight,** the mass of one equivalent. In terms of the integer n defined in Equation 5.1

$$\text{equivalent weight (eq wt)} = \frac{\text{mass of 1 mol}}{n} \qquad (5.2)$$

If there are two equivalents per mole, then each equivalent must be half a mole. The mass of one equivalent, therefore, is half the mass of a mole.

For example, for the complete neutralization of H_2SO_4, $n = 2$. The equivalent weight of H_2SO_4 is its formula mass divided by 2.

$$\text{eq wt } H_2SO_4 = \frac{98.0 \text{ g}}{2} = 49.0 \text{ g}$$

In other words, 1 eq $H_2SO_4 = 49.0$ g H_2SO_4.

EXAMPLE 5.20. USING EQUIVALENTS IN AN ACID-BASE CALCULATION

PROBLEM: How many grams of H_3PO_4 will be completely neutralized to give PO_4^{3-} by 20.0 g NaOH?

SOLUTION: Let's calculate the mass of one equivalent for each reactant. These masses will exactly react with each other because 1 eq H_3PO_4 reacts with exactly 1 eq NaOH.

$$1 \text{ eq NaOH} \Longleftrightarrow 1 \text{ eq } H_3PO_4$$

Sodium hydroxide has a formula mass of 40.0. Therefore,

There is 1 OH^- per NaOH.

$$1 \text{ eq NaOH} = \frac{40.0 \text{ g NaOH}}{1} = 40.0 \text{ g NaOH}$$

The formula mass of phosphoric acid is 98.0.

All three H^+ of H_3PO_4 are being neutralized.

$$1 \text{ eq } H_3PO_4 = \frac{98.0 \text{ g } H_3PO_4}{3} = 32.7 \text{ g } H_3PO_4$$

Therefore, we can say that

$$40.0 \text{ g NaOH} \Longleftrightarrow 32.7 \text{ g H}_3\text{PO}_4$$

Now we use this in a conversion factor to find our answer.

$$20.0 \text{ g NaOH} \times \left(\frac{32.7 \text{ g H}_3\text{PO}_4}{40.0 \text{ g NaOH}} \right) \Longleftrightarrow 16.4 \text{ g H}_3\text{PO}_4$$

Thus, 20.0 g NaOH will neutralize 16.4 g H_3PO_4. Notice that we did the problem without a chemical equation. This is the only real usefulness of the concept of equivalents—for certain types of reactions we don't need to write an equation to establish the stoichiometry.

Oxidation–reduction reactions

An equivalent of an oxidizing agent or reducing agent in a redox reaction is the amount of substance that gains or loses one mole of electrons. This definition ensures that one equivalent of oxidizing agent reacts exactly with one equivalent of reducing agent.

Equations 5.1 and 5.2 also apply to redox reactions. *The value of n for an oxidizing or reducing agent is just the total number of electrons gained or lost by one formula unit of the substance.* For example, suppose we were going to use calcium iodate, $Ca(IO_3)_2$, in a reaction in which the iodate ion was going to give I^- as a product.

$$\underset{+5}{Ca(IO_3)_2} \longrightarrow \underset{-1}{I^-}$$

For our purposes here, it's convenient to think about the change in oxidation number as a gain or loss of electrons. If we do this, we see that the oxidation number of iodine changes from $+5$ in $Ca(IO_3)_2$ to -1 in the I^- ion, which corresponds to a gain of $6e^-$ for each iodine. But one formula unit of $Ca(IO_3)_2$ has two iodines, so the total number of electrons transferred by one formula unit is 12. Therefore, in this reaction, $n = 12$ for $Ca(IO_3)_2$, so 1 mol $Ca(IO_3)_2$ is equal to 12 eq $Ca(IO_3)_2$.

$$\text{number eq } Ca(IO_3)_2 = \text{number mol } Ca(IO_3)_2 \times 12$$

The equivalent weight of $Ca(IO_3)_2$ is $\frac{1}{12}$ of the mass of a mole.

$$\text{eq wt } Ca(IO_3)_2 = \frac{\text{mass of 1 mol } Ca(IO_3)_2}{12}$$

Notice that an equivalent in acid–base reactions is defined differently than an equivalent in redox reactions.

To calculate equivalent weights in redox reactions, you must know the *changes* in oxidation numbers that the *reactants* undergo.

EXAMPLE 5.21. DETERMINING THE NUMBER OF EQUIVALENTS PER MOLE FOR A REACTANT IN A REDOX REACTION

PROBLEM: In a reaction between nitric acid and zinc, the NO_3^- ion can be reduced to NH_4^+ ion. How many equivalents are there per mole of HNO_3 in this reaction?

SOLUTION: In the reaction, the following change occurs in the oxidation number of the nitrogen:

$$\underset{+5}{NO_3^-} \longrightarrow \underset{-3}{NH_4^+}$$

This corresponds to a gain of 8 electrons per NO_3^-, so in the reaction, there must be 8 eq per mole

$$1 \text{ mol } HNO_3 = 8 \text{ eq } HNO_3$$

Note that if HNO_3 were to function as an acid instead of an oxidizing agent, there would be one equivalent per mole.

EXAMPLE 5.22. USING EQUIVALENTS IN A REDOX CALCULATION

PROBLEM: $FeSO_4$ reacts with $KMnO_4$ in H_2SO_4 to produce $Fe_2(SO_4)_3$ and $MnSO_4$. How many grams of $FeSO_4$ react with 3.71 g $KMnO_4$?

SOLUTION: In this reaction,

For convenience, we are thinking of changes in oxidation numbers as changes in the numbers of electrons.

$$1 \text{ mol of } FeSO_4 \text{ loses 1 mol of electrons:} \quad \underset{+2}{Fe} \xrightarrow{-1e^-} \underset{+3}{Fe}$$

$$1 \text{ mol of } KMnO_4 \text{ gains 5 mol of electrons:} \quad \underset{+7}{Mn} \xrightarrow{+5e^-} \underset{+2}{Mn}$$

The equivalent weight of $FeSO_4$ in this reaction is the same as the mass of one mole because each mole of $FeSO_4$ loses 1 mol of electrons.

$$1 \text{ eq } FeSO_4 = \frac{1 \text{ mol } FeSO_4}{1} = 152 \text{ g } FeSO_4$$

The equivalent weight of $KMnO_4$ is $\frac{1}{5}$th the mass of 1 mol because each mole of $KMnO_4$ gains 5 mol of electrons.

$$1 \text{ eq } KMnO_4 = \tfrac{1}{5} \text{ mol } KMnO_4 = \tfrac{1}{5}(158 \text{ g } KMnO_4)$$

$$1 \text{ eq } KMnO_4 = 31.6 \text{ g } KMnO_4$$

Since 1 eq of $FeSO_4$ must react with precisely 1 eq of $KMnO_4$,

$$152 \text{ g } FeSO_4 \Leftrightarrow 31.6 \text{ g } KMnO_4$$

The solution to the problem can then be obtained as

$$3.71 \text{ g } \cancel{KMnO_4} \times \left(\frac{152 \text{ g } FeSO_4}{31.6 \text{ g } \cancel{KMnO_4}} \right) \Leftrightarrow 17.8 \text{ g } FeSO_4$$

Normality

Another concentration unit that can be used in solving problems in solution stoichiometry is **normality,** defined as *the number of equivalents per liter of solution.*

$$\text{normality} = \frac{\text{number of equivalents}}{1 \text{ L of solution}}$$

Once again we have a ratio, and like molarity, it relates the amount of solute to the total volume of the solution. Normality also can be used as a conversion factor, which involves "translating the label." Thus, a concentration of 1.00 N (a 1.00 *normal* solution) can be translated to mean

$$\frac{1.00 \text{ eq solute}}{1000 \text{ mL solution}} \quad \text{or} \quad \frac{1000 \text{ mL solution}}{1.00 \text{ eq solute}}$$

In the preceding discussion, an equivalent of an oxidizing or reducing agent was defined as the quantity that either gains or loses 1 mol of electrons. We saw in Example 5.22 that when the permanganate ion in $KMnO_4$ is reduced to Mn^{2+}, five electrons are gained by each formula unit of $KMnO_4$. One mole of $KMnO_4$, therefore, consumes 5 mol of electrons, and therefore, 1 mol $KMnO_4$ is equal to 5 eq $KMnO_4$. This means that a 0.10 M $KMnO_4$ solution, which contains 0.10 mol $KMnO_4$ per liter, can also be considered to contain 0.50 eq $KMnO_4$ per liter. We would say that the concentration of this solution is 0.50 normal, or 0.50 N. Notice that there is a very simple relationship between normality and molarity — the normality is always an integral multiple of the molarity

$$N = n \times M$$

where n is the same integer as before. For substances undergoing oxidation or reduction, the integer n corresponds to the number of electrons transferred per formula unit. When $KMnO_4$ is reduced to Mn^{2+}, $n = 5$. If the product of the reduction is MnO_2, only three electrons are acquired by each $KMnO_4$ formula unit, and n would equal 3. Similarly, when oxalic acid, $H_2C_2O_4$, is oxidized to produce CO_2, two electrons are lost by each $H_2C_2O_4$ formula unit and the normality of a solution of $H_2C_2O_4$ is twice its molarity.

$$H_2Cr_2O_4 \longrightarrow 2CO_2 + 2H^+ + 2e^-$$

For solutions of acids and bases, n is once again the number of H^+ provided by a formula unit of the acid or the number of OH^- made available by a formula unit of the base. For complete neutralization, a $1.00\ M\ H_3PO_4$ solution could also be labeled $3.00\ N\ H_3PO_4$; a $1.00\ M\ Ba(OH)_2$ solution is also $2.00\ N\ Ba(OH)_2$.

> The normality is never less than the molarity.

EXAMPLE 5.23. PREPARING A SOLUTION OF A GIVEN NORMALITY

PROBLEM: How many grams of $K_2Cr_2O_7$ are needed to prepare 250 mL of a $0.100\ N$ $K_2Cr_2O_7$ solution if the solution is to be used in a reaction in which the $Cr_2O_7^{2-}$ ion is to be reduced to Cr^{3+}?

SOLUTION: First, let's calculate the number of equivalents of $Cr_2O_7^{2-}$ that must be in the solution to be prepared. The normality gives us these conversion factors

$$\frac{0.100\ eq\ Cr_2O_7^{2-}}{1000\ mL\ soln} \quad and \quad \frac{1000\ mL\ soln}{0.100\ eq\ Cr_2O_7^{2-}}$$

We will use the first.

$$250\ mL\ soln \times \left(\frac{0.100\ eq\ Cr_2O_7^{2-}}{1000\ mL\ soln}\right) \Leftrightarrow 0.0250\ eq\ Cr_2O_7^{2-}$$

When $Cr_2O_7^{2-}$ is reduced to Cr^{3+}, each Cr gains 3 electrons (each Cr changes oxidation state from $+6$ to $+3$). A formula unit of $Cr_2O_7^{2-}$ has 2 chromium atoms that undergo this change, so a formula unit of $Cr_2O_7^{2-}$ gains 6 electrons. This means that there are 6 equivalents per mole. Stated mathematically

$$1\ mol\ K_2Cr_2O_7 = 6\ eq\ K_2Cr_2O_7$$

The formula mass of $K_2Cr_2O_7$ is 294.2, so the number of grams of this salt needed is

$$0.0250\ eq\ K_2Cr_2O_7 \times \left(\frac{1\ mol\ K_2Cr_2O_7}{6\ eq\ K_2Cr_2O_7}\right) \times \left(\frac{294.2\ g\ K_2Cr_2O_7}{1\ mol\ K_2Cr_2O_7}\right) = 1.23\ g\ K_2Cr_2O_7$$

To prepare the solution, we would dissolve $1.23\ g\ K_2Cr_2O_7$ in enough water to make exactly 250 mL of solution.

> You learned to work similar problems dealing with molar concentrations in Chapter 3.

Normality is useful because mixing equal volumes of solutions having the same normality will lead to complete reaction between their solutes; 1 liter of 1 N acid will completely neutralize 1 liter of 1 N base because 1 equivalent of acid reacts with 1 equivalent of base. In the same manner, 1 liter of 1 N oxidizing agent will completely oxidize the contents of 1 liter of 1 N reducing agent.

EXAMPLE 5.24. USING NORMALITY IN A SOLUTION STOICHIOMETRY CALCULATION

PROBLEM: How many milliliters of 0.150 N $K_2Cr_2O_7$ are required to react with 75.0 mL of 0.400 N $H_2C_2O_4$?

SOLUTION: In 75.0 mL of 0.400 N $H_2C_2O_4$, there are

$$75.0 \text{ mL soln} \times \left(\frac{0.400 \text{ eq } H_2C_2O_4}{1000 \text{ mL soln}} \right) \Longleftrightarrow 3.00 \times 10^{-2} \text{ eq } H_2C_2O_4$$

This requires 3.00×10^{-2} eq of $K_2Cr_2O_7$ for complete oxidation. The volume of potassium dichromate solution needed for this is

$$3.00 \times 10^{-2} \text{ eq } K_2Cr_2O_7 \times \left(\frac{1000 \text{ mL soln}}{0.150 \text{ eq } K_2Cr_2O_7} \right) \Longleftrightarrow 200 \text{ mL soln}$$

Titrations

In Example 5.24 we used the idea that equal numbers of equivalents of oxidizing and reducing agent must react. Since the product of volume (V) $\times$ normality (N) gives equivalents, when any two substances in solution (let us call them A and B) undergo complete reaction

$$V_A N_A = V_B N_B$$

which means

$$\text{(number of equivalents of } A) = \text{(number of equivalents of } B)$$

It is important to remember that this equation only applies if the concentration is expressed in normality. If you use molarity instead, you may get wrong answers.

In a titration, if we have combined equal numbers of equivalents of reactants, we have reached the **equivalence point.** If we have chosen the right indicator, we also will have reached the endpoint and the titration will be halted. Ideally, of course, the endpoint should occur at the equivalence point, but how close they come depends on the choice of indicators, especially in acid–base titrations. We will examine this in greater detail in Chapter 16.

The use of equivalents and normality in a volumetric analysis is illustrated in Example 5.25.

EXAMPLE 5.25. USING NORMALITY AND EQUIVALENTS IN A TITRATION CALCULATION

PROBLEM: A 1.000-g sample of an iron ore containing Fe_2O_3 was dissolved in acid and all the iron converted to Fe^{2+}. The solution was titrated with 90.40 mL of 0.1000 N $KMnO_4$ to give Mn^{2+} and Fe^{3+} as products. What percentage of the ore is Fe_2O_3?

SOLUTION: The number of equivalents of $KMnO_4$ reduced was

$$90.40 \text{ mL soln} \times \left(\frac{0.1000 \text{ eq } KMnO_4}{1000 \text{ mL soln}} \right) \Longleftrightarrow 9.040 \times 10^{-3} \text{ eq } KMnO_4$$

This amount of $KMnO_4$ has oxidized 9.040×10^{-3} eq Fe^{2+} to Fe^{3+}. Since 1 mol of Fe^{2+} loses one mole of electrons,

$$1 \text{ mol } Fe^{2+} = 1 \text{ eq } Fe^{2+}$$

Therefore, 9.040×10^{-3} eq $Fe^{2+} = 9.040 \times 10^{-3}$ mol Fe^{2+}. Each mole of Fe_2O_3 contains 2 mol of Fe.

$$1 \text{ mol } Fe_2O_3 \Longleftrightarrow 2 \text{ mol } Fe$$

The number of moles of Fe_2O_3 in the sample was

$$9.040 \times 10^{-3} \text{ mol Fe} \times \left(\frac{1 \text{ mol } Fe_2O_3}{2 \text{ mol Fe}} \right) \Longleftrightarrow 4.520 \times 10^{-3} \text{ mol } Fe_2O_3$$

The number of grams of Fe_2O_3 was

$$4.520 \times 10^{-3} \text{ mol } Fe_2O_3 \times \left(\frac{159.7 \text{ g } Fe_2O_3}{1 \text{ mol } Fe_2O_3} \right) \Longleftrightarrow 0.7218 \text{ g } Fe_2O_3$$

Since the sample weighed 1.000 g,

$$\text{percent } Fe_2O_3 = \frac{0.7218 \text{ g}}{1.000 \text{ g}} \times 100\% = 72.18\%$$

REVIEW QUESTIONS AND PROBLEMS

Problems whose numbers are in blue have their answers in Appendix D at the back of the book. The more difficult problems are marked with asterisks.

Reactions in Solution — General

5.1 Why are solutions usually employed for carrying out chemical reactions?

5.2 Give the meaning of the terms solvent, solute, concentrated, dilute, saturated, supersaturated, and unsaturated.

5.3 What is the meaning of the term solubility?

5.4 Why do we specify the temperature when giving the solubility of a solute in a particular solvent?

5.5 Can a saturated solution be dilute? Explain your answer.

5.6 How can a supersaturated solution be changed to a saturated solution without changing its temperature?

5.7 The solubility of sugar in water increases with increasing temperature. Describe how you would prepare a supersaturated solution of sugar in water.

5.8 At 25 °C, the solubility of $PbSO_4$ in water is 0.043 g/L. How would you describe (concentrated, dilute; saturated, unsaturated, supersaturated) a solution that contains 0.050 g of $PbSO_4$ per liter?

Electrolytes

5.9 What is an electrolyte? What is a nonelectrolyte?

5.10 How can one distinguish between a strong and a weak electrolyte? What are present in solutions of electrolytes that are absent in solutions of nonelectrolytes?

5.11 Write chemical equations for the dissociation of the following strong electrolytes: KCl, $(NH_4)_2SO_4$, Na_3PO_4, $NaOH$, HCl.

5.12 Why is the hydronium ion often abbreviated H^+?

5.13 What is a dynamic equilibrium?

5.14 Cadmium sulfate, $CdSO_4$, is an exception to the rule that salts are strong electrolytes. In a $1.00 \; M$ solution, $CdSO_4$ is only 7% dissociated. Write an equation to represent the equilibrium dissociation of this compound.

5.15 Write an equation to represent the equilibrium dissociation of water.

5.16 What is the meaning of the term *position of equilibrium*?

5.17 What can be said about the position of equilibrium (a) in the dissociation of water; (b) in the dissociation of HCl?

* 5.18 In the equilibrium ionization of acetic acid in water, how would you expect the rate of the reverse reaction (i.e., $C_2H_3O_2^- + H_3O^+ \rightarrow HC_2H_3O_2 + H_2O$) to change as the solution is diluted by adding more solvent? How should the rate of the forward reaction be affected? What should happen to the percentage ionization of $HC_2H_3O_2$ as the solution is made more dilute?

Ionic Reactions

5.19 What is the term used to describe a solid that is formed in a solution during a chemical reaction?

5.20 What is a metathesis reaction? What other name is often applied to this kind of reaction?

5.21 What is meant by the term *spectator ion*?

5.22 How can a precipitate be separated from a reaction mixture?

5.23 What are the differences between a molecular equation, an ionic equation, and a net ionic equation?

5.24 Of what value is a net ionic equation?

5.25 The compounds K_2CO_3, Na_2CO_3, KCl, and NaCl are soluble in water, but $CaCO_3$ is not. Given the following molecular equation,

$$CaCl_2(aq) + Na_2CO_3(aq) \longrightarrow CaCO_3(s) + 2NaCl(aq)$$

write the ionic and net ionic equations for the reaction between solutions of $CaCl_2$ and K_2CO_3.

5.26 Write ionic and net ionic equations for the following:
(a) $CuCl_2(aq) + Pb(NO_3)_2(aq) \rightarrow$
$$Cu(NO_3)_2(aq) + PbCl_2(s)$$
(b) $FeSO_4(aq) + 2NaOH(aq) \rightarrow$
$$Fe(OH)_2(s) + Na_2SO_4(aq)$$
(c) $ZnSO_4(aq) + BaCl_2(aq) \rightarrow ZnCl_2(aq) + BaSO_4(s)$
(d) $2AgNO_3(aq) + K_2SO_4(aq) \rightarrow$
$$Ag_2SO_4(s) + 2KNO_3(aq)$$
(e) $(NH_4)_2CO_3(aq) + CaCl_2(aq) \rightarrow$
$$2NH_4Cl(aq) + CaCO_3(s)$$

5.27 Write *balanced* ionic and net ionic equations for the following:
(a) $Cu(NO_3)_2(aq) + NaOH(aq) \rightarrow$
$$Cu(OH)_2(s) + NaNO_3(aq)$$
(b) $BaCl_2(aq) + Al_2(SO_4)_3(aq) \rightarrow BaSO_4(s) + AlCl_3(aq)$
(c) $Hg_2(NO_3)_2(aq) + HCl(aq) \rightarrow$
$$Hg_2Cl_2(s) + HNO_3(aq)$$
(d) $Bi(NO_3)_3(aq) + Na_2S(aq) \rightarrow Bi_2S_3(s) + NaNO_3(aq)$
(e) $CaCl_2(aq) + Na_2SO_4(aq) \rightarrow CaSO_4(s) + NaCl(aq)$

Acid–Base Reactions

5.28 What are some properties of acids and bases?

5.29 For aqueous solutions, how are acids and bases defined?

5.30 How can you test a solution in the laboratory to find out whether it is basic?

5.31 What is the difference between a strong acid and a weak acid?

5.32 Make a list of the names and chemical formulas of the strong acids. (Refer to Table 5.2, if necessary.) Write equations for their ionization in water.

5.33 Write the equations for the stepwise ionization of the following acids in water: (a) H_2SO_3 and (b) H_3AsO_4.

5.34 Define these terms: (a) monoprotic acid, (b) diprotic acid, (c) triprotic acid, and (d) polyprotic acid.

5.35 What is an acidic anhydride? What is a basic anhydride?

5.36 For each of the following, state whether their aqueous solutions would be expected to be acidic or basic: (a) N_2O_3 (b) CaO (c) P_4O_{10} (d) SeO_2 (e) Cs_2O

5.37 The anhydride of nitric acid is N_2O_5. Write a balanced equation showing the reaction of N_2O_5 with water to give HNO_3.

5.38 What acid is formed by the reaction of P_4O_{10} with H_2O?

5.39 A newly discovered element is found to react with oxygen to form an oxide that, when dissolved in water, causes litmus to turn blue. Would the element be classed as a metal or a nonmetal? Why?

5.40 What are the two classes of substances that are bases in an aqueous solution?

5.41 Write an equation for the reaction of the oxide ion with H_2O.

5.42 What is the name of the base formed by the reaction of K_2O with water?

5.43 Write an equation for the ionization of ammonia in water. Is ammonia a strong or a weak base?

5.44 Hydrazine, N_2H_4, is a weak base in water. Write an equation that shows how this molecule reacts to produce a basic solution.

5.45 What is the net ionic equation for the reaction of a strong acid with a strong base?

5.46 Complete and balance the following equations:
(a) $KOH + HCl \rightarrow$ (d) $CuO + HBr \rightarrow$
(b) $NaOH + HC_2H_3O_2 \rightarrow$ (e) $Fe_2O_3 + H_2SO_4 \rightarrow$
(c) $NH_3(aq) + HCl \rightarrow$

5.47 What is an acid salt? Write the formulas of all the salts that can be formed by the reaction of KOH with H_3PO_4. Name each of these salts.

Metathesis Reactions

5.48 What are the three driving forces for metathesis reactions?

5.49 Without looking at the solubility rules, indicate whether the following are soluble or insoluble: KCl, $(NH_4)_2SO_4$, $AgNO_3$, $PbSO_4$, $Mn(OH)_2$, $FePO_4$, $CaCO_3$, $Zn(ClO_4)_2$, $Ba(C_2H_3O_2)_2$, NiO.

5.50 Without looking at the solubility rules, indicate whether the following are soluble or insoluble: KNO_3, $FeCl_2$, $NiCO_3$, $(NH_4)_2HPO_4$, Hg_2Cl_2, $Al(OH)_3$, PbI_2, CuI_2, $SrBr_2$, CoS.

5.51 Write ionic and net ionic equations for the following:
(a) $Al(OH)_3(s) + 3HCl(aq) \rightarrow AlCl_3(aq) + 3H_2O$
(b) $CuCO_3(s) + H_2SO_4(aq) \rightarrow$
$$CuSO_4(aq) + H_2O + CO_2(g)$$
(c) $Cr_2(CO_3)_3(s) + 6HNO_3(aq) \rightarrow$
$$2Cr(NO_3)_3(aq) + 3H_2O + 3CO_2(g)$$

5.52 Write net ionic equations for each of the following:
(a) $NaBr + AgNO_3 \rightarrow AgBr + NaNO_3$
(b) $CoCO_3 + 2HNO_3 \rightarrow Co(NO_3)_2 + CO_2 + H_2O$
(c) $NaC_2H_3O_2 + HNO_3 \rightarrow NaNO_3 + HC_2H_3O_2$
(d) $Pb(NO_3)_2 + (NH_4)_2SO_4 \rightarrow PbSO_4 + 2NH_4NO_3$
(e) $H_2S + Cu(NO_3)_2 \rightarrow 2HNO_3 + CuS$
(f) $NaOH + NH_4Cl \rightarrow NH_3 + H_2O + NaCl$

5.53 Write balanced net ionic equations for each of the following:
(a) $CoS + HCl \rightarrow H_2S + CoCl_2$
(b) $PbCO_3 + HNO_3 \rightarrow H_2O + CO_2 + Pb(NO_3)_2$
(c) $PbCO_3 + H_2SO_4 \rightarrow PbSO_4 + H_2O + CO_2$
(d) $SnCl_2 + NaOH \rightarrow Sn(OH)_2 + NaCl$

(e) $Ag_2O + HCl \rightarrow AgCl + H_2O$

(f) $MgSO_4 + NiCl_2 \rightarrow MgCl_2 + NiSO_4$

5.54 Write molecular, ionic, and net ionic equations for the reaction (if any) between
(a) Na_2SO_4 and $BaCl_2$
(b) $Ca(NO_3)_2$ and $(NH_4)_2CO_3$
(c) $NaC_2H_3O_2$ and HNO_3
(d) $NaOH$ and $CuCl_2$
(e) $(NH_4)_2CO_3$ and HNO_3

5.55 Write molecular, ionic, and net ionic equations for the reaction (if any) between
(a) H_2SO_4 and $BaSO_4$
(b) NH_4Br and $MnSO_4$
(c) K_2S and $Ni(C_2H_3O_2)_2$
(d) $MgSO_4$ and $LiOH$
(e) $AgC_2H_3O_2$ and KCl

5.56 Write molecular, ionic, and net ionic equations for any reaction that would occur between
(a) $AgBr$ and KI
(b) $BaCl_2$, SO_2, and H_2O
(c) $Na_2C_2O_4$ and HCl
(d) K_2SO_3 and HCl
(e) $BaCO_3$ and H_2SO_4

* 5.57 Based on what you've learned in this chapter, predict what will occur if CO_2 is bubbled into a solution of $NaOH$. Use chemical equations to describe any reactions that will take place.

Ion–Electron Method

5.58 Balance the following equations by the ion–electron method. All reactions occur in an acidic solution.
(a) $Cu + NO_3^- \rightarrow Cu^{2+} + NO$
(b) $Zn + NO_3^- \rightarrow Zn^{2+} + NH_4^+$
(c) $Cr + H^+ \rightarrow Cr^{3+} + H_2$
(d) $Cr_2O_7^{2-} + H_3AsO_3 \rightarrow Cr^{3+} + H_3AsO_4$
(e) $I^- + SO_4^{2-} \rightarrow I_2 + H_2S$
(f) $Ag^+ + AsH_3 \rightarrow H_3AsO_4 + Ag$
(g) $S_2O_8^{2-} + HNO_2 \rightarrow NO_3^- + SO_4^{2-}$
(h) $MnO_2 + Br^- \rightarrow Br_2 + Mn^{2+}$
(i) $S_2O_3^{2-} + I_2 \rightarrow I^- + S_4O_6^{2-}$
(j) $IO_3^- + HSO_3^- \rightarrow I^- + SO_4^{2-}$

5.59 Balance the following equations by the ion–electron method. All reactions take place in an acidic solution.
(a) $Cr_2O_7^{2-} + CH_3CH_2OH \rightarrow Cr^{3+} + CH_3CHO$
(b) $PbO_2 + Cl^- \rightarrow Pb^{2+} + Cl_2$
(c) $Mn^{2+} + BiO_3^- \rightarrow MnO_4^- + Bi^{3+}$
(d) $ClO_3^- + HAsO_2 \rightarrow H_3AsO_4 + Cl^-$
(e) $PH_3 + I_2 \rightarrow H_3PO_2 + I^-$
(f) $MnO_4^- + S_2O_3^{2-} \rightarrow S_4O_6^{2-} + Mn^{2+}$
(g) $Mn^{2+} + PbO_2 \rightarrow MnO_4^- + Pb^{2+}$
(h) $As_2O_3 + NO_3^- \rightarrow H_3AsO_4 + N_2O_3$
(i) $P + Cu^{2+} \rightarrow Cu + H_2PO_4^-$
(j) $MnO_4^- + H_2S \rightarrow Mn^{2+} + S$

5.60 Balance the following equations by the ion–electron method. All reactions take place in a basic solution.
(a) $CN^- + AsO_4^{3-} \rightarrow AsO_2^- + CNO^-$
(b) $CrO_2^- + HO_2^- \rightarrow CrO_4^{2-} + OH^-$
(c) $Zn + NO_3^- \rightarrow Zn(OH)_4^{2-} + NH_3$
(d) $Cu(NH_3)_4^{2+} + S_2O_4^{2-} \rightarrow SO_3^{2-} + Cu + NH_3$

(e) $N_2H_4 + Mn(OH)_3 \rightarrow Mn(OH)_2 + NH_2OH$
(f) $MnO_4^- + C_2O_4^{2-} \rightarrow MnO_2 + CO_3^{2-}$
(g) $ClO_3^- + N_2H_4 \rightarrow NO_3^- + Cl^-$

* 5.61 Balance the following equations by the ion–electron method:
(a) $P_4 \rightarrow PH_3 + H_2PO_2^-$ (basic solution)
(b) $Cu + Cl^- + As_4O_6 \rightarrow CuCl + As$ (acidic solution)
(c) $IPO_4 \rightarrow I_2 + IO_3^- + H_2PO_4^-$ (acidic solution)
(d) $NO_2 \rightarrow NO_3^- + NO$ (acidic solution)
(e) $Br_2 \rightarrow Br^- + BrO_3^-$ (basic solution)
(f) $HSO_2NH_2 + NO_3^- \rightarrow N_2O + SO_4^{2-}$ (acidic solution)
(g) $ClO_3^- + Cl^- \rightarrow Cl_2 + ClO_2$ (acidic solution)
(h) $ClO_2 \rightarrow ClO_2^- + ClO_3^-$ (basic solution)
(i) $Se \rightarrow Se^{2-} + SeO_3^{2-}$ (basic solution)
(j) $ICl \rightarrow IO_3^- + I_2 + Cl^-$ (acidic solution)
(k) $FNO_3 \rightarrow O_2 + F^- + NO_3^-$ (basic solution)
(l) $Fe(OH)_2 + O_2 \rightarrow Fe(OH)_3$ (basic solution)

Oxidizing and Reducing Agents

5.62 Chlorine, Cl_2, is a powerful oxidizing agent, but it is seldom used in that capacity in the lab. Why?

5.63 What color is characteristic of solutions that contain (a) CrO_4^{2-}, (b) $Cr_2O_7^{2-}$, (c) MnO_4^-.

5.64 Complete and balance net ionic equations for reactions of the following:
(a) $SO_3^{2-} + CrO_4^{2-}$ (slightly acidic solution)
(b) $S_2O_3^{2-} + I_2$ (acidic solution)
(c) $S_2O_3^{2-} + Cl_2$ (acidic solution)

5.65 What is the correct balanced net ionic equation for the reaction between sodium sulfite and sodium chromate if the reaction takes place in an acidic solution?

5.66 What is the net ionic equation for the reaction of Na_2SO_3 and $KMnO_4$ in a basic solution?

5.67 What reaction will occur if SO_3^{2-} is oxidized by CrO_4^{2-} in a very basic solution?

Stoichiometry of Ionic Reactions

5.68 What is the meaning of the concentration unit *percent by mass (weight/weight percent)*?

5.69 What is the meaning of the concentration unit *parts per million*?

5.70 Modern analytical techniques have achieved such sensitivities that pollutants and potential cancer-causing agents can be detected at the level of *parts per billion, ppb*. What does this concentration term mean?

5.71 The concentration of fluoride ion in sea water is approximately 0.001 g per 1000 g of sea water. Express this concentration in
(a) percent by mass
(b) parts per million
(c) parts per billion

5.72 Mercury is an extremely toxic substance that deactivates enzyme molecules that promote biochemical reactions. A 25.0-g sample of tuna fish taken from a large shipment was analyzed for this substance and found to contain 2.1×10^{-5} mol of Hg. By law, foods having a mercury content above 0.50 ppm cannot be sold (they cannot even be given away). Determine whether this shipment of tuna must be confiscated.

5.73 What is the molarity of each of the following?
(a) 1.50 mol of NaCl in 2.00 L of solution
(b) 0.248 mol of KCN in 250 mL of solution
(c) 0.750 mol of H_2SO_4 in 1.35 L of solution
(d) 85.5 g of HNO_3 in 1.00 L of solution
(e) 44.5 g of $NH_4C_2H_3O_2$ in 600 mL of solution

5.74 How many moles of solute are in (a) 250 mL of 0.100 M KCl; (b) 1.65 L of 1.40 M $HClO_4$; (c) 0.0250 L of 0.0100 M $HC_2H_3O_2$?

5.75 How many grams of Na_2CO_3 are required to prepare 300 mL of a 0.150 M solution?

5.76 How many grams of $Ba(OH)_2$ are required to prepare 250 mL of a solution that has a hydroxide ion concentration of 0.300 M?

5.77 Pure nitric acid has a density of 1.513 g/mL. What is its molar concentration?

5.78 A solution of $MgSO_4$ contains 22.0% $MgSO_4$ by mass and contains 273.8 g of the salt per liter. What is the density of the solution in g/mL? What is the molarity of the solution?

5.79 What are the molar concentrations of the ions in each of the following salt solutions?
(a) 0.100 M LiCl   (d) 0.600 M $NaHSO_4$
(b) 0.250 M $CaCl_2$ (e) 0.400 M $Fe_2(SO_4)_3$
(c) 1.20 M $(NH_4)_2SO_4$

5.80 What are the molar concentrations of the ions in each of these salt solutions?
(a) 0.0250 M $Ba(OH)_2$
(b) 0.300 M $Cd(NO_3)_2$
(c) 0.400 M Na_2HPO_4
(d) 0.100 M $Cr_2(SO_4)_3$
(e) 0.0450 M $Hg_2(NO_3)_2$

5.81 The concentration of SO_4^{2-} is 0.100 M in a solution of Na_2SO_4. What is the molar concentration of Na_2SO_4?

5.82 The concentration of Cl^- is 0.160 M in a solution of $FeCl_3$. What is the molar concentration of $FeCl_3$?

5.83 The concentration of Al^{3+} is 0.140 M in a solution of $Al_2(SO_4)_3$. What is the molar concentration of $Al_2(SO_4)_3$?

5.84 How many moles of each kind of ion are present in
(a) 50.0 mL of 0.200 M NaCl
(b) 30.0 mL of 0.160 M $CaCl_2$
(c) 27.0 mL of 0.650 M Na_2SO_4
(d) 135 mL of 0.820 M $(NH_4)_2SO_4$
(e) 75.0 mL of 0.250 M $Al_2(SO_4)_3$

5.85 What are the molar concentrations of the ions in a solution prepared by dissolving 10.45 g $CuSO_4 \cdot 5H_2O$ in a volume of 150.0 mL of solution?

5.86 How many milliliters of 0.100 M NaOH are needed to react with 5.00×10^{-3} mol of H_2SO_4 to give Na_2SO_4?

5.87 How many milliliters of 1.250 M HNO_3 contain enough nitric acid to dissolve a copper penny weighing 3.22 g? The net ionic equation is

$$3Cu(s) + 8H^+(aq) + 2NO_3^-(aq) \longrightarrow \\ 3Cu^{2+}(aq) + 2NO(g) + 4H_2O$$

5.88 Copper(II) carbonate dissolves in perchloric acid ($HClO_4$).
(a) Write a net ionic equation for the reaction.
(b) How many milliliters of 1.35 M $HClO_4$ are needed to prepare 5.25 g of $Cu(ClO_4)_2$?
(c) How many grams of $CuCO_3$ are needed to prepare 5.25 g $Cu(ClO_4)_2$?

5.89 How many milliliters of 0.300 M NaOH are required to react with 500 mL of 0.170 M H_3PO_4 to yield (a) Na_3PO_4, (b) Na_2HPO_4, (c) NaH_2PO_4?

5.90 How many milliliters of 0.100 M $BaCl_2$ are required to react with 25.0 mL of 0.200 M H_2SO_4?

5.91 How many milliliters of 0.1000 M $BaCl_2$ are required to react completely with 25.0 mL of 0.200 M $Fe_2(SO_4)_3$?

5.92 A 0.244-g sample of benzoic acid (a monoprotic acid) requires 20.0 mL of 0.100 M NaOH for complete neutralization. Calculate the molecular mass of the acid.

5.93 In an experiment 20.0 mL of 0.200 M $AgNO_3$ is added to 30.0 mL of 0.200 M NaCl.
(a) Write the net ionic equation for the reaction that occurs.
(b) How many moles of precipitate are formed?
(c) How many grams of precipitate are formed?
(d) What are the molar concentrations of each of the ions remaining in the solution after the reaction is complete?

5.94 How many grams of AgCl will be formed if 25.0 mL of 0.050 M HCl is added to 100 mL of 0.50 M $AgNO_3$?

5.95 Caproic acid is a foul-smelling substance found in certain excretions of male goats during the breeding season. The acid has an empirical formula of C_3H_6O. A 0.100-g sample of the acid required 17.2 mL of 0.0500 M NaOH for complete reaction. Assuming that the acid is monoprotic, calculate (a) its molecular mass; (b) its molecular formula.

5.96 If 380 mL of 0.273 M $Ba(OH)_2$ is added to 500 mL of 0.520 M HCl, will the mixture be acidic or basic? Calculate the molar concentration of H^+ (or OH^- if the solution is basic) in the final mixture. Assume that volumes are additive.

5.97 In an experiment 40.0 mL of 0.270 M $Ba(OH)_2$ is added to 25.0 mL of 0.330 M $Al_2(SO_4)_3$.

(a) Write the net ionic equation for the reaction that occurs.

(b) What total mass (in grams) of precipitate is formed?

(c) What is the molar concentration of each of the ions remaining in solution?

Chemical Analysis and Titrations

5.98 What is the purpose of a *chemical analysis*?

5.99 A 0.249-g sample of a compound containing titanium and chlorine was dissolved in water and treated with silver nitrate solution. the silver chloride that formed was found to weigh 0.694 g after being filtered, washed, and dried. What is the empirical formula of the original compound?

5.100 A certain lead ore contains the compound $PbCO_3$. A sample of this ore weighing 1.526 g was treated with nitric acid, which dissolved the $PbCO_3$, and the resulting solution was then treated with Na_2SO_4. This gave a precipitate of $PbSO_4$ which was dried and found to weigh 1.081 g. What is the percentage by mass of $PbCO_3$ in the ore?

* 5.101 A 1.850-g sample of a mixture of $CuCl_2$ and $CuBr_2$ was dissolved in water and mixed thoroughly with a 1.800-g portion of AgCl. After the reaction, the solid, which now consisted of a mixture of AgCl and AgBr, was filtered, washed, and dried. Its mass was found to be 2.052 g. What percent by mass of the original mixture was $CuBr_2$?

5.102 Describe the following: (a) buret, (b) titration, (c) titrant, and (d) endpoint.

5.103 What is the function of an indicator? What color is phenolphthalein in (a) an acidic solution and (b) a basic solution?

5.104 A 15.00-mL portion of a solution of H_2SO_4 of unknown concentration was titrated with 0.1500 M NaOH. The titration required 21.30 mL of the base. Assuming complete neutralization of the H_2SO_4, what was the acid's molarity?

5.105 A 1.030-g portion of a mixture containing $CaCO_3$, $CaSO_4 \cdot 2H_2O$, and $BaSO_4$ was heated, which caused the $CaCO_3$ to decompose into CaO and CO_2. The resulting solid was treated with water, which caused the CaO to dissolve, forming $Ca(OH)_2$. This solution required 37.25 mL of 0.120 M HCl for complete neutralization in a titration. What was the percentage by mass of $CaCO_3$ in the original sample?

* 5.106 A mixture of the two monoprotic acids—lactic acid, $HC_3H_5O_3$ (found in sour milk), and caproic acid, $HC_6H_{11}O_2$ (found in excretions from the goat)—was titrated with 0.0500 M NaOH. A 0.1000-g sample of the mixture required 20.4 mL of the base. What was the mass in grams of each acid in the sample?

* 5.107 An amino acid isolated from a piece of animal tissue was

believed to be glycine, $NH_2CH_2CO_2H$. A 0.0500-g sample was treated in such a way that all the nitrogen in it was converted to ammonia. This NH_3 was added to 50.0 mL of 0.05000 M HCl, neutralizing part of the acid.

$$NH_3 + HCl \longrightarrow NH_4Cl$$

The acid remaining in the solution was titrated with 0.0600 M NaOH, which required 30.57 mL of the base for neutralization.

(a) How many moles of HCl were neutralized by the NH_3?

(b) How many grams of nitrogen were in the 0.0500-g sample?

(c) What was the percentage by mass of nitrogen in the sample? How does it compare with the percentage nitrogen calculated for glycine?

Equivalent Weights and Normality

5.108 How are equivalents defined for acids and bases? How are they defined for oxidation–reduction?

5.109 If two substances A and B react, what relationship exists between the number of equivalents of A and B that react?

5.110 How many equivalents are there in 0.200 mol $Ba(OH)_2$?

5.111 How many moles of H_3PO_4 must be used to give 5.00 eq H_3PO_4 if the acid is to be neutralized to give PO_4^{3-}?

5.112 How many equivalents are there in 0.140 mol H_3AsO_4 if it is neutralized to give $HAsO_4^{2-}$?

5.113 What is the equivalent weight of $MnSO_4$ when it is oxidized to produce (a) Mn_2O_3, (b) MnO_2, (c) K_2MnO_4, (d) $KMnO_4$?

5.114 How many grams of Na_2CrO_4 are needed to give 0.400 eq Na_2CrO_4 if it will be reduced to Cr^{3+}?

5.115 How many grams of $MnSO_4 \cdot 6H_2O$ are needed to prepare 300 mL of 0.100 N $MnSO_4$ if the Mn^{2+} will be oxidized to MnO_4^-?

5.116 How many grams of $NaBiO_3$ are required to react with 0.500 g of $Mn(NO_3)_2$ to produce $NaMnO_4$ and $Bi(NO_3)_3$? Solve the problem using equivalent weights.

5.117 What is the equivalent weight of
(a) H_3PO_4 when neutralized to HPO_4^{2-}
(b) $HClO_4$ when it reacts as an acid
(c) $NaIO_3$ when reduced to I^-
(d) $NaIO_3$ when reduced to I_2
(e) $Al(OH)_3$

5.118 What is the normality of each of the following solutions:
(a) 22.0 g of $Sr(OH)_2$ in 800 mL of solution
(b) 500 mL of 0.25 M H_2SO_4, for complete neutralization
(c) 0.150 M H_3PO_4, when neutralized to HPO_4^{2-}
(d) 41.7 g of $K_2Cr_2O_7$ in 600 mL of solution when used in a reaction in which one product is Cr^{3+}

(e) 25.0 g of Na_2O dissolved in sufficient water to give 1.50 L of solution

(f) 0.135 eq of H_2SO_4 in 400 mL of solution

5.119 A volume of 129 mL of 0.850 N $Ba(OH)_2$ is required to completely neutralize a 4.93-g sample of an acid. What is the equivalent weight of the acid?

* 5.120 In acid solution, 45.0 mL of $KMnO_4$ solution is required to react with 50.0 mL of 0.250 N $H_2C_2O_4$ to give Mn^{2+} and CO_2 as products. How many milliliters of this same $KMnO_4$ solution are required to oxidize 25.0 mL of 0.250 N $K_2C_2O_4$ to yield, in basic solution, MnO_2 and CO_2 as products?

* 5.121 A 10.0-mL portion of a solution of HCl was diluted to exactly 50.0 mL. If 5.00 mL of this solution required 41.0 mL of 0.255 N NaOH for complete neutralization, what was the concentration of the original HCl solution before dilution?

5.122 How many milliliters of 0.500 N $K_2Cr_2O_7$ must be used to completely oxidize the contents of 120 mL of 0.850 N $H_2C_2O_4$?

5.123 How many milliliters of water must be added to 85.0 mL of 1.00 N H_3PO_4 to give a solution that is 0.650 N H_3PO_4? Assume that the volumes of the liquids are additive.

* 5.124 To what volume must 250 mL of 1.40 M H_2SO_4 be diluted to give a solution, 25.0 mL of which is able to be completely neutralized by 15.0 mL of 0.750 M NaOH?

Additional Exercises

5.125 A particular pollutant was present in water at a concentration of 825 ppm. What was the concentration in percent by mass? If the pollutant was benzene, C_6H_6, what was its molar concentration (assume that 1000 mL of the solution has a mass of essentially 1000 g)?

* 5.126 Silver iodide is less soluble than silver bromide, which in turn is less soluble than silver chloride. A 0.2000-g sample that was known to be a mixture of NaCl, NaBr, and NaI was dissolved in water and an excess of $AgNO_3$ was added. The precipitate containing AgCl, AgBr, and AgI was filtered, dried, and found to weigh 0.4120 g. This solid was then placed in water and treated with NaBr solution, which converted any AgCl to AgBr (but it did not affect the AgI). The precipitate that now consisted of AgBr and AgI was filtered, dried, and found to weigh 0.4881 g. It was then placed in water and treated with NaI, which converted the AgBr to AgI. When the solid (now consisting entirely of AgI) was filtered and dried, it weighed 0.5868 g. What were the percentages by mass of NaCl, NaBr, and NaI in the original sample?

* 5.127 A 0.1000-g sample of a mixture of $FeSO_4$ and $Fe_2(SO_4)_3$ was dissolved in an H_2SO_4 solution and titrated with 0.00400 M $KMnO_4$. The titration required 15.8 mL of

the $KMnO_4$ solution. What percent (by mass) of the mixture was $FeSO_4$?

5.128 A sample of $Cr_2(SO_4)_3$ weighing 0.500 g was dissolved in water and an excess of dilute NaOH was added, precipitating $Cr(OH)_3$. This precipitate was collected by filtration. How many milliliters of 0.400 M HNO_3 are needed to dissolve the precipitate?

* 5.129 A mixture of $MgCO_3$ and $CaCO_3$ (a dolomitic limestone used in agriculture) was heated to produce MgO and CaO. A 2.000-g sample of this oxide mixture was allowed to react with 100 mL of 1.00 M HCl. The excess HCl required 19.6 mL of 1.00 M NaOH for complete neutralization. What were the percentages by mass of $CaCO_3$ and $MgCO_3$ in the original limestone sample?

* 5.130 A sample of rock containing limestone ($CaCO_3$) was heated, converting the $CaCO_3$ to CaO. This was treated with H_2O to give $Ca(OH)_2$, which was then titrated with HCl. In one analysis, a 0.2000-g sample, taken *after* converting the $CaCO_3$ to CaO as described above, required 30.30 mL of 0.1000 M HCl for complete neutralization. What was the percentage by mass of $CaCO_3$ in the original rock?

5.131 Ascorbic acid (vitamin C) is a diprotic acid having the formula $H_2C_6H_6O_6$. A sample of a vitamin supplement was analyzed by titrating a 0.1000-g sample dissolved in water with 0.0200 M NaOH. A volume of 15.2 mL of the base was required to completely neutralize the ascorbic acid. What was the percentage by mass of ascorbic acid in the sample?

5.132 How many equivalents are there in 1 mol of HIO_3 if it is to be used in a reaction in which
(a) it is neutralized by a base?
(b) the iodine is reduced to I_2?

5.133 A 1.362-g sample of an iron ore that contains Fe_3O_4 was dissolved in acid and all the iron was reduced to Fe^{2+}. The solution was then acidified with H_2SO_4 and titrated with 39.42 mL of 0.0281 M $KMnO_4$ solution, which oxidized the iron to Fe^{3+}.
(a) What is the net ionic equation for the reaction between MnO_4^- and Fe^{2+}?
(b) What was the percentage of Fe_3O_4 in the sample?

5.134 A mixture of $CaCl_2$ and NaCl weighing 2.385 g was dissolved in water and treated with a solution of sodium oxalate, $Na_2C_2O_4$, which produced a precipitate of calcium oxalate, CaC_2O_4. This precipitate was filtered from the mixture and then dissolved in HCl to give $H_2C_2O_4$.

$$CaC_2O_4(s) + 2H^+(aq) \longrightarrow Ca^{2+}(aq) + H_2C_2O_4(aq)$$

The $H_2C_2O_4$ was titrated with $KMnO_4$, giving CO_2 and Mn^{2+} as products. The titration required 19.64 mL of 0.2000 M $KMnO_4$.
(a) How many moles of CaC_2O_4 had precipitated?
(b) What was the percentage by mass of $CaCl_2$ in the original sample?

THE STRUCTURE OF MATTER AND THE ORIGIN OF CHEMICAL PERIODICITY

6

ENERGY AND ENERGY CHANGES: THERMOCHEMISTRY

In the preceding chapters you've learned how some of the elements react and the kinds of compounds they form, and you've learned how to deal quantitatively with the amounts of chemicals in chemical reactions. In this chapter we turn our attention to another important aspect of chemicals and chemical reactions—the energy chemicals possess and the energy changes that take place when chemical reactions occur.

Energy is a term with which you're surely familiar. You wake up in the morning "full of energy," and gasoline is a "source of energy" for your car. The importance of energy and energy resources is something that is generally recognized by nearly everyone, yet few people outside science really understand what energy is. Learning the meaning of energy, therefore, should be one of your principal goals as you study this chapter, not only because it will make you a better-informed citizen, but also because understanding energy is vital to understanding the fundamental reasons for chemical and physical properties.

6.1 ENERGY AND ENERGY TRANSFER

Energy is a more difficult concept to understand than *matter*, because energy and matter are so different. You can't see and touch energy or put energy in a bottle to study it. All you can do is study the effects energy has on objects.

Energy is usually defined as the capacity to do work; it is something matter has that can make things happen. When an object has energy, it can affect other objects by doing work on them. For example, a moving car has energy because it can "do work" on another car by moving it some distance in a collision. Similarly, coal and oil have energy that is liberated during combustion as heat, which can be harnessed to make machinery do work.

An object can possess energy in just two ways: as kinetic energy and as potential energy. The total amount of energy an object has is therefore the sum of its kinetic energy and its potential energy.

Kinetic energy (K.E.) is energy an object has when it is moving and can be calculated by the equation

$$K.E. = \tfrac{1}{2}mv^2$$

where m is the object's mass and v is its velocity (or speed). Thus, kinetic energy depends on both an object's mass and its speed, which is really something you already know. For example, you know that a truck moving at 20 mi/hr can do more "work" on the rear end of a car than a bicycle moving at the same speed, because the truck has a larger mass. Similarly, a truck moving at 80 mi/hr can do more work on a car than one traveling at just 5 mi/hr.

Potential energy (P.E.) is stored energy, and it is the energy an object has because it is either attraced to or repelled by some other object. In fact, *if an object feels no attractive or repulsive forces, then it has no potential energy.*

It is easy to see how potential energy is related to attractions and repulsions by considering some examples. As you know, objects are attracted to the earth by gravity. If you want to raise something from the floor to a desk top, you have to *do work* to lift it. The work that you do is stored as potential energy when the object's distance from the earth increases, and it can be recovered and harnessed by letting the object fall back to earth. That's how an old-time grandfather clock works (Figure 6.1). To provide the energy the clock needs to operate, you pull a chain that lifts a weight. This gives the weight potential energy. Then, as the weight slowly drops, controlled by the swinging of the clock's pendulum, it pulls the chain over a gear, causing the clock mechanism to move. These workings of the clock illustrate an important relationship between potential energy and the distance between objects that attract each other.

> When objects that attract each other are pulled apart, their potential energy increases, and when they move toward each other, their potential energy decreases.

A somewhat similar situation exists when there are repulsions between objects. Consider, for example, the magnets illustrated in Figure 6.2. As you probably know from playing with magnets, when they're held in a certain way, the end of one magnet repels the end of another; we say the "north pole" of one repels the "north pole" of the other. Because the repulsions push the magnets apart, we have to do some work to move them closer together. This causes their potential energy to increase; the energy we expended pushing them together is now stored by the magnets. If we release them, they will push each other apart, and this "push" could also be made to do work for us. Here we see another important fact about potential energy.

> When objects that repel are pushed toward each other, their potential energy increases, and when they are moved apart, their potential energy decreases.

The relationships between potential energy and the distances between objects that attract or repel is very important to understand. As you will see, all sorts of chemical and physical properties of substances are related to potential energy changes, and they can be understood and explained by examining how atoms and subatomic particles attract and repel each other.

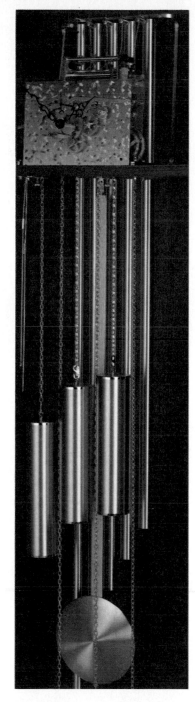

Figure 6.1. *The weights in a grandfather clock provide the energy needed to make the clock work. Raising the weights gives them potential energy, which is converted to kinetic energy as they gradually descend, pulling the chain over a gear and driving the clock's mechanism.*

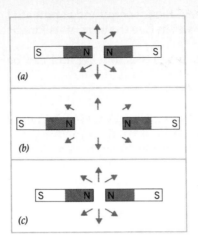

Figure 6.2. (a) *The north poles of two magnets repel when placed near each other. (b) When the magnets are moved apart, their potential energy decreases. (c) When they are moved together, their potential energy increases.*

Conservation of energy

Scientists have found that in any isolated system (including our universe, we believe), the total amount of energy is constant. This leads to a major physical law, called the **law of conservation of energy,** which states that *energy can be neither created nor destroyed. It can only be transformed from one kind to another.* The law of conservation of energy controls what happens to energy during chemical and physical changes. For example, we've seen that kinetic energy can change to potential energy when we raise a clock weight, and potential energy changes to kinetic energy when the weight descends and moves the gears of the clock. But we can't destroy energy and we can't make it from nothing. If we want to increase the energy of something by a certain amount, it has to come from something else.

How atoms and molecules possess energy

Atoms and molecules (and even ions) are objects just like clock weights and magnets, and therefore they can have energy in exactly the same way: as kinetic energy and as potential energy. The way they do this, and how we observe their energy, is explained as follows.

This explanation is part of the kinetic theory of matter, which we will explore further in Chapter 11.

Kinetic Energy In any substance, whether it's a solid, liquid, or gas, the individual atomic-sized particles are in constant motion, colliding with and bouncing off each other. If you study an object such as a pencil, however, you don't notice this because the atoms are too small to see and because the motions are random. This randomness causes the motions to cancel out, so the pencil itself doesn't jump around, even though the tiny particles within it are moving.

Because of collisions, the speeds and kinetic energies of the atoms and molecules are constantly changing.

Because atoms and molecules are moving, they possess kinetic energy. At any particular instant, some particles might be moving slowly and therefore have small kinetic energies. Other faster-moving particles have larger kinetic energies. For any collection of these particles, however, there is some average value for the kinetic energy. Physicists have shown that *for any object, the average kinetic energy of its atomic-sized particles is directly proportional to the absolute temperature (Kelvin temperature) of the object.* This means that if something is hot, its atoms and molecules have a larger average kinetic energy, and are moving faster (on average), than the atoms and molecules in something that is cool. (Interestingly, a thermometer is therefore a device that monitors the average kinetic energies of atoms, molecules, and ions!)

Potential Energy Potential energy arises from attractions and repulsions, and as you learned in Chapter 4, atoms are made of electrically charged particles (nuclei and electrons) that attract and repel each other. Therefore, electrons and nuclei have potential energy, which changes whenever their distances change. There are potential energy changes when electrons are transferred between atoms during the formation of ions, and there are potential energy changes when atoms share electrons during the formation of molecular substances.

The potential energy that substances have because of the attractions and repulsions between their subatomic particles is sometimes called **chemical energy.** When chemicals react, there are changes in the nature of the attractions (the *chemical bonds*) between their atoms, so there are changes in chemical energy (potential energy) that we observe as energy being given off or absorbed during the reaction.

> The nature of chemical bonds will be discussed in Chapters 8 and 9.

Heat energy

One of the most common ways that we encounter energy is in the form of heat. When you think of the combustion of a fuel, such as the butane in a cigarette lighter, you think immediately of the heat that's generated. When something hot is placed next to something cold, *heat flows* into the cold object from the hot one. How is heat and the transfer of heat related to the kinds of energy discussed above?

Heat is really kinetic energy — the kinetic energy of atoms and molecules. When something is hot, the average kinetic energy of its molecules is large and it has a lot of heat in it. If it is cold, the average kinetic energy is small and the object has less heat in it.

> Heat isn't another kind of energy; it's just the kinetic energy of atoms and molecules.

When a hot object is placed in contact with a cold one, heat flows from hot to cold until both eventually come to the same temperature. If we could view the interface between the hot and cold objects at an atomic level, we would see rapidly moving molecules in the hot object and slow-moving molecules in the cold one (Figure 6.3). As we watched, we would see the faster molecules on one side collide with the slower ones on the other. These collisions would cause the faster ones to slow down and the slower ones to speed up, and we would witness kinetic energy being transferred from the hotter object to the cooler one through the collisions between their molecules. Eventually, the average kinetic energies of the particles in both objects would become the same and they would have come to the same temperature.

 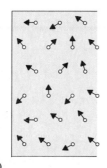
(a)

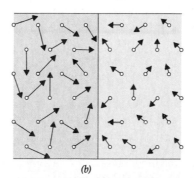
(b)

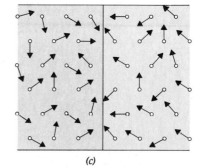
(c)

Figure 6.3. (a) *Two objects at different temperatures. The molecules in the hot object on the left are moving faster than those in the cool object on the right.* (b) *When the objects are placed in contact, collisions between the faster-moving molecules in the hot object and the slower ones in the cool object cause the fast molecules to slow down and the slow molecules to speed up.* (c) *Eventually, thermal equilibrium is reached and the average kinetic energies in both objects become the same.*

6.2 ENERGY CHANGES IN CHEMICAL REACTIONS

Of course, we can't really "produce" energy. The energy we take from a reaction must come at the expense of the chemicals taking part in the reaction.

One of civilization's principal uses for chemical reactions is the "production" of energy — the energy that we need for all the common tasks we perform. The combustion of fuels such as oil and coal is used to generate electricity. Gasoline is burned in your car engine to provide the power to make the car go. If you have a gas stove in your home, you burn methane, the principal component of natural gas, to provide the heat for cooking. And through the sequence of reactions called metabolism, the food you eat provides you with the energy you need to function.

In almost every chemical reaction, there is either an absorption or a release of energy. Let's examine how this comes about and how the energy change makes itself known to us.

Suppose we were to carry out a chemical change in an insulated container, so that no heat could either enter or leave the reaction mixture. Stated another way, suppose we carry out the reaction so that the *total* energy stays the same. Let's also suppose that the potential energy of the products is lower than the potential energy of the reactants, so when the reaction occurs, there is a *decrease* in the potential energy. This energy can't just disappear, because the total energy (kinetic plus potential) has to stay constant. Therefore, if the potential energy goes down, the kinetic energy must go up; potential energy is changed to kinetic energy. This increase in the amount of kinetic energy causes the average kinetic energy of the molecules to increase, which we observe as a rise in the temperature of the reaction mixture. The reaction mixture becomes hot.

Most of the time, chemical reactions are not insulated from the rest of the world. If the reaction mixture becomes hot, as described above, heat can flow to the surroundings. Any change that is able to release energy to the surroundings like this is said to be an **exothermic change.** Notice that *when an exothermic reactions occurs, the temperature of the reaction mixture rises and the potential energy of the chemicals involved in the reaction decreases.*

Sometimes chemical changes occur in which there is an increase in the potential energy of the substances involved. When this happens, there is a corresponding decrease in the kinetic energy and the temperature of the reaction mixture drops. If the reaction mixture isn't insulated from the surroundings, heat can then flow into the reaction mixture and the change is said to be an **endothermic change.** Notice that *when an endothermic reaction occurs, the temperature of the reaction mixture drops and the potential energy of the chemicals involved in the reaction increases.*

The combustion reactions taking place in this bonfire are exothermic and release their energy to the surroundings.

EXAMPLE 6.1. ANALYZING ENERGY CHANGES

PROBLEM: When gasoline vapors are mixed with air and a spark is passed through the mixture, a rapid reaction occurs that can be represented by the chemical equation

$$2C_8H_{18} + 25O_2 \longrightarrow 16CO_2 + 18H_2O$$

The reaction mixture becomes very hot, and the heat generated is able to power a car. How should this reaction be described, as exothermic or endothermic? Which has the higher potential energy, $2C_8H_{18} + 25O_2$ or $16CO_2 + 18H_2O$?

SOLUTION: Since the reaction mixture becomes hot, the reaction must be exothermic. In an exothermic reaction, the potential energy decreases, so the products of the reaction ($16CO_2 + 18H_2O$) must have a lower potential energy than the reactants.

6.3 MEASURING ENERGY IN CHEMICAL REACTIONS

In discussing energy changes, we have looked at only one of the ways that energy can be exchanged with the surroundings — as heat. Some reactions also are able to give off light, and we will discuss light as a form of energy in the next chapter. Some reactions can also supply their energy as electricity, which is what happens in batteries like the one you use to start your car. We will look at these reactions, too, later in the book. These various forms of energy can be converted from one to the other and they are therefore ultimately equivalent. This makes expressing amounts of energy easy because we don't need a multitude of units to do the job.

The SI unit for energy, the **joule (J)**, is derived from the definition of kinetic energy. One joule is 1 kg m^2/s^2 and corresponds to the amount of energy possessed by an object with a mass of 2 kg traveling at a velocity of 1 m/s (in English units, an object weighing 4.4 lb traveling at a speed of 197 ft/min, or about 2.2 mi/hr).[1]

$$1 \text{ J} = 1 \text{ kg m}^2/\text{s}^2$$

A smaller energy unit sometimes used in physics is the *erg*, which is 1×10^{-7} J. In referring to energies involved in chemical reactions between mole-sized amounts of reactants, we often use a larger unit, the **kilojoule (kJ)**. One kilojoule = 1000 joules (1 kJ = 1000 J).

All forms of energy can be changed completely to heat, and when chemists measure energy, it is usually in the form of heat. The traditional unit used to express heat is called the **calorie** (abbreviated **cal**) and its original definition relied on the effect that heat has on the temperature of a substance. Originally, the calorie was defined as the amount of heat needed to raise the temperature of 1 g of water, initially at 15 °C, by 1 °C.[2] The **kilocalorie (kcal)**, like the kilojoule, is a more appropriately sized unit for dealing with energy changes in chemical reactions. It is also the unit used in nutrition for reporting the energy content of foods; the Calorie (with a capital C) is the same as a kilocalorie. Thus, when we read that a serving of mashed potatoes contains 230 Cal, we are being told that 230 kcal of energy is liberated when the body metabolizes this food.

Until relatively recently, all the chemical scientific literature has used the calorie (or kilocalorie) for reporting energy changes. With the acceptance of the SI, the joule (or kilojoule) is now preferred and the calorie has been redefined in terms of this SI unit. The calorie and kilocalorie are now defined *exactly* by the following relationships:

$$1 \text{ cal} = 4.184 \text{ J}$$
$$1 \text{ kcal} = 4.184 \text{ kJ}$$

In this book, we will use joules and kilojoules almost exclusively. However, there is much valuable data contained in the older scientific literature. To use it effectively in modern science requires translating it (or comparing it) to joules or kilojoules, so you should learn these conversion factors.

Heat capacity and specific heat

The property of water that permitted the original definition of the calorie is the size of the temperature change that water undergoes when it absorbs or loses a given

Objects can have only K.E. and P.E., but energy can be transferred between objects as heat, light, electricity, or even as sound.

You should learn these conversion factors.

---

[1] Remember, K.E. = $\frac{1}{2}mv^2$. In this case, K.E. = $\frac{1}{2}(2 \text{ kg})(1 \text{ m/s})^2 = 1 \text{ kg} \cdot \text{m}^2/\text{s}^2 = 1$ J.

[2] It is necessary to specify the temperature of the water because the amount of heat required to raise the temperature of 1 g of water by 1 °C varies slightly with the temperature of the water. For example, at 25 °C (approximately room temperature), it takes 0.998 cal to raise the temperature of 1 g of H_2O by 1 °C.

amount of heat. The general term for this property is **heat capacity,** which is defined as the amount of heat needed to change the temperature of something by 1 °C.

Heat capacity is an extensive property, which means that its magnitude depends on the size of the sample. For instance, to raise the temperature of 1 g of water by 1 °C requires 4.18 J (1 cal), but to change the temperature of 100 g of water by 1 °C requires 100 times as much energy, or 418 J. The 1-g sample has a heat capacity of 4.18 J/°C, whereas the 100-g sample has a heat capacity of 418 J/°C.

An intensive property related to heat capacity is **specific heat,** which is defined as the amount of heat needed to raise the temperature of 1 g of a substance by 1 °C. For water, the specific heat is 4.18 J g^{-1} $°C^{-1}$. Most substances have much smaller specific heats than water. Iron, for instance has a specific heat of only 0.452 J g^{-1} $°C^{-1}$. This means that it takes less heat to raise the temperature of 1 g of iron by 1 °C than it does to cause the same temperature change for 1 g of water. It also means that a given amount of heat will raise the temperature of 1 g of iron more than it will raise the temperature of 1 g of water.

The large specific heat of water is responsible for the moderating effects the oceans have on weather. These large bodies of water cool much more slowly in winter than do large landmasses, so air moving onto land after passing over the oceans is never as cold as air that has traveled long distances over the much colder landmasses. Similarly, in summer the air near the oceans is never quite as hot as air over the central parts of continents because the oceans warm much more slowly than the landmasses.

The following examples illustrate how we can calculate and use heat capacity and specific heat.

> Heat capacity has units of J/°C or cal/°C, or J/K or cal/K.

> Specific heat has units of
>
> $$\frac{J}{g \, °C} \quad \text{or} \quad \frac{cal}{g \, °C},$$
>
> which can also be written J g^{-1} $°C^{-1}$ and cal g^{-1} $°C^{-1}$. Because the temperature unit in the denominator is for a temperature change and because a Celsius degree is the same size as a kelvin, these units can also be written as J g^{-1} K^{-1} or cal g^{-1} K^{-1}.

EXAMPLE 6.2. CALCULATING THE HEAT CAPACITY OF AN OBJECT

PROBLEM: What is the heat capacity expressed in kJ/°C of a 2.00-kg copper bar, given that the specific heat of copper is 0.387 J g^{-1} $°C^{-1}$?

ANALYSIS: The units here give us the method we need to solve the problem. Heat capacity has units of (energy)/(temperature) whereas specific heat has units of (energy)/(mass × temperature). Multiplying the specific heat by the mass of the copper bar will give us the heat capacity in units of J/°C, which we can then convert to kJ/°C.

$$\text{specific heat} \times \text{mass} = \text{heat capacity}$$

$$\left(\frac{J}{\cancel{g} \, °C}\right) \times \cancel{g} = \frac{J}{°C}$$

SOLUTION: The analysis has given us the method. Now all we need to do is the arithmetic.

$$(0.387 \text{ J } \cancel{g^{-1}} \, °C^{-1}) \times 2000 \, \cancel{g} = 774 \text{ J } °C^{-1}$$

Changing joules to kilojoules, we find that the heat capacity is 0.774 kJ/°C.

> The mass was given to three significant figures, so the answer is rounded to three significant figures.

EXAMPLE 6.3. CALCULATIONS INVOLVING SPECIFIC HEAT

PROBLEM: How many joules are needed to raise the temperature of an iron nail that has a mass of 7.05 g from 25 °C to 100 °C? The specific heat of iron is 0.452 J g^{-1} $°C^{-1}$.

ANALYSIS: Once again, the units come to the rescue. If we multiply the specific heat by the mass and the temperature change, the units g and °C will cancel, giving us the desired units of energy.

specific heat × mass × temperature change = heat energy

$$\left(\frac{J}{g\,°C}\right) \times g \times °C = J$$

Heat affects a body by *changing* its temperature. That's why we use the temperature change here.

SOLUTION: All we need to do is the arithmetic. The temperature change is 75 °C (100 °C − 25 °C). Therefore, the answer to the problem is

$$(0.452\ J\ g^{-1}\ °C^{-1}) \times 7.05\ g \times 75\ °C = 240\ J\ \text{(rounded)}$$

Notice that the answer has been rounded to two significant figures.

Figure 6.4. *A bomb calorimeter.*

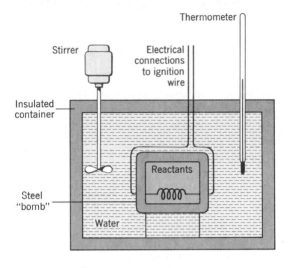

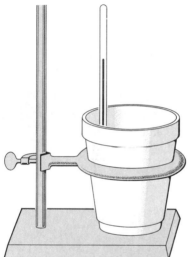

Figure 6.5. *A coffee cup calorimeter, consisting of a Styrofoam coffee cup supported by a metal ring. A thermometer is used to monitor the temperature of the reaction mixture.*

Calorimetry: measuring energy changes in chemical reactions

The energy change in a chemical reaction can always be made to appear as heat, so it is conveniently termed the **heat of reaction.** An apparatus that's used to measure the heat of reaction is called a **calorimeter** (literally, a *calorie meter,* although we now express the heat in joules rather than calories). There are various designs for this apparatus, and one called a **bomb calorimeter** is shown in Figure 6.4. This kind of calorimeter is normally used to study exothermic reactions that don't begin until they are initiated by applying heat—reactions such as the combustion of CH_4 with O_2 or the reaction of H_2 with O_2. The apparatus consists of a strong steel container (the *bomb*) into which the reactants are placed. The bomb is then immersed in an insulated bath that's fitted with a stirrer and a thermometer. The initial temperature of the bath is measured and then the reaction is set off by a small heater wire within the bomb. The heat given off in the reaction is absorbed by the bomb and the bath, causing the temperature of the entire apparatus to rise. From the temperature change and the previously measured heat capacity of the apparatus, the amount of heat given off in the reaction can be calculated.

A bomb calorimeter is capable of very precise measurements. A simpler apparatus that you might encounter in the lab is shown in Figure 6.5. This is sometimes called a *coffee cup calorimeter* because it uses a Styrofoam coffee cup to contain the reaction mixture. Its use follows the same principle as the bomb calorimeter. First the temperature of the reactants is measured, and then they are combined in the coffee cup where the heat of reaction changes the temperature of the reaction mixture. After the reaction is complete (usually just a few seconds), the final temperature is measured. From the temperature change and an estimate of the heat capacity of the reaction mixture, a reasonably good estimate of the heat of reaction can be made.

A bomb calorimeter is nearly always used to measure the energy given off in exothermic reactions.

A Styrofoam coffee cup is a good insulator and is inexpensive.

EXAMPLE 6.4. MEASURING THE HEAT OF REACTION WITH A BOMB CALORIMETER

PROBLEM: In an experiment, 0.100 g of H_2 and 0.800 g of O_2 were compressed into a 1.00-L bomb, which was then placed into a calorimeter that has a heat capacity of 9.08×10^4 J/°C. The initial temperature of the calorimeter was measured to be 25.000 °C and after the reaction had taken place, the final temperature of the calorimeter was 25.155 °C. Calculate the amount of heat given off in the reaction of H_2 and O_2 to form H_2O, expressed (a) in kilojoules and (b) in kilojoules per mole of H_2O formed.

SOLUTION: (a)   As described in the text, we can calculate the heat evolved by multiplying the heat capacity (9.08×10^4 J/°C) by the temperature change in °C. The temperature change is 0.155 °C. Therefore,

$$\left(\frac{9.08 \times 10^4 \text{ J}}{1 \text{ °C}}\right) \times (0.155 \text{ °C}) = 1.41 \times 10^4 \text{ J}$$

Changing joules to kilojoules gives an answer of 1.41×10^1 kJ (14.1 kJ). This is the amount of heat given off by the reactants in this experiment.

(b)   The amount of heat liberated in a reaction depends on the amounts of the reactants that are used. Large amounts of reactants produce large energy changes. When we compare energy changes for different reactions, heats of reaction are often expressed on a "per mole" basis.

The reaction between H_2 and O_2 to form water is

$$2H_2 + O_2 \longrightarrow 2H_2O$$

The amount of H_2O produced in the reaction, expressed in moles, is

$$0.100 \text{ g } H_2 \times \frac{1 \text{ mol } H_2}{2.016 \text{ g } H_2} \times \frac{2 \text{ mol } H_2O}{2 \text{ mol } H_2} = 0.0496 \text{ mol } H_2O$$

To get units of kilojoules per mole of water, we just take the ratio of the number of kilojoules to the number of moles.

$$\frac{14.1 \text{ kJ}}{0.0496 \text{ mol } H_2O} = 284 \text{ kJ/mol } H_2O$$

6.4 HEATS OF REACTION AND THERMOCHEMISTRY

We will study some of the principles of thermodynamics in Chapter 14.

The study of heats of reaction is called **thermochemistry,** which is part of a larger branch of science called thermodynamics. Before we can deal further with thermochemical principles, however, we have to define some terms. One of these is the term **system,** which simply is that part of the universe we happen to be studying. It might be a chemical reaction taking place in a beaker, for example. Outside the system are the **surroundings.** In describing a system, we must specify its properties very precisely. We give the temperature, pressure, number of moles of each component, and whether the components are liquids, solids, or gases. Once these variables are set, all the properties of the system are fixed and we have described the **state** of the system.

The reason we specify the state of the system so precisely is because the energy of the system depends on these variables. At a higher temperature, for example, the system has a larger amount of kinetic energy.

When a change takes place in a system, we say that the system goes from one state to another. If the system is insulated from the surroundings, so no heat can flow between them, then any change that takes place within the system is said to be an **adiabatic change.** During an adiabatic change, the temperature of the system shifts; it increases if the reaction is exothermic and it decreases if the reaction is endothermic. If the system is not insulated from the surroundings, then heat can flow between them and it is possible for the system to be maintained at a constant temperature when a reaction occurs. A change that occurs at constant temperature is called an **isothermal change.**

As we noted earlier, when an exothermic or endothermic reaction occurs, there is a change in the potential energy of the chemicals involved. The heat of reaction that we measure is therefore equal to this potential energy change. We will be dealing with changes in quite a few quantities from here on, so let's establish some rules for expressing changes in general.

The symbol Δ (Greek letter delta) is normally used to signify a change in a quantity. For instance, a change in temperature can be represented by Δt, where t stands for temperature. The usual practice in specifying a change is to subtract the initial value of the quantity from the final value. For example, a change in temperature Δt is defined as

$$\Delta t = t_{final} - t_{initial}$$

Similarly, a change in potential energy, $\Delta(P.E.)$, is

$$\Delta(P.E.) = (P.E.)_{final} - (P.E.)_{initial}$$

One of the things that this definition does is estabish an algebraic sign convention for exothermic and endothermic changes. In an exothermic change, the potential energy of the products is lower than the potential energy of the reactants; the final P.E. is less than the initial P.E. and $\Delta(P.E.)$ is a negative quantity. Just the opposite is true for an endothermic change, for which $\Delta(P.E.)$ is positive.

Exothermic change	$\Delta(P.E.)$ is negative
Endothermic change	$\Delta(P.E.)$ is positive

Enthalpy and enthalpy changes

The two calorimeters described in Figures 6.4 and 6.5 differ in a significant way. A reaction in a bomb calorimeter occurs at constant volume, because the bomb cannot expand or contract. This means that if gases are formed in the reaction, their pressures can build up and the pressure of the system can change. Because of the constant volume conditions, the heat of reaction measured with a bomb calorimeter is called the *heat of reaction at constant volume.* The coffee cup calorimeter is open to the atmosphere, and if a reaction occurs in it that produces a gas, the gas can expand into the atmosphere and the pressure on the system can remain constant. The energy change measured with a coffee cup calorimeter is therefore the *heat of reaction at constant pressure.*

Heats of reaction measured at constant volume and at constant pressure do not differ by much, but they are not identical (for reasons we will discuss in Chapter 14). Since most of the reactions that are of interest to us occur in open containers, exposed to a constant pressure from the atmosphere, we will devote the remainder of our discussions to heats of reaction at constant pressure.

The heat of reaction at constant pressure is called the **enthalpy change** for the reaction and is represented symbolically as ΔH. It is defined as

$$\Delta H = H_{final} - H_{initial} \tag{6.1}$$

Enthalpy is from the Greek word *enthalpein* meaning *to heat in.*

Although this is the formal definition of ΔH, the enthalpies, H, of the initial and final states (which are actually related to the total amounts of energy that these states have) can't really be measured. This is because the total energy of a system includes the sum of all of its kinetic energy and all of its potential energy. These total energies can't be obtained because we really don't know exactly how fast the molecules of the system are moving[3] and because we can't possibly take into account all the attractions and repulsions felt by all the molecules in a system. Nevertheless, the definition given by Equation 6.1 is important because it does establish the algebraic signs that ΔH has for exothermic and endothermic changes. An exothermic change has H_{final} less than $H_{initial}$, so ΔH has a negative value. By a similar analysis, we find that ΔH for an endothermic change has a positive value.

The magnitude of ΔH for any particular reaction depends to some degree on what the pressure on the system happens to be, so if we wish to compare ΔH values for different reactions, they have to be measured at the same pressure. Scientists have defined a unit of pressure called *1 standard atmosphere*, abbreviated *1 atm*, which corresponds roughly to the average atmospheric pressure at sea level. This is usually chosen as a reference pressure for the measurement of enthalpy changes, and most of the ΔH values given in this book are for a pressure of 1 atm.

State functions

A **state function** (or **state variable**) is a quantity whose value depends only on the current state of a system, and does not depend on the prior history of the system. An example is temperature. A sample of water, for instance, might have a temperature of 25 °C. This temperature doesn't depend on what the temperature of the water might have been at some previous time; its value is just the current temperature.

One of the facts about state functions is that when one changes, *how* the change comes about doesn't affect the magnitude of the change. For example, the temperature of the water could be changed to 60 °C. The temperature change, Δt, is just the difference between the two temperatures.

$$\Delta t = t_{final} - t_{initial}$$
$$\Delta t = 60 \text{ °C} - 25 \text{ °C} = 35 \text{ °C}$$

Since Δt depends only on the initial and final states, it is referred to as a state function, too.

The water could have been cooled first to 10 °C and then raised to 60 °C, or it could have been heated to 95 °C and then cooled to 60 °C, or the temperature change could have followed any other path. When the final temperature of 60 °C is reached, however, the net temperature change is the same; $\Delta t = 35$ °C.

Enthalpy, like temperature, is a state function. Therefore, no matter how we go from one state to another, the net energy change, and therefore the net enthalpy change, is the same. If this weren't true, then it would be possible to go from some state 1 to some state 2 by one path that requires the input of a certain amount of energy, and then from state 2 back to state 1 by a *different* path that releases even more energy. If this were possible, we could repeat the cycle over and over again, and each time have some energy to spare. We would have invented a perpetual motion machine — a device that creates energy. But because the law of conservation of energy is a *fact*, perpetual motion schemes such as this are impossible. (In fact, the U.S. Patent Office refuses to issue patents on them unless the inventor is able to supply a working model.)

After this rule was adopted, the U.S. Patent Office stopped receiving patent applications for perpetual motion machines.

[3] It's impossible to determine in an absolute sense how fast anything is moving. An object on a table may seem to be motionless, but it is on the surface of a planet that rotates around its axis once every 24 hours. The earth is moving through space around a star (our sun) that is moving through a galaxy that is moving through the universe. Since we have no fixed motionless point of reference, we can't possibly determine exactly how fast anything is moving, so we can't figure out exactly how much kinetic energy something has.

6.5 HESS'S LAW OF HEAT SUMMATION

Because enthalpy is a state function, the magnitude of ΔH for a chemical reaction does not depend on the path taken by the reactants as they proceed to form the products. To see the significance of this in the study of heats of reaction, let's examine a change that's familiar to you — the vaporization of water at its boiling point. Specifically, let's consider the conversion of 1 mol of liquid water, $H_2O(l)$, to 1 mol of gaseous water, $H_2O(g)$, at 100 °C and a pressure of 1 atm. This process absorbs 41 kJ, so $\Delta H = +41$ kJ. The overall change can be represented by the equation

$$H_2O(l) \longrightarrow H_2O(g) \qquad \Delta H = +41 \text{ kJ}$$

The positive sign of ΔH tells us that the change is endothermic.

An equation written in this manner, in which the energy change is also shown, is called a **thermochemical equation**. In a thermochemical equation, the coefficients are taken to represent the numbers of moles of the reactants and products. This thermochemical equation states that 1 mol of liquid water is changed to 1 mol of gaseous water by the absorption of 41 kJ.

Changing 1 mol of liquid water to 1 mol of gaseous water will *always* absorb this same amount of energy, provided we refer to this same pair of initial and final states. It doesn't matter how we carry out the change. We could even go as far as to decompose the mole of liquid water into gaseous H_2 and O_2 and then recombine the elements to give a mole of gaseous water. The *overall* enthalpy change would still be the same, $+41$ kJ. Thus, *it is possible to look at some overall change as the net result of a sequence of steps, and the net value of ΔH for the overall process is merely the sum of all the enthalpy changes that take place along the way.* This last statement constitutes **Hess's law of heat summation.**

Hess's law is really just another way of stating the law of conservation of energy.

Thermochemical equations serve as a useful tool for applying Hess's law. For example, the thermochemical equations that correspond to the indirect path just described for the vaporization of water at 100 °C are

$$H_2O(l) \longrightarrow H_2(g) + \tfrac{1}{2}O_2(g) \qquad \Delta H = +283 \text{ kJ}$$

$$H_2(g) + \tfrac{1}{2}O_2(g) \longrightarrow H_2O(g) \qquad \Delta H = -242 \text{ kJ}$$

Notice that fractional coefficients are allowed in thermochemical equations. This is because a coefficient of $\tfrac{1}{2}$ is taken to mean $\tfrac{1}{2}$ mol. (In ordinary chemical equations, however, fractional coefficients are avoided because they are meaningless on a molecular level; one cannot have half an atom or molecule and still retain the chemical identity of the species.)

The two equations above tell us that 283 kJ are required to decompose 1 mol of $H_2O(l)$ into its elements, and that 242 kJ are evolved when they recombine to produce 1 mol of $H_2O(g)$. The net change (the vaporization of one mole of water) is obtained by adding the two chemical equations together and then canceling any quantities that appear on both sides of the arrow.

$$H_2O(l) + \cancel{H_2(g)} + \cancel{\tfrac{1}{2}O_2(g)} \longrightarrow H_2O(g) + \cancel{H_2(g)} + \cancel{\tfrac{1}{2}O_2(g)}$$

or

$$H_2O(l) \longrightarrow H_2O(g)$$

This is similar to canceling spectator ions from an ionic equation.

We also find that the heat of the overall reaction is equal to the algebraic sum of the heats of reaction for the two steps.

$$\Delta H = +283 \text{ kJ} + (-242 \text{ kJ})$$

$$\Delta H = +41 \text{ kJ}$$

Thus, *when we add thermochemical equations to obtain some net change, we also add their corresponding heats of reaction.*

Figure 6.6. *An enthalpy diagram for the "reaction"*

$$H_2O(l) \longrightarrow H_2O(g)$$

The enthalpy of the free elements has arbitrarily been chosen as the zero point on the enthalpy scale. Notice that the net enthalpy change following the two-step path involving the free elements as intermediates is the same as the enthalpy change for the direct conversion of the liquid to vapor.

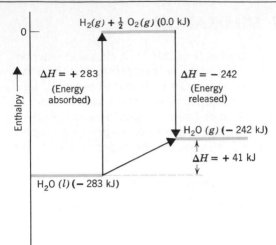

To describe the nature of these thermochemical changes, we can also illustrate them graphically (Figure 6.6). This type of figure is frequently called an **enthalpy diagram.** Notice that we have chosen the enthalpy of the free elements as the zero point on the energy scale. This choice is entirely arbitrary because we are interested only in determining differences in H. In fact, we have no way at all of knowing absolute enthalpies. We can only measure ΔH.

Manipulating thermochemical equations

In the preceding discussion, you learned that when thermochemical equations can be added to give some final equation, we just add up their ΔHs to get the ΔH corresponding the the final equation. Example 6.5 below illustrates this by combining three equations to get a fourth. As you will see, however, sometimes the thermochemical equations have to be manipulated and rearranged before they are added. These manipulations affect the values of ΔH, so before going on to Example 6.5, let's look at the kinds of operations you can perform on thermochemical equations and how these operations affect the value of ΔH.

Multiplying or dividing the coefficients by some factor

We can perform these manipulations because ΔH is a state function.

If we multiply or divide the coefficients of a thermochemical equation by some factor, we must do the same to the value of ΔH. For example, the decomposition of liquid water has the thermochemical equation

$$H_2O(l) \longrightarrow H_2(g) + \tfrac{1}{2}O_2(g) \qquad \Delta H = +283 \text{ kJ}$$

If we multiply the equation through by 2, this doubles the number of moles of everything, and as you might expect, this doubles the amount of heat absorbed. When we double the coefficients, we double the value of ΔH.

$$2H_2O(l) \longrightarrow 2H_2(g) + O_2(g) \qquad \Delta H = +566 \text{ kJ}$$

Changing the direction of the thermochemical equation

If a reaction is exothermic in one direction, it must be endothermic if it occurs in the opposite direction. Therefore, if we reverse the direction of a thermochemical equation, we just change the sign of its ΔH.

Reversing this equation,

$$H_2O(l) \longrightarrow H_2(g) + \tfrac{1}{2}O_2(g) \qquad \Delta H = +283 \text{ kJ}$$

gives this equation:

$$H_2(g) + \tfrac{1}{2}O_2(g) \longrightarrow H_2O(l) \qquad \Delta H = -283 \text{ kJ}$$

Now let's look at how we apply these kinds of operations in solving a Hess's law problem.

EXAMPLE 6.5. APPLYING HESS'S LAW BY COMBINING THERMOCHEMICAL EQUATIONS

PROBLEM: The thermochemical equation for the combustion of acetylene, the fuel used in torches, is given by Equation (1)

(1) $\qquad 2C_2H_2(g) + 5O_2(g) \longrightarrow 4CO_2(g) + 2H_2O(l) \qquad \Delta H_1 = -2602 \text{ kJ}$

Ethane, another hydrocarbon fuel, reacts as follows:

(2) $\qquad 2C_2H_6(g) + 7O_2(g) \longrightarrow 4CO_2(g) + 6H_2O(l) \qquad \Delta H_2 = -3123 \text{ kJ}$

Finally, hydrogen and oxygen combine by this equation:

(3) $\qquad H_2(g) + \tfrac{1}{2}O_2(g) \longrightarrow H_2O(l) \qquad \Delta H_3 = -286 \text{ kJ}$

> The ΔHs are numbered simply for ease of identification in the solution of the problem.

All these data are for the same temperature and pressure: 25 °C and 1 atm. Use these thermochemical equations to calculate ΔH for the reaction

(4) $\qquad C_2H_2(g) + 2H_2(g) \longrightarrow C_2H_6(g) \qquad \Delta H_4 = ?$

ANALYSIS: To solve this problem, we have to combine Equations (1), (2) and (3) in such a way that when they are added, everything cancels except the formulas in Equation (4). (Let's call this the *target equation.*) To get this to happen, we have to rearrange the given equations. For example, in the target equation, one C_2H_6 appears on the right, but in Equation (2) it is on the left with a coefficient of 2. This means we have to reverse Equation (2) and divide its coefficients by 2; this will bring C_2H_6 to the right with a coefficient of 1. Of course, this means we have to change the sign of the ΔH and divide it by 2 also. As we go about rearranging the given equation, we have to keep our eye on the target equation, so we can be sure that we end up with the correct formulas with the correct coefficients on the correct side of the arrow.

The greatest difficulty students have with problems of this type is getting started. Practice will help you learn, but there are some rules that are useful to keep in mind. When you scan the given equations, look for formulas that appear only once among them. For example, C_2H_2 only appears in Equation (1), so we know exactly what we have to do to get it in the right place for the target equation. Avoid formulas that occur in more than one of the given equations. In this problem we don't want to begin working with O_2, because we can't tell which of the given equations have to be arranged to get the O_2 to cancel.

SOLUTION: In the target equation, $1C_2H_2$ is on the left, so we use Equation (1) with its coefficients divided by 2, which gives Equation (5) below. Notice that we also divide the ΔH for Equation (1) by 2 to get the ΔH for Equation (5). In the target equation there is $2H_2$ on the left, so we multiply the coefficients of Equation (3) and its ΔH by 2 to give Equation (6) and its ΔH. Finally, as we saw in the analysis above, we reverse Equation (2) and divide its coefficients by 2 and make the appropriate adjustments in the ΔH. This gives Equation (7).

(5) $\quad C_2H_2(g) + \tfrac{5}{2}O_2(g) \longrightarrow 2CO_2(g) + H_2O(l) \qquad \Delta H_5 = -\dfrac{2602 \text{ kJ}}{2} = -1301 \text{ kJ}$

(6) $\quad 2H_2(g) + O_2(g) \longrightarrow 2H_2O(l) \qquad \Delta H_6 = 2(-286 \text{ kJ}) = -572 \text{ kJ}$

> On the left, $\tfrac{5}{2}O_2 + O_2$ gives $\tfrac{7}{2}O_2$

(7) $\quad 2CO_2(g) + 3H_2O(l) \longrightarrow C_2H_6(g) + \tfrac{7}{2}O_2(g) \qquad \Delta H_7 = +\dfrac{3123 \text{ kJ}}{2} = +1561 \text{ kJ}$

Adding Equations (5), (6), and (7) gives

$$C_2H_2(g) + 2H_2(g) + \tfrac{7}{2}O_2(g) + 2CO_2(g) + 3H_2O(l) \longrightarrow 2CO_2(g) + 3H_2O(l) + C_2H_6(g) + \tfrac{7}{2}O_2(g)$$

Canceling elements and compounds that are the same on both sides, we get

$$C_2H_2(g) + 2H_2(g) \longrightarrow C_2H_6(g)$$

which is the equation we want. Since it is obtained by adding Equations (5), (6) and (7), its ΔH (ΔH_4 in the statement of the problem) is obtained by adding the ΔHs of (5), (6) and (7).

$$\Delta H_4 = \Delta H_5 + \Delta H_6 + \Delta H_7$$

$$\Delta H_4 = (-1301 \text{ kJ}) + (-572 \text{ kJ}) + (1561 \text{ kJ})$$

$$= -312 \text{ kJ}$$

The enthalpy change for Equation (4) is therefore -312 kJ.

Heats of formation

A particularly useful type of thermochemical equation corresponds to the formation of *one mole* of a substance from its elements. The enthalpy changes associated with these reactions are called **heats of formation** or **enthalpies of formation** and are denoted as ΔH_f. For example, thermochemical equations for the formation of liquid and gaseous water at 100 °C and 1 atm are, respectively,

$$H_2(g) + \tfrac{1}{2}O_2(g) \longrightarrow H_2O(l) \qquad \Delta H_f = -283 \text{ kJ/mol}$$

$$H_2(g) + \tfrac{1}{2}O_2(g) \longrightarrow H_2O(g) \qquad \Delta H_f = -242 \text{ kJ/mol}$$

How can we use these equations to obtain the heat of vaporization of water? Clearly, we must reverse the first equation and then add it to the second. When we reverse this equation, we must also remember to change the sign of ΔH (If the formation of $H_2O(l)$ is exothermic, as indicated by a negative ΔH_f, the reverse process must be endothermic.)

(Exothermic) $\qquad H_2(g) + \tfrac{1}{2}O_2(g) \longrightarrow H_2O(l) \qquad \Delta H = \Delta H_f = -283 \text{ kJ}$

(Endothermic) $\qquad H_2O(l) \longrightarrow H_2(g) + \tfrac{1}{2}O_2(g) \qquad \Delta H = -\Delta H_f = +283 \text{ kJ}$

When this last equation is added to that for the formation of $H_2O(g)$, we obtain

$$H_2O(l) \longrightarrow H_2O(g)$$

and the heat of reaction is

$$\Delta H = \Delta H_{f\,H_2O(g)} - \Delta H_{f\,H_2O(l)}$$

$$\Delta H = -242 \text{ kJ} - (-283 \text{ kJ}) = +41 \text{ kJ}$$

Notice that the heat of reaction for the overall change is equal to the heat of formation of the product *minus* the heat of formation of the reactant. In general, we can write that for any overall reaction

Equation 6.2 is a particularly useful form of Hess's law.

$$\Delta H_{\text{reaction}} = \left(\begin{array}{c} \text{sum of the } \Delta H_f \\ \text{of the products} \end{array} \right) - \left(\begin{array}{c} \text{sum of the } \Delta H_f \\ \text{of the reactants} \end{array} \right) \qquad (6.2)$$

6.6 STANDARD STATES

The magnitude of ΔH_f depends on the conditions of temperature, pressure, and the physical state (gas, liquid, solid, crystalline form) of the reactants and products. For instance, at 100 °C and 1 atm, the heat of formation of liquid water is -283 kJ/mol, while at 25 °C and 1 atm, ΔH_f for $H_2O(l)$ is -286 kJ/mol. To avoid the necessity of always having to specify the conditions for which ΔH_f is recorded and to permit comparisons between ΔH_f for various compounds, a standard set of conditions is chosen, usually 25 °C and a pressure of 1 atm. Under these conditions a substance is said to be in its **standard state.** Heats of formation of substances in their standard states are indicated as ΔH_f°. For example, the standard heat of formation of liquid water $\Delta H_{f\,H_2O(l)}^\circ = -286$ kJ/mol and is the heat liberated when H_2 and O_2, each in their natural form at 25 °C and 1 atm, react to produce $H_2O(l)$ at 25 °C and 1 atm.

Table 6.1 contains standard heats of formation for a variety of different substances. Such a table is very useful because it allows us to use Equation 6.2 to calculate the **standard heats of reaction,** ΔH°, for a large number of different chemical changes. In performing these calculations, we take the ΔH_f° of any element in its natural, most stable form at 25 °C and 1 atm to be equal to zero.

A more complete table of standard heats of formation is given in Appendix C.

For an element in its standard state, $\Delta H_f^\circ = 0$

This is reasonable, because there is no change if we are forming the most stable form of an element from itself. We are using the elements as a reference point, and to get to this reference point *from* the reference point involves no energy change.

The following are some examples of how we use ΔH_f° values in calculations.

EXAMPLE 6.6. CALCULATING ΔH° FOR A REACTION FROM STANDARD HEATS OF FORMATION

PROBLEM: Many careful cooks keep sodium bicarbonate (baking soda) handy because it is a good extinguisher for oil or grease flames. Its decomposition products help smother the flames. The reaction for the decomposition is

$$2NaHCO_3(s) \longrightarrow Na_2CO_3(s) + H_2O(g) + CO_2(g)$$

Calculate ΔH° for the reaction in units of kilojoules from the standard heats for formation of the reactants and products.

ANALYSIS: We will use Equation 6.2, which states in effect

$$\Delta H^\circ = (\text{sum of } \Delta H_f^\circ \text{ products}) - (\text{sum of } \Delta H_f^\circ \text{ reactants})$$

This means we have to add up all the heat evolved or absorbed during the formation of the products from their elements and then subtract the heat evolved or absorbed by the formation of the reactants from their elements. We have to be careful about the units here, however. A heat of formation is the energy associated with the formation of one mole of a compound and has units of kJ/mol. The coefficients in the equation specify the numbers of moles of each reactant, so to get the total amount of energy contributed by each substance, we multiply its heat of formation by its coefficient in the equation, as shown below.

SOLUTION: Using the data in Table 6.1, we obtain for the products

$$1 \text{ mol } Na_2CO_3(s) \times \left[\frac{-1131 \text{ kJ}}{1 \text{ mol } Na_2CO_3(s)} \right] = -1131 \text{ kJ}$$

$$1 \text{ mol } H_2O(g) \times \left[\frac{-242 \text{ kJ}}{1 \text{ mol } H_2O(g)} \right] = -242 \text{ kJ}$$

$$1 \text{ mol } CO_2(g) \times \left[\frac{-394 \text{ kJ}}{1 \text{ mol } CO_2(g)} \right] = -394 \text{ kJ}$$

Total of ΔH_f° of the products $= -1767$ kJ

Table 6.1. Standard heats of formation of some substances at 25 °C and 1 atm

Substance	$\Delta H_f°$ (kJ/mol)	Substance	$\Delta H_f°$ (kJ/mol)	Substance	$\Delta H_f°$ (kJ/mol)
Al(s)	0	$HCHO_2(g)$	−363	LiCl(s)	−408.8
$AlCl_3(s)$	−704	(formic acid)		Mg(s)	0
$Al_2O_3(s)$	−1676	$HC_2H_3O_2(l)$	−487.0	$MgCl_2(s)$	−641.8
$Al_2(SO_4)_3(s)$	−3441	(acetic acid)		$MgCl_2 \cdot 2H_2O(s)$	−1280
As(s)	0	HCHO(g)	−108.6	$Mg(OH)_2(s)$	−924.7
$AsH_3(g)$	+66.4	(formaldehyde)		$KMnO_4(s)$	−813.4
$As_4O_6(s)$	−1314	$CH_3CHO(g)$	−167	$MnSO_4(s)$	−1064
$As_2O_5(s)$	−925	(acetaldehyde)		$N_2(g)$	0
Ba(s)	0	$(CH_3)_2CO(l)$	−248.1	$NH_3(g)$	−46.0
$BaCO_3(s)$	−1219	(acetone)		$NH_4Cl(s)$	−314.4
$BaCl_2(s)$	−860.2	$C_6H_5CO_2H(s)$	−385.1	NO(g)	+90.4
$Ba(OH)_2$	−998.22	(benzoic acid)		$NO_2(g)$	+34
$BaSO_4(s)$	−1465	$CO(NH_2)_2(s)$	−333.5	$N_2O(g)$	+81.5
$Br_2(l)$	0	(urea)		$HNO_3(l)$	−174.1
$Br_2(g)$	+30.9	$Cl_2(g)$	0	$O_2(g)$	0
HBr(g)	−36	HCl(g)	−92.5	$O_3(g)$	+143
Ca(s)	0	HCl(aq)	−167.2	P(s, white)	0
$CaCO_3(s)$	−1207	$Cr_2O_3(s)$	−1141	$P_4O_{10}(s)$	−2984
$CaCl_2(s)$	−795.8	$(NH_4)_2Cr_2O_7(s)$	−1807	$H_3PO_4(s)$	−1279
CaO(s)	−635.5	$K_2Cr_2O_7(s)$	−2033.01	K(s)	0
$Ca(OH)_2(s)$	−986.6	Cu(s)	0	KCl(s)	−436.8
$Ca_3(PO_4)_2(s)$	−4119	$CuCl_2(s)$	−172	$SiH_4(g)$	+33
$CaSO_3(s)$	−1156	CuO(s)	−155	$SiO_2(s, alpha)$	−910.0
$CaSO_4(s)$	−1433	$Cu_2S(s)$	−79.5	Na(s)	0
$CaSO_4 \cdot \frac{1}{2}H_2O(s)$	−1573	CuS(s)	−53.1	NaF(s)	−571
$CaSO_4 \cdot 2H_2O(s)$	−2020	$CuSO_4(s)$	−771.4	NaCl(s)	−413
C(s, graphite)	0	$CuSO_4 \cdot 5H_2O(s)$	−2279.7	NaBr(s)	−360
C(s, diamond)	+1.88	$F_2(g)$	0	NaI(s)	−288
$CCl_4(l)$	−134	HF(g)	−271	$NaHCO_3(s)$	−947.7
CO(g)	−110	$H_2(g)$	0	$Na_2CO_3(s)$	−1131
$CO_2(g)$	−394	$H_2O(l)$	−286	$Na_2O_2(s)$	−504.6
$CO_2(aq)$	−413.8	$H_2O(g)$	−242	NaOH(s)	−426.8
$H_2CO_3(aq)$	−699.65	$H_2O_2(l)$	−187.8	$Na_2SO_4(s)$	−1384.49
$CS_2(l)$	+89.5	$I_2(s)$	0	S(s, rhombic)	0
$CS_2(g)$	+117	$I_2(g)$	+62.4	$SO_2(g)$	−297
$CH_4(g)$	−74.9	HI(g)	+26	$SO_3(g)$	−396
$C_2H_2(g)$	+227	Fe(s)	0	$H_2SO_4(l)$	−813.8
$C_2H_4(g)$	+51.9	$Fe_2O_3(s)$	−822.2	$SnCl_4(l)$	−511.3
$C_2H_6(g)$	−84.5	$Fe_3O_4(s)$	−1118.4	$SnO_2(s)$	−580.7
$C_3H_8(g)$	−104	Pb(s)	0	Zn(s)	0
$C_4H_{10}(g)$	−126	PbO(s)	−217.3	ZnO(s)	−348
$C_6H_6(l)$	+49.0	$PbO_2(s)$	−277	$ZnSO_4(s)$	−982.8
$CH_3OH(l)$	−238	$Pb(OH)_2(s)$	−515.9		
$C_2H_5OH(l)$	−278	$PbSO_4(s)$	−920.1		
		Li(s)	0		

For the single reactant,

$$2 \text{ mol NaHCO}_3(s) \times \left[\frac{-947.7 \text{ kJ}}{1 \text{ mol NaHCO}_3(s)} \right] = -1895 \text{ kJ}$$

Now we substitute these values into the equation

$$\Delta H^\circ = (\text{sum of } \Delta H_f^\circ \text{ products}) - (\text{sum of } \Delta H_f^\circ \text{ reactants})$$

which gives

$$\Delta H^\circ = (-1767 \text{ kJ}) - (-1895 \text{ kJ})$$
$$= +128 \text{ kJ}$$

EXAMPLE 6.7. CALCULATING ΔH° FOR A REACTION FROM STANDARD HEATS OF FORMATION

PROBLEM: Calculate ΔH° for the reaction

$$2\text{Na}_2\text{O}_2(s) + 2\text{H}_2\text{O}(l) \longrightarrow 4\text{NaOH}(s) + \text{O}_2(g)$$

How many kilojoules are liberated when 25.0 g Na_2O_2 react according to this equation?

SOLUTION: We use Equation 6.2.

$$\Delta H^\circ = [4 \, \Delta H_{f\text{NaOH}(s)}^\circ + \Delta H_{f\text{O}_2(g)}^\circ] - [2 \, \Delta H_{f\text{Na}_2\text{O}_2(s)}^\circ + 2 \, \Delta H_{f\text{H}_2\text{O}(l)}^\circ]$$

All the data are available in Table 6.1. Therefore,

$$\Delta H^\circ = \left[4 \text{ mol} \times \left(\frac{-426.8 \text{ kJ}}{1 \text{ mol NaOH}} \right) + 0.00 \right] - \left[2 \text{ mol} \times \left(\frac{-504.6 \text{ kJ}}{1 \text{ mol Na}_2\text{O}_2} \right) + 2 \text{ mol} \times \left(\frac{-286 \text{ kJ}}{1 \text{ mol H}_2\text{O}(l)} \right) \right]$$

$$\Delta H^\circ = (-1707 \text{ kJ}) - (-1581 \text{ kJ})$$
$$= -126 \text{ kJ}$$

To calculate the number of kilojoules liberated by 25.0 g Na_2O_2, we have to realize that the ΔH° that we calculated is the energy given off when 2 mol Na_2O_2 react. Therefore,

$$2 \text{ mol Na}_2\text{O}_2 \Longleftrightarrow -126 \text{ kJ}$$

The formula mass of Na_2O_2 is 78.0, so

$$25.0 \text{ g Na}_2\text{O}_2 \times \left(\frac{1 \text{ mol Na}_2\text{O}_2}{78.0 \text{ g Na}_2\text{O}_2} \right) \times \left(\frac{-126 \text{ kJ}}{2 \text{ mol Na}_2\text{O}_2} \right) \Longleftrightarrow -20.2 \text{ kJ}$$

The reaction of 25.0 g Na_2O_2 liberates 20.2 kJ.

It is often impossible to measure directly the heat of formation of a compound. For example, no one has yet found a way to cause gaseous hydrogen and graphite (the most stable crystalline form of carbon) to react to form methane, CH_4, in a way that allows us to measure the energy change for the reaction. Therefore, in order to determine ΔH_f° for compounds such as methane, indirect methods must be used. One technique that works well with many carbon-containing compounds is to burn the compound in a bomb calorimeter and measure the heat of combustion. In these reactions, the heats of formation of the products are generally known, so the unknown value of ΔH_f° for the reactant can be calculated as illustrated in Example 6.8.

EXAMPLE 6.8. USING THE HEAT OF COMBUSTION TO CALCULATE ΔH_f°

PROBLEM: The combustion of 1 mol of benzene, $C_6H_6(l)$, to produce $CO_2(g)$ and $H_2O(l)$ liberates 3271 kJ when the products are returned to 25 °C and 1 atm. What is the standard heat of formation of $C_6H_6(l)$ expressed in kilojoules per mole?

SOLUTION: The equation for the combustion of 1 mol of C_6H_6 is

$$C_6H_6(l) + 7\tfrac{1}{2}O_2(g) \longrightarrow 6CO_2(g) + 3H_2O(l)$$

The standard heat of reaction $\Delta H^\circ = -3271$ kJ. From Equation 6.2, we know that

$$\Delta H^\circ = [6\,\Delta H^\circ_{f\,CO_2(g)} + 3\,\Delta H^\circ_{f\,H_2O(l)}] - [\Delta H^\circ_{f\,C_6H_6(l)}]$$

Solving for the heat of formation of benzene, we obtain

$$\Delta H^\circ_{f\,C_6H_6(l)} = 6\,\Delta H^\circ_{f\,CO_2(g)} + 3\,\Delta H^\circ_{f\,H_2O(l)} - \Delta H^\circ$$

From Table 6.1 we can obtain the heats of formation of CO_2 and H_2O. Therefore,

$$\Delta H^\circ_{f\,C_6H_6(l)} = 6(-394)\text{ kJ} + 3(-286)\text{ kJ} - (-3271\text{ kJ})$$

$$\Delta H^\circ_{f\,C_6H_6(l)} = +49\text{ kJ}$$

Since 1 mol of C_6H_6 is involved,

$$\Delta H_f^\circ = +49\text{ kJ/mol}$$

REVIEW QUESTIONS AND PROBLEMS

Problems whose numbers are in blue have their answers in Appendix D at the back of the book. The more difficult problems are marked with asterisks.

Energy and Energy Transfer

6.1 Define *energy*. In what two ways can matter have energy?

6.2 If the speed of an object is doubled, by what factor does its kinetic energy increase?

6.3 If the potential energy of an object decreases as it is moved away from another object, what kind of force (attractive or repulsive) must exist between the two?

6.4 State the law of conservation of energy.

6.5 What name is sometimes given to the kind of potential energy that chemicals have because of the attractions and repulsions between their subatomic particles?

6.6 A hydrogen atom consists of one proton and one electron. If the electron is removed from a hydrogen atom, does its potential energy increase or decrease? Explain your answer.

6.7 When you throw a ball into the air, it slows as it rises, comes to a stop, and then speeds up as it falls. Analyze this phenomenon in terms of the K.E. and P.E. of the ball.

6.8 What property that we can measure in the laboratory is related to the speed at which the molecules of a substance are moving? Specifically, what is this relationship?

6.9 What is the origin of the potential energy possessed by atoms and other atomic-sized particles?

6.10 If the temperature of an object increases from 100 °C to 200 °C, by what factor does the average kinetic energy of its molecules increase? How does the average kinetic energy change if the temperature is changed from 27 °C to 327 °C?

6.11 What is *heat energy*? By what mechanism does heat flow from a hot object into a cool object?

6.12 When something hot is placed in contact with something cool, the hot object becomes cooler while the cool one becomes warmer. How does the amount of heat lost by the hot object compare with the amount of heat absorbed by the cool one?

6.13 Magnesium is used in fireworks and incendiary bombs because it burns fiercely, liberating lots of heat. The reaction of magnesium with oxygen is $2Mg + O_2 \rightarrow 2MgO$.
(a) Is this reaction exothermic or endothermic?
(b) Which has the larger potential energy, $2Mg + O_2$ or $2MgO$?
(c) As this reaction takes place, what happens to the kinetic energy of the particles (atoms, molecules, etc.)?

6.14 Instant cold packs can be purchased in a pharmacy. They contain plastic pouches filled with water and a vial filled with solid ammonium nitrate. When the vial is crushed, the ammonium nitrate dissolves in the water and as the solution forms, it becomes quite cool. When this solution forms, does the potential energy of the ions (NH_4^+ and NO_3^-) and the water increase or decrease? Explain your answer.

Energy Units, Heat Capacity, and Specific Heat

6.15 What is the definition of the *joule?* How many joules equal 1 calorie? How many kilojoules equal 1 kilocalorie?

6.16 What is an *erg?* How many ergs equal (a) 1 joule and (b) 1 calorie?

6.17 Perform the following conversions:
(a) 345 J to cal
(b) 546 cal to J
(c) 234 kJ to kcal
(d) 1.257 kcal to kJ

6.18 A truck having a mass of 4500 kg is moving at a speed of 1.79 m/s. Calculate its kinetic energy in joules. How many calories does this correspond to?

6.19 Calculate the kinetic energy, in joules, of a 145-lb athlete running at a speed of 15.3 mi/hr.

6.20 Calculate the kinetic energy, in joules and in kilojoules, of a 225-kg motorcycle traveling at 80.0 km/hr.

6.21 A car weighing 2.40 tons (1 ton = 2000 lb) is traveling at 35.0 mi/hr. What is its kinetic energy in joules? If all this energy were converted to heat, how many kilograms of water could have their temperature raised by 12.0 °C?

6.22 A car with a mass of 1.5×10^3 kg skids to a halt from a speed of 60.0 m/s. How many joules of heat are generated by friction between the car and the pavement?

6.23 Work (which has the same units as energy) is accomplished by moving an opposing force some distance. The amount of work is calculated as the product of force times distance (work = force × distance). Now, suppose a packing crate filled with machinery has a mass of 1.4×10^3 kg and is moved a distance of 40.0 m by a person exerting a force of 515 N. (The newton, N, is the SI unit of force: 1 N = 1 kg m/s².) How many joules of work have been accomplished?

6.24 What is the definition of *heat capacity?* What is the definition of *specific heat?* Be sure to specify the units for each.

6.25 Why do the oceans have a moderating influence on the summer and winter temperatures of landmasses along their shores?

6.26 What is the specific heat of water expressed in both calorie and joule energy units? What is the heat capacity of 1.87×10^3 g of water expressed in J/°C?

6.27 The *molar heat capacity* is the amount of heat needed to raise the temperature of 1 mol of a substance by 1 °C. What is the molar heat capacity of water expressed in J/mol °C?

6.28 How much will the temperature of 150 g of liquid water change if
(a) it gains 40.0 J?   (b) it loses 35.0 cal?

6.29 How many joules are required to raise the temperature of 0.500 kg of liquid water by 24.0 °C? How many calories are needed?

6.30 If 250 g of water at 30.0 °C is heated to 45.0 °C, by how much does the kinetic energy of the water molecules increase? (Express your answer in joules.)

6.31 A copper penny with a mass of 3.14 g is heated to 100.0 °C in boiling water. It is quickly dried and placed into a Styrofoam cup containing 10.0 g of water at 25.0 °C. Copper has a specific heat of 0.387 J/g °C. What will be the final temperature of the penny and the water? (Assume that a negligible amount of heat is absorbed by the cup.) *Hint:* The law of conservation of energy requires that the amount of heat lost by the penny be equal to the amount of heat absorbed by the water.

6.32 A 22.5-g copper bar is heated to a temperature of 100.0 °C in boiling water. After being quickly dried, it was placed into 20.0 g of liquid methyl alcohol (CH_3OH) with a temperature of 10.0 °C. The temperature of the mixture rose to 23.1 °C. The specific heat of copper is 0.387 J/g °C. Calculate the specific heat of the methyl alcohol. What is the molar heat capacity of the methyl alcohol?

6.33 A metal specimen with a mass of 25.467 g was heated to 100.0 °C in boiling water. The sample was quickly dried and placed in a Styrofoam cup that contained 15.0 g of H_2O having a temperature of 24.3 °C. The mixture was stirred quickly and the temperature of the water rose to 31.2 °C. Calculate the specific heat of the metal. (The hint given in the Question 6.31 applies here, too.)

Calorimetry

6.34 What is a calorimeter? What is a bomb calorimeter?

6.35 An electric heater was immersed in a calorimeter and electricity was used to supply 1347 J of heat. The temperature of the calorimeter rose from 25.000 °C to 26.135 °C. Calculate the heat capacity of the calorimeter in J/°C.

6.36 A certain exothermic reaction in a bomb calorimeter liberated 14.3 kJ. If the initial temperature of the calorimeter was 25.000 °C and the heat capacity of the calorimeter was 1.78×10^4 J/°C, what was the final temperature of the calorimeter?

6.37 A 1.00-mol sample of propane, a gas used for cooking in many rural areas, was placed in a bomb calorimeter with excess oxygen and ignited. The reaction was

$$C_3H_8(g) + 5O_2(g) \longrightarrow 3CO_2(g) + 4H_2O(l)$$

The initial temperature of the calorimeter was 25.000 °C and its total heat capacity was 97.1 kJ/°C. The reaction raised the temperature of the calorimeter to 27.282 °C.
(a) How many joules were liberated in this reaction?
(b) What is the heat of reaction of propane with oxygen expressed in kilojoules per mole of C_3H_8 burned?

6.38 The reaction of $OF_2(g)$ with water vapor follows the equation

$$OF_2(g) + H_2O(g) \longrightarrow O_2(g) + 2HF(g)$$

The reaction of 0.500 mol of OF_2 at 25 °C and a pressure of 1 atm liberated 162 kJ. Calculate the value of $\Delta H°$ for this reaction in units of kJ per mole of OF_2.

6.39 Toluene, C_7H_8, is used in the manufacture of explosives such as TNT (trinitrotoluene). A 1.500-g sample of liquid toluene was placed in a bomb calorimeter along with excess oxygen. When the combustion of the C_7H_8 was initiated, the temperature of the calorimeter rose from 25.000 °C to 26.413 °C. The products of the combustion are $CO_2(g)$ and $H_2O(l)$, and the heat capacity of the calorimeter was 45.06 kJ/°C. The reaction is

$$C_7H_8(l) + 9O_2(g) \longrightarrow 7CO_2(g) + 4H_2O(l)$$

(a) How many joules were liberated in this experiment?
(b) How many joules would be liberated under similar conditions if 1 mol of toluene were burned?

Thermochemistry

6.40 Define the following terms: (a) system, (b) surroundings, (c) state of a system, (d) isothermal change, and (e) adiabatic change.

6.41 If there is a decrease in the total potential energy of a system of chemical substances when they undergo a reaction, how will the average kinetic energy of the particles of the system change if the reaction is carried out (a) isothermally and (b) adiabatically?

6.42 When a "coffee cup" calorimeter is used to measure the heat of a reaction, is the reaction being carried out isothermally or adiabatically?

6.43 What is a state function? Why must the energy change for a reaction be a state function, provided the change occurs between the same two states.

6.44 How would we represent a change in the volume of a system symbolically? (Use V to stand for volume.)

6.45 If the temperature of 1 mol of liquid H_2O is increased from 25 °C to 45 °C, calculate the increase in its total kinetic energy. (Assume the potential energy of the water molecules stays the same.) Suppose this same mole of liquid water at 25 °C was first heated to 80 °C and then cooled to 45 °C. How much would the kinetic energy increase when the water is heated to 80 °C and how much would it decrease when the water is cooled from 80 °C to 45 °C? Calculate the net change in K.E. for the 25 °C → 80 °C → 45 °C path? How does this value compare with the K.E. change for the single step 25 °C → 45 °C path?

6.46 Why can't we calculate or measure the total kinetic energy of an object? What would we need to know in order to determine the total potential energy of an object?

6.47 What is a *perpetual motion machine?* Why are they impossible?

Hess's Law

6.48 State *Hess's law of heat summation.*

6.49 Under what conditions do we consider a substance to be in its *standard state?*

6.50 What is a thermochemical equation? Why are fractional coefficients permitted in thermochemical equations?

6.51 The following are two reactions showing the formation of 1 mol of SO_3:

$$SO_2(g) + \tfrac{1}{2}O_2(g) \longrightarrow SO_3(g)$$

$$S(s) + \tfrac{3}{2}O_2(g) \longrightarrow SO_3(g)$$

Should their enthalpy changes both be labeled $\Delta H_f°$ if they occur at 25 °C and 1 atm? Explain your answer.

6.52 The reaction

$$2NO(g) + Cl_2(g) \longrightarrow 2NOCl(g)$$

has $\Delta H° = -77.4$ kJ. What are $\Delta H°$ corresponding to the following?
(a) $NO(g) + \tfrac{1}{2}Cl_2(g) \rightarrow NOCl(g)$
(b) $6NOCl(g) \rightarrow 6NO(g) + 3Cl_2(g)$

6.53 Consider the following three thermochemical equations:

(1) $CH_3OH(l) + O_2(g) \longrightarrow HCHO_2(l) + H_2O(l)$
$$\Delta H° = -411 \text{ kJ}$$

(2) $CO(g) + 2H_2(g) \longrightarrow CH_3OH(l) \quad \Delta H° = -128 \text{ kJ}$

(3) $HCHO_2(l) \longrightarrow CO(g) + H_2O(l) \quad \Delta H° = -33 \text{ kJ}$

Suppose Equation (1) is reversed and divided by 2, Equations (2) and (3) are multiplied by $\tfrac{1}{2}$, and then the three adjusted equations are added. What is the net reaction and what is the value of $\Delta H°$ for the net reaction?

6.54 Chlorofluoromethanes (CFMs) are carbon compounds of chlorine and fluorine and are also known as Freons. Examples are Freon-11 ($CFCl_3$) and Freon-12 (CF_2Cl_2), which have been used as aerosol propellants. Freons are also used in refrigeration and air conditioning systems. It is feared that continued use of the Freons may result in the depletion of the atmospheric ozone layer that protects the earth's inhabitants from harmful ultraviolet radiation. In the stratosphere CFMs absorb high-energy radiation and produce Cl atoms that hasten the removal of ozone, O_3, from the air. Possible reactions are

(1) $O_3 + Cl \longrightarrow O_2 + ClO \qquad \Delta H° = -126 \text{ kJ}$
(2) $ClO + O \longrightarrow Cl + O_2 \qquad \Delta H° = -268 \text{ kJ}$
(3) $O_3 + O \longrightarrow 2O_2$

The O atoms come from the breaking apart of O_2 molecules caused by radiation from the sun. Use Equations (1) and (2) to calculate the value of $\Delta H°$ (in kilojoules) for

Equation (3), the net reaction for the removal of O_3 from the atmosphere.

6.55 Given the following thermochemical equations,

$$CaO(s) + Cl_2(g) \longrightarrow CaOCl_2(s) \qquad \Delta H° = -110.9 \text{ kJ}$$
$$H_2O(l) + CaOCl_2(s) + 2NaBr(s) \longrightarrow$$
$$2NaCl(s) + Ca(OH)_2(s) + Br_2(l)$$
$$\Delta H° = -60.2 \text{ kJ}$$

$$Ca(OH)_2(s) \longrightarrow CaO(s) + H_2O(l) \qquad \Delta H° = +65.1 \text{ kJ}$$

calculate the value of $\Delta H°$ (in kilojoules) for the reaction

$$\tfrac{1}{2}Cl_2(g) + NaBr(s) \longrightarrow NaCl(s) + \tfrac{1}{2}Br_2(l)$$

6.56 Given the following thermochemical equations,

$$2Cu(s) + S(s) \longrightarrow Cu_2S(s) \qquad \Delta H° = -79.5 \text{ kJ}$$

$$S(s) + O_2(g) \longrightarrow SO_2(g) \qquad \Delta H° = -297 \text{ kJ}$$

$$Cu_2S(s) + 2O_2(g) \longrightarrow 2CuO(s) + SO_2(g)$$
$$\Delta H° = -527.5 \text{ kJ}$$

calculate the value of the standard enthalpy of formation (in kilojoules) of $CuO(s)$.

6.57 Given the following thermochemical equations,

$$4NH_3(g) + 7O_2(g) \longrightarrow 4NO_2(g) + 6H_2O(g)$$
$$\Delta H° = -1132 \text{ kJ}$$

$$6NO_2(g) + 8NH_3(g) \longrightarrow 7N_2(g) + 12H_2O(g)$$
$$\Delta H° = -2740 \text{ kJ}$$

calculate the value of $\Delta H°$ (in kilojoules) for the reaction

$$4NH_3(g) + 3O_2(g) \longrightarrow 2N_2(g) + 6H_2O(g)$$

6:58 Given the following thermochemical equations,

$$3Mg(s) + 2NH_3(g) \longrightarrow Mg_3N_2(s) + 3H_2(g)$$
$$\Delta H° = -371 \text{ kJ}$$

$$\tfrac{1}{2}N_2(g) + \tfrac{3}{2}H_2(g) \longrightarrow NH_3(g) \qquad \Delta H° = -46 \text{ kJ}$$

calculate $\Delta H°$ (in kilojoules) for the reaction

$$3Mg(s) + N_2(g) \longrightarrow Mg_3N_2(s)$$

6.59 Use the following thermochemical equations,

$$8Mg(s) + Mg(NO_3)_2 \longrightarrow Mg_3N_2(s) + 6MgO(s)$$
$$\Delta H° = -3884 \text{ kJ}$$

$$Mg_3N_2(s) \longrightarrow 3Mg(s) + N_2(g) \qquad \Delta H° = +463 \text{ kJ}$$

$$2MgO(s) \longrightarrow 2Mg(s) + O_2(g) \qquad \Delta H° = +1203 \text{ kJ}$$

to calculate the standard heat of formation (in kilojoules) of $Mg(NO_3)_2$.

6.60 Important reactions in the production of ozone, O_3, in polluted air are

$$(1) \quad 2NO(g) + O_2(g) \longrightarrow 2NO_2(g)$$

$$(2) \quad NO_2(g) \longrightarrow NO(g) + O(g)$$

$$(3) \quad O_2(g) + O(g) \longrightarrow O_3(g)$$

Given that the reaction $O_2(g) \rightarrow 2O(g)$ has $\Delta H° = +498$ kJ, use the data in Table 6.1 to calculate $\Delta H°$ (in kilojoules) for reactions (1), (2), and (3).

6.61 Given the following thermochemical equations,

$$2H_2(g) + O_2(g) \longrightarrow 2H_2O(l) \qquad \Delta H° = -571.5 \text{ kJ}$$

$$N_2O_5(g) + H_2O(l) \longrightarrow 2HNO_3(l) \qquad \Delta H° = -76.6 \text{ kJ}$$

$$\tfrac{1}{2}N_2(g) + \tfrac{3}{2}O_2(g) + \tfrac{1}{2}H_2(g) \longrightarrow HNO_3(l)$$
$$\Delta H° = -174 \text{ kJ}$$

calculate $\Delta H°$ for the reaction

$$2N_2(g) + 5O_2(g) \longrightarrow 2N_2O_5(g)$$

6.62 When plaster of Paris, $CaSO_4 \cdot \tfrac{1}{2}H_2O$, is mixed with water, it combines with some H_2O to produce gypsum, $CaSO_4 \cdot 2H_2O$. The reaction is exothermic, which explains why a plaster cast on a broken arm or leg becomes warm as the cast hardens. Given that for $CaSO_4 \cdot \tfrac{1}{2}H_2O(s)$, $\Delta H_f° = -1573$ kJ/mol, and for $CaSO_4 \cdot 2H_2O(s)$, $\Delta H_f° = -2020$ kJ/mol, calculate $\Delta H°$ (in kilojoules) for the reaction

$$CaSO_4 \cdot \tfrac{1}{2}H_2O(s) + \tfrac{3}{2}H_2O(l) \longrightarrow CaSO_4 \cdot 2H_2O(s)$$

6.63 The evaporation of perspiration is one way the body disposes of excess thermal energy produced during exercise and thereby is able to maintain a constant body temperature. How many kilojoules are removed from the body by the evaporation of 10.0 g of H_2O at 25 °C? [*Hint*: The "reaction" is $H_2O(l) \rightarrow H_2O(g)$.]

6.64 The combustion of ethanol, $C_2H_5OH(l)$, to give $CO_2(g)$ and $H_2O(l)$ evolves 1.37×10^3 kJ per mole of $C_2H_5OH(l)$ when the products are returned to 25 °C and 1 atm. Use this information plus the data in Table 6.1 to calculate $\Delta H_f°$ for $C_2H_5OH(l)$. The equation for the combustion of 1 mol of ethanol is

$$C_2H_5OH(l) + 3O_2(g) \longrightarrow 2CO_2(g) + 3H_2O(l)$$

6.65 Use the data in Table 6.1 to calculate $\Delta H°$ in kilojoules for the following reactions:
(a) $2Al(s) + Fe_2O_3(s) \rightarrow Al_2O_3(s) + 2Fe(s)$
(b) $SiH_4(g) + 2O_2(g) \rightarrow SiO_2(s) + 2H_2O(g)$
(c) $CaO(s) + SO_3(g) \rightarrow CaSO_4(s)$
(d) $CuO(s) + H_2(g) \rightarrow Cu(s) + H_2O(g)$
(e) $C_2H_4(g) + H_2(g) \rightarrow C_2H_6(g)$

6.66 Use the data in Table 6.1 to calculate $\Delta H°$ in kilojoules for each of the following reactions:
(a) $C_2H_2(g) + H_2(g) \rightarrow C_2H_4(g)$
(b) $SO_3(g) + H_2O(l) \rightarrow H_2SO_4(l)$
(c) $Mg(OH)_2(s) + 2HCl(g) \rightarrow MgCl_2 \cdot 2H_2O(s)$
(d) $CO_2(g) + H_2(g) \rightarrow CO(g) + H_2O(g)$
(e) $10N_2O(g) + C_3H_8(g) \rightarrow$
$$10N_2(g) + 3CO_2(g) + 4H_2O(g)$$

Additional Exercises

6.67 When 200 mL of 1.00 M NaOH at 25.00 °C was mixed with 150 mL of 1.00 M HCl, also at 25.00 °C, in a Styrofoam "coffee cup" calorimeter, the temperature of the reaction mixture rose to 30.00 °C. Calculate the value of ΔH in kilojoules for the neutralization of 1 mol of H^+ by 1 mol of OH^- [i.e., $H^+ + OH^- \rightarrow H_2O(l)$]. Assume that the specific heat of each solution is 4.18 J/g °C.

6.68 Given the following thermochemical equations:

$2Cu(s) + O_2(g) \longrightarrow 2CuO(s)$ $\Delta H° = -155$ kJ

$Cu(s) + S(s) \longrightarrow CuS(s)$ $\Delta H° = -53.1$ kJ

$S(s) + O_2(g) \longrightarrow SO_2(g)$ $\Delta H° = -297$ kJ

$4CuS(s) + 2CuO(s) \rightarrow 3Cu_2S(s) + SO_2(g)$
$\Delta H° = -13.1$ kJ

calculate the value of $\Delta H°$ (in kilojoules) for the reaction

$$CuS(s) + Cu(s) \longrightarrow Cu_2S(s)$$

6.69 It is estimated that the body generates up to 5900 kJ of thermal energy per hour during heavy physical exercise, which the body eliminates by perspiring. How many grams of glucose, $C_6H_{12}O_6$, would have to be metabolized per hour (to give CO_2 and H_2O) in order to generate this excess thermal energy if it is assumed that 60% of the available energy from the metabolism appears as excess heat (the remaining 40% is the energy we use to move limbs, pump blood, etc.)? For $C_6H_{12}O_6$, $\Delta H_{combustion} = -2820$ kJ/mol of $C_6H_{12}O_6$.

6.70 Given the following thermochemical equations,

$Fe_2O_3(s) + 3CO(g) \longrightarrow 2Fe(s) + 3CO_2(g)$
$\Delta H = -28$ kJ

$3Fe_2O_3(s) + CO(g) \longrightarrow 2Fe_3O_4(s) + CO_2(g)$
$\Delta H = -59$ kJ

$Fe_3O_4(s) + CO(g) \longrightarrow 3FeO(s) + CO_2(g)$
$\Delta H = +38$ kJ

calculate ΔH for the reaction

$$FeO(s) + CO(g) \longrightarrow Fe(s) + CO_2(g)$$

without referring to the data in Table 6.1.

6.71 Use the results of Problem 6.70 and the data in Table 6.1 to calculate the standard heat of formation of FeO. Express the answer in kilojoules per mole.

6.72 Acetylene, C_2H_2, a gas used in welding torches, is produced by the action of water on calcium carbide, CaC_2. Given the thermochemical equations below, calculate $\Delta H_f°$ for acetylene. Express your answer in kilojoules per mole.

$CaO(s) + H_2O(l) \longrightarrow Ca(OH)_2(s)$ $\Delta H° = -65.3$ kJ

$CaO(s) + 3C(s, graphite) \longrightarrow CaC_2(s) + CO(g)$
$\Delta H° = +462.3$ kJ

$CaC_2(s) + 2H_2O(l) \longrightarrow Ca(OH)_2(s) + C_2H_2(g)$
$\Delta H° = -126$ kJ

$2C(s, graphite) + O_2(g) \longrightarrow 2CO(g)$
$\Delta H° = -220$ kJ

$2H_2O(l) \longrightarrow 2H_2(g) + O_2(g)$ $\Delta H° = +572$ kJ

ELECTRONIC STRUCTURE AND THE PERIODIC TABLE

Chemical facts, such as those you learned in Chapters 4 and 5, are a vital part of chemistry. You certainly can't say that you know chemistry if you don't know how certain substances behave. But facts are just a part of chemistry. We are not satisfied just knowing *how* substances behave; we want to also know *why*. Why are some elements metals and others nonmetals? Why are nonmetal–nonmetal compounds molecular, whereas metal–nonmetal compounds tend to be ionic? And in the periodic table (see Figure 4.8), why are there so many empty spaces? Why aren't there elements to fill them? These are examples of the kinds of questions that arise naturally as we learn facts. The search for answers is an important part of chemical research, and it is on its way to providing us with a fundamental understanding of the basic processes that control our world.

In Chapter 4 you learned that an atom is composed of protons and neutrons located in its nucleus at the center of the atom, surrounded by enough electrons to make the atom electrically neutral. This picture of atomic structure is sufficient to explain some properties of the elements — the existence of isotopes, for example — but it still doesn't explain their chemical and physical properties.

When atoms react with each other, only their outer parts come into contact. Their nuclei are so tiny and buried so deeply inside the atoms that they never come close to each other. Therefore, chemical similarities and differences among atoms of different elements must be controlled by how the electrons are arranged in the outer regions of the atoms. This arrangement of electrons is called an atom's **electronic structure.**

We begin this chapter with a discussion of the experimental facts that helped solve the mystery of electronic structure and led ultimately to our currently accepted theory. Our goal is to be able to describe the electronic structure of an atom and relate it to the atom's position in the periodic table. From this knowledge we can begin to understand some of the trends in the chemical and physical properties of the elements.

7.1 ELECTROMAGNETIC RADIATION AND ATOMIC SPECTRA

The key that permitted the deduction of the electronic structures of the elements is an analysis of the light emitted by atoms when they've been energized, or *excited*, by being heated in a flame or by having an electric discharge passed through them. However, before we can discuss this, you first must learn what light is.

In Chapter 6 you learned that matter can *possess* energy in two ways: as kinetic energy and as potential energy. You also learned that energy can be *transferred* between objects. One way for this to happen is the transfer of heat, which is really a transfer of kinetic energy through the collisions of atoms and molecules. Another way that energy can be transferred between samples of matter is as light, which is really a form of energy.

Light in all its varied forms—X rays, visible light, infrared and ultraviolet radiation, and radio and TV waves—is called **electromagnetic radiation.** It travels through space as a wave at a constant speed called the speed of light, 3.00×10^8 m/s. Light waves, like waves in general, are characterized by certain properties. One is the intensity of the wave, which is measured by its **amplitude,** the maximum height of its peaks, as shown in Figure 7.1. Another characteristic of the wave is its **wavelength,** λ, the distance between consecutive peaks or troughs in the wave (see Figure 7.1). A third quantity that characterizes a wave is its **frequency,** ν, which is the number of peaks that pass a given point in one second as the wave moves by. For any wave, a relationship between its wavelength and its frequency exists.

$$(\text{wavelength}) \times (\text{frequency}) = \text{speed of the wave}$$

For light waves, the speed is given by the symbol c. This gives the useful equation

$$\lambda \cdot \nu = c \tag{7.1}$$

Light waves come in all sorts of wavelengths, giving rise to the **electromagnetic spectrum** that is illustrated in Figure 7.2. The units used to express these wavelengths depend on the region of the spectrum in which the radiation occurs. CB radios, for example, broadcast waves having wavelengths of about 11 meters, so meters are convenient units to express these wavelengths. Visible light, which encompasses only the very narrow portion of the electromagnetic spectrum to which our eyes are sensitive, has wavelengths of the order of several hundred nanometers, so this unit is convenient for reporting the lengths of visible light waves (Figure 7.2b).

Frequency, in general, describes the number of events *per unit of time*. If we are discussing family gatherings, for instance, they might take place *three times per year*. If we are speaking of an alternating electric current, we find that the current reverses direction *120 times per second*. In each case, the notion of frequency is in the *per unit of time*, and the unit that we associate with frequency has the dimension of 1/time or time⁻¹. In the SI, the unit of time is the second, s. Therefore, the SI unit of frequency is 1/s or s⁻¹. This unit is given the name **hertz,** and its symbol is **Hz.**

$$1 \text{ Hz} = 1 \text{ s}^{-1}$$

We can also write the speed of light as 3.00×10^8 m s⁻¹, since 1 s⁻¹ = $1/$s.

Light is called electromagnetic radiation because it causes disturbances in electric and magnetic fields.

The speed of light c is the same for all "kinds" of electromagnetic radiation: X rays, ultraviolet light, visible light, and others.

The visible spectrum ranges from approximately 400 to 700 nm.

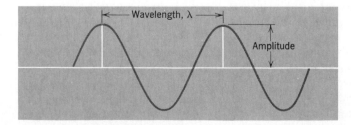

Figure 7.1. *Properties of a wave.*

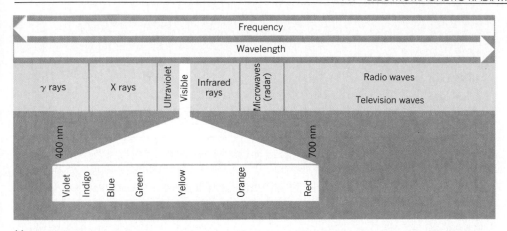

(a)

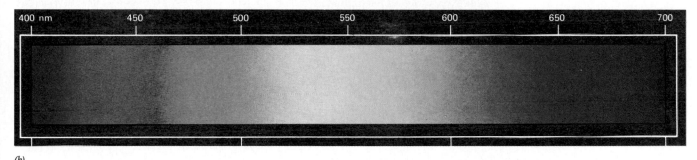

(b)

Figure 7.2. (a) *The electromagnetic spectrum.* (b) *The colors in the visible portion of the spectrum.*

EXAMPLE 7.1. CONVERTING FREQUENCY TO WAVELENGTH

PROBLEM: A ham radio operator broadcasts at a frequency of 14.2 MHz (megahertz). What is the wavelength of the radio waves put out by the transmitter?

SOLUTION: Since we wish to calculate the wavelength, let's solve Equation 7.1 for λ.

$$\lambda = \frac{c}{v}$$

The speed of light can be expressed as 3.00×10^8 m s^{-1}, and the frequency can be written $v = 14.2 \times 10^6$ Hz or 14.2×10^6 s^{-1} (remember, *mega* means $\times 10^6$). Substituting these values gives

$$\lambda = \frac{3.00 \times 10^8 \text{ m s}^{-1}}{14.2 \times 10^6 \text{ s}^{-1}} = 21.1 \text{ m}$$

EXAMPLE 7.2. CONVERTING FREQUENCY TO WAVELENGTH

PROBLEM: What is the wavelength, in nanometers, of green light having a frequency of 6.67×10^{14} Hz?

SOLUTION: Again we solve Equation 7.1 for λ ($\lambda = c/v$). Using $c = 3.00 \times 10^8$ m s^{-1} and $v = 6.67 \times 10^{14}$ s^{-1}, we get

$$\lambda = \frac{3.00 \times 10^8 \text{ m s}^{-1}}{6.67 \times 10^{14} \text{ s}^{-1}} = 4.50 \times 10^{-7} \text{ m}$$

Since 1 nm = 10^{-9} m,

The SI prefix *nano* means "$\times 10^{-9}$."

$$\lambda = 4.50 \times 10^{-7} \text{ m} \times \left(\frac{1 \text{ nm}}{10^{-9} \text{ m}}\right)$$

$$= 450 \text{ nm}$$

EXAMPLE 7.3. CONVERTING WAVELENGTH TO FREQUENCY

PROBLEM: What is the frequency of infrared radiation that has a wavelength of 1.25×10^3 nm?

SOLUTION: First we solve Equation 7.1 for ν ($\nu = c/\lambda$). If we use $c = 3.00 \times 10^8$ m s^{-1}, we must express the wavelength in meters.

$$\lambda = 1.25 \times 10^3 \text{ nm} \times \left(\frac{10^{-9} \text{ m}}{1 \text{ nm}}\right)$$

$$= 1.25 \times 10^{-6} \text{ m}$$

Now we can solve for the frequency.

$$\nu = \frac{3.00 \times 10^8 \text{ m s}^{-1}}{1.25 \times 10^{-6} \text{ m}}$$

$$= 2.40 \times 10^{14} \text{ s}^{-1}$$

The frequency is 2.40×10^{14} Hz.

Atomic spectra

If sunlight or the light from an ordinary electric light bulb is formed into a narrow beam and then passed through a prism onto a screen, a rainbow of colors is produced (Figure 7.3). By refraction, the prism splits the white light into a spectrum that is composed of light of all colors. This "rainbow" is called a **continuous spectrum**

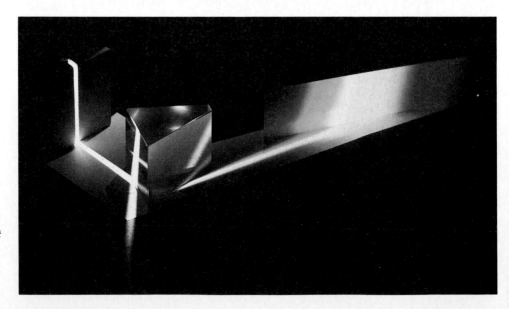

Figure 7.3. *Refraction of white light by a prism separates the light into a continuous spectrum, which contains light of all colors (all frequencies).*

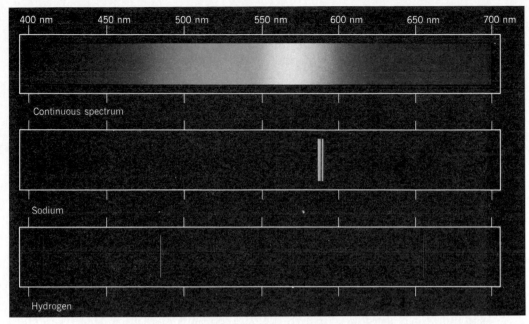

(a)

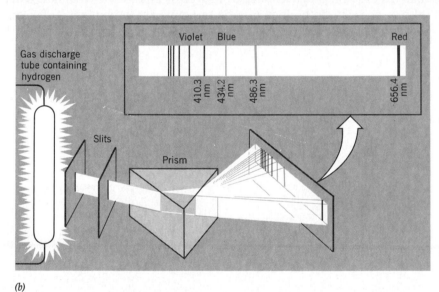

(b)

Figure 7.4. (a) *The emission spectra of sodium and hydrogen in the visible region compared to a continuous spectrum. (b) The formation of the emission spectrum of hydrogen. An electric spark is passed through hydrogen gas in a glass tube. The light emitted by hydrogen atoms is formed into a narrow beam and passed through a prism, which divides the beam into several beams that form the line spectrum when they strike the screen. The four lines whose wavelengths are listed are those that can be seen in the visible portion of the hydrogen spectrum.*

because all the wavelengths of visible light are present. However, if the light passed through the prism comes from an electric spark passing through a gas such as hydrogen or sodium vapor, or if the light comes from a compound heated in a flame, the spectrum produced looks quite different, as shown in Figure 7.4. Instead of a rainbow of colors, only a few colored lines are seen. These lines are the image of the slit that is used to create the narrow beam of light, and because of the appearance of the spectrum, it is often called a **line spectrum.** Since the light is emitted by *atoms* that have been excited (energized), it is also called an **atomic emission spectrum,** or simply an **emission spectrum** or **atomic spectrum.**

It is quite obvious that the light given off by an element when it produces an emission spectrum does not consist of radiation of all wavelengths. The dark regions of the spectrum correspond to wavelengths for which light is *not* emitted. You can also see from Figure 7.4 that the lines in the emission spectrum of hydrogen have different wavelengths than the lines in the spectrum of sodium. In fact, each element has its own characteristic emission spectrum that can be used to identify it. For example, you have probably seen crime shows on TV in which the police take paint chips from the clothing of a hit-and-run victim for analysis. If the elements contained in the paint, along with their relative amounts, match a similar analysis of paint scrapings taken from a suspect's car, there is fairly strong evidence that the suspect's car was involved in the crime. Such an analysis can be done easily by using atomic spectra, because each element in the paint produces its own characteristic set of emission lines.

The atomic spectra of certain elements are also used in modern street lighting. In older incandescent lighting, in which a tungsten filament is heated white-hot by an electric current, much of the emitted energy is in the form of infrared light. Unfortunately, our eyes are not sensitive to this form of radiation, so this energy is wasted — only a relatively small fraction of the electrical energy actually appears as visible light. Modern streetlights consist of high-intensity mercury or sodium vapor lamps in which the light is produced by the atomic emission spectra of these elements. In these lamps, most of the electrical energy appears as light in the visible portion of the spectrum, and they are therefore much more efficient. For example, sodium emits intense light at a wavelength of 589 nm, which is yellow. The golden glow of streetlights in scenes like that shown in Figure 7.5 is from this emission line of sodium, which is produced in high-pressure sodium vapor lamps. There is another, much more yellow light emitted by low-pressure sodium vapor lamps that you may also have seen. Both kinds of sodium lamps are very energy efficient, and most communities have switched to this kind of lighting to save money on the costs of electricity.

Mercury vapor lamps, which are nearly as efficient as sodium vapor lamps, give a bluish-white light.

7.2 ATOMIC STRUCTURE AND THE BOHR THEORY

The existence of atomic spectra was discovered in the latter part of the nineteenth century. Although measurements could be made at that time of the wavelengths of

Figure 7.5. *Energy-efficient high-pressure sodium lamps light Park Avenue in New York City. The golden-yellow tone of the lighting is caused by the intense emission of yellow light by sodium at 589 nm.*

the spectral lines, no one was able to explain why such spectra were produced. One thing was clear, however. Heavy elements produce complex spectra, while the lightest of the elements, hydrogen, produces the simplest spectrum with the fewest lines.

Progress in understanding atomic spectra began when, almost by trial and error, an equation was discovered that could be used to calculate the wavelengths of all the spectral lines of hydrogen. This is called the Rydberg equation

$$\frac{1}{\lambda} = 109{,}678 \text{ cm}^{-1} \left(\frac{1}{n_1{}^2} - \frac{1}{n_2{}^2} \right) \tag{7.2}$$

Notice how remarkably simple this equation is. On the right side there is just one constant, 109,678 cm^{-1}, and two variables, n_1 and n_2. The variables n_1 and n_2 stand for integers (whole numbers) and can have values of 1, 2, 3, . . . , ∞, with the requirement that n_2 must always be larger than n_1. Thus, if we choose $n_1 = 1$, then n_2 can have a value of 2 or 3 or 4, . . . , and so on, and if we take $n_1 = 2$, then n_2 can be given a value of 3 or 4 or 5, . . . , and so on. Whichever pair of integers we choose, however, we always obtain the wavelength of a line in the hydrogen spectrum. The following example illustrates how easy this is to accomplish.

Many of the lines in the emission spectrum of hydrogen and the other elements lie outside the visible portion of the electromagnetic spectrum.

EXAMPLE 7.4. USING THE RYDBERG EQUATION TO CALCULATE THE WAVELENGTH OF A LINE IN THE SPECTRUM OF HYDROGEN

PROBLEM: Calculate the wavelength, in nanometers, of the line in the spectrum of hydrogen corresponding to $n_1 = 2$ and $n_2 = 4$ in the Rydberg equation.

SOLUTION: All we have to do is substitute the values of n_1 and n_2 into the Rydberg equation as follows:

$$\frac{1}{\lambda} = 109{,}678 \text{ cm}^{-1} \left(\frac{1}{n_1{}^2} - \frac{1}{n_2{}^2} \right)$$

$$= 109{,}678 \text{ cm}^{-1} \left(\frac{1}{2^2} - \frac{1}{4^2} \right)$$

$$= 2.056 \times 10^4 \text{ cm}^{-1}$$

We now have the reciprocal of the wavelength in units of cm^{-1}. Taking the reciprocal of each side gives the wavelength in centimeters.

$$\lambda = \frac{1}{2.056 \times 10^4 \text{ cm}^{-1}}$$

$$= 4.864 \times 10^{-5} \text{ cm}$$

To change to nanometers, we use the relationships

$$1 \text{ cm} = 10^{-2} \text{ m}$$

$$1 \text{ nm} = 10^{-9} \text{ m}$$

Applying these as conversion factors gives

$$\lambda = 4.864 \times 10^{-5} \text{ cm} \left(\frac{10^{-2} \text{ m}}{1 \text{ cm}} \right) \left(\frac{1 \text{ nm}}{10^{-9} \text{ m}} \right)$$

$$= 486.4 \text{ nm}$$

You might note that this is the wavelength of the green line in the visible spectrum of hydrogen shown in Figure 7.4 on page 195.

What is amazing about the Rydberg equation is its simplicity. How remarkable it is to be able to calculate the wavelength of a spectral line by just choosing two whole numbers, and no matter which pair of whole numbers we choose, we always obtain the wavelength of one of the spectral lines. It is no wonder that scientists saw the Rydberg equation as a fundamental clue to the internal structure of the atom.

The energy of light waves

By analyzing the light given off by heated objects, a German physicist named Max Planck concluded in 1900 that light sometimes behaves as though it were composed of tiny particles or **quanta** of energy (later called **photons**). Planck also found that the energy of a photon is directly proportional to the frequency of the light,

$$E_{photon} = h\nu \tag{7.3}$$

where h is a proportionality constant known as **Planck's constant.** Its value is 6.63×10^{-34} J $\cdot$ s (the units are thus a product of energy $\times$ time).

Another phenomenon that illustrates the particlelike behavior of light is the photoelectric effect, in which it is observed that light is able to knock electrons off the surfaces of certain metals (most notably, the alkali metals). These ejected electrons can be detected electronically, and the photoelectric effect has been routinely used in door-opening devices and alarm systems to detect any interruption that might occur in a beam of light.

In studying the photoelectric effect, scientists discovered that there are some interesting restrictions on the kind of light able to cause the phenomenon. If light of low frequency (long wavelength) is used, no electrons are ejected at all, regardless of how intense the light is. As the frequency of the light shining on the surface is gradually increased, electrons begin to be ejected when a certain minimum frequency is finally reached. Then, as the frequency of the light is increased further, the kinetic energy of the ejected electrons increases.

The brightness or intensity of the light is a measure of the *number* of photons that are striking the surface per second. However, each photon has an energy that depends on its frequency.

Albert Einstein explained this behavior in 1905 by also suggesting that light is composed of particlelike photons whose energy is directly proportional to the frequency of the light. When the frequency is low, each photon has a very small amount of energy and is incapable of knocking an electron off the surface of the metal. As the frequency (and energy) increases, eventually the photon does pack enough energy to eject an electron, and as the frequency (and energy) is increased further, the *extra* energy, beyond the minimum needed to eject the electron, appears as the electron's kinetic energy.

Energy levels in atoms

The relationship between the frequency and the energy of light opened the door to the understanding of the internal structure of the atom. When an atom emits a photon having a specific energy, the atom loses that specific amount of energy. In other words, the photons emitted by atoms of an element tell us what energy changes are occurring in those atoms. Now, in an atomic spectrum we see light of only certain discrete frequencies being emitted. This suggests that certain energy changes are possible, but others (corresponding to frequencies that are *not* part of the spectrum) are impossible.

How is it that for a given atom, we always find the same allowed energy changes giving rise to the same atomic spectrum? And why are these energy changes different for different atoms? The preliminary answer to these questions comes in the form of a hypothesis: Suppose that in an atom, the electrons are not allowed to have random amounts of energy, but instead are restricted to certain specific energies. Stated differently, suppose the electrons in an atom are restricted to certain

energy levels. If this is true, then when an electron loses energy, it can "fall" from one energy level to a lower one, but it can't come to some intermediate energy. Therefore, whenever an electron goes between these two energy levels, it will always change energy by the same amount and always emit a photon of the same frequency. Within any given atom, there are many energy levels and many different changes possible, but the energy changes are always the same and the photons emitted by a collection of these atoms always correspond to the same set of frequencies. Thus, the emission spectrum is always the same for any given element, and it is different for different elements simply because each element has its own characteristic set of energy levels for its electrons.

The Bohr theory of hydrogen

Niels Bohr, a Danish physicist, was the first to develop a theory of atomic structure employing the notion of electron energy levels to explain atomic spectra. He chose hydrogen for his theoretical model, simply because it has the simplest atom (one proton and one electron) and because it produces the simplest atomic spectrum.

Bohr's theory proposed a "solar system" model of the hydrogen atom, with the electron able to travel around the nucleus in orbits of fixed size and energy (see Figure 7.6). Because the electron and proton attract each other, the larger the orbit, the greater the energy of the electron. From this model of the atom, Bohr derived mathematically an equation for the energy of the electron that has the form

$$E = -A/n^2$$

in which A is a constant that can be evaluated from Planck's constant and the known mass and charge of the electron. The value of A is 2.18×10^{-18} J. The quantity n is an integer that identifies the orbit of the electron; it can have values of 1, 2, 3, . . . , ∞ and is called a **quantum number** because its value determines the "quantity" of energy that the electron has in a particular orbit.

Because of the negative sign in the equation, the orbit of lowest energy has $n = 1$ and that of highest energy has $n = \infty$. The size of the orbit also increases with increasing n.

The atomic spectrum of hydrogen was explained by Bohr as follows. Ordinarily, the electron in a hydrogen atom is in its lowest energy orbit (its lowest energy level)

The zero point on the energy scale is arbitrarily chosen to be the energy the electron has when it is in its highest energy level. Therefore, when the electron has a lower energy, the value for its energy is negative.

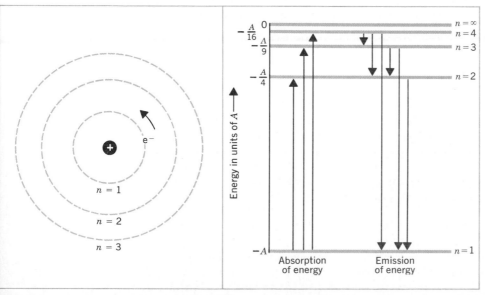

(a)   (b)

Figure 7.6. *Bohr's model of the atom. (a) The electron is able to travel along certain specific orbits of fixed energy. This part of Bohr's theory was later shown to be wrong. (b) The energy of the electron changes by specific amounts when the electron goes from one energy level to another. This part of Bohr's theory is correct and is incorporated in our currently accepted theory.*

with $n = 1$. When the atom is excited, the electron is raised to a higher-energy orbit—say, with $n = 2, 3,$ or 4. The atom is not stable in this state, however, so the electron drops to a lower-energy orbit, as illustrated in Figure 7.6b. For each drop, the electron changes energy by a specific amount, an amount equal to the difference in energy between the two orbits. This amount of energy is released by the atom in the form of a photon having an appropriate frequency.

Bohr used his equation for the energy and his concept of how an atomic spectrum is formed to derive an equation for the reciprocal of the wavelength of the spectral lines for hydrogen. Amazingly, his mathematically derived equation matched the Rydberg equation almost exactly.

$$\text{Derived by Bohr:} \qquad \frac{1}{\lambda} = 109{,}730 \text{ cm}^{-1} \left(\frac{1}{n_1^2} - \frac{1}{n_2^2} \right)$$

The experimentally derived equation is the Rydberg equation.

$$\text{Determined experimentally:} \qquad \frac{1}{\lambda} = 109{,}678 \text{ cm}^{-1} \left(\frac{1}{n_1^2} - \frac{1}{n_2^2} \right)$$

Bohr's success delighted physicists, because they believed that the secret of the atom had finally been revealed. Unfortunately, however, Bohr's model only worked for hydrogen. When applied to other atoms, it failed to predict their atomic spectra correctly, suggesting a basic flaw in the model. Nevertheless, Bohr's introduction of the concept of quantum numbers and the success of the energy level concept at explaining the origin of atomic spectra were important steps forward.

EXAMPLE 7.5. CALCULATING ENERGY CHANGES IN THE HYDROGEN ATOM

PROBLEM: Calculate the energy of the photon emitted when an electron drops from the fifth to the second energy level in a hydrogen atom. Calculate the frequency and wavelength (in nanometers) of this photon.

ANALYSIS: According to the Bohr theory, the energy of the hydrogen atom is given by the equation

$$E = -A/n^2$$

where $A = 2.18 \times 10^{-18}$ J. Let's use this to derive an equation for an energy change between two states with different values for their quantum number. For two energy levels having quantum numbers n_1 and n_2, the change in energy going from one to the other is

$$\Delta E = E_1 - E_2$$

$$= \left(\frac{-A}{n_1^2} \right) - \left(\frac{-A}{n_2^2} \right)$$

$$= A \left(\frac{1}{n_2^2} - \frac{1}{n_1^2} \right)$$

This energy change is the energy of the photon. The frequency can then be calculated using the Planck relationship, and finally, the wavelength is obtained as in Example 7.1.

SOLUTION: In this problem, we can take $n_1 = 5$ and $n_2 = 2$. Substituting these into the equation above gives

$$\Delta E = 2.18 \times 10^{-18} \text{ J} \left(\frac{1}{2^2} - \frac{1}{5^2} \right)$$

$$= 2.18 \times 10^{-18} \text{ J} \left(\frac{1}{4} - \frac{1}{25} \right)$$

$$= 4.58 \times 10^{-19} \text{ J}$$

The energy of a photon is related to its frequency by the equation

$$E = h\nu$$

Solving for the frequency gives

$$\nu = \frac{E}{h}$$

$$= \frac{4.58 \times 10^{-19}\ \cancel{J}}{6.63 \times 10^{-34}\ \cancel{J} \cdot s}$$

$$= 6.91 \times 10^{14}\ Hz$$

This is the frequency of the photon that's emitted.

Wavelength and frequency are related by the equation

$$\lambda\nu = c$$

Solving for the wavelength yields

$$\lambda = \frac{c}{\nu}$$

$$= \frac{3.00 \times 10^{8}\ \cancel{m}\ \cancel{s^{-1}}}{6.91 \times 10^{14}\ \cancel{s^{-1}}} \times \frac{1\ nm}{10^{-9}\ \cancel{m}}$$

$$= 434\ nm$$

This is the violet line that you can see in the visible spectrum of hydrogen in Figure 7.4 on page 195.

7.3 THE WAVE NATURE OF MATTER: WAVE MECHANICS

The currently accepted theory explaining the behavior of electrons in atoms is called **wave mechanics,** which has its roots in a hypothesis put forward by Louis de Broglie in 1924. De Broglie suggested that if light can behave in some instances as though it were composed of particles, perhaps particles, at times, exhibit properties that we normally associate with waves.

De Broglie's argument proceeded as follows. Einstein had shown that the energy equivalent, E, of a particle of mass, m, is equal to

$$E = mc^2 \qquad (7.4)$$

where c is the speed of light. A photon whose energy is E could thus be said to have an effective mass equal to m. Max Planck had shown that the energy of a photon is given by Equation 7.3,

$$E = h\nu = \frac{hc}{\lambda}$$

Equating these two gives

$$\frac{hc}{\lambda} = mc^2$$

When we solve for λ, the wavelength, we obtain

$$\lambda = \frac{h}{mc}$$

Figure 7.7. *The production of a diffraction pattern by constructive and destructive interference of light waves.*

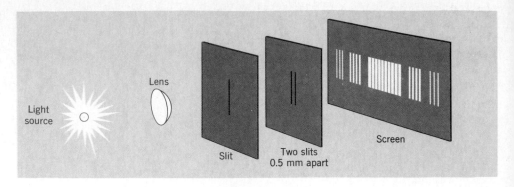

If this equation also applies to particles, such as the electron, the equation can be written as

$$\lambda = \frac{h}{mv} \tag{7.5}$$

where we have replaced c, the speed of light, by v, the speed of the particle.

Experimental evidence for this dual wave–particle nature of matter exists in the form of a phenomenon called diffraction, a property that can only be explained by wave motion. If light is allowed to pass through a very thin slit whose width is about the same as the wavelength of the light, the slit behaves as if it were a tiny light source, scattering light in all directions. This phenomenon is called **diffraction.** If two of these slits are placed alongside each other, each behaves as a separate source of light. When a screen is placed so that this light falls on it, we observe a pattern, called a **diffraction pattern,** that consists of light and dark areas as shown in Figure 7.7. In the bright areas, light waves that arrive from each hole are *in phase;* that is, the peaks and troughs of the two waves are lined up so that the amplitudes of the waves add together to produce a resultant wave of greater intensity. This is illustrated in Figure 7.8a. In the darkened areas, waves that arrive from the two slits are *out of phase* with each other, which means that the peaks of one wave coincide with the troughs of the other. When this happens, the amplitudes of the waves cancel (Figure 7.8b), so that zero intensity (darkness) is observed. Similar diffraction patterns can be produced with certain particles, including electrons, protons, and neutrons. Since diffraction can only be explained as a property of waves, it is clear that under certain circumstances matter does have wave properties. The reason that the wave nature of matter was not discovered earlier is that objects large enough to see, either with the naked eye or a microscope, possess so much mass that their wavelengths are much too short to be observed.

When the waves are in phase, they undergo **constructive interference** and their intensities add. When out of phase, the waves undergo **destructive interference** and their intensities cancel.

Figure 7.8. *Constructive and destructive interference of waves.*

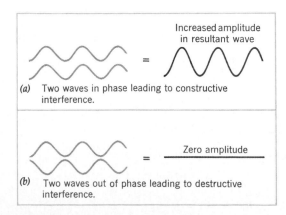

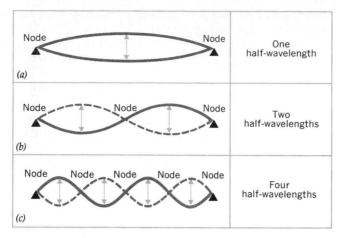

Node .... Node	One half-wavelength
(a)	
Node ... Node ... Node	Two half-wavelengths
(b)	
Node ... Node ... Node ... Node ... Node	Four half-wavelengths
(c)	

Figure 7.9. *Vibrations on a guitar string. (a) An open string is plucked. Only half a wave is formed, so the wavelength is twice the length of the string. (b) A harmonic is produced by briefly touching the string at its midpoint as it is plucked. (c) A still higher harmonic with a shorter wavelength and more nodes along the string.*

Standing waves and quantum numbers

One of the significant contributions of the Bohr theory was the introduction of an integer quantum number. If we consider the electron to be a wave that surrounds the nucleus, we find that the appearance of integers occurs in a very natural way. To see how this might happen, let's first consider some simpler waves, those that occur on a guitar string. When a string is plucked, it moves up and down in the center, but its ends remain motionless as shown in Figure 7.9*a*. At the ends, the height of the wave—its amplitude—is zero. These points of zero amplitude are called **nodes,** and any waves that have stationary nodes, whether on a guitar string or not, are called **standing waves.**

If you've ever played the guitar, you know that even without using your finger to press the string at the frets along the neck of the instrument, it is possible to play a variety of notes called harmonics. For example, if the string is briefly touched at its midpoint as it is plucked, a note an octave higher is produced. Figure 7.9*b* shows that when this harmonic is played, there are three nodes along the string. Still other harmonics produce even more nodes.

In examining the waves in Figure 7.9, we can see that there are certain restrictions on their allowed wavelengths. Because each end of the string must always be a node, there must be a *whole number* of *half-wavelengths* repeated along the length of string. Waves for which this is not true simply cannot exist on the string. What we see, therefore, is that a whole number, sort of a "musical quantum number", arises automatically when we consider the standing waves possible on a guitar string.

A similar kind of situation exists in atoms because the electron's matter waves are standing waves, too. The shapes of the matter waves are much different than those on a guitar string, of course, because the atom is three-dimensional and the conditions restricting the locations of the nodes are also much different. Nevertheless, the restrictions that determine which electron waves can exist give rise quite naturally to integer quantum numbers. In three dimensions, however, there are three quantum numbers instead of only one.

In 1926, Erwin Schrödinger (1887–1961) applied mathematics to the investigation of the standing waves in the hydrogen atom and began a field of study called **wave mechanics** or **quantum mechanics.** The mathematics here is quite advanced, so we will avoid it completely and look only at the results of the theory.

Schrödinger solved a mathematical equation, called a wave equation, for the

If the length of the string is L, then $n(\lambda/2) = L$ where n is a whole number. The allowed wavelengths along the string, therefore, are $\lambda = 2L/n$.

Schrödinger shared the Nobel prize for physics in 1933 with Paul Dirac, another physicist, for their pioneering work in quantum mechanics.

hydrogen atom.[1] He obtained a set of mathematical functions called **wave functions** (usually represented by the Greek letter psi, ψ) that describe the allowed shapes and energies of the electron waves. Each of these different possible waves is called an **orbital** (to distinguish it from Bohr's orbits). *Each orbital in an atom has a characteristic energy and is viewed as describing a region around the nucleus where the electron can be expected to be found.* The wave functions describing the orbitals are themselves characterized by the values of three quantum numbers (as we hinted earlier).

According to wave mechanics, each of the energy levels in an atom is associated with one or more orbitals. In atoms that contain more than one electron, the distribution of the electrons about the nucleus is determined by the number and kind of orbitals occupied. Therefore, in order to investigate the way the electrons are arranged in space, we must first examine the energy levels in the atom. This is best accomplished through a discussion of the quantum numbers.

1. The **principal quantum number, n.** The energy levels in an atom are arranged roughly into main levels, or **shells,** as determined by the principal quantum number, n. The larger the value of n, the greater the average energy of the levels belonging to the shell. We will also see that n determines the size of the orbitals. As in the Bohr theory, n may have values of 1, 2, 3, . . . , and so on up to infinity. Letters are also associated with these shells as shown below.

> The larger the value of n, the greater the average distance of the electron from the nucleus.

Principal quantum number	1	2	3	4 . . .
Letter designation		K	L	M	N . . .

For example, sometimes the shell with $n = 1$ is referred to as the K shell.

2. The **azimuthal quantum number, l.** Wave mechanics predicts that each main shell is composed of one or more **subshells,** or sublevels, each of which is specified by a secondary quantum number, l, called the azimuthal quantum number. As we will see, this quantum number determines the shape of an orbital and, to a certain degree, its energy. For any given shell, l may have values of 0, 1, 2, and so on, up to a maximum of $n - 1$ for that shell. Thus, when $n = 1$, the largest (and only) value of l that is allowed is $l = 0$. Therefore, the K shell consists of only one subshell. When $n = 2$, two values of l occur, $l = 0$ and $l = 1$; hence, the L shell is composed of two subshells. The values of l that occur for each value of n are summarized in the table in the margin.

Notice that *the number of subshells in any given shell is simply equal to the value of n for that shell.*

For the purposes of discussing the distribution of electrons in an atom, it is common practice to associate letters with the various values of l

n	l
1	0
2	0, 1
3	0, 1, 2
4	0, 1, 2, 3
.	.
.	.
.	.
n	0, 1, 2, . . . , $n - 1$

Value of l	0	1	2	3	4	5	6 . . .
Subshell designation	s	p	d	f	g	h	i . . .

The first four letters find their historical origin in the study of the atomic spectra of the alkali metals (lithium through cesium). In these spectra four series of lines are observed and they are termed the *s*harp, *p*rincipal, *d*iffuse, and *f*undamental series, hence the letters *s*, *p*, *d*, and *f*. For $l = 4, 5, 6$, and so on, we just continue with the alphabet. For our purpose, however, we will be interested only in *s*, *p*, *d*, and *f* subshells, because they are the only ones populated by electrons in atoms in their **ground state** (state of lowest energy).

---

[1] The wave equation can really be solved only for a very few species. Fortunately, one of them is the hydrogen atom, and the results obtained for hydrogen, as it turns out, can be extended quite successfully to the other elements in the periodic table.

Table 7.1. Summary of quantum numbers

Principal Quantum Number, n (shell)	Azimuthal Quantum Number, l (subshell)	Subshell Designation	Magnetic Quantum Number, m_l (orbital)	Number of Orbitals in Subshell
1	0	1s	0	1
2	0	2s	0	1
	1	2p	−1 0 +1	3
3	0	3s	0	1
	1	3p	−1 0 +1	3
	2	3d	−2 −1 0 +1 +2	5
4	0	4s	0	1
	1	4p	−1 0 +1	3
	2	4d	−2 −1 0 +1 +2	5
	3	4f	−3 −2 −1 0 +1 +2 +3	7

To specify a subshell within a given shell, we write the value of n for the shell followed by the letter designation of the subshell. For example, the s subshell of the second shell ($n = 2$, $l = 0$) would be called the 2s subshell. Similarly, the p subshell of the second shell ($n = 2$, $l = 1$) would be the 2p subshell.

3. The **magnetic quantum number, m_l.** Each subshell is composed of one or more orbitals. An orbital within a particular subshell is distinguished by its value of m_l, which serves to determine its orientation in space relative to the other orbitals. The magnetic quantum number derives its name from the fact that it can be used to explain the appearance of additional lines in atomic spectra produced when atoms are caused to emit light while in a magnetic field. It has integer values that range between $-l$ and $+l$. When $l = 0$, only one value of m_l is permitted, $m_l = 0$; therefore, an s subshell consists of only one orbital (we call it an s orbital). A p subshell ($l = 1$) contains three orbitals corresponding to m_l equal to -1, 0, and $+1$. In a similar fashion, we find that a d subshell ($l = 2$) is composed of five orbitals and an f subshell ($l = 3$), seven. This is summarized in Table 7.1. Notice the simple progression in the numbers of orbitals per subshell: 1, 3, 5, 7, and so on.

In describing the orbitals (electron waves) in an atom, we can assign each one a set of values for n, l, and m_l. For example, one wave will have $n = 1$, $l = 0$, and $m_l = 0$. This lets us identify it as a 1s orbital, and we think of the 1s orbital as being *occupied* by that electron. In a certain sense, it is as though we viewed the atom as a sort of parking garage for electrons, in which each orbital is a possible parking space corresponding to a particular wave shape and energy. When an electron is in one of these parking spaces, it has the wave shape and energy of that particular orbital.

Energy levels in atoms

Bohr was fortunate to have chosen hydrogen when he formulated his theory of electronic structure. Because of its simplicity, hydrogen is the only element in which all the orbitals having the same value of n have the same energy. For hydrogen, therefore, only one quantum number is needed to describe the energy. For atoms with more than one electron, however, the energies of the orbitals also depend on

Figure 7.10. *An approximate energy level diagram for atomic orbitals in multielectron atoms. We use this diagram to determine which orbitals in an atom are populated by electrons.*

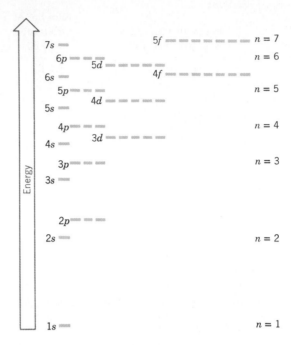

the quantum number l, and Figure 7.10 illustrates how the energies of the shells, subshells, and orbitals vary for multielectron atoms. There are several points about this diagram worth noting. First, we see that the average energies of the shells increase with increasing value of the principal quantum number, n. Thus, the shell with $n = 1$ lies lowest in energy; above that there is the shell with $n = 2$, consisting of the $2s$ and $2p$ subshells; higher still we find the shell with $n = 3$ (composed of $3s$, $3p$, and $3d$ subshells), and so on.

Also note that as n becomes larger, the spacing between successive shells becomes less, as illustrated on the right side of Figure 7.10. Because of this narrowing energy separation, we begin to observe overlap among the subshells of the third and higher shells. The $4s$ subshell, for example, lies lower in energy than the $3d$ subshell. This overlap is even more pronounced in higher shells, where the $5s$ subshell lies below the $4d$, the $6s$ and $4f$ below the $5d$, and the $7s$ and $5f$ below the $6d$.

In Figure 7.10 we have indicated each orbital by means of a dash. Each s subshell is shown as a single dash to stress that it is composed of only one orbital. Likewise, p subshells are shown as three dashes, d subshells as five dashes, and f subshells as seven dashes. Observe that all the orbitals of a given subshell are shown to have the same energy.

The sequence of energy levels described by Figure 7.10 turns out to be of critical importance in determining the arrangement of electrons in an atom. Before discussing this, however, we must look at yet another quantum number.

7.4 ELECTRON SPIN AND THE PAULI EXCLUSION PRINCIPLE

In addition to the three quantum numbers, n, l, and m_l, which come directly from the solution of the wave equation, there is yet another, called the **spin quantum number, m_s**. This quantum number arises because the electron behaves as if it were spinning (in much the same manner as the earth spins about its axis). The circular motion of electric charge that results causes the electron to act as a tiny electromagnet, just as passing an electric current through a wire wrapped about a nail causes the nail to become magnetic (see Figure 7.11). Since the electron can only spin in either of two directions, m_s can have only two values. These turn out to be $+\frac{1}{2}$ and $-\frac{1}{2}$, although the actual values are not really important to us.

We find, therefore, that each electron in an atom can be assigned a set of values for its four quantum numbers, n, l, m_l, and m_s, which determine the orbital in which the electron will be found and the direction in which the electron will be spinning. There is a restriction, however, on the values that these quantum numbers can have. This is expressed as the **Pauli exclusion principle,** which states that *no two electrons in any one atom can have all four quantum numbers the same.* This means that if we choose a particular set of values for n, l, and m_l corresponding to a particular orbital (for example, $n = 1$, $l = 0$, $m_l = 0$; the 1s orbital), we are able to have only two electrons with different values of the spin quantum number, m_s (that is, either $m_s = +\frac{1}{2}$ or $m_s = -\frac{1}{2}$). In effect, *this limits the number of electrons in any given orbital to two, and it also requires that the spins of these two electrons be in opposite directions.*

Because the Pauli exclusion principle restricts the number of electrons in any orbital to two, the maximum number of electrons that can be accommodated in s, p, d, and f subshells can be summarized as follows:

Subshell	Number of Orbitals	Maximum Number of Electrons
s	1	2
p	3	6
d	5	10
f	7	14

The maximum number of electrons permitted in any shell is equal to $2n^2$. For example, the K shell ($n = 1$) can hold up to two electrons and the L shell ($n = 2$) can hold a maximum of eight.

The spin of the electron is also responsible for most of the magnetic properties that we find associated with atoms and molecules. Materials that are **diamagnetic** experience no attraction for another magnet.[2] In these substances there are the same number of electrons of each spin, so the magnetic effects caused by their spins cancel. **Paramagnetic** substances, on the other hand, are weakly attracted to a magnetic field. In these materials there are more electrons of one spin than the other (as will always be true when an atom or molecule has an odd number of electrons) and total cancellation does not occur. The extra electrons of one spin cause the atom or molecule, as a whole, to behave as if it were itself a tiny magnet. **Ferromagnetic** substances, of which iron is the most common example, owe their very strong magnetic behavior to interactions among paramagnetic atoms in the solid state. Ferromagnetism is about one million times stronger than paramagnetism. This phenomenon is discussed in more detail in Chapter 21.

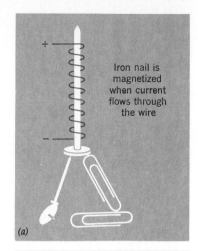

(a) Iron nail is magnetized when current flows through the wire

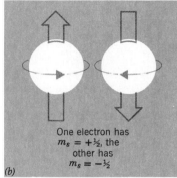

(b) One electron has $m_s = +\frac{1}{2}$, the other has $m_s = -\frac{1}{2}$

Figure 7.11. *The spin of the electron. (a) In an electromagnet, circulation of electric charge through a wire wrapped around the nail produces a magnetic field. (b) Because the electron behaves like a little electromagnet, we imagine that its negative charge also moves in a circular path. This is accomplished when the electron spins about an axis through its center.*

Ferromagnetism is a property of the solid state. If iron is melted, it loses its ferromagnetism and becomes just paramagnetic.

7.5 THE ELECTRON CONFIGURATIONS OF THE ELEMENTS

The way the electrons are distributed among the orbitals of an atom is its **electronic structure** or **electron configuration.** As we've suggested earlier, this is determined

---

[2] They are, in fact, repelled slightly by a magnetic field. This is a result of the motion of the electrons in the atom and is not associated with the electron's spin.

by the order in which the subshells occur on the scale of increasing energy. The reason is that in an atom in its ground state, the electrons will be found in the lowest energy levels available. In hydrogen, for instance, the single electron will be located in the 1s subshell because it is this level that has the lowest energy. To indicate that the 1s subshell is populated by one electron, we use a superscript (in this case 1) on the subshell designation. Thus, we denote the electron configuration of hydrogen as $1s^1$. As we proceed in this discussion, it will also be necessary to keep tabs on the electron spins. One method that is often employed is to symbolize an electron with its spin in one direction by an arrow pointing up, ↑, and an electron with opposite spin as an arrow pointing down, ↓. To indicate the distribution of electrons among the orbitals of the atom, we will place the arrows over bars that symbolize orbitals. Hydrogen, for example, is represented as

The lowest energy distribution of electrons is an atom's **ground state.**

$$\text{H} \quad \underline{\uparrow} \atop 1s$$

This kind of representation of the electron configuration is usually called an **orbital diagram.**

To obtain the electron configurations of the other elements in the periodic table, let us imagine that we are able to proceed from one atom to the next by adding a proton and any necessary neutrons to the nucleus, followed by an electron that we place into the lowest available energy level. As you follow this discussion, refer both to the periodic table and the energy-level diagram in Figure 7.10. A complete table of the electron configurations of the elements is contained in Table 7.2.

Hydrogen is the simplest element, consisting of just a single proton and one electron. The next element, atomic number, 2, is helium. Here there are two electrons to consider, and because the 1s orbital can accommodate them both, the electronic structure of helium is $1s^2$ and its orbital diagram is

$$\text{He} \quad \underline{\uparrow\downarrow} \atop 1s$$

Note that in placing the electrons in the same orbital, we have indicated that their spins are in opposite directions as required by the Pauli exclusion principle. We refer to this by saying that their *spins are paired*, or simply that the electrons are paired.

When two electrons have opposite spins, they are said to be *paired*.

The next two elements following He are Li and Be, which have three and four electrons, respectively. In each element, the first two electrons will enter the 1s subshell, and since no more than two electrons can occupy an s subshell, the remaining electron(s) must occupy the 2s subshell. The electron configurations of Li and Be, then, are Li, $1s^2 2s^1$ and Be, $1s^2 2s^2$. We could also show this as

$$\text{Li} \quad \underline{\uparrow\downarrow} \qquad \underline{\uparrow}$$

$$\text{Be} \quad \underline{\uparrow\downarrow} \qquad \underline{\uparrow\downarrow} \atop 1s \qquad 2s$$

Frequently we abbreviate the electron configuration, especially for atoms with many electrons. Notice that for Li and Be, both have completed 1s subshells. This $1s^2$ configuration is the same as that for the element helium, which we abbreviate by replacing $1s^2$ by the symbol for He in brackets.

We can also write these configurations as

Li [He] $2s^1$

Be [He] $2s^2$

$$\text{Li} \quad [\text{He}] \quad \underline{\uparrow}$$

$$\text{Be} \quad [\text{He}] \quad \underline{\uparrow\downarrow} \atop 2s$$

Table 7.2. The electron configurations of the elements

Atomic Number				Atomic Number				Atomic Number			
1	H	$1s^1$		36	Kr	[Ar]	$4s^23d^{10}4p^6$	71	Lu	[Xe]	$6s^24f^{14}5d^1$
2	He	$1s^2$		37	Rb	[Kr]	$5s^1$	72	Hf	[Xe]	$6s^24f^{14}5d^2$
3	Li	[He]	$2s^1$	38	Sr	[Kr]	$5s^2$	73	Ta	[Xe]	$6s^24f^{14}5d^3$
4	Be	[He]	$2s^2$	39	Y	[Kr]	$5s^24d^1$	74	W	[Xe]	$6s^24f^{14}5d^4$
5	B	[He]	$2s^22p^1$	40	Zr	[Kr]	$5s^24d^2$	75	Re	[Xe]	$6s^24f^{14}5d^5$
6	C	[He]	$2s^22p^2$	41	Nb	[Kr]	$5s^14d^4$	76	Os	[Xe]	$6s^24f^{14}5d^6$
7	N	[He]	$2s^22p^3$	42	Mo	[Kr]	$5s^14d^5$	77	Ir	[Xe]	$6s^24f^{14}5d^7$
8	O	[He]	$2s^22p^4$	43	Tc	[Kr]	$5s^24d^5$	78	Pt	[Xe]	$6s^14f^{14}5d^9$
9	F	[He]	$2s^22p^5$	44	Ru	[Kr]	$5s^14d^7$	79	Au	[Xe]	$6s^14f^{14}5d^{10}$
10	Ne	[He]	$2s^22p^6$	45	Rh	[Kr]	$5s^14d^8$	80	Hg	[Xe]	$6s^24f^{14}5d^{10}$
11	Na	[Ne]	$3s^1$	46	Pd	[Kr]	$4d^{10}$	81	Tl	[Xe]	$6s^24f^{14}5d^{10}6p^1$
12	Mg	[Ne]	$3s^2$	47	Ag	[Kr]	$5s^14d^{10}$	82	Pb	[Xe]	$6s^24f^{14}5d^{10}6p^2$
13	Al	[Ne]	$3s^23p^1$	48	Cd	[Kr]	$5s^24d^{10}$	83	Bi	[Xe]	$6s^24f^{14}5d^{10}6p^3$
14	Si	[Ne]	$3s^23p^2$	49	In	[Kr]	$5s^24d^{10}5p^1$	84	Po	[Xe]	$6s^24f^{14}5d^{10}6p^4$
15	P	[Ne]	$3s^23p^3$	50	Sn	[Kr]	$5s^24d^{10}5p^2$	85	At	[Xe]	$6s^24f^{14}5d^{10}6p^5$
16	S	[Ne]	$3s^23p^4$	51	Sb	[Kr]	$5s^24d^{10}5p^3$	86	Rn	[Xe]	$6s^24f^{14}5d^{10}6p^6$
17	Cl	[Ne]	$3s^23p^5$	52	Te	[Kr]	$5s^24d^{10}5p^4$	87	Fr	[Rn]	$7s^1$
18	Ar	[Ne]	$3s^23p^6$	53	I	[Kr]	$5s^24d^{10}5p^5$	88	Ra	[Rn]	$7s^2$
19	K	[Ar]	$4s^1$	54	Xe	[Kr]	$5s^24d^{10}5p^6$	89	Ac	[Rn]	$7s^26d^1$
20	Ca	[Ar]	$4s^2$	55	Cs	[Xe]	$6s^1$	90	Th	[Rn]	$7s^26d^2$
21	Sc	[Ar]	$4s^23d^1$	56	Ba	[Xe]	$6s^2$	91	Pa	[Rn]	$7s^25f^26d^1$
22	Ti	[Ar]	$4s^23d^2$	57	La	[Xe]	$6s^25d^1$	92	U	[Rn]	$7s^25f^36d^1$
23	V	[Ar]	$4s^23d^3$	58	Ce	[Xe]	$6s^24f^15d^1$	93	Np	[Rn]	$7s^25f^46d^1$
24	Cr	[Ar]	$4s^13d^5$	59	Pr	[Xe]	$6s^24f^3$	94	Pu	[Rn]	$7s^25f^6$
25	Mn	[Ar]	$4s^23d^5$	60	Nd	[Xe]	$6s^24f^4$	95	Am	[Rn]	$7s^25f^7$
26	Fe	[Ar]	$4s^23d^6$	61	Pm	[Xe]	$6s^24f^5$	96	Cm	[Rn]	$7s^25f^76d^1$
27	Co	[Ar]	$4s^23d^7$	62	Sm	[Xe]	$6s^24f^6$	97	Bk	[Rn]	$7s^25f^9$
28	Ni	[Ar]	$4s^23d^8$	63	Eu	[Xe]	$6s^24f^7$	98	Cf	[Rn]	$7s^25f^{10}$
29	Cu	[Ar]	$4s^13d^{10}$	64	Gd	[Xe]	$6s^24f^75d^1$	99	Es	[Rn]	$7s^25f^{11}$
30	Zn	[Ar]	$4s^23d^{10}$	65	Tb	[Xe]	$6s^24f^9$	100	Fm	[Rn]	$7s^25f^{12}$
31	Ga	[Ar]	$4s^23d^{10}4p^1$	66	Dy	[Xe]	$6s^24f^{10}$	101	Md	[Rn]	$7s^25f^{13}$
32	Ge	[Ar]	$4s^23d^{10}4p^2$	67	Ho	[Xe]	$6s^24f^{11}$	102	No	[Rn]	$7s^25f^{14}$
33	As	[Ar]	$4s^23d^{10}4p^3$	68	Er	[Xe]	$6s^24f^{12}$	103	Lr	[Rn]	$7s^25f^{14}6d^1$
34	Se	[Ar]	$4s^23d^{10}4p^4$	69	Tm	[Xe]	$6s^24f^{13}$				
35	Br	[Ar]	$4s^23d^{10}4p^5$	70	Yb	[Xe]	$6s^24f^{14}$				

Here we focus our attention on the electronic structure of the outermost shell — the shell with highest n — that, in chemical reactions, is responsible for chemical changes. Electrons in shells below the outer shell are said to be **core electrons.** In this example, the inner filled $1s$ subshell is called the helium core. We will frequently find it useful to consider only those electrons that occur outside a core of electrons that corresponds to the electron configuration of one of the noble gases. In writing an abbreviated configuration, therefore, we always select the electron configuration corresponding to the preceding noble gas to be replaced by the bracketed chemical symbol.

Z stands for atomic number.

At beryllium, which has four electrons, the 2s subshell is completed. The fifth electron of boron ($Z = 5$) must then enter the next lowest available subshell, which is the 2p. This gives boron the configuration $1s^2 2s^2 2p^1$. Similarly, the fifth and sixth electrons of carbon must also enter the 2p subshell; thus, we represent carbon as $1s^2 2s^2 2p^2$. However, if we examine the distribution of the electrons over the various orbitals, we face a choice; the electrons could be arranged in the following three ways:[3]

Notice that all the orbitals of the p subshell are shown, even though not all of them are occupied by electrons.

or

C   [He]   ⇅   ⇅   __   __

or

C   [He]   ⇅   ↑   ↓   __

or

C   [He]   ⇅   ↑   ↑   __
         $2s$ $2p$

The last two electrons can be paired in the same orbital, paired in different orbitals, or arranged so that their spins are in the same direction (unpaired).

As it turns out, experiments show that the last diagram gives the electron configuration that is lowest in energy. **Hund's rule** summarizes this experimental evidence: *Electrons entering a subshell containing more than one orbital will be spread out over the available equal-energy orbitals with their spins in the same direction.* For nitrogen ($Z = 7$), therefore, the electron configuration is written as $1s^2 2s^2 2p^3$, and its ground state has the orbital diagram

N   [He] $2s^2 2p^3$

N   [He]   ⇅   ↑   ↑   ↑
         $2s$ $2p$

Finally, the elements oxygen, fluorine, and neon ($Z = 8, 9,$ and 10, respectively) lead to the completion of the 2p subshell.

O   [He] $2s^2 2p^4$

F   [He] $2s^2 2p^5$

Ne   [He] $2s^2 2p^6$

O   [He]   ⇅   ⇅   ↑   ↑

F   [He]   ⇅   ⇅   ⇅   ↑

Ne   [He]   ⇅   ⇅   ⇅   ⇅
          $2s$ $2p$

After the 2p subshell is filled at Ne, the next lowest available energy level is the 3s. This becomes populated with Na and Mg ($Z = 11$ and 12). After this the 3p subshell is gradually filled by the next six electrons as we complete the configurations of the atoms Al through Ar ($Z = 13$ through 18). Then since the 4s subshell lies at lower energy than the 3d, it is occupied next by the nineteenth and twentieth electrons of K and Ca ($Z = 19$ and 20).

---

[3] These are the only three possibilities that we have to consider because in an isolated atom, each of the p orbitals is equivalent in energy. Thus, the arrangements

⇅   __   __   __   ⇅   __   __   __   ⇅
   $2p$ $2p$ $2p$

are indistinguishable from one another experimentally.

Examination of Figure 7.10 reveals that after the $4s$ subshell is completed, additional electrons begin to populate the $3d$ subshell. Scandium, therefore, will have the electron configuration $1s^2 2s^2 2p^6 3s^2 3p^6 4s^2 3d^1$. Normally, we order the subshells to place all those with the same value of the principal quantum number together. For scandium, this gives

$$\text{Sc} \quad 1s^2 2s^2 2p^6 3s^2 3p^6 3d^1 4s^2$$

and the orbital diagram is

Sc   [Ar]   ↑   __   __   __   __   ↑↓
                 $3d$ $4s$

As we proceed through Ti and V (Z = 22 and 23), two more electrons are added to the $3d$ subshell; however, when we get to Cr (Z = 24), we find the structure

Cr   [Ar]   ↑   ↑   ↑   ↑   ↑   ↑
                 $3d$ $4s$

instead of

Cr   [Ar]   ↑   ↑   ↑   ↑   __   ↑↓
                 $3d$ $4s$

This unexpected result occurs because a half-filled or completely filled subshell possesses an extra, added stability. The origin of this extra stability is very complex, so we cannot discuss it here. Nevertheless, the phenomenon is quite important and should be kept in mind. We see it again in period 4, for example, when we get to copper. On the basis of our energy-level diagram in Figure 7.10, we would predict copper to have the electron configuration

Cu   [Ar]   ↑↓   ↑↓   ↑↓   ↑↓   ↑   ↑↓
                 $3d$ $4s$

The actual structure of the ground state is given by

Cu   [Ar]   ↑↓   ↑↓   ↑↓   ↑↓   ↑↓   ↑
                 $3d$ $4s$

Similar exceptions occur elsewhere. For example, Ag and Au have filled d subshells just as copper does.

Transferring an electron from the $4s$ to $3d$ subshell of copper produces one filled and one half-filled subshell, instead of the filled $4s$ and the neither filled nor half-filled $3d$ subshell that we initially would predict. Because the electron configurations of Cr and Cu are not predictable by our rules, they must be remembered as exceptions.

After the $3d$ subshell is completed at atomic number 30 (zinc), the $4p$ subshell is gradually filled as we proceed from Ga to Kr (Z = 31 to 36). This is followed by the completion of the $5s$ subshell from Rb to Sr (Z = 37, 38); the $4d$ subshell as we progress across the second row of transition elements (Z = 39 to 48); the $5p$ from In to Xe (Z = 49 to 54); and the $6s$ with Cs and Ba (Z = 55, 56).

It should be noted that this explanation does not account for all the irregularities that actually occur among the electron configurations of the elements.

Based on the energy-level sequence in Figure 7.10, we would expect that after the $6s$ subshell had been filled, the $4f$ subshell should be populated next. Actually, at La (Z = 57) the last electron enters the $5d$ subshell instead. The $4f$ subshell is filled afterward, with a few minor irregularities. As we go to higher and higher shells, these irregularities become more frequent because the spacing between subshells becomes smaller and smaller. As we proceed from atom to atom, the energies of the various subshells shift about somewhat as the nuclear charge increases and as the electron populations of the subshells change. The result is that it is difficult to predict accurately the electron configuration of elements of very high atomic number. Nevertheless, we can account for the occurrence of the lanthanide elements by the filling of the $4f$ subshell (an f subshell can accommodate 14 electrons and the lanthanide series consists of 14 elements). Likewise, we can account for the actinide elements by the filling of the $5f$ subshell.

7.6 THE PERIODIC TABLE AND ELECTRON CONFIGURATIONS

In the last section we saw that the results of wave mechanics could be used to predict the electron configurations of the elements. These electron configurations are based on theory. But what *experimental* evidence is there, other than atomic spectra, that the theory is correct?

One of the strongest supports for the assignment of electron configurations is the periodic table itself. Recall that this table was created purely on the basis of the *observed* chemical and physical properties of the elements. For example, the elements in Group IA are placed where they are because they are all very reactive metals that form compounds with the same general formulas (e.g., MCl, M_2O). Let's examine the basic layout of the periodic table on the inside front cover of the book. Notice that Groups IA and IIA form a block of 2 columns of elements. Groups IIIA through the noble gases form a block of 6 columns, the transition elements form a block of 10 columns, and the two rows of inner transition elements can be viewed as a block of 14 columns of elements. Isn't it amazing that the periodic table is divided into blocks of 2, 6, 10, and 14 columns, which are exactly the numbers of electrons that the theory tells us can occupy s, p, d, and f subshells?

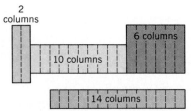

The periodic table is divided into four regions having 2, 6, 10, and 14 columns.

Now let's examine the electron configurations of the elements in a group. For the elements in Group IA we have

Li	[He]	$2s^1$	Rb	[Kr]	$5s^1$
Na	[Ne]	$3s^1$	Cs	[Xe]	$6s^1$
K	[Ar]	$4s^1$			

Notice that each of these elements has one electron in an s orbital outside a noble gas core. If we make the reasonable assumption that it is the outer electrons of atoms that come into contact when they react, it is not at all surprising that the Group IA elements have similar chemical and physical properties. Similar likenesses occur among the electron configurations of the atoms in other groups as well. For example, all the elements in Group IIA have outer-shell configurations that we might generalize as ns^2.

Be	[He]	$2s^2$	Sr	[Kr]	$5s^2$
Mg	[Ne]	$3s^2$	Ba	[Xe]	$6s^2$
Ca	[Ar]	$4s^2$			

Because of observations such as these, chemists today are confident that the chemical properties of the elements are controlled by their electron configurations, so it is important that you develop the ability to write them down. The best aid for this is the periodic table itself.

As we have just seen, the order of filling of the energy levels can be used to account for the structure of the periodic table and the similarities among the properties of the elements. We can also work in the other direction and use the periodic table to deduce electronic structures. If we look back over the procedure for determining electronic structures, we find that for any element in Groups IA and IIA, the final electron is added to an s subshell, and that the principal quantum number of that subshell is the same as the period number. Sodium, for instance, is a period 3 element and has its outer electron in the $3s$ subshell. For elements in Group IIIA to Group 0, the last electron is added to a p subshell whose value of n is also the same as the period number. For example, as we complete the electron configurations of the period 2 elements boron through neon, the final electron is placed in a $2p$ subshell. In

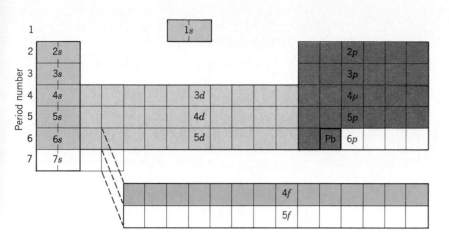

Figure 7.12. *The use of the periodic table to predict electron configurations. The shaded areas identify the subshells that are filled to obtain the electron configuration of lead (see Example 7.7).*

the case of the transition elements, the final electron that we add is placed into a *d* subshell with *n* equal to *one less* than the period number. For example, with iron (a fourth-period element), the last electron enters a 3*d* subshell. Finally, notice that the electronic structure of an inner transition element (i.e., one from the lanthanide or actinide series) is completed by placing an electron in an *f* subshell whose principal quantum number is *two* less than the period number.

We can use these observations by making the periodic table tell us which subshells are filled as we build up the electron configuration of an atom. As before, we begin with hydrogen and we proceed through the elements in the periodic table in order of increasing atomic number until we arrive at the element in which we are interested. As we move across a given period, we add electrons to an *s* subshell when we pass through Groups IA and IIA, and to a *p* subshell when we pass through Groups IIIA through Group 0. The value of *n* for these subshells is the same as the period number. As we pass through a row of transition elements, we fill a *d* subshell with *n* equal to the period number minus 1, and as we move across a row of inner transition elements, we fill an *f* subshell with *n* equal to the period number minus two. This is summarized in Figure 7.12, and the following examples illustrate how this method works.

EXAMPLE 7.6. PREDICTING ELECTRON CONFIGURATIONS

PROBLEM: What is the electron configuration of antimony (Sb)?

SOLUTION: Antimony has atomic number 51. To reach this element we have to pass completely through periods 1, 2, 3, and 4, and part of the way across period 5. As we do this, here is what we get:

Period 1 — fill the 1*s* subshell, which gives $1s^2$

Period 2 — fill the 2*s* and 2*p* subshells $2s^22p^6$

Period 3 — fill the 3*s* and 3*p* subshells $3s^23p^6$

Period 4 — fill the 4*s*, 3*d*, and 4*p* subshells (in that order) $4s^23d^{10}4p^6$

Period 5 — to get as far as cadmium ($Z = 48$), we have to fill the 5*s* and 4*d* subshells. Then we have to move three spaces into the *p* region to get to Sb. This gives $5s^24d^{10}5p^3$

Putting this all together, we have

$$1s^22s^22p^63s^23p^64s^23d^{10}4p^65s^24d^{10}5p^3$$

If we collect all the subshells of a given shell together we obtain

$$1s^2 2s^2 2p^6 3s^2 3p^6 3d^{10} 4s^2 4p^6 4d^{10} 5s^2 5p^3$$

We can also write the configuration by showing the noble gas core plus the electrons outside of it

$$[Kr] \quad 4d^{10} 5s^2 5p^3$$

EXAMPLE 7.7. PREDICTING ELECTRON CONFIGURATIONS

PROBLEM: What is the electron configuration of lead?

SOLUTION: Lead has atomic number 82. This means that in building up the lead atom, we cross through periods 1 to 5 and part of 6. Proceeding from left to right across one period after another, we fill, in order, the subshells $1s$, $2s$, $2p$, $3s$, $3p$, $4s$, $3d$, $4p$, $5s$, $4d$, $5p$, $6s$, $4f$, $5d$, and finally end by placing two electrons into the $6p$ subshell. Taking into account the maximum population of each subshell, we obtain as the electron configuration of lead

$$1s^2 2s^2 2p^6 3s^2 3p^6 4s^2 3d^{10} 4p^6 5s^2 4d^{10} 5p^6 6s^2 4f^{14} 5d^{10} 6p^2$$

As before, we might prefer to write all subshells of a given shell together. Thus, for lead we would have

$$1s^2 2s^2 2p^6 3s^2 3p^6 3d^{10} 4s^2 4p^6 4d^{10} 4f^{14} 5s^2 5p^6 5d^{10} 6s^2 6p^2$$

When we begin to discuss the chemical bonds formed between atoms, we often will be interested only in the electron population of the outer shell, or **valence shell,** of an atom. (Valence is a word that relates to the number of chemical bonds an atom is able to form.) This is because it is only the outer parts of atoms that come into contact with each other when the atoms combine to form compounds. If all we are interested in is the configuration of an atom's valence shell, we do not have to work through the entire electron configuration. Instead, we can locate the element in the periodic table and immediately find the information we need. For example, suppose we wished to know only the outer-shell electron configuration for tin, Sn. We find this element in Group IVA and period 5. The period number tells us that the outer shell has $n = 5$, so we are only interested in populated orbitals that have 5 as their principal quantum number. Crossing period 5 as far as Sn, we place two electrons in the $5s$, ten in the $4d$, and finally two in the $5p$.

$$Sn \quad [Kr] \quad 4d^{10} 5s^2 5p^2$$

We can ignore the electrons in the $4d$ because this subshell is not part of the valence shell, so the valence shell configuration of tin is

$$5s^2 5p^2$$

EXAMPLE 7.8. PREDICTING OUTER-SHELL ELECTRON CONFIGURATIONS

PROBLEM: What is the outer-shell configuration of silicon?

SOLUTION: Silicon (Si) is in period 3; therefore, the outer shell is the third shell. To get to Si in period 3, we fill the $3s$ subshell and place two electrons in the $3p$ subshell. This gives us

$$Si \quad 3s^2 3p^2$$

EXAMPLE 7.9. CONSTRUCTING ORBITAL DIAGRAMS BY USE OF THE PERIODIC TABLE

PROBLEM: Use the periodic table to construct the orbital diagram of the valence shell of tellurium, Te.

SOLUTION: First, let's determine the electron configuration of the valence shell. Then, we can translate that into the appropriate orbital diagram. We begin by noting that Te is in period 5, which means that its valence shell is the 5th shell—we are only interested in orbitals that are part of the 5th shell. As we cross the 5th row, we first fill the $5s$ orbital (which gives $5s^2$); then we fill the $4d$ orbital (which gives $4d^{10}$), and then we place 4 electrons into the $5p$ subshell (which gives $5p^4$). The electron configuration of Te is therefore

$$\text{Te} \quad [\text{Kr}] \quad 4d^{10}5s^25p^4$$

and the valence shell configuration is

$$\text{Te} \quad 5s^25p^4$$

The next step is to translate this into an orbital diagram. We begin by drawing one dash for the $5s$ orbital and *three* dashes for the $5p$ orbitals. Don't forget to show all the orbitals of a given subshell, even if not all of them have electrons. The best way to start is to draw the dashes for the orbitals and then fill in the electrons. For Te, we get

$$\text{Te} \quad \underline{\quad} \quad \underline{\quad}\;\underline{\quad}\;\underline{\quad}$$
$$\qquad\quad 5s \qquad\quad 5p$$

Now we place 2 electrons in the $5s$ orbital, being sure to indicate that they are paired. Next, we place electrons in the $5p$ orbitals, one orbital at a time until they are each half-filled (which accounts for 3 of the 4 electrons); then we add the 4th electron to one of the half-filled $5p$ orbitals with its spin paired. This gives the completed orbital diagram

$$\text{Te} \quad \underline{\uparrow\downarrow} \quad \underline{\uparrow\downarrow}\;\underline{\uparrow}\;\underline{\uparrow}$$
$$\qquad\quad 5s \qquad\quad 5p$$

Notice that the $4d$ subshell is *not* part of the valence shell. Also, note that the valence shell includes both $5s$ and $5p$ subshells.

7.7 THE SHAPES OF ATOMIC ORBITALS

In the previous sections you learned how to determine the distribution of an atom's electrons among its atomic orbitals, which is controlled by the energies of the orbitals. Another facet of electronic structure that is of interest to us is how the electrons are distributed in space around the nucleus, and this is determined by the shapes of the electron waves.

One of the most difficult concepts to comprehend is the way the wave nature of the electron spreads the electron out around the nucleus of an atom, so let's begin our discussion with an analogy.

Suppose you gave a piece of paper and a crayon to a child, walked away, and then returned after the child had used up the entire crayon by creating a drawing of some kind. The crayon hasn't actually disappeared, of course. It is now spread out over the paper. You can point to different parts of the drawing and say that more of the crayon is here than there, but to speak of the whole crayon and how it is distributed, it is necessary to describe the entire drawing.

The wave nature of matter does to an electron what a child does to a crayon. In an orbital, the electron wave spreads the electron's mass and charge through the space around the nucleus of the atom. In some places the concentration of mass and charge is large; in other places the concentration is small, but to speak of the entire electron and how its charge and mass are distributed, it is necessary to describe the entire electron wave.

Frequently the term *electron cloud* is used to describe the space-filling nature of the electron wave. Where there is a large concentration of the electron's charge (a large amount of charge per unit volume), we say that the **electron density** is high. In those places where the electron's charge is thinly spread (where the amount of charge per unit volume is small), we say there is a low electron density.

One way to illustrate the way the electron density varies from place to place in an orbital is shown in Figure 7.13 for a 1s orbital. This is sometimes called a dot density diagram, in which the number of dots per unit area (the dot density) is proportional to the electron density. Figure 7.13 actually represents a cross section of a 1s orbital, and as you can see, the electron density is large very close to the nucleus and gradually decreases as the distance from the nucleus increases.

The shapes of s orbitals

One of the properties of the 1s orbital (or any s orbital, for that matter) is that at any particular distance from the nucleus, the electron density is the same, regardless of which direction we go. Stated another way, points of equal electron density lie on the surface of a sphere with the nucleus at the center. Because of this, we say that s orbitals are spherical.

Higher-energy s orbitals differ from the 1s in some respects. Figure 7.14 compares the way the electron density varies for 1s, 2s, and 3s orbitals. The diagrams have been drawn so that the size of the sphere defining the shape of the orbital is large enough to encompass most (say, 90%) of the electron density in the orbital.

First, for the 2s orbital, notice that as we move out from the nucleus, the electron density gradually drops to zero and then increases again before gradually decreasing once more. You learned earlier that those places where the amplitude or intensity of the wave drops to zero are called nodes. Thus, the electron wave for a 2s orbital has a node just like the standing waves on a guitar string. Similarly, the electron wave for a 3s orbital has two nodes, as you can see in Figure 7.14.

Another thing to notice in Figure 7.14 is the size of the orbital. As mentioned earlier, the larger the value of the principal quantum number of an orbital, the larger is its size. In Figure 7.14 you see that as the value of n increases, the size of the sphere needed to enclose 90% of the electron density also increases.

The electron density in an orbital doesn't end abruptly at some particular distance from the nucleus. Instead, it gradually decreases, approaching zero at very large distances.

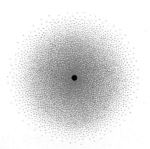

Figure 7.13. *A dot density diagram that illustrates the way the electron density varies from place to place in a 1s orbital. Where there are many dots per unit area, the electron density is high.*

The nodes in the 2s and 3s orbitals are actually spheres on which the electron density is zero.

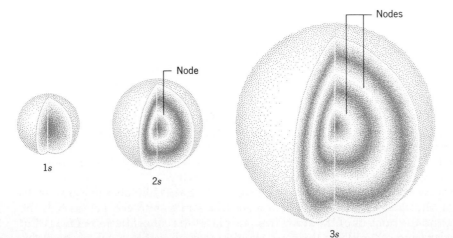

Figure 7.14. *Electron density distributions in the 1s, 2s, and 3s orbitals of an atom.*

The shapes of p orbitals

The shape of the electron cloud characteristic of a 2p orbital is illustrated in Figure 7.15. We see that for this orbital the electron density is not distributed in a spherically symmetrical way. Instead, it is concentrated in two equal-size regions that are located on opposite sides of the nucleus with their centers along a straight line that passes through the center of the atom. An electron in this orbital is actually divided between these two regions, which are separated from each other by a nodal plane (a plane on which the electron density is zero) that passes through the nucleus and is perpendicular to the axis of the orbital.

The p orbitals with larger values of their principal quantum number are larger (extend farther from the nucleus) than a 2p orbital and also contain nodes in addition to the nodal plane passing through the nucleus. A cross section of a 3p orbital is illustrated in Figure 7.16. Despite these differences, there is a concentration of electron density along certain specific directions in all p orbitals, and we say that p orbitals have definite *directional properties*. As you will see, this allows us to understand why molecules have the shapes they do. Because we are interested primarily in the directional properties of the orbitals, we will represent all s orbitals as spheres and all p orbitals as a pair of dumbbell-shaped lobes pointing in opposite directions from the nucleus.

A p subshell consists of three p orbitals, each with the same shape. They differ from one another only in the directions in which their electron densities are concentrated. These directions lie at right angles to one another as shown in Figure 7.17. Because the p orbitals can be drawn on a set of *xyz* axes, it is common practice to identify the orbitals as p_x, p_y, and p_z.

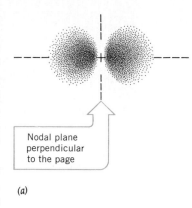

Nodal plane
perpendicular
to the page

(a)

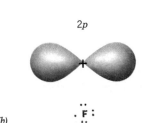

2p

(b)

Figure 7.15. *The shape of a* 2p *orbital.* (a) *A dot density diagram shows that the electron density is concentrated in two regions that lie on opposite sides of the nucleus.* (b) *The kind of drawing we will use in this book to represent a p orbital.*

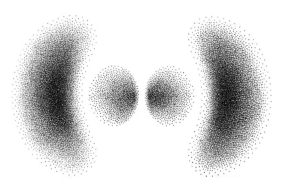

Figure 7.16. *A cross section of a* 3p *orbital. Notice that there is a node between the two regions of electron density on each side of the nucleus.*

The shapes of d orbitals

The d orbitals are a bit more complicated than p orbitals. Because of this, it is difficult to draw all of them together on the same set of axes, as we have done for the p orbitals. In Figure 7.18, they are shown separately, but in an atom, they are really superimposed on the same nucleus.

The first thing to notice about the d orbitals is that they do not all look alike. Four of them, labeled d_{xy}, d_{xz}, d_{yz}, and $d_{x^2-y^2}$, have the same shape, and each is composed of four lobes of electron density. The difference is that they point in different directions. The fifth, labeled d_{z^2}, consists of two large lobes of electron density pointing in opposite directions along the z axis, plus a donut of electron density in the xy plane. The labels for the d orbitals are not as descriptive as those for the p orbitals and actually have their origin in the mathematics of wave mechanics. Although we will use these labels when we wish to refer to the various d orbitals, you need not worry about where they come from.

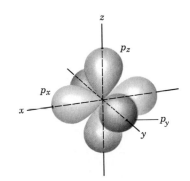

Figure 7.17. *The three* p *orbitals of a* p *subshell point in directions that are mutually perpendicular. By imagining that they point along a set of xyz axes, we can label them* p_x, p_y, *and* p_z.

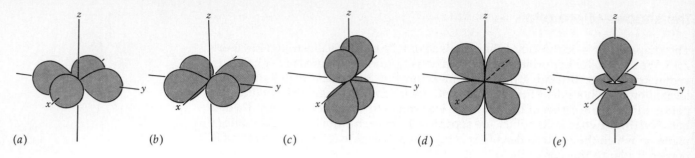

Figure 7.18. *Directional properties of the* d *orbitals.* (a) $d_{x^2-y^2}$, (b) d_{xy}, (c) d_{xz}, (d) d_{yz}, *and* (e) d_{z^2}.

The shapes of the *f* orbitals are even more complex than the *d* orbitals. Fortunately, they are only needed in discussions of the chemistry of the inner transition elements, which is not included in this book. Therefore, we will not burden you with the shapes of the *f* orbitals.

7.8 THE VARIATION OF PROPERTIES WITH ATOMIC STRUCTURE

Within the periodic table, many of the properties of the elements vary in a more or less regular fashion as we proceed from left to right across a period or from top to bottom in a group. We saw this in Chapter 4 where we noted a change in the metallic character of the elements within a period and within a group. Most of these variations can be accounted for directly in terms of variations in the electronic structures of the elements. In this section, we will look briefly at three important properties whose variations are easily seen and which are related to each other.

Atomic size

As we've mentioned previously, atoms are very tiny — they have diameters of about 10^{-10} m. A unit that has been used for many years to report the sizes of atoms, ions, or molecules is the **angstrom (Å).**

$$1 \text{ Å} = 10^{-8} \text{ cm} = 10^{-10} \text{ m}$$

This is not a recognized SI unit and in current scientific literature these dimensions are usually expressed either in nanometers (10^{-9} m) or picometers (10^{-12} m). It is useful to remember the conversions.

$$1 \text{ Å} = 0.1 \text{ nm}$$

$$1 \text{ Å} = 100 \text{ pm}$$

To conform with current practice, we will express atomic and molecular dimensions in picometers.

In any discussion of atomic size, we immediately face a basic problem of definition: Exactly what do we mean by the size or radius of an atom? We have seen that the electron density in an atom does not end abruptly at some particular distance from the nucleus. Instead it trails off gradually, approaching zero at very large distances from the center of the atom. Because of this, it is difficult to define precisely what we mean by the size of an atom. One attempt to solve this problem might be to take the radius of an atom as half the distance between neighboring atoms when the element is present in its most dense form (i.e., most highly compacted form, which is usually the solid). Even this definition, however, is complicated because when atoms are attached to each other by chemical bonds, as they are in molecules like H_2 or Cl_2, they approach each other more closely than nonbonded atoms do (e.g., the noble gases when they are frozen). Also, the atomic radius that we measure for atoms of a pure element will not necessarily be the same in compounds. For example, carbon atoms in diamond (pure carbon) are separated by a distance of 154 pm and we would thus assign carbon a radius of 77 pm. In the ethane molecule, C_2H_6, the carbon–carbon distance is also 154 pm; however, in ethylene, C_2H_4, and acetylene, C_2H_2, we find carbon–carbon distances equal to 137 and 120 pm, respectively. These lead to atomic radii for carbon of 69 and 60 pm, both considerably smaller than 77 pm.

The way atomic size can be measured is described in Chapter 12.

Despite this difficulty of definition, we can compare the atomic radii of the elements if they are measured under circumstances that lead to essentially similar kinds of bonds between their atoms. Figure 7.19 illustrates the variation of atomic radius with atomic number. We see that as we proceed down within a group, the sizes of atoms generally increase, and that as we proceed from left to right across a period, a gradual decrease in size is observed.

In order to interpret these trends within the periodic table in terms of electronic structure, we must look at the factors that determine the size of the outer shell of an atom, that is, the average distance at which electrons in the outer shell occur. As we have discussed in the preceding section, one factor is the principal quantum number of the outer shell (recall that the electron occurs at increasingly larger distances from the nucleus with increasing value of n).

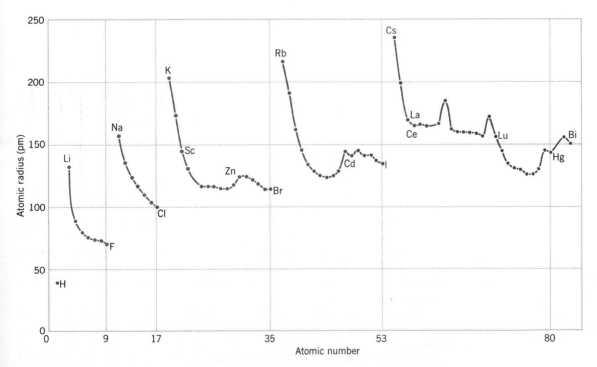

Figure 7.19. *A graph of atomic radius versus atomic number. In general, the sizes of atoms decrease from left to right across a period and increase from top to bottom in a group.*

Effective nuclear charge

The size of the outer shell also depends on the amount of positive charge that the outer electrons feel. The higher the positive nuclear charge, the more the outer electrons are pulled toward the center of the atom. In multielectron atoms, the positive charge felt by the outer electrons is always less than the full nuclear charge. This is because electrons in inner shells partially shield those in the outer shell. Stated another way, the negative charge of the electrons in the core, whose electron density lies partially between the nucleus and the valence shell, partly offsets the positive charge of the nucleus. *The residual net charge felt by the outer valence electrons is called the* **effective nuclear charge.** In a sodium atom, for example, there are 10 electrons in the neon core. These partially shield the outer 3s electron of Na from the positive charge of the 11 protons in the nucleus. As a result, the 3s electron of sodium feels an effective nuclear charge that is much less than +11. (Actually, it is only about +2.8.)

If the neon core of the sodium atom were 100% effective at shielding the outer shell, the 3s electron would feel an effective nuclear charge of only about 1+.

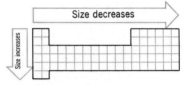

The variation of atomic size in the periodic table.

The variation in atomic size across a period is not always a smooth one, as demonstrated by the irregularities seen for period 6 in Figure 7.19.

As we proceed downward from one atom to another within a group, each successive element has its outer electrons in a shell with a larger value of *n*. The effective nuclear charge felt by these electrons stays nearly the same, so the dominant effect is an increase in size that accompanies an increase in the value of the principal quantum number of the outer-shell orbitals.

For the representative elements, as we move from left to right across a period, we add electrons to the same shell and simultaneously increase the nuclear charge. Because they are in the *same* shell, the outer-shell electrons do not shield each other from the nucleus very well, so the effective nuclear charge experienced by any one electron in the outer shell increases. This increase in effective nuclear charge leads to a greater attraction for the outer-shell electrons, so they are pulled in closer to the nucleus, and the sizes of the atoms decrease.

The variation in size as we pass through a row of transition elements is much less than among the representative elements. This is because electrons are being added to an inner shell as the nuclear charge gets larger. In the first row of the transition elements, for instance, the outer electrons occur in a 4s subshell, but each successive electron is added to the inner 3d subshell as we proceed across the table. The inner-shell electrons are almost completely effective at shielding the outer shell from the nuclear charge, so the outer 4s electrons experience only a small, gradual increase in effective nuclear charge across this region of the periodic table. Therefore, only small changes in size occur.

The lanthanide contraction

As we move from left to right across a row of transition elements, there is a gradual decrease in size because the inner electrons that are being added are not completely effective at shielding the outer electrons. A similar phenomenon also occurs among the inner transition elements—for example, the lanthanides. For these elements, the electrons being added go into the 4f subshell. Although they shield the outer 6s electrons quite well, the shielding isn't perfect, so the outer electrons experience a very small gradual increase in effective nuclear charge, and the sizes of the atoms decrease on going from Ce to Lu.

If you look at the periodic table, you see that the lanthanide elements fall between La (in Group IIIB) and Hf (in Group IVB). In the fifth period, however, there is no similar row of elements. On moving from Y to Zr, there is an increase of one proton to the nucleus and only one electron to the atom, so only a small decrease in size occurs. However, in the sixth period, a much larger decrease in size occurs between La and Hf because of the intervening lanthanide elements. *This additional decrease in size is known as the* **lanthanide contraction;** it causes Hf to be the same size as Zr, even though Hf is below Zr in Group IVB.

The effects of the lanthanide contraction extend beyond Hf in the sixth period. All the rest of the transition elements in the sixth period are nearly the same size as the elements above them in the fifth period. One result of this is that the sixth-period transition metals are all extremely dense. Their atoms are the same size as the atoms above them, but they pack a lot more mass. Even beyond the transition elements, we see the influence of the lanthanide contraction with unusually dense elements such as lead and bismuth.

The sizes of ions

You learned earlier that when metals and nonmetals react with each other, they tend to form ions. In general, positive ions are smaller than the neutral atoms from which they are formed, while negative ions are larger than the neutral atoms (Table 7.3).

When an atom loses electrons to form a positive ion, the electrons that are lost always come from the shell with the largest value of the principal quantum number. An atom of sodium, for example, has the electron configuration

$$\text{Na} \quad [\text{Ne}] \quad 3s^1$$

When it forms the Na^+ ion, the higher-energy $3s$ electron is lost, exposing the smaller neon core beneath. Many positive ions are formed by the emptying of the outer shells of their neutral atoms, and their sizes reflect the smaller sizes of the noble gas cores that are exposed when the ions are produced.

When a transition metal forms a positive ion, the outer s subshell is emptied

Table 7.3. Atomic and ionic radii (in picometers)

		Positive Ions					Negative Ions		
		Atomic Radius	Ionic Radius	Charge			Atomic Radius	Ionic Radius	Charge
Group IA	Li	135	60	(+1)	Group VIIA	F	64	136	(−1)
	Na	154	95	(+1)		Cl	99	181	(−1)
	K	196	133	(+1)		Br	114	195	(−1)
	Rb	211	148	(+1)		I	133	216	(−1)
	Cs	225	169	(+1)	Group VIA	O	66	140	(−2)
Group IIA	Be	90	31	(+2)		S	104	184	(−2)
	Mg	130	65	(+2)		Se	117	198	(−2)
	Ca	174	99	(+2)		Te	137	221	(−2)
	Sr	192	113	(+2)	Group VA	N	70	171	(−3)
	Ba	198	135	(+2)		P	110	212	(−3)
Group IIIA	Al	143	50	(+3)					
	Ga	122	62	(+3)					
	In	162	81	(+3)					

Elements That Form More Than One Ion					
Fe	126	Fe^{2+}	76	Fe^{3+}	64
Co	125	Co^{2+}	78	Co^{3+}	63
Cu	128	Cu^+	96	Cu^{2+}	69

before the underlying d *subshell is disturbed.* For example, iron has the electron configuration

$$\text{Fe} \quad [\text{Ar}] \quad 3d^6 4s^2$$

Removal of the two $4s$ electrons gives the Fe^{2+} ion, which has the configuration

$$Fe^{2+} \quad [\text{Ar}] \quad 3d^6$$

The Fe^{3+} ion is produced when another electron is removed from the $3d$ subshell.

$$Fe^{3+} \quad [\text{Ar}] \quad 3d^5$$

For the change $Fe \rightarrow Fe^{2+}$, the outer shell is emptied, which leaves the smaller core beneath. For the change $Fe^{2+} \rightarrow Fe^{3+}$, an electron is removed from the $3d$ subshell, which decreases the repulsions between the electrons that remain. This allows the nucleus to pull them closer together and the size of the ion decreases.

The rule is as follows: Particles become smaller when electrons are removed from them and larger when electrons are added.

When negative ions are formed from neutral atoms, electrons are added to the outer shell, which increases the electron–electron repulsions. These repulsions cause the electrons to spread out and thereby cause the size of the outer shell to increase, so the negative ion is larger than the neutral atom.

Ionization energy

The **ionization energy** *is defined as the energy required to remove an electron from an isolated gaseous atom or ion in its ground state.* We can represent this process by an equation such as

$$\text{Na}(g) \longrightarrow \text{Na}^+(g) + e^-$$

The potential energy of an electron drops as it becomes attached to an atom to which it is attracted. The amount that the energy is lowered equals the energy that's released.

This is an endothermic process; because the electron is attracted to the positive nucleus, energy must be supplied to remove it. Since all atoms other than hydrogen possess more than one electron, they also have more than one ionization energy. The amount of energy required to remove the first electron is called the first ionization energy, that required to remove the second electron is called the second ionization energy, and so forth. As we might expect, successive ionization energies increase in magnitude because the species from which the electron is removed becomes progressively more positively charged. For example, the first ionization energy involves the removal of an electron from a neutral atom, but the second ionization energy involves the removal of an electron from an ion whose charge is $1+$.

The noble gas core is so stable because a great deal of energy is required to disrupt it.

Table 7.4 contains successive ionization energies for the first 20 elements in the periodic table. They are given in units of kilojoules per mole (kJ/mol) and represent energies needed to remove electrons from 1 mol of gaseous atoms. An examination of the data in this table points out the great stability associated with an electron core having a noble gas electron configuration. We see, for example, that for a Group IA element the first ionization energy is relatively low and that the second ionization energy is very much greater. For the Group IIA elements a very large increase in ionization energy occurs after two electrons have been removed, while for Group IIIA elements the break occurs after the third electron has been lost. In fact, we see that in general a very large jump in ionization energy always occurs after an atom has lost a number of electrons that is numerically equal to its group number. Since a Group IA element contains one electron outside a noble gas electron configuration, a Group IIA element, two, and so on, these larger increases in ionization energy must reflect the extreme difficulty encountered in trying to break into the noble gas structure that lies below the outer shell.

The variation of the first ionization energy across periods and down groups, illustrated in Figure 7.20, quite closely parallels the trends in atomic size. This really

Table 7.4. Ionization energies of the first 20 elements (kJ/mol)

	First	Second	Third	Fourth	Fifth	Sixth	Seventh	Eighth
H	1,312							
He	2,371	5,247						
Li	520	7,297	11,810					
Be	900	1,757	14,840	21,000				
B	800	2,430	3,659	25,020	32,810			
C	1,086	2,352	4,619	6,221	37,800	47,300		
N	1,402	2,857	4,577	7,473	9,443	53,250	64,340	
O	1,314	3,391	5,301	7,468	10,980	13,320	71,300	84,050
F	1,681	3,375	6,045	8,418	11,020	15,160	17,860	92,000
Ne	2,080	3,963	6,276	9,376	12,190	15,230	—	—
Na	495.8	4,565	6,912	9,540	13,360	16,610	20,110	25,490
Mg	737.6	1,450	7,732	10,550	13,620	18,000	21,700	25,660
Al	577.4	1,816	2,744	11,580	15,030	18,370	23,290	27,460
Si	786.2	1,577	3,229	4,356	16,080	19,790	23,780	29,250
P	1,012	1,896	2,910	4,954	6,272	21,270	25,410	29,840
S	999.6	2,260	3,380	4,565	6,996	8,490	28,080	31,720
Cl	1,255	2,297	3,850	5,146	6,544	9,330	11,020	33,600
Ar	1,520	2,665	3,947	5,770	7,240	8,810	11,970	13,840
K	418.8	3,069	4,600	5,879	7,971	9,619	11,380	14,950
Ca	589.5	1,146	4,941	6,485	8,142	10,520	12,350	13,830

should not be too surprising, since the energy required to remove an electron from an atom completely should depend in part on how far away it is from the nucleus. In addition, the same factors responsible for causing an outer shell to contract in size as we proceed across a period will also lead to the electron being held more tightly. Thus, as we proceed down within a group (e.g., the alkali metals), the increase in size that occurs is accompanied by a decrease in ionization energy. As we move across a

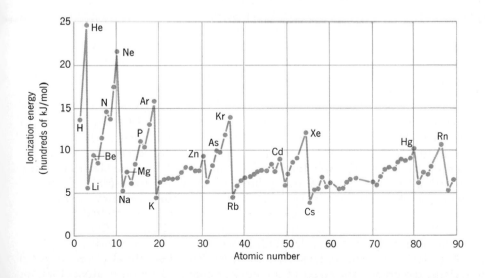

Figure 7.20. *The variation of first ionization energy with atomic number.*

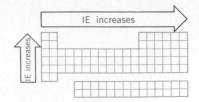

The variation of ionization energy in the periodic table.

period, from left to right, the increased effective nuclear charge experienced by the outer-shell electrons causes the shell to shrink in size and also makes it more difficult to remove an electron.

Irregularities in trends across a period

If we examine more closely the trend in ionization energy across a period, we note some irregularities. In period 2, for example, we might expect a uniform increase in ionization energy as we go from Li to Ne. What we actually find, however, is that the ionization energy of beryllium is larger than that of boron and the ionization energy of nitrogen is larger than that of oxygen. These reversals can also be explained by the electronic structures of the elements.

In the case of beryllium, the first electron to be removed comes from the filled $2s$ subshell, whereas the electron removed first from boron is in the singly occupied $2p$ subshell. The $2p$ subshell is higher in energy than the $2s$, so the $2p$ electron of boron is more easily removed than a $2s$ electron of beryllium.

When we get to nitrogen, we find that we have a half-filled $2p$ subshell (electronic structure of nitrogen is $1s^2 2s^2 2p^3$), while in oxygen the $2p$ subshell is occupied by four electrons. The fourth electron in this $2p$ subshell is in an orbital already occupied by another electron, so it experiences considerable electron–electron repulsion. As a result, this electron is more easily removed than one of the electrons in a singly occupied orbital in the nitrogen atom. Note that the same inverted order of values for the ionization energy also occurs in periods 3 and 4, where the ionization energy of phosphorus is greater than for sulfur and that for arsenic is greater than for selenium.

Electron affinity

The **electron affinity** *is the energy that is released or absorbed when an electron is added to a gaseous atom or ion in its ground state.* Such a process occurs, for example, when a chlorine atom picks up an electron to become a negative ion

$$Cl(g) + e^- \longrightarrow Cl^-(g)$$

The potential energy of an electron increases as it is pulled away from an atom, so the process is endothermic. (Remember, an endothermic change is accompanied by a potential energy increase.)

The electron affinity (like the ionization energy) applies to isolated atoms and usually represents an exothermic process. This is so because we are placing the electron into an environment where it experiences the attraction of the nucleus. We can see how the addition of an electron to an atom would release energy by considering the reverse process, pulling the electron away from the attractive force of the nucleus. If removing the electron requires work (i.e., is endothermic), the opposite process would release energy.

There are instances when more than one electron is added to the outer shell of the atom. For example, oxygen reacts to form the ion, O^{2-}, which requires that an oxygen atom pick up two electrons. The first electron enters a neutral atom, but the second electron, which gives the ion a charge of $2-$, must be forced onto an already negative ion. This requires work; therefore, we find that the second electron affinity of an atom is an endothermic quantity. Table 7.5 contains electron affinities for some representative elements. The table is not complete because electron affinities are difficult to measure and, for many elements, have not been accurately determined.

Adding more than one electron to an isolated atom is always endothermic overall.

As with the ionization energy, the variations in electron affinity generally parallel the variations in atomic size. This is because we are considering the placement of an electron into the outer shell of the atom. The closer the electron can get to the nucleus, the greater the effect of the nuclear charge. Therefore, atoms that are very small and have outer shells that experience a high effective nuclear charge (i.e.,

Table 7.5. Electron affinities^a for the representative elements (kJ/mol)

IA	IIA	IIIA	IVA	VA	VIA	VIIA
H						
-73						
Li	Be	B	C	N	O	F
-60	$\approx +100$	-27	-122	$\approx +9$	-141	-328
Na	Mg	Al	Si	P	S	Cl
-53	$\approx +30$	-44	-134	-72	-200	-348
K	Ca	Ga	Ge	As	Se	Br
-48	—	-30	-120	-77	-195	-325
Rb	Sr	In	Sn	Sb	Te	I
-47	—	-30	-121	-101	-190	-295
Cs	Ba	Tl	Pb	Bi	Po	At
-45	—	-30	-110	-110	-183	-270

^{a} Negative values mean that the process $M + e^- \rightarrow M^-$ is exothermic.

elements in the upper right of the periodic table) have very large exothermic electron affinities. On the other hand, atoms that are large and whose outer shells feel the effect of a small effective nuclear charge (such as the elements in Groups IA and IIA) have less exothermic electron affinities.

If we examine Table 7.5, we find that the nonmetals of the second period have electron affinities that are less exothermic than the elements of the same group just below them in period 3. This is a bit of a surprise because in a given group, size increases going down. Apparently, the crowding of electrons in the outer shell of a period 2 element makes the mutual repulsions of the electrons substantially greater than in the outer shell of a period 3 element. Therefore, even though the electron that is added to a period 2 element gets closer to the nucleus than one added to a period 3 element, the greater repulsions in the smaller outer shell lead to a lesser *net* amount of energy being evolved.

Finally, carbon has a rather substantial exothermic electron affinity, while that of nitrogen is actually endothermic. With carbon, the entering electron can occupy a vacant $2p$ orbital and therefore experiences only minimal electron–electron repulsions. With nitrogen, however, an additional electron must be placed into an orbital that is already occupied by an electron. The greater electron–electron repulsions that result, as well as the loss of the extra stability that a nitrogen atom has because of its half-filled p subshell, cause the electron affinity to be an endothermic quantity for nitrogen.

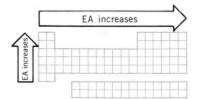

The variation of electron affinity in the periodic table.

REVIEW QUESTIONS AND PROBLEMS

Problems whose numbers are in blue have their answers in Appendix D at the back of the book. The more difficult problems are marked with asterisks.

Electromagnetic Radiation

7.1 Sketch a wave and label its wavelength and its amplitude. What relationship exists between the wavelength and the frequency of a light wave?

7.2 What is the value of the speed of light in meters per second?

7.3 What are the units of frequency in the SI? Why are the wavelengths of radio waves given in meters and those of visible light given in nanometers? What wavelength limits (in nanometers) encompass the visible portion of the electromagnetic spectrum?

7.4 How do the wavelengths of infrared light and ultraviolet light compare to the wavelengths of visible light?

7.5 Arrange the following in order of increasing wavelength: visible light, microwaves, ultraviolet light, X rays, infrared light, TV waves, gamma rays.

7.6 Which type of radio broadcasts have a higher frequency, FM or AM? (If necessary, refer to the local radio listings in a newspaper.)

7.7 Calculate the wavelength of light (in nanometers) whose frequency is 8.0×10^{15} Hz. Calculate the frequency of light (in hertz) whose wavelength is 200.0 nm.

7.8 A radar transmitter broadcasts electromagnetic radiation in the microwave region of the spectrum. A typical radar wave has a frequency of 9.40×10^9 Hz. What is its wavelength (a) in meters and (b) in centimeters?

7.9 Radio station WCBS in New York broadcasts its FM signal at a frequency of 101.1 megahertz (MHz). The AM signal is broadcast at 880 kilohertz (kHz). What are the wavelengths of these signals expressed in meters?

7.10 Calculate the wavelength of WCBS-AM (Question 7.9) in feet. How does this compare to the length of a football field?

Atomic Spectra

7.11 What is the difference between a continuous spectrum and an emission line spectrum? Describe how you could produce each of them.

7.12 From the point of view of atomic structure, what is the significance of emission line spectra?

7.13 What other terms are used to describe line spectra?

7.14 Mercury vapor lamps, used for highway and street lights, produce light corresponding to the atomic spectrum of mercury. One of the lines in the visible spectrum is green and has a wavelength of 546 nm. What is the frequency of this line in hertz?

7.15 Use the data in Figure 7.4b to calculate the frequencies (in hertz) of the visible lines in the atomic spectrum of hydrogen.

7.16 What is the wavelength in nanometers of the light that gives sodium vapor lamps their characteristic yellow color? What is the frequency of this light in hertz?

7.17 Why is it possible to use atomic emission spectra to determine the elemental composition of a substance?

Bohr Theory and the Energy of Light Waves

7.18 Why does Planck's constant have units of *energy × time?*

7.19 What is a photon?

7.20 Describe the photoelectric effect. How does it support the concept that light is composed of photons, each with an energy that depends on the frequency of the light?

7.21 Calculate the energy in joules of a photon having a frequency of 3×10^{15} Hz. If a photon has an energy of 2×10^{-20} J, what is its wavelength in meters?

7.22 Calculate the energy contained in one photon of yellow light emitted by sodium with a wavelength of 589 nm. What is the energy of a mole of these photons, expressed in kilojoules? By how many Celsius degrees could the temperature of 10.0 kg of water be raised by the energy in one mole of these photons?

7.23 Calculate the energy in joules of one mole of photons that have
(a) a frequency of 2.6×10^{14} Hz.
(b) a wavelength of 546 nm.

7.24 Why does Planck's relationship [Equation 7.3], along with atomic emission spectra, suggest the existence of energy levels in atoms?

7.25 Describe Bohr's model of the atom. What initial evidence was there that Bohr's theory might be correct? Why was his theory eventually discarded?

7.26 How does Bohr's theory explain the emission and absorption of light by a hydrogen atom?

7.27 Use the Rydberg equation to calculate the wavelength (in nanometers) of the spectral line of hydrogen that would result when an electron drops from the fourth Bohr orbit to the second, and from the sixth Bohr orbit to the third.

7.28 How many joules must be supplied to raise an electron from the first Bohr orbit to the third?

7.29 What is the energy in joules of an electron in the third energy level ($n = 3$) of a hydrogen atom? What is the energy in joules of an electron in the second energy level? How many joules does an electron lose if it falls from the third to the second energy level in a hydrogen atom? What is the frequency (in hertz) of the photon that's emitted? What is the wavelength of this photon in nanometers? What color is the light that's given off?

Wave Mechanics

7.30 State the de Broglie relationship.

7.31 What is the effective mass in grams of a mole of photons that have a wavelength of 589 nm?

*7.32 Calculate the kinetic energy in joules of an electron with a wavelength of 0.10 nm.

7.33 Why don't we observe the wave properties of large objects such as baseballs or airplanes?

7.34 Describe the constructive and destructive interference of two waves. How is this phenomenon related to the production of a diffraction pattern?

7.35 What experimental evidence is there for the wave properties of the electron?

7.36 What is a standing wave? What are nodes?

Quantum Numbers

7.37 What is the symbol for the principal quantum number? What are its allowed values? What physical characteristic of the electron wave is associated with the value of its principal quantum number?

7.38 What is the symbol for the azimuthal quantum number? What are its allowed values? What characteristic of an electron wave is associated with the value of its azimuthal quantum number?

7.39 What is the symbol for the magnetic quantum number? What are its allowed values? What characteristic of an electron wave is associated with the value of its magnetic quantum number?

7.40 What is an orbital?

7.41 How many subshells are in the fourth shell?

7.42 If the value of n for an electron is 5, what subshells are possible for this electron?

7.43 If the largest value of m_l for an electron is 3, what kind of subshell must it be in?

7.44 How many orbitals are there in the shell with $n = 5$?

7.45 What does the term *ground state of an atom* mean?

7.46 How many electrons can be accommodated in each of the following types of subshells: s, p, d, f, g, h? What is the lowest value of n for a shell that has an h subshell? What are the allowed values of m_l for an h subshell?

7.47 Within a given shell, how do the energies of the s, p, d, and f subshells compare?

7.48 Why did Bohr's theory work for hydrogen, but fail for atoms with two or more electrons?

Electron Spin

7.49 What property of an electron is associated with its "spin"?

7.50 What are the only values that m_s can have?

7.51 What is the Pauli exclusion principle? What significance does it have in determining the electronic structure of an atom?

7.52 Give the values of n, l, m_l, and m_s for each electron in a filled L shell.

7.53 How many electrons can be placed into the M shell of an atom?

7.54 What do we mean when we say that two electrons are paired?

7.55 What magnetic property is associated with unpaired electrons in an atom, molecule, or ion? What is the magnetic property associated with the complete pairing of all the electrons in an atom, molecule, or ion?

Electron Configurations

7.56 What is Hund's rule?

7.57 Use the periodic table as a guide in predicting the complete electron configurations of these elements: P, Ni, As, Ba, Rh, Ho, Ge.

7.58 Write the complete electron configuration for Rb, Sn, Br, Cr, and Cu.

7.59 Give the outer-shell electron configuration for K, Al, F, S, Tl, and Bi.

7.60 Use the periodic table to arrive at the electronic structure of the outer shells of the atoms of Si, Se, Sr, Cl, O, S, As, and Ga.

7.61 Why are there no elements in period 3 between Mg and Al?

7.62 How many electrons are in p orbitals in (a) As, (b) Si, (c) Ru?

7.63 Draw the orbital diagram for (a) phosphorus and (b) calcium.

7.64 Draw the orbital diagram for the valence shell of (a) Sn, (b) Br, (c) Ba.

7.65 Which of the following atoms are diamagnetic: (a) Cd, (b) Ge, (c) Pt, (d) Sr, (e) Kr?

7.66 Construct the orbital diagram for each of the elements in the first row of transition elements ($Z = 21$ to $Z = 30$). How many unpaired electrons are there in each of these atoms? Which of these elements are paramagnetic and which are diamagnetic?

Shapes of Orbitals

7.67 How does the shape of an orbital depend on its value of l? How does the size of an orbital depend on its value of n?

7.68 On a single set of Cartesian coordinate axes, sketch the shapes of the three p orbitals. Label them p_x, p_y, and p_z.

7.69 How do the $1s$ and $2s$ orbitals differ? How are they alike?

7.70 How does the shape of an s orbital differ from that of a p orbital? Sketch the shapes of these kinds of orbitals.

7.71 On separate sets of axes, sketch the shapes of the following orbitals: (a) d_{xy}, (b) $d_{x^2-y^2}$, (c) d_{z^2}

7.72 On the basis of the mutual repulsion of electrons and the spatial orientations of the p orbitals in a given subshell, the correct orbital diagram for the ground state of nitrogen seems quite reasonable. Why?

Periodic Trends in Properties

7.73 What difficulties are there in defining the size of an atom or an ion? What units are used for expressing atomic size?

7.74 A potassium atom has a diameter of about 4.06 Å. Express this in meters, nanometers, and picometers.

7.75 What is meant by the term *effective nuclear charge*?

7.76 If the core electrons in the following atoms were 100% effective at shielding the outer-shell electrons, what would be the effective nuclear charge felt by the outer-shell electrons? (Assume electrons in the same shell provide no shielding for each other.) (a) Mg, (b) Al, (c) Si, (d) S, (e) Cl.

7.77 If the core electrons in the following atoms were 100% effective at shielding the outer-shell electrons, what would be the effective nuclear charge felt by the outer-shell electrons? (Assume electrons in the same shell provide no shielding for each other.) (a) Be, (b) Mg, (c) Ca, (d) Sr, (e) Ba.

7.78 Explain why atomic size decreases from left to right across a period and why it increases from top to bottom in a group. (*Hint:* First work Questions 7.76 and 7.77.)

7.79 Give the electron configurations of the following ions: (a) Ca^{2+}, (b) S^{2-}, (c) Cl^-, and (d) K^+.

7.80 Give the electron configurations and orbital diagrams for the following ions: (a) Cr^{3+}, (b) Mn^{2+}, (c) Mn^{3+}, (d) Co^{2+}, (e) Co^{3+}, (f) Ni^{2+}.

7.81 Choose the largest atom: Ge, Sb, Sn, As.

7.82 Choose the larger species in each pair:

(a) S or Se   (d) O^+ or O^-
(b) C or N   (e) S or S^{2-}
(c) Fe^{2+} or Fe^{3+}

7.83 Explain the variation in ionic size observed for the series, N^{3-}, O^{2-}, and F^- (Table 7.3) in terms of the effective nuclear charge and electron–electron repulsions experienced by their outer-shell electrons.

7.84 What is the lanthanide contraction? How might this be used to explain why the elements in the sixth period following the lanthanides have higher ionization energies than the elements directly above them in the fifth period (e.g., the ionization energy of Pt = 870 kJ/mol, while that of Pd = 805 kJ/mol)?

7.85 In Table 7.3, we find some elements that form more than one positive ion. In each case the ion with the greater positive charge is smaller. Why is this so?

7.86 Define *ionization energy* and *electron affinity*.

7.87 How can we explain the variation in ionization energy across a period in the periodic table?

7.88 Choose the species with the larger ionization energy.
(a) Li or Be   (c) C or N
(b) Be or B   (d) N or O

(e) Ne or Na   (g) Na^+ or Mg^+
(f) S or S^+
Check your answers by referring to Table 7.4.

7.89 Choose the species with the more exothermic electron affinity:
(a) S or Cl   (c) P or As
(b) S or S^- (d) O or S

7.90 Why is the second electron affinity for an element always an endothermic quantity?

Additional Exercises

7.91 Use the Rydberg equation [Equation 7.2] to calculate the ionization energy of hydrogen.

7.92 If an electron in a hydrogen atom falls from the energy level with $n = 4$ to the energy level with $n = 1$, what is the energy (in joules), the wavelength (in nanometers), and the frequency (in hertz) of the photon that is emitted?

7.93 Can an atom with an odd atomic number be diamagnetic? Explain.

7.94 How many unpaired electrons would there be in the ion Gd^{3+}?

7.95 Why are the elements in the center of period 6 so dense?

7.96 Predict the electron configuration of element 106.

7.97 If the element with $Z = 114$ was discovered, in which group would you expect to find it? What would be its electron configuration?

7.98 If the ion Ar^- were to be formed, what would be its electron configuration?

7.99 Draw a graph, on a set of axes like that below, of the ionization energy versus the number of electrons removed from the atom for each of the elements Li, C, O, S, and Ne.

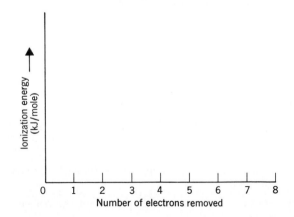

* 7.100 If all the energy required to remove the electrons from 1 mole of H atoms was used instead to heat water, how many grams of water could have their temperature increased by 25 °C?

CHAPTER

8

CHEMICAL BONDING:
GENERAL CONCEPTS

Chemical bonds are the attractions between atoms that hold compounds together. Their strengths determine the chemical properties of substances, and the way chemical bonds change when substances react is what controls the amount of energy released or absorbed during the reaction.

The kinds of chemical bonds formed by an atom depend on the atom's electronic structure. For example, ionization energy and electron affinity control the ease with which an atom gains or loses electrons, and as you learned in the last chapter, these properties depend on electronic structure and the position of the element in the periodic table.

Chemical bonds can be divided into two broad categories: **ionic bonds** and **covalent bonds.** You should recall that ionic compounds are formed by the transfer of electrons between atoms to give electrically charged particles that attract each other. This attraction between oppositely charged ions is what constitutes an ionic bond. Covalent bonds are formed by the sharing of electrons between atoms. In a very simple sense, the mutual attraction of the nuclei of the atoms for the electrons shared between them is what constitutes a covalent bond.

In this chapter we will examine how ionic and covalent bonds are formed between atoms, and the conditions that tend to lead to these kinds of bonds. Toward the end of the chapter, you will also learn that the simple division of chemical bonds between ionic and covalent is really a bit of oversimplification. To understand the properties of substances, we must really take a more sophisticated look at the nature of chemical bonds.

8.1 BONDING IN IONIC COMPOUNDS

In Chapter 4 we saw that metals tend to react with nonmetals to form ionic compounds. We also discussed how to use the periodic table to help remember the charges on the ions formed by the representative elements. What we could not do in Chapter 4, however, is explain *why* ionic compounds form and *why* the elements form the ions they do. That's our goal now, so let's examine the formation of a typical ionic compound to see how the electronic structures of the atoms influence the number of electrons transferred, as well as the abilities of the atoms to form ionic compounds.

When lithium and fluorine react, they form the ionic compound, LiF, which contains Li^+ and F^- ions. The electron configurations of the Li and F atoms are

$$Li \quad 1s^2 2s^1$$

and

$$F \quad 1s^2 2s^2 2p^5$$

The loss of an electron by lithium and the gain of an electron by fluorine produces the following changes in their electron configurations:

$$Li \, (1s^2 2s^1) \longrightarrow Li^+ \, (1s^2) + e^-$$
$$F \, (1s^2 2s^2 2p^5) + e^- \longrightarrow F^- \, (1s^2 2s^2 2p^6)$$

Recall that when an atom loses electrons, they come from the shell with largest n.

Notice that each of the ions formed in this reaction has the same electron configuration as one of the noble gases. Lithium has the configuration of helium and fluorine has the configuration of neon.

In a similar reaction, calcium and oxygen atoms combine to form the ionic compound CaO. The changes in electron configurations for the atoms in this reaction are

$$Ca \, (1s^2 2s^2 2p^6 3s^2 3p^6 4s^2) \longrightarrow Ca^{2+} \, (1s^2 2s^2 2p^6 3s^2 3p^6) + 2e^-$$
$$O \, (1s^2 2s^2 2p^4) + 2e^- \longrightarrow O^{2-} \, (1s^2 2s^2 2p^6)$$

Once again, the ions formed have the same electron configuration as a noble gas, argon in the case of Ca^{2+} and neon in the case of O^{2-}.

In attempting to explain these reactions, we need answers to several questions. First, why do metals such as Li and Ca lose electrons and why do nonmetals such as F and O gain them? Second, why do electron loss and electron gain cease once noble gas electron configurations are reached?

Primarily, it is the change in the potential energy that controls the formation of the ions in compounds such as LiF and CaO. For an ionic compound to be stable, its formation from the elements must be exothermic, which means that the potential energy of the compound must be lower than that of the elements. In turn, this means that any endothermic contributions to the net energy change must be smaller than the exothermic ones.

In general, the lower the potential energy, the more stable the compound.

Metals form cations because they lose electrons relatively easily. As you learned in the last chapter, the ionization energies of metals are much less than those of the nonmetals. For the representative metals, loss of electrons ceases once the valence shell has been emptied because breaking into the inner noble gas core is extremely difficult (i.e., it requires a very large amount of energy).

For most of the nonmetals, addition of an electron to the atom is exothermic, so this favors the formation of anions by the nonmetals. However, formation of an anion with a charge of $2-$ or larger is always endothermic.

If ionization energy and electron affinity were the only factors affecting the formation of an ionic compound, very few would ever form at all. In almost all cases, the energy needed to remove the electrons from the metal is larger than the energy released on the formation of the anion, so the formation of the ions from the neutral atoms is nearly always endothermic. What, then, allows ionic compounds to exist?

The lattice energy

The principal reason that ionic compounds are stable is because the attractions between the ions, which occur when the compound is formed, produce a large decrease in the potential energy. To see this, let's examine the potential energies of two situations: one in which the neutral atoms are alongside each other and the other in which the ions are next to each other. Let's see how the energy changes if we separate the neutral atoms and then bring the particles back together as ions.

Neutral atoms attract each other only very weakly. (In fact, from what you've learned so far, you would have no reason to believe that there are any attractions at all between the neutral atoms.) Since the attractive forces are so weak, separating the atoms from each other requires just a small increase in potential energy. On the other hand, if the particles are brought back together as ions, which attract each other strongly, a large decrease in potential energy occurs. The net result is that the ions in a crystal end up at a much lower potential energy than the neutral atoms. This potential energy lowering is called the **lattice energy,** and its size more than makes up for the increase in potential energy associated with the creation of the ions. As a result, the formation of the ionic compound is exothermic.

Now we are able to understand why so many ions tend to have a noble gas electron configuration. It doesn't take too much energy to empty the valence shell of a metal, so the lattice energy is exothermic enough to compensate for this endothermic contribution to the overall energy change. However, dipping into the noble gas core beneath the outer shell requires a tremendous input of energy — more than can be made up by the exothermic lattice energy. As a result, electron loss stops once a noble gas core is exposed.

For a nonmetal, adding electrons to the valence shell is either exothermic or somewhat endothermic. However, once the valence shell is completed, any additional electrons must enter the next higher shell. This also requires a tremendous input of energy to accomplish — more than can be compensated by the lattice energy. As a result, nonmetals never gain more than enough electrons to give them a complete ns^2np^6 "noble gas" configuration.

The tendency of the ions of many of the representative elements to have a noble gas configuration, with 8 electrons in the outer shell, is the basis of the **octet rule:** *When metals and nonmetals of the A Groups react, they often tend to gain or lose electrons until there are eight electrons in the outer shell.* As we will see, this rule is really more useful in its application to covalent bonds between atoms.

> Ionic compounds are favored when atoms of low ionization energy (metals) combine with atoms with exothermic electron affinities (nonmetals).

> Except for helium, the noble gases have eight electrons in their outer shell.

Failure of the octet rule

Except for the metals in Groups IA and IIA and aluminum, the octet rule doesn't work well for cations. When a transition element or **post-transition element** (an element to the right of a row of transition elements, such as tin or lead) forms a positive ion, the outer-shell electron configuration is generally not the same as that of a noble gas.

Recall that when a positive ion is formed from an atom, electrons are *always* lost first from the shell with the largest value of n. This means that a transition element always loses electrons from its outer s subshell before any electrons are lost from its outermost d subshell. For example, the zinc ion, Zn^{2+}, is formed when a zinc atom loses its outer $4s$ electrons.

$$Zn\ ([Ar]\ 3d^{10}4s^2) \longrightarrow Zn^{2+}\ ([Ar]\ 3d^{10}) + 2e^-$$

The electron configuration of the Zn^{2+} ion also can be rewritten as

$$Zn^{2+}\ [Ne]\ 3s^23p^63d^{10}$$

and we see that its outer shell, $3s^23p^63d^{10}$, does not have the usual noble gas configuration, ns^2np^6. It does have one thing in common with a noble gas configuration, however — all the subshells in the outer shell are complete. Because of this similarity, the $ns^2np^6nd^{10}$ configuration is called a **pseudo-noble-gas configuration.**

Many of the transition and post-transition metals form ions with neither a noble gas configuration nor a pseudo-noble-gas configuration. An example is iron, which forms Fe^{2+} and Fe^{3+} ions. Depending on the particular circumstances, iron atoms lose electrons until the extra energy needed to take one more is greater than can be made up by the lattice energy. For iron, this sometimes gives Fe^{2+} and sometimes Fe^{3+}. (Although we can understand why this occurs for elements like iron, predicting what will happen for any particular metal isn't possible, so the formulas of the ions of the transition elements just have to be learned, as we said in Chapter 4.)

8.2 LEWIS SYMBOLS

G. N. Lewis received his Ph.D. from Harvard and later served as Professor of Chemistry and Dean at the University of California.

Often it is useful to keep tabs on the valence shell[1] electrons of atoms when they become joined by chemical bonds. A system to accomplish this was introduced by Gilbert N. Lewis (1875–1946), one of America's best-known chemists. The system employs a special type of notation, called **Lewis symbols.**

To construct the Lewis symbol for an element, we write its atomic symbol surrounded by a number of dots (or X's or circles, etc.), each of which represents one electron in the atom's valence shell. For example, hydrogen, which has one electron in its valence shell, is given the Lewis symbol, H·. Any atom, in fact, with one electron in its outer shell has a similar Lewis symbol. This includes any element in Group IA of the periodic table, so that each of the elements Li, Na, K, Rb, Cs, and Fr has a Lewis symbol that we might generalize as X· (where X = Li, Na, etc.). Generalized Lewis symbols for the representative elements are given in Table 8.1.

Table 8.1. Lewis symbols for A-Group elements

Group	IA	IIA	IIIA	IVA	VA	VIA	VIIA	O
Symbol	X·	·X·	·X̣·	·X̣·̣	·X̣⋮	·X̤⋮	·X̤⋮	⋮X̤⋮

Lewis symbols usually are *not* used for the transition elements.

In general, the number of valence electrons that an atom of a representative element has is equal to its group number. Therefore, we see that the group number is also equal to the number of dots in the atom's Lewis symbol. This is useful to remember because it makes writing the Lewis symbol for an element very simple. Notice that in Table 8.1 the number of unpaired electrons for atoms in Groups IIA, IIIA, and IVA doesn't agree with the predictions you would make by writing their electron configurations. The Lewis symbols are written this way only because when the atoms form bonds, they *behave* as though they have the number of unpaired electrons shown by their Lewis symbols.

EXAMPLE 8.1. WRITING THE LEWIS SYMBOL FOR AN ATOM

PROBLEM: What is the Lewis symbol for germanium ($Z = 32$)?

SOLUTION: Germanium is in Group IVA and therefore has four valence electrons. Its Lewis symbol has four dots that we arrange symmetrically around the chemical symbol.

$$·\overset{\cdot}{\underset{\cdot}{Ge}}·$$

[1] As noted in the previous chapter, *valence* is a term sometimes associated with chemical bonding. It normally describes an atom's bond-forming capacity.

Lewis symbols are used to describe chemical bonds between atoms. The formulas that we construct using Lewis symbols are called **Lewis structures** or **electron dot formulas.** They are most useful for describing covalent bonds, but they can also be used to diagram what happens when atoms combine to form ionic compounds. For example, the reaction between atoms of lithium and fluorine can be shown as follows:

$$Li \cdot + \cdot \overset{\displaystyle \cdot \cdot}{\underset{\displaystyle \cdot \cdot}{F}} : \longrightarrow Li^+ \left[: \overset{\displaystyle \cdot \cdot}{\underset{\displaystyle \cdot \cdot}{F}} : \right]^-$$

$Li^+ \left[: \overset{\cdot \cdot}{\underset{\cdot \cdot}{F}} : \right]$ is the Lewis structure for LiF.

The brackets around the fluorine on the right are used to indicate that all four pairs of electrons are the exclusive property of the fluoride ion. Notice that with one electron transferred from lithium to fluorine, the valence shell of lithium is emptied and the fluoride ion ends with a Lewis symbol that's the same as for one of the noble gases. In a similar way, we can also diagram the reactions for the formation of $CaCl_2$ and Li_2O.

$$Ca : + \begin{array}{c} \nearrow \cdot \overset{\cdot \cdot}{\underset{\cdot \cdot}{Cl}} : \\ \\ \searrow \cdot \overset{\cdot \cdot}{\underset{\cdot \cdot}{Cl}} : \end{array} \longrightarrow Ca^{2+}, 2 \left[: \overset{\cdot \cdot}{\underset{\cdot \cdot}{Cl}} : \right]^-$$

or

$CaCl_2$

$$\begin{array}{c} Li \cdot \\ \\ Li \cdot \end{array} + \cdot \overset{\cdot \cdot}{\underset{\cdot}{O}} : \longrightarrow 2Li^+, \left[: \overset{\cdot \cdot}{\underset{\cdot \cdot}{O}} : \right]^{2-}$$

8.3 THE COVALENT BOND

In many instances, the formation of an ionic substance is not energetically favorable. For example, the creation of a cation may require too large an energy input (ionization energy) to be recovered by the energy released when the anion is formed and the ionic solid is produced (electron affinity and lattice energy). In these situations, a covalent bond is formed.

A **covalent bond** *results from the sharing of a pair of electrons between atoms.* The binding force results from the attraction between the shared electrons and the positive nuclei of the atoms entering into the bond. In this sense the electrons serve as a sort of glue cementing the atoms together. Consider, for example, the formation of the H_2 molecule from two hydrogen atoms. As the atoms approach each other, the single $1s$ electron on each of them begins to feel the attraction of both nuclei. The electron density therefore begins to shift to the region between the nuclei, as shown in Figure 8.1.

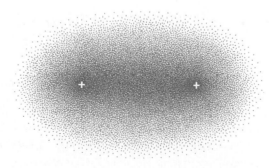

Figure 8.1. *The electron distribution in the H_2 molecule.*

Figure 8.2. *An energy diagram for the formation of H_2 from two hydrogen atoms. The bond energy is 435 kJ/mol of bonds, and the bond length is 75 pm (0.75 Å).*

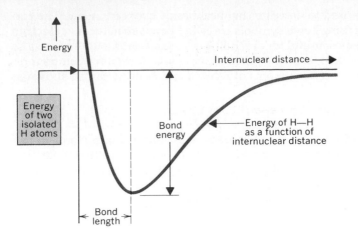

If we examine the energy changes accompanying the formation of the bond, we find that as the atoms come together, the energy begins to decrease. This is because the electrons are coming closer to another positive nucleus to which they are also attracted (remember how potential energy changes between particles that attract each other). The energy curve for the molecule is shown in Figure 8.2. Notice that at small internuclear distances the energy rises steeply. This is caused by the repulsions between the nuclei. The most stable (lowest energy) distance of separation between the two nuclei occurs when the energy is a minimum. At this point the attractions and repulsions are balanced. The depth of this minimum is the amount of energy that must be supplied to separate the atoms and is called the **bond energy.** The distance between the nuclei when the energy is a minimum is called the **bond length** or **bond distance.**

At small distances, the repulsions between the nuclei outweigh the attractions that the nuclei have for the electron cloud of the shared electron pair between them.

When two atoms such as hydrogen share a pair of electrons, the spins of the electrons become paired. This is an important aspect of the creation of a covalent bond. Each H atom completes its valence shell by acquiring a share of an electron from another atom. We can indicate the formation of H_2, using Lewis symbols, as

A covalent bond is sometimes called an **electron pair bond.**

$$H\cdot + H\cdot \longrightarrow H\!:\!H$$

in which the pair of electrons in the bond is shown as a pair of dots between the two H atoms. Often a dash is used instead of the pair of dots, so the H_2 molecule can be represented as H—H.

One dash stands for two electrons with their spins paired.

Often we will find it necessary to count the number of electrons belonging to each atom in a molecule held together by covalent bonds. For H_2, the electron pair in the bond is shared by both atoms, so we can count both electrons as belonging to both atoms. Notice that by forming the covalent bond, both H atoms have, in effect, achieved a noble gas configuration.

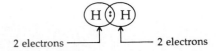

The number of covalent bonds that an atom forms can often be explained simply by counting the number of electrons required to achieve a noble gas configuration. For example, a carbon atom has four electrons in its valence shell. To attain a noble gas configuration, it usually acquires, through sharing, four additional electrons. Therefore, a carbon atom usually forms four covalent bonds, and with hydrogen it forms the molecule CH_4, which is called methane (the chief component of natural gas).

$$\cdot \overset{\displaystyle \cdot}{\underset{\displaystyle \cdot}{C}} \cdot \; + \; 4H\times \; \longrightarrow \; H \overset{\displaystyle \overset{H}{\times}}{\underset{\displaystyle \underset{H}{\overset{\times}{\times}}}{\times C \times}} H$$

Nitrogen, which has five valence electrons, has to gain only three electrons through sharing to complete an octet; therefore, nitrogen forms three covalent bonds with hydrogen to form the ammonia molecule, NH_3. In a similar fashion it is easy to see why the formula for water is H_2O and that for hydrogen fluoride is HF.

$$H \overset{\displaystyle \cdot\cdot}{\underset{\displaystyle \underset{H}{\times}}{\times N \times}} H \qquad H \overset{\displaystyle \cdot\cdot}{\underset{\displaystyle \underset{H}{\times}}{\times O \colon}} \qquad H \overset{\displaystyle \cdot\cdot}{\underset{\displaystyle \cdot\cdot}{\times F \colon}}$$

In the Lewis structures above, we have used ×'s to represent the hydrogen electrons just to make it easier to see what happens when the bonds are formed. Ordinarily, however, we don't distinguish among the various electrons in the bonds because all electrons are really alike.

In each of the molecules that we've discussed up to this point, the atoms have been joined by covalent bonds that each consist of a single pair of electrons. These are called **single bonds.** It is also possible for a pair of atoms to share two or even three pairs of electrons. For example, in carbon dioxide the carbon shares two pairs of electrons with each oxygen atom. These are **double bonds.**

$$\overset{\displaystyle \cdot\cdot}{\underset{\displaystyle \cdot\cdot}{O}} \colon\colon C \colon\colon \overset{\displaystyle \cdot\cdot}{\underset{\displaystyle \cdot\cdot}{O}}$$

The electrons in the bonds are counted as belonging to each atom, so we can count eight electrons around carbon and around each oxygen.

8 electrons

Using dashes to represent a pair of electrons, we can also draw this as

$$\overset{\displaystyle \cdot\cdot}{\underset{\displaystyle \cdot\cdot}{O}} = C = \overset{\displaystyle \cdot\cdot}{\underset{\displaystyle \cdot\cdot}{O}}$$

Nitrogen is an example of a molecule with a **triple bond.**

$$\colon N \colon\colon\colon N \colon \qquad \text{or} \qquad \colon N \equiv N \colon$$

If we count all six electrons between the atoms as belonging to both, each atom has an octet.

Exceptions to the octet rule

There are many examples of covalent compounds that fail to obey the octet rule. For instance, the molecule $BeCl_2$, which exists in the gas phase at high temperatures, is formed by the pairing of the two Be valence electrons with electrons on two chlorine atoms[2]

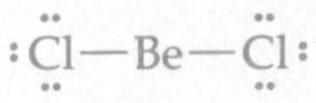

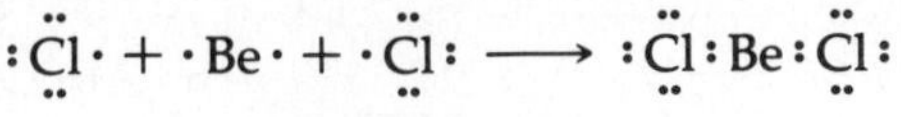

$$:\ddot{C}l-Be-\ddot{C}l:$$

$$:\ddot{C}l\cdot + \cdot Be\cdot + \cdot\ddot{C}l: \longrightarrow :\ddot{C}l:Be:\ddot{C}l:$$

In this molecule the Be atom has only four electrons in its valence shell. Another example is BCl_3

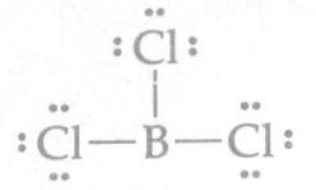

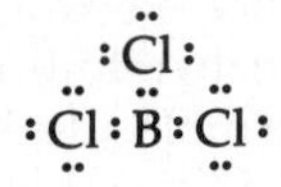

$$:\ddot{C}l:$$
$$:\ddot{C}l:B:\ddot{C}l:$$

$$:\ddot{C}l:$$
$$\,\, |$$
$$:\ddot{C}l-B-\ddot{C}l:$$

Molecules with atoms that are surrounded by less than an octet of electrons are rare and must simply be learned as exceptions. The method we will learn later for drawing Lewis structures doesn't work for them.

There are many molecules in which the central atom has more than eight electrons in its valence shell. Two typical examples are PCl_5 and SF_6. To form covalent bonds between the central atom and each of the surrounding atoms, more than four pairs of electrons are needed. In PCl_5, for example, there are five covalent bonds; in SF_6 there are six. The central atom in each of these molecules uses all its valence electrons to form covalent bonds.

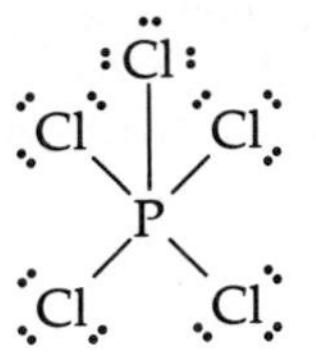

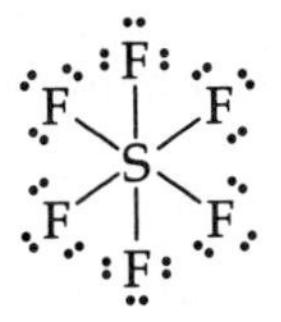

In these compounds, both phosphorus and sulfur have exceeded the number of electrons required for a noble gas electron configuration. This can occur with these elements because, in each case, the valence shell can accommodate more than eight electrons (both P and S are in the third period and the third shell can contain up to 18 electrons because of the availability of the relatively low-energy $3d$ subshell). Elements in the second period (Li to Ne) never form compounds with more than eight electrons in their valence shell because the second shell cannot have more than an octet.

8.4 DRAWING LEWIS STRUCTURES

Lewis structures for covalently bonded molecules and polyatomic ions (which are held together by covalent bonds, too) are very useful. One reason, as we will see in the next chapter, is that they allow us to predict the shapes of molecules and polyatomic ions. Therefore, you should learn how to draw them.

The first step in drawing a Lewis structure is deciding which atoms are attached to each other. For example, in CO_2, we must know that there are two O atoms bonded to the C atom, and that the molecule does not have a structure such as O—O—C. The only way of being sure, of course, is to obtain the structural infor-

---

[2] Even though $BeCl_2$ is formed from elements in Groups IIA and VIIA, it is covalent instead of ionic. The reason is discussed in Chapter 20.

mation experimentally, and this is what a chemist would do if a new compound has been discovered. (There are methods for determining molecular structures, but they are beyond the scope of this book and we won't attempt to discuss them.)

Even in the absence of concrete structural data, the arrangement of atoms often can be inferred from the formula of the species. In simple binary molecules and polyatomic ions, such as CO_2 or CO_3^{2-}, the central atom usually appears first in the formula. For example, carbon is the central atom in both CO_2 and CO_3^{2-}. This generalization also holds for NH_3, NO_2, NO_3^-, SO_3, and SO_4^{2-}. Unfortunately, it is not true for H_2O and H_2S (in which H atoms are bound to O and S, respectively). Nor is it true for molecules such as HClO (in which the O is the central atom) or ions like SCN^- (in which C is central). The structure of the molecule is therefore not always obvious. If you must guess, the most symmetrical arrangement of atoms has the greatest chance of being correct. Once we know the arrangement of atoms in the molecule, however, we can then go about distributing the valence electrons. The procedure for doing this can be summarized in the following steps, which we will illustrate by some examples.

It's useful to remember that hydrogen can never be a central atom because it forms only one covalent bond.

1. *Count all the valence electrons of the atoms.* If the species is an ion, add an additional electron for each negative charge or subtract an electron for each positive charge.
2. *Place one pair of electrons in each bond.*
3. *Complete the octets of the atoms bonded to the central atom.* (Remember, however, that the valence shell of any hydrogen atom is complete with only two electrons.)
4. *Place any additional electrons on the central atom in pairs.*
5. *If the central atom still has less than an octet, you must form multiple bonds so that each atom has an octet.*

Now let's look at some examples that show how this procedure works.

EXAMPLE 8.2. DRAWING LEWIS STRUCTURES

PROBLEM: What is the Lewis structure for CCl_4 (carbon tetrachloride, a substance once used as a dry cleaning solvent until it was found to be very toxic)?

SOLUTION: First we need the arrangement of atoms. The formula suggests that it is

$$
\begin{array}{ccc}
 & Cl & \\
Cl & C & Cl \\
 & Cl &
\end{array}
$$

This arrangement of atoms, which only shows which atoms are bonded to each other, is sometimes called the *skeletal structure*.

Now we count the valence electrons

$$
\begin{array}{lr}
\text{carbon (Group IVA) contributes } 4e^- & 4e^- \\
\text{chlorine (Group VIIA) contributes } 7e^- \text{ each} & \underline{28e^-} \\
\text{Total} & 32e^-
\end{array}
$$

We begin distributing the electrons by placing a pair in each bond. This gives

$$
\begin{array}{ccc}
 & \overset{\cdot\cdot}{Cl} & \\
Cl & :C: & Cl \\
 & \overset{\cdot\cdot}{Cl} &
\end{array}
$$

This has used $8e^-$, so there are $32e^- - 8e^- = 24e^-$ left. We now complete the valence shells of the Cl atoms.

$$:\ddot{C}l: \quad \quad \quad :\ddot{C}l:$$
$$:\ddot{C}l:C:\ddot{C}l: \quad \text{or} \quad :\ddot{C}l-C-\ddot{C}l:$$
$$:\ddot{C}l: \quad \quad \quad :\ddot{C}l:$$

This has used all $24e^-$, so none are left. Each atom has an octet and, therefore, we can stop. This is the Lewis structure for CCl_4.

EXAMPLE 8.3. DRAWING LEWIS STRUCTURES

PROBLEM: What is the Lewis structure for SF_4?

SOLUTION: We begin by choosing an arrangement of atoms. Once again, the formula provides a clue to the skeletal structure

$$F$$
$$F \quad S \quad F$$
$$F$$

Next, we count valence electrons: 6 from sulfur and 7 from each fluorine give a total of $6 + 28 = 34$ electrons. As before, we start by placing a pair in each bond.

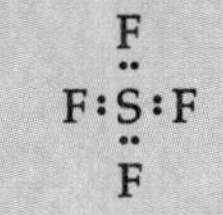

$$F$$
$$F:\ddot{S}:F$$
$$F$$

This leaves us with $34 - 8 = 26$ electrons. Next we complete the valence shells of fluorine

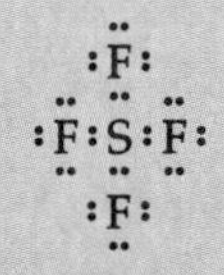

$$:\ddot{F}:$$
$$:\ddot{F}:\ddot{S}:\ddot{F}:$$
$$:\ddot{F}:$$

This has used $24e^-$, so there are still $2e^-$ left. Step 4 tells us to place them on the sulfur (the central atom). Rearranging the fluorines to make room for the extra dots on sulfur gives us the dot structure for SF_4.

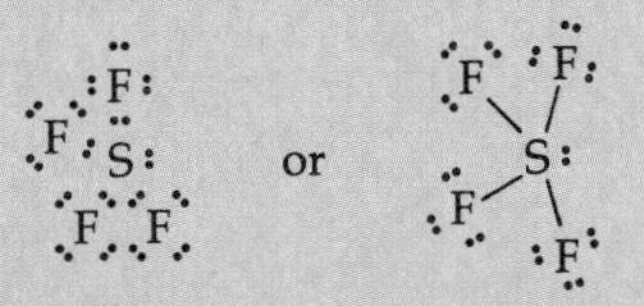

$$\text{or}$$

EXAMPLE 8.4. DRAWING LEWIS STRUCTURES

PROBLEM: What is the Lewis structure for the carbonate ion, $CO_3{}^{2-}$?

SOLUTION: This time we have an ion whose formula suggests the atom arrangement

$$O$$
$$O \quad C \quad O$$

Next we count valence electrons; carbon supplies $4e^-$, each oxygen supplies $6e^-$, and the negative charge adds another $2e^-$. The total is $4 + 18 + 2 = 24$ electrons. Placing a pair in each bond gives

$$\begin{matrix} & \overset{\textstyle O}{} & \\ O & : C : & O \end{matrix}$$

This has used $6e^-$, leaving us with $18e^-$. This are used to complete the octets around oxygen.

$$\begin{matrix} & : \overset{..}{\underset{..}{O}} : & \\ : \overset{..}{\underset{..}{O}} & : C : & \overset{..}{\underset{..}{O}} : \end{matrix}$$

This time we've run out of electrons, but carbon doesn't have an octet. What we must do now is move one of the unshared pairs on an oxygen into one of the bonds (it doesn't matter which one).[3]

$$\begin{matrix} & : \overset{..}{\underset{..}{O}} : & \\ : \overset{..}{\underset{..}{O}} : & C : & \overset{..}{\underset{..}{O}} : \end{matrix} \quad \text{gives} \quad \begin{matrix} & : \overset{..}{\underset{..}{O}} : & \\ : \overset{..}{\underset{..}{O}} :: & C : & \overset{..}{\underset{..}{O}} : \end{matrix}$$

Notice that the oxygen hasn't lost the pair, but carbon completes its octet by sharing it. Finally, we should be sure to indicate the charge on the ion by enclosing the Lewis structure in brackets with the charge outside.

By creating the double bond, we increase the number of electrons around the carbon without actually taking them away from the oxygen atom.

$$\left[\begin{matrix} & : \overset{..}{O} : & \\ & | & \\ : \overset{..}{O} = & C - & \overset{..}{O} : \end{matrix} \right]^{2-}$$

EXAMPLE 8.5. DRAWING LEWIS STRUCTURES

PROBLEM: Draw the electron-dot formula for the poisonous gas hydrogen cyanide, HCN (C is the central atom).

SOLUTION: There are 10 valence electrons to distribute (1 from H, 4 from C, 5 from N). First we place a pair in each bond.

$$H : C : N$$

This accounts for 4 electrons. As we add the remaining $6e^-$, we must keep in mind that the valence shell of H can hold only $2e^-$. No more electrons can be placed around H because it already has 2. Completing the octet of nitrogen gives

$$H : C : \overset{..}{\underset{..}{N}} :$$

---

[3] Applying this approach to drawing the Lewis structures of BCl_3 or $BeCl_2$ would give

$$\begin{matrix} Cl - B = Cl \\ | \\ Cl \end{matrix} \quad \text{and} \quad Cl = Be = Cl$$

However, in many of their chemical properties, these molecules *behave* as though they only contain single bonds, so the structures with double bonds are not acceptable, even though they satisfy the octet rule. As mentioned earlier, BCl_3 and $BeCl_2$ are exceptions to the octet rule and must be learned as special cases.

Once again the central atom doesn't have an octet. This can be corrected as follows:

$$H:C:N: \longrightarrow H:C:::N:$$

The HCN molecule contains a triple bond.

8.5 BOND ORDER AND SOME BOND PROPERTIES

At the beginning of Section 8.3, we described two features that characterize a covalent bond: bond length and bond energy. The bond length, you recall, is the distance between the nuclei of the two atoms joined by the bond. The bond energy is the energy needed to pull the atoms apart and produce neutral fragments. For a diatomic molecule such as H_2, this represents the process

$$H:H(g) \longrightarrow H\cdot(g) + H\cdot(g)$$

while in a molecule such as C_2H_6 the carbon–carbon bond energy represents the energy needed to cause the reaction

$$H-\underset{\underset{H}{|}}{\overset{\overset{H}{|}}{C}}-\underset{\underset{H}{|}}{\overset{\overset{H}{|}}{C}}-H(g) \longrightarrow H-\underset{\underset{H}{|}}{\overset{\overset{H}{|}}{C}}\cdot(g) + \cdot\underset{\underset{H}{|}}{\overset{\overset{H}{|}}{C}}-H(g)$$

It should not be surprising to learn that the magnitudes of the bond length and the bond energy differ for bonds between different atoms. Some bonds are strong and some are weak; some are long and some are short.

One factor that affects the bond length and the bond energy is the amount of electron density between the nuclei. A convenient way of expressing this is by giving the **bond order,** *the number of covalent bonds that exist between a pair of atoms.* Consider, for example, the following molecules:

$$H-\underset{\underset{H}{|}}{\overset{\overset{H}{|}}{C}}-\underset{\underset{H}{|}}{\overset{\overset{H}{|}}{C}}-H \qquad H-\overset{\overset{H}{|}}{C}=\overset{\overset{H}{|}}{C}-H \qquad H-C\equiv C-H$$

ethane   ethylene   acetylene

The carbon–carbon bond order in ethane is 1, in ethylene it is 2, and in acetylene it is 3.

As long as we are dealing with bonds between the same elements, we can relate bond length and bond energy to the bond order. As the bond order between a pair of atoms increases, additional electron density is placed between the two nuclei, which causes them to be pulled together. Therefore, *the bond length decreases as the bond order increases.* Increasing the bond order also makes it more difficult to pull the bonded atoms apart. Therefore, *the bond energy increases as the bond order increases.* Data that illustrate this are shown in Table 8.2.

Another bond property related to the bond order is the **vibrational frequency** of the atoms joined by the bond. The atoms within a molecule are not stationary; they are in constant motion. This motion can be resolved into two basic types: vibration in which a pair of atoms move toward and away from each other along a line joining their centers, much as two balls connected by a spring (Figure 8.3a); and

There are ways of experimentally determining bond lengths and bond energies.

Table 8.2. Variation of bond properties with bond order

Bond	Bond Order	Average Bond Length (pm)	Average Bond Energy (kJ/mol)	Average Vibrational Frequency (Hz)
C—C	1	154	348	3.0×10^{13}
C—C	2	137	607	4.9×10^{13}
C≡C	3	120	833	6.6×10^{13}
C—O	1	143	356	3.2×10^{13}
C=O	2	123	724	5.2×10^{13}
C—N	1	147	292	3.7×10^{13}
C≡N	3	116	879	6.8×10^{13}

bending in which the angle between the three atoms alternately increases and decreases (Figure 8.3b). For simplicity we will restrict our discussion to vibrational motion.

There are two factors that affect the frequency of vibration (i.e., the number of vibrations per second). One is the masses of the atoms bonded together and the other is the bond order. *For a given pair of atoms, as the bond order increases, so does the vibrational frequency.* This is because increasing the bond order increases the attractive forces holding the nuclei together, in effect, stiffening the "spring" between the two atoms.

Today the measurement of the vibrational frequencies of bonds is really quite simple. It happens that these vibrational frequencies are about the same as the frequency of infrared radiation, and when infrared light shines on a substance,

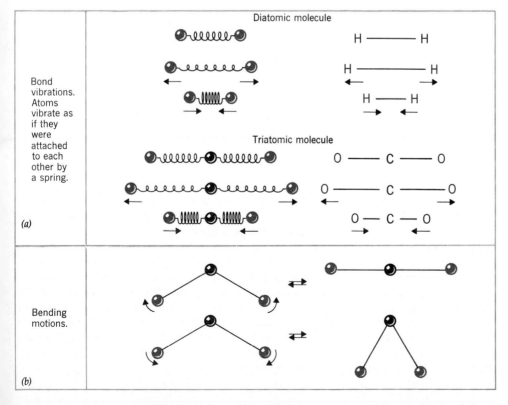

Figure 8.3. *The motions of atoms within molecules.*

radiation is absorbed if it has the same frequency as the vibrational frequency of a bond. By observing which frequencies are selectively removed from a continuous infrared spectrum, we can deduce these vibrational frequencies. The data recorded in the right hand column of Table 8.2 were obtained in this way.

In complex molecules there are many different vibrational modes available to the atoms and many different frequencies are absorbed from the infrared "rainbow." The infrared absorption spectra of most molecules are therefore quite complicated. Nevertheless, an experienced chemist often finds such absorption spectra extremely valuable as an aid in deducing molecular structure. In addition, each molecule, because of its unique structure, gives rise to its own characteristic absorption spectrum, which can be used to identify the compound and thus serves as a sort of fingerprint. Examples of infrared absorption spectra of some drugs are shown in Figure 8.4.

Virtually every modern chemistry laboratory has one or more instruments to measure and record infrared absorption spectra.

EXAMPLE 8.6. RELATING BOND PROPERTIES TO BOND ORDER

PROBLEM: Compare the carbon–oxygen bond lengths, bond energies, and bond vibrational frequencies in the molecules

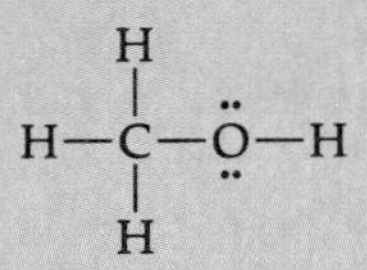

methyl alcohol 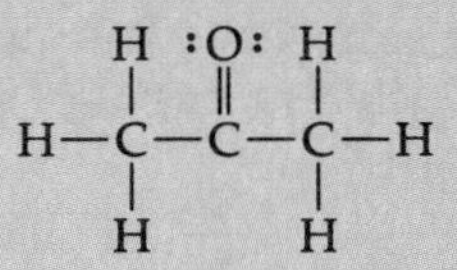   acetone

SOLUTION: The C—O bond order in methyl alcohol is 1, and in acetone it is 2. This means that compared to methyl alcohol, the C—O bond in acetone will be shorter, have a larger bond energy, and have a higher vibrational frequency.

8.6 RESONANCE

If we draw the Lewis structure for the NO_2^- ion following the rules given in Section 8.3, we obtain

$$\left[\ddot{O} - \ddot{N} = \ddot{O} \right]^-$$

Based on our discussion in the last section, we would expect the two nitrogen–oxygen bonds to be different. The Lewis structure suggests that one should be shorter, stronger, and have a larger vibrational frequency than the other. Surprisingly, however, all experimental evidence indicates that the two nitrogen–oxygen bonds are identical. In fact, it appears that they have properties that are about midway between what would be expected for a N—O single bond and a N—O double bond. Quite clearly, there is something wrong with the Lewis structure we've drawn. We might attempt to remedy the situation by dividing one of the pairs of electrons in the double bond, placing one in each bond, but this isn't satisfactory either because it would suggest that there are unpaired electrons in the ion, which there are not. Experimentally, it can be demonstrated that the NO_2^- ion is diamagnetic, which means that all the electrons are paired.

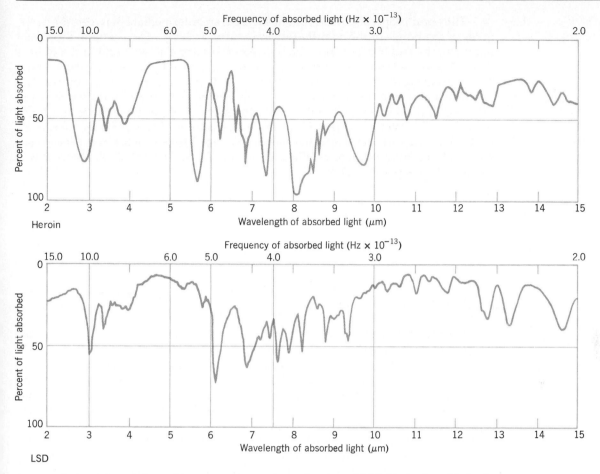

Figure 8.4. *Infrared absorption spectra for two drugs, heroin and LSD. Notice how distinctly different the spectra are.*

The way we get around this problem is as follows. When we draw the Lewis structure for NO_2^-, we reach this stage

$$\left[:\ddot{O}-\ddot{N}-\ddot{O}:\right]^-$$

Next, we realize that the nitrogen has less than an octet, so it is necessary to create a double bond in the structure. However, we have two choices — we can place it in the bond at the left, or in the bond at the right. Let's draw both structures

$$\left[:\ddot{O}=\ddot{N}-\ddot{O}:\right]^- \qquad \left[:\ddot{O}-\ddot{N}=\ddot{O}:\right]^-$$

Obviously, neither structure is satisfactory, but scientists have found that they can use both of these structures *together* to represent the actual structure of the ion. The solution to the problem is called **resonance.** We say that the actual structure of the ion does not correspond to either of the two structures we have drawn, but instead, to a structure somewhere in between that has the properties of both. This true structure, which we cannot actually draw with Lewis symbols, is said to be a **resonance hybrid,** and the two structures that we *can* draw are said to be **resonance structures** or **contributing structures.**

In the NO_2^- ion, there is nothing physically that makes any one N—O bond special. Nature, therefore, treats them equally by distributing the double bond equally between the two bonds.

Because the N—O bonds are the same, we say they are *equivalent*.

Fractional bond orders not only are possible but actually occur in many different compounds.

The bond properties of a resonance hybrid are intermediate between the properties of the bonds of the contributing structures. In NO_2^-, for example, the actual bond lengths and bond energies lie between those for a single bond and a double bond. A rough estimate of where they lie can be obtained by calculating the average N—O bond order. This is done by obtaining the total number of bonds between the nitrogen and one of the oxygens in all the resonance structures, and then dividing this total by the number of resonance structures. For example, if we choose the bond to the left of the nitrogen in our two resonance structures, we see there are two bonds in the first structure and one in the second. The total therefore is three. Since there are two contributing structures, the average bond order is $\frac{3}{2} = 1.5$. As noted, this is suggested by experimental evidence.

$$\left[:\ddot{O}=\ddot{N}-\ddot{O}:\right]^- \qquad \left[:\ddot{O}-\ddot{N}=\ddot{O}:\right]^-$$

two bonds ⌐_____⌐ one bond

$$\begin{array}{r} 3 \text{ bonds} \rightarrow \\ 2 \text{ structures} \rightarrow \end{array} \frac{3}{2} = 1.5 \text{ (average bond order)}$$

It is really quite unfortunate that the word resonance was used to describe this phenomenon, because the impression often is that the actual structure of a molecule or ion fluctuates back and forth between the contributing structures. This definitely is not so! The actual structure of the NO_2^- ion, for example, never corresponds to the one on the left and it never corresponds to the one on the right. The nitrite ion has only one, single electronic structure that is intermediate between the two contributing structures. Our difficulty is that we just can't draw it adequately with Lewis symbols. The problem is somewhat like attempting to describe the beast you would get if you were to crossbreed a donkey with a zebra. When you try to picture the offspring (a zonkey?), you visualize it with characteristics of both parents — perhaps with long ears, stripes, and stubborn. But you know that it wouldn't be a zebra one minute and a donkey the next.

Sometimes, we need more than two resonance structures to explain the structure of a molecule or ion. For instance, when we draw the Lewis structure of the nitrate ion, NO_3^-, we get to this point.

$$\left[\begin{array}{c} :\ddot{O}: \\ | \\ N \\ \diagup \quad \diagdown \\ :\ddot{O}. \qquad .\ddot{O}: \end{array}\right]^-$$

The nitrogen doesn't have an octet, so we need to create a double bond. However, this time there are three choices for its location, so we must draw three resonance structures

$$\left[\begin{array}{c} \ddot{O}: \\ \| \\ N \\ \diagup \quad \diagdown \\ :\ddot{O}. \quad .\ddot{O}: \end{array}\right]^- \longleftrightarrow \left[\begin{array}{c} :\ddot{O}: \\ | \\ N \\ \diagup \quad \diagdown \\ :\ddot{O} \quad .\ddot{O}: \end{array}\right]^- \longleftrightarrow \left[\begin{array}{c} :\ddot{O}: \\ | \\ N \\ \diagup \quad \diagdown \\ :\ddot{O}. \quad \ddot{O}: \end{array}\right]^-$$

The double-headed arrow here is used to indicate "resonance" among these three structures. In the true structure, we would expect an average bond order of $\frac{4}{3} = 1.33$ for each of the three N—O bonds.

Resonance is not restricted just to inorganic substances. For example, benzene (C_6H_6) consists of six carbon atoms in a ring. Each carbon is bonded to two other carbons and a hydrogen atom. Two Lewis structures can be drawn that satisfy the octet rule, as shown below.

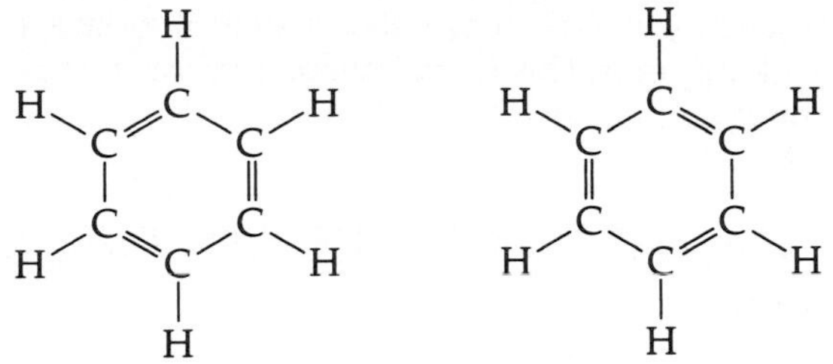

Notice that the only difference between the two structures is the placement of the double bonds. In the actual benzene molecule, all the carbon–carbon bonds are identical, so neither of these structures is acceptable individually. Instead, we view the molecule as a resonance hybrid of both with an average C—C bond order of 1.5.

The benzene molecule is often drawn as simply a hexagon with alternating double bonds, with the vertices assumed to contain carbon atoms, each attached to one hydrogen atom.

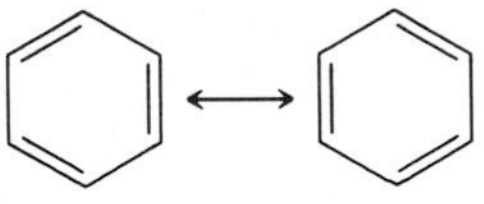

Modified Lewis structures for resonance hybrids

In each of the resonance hybrids we've described, resonance distributes additional electron density into each of the bonds involved. In the NO_2^- ion, for example, the equivalent of half a bond is added to each N—O "single bond." The same happens for each C—C bond in benzene. In the NO_3^- ion, one-third of a bond is added to each N—O "single bond." In an attempt to draw a single structure for these species, the fractional bonds are sometimes represented by broken dashes. Thus, for NO_2^- and NO_3^-, we have

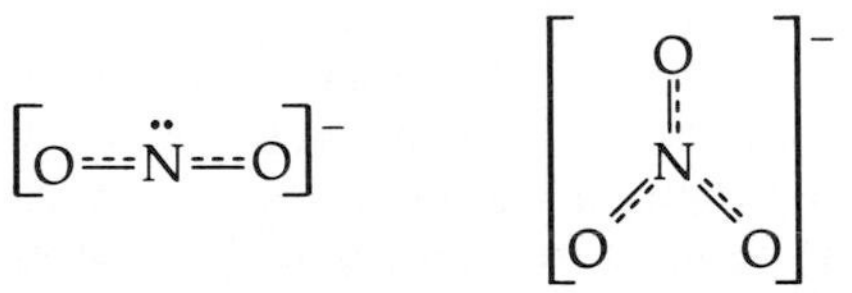

$$\left[O \cdots \ddot{N} \cdots O \right]^- \qquad \left[\begin{array}{c} O \\ \| \\ N \\ O \quad\quad O \end{array} \right]^-$$

In these structures, we do not attempt to indicate the locations of the unshared pairs of electrons. The purpose here is to show the equal distribution of electron density among the bonds.

The benzene molecule is a very important one among organic compounds and is often represented by the following symbol to indicate the uniform distribution of the "extra" bonds around the carbon ring:

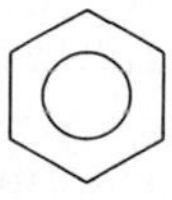

Bond energy and resonance

One of the interesting consequences of resonance is that molecules in which it occurs are more stable than might otherwise have been predicted, which suggests that bonds involving resonance are stronger than "ordinary" bonds. For example, it has been calculated that the standard heat of formation of one of the two resonance structures of C_6H_6 (really a hypothetical molecule) would be $+230$ kJ/mol. However, the measured standard heat of formation of benzene is $+84$ kJ/mol, which means that the actual molecule is more stable than any one of its contributing structures by about 146 kJ/mol. This extra lowering of the energy of the molecule is called **resonance energy.**

8.7 FORMAL CHARGE AND THE SELECTION OF LEWIS STRUCTURES

If we construct the Lewis structure for the sulfuric acid molecule, H_2SO_4, we obtain

$$
\begin{array}{c}
\ddot{\,}\ddot{\,} \\
:\ddot{O}: \\
| \\
H\!-\!\ddot{\underset{\cdot\cdot}{O}}\!-\!S\!-\!\ddot{\underset{\cdot\cdot}{O}}\!-\!H \\
| \\
:\ddot{O}: \\
\ddot{\,}\ddot{\,}
\end{array}
$$

It obeys the octet rule, and there doesn't seem to be any need to attempt to write any other structures for it. However, in sulfuric acid, it is found experimentally that the bonds between the sulfur and the oxygens not also bonded to hydrogen (those at the top and bottom in this structure) are shorter than we expect for S—O single bonds. Therefore, we need to consider Lewis structures that place additional electron density in these bonds.

Because sulfur is in period 3, its valence shell can expand beyond 8, and this allows us to write the following Lewis structure for H_2SO_4:

The valence shell of sulfur has $3s$, $3p$, and $3d$ subshells, which can hold up to 18 electrons all together.

$$
\begin{array}{c}
\ddot{\,}\ddot{\,} \\
O: \\
\| \\
H\!-\!\ddot{\underset{\cdot\cdot}{O}}\!-\!S\!-\!\ddot{\underset{\cdot\cdot}{O}}\!-\!H \\
\| \\
O: \\
\ddot{\,}\ddot{\,}
\end{array}
$$

The double bonds in the structure account for the shorter S—O bonds that are observed, and since this Lewis structure agrees better with the experimental evidence, it is the one that's preferred. The question is, however, are there any criteria that we can apply that would have allowed us to predict that this structure is better than the one with only single bonds, even though it violates the octet rule unnecessarily? To answer this question, let's take a closer look at the Lewis structures we've drawn.

In the first Lewis structure drawn for H_2SO_4, the sulfur has four single bonds to oxygen atoms. If the electrons in a bond are shared equally by S and O, then each atom "owns" half of the electron pair, or the equivalent of one electron. In other words, the four bonds place the equivalent of four electrons in the valence shell of the sulfur. A single atom of sulfur by itself, however, has six valence electrons. This means that in this Lewis structure, the sulfur has two electrons *fewer* than it does as just an isolated atom. Thus, at least in a bookkeeping sense, it would appear that if sulfur obeyed the octet rule in H_2SO_4, it would have a charge of $2+$. This apparent charge on the sulfur atom is called its **formal charge.**

Formal charges are a bookkeeping system that we apply to Lewis structures.

We can also calculate the formal charges on the hydrogen and oxygen atoms. Each hydrogen is held by one bond, which places one electron on it. An isolated H atom has just one electron, so the hydrogens have no formal charge (i.e., their formal charges are zero).

For the oxygens, we have two kinds to consider. Each oxygen that is also bonded to a hydrogen has two single bonds, which each contribute one electron to the O atom, as well as two unshared electron pairs that it owns exclusively. That comes to a total of 6 electrons, and since an isolated O atom has 6 valence electrons, these oxygens carry no formal charge. Now let's look at the oxygens *not* bonded to hydrogen. Each one has three unshared electron pairs and therefore owns these 6 electrons exclusively. Each also has a share of the electron pair in the bond to sulfur, and that contributes one more. This gives each of these oxygens a total of 7 electrons, which is one more than is found in the valence shell of an isolated atom, so the formal charge on each of these two oxygens is 1−.

Formal charges are indicated in a Lewis structure by placing them in circles alongside the atoms, as illustrated below

$$H-\overset{..}{\underset{..}{O}}-\overset{\overset{:\overset{..}{O}:^{\ominus}}{|}}{\underset{\underset{:\overset{..}{O}:_{\ominus}}{|}}{S^{2+}}}-\overset{..}{\underset{..}{O}}-H$$

The sum of the formal charges must equal the electrical charge on the particle.

Notice that the sum of the formal charges is zero, which it must be because the H_2SO_4 molecule is electrically neutral. As a rule, the sum of the formal charges in a Lewis structure must equal the actual electrical charge on the species.

In general, the formal charge on an atom in a Lewis structure is calculated according to the formula

$$\begin{pmatrix}\text{formal}\\\text{charge}\end{pmatrix}=\begin{pmatrix}\text{number of }e^-\text{ in valence}\\\text{shell of isolated atom}\end{pmatrix}-\begin{pmatrix}\text{number of bonds}\\\text{to the atom}\end{pmatrix}-\begin{pmatrix}\text{number of}\\\text{unshared }e^-\end{pmatrix}$$

Thus, for the sulfur in this Lewis structure

$$\text{formal charge}=6-4-0=+2$$

and for each oxygen not bonded to hydrogen

$$\text{formal charge}=6-1-6=-1$$

Now let's calculate the formal charges on the atoms in the second Lewis structure for H_2SO_4 (the one with the two S—O double bonds). The sulfur is now attached to other atoms by 6 bonds, so its formal charge is zero.

For sulfur

$$\text{formal charge}=6-6-0=0$$

For the oxygens not also bonded to hydrogen

$$\text{formal charge}=6-2-4=0$$

Keep in mind that formal charges do not necessarily equal the actual charges on the atoms in a molecule or ion.

Thus, in this second structure, none of the atoms carries either a positive or a negative formal charge. Now, let's ask the question, "How does the energy of this Lewis structure, with its double bonds, compare to the energy of the Lewis structure with all single bonds?" Imagine changing the one with the double bonds to the one with the single bonds. This involves creating two pairs of positive–negative charge from something electrically neutral. Stated another way, it involves separating negative charges from positive charges, and as you know by now, this involves an increase in the potential energy. Our conclusion, therefore, is that the singly bonded structure has a higher potential energy than the one with the double bonds. This

In general, the lower the energy of a structure, the more stable it is and the more likely it will be formed.

makes the structure with the double bonds preferable to the one with only single bonds, and it gives us a rule that we can use in selecting the best Lewis structures for a molecule or ion

> When several Lewis structures are possible, those with the smallest formal charges are the most stable and are preferred.

Let's look as some examples that illustrate the calculation of formal charges and the use of them in selecting Lewis structures.

EXAMPLE 8.7. CALCULATING FORMAL CHARGES

PROBLEM: Determine the formal charges on the atoms in the two structures shown below for the molecule $POCl_3$. Which of these two structures is the better structure (i.e., which is preferred)?

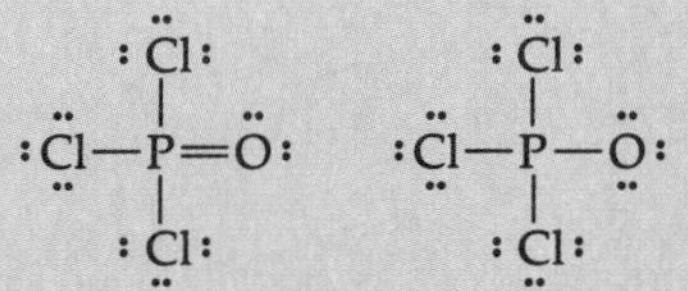

ANALYSIS: Calculating the formal charges is straightforward; we just apply the formula given above. Then, to choose the better structure, we look for the one with the smaller formal charges.

SOLUTION: We begin by calculating the formal charges. For the chlorines, which have 7 valence electrons as isolated neutral atoms, we have

$$\text{formal charge} = 7 - 1 - 6 = 0$$

Each chlorine has a formal charge of zero in both structures. Now let's examine the other atoms.

For structure I

Phosphorus: formal charge $= 5 - 5 - 0 = 0$

Oxygen: formal charge $= 6 - 2 - 4 = 0$

For structure II

Phosphorus: formal charge $= 5 - 4 - 0 = +1$

Oxygen: formal charge $= 6 - 1 - 6 = -1$

Placing these formal charges on the respective Lewis structures gives

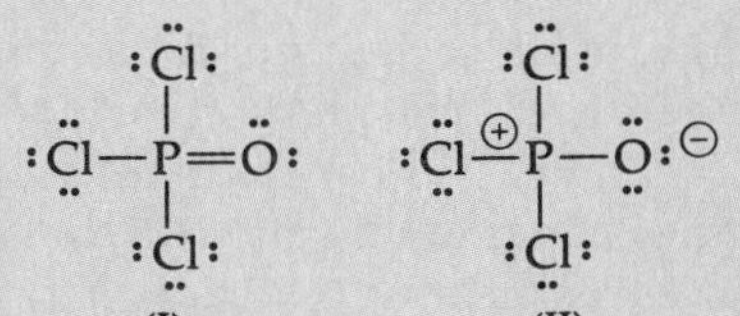

Finally, since structure I has the smallest formal charges, it is lower in energy and is the better Lewis structure for this molecule.

EXAMPLE 8.8. USING FORMAL CHARGES TO SELECT LEWIS STRUCTURES

PROBLEM: Select the best Lewis structure for chloric acid, $HClO_3$, in which the chlorine atom is surrounded by three oxygen atoms, one of which is also bonded to a hydrogen atom.

ANALYSIS: First, we have to construct a Lewis structure that obeys the octet rule, and this is done as described earlier. Next, we have to decide whether it is the most stable Lewis structure that can be drawn. To do this, we will calculate the formal charges on the atoms. If they are zero, we need look no further, but if they are nonzero, we will look for a way to reduce the formal charges.

SOLUTION: First, let's look at the Lewis structure obtained by following the rules given earlier. This is

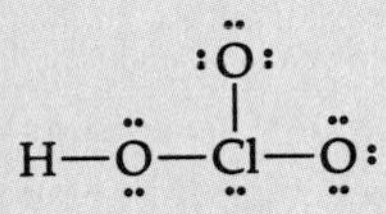

$$
\begin{array}{c}
:\ddot{O}: \\
| \\
H-\ddot{O}-Cl-\ddot{O}:
\end{array}
$$

Now let's assign formal charges to each of the atoms.

Hydrogen

$$\text{formal charge} = 1 - 1 - 0 = 0$$

Oxygen (bonded to H and Cl)

$$\text{formal charge} = 6 - 2 - 4 = 0$$

Other oxygens

$$\text{formal charge} = 6 - 1 - 6 = -1$$

Chlorine

$$\text{formal charge} = 7 - 3 - 2 = +2$$

Therefore, our Lewis structure looks like this:

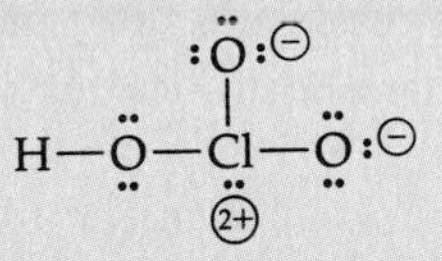

$$
\begin{array}{c}
:\ddot{O}:^{\ominus} \\
| \\
H-\ddot{O}-Cl-\ddot{O}:^{\ominus} \\
{}_{\textcircled{2+}}
\end{array}
$$

Now let's see whether we can reduce the number of formal charges in the structure by moving electrons around. If we move an unshared pair of electrons from one of the oxygens not bonded to the hydrogen into the Cl—O bond, thereby creating a double bond, this makes the formal charge on the oxygen less negative and that on the chlorine less positive. (In effect, it takes an electron from oxygen and moves it onto the chlorine.) Forming two such double bonds gives

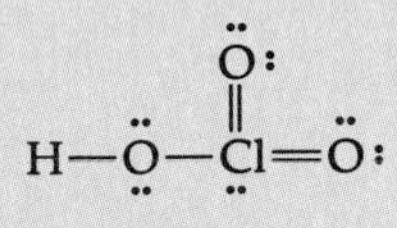

$$
\begin{array}{c}
\ddot{O}: \\
\| \\
H-\ddot{O}-Cl=\ddot{O}:
\end{array}
$$

Now there are no formal charges on the atoms, so we select it as the preferred structure.

EXAMPLE 8.9. USING FORMAL CHARGES TO SELECT RESONANCE STRUCTURES

PROBLEM: Select the best Lewis structure(s) for the sulfate ion, SO_4^{2-}.

SOLUTION: First, let's draw a Lewis structure that obeys the octet rule.

$$\begin{bmatrix} \ddot{:O}: \\ | \\ :\ddot{O}-S-\ddot{O}: \\ | \\ :\ddot{O}: \end{bmatrix}^{2-}$$

Now let's calculate formal charges.

For sulfur

$$\text{formal charge} = 6 - 4 - 0 = +2$$

For oxygen

$$\text{formal charge} = 6 - 1 - 6 = -1$$

Adding the formal charges to the Lewis structure gives

$$\begin{bmatrix} \ddot{:O}: \ominus \\ | \\ \ominus: \ddot{O} \underset{\textcircled{2+}}{-} S - \ddot{O}: \ominus \\ | \\ :\ddot{O}: \ominus \end{bmatrix}^{2-}$$

Next, we try to reduce these formal charges. As before, moving an unshared pair of electrons from oxygen to the S—O bond makes the charge on the oxygen less negative and the charge on the sulfur less positive. Since there are two positive formal charges on the sulfur, we need to do this twice to give sulfur a zero formal charge. The result is

All the formal charges cannot be reduced to zero in this formula because the sum of the charges must equal the charge on the sulfate ion, 2 −.

$$\begin{bmatrix} \ddot{:O}: \ominus \\ | \\ :\ddot{O}=S=\ddot{O}: \\ | \\ :\ddot{O}: \ominus \end{bmatrix}^{2-}$$

Notice, however, that in creating the double bonds, we had choices as to where to place them. There are six different arrangements for these double bonds, as shown below, so there are six resonance structures.

$$\begin{bmatrix} \ddot{:O}: \\ | \\ :\ddot{O}=S=\ddot{O}: \\ | \\ :\ddot{O}: \end{bmatrix}^{2-} \longleftrightarrow \begin{bmatrix} \ddot{O}: \\ \| \\ :\ddot{O}-S-\ddot{O}: \\ | \\ \ddot{O}: \end{bmatrix}^{2-} \longleftrightarrow \begin{bmatrix} \ddot{O}: \\ \| \\ :\ddot{O}-S=\ddot{O}: \\ | \\ :\ddot{O}: \end{bmatrix}^{2-} \longleftrightarrow$$

$$\begin{bmatrix} \ddot{:O}: \\ | \\ :\ddot{O}=S-\ddot{O}: \\ \| \\ \ddot{O}: \end{bmatrix}^{2-} \longleftrightarrow \begin{bmatrix} \ddot{O}: \\ | \\ :\ddot{O}=S-\ddot{O}: \\ | \\ :\ddot{O}: \end{bmatrix}^{2-} \longleftrightarrow \begin{bmatrix} \ddot{:O}: \\ | \\ :\ddot{O}-S=\ddot{O}: \\ \| \\ \ddot{O}: \end{bmatrix}^{2-}$$

Thus, we consider the sulfate ion to be a resonance hybrid of these six contributing structures.

8.8 COORDINATE COVALENT BONDS

When a nitrogen atom combines with three hydrogen atoms to form the molecule NH_3, the nitrogen atom has completed its octet. We might expect, therefore, that the maximum number of covalent bonds that we would observe a nitrogen atom to form is three. There are instances, however, when nitrogen may have more than three covalent bonds. In the ammonium ion, NH_4^+, which is formed in the reaction

$$
\begin{array}{c}
\text{H} \\
\text{H:}\overset{\displaystyle\cdot\cdot}{\text{N}}\text{:} + \text{H}^+ \\
\text{H}
\end{array}
\longrightarrow
\left[
\begin{array}{c}
\text{H} \\
\text{H:}\overset{\displaystyle\cdot\cdot}{\text{N}}\text{:H} \\
\text{H}
\end{array}
\right]^+
$$

the nitrogen is covalently bound to four hydrogen atoms. When the additional bond between the H^+ and the N atom is created, both of the electrons in the bond come from the nitrogen. *This type of bond, in which a pair of electrons from one atom is shared by two atoms, is called a* **coordinate covalent bond.** It is important that you remember that the coordinate covalent bond is really no different, once formed, than any other covalent bond and that our distinction is primarily aimed at keeping track of electrons; that is, it is "bookkeeping."

When Lewis structures are written using dashes to represent electron pairs, the coordinate covalent bond is sometimes indicated by means of an arrow pointing away from the atom supplying the electron pair. For example, the product of the reaction of boron trichloride, BCl_3, and ammonia, NH_3, is a substance known as an **addition compound** because it is formed by simply adding one molecule to another.

$$
\begin{array}{c}
\text{H} \quad \text{Cl} \\
| \qquad | \\
\text{H—N: + B—Cl} \\
| \qquad | \\
\text{H} \quad \text{Cl}
\end{array}
\longrightarrow
\begin{array}{c}
\text{H} \; \text{Cl} \\
| \;\; | \\
\text{H—N:B—Cl} \\
| \;\; | \\
\text{H} \; \text{Cl}
\end{array}
$$

To show that the electron pair shared between the B and N originates on the nitrogen, the Lewis structure of this addition compound can be written

$$
\begin{array}{c}
\text{H} \quad \text{Cl} \\
| \qquad | \\
\text{H—N}\rightarrow\text{B—Cl} \\
| \qquad | \\
\text{H} \quad \text{Cl}
\end{array}
$$

Using this type of notation, we are tempted to write the structure of the NH_4^+ ion as

$$
\left[
\begin{array}{c}
\text{H} \\
| \\
\text{H—N}\rightarrow\text{H} \\
| \\
\text{H}
\end{array}
\right]^+
$$

This gives the impression that one of the N—H bonds is different from the other three. It has been shown experimentally, however, that all four N—H bonds are identical. Therefore, to avoid conveying false impressions, we write the NH_4^+ ion simply as

$$
\left[
\begin{array}{c}
\text{H} \\
| \\
\text{H—N—H} \\
| \\
\text{H}
\end{array}
\right]^+
$$

Remember, once the bond is formed, it doesn't matter where the electrons in it came from.

8.9 POLAR MOLECULES AND ELECTRONEGATIVITY

When we realize that each element has a different nuclear charge and electron configuration, it is not unreasonable to expect that atoms of different elements have different abilities to attract electrons when they form chemical bonds. **Electronegativity** *is the term used to describe an atom's attraction for the electrons in a bond.* It is important not to confuse this term with electron affinity, which is an energy term and refers to an isolated atom.

When two identical atoms combine, as for instance in H_2, both have the same electronegativity. Because each atom is equally capable of attracting the electron pair in the bond, the pair will be distributed equally over both atoms. This means that each atom has half of the electron pair around it, which is equivalent to one electron. Therefore, the electron the atom has donated to the covalent bond has not been lost at all and the atom carries a net charge of zero both before and after the bond has been formed.

If the electronegativities of the two atoms joined by a bond are different, the electron pair will be pulled more toward the atom with the higher electronegativity. For example, consider the HCl molecule. It happens that chlorine is more electronegative than hydrogen, so the pair of electrons in the HCl bond are not shared equally by the two atoms. Instead, more than half of the electron density of the bond pair becomes concentrated around the chlorine. As a result, the chlorine end of the molecule becomes partially negative, while the hydrogen end becomes partially positive. We indicate this as follows for the HCl molecule,

$$\overset{\delta+}{H}\!-\!\overset{\delta-}{\underset{..}{\overset{..}{Cl}}}\!:$$

δ is the lowercase Greek letter delta.

where $\delta+$ and $\delta-$ are meant to represent the partial positive and negative charges. These symbols are used because the charges are not full $1+$ and $1-$ charges; electron transfer from H to Cl is not complete, and HCl is far from being ionic. (In fact, measurements suggest that these charges are really only about $+0.17$ on the hydrogen and -0.17 on the chlorine.)

Within a molecule, equal positive and negative charges separated by a distance constitute a **dipole**. Therefore, the HCl molecule, with its centers of positive and negative charge, is a dipole and is said to be **polar**. In fact, any diatomic molecule (a molecule formed from two atoms) formed from two elements of different electronegativity will be polar.

A dipole is defined quantitatively by its **dipole moment,** the product of the charge on either end of the dipole times the distance between the charges. A very polar molecule is one with a large dipole moment, while a nonpolar molecule will have no dipole moment at all.

When three or more atoms are bonded together, it is possible to have a nonpolar molecule even though there are polar bonds present. Carbon dioxide is an example. The CO_2 molecule is linear and may be represented as

$$:\!\overset{..}{\underset{\delta-}{O}}\!=\!\overset{\delta+}{\underset{\delta+}{C}}\!=\!\overset{\delta-}{\underset{..}{O}}\!:$$

showing that oxygen is more electronegative than carbon.

A polar bond is a bond dipole.

The overall dipole moment of a molecule arises as a sum of the individual **bond dipoles** within the molecule, which add together like vectors. In CO_2 these bond dipoles are oriented in opposite directions, and they exactly cancel each other.

$$:\!\overset{..}{O}\!=\!C\!=\!\overset{..}{\underset{..}{O}}\!:$$
$$\longleftarrow \quad \longrightarrow$$

(An arrow with a plus sign on one end is used to represent the bond dipole.)

In the water molecule, which happens to have a bent or angular shape, the two bond dipoles do not cancel each other entirely, but rather are partially additive. As a result, the H_2O molecule does have a net dipole moment (indicated by the heavy arrow) and is polar.

We will have more to say about the effect of molecular structure on the polarity of molecules in the next chapter.

We would like to have some quantitative measure of electronegativity so that we can make predictions concerning the polarity of bonds. One approach toward this, taken by R. S. Mulliken in 1934, uses the average of the ionization energy and electron affinity. A very electronegative element has a very high ionization energy, so it is difficult to remove its electrons. It also has a very high electron affinity, so a very stable species results when electrons are added. On the other hand, an element of low electronegativity will have a low ionization energy and low electron affinity so that it loses electrons readily and has little tendency to pick them up. Unfortunately, it is very difficult to measure the electron affinity of an element. Therefore, this method of assigning electronegativities is not universally applicable.

The most widely used scale of electronegativities is based on one developed by Linus Pauling. He observed that when atoms of different electronegativities are combined, their bonds are stronger than expected. Presumably two factors contribute to the bond strength. One of them is the covalent bonding between the atoms. The other is an additional binding produced by an attraction between the oppositely charged ends of the bond dipole. The extra bond strength, then, was attributed to this additional binding and Pauling used this concept to develop his table of electronegativities. Figure 8.5 is a table of the electronegativities of the elements that are quite close to Pauling's original values.

When we are interested in knowing something about the polarity of a bond, we are mainly interested in the *difference* in electronegativity between the two atoms joined by the bond. If this difference is small, the bond will be relatively nonpolar, but if it is large, the bond will be polar. And if the difference in electronegativity is very large, the electron pair will be concentrated almost exclusively around the more

Elements with low electronegativities are often said to be **electropositive** because they tend to give away their electron rather easily and acquire partial positive charges.

Linus Pauling is an American chemist. He won the 1954 Nobel prize in chemistry for his work on the nature of matter and the 1962 Nobel peace prize for his efforts to ban nuclear testing worldwide. He has become known for advocating large doses of vitamin C to ward off the common cold.

H 2.1																
Li 1.0	Be 1.5											B 2.0	C 2.5	N 3.1	O 3.5	F 4.1
Na 1.0	Mg 1.3											Al 1.5	Si 1.8	P 2.1	S 2.5	Cl 2.9
K 0.9	Ca 1.1	Sc 1.2	Ti 1.3	V 1.5	Cr 1.6	Mn 1.6	Fe 1.7	Co 1.7	Ni 1.8	Cu 1.8	Zn 1.7	Ga 1.8	Ge 2.0	As 2.2	Se 2.4	Br 2.8
Rb 0.9	Sr 1.0	Y 1.1	Zr 1.2	Nb 1.3	Mo 1.3	Tc 1.4	Ru 1.4	Rh 1.5	Pd 1.4	Ag 1.4	Cd 1.5	In 1.5	Sn 1.7	Sb 1.8	Te 2.0	I 2.2
Cs 0.9	Ba 0.9	La 1.1	Hf 1.2	Ta 1.4	W 1.4	Re 1.5	Os 1.5	Ir 1.6	Pt 1.5	Au 1.4	Hg 1.5	Tl 1.5	Pb 1.6	Bi 1.7	Po 1.8	At 2.0
Fr 0.9	Ra 0.9	Ac 1.0	Lanthanides: 1.0–1.2 Actinides: 1.0–1.2													

Figure 8.5. *A table of electronegativities based on Pauling's original electronegativity scale.*

electronegative atom and the bond will be essentially ionic. We see, therefore, that the degree of ionic character of the bond, as measured by the amount of charge carried by the atoms at each end, can vary from zero (in H_2, for example) to essentially 100%, depending on the electronegativities of the bonded atoms. There really is no sharp dividing line between ionic and covalent substances, and even in compounds that we think of as being ionic, such as NaCl, there is still some degree of covalent character to the bonds between the atoms. As a very rough guide, bonds become more than 50% ionic when the electronegativity difference between the atoms is larger than about 1.7, and we normally think of substances with these bonds as being ionic compounds.

If the bonds are more ionic than covalent, we tend to think of the substance as an ionic compound.

Finally, electronegativity trends within the periodic table are worth noting. We see that the most electronegative elements are located in the upper right portion of the table; the least electronegative are found in the lower left. This is consistent with the trends in ionization energy (IE) and electron affinity (EA) discussed in Chapter 7, where we saw that elements with the largest IE and EA are in the upper right region of the periodic table and those with the smallest IE and EA are in the lower left. It is also consistent with our observations that atoms from opposite ends of the periodic table — lithium and fluorine, for example — form bonds that are essentially 100% ionic, and that atoms such as carbon and oxygen form covalent bonds that are only somewhat polar.

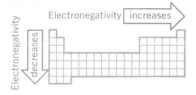

REVIEW QUESTIONS AND PROBLEMS

Problems whose numbers are in blue have their answers in Appendix D at the back of the book. The more difficult problems are marked with asterisks.

Ionic Bonding

8.1 What is an *ionic bond*?

8.2 Write the electron configurations of the following ions: (a) Ba^{2+}, (b) Se^{2-}, (c) Al^{3+}, (d) Na^+, and (e) Br^-.

8.3 Which noble gases have the same electron configurations as the ions in the preceding question?

8.4 In which of the following compounds does the cation and anion have *exactly* the same electron configuration?
(a) NaCl   (c) CaO   (e) $SnBr_2$
(b) Na_2O (d) Mg_3N_2 (f) KBr

8.5 Why doesn't aluminum form the compound $AlCl_4$? Why doesn't oxygen form the compound Na_3O?

8.6 What is the meaning of the term *lattice energy*? In what way is the lattice energy involved in determining the stability of an ionic compound?

8.7 Use the tables of ionization energy and electron affinity on pages 223 and 225 to calculate the energy changes for the following reactions:
(a) $Na(g) + Cl(g) \rightarrow Na^+(g) + Cl^-(g)$
(b) $Na(g) + 2Cl(g) \rightarrow Na^{2+}(g) + 2Cl^-(g)$
Approximately how many times larger would the lattice energy of $NaCl_2$ have to be compared to the lattice energy of NaCl if $NaCl_2$ is to be more stable than NaCl?

8.8 What is the octet rule? What is its origin? Give an example of one ionic compound that obeys the octet rule and one that does not.

8.9 Does the compound LiF obey the octet rule?

8.10 Why are many of the transition elements able to form more than one positive ion?

8.11 What is the principal reason that KF(s) is more stable than K(s) and $F_2(g)$?

8.12 Give the electron configuration of the following ions: Pb^{2+}, Pb^{4+}, Mn^{2+}, Mn^{3+}, Sb^{3+}, Sc^{3+}, Ti^{2+}.

8.13 Give the electron configurations of these ions: (a) Zn^{2+}, (b) Sn^{2+}, (c) Bi^{3+}, (d) Cr^{2+}, (e) Fe^{3+}, and (f) Ag^+.

8.14 What is the *pseudonoble-gas* configuration? Give *three* ions that have such a configuration in their outer shells.

8.15 Which of the following pairs of ions have *exactly* the same electron configurations: (a) K^+ and Cl^-, (b) Na^+ and Br^-, (c) Li^+ and F^-, (d) Sr^{2+} and Br^-, (e) Mg^{2+} and C^{4-}?

8.16 Indicate how the electron configuration changes for each atom when the following ionic compounds are formed from the elements: K_2O, Mg_3N_2, Na_2S, and $BaBr_2$.

8.17 On the basis of their electron configurations, why do many of the transition elements form ions with a charge of $2+$?

Lewis Symbols

8.18 What is the purpose of Lewis symbols?

8.19 Write Lewis symbols for (a) selenium (b) bromine, (c) aluminum, (d) barium, (e) germanium, and (f) phosphorus.

8.20 Why do all the Lewis symbols of the elements in a given group have the same number of dots?

8.21 Use Lewis symbols and the octet rule to explain why the formula for the ionic compound formed from Mg and C is Mg_2C.

8.22 Draw Lewis structures for the following ionic compounds: (a) BaO, (b) Na_2O, (c) KF, (d) CaS, and (e) Li_3N.

8.23 (a) How many unpaired electrons does a carbon atom have in its ground state?
(b) Draw the Lewis symbol for carbon. How many unpaired electrons does it suggest?
(c) Why is the discrepancy between the answers to parts (a) and (b) allowed?

Covalent Bonding

8.24 What is the difference between an ionic bond and a covalent bond?

8.25 Fluorine and chlorine atoms can combine to form the covalently bonded molecule ClF. Why, energetically, is ionic bonding *not* favored for ClF?

8.26 How many hydrogen atoms would be expected to bond covalently to each of the following atoms: (a) Ge, (b) S, (c) Br, (d) Si, (e) P?

8.27 Use Lewis symbols to diagram the reaction for the formation of the covalently bonded molecules NH_3, H_2O, and HF from their atoms.

8.28 What is a double bond? What is a triple bond?

8.29 How many unpaired electrons are there in a molecule of (a) NH_3, (b) H_2O, (c) CH_4, (d) HCl?

8.30 Define *bond energy* and *bond length*.

8.31 Write the Lewis structures of $BeCl_2$ and BCl_3. In each molecule, how many electrons are in the valence shell of the central atom?

8.32 Why do elements from period 2 never exceed an octet in their valence shells?

8.33 How many orbitals are in the valence shell of (a) carbon, (b) nitrogen, (c) phosphorus, (d) chlorine?

8.34 What is the maximum number of covalent bonds that can be formed by an atom of nitrogen? Could a phosphorus atom form more covalent bonds than a nitrogen atom? Explain your answer.

Drawing Lewis Structures

8.35 Which atoms are bonded to which other atoms in the following molecules or ions? (Sketch the *skeletal structure* of each of them.) (a) NH_3, (b) PCl_3, (c) SCl_2, (d) NO_2^-, (e) BrF_5, (f) PCl_4^+.

8.36 How many valence electrons are there in an atom of (a) arsenic, (b) iodine, (c) silicon, (d) tin, and (e) sulfur?

8.37 How many valence electrons are in each of the species in Question 8.35?

8.38 Why, in general, do we not choose hydrogen as the central atom in a molecule?

8.39 What is the maximum number of bonds formed by a hydrogen atom?

8.40 Draw Lewis structures that obey the octet rule for the molecules PCl_3, SiH_4, BCl_3, H_2S, C_3H_8, and CO.

8.41 Draw Lewis structures that obey the octet rule for the molecules Cl_2, SO_2, OF_2, SnH_4, C_2H_4, and SCl_2.

8.42 Draw Lewis structures that obey the octet rule for the ions Cl^-, S^{2-}, ClO^-, ClO_4^-, SO_3^{2-}, and SeO_4^{2-}.

8.43 Draw Lewis structures that obey the octet rule for the ions NO_3^-, NO^+, NO_2^-, and CO_3^{2-}.

8.44 Draw Lewis structures for SeF_6, SeF_4, ICl_3, $AsCl_5$, ICl_2^-, ICl_4^-, and XeF_4.

8.45 Which of the following compounds do not obey the octet rule: ClF_3, OF_2, SF_4, SO_2, IF_7, NO_2, BCl_3?

Bond Order and Bond Properties

8.46 Define *bond order*.

8.47 What are the various bond orders in the following molecules?
(a) CCl_4
(b) HCN (See Example 8.5.)
(c) CO_2
(d) NO^+
(e)

$$H-\overset{\overset{\displaystyle H}{|}}{\underset{\underset{\displaystyle H}{|}}{C}}-\overset{..}{N}=C=\overset{..}{\overset{..}{O}}:$$

(Methyl isocyanate, the chemical that caused the tragedy in Bhopal, India, in December 1984.)

8.48 How does bond energy vary with bond order? Why?

8.49 How does bond length vary with bond order? Why?

8.50 Describe the bending and vibrational motions within the SO_2 molecule.

8.51 How does the bond vibrational frequency vary with bond order? Why? How are bond vibrational frequencies measured?

8.52 The C—C bond length in a series of compounds was found to be as follows: compound 1, 154 pm; compound 2, 137 pm; compound 3, 146 pm; and compound 4, 140 pm. Arrange these in order of increasing C—C bond order. How would you expect the C—C bond energies to vary?

Resonance

8.53 What is a resonance hybrid? Why is the concept of resonance used?

8.54 Draw resonance structures that obey the octet rule for SO_3, NO_3^- and CO_3^{2-}; SO_2 and NO_2^-.

8.55 Draw resonance structures that obey the octet rule for HNO_3, SeO_2, SeO_3, and N_2O_4.

Skeletal structures

8.56 Draw resonance structures for $C_2O_4^{2-}$, CH_3COO^-, and N_3^-.

Skeletal structures

8.57 How would you expect the N—O bond length in NO_2^- to compare with that in NO_3^-?

8.58 Compare the C—O bond properties—bond order, bond energy, bond length, and vibrational frequencies—in the following. (*Hint:* Draw resonance structures when necessary.)

8.59 Draw resonance structures that obey the octet rule for SO_2 and SO_3. How would you expect their S—O bond lengths, bond energies, and bond vibrational frequencies to compare?

8.60 Draw the two resonance structures for benzene. Draw the single structure that is usually used to represent the resonance hybrid.

8.61 Draw single structures to represent the resonance hybrids of SO_2 and SO_3.

8.62 For a molecule that exists as a hybrid of two or more resonance structures, how does the energy of the actual molecule compare to the energy we would calculate for any of the individual resonance structures?

8.63 What is meant by the term *resonance energy*?

Formal Charge

8.64 Give the formula used to calculate the formal charge on an atom in a Lewis structure.

8.65 What is the sum of the formal charges in the ion $S_2O_3^{2-}$ in which one of the sulfur atoms is in the center surrounded by three oxygens and the other sulfur?

8.66 Draw Lewis structures that obey the octet rule for the following ions and assign formal charges to the atoms in the structures:
(a) ClO_3^-
(c) PO_4^{3-}
(e) ClO_4^-
(b) BrO_2^-
(d) SeO_3^{2-}

8.67 Draw Lewis structures that obey the octet rule for the following molecules and assign formal charges to the atoms in the structures: (a) $HClO_4$, (b) $HClO_2$, (c) H_2SO_3, (d) H_3PO_4.

8.68 For each of the ions in Question 8.66, select a better Lewis structure based on the assignment of formal charges. Where necessary, indicate resonance structures.

8.69 For each of the molecules in Question 8.67, select a better Lewis structure based on the assignment of formal charges. Where necessary, indicate resonance structures.

8.70 A student proposed the following two structures for the nitric acid molecule and suggested that structure 2 is preferred because of the assignment of formal charges. Why must structure 2 be rejected, however?

8.71 Draw resonance structures for the azide ion N_3^-, which has the skeletal structure $N \cdots N \cdots N$. Use formal charges to identify the highest and lowest energy structures.

8.72 Why is the Lewis structure for CO_2 usually written as $\ddot{O}=C=\ddot{O}$ rather than $:O\equiv C-\ddot{O}:$?

8.73 Why do we want to reject the following structure for nitric acid as unacceptable compared to structure 1 in Question 8.70?

8.74 Fluorous acid, HFO_2, has not been isolated. If its skeletal structure must be $H \cdots O \cdots F \cdots O$, draw the Lewis structure for the molecule and explain why it is a high-energy structure, based on electronegativities and formal charges, and why it is therefore not surprising that HFO_2 cannot be made.

8.75 What are the bond orders of the bonds in the ions whose structures you chose in Question 8.68?

8.76 What are the average bond orders of the P—O bonds in the acid H_3PO_4? [Be sure to choose the best Lewis structure(s) based on the assignment of formal charges.]

8.77 In the compound SO_2Cl_2, the sulfur is surrounded by two chlorine atoms and two oxygen atoms. Draw the best Lewis structure for this substance.

Coordinate Covalent Bonds

8.78 What is a coordinate covalent bond? How does it differ from other covalent bonds?

8.79 Use Lewis symbols to show the formation of a coordinate covalent bond in the reaction

$$AlCl_3 + Cl^- \longrightarrow AlCl_4^-$$

8.80 The molecule CH_4 would not be expected to participate in the formation of a coordinate covalent bond. Why?

8.81 The BF_3 molecule reacts with F^- to form BF_4^-. Use Lewis symbols to explain this reaction in terms of coordinate covalent bonding.

Electronegativity

8.82 Define electronegativity. What is the difference between electronegativity and electron affinity?

8.83 Define *polar*, *dipole*, and *dipole moment*.

8.84 What trends in electronegativity occur in the periodic table? What correlation, if any, exists between ionization energy and electronegativity?

8.85 Use the periodic table to decide which of the following bonds should be more polar:
(a) P—F or P—O
(b) Al—S or Al—Cl
(c) Se—Cl or Se—Br

8.86 Which of the following contain bonds that are more than 50% ionic: $AlCl_3$, MgO, Al_2O_3, NF_3, CsF, $FeCl_2$, SO_2, Ca_3P_2, Mg_2Si? (Use the data in Figure 8.5.)

8.87 Which of the following have bonds that are less than 50% ionic: NH_3, MnF_2, BCl_3, $MgCl_2$, BeI_2, NaH? (Use the data in Figure 8.5.)

8.88 Arrange the following compounds in order of increasing ionic character of their bonds: SO_2, H_2S, SF_2, OF_2, ClF_3, H_2Se, F_2. (Use the data in Figure 8.5.)

8.89 Which one of the following is most electropositive: Al, Rb, Mg, N?

8.90 The O—S—O bond angle in SO_2 is less than 180°. Explain why this is a polar molecule. Which end of the molecule carries the positive charge?

8.91 The XeF_4 molecule has a square planar structure. The fluorine atoms are located at the corners of a square, with the Xe located in the center of the square. Are the bonds polar? Is the molecule polar?

Additional Exercises

8.92 Given the following thermochemical equations:

$$K(s) \longrightarrow K(g) \qquad \Delta H = +90.0 \text{ kJ}$$
$$\tfrac{1}{2}Cl_2(g) \longrightarrow Cl^-(g) \qquad \Delta H = +119 \text{ kJ}$$

$$K(g) \longrightarrow K^+(g) + e^- \qquad \Delta H = +419 \text{ kJ}$$
$$e^- + Cl(g) \longrightarrow Cl^-(g) \qquad \Delta H = -348 \text{ kJ}$$
$$K^+(g) + Cl^-(g) \longrightarrow KCl(s) \qquad \Delta H = -704 \text{ kJ}$$

use Hess's law to calculate ΔH for the reaction

$$K(s) + \tfrac{1}{2}Cl_2(g) \longrightarrow KCl(s)$$

Which of the energy terms permits the formation of KCl to be exothermic overall?

* 8.93 The lattice energy of an ionic compound can be defined quantitatively as the energy evolved when the isolated gaseous ions are brought together to give the ionic solid. Use Hess's law and the information below to calculate the lattice energy for $CaCl_2$ in units of kJ/mol of $CaCl_2$ formed. Energy needed to vaporize 1 mol of $Ca(s)$ = 192 kJ; first ionization energy of Ca = 589.5 kJ/mol; second ionization energy of Ca = 1146 kJ/mol; electron affinity of Cl = −348 kJ/mol; bond energy of Cl_2 = 238 kJ/mol of Cl—Cl bonds; energy change for the reaction $Ca(s) + Cl_2(g) \rightarrow CaCl_2(s)$, −795 kJ/mol of $CaCl_2(s)$ formed.

* 8.94 Given the following data use Hess's law to calculate the electron affinity of Br. The energy change for the reaction $Na(s) + \tfrac{1}{2}Br_2(l) \rightarrow NaBr(s)$ is −360 kJ. The energy needed to vaporize 1 mol of $Na(s)$ is 109 kJ. The energy needed to vaporize 1 mol of $Br_2(l)$ is 31 kJ. The ionization energy of $Na(g)$ is 495.8 kJ/mol. The bond energy of Br_2 is 192 kJ/mol of Br—Br bonds. The lattice energy of NaBr is −734.3 kJ/mol.

* 8.95 Below are calculated and experimental bond energies for the hydrogen halides. Use these data to show that the electronegativities of the halogens decrease from F to I.

Calculated and experimental bond energies

	Calculated (kJ/mol)	Experimental (kJ/mol)
HF	295	565
HCl	337	431
HBr	310	360
HI	290	300

* 8.96 Hydrogen is more electronegative than any of the Group IA elements. Based on this statement and the data below, show that the electronegativities of the alkali metals decrease from Li to Rb.

Calculated and experimental bond energies

	Calculated (kJ/mol)	Experimental (kJ/mol)
Li—H	272	238
Na—H	256	200
K—H	244	180
Rb—H	242	160

COVALENT BONDING AND MOLECULAR STRUCTURE

In Chapter 8 we found that chemical bonds can be broadly classified into two main categories: ionic bonds and covalent bonds. The ionic bond arises as a purely electrostatic attraction between oppositely charged particles and is therefore nondirectional. This means that the arrangement of ions in a cluster is determined simply by the balancing of attractive and repulsive forces between the ions, not by their electronic structures. The covalent bond, on the other hand, has very definite directional properties. Covalently bound substances, such as molecules or polyatomic ions, have characteristic shapes that are usually retained when these substances undergo physical changes such as melting or vaporization.

The *shapes* of molecules, which means the way their atoms are arranged in space, affect many of their physical and chemical properties. In Chapter 8, for example, we learned that molecular shape can determine whether or not a molecule is polar — a phenomenon that will be explored further in this chapter. Later you will learn that molecular polarity has a very strong influence on such physical properties as melting point and boiling point. Molecular shape can also affect chemical properties. In biological systems, such as our own bodies, the chemical reactions that keep us alive (and even allow us to study chemistry) depend on the very precise fitting together of molecules. If this fit is destroyed, which generally is what happens in cases of poisoning, the organism dies. Thus, an understanding of molecular geometry and the factors that affect it are critical to our understanding of chemistry.

So far, the simple picture of a covalent bond as a pair of dots shared between two atoms has given us no information about molecular structure. In this chapter we will first see how Lewis structures can be used to predict molecular shapes with a surprisingly high degree of accuracy. Then we will examine how modern theories based on quantum mechanics attempt to answer the "why" and "how" questions — *why* molecules have the shapes they do, and *how* atoms are able to share electrons with each other. As you will see, we will examine some rather different approaches to understanding molecular structure, and you should keep in mind that none of the theories about bonding is perfect. Otherwise we would only have to consider one of them. Each theory has its usefulness for particular purposes, and the model a chemist chooses in a particular circumstance depends on what aspect of bonding and structure is being examined.

9.1 MOLECULAR SHAPES

Although the number of different molecules is enormous, the number of different ways in which atoms arrange themselves around one another is rather limited. Because of this, understanding and describing molecular shapes are not as complicated as they might otherwise be.

Most molecules have shapes that can be considered to be derived from a basic set of just five different geometries. Our goal in this section is to make you familiar with them so that you can visualize the structures of molecules in three dimensions. You should learn how these molecular shapes are characterized by the various angles between bonds, and even though you may find it difficult at first, you should begin to attempt to sketch the structures. (For some of the more complex ones, some instruction is provided.) Practice until you are able to sketch the shapes so that they convey three-dimensional information to you. If you do this, you will find the study of molecular structures much easier and much more interesting. Let's look now at these five basic structures, starting with the simplest one.

The names for some of these structures may be unfamiliar to you. Nevertheless, you should learn to associate the name with the three-dimensional figures they represent.

Linear molecules

An arrangement of atoms is linear when they are all in a straight line. The angle formed between two bonds that go to the same central atom, which we call the **bond angle,** is 180°.

Planar triangular molecules

A planar triangular arrangement of four atoms has them all in the same plane. The central atom is surrounded by three others located at the corners of a triangle. The bond angles are all 120°.

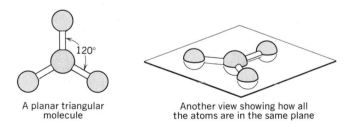

A planar triangular
molecule

Another view showing how all
the atoms are in the same plane

Tetrahedral molecules

A tetrahedron is a four-sided pyramid having equilateral triangles as faces. In a tetrahedral molecule, the central atom is located in the center of this tetrahedron and four other atoms are located at the corners. The bond angles are all equal and have values of 109.5°.

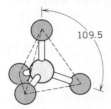

A tetrahedron   A tetrahedral molecule

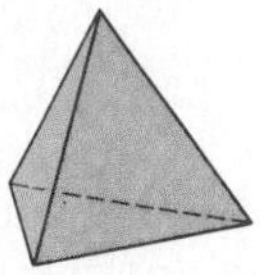

To draw a tetrahedron, begin by making a check mark as illustrated below (step 1). Then pick a point above the check mark (step 2) and draw three straight lines to the ends and the center of the check mark (step 3). Finally, draw a dotted line across the back to complete the tetrahedron (step 4).

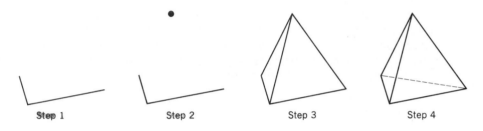

Step 1   Step 2   Step 3   Step 4

Trigonal bipyramidal molecules

A trigonal bipyramid consists of two trigonal pyramids (pyramids with triangular bases similar to tetrahedrons) that share a common face.

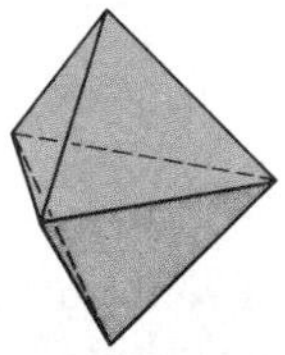

A trigonal bipyramid

Imagine the trigonal bipyramid centered inside a sphere with the atoms in the triangular plane around the equator and the vertical bonds pointing north and south. The three bonds in the triangular plane are then referred to as *equatorial bonds* and those pointing to the top and bottom are said to be *axial bonds.*

In a trigonal bipyramidal molecule, a central atom is surrounded by five others. The central atom is located in the center of the triangular plane shared by the upper and lower pyramids. The five atoms attached to the central atom are located at the five corners. In this kind of molecule, the bond angles are not all the same. Between any two bonds that lie in the central triangular plane, the bond angle is 120°. The angle is only 90° between a bond in the central triangular plane and a bond that points to the top or bottom of the trigonal bipyramid.

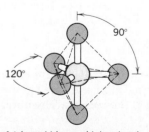

A trigonal bipyramidal molecule

When we draw a trigonal bipyramidal molecule, we normally sketch a triangle as it would look tilted backward so that we would be looking at it from its edge. Then we draw lines to the top and bottom of the trigonal bipyramid.

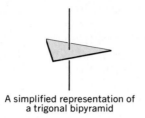

A simplified representation of
a trigonal bipyramid

Octahedral molecules

An octahedron is a geometrical figure that has eight faces. We can think of it as two square pyramids sharing a common square base. Notice that the figure has only six corners even though it has eight faces.

An octahedron

In an octahedral molecule, the central atom is surrounded by six others. The central atom is located in the center of the square plane that passes through the middle of the octahedron. The six atoms bonded to it are at the six corners of the octahedron. The angle between any pair of adjacent bonds is the same and has a value of 90°.

Remember, an octahedral molecule has only six atoms joined to the central atom.

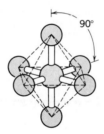

A simplified drawing of an octahedron generally shows the square plane in the center, as it would look tilted back, with lines running to the top and bottom of the octahedron.

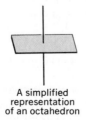

A simplified
representation
of an octahedron

The square plane, tilted back and viewed in perspective, resembles a parallelogram. Draw this first. Then draw a line from the center upward to the top apex of the octahedron. Finally, draw a line from the center to the bottom apex, but don't show the part hidden by the square plane in the center.

Before going on to the next section, you should practice drawing each of the five structures described here. If you understand them well, it will make the rest of this chapter easier to understand, too.

9.2 VALENCE SHELL ELECTRON PAIR REPULSION THEORY

One of the primary goals of chemical bonding theory is to explain and (we hope) to predict molecular structure. A theory that is exceedingly simple in its concept and remarkably successful in its ability to predict molecular geometry accurately is called the **valence shell electron pair repulsion theory (VSEPR theory).** It is not necessary to employ the notion of atomic orbitals at all. We will see, instead, that if an electron dot structure can be drawn for a molecule, its general shape can be predicted.

If necessary, review the procedures for writing Lewis structures on page 237.

When we seek to predict the shape of a molecule, we look for a way to decide how atoms or groups of atoms (for which we will use the general term **ligand**) are arranged around a central atom. For example, in a molecule such as SO_2, how are the oxygen atoms (the ligands) arranged around the sulfur atom? Are the three atoms in a straight line (i.e., is the molecule linear), or is the bond angle something less than 180°? To decide such questions, *the VSEPR theory proposes that the geometric arrangement of ligands around the central atom is determined **solely** by the repulsions among the electron pairs in the valence shell of the central atom.* According to the theory, these electron pairs assume positions such that the repulsions between them are a minimum, and the ligands follow along. To see how this works, let's begin by considering the simple molecule $BeCl_2$. Its electron dot structure is given as

$$: \overset{\displaystyle ..}{Cl} : Be : \overset{\displaystyle ..}{Cl} :$$

This particular molecule, you recall, violates the octet rule, and there are only two pairs of electrons in the valence shell of Be. According to the VSEPR theory, these electron pairs will arrange themselves to be as far apart as possible, so that the repulsion between them is at a minimum. When there are two electron pairs in the valence shell, this minimal repulsion occurs when they are located on opposite sides of the nucleus, which we can represent as

The diagram suggests the approximate locations of the electron clouds containing the electron pairs.

In the $BeCl_2$ molecule, the ligands (i.e., the chlorine atoms) are attached to the Be by the sharing of these electron pairs. This means that the chlorines must be placed where the electron pairs are, and the molecule should therefore have the *linear* structure

$$Cl \overset{\overset{\displaystyle 180°}{\frown}}{—Be—} Cl$$

This is, in fact, the shape of the $BeCl_2$ molecule in the gas phase.

We can also extend this reasoning to situations involving double or triple bonds. For instance, the CO_2 molecule has the dot structure

$$: \overset{\displaystyle ..}{O} :: C :: \overset{\displaystyle ..}{O} :$$

where we see that there are double bonds between C and O. Both pairs of electrons in a double bond must be confined to the same general region in the valence shell of an atom — otherwise, it wouldn't be a *double* bond. Therefore, in terms of their effect on determining molecular geometry, a group of four electrons in a double bond behaves much like a group of two electrons in a single bond. In the valence shell of carbon, therefore, we have *two* groups of four, and these groups will locate themselves on opposite sides of the carbon nucleus so that the repulsions between them are a minimum.

As before, the ligands (in this case, oxygen) are attached to the central atom through these electron pairs and we again have a *linear* structure.

$$:\overset{\cdot\cdot}{O}=C=\overset{\cdot\cdot}{O}:$$

The actual shape of the electron density distribution in a double bond will be discussed later. This diagram is just meant to show that the double bonds point in directions that are 180° apart.

When there are more than two pairs (or groups of pairs) of electrons in the valence shell, we find other geometric arrangements as shown in Figure 9.1. Electron pairs arranged in the valence shell in this manner lead to minimum repulsions. Let us see how we can use these electron pair arrangements to predict molecular structure.

Notice that these geometries are the same five that we discussed in Section 9.1

Three groups of electrons in the valence shell

In Chapter 8 we saw that the molecule BCl_3 has the dot structure

$$:\overset{\cdot\cdot}{\underset{\cdot\cdot}{Cl}}:$$
$$:\overset{\cdot\cdot}{\underset{\cdot\cdot}{Cl}}:\overset{\cdot\cdot}{B}:\overset{\cdot\cdot}{\underset{\cdot\cdot}{Cl}}:$$

Thus, there are three electron pairs around boron. According to Figure 9.1, we therefore expect the three chlorine atoms to be arranged around the boron atom at the corners of a planar equilateral triangle. Experimentally, this is the structure that is found for BCl_3, which is said to be a planar triangular molecule.

Now let's consider the molecule SO_2. The electron dot structure for one of its two resonance structures is

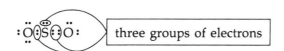 three groups of electrons

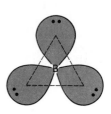

Around the sulfur there are three groups of electrons—two groups each with one pair, and one group with two pairs (the double bond). To have minimum repulsions, these groups of electrons are situated at the corners of a planar triangle with the sulfur in the center.

We count the unshared pair, too, because it is in the valence shell just like the electrons in the bonds.

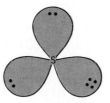

Attaching the oxygen atoms, one to a single pair and one to the double pair, we have

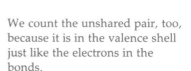

Number of Electron Pairs	Shape		Example
Two pairs	linear		$BeCl_2$
Three pairs	planar triangular		BCl_3
Four pairs	tetrahedral (A tetrahedron is pyramid-shaped. It has four triangular faces and four corners.)		CH_4
Five pairs	trigonal bipyramidal This figure consists of two three-sided pyramids joined by sharing a common face—the triangular plane through the center.)		PCl_5
Six pairs	octahedral (An octahedron is an eight sided figure with *six* corners. It consists of two square pyramids that share a common square base.)		SF_6

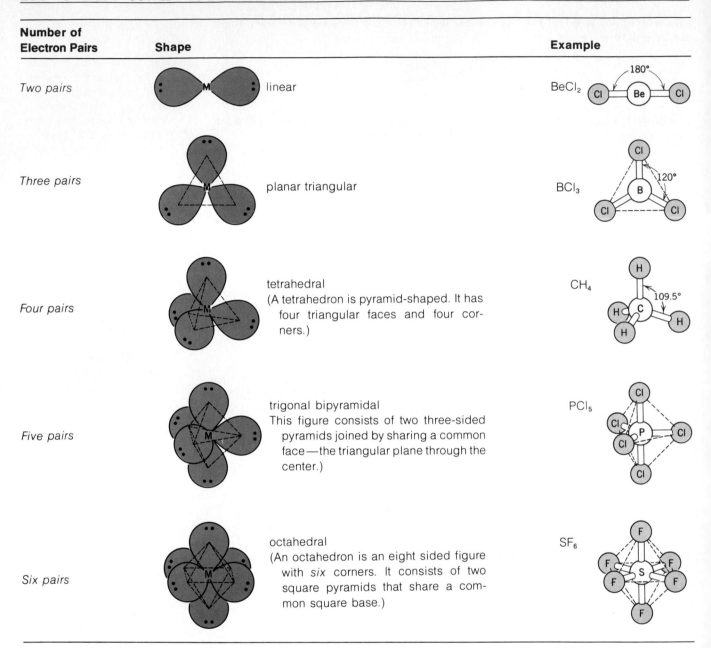

Figure 9.1. *Arrangements of electron pairs that give minimum repulsions.*

We use the terms *molecular shape, molecular structure, molecular geometry,* and *arrangement of atoms* interchangeably. They all refer to the way the atoms in a molecule are arranged in space relative to each other.

Thus, the theory predicts that the two oxygen atoms and the sulfur do not lie in a straight line. How then should we describe the structure?

When we give the shape of the SO_2 molecule, or of any other molecule, we always describe how the *atoms* in the molecule are arranged around the central atom, *not* how the electrons are arranged. Therefore, even though the electron groups in the valence shell of sulfur are presumed to be in a triangle, we *do not* describe the SO_2 molecule as triangular. Instead, we say that it is nonlinear, bent, angular, or V-shaped, or use some other description that simply says the three atoms are not arranged in a straight line.

An important aspect of the structure of the SO_2 molecule is the unshared pair of electrons in the valence shell of the sulfur. Such an unshared pair of electrons is

Number of Pairs in Bonds	Number of Lone Pairs	Structure	
3	0	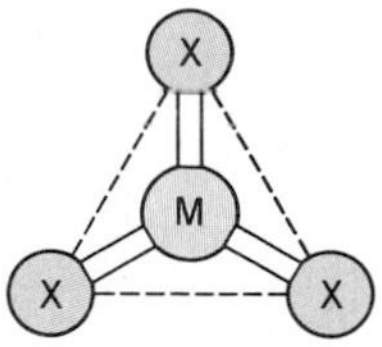	Planar triangular (Example, BCl_3)
2	1	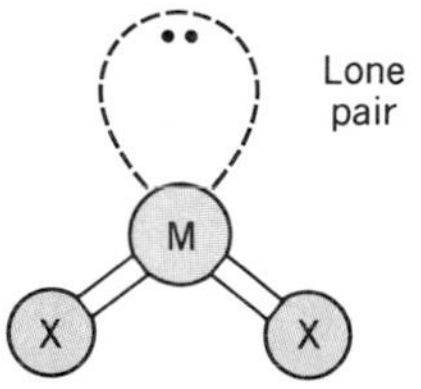	V-Shaped (Example, SO_2)

Figure 9.2. *Geometries of molecules or ions in which the central atom has three electron pairs (or groups of electron pairs) in its valence shell.*

called a **lone pair,** and lone pairs on the central atom have a strong influence on the shapes of molecules. In the case of SO_2, we see that it forces the bond pairs away from a linear arrangement and squeezes them toward each other. We will see similar effects in other molecules.

In summary, when there are three groups of electrons around an atom, they are arranged at the corners of a triangle. If they are all bonded to ligands, we have a molecule that we might generalize as MX_3, in which M stands for the central atom and X for a ligand. The structure of an MX_3 molecule is planar triangular. If only two groups are bonded, leaving one lone pair, we have a species MX_2E, where we use E to represent the lone pair. In an MX_2E molecule, the atomic nuclei are situated so as to give a nonlinear structure. Figure 9.2 illustrates the structures that we find for these kinds of molecules or polyatomic ions. In the figure, the lone pair on the MX_2E species is shown as an electron cloud. Notice that we have omitted from this figure, as well as from our discussion, molecules with the formula MXE_2. This would be a diatomic molecule, and when only two atoms are bonded to each other, there's only one way for them to be connected. Only when there are three or more atoms in a molecule or ion do we have a choice of geometries.

We can determine the arrangement of atoms in a molecule experimentally, but not the geometrical arrangement of their electrons.

Four groups of electrons in the valence shell

If an atom has four pairs of electrons in its valence shell, minimum repulsions occur if they are arranged tetrahedrally. We have just seen that when there are three electron pairs (or groups of pairs) in the valence shell of a central atom, two possible molecular shapes can occur depending on whether or not one of the pairs is a lone pair. For molecules in which the central atom has four pairs in its valence shell, there are three possible molecular shapes — all of which are derived from the tetrahedral arrangement of the electrons. Once again using M for the central atom, X for a ligand, and E for a lone pair, we can represent these as follows (see also Figure 9.3):

MX_4 These are tetrahedral molecules with ligands bonded by all four electron pairs. An example is methane, CH_4.

Figure 9.3. *Geometries of molecules in which the central atom has four pairs of electrons in its valence shell.*

Number of Pairs in Bonds	Number of Lone Pairs	Structure	
4	0		Tetrahedral (Example, CH_4) All bond angles are 109.5°.
3	1		Trigonal pyramidal (Pyramid-shaped) (Example, NH_3)
2	2		Nonlinear, bent (Example, H_2O)

MX_3E When one lone pair is present, a *trigonal pyramidal* molecule is formed. This is a molecule shaped like a pyramid with a triangular base. An example is ammonia, NH_3, that has the nitrogen atom at the top of the pyramid and the three hydrogens around the base. Notice that we describe the shape of the ammonia molecule according to how the atoms are arranged, not by how the electrons are arranged.

MX_2E_2 Two lone pairs gives a nonlinear or angular structure   (e.g., H_2O).

Five electron pairs

Five electron pairs will have minimum repulsions if they are arranged at the corners of a trigonal bipyramid, as shown in Figure 9.1. This gives four possible structures, depending on the number of lone pairs, as illustrated in Figure 9.4.

MX_5 All electron pairs are used in bonds and a trigonal bipyramidal molecule is formed.

MX_4E If one of the five pairs is a lone pair, one might suspect that there are two possible molecular structures, one with the lone pair in the central triangular plane (structure I in the margin) and the other with the lone pair perpendicular to this plane (structure II).

It turns out that the repulsions are less in structure I than in II. To understand this, we have to first realize that a lone pair is larger in volume than a pair of electrons in a bond. In a bond, the electrons are under the influence of two positive nuclei, but in a lone pair the electrons are attracted to only one nucleus. The greater total nuclear charge felt by the bonding pair causes it to be pulled into a smaller volume, so effectively it takes up less space than a lone pair. Because the lone pair is larger than the bonding pair, it exerts a greater repulsion toward other pairs in the valence shell.

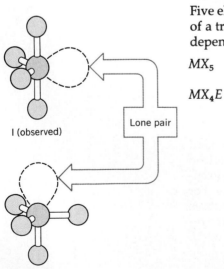

I (observed)

Lone pair

II (not observed)

Number of Pairs in Bonds	Number of Lone Pairs	Structure	
5	0		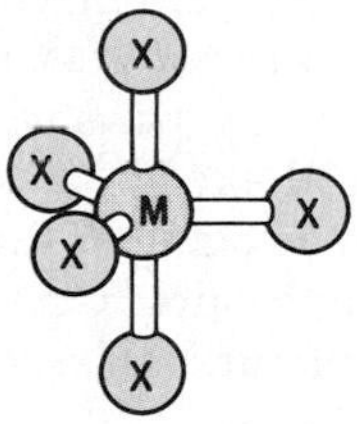 Trigonal bipyramidal (Example, PCl_5)
4	1		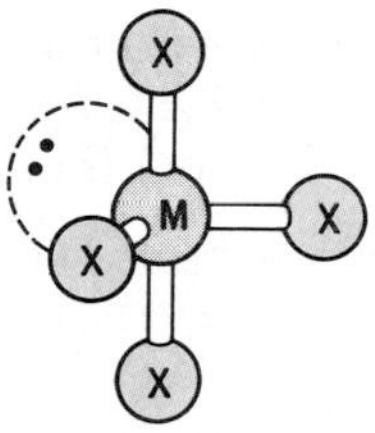 Unsymmetrical tetrahedron 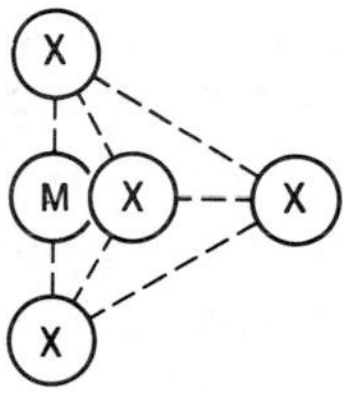 (Example, SF_4)
3	2		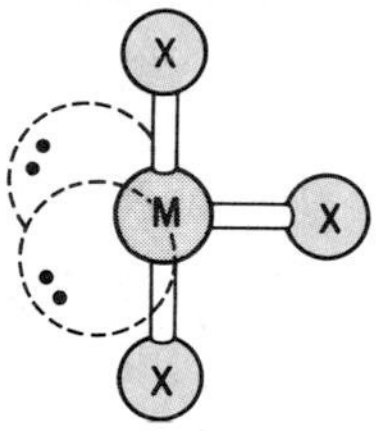 T-Shaped 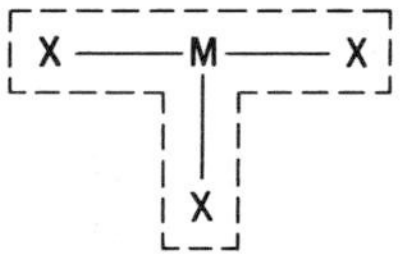 (Example, ClF_3)
2	3		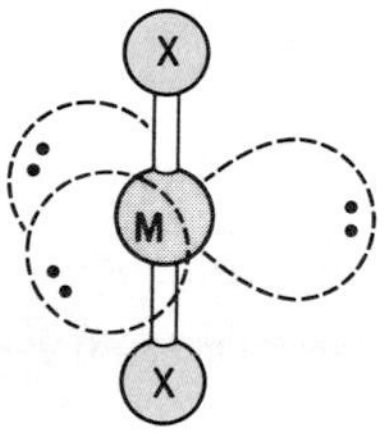 Linear (Example, I_3^-)

Figure 9.4. *Molecular structures that result when the central atom has five electron pairs in its valence shell.*

Now that we realize that a lone pair creates a greater amount of repulsion than a bonding pair, let's examine structures I and II to see which will give the least repulsions. We do this by examining how the bonds in the structures are oriented relative to the lone pair. In structure I, the lone pair is alongside two bonds that are at angles of 90° (those that point up and down) and two bonds that are at angles of 120° (those that are in the triangular plane of the trigonal bipyramid along with the lone pair). In structure II, the lone pair is alongside *three* bonds at angles of 90°, and the fourth bond is at an angle of 180°. In terms of their influence on structure, it is only the repulsions due to the nearest neighbors that are important. In structure I, the lone pair has only two nearest-neighbor bonds, but in structure II it has three. Structure I, therefore, has the lesser

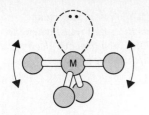

The MX$_4$E structure is sometimes said to be a "seesaw" structure because, when tilted on its side, it resembles the children's apparatus of the same name.

total amount of repulsion, so it is preferred. In fact, it is *always* found that the lone pairs prefer to be in the triangular plane of the trigonal bipyramid, even when there are two or three lone pairs.

The shape of the MX_4E molecule is difficult to describe—the term we will use is *unsymmetrical tetrahedral* or *distorted tetrahedral*.

MX_3E_2 This structure has two lone pairs in the central triangular plane and the atoms are arranged in the form of the letter T drawn on its side (⊣). The molecule is described as being T-shaped.

MX_2E_3 Now there are three lone pairs in the triangular plane and the atoms are in a straight line; the structure is described as *linear*.

Six electron pairs

These have minimum repulsions when arranged octahedrally (Figure 9.1). This gives *five* possibilities: MX_6, MX_5E, MX_4E_2, MX_3E_3, and MX_2E_4. Only the first three are actually observed to occur, however. These are shown in Figure 9.5.

MX_6 When all electron pairs are used for bonding, an *octahedral* structure is formed.

Figure 9.5. *Molecular structures that result when the central atom has six electron pairs in its valence shell.*

Number of Pairs in Bonds	Number of Lone Pairs	Structure	
6	0		Octahedral (Example, SF$_6$) All bond angles are 90°.
5	1		Square pyramidal (Example, BrF$_5$)
4	2		Square planar (Example, XeF$_4$)

Table 9.1. Summary of molecular shapes

Type of Molecule or Ion	Shape
MX_2	Linear
MX_3	Planar triangular
MX_2E	Nonlinear (angular, bent)
MX_4	Tetrahedral
MX_3E	Trigonal pyramidal
MX_2E_2	Nonlinear (angular, bent)
MX_5	Trigonal bipyramidal
MX_4E	Distorted tetrahedral
MX_3E_2	T-shaped
MX_2E_3	Linear
MX_6	Octahedral
MX_5E	Square pyramidal
MX_4E_2	Square planar

MX_5E The atoms in this structure are at the corners of a pyramid having a square base, so the structure is described as being *square pyramidal*.

MX_4E_2 With two lone pairs, minimum repulsions occur if they are as far apart as possible. This gives an arrangement of atoms that we describe as *square planar*.

Let's look now at some examples that illustrate how we use the VSEPR theory to predict the shapes of molecules and ions. It is best if you try to visualize the structure and identify it by its name. If this is an overwhelming problem for you, the structure can be found by identifying the number of ligands and lone pairs around the central atom, and then referring to Table 9.1.

EXAMPLE 9.1. PREDICTING THE SHAPE OF AN ION BY THE VSEPR THEORY

PROBLEM: What is the shape of the sulfate ion, SO_4^{2-}?

SOLUTION: We begin by drawing a Lewis structure for the ion following the procedure given in Chapter 8. This gives

$$\left[\begin{array}{c} :\ddot{O}: \\ \parallel \\ :\ddot{O}-S-\ddot{O}: \\ \parallel \\ :\ddot{O}: \end{array} \right]^{2-}$$

We could also use the Lewis structure that obeys the octet rule here. It isn't necessary to take into account formal charges when we apply the VSEPR theory.

This is one of four resonance structures, but we just need to know one of them to predict the shape of the ion. We treat the double bonds just as a single bond for the purposes of determining the structure, and we see that the sulfur has four groups of electrons in its valence shell. These must be arranged tetrahedrally, as illustrated below.

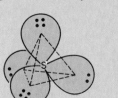

Attaching the oxygen atoms gives a tetrahedral ion.

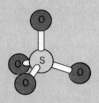

We can get the same answer, of course, by recognizing (from the Lewis structure) that SO_4^{2-} is an MX_4 species, which is tetrahedral.

EXAMPLE 9.2. PREDICTING THE SHAPE OF AN ION BY THE VSEPR THEORY

PROBLEM: Formate ion, HCO_2^-, comes from formic acid, the substance that produces the stinging sensation in bites from fire ants. The dot structure of formate ion is

$$\left[\begin{array}{c} \overset{\displaystyle :\overset{..}{O}:}{\underset{\displaystyle H-C-\overset{..}{\underset{..}{O}}:}{\|}} \end{array} \right]^-$$

What is its shape?

SOLUTION: The double bond behaves just like a single bond for purposes of predicting molecular shape. This ion has three groups of electrons around the carbon and they are arranged in a planar triangular fashion. Since all groups are used in bonding, the ion (an MX_3 type) has a planar triangular shape.

EXAMPLE 9.3. PREDICTING THE SHAPE OF A MOLECULE BY THE VSEPR THEORY

PROBLEM: Arsenic, a well-known poison, can be detected by converting its compounds to the unstable substance AsH_3 (arsine), which decomposes easily on a clean hot glass surface where it deposits a mirrorlike coating of pure arsenic. What is the shape of an AsH_3 molecule?

SOLUTION: The first step is always to draw the Lewis structure. This is

$$H-\overset{..}{\underset{|}{As}}-H$$
$$H$$

The four electron pairs are arranged tetrahedrally, but only three of them are used in bonds; one is a lone pair. The molecular structure that we get is

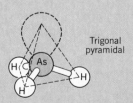

Trigonal pyramidal

We get the same answer recognizing AsH_3 as an MX_3E molecule.

9.3 THE POLARITY OF MOLECULES AND MOLECULAR STRUCTURE

In Section 8.9, we discussed very briefly how shape can influence whether or not a molecule with polar bonds is a polar molecule. Molecular polarity is an important property because many of the physical properties of substances depend on it. This is because polar molecules attract each other. When a large number of polar molecules are together, they tend to orient themselves so that the partially positive end of one is near the partially negative end of another. The electrostatic attraction of opposite charges leads to a net attraction between the molecules. In Chapter 12 we will see how these attractions influence such properties as boiling point and ease of evaporation. But first, let's look more closely at the way molecular polarity is related to molecular structure. We'll begin with a quick review of some of the principles developed in Section 8.9.

A covalent bond will be polar when the two atoms joined by the bond differ in electronegativity, and you should recall that the atom with the larger electronegativity is the one that carries the partial negative charge. The other carries an equal partial positive charge.

In order for a molecule to be polar, it must be a dipole. This means that opposite ends of the molecule must carry *opposite* electrical charges. If there are no charges on opposite ends, or if the charges are of the same sign, the molecule is not a dipole, and it is not considered polar.

When a molecule consists of only two atoms, it must be polar if the bond is polar. This has to be true because the two atoms in the molecule have equal and opposite partial charges. However, if there are three or more atoms in a molecule, it is possible for it to be nonpolar, even when the bonds are polar. In Section 8.9 we discussed two simple cases illustrating this, carbon dioxide and water.

Carbon dioxide has the Lewis structure

$$:\overset{..}{O}=C=\overset{..}{O}:$$

and a linear shape. The bonds in CO_2 are polar, with the more electronegative oxygen carrying the negative charge in each case. Because the molecule is linear, these partially negative oxygens are found at opposite ends of the molecule, so the opposite ends carry the same electrical charge. This means, of course, that the molecule is nonpolar.

As we noted in Section 8.9, the overall polarity of the molecule also can be considered to result from the interaction of the various polar bonds within the molecule. To analyze these effects, we view the individual **bond dipoles** — the dipoles *within* the molecule produced by the unequal sharing of electrons in the bonds — as vectors that can either reinforce or cancel each other. As you saw on page 252, the bond dipoles in CO_2 point in opposite directions and cancel completely, which tells us that CO_2 is nonpolar.

$$:\overset{..}{O}=C=\overset{..}{O}:$$

(Bond dipoles, ↔, cancel because they point in opposite directions.)

On page 253, you saw a similar analysis for water. In this case, the individual bond dipoles do not cancel entirely, because the molecule is nonlinear. Instead, they partially add in one direction to give a net dipole for the molecule.

(Heavy arrow is the net dipole for the molecule.)

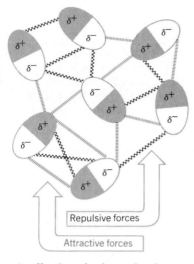

A collection of polar molecules orient themselves so the attractions are greater than the repulsions.

The differences between water and carbon dioxide illustrate how important molecular structure is in determining molecular polarity. This is one of the reasons why the VSEPR theory is so valuable: Its ability to predict molecular structure enables us to also make predictions about the polarity of molecules. To do this, let's begin by analyzing how bond dipoles are expected to interact in some simple arrangements, and then we can extend these simple cases to the study of more complex structures. In each of these discussions, we assume that all the atoms bonded to the central atom are the *same*.

> If the atoms bonded to the central atom are not the same, the individual bonds will differ in their polarities and cancellation of the bond dipoles usually can't occur.

A linear arrangement of bond dipoles

We have already examined this arrangement of bond dipoles in discussing the CO_2 molecule. As we will see, other molecular structures have this arrangement of dipoles within them, so it is useful to remember that a pair of equivalent bond dipoles pointing in opposite directions cancel each other.

$$\longleftarrow \cdot \longrightarrow \qquad \text{(Cancellation of the bond dipoles is obvious here.)}$$

A planar triangular arrangement of bond dipoles

Here we have bond dipoles arranged as shown below.

Although not as obvious as with the linear arrangement, complete cancellation of the bond dipoles occurs here, too. The two dipoles pointing downward at an angle cancel the one pointing up, and so forth. Because of this, we expect that planar triangular molecules such as BCl_3 and SO_3 should be nonpolar, and indeed, they are. In fact, any planar triangular molecule will be nonpolar, if the central atom is bonded to three other identical atoms. (Of course, if the three peripheral atoms are not the same, then the bonds will differ in their polarities and complete cancellation cannot occur. A molecule MX_2Y, in which atoms X and Y are different, will therefore be polar.)

A tetrahedral arrangement of bond dipoles

In this structure the bond dipoles point toward the corners of a tetrahedron.

This arrangement also leads to complete cancellation of the dipoles. Each individual bond dipole is canceled by the effects of the other three that point partially in the opposite direction. As a result, tetrahedral molecules such as CCl_4 or CF_4 are nonpolar, even though they have polar bonds.

Trigonal bipyramidal and octahedral arrangements of bond dipoles

To analyze these structures, we can make use of what we've learned so far about simpler ones. As you can see in Figure 9.6a, the trigonal bipyramidal structure actually consists of a triangular arrangement of bonds in a plane through the center, plus a pair of bonds pointing in opposite directions perpendicular to this plane. We have seen that a planar triangular distribution of bond dipoles leads to cancellation, so the three bonds in the triangular plane in the center of the trigonal bipyramid contribute nothing to the polarity of the molecule. We have also seen that a pair of bond dipoles pointing in opposite directions cancel each other, so the two bonds that are perpendicular to the triangular plane in the trigonal bipyramid also cancel. Overall, then, we get complete cancellation of all the bond dipoles, which means that trigonal bipyramidal molecules such as PCl_5 will be nonpolar, regardless of how polar their bonds happen to be.

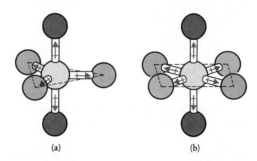

(a) (b)

Figure 9.6. (a) *A molecule with the general formula* MX_5 *and a trigonal bipyramidal structure is nonpolar. The set of three bond dipoles in the triangular plane in the center (in blue) cancel, as do the linear set of bond dipoles (red).* (b) *In an octahedral* MX_6 *molecule there are three sets of linear pairs of dipoles shown in red, blue, and green. Cancellation occurs in each set, so the molecule is nonpolar.*

Symmetrical octahedral molecules such as SF_6 are nonpolar, too. In the octahedral structure (Figure 9.6b), there are three pairs of bonds pointing in opposite directions. Total cancellation occurs for each pair, so all the individual bond dipoles in the octahedral structure are canceled and a nonpolar molecule is the result.

Molecules that have lone pairs on the central atom

In most cases, when there are lone pairs in the valence shell of the central atom, the bond dipoles do not completely cancel. Consider, for example, the case of SO_2, which has a V-shaped structure and one lone pair on the central sulfur atom.

For SO_2, the two bond dipoles that point partially in one direction are not offset by a third one in the opposite direction. As with the water molecule, the bond dipoles are partially additive and the molecule as a whole is polar.

A similar imbalance is found in most of the other structures that have lone pairs on the central atom. There are two exceptions, however. One is the linear MX_2E_3 structure and the other is the square planar MX_4E_2 structure, which are shown in

Figure 9.7. *In a linear* MX_2E_3 *molecule and a square planar* MX_4E_2 *molecule, bond dipoles cancel to give nonpolar molecules.*

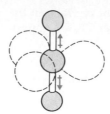
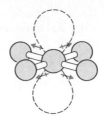

Figure 9.7. For these two, the bond dipoles are symmetrically arranged and their effects cancel completely to give nonpolar molecules.

EXAMPLE 9.4. PREDICTING WHETHER OR NOT A MOLECULE IS POLAR

PROBLEM: Is the SiF_4 molecule polar?

SOLUTION: The first step is to draw the Lewis structure and then determine the shape of the molecule by using the VSEPR theory. By the rules we developed in Chapter 8, the Lewis structure of the molecule is

$$:\ddot{F}:$$
$$|$$
$$:\ddot{F}—Si—\ddot{F}:$$
$$|$$
$$:\ddot{F}:$$

The VSEPR theory tells us that the molecule is tetrahedral. We've seen that in this structure all the bond dipoles cancel, so we can say with confidence that SiF_4 is a nonpolar molecule.

EXAMPLE 9.5. PREDICTING WHETHER OR NOT A MOLECULE IS POLAR

PROBLEM: Is the chloroform molecule, $CHCl_3$, polar?

SOLUTION: The Lewis structure for chloroform is

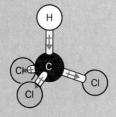

$$:\ddot{Cl}:$$
$$|$$
$$:\ddot{Cl}—C—H$$
$$|$$
$$:\ddot{Cl}:$$

and it is a tetrahedral molecule. However, not all the bonds in the molecule are the same. Chlorine is a more electronegative atom than hydrogen, so the C—Cl bonds are more polar than the C—H bond. In addition, carbon is slightly more electronegative than hydrogen, so the C—H bond dipole actually adds to the effects of the C—Cl bond dipoles. As a result, we expect the $CHCl_3$ molecule to be polar.

EXAMPLE 9.6. PREDICTING WHETHER OR NOT A MOLECULE IS POLAR

PROBLEM: Is the PCl_3 molecule polar?

SOLUTION: As before, we begin with the Lewis structure.

$$:\ddot{C}l-P-\ddot{C}l:$$
$$|$$
$$:\ddot{C}l:$$

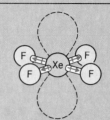

VSEPR theory tells us that the molecule has a trigonal pyramidal shape, and we note that there is a lone pair of electrons on the phosphorus atom. In this structure, the bond dipoles do not cancel, and we can expect that the molecule will be polar.

EXAMPLE 9.7. PREDICTING WHETHER OR NOT A MOLECULE IS POLAR

PROBLEM: Is the XeF$_4$ molecule polar?

SOLUTION: First we draw the Lewis structure.

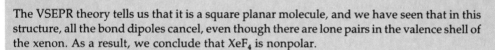

$$:\ddot{F}:$$
$$|$$
$$:\ddot{F}-Xe-\ddot{F}:$$
$$|$$
$$:\ddot{F}:$$

The VSEPR theory tells us that it is a square planar molecule, and we have seen that in this structure, all the bond dipoles cancel, even though there are lone pairs in the valence shell of the xenon. As a result, we conclude that XeF$_4$ is nonpolar.

9.4 ORBITAL OVERLAP AND THE COVALENT BOND

The electron pair repulsion theory is a useful device for predicting molecular geometry, but it still doesn't answer the basic question: How do atoms share electrons between their valence shells? To find the answer, we must look to the results of quantum mechanics to see how the orbitals of atoms interact with each other when bonds are formed.

There are two important approaches to chemical bonding based on the results of quantum mechanics. One of these, called the **valence bond theory,** permits us to retain our picture of individual atoms coming together to form a covalent bond, and it is the theory that we will discuss here. The other, which conceptually is somewhat more difficult to deal with, is called the **molecular orbital theory.** The molecular orbital theory views a molecule as a set of positive nuclei with orbitals that extend over the entire molecule. The electrons that populate these orbitals do not necessarily belong to any individual atoms but, instead, to the molecule as a whole.

The basic postulates of the valence bond theory can be stated as follows.

1. When two atoms form a covalent bond, an atomic orbital on one of them *overlaps* with an atomic orbital of the other. By **overlap,** we mean that the two orbitals share some common region in space.
2. Two electrons with their spins paired can be shared between the two overlapping orbitals, the electron density being concentrated between the nuclei of the bonded atoms.
3. The strength of the covalent bond, as measured by the amount of energy needed to break it, is proportional to the amount of overlap of the orbitals — the greater the degree of overlap, the stronger the bond and the more the

Molecular orbital theory is discussed in Appendix B. When the valence bond and molecular orbital theories are extended and refined, both give essentially the same results.

potential energy of the atoms is lowered when the bond is formed. As a result, atoms tend to position themselves so that a maximum amount of orbital overlap is achieved.

Let's see how this theory explains the bonding in some familiar compounds. The simplest of these is the hydrogen molecule, which is formed from two hydrogen atoms, each of which has a single electron in a 1s orbital. According to valence bond theory, we would view the H—H bond as resulting from the overlap of the two 1s orbitals, as shown in Figure 9.8. The electron density that this gives in the molecule is the same as that described in Chapter 8 (Figure 8.1).

Figure 9.8. *Formation of the bond in the H_2 molecule by the overlap of the 1s orbitals of the hydrogen atoms.*

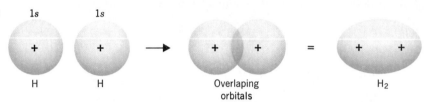

In the HF molecule we have a somewhat different state of affairs. Fluorine has the valence shell electron configuration

$$F \quad \frac{\uparrow\downarrow}{2s} \quad \frac{\uparrow\downarrow}{} \frac{\uparrow\downarrow}{} \frac{\uparrow}{2p}$$

where we find one of the 2p orbitals occupied by a single electron. It is with this partially occupied 2p orbital that the hydrogen 1s orbital overlaps, as illustrated in Figure 9.9. In this case the hydrogen electron and the fluorine electron can pair up and be shared between the two nuclei. Note that the 1s orbital of the hydrogen atom does not overlap with an already filled atomic orbital on fluorine because then there would be three electrons in the bond (two from the fluorine 2p orbital and one from the hydrogen 1s orbital). This situation is not permitted. In the valence bond theory, *only two electrons with their spins paired may be shared in one set of overlapping orbitals.*

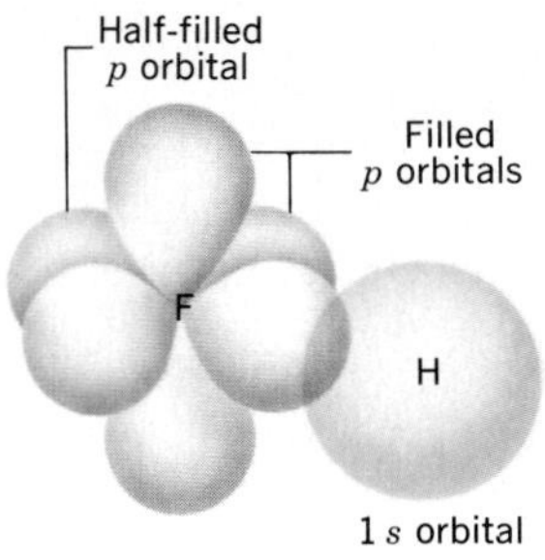

Suppose that we now consider the molecule H_2O. Here we have two hydrogen atoms bound to a single oxygen atom. The outer-shell electron configuration of oxygen is

$$O \quad \frac{\uparrow\downarrow}{2s} \quad \frac{\uparrow\downarrow}{} \frac{\uparrow}{} \frac{\uparrow}{2p}$$

and it shows that there are two unpaired electrons in p orbitals. This allows the two hydrogen atoms, with their electrons in 1s orbitals, to bond to the oxygen by means of the overlap of their 1s orbitals with these partially filled oxygen p orbitals (Figure 9.10). We can represent this using the following orbital diagram:

$$O \text{ (in } H_2O) \quad \frac{\uparrow\downarrow}{2s} \quad \frac{\uparrow\downarrow}{} \frac{\uparrow\downarrow}{} \frac{\uparrow\downarrow}{2p}$$

The colored arrows represent the electrons from the hydrogen atoms. Since the p orbitals are oriented at 90° to one another, we expect the H—O—H bond angle in water to also be 90°. Actually, this angle is 104.5°. One explanation for this discrepancy (we will see another later) is that because the oxygen atom is so small, the electron clouds of the hydrogen atoms would be forced to overlap if the bond angle were just 90° (see Figure 9.11). But because of bond formation with oxygen, each of

Figure 9.9. *Formation of the bond in HF by the overlap of the partially filled fluorine 2p orbital with the 1s orbital of hydrogen.*

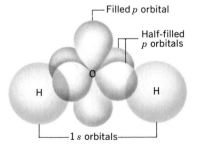

Figure 9.10. *Bonding in the H_2O molecule. Overlap of two half-filled oxygen 2p orbitals with hydrogen 1s orbitals give two O—H bonds at a predicted angle of 90°.*

these hydrogen orbitals is effectively occupied by a pair of electrons, so their overlap would place two electron pairs in the same region of space. This is not allowed; in fact, it is part of the reason for the Pauli exclusion principle, which states that only two electrons with paired spins can occupy the same orbital. As a result, strong repulsions between the electron clouds of the hydrogen atoms push them apart and cause the bond angle to increase to its 104.5° value.

We can also apply the valence bond theory to the ammonia molecule. Nitrogen, being in Group VA, has three unpaired electrons in its *p* subshell.

$$N \quad \underset{2s}{\underline{\uparrow\downarrow}} \quad \underset{2p}{\underline{\uparrow}\ \underline{\uparrow}\ \underline{\uparrow}}$$

Three hydrogen atoms can form bonds to nitrogen by overlapping their 1*s* orbitals with the half-filled *p* orbitals as shown in Figure 9.12*a*. The orbital diagram shows how the nitrogen completes its valence shell by this process.

$$N \text{ (in NH}_3) \quad \underset{2s}{\underline{\uparrow\downarrow}} \quad \underset{2p}{\underline{\uparrow\downarrow}\ \underline{\uparrow\downarrow}\ \underline{\uparrow\downarrow}} \quad \text{(Colored arrows = H electrons.)}$$

As in the water molecule, the H—N—H bond angles are larger than the predicted 90°, having values of 107°. The resulting picture provided by valence bond theory therefore explains the trigonal pyramidal shape of the NH_3 molecule, Figure 9.12*b*.

9.5 HYBRID ORBITALS AND MOLECULAR STRUCTURE

The very simple picture of the overlap of half-filled atomic orbitals that we have just developed cannot be used to explain the geometry of all molecules. It works well for H_2 and HF, and with some "fudging" it also explains the structures of water and ammonia molecules. If we examine the methane (CH_4) molecule, however, this simple model breaks down completely. With carbon, we would initially expect only two bonds to be formed with hydrogen since the valence shell of carbon contains only two unpaired electrons in half-filled orbitals.

$$C \quad \underset{2s}{\underline{\uparrow\downarrow}} \quad \underset{2p}{\underline{\uparrow}\ \underline{\uparrow}\ \underline{}}$$

The species CH_2, however, does not exist as a stable molecule. Instead, the simplest compound between carbon and hydrogen is methane, which has the formula CH_4. Attempting to explain the structure of this molecule by spreading the electrons out to give

$$\underset{2s}{\underline{\uparrow}} \quad \underset{2p}{\underline{\uparrow}\ \underline{\uparrow}\ \underline{\uparrow}}$$

suggests that three of the C—H bonds will be formed by the overlap of hydrogen 1*s* orbitals with carbon 2*p* orbitals, while the remaining bond would result from the overlap of the carbon 2*s* orbital with a hydrogen 1*s* orbital. This fourth C—H bond should certainly be different from the other three bonds, because it is formed from a different kind of orbital. It has been found experimentally, however, that *all* four C—H bonds are identical and the molecule has a structure in which the carbon atom lies at the center of a tetrahedron with the hydrogen atoms located at the four corners (Figure 9.13). Apparently, the orbitals that carbon uses to form bonds in molecules

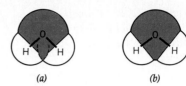

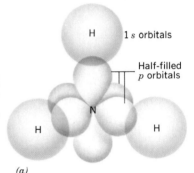

Figure 9.11. (a) *Because of the small size of the oxygen atom, the* 1s *orbitals of hydrogen would overlap if the bond angle were 90°, causing very strong repulsions between the electron pairs.* (b) *Opening up the bond angle to* 104.5° *relieves this strong repulsion between the hydrogens.*

Figure 9.12. *Bonding in* NH_3 *gives a trigonal pyramidal molecule.* (a) *Overlap of the* 2p *orbitals of nitrogen with* 1s *orbitals of hydrogen.* (b) *The trigonal pyramidal shape of the ammonia molecule.*

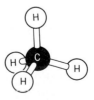

Figure 9.13. *The structure of methane,* CH_4.

Figure 9.14. *Formation of a pair of* sp *hybrid orbitals by mixing one* s *and one* p *atomic orbital.*

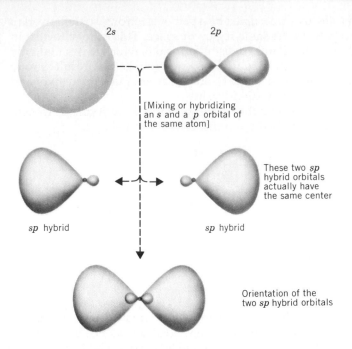

$2s$ 　　　　 $2p$

[Mixing or hybridizing an *s* and a *p* orbital of the same atom]

These two *sp* hybrid orbitals actually have the same center

sp hybrid 　　　　 *sp* hybrid

Orientation of the two *sp* hybrid orbitals

There are no simple atomic orbitals arranged in a tetrahedral fashion.

An atom forms hybrid orbitals by mixing appropriate combinations of its atomic orbitals.

It takes energy to form hybrid orbitals, but this energy is more than paid back by the much stronger bonds that hybrid orbitals form.

In the labels used for the various hybrid orbitals, the superscripts specify the numbers of each kind of atomic orbital incorporated in the hybrid set.

like CH_4, and that other atoms use to form bonds in the more complex trigonal bipyramidal and octahedral structures, are not just simple atomic orbitals. The question is, "What kinds of orbitals are they?"

The solution to this apparent dilemma is found in the behavior of waves. Scientists have discovered that the electron waves corresponding to the various atomic orbitals can be considered to combine with each other by constructive and destructive interference to form new **hybrid atomic orbitals** that have different shapes and different directional properties. This is illustrated in Figure 9.14 for the mixing or **hybridization** of an *s* orbital and a *p* orbital. The result is a *pair* of orbitals that we label *sp hybrid orbitals.*

Hybrid orbitals such as these have some interesting properties that help explain why atoms often seem to use them in bond formation. First, notice that the shape of a hybrid orbital is quite different than that of either an *s* or a *p* orbital. It consists of two "lobes" that point in opposite directions, but one lobe is much larger than the other. This large lobe extends farther from the nucleus than either the *s* or *p* orbital and is able to overlap with orbitals on other atoms much more effectively. Therefore, bonds formed by hybrid orbitals will be stronger than bonds formed by just simple *s* or *p* orbitals. Second, notice that by hybridization we have formed two orbitals that are identical, except for the directions in which they point. This allows for the formation of two equivalent bonds that point in opposite directions, and we will examine a molecule shortly for which *sp* hybrid orbitals are needed to explain the bonding and structure.

A pair of *sp* hybrid orbitals is formed by mixing one *s* and one *p* orbital. Other sets of hybrids can be formed by mixing additional simple atomic orbitals, and these are described in Table 9.2 and illustrated in Figure 9.15. The notation used to identify the hybrid orbitals specifies by superscripts the number of atomic orbitals of each type that are mixed. For example, the hybrid orbitals labeled sp^3d^2 are formed by mixing *one s* orbital, *three p* orbitals, and *two d* orbitals. You might also notice that the directional properties of these hybrid orbitals are identical to the basic geometries predicted by the VSEPR theory based on the mutual repulsions of electron pairs in the valence shell of an atom.

Table 9.2. Hybrid orbitals

Hybrid Orbitals	Number of Orbitals	Orientation
sp	2	Linear
sp^2	3	Planar triangle
sp^3	4	Tetrahedral
sp^3d	5	Trigonal bipyramidal
sp^3d^2	6	Octahedral

Using hybrid orbitals to explain bonding and structure

Let us see how we can use the information contained in Table 9.2 and Figure 9.15 to account for the structures of some typical molecules. We might begin with a mole-

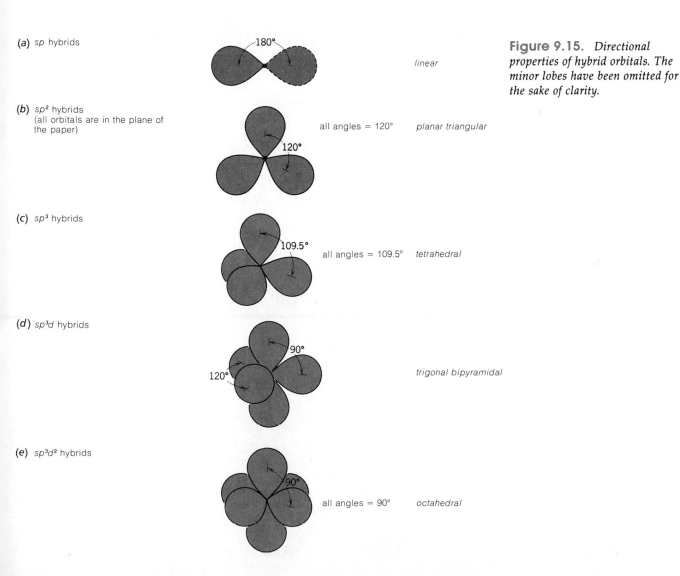

(a) *sp* hybrids

180°

linear

(b) *sp²* hybrids
(all orbitals are in the plane of the paper)

all angles = 120°   *planar triangular*

120°

(c) *sp³* hybrids

109.5°

all angles = 109.5°   *tetrahedral*

(d) *sp³d* hybrids

90°

120°

trigonal bipyramidal

(e) *sp³d²* hybrids

90°

all angles = 90°   *octahedral*

Figure 9.15. *Directional properties of hybrid orbitals. The minor lobes have been omitted for the sake of clarity.*

cule such as BeH_2, which in the gas phase would be a linear molecule with the dot structure

VSEPR theory predicts that BeH_2 should be linear.

$$H \colon Be \colon H$$

The electronic structure of beryllium's valence shell is

$$Be \quad \underset{2s}{\underline{\text{⇅}}} \quad \underset{2p}{\underline{\quad} \; \underline{\quad} \; \underline{\quad}}$$

In order to form two covalent bonds with H atoms, the Be atom must provide two half-filled (i.e., singly occupied) orbitals. This can be accomplished by creating a pair of sp hybrids and placing one electron in each of them.

An atom's valence electrons are distributed among the hybrid orbitals following Hund's rule (page 210).

$$Be \quad \underset{sp}{\underbrace{\underline{\uparrow} \; \underline{\uparrow}}} \quad \underset{\text{unhybridized } 2p \text{ orbitals}}{\underbrace{\underline{\quad} \; \underline{\quad}}}$$

The two H atoms can then bond to the beryllium atom by the overlap of their respective singly occupied s orbitals with the singly occupied Be sp hybrids as shown in Figure 9.16. The orbital diagram for the molecule is

$$Be \text{ (in } BeH_2) \quad \underset{sp}{\underbrace{\underline{\text{⇅}} \; \underline{\text{⇅}}}} \quad \underset{p}{\underbrace{\underline{\quad} \; \underline{\quad}}} \qquad \text{(Colored arrows = H electrons.)}$$

Because of the orientation of the sp hybrid orbitals, the H atoms are forced to lie on opposite sides of the Be, and a linear H—Be—H molecule results.

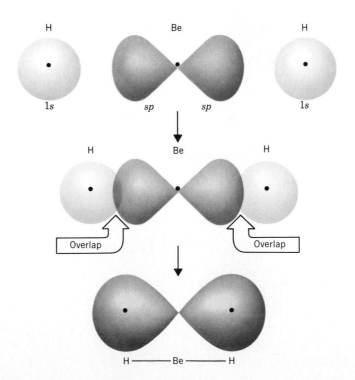

Figure 9.16. *Bonding in the BeH_2 molecule according to valence bond theory. Only the larger lobe of each sp hybrid is shown.*

Let us now return to our problem of the structure of CH_4. If we use hybrid orbitals on the carbon atom, we find that in order to provide four orbitals with which hydrogen $1s$ orbitals can overlap, we must use a set of sp^3 hybrids.

C   ⇅   ↑   ↑   ___   (Unhybridized)
     2s   2p

   ↑   ↑   ↑   ↑   (Hybridized)
      sp^3

$(s + p + p + p)$ gives a set of four equivalent sp^3 hybrid orbitals.

In Figure 9.15 we see that these orbitals point toward the vertices of a tetrahedron. Therefore, when the four hydrogen atoms are attached to the carbon by orbital overlap with these sp^3 hybrids,

$$C \text{ (in } CH_4) \quad \underline{⇅} \;\; \underline{⇅} \;\; \underline{⇅} \;\; \underline{⇅}$$
$$sp^3$$

a tetrahedral molecule results, as shown in Figure 9.17. This is in agreement with the structure that is found by experiment.

As another example, consider the molecule SF_6. Sulfur, being in Group VIA, has six valence electrons distributed over the $3s$ and $3p$ subshells

S   ⇅   ⇅   ↑   ↑   ___ ___ ___ ___ ___
    3s   3p   3d

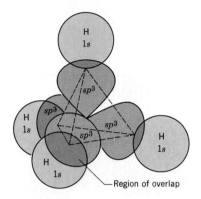

Figure 9.17. *Formation of methane by overlap of hydrogen* 1s *orbitals with carbon* sp³ *hybrid orbitals.*

Until now, we've omitted the $3d$ subshell when writing the electron configuration of a period 3 element.

Here we have shown the empty $3d$ subshell as well as the $3s$ and $3p$ subshells that contain electrons. In order for sulfur to form six covalent bonds to fluorine, six half-filled orbitals must be created. This can be accomplished by using two of the unoccupied $3d$ orbitals, and an sp^3d^2 hybrid set is formed.

S   ⇅   ⇅   ↑   ↑   ___ ___ ___ ___ ___
    3s   3p   3d

   ↑   ↑   ↑   ↑   ↑   ↑   ___ ___ ___   (Hybridized)
       sp^3d^2 3d   (Not used in hybrids)

$(s + p + p + p + d + d)$ gives a set of six equivalent sp^3d^2 hybrid orbitals.

S (in SF_6)   ⇅ ⇅ ⇅ ⇅ ⇅ ⇅   ___ ___ ___   (Colored arrows =
          sp^3d^2 3d   F electrons.)
          (Not used
          in hybrids)

The sp^3d^2 orbitals point toward the corners of an octahedron, which explains the octahedral geometry of SF_6.

Molecules in which there are lone pairs on the central atom

In our earlier discussion we viewed the structure of H_2O and NH_3 as resulting from the use of half-filled p atomic orbitals on the oxygen and nitrogen atoms, respectively. An alternative view of the bonding in these molecules employs sp^3 hybrid orbitals on the central atom. In the tetrahedral set of hybrids, the orbitals are oriented at angles of 109.5° to each other. The bond angles in water (104.5°) and ammonia (107°) are not too different from the tetrahedral angle and, using water as an example, we might consider the molecule to result from the overlap of hydrogen 1s orbitals with two partially occupied sp^3 orbitals on the oxygen atom.

$$\text{O} \quad \underset{2s}{\underline{\uparrow\downarrow}} \quad \underset{2p}{\underline{\uparrow\downarrow} \quad \underline{\uparrow} \quad \underline{\uparrow}} \qquad \text{(Unhybridized)}$$

$$\underset{sp^3}{\underline{\uparrow\downarrow} \quad \underline{\uparrow\downarrow} \quad \underline{\uparrow} \quad \underline{\uparrow}} \qquad \text{(Hybridized)}$$

$$\text{O (in H}_2\text{O)} \quad \underset{sp^3}{\underline{\uparrow\downarrow} \quad \underline{\uparrow\downarrow} \quad \underline{\uparrow\downarrow} \quad \underline{\uparrow\downarrow}} \qquad \text{(Colored arrows = H electrons.)}$$

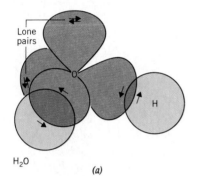

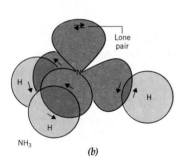

H₂O   *(a)*

NH₃   *(b)*

Figure 9.18. *The use of* sp³ *hybrids for bonding in* (a) *H₂O and* (b) *NH₃.*

Notice that only two of the hybrid orbitals are involved in bond formation while the other two contain nonbonded lone pairs of electrons. In the case of ammonia, three of the sp^3 orbitals are used in bonding and the fourth orbital contains a lone pair of electrons (Figure 9.18). There is, in fact, rather strong experimental evidence to indicate that this lone pair does indeed project out from the nitrogen atom as implied in this picture of the NH₃ molecule. It is worth noting that in our previous description of NH₃, we found this lone pair of electrons in an s orbital that would have spread the electron pair symmetrically about the nucleus.

In the case of H₂O and NH₃, the H—X—H bond angles (104.5° and 107°, respectively) are less than the tetrahedral angle of 109.5° observed in the molecule, CH₄. One way to account for this is to consider the influence of the lone-pair electrons present in hybrid orbitals of the central atom. As mentioned earlier, a pair of electrons in a bond is attracted to two nuclei and, therefore, might be expected to occupy a smaller effective volume than a pair of electrons in a nonbonded orbital, which experience the attraction of only one nucleus. The lone-pair electrons, then, because of their greater space requirement, tend to crowd together the electron pairs located in the bonds and hence reduce the bond angle to something less than 109°. On this basis we anticipate a greater reduction in bond angle for water than for ammonia, since water has two lone pairs while ammonia has only one.

Valence bond theory and the VSEPR theory

As we noted earlier, the orientations of the hybrid orbitals in Figure 9.15 are the same as the orientations that give minimum repulsions between electron pairs described in the valence shell electron pair repulsion theory, and that the two theories give identical results. For example, the shapes of methane, water, and ammonia predicted by the VSEPR theory are the same as those accounted for by using sp^3 hybrids in valence bond theory. Both theories use a tetrahedral arrangement of electron pairs in these molecules. This useful correlation gives us a simple way of anticipating the kinds of hybrid orbitals that an atom will use in a particular molecule. For instance, let's look again at SF₆. If we draw a dot structure for the molecule, we get

The VSEPR theory predicts that the six electron pairs around sulfur should be arranged octahedrally. Now we can ask ourselves, "What kinds of hybrid orbitals have an octahedral geometry?" The answer, of course, is sp^3d^2, and that is exactly what we used in our explanation of the structure of SF_6 by valence bond theory. We see, therefore, that the VSEPR theory can be used to help us choose the right kinds of hybrid orbitals to use in the valence bond theory. The two theories complement one another very nicely in explaining the bonding in molecules.

EXAMPLE 9.8. USING THE VSEPR THEORY TO PREDICT HYBRIDIZATION

PROBLEM: How can valence bond theory explain the bonding in the PCl_3 molecule?

SOLUTION: Since we aren't told what the structure of the molecule is, let's begin by using the VSEPR theory. This requires that we generate the Lewis structure of the molecule, which is

$$:\ddot{C}l—\ddot{P}—\ddot{C}l:$$
$$|$$
$$:\ddot{C}l:$$

The VSEPR theory predicts that the electron pairs should be arranged tetrahedrally around the phosphorus atom. This would require that the phosphorus atom use sp^3 hybrid orbitals. We next examine the electronic structure of phosphorus. The orbital diagram for the atom is

$$P \quad \underline{\uparrow\downarrow} \quad \underline{\uparrow} \; \underline{\uparrow} \; \underline{\uparrow}$$
$$3s \qquad\quad 3p$$

Creating sp^3 hybrids would give the following before bond formation:

$$P \quad \underline{\uparrow\downarrow} \; \underline{\uparrow} \; \underline{\uparrow} \; \underline{\uparrow}$$
$$sp^3$$

Each chlorine has the following electronic structure:

$$Cl \quad \underline{\uparrow\downarrow} \quad \underline{\uparrow\downarrow} \; \underline{\uparrow\downarrow} \; \underline{\uparrow}$$
$$3s \qquad\quad 3p$$

The P—Cl bonds would therefore be formed by the overlap of an unpaired electron in a chlorine $3p$ orbital with an unpaired electron in an sp^3 orbital on the phosphorus atom. We also see that the lone pair on the phosphorus atom occupies one of the sp^3 orbitals.

$$P \text{ (in } PCl_3\text{)} \quad \underline{\uparrow\downarrow} \; \underline{\uparrow\downarrow} \; \underline{\uparrow\downarrow} \; \underline{\uparrow\downarrow} \quad \text{(Colored arrows are chlorine electrons.)}$$
$$sp^3$$

EXAMPLE 9.9. USING THE VSEPR THEORY TO PREDICT HYBRIDIZATION

PROBLEM: Determine the kind of hybrid orbitals used by sulfur in SF_4, and explain the bonding in this molecule according to valence bond theory.

SOLUTION: Let's use the VSEPR theory to help us choose hybrid orbitals. This means that we first need the Lewis structure for SF_4. Following our usual procedure, we get

Notice that there are five electron pairs around the sulfur. The VSEPR theory tells us that they should be located at the corners of a trigonal bipyramid. The hybrid orbital set that is trigonal bipyramidal is sp^3d.

We have included the vacant $3d$ subshell because we know we will need a d orbital to form the sp^3d hybrid set.

Now we examine the electronic structure of sulfur.

Forming the sp^3d hybrid gives

Notice that we have enough half-filled orbitals to form the four bonds to fluorine.

S (in SF_4)   (Colored arrows = F electrons.)

This gives us our bonding picture for SF_4, in which a lone pair occupies one of the hybrid orbitals. The structure of SF_4 is shown in Figure 9.19.

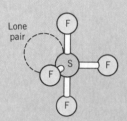

Figure 9.19. *The structure of SF_4. Notice the lone pair of electrons in the sp³d hybrid orbital. The VSEPR theory, by predicting the structure, even tells us which of the hybrid orbitals houses the lone pair.*

EXAMPLE 9.10. USING THE VALENCE BOND THEORY TO EXPLAIN BONDING

PROBLEM: The H_2S molecule is nonlinear, with an H—S—H bond angle of approximately 92°. Explain the structure of this molecule by using the valence bond theory.

SOLUTION: In this example we are given actual data about the molecule. Whatever explanation we devise, it must adequately explain the facts. With this in mind, let's begin as before. The Lewis structure for H_2S is

$$H—\overset{\cdot\cdot}{\underset{\cdot\cdot}{S}}—H$$

The VSEPR theory would predict that the electron pairs are arranged tetrahedrally, and that in turn would suggest sp^3 hybrids. However, there is a problem with this. The angle between bonds in a tetrahedral structure is 109.5°, but the bond angle in H_2S is only 92°. Is there a better explanation for the kinds of orbitals used by the sulfur? Let's look once again at the electronic structure of sulfur.

This time we haven't included the $3d$ subshell because it isn't needed.

Notice that the two unpaired electrons sulfur needs for bonding are in p orbitals. By now you should know that p orbitals are at 90° angles to each other, which is very close to the bond angle that we are trying to explain. Therefore, it appears that in this molecule, the sulfur atom *does not* use hybrid orbitals. The bonding picture for the molecule is simply

S $\underset{3s}{\underline{\uparrow\downarrow}}$ $\underset{3p}{\underline{\uparrow\downarrow}\;\underline{\uparrow\downarrow}\;\underline{\uparrow\downarrow}}$ (Colored arrows are hydrogen electrons.)

This example illustrates how we must be careful in applying the various bonding theories. In the absence of actual data about the structure of the molecule, we can use the VSEPR theory to anticipate the structure and then derive a valance bond explanation of the bonding. Usually this explanation will be correct. The case of H_2S, however, illustrates that the VSEPR theory can be wrong about bond angles. When there are actual data about the structure of the molecule, the data must be explained properly, regardless of what the VSEPR theory may predict.

Coordinate covalent bonds

Before we move on, a word should be said about the coordinate covalent bond. An example of this, you remember, is provided by the ammonium ion.

$$\begin{bmatrix} \text{H} \\ \overset{\bullet\times}{\underset{\bullet\times}{\text{H}\,\overset{\bullet}{\times}\,\text{N}\,\overset{\bullet}{\times}\,\text{H}}} \\ \text{H} \end{bmatrix}^{+} \quad \text{or} \quad \begin{bmatrix} \text{H} \\ | \\ \text{H}-\text{N}\!\rightarrow\!\text{H} \\ | \\ \text{H} \end{bmatrix}^{+}$$

According to valence bond theory, two electrons are shared between two overlapping orbitals. It doesn't matter, however, where the electrons come from. If one comes from each of the overlapping orbitals, an ordinary covalent bond is formed. If one orbital is empty and the other is filled, both electrons can come from the filled orbital and a coordinate covalent bond is formed. Thus, we can imagine the coordinate covalent bond in the ammonium ion to be formed by the overlap of an *empty 1s* orbital centered on a proton (a hydrogen ion, H^+) with the *completely filled* lone-pair orbital on the nitrogen of an ammonia molecule as shown in Figure 9.20. The

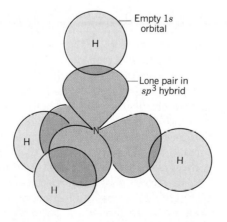

Empty $1s$ orbital

Lone pair in sp^3 hybrid

Figure 9.20. *Formation of a coordinate covalent bond by the overlap of an empty orbital on hydrogen with a filled orbital on nitrogen.*

electron pair is then shared in the region of orbital overlap. Once the bond is formed, of course, it is a full-fledged covalent bond whose properties do not depend on its origin. Therefore, the four N—H bonds in NH_4^+ are identical and the ion is usually represented as simply

$$\begin{bmatrix} & H & \\ & | & \\ H - & N & - H \\ & | & \\ & H & \end{bmatrix}^+$$

This same argument can be extended to other coordinate covalent bonds as well.

9.6 MULTIPLE BONDS

Double and triple bonds occur when two and three pairs of electrons, respectively, are shared between two atoms. As examples, we have seen the molecules ethylene, C_2H_4, and acetylene, C_2H_2.

$$\begin{array}{c} H \\ \diagdown \\ \\ \diagup \\ H \end{array} C = C \begin{array}{c} H \\ \diagup \\ \\ \diagdown \\ H \end{array} \qquad\qquad H - C \equiv C - H$$

ethylene   acetylene

The bonding in ethylene is usually interpreted in the following way. In order to form bonds to three other atoms (two hydrogens and one carbon), each carbon atom employs a set of sp^2 hybrids.

$$C \quad \underset{2s}{\underline{\uparrow\downarrow}} \quad \underset{2p}{\underline{\uparrow}\;\underline{\uparrow}\;\underline{}}$$

gives

$$C \quad \underbrace{\underline{\uparrow}\;\underline{\uparrow}\;\underline{\uparrow}}_{sp^2} \quad \underset{p}{\underline{\uparrow}} \quad \text{(Unhybridized)}$$

Two of these hybrid orbitals are used for overlap with hydrogen $1s$ orbitals, while the third sp^2 orbital overlaps with a similar sp^2 orbital on the other carbon atom, as shown in Figure 9.21. This, then, accounts for all of the C—H bonds in C_2H_4, as well as *one* of the electron pairs shared between the two carbons.

Because of the way the sp^2 orbitals are created, each carbon atom also has an unhybridized p atomic orbital that is perpendicular to the plane of the sp^2 orbitals and that projects above and below the plane of these hybrids (Figure 9.21). When the two carbon atoms are joined together, these p orbitals approach each other sideways and, in addition to the bond formed from the overlap of sp^2 orbitals, a second bond is formed in which the electron cloud is concentrated above and below the carbon – carbon axis. This is also illustrated in Figure 9.21.

In terms of this interpretation, the double bond in ethylene consists of two distinctly different kinds of bonds, and to differentiate between them a specific notation is employed. A bond that concentrates electron density along the line joining the bound nuclei is called a σ **bond (sigma bond)**. The overlap of the sp^2 orbitals of adjacent carbons therefore gives rise to a σ bond. The bond that is formed by the sideways overlap of two p orbitals, and that provides electron density above and below a plane containing both of the bound nuclei, is called a π **bond (pi bond)**.

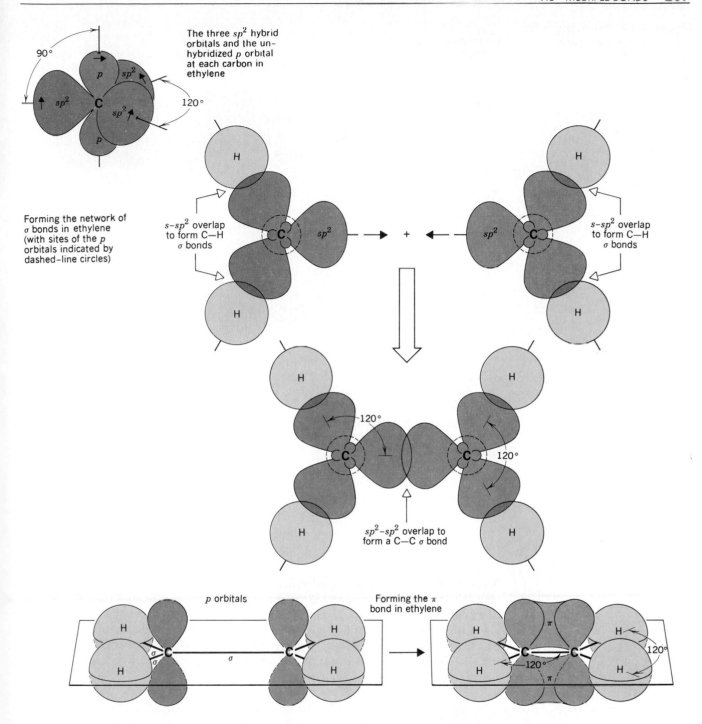

The three sp^2 hybrid orbitals and the un-hybridized p orbital at each carbon in ethylene

Forming the network of σ bonds in ethylene (with sites of the p orbitals indicated by dashed-line circles)

$s-sp^2$ overlap to form C—H σ bonds

$s-sp^2$ overlap to form C—H σ bonds

sp^2-sp^2 overlap to form a C—C σ bond

p orbitals

Forming the π bond in ethylene

Figure 9.21. *Formation of the carbon–carbon double bond in ethylene, C_2H_4. (Adapted from J. R. Holum, Organic and Biological Chemistry, 2nd edition, 1986, John Wiley & Sons, New York.)*

Thus, in ethylene we find the double bond consists of one σ bond and one π bond. Notice that in this double bond the two electron pairs manage to avoid one another by occupying different regions in space.

Another point to note is that bonds formed by the overlap of the hydrogen 1s

The σ bond is sandwiched between the two halves of the π bond like a frankfurter between the two halves of a bun.

Figure 9.22. *Formation of the carbon–carbon triple bond in acetylene, C₂H₂.*

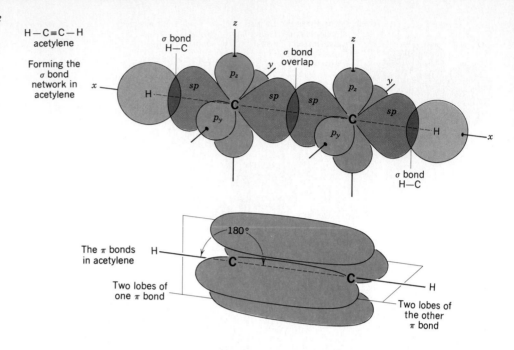

orbitals with carbon sp^2 hybrid orbitals (Figure 9.21) also concentrate electron density along a line joining bound atoms. Therefore, these C—H bonds would also be termed σ bonds.

In acetylene each carbon is bound to only two other atoms, a hydrogen and a carbon atom. Two orbitals are needed for this purpose, and a pair of sp hybrid orbitals are used.

C $\underset{2s}{\uparrow\downarrow}$ $\underset{2p}{\uparrow \quad \uparrow \quad \underline{}}$

$\underbrace{\uparrow \quad \uparrow}_{sp}$ $\underbrace{\uparrow \quad \uparrow}_{p}$ (Unhybridized)

This leaves two singly occupied unhybridized p orbitals on each carbon that are mutually perpendicular as well as perpendicular to the sp hybrids. When the carbon atoms join by way of σ bond formation between an sp hybrid orbital on each carbon, the p orbitals can also overlap to yield two π bonds that surround the axis joining the carbon nuclei (Figure 9.22). A triple bond therefore consists of one σ and two π bonds. The two π bonds in acetylene (or in any triple bond) give a total electron distribution that is cylindrical about the bond axis.

Summary

In arriving at the structure of a molecule, such as ethylene or acetylene, *the shape of the molecular framework is determined by the σ bonds that arise from the overlap of the hybrid orbitals.* Double and triple bonds in a structure result from additional π bonds. In summary, we find the following:

single bond—one σ bond
double bond—one σ + one π bond
triple bond—one σ + two π bonds

EXAMPLE 9.11. IDENTIFYING SIGMA AND PI BONDING, AND THE KINDS OF HYBRID ORBITALS USED BY ATOMS IN A MOLECULE

PROBLEM: Identify the kinds of hybrid orbitals used by the atoms in acetic acid whose structure is

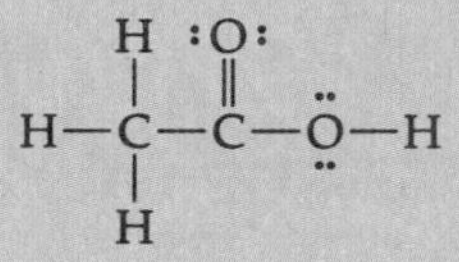

What kinds of bonds (σ, π) exist between the atoms?

SOLUTION: To identify the kinds of hybrid orbitals that an atom uses, we simply count the number of groups of electrons around the atom. Then we choose a hybrid set that has that number of orbitals. For example, the carbon on the left has four bonds (four pairs) and uses sp^3 orbitals. The carbon next to it has three groups of electrons, which means it uses sp^2 orbitals. The doubly bonded oxygen has three groups of electrons, so it uses sp^2 orbitals, too. Finally, the singly bonded oxygen has four pairs around it and must use sp^3 hybrids. (No doubt you've noticed that we haven't said anything about hydrogen—hydrogen always uses only its $1s$ orbital for bonding.)

Now we can identify the kinds of bonds in the molecule.

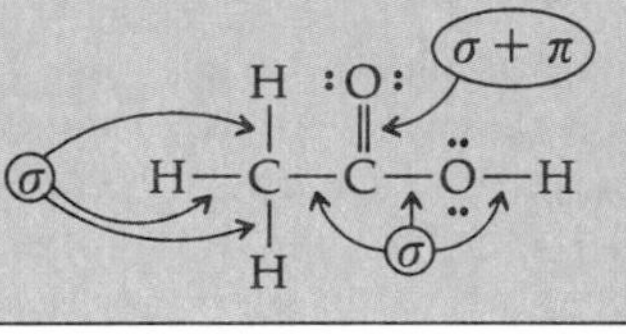

Hydrogen only forms one σ bond.

9.7 RESONANCE STRUCTURES

In Chapter 8 we saw that there are instances in which we cannot draw a single satisfactory Lewis dot formula for a molecule or ion. Some examples, you might remember, are SO_2, SO_3, and NO_2^-. Sulfur dioxide, for instance, is represented as

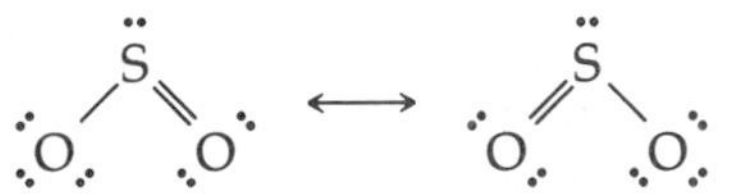

and it was stated that the actual electronic structure of this molecule corresponds to a resonance hybrid of these two structures.

Electron dot formulas, as we have drawn them, closely correspond to the valence bond pictures developed in the preceding sections. Each pair of dots drawn between two atoms represents a pair of electrons shared in a region where atomic orbitals of the bonded atoms overlap. When we draw one of the resonance structures of SO_2, we are therefore referring to a bonding picture in which one of the S—O bonds consists of a single σ bond while the other is composed of one σ and one π bond.

When the valence bond theory was developed, it was recognized that there were numerous instances in which a single valence bond structure was inadequate in accounting for the molecular structure, so the concept of resonance evolved. The inability in these instances to draw a single picture to describe the electron density in the molecule is one drawback of valence bond theory. Nevertheless, the correspondence between the valence bond structures that are based on orbital overlap and the simple electron dot formulas makes the valence bond concept a very useful one.

9.8 SINGLE BONDS VERSUS MULTIPLE BONDS: THE MOLECULAR STRUCTURES OF THE ELEMENTAL NONMETALS

We have seen that atoms are able to share electrons in two basic ways. One is by the formation of σ bonds, which frequently involves the overlap of hybrid orbitals. The second is by the formation of π bonds, which normally requires the sideways overlap of unhybridized p orbitals. (Pi bonds can also be formed by d orbitals, but we do not discuss these in this book.)

Not all atoms have the same tendencies to form π bonds. The ability of an atom to form π bonds determines its ability to form multiple bonds, and this in turn greatly affects the kinds of molecular structures that the element produces. One of the most striking illustrations of this is the molecular structures of the elemental nonmetals and metalloids.

If we take a brief look at the electronic structures of the nonmetals and metalloids, we see that their atoms generally contain valence shells that are only partially filled. Exceptions to this, of course, are the noble gases, which have electronically complete valence shells. We have also seen that atoms tend to complete their valence shells by bond formation. Therefore, in their elemental states, most of the nonmetallic elements do not exist as single atoms, but instead form bonds to each other and produce more complex molecular species. (The only nonmetals that actually occur in nature as collections of single atoms are the noble gases.)

One of the controlling factors in determining the complexity of the molecular structures of the nonmetals and metalloids is their abilities to form multiple bonds. Small atoms, such as those in period 2, are able to approach each other closely. As a result, effective sideways overlap of their p orbitals can occur, and these atoms form strong π bonds. Therefore, carbon, nitrogen, and oxygen are able to form multiple bonds about as easily as they are able to form single bonds. On the other hand, when the atoms are large — which is the case for atoms from periods 3, 4, and so on — π-type overlap between their p orbitals is relatively ineffective. Therefore, rather than form a double bond consisting of one σ bond and one π bond, these elements prefer to use two separate σ bonds to bond their atoms together. This leads to a useful generalization: *Elements in period 2 are able to form multiple bonds fairly readily, while elements below them in periods 3, 4, 5, and 6 have a tendency to prefer single bonds.* Let's look at some of the consequences of this.

Oxygen and nitrogen have 6 and 5 electrons, respectively, in their valence shells. This means that an oxygen atom needs two electrons to complete its valence shell, and a nitrogen atom needs three. Although a perfectly satisfactory Lewis structure for O_2 can't be drawn, experimental evidence suggests that the oxygen molecule does possess a double bond. The nitrogen molecule, which we discussed earlier, contains a triple bond. Oxygen and nitrogen, because of their small size, are capable of multiple bonding because they are able to form strong π bonds. This allows them to form a sufficient number of bonds with just a *single* neighbor to complete their valence shells, so they are able to form diatomic molecules.

Oxygen, in addition to forming the stable species O_2, also can exist in another exceedingly reactive molecular form called **ozone,** which has the formula O_3. The structure of ozone can be represented as a resonance hybrid

In ozone, oxygen is also participating in π bonding.

This unstable molecule can be generated by the passage of an electric discharge through ordinary O_2, and the pungent odor of ozone can often be detected in the vicinity of high-voltage electrical equipment. It is also formed in limited amounts in the upper atmosphere by the action of ultraviolet radiation from the sun on O_2. The presence of ozone in the upper atmosphere shields the earth and its creatures from exposure to intense and harmful ultraviolet light.

The existence of an element in more than one form, either as the result of differences in molecular structure as with O_2 and O_3, or as the result of differences in the packing of molecules in the solid, is a phenomenon called **allotropy.** The different forms of the element are called **allotropes** — O_2 is one allotrope of oxygen, and O_3 is another.

Let's turn our attention now to another period 2 nonmetal, carbon. This element has four electrons in its valence shell, so it must share four electrons to complete its octet. There is no way for carbon to form a quadruple bond, so a simple C_2 species is not stable under ordinary conditions. Instead, carbon completes its octet in either of two different ways, and two allotropic forms of elemental carbon are found. One of these is **diamond.** In diamond, each carbon atom uses sp^3 hybrid orbitals to form covalent bonds to four other carbon atoms at the corners of a tetrahedron. Each of those atoms, in turn, is bonded through sp^3 hybrid orbitals to three more, and so on, as illustrated in Figure 9.23a. This produces a gigantic three-dimensional network of interconnected covalent bonds. A diamond crystal of gem quality, such as one of those in Figure 9.23b, therefore consists of a huge number of carbon atoms covalently bonded together in one enormous molecule. Since breaking a diamond crystal involves rupturing a very large number of covalent bonds, diamond is extremely hard.

The second allotrope of carbon is called **graphite.** This is the form that probably is more familiar to you. In graphite, the carbon atoms are arranged in the form of hexagonal rings connected together in large planar sheets, perhaps somewhat remi-

The carbon in soot, coal, and charcoal briquettes is composed of graphite fragments.

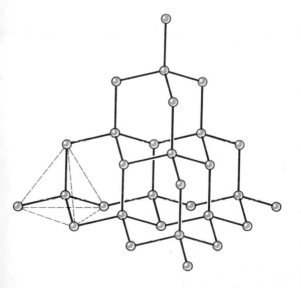

Figure 9.23. (a) *In diamond, each carbon atom is covalently bonded to four other carbon atoms.* (b) *A gem-quality diamond.*

(a)   *(b)*

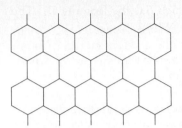

Figure 9.24. *Sigma-bond framework of a graphite sheet.*

Figure 9.25. *One of the resonance structures for graphite.*

C $\frac{\uparrow\downarrow}{2s}$ $\frac{\uparrow\;\;\uparrow}{2p}$

hybridization

$\frac{\uparrow\;\;\uparrow\;\;\uparrow}{sp^2}$ $\frac{\uparrow}{p}$

The unhybridized *p* orbitals are perpendicular to the plane of the carbon atoms.

It has been suggested that small molecules such as O_2 or H_2O trapped between layers of carbon atoms in graphite serve as microscopic ball bearings that help the layers slide over each other easily.

niscent of chicken wire (Figure 9.24). Each carbon is surrounded by three nearest neighbors at angles of 120° from one another. This means that the molecular framework is based on σ bonds produced by the overlap of *sp²* hybrid orbitals on the carbon atoms. On each of the carbons throughout the entire structure, there remains an unhybridized *p* orbital containing an unpaired electron. These *p* orbitals are ideally situated to form π bonds with their neighbors, and one of the many resonance forms of a graphite sheet is illustrated in Figure 9.25.

In the total graphite structure, shown in Figure 9.26, the planes of carbon atoms are stacked in layers so that each carbon atom lies above another in every second layer. Within any given layer, adjacent carbon atoms are fairly close together (141 pm), while the spacing between successive planes is much greater (335 pm). The different planes of carbon atoms are not held together by covalent bonds, but instead by much weaker forces. As a result, the layers are able to slide over one another with relative ease, and as you may know, graphite has many applications as a dry lubricant. As every schoolchild is taught, the lead in a lead pencil is actually graphite, and when one writes with a pencil, the layers of graphite slide off one another and onto the paper.

In graphite, carbon exhibits multiple bonding, as do nitrogen and oxygen in their molecular forms. As we noted earlier, their abilities to do this reflects their

Figure 9.26. *The stacking of sheets of carbon atoms in the graphite structure. The individual layers are weakly attracted to each other and can slide over each other relatively easily.*

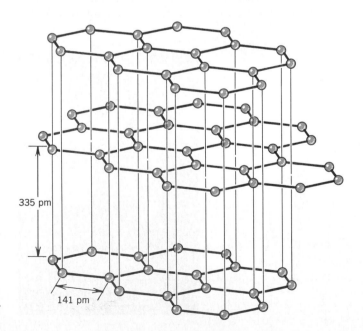

335 pm

141 pm

ability to form strong π bonds — a requirement for the formation of a double or triple bond. When we move to the third and successive periods, a different state of affairs exists. Here, we have much larger atoms that are not able to form strong π bonds. These elements prefer to form only single bonds, and the molecular structures of the free elements reflect this.

Elements of Group VIIA

Each of the elements in Group VIIA are diatomic in the free state. This is so because π bonding is not necessary in any of their molecular structures. Chlorine, for example, requires just one electron to complete its octet, so it only needs to form one covalent bond with another atom. Therefore, it forms one σ bond to another chlorine atom and is able to exist as diatomic Cl_2. Bromine and iodine form diatomic Br_2 and I_2 molecules for the same reason. The structures of the remaining nonmetals and metalloids are considerably more complex, however.

Elements of Group VIA

Below oxygen in Group VIA is sulfur, which has the Lewis symbol

$$:\overset{\displaystyle\cdot}{\underset{\displaystyle\cdot\cdot}{S}}:$$

It requires two electrons to obtain an octet, so it must form two covalent bonds. But sulfur doesn't form π bonds well to other sulfur atoms, so it prefers to form two single bonds to different sulfur atoms. Each of these also prefers to bond to two different sulfur atoms and this gives rise to a

$$-\overset{\displaystyle\cdot\cdot}{\underset{\displaystyle\cdot\cdot}{S}}-\overset{\displaystyle\cdot\cdot}{\underset{\displaystyle\cdot\cdot}{S}}-\overset{\displaystyle\cdot\cdot}{\underset{\displaystyle\cdot\cdot}{S}}-\overset{\displaystyle\cdot\cdot}{\underset{\displaystyle\cdot\cdot}{S}}-$$

sequence. Actually, in sulfur's most stable form, the sulfur atoms are arranged in an 8-member ring to give a molecule with the formula S_8. The S_8 ring has a puckered crownlike shape, which is illustrated in Figure 9.27. Selenium, below sulfur in Group VIA, also forms Se_8 rings in one of its allotropic forms. And selenium and tellurium also exist in a gray form in which there are long Se_x and Te_x chains (the subscript x is a large number).

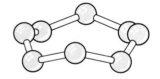

Figure 9.27. *The structure of the S_8 ring.*

Elements of Group VA

The elements below nitrogen in Group VA all have five valence electrons. Phosphorus is an example

$$\cdot\overset{\displaystyle\cdot}{\underset{\displaystyle\cdot\cdot}{P}}\cdot$$

To achieve a noble gas structure, it must acquire three more electrons. Since there is little tendency for phosphorus to form multiple bonds, as nitrogen does when it forms N_2, the octet is completed by the formation of three single bonds to three different phosphorus atoms.

   The simplest elemental form of phosphorus is a waxy solid called **white phosphorus.** It consists of P_4 molecules in which each phosphorus atom lies at a corner of a tetrahedron, as illustrated in Figure 9.28. Notice that in this structure each phosphorus is bound to three others. This allotrope of phosphorus is very reactive partly

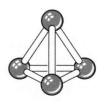

Figure 9.28. *The structure of white phosphorus, P_4.*

Figure 9.29. *Streamers of white smoke arc through the air as a phosphorus bomb explodes. The burning particles are chunks of white phosphorus.*

By contrast, the strong triple bond in N_2 makes elemental nitrogen very unreactive except at high temperatures.

because of the very small P—P—P bond angle of 60°. At this small angle, the orbitals of the phosphorus atoms don't overlap very well, so the bonds are weak. This means that breaking a P—P bond occurs easily, and this is the first step in causing this molecule to react with something. As a result, P_4 molecules are readily attacked by other chemicals, especially oxygen. White phosphorus is so reactive toward oxygen that it ignites and burns spontaneously in air. For this reason, white phosphorus is used in military incendiary devices, and you've probably seen movies in which exploding phosphorus shells produce arching showers of smoking particles (Figure 9.29).

A second allotrope of phosphorus that is much less reactive is called **red phosphorus.** At the present time, its structure is unknown, although it has been suggested that it contains P_4 tetrahedra joined at the corners as shown in Figure 9.30. Red phosphorus is also used in explosives and fireworks, and it is mixed with fine sand and used on the striking surfaces of matchbooks. As a match is drawn across the surface, friction ignites the phosphorus, which then ignites the ingredients in the tip of the match.

A third allotrope of phosphorus is called **black phosphorus,** which is formed by heating white phosphorus at very high pressures. This variety has a layer structure in which each phosphorus atom in a layer is covalently bonded to three others in the same layer. As in graphite, these layers are stacked one atop another, with only weak forces between the layers. As you might expect, black phosphorus has many similarities to graphite.

The elements arsenic and antimony, which are just below phosphorus in the periodic table, are also able to form somewhat unstable yellow allotropic forms containing As_4 and Sb_4 molecules, but their most stable forms have a metallic appearance with structures similar to black phosphorus.

Figure 9.30. *Red phosphorus is believed to be composed of long chains of phosphorous tetrahedra connected at their corners.*

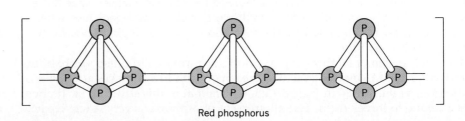

Red phosphorus

Elements of Group IVA

Finally, we look at the heavier elements in Group IVA: silicon and germanium. To complete their octets, each must form four covalent bonds. Unlike carbon, however, they have no apparent tendency to form multiple bonds, so they don't form allotropes that have a graphite structure. Instead, each of them forms a solid with a structure similar to diamond.

Summary

In this section we have examined the structures of the elemental forms of the nonmetallic elements. Some of these are simply diatomic molecules, but others are quite complex. This complexity can be traced to two factors. One is the number of electrons that the atom of the element needs to complete its octet. The second is the preference of the heavier (and larger) nonmetals and metalloids for single bonds. This preference prevents them from forming diatomic molecules and causes them to form more complex structures.

REVIEW QUESTIONS AND PROBLEMS

Problems whose numbers are in blue have their answers in Appendix D at the back of the book.

Molecular Shapes

9.1 Make sketches of the five common molecular shapes. (If necessary, refer to the drawings in Section 9.1.)

9.2 What is the bond angle in a linear molecule?

9.3 What are the bond angles in planar triangular, tetrahedral, and octahedral molecules?

9.4 What are the bond angles in a trigonal bipyramidal molecule?

9.5 How many atoms are bonded to the central atom in a molecule that has
(a) a tetrahedral shape?
(b) an octahedral shape?
(c) a trigonal bipyramidal shape?

VSEPR Theory

9.6 What is the basic postulate of the valence shell electron pair repulsion theory?

9.7 Use the valence shell electron pair repulsion theory to predict the geometry of each of the following (in each case, the central atom is written first):
(a) NF_3
(b) PH_4^+
(c) CO_3^{2-}
(d) NO_2^-
(e) $SeCl_4$
(f) ICl_2^-
(g) BrF_5
(h) CCl_4
(i) $AlCl_6^{3-}$
(j) $SbCl_5$
(k) $SnCl_2$

9.8 Make sketches of the structures of the molecules and ions in Question 9.7.

9.9 For each of the following, use the valence shell electron pair repulsion theory to predict (1) the arrangement of electron pair groups around the central atom (written first in the formula) and (2) the molecular shape.
(a) CH_2O
(b) ClO_3^-
(c) SiF_6^{2-}
(d) PCl_3
(e) $SOCl_2$
(f) $SnCl_4$
(g) SO_3
(h) BF_4^-
(i) OF_2
(j) XeF_4

9.10 Use the valence shell electron pair repulsion theory to predict for each of the following: (1) the geometric arrangement of electron pairs around the central atom (written first in the formulas), (2) the molecular shape.
(a) ClO_2^-
(b) SCl_2
(c) $SbCl_6^-$
(d) PCl_4^+
(e) IF_4^-
(f) PO_4^{3-}
(g) CH_3^+
(h) ICl_3
(i) NO_3^-
(j) AsH_3
(k) $POCl_3$

9.11 Describe the changes in molecular geometry that take place during the following reactions:
(a) $BF_3 + F^- \rightarrow BF_4^-$
(b) $PCl_5 + Cl^- \rightarrow PCl_6^-$
(c) $ICl_3 + Cl^- \rightarrow ICl_4^-$
(d) $SF_2 + F_2 \rightarrow SF_4$
(e) $C_2H_2 + H_2 \rightarrow C_2H_4$
(f)
$$Cl-\overset{\overset{O}{\|}}{C}-Cl \longrightarrow COCl^+ + Cl^-$$

9.12 On p. 266, it was suggested that lone electron pairs tend to repel rather strongly electron pairs between bonded atoms; in other words, lone-pair–bond-pair repulsions are greater than bond-pair–bond-pair repulsions. This will lead to structural distortions of some of the idealized molecular geometries pictured in Figures 9.2 through 9.5. Predict the nature of these distortions and sketch the shapes of the resulting molecules.

Polarity of Molecules

9.13 Why are some covalent bonds polar? What is a dipole?

9.14 Why can a molecule that has polar bonds be a nonpolar molecule?

9.15 The molecule SO_2 is polar, but SO_3 is nonpolar. Why?

9.16 Which of the following are polar molecules? (The central atom is written first in each of them.)
(a) SCl_2 (f) XeF_2
(b) BF_3 (g) SF_4
(c) ICl_3 (h) XeO_3
(d) $POCl_3$ (i) $SnCl_4$
(e) PCl_5 (j) BrF_5

9.17 A certain molecule with the formula MX_3 is nonpolar, even though M and X differ substantially in electronegativity. What is the likely geometry of the MX_3 molecule?

Valence Bond Theory and Hybrid Orbitals

9.18 What are the basic postulates of the valence bond theory?

9.19 What is meant by orbital overlap? What is its importance in covalent bond formation?

9.20 Use valence bond theory to explain the bonding in the Cl_2 molecule. Make a sketch that shows the orbital overlap involved in the formation of the Cl—Cl bond. What happens to the spins of the electrons as the Cl—Cl bond is formed?

9.21 Use valence bond theory to explain the bonding in the HCl molecule. Use a sketch to show the overlap of orbitals that leads to the formation of the bond.

9.22 Based on previous discussions in Chapter 8, describe what happens to the potential energy of the H and Cl atoms as they come together to form the H—Cl bond. Sketch a graph showing how the potential energy varies with internuclear distance. In HCl, the average bond length is 127 pm, and the bond energy is 431 kJ/mol. Indicate these values on your sketch.

9.23 What is a hybrid orbital? How is the shape of a hybrid orbital different than the shape of a "pure" atomic orbital? Why do atoms usually prefer to use hybrid orbitals in bond formation?

9.24 What are the geometries of the following kinds of hybrid orbitals?
(a) sp (b) sp^2 (c) sp^3 (d) sp^3d (e) sp^3d^2

9.25 What angles exist between the orbitals in
(a) sp^3 hybrids   (c) sp hybrids
(b) sp^2 hybrids   (d) sp^3d^2 hybrids

9.26 Make a sketch that shows how the bonds are formed by orbital overlap in the planar triangular BCl_3 molecule. Diagram the electron configuration of boron in BCl_3.

9.27 Why is it necessary to suggest carbon's use of hybrid orbitals in attempting to explain the structure of CH_4?

9.28 Give *two* explanations for the 104.5° bond angle in H_2O.

9.29 What evidence suggests that nitrogen uses sp^3 hybrid orbitals for bonding in NH_3?

9.30 In arsine, AsH_3, the H—As—H bond angles are very close to 90°. Based on this information, explain the bonding in AsH_3 according to the valence bond theory.

9.31 The H—Se—H bond angle in H_2Se is 91°. According to valence bond theory, what kinds of orbitals does Se use for bonding in this molecule?

9.32 In PF_3, the F—P—F bond angle is approximately 104°. Why is it more reasonable to suggest that phosphorus uses sp^3 hybrids for bonding, rather than pure p orbitals? Diagram the outer-shell electron configuration of P and F. Show how hybridization takes place and how the bonds are formed.

9.33 In PF_3, is there any way to tell *for sure* whether fluorine uses a pure p orbital for bonding or some kind of hybrid orbital? Explain your answer.

9.34 Describe the bonding in the molecule $SnCl_2$. (Begin by using the VSEPR theory to predict its shape.)

9.35 Use the VSEPR theory to predict the shapes of each of the following. Then, use the orbital diagrams for the central atom to explain the bonding in each species.
(a) BCl_3 (f) $SbCl_6^-$
(b) NH_4^+ (g) PCl_3
(c) PCl_5 (h) TeF_4
(d) $AlCl_6^{3-}$ (i) ClO_4^-
(e) $BeCl_2$

9.36 Use your answers to Question 9.7 to suggest the types of hybrid orbitals that would be used in valence bond theory to account for these geometries.

9.37 What kind of hybrid orbitals would be used by the central atom in each species in Question 9.9?

9.38 Diagram the bonding in $SbCl_5$. What kind of hybrid orbitals are involved in the bonding?

9.39 It is possible for SiF_4 to react with F^- to give SiF_6^{2-} but it is not possible for CF_4 to form CF_6^{2-}. Why?

9.40 We have discussed the reaction $BCl_3 + NH_3 \rightarrow Cl_3BNH_3$ earlier (Chapter 8). What kind of hybrid orbitals are used by B and N before and after reaction? How does the geometry change about B and N as the reaction occurs?

9.41 Which of the species in Question 9.35 have one or more bonds that would be considered to have been formed by way of coordinate covalent bonding? How, if at all, does a coordinate covalent bond differ from a normal covalent bond once it has been formed?

9.42 $SnCl_4$ is a volatile liquid composed of individual $SnCl_4$ molecules. Describe the bonding that is expected in this molecule.

9.43 Hybrid orbitals are not symmetrical about the nucleus. They concentrate electron density on the side of the nucleus where the orbital is large. Lone electron pairs in hybrid orbitals therefore are expected to contribute to the dipole moment of the molecule. It is observed experimentally that NF_3 is nearly a nonpolar molecule; NH_3 is very polar. The electronegativity difference between N and F is nearly the same as that between N and H. How does this support the view that in both NF_3 and NH_3 the nitrogen uses sp^3 hybrid orbitals?

Multiple Bonds and Resonance

9.44 Describe a σ bond, a π bond. What constitutes a double bond? A triple bond?

9.45 What kind of hybrid orbitals are used by each atom in the molecule below? What kinds of bonds (σ, π) occur between the atoms?

$$H-C\equiv C-\overset{\underset{|}{H}}{C}=\overset{\underset{|}{H}}{C}-\overset{\overset{:O:}{\|}}{C}-\overset{..}{\underset{..}{O}}-\overset{\underset{|}{H}}{\underset{\underset{H}{|}}{C}}-H$$

9.46 Describe the bonding in the N_2 molecule according to the valence bond theory.

9.47 Use valence bond theory to explain the bonding in the cyanide ion, CN^-.

9.48 Make a sketch showing the overlap of orbitals to form the σ and π bonds in the formaldehyde molecule.

$$\overset{H}{\underset{H}{>}}C=\overset{..}{O}:$$

9.49 Draw resonance structures for the nitrite ion, NO_2^-. What kind of hybrid orbitals are used by the nitrogen atom in this ion?

Molecular Structures of the Nonmetals

9.50 Why are the period 2 nonmetals able to form much stronger π bonds than the nonmetals of period 3? Why does a period 3 nonmetal prefer to form several σ bonds instead of one σ bond and several π bonds?

9.51 Even though the nonmetals of periods 3, 4, and 5 do not tend to form π bonds, each of the halogens is able to exist as a diatomic molecule (Cl_2, Br_2, I_2). Why?

9.52 What are *allotropes*?

9.53 What are the two allotropes of oxygen?

9.54 Describe the molecular structure of elemental sulfur.

9.55 What is the molecular structure of white phosphorus?

9.56 Why is white phosphorus so chemically reactive?

9.57 Compare the molecular structures of white and red phosphorus.

9.58 Compare the structures of graphite and diamond. What kinds of hybrid orbitals are used by carbon in each of these?

9.59 Silicon only forms a diamondlike molecular structure. Why doesn't it form a structure similar to graphite?

10

CHEMICAL REACTIONS
AND THE PERIODIC
TABLE

Now that you have learned something about the structures of atoms and the way atoms attract each other by chemical bonds, we can take a fresh look at chemical reactivity to learn how the chemical properties of substances are related to electronic structure and bonding. As in Chapter 4, one of our principal goals will be to correlate the chemical properties and reactions of the elements with their positions within the periodic table. In this way, chemical facts become simpler to remember and the periodic table becomes our guide in following trends in chemical reactivity.

10.1 THE REACTIONS OF METALS AS REDUCING AGENTS

You have learned that metals are elements with low ionization energies and low electronegativities: They lose electrons easily and have little tendency to gain them. As a result, in their reactions with the nonmetallic elements, they tend to form positive ions (cations), and in the process they are oxidized. As discussed in Chapter 4, this means that they serve as reducing agents. A reaction that we've examined before that illustrates this is the reaction of sodium with chlorine to give sodium chloride.

$$2Na(s) + Cl_2(g) \longrightarrow 2NaCl(s)$$

Recall that oxidation is a loss of electrons or an increase in oxidation number. Something that is oxidized causes something else to be reduced and is said to be a reducing agent.

The chlorine oxidizes the sodium to give Na^+ ions, and in the process we say that the sodium reduces the chlorine to Cl^- (an anion); chlorine is the oxidizing agent and sodium is the reducing agent.

The ability of metals to serve as reducing agents is not restricted to their reactions with the nonmetallic elements. Many other substances can oxidize metals and therefore cause metals to be reducing agents. By studying these reactions, we are able to rank the metals according to their reducing abilities.

The reaction of metals with acids

One of the characteristic ways that metals behave as reducing agents is in their reactions with acids. A typical example is the reaction of zinc with hydrochloric acid or sulfuric acid.

$$Zn(s) + 2HCl(aq) \longrightarrow ZnCl_2(aq) + H_2(g)$$

$$Zn(s) + H_2SO_4(aq) \longrightarrow ZnSO_4(aq) + H_2(g)$$

The net ionic equation for both of these is the same.

$$Zn(s) + 2H^+(aq) \longrightarrow Zn^{2+}(aq) + H_2(g)$$

In the reaction, the substance oxidized is zinc and the substance reduced is hydrogen ion. Therefore, zinc is the reducing agent and hydrogen ion is the oxidizing agent. (Recall that H^+ is actually attached to H_2O in solution, so the reacting species is the hydronium ion, H_3O^+. For simplicity, however, we abbreviate H_3O^+ as H^+ because the hydrogen ion is the "active ingredient" in the hydronium ion.)

In aqueous solutions of HCl or H_2SO_4 the only oxidizing agent is H^+; under ordinary conditions neither Cl^- nor $SO_4{}^{2-}$ is reduced. Acids such as HCl and H_2SO_4, in which the only effective oxidizing agent is H^+, are called **nonoxidizing acids.** (This may seem strange, since the acid does attack metals by oxidation, but the term is used to differentiate these acids from others in which the anion of the acid is also an oxidizing agent.)

Some other common metals that also react with nonoxidizing acids are iron, magnesium, and aluminum. In each case, the reaction gives hydrogen and the metal ion in solution.

$$Fe(s) + 2H^+(aq) \longrightarrow Fe^{2+}(aq) + H_2(g)$$

$$Mg(s) + 2H^+(aq) \longrightarrow Mg^{2+}(aq) + H_2(g)$$

$$2Al(s) + 6H^+(aq) \longrightarrow 2Al^{3+}(aq) + 3H_2(g)$$

If you spill battery acid, H_2SO_4, on the metal parts of your car, the acid will attack the metal unless it is washed off with water.

In general, then, for metals that react with nonoxidizing acids

$$metal + H^+ \longrightarrow metal\ ion + H_2(g)$$

As implied in the preceding paragraph, not all metals are able to be oxidized by hydrogen ion. Two common metals that fit into this category are copper and silver; if either one of these metals is placed in a solution of hydrochloric acid, nothing happens. This simply reflects the fact that some metals, such as copper and silver, are more difficult to oxidize than others, and that H^+ simply is not up to the job. A stronger oxidizing agent than H^+ is needed to oxidize these metals.

One acid capable of dissolving copper and silver is nitric acid, HNO_3. This is an example of an **oxidizing acid;** in addition to H^+, a solution of this acid also contains the nitrate ion, which is a stronger oxidizing agent than H^+. The vigorous reaction of copper with concentrated HNO_3 is shown in Figure 10.1. The reddish-brown gas that is produced is nitrogen dioxide, NO_2, formed in the reaction

$$Cu(s) + 2NO_3{}^-(aq) + 4H^+(aq) \longrightarrow Cu^{2+}(aq) + 2NO_2(g) + 2H_2O$$

An iron nail in a solution of HCl gradually evolves hydrogen as the metal is oxidized by H^+ ion. The small bubbles contain H_2 gas.

In this reaction, it is the $NO_3{}^-$ ion that is reduced to NO_2. No H_2 is formed because H^+ is not reduced; the hydrogen ions become incorporated in the H_2O that is also formed among the products.

When HNO_3 serves as an oxidizing agent, the products depend, to a degree, on how concentrated the acid is. For example, with copper there are the following reactions:

With dilute HNO_3

$$3Cu(s) + 2NO_3{}^-(aq) + 8H^+(aq) \longrightarrow 3Cu^{2+}(aq) + 2NO(g) + 4H_2O$$

Figure 10.1. *A pure copper penny reacts violently with concentrated nitric acid, as this sequence of photographs shows. The dark red-brown vapors are nitrogen dioxide, NO_2, which is the same gas that gives smog its characteristic color.*

With concentrated HNO_3

$$Cu(s) + 2NO_3^-(aq) + 4H^+(aq) \longrightarrow Cu^{2+}(aq) + 2NO_2(g) + 2H_2O$$

Similar reactions occur with silver. Once again regardless of the concentration of the HNO_3, H_2 is not among the products. Instead, nitrate ion is reduced to either NO or NO_2.

Earlier we mentioned sulfuric acid as an example of a nonoxidizing acid, and this is indeed how sulfuric acid behaves in dilute aqueous solutions. However, when concentrated and hot, sulfuric acid is also an oxidizing agent. For example, hot concentrated sulfuric acid reacts with copper as follows:

$$Cu + 2H_2SO_4 \xrightarrow{\text{heat}} CuSO_4 + SO_2 + 2H_2O$$

Hot concentrated sulfuric acid is a dangerous substance.

The net ionic equation is

$$Cu(s) + 4H^+(aq) + SO_4^{2-}(aq) \longrightarrow Cu^{2+}(aq) + SO_2(g) + 2H_2O$$

Notice that in this case it is the sulfate ion, SO_4^{2-}, that is reduced to SO_2 rather than H^+.

The tendency of metals to react with acids provides a very crude method for sorting metals according to their abilities to serve as reducing agents. Metals such as zinc, iron, magnesium, and aluminum, which are able to react with H^+, are more easily oxidized and are better reducing agents than metals such as copper and silver, which are unaffected by nonoxidizing acids. But how can we differentiate among zinc, iron, magnesium, and aluminum, and how do copper and silver compare in their abilities to be reducing agents?

The activity series for metals

The reaction of an acid with a metal is characteristic of a much broader class of chemical reactions in which one element displaces another from a compound. Some people call them **single displacement reactions.** Another example of this kind of reaction is the change that occurs when a strip of metallic zinc is dipped into a solution containing copper sulfate (Figure 10.2). After a while, the zinc strip has acquired a heavy deposit of reddish-brown copper, and the blue color of the copper

Figure 10.2. *The reaction of zinc with copper ion. (Left) A piece of shiny zinc next to a beaker containing a copper sulfate solution. (Center) When the zinc is placed into the solution, copper ions are reduced to the free metal as the zinc dissolves. (Right) After a while the zinc becomes coated with a thick red-brown layer of copper. Notice that the blue color of the solution has faded somewhat.*

ion in the solution has faded. If we were to analyze the solution, it would be found to contain zinc ion. The net ionic reaction that has occurred is

$$Zn(s) + Cu^{2+}(aq) \longrightarrow Cu(s) + Zn^{2+}(aq)$$

Notice the similarity to the reaction of zinc with hydrogen ion.

$$Zn(s) + 2H^+(aq) \longrightarrow H_2(g) + Zn^{2+}(aq)$$

Reactions like that of zinc with copper ion also allow us to rank metals according to their ease of oxidation. For example, we have just seen that zinc is able to reduce copper ion in a solution, but if we dip a copper strip into a solution that contains Zn^{2+}, nothing happens (as we see in Figure 10.3).

$$Cu(s) + Zn^{2+}(aq) \longrightarrow \text{No reaction}$$

Thus, zinc can displace copper from copper compounds, but copper can't displace zinc from zinc compounds. Stated another way, zinc willingly gives up electrons to copper ion, but copper will not give up electrons to zinc ion. What this means is that zinc is more easily oxidized than copper (a fact that we've already established by comparing the effects of H^+ on zinc and copper metals.)

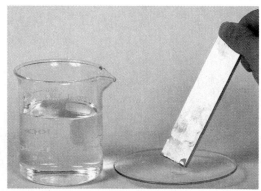

Figure 10.3. *Although metallic zinc will displace copper from a solution that contains Cu^{2+} ion, metallic copper will not displace Zn^{2+} from its solutions. Here we see that the copper bar is unaffected by being dipped into a solution of zinc sulfate.*

By comparing the abilities of metals to displace others from compounds, we can place the metals in order of decreasing ease of oxidation. For example, consider the following reactions that are observed to occur experimentally:

$$Fe(s) + Pb^{2+}(aq) \longrightarrow Fe^{2+}(aq) + Pb(s)$$

$$Mg(s) + Fe^{2+}(aq) \longrightarrow Mg^{2+}(aq) + Fe(s)$$

$$Pb(s) + Cu^{2+}(aq) \longrightarrow Pb^{2+}(aq) + Cu(s)$$

Let's see what these reactions tell us. The first shows that iron is more easily oxidized than lead. The second indicates that magnesium is more easily oxidized than iron, which means that it is also more easily oxidized than lead. Finally, the third reaction tells us that lead is more easily oxidized than copper. Placed in order of decreasing ease of oxidation, they are

$$Mg > Fe > Pb > Cu$$

A list of metals arranged this way is called an **activity series,** and a more extensive version is given in Table 10.1. Those metals at the top of the table are most easily oxidized, and those at the bottom are least easily oxidized. Notice that the alkali and alkaline earth metals are high on the list, indicating their extreme ease of oxidation. Also notice that the noble metals are at the bottom of the list, which is indicative of their high resistance to oxidation.

Table 10.1. The activity series of metals

Metal	Oxidation Reaction
Lithium	$Li \longrightarrow Li^+ + e^-$
Cesium	$Cs \longrightarrow Cs^+ + e^-$
Rubidium	$Rb \longrightarrow Rb^+ + e^-$
Potassium	$K \longrightarrow K^+ + e^-$
Barium	$Ba \longrightarrow Ba^{2+} + 2e^-$
Strontium	$Sr \longrightarrow Sr^{2+} + 2e^-$
Calcium	$Ca \longrightarrow Ca^{2+} + 2e^-$
Sodium	$Na \longrightarrow Na^+ + e^-$
Magnesium	$Mg \longrightarrow Mg^{2+} + 2e^-$
Aluminum	$Al \longrightarrow Al^{3+} + 3e^-$
Manganese	$Mn \longrightarrow Mn^{2+} + 2e^-$
Zinc	$Zn \longrightarrow Zn^{2+} + 2e^-$
Chromium	$Cr \longrightarrow Cr^{3+} + 3e^-$
Iron	$Fe \longrightarrow Fe^{2+} + 2e^-$
Cadmium	$Cd \longrightarrow Cd^{2+} + 2e^-$
Cobalt	$Co \longrightarrow Co^{2+} + 2e^-$
Nickel	$Ni \longrightarrow Ni^{2+} + 2e^-$
Tin	$Sn \longrightarrow Sn^{2+} + 2e^-$
Lead	$Pb \longrightarrow Pb^{2+} + 2e^-$
Hydrogen	$H_2 \longrightarrow 2H^+ + 2e^-$
Copper	$Cu \longrightarrow Cu^{2+} + 2e^-$
Silver	$Ag \longrightarrow Ag^+ + e^-$
Mercury	$Hg \longrightarrow Hg^{2+} + 2e^-$
Platinum	$Pt \longrightarrow Pt^{2+} + 2e^-$
Gold	$Au \longrightarrow Au^{3+} + 3e^-$

Ease of Oxidation of Metal Decreases

Ease of Reduction of Metal Ion Increases

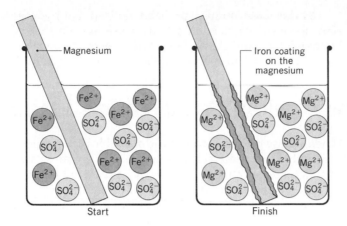

Figure 10.4. *At the start of the reaction between magnesium and iron(II) sulfate, the solution contains the solute $FeSO_4$. During the reaction, Mg is oxidized and its ions replace Fe^{2+} ions in the solution. After the reaction is over, all the Fe^{2+} ions have been replaced by Mg^{2+} ions. Iron coats the magnesium bar and the solution contains the ions of the compound $MgSO_4$.*

The activity series also lets us compare the ease of reduction of metal ions. When a metal is difficult to oxidize, its cation is easily reduced. Metallic gold, for example, is very difficult to oxidize but its ion, Au^{3+}, is quite easy to reduce.

One of the benefits of the activity series is that we can use it to determine the outcome of single displacement reactions. Any metal on the list is able to displace a metal below it from its compounds. For example, magnesium is above iron in the series. This means that magnesium is more easily oxidized than iron, and it is a better reducing agent than iron. If magnesium metal is placed into a solution of an iron compound, the magnesium will be oxidized and the iron ion will be reduced. After the reaction is over, the solution will contain a magnesium compound instead of the corresponding iron compound (see Figure 10.4)

EXAMPLE 10.1. USING THE ACTIVITY SERIES TO PREDICT CHEMICAL REACTIONS

PROBLEM: What will happen if a piece of chromium is dipped into a solution of silver nitrate?

SOLUTION: If we refer to the activity series in Table 10.1, we see that chromium is above silver in the table. Therefore, chromium is able to reduce silver ion to metallic silver. The net ionic equation for the reaction is

$$Cr(s) + 3Ag^+(aq) \longrightarrow Cr^{3+}(aq) + 3Ag(s)$$

Notice that three silver ions react for each chromium atom that reacts. This is necessary to make the net charge on both sides of the equation the same—a requirement for a balanced net ionic equation.

In an ionic equation, both charge and *mass must be balanced.*

EXAMPLE 10.2. USING THE ACTIVITY SERIES TO PREDICT CHEMICAL REACTIONS

PROBLEM: What will happen if a piece of lead is dipped into a solution that contains aluminum sulfate?

SOLUTION: As before, we examine the activity series. This time we find that lead is *below* aluminum in the series, which means that metallic lead is unable to reduce aluminum ion. As a result, there is no reaction in this mixture.

$$Pb(s) + Al^{3+}(aq) \longrightarrow \text{No reaction}$$

By this time you have probably noticed that hydrogen, H_2, is also listed in the activity series. Its location sets off those metals that are able to be oxidized by hydrogen ion. Any metal above hydrogen in the series is able to reduce H^+ to give H_2, so any metal above hydrogen will react with nonoxidizing acids such as HCl.

10.2 PERIODIC TRENDS IN THE REACTIVITY OF METALS

When we use the term *reactivity* in describing the properties of metals, we mean the ease with which a metal loses electrons to become a cation. A very reactive metal is one that loses electrons easily and so is oxidized readily.

The activity series discussed in the preceding section ranks the metals according to their reactivity. Although this series is valuable for solving problems such as Examples 10.1 and 10.2, often it is just useful to know how the reactivities of the metals vary within the periodic table — to know the general locations of the very reactive metals and the very unreactive ones. These kinds of periodic trends are depicted in Figure 10.5.

One might expect that the lower the ionization energy of a metal, the more easily oxidized it would be.

Examining Figure 10.5, you will see that the trends in reactivity roughly parallel the variations in ionization energy, which should not be too surprising since metals lose electrons when they react. The parallel is only approximate, however, because the ionization energy applies to isolated gaseous atoms forming isolated gaseous ions. In chemical reactions, metals normally react as solids to give ions in solution, so the ionization energy is only one factor involved.

Notice that the most reactive elements are located in Groups IA and IIA. These are the elements at the top of the activity series in Table 10.1. Also notice that the least reactive metals are located near each other in period 6 just to the right of center in the block of transition metals.

The ease of oxidation of metals in an important property. Many of the practical uses of these elements depend on it (or rather, the *lack* of it). This is so because the air oxidation of metals, which we often call *corrosion*, produces products that lack metallic properties. Corrosion, therefore, wipes out the desirable properties for which metals are often chosen. For this reason, extremely reactive metals such as those in Group IA have very few practical uses, and none that would require exposure to the atmosphere.

We are able to tolerate a moderate ease of oxidation in some metals — iron, for example — because they have such desirable physical properties. But if severely corrosive conditions are to be encountered, the metal must be protected. Huge sums of money are spent annually, for instance, to paint steel structures such as bridges.

For metals that are able to reduce hydrogen ion to H_2 (i.e., those that react with nonoxidizing acids), there is an interesting parallel between ease of oxidation and the vigor with which they react with hydrogen ions. In general, it is usually found

Figure 10.5. *General trends in the ease with which metals are oxidized. Elements with generally low ionization energies are easily oxidized.*

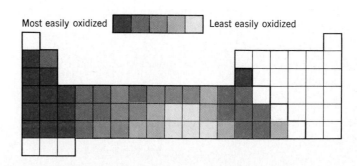

Most easily oxidized   Least easily oxidized

The tendency of steel to react with oxygen and moisture to form rust is a problem often solved by applying a protective coating of paint. Here we see a bridge receiving its periodic paint job. The red color is a rust-inhibiting undercoat.

that the more easily oxidized the metal, the faster it evolves H_2 (temperature and acid concentration being constant). The general reaction is the same; the metal loses electrons to become a cation while the H^+ ion is reduced to H_2. For example,

$$M(s) + 2H^+(aq) \longrightarrow M^{2+}(aq) + H_2(g)$$

where M stands for a metal such as Fe, Zn, or Mg. Although the products are essentially the same in each reaction, the speeds of the reactions differ by quite a lot, as illustrated in Figure 10.6. These differences reflect the fact that magnesium is more easily oxidized than zinc, and that zinc is more easily oxidized than iron.

Of all the metals, those of Group IA are the most easily oxidized. In fact, it would be extremely dangerous to place an alkali metal such as sodium or potassium in hydrochloric acid because the reaction is explosively violent. These metals, due to their low ionization energies, are so easily oxidized by sources of protons that they all

This parallel is only approximate, so we can't really use it in place of the activity series to rank metals according to their ease of oxidation.

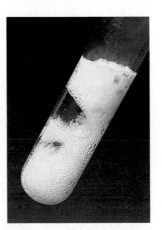

Figure 10.6. *These three test tubes contain HCl at the same concentration and allow us to compare the rates of reaction of HCl with (a) iron, (b) zinc, and (c) magnesium.*

react vigorously with water to liberate hydrogen gas, as shown in Figure 1.12 on page 23. For sodium, the reaction is

$$2Na(s) + 2H_2O(l) \longrightarrow 2Na^+(aq) + 2OH^-(aq) + H_2(g)$$

The violence of this reaction increases going down the group. Rubidium, for example, reacts explosively with water.

The Group IIA elements have slightly larger ionization energies than the elements in Group IA, so they generally are not as easily oxidized. Calcium and the elements below it react with water, but less vigorously than the alkali metals.

$$Ca(s) + 2H_2O(l) \rightarrow Ca^{2+}(aq) + 2OH^-(aq) + H_2(g)$$

Mg does react slowly with boiling water to release H_2.

At room temperature, magnesium doesn't react with water because of an insoluble oxide coating that forms on its surface when the metal is exposed to air. Beryllium doesn't react with water at all; it is even less reactive than magnesium.

Most of the transition metals of period 4 react with dilute nonoxidizing acids such as HCl and H_2SO_4. Iron and zinc are examples that were discussed earlier. An exception to this general rule is copper, which requires an oxidizing acid such as HNO_3. The post-transition metals (the metals in Groups IIIA, IVA, and VA) are also reasonably reactive and react with nonoxidizing acids (although the reaction of lead is slow).

As noted earlier, the metals that are least reactive are located in the lower center portion of the block of transition metals (see Figure 10.5.). Here we find platinum (Pt), iridium (Ir), and gold (Au)—metals so unreactive that they do not even react with oxidizing acids such as concentrated HNO_3. Their superior resistance to oxidation led them to be called **noble metals,** and many of their industrial applications are based on this property. Gold, for instance, is used to coat electrical contacts in low-voltage computer circuits because it doesn't corrode and build up a surface coating that would otherwise restrict the flow of electricity. Although these metals resist attack by HNO_3, they do dissolve slowly in a mixture of concentrated HNO_3 and concentrated HCl known as **aqua regia.** (Alchemists, using Latin, named this mixture *royal water* because it alone attacked gold, a *noble* metal.)

Aqua regia is made by mixing three parts of concentrated HCl with one part concentrated HNO_3 by volume (e.g., 30 mL HCl and 10 mL HNO_3).

10.3 THE REACTIONS OF NONMETALS AS OXIDIZING AGENTS

In the preceding sections, you learned about the abilities of the metals to function as reducing agents. Nonmetals often behave in just the opposite way by functioning as oxidizing agents. In these instances, the nonmetal atom gains one or more electrons (or its oxidation number decreases by one or more units).

We have already mentioned a number of reactions between metals and nonmetals in which the nonmetal is the oxidizing agent. Examples are the reaction of sodium with chlorine (Figure 1.2, page 4), and the reactions of sulfur with copper and zinc (Figures 1.14 and 1.19, respectively). Another example is the reaction of elemental aluminum with bromine, Br_2, shown in Figure 10.7. As you can see, many of these reactions are highly exothermic.

The tendency of the nonmetals to be oxidizing agents follows their periodic trends in electronegativity, which is just what you might expect. A very electronegative element has a strong tendency to attract electrons to itself and so should be able to remove electrons readily from other substances, thereby causing these substances to be oxidized. Since electronegativity increases from left to right across a period and from bottom to top within a group, the best oxidizing agents are located in the upper right of the periodic table.

It is useful to know the specific trends in oxidizing ability across a period and down a group. Within a given period, the strongest oxidizing agents are located at

Figure 10.7. *Bromine, the red liquid in the beaker, acts as an oxidizing agent when it combines vigorously and exothermically with aluminum to form the compound $AlBr_3$. The smoke contains unreacted Br_2 as well as some of the product.*

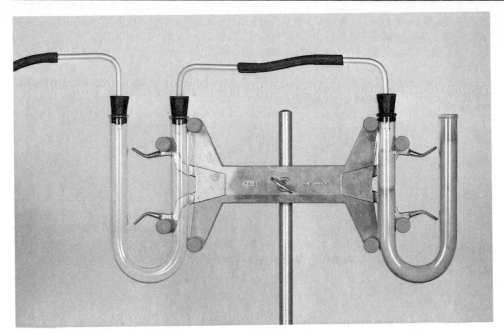

The faint pale yellow-green color of chlorine gas can barely be seen in the U-shaped tube on the left, but on contact with solid NaBr in the tube on the right, its reaction with Br^- liberates reddish-orange vapors of Br_2.

the right. For example, in period 2 the strongest oxidizing agent is fluorine, and the second best is oxygen. Similarly, in period 3, the strongest oxidizing agent is chlorine. Within a group, the elements that are the best oxidizing agents are at the tops of their respective groups. Among the halogens, for example, the best oxidizing agent is fluorine, followed by chlorine, and so forth.

$$\text{------ Oxidizing strength decreases} \longrightarrow$$

$$F_2 > Cl_2 > Br_2 > I_2$$

In Group VIA, the best oxidizing agent is oxygen, followed by sulfur, and so on.

In the previous section you saw that a metal that is easily oxidized is able to displace one that is less easily oxidized from compounds. In other words, a stronger reducing agent can displace a weaker one. A similar phenomenon is observed among the nonmetals. An element that is a better oxidizing agent is able to displace a weaker oxidizing agent from compounds. For example, if fluorine is passed over sodium chloride, the more powerful oxidizing agent, fluorine, displaces the weaker one, chlorine, by oxidizing Cl^- to Cl_2.

$$F_2(g) + 2NaCl(s) \longrightarrow Cl_2(g) + 2NaF(s)$$

In a similar way, chlorine is able to oxidize bromide ion and iodide ion to molecular bromine and iodine, respectively.

$$Cl_2 + 2Br^- \longrightarrow 2Cl^- + Br_2$$

$$Cl_2 + 2I^- \longrightarrow 2Cl^- + I_2$$

In addition, because Br_2 is a better oxidizing agent than I_2 and is able to oxidize I^- to the free element,

$$Br_2 + 2I^- \longrightarrow I_2 + 2Br^-$$

10.4 MOLECULAR OXYGEN AS AN OXIDIZING AGENT

One of the most common and powerful oxidizing agents in the laboratory, in commerce, and in our everyday existence is molecular oxygen, O_2. As we discussed earlier, scientists realized very early that oxygen combines with many substances in reactions that became known in general as *oxidation.* The removal of oxygen from a compound was similarly termed *reduction,* and it wasn't until some time later that the general nature of redox reactions was appreciated. Then the term oxidation was extended to cover any change in which an element undergoes an increase in oxidation number as a result of reaction with an oxidizing agent.

Although the reactions of oxygen are varied, there are some that are typical of specific classes of reactants.

Reactions of metals with oxygen

The formation of metal oxides by the direct reaction of metals with oxygen is a common event. In fact, in the form of corrosion, it represents both a nuisance and a source of economic waste in our modern society. Iron reacts with oxygen in the presence of moisture to form rust, an iron oxide whose crystals contain water molecules in variable amounts (as indicated by the variable coefficient x in the following equation).

Rust is a hydrated iron(III) oxide.

$$2Fe(s) + \tfrac{3}{2}O_2(g) + xH_2O(l) \longrightarrow Fe_2O_3 \cdot xH_2O(s)$$
$$\text{rust}$$

Aluminum, another common structural metal, also forms an oxide by direct reaction with oxygen in the air.

$$2Al(s) + \tfrac{3}{2}O_2(g) \longrightarrow Al_2O_3(s)$$

As you learned earlier, aluminum is even more easily oxidized than iron, so a fresh aluminum surface reacts rapidly with O_2 and quickly becomes covered by a thin oxide layer. Unlike rust on iron, however, the Al_2O_3 adheres tightly to the aluminum surface, effectively protecting it from further attack. In fact, because of the extreme ease of oxidation of aluminum, it is this protective oxide coating that makes it at all possible to use aluminum as a structural metal.

The corrosion of iron and aluminum in the air are relatively slow reactions. However, sometimes the reactions of metals with oxygen can be much more rapid and can evolve lots of heat and light. Such reactions with oxygen are usually identified by the term **combustion.** An example is the combustion of magnesium.

The combustion of magnesium metal is shown in the photograph on page 103.

$$2Mg(s) + O_2(g) \longrightarrow 2MgO(s)$$

This chemical change produces not only a great deal of heat, but also intense light—a fact that accounts for one of the important uses of magnesium—in flares and flashbulbs. A flashbulb (Figure 10.8) contains a fine magnesium wire in an atmosphere of pure oxygen. The flashbulb is fired by passing a small electric current through the wire, which heats it and sets off the combustion reaction. Afterwards, the interior of the bulb is coated with a thin deposit of the white powder MgO.

Although the corrosion of iron is a slow reaction, iron can be made to react much more rapidly by raising the temperature and increasing the O_2 concentration. For example, the cutting of steel by an acetylene torch is accomplished by first heating the steel to a high temperature with an oxygen–acetylene flame. Once the metal is very hot, the acetylene is turned off and the steel is bathed in a stream of pure oxygen. At this high temperature, the iron burns rapidly in the pure oxygen, and in the process a large amount of heat is evolved that melts the steel and sends showers of sparks flying (see Figure 10.9).

Figure 10.8. *Combustion of the fine magnesium wire inside a flashbulb produces a burst of light and leaves the interior of the bulb coated with white magnesium oxide powder.*

Reactions of nonmetals with oxygen

Oxygen also combines directly with most of the nonmetals to form covalent oxides. An example, with which you are surely familiar, is the reaction of O_2 with carbon (in

Figure 10.9. *Steel is cut by a stream of pure oxygen whose reaction with red-hot iron produces enough heat to melt the metal and send a shower of burning steel sparks flying.*

the form of charcoal briquets or coal). In the presence of an excess amount of O_2, the product is carbon dioxide

$$C(s) + O_2(g) \longrightarrow CO_2(g)$$

but if only a limited supply of oxygen is available, significant amounts of carbon monoxide are formed

$$2C(s) + O_2(g) \longrightarrow 2CO(g)$$

Because of this second reaction, bags of charcoal contain the warning shown in the photograph in Figure 10.10.

Carbon monoxide itself is able to react with oxygen and form CO_2.

$$2CO(g) + O_2(g) \longrightarrow 2CO_2(g)$$

The reaction is quite exothermic ($\Delta H = -284$ kJ per mole of CO burned), and CO is used industrially as a fuel because it can be made cheaply from coal and because it is easily transported through pipelines.

Two other nonmetals that react readily with oxygen are sulfur and phosphorus. Sulfur burns in the air with a blue flame and produces sulfur dioxide, a gas with an irritating choking odor.

$$S(s) + O_2(g) \longrightarrow SO_2(g)$$

This reaction is the first step in the synthesis of sulfuric acid, which is discussed in detail later in this book.

In Chapter 9 we discussed the allotropes of phosphorus: red phosphorus and white phosphorus. Both burn in oxygen to give the oxide P_4O_{10}, although the

$$C + O_2 \rightarrow CO_2$$
$$0 \quad\;\; 0 \quad\;\; +4-2$$

By analyzing oxidation numbers, we see that oxygen is an oxidizing agent in these reactions, too.

Figure 10.10. *The combustion of carbon in a limited supply of oxygen can produce dangerous carbon monoxide. This is why bags of charcoal briquettes contain a warning about the hazards of using them indoors.*

White phosphorus burns under water when O_2 is bubbled into the solution at temperatures above about 70°C, illustrating the extreme reactivity of this allotrope of phosphorus.

reaction of white phosphorus is spontaneous; it ignites spontaneously when the P_4 is exposed to O_2.

$$P_4(s) \quad + 5O_2(g) \longrightarrow P_4O_{10}(s)$$
white
phosphorus

Not all the nonmetals react directly with oxygen to form oxides. Nitrogen is an example of one that doesn't, which explains why our atmosphere, a mixture of nitrogen and oxygen, is stable. Attempts to ignite mixtures of N_2 and O_2 are unsuccessful because reactions between them are endothermic. However, if air is heated to very high temperatures, such as in an automobile engine, a small amount of the oxide NO is produced. When released to the atmosphere through the auto exhaust, this compound begins a chain of reactions eventually causing smog.

Reactions of organic compounds with oxygen

Organic compounds are generally considered to be the compounds of carbon. Their number is enormous, and it is certainly not our intention to discuss them at length here. Our goal is merely to look at the typical kinds of reactions that they undergo when they are burned in O_2.

The simplest organic compounds are called *hydrocarbons*, compounds composed only of carbon and hydrogen. The simplest hydrocarbon is methane, CH_4, which is the chief component of natural gas. Hydrocarbons are also the main constituents of gasoline, kerosene, diesel oil, and candle wax.

Methane and other hydrocarbons burn readily in air. When sufficient oxygen is available, the products of combustion are carbon dioxide and water.

$$CH_4 \quad + \quad 2O_2 \longrightarrow CO_2 + 2H_2O$$
methane

$$2C_4H_{10} \quad + \quad 13O_2 \longrightarrow 8CO_2 + 10H_2O$$
butane
(in disposable
cigarette lighters)

$$2C_8H_{18} \quad + \quad 25O_2 \longrightarrow 16CO_2 + 18H_2O$$
octane
(in gasoline)

However, in a more limited supply of oxygen, the products can include carbon monoxide. For example,

$$2CH_4 + 3O_2 \longrightarrow 2CO + 4H_2O$$

With an extremely limited supply of oxygen, only the hydrogen combines to form water, and a sooty flame containing elemental carbon is produced.

$$CH_4 + O_2 \longrightarrow C + 2H_2O$$

Carbon formed by this reaction is called *lampblack* and its production has been one of the commercial uses of methane.

Organic compounds often contain elements in addition to carbon and hydrogen. If they contain oxygen, this is incorporated in the CO_2 and H_2O formed in the combustion reaction. For example, the combustion of methyl alcohol, CH_3OH, the fuel in Sterno, follows the equation

$$2CH_3OH + 3O_2 \longrightarrow 2CO_2 + 4H_2O$$

Similarly, ethyl alcohol, C_2H_5OH, which is also known as grain alcohol and is the alcohol used to prepare the automotive fuel gasohol, burns according to the equation

$$C_2H_5OH + 3O_2 \longrightarrow 2CO_2 + 3H_2O$$

A common material whose combustion has been used since the discovery of fire to warm bodies and cook foods is wood. The major combustible component of wood is cellulose, a fibrous substance that gives plants their structural strength. Cellulose is also composed of carbon, hydrogen, and oxygen. Its molecules are *polymers,* meaning they are made up of many similar smaller units linked together to give large molecules. The lengths of the molecules can vary somewhat, so we usually specify their compositions by giving the formula of the repeating unit in parentheses and indicating by a subscript the number of such units in the molecule. When the exact number is unknown or unimportant, we use the subscript x or n. For cellulose, the repeating unit has the formula $C_6H_{10}O_5$, so the formula for a large cellulose molecule is $(C_6H_{10}O_5)_x$, where x is a large number. The equation for the combustion of cellulose, however, can be given as just the equation for the combustion of its empirical formula unit, $C_6H_{10}O_5$.

$$C_6H_{10}O_5 + 6O_2 \longrightarrow 6CO_2 + 5H_2O$$

An important source of the sulfur dioxide found in polluted air is the presence of sulfur-containing organic compounds in coal and fuels made from high-sulfur crude oils. When these organic compounds burn, the sulfur in them is oxidized to SO_2. An example is the combustion of a compound called propyl mercaptan, C_3H_7SH, which is a strong-smelling compound that evolves from freshly chopped onions.

$$C_3H_7SH + 6O_2 \longrightarrow 3CO_2 + 4H_2O + SO_2$$

Other sulfur-containing compounds react is a similar way to yield the same products.

10.5 CHEMICAL REACTIONS OF THE HYDROGEN ION: BRØNSTED–LOWRY ACIDS AND BASES

The concept of acids and bases is not new to you. You learned how to name them in Chapter 4 and you studied acid–base reactions in aqueous solutions in Chapter 5. The Arrhenius definition of acids and bases that you learned at that time—that acids produce H_3O^+ ion in water and bases produce OH^- ion in water—was sufficient to get you this far. However, you are now ready to see that the acid–base concept extends beyond the limits of aqueous solutions and that these substances can be defined in more general ways.

Figure 10.11 is a photo of open bottles of concentrated ammonia and concentrated hydrochloric acid beside each other. Above the bottles is a smoke consisting of microscopic crystals of ammonium chloride, NH_4Cl, formed by the reaction of gaseous HCl with gaseous NH_3. (The HCl escapes from the concentrated hydrochloric acid, and the NH_3 escapes from the concentrated aqueous ammonia solution.) The equation for the reaction is

$$NH_3(g) + HCl(g) \longrightarrow NH_4Cl(s)$$

The product of this reaction, NH_4Cl, is the same salt that is formed if the solutions in the bottles are mixed.

Since the reaction of HCl with NH_3 gives the same product regardless of whether or not a solvent is present, it seems reasonable to consider both to be acid–base reactions. However, the Arrhenius definition is inadequate for the job. In the gas phase there are no hydronium ions or hydroxide ions, so we can't interpret the gaseous reaction as an acid–base reaction in the same way that we analyzed others in Chapter 5. We obviously need a more general definition of acids and bases.

A somewhat more general approach to acids and bases was proposed independently in 1923 by the Danish chemist J. N. Brønsted and the British chemist T. M.

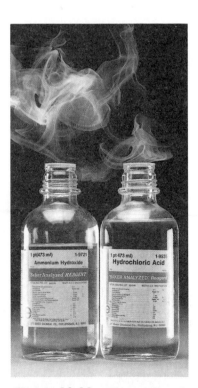

Figure 10.11. *The reaction of gaseous hydrogen chloride with gaseous ammonia, each coming from its concentrated aqueous solution, produces a white cloud of ammonium chloride.*

Lowry. The Brønsted–Lowry definitions of acids and bases are as follows:

*An **acid** is a substance that donates a proton (a hydrogen ion, H^+) to some other substance.*

*A **base** is a substance that accepts a proton from an acid.*

Stated more briefly, an acid is a proton donor and a base is a proton acceptor.

Based on this definition, we can now analyze the reaction between gaseous HCl and gaseous NH_3. This reaction can be diagrammed using Lewis structures as follows:

$$\begin{array}{c} H \\ \vdots \\ H\!:\!N\!:\overset{\curvearrowleft}{+\textcircled{H}}\!:\!\overset{..}{\underset{..}{Cl}}\!: \end{array} \longrightarrow \left[\begin{array}{c} H \\ \vdots \\ H\!:\!N\!:\!H \\ \vdots \\ H \end{array}\right]^{+} \left[:\overset{..}{\underset{..}{Cl}}\!:\right]^{-}$$

Since a proton is transferred from the HCl to the NH_3, this is clearly an acid–base reaction in the Brønsted–Lowry sense. In the reaction, HCl is the acid because it is the proton donor, and NH_3 is the base because it is the proton acceptor. Thus, under both the Arrhenius and Brønsted–Lowry definitions, HCl is an acid and NH_3 is a base.

The reaction between HCl and NH_3 illustrates the more general nature of the Brønsted–Lowry definition of acids and bases. Any reaction in which a proton is transferred from one particle to another is an acid–base reaction; hydronium ion and hydroxide ion do not have to be involved. In fact, acid–base reactions do not even require a solvent.

Conjugate acids and bases

Let us consider a reaction similar to others that we discussed earlier, but this time we will view it in terms of the Brønsted–Lowry definition. Hydrogen fluoride, HF, is a weak acid in water and undergoes ionization that proceeds by the reaction:

$$HF + H_2O \longrightarrow H_3O^+ + F^-$$

The reason why HCl is a strong acid, while HF is weak, will be discussed in the next section.

In this reaction, the HF is functioning as an acid because it is donating a proton to the water molecule. Water, on the other hand, is functioning as a *base* because it is accepting a proton from the HF.

Unlike HCl, which is a strong acid, HF is a weak acid and is incompletely ionized in water. Because of this, there is an equilibrium involved in the ionization, which properly should be written

$$HF + H_2O \rightleftharpoons H_3O^+ + F^-$$

Let's examine the reverse reaction, the reaction of the hydronium ion with fluoride ion, more closely.

$$H_3O^+ + F^- \longrightarrow HF + H_2O$$

This reverse reaction is also a Brønsted–Lowry acid–base reaction, with the hydronium ion serving as the acid and the fluoride ion as the base. Therefore, in the equilibrium there are two acids and two bases, one of each on either side of the arrow.

$$\underset{\text{acid}}{HF} + \underset{\text{base}}{H_2O} \rightleftharpoons \underset{\text{acid}}{H_3O^+} + \underset{\text{base}}{F^-}$$

An acid differs from its conjugate base by *only* one proton. All the other atoms are the same.

When the acid HF reacts, it forms the base F^-. These two substances are related to each other by the loss or gain of a *single* proton and constitute a **conjugate acid–base pair.** We say that F^- is the **conjugate base** of HF, and HF is the **conjugate**

acid of F^-. In this equilibrium, we also can identify H_2O and H_3O^+ as a conjugate pair. Water is the conjugate base of H_3O^+, and H_3O^+ is the conjugate acid of H_2O.

Another example of a Brønsted–Lowry acid–base reaction occurs in aqueous solutions of ammonia.

$$NH_3 + H_2O \rightleftharpoons NH_4^+ + OH^-$$

In this case water serves as an acid by giving up a proton to a molecule of NH_3, which thereby acts as a base. In the reverse reaction, on the other hand, NH_4^+ is the acid and OH^- is the base. Again we have two acid–base conjugate pairs: NH_3 and NH_4^+ is one of them, and H_2O and OH^- is the other.

In general, ignoring the charges on the species involved, we can represent any Brønsted–Lowry acid–base reaction as

$$\text{acid (HX)} + \text{base (Y)} \rightleftharpoons \text{base (X)} + \text{acid (HY)}$$

where acid (HX) and base (X) represent one conjugate pair and acid (HY) and base (Y) the other. The members of a conjugate pair differ *only by one proton*. They are otherwise the same. Also, within a conjugate pair, the acid has one more hydrogen (actually, H^+) than the base.

EXAMPLE 10.3. WRITING THE FORMULAS FOR CONJUGATE ACIDS AND BASES

PROBLEM: (a) What is the formula for the conjugate base of $N_2H_5^+$? (b) What is the formula for the conjugate acid of NH_2^-?

SOLUTION: (a) The conjugate base has *one* less proton than the acid, so the conjugate base of $N_2H_5^+$ must have only 4 hydrogens and it must have *one* less positive charge. Therefore, the conjugate base is N_2H_4.

(b) We have to add a proton to NH_2^- to get its conjugate acid. The acid will therefore have *one* more hydrogen and *one* more positive charge. Since the charge on the base is -1, adding a positive charge gives a net charge of zero. The conjugate acid is NH_3.

EXAMPLE 10.4. IDENTIFYING CONJUGATE ACIDS AND BASES

PROBLEM: Identify the acid–base conjugate pairs in the reaction

$$H_2S + (CH_3)_2NH \rightleftharpoons (CH_3)_2NH_2^+ + HS^-$$

For each pair, which is the acid and which is the base?

SOLUTION: Our first task is to identify the conjugate pairs. We know that the *only* difference between them is the number of hydrogens and the charge; everything else must be the same. Scanning the equation, we see that H_2S and HS^- both have sulfur in them, and the other substances in the equation do not have sulfur. Therefore, H_2S and HS^- must represent one of the conjugate pairs. We also see that the HS^- ion can be formed from H_2S by the loss of a proton, so H_2S is the conjugate acid and HS^- is the conjugate base. Comparing the other two substances, $(CH_3)_2NH$ and $(CH_3)_2NH_2^+$, we see that they are identical except for the additional H^+ possessed by $(CH_3)_2NH_2^+$. Therefore, $(CH_3)_2NH_2^+$ is the conjugate acid and $(CH_3)_2NH$ is the conjugate base.

$$H_2S + (CH_3)_2NH \rightleftharpoons (CH_3)_2NH_2^+ + HS^-$$
$$\text{acid} \qquad \text{base} \qquad\qquad \text{acid} \qquad \text{base}$$

Perhaps you noticed that in the reactions of HF and NH_3 with water, water functions as a base in one instance and as an acid in the other. A substance, such as water, that can serve as either an acid or a base, depending on conditions, is said to be **amphiprotic** or **amphoteric**. Water is not the only substance that behaves this way. For example, water, acetic acid, and ammonia all undergo reactions with themselves to produce ions.

$$H_2O + H_2O \rightleftharpoons H_3O^+ + OH^-$$

$$HC_2H_3O_2 + HC_2H_3O_2 \longrightarrow H_2C_2H_3O_2^+ + C_2H_3O_2^-$$

$$NH_3 + NH_3 \rightleftharpoons NH_4^+ + NH_2^-$$

$$\textbf{acid} + \textbf{base} \rightleftharpoons \textbf{acid} + \textbf{base}$$

When a substance reacts with itself to form ions, it is called an **autoionization reaction.** These particular reactions can be diagrammed with Lewis structures that illustrate how one molecule functions as an acid while the other behaves as a base.

In each reaction. the substance on the left is both an acid and a base.

Solutions of metal ions in water

An interesting example of a Brønsted–Lowry acid–base reaction occurs in aqueous solutions that contain metal ions with high positive charges. For example, solutions of the salt $AlCl_3$ are acidic, as are solutions of salts that contain Cr^{3+} and Fe^{3+}.

When a salt is dissolved in water, it dissociates completely and the positive and negative ions become surrounded by water molecules—they become hydrated. Each negative ion has a cluster of water molecules around it, with the positive ends of the water dipoles pointing at the negative ion. Similarly, each positive ion is surrounded by a number of water molecules arranged so that the negative ends of their dipoles point at the positive ion. If the salt contains a metal ion with a large positive charge, such as the aluminum ion, Al^{3+}, the water molecules that surround the cation are held especially tightly, and the positive ion with its layer of water molecules moves around in the solution as a single unit. In fact, when salts of Al^{3+} are crystallized from aqueous solutions, their crystals contain the octahedral $Al(H_2O)_6^{3+}$ ion, and it is believed that this ion exists in aqueous solutions of aluminum salts as well.

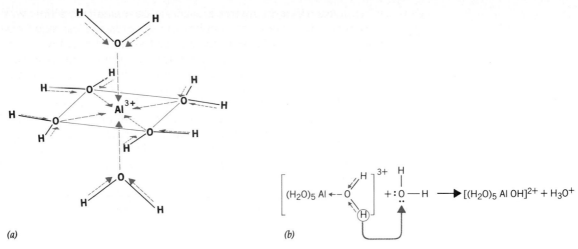

Figure 10.12. (a) *Dotted lines in color indicate how electron density is drawn toward the highly charged Al³⁺ ion and away from the O—H bonds. This increases the polarity of the O—H bonds and makes it easier to remove the hydrogens as H⁺. (b) A proton is transferred from one of the water molecules surrounding the Al³⁺ ion to a water molecule in the solvent. In this reaction the Al(H₂O)₆³⁺ ion is serving as a Brønsted acid.*

The acidity of solutions that contain ions such as Al^{3+}, Cr^{3+}, and Fe^{3+} is explained as follows. The high charge on the metal ion draws electron density toward itself and away from the oxygen atoms of the water molecules that surround it. These oxygen atoms, in turn, draw electron density from the O—H bonds (Figure 10.12). This makes the O—H bonds even more polar than they are in an ordinary water molecule. In other words, the hydrogen atoms of the water molecules surrounding the metal ion carry an even greater partial positive charge than the hydrogen atoms in an ordinary water molecule. As a result, a hydrogen is rather easily removed as H^+ from an ion such as $Al(H_2O)_6^{3+}$ and the ion is acidic. Its reaction with water can be viewed as a Brønsted–Lowry acid–base reaction.

$$Al(H_2O)_6^{3+} + H_2O \rightleftharpoons Al(H_2O)_5OH^{2+} + H_3O^+$$

$$\text{acid} \qquad \text{base} \qquad\qquad \text{base} \qquad\qquad \text{acid}$$

10.6 THE STRENGTHS OF ACIDS AND BASES: PERIODIC TRENDS

At various times in this chapter we've mentioned that some acids are strong and others are weak. When we talk of the strength of an acid in Brønsted–Lowry terms, we mean the ability of the acid to donate a proton to another species. And when we speak of the strength of a base, we mean its relative ability to capture a proton from an acid. But how do we compare these abilities among a series of acids or bases?

Any Brønsted–Lowry acid–base reaction can be viewed as two opposing or competing reactions between pairs of acids and bases. In one sense, the two conjugate bases can be considered to be competing for a proton. Consider, for example, the reaction of HCl with water, a reaction that we've discussed several times before. If, for the moment, we write the reaction as an equilibrium

$$HCl + H_2O \rightleftharpoons H_3O^+ + Cl^-$$

we can imagine that the two bases, H_2O and Cl^-, are in competition for the proton. In the forward direction, the H_2O captures the proton, and in the reverse reaction the Cl^- ion captures it. You've already learned that HCl is completely ionized in water, which means that the forward reaction goes essentially all the way toward completion, while the reverse reaction hardly proceeds at all. In equilibrium terms, we say that the position of equilibrium lies far to the right, in favor of the products. This must mean that the H_2O has a far greater affinity for protons than the Cl^- ion, because in the competition the H_2O wins — the water is capable of capturing essentially all the available H^+. We express this relative ability to pick up a proton by stating that water is a stronger base than the chloride ion.

We can also compare the relative strengths of the two acids in this reaction, HCl and H_3O^+. Because the position of equilibrium lies so far to the right, HCl must be far more effective at donating protons than H_3O^+ — after all, there are no HCl molecules left in the solution. This means that hydrogen chloride is a stronger acid than hydronium ion.

It is often desirable to compare the strengths of two or more acids relative to the same base. This is what we do when we compare the strengths of HCl and HF in water. In water, HCl is essentially 100% ionized, but in a 1 M solution of HF only about 3% of the HF is ionized.

The larger the percentage ionized, the stronger the acid.

$$HCl + H_2O \longrightarrow H_3O^+ + Cl^- \quad (100\%)$$

$$HF + H_2O \rightleftharpoons H_3O^+ + F^- \quad (\text{about } 3\% \text{ in } 1\ M\ \text{HF})$$

This information suggests that in water, HCl is a stronger acid than HF, because the HCl is more effective at giving its protons to water molecules. You can also see that the conclusion arrived at here is in agreement with our previous definitions of strong and weak acids: Strong acids are completely ionized in water, and weak acids are incompletely ionized.

Trends in the strengths of acids

Often, for simplicity, acids described in Brønsted–Lowry terms are just called Brønsted acids.

If we begin to compare the strengths of Brønsted acids, we begin to see certain patterns. For oxoacids of the same element, for instance, we find that the greater the number of oxygen atoms in the acid, the stronger the acid. Thus, H_2SO_4 is stronger than H_2SO_3 and HNO_3 is stronger than HNO_2. This trend is even more pronounced among the oxoacids of the halogens, which except for fluorine form four different oxoacids each. For chlorine, we find that $HClO_4$ is stronger than $HClO_3$, which is stronger than $HClO_2$, which is stronger than HClO. If we examine the oxoacids of different nonmetals, we also observe variations in acid strength that we can associate with the position of the nonmetal in the periodic table. And periodic trends also exist among the binary acids — acids such as HF, HCl, and H_2S.

To study trends in acid strength and to learn the reasons behind the trends, we find it convenient to divide acids into two categories, oxoacids and binary acids. As you will see, the factors that ultimately control their proton-donating capacity are somewhat different.

Oxoacids

As just noted, the strength of an oxoacid H_nXO_m depends on two factors. One is the number of oxygen atoms, n, in its structure and the other is the location of the nonmetal atom X within the periodic table. Before examining periodic trends, let's see why and how the number of oxygens in the acid affects its acidity.

Oxoacids with the Same Central Nonmetal As mentioned above, the following is observed experimentally. *For oxoacids of the same nonmetal, the greater the number of oxygen atoms in its structure, the stronger the acid.* To understand why this is so, let's compare sulfuric acid and sulfurous acid.

An oxoacid is acidic in water because a hydrogen from one of its O—H bonds is transferred to a water molecule in the surrounding solvent to yield a hydronium ion, H_3O^+. Using Lewis formulas, we diagram this reaction as follows for H_2SO_4:

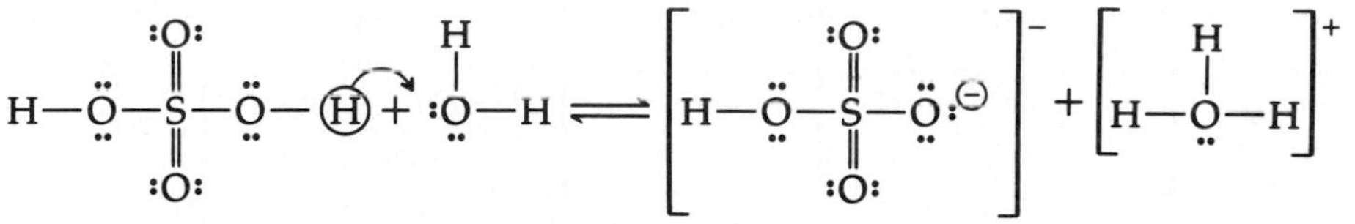

Notice that we have chosen structures with the smallest formal charges, so two of the oxygens attached to the sulfur have double bonds.

Formal charge was discussed in Section 8.7 beginning on page 246.

For H_2SO_4, this ionization reaction actually proceeds essentially to completion, but if we think of it as an equilibrium, the extent to which it proceeds to the right is controlled by two factors. One is the tendency of the oxoacid molecule (H_2SO_4 in this case) to give up a proton. The other is the tendency of the anion (HSO_4^- here) to acquire a proton to reform the acid molecule. What we need to know, then, is how the number of oxygens affects these two opposing tendencies.

The tendency of the acid molecule to give up a proton is controlled by the polarity of the O—H bond; anything that can increase the polarity of the bond, and thereby give the hydrogen a greater partial positive charge, should make the acid a better proton donor. This is so because the larger the partial positive charge on the hydrogen, the easier it is for the hydrogen to separate from the rest of the molecule as an H^+ ion. Now let's compare the structure of H_2SO_4, a strong acid in water, with the structure of H_2SO_3, which is a weak acid (about 11% ionized in a 1 M solution).

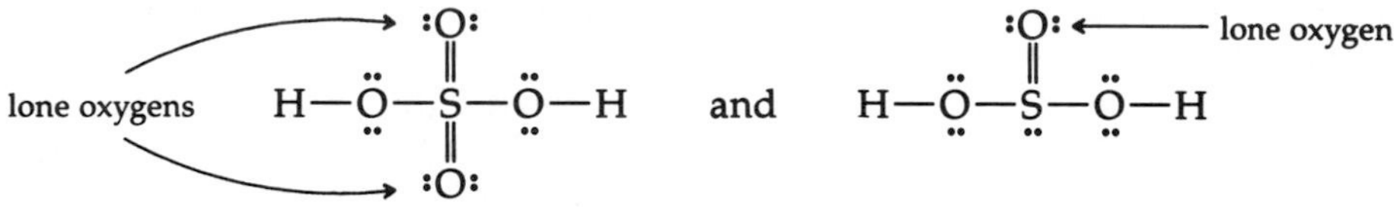

Notice that we have called attention to the oxygen atoms that are not attached to hydrogen atoms by calling them *lone oxygens*.

In each of these acids, sulfur is bonded to two O—H groups. In H_2SO_4, however, the sulfur is also attached to two lone oxygens, whereas in H_2SO_3 it is bonded to only one. We know that oxygen is a very electronegative element and we expect these S—O bonds to the lone oxygens to be polar with the negative charge concentrated around the oxygen end of the bond dipole, as shown below. Notice that the lone oxygens have the effect of drawing electron density from the sulfur atoms, leaving the latter with something of a positive charge, the magnitude of which is greater in H_2SO_4 than in H_2SO_3.

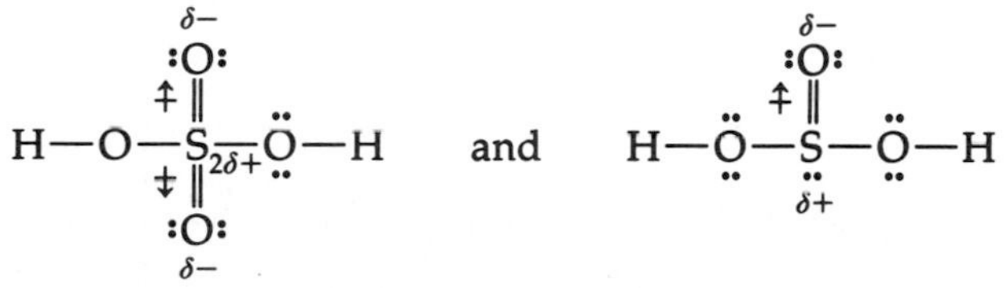

The positive charge on the sulfur tends to draw electron density from the other S—O bonds and, in turn, from the O—H bonds, thereby increasing the partial positive charge on the hydrogens. The effect is more pronounced in H_2SO_4 than in H_2SO_3 because of the greater positive charge on the S in H_2SO_4, so the hydrogens in sulfuric acid carry a more positive partial charge than those in sulfurous acid. This makes H_2SO_4 a better donor of H^+ ions than H_2SO_3, so taken alone, the effects of the lone oxygens suggest that sulfuric acid is stronger than sulfurous acid.

Now let's look at how the structure of the anion affects the strength of the oxoacid. The ions formed when H_2SO_4 and H_2SO_3 lose a proton are

In the assignment of formal charges, the negative charge of the anion is placed on one of the oxygen atoms. But these are not the only structures that we should draw for these anions. For the HSO_4^- ion, there are three equivalent resonance structures

which means that the single negative charge is spread over all three oxygen atoms. Each oxygen therefore carries only a $-\frac{1}{3}$ charge. If we draw similar resonance structures for the HSO_3^- ion, however, only two are possible and each oxygen carries a charge of $-\frac{1}{2}$.

Now we can see why the HSO_4^- ion has less of a tendency to pick up protons than the HSO_3^- ion. Comparing these two, the greater negative charge on the oxygens in the HSO_3^- ion makes it more likely that this anion will capture a proton and reform H_2SO_3.

So, let's summarize what we have found. Because H_2SO_4 has the greater number of lone oxygens, it loses protons more easily than H_2SO_3. Because each oxygen in the HSO_4^- ion carries less of a negative charge than each oxygen in the HSO_3^- ion, the HSO_4^- ion has the lesser tendency to regain protons. Both of these taken together tell us that sulfuric acid will be more completely ionized in solution than sulfurous acid and that H_2SO_4 is the stronger acid.

We can apply similar arguments to the oxoacids of the other nonmetals. We mentioned those of chlorine

Hypochlorous acid	HClO	
Chlorous acid	$HClO_2$	
Chloric acid	$HClO_3$	
Perchloric acid	$HClO_4$	

As we go from HClO to $HClO_4$, the number of lone oxygens increases, so the O—H bonds become more polar. This means the molecules become better donors of protons. Moreover, as we go from HClO to $HClO_4$, the number of oxygens over which the negative charge can be spread in the anion increases, so the abilities of the anions to regain protons decrease from ClO^- to ClO_4^-. Both of these factors taken together suggest that the tendency to donate a proton and not gain it back increases from HClO to $HClO_4$, so the acidity increases from HClO to $HClO_4$.

The smaller the negative charge carried by each oxygen, the less able the anion is to capture a proton and become the original acid molecule.

Compared to H_2SO_3, H_2SO_4 loses protons more easily and its anion captures them less easily, so the overall tendency for H_2SO_4 to ionize and stay that way is greater.

Oxoacids with Different Central Nonmetal Atoms *For acids with the same number of oxygens around the nonmetal, the acidity is controlled by the electronegativity of the nonmetal and therefore the location of the nonmetal in the periodic table.* Consider, for example, the oxoacids of the halogens with the general formula HXO_3 whose molecules have the structure

$$
\begin{array}{c}
\ddot{\text{:}}\text{O}\text{:} \\
\parallel \\
\ddot{\text{O}}\!\!=\!\!\ddot{\text{X}}\!-\!\ddot{\text{O}}\!-\!\text{H}
\end{array}
$$

As the electronegativity of X increases, electron density is drawn from the oxygens, including the one attached to the hydrogen. This also leads to some pulling of the electron density from the O—H bond and an increase in the partial positive charge carried by the hydrogen atom. As discussed earlier, this makes it easier for the molecule to lose the hydrogen as an H^+ ion.

In the periodic table, electronegativity increases from bottom to top in a group. This means that the acidity of these oxoacids increases from iodine at the bottom of Group VIIA to chlorine near the top.

$$\xrightarrow{\hspace{1cm}\text{Increasing acid strength}\hspace{1cm}}$$

$$HIO_3 < HBrO_3 < HClO_3$$

Similar trends are observed for other oxoacids in other groups. For example, H_3PO_4 is stronger than H_3AsO_4 and H_2SO_4 is stronger than H_2SeO_4.

You also learned that electronegativity increases from left to right in a period, and it is observed that the strengths of the oxoacids increase in the same direction, from phosphorus to chlorine, for example.

$$H_3PO_4 \qquad\qquad H_2SO_4 \qquad\qquad HClO_4$$

$$
\begin{array}{ccc}
\begin{array}{c}
\ddot{\text{:}}\text{O}\text{:} \\
\parallel \\
\text{H}\!-\!\ddot{\text{O}}\!-\!\text{P}\!-\!\ddot{\text{O}}\!-\!\text{H} \\
| \\
\ddot{\text{:}}\text{O}\text{:} \\
| \\
\text{H}
\end{array}
&
\begin{array}{c}
\ddot{\text{:}}\text{O}\text{:} \\
\parallel \\
\text{H}\!-\!\ddot{\text{O}}\!-\!\ddot{\text{S}}\!-\!\ddot{\text{O}}\!-\!\text{H} \\
\parallel \\
\ddot{\text{:}}\text{O}\text{:}
\end{array}
&
\begin{array}{c}
\ddot{\text{:}}\text{O}\text{:} \\
\parallel \\
\text{H}\!-\!\ddot{\text{O}}\!-\!\ddot{\text{Cl}}\!\!=\!\!\ddot{\text{O}} \\
\parallel \\
\ddot{\text{:}}\text{O}\text{:}
\end{array}
\end{array}
$$

The molecules become better H^+ donors and their anions become poorer H^+ acceptors going from H_3PO_4 to $HClO_4$.

$$\xrightarrow{\hspace{0.5cm}\text{Electronegativity of central atom increases}\hspace{0.5cm}}$$
$$\xrightarrow{\hspace{1cm}\text{Acidity of oxoacid increases}\hspace{1cm}}$$

Notice that as we go from left to right, not only does the electronegativity increase for the central atom, but the number of lone oxygens also increases.

Periodic trends in the strengths of the oxoacids are shown in Figure 10.13.

Binary acids

We can also find periodic trends in the strengths of the binary acids of the nonmetals. Recall that these acids are water solutions of nonmetal hydrides—for example, HF, HCl, and H_2S. They produce acidic solutions by reacting with water. For instance,

$$
\begin{array}{c}
\text{H}\!-\!\ddot{\text{O}}\text{:} \\
| \\
\text{H}
\end{array}
+ \;\text{H}\text{:}\ddot{\text{Cl}}\text{:} \;\longrightarrow\;
\left[
\begin{array}{c}
\text{H}\!-\!\ddot{\text{O}}\!-\!\text{H} \\
| \\
\text{H}
\end{array}
\right]^{+}
+ \left[\text{:}\ddot{\text{Cl}}\text{:}\right]^{-}
$$

When we consider binary acids derived from elements belonging to the same group, an increase in acidity down the group is observed experimentally. Thus, the hydrogen halides increase in acidity in the order HF < HCl < HBr < HI.

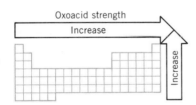

Figure 10.13. *Variations in the strengths of oxoacids with similar formulas.*

At first glance this order of acidity seems opposite to what we might predict. We know, for instance, that fluorine is more electronegative than chlorine; therefore, the HF bond is more polar than the HCl bond. Consequently, the H in HF is more positively charged than the H in HCl. Thus, we are tempted to predict that HF should lose a proton more readily than HCl. This, however, is precisely the *reverse* of their acid strengths in water, where HF is weak and HCl is essentially 100% ionized.

The solution to this puzzle can be understood by realizing that there are actually two opposing factors contributing to the acidity of these compounds. One is, indeed, the ionic character of the H—X bond; the other is the H—X bond strength. As we proceed down within a group, the nonmetal becomes progressively larger and there is an accompanying rapid decrease in the strength of the H—X bond. This weakening of the H—X bond turns out to be more than sufficient to compensate for the decrease in the polarity of the bonds, and a net increase in acid strength is observed.

When we look at the acidity of the binary hydrogen compounds of elements in the same period, for example, NH_3, H_2O, and HF, the dominant factor becomes the polarity of the H—X bond. As we go from left to right within the period, there is little change in the size of the nonmetal and relatively little change in the H—X bond energy. There is, however, a very dramatic increase in the ionic character of the H—X bond, which is reflected in a rapid increase in acidity from ammonia to hydrogen fluoride. This is summarized in Figure 10.14.

In general, bond strengths tend to decrease as the bonded atoms become larger.

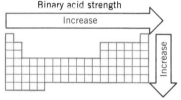

Figure 10.14. *Variations in the strengths of the binary acids of the nonmetals.*

10.7 LEWIS ACIDS AND BASES; COMPLEX IONS OF METALS

The Brønsted–Lowry definition of acids and bases is more general than the Arrhenius definition because it removes the restriction of having to deal with reactions in aqueous solution. However, even the Brønsted–Lowry concept is restricted in scope, since it limits discussion of acid–base phenomena to proton-transfer reactions. There are many reactions that have all the earmarks of acid–base reactions but that do not fit the Brønsted–Lowry mold. The approach taken by the American chemist Gilbert N. Lewis further extends the acid–base concept to cover these cases.

This is the same G. N. Lewis who originated the Lewis symbols and structures you learned to draw in Chapter 8.

In the Lewis definition of acids and bases: *A* **base** *is defined as a substance that can donate a* **pair** *of electrons to the formation of a covalent bond. An* **acid** *is a substance that can accept a pair of electrons to form the bond.*

A simple example of a Lewis acid–base reaction is the reaction of a hydrogen ion with a hydroxide ion

Lewis base: electron pair donor.
Lewis acid: electron pair acceptor.

$$\left(H^+\right) + \left[:\ddot{O}-H\right]^- \longrightarrow \overset{H \qquad H}{\underset{\ddot{O}}{\diagdown\diagup}}$$

The hydroxide ion is the Lewis base because it furnishes the pair of electrons that become shared with the hydrogen. The hydrogen ion, on the other hand, is the Lewis acid because it accepts a share of the pair of electrons when the O—H bond is created.

Another example is the reaction between BF_3 and ammonia,

Recall that NH_3BF_3 is an example of an addition compound.

$$\begin{array}{ccc} H & F \\ | & | \\ H-N: + & B-F \\ | & | \\ H & F \end{array} \longrightarrow \begin{array}{ccc} H & F \\ | & | \\ H-N \rightarrow & B-F \\ | & | \\ H & F \end{array}$$

In this case, the NH_3 functions as the base and BF_3 serves as the acid. Compounds containing elements with incomplete valence shells, such as BF_3 or $AlCl_3$, tend to be Lewis acids, while compounds or ions that have lone pairs of electrons can behave as

Lewis bases. When the Lewis acid–base reaction occurs, a coordinate covalent bond is formed.

Still other examples of Lewis acid–base reactions are provided by the reactions of metal oxides with nonmetal oxides. Recall that metal oxides, in water, produce hydroxides. For instance,

$$Na_2O + H_2O \longrightarrow 2NaOH$$

Nonmetal oxides react to form acids as illustrated by the reaction

$$SO_3 + H_2O \longrightarrow H_2SO_4$$

When these two solutions are mixed, neutralization occurs with the production of the solvent plus a salt,

$$2NaOH + H_2SO_4 \longrightarrow 2H_2O + Na_2SO_4$$

The production of Na_2SO_4 from Na_2O and SO_3 can take place directly without the introduction of any water whatsoever, as shown by the equation

$$Na_2O(s) + SO_3(g) \longrightarrow Na_2SO_4(s)$$

According to the Lewis definition, this also is a neutralization reaction between a Lewis base (oxide ion) and a Lewis acid (sulfur trioxide):

base   acid   sulfate ion

A pair of electrons in the double bond moves to the oxygen on top, allowing the sulfur to accept the share of a pair of electrons from the oxide ion.

In this case, we find that some electronic rearrangement must take place as the oxygen becomes attached to the sulfur. Nevertheless, the overall change can be viewed as a neutralization reaction.

The reaction between Na_2O and SO_3 illustrates the limitations of the Brønsted–Lowry concept. Since no protons are involved in the reaction, it would never be classified as an acid–base reaction under the Brønsted–Lowry definition.

Reactions of this type, between an oxide such as CaO and SO_2 or SO_3, are important for the removal of sulfur oxides from the gases produced by combustion of high-sulfur fuels. For example,

$$CaO(s) + SO_2(g) \longrightarrow CaSO_3(s)$$

A reaction quite analogous to that just described has been used on spacecraft to remove carbon dioxide from the air breathed by the astronauts. In this case, carbon dioxide reacts with LiOH,

$$CO_2(g) + LiOH(s) \longrightarrow \underset{\text{lithium bicarbonate}}{LiHCO_3(s)}$$

(Lithium hydroxide is used because of the very low atomic mass of Li. This results in many moles of LiOH per pound.) This reaction also can be viewed as a Lewis acid–base reaction

bicarbonate ion

Thus far we have looked only at simple acid–base neutralizations. The acid–base reactions discussed in Section 10.5 can also be treated from the Lewis point of view. In the Brønsted–Lowry theory these reactions were regarded as competitions in which the stronger acid prevails by losing its proton. Under the Lewis definition these reactions are looked on as constituting the displacement of one base (the weaker one) by another. Referring to the reaction of HCl with H_2O, for example,

$$HCl + H_2O \longrightarrow H_3O^+ + Cl^-$$

we see that the Lewis theory interprets the change as the result of replacing the Cl^- ion in HCl by the stronger base, H_2O.

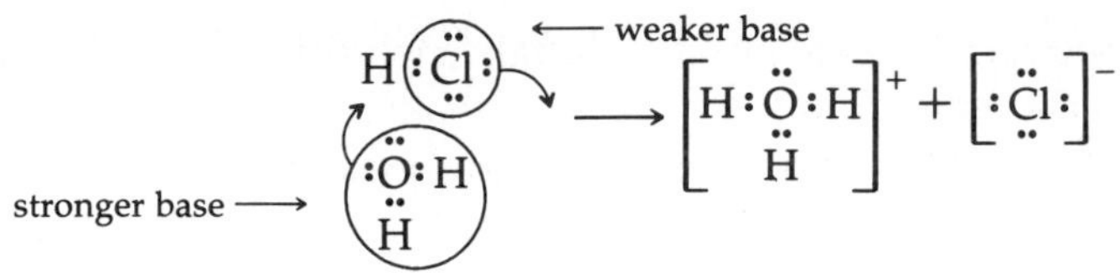

In other words, the stronger base, H_2O, pushes out the weaker one, Cl^-. Here we interpret the acid to be the H^+ ion instead of the entire HCl molecule; the H^+ is changing partners as it moves from the weaker base to the stronger one.

Complex ions

In our previous discussions of the compounds of metals, you have been led to believe that metals form only ionic bonds. For some metals, such as those in Group IA, this is essentially correct. But for many others, especially the transition metals and post-transition metals, this is not the case. The ions of these metals often become attached rather strongly to anions or neutral molecules to form more complicated species that we call **complex ions,** or simply **complexes.** In these substances the metal ion behaves as a Lewis acid and becomes bonded covalently to other species that serve as Lewis bases. A common example is the copper(II) ion.

When the Cu^{2+} ion is formed in an aqueous solution, it quickly becomes bound to four water molecules to give a pale blue complex ion with the formula $Cu(H_2O)_4{}^{2+}$.

$$Cu^{2+} + 4H_2O \longrightarrow Cu(H_2O)_4{}^{2+}$$

The formation of this complex can be diagrammed as follows by using Lewis formulas:

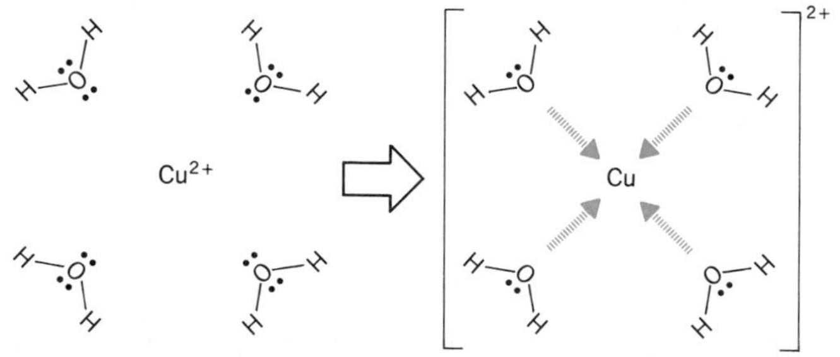

Notice that water molecules provide the electron pairs for the bonds to the Cu^{2+} ion, so H_2O here is behaving as a Lewis base. By accepting a share of an electron pair from each of the water molecules, the Cu^{2+} ion is behaving as a Lewis acid.

Complex ions are very important species. Many biological substances, such as hemoglobin, vitamin B_{12}, and chlorophyll, contain complexes as part of their overall structure, and their biological functions depend on the presence of a metal ion. Complexes are also found in household products and even foods. It is not surprising, therefore, that the study of complexes is an important specialty within the science of chemistry. As with any specialty, it has developed its own terminology.

The Lewis bases that attach themselves to metal ions are called **ligands,** from the Latin *ligare,* meaning "to bind." [We have used this term before in our discussion of the VSEPR theory when referring to the groups of atoms attached to (bound to) some central atom in a molecule.] The atom in the ligand that donates the electron pair to the metal ion is called the **donor atom** and the metal ion itself is said to be an **acceptor.** Because we can view the formation of the complex as occurring by the formation of coordinate covalent bonds from the ligands to the metal ion, compounds that contain complexes are called **coordination compounds.**

We will study the structures of complexes and more about the bonding in them when we get to Chapter 21. For now, however, let's look at some of the kinds of substances that serve as ligands and how we arrive at the formulas for complexes.

Ligands Ligands in complexes are either anions or neutral molecules that contain one or more atoms with at least one unshared pair of electrons that can be donated to a metal ion.

Many anionic ligands are simple ions, such as the halides (F^-, Cl^-, etc.) or sulfide ion (S^{2-}). Others are polyatomic anions such as nitrite ion (NO_2^-), hydroxide ion (OH^-), cyanide ion (CN^-), thiocyanate ion (SCN^-), carbonate ion (CO_3^{2-}), and thiosulfate ion ($S_2O_3^{2-}$). This is just a small sample of them; the list is long.

Among the neutral molecules that behave as ligands, the most common is water. In fact, when almost any metal ion is placed in water, it forms a complex with H_2O, and the chemistry of metal ions in aqueous solution tends to be the chemistry of their complexes with water. The number of water molecules that become attached to a metal ion depends on the metal. Copper(II) ion is one that binds four water molecules in $Cu(H_2O)_4^{2+}$, but many metals become bonded to six water molecules. The acidic $Al(H_2O)_6^{3+}$ ion described in Figure 10.12 is one example.

Many complexes of the transition metals are colored (the origin of the color will be discussed in Chapter 21), and the colors that we often associate with metal ions in solution and their salts in the solid state are the colors of their complexes with water. Copper salts, for example, are often blue because of the presence of the $Cu(H_2O)_4^{2+}$ ion. When copper sulfate crystallizes, it forms the hydrate $CuSO_4 \cdot 5H_2O$, which would really be better written as $[Cu(H_2O)_4]SO_4 \cdot H_2O$ because it contains the complex. Similarly, hydrates such as green $NiCl_2 \cdot 6H_2O$ and reddish-pink $CoCl_2 \cdot 6H_2O$ contain the complexes $Ni(H_2O)_6^{2+}$ and $Co(H_2O)_6^{2+}$, respectively. These give their characteristic colors to both the solid salts and to solutions that contain them, as shown in Figure 10.15. Even though most hydrates of metal salts contain similar

(a) (b)

Figure 10.15. (a) *Cobalt chloride crystals have the formula $CoCl_2 \cdot 6H_2O$. They actually contain the complex ion $Co(H_2O)_6^{2+}$, which gives its reddish-pink color to both the crystals and the aqueous solution. (b) Nickel chloride crystals with the formula $NiCl_2 \cdot 6H_2O$ contain the ion $Ni(H_2O)_6^{2+}$, which gives its green color to the crystals and the aqueous solution of the salt.*

Figure 10.16. (a) *An aqueous solution of $CuSO_4$ before the addition of aqueous ammonia has the pale blue color of the $Cu(H_2O)_4{}^{2+}$ ion. (b) After ammonia is added, the solution turns dark blue because the $Cu(NH_3)_4{}^{2+}$ complex forms.*

(a)

(b)

complexes, we often don't bother to make the distinction and we usually write their formulas as hydrates in the usual way.

Another important neutral ligand is ammonia, which also forms complexes with many metal ions. When ammonia is added to a solution containing the pale blue $Cu(H_2O)_4{}^{2+}$ ion, for example, the water molecules are displaced and the deep blue complex $Cu(NH_3)_4{}^{2+}$ is formed (see Figure 10.16). The reaction is

$$\underset{\text{pale blue}}{Cu(H_2O)_4{}^{2+}(aq)} + 4NH_3(aq) \longrightarrow \underset{\text{deep blue}}{Cu(NH_3)_4{}^{2+}(aq)} + 4H_2O$$

We described a similar reaction on page 322 for the displacement of Cl^- from HCl by H_2O.

Once again, we have a case of a stronger Lewis base (NH_3) displacing a weaker one (H_2O) from a Lewis acid (the Cu^{2+} ion).

Polydentate Ligands Ligands are often classified according to the number of donor atoms they have that are able to bind to a metal ion. Ligands such as NH_3, Cl^-, and CN^- can bind through only one donor atom and are called **monodentate** ligands ("one-toothed" ligands). There also are many ligands that have two or more donor atoms that are able to simultaneously bind to a metal ion. As a class, they are called **polydentate ligands.** Within this class are **bidentate ligands,** which have two donor atoms through which they can become attached to a metal. Two common examples are ethylenediamine and oxalate ion, whose structures are shown below.

$$\underset{\text{ethylenediamine}}{H_2\overset{\cdot\cdot}{N}-CH_2-CH_2-\overset{\cdot\cdot}{N}H_2} \qquad \underset{\text{oxalate ion}}{\left[\overset{\cdot\cdot}{\underset{\cdot\cdot}{:}}O-\overset{\overset{:O::O:}{\|\;\|}}{C-C}-\overset{\cdot\cdot}{\underset{\cdot\cdot}{O}}: \right]^{2-}}$$

They become attached to a metal ion as follows:

One of the most commercially useful polydentate ligands is a compound called ethylenediaminetetraacetic acid, which is usually abbreviated EDTA.

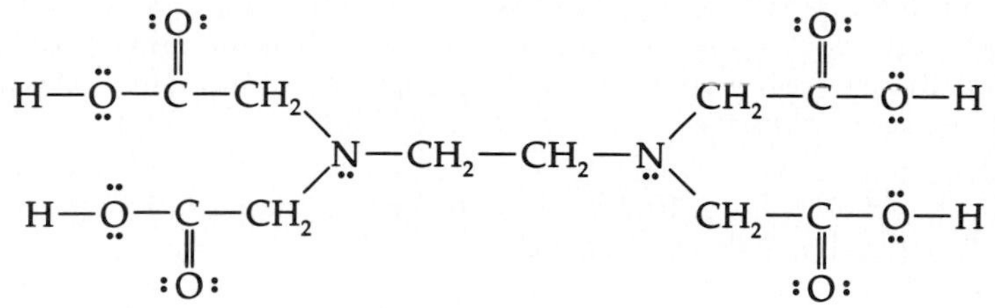

The H atoms attached to the oxygens in the structure above are acidic and can be removed to yield the anion EDTA^{4-}. This anion, shown below, has *six* donor atoms (printed in color) that it can use to hold a metal ion, and it forms very stable complexes by wrapping itself around the metal ion.

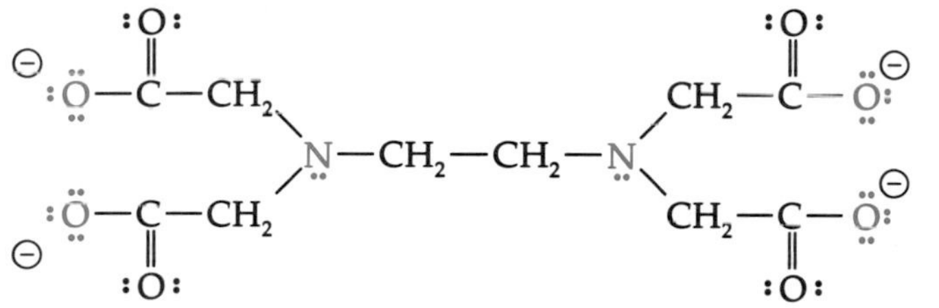

One reason for EDTA's importance as a ligand is its relative nontoxicity. It is used in small amounts in foods such as salad dressings where, by forming complexes, the EDTA effectively prevents trace amounts of certain metal ions from taking part in reactions that lead to spoilage. You will often find listed among the ingredients in such food products CaNa$_2$EDTA; the calcium salt is used because the EDTA^{4-} anion has such a strong affinity for calcium ions that it will remove Ca^{2+} from bones unless some calcium is already present. Many shampoos contain Na$_4$EDTA as a "water softener" because the EDTA^{4-} ion binds Ca^{2+}, Mg^{2+}, and Fe^{2+} ions, which are present in "hard water" and interfere with the action of soap. EDTA is also sometimes added in small amounts to whole blood to prevent clotting. It ties up Ca^{2+} ions that are required for clotting. EDTA is even used as an antidote for poisoning by "heavy metals" such as lead because it fastens itself to the lead ions and permits them to be expelled by the body.

Formulas for Complexes Writing the formula for a complex is very simple. The metal ion is written first, followed by the ligands, and the charge is just the algebraic sum of the charges on the metal ion and on the ligands. For instance, copper(II) ion forms a complex with chloride ion that contains the Cu^{2+} ion and four Cl$^-$ ions. The charge on the complex is the sum of the 2+ charge of the metal ion and the four 1− charges provided by the four chloride ions. The net charge is $[(2+) + 4(1-)] = 2-$, so the formula is CuCl$_4{}^{2-}$. Sometimes when we write the formula of a compound and wish to identify clearly the ligands around the metal ion, we enclose the formula for the complex within square brackets. That's what we did when we wrote the formula for copper sulfate to show that four water molecules are attached to each copper ion in the solid. Similarly, the hydrate of cobalt chloride illustrated in Figure 10.15 can be written [Co(H$_2$O)$_6$]Cl$_2$, which tells us that the solid contains two kinds of particles, [Co(H$_2$O)$_6$]$^{2+}$ ions and Cl$^-$ ions.

EXAMPLE 10.5. WRITING THE FORMULA FOR A COMPLEX

PROBLEM: In photography, unexposed silver bromide (AgBr) is removed from photographic film by causing it to dissolve in a solution of sodium thiosulfate, $Na_2S_2O_3$. The thiosulfate ion, $S_2O_3^{2-}$, forms a complex with the silver ion in which there are two thiosulfate ions per silver ion. Write the formula for this complex.

SOLUTION: We write the silver first, followed by the two thiosulfate ions in parentheses. Let's do that much and then figure out the charge.

$$Ag(S_2O_3)_2^{(?)}$$

The silver ion is Ag^+, so the net charge on the complex is

$$\text{net charge} = (1+) + 2(2-)$$
$$= 3-$$

The formula of the complex, therefore, is $Ag(S_2O_3)_2^{3-}$.

REVIEW QUESTIONS AND PROBLEMS

Problems whose numbers are in blue have their answers in Appendix D at the back of the book; the more difficult problems are marked with asterisks.

Metals as Reducing Agents

10.1 What kinds of oxidation states do metals have in virtually all their compounds?

10.2 Why do metals nearly always function as reducing agents when they react?

10.3 Why is HCl called a nonoxidizing acid?

10.4 Manganese reacts with HCl to give H_2 gas and Mn^{2+} in solution. Write a balanced net ionic equation for the reaction.

10.5 Write an equation for the reaction of HBr(aq) with aluminum.

10.6 Why is HNO_3 referred to as an oxidizing acid?

10.7 Write balanced net ionic equations for the reaction of silver with (a) concentrated HNO_3 and (b) dilute HNO_3.

10.8 Why is nitric acid used to dissolve silver instead of hydrochloric acid?

10.9 If zinc is allowed to react with very dilute nitric acid, zinc is oxidized to Zn^{2+}, while the NO_3^- ion can be reduced to NH_4^+. Write a balanced net ionic equation for the reaction.

10.10 When hot and concentrated, H_2SO_4 is a potent oxidizing agent. It oxidizes zinc to Zn^{2+} and the sulfate can be reduced to H_2S. Write a balanced net ionic equation for the reaction.

10.11 Write a balanced molecular equation for the reaction of hydrochloric acid with (a) magnesium and (b) aluminum.

10.12 Write balanced equations for these reactions:
(a) chromium + hydrochloric acid →
$\qquad$ chromium(III) chloride + hydrogen
(b) nickel + sulfuric acid →
$\qquad$ nickel(II) sulfate + hydrogen

10.13 Write net ionic equations for the reactions in Questions 10.11 and 10.12. What is the oxidizing agent in each of these reactions?

10.14 What is a *single displacement reaction?*

10.15 The following reactions are observed to occur:

$$Mg(s) + Cu^{2+}(aq) \longrightarrow Mg^{2+}(aq) + Cu(s)$$
$$Sn(s) + Cu^{2+}(aq) \longrightarrow Sn^{2+}(aq) + Cu(s)$$
$$Cu(s) + 2Ag^+(aq) \longrightarrow Cu^{2+}(aq) + Ag(s)$$
$$Mg(s) + Cd^{2+}(aq) \longrightarrow Mg^{2+}(aq) + Cd(s)$$
$$Cd(s) + Sn^{2+}(aq) \longrightarrow Cd^{2+}(aq) + Sn(s)$$

On the basis of the results of these reactions and without referring to Table 10.1, arrange the metals here in order of increasing ease of oxidation.

10.16 Using the activity series in Table 10.1, predict the outcome of the following reactions:
(a) $Al(s) + Zn^{2+}(aq) \rightarrow$ $\quad$ (f) $Hg(l) + H^+(aq) \rightarrow$
(b) $Sn(s) + Cu^{2+}(aq) \rightarrow$ $\quad$ (g) $Ni(s) + H^+(aq) \rightarrow$
(c) $Ag(s) + Co^{2+}(aq) \rightarrow$ $\quad$ (h) $Cd(s) + H_2O \rightarrow$
(d) $Mn(s) + Pb^{2+}(aq) \rightarrow$ $\quad$ (i) $Ba(s) + H_2O \rightarrow$
(e) $Cu(s) + Mg^{2+}(aq) \rightarrow$ $\quad$ (j) $H_2(g) + Pt^{2+}(aq) \rightarrow$

10.17 Where, in general, are the easily oxidized metals located in the periodic table? Where do we find the metals that are least easily oxidized?

10.18 According to the trends in ionization energy, which element in each of the following pairs should be more easily oxidized?
(a) Rb or Sr
(b) Na or Rb
(c) Na or Al
(d) Al or Ca

10.19 Why do the metals below magnesium in Group IIA have very few practical applications?

10.20 Why are gold and platinum known as noble metals?

10.21 Would you expect calcium to react with HCl more or less rapidly than magnesium? What is the basis for your answer?

10.22 What is *aqua regia*?

10.23 Write a balanced ionic equation for the reaction of water with (a) sodium, (b) rubidium, (c) strontium.

Nonmetals as Oxidizing Agents

10.24 In general, how does the oxidizing ability of the nonmetals vary (a) across a period and (b) down a group?

10.25 Arrange the following lists of nonmetals in order of increasing strength of the nonmetals as oxidizing agents:
(a) O, F, N, C
(b) I, Cl, Br, F

10.26 Complete the following equations. If no reaction occurs, write N.R.
(a) $F_2 + Cl^- \rightarrow$
(b) $Br_2 + Cl^- \rightarrow$
(c) $I_2 + Cl^- \rightarrow$
(d) $Br_2 + I^- \rightarrow$

10.27 Write the balanced chemical equation for the reaction of bromine with aluminum. Predict the reaction of bromine with zinc.

Oxygen as an Oxidizing Agent

10.28 What is combustion?

10.29 What is rust? Write a chemical equation for its formation.

10.30 What is the product of corrosion of aluminum? Why does this cause less of a problem than the rusting of iron?

10.31 What chemical reaction takes place inside a flashbulb?

10.32 Write a chemical equation for the reaction of oxygen with (a) iron, (b) lithium, (c) calcium, (d) magnesium, (e) aluminum.

10.33 Write a chemical equation for the reaction of oxygen (present in plentiful supply) with (a) carbon, (b) sulfur, (c) phosphorus.

10.34 Write a chemical equation for the combustion of carbon in a limited supply of oxygen.

10.35 Write chemical equations for the complete combustion of (a) C_9H_{20}, (b) $C_2H_4(OH)_2$, (c) $(CH_3)_2S$.

10.36 If CH_4 is burned in a limited supply of O_2, what are the products of combustion?

10.37 One of the ingredients in candle wax is eicosane, $C_{20}H_{42}$. When candles burn, they usually produce a yellow sooty flame. Write the various chemical equations that illustrate the combustion of eicosane under these conditions.

10.38 Write the balanced chemical equation for the complete combustion of cellulose, $(C_6H_{10}O_5)_x$, in oxygen.

Brønsted–Lowry Acids and Bases

10.39 What are the Brønsted–Lowry definitions of an acid and a base?

10.40 What is the conjugate base of (a) NH_3, (b) NH_4^+, (c) $HC_2H_3O_2$, (d) H_3PO_4, (e) HNO_3?

10.41 What is the conjugate acid of (a) HSO_4^-, (b) SO_4^{2-}, (c) H_2O, (d) Cl^-, (e) CHO_2^-?

10.42 Why is the Brønsted–Lowry definition of acids and bases less restrictive than the Arrhenius definition?

10.43 Identify the two acid–base conjugate pairs in each of the following reactions:
(a) $C_2H_3O_2^- + H_2O \rightleftharpoons OH^- + HC_2H_3O_2$
(b) $HF + NH_3 \rightleftharpoons NH_4^+ + F^-$
(c) $HSO_4^- + HPO_4^{2-} \rightleftharpoons SO_4^{2-} + H_2PO_4^-$
(d) $Al(H_2O)_6^{3+} + OH^- \rightleftharpoons Al(H_2O)_5OH^{2+} + H_2O$
(e) $N_2H_4 + H_2O \rightleftharpoons N_2H_5^+ + OH^-$
(f) $NH_2OH + HCl \rightleftharpoons NH_3OH^+ + Cl^-$
(g) $O^{2-} + H_2O \rightleftharpoons 2OH^-$
(h) $H^- + H_2O \rightleftharpoons H_2 + OH^-$
(i) $NH_2^- + N_2H_4 \rightleftharpoons NH_3 + N_2H_3^-$
(j) $HNO_3 + H_2SO_4 \rightleftharpoons H_3SO_4^+ + NO_3^-$

10.44 Identify the acid–base conjugate pairs in each of the following reactions:
(a) $HClO_4 + N_2H_4 \rightleftharpoons N_2H_5^+ + ClO_4^-$
(b) $HSO_3^- + H_3PO_3 \rightleftharpoons H_2SO_3 + H_2PO_3^-$
(c) $C_5H_5NH^+ + (CH_3)_3N \rightleftharpoons C_5H_5N + (CH_3)_3NH^+$
(d) $CO_3^{2-} + H_2O \rightleftharpoons HCO_3^- + OH^-$
(e) $HCHO_2 + C_7H_5O_2^- \rightleftharpoons C_7H_5O_2H + CHO_2^-$
(f) $H_2C_2O_4 + CH_3NH_2 \rightleftharpoons HC_2O_4^- + CH_3NH_3^+$
(g) $H_2CO_3 + H_2O \rightleftharpoons HCO_3^- + H_3O^+$
(h) $C_2H_5OH + NH_2^- \rightleftharpoons C_2H_5O^- + NH_3$
(i) $NO_2^- + N_2H_5^+ \rightleftharpoons HNO_2 + N_2H_4$
(j) $HCN + H_2SO_4 \rightleftharpoons H_2CN^+ + HSO_4^-$

10.45 Write autoionization reactions for the following solvents:
(a) $H_2O(l)$ (b) $NH_3(l)$ (c) $HCN(l)$

10.46 What would be the formula of the conjugate acid of dimethylamine, $(CH_3)_2NH$? What would be the formula of its conjugate base?

10.47 In water, the HCO_3^- ion is amphiprotic. Write equilibria to illustrate this.

10.48 Explain why a solution that contains the ion $Cr(H_2O)_6^{3+}$ is acidic.

Strengths of Acids

10.49 Which is the stronger acid?
(a) H_2SO_3 or $HClO_3$
(b) HNO_3 or H_2CO_3
(c) H_3AsO_4 or H_3PO_4

(d)

$$H-O-\overset{\overset{\displaystyle O}{\|}}{\underset{\underset{\displaystyle O}{\|}}{S}}-O-H \quad \text{or} \quad H-O-\overset{\overset{\displaystyle O}{\|}}{\underset{\underset{\displaystyle O}{\|}}{Se}}-O-H$$

(e) $HClO_3$ or $HBrO_3$
(f) $HOBr$ or $HBrO_3$
(g) H_2S or H_2Se
(h) H_2Se or HBr
(i) PH_3 or NH_3

*10.50 For each of the following pairs of substances, choose the one in which the hydrogen printed in color is the more acidic:

(a)

$$H-\overset{\overset{\displaystyle Cl}{|}}{\underset{\underset{\displaystyle Cl}{|}}{C}}-C\overset{\displaystyle O}{\underset{\displaystyle OH}{\diagup}} \quad \text{or} \quad H-\overset{\overset{\displaystyle Cl}{|}}{\underset{\underset{\displaystyle H}{|}}{C}}-C\overset{\displaystyle O}{\underset{\displaystyle OH}{\diagup}}$$

(b)

$$H-\overset{\overset{\displaystyle Cl}{|}}{\underset{\underset{\displaystyle H}{|}}{C}}-C\overset{\displaystyle O}{\underset{\displaystyle OH}{\diagup}} \quad \text{or} \quad H-\overset{\overset{\displaystyle F}{|}}{\underset{\underset{\displaystyle H}{|}}{C}}-C\overset{\displaystyle O}{\underset{\displaystyle OH}{\diagup}}$$

*10.51 Which compound would you expect to be more acidic, CH_3OH or CH_3SH? Explain your answer.

10.52 Why is HNO_3 a stronger acid than HNO_2?

10.53 Why is HCl a stronger acid than HF?

10.54 Why is HCl a stronger acid than H_2S?

10.55 Explain using Lewis structures how the electronic structure of the anions affect the trend toward increasing acidity for H_3PO_4, H_2SO_4, and $HClO_4$.

10.56 Explain *why* $HClO_4$ is a stronger acid than $HClO_3$.

10.57 Use the arguments that we presented to explain relative acid strengths of oxoacids to compare the strengths of the HSO_3^- and HSO_4^- ions as Brønsted *bases*.

Lewis Acids and Bases

10.58 What is the Lewis definition of a base? What is a Lewis acid?

10.59 Explain why the reaction $H^+ + NH_3 \rightarrow NH_4^+$ is a Lewis acid–base reaction. Which is the Lewis acid and which is the Lewis base?

10.60 Boron trichloride, BCl_3, reacts with diethyl ether, $(C_2H_5)_2O$, to form an *addition compound*, which we can write as $Cl_3B \leftarrow O(C_2H_5)_2$. Use electron dot formulas to

interpret this reaction as a Lewis acid–base neutralization.

10.61 Explain why the reaction of CO_2 with H_2O to produce H_2CO_3, which we can also write as $CO(OH)_2$, can be viewed as a Lewis acid–base neutralization.

Complexes of Metal Ions

10.62 Silver ion reacts with ammonia to form an ion having the formula $Ag(NH_3)_2^+$ in which two ammonia molecules are bound to the Ag^+ ion. Explain how this can be viewed as a Lewis acid–base reaction.

10.63 Use Lewis symbols to diagram the formation of the $CuCl_4^{2-}$ ion from Cu^{2+} and Cl^- ions. In the reaction, which is a Lewis acid and which is a Lewis base?

10.64 Define the following terms: (a) ligand, (b) complex, (c) donor atom, (d) acceptor atom, (e) polydentate ligand, (f) bidentate ligand, and (g) monodentate ligand.

10.65 Why are compounds that contain complex ions called coordination compounds?

10.66 In general, what kinds of substances serve as ligands in the formation of complex ions with metals?

10.67 Explain how the following reactions involve the displacement of one Lewis base by another. Identify the Lewis acid in each reaction.
(a) $Zn(H_2O)_4^{2+} + 4CN^- \longrightarrow Zn(CN)_4^{2-} + 4H_2O$
(b) $Pt(NH_3)_4^{2+} + Cl^- \rightarrow Pt(NH_3)_3Cl^+ + NH_3$

10.68 Make a sketch to show how the oxalate ion behaves as a bidentate ligand.

10.69 Draw the structure of the $EDTA^{4-}$ anion. Identify its donor atoms. Why is EDTA an important ligand? Give three of its uses.

10.70 Why do so many copper salts give blue-colored solutions?

10.71 What is the color of each of the following: (a) $Cu(NH_3)_4^{2+}$, (b) $Cu(H_2O)_4^{2+}$, (c) $Co(H_2O)_6^{2+}$, (d) $Ni(H_2O)_6^{2+}$?

10.72 Manganous chloride has the formula $MnCl_2 \cdot 6H_2O$. On the basis of what you have learned in this chapter, how are the water molecules probably held within crystals of this salt?

10.73 Write the formulas of the complexes formed from the following:
(a) Ag^+ and two I^- ions
(b) Ag^+ and two NH_3 molecules
(c) Co^{3+} and three oxalate ions
(d) Co^{3+} and six NO_2^- ions
(e) Cr^{3+} and one $EDTA^{4-}$ ion
(f) Ni^{2+} and four CN^- ions
(g) Fe^{3+} and six SCN^- ions

10.74 What is the oxidation state of the metal in each of the following complexes? (a) $Cd(CN)_4^{2-}$, (b) $Fe(CN)_6^{3-}$, (c) $Fe(CN)_6^{4-}$, (d) $[Cr(NH_3)_4Cl_2]^+$, (e) $Pt(NH_3)_2Cl_2$

PHYSICAL PROPERTIES AND THE STATES OF MATTER

CHAPTER

11

PROPERTIES OF GASES

Matter is capable of existing in three different physical forms or **states:** solid, liquid, and gas. In this chapter and the next, we will examine the physical and chemical characteristics of these states and the transformations that occur among them. This chapter deals with the gaseous state, in which the **intermolecular forces of attraction**—the attractions that one molecule experiences toward others around it—are weak. These weak forces allow for rapid, independent movement of the molecules and cause the physical behavior of a gas to be nearly independent of its chemical composition. Instead, the behavior of a gas is controlled by its volume, pressure, temperature, and number of moles. Since these variables are of paramount importance, we will begin our discussion by taking a close look at them as they apply to the gaseous state.

11.1 VOLUME AND PRESSURE

When a gas is introduced into a container, the molecules move freely within it and occupy the container's entire volume. As a result, the volume of a gas is given simply by specifying the volume of the vessel in which it is held. Because gases mix freely with one another, when there are several gases in a mixture, the volume of each component is the same as the volume occupied by the entire mixture.

Pressure is defined as force per unit area; it is an intensive quantity formed as a ratio of two extensive quantities: force and area. For example, if a 100-lb force is exerted on a piston whose total area is 100 in.^{2}, the pressure acting on each square inch is 100 lb/100 in.^{2} = 1 lb/in.^{2} (1 pound per square inch or 1 psi). A force of only 1 lb pressing on an area of 1 in.^{2} would exert exactly the same pressure because the ratio of force to area is the same.

The *unit area* in this discussion is 1 in.^{2}.

$$\frac{100 \text{ lb}}{100 \text{ in.}^2} = \frac{1 \text{ lb}}{1 \text{ in.}^2}$$

If the 100-lb force is exerted on a smaller area, for example, 1 in.^{2}, the pressure is greatly magnified (Figure 11.1). Now the pressure is 100 lb/in.^{2} = 100 psi. The dependence of pressure on both force and the area over which it is spread has been experienced firsthand by anyone who has ever stepped on a nail. A 110-lb person

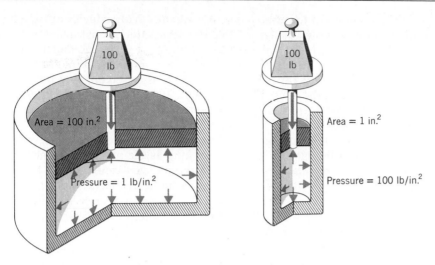

Figure 11.1. *Pressure is the ratio of force to area. The same force is exerted by the 100-lb weight on both pistons, but the smaller area of the piston on the right gives a larger pressure. Within each cylinder the same pressure is exerted on all the walls.*

stepping on even a dull nail having a point area of 0.01 in.^{2} will experience a pressure of *11,000 psi!* This is more than enough to cause the nail to puncture the skin.

If the pressure generated by a piston is applied to a fluid (a gas or liquid), as illustrated in Figure 11.1, it is transmitted uniformly in all directions so that all the walls of the container experience the same pressure. If the piston is supported by the fluid, then the fluid also exerts an equal pressure on the piston as well as the other walls of the container.

The pressure of the atmosphere

The atmosphere that surrounds the earth is a mixture of gases. It exerts a pressure that is easily demonstrated by the experiment illustrated in Figure 11.2, which shows what happens if the air is pumped out of a steel can. Before the air is removed, the pressure inside and outside the can is the same. However, when the air is removed, the pressure of the atmosphere pushes only on the outside, and it is large enough that the can is crushed.

 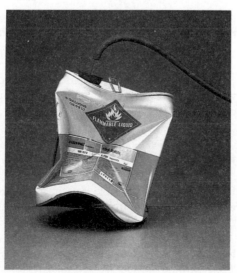

Figure 11.2. *On the left, air has not been removed from the can and the atmospheric pressure is exerted equally on its inner and outer surfaces. On the right, we see what happens when the air is removed from the inside of the can. The atmospheric pressure is so great that it crushes the can.*

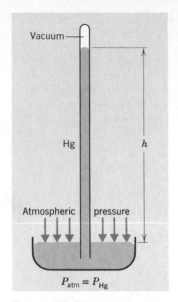

Figure 11.3. *A mercury barometer. The atmospheric pressure supports a column of mercury of height* h, *which changes as the atmospheric pressure changes.*

The symbol for the newton is N.

Normal barometric pressure at sea level is approximately 29.9 in. of mercury.

The pressure of the atmosphere is measured with a device called a **barometer,** and one kind is shown in Figure 11.3. The apparatus is constructed by filling a glass tube about 1 meter long with mercury, capping the end so no mercury can spill out, and then inverting the filled tube into a dish of mercury so that the capped end is submerged. When the cap is removed, some of the mercury runs out, but not much of it. Instead, the tube is left nearly full, with just a relatively small space (essentially a vacuum) above the mercury. The mercury is kept in the tube by the atmospheric pressure pushing down on the surface of the liquid in the dish.

The height of the mercury column, h, in the tube turns out to be independent of the diameter of the tube or its length, as long as the tube is long enough to leave a space above the mercury. The height of the mercury does change, however, when the atmospheric pressure changes. For example, when a storm approaches, the atmospheric pressure drops and the column becomes shorter. In fact, it is the height of such a column of mercury that is reported, usually in "inches of mercury," when weather reports on TV or radio give the barometric pressure.

When the height of the mercury column in the tube isn't changing, the pressure exerted by the mercury column exactly balances the pressure of the atmosphere. We can express the atmospheric pressure, therefore, in terms of the height of this column. At sea level, the height fluctuates around a value of approximately 760 mm, which led to the original definition of one of our standard units of pressure called the **standard atmosphere (atm).** As first defined, the standard atmosphere was equal to the pressure exerted by a column of mercury 760 mm high at sea level and at a temperature of 0 °C.[1]

$$1 \text{ atm} = 760 \text{ mm Hg}$$

In English units, this corresponds to a pressure of 14.7 lb/in.[2].

The SI unit of pressure is the **pascal (Pa),** defined as 1 newton (the SI unit of force) per square meter.

$$1 \text{ Pa} = 1 \text{ N/m}^2$$

Since the introduction of the SI, the standard atmosphere has been redefined in terms of the pascal, so its current definition is

$$1 \text{ atm} = 101,325 \text{ Pa} = 101.325 \text{ kPa}$$

A smaller unit of pressure, which we often use in experimental work, is the **torr,** named after Evangelista Torricelli, the inventor of the barometer. It is defined so that 760 torr equals 1 atm.

$$1 \text{ atm} = 760 \text{ torr}$$

For all but the most precise measurements of gas pressures, 1 torr can be taken to be the pressure exerted by a column of mercury 1 mm high.

$$1 \text{ torr} = 1 \text{ mm Hg}$$

In the chemistry lab, we find torr and atm to be more convenient units to work with than the pascal. Therefore, for simplicity, we will not use this SI unit in our computations.

Measuring pressures of trapped gases

It is often desirable to know the pressure of a gas present in a closed system (for example, the pressure of gases produced during a chemical reaction). The instru-

---

[1] The height of the column of mercury, which is supported by atmospheric pressure, varies with both the density of the mercury and the pull of gravity on the mercury in the column. Since density varies with temperature and the pull of gravity varies with altitude, in this definition of the standard atmosphere, it was necessary to specify a reference temperature (0 °C) as well as a reference altitude (sea level).

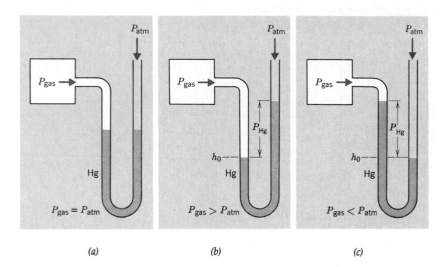

Figure 11.4. *An open-end manometer. (a) The pressure of the trapped gas is equal to the atmospheric pressure. (b) The pressure of the trapped gas is greater than the atmospheric pressure. (c) The pressure of the trapped gas is less than the atmospheric pressure.*

ment normally used for these pressure measurements is called a **manometer.** An open-end manometer (Figure 11.4) is simply a U-shaped tube containing some liquid, such as mercury. One arm of the tube is connected to a system whose pressure is to be measured, while the other arm remains open to the atmosphere. When the pressure of the gas inside the system (P_{gas}) is equal to P_{atm}, the level of the liquids in both arms will be the same, as shown in Figure 11.4a. If the pressure of the gas is greater than P_{atm}, the mercury in the left arm will be forced downward, causing the mercury in the right arm to rise (Figure 11.4b). We obtain the pressure of the gas in this system by comparing the pressures exerted in both arms at a reference level, h_0, which is chosen to be the height of the shorter column. The pressure exerted on the left column when $P_{gas} > P_{atm}$ is simply P_{gas}, while at the same level in the right arm the pressure is P_{atm} plus the pressure exerted by the column of mercury that rises above the reference level, P_{Hg}. When the levels are stationary, the pressures at the reference level on both sides are equal and

$$P_{gas} = P_{atm} + P_{Hg}$$

The atmospheric pressure (P_{atm}) is found with a barometer, and P_{Hg} is simply the difference in the heights of the two mercury columns. Similarly, when $P_{gas} < P_{atm}$, shown in Figure 11.4c, the pressure in the left arm at the reference level is $P_{gas} + P_{Hg}$, while in the right column the pressure is P_{atm}. In this case, when the columns are stationary

$$P_{gas} + P_{Hg} = P_{atm}$$

so that

$$P_{gas} = P_{atm} - P_{Hg}$$

Therefore, when $P_{gas} < P_{atm}$, the pressure of the gas in the system is found by subtracting the difference in the heights of the columns from atmospheric pressure.

A closed-end manometer, shown in Figure 11.5, is often used for measuring low pressures because it can be of compact size. It consists of a U-shaped tube with one arm sealed at the top. If the sealed tube on the right is short and the arm on the left is open to the atmosphere, mercury will fill the tube on the right completely. When the manometer is connected to an apparatus filled with a gas at a low pressure, the level on the left can rise and that on the right can drop, as shown in Figure 11.5b. At the reference level, the pressure exerted on the left side is just the pressure of the gas, P_{gas}, while on the right at this same level the pressure is P_{Hg}, because the space above

In problems dealing with open-end manometers, it is usually best to draw a picture of the apparatus and then decide whether to add or subtract the pressure difference from the atmospheric pressure.

Figure 11.5. *A closed-end manometer can be of compact design when low pressures are to be measured. (a) When the gas pressure equals atmospheric pressure, the mercury is forced to the top of the right arm of the manometer if the tube length is short. (b) The pressure of the gas can be read directly if it is considerably smaller than atmospheric pressure.*

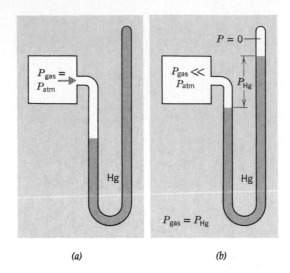

(a)   (b)

the mercury is a vacuum. When the columns are stationary, these pressures are equal, so $P_{gas} = P_{Hg}$. Thus, the pressure of the gas is measured by taking the difference in the heights of the liquid in the two arms of the manometer. One of the advantages of using a closed-end manometer, therefore, is that there is no need to make a separate measurement of the barometric pressure.

Using liquids other than mercury in manometers

Mercury is not the only fluid ever used in manometers, especially if low pressures are to be measured. This is because mercury, with its large density, gives a very small difference in the heights of the columns in a manometer when the pressure is small. On the other hand, if a liquid of lower density is used, the difference in heights can be much larger. For example, a pressure of 1 torr gives a mercury height of 1 mm, but because water is only 1/13.6 as dense as mercury, a water column must be 13.6 times as high to exert the same pressure. A similar effect would be produced if we used a fluid that was half as dense as mercury: A pressure of 1 torr would give a mercury height of 1 mm, but a height of 2 mm for the other fluid.

There is a very simple relationship between the heights of columns of fluids in manometers (or barometers) and their densities. Consider two liquids A and B. If, for a given pressure, the height of the column of liquid A is h_A and the height of the column of liquid B is h_B, then

Here d_A and d_B are the densities of liquids A and B.

$$h_B = h_A \times \frac{d_A}{d_B}$$ (11.1)

EXAMPLE 11.1. USING FLUIDS OTHER THAN MERCURY IN A MANOMETER

PROBLEM: A liquid with a density of 1.15 g/mL was used in an open-end manometer. In a particular experiment, a difference in the heights of the levels in the two arms was measured to be 14.7 mm, with the level in the arm connected to an apparatus containing a trapped gas lower than the level open to the atmosphere. The barometric pressure had been measured to be 756.00 torr. What was the pressure of the trapped gas?

ANALYSIS: The first step is to visualize the experimental setup. The condition described in the problem corresponds to the one shown in Figure 11.4a, so we know we have to *add* the difference in pressures, which we might call P_{liquid}, to the atmospheric pressure.

$$P_{gas} = P_{atm} + P_{liquid}$$

However, before we can do this, we have to convert the difference in heights to a pressure difference in torr, which means we need to use Equation 11.1.

SOLUTION: First, let's convert the difference in heights of the liquid levels from "mm liquid" to "mm Hg," which we can then equate to torr. The density of mercury is 13.6 g/mL. Therefore,

$$h_{Hg} = h_{liquid} \times \frac{d_{liquid}}{d_{Hg}}$$

$$= 14.7 \text{ mm liquid} \times \frac{1.15 \text{ g/mL}}{13.6 \text{ g/mL}}$$

$$= 1.24 \text{ mm Hg}$$

Since we can take 1 mm Hg to be equal to 1 torr, a height difference of 1.24 mm Hg corresponds to a pressure difference of 1.24 torr. The pressure of the gas, therefore, is

$$P_{gas} = 756.00 \text{ torr} + 1.24 \text{ torr}$$
$$= 757.24 \text{ torr}$$

We could do a calculation similar to that in Example 11.1 to show how high a column of water would be if it were used in a barometer instead of mercury. At an atmospheric pressure of exactly 1 atm, the water column in such a barometer would be 1.03×10^4 mm high. In English units, this translates to a height of 33.8 ft, which explains why Hg is used instead of H_2O—if H_2O were used, the barometer would have to be nearly three stories high. Of more practical interest, however, is that this calculation explains why pumps that work by suction can't possibly lift water more than about 33 ft high, no matter how hard they work. Even if the pump were able to create a perfect vacuum, an atmospheric pressure of 1 atm could only push water up a pipe 33.8 ft.

11.2 PRESSURE–VOLUME–TEMPERATURE RELATIONSHIPS FOR A FIXED AMOUNT OF GAS

You are already familiar with many of the ways gases behave, simply because of everyday experiences. For example, you know that if you squeeze air into a smaller volume by pushing down on the handle of a bicycle pump, without letting any of the air escape into a tire (Figure 11.6), the pressure of the air increases—you already know that *the pressure of a gas increases when its volume is decreased*. If you are a hot-air balloonist, you also know how air behaves when it's heated. The air expands. By expanding, a given mass of it occupies more space, so the mass per unit volume (its density) decreases. That is what allows the hot air (and the balloon that holds it) to float in the more dense cool air surrounding the balloon (Figure 11.7). Thus, you already know that *the volume of a gas expands when its temperature increases*. And you probably know that if you place an aerosol can in a fire, it will explode. You know that *the pressure of a confined gas increases when its temperature increases*.

Observations somewhat similar to these are what led scientists a long time ago to study the behavior of gases systematically. Generally, their experiments involved trapping a sample of a gas and then observing what happens when the pressure,

Aerosol cans have warnings printed on them such as "Do not incinerate" or "Do not store over 120 °F."

Figure 11.6. *Pushing down on the handle of the pump decreases the volume of the air and increases its pressure if the gas can't escape through the hose.*

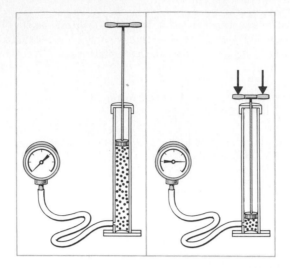

volume, or temperature is varied. Because three variables are involved, one was held constant while the relationship between the other two was determined. From these studies, several *gas laws* were established. For example, in 1662 Robert Boyle, an Irish chemist and physicist, reported that if the temperature of a gas sample is held constant, its volume is inversely proportional to the pressure exerted on it. This relationship became known as Boyle's law and can be expressed in equation form as

$$V \propto \frac{1}{P} \quad \text{(at constant temperature)}$$

The proportionality can be changed to an equality by inserting a proportionality constant,

$$V = \frac{\text{constant}}{P}$$

Figure 11.7. *A propane gas burner just inside the lower throat of a hot-air balloon heats the air in the balloon. The air expands in volume, and because the balloon has an open end, some air spills out. This reduces the density of the remaining air, and the balloon rises in the denser outside air.*

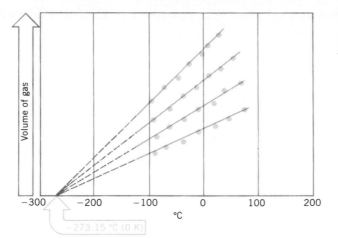

Figure 11.8. *A graph of the volume of various gas samples as a function of temperature.*

or, clearing the fraction,

$$PV = \text{constant}$$

In 1787 a French chemist named Jacques Alexander Charles became interested in hot-air ballooning, which at that time was beginning to be popular in France. He studied what happens to the volume of a sample of gas when its temperature is varied, keeping the pressure constant. When data from experiments like his are plotted, a graph like that shown in Figure 11.8 is obtained. Here the volume of the gas is plotted against the temperature in degrees Celsius. The colored points correspond to typical data, and the lines are drawn to most closely fit the data. Each line represents data collected for a different sample. Because all real gases eventually condense if cooled sufficiently, the solid portions of the lines correspond to temperatures at which measurements are possible; at lower temperatures the gas condenses. However, if the lines are extrapolated (extended) back to a hypothetical volume of zero, they all meet at the same temperature, $-273.15\ ^{\circ}\text{C}$. What is especially significant is that this same behavior is exhibited by all gases; when plots of volume versus temperature are extrapolated to zero volume, the temperature axis is always crossed at $-273.15\ ^{\circ}\text{C}$. This point represents the temperature at which all gases, if they did not condense, would have a volume of zero, and below which they would have a negative volume. Negative volumes are impossible, of course, so it was reasoned that this must be the coldest possible temperature and it was called **absolute zero.**

As you learned earlier, absolute zero corresponds to the zero point on the Kelvin temperature scale (also called the **absolute temperature scale**). Recall that to obtain the Kelvin temperature, we add 273.15 to the Celsius temperature.

$$T_K = T._C + 273.15$$

For most purposes we will need only three significant figures, so 0 °C = 273 K and we can use the approximate relationship

$$T_K = T._C + 273$$

The straight lines in Figure 11.8 suggest that at constant pressure the volume of a gas is directly proportional to its temperature, provided the temperature is expressed in kelvins. This became known as Charles's law and is expressed mathematically as

$$V \propto T \quad \text{(at constant pressure)}$$

where T stands for the absolute temperature. (As you will see, the absolute temperature scale is always used whenever temperature enters numerically into a calculation

When Jacques Charles became interested in hot-air balloons, they were inflated with air heated by burning straw.

A straight-line graph results when there is a direct proportionality between the two quantities being plotted.

When you encounter a capital letter T in an equation, it almost always stands for the absolute temperature.

involving pressures or volumes of a gas.) Charles's law can also be expressed as an equality by adding a proportionality constant.

$$\frac{V}{T} = \text{constant}$$

Taking a somewhat different approach, Joseph Gay-Lussac, a contemporary of Charles, studied how the pressure of a gas depends on its temperature when the volume is held constant. He also found a proportionality involving the absolute temperature. The pressure of a gas held at constant volume is directly proportional to the absolute temperature. Gay-Lussac's law in equation form is

$$P \propto T \quad \text{(at constant volume)}$$

or

$$\frac{P}{T} = \text{constant}$$

The combined gas law

The three gas laws that we've mentioned above can be brought together into a single law known as the **combined gas law**, which can be expressed as

$$\frac{PV}{T} = \text{constant}$$

Usually, we use the combined gas law in problems in which we know some given set of conditions of temperature, pressure and volume for a fixed amount of gas, and we want to find out how one of these variables will change when we change the others. If we label the initial conditions of P, V, and T with a subscript i and the final conditions with the subscript f, the combined gas law can be written in the following useful form:

The combined gas law only applies when the amount of gas remains constant.

$$\frac{P_i V_i}{T_i} = \frac{P_f V_f}{T_f} \tag{11.2}$$

Let's look at some examples that illustrate how we use it.

EXAMPLE 11.2. USING THE COMBINED GAS LAW

PROBLEM: A sample of a gas exerts a pressure of 625 torr in a 300-mL vessel at 25 °C. What pressure would this gas sample exert if it were placed in a 500-mL container at 50 °C?

ANALYSIS: The first thing to notice about this problem is that it deals with a fixed amount of gas being transferred from one container to another, with changes in conditions of pressure, temperature, and volume. This is the clue that tells us we need to apply the combined gas law.

As you will see below, the best way to organize the solution to the problem is to make a small table of the data corresponding to the initial and final conditions by extracting the information from the statement of the problem. This will avoid errors and help in identifying the unknown quantity being solved for.

To solve the problem, we need to solve Equation 11.2 for the desired quantity, in this case the final pressure. When this is done, the resulting equation can be arranged to look like the following:

Ratio of temperatures

$$P_f = P_i \times \left(\frac{V_i}{V_f}\right) \times \left(\frac{T_f}{T_i}\right)$$

Ratio of volumes

Notice that the final pressure is equal to the initial pressure multiplied by a volume ratio and a temperature ratio. Of course, we can just substitute the proper quantities from the table of data and solve for the answer. However, we can also use reasoning as a check on our having solved Equation 11.2 correctly. To do this, we use what we know about the effects of volume changes and temperature changes on the pressure of a gas to determine which quantities go in the numerators and denominators of the ratios. To help you in your thinking, this is how we will work out the solutions to this and other problems in this section.

SOLUTION: Let's begin by tabulating the data. In doing this, we have to be especially careful about the temperatures, which must be expressed in kelvins before we insert them into the solution to the problem. Thus, the initial temperature of 25 °C becomes 298 K (i.e., 25 + 273), and the final temperature of 50 °C becomes 323 K.

	Initial (i)	Final (f)
P	625 torr	?
V	300 mL	500 mL
T	298 K	323 K

It is simple enough to substitute these into the equation obtained in the analysis section to calculate the answer. However, let's see whether we can use reasoning to decide how the arithmetic should be set up. We need to establish the volume and temperature ratios to solve for the final pressure.

$$P_f = P_i \times (\text{volume ratio}) \times (\text{temperature ratio})$$

Let's do this by imagining that we carry out the net change in two steps, first holding the temperature constant while we change the volume, and then in the second step holding the volume constant while we change the temperature.

Notice that in this problem the volume is increasing from 300 mL to 500 mL. We know that this should lead to a decrease in the pressure, because the pressure of a gas is inversely proportional to its volume; in other words, when the volume goes up, the pressure goes down. The volume change favors a pressure decrease, so the volume ratio should have a value that tends to make the final pressure *smaller* than the initial pressure, and this requires that the smaller volume be in the numerator of the volume ratio. The smaller volume is the initial volume, so the volume ratio is

$$\left(\frac{300 \text{ mL}}{500 \text{ mL}}\right)$$

and we have now gotten this far:

$$P_f = P_i \times \left(\frac{300 \text{ mL}}{500 \text{ mL}}\right) \times (\text{temperature ratio})$$

Notice that the ratio of volumes is exactly what we get by blindly substituting values from the table into the solved equation, but we got the proper ratio by reasoning instead of worrying about keeping the subscripts i and f straight.

Now let's look at the temperature ratio. The temperature of the gas is increasing as we go from the initial to the final conditions. You already know that the pressure of a gas increases when its temperature rises, so the temperature change works to make the final pressure larger than the initial pressure. In this case, we want to multiply by a temperature ratio that has the larger value in the numerator, so the temperature ratio must be

$$\left(\frac{323 \text{ K}}{298 \text{ K}}\right)$$

Therefore, the final pressure is given by the expression

$$P_f = P_i \times \left(\frac{300 \text{ mL}}{500 \text{ mL}}\right) \times \left(\frac{323 \text{ K}}{298 \text{ K}}\right)$$
$$= 406 \text{ torr}$$

Notice that the units for volume and temperature cancel in the ratios.

Scientists who first studied gases soon realized that the pressure and temperature controlled the volume observed for a gas sample. Therefore, to compare different gas samples, they defined a set of reference conditions. These conditions are known as **standard temperature and pressure,** or simply **STP,** and are 0 °C (273 K) and 1 atm (760 torr).

EXAMPLE 11.3. USING THE COMBINED GAS LAW

PROBLEM: What would be the volume of a gas at STP if it was found to occupy a volume of 255 mL at 25 °C and 650 torr?

SOLUTION: Once again, the problem involves the combined gas law.

$$\frac{P_i V_i}{T_i} = \frac{P_f V_f}{T_f}$$

It is absolutely essential that you express the temperatures in kelvins in these calculations.

First, let's tabulate the data as in the last example. Once again, we have to change Celsius temperatures to kelvins. We also have to realize that STP translates to mean 760 torr and 273 K.

	Initial (i)	Final (f)
P	650 torr	760 torr
V	255 mL	?
T	298 K	273 K

Now we solve the combined gas law for the final volume.

$$V_f = V_i \times \left(\frac{P_i}{P_f}\right) \times \left(\frac{T_f}{T_i}\right)$$

Notice that we've arranged the solved equation so that the final volume is equal to the initial volume multiplied by a ratio of pressures and a ratio of temperatures.

$$V_f = V_i \times \text{(ratio of pressures)} \times \text{(ratio of temperatures)}$$

Now, we look at the data to decide how the ratios should be set up. First, we see that the pressure is increasing from 650 torr to 760 torr. From the properties of gases we know that when the pressure goes up, the gas volume goes down, so we need a ratio that will make the volume smaller. This requires the smaller pressure in the numerator.

Examining the temperature, we see that it is falling from 298 K to 273 K. When a gas cools, its volume shrinks, so the temperature ratio should tend to make the final volume smaller than the initial volume. This means that the smaller temperature must be in the numerator.

The result of this is as follows:

$$V_f = 255 \text{ mL} \times \left(\frac{650 \text{ torr}}{760 \text{ torr}} \right) \times \left(\frac{273 \text{ K}}{298 \text{ K}} \right)$$

$$= 200 \text{ mL}$$

Thus, the volume at STP is 200 mL.

EXAMPLE 11.4. USING THE COMBINED GAS LAW

PROBLEM: A sample of a gas occupies 250 mL at 27 °C. What volume will it occupy at 35 °C if there is no change in pressure?

SOLUTION: When one of the variables in the combined gas law remains constant, it allows us to simplify the equation. In this case the initial and final pressures are the same, so let's just represent the pressure by the symbol P. Substituting this into the combined gas law gives

$$\frac{PV_i}{T_i} = \frac{PV_f}{T_f}$$

Dividing both sides by P eliminates it from the equation, which then becomes

$$\frac{V_i}{T_i} = \frac{V_f}{T_f}$$

(Actually, this is a statement of Charles's law, but we don't need to know that to solve the problem.)

Let's tabulate the data before proceeding further. As usual, we convert °C to K.

	Initial (i)	Final (f)
V	250 mL	?
T	300 K	308 K

27 °C + 273 = 300 K

35 °C + 273 = 308 K

Solving for the final volume, we get

$$V_f = V_i \times \left(\frac{T_f}{T_i} \right)$$

or

$$V_f = V_i \times \text{(ratio of temperatures)}$$

We see that the temperature is increasing, and because a gas expands when heated, we expect the volume to become larger. This requires that the larger temperature be in the numerator of the ratio. Therefore,

$$V_f = 250 \text{ mL} \times \left(\frac{308 \text{ K}}{300 \text{ K}} \right)$$

$$= 257 \text{ mL}$$

The preceding example illustrates an important principle to remember. If one of the variables (pressure, temperature, volume) remains constant in a problem, you can just drop it out of the combined gas law.

The ideal gas

The combined gas law tells us that for a gas sample, the ratio PV/T is a constant.

$$\frac{PV}{T} = \text{constant} \tag{11.3}$$

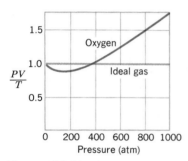

Figure 11.9. *A graph of* PV/T *versus pressure for an ideal gas and for one mole of oxygen at 0 °C.*

However, if very precise measurements are made for real gases—gases such as methane (CH_4) and oxygen—it is found that this is not quite true. For example, if a graph is made of the value of the PV/T ratio as a function of pressure for a real gas, it looks like the one shown in Figure 11.9. The horizontal line is what would be expected if Equation 11.3 was obeyed exactly, but as you can see, the real values for this ratio drop below the line at low pressures and rise above it at higher pressures.

A hypothetical gas that would obey the combined gas law over all ranges of temperature and pressure is called an **ideal gas.** Real gases are said to deviate from ideal behavior, and the significance of this deviation will be discussed later. Fortunately, at relatively low pressures, which includes atmospheric pressure, and at relatively high temperatures, many gases come quite close to being ideal, so the combined gas law works satisfactorily for most computations of the kind we have just seen.

11.3 THE IDEAL GAS LAW

We have seen that for a fixed amount of a gas (assuming it to behave as an ideal gas)

$$\frac{PV}{T} = \text{constant}$$

At a constant temperature, the pressure and volume depend on the amount of gas in the container. If you blow more air into a balloon, which is subjected to the constant pressure of the atmosphere, its size (volume) increases. If you pump more air into a tire, whose volume remains essentially constant, the pressure increases.

The constant in this equation, however, is only a constant if we keep the amount of gas the same. In fact, the value of the constant is directly proportional to the number of moles of gas; if we double the number of moles, the value of the constant doubles. If we represent the number of moles of gas by the symbol n, the equation becomes

$$\frac{PV}{T} = n \times (\text{other constant})$$

This "other constant" is represented by the symbol R and is called the **universal gas constant.** When we use this symbol, the equation becomes

$$\frac{PV}{T} = nR$$

If we clear fractions by multiplying both sides by the temperature, we obtain the equation as it is normally written.

$$PV = nRT \qquad (11.4)$$

Equation 11.4 embodies all the other gas laws we have discussed and is known as the **ideal gas law.** It is also called the **equation of state for an ideal gas** because it relates the variables (P, V, T, n) that specify the properties of the gas. If the values of any three of these variables are known, the fourth can have only one value as determined by Equation 11.4.

Molar volume

To use Equation 11.4, we need the value of R, which we can obtain by substituting measured values of P, V, T, and n into the equation. If we choose STP (1 atm and 0 °C or 273 K), and take one mole of gas, we can measure the volume, which we call the **molar volume at STP** because it is the volume of one mole under these conditions. If we do this for many gases, we find that the values fluctuate somewhat because real gases are not "ideal." Some typical values are given in Table 11.1. From the results of many measurements, the average volume occupied by one mole of a gas at STP is 22.4 L, and this value is therefore taken to be the molar volume of an ideal gas at STP. Using this value, we can calculate what R is.

$$R = \frac{PV}{nT}$$

$$= \frac{(1\ atm)(22.4\ L)}{(1\ mol)(273\ K)}$$

$$= 0.0821 \frac{L\ atm}{mol\ K}$$

or

$$R = 0.0821\ L\ atm\ mol^{-1}\ K^{-1}$$

More precise measurements give $R = 0.082057\ L\ atm\ mol^{-1}\ K^{-1}$.

The constant R can have other numerical values depending on the units used to express pressure and volume. Example 11.5 illustrates how we can convert R from one set of units to another. The most useful values of R, with their corresponding units, are included on the inside back cover of this book.

Table 11.1. Molar volumes of several real gases at STP

Substance	Molar Volume (liters)
Oxygen, O_2	22.397
Nitrogen, N_2	22.402
Hydrogen, H_2	22.433
Helium, He	22.434
Argon, Ar	22.397
Carbon dioxide, CO_2	22.260
Ammonia, NH_3	22.079

EXAMPLE 11.5. CHANGING THE UNITS OF THE GAS CONSTANT

PROBLEM: What is the value of R when pressure is expressed in torr and volume is expressed in milliliters?

SOLUTION: This is simply a unit conversion problem making use of the relationships

$$1 \text{ L} = 1000 \text{ mL}$$

$$1 \text{ atm} = 760 \text{ torr}$$

$$R = \left(\frac{0.0821 \text{ L atm}}{\text{mol K}} \right) \times \left(\frac{1000 \text{ mL}}{1 \text{ L}} \right) \times \left(\frac{760 \text{ torr}}{1 \text{ atm}} \right)$$

$$= \frac{6.24 \times 10^4 \text{ mL torr}}{\text{mol K}}$$

When you use R in a computation, be sure the units cancel correctly.

The choice of the value of R to be used in a given computation is governed by the units of P and V. Most people find it best, however, to learn one value of R and to convert P and V to units that can be used with that value of R.

The following are some examples of the application of the ideal gas law.

EXAMPLE 11.6. USING THE IDEAL GAS LAW

PROBLEM: What volume will 25.0 g of O_2 occupy at 20°C and a pressure of 0.880 atm?

SOLUTION: From the ideal gas law,

$$V = \frac{nRT}{P}$$

We will use $R = 0.0821$ L atm mol^{-1} K^{-1}. Tabulating our data, we get

P	0.880 atm
V	?
n	25.0 g of $O_2 \times \dfrac{1 \text{ mol of } O_2}{32.0 \text{ g of } O_2} = 0.781$ mol O_2
T	$20 + 273 = 293$ K

Substituting, we obtain

$$V = \frac{(0.781 \text{ mol}) \times (0.0821 \text{ L atm mol}^{-1} \text{ K}^{-1}) \times (293 \text{ K})}{(0.880 \text{ atm})}$$

$$= 21.3 \text{ L}$$

EXAMPLE 11.7. DETERMINING THE MOLECULAR MASS OF A GAS

PROBLEM: A student collected natural gas from a laboratory gas jet at 25 °C in a 250-mL flask until the pressure of the gas was 550 torr. The gas sample weighed 0.118 g at a temperature of 25 °C. From these data, calculate the molecular mass of the gas.

ANALYSIS: To measure the molecular mass of a substance, we need to establish experimentally a relationship between moles and grams. If we can find out how many moles correspond to some number of grams, all we need to do is take the ratio of grams to moles. The numerical value of this ratio is the molecular mass.

In this problem, we are given the number of grams of gas, so we have one of the pieces of information we need. We are also given the pressure, volume, and temperature of the gas sample, from which we can calculate the number of moles in the sample by using the ideal gas law. Once we have the number of moles, we set up the ratio of grams/moles to calculate the molecular mass.

SOLUTION: Let's solve the ideal gas law for the number of moles, n.

$$n = \frac{PV}{RT}$$

Again, we use $R = 0.0821$ L atm mol^{-1} K^{-1}. Our data are

P	$550 \text{ torr} \times \dfrac{1 \text{ atm}}{760 \text{ torr}} = 0.724 \text{ atm}$
V	$250 \text{ mL} \times \dfrac{1 \text{ L}}{1000 \text{ mL}} = 0.250 \text{ L}$
n	?
T	$25 + 273 = 298 \text{ K}$

Substituting, we obtain

$$n = \frac{(0.724 \text{ atm}) \times (0.250 \text{ L})}{(0.0821 \text{ L atm mol}^{-1} \text{ K}^{-1}) \times (298 \text{ K})}$$

$$= \frac{0.00740}{\text{mol}^{-1}} = 0.00740 \text{ mol}$$

Now we can take the ratio of grams to moles for this sample.

$$\frac{0.118 \text{ g}}{0.00740 \text{ mol}} = 15.9 \text{ g/mol}$$

A molecular mass can be calculated if you know the number of moles in a given mass of the compound. Simply take the ratio of grams to moles.

If there are 15.9 grams per mole, the molecular mass must be 15.9. (Natural gas is methane, CH_4, which has a molecular mass of 16.0.)

EXAMPLE 11.8. DETERMINING THE MOLECULAR MASS OF A GAS FROM ITS DENSITY

PROBLEM: A student measured the density of a gaseous compound to be 1.34 g/L at 25 °C and 760 torr and was told that the compound was composed of 79.8% carbon and 20.2% hydrogen.

(a) What is the empirical formula of the compound?

(b) What is its molecular mass?

(c) What is the molecular formula of the compound?

SOLUTION: (a) Following the procedure outlined in Section 2.8, we find the empirical formula of the carbon–hydrogen compound. If we assume a 100-g sample,

$$79.8 \text{ g C} \times \left(\frac{1 \text{ mol C}}{12.0 \text{ g C}} \right) = 6.65 \text{ mol C}$$

$$20.2 \text{ g H} \times \left(\frac{1 \text{ mol H}}{1.01 \text{ g H}} \right) = 20.0 \text{ mol H}$$

The empirical formula is $C_{\frac{6.65}{6.65}} H_{\frac{20.0}{6.65}}$ or CH_3, which would give an empirical formula mass of $CH_3 = 15.0$.

(b) The density gives the weight of 1 L of the gas.

$$1.00 \text{ L} \Longleftrightarrow 1.34 \text{ g}$$

To calculate molecular mass we need a relationship between mass and moles. Following the procedure in Example 11.7, we get

P	760 torr $= 1$ atm
V	1.00 L
n	?
T	$25 + 273 = 298$ K

$$n = \frac{PV}{RT} = \frac{(1.00 \text{ atm}) \times (1.00 \text{ L})}{(0.0821 \text{ L atm mol}^{-1} \text{ K}^{-1}) \times (298 \text{ K})}$$

$$= 0.0409 \text{ mol}$$

Thus,

$$0.0409 \text{ mol} = 1.34 \text{ g}$$

To obtain the molecular mass, we take the ratio of grams to moles.

$$\frac{1.34 \text{ g}}{0.0409 \text{ mol}} = 32.8 \text{ g/mol}$$

The molecular mass is 32.8.

(c) Recall that the molecular formula must have subscripts that are whole-number multiples of the subscripts in the empirical formula. Thus, possibilities are CH_3, C_2H_6, C_3H_9, and so on. You also learned that the molecular mass is a multiple of the mass of the empirical formula unit, so the actual molecular mass must be a multiple of the mass of CH_3, which is 15.0. The measured value, 32.8, is approximately twice 15.0, so the molecular formula must be C_2H_6. (This is the hydrocarbon ethane.)

The fact that the actual molecular mass (30.0) and the measured value (32.8) differ is not uncommon and would be attributed to *experimental error,*—error that creeps into the results of an experiment in the natural course of making measurements.

An equation relating the molecular mass of a gas to its density directly can easily be derived from the ideal gas law and used to solve problems such as part (b) of this last example. We know, for example, that the number of moles of a substance is obtained by dividing its mass, in grams, by the molecular mass.

$$\text{number of moles } (n) = \frac{\text{number of grams } (g)}{\text{molecular mass } (M)}$$

or simply,

$$n = \frac{g}{M}$$

Substituting for n in the ideal gas law gives

$$PV = \frac{g}{M} RT$$

which can be rearranged to solve for M.

$$M = \frac{g}{V} \frac{RT}{P} \tag{11.5}$$

Density (d) is a ratio of mass to volume, g/V. This allows us to write Equation 11.5 as

$$M = d \frac{RT}{P} \tag{11.6}$$

To solve part (b) of Example 11.8 we can substitute the given values for d, R, T, and P into this equation.

$$M = \frac{1.34 \text{ g } (0.0821 \text{ } \cancel{L} \text{ atm mol}^{-1} \text{ } \cancel{K}^{-1})(298 \text{ } \cancel{K})}{(1.00 \text{ } \cancel{L})(1 \text{ } \cancel{atm})}$$

$$= 32.8 \text{ g/mol}$$

Some people find it easy to remember equations like (11.5) and (11.6). Others find it easier to approach problems as shown in Examples 11.7 and 11.8. The choice is up to you.

11.4 CHEMICAL REACTIONS BETWEEN GASES

Many gases are able to undergo reactions with each other. Hydrogen and oxygen combine explosively to form water. Nitrogen and hydrogen, under appropriate conditions, combine to form ammonia.

One of the early observations about gaseous reactions was that if they are carried out so that the volumes of the gases involved are measured at the same temperature and pressure, the volumes are related in a very simple fashion. For instance, in the reaction of hydrogen with oxygen to form water vapor, it is found that for every two volumes of hydrogen that react, one volume of oxygen is consumed and two volumes of water vapor are formed, *provided the volumes are measured at the same temperature and pressure.* Expressed in equation form, this is

2 volumes hydrogen + 1 volume oxygen $\longrightarrow$ 2 volumes gaseous water

If we compare this to the chemical equation that we would now write for this reaction, it is easy to see why this relationship exists.

$$2H_2(g) + O_2(g) \longrightarrow 2H_2O(g)$$

Because the volume of a gas at constant temperature and pressure is proportional to the number of moles, when there is a 2-to-1 mole ratio, there must be a 2-to-1 volume ratio.

The whole-number volume ratios that we observe here are easy to understand today, but when first observed, the chemical formulas for gases such as hydrogen and oxygen had not been established. Therefore, the correspondence between volume ratios and mole ratios was not obvious. In fact, it was these whole-number volume ratios that led Amadeo Avogadro to propose what became known as **Avogadro's principle.** *Equal volumes of gases at the same temperature and pressure contain equal numbers of molecules.* Amazingly, this simple idea was what permitted chemists to establish the chemical formulas of some critically important gases and led ulti-

Gay-Lussac's *law of combining volumes* states that at constant T and P, the volumes of gases that react or are produced in a chemical reaction are in whole-number ratios.

mately to our current table of atomic masses.[2] So great was the importance of Avogadro's principle that the number of particles in one mole was named after him.

When volumes are measured at the same T and P, stoichiometry problems involving gases are simple, as illustrated below.

EXAMPLE 11.9. STOICHIOMETRY OF GASEOUS REACTIONS

PROBLEM: How many liters of O_2, at STP, are required for the complete combustion of 4.50 L of butane, C_4H_{10}, at STP? Butane is the fuel in disposable cigarette lighters.

SOLUTION: As with any problem in stoichiometry, we should first write a balanced equation for the reaction. This is

$$2C_4H_{10} + 13O_2 \longrightarrow 8CO_2 + 10H_2O$$

To solve the problem we can compute the number of moles of C_4H_{10} using the molar volume at STP.

$$4.50 \text{ L } C_4H_{10} \times \left(\frac{1 \text{ mol } C_4H_{10}}{22.4 \text{ L } C_4H_{10}} \right) \Longleftrightarrow 0.201 \text{ mol } C_4H_{10}$$

Next we calculate the number of moles of O_2 required, using the coefficients in the equation

$$0.201 \text{ mol } C_4H_{10} \times \left(\frac{13 \text{ mol } O_2}{2 \text{ mol } C_4H_{10}} \right) \Longleftrightarrow 1.31 \text{ mol } O_2$$

Finally we can calculate the volume of O_2, again using the molar volume of a gas at STP.

$$1.31 \text{ mol } O_2 \times \left(\frac{22.4 \text{ L } O_2}{1 \text{ mol } O_2} \right) \Longleftrightarrow 29.3 \text{ L } O_2$$

The volume of O_2 required is 29.3 L.

In problems involving gaseous reactants or products at the *same temperature and pressure*, we can take a shortcut to the answer. If we set up the calculation above with all the conversion factors strung together, we have

$$4.50 \text{ L } C_4H_{10} \times \left(\frac{1 \text{ mol } C_4H_{10}}{22.4 \text{ L } C_4H_{10}} \right) \times \left(\frac{13 \text{ mol } O_2}{2 \text{ mol } C_4H_{10}} \right) \times \left(\frac{22.4 \text{ L } O_2}{1 \text{ mol } O_2} \right) \Longleftrightarrow 29.3 \text{ L } O_2$$

Remember, this only works if the volumes are compared at the same T and P.

The number 22.4 appears in both numerator and denominator and therefore cancels. The volumes of reactants (or products) are simply related by the coefficients in the equation. Thus, in this problem we could state that

$$2 \text{ L } C_4H_{10} \Longleftrightarrow 13 \text{ L } O_2$$

Realizing this, we see that the solution to the problem could have been obtained as

$$4.50 \text{ L } C_4H_{10} \times \left(\frac{13 \text{ L } O_2}{2 \text{ L } C_4H_{10}} \right) \Longleftrightarrow 29.3 \text{ L } O_2$$

[2] Avogadro's principle was used to show that gases such as H_2, O_2, and Cl_2 must be at least diatomic. For example, if Avogadro's principle is correct, then two volumes of gaseous water must contain twice as many molecules as one volume of gaseous oxygen. This can be so only if each molecule of oxygen has twice as many O atoms as a molecule of water. If there is one O atom in a water molecule, then oxygen must be O_2. Similar arguments led to the formulas of the other diatomic gases as well.

EXAMPLE 11.10. STOICHIOMETRY OF GASEOUS REACTIONS

PROBLEM: The drain cleaner Drano contains small bits of aluminum, which react with NaOH (the main ingredient in this product) to produce bubbles of hydrogen. These bubbles presumably are designed to stir the mixture and hasten its action. How many milliliters of H_2, measured at STP, will be released when 0.150 g of Al is dissolved? The chemical equation is

$$2Al + 2OH^- + 2H_2O \longrightarrow 3H_2 + 2AlO_2^-$$

SOLUTION: First we calculate the number of moles of Al that react.

$$0.150 \text{ g Al} \times \left(\frac{1 \text{ mol Al}}{27.0 \text{ g Al}} \right) = 0.00556 \text{ mol Al}$$

Next we calculate the number of moles of H_2 produced.

$$0.00556 \text{ mol Al} \times \left(\frac{3 \text{ mol } H_2}{2 \text{ mol Al}} \right) \Longleftrightarrow 0.00834 \text{ mol } H_2$$

Since 1 mol H_2 = 22.4 L H_2 at STP,

$$0.00834 \text{ mol } H_2 \times \left(\frac{22.4 \text{ L } H_2}{1 \text{ mol } H_2} \right) = 0.187 \text{ L } H_2$$

Expressed in milliliters, the answer is 187 mL H_2.

In this case, we could use the molar volume at STP to change moles of H_2 to milliliters of H_2. If the conditions of T and P were other than STP, we would have needed to use the ideal gas law to relate moles to volume.

11.5 DALTON'S LAW OF PARTIAL PRESSURES

When two or more gases that do not react chemically are placed in the same container, the pressure exerted by each gas in the mixture is the same as it would be if it were the only gas in the container. The pressure exerted by each gas in a mixture is called its **partial pressure** and, as observed by John Dalton, the total pressure is equal to the sum of the partial pressures of each gas in the mixture. This statement, known as **Dalton's law of partial pressures,** can be expressed as

$$P_T = p_a + p_b + p_c + \cdots \qquad (11.7)$$

where P_T is the total pressure of the mixture (which could be measured with a manometer) and p_a, p_b, and p_c are the partial pressures of gases a, b, and c, respectively. For example, if nitrogen, oxygen, and carbon dioxide were placed in the same vessel, the total pressure of the mixture would be

$$P_T = p_{N_2} + p_{O_2} + p_{CO_2}$$

Thus, if the partial pressure of nitrogen were 200 torr, that of oxygen 250 torr, and that of carbon dioxide 300 torr, the total pressure of the mixture would be

$$P_T = 200 \text{ torr} + 250 \text{ torr} + 300 \text{ torr}$$

$$= 750 \text{ torr}$$

Dalton's law can be useful in determining the pressure resulting from the mixing of two gases that were originally in separate containers, as shown by the following example.

EXAMPLE 11.11. USING DALTON'S LAW OF PARTIAL PRESSURES

PROBLEM: If 200 mL of N_2 at 25 °C and a pressure of 250 torr are mixed with 350 mL of O_2 at 25 °C and a pressure of 300 torr, so that the resulting volume is 300 mL, what would be the final pressure in torr of the mixture at 25 °C?

SOLUTION: From Dalton's law we know that we can treat each gas in the mixture as if it were the only gas present. Therefore, we can calculate *independently* the new pressures of N_2 and O_2 when placed in the 300-mL container. We will use the combined gas law, eliminating the temperature because it stays the same. As in earlier examples, let's first tabulate the data.

For N_2	(i)	(f)		For O_2	(i)	(f)
p	250 torr	?		p	300 torr	?
V	200 mL	300 mL		V	350 mL	300 mL

For each calculation we can write

$$p_f = p_i \times \text{(ratio of volumes)}$$

Since the volume of the N_2 is increasing, its pressure must decrease; p_f must be less than p_i. This requires a volume ratio smaller than one, which means that the larger volume must be in the denominator. Thus,

$$p_{N_2} = 250 \text{ torr} \times \left(\frac{200 \text{ mL}}{300 \text{ mL}}\right)$$

$$= 167 \text{ torr}$$

For O_2 the volume is decreasing; p_f must be greater than p_i. This requires a volume ratio larger than one.

$$p_{O_2} = 300 \text{ torr} \times \left(\frac{350 \text{ mL}}{300 \text{ mL}}\right)$$

$$= 350 \text{ torr}$$

The total pressure of the mixture is the sum of the partial pressures.

$$P_T = p_{N_2} + p_{O_2} = 167 \text{ torr} + 350 \text{ torr}$$

$$P_T = 517 \text{ torr}$$

Collecting gases over water

This can only be done if the gas has a low solubility in water.

Gases prepared in the laboratory are quite often collected by the displacement of water, as shown in Figure 11.10. A gas collected in this manner becomes "contaminated" with water molecules that evaporate into the gas. These water molecules also exert a pressure called the **vapor pressure**. For reasons that we will discuss in Chapter 12, the vapor pressure of water depends *only* on the temperature of the liquid water (Table 11.2). The pressure of the water vapor contributes to the total pressure of the "wet" gas, so we can write

$$P_T = p_{\text{gas}} + p_{H_2O}$$

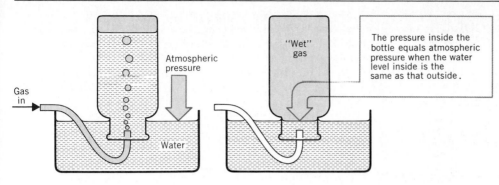

Figure 11.10. *Collection of a gas by the displacement of water. As the gas bubbles through the water, evaporation adds water vapor to the gas. The total pressure of the gas mixture is the sum of the partial pressure of the gas plus the partial pressure of the water vapor.*

The partial pressure of the gas is therefore obtained by subtracting the vapor pressure of water from the total pressure, P_T, exerted by the "wet" gas.

$$p_{gas} = P_T - p_{H_2O}$$

Table 11.2. Vapor pressure of water as a function of temperature

Temp. (°C)	Pressure (torr)	Temp. (°C)	Pressure (torr)	Temp. (°C)	Pressure (torr)
0	4.6	18	15.5	40	55.3
1	4.9	19	16.5	45	71.9
2	5.3	20	17.5	50	92.5
3	5.7	21	18.7	55	118.0
4	6.1	22	19.8	60	149.4
5	6.5	23	21.1	65	187.5
6	7.0	24	22.4	70	233.7
7	7.5	25	23.8	75	289.1
8	8.0	26	25.2	80	355.1
9	8.6	27	26.7	85	433.6
10	9.2	28	28.3	90	525.8
11	9.8	29	30.0	95	634.1
12	10.5	30	31.8	96	657.6
13	11.2	31	33.7	97	682.1
14	12.0	32	35.7	98	707.3
15	12.8	33	37.7	99	733.2
16	13.6	34	39.9	100	760.0
17	14.5	35	42.2	101	787.6

EXAMPLE 11.12. COLLECTING A GAS OVER WATER

PROBLEM: A student generates oxygen gas in the laboratory and collects it in a manner similar to that shown in Figure 11.10. The gas is collected at 25 °C until the levels of the water inside and outside the flask are equal. If the volume of the gas is 245 mL and the atmospheric pressure is 758 torr,

  (a) What is the pressure, in torr, of O_2 gas in the "wet" gas mixture at 25 °C?
  (b) What would be the volume in liters of dry oxygen at STP?

SOLUTION: (a) Since the water level inside the collection bottle is the same as outside, the pressures inside and outside must be the same. This means that the total pressure of the "wet" gas must be equal to the atmospheric pressure, which is 758 torr.

The partial pressure of the O_2 is obtained by subtracting the vapor pressure of water at 25 °C (the temperature of the water through which the O_2 bubbled) from the total pressure of the mixture. Thus,

$$P_T = p_{gas} + p_{H_2O}$$

Substituting p_{O_2} for p_{gas} and rearranging, we have

$$p_{O_2} = P_T - p_{H_2O}$$

According to Table 11.2, the partial pressure of water vapor at 25 °C is 23.8 torr. The atmospheric pressure was given as 758 torr. Therefore, the partial pressure of O_2 is

$$p_{O_2} = 758 \text{ torr} - 23.8 \text{ torr}$$
$$= 734 \text{ torr} \text{(rounded to the correct number of significant figures)}$$

This is the pressure exerted by the oxygen alone (i.e., the "dry" oxygen).

(b) This part of the problem is a combined gas law calculation. First, we tabulate the data as follows:

	(i)	(f)	
V	245 mL	?	
P	734 torr	760 torr	STP
T	298 K	273 K	

Next, recall that

$$V_f = V_i \times \text{(ratio of pressures)} \times \text{(ratio of temperatures)}$$

The pressure change should tend to decrease the volume; therefore, the pressure ratio should be less than one. The temperature decrease should also decrease the volume; therefore, the temperature ratio should be less than one. Thus,

$$V_f = 245 \text{ mL} \times \left(\frac{734 \text{ torr}}{760 \text{ torr}}\right) \times \left(\frac{273 \text{ K}}{298 \text{ K}}\right)$$

$$V_f = 217 \text{ mL at STP}$$

11.6 GRAHAM'S LAW OF EFFUSION

If two gases are placed in the same container, their molecules gradually mix until the composition of the gas becomes uniform. This process of mixing is called **diffusion** and is something everyone has experienced at one time or another. For example, if someone wearing perfume sits near you in a movie theater, it isn't long before you smell the perfume's aroma. The molecules of the perfume diffuse through the air and reach you rather quickly.

A process somewhat similar to diffusion is called **effusion**. This is a process by which a gas, under pressure, escapes from a vessel by passing through a very small opening, as illustrated in Figure 11.11. Effusion is also responsible for the fate of helium-filled rubber balloons. Perhaps as a child you brought home such a balloon and found that by the next morning much of the helium had escaped. Actually, the helium had effused through tiny pores in the rubber.

Effusion.

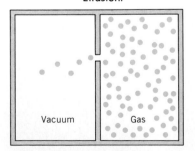

Vacuum   Gas

Figure 11.11. *Effusion of a gas into a vacuum.*

Thomas Graham (1805–1869), a British chemist, studied the rates of effusion of various gases through porous plugs of plaster of Paris. He found that if he measured these rates under the same conditions of temperature and pressure, the rates were inversely proportional to the *square roots* of the densities of the gases. This statement, known as **Graham's law,** can be expressed mathematically as

$$\text{rate of effusion} \propto \sqrt{\frac{1}{d}}$$

The rate of effusion of two gases (labeled simply A and B) can be compared by dividing the rate of one by the other; that is,

$$\frac{\text{rate of effusion } (A)}{\text{rate of effusion } (B)} = \sqrt{\frac{d_B}{d_A}}$$

By taking the ratio, the proportionality constant (whatever it may be) cancels from numerator and denominator.

Looking back at Equation (11.6), we see that the density of a gas is directly proportional to its molecular mass. This allows us to rewrite the ratio of effusion rates as

$$\frac{\text{rate of effusion } (A)}{\text{rate of effusion } (B)} = \sqrt{\frac{d_B}{d_A}} = \sqrt{\frac{M_B}{M_A}} \qquad (11.8)$$

Graham's law works for diffusion, too.

where M_A and M_B are the molecular masses of gases A and B, respectively. One of the things Equation 11.8 tells us is that lightweight gases effuse and diffuse rapidly, whereas those with large molecular masses effuse and diffuse more slowly.

EXAMPLE 11.13. CALCULATING RELATIVE RATES OF EFFUSION

PROBLEM: Which gas will effuse faster, ammonia or carbon dioxide? What are their relative rates of effusion?

SOLUTION: The molecular mass of CO_2 is 44 and that of NH_3 is 17. Therefore NH_3 will effuse faster. We can calculate how much faster with Equation 11.8.

$$\frac{\text{rate of effusion } (NH_3)}{\text{rate of effusion } (CO_2)} = \sqrt{\frac{M_{CO_2}}{M_{NH_3}}} = \sqrt{\frac{44}{17}} = 1.6$$

Therefore, the rate of effusion of NH_3 is 1.6 times larger than the rate of CO_2.

11.7 KINETIC MOLECULAR THEORY AND THE GAS LAWS

We have seen that the study of gases produced a variety of different gas laws. The people who discovered these laws could not help wondering what gases must be composed of to behave as they do. What is the origin of the pressure of a gas? Why are gases so compressible? Why do they expand when heated? And why do light gases effuse more rapidly than heavy ones? These questions, among others, illustrate the need for a theoretical model of a gas.

The **kinetic molecular theory** evolved in an attempt to explain the physical behavior of substances. According to this theory, any given sample of matter is composed of a huge number of small particles (molecules or individual atoms) that are in constant, random motion. In addition, the theory proposes that the average kinetic energy that these particles possess is proportional to the absolute temperature of the sample, a concept that you were introduced to in Chapter 6. As we will see, these two basic postulates apply to all three states of matter: solids, liquids, and gases. What we want to do now is to see how they explain the properties of a gas.

Although the postulates can be used to actually derive the ideal gas law mathematically, we will only use them qualitatively.

The pressure–volume relationship: Boyle's law

At 0 °C, the average speed of an oxygen molecule is about 1000 mi/hr.

The most striking quality of a gas is its compressibility. The molecules envisioned in the kinetic molecular theory must therefore be very tiny and very far apart in a gas so that there is plenty of empty space between them. Only in this way could they be so easily crowded together. As these tiny particles fly about, they collide with each other and the walls of the container. Each impact with a wall exerts a tiny push, and the cumulative effects of enormous numbers of such impacts each second on each square centimeter of the wall give rise to the pressure exerted on the wall by the gas.

With this model of a gas in mind, we can now explain Boyle's law. If we halve the volume of a gas, we pack twice as many molecules into each cubic centimeter. Over each square centimeter of wall there are now twice as many molecules as before, and therefore there must be twice as many molecule–wall collisions each second. This means that the pressure has doubled, as illustrated in Figure 11.12. If halving the volume doubles the pressure, then pressure and volume are inversely proportional to each other, which is a statement of Boyle's law.

An ideal gas, you recall, would obey Boyle's law exactly under all conditions. This means that no matter how tightly packed the molecules were, it would always be possible to halve their volume by doubling the pressure. This could happen over and over again, of course, only if the gas were composed of particles having no volume themselves, so that the entire volume would be empty space. Real molecules do have finite volumes, however, so no real gas could obey Boyle's law perfectly, especially at high pressure. We will discuss the consequences of this further in the next section.

Distribution of molecular speeds

The second postulate of the kinetic molecular theory is that the average kinetic energy of a collection of molecules is proportional to their absolute temperature. Let's take a close look at the distribution of molecular energies implied by the use of the term *average kinetic energy* because it will be of great value to us later on.

In a gas (or a liquid or a solid), molecules are in constant motion. Because they

Figure 11.12. *When the volume of a gas is halved on going from* (a) *to* (b) *twice as many molecules are crowded into each tiny unit of volume. This produces twice as many molecule–wall collisions each second, which doubles the pressure.*

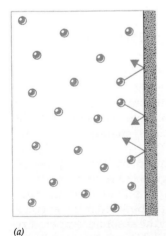
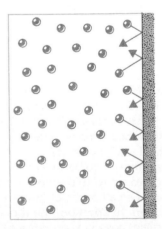

(a)　　　　　　(b)

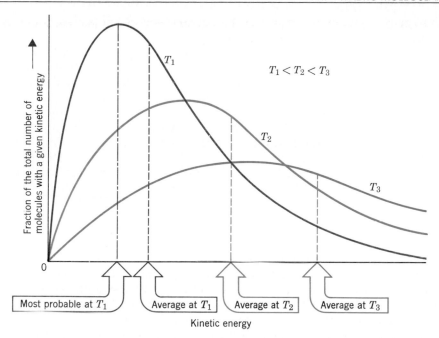

Figure 11.13. *The distribution of kinetic energies in a collection of molecules at three different temperatures.*

$T_1 < T_2 < T_3$

Fraction of the total number of molecules with a given kinetic energy

T_1

T_2

T_3

0

Most probable at T_1 | Average at T_1 | Average at T_2 | Average at T_3

Kinetic energy

continually collide with each other, there is a wide range of different molecular speeds. For example, occasionally a collision might leave a particular molecule almost motionless, until it suffers another collision and is sent on its way again an instant later. Another molecule might receive several "rear-end" collisions in a row that give it a very high speed. In this way the speeds of molecules are forever changing through collisions with others. At any instant, some molecules will be moving slowly and some will be going very fast, although most will have speeds in between.

Because each molecule has a kinetic energy equal to $\frac{1}{2}mv^2$, where m is its mass and v is its speed, there is also a distribution of kinetic energies associated with the distribution of molecular speeds. Figure 11.13 illustrates the shape of this kinetic energy distribution at three different temperatures. Plotted vertically are the fractions of all the molecules that have the particular kinetic energies given along the horizontal axis. For example, at zero K.E., which corresponds to molecules standing still, this fraction is essentially zero since very few, if any, molecules are motionless at any instant. The fraction having a particular K.E. increases as we move to higher energies (higher speeds) and eventually becomes a maximum. At still higher kinetic energies the fraction decreases and gradually approaches zero again at kinetic energies corresponding to very fast-moving molecules. The curve does not go all the way to zero, however, because there is virtually no upper limit to molecular speeds, other than the speed of light.

This is called a *Maxwell–Boltzmann* distribution.

The maximum on this curve represents the kinetic energy possessed by the largest fraction of molecules. This kinetic energy would be found most frequently (that is, with the greatest probability) if we were able to examine molecules at random; hence, it is called the *most probable kinetic energy*. The average kinetic energy occurs at a higher value than the most probable kinetic energy because the curve is not symmetrical. Just as a few "curve breakers" in a chemistry class tend to raise the class average on exams, the high-velocity molecules shift the average K.E. above the most probable K.E.

When the temperature of a substance is raised, the curve changes so that the average K.E. increases, as shown in Figure 11.13. More molecules have high speeds and fewer have low ones and, on the average, the molecules move faster. Thus, as we described earlier, when heat is added to a substance to raise its temperature, the energy goes into increasing the average kinetic energy and increases the speeds of the particles.

The relationship between kinetic energy and temperature also leads quite naturally to the concept of an absolute zero. As kinetic energy is removed from a substance, its molecules move more and more slowly. If all the molecules were to cease moving, their average kinetic energy would be zero and, since negative kinetic energies are impossible (a molecule cannot be going slower than when it is standing still), the temperature of the substance would also be at its lowest point. It is this temperature that we refer to as absolute zero — the temperature at which all molecular motion has ceased.[3] It should be understood, however, that electronic motion would still continue at absolute zero. Even though the molecules would be motionless, the electrons would still be "whizzing" about their respective nuclei.

Temperature and the gas laws

You've probably begun to wonder what all of this has to do with the gas laws. Let's look at Gay-Lussac's pressure–temperature law first. This law states that if we keep the volume constant, the pressure is directly proportional to the absolute temperature. In other words, when the temperature goes up, so does the pressure. How can we explain this? According to kinetic theory, raising the temperature increases the average kinetic energy of the molecules, so the molecules move faster. This means that they will strike the wall more frequently, and that when they hit the wall, the average force of impact will be greater. These factors make the pressure increase.

We can also explain Charles's law, which says that the volume increases when we raise the temperature, provided we keep the pressure constant. We have just seen that raising the temperature causes more molecules to hit the wall each second and also causes the force of impact of the molecules with the walls to increase. The only way to keep the pressure constant is to simultaneously decrease the number of collisions per second with each square centimeter of wall. This can be accomplished by allowing the gas to expand so fewer molecules are over each square centimeter of wall. In other words, to keep the pressure of a gas constant as we raise its temperature, we have to allow it to expand and occupy a larger volume.

Now let's look at what happens if we cool a gas. Following the reasoning in the preceding paragraph, to keep the pressure constant as the gas cools, we must decrease the volume. The molecules move more and more slowly and the space between them becomes less and less. All real gases eventually condense to a liquid when they are cooled because attractive forces between the molecules eventually cause "sticky" collisions. An ideal gas, however, would not condense regardless of how much we cooled it, so another property of the molecules of an ideal gas is that they have no intermolecular attractions. Consequently, *an ideal gas is a hypothetical substance whose molecules have no volume and no intermolecular attractive forces.*

When the molecules stick together, they settle to the bottom of the container as a liquid.

Graham's law

The postulate of the kinetic theory relating average kinetic energy to temperature can be used very simply to derive Graham's law. Suppose we have two different gases, A and B. If they are both at the same temperature, the average kinetic energies

---

[3] In fact, even at absolute zero there is still a residual molecular motion that is required by the *Heisenberg uncertainty principle.* This principle states that we cannot simultaneously know exactly the position and momentum of a particle. We will see in the next chapter that the average positions of particles in the solid state can be determined, and if the particles were motionless, we would also know that their momentum was zero, thereby violating the uncertainty principle.

of their molecules must be the same. This means that

$$\overline{K.E._A} = \overline{K.E._B}$$

or

$$\tfrac{1}{2}m_A\overline{v_A^2} = \tfrac{1}{2}m_B\overline{v_B^2} \qquad (11.9)$$

m_A and m_B are the masses of molecules A and B.

where $\overline{v^2}$ is called the **mean square speed** of the molecules and is the average of the speeds squared of all the molecules; that is,

$$\overline{v^2} = \frac{v_1^2 + v_2^2 + v_3^2 + \cdots}{n_T}$$

where v_1, v_2, v_3, and so on represent the speeds of molecules 1, 2, 3, and so on, and n_T is the total number of molecules present. Equation 11.9 can be rearranged to give

$$\frac{\overline{v_A^2}}{\overline{v_B^2}} = \frac{m_B}{m_A}$$

Taking the square root of both sides, we obtain

$$\frac{\overline{v_A}}{\overline{v_B}} = \sqrt{\frac{m_B}{m_A}} \qquad (11.10)$$

where $\overline{v}$ is called the root-mean-square speed. For any molecule, its actual mass is proportional to its molecular mass — that's the principle upon which the atomic mass table is based. This means that

$$M \propto m$$

Therefore, we can substitute M_A and M_B for m_A and m_B into Equation 11.10 and arrive at

$$\frac{\overline{v_A}}{\overline{v_B}} = \sqrt{\frac{M_B}{M_A}}$$

We can make these substitutions because whatever the proportionality constants are, they cancel from numerator and denominator in the ratio.

The rate at which gases effuse should be directly proportional to the velocity of their molecules, with faster molecules effusing at a higher rate. Thus, we are led to conclude that

$$\frac{\text{rate of effusion } (A)}{\text{rate of effusion } (B)} = \frac{\overline{v_A}}{\overline{v_B}} = \sqrt{\frac{M_B}{M_A}}$$

or simply

$$\frac{\text{rate of effusion } (A)}{\text{rate of effusion } (B)} = \sqrt{\frac{M_B}{M_A}}$$

which is Graham's law.

Avogadro's principle

Equal volumes of gases at the same temperature and pressure have equal numbers of molecules. That is how we previously stated Avogadro's principle. But we could have also stated this principle as follows. Equal numbers of gas molecules occupying the same volume at the same temperature exert the same pressure. We can explain this by noting that the average force of impact of the molecules colliding with a given area of the wall depends on their average kinetic energy, and therefore on their temperature. If the temperature of two gas samples is the same, the average kinetic energy of their molecules must be equal, and if the number of molecules per unit volume is the same, it follows that their pressures must also be the same.

Dalton's law of partial pressures

Molecules of an ideal gas are unaware of each other's existence, except when they collide, because they have no attractions for each other. In a mixture of nonreacting gases, each behaves independently and exerts a pressure that is the same as it would exert if it were alone. The cumulative effect of the individual partial pressures is the total pressure.

Avogadro's principle can also be applied to gas mixtures. For example, consider the earth's atmosphere, where about 1 of every 5 molecules is O_2 and 4 of every 5 are N_2. Since one-fifth of the molecules are oxygen, only one-fifth of the pressure is contributed by O_2; the other four-fifths of the pressure is contributed by N_2.

The partial pressure of a gas is related quantitatively to the total pressure by its **mole fraction** (usually given by the symbol X), which is the number of moles of the gas in question divided by the total number of moles of gas in the mixture. For some gas A,

$$X_A = \frac{\text{number of moles of } A}{\text{total number of moles of gas in the mixture}} \tag{11.11}$$

If p_A is the partial pressure of A and P_T is the total pressure, then

$$p_A = X_A P_T \tag{11.12}$$

In the atmosphere, for example, in each 5 mol of air there are 1 mol O_2 and 4 mol N_2. Therefore,

$$X_{O_2} = \frac{1 \text{ mol}}{5 \text{ mol}} = 0.2$$

$$X_{N_2} = \frac{4 \text{ mol}}{5 \text{ mol}} = 0.8$$

If the total pressure of a sample of air were 500 torr, then

$$p_{O_2} = 0.2 \,(500 \text{ torr})$$
$$= 100 \text{ torr}$$
$$p_{N_2} = 0.8 \,(500 \text{ torr})$$
$$= 400 \text{ torr}$$

11.8 REAL GASES

The differences between a real gas and a hypothetical ideal gas became apparent in the last section. Molecules of an ideal gas are abstract points in space and have no volume, whereas a real gas is composed of actual molecules whose atoms occupy some space. The effects of this are seen when we compress a gas to a very high pressure. Its volume is larger than would be expected for an ideal gas under the same conditions.

You also learned in the last section that molecules of an ideal gas would have no attractive forces between them and they could be cooled to absolute zero without condensing to a liquid. Molecules of a real gas do attract each other, however. Their behavior is similar to that shown in Figure 11.14. As the gas is cooled, its volume begins to fall below the Charles's law value. Then, suddenly, the substance condenses to a liquid with a much smaller volume. At still lower temperatures it freezes to a solid. Another manifestation of the attractive forces between gas molecules is the cooling that occurs when a compressed gas is allowed to expand freely into a vacuum. As the gas expands, the average distance of separation between the molecules increases. Since there are forces of attraction between them, moving the mole-

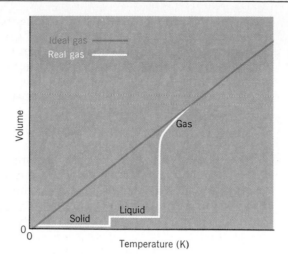

Figure 11.14. *The variation of volume with temperature for an ideal gas and for a real gas.*

cules apart increases their potential energy and causes a corresponding decrease in their kinetic energy. (Remember, the total energy must stay the same according to the law of conservation of energy.) This lowering of the kinetic energy is observed as a lowering of the temperature of the gas.

An equation of state for real gases

Because real gases deviate from ideal behavior, especially at high pressure and low temperature, the ideal gas law can't be used to make highly accurate calculations. One way to improve the accuracy is to modify the ideal gas law to take into account the factors that cause a real gas to differ from an ideal gas.

Suppose that the gas molecules in a container could be stopped and allowed to settle to the bottom. We would see that part of the volume of the container is occupied by the gas molecules. The remaining free space is somewhat less than the volume of the container. If, in this hypothetical case, another gas molecule were added, it could move in the free space, but not in the entire volume of the container. This same situation exists when the molecules are moving, and the volume within which the molecules *cannot* move because of their own finite volume is called the **excluded volume.**

In an ideal gas the molecules would have no volume themselves, so the ideal gas would be entirely empty space into which other molecules could be squeezed when the gas is compressed. If we associate the empty space available in a real gas with this *ideal volume*, V_{ideal}, then the measured volume of the real gas, V_{meas}, is actually slightly larger than V_{ideal} by an amount that is related to the size of the real molecules. According to J. D. van der Waals (1837–1923), a Dutch physicist, the measured volume is

$$V_{meas} = V_{ideal} + nb$$

where *b* is the correction due to the excluded volume per mole and *n* is the number of moles of gas. Solving for the ideal gas volume, we have

$$V_{ideal} = V_{meas} - nb \tag{11.13}$$

Van der Waals also included a correction to the pressure that takes into account the attractive forces between the molecules of a real gas. In an ideal gas, the molecules would travel in straight lines because there are no attractive forces to cause their paths to curve. But in a real gas, the attractive forces do cause molecules to curve and change directions as they pass by each other, as shown in Figure 11.15. As a

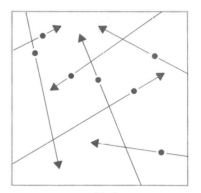

(a)   Ideal gas

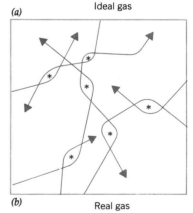

(b)   Real gas

Figure 11.15. (a) *In an ideal gas the molecules travel in straight lines.* (b) *In a real gas the paths curve as one molecule passes close to another because the molecules attract each other. Asterisks indicate the points at which molecules pass close to each other.*

result, the molecules of a real gas travel longer distances to reach the walls, and this takes more time. This means that the real gas molecules don't strike the walls as frequently as the molecules of an ideal gas. A lower collision frequency means, of course, that the real gas will exert a smaller pressure than the ideal gas. Van der Waals corrected for this with the term n^2a/V^2, which he added to the measured pressure to bring it up to the pressure that an ideal gas would exert.

In a real gas, the attractive forces cause the molecules to linger somewhat within the body of the gas, so the molecules don't strike the walls as often.

$$P_{ideal} = P_{meas} + \frac{n^2a}{V^2} \tag{11.14}$$

The quantity a is proportional to the strengths of the attractive forces, and we see that the stronger the attractions, the more must be added to the measured pressure to make it as large as the ideal gas pressure. This is reasonable if we consider that when the attractive forces are very large, the molecules will undergo large deflections in their paths. This means that they will tarry longer within the body of the gas, thereby striking the walls less frequently.

Substituting these corrected pressures and volumes (from Equations 11.14 and 11.13, respectively) into the ideal gas equation gives us

Van der Waals received the 1910 Nobel prize in physics for his development of this equation.

$$\left(P + \frac{n^2a}{V^2}\right)(V - nb) = nRT \tag{11.15}$$

in which all symbols stand for measured quantities. This equation is called the **van der Waals equation of state for a real gas.** It is more complex than the ideal equation, but it does work well for many gases over fairly wide ranges of temperature and pressure.

The values of the constants a and b depend on the nature of the gas because the molecular volumes and the molecular attractions vary from gas to gas. Some typical values of a and b are found in Table 11.3. We see that molecules containing many atoms, such as C_2H_5OH, have large values of b. This is not surprising, since such molecules would be expected to be larger than molecules containing only a few atoms.

The variation among the values of a reflects variations in the strengths of the intermolecular attractions. It is easy to understand why polar molecules such as NH_3, H_2O, CH_3OH, and C_2H_5OH attract each other. These molecules are dipoles that tend to align themselves so that the partial positive charge on one attracts the partial negative charge on another (Figure 11.16). The attractions between nonpolar molecules, such as O_2, CH_4, and C_2H_6, or between isolated atoms such as helium and the other noble gases, are more difficult to explain. We will study all these intermolecular attractions in detail in the next chapter.

From the discussion above, we see that the values of a and b enable us to expand our knowledge about the molecules of which a real gas is composed. The van der

Table 11.3. The van der Waals constants for real gases

	a (L² atm/mol²)	b (L/mol)
He	0.034	0.0237
O_2	1.36	0.0318
NH_3	4.17	0.0371
H_2O	5.46	0.0305
CH_4	2.25	0.0428
C_2H_6	5.489	0.06380
CH_3OH	9.523	0.06702
C_2H_5OH	12.02	0.08407

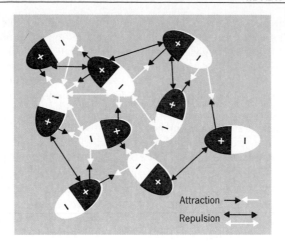

Figure 11.16. *Electrostatic attractions between dipoles. The attractions outweigh the repulsions, so the molecules feel a net attraction toward each other.*

Waals constants for a gas are obtained by making careful measurements of *P*, *V*, and *T* and then choosing values of *a* and *b* that make the van der Waals equation produce the best match with the experimental data. In this sense, *a* and *b* are experimentally determined quantities that enable us to check our theories on molecular size and attractions.

REVIEW QUESTIONS AND PROBLEMS

Problems whose numbers are in blue have their answers in Appendix D at the back of the book. The more difficult problems are marked with asterisks.

Pressures of Gases; Barometers and Manometers

11.1 Define pressure. Explain why the height of mercury in a barometer is independent of the cross-sectional area of the tube.

11.2 What is the SI unit of pressure and how is it related to the standard atmosphere?

11.3 Perform the following conversions:
  (a) 1.50 atm to torr
  (b) 785 torr to atm
  (c) 3.45 atm to Pa
  (d) 3.45 atm to kPa
  (e) 165 torr to Pa
  (f) 342 kPa to atm
  (g) 11.5 kPa to torr

*11.4 Mercury has a density of 13.6 g/mL. Calculate the value of the standard atmosphere in the units, lb/in.^{2}.

11.5 An open-end manometer containing mercury was connected to a vessel containing a gas at a pressure of 740 torr. The atmospheric pressure was 765 torr. Sketch a diagram showing the relative heights of the mercury in each arm of the manometer.

11.6 What is the advantage of using a closed-end manometer for pressure measurements?

11.7 Why is mercury a useful substance for use in barometers and manometers?

11.8 Suppose that in the closed-end manometer shown in Figure 11.5 the mercury in the closed arm was 15.8 cm higher than in the arm connected to the container full of gas. What was the pressure of the gas in the container expressed in torr?

11.9 An open-end manometer was connected to a flask containing a gas at an unknown pressure. The mercury in the arm open to the atmosphere was 65 mm higher than in the closed end. The atmospheric pressure was 733 torr. What was the pressure of the gas in the flask expressed in torr?

11.10 An open-end manometer was connected to a vessel containing a gas at a pressure of 535 torr. The atmospheric pressure was 774 torr. What was the difference in height (in millimeters) between the mercury levels in the manometer?

11.11 A flask containing a gas was connected to both a closed-end and an open-end manometer. In the closed-end manometer, the mercury in the sealed arm was 755 mm above the level in the arm connected to the flask. In the open-end manometer, the arm connected to the gas was 17 mm higher than the side open to the air. What was the atmospheric pressure in torr?

11.12 A manometer connecting two flasks (labeled *A* and *B*) contains an oil having a density of 0.847 g/mL. The oil in the arm connected to flask *A* is 74.0 cm higher than the oil in the arm connected to flask *B*. The gas in flask *A* has a pressure of 836 torr. What is the pressure of the gas in flask *B* in torr?

11.13 A person digs a well and finds water 35 ft below the surface. If the average atmospheric pressure at the well is 1 atm, will the person be able to draw water from the well using a pump, mounted at ground level, that works by suction? Explain your answer.

Combined Gas Law

11.14 What is the combined gas law?

11.15 Show how Boyle's law, Charles's law, and Gay-Lussac's law can be obtained from the combined gas law.

11.16 In terms of Charles's law, why is −273.15 °C the lowest possible temperature?

11.17 Why is it dangerous to incinerate an aerosol can?

11.18 If you want to be sure that the tires of your car are properly inflated, you should check them before beginning a trip rather than after having driven on them for a long period. Why?

11.19 Turbocharging is a way of increasing the power output from a gasoline or diesel engine. The exhaust gases are used to turn a turbine connected to a compressor. The compressor forces more air into the cylinders so that more fuel can be burned and more power can be produced. Compressing the air, however, heats it, and even more power can be produced if the air from the compressor is cooled before it is let into the cylinders. Why?

11.20 What makes a hot-air balloon rise?

11.21 State Boyle's law in words. Is the law obeyed exactly by all gases? What is a gas that exactly obeys the gas laws called?

11.22 A gas has a volume of 350 mL at 740 torr. How many milliliters will the gas occupy at 900 torr if the temperature remains constant?

11.23 A sample of SO_2 occupies 1.45 L at 2.75 atm. If we assume no temperature change, how many liters will this gas occupy at 800 torr?

11.24 A gas is compressed at constant temperature from a volume of 540 mL to 320 mL. If the initial pressure was 475 torr, what is the final pressure in torr?

11.25 A gas exerts a pressure of 350 torr at 20 °C. How many torr will it exert if its temperature is raised to 40 °C without a change in volume?

11.26 A gas has a pressure of 655 torr at 25 °C. To what Celsius temperature must it be heated to raise its pressure to 825 torr?

*11.27 An automobile tire is inflated to a *gauge pressure* of 29 lb/in.² at 65 °F. (The actual pressure is 29 lb/in.² *above* atmospheric pressure.) After a trip the temperature of the tire has risen to 130 °F. What will be the gauge pressure of the air in the tire (in lb/in.²), if we assume that no air has leaked out and that the volume of the tire hasn't changed? Assume the atmospheric pressure is 14.7 lb/in.².

11.28 At 25 °C and 1 atm a gas occupies a volume of 1.50 L. How many liters will it occupy at 100 °C and 1 atm?

11.29 A balloon has a volume of 2.0 L indoors at a temperature of 25 °C (77 °F). If it is taken outdoors on a very cold winter day when the temperature is −28.9 °C (−20 °F), what will its volume be in liters? Assume constant air pressure within the balloon.

11.30 What would be the final volume of a 2.00-L sample of a gas that is heated from 26 °C to 100 °C at constant pressure?

11.31 A sample of O_2 occupies 285 mL at 25 °C. At what Celsius temperature will it occupy 350 mL if the pressure remains constant?

11.32 At what Celsius temperature will a gas sample occupy 0.850 L at 1.00 atm pressure if it occupies 400 mL at 32 °C and 1.00 atm?

11.33 A gas exerts a pressure of 200 kPa in a container having a volume of 350 cm³. What will its pressure be in kPa if the gas is transferred to a 400 cm³ container at the same temperature?

11.34 A bicycle pump has a barrel that is 75.0 cm long (about 30 in.). If we assume that on the upstroke air is drawn in at a pressure of 1.00 atm, how long must the downstroke be (in centimeters) to raise the pressure of the air to 5.50 atm, if the temperature of the air remains constant? (That's about the pressure in the tire of a 10-speed bike.)

11.35 What is STP?

11.36 If a gas, originally in a 50.0-mL container at a pressure of 645 torr, is transferred to another container whose volume is 65.0 mL, what would be its new pressure in torr if
(a) there was no temperature change?
(b) the temperature of the first container was 25 °C and that of the second was 35 °C?

11.37 A 300-mL sample of a gas exerts a pressure of 450 torr at 27 °C. What pressure (in torr) would it exert in a 200-mL container at 20 °C?

11.38 A 2.00-L sample of a gas originally at 25 °C and a pressure of 700 torr is allowed to expand to a volume of 5.00 L. If the final pressure of the gas is 585 torr, what is its final temperature in degrees Celsius?

11.39 The density of CO_2 is 1.96 g/L at 0 °C and 1 atm. Determine its density in grams per liter at 650 torr and 25 °C.

11.40 If a 50.0-mL sample of gas exerts a pressure of 450 torr at 35 °C, how many milliliters will it occupy at STP?

Ideal Gas Law

11.41 Suppose that the ideal gas law were $PV^2 = nR/T^2$. What would be the units of R if P is in atm and V in liters?

11.42 What determines the choice of the value of R that you should use in an ideal gas law calculation?

11.43 What is the value of the gas constant in the units Pa m³/mol K?

11.44 Calculate the number of liters occupied, *at STP*, by
(a) 0.200 mol O_2
(b) 12.4 g Cl_2
(c) a mixture of 0.100 mol N_2 and 0.050 mol O_2

11.45 Calculate the mass, in grams, of 245 mL of SO_2 at STP.

11.46 What is the density of butane, C_4H_{10}, at STP expressed in grams per liter?

11.47 The density of a gas was found to be 1.96 g/L at STP. What is its molecular mass?

11.48 In the laboratory a student filled a 250-mL container with an unknown gas until a pressure of 760 torr was obtained. The sample of gas was found to weigh 0.164 g. Calculate the molecular mass of the gas if the temperature in the laboratory was 25 °C.

11.49 Calculate the pressure, in torr and atmospheres, that would be exerted by 25.0 kg of steam (H_2O) in a 1000-L boiler at 200°C if we assume ideal gas behavior.

11.50 The density of a gas was found to be 1.81 g/L at 30 °C and 760 torr. What is its molecular mass?

11.51 How many milliliters would be occupied by 0.234 g of NH_3 at 30 °C and a pressure of 0.847 atm?

11.52 A chemist observed a gas being evolved in a chemical reaction and collected some of it for analyses. It was found to contain 80.0% carbon and 20.0% hydrogen. It was also observed that 500 mL of the gas at 760 torr and 0 °C weighed 0.6695 g.
(a) What is the empirical formula of the gaseous compound?
(b) What is its molecular mass?
(c) What is its molecular formula?

*11.53 A 0.2000-g sample of a fishy smelling liquid known to contain only carbon, hydrogen, and nitrogen was burned and produced 0.482 g CO_2 and 0.271 g H_2O. A second sample weighing 0.2500 g was treated in such a way that all the nitrogen in the substance was converted to N_2. This gas was collected and found to occupy 42.3 mL at 26.5 °C and 755 torr.
(a) What are the percentages of carbon, hydrogen, and nitrogen in the compound?
(b) What is the empirical formula of the compound?

11.54 A sample of N_2 in a flask at 25 °C exerts a pressure of 525 torr. When 0.100 g of O_2 is added to this flask, the pressure rises to 760 torr. The temperature remains constant and there is no reaction between the N_2 and O_2. How many grams of N_2 are in the flask?

*11.55 The product PV has the dimensions of energy. Given the data, $1 J = 1 N \cdot m$, 1 atm $= 101,325$ Pa, and 1 Pa $= 1 N/m^2$, calculate the number of joules equal to 1 liter atmosphere. What is the value of the gas constant, R, in $J \cdot mol^{-1} K^{-1}$ and $cal \cdot mol^{-1}K^{-1}$?

Gas Stoichiometry

11.56 In the reaction

$$N_2(g) + 3H_2(g) \longrightarrow 2NH_3(g)$$

how many milliliters of N_2, measured at STP, are required to produce 400 mL of NH_3, measured at STP? How many milliliters of H_2 at STP are required?

11.57 In the reaction

$$2NO(g) + 2H_2(g) \longrightarrow 2H_2O(g) + N_2(g)$$

how many milliliters of N_2, measured at STP, would be produced from (a) 0.00140 mol NO,
(b) 1.3×10^{-3} g H_2?

11.58 Oxygen gas, generated in the reaction $KClO_3 \rightarrow KCl + O_2$ (unbalanced), was collected over water at 30 °C in a 150-mL vessel until the total pressure was 600 torr.
(a) How many grams of dry O_2 were produced?
(b) How many grams of $KClO_3$ were consumed in the reaction?

11.59 Nitric acid is produced by dissolving NO_2 in water according to the equation

$$3NO_2(g) + H_2O(l) \longrightarrow 2HNO_3(l) + NO(g)$$

How many milliliters of NO_2 at 25 °C and 770 torr are required to produce 10.0 g of HNO_3?

*11.60 120 mL of NH_3 at 25 °C and 750 torr was mixed with 165 mL of O_2 at 50 °C and 635 torr and transferred to a 300-mL reaction vessel where they were allowed to react according to the equation

$$4NH_3(g) + 5O_2(g) \longrightarrow 4NO(g) + 6H_2O(g)$$

What was the total pressure (in torr) in the reaction vessel at 150 °C after the reaction is over? Assume the reaction goes to completion.

11.61 Calculate the maximum number of milliliters of CO_2, at 750 torr and 28 °C, that could be produced by reacting 500 mL of CO, at 760 torr and 15 °C, with 500 mL of O_2 at 770 torr and 0 °C.

*11.62 Ozone, O_3, is an important species in the chain of reactions that lead to the production of smog. In an ozone analysis, 2.0×10^5 L of air at STP was drawn through a solution of NaI where the O_3 undergoes the reaction

$$O_3 + 2I^- + H_2O \longrightarrow O_2 + I_2 + 2OH^-$$

The I_2 formed was titrated with $0.0100\ M$ $Na_2S_2O_3$ with which it reacts.

$$I_2 + 2S_2O_3{}^{2-} \longrightarrow 2I^- + S_4O_6{}^{2-}$$

In the analysis, 0.420 mL of the $Na_2S_2O_3$ solution was required to completely react with all of the I_2.
(a) Calculate the number of moles of I_2 that reacted with the $S_2O_3{}^{2-}$ solution.
(b) How many moles of I_2 were produced in the first reaction?
(c) How many moles of O_3 were contained in the 200,000 L of air?
(d) How many milliliters would the O_3 occupy at STP?
(e) What is the concentration of O_3, in parts per million by volume, in the air sample?

11.63 An important reaction in the production of nitrogen fertilizers is the oxidation of ammonia

$$4NH_3(g) + 5O_2(g) \xrightarrow{500\ °C} 4NO(g) + 6H_2O(g)$$

How many liters of O_2, measured at 25 °C and 0.895 atm, must be used to produce 100 L of NO at 500 °C and 750 torr?

11.64 A student collected 35.0 mL of O_2 over water at 25 °C and a total pressure of 745 torr from the decomposition of a 0.2500-g sample known to contain a mixture of KCl and $KClO_3$. The reaction that produced the oxygen was

$$2KClO_3 \longrightarrow 2KCl + 3O_2$$

(a) How many moles of O_2 were collected?
(b) How many grams of $KClO_3$ were decomposed?
(c) What percentage by mass of the sample was $KClO_3$?

Dalton's Law

11.65 What is Dalton's law of partial pressures?

11.66 A 1.00-L mixture of gases is produced from 1.00 L of N_2 at 200 torr, 1.00 L of O_2 at 500 torr, and 1.00 L of Ar at 150 torr. What is the pressure of the mixture in torr?

11.67 A 1.00-L flask is filled by placing in it the contents of a 2.00-L flask of N_2 at 300 torr and a 2.00-L flask of H_2 at 80 torr. What is the pressure (expressed in atm) of the mixture in the 1.00-L flask?

11.68 What would be the total pressure in torr of a mixture prepared by adding 20.0 mL of N_2 at 0 °C and 740 torr plus 30.0 mL of O_2 at 0 °C and 640 torr to a 50.0-mL container at 0 °C?

11.69 A mixture of N_2 and O_2 has a volume of 100 mL at a temperature of 50 °C and a pressure of 800 torr. It was prepared by adding 50.0 mL of O_2 at 60 °C and 400

torr with X mL of N_2 at 40 °C and 400 torr. What is the value of X?

11.70 A gas is collected by the displacement of water until the total pressure inside a 100-mL flask is 700 torr at 25 °C. How many milliliters would the dry gas occupy at STP?

11.71 A mixture of N_2 and O_2 in a 200-mL vessel exerts a pressure of 720 torr at 35 °C. If there is 0.0020 mol of N_2 present,
(a) What is the mole fraction of N_2 (see Equation 11.12)?
(b) What is the partial pressure of N_2 in torr?
(c) What is the partial pressure of O_2 in torr?
(d) How many moles of O_2 are present?

11.72 The air exhaled by an average human being might have the following typical composition, expressed in terms of partial pressures: N_2, 569 torr; O_2, 116 torr; CO_2, 28 torr; water vapor, 47 torr. What are the mole fractions of each gas?

11.73 How many milliliters of CO_2 at 30 °C and 700 torr must be added to a 500-mL container of N_2 at 20 °C and 800 torr to give a mixture having a pressure of 900 torr at 20 °C?

11.74 How many milliliters would be occupied by 0.0244 g of O_2 if it were collected over water at 23 °C and at a total pressure of 740 torr?

*11.75 Three gases were added to the same 10.0-L container to give a total pressure of 800 torr at 30 °C. If the mixture contained 8.00 g of CO_2, 6.00 g of O_2, and an unknown amount of N_2, calculate the following:
(a) the total number of moles of gas in the container
(b) the mole fraction of each gas
(c) the partial pressure of each gas, in torr
(d) the number of grams of N_2 in the container

*11.76 A gas, at a total pressure of 800 torr and a volume of 500 mL over water at 35 °C, is compressed to a volume of 250 mL, also over water at 35 °C. Calculate the final pressure of the wet gas in torr.

*11.77 During a rainstorm in July in New York City, the humidity was found to be 100%, which means that the air was saturated with water vapor. The atmospheric pressure was 740 torr and the temperature was 31°C. Dry air has an average molecular mass of 28.8. Calculate the number of grams of water in 1.00 L of the air during the storm.

11.78 When 280 mL of gas is collected over water at 20 °C, the water level inside the collection bottle is 28.4 mm higher than the water level outside. The atmospheric pressure is 763 torr. How many milliliters would the dry gas occupy at STP?

Graham's Law

11.79 What is the difference between *diffusion* and *effusion*?

11.80 What is Graham's law of effusion?

11.81 Compare the rates of effusion of He and Ne. Which gas effuses faster, and how much faster?

11.82 If, at a particular temperature, the average speed of CH_4 molecules is 1000 mi/hr, what would be the average speed of CO_2 molecules at the same temperature expressed in miles per hour?

11.83 The rate of effusion of an unknown gas was determined to be 2.92 times greater than that of NH_3. What is the approximate molecular mass of the unknown gas?

Kinetic Theory

11.84 What are the postulates of the kinetic molecular theory?

11.85 In terms of kinetic theory, what is the origin of the pressure of a gas?

11.86 In qualitative terms, why do molecules of gases with low molecular masses diffuse faster than gases with high molecular masses, provided they are at the same temperature?

11.87 In terms of the kinetic theory, how should the rate of diffusion of a gas be affected by an increase in temperature? Explain.

11.88 Why does the pressure of a gas increase when the volume is decreased (at constant temperature)?

11.89 Why does the pressure of a gas increase when the temperature is increased, provided the volume is constant?

11.90 If a warm object is placed in contact with a cool object, heat transfer occurs until they both come to the same temperature. How can the kinetic molecular theory account for this heat transfer and these temperature changes that occur?

11.91 Sketch a graph showing the distribution of kinetic energies for a gas at two different temperatures. For each temperature indicate the most probable K.E. and the average K.E. Why is the curve not symmetrical?

Real Gases

11.92 What is meant by nonideal behavior of a gas? Under what conditions is this behavior most evident?

11.93 Explain why most gases cool on expansion into a vacuum.

11.94 What physical significance do the constants a and b have in the van der Waals equation of state for real gases?

11.95 Why did van der Waals subtract a correction from the measured volume? Why did he add a correction to the measured pressure?

11.96 Use the van der Waals equation to calculate the pressure, in atm, exerted by 1.000 mol of He at 0.00 °C in a volume of 22.400 L. Use $R = 0.082057$ L atm/mol K. Compare this to the pressure an ideal gas would exert under these same conditions.

11.97 Use the van der Waals equation to calculate the pressure, in atm, exerted by 1.000 mol of C_2H_6 at 0.00 °C in a volume of 22.400 L. Use $R = 0.082057$ L atm/mol K. Compare this to the pressure of an ideal gas under these same conditions.

*¤11.98** Calculate the molar volume (in liters) of O_2 at STP from the van der Waals equation and compare it to the value in Table 11.1. Use $R = 0.082057$ L atm/mol K. (*Hint:* The volume can be obtained by successive approximations if you solve for the V in the term, $V - nb$).

Additional Exercises

*¤11.99** When 0.230 g of a metal was dissolved in HCl, 348 mL of H_2 at 24 °C and a pressure of 680 torr was evolved. Calculate the equivalent weight of the metal acting as a reducing agent. From what you have been taught about the ions formed by the metals and the possible atomic mass of the metal described in this problem, identify the metal.

*¤11.100** A 0.4500-g sample of a compound known to contain only hydrogen, carbon, nitrogen, and oxygen was burned in O_2, yielding 0.671 g of CO_2 and 0.345 g of H_2O. In a separate experiment, the nitrogen in 0.3600 g of the compound was changed to N_2 and collected to give 153 mL of the gas at 25 °C and 740 torr. What is the empirical formula of the compound?

*¤11.101** If 250 mL of O_2 over water at 30 °C and a total pressure of 740 torr is mixed with 300 mL of N_2 over water at 25 °C and a total pressure of 780 torr, what will the total pressure be if the mixture is in a 500-mL container over water at 35 °C?

12 STATES OF MATTER AND INTERMOLECULAR FORCES

The physical properties of gases, which we studied in the last chapter, are vastly different from those of the other two states of matter, liquids and solids. For one thing, we can't even see a gas unless it is colored. We would be totally unaware that gases such as those in our atmosphere exist if we couldn't feel a breeze and see it rustle the leaves on a tree, or if we hadn't discovered that gases exert a pressure. Yet who would fail to recognize a lake full of water or a gigantic icy glacier?

The water vapor in the air (which we recognize as humidity), the water of a lake, and the water frozen in a glacier are all forms of the same chemical substance. They are each composed of molecules of water and have the same set of chemical properties. It is their physical properties that make them seem so different. In this chapter we will study why gases, liquids, and solids are so different, and what important properties liquids and solids have. We will also study how the three states of matter change from one to another and what this tells us about them.

12.1 THE IMPORTANCE OF INTERMOLECULAR ATTRACTIONS

One of the most interesting things about gases is that their behavior is nearly independent of their chemical composition. Without too much error, we can treat all gaseous substances as "ideal gases," provided their temperatures are reasonably high. In fact, it is this very lack of dependence of physical properties on chemical composition that permits the existence of the gas laws that we find so useful. For liquids and solids, on the other hand, there are no universal laws comparable to the gas laws. The physical properties of a liquid or solid depend quite strongly on the kinds of particles of which a substance is composed.

This fundamental difference in the dependence of properties on chemical makeup is not difficult to understand if we consider the major way that gases differ from liquids and solids — in the closeness of the packing of the particles — and the way the distances between the particles influence the effectiveness of intermolecular attractive forces.

If you have ever played with magnets, you already know something about how distance affects the strengths of attractions. When two magnets are close together, the north pole of one attracts the south pole of another quite strongly. On the other hand, if you separate the magnets by several inches, you can hardly feel any attraction between them at all. The strengths of the attractions decrease very quickly as the distance between the magnets increases, and the same applies to the strengths of the attractions between molecules.

In a gas, where the particles are far apart, the attractions are uniformly very weak simply because the distances between the particles are so great. Therefore, when several gases are compared, the differences that do exist among the abilities of the molecules to attract each other are essentially leveled out; the differences are so small that they are almost negligible, unless very precise measurements are made. As a result, all gases behave pretty much alike.

When the particles of a gas are condensed to a liquid or a solid, they approach each other very closely, so there is hardly any distance between them. Because the particles are so close, the intermolecular attractions are greatly magnified, and as a result, the behavior of liquids and solids is very strongly influenced by the nature of the attractive forces. Furthermore, because the intermolecular distances are so small and the attractions so strong in these condensed phases, the differences that are negligible in the gas phase are now much more important. Therefore, the way in which chemical composition influences intermolecular attractions becomes a very significant factor in determining the differences among the physical properties of liquids and solids. So, we see that to understand liquids and solids, we need to know about intermolecular attractions and the factors that control their strengths.

> The closer two molecules are, the more strongly they attract each other.

12.2 TYPES OF INTERMOLECULAR ATTRACTIVE FORCES

Before we discuss the attractive forces between molecules, it is important that you understand that *intermolecular attractions* are much weaker than *intramolecular attractions* — the attractions between atoms *within* molecules. Intramolecular attractions are chemical bonds and they hold molecules together. When these attractions are disrupted, chemical reactions occur, so the strengths of the chemical bonds between atoms principally determine chemical properties.

Physical properties of substances are observed without changing the chemical composition of the substance being studied, so chemical bonds are not affected. Instead, it is the strengths of the intermolecular attractions that determine what happens. With that in mind, let's now examine how intermolecular attractions arise.

> Attractions *within* molecules (chemical bonds) determine chemical properties; attractions *between* molecules determine physical properties.

Dipole–dipole attractions

Polar molecules have ends that are oppositely charged. In a collection of these molecules, the individual dipoles tend to orient themselves so that the partial positive charge on one is near the partial negative charge on others. Because the molecules are constantly moving and colliding with each other, this alignment is far from perfect, particularly in liquids and gases. Nevertheless, the attractions between the oppositely charged ends of the dipoles outweigh the repulsions between like-charged ends, and a net overall attraction exists between them (Figure 12.1).

Dipole–dipole attractions are normally considerably weaker than ionic or covalent bonds — they are only about 1% as strong. Their strength also decreases very rapidly as the distance between the dipoles increases, so their effects between the widely spaced molecules in a gas are very much less than between tightly packed molecules in a liquid or solid. This is why the molecules of a gas behave almost as though there were no attractive forces at all.

Figure 12.1. *Electrostatic attractions between dipoles.*

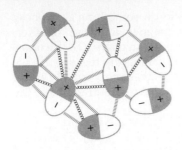

The energy required to separate a pair of dipoles is proportional to $1/d^3$, where d is the distance between the dipoles.

The strengths of dipole–dipole attractions are greatest, naturally, between the most polar molecules. You have already learned how to compare the relative polarities of bonds as well as how to use the structures of molecules predicted by the VSEPR theory to predict which molecules are polar and which are not.

Hydrogen bonding

Remember, the closer the attracting particles, the stronger the attraction. The small sizes of N, O, and F make these dipole–dipole attractions exceptionally strong.

A particularly strong dipole–dipole attraction occurs between molecules in which hydrogen is covalently bonded to a very small, highly electronegative element such as fluorine, oxygen, or nitrogen. In these instances extremely polar molecules are formed in which the small hydrogen atom carries a substantial partial positive charge. Because the positive end of this dipole can closely approach the negative end of a neighboring dipole, the force of attraction between the two is quite large. This special kind of dipole–dipole interaction is called a **hydrogen bond** and is about 5 to 10% as strong as an ordinary covalent bond.

Hydrogen bonding is a very important type of attractive force. In water, for example, the molecules interact strongly with each other by hydrogen bonding (Figure 12.2). This produces attractive forces that are much stronger than those between other molecules of similar size and mass and is what makes water a liquid at room temperature. Hydrogen bonding is also responsible for controlling the orientation of water molecules in ice (Figure 12.3). Each water molecule is surrounded tetrahedrally by four others to which it is held by hydrogen bonds. This causes ice to have a very "open" structure and makes ice less dense than liquid water. That's why ice cubes and icebergs float (much to the distress of the captain of the *Titanic*).

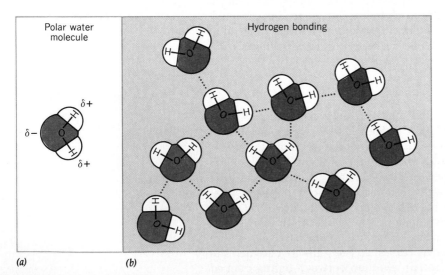

(a)   *(b)*

Figure 12.2. *Hydrogen bonding in water. (a) The water molecule is very polar. (b) Hydrogen bonding produces strong attractions between water molecules.*

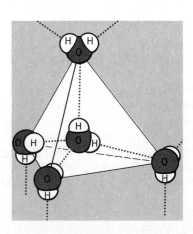

Figure 12.3. *Hydrogen bonding (dotted lines) between water molecules in ice.*

London forces

Even uncombined atoms and nonpolar molecules experience weak attractions. We know this because substances like helium and hydrogen can be condensed to liquids if they are cooled to a sufficiently low temperature. Attractive forces must exist to hold them together in the liquid state. These attractions are called **London forces,** after the German physicist Fritz London who provided an explanation of them.

According to this explanation, when electrons move around in an atom or molecule, their motion is somewhat random, so that at any instant there is a chance that more electrons will be on one side of the particle than the other. At the particular instant that one side has more electrons, the particle will be a dipole. We call it an *instantaneous dipole* because its existence is only momentary. In a collection of atoms or molecules the electron motions in nearby neighboring particles are not entirely independent. As the negative end of an instantaneous dipole begins to form in one of them, it pushes electrons away in the particle alongside, as shown in Figure 12.4. We say that the instantaneous dipole *induces* a dipole in its neighbor. As you can see, because of the way the dipoles are formed, they attract each other, and this attraction produces a momentary tug that helps hold them together.

An important thing about London forces is the way in which they oscillate on and off very rapidly. The momentary dipoles that are responsible for the attractions exist one moment and are gone the next as the electrons continue their journeys. This flickering on and off of the dipoles is illustrated in Figure 12.5, and because of their fleeting existence, London forces are rather weak on average. Nevertheless, they are present between all particles — ions as well as polar and nonpolar molecules. They play only a minor role in the attractions between ions because the forces of attractions between ions are so strong. But London forces do play a significant role in the attractions between molecules of all kinds, especially nonpolar ones.

London forces are often called *London dispersion forces.*

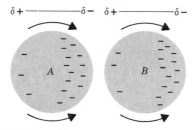

Figure 12.4. *London forces. As the instantaneous dipole forms in atom A, it induces a dipole in atom B.*

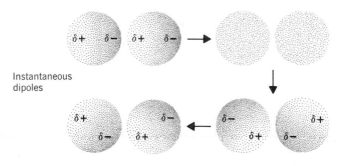

Instantaneous dipoles

Figure 12.5. *Instantaneous views of how the electron density fluctuates in two neighboring atoms, giving rise to fleeting attractions between the momentary dipoles.*

The strengths of London forces depend on several factors. One of these is the complexity of the molecule. Consider, for example, the hydrocarbon molecules propane, C_3H_8, and hexane, C_6H_{14}, whose structures are shown below.

The strengths of London forces, as measured by the energy needed to separate the particles, is proportional to $1/d^6$.

propane

hexane

Each of these molecules contains relatively nonpolar C—H bonds and overall they are essentially nonpolar molecules. In their respective liquids, they attract each other by similar London forces. Yet the attractive forces that hold hexane molecules within liquid hexane are considerably stronger than those that hold propane molecules in

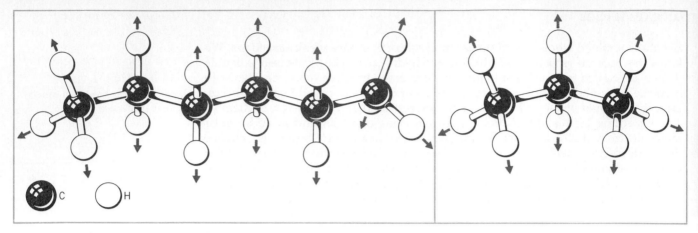

Figure 12.6. *The C_6H_{14} molecule has more places along its chain where it can be held to other molecules than the shorter C_3H_8 molecule has.*

its liquid. The reason seems to be the number of places along a molecular chain that can be attracted to other molecules that surround it. In each of these substances, the attractions are essentially between the hydrogens of one molecule and the hydrogens of other molecules around it; the carbons are buried beneath an outer covering of hydrogen atoms. In propane, C_3H_8, there are 8 hydrogen atoms, so there are potentially 8 places that this molecule can attach itself to other molecules (Figure 12.6). In hexane, C_6H_{14}, however, there are 14 hydrogen atoms, so there are potentially 14 places that it can bind to other molecules. Even though the binding force may be the same at each hydrogen, because hexane has more hydrogens, it feels the greater *overall* attractions to its neighbors.

In complex molecules with many atoms, the cumulative effects of the London forces can be quite substantial.

Another factor that influences the strengths of London forces is molecular size. If we examine molecules with the same general formula, such as the halogens (F_2, Cl_2, Br_2), we find that large molecules attract each other more strongly than small molecules. This is explained as follows.

As we proceed from F_2 to Br_2, the atoms that make up the molecules become larger. As the size increases, the outer electrons become farther from the nuclei and are not held as tightly. Because of this, the electron cloud of a large molecule is more easily distorted, or *polarized,* and it is easier to create the instantaneous dipoles that are responsible for the London forces. (The ease of distortion of the electron cloud is referred to as **polarizability**.) The result is that the London forces are stronger between molecules composed of large, easily polarized atoms such as bromine than between molecules composed of small atoms such as fluorine.

12.3 COMPARING SOME GENERAL PROPERTIES OF GASES, LIQUIDS, AND SOLIDS

In Section 12.1, we noted that liquids and solids owe their physical properties to the closeness of the molecules within them, which causes the intermolecular forces to be relatively strong. Although both factors—closeness of packing and intermolecular forces—are related, certain physical properties are influenced more by one than the other. Two properties particularly affected by how tightly the particles are packed are compressibility and the rates of diffusion. Properties particularly influenced by the strengths of the intermolecular attractive forces are shapes, volumes, and the ability to flow; surface tension; and the rate of evaporation.

The hydraulic cylinders on this earth-moving machine use the incompressibility of liquids to transmit the large forces needed to make it operate.

Compressibility

Compressibility is a term that refers to the degree to which the volume of a substance is decreased by the application of pressure. If the volume changes a lot when the pressure is increased, the substance is said to be very *compressible*.

In a gas the molecules are widely separated, so there is a lot of empty space into which they can be crowded. As a result, gases are very compressible. The molecules in a liquid or a solid, however, are tightly packed and there is very little empty space between them. For this reason, increasing the pressure has hardly any effect on their volume, and they are virtually incompressible.

The incompressibility of liquids is an important and useful property. Many types of hydraulic machinery depend on it to transmit enormous forces that lift and move heavy things. You depend on it yourself when you step on the brakes of a car. The force exerted by your foot is first magnified and then transmitted by an oil through the brake lines to the wheels, where it causes the brake shoes to rub against a surface and stop the car. If air gets into the brake lines, the force you exert with your foot simply compresses the air, and the car won't stop very quickly at all. (Disaster!)

There are repulsions between molecules, too. When molecules are pushed *very* close together, the electron clouds of the atoms repel each other strongly, so the molecules can't just meld together under pressure.

Compressed gases are used in many practical applications, too.

Diffusion

Molecules diffuse rapidly in a gas, compared to a liquid or a solid, because they move relatively long distances between collisions. They get where they're going quickly because there are relatively few interruptions along their paths, as shown in Figure 12.7. When two liquids mix, however, the molecules of one diffuse throughout the molecules of the other at a much slower rate than when two gases are mixed. We can

Rates of diffusion increase with increasing temperature because molecules move faster at higher temperatures.

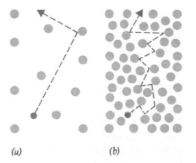

(a)   *(b)*

Figure 12.7. *Diffusion can occur more rapidly (a) in a gas than (b) in a liquid because a molecule moves farther between collisions in a gas.*

Figure 12.8. *Diffusion in liquids.* (a) *An ink drop is placed into water.* (b) *The ink has diffused throughout the water, giving a uniform solution.*

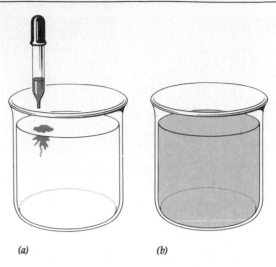

(a)   (b)

observe the diffusion of two liquids by dropping a small quantity of ink into some water. As shown in Figure 12.8, when the ink drop strikes the water, we see it as a concentrated dot, which slowly spreads throughout the liquid. Diffusion takes place because the molecules in both liquids are able to move throughout the container. However, since the molecules in a liquid are so close together, the average distance that they travel between collisions—their *mean free path*—is very short. The molecules undergo billions of collisions before going very far, and these constant interruptions in their paths keep them from spreading rapidly throughout the liquid.

Diffusion within solids is even much slower than in liquids. Not only are the molecules very tightly packed, but they are also held quite rigidly in place. They are not free to roam about, even though each may vibrate and bounce around inside the small cavity that it occupies within a crystal. Only molecules (or ions, if the solid is ionic) that have very large kinetic energies are able to muscle their way past their neighbors, so diffusion is extremely slow at ordinary temperatures.

High-temperature solid-state diffusion is important in manufacturing electronic components such as those found in computers.

Volume and shape

The most obvious property of gases, liquids, and solids is the way they behave when transferred from one container to another. A gas, as we learned, expands to completely fill whatever container it is placed in. A liquid, however, retains a constant volume, but conforms to the shape of its container. Gases and liquids are both fluids; they flow and can be pumped from place to place. A solid, however, is not a fluid and maintains both its shape and volume.

In a gas the intermolecular attractive forces are so weak that the rapidly moving molecules can easily overcome them and expand to fill a container. The strengths of these forces are much larger in a liquid, however, and are responsible for holding the molecules close together. In a solid the attractive forces are still larger and hold the molecules more or less firmly in place so that they cannot move over and around each other. This prevents solids from flowing like gases and liquids.

Surface tension

Have you ever noticed how raindrops form beads of water on a freshly waxed car? Have you ever wondered why moist grains of sand stick together, but fall apart if either dry or completely submerged in water? These are phenomena that are caused by a property of liquids called surface tension.

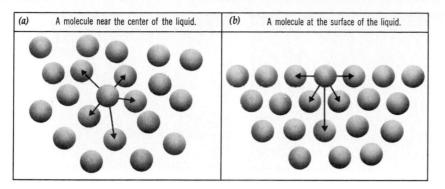

Figure 12.9. *Intermolecular attractive forces in a liquid. (a) A molecule within the bulk of the liquid is attracted by other molecules on all sides. (b) A molecule at the surface feels the attractions of those alongside and below it, but no attractions from above the surface.*

In a liquid, each molecule moves about always under the influence of its neighbors. A molecule within the liquid is completely surrounded by others to which it is attracted (Figure 12.9a). However, a molecule at the surface is not completely surrounded and feels attractions only to molecules below and beside it (Figure 12.9b). Molecules at the surface thus feel a net attraction in a direction toward the interior of the liquid. For a molecule to come to the surface, it must overcome this attraction. In other words, its potential energy must increase. Work must be done to pull it to the surface. Making the surface of a liquid larger, therefore, requires an input of energy, and the amount of energy needed is proportional to the liquid's **surface tension.**

The lowest energy (most stable) state for a given volume of liquid is when its surface area is a minimum. This gives the fewest high-energy surface molecules. The shape that satisfies this condition is a sphere, which is why raindrops are nearly spherical. All liquids strive to minimize their surface areas and tend toward spherical shapes as much as possible. This natural tendency is what allows you to fire polish glass tubing in the lab. As the glass softens, the sharp edges become rounded because the attractive forces within the glass tend to reduce the surface area. In Figure 12.10 we see why moist grains of sand are drawn together. The film of water between them achieves a lower energy by reducing its surface area and, in doing so, pulls the grains together. Pulling the grains of sand apart requires that the surface tension be overcome, so the particles tend to stay together.

The magnitude of a liquid's surface tension depends on the strengths of the attractive forces between its molecules. When the attractive forces are large, as in H_2O, the surface tension is large. On the other hand, liquids such as gasoline that are composed of nonpolar molecules have small surface tensions because the intermolecular attractions are much weaker.

A property that we often think of when a liquid is mentioned is its ability to **wet** things, to spread across their surfaces as a thin film. Water, for example, spreads smoothly across a clean glass surface. This ability reflects a similarity among the strengths of the attractive forces between the individual water molecules and between the water molecules and the glass surface.

The units of surface tension are energy per unit area. The surface tension is a measure of the energy needed to create a unit area of surface.

Liquids tend to form spherical drops because a sphere has the smallest surface area for a given volume.

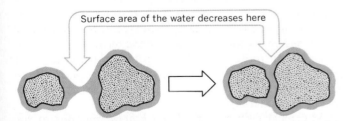

Surface area of the water decreases here

Figure 12.10. *Grains of sand such as these are drawn together when wet as the film of water between them reduces its surface area, thereby achieving a lower energy.*

Figure 12.11. (a) *When the glass is clean, water wets it easily and spreads smoothly across the surface as it leaves the dropper.* (b) *When the glass is coated with grease, the water refuses to spread and forms a nearly spherical bead instead.*

(a)

(b)

Detergents contain substances, called surfactants, that lower the surface tension of water and allow the detergent solution to spread more easily over greasy surfaces.

Glass is made from SiO_2 (sand) and other chemicals, and its surface contains lots of oxygen atoms. In water, as you learned, the molecules attract each other by hydrogen bonding between the hydrogen of one O—H bond and the oxygen of another H_2O molecule. But the O—H bonds of water can also hydrogen bond to oxygens in other molecules, including the oxygens in the surface of glass. Because of these similar attractions, it takes little effort to spread water molecules across the glass surface, so the surface tension of water is easily overcome.

If a glass surface has a film of grease on it, which is made of nonpolar molecules, water doesn't spread across it easily. In this case, there are no polar substances in the grease to which the water molecules can hydrogen bond, so it takes a lot of work to increase the surface area of the H_2O. As a result, the water resists increasing its surface area and tends to form beads that are as nearly spherical as possible (see Figure 12.11).

Substances with very low surface tension wet surfaces very easily, regardless of what the surface happens to be. Hydrocarbon solvents such as naphtha or gasoline spread across glass and greasy surfaces with about equal ease because the hydrocarbon molecules attract each other so weakly. It hardly takes any effort to expand the surfaces of their liquids, so they can easily spread out across a surface.

Evaporation

Naphthalene sublimes when heated, and the vapor condenses to give beautiful flaky crystals on the underside of a watch glass cooled by ice.

In a liquid or a solid, just as in a gas, the molecules are constantly undergoing collisions, giving rise to a distribution of individual molecular velocities and, of course, kinetic energies. Even at room temperature, a small percentage of the molecules are moving with relatively high kinetic energies. If some of these faster-moving molecules possess enough kinetic energy to overcome the attractive forces within the liquid or solid, they can escape through the surface into the gaseous state—they **evaporate.**

The evaporation of a liquid is something everyone has seen. A small amount of spilled gasoline or nail polish remover quickly disappears, and rain puddles evaporate after a summer shower. Solids can also evaporate, although most of them that we encounter under normal conditions don't seem to disappear rapidly, if at all. But have you ever seen dry ice? It is composed of solid carbon dioxide. It is called *dry* ice because the solid doesn't melt; it simply evaporates. Have you ever wondered what happens to the crystals of naphthalene (moth flakes) that you sprinkle in a drawer or garment bag? They evaporate, too. Direct conversion of a solid to a gas, without melting, is called **sublimation.** Solid carbon dioxide and naphthalene are two substances that readily sublime at atmospheric pressure.

Figure 12.12 represents a typical distribution of the kinetic energies of the molecules in a liquid. The shaded area corresponds to the fraction of the total number of molecules that possess sufficient kinetic energy to evaporate. Just as

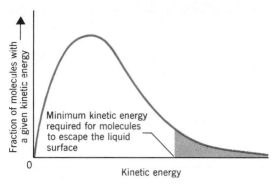

Figure 12.12. *The kinetic energy distribution in a liquid. The shaded area represents the total fraction of molecules that have kinetic energies equal to or greater than the minimum needed to escape the surface.*

removing the smart students from a chemistry class will lower the class average on exams, the loss of the higher-energy fraction because of evaporation leads to a lowering of the average kinetic energy of the remaining molecules. Since the temperature is directly proportional to the average kinetic energy, this results in a decrease in the temperature of a liquid as it evaporates. For example, we have all felt cool after a bath, because the evaporation of water from the body has drawn heat from us. In fact, evaporation of perspiration provides the body with a mechanism for controlling body temperature.

Several things affect the rates at which liquids evaporate. One is the surface area. Increasing the surface area brings more high-energy molecules close to the surface, so the total number that can escape each second is larger.

Another factor is the temperature. You know that water evaporates more rapidly on a warm day than on a cool one. The reason can be seen in Figure 12.13. Notice that the shaded area under the curve, which is the *total* fraction of the molecules that have enough energy to escape the liquid, is larger at the higher temperature. At the higher temperature, more molecules evaporate per second simply because more molecules have the necessary energy to do so.

A third factor is the strengths of the intermolecular attractions. Figure 12.14 compares two liquids at the same temperature. In one (liquid *A*) the attractive forces are very strong, so only very high-speed molecules have enough kinetic energy to overcome the attractions and escape. There are relatively few of these molecules, so the rate of evaporation is slow. In the other liquid the attractive forces are weak, and a much larger fraction have the energy needed to escape. Therefore, this liquid evaporates quickly. The attractive forces in most solids are much greater than in liquids, which explains why solids usually do not evaporate (sublime) easily.

Steam issues from the cooling tower of a nuclear power plant in Arkansas. The temperature-lowering effect produced by the evaporation of water is often used commercially for cooling purposes.

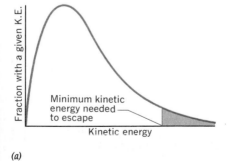

(a)

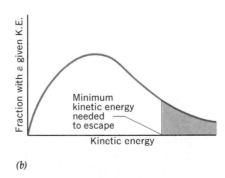

(b)

Figure 12.13. *Kinetic energy distributions* **(a)** *at low temperature and* **(b)** *at high temperature. The same minimum kinetic energy is needed to escape at both temperatures, but the total fraction of molecules having at least this much energy (the shaded area) is larger at the higher temperature.*

Figure 12.14. *The kinetic energy distribution is the same for two different liquids at the same temperature, but the kinetic energy needed by A molecules to escape from the liquid is greater than that needed by B molecules. The total fraction of molecules with sufficient kinetic energy to escape is greater for liquid B than for liquid A.*

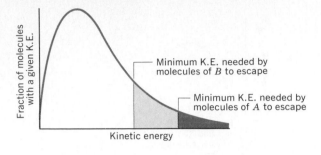

12.4 HEATS OF VAPORIZATION

In the preceding section we noted that liquids become cool when they evaporate because the departure of high-energy molecules lowers the average kinetic energy of those that remain behind. If we wished to keep the temperature of the liquid constant during evaporation, we would have to add heat. This is to replace the energy taken away by the molecules that evaporate, and we can think of it as the energy needed to cause the evaporation, at a constant temperature, of a given amount of the liquid. Normally, the amount of liquid is taken to be a mole, and *the amount of energy needed to cause the evaporation of one mole of the liquid at a constant pressure is called its* **molar heat of vaporization,** or simply its **heat of vaporization** (the word *molar* is understood).

The term *heat of vaporization* is usually taken to mean *molar heat of vaporization.*

Because the molar heat of vaporization is measured at constant pressure, it is really an enthalpy change. It is represented symbolically as $\Delta H_{\text{vaporization}}$ (or simply ΔH_{vap}). In a formal sense, it is defined as the difference between the enthalpy of the vapor and the enthalpy of the liquid.

$$\Delta H_{\text{vap}} = H_{\text{vapor}} - H_{\text{liquid}}$$

H_{vapor} is the enthalpy of the vapor and H_{liquid} the enthalpy of the liquid.

You should recall that we have used similar definitions before. As before, H_{vapor} and H_{liquid} cannot actually be measured; the definition simply serves to define the algebraic sign of ΔH_{vap}, which is positive (endothermic) for evaporation. This is what we expect, because pulling the molecules of a liquid apart to form the gas increases their potential energy. If the liquid is isolated from the surroundings, then forming the vapor will lead to a corresponding lowering of the kinetic energy and a lowering of the temperature. If the temperature is to be kept from falling, heat must be allowed to flow in from the surroundings. Therefore, at constant temperature, evaporation absorbs heat from the surroundings and is an endothermic process.

The molar heat of vaporization is an important physical property of a substance. Engineers have to know about it when they design chemical plants because they must know how much energy will be needed to vaporize the solvents used in various processes. The heat of vaporization of water is important to meteorologists because much of the solar energy absorbed by the earth goes to evaporating water from the oceans. In a sense, this energy becomes stored in the water vapor, and when the water condenses, it is released. This is the origin of the enormous amounts of energy contained in violent hurricanes and other rain storms.

EXAMPLE 12.1. USING THE HEAT OF VAPORIZATION

PROBLEM: The molar heat of vaporization of water is 40.6 kJ/mol. How many kilojoules of heat energy are required to convert 1.00 L of water to steam?

SOLUTION: Water has a density of 1.00 g/mL. Therefore, 1.00 L of water weighs 1.00×10^3 g. The problem, as is generally the case, is one of unit conversion.

$$1.00 \times 10^3 \text{ g H}_2\text{O} \Longleftrightarrow (?) \text{ kJ}$$

This can be set up as

$$1.00 \times 10^3 \text{ g H}_2\text{O} \times \left(\frac{1 \text{ mol H}_2\text{O}}{18.0 \text{ g H}_2\text{O}} \right) \times \left(\frac{40.6 \text{ kJ}}{1 \text{ mol H}_2\text{O}} \right) \Longleftrightarrow 2260 \text{ kJ (to three significant figures)}$$

Thus, 2260 kJ of heat is required.

A useful feature of the heat of vaporization is that its magnitude provides a good measure of the strengths of the attractive forces in a liquid. This is easy to understand. When the intermolecular attractions are strong, a lot of energy must be supplied to pull the molecules apart as the liquid is changed to a gas.

Table 12.1 lists some values of ΔH_{vap} for various substances. We can use these data to gain some insight into the way chemical composition and structure influence intermolecular attractions. For example, if we look at the series of hydrocarbons, CH_4 through $C_{10}H_{22}$, we observe a steady increase in ΔH_{vap} with an increase in the number of atoms. As we mentioned earlier, when comparing molecules composed of the same or similar sized atoms, the more complex the molecule, the stronger are the cumulative effects of the London forces. In $C_{10}H_{22}$ these cumulative effects are quite impressive; this substance, with its nonpolar hydrocarbon molecules, has a heat of vaporization more than twice as large as that of HCl, which is made up of very polar molecules.

In Table 12.1, we can also see the effects of molecular size. Notice that the heats of vaporization increase from F_2 to Br_2. As noted in Section 12.2, as the atoms in the

For molecules composed of the same or similar sized atoms, London forces increase as the number of atoms increases.

$C_{10}H_{22}$ has 22 hydrogens through which it can be attracted to its neighbors by London forces.

Table 12.1. Heats of vaporization and boiling points of various substances

Compound	ΔH_{vap} (kJ/mol)	Boiling Point (°C)
CH_4	9.20	−161
C_2H_6	14	−89
C_3H_8	18.1	−30
C_4H_{10}	22.3	0
C_6H_{14}	28.6	68
C_8H_{18}	33.9	125
$C_{10}H_{22}$	35.8	160
F_2	6.52	−188
Cl_2	20.4	−34.6
Br_2	30.7	59
HF	30.2	17
HCl	15.1	−84
HBr	16.3	−70
HI	18.2	−37
H_2O	40.6	100
H_2S	18.8	−61
NH_3	23.6	−33
PH_3	14.6	−88
SiH_4	12.3	−112

For molecules with the same number of atoms, London forces increase as the size of the atoms increases.

molecules become larger, they become more easily polarized and the London forces become stronger. We also see this effect going from HCl to HI, where the halogen atom becomes progressively larger from Cl to I.

Evidence for hydrogen bonding is seen if we compare the heats of vaporization of HF and HCl. Even though Cl is larger than HF, the ΔH_{vap} of HF is larger than that of HCl, which is attributed to hydrogen bonding in HF. We see the same inverted order in ΔH_{vap} for H_2O and H_2S, and for NH_3 and PH_3, where, again, hydrogen bonding is significant for H_2O and NH_3 but not for H_2S and PH_3. Oxygen, fluorine, and nitrogen are all very small and are the most electronegative elements in the periodic table, while the elements below them are much larger and much less electronegative. It is not surprising, therefore, that hydrogen bonding is important only for H_2O, HF, and NH_3. Normal behavior is reached in Group IVA hydrides, where ΔH_{vap} for CH_4 is less than ΔH_{vap} for SiH_4. Here, neither CH_4 nor SiH_4 has any tendency to hydrogen-bond because they are nonpolar.

12.5 VAPOR PRESSURES OF LIQUIDS AND SOLIDS

When a liquid evaporates from an open container, all the liquid will eventually disappear because the molecules that enter the vapor simply diffuse away into the atmosphere. However, if the liquid is placed into a closed container, the molecules that evaporate cannot escape and accumulate in the vapor space above the liquid. There they exert a pressure, just like any other gas molecules, and we call this pressure the **vapor pressure.**

If we study how the vapor pressure changes when a liquid is placed into an evacuated container, we find that it initially rises and then gradually becomes constant. In addition, for a given liquid at a given temperature, the final value of the vapor pressure is always the same, regardless of the size of the container or the amount of liquid in it, just as long as some liquid is still present when the limiting pressure is reached.

The behavior of the vapor pressure just described is explained in the following way. When the liquid is introduced into the container, it begins to evaporate and molecules of the substance start to accumulate above the liquid in the vapor space. These molecules collide with the walls of the container and produce the pressure. As time passes, more and more molecules of the substance enter the vapor phase, and the pressure rises. But during this time something else is happening. One of the walls surrounding the gas space is the liquid itself, and when molecules from the gas strike the liquid's surface, there is a high probability that they will become trapped there. This is because the kinetic energy of an incoming molecule tends to become scattered among the liquid molecules at the surface, and there is little likelihood that the incoming molecule will retain enough kinetic energy to escape again. As a result, gaseous molecules of the substance that strike the liquid's surface return to the liquid state. Thus, we not only have evaporation taking place, but condensation as well.

The rate at which molecules evaporate from the liquid is determined by the liquid's temperature, which we will assume is being kept constant. Therefore, the rate of evaporation is constant. This means that molecules are moving into the vapor at a constant rate. The rate of condensation, however, depends on the concentration of molecules in the vapor. The more molecules per unit volume, the more collisions they will make with the surface of the liquid, and the faster they will condense. When the liquid is first placed in the container, none of its molecules is in the vapor, so no condensation is taking place. As a result, initially there will be a net influx of molecules into the gas phase as the liquid evaporates. However, as the number of molecules in the vapor increases, the rate of condensation will also increase, and this will continue until the liquid is evaporating as fast as its vapor is condensing (Figure

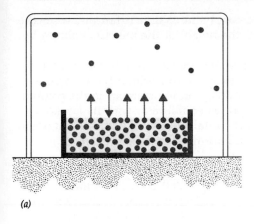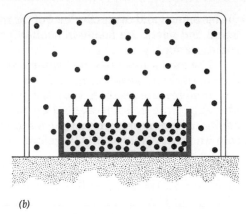

(a)

(b)

Figure 12.15. (a) *A liquid shortly after it has begun to evaporate into an evacuated container. The rate of evaporation is greater than the rate of condensation.* (b) *When equilibrium is reached, the rates of evaporation and condensation are equal.*

12.15). Once this happens, there will no longer be any change with time in the number of molecules in the gas. (For each hundred molecules that leave the liquid, one hundred return.) Since the number of molecules in the vapor has become constant, so has the pressure. At this point, the liquid is in *dynamic equilibrium* with its vapor, and the pressure exerted by the vapor once equilibrium has been established is called the **equilibrium vapor pressure** of the liquid.

The concept of a dynamic equilibrium was introduced earlier in our discussion of weak electrolytes.

One way that we can measure the vapor pressure of a liquid (usually, the term *vapor pressure* is used to mean *equilibrium vapor pressure;* the word "equilibrium" is understood) is by the use of a barometer, as shown in Figure 12.16. First the height of the mercury in the barometer is measured accurately. Next, the liquid whose vapor pressure is to be determined is carefully added to the barometer by means of an eye dropper and allowed to rise to the top of the mercury in the column, as shown in Figures 12.16*b*, 12.16*c*, and 12.16*d* (most liquids are less dense than mercury and will therefore float on the mercury surface). The space above the mercury column in Figure 12.16*a* is for all practical purposes a vacuum,[1] so there is no pressure exerted on the top of the mercury column. The space above the mercury in Figures 12.16*b*, 12.16*c*, and 12.16*d*, however, contains a very small amount of liquid plus its vapor. As the liquid evaporates, the pressure of the trapped vapor pushes mercury out of

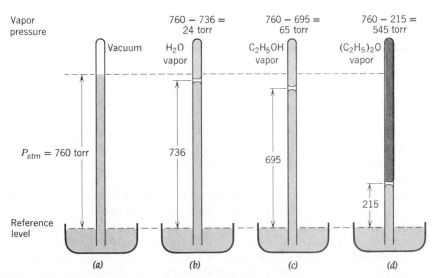

Figure 12.16. *Estimating the vapor pressure of a liquid. When a small amount of liquid is allowed to rise to the top of a barometer, its evaporation produces a pressure (its vapor pressure) that forces mercury out of the tube.* (a) *No liquid above the mercury;* (b) *H_2O;* (c) *ethyl alcohol, C_2H_5OH;* (d) *diethyl ether, $(C_2H_5)_2O$.*

[1] Mercury itself does have a finite vapor pressure (about 10^{-3} torr at room temperature) and, therefore, should never be left in an open container because of its high toxicity.

the barometer and causes the level of the mercury in the column to fall, and when the liquid and vapor are finally in equilibrium, the height of the mercury column becomes stationary.

The total pressure exerted at the reference level outside each barometer will be the atmospheric pressure, P_{atm}. The total pressure exerted within the barometer is P_{Hg}, the pressure resulting from the pull of gravity on the mercury in the column, plus P_{vapor}, the pressure exerted by the vapor in equilibrium with its liquid. The additional pressure exerted by the weight of the small amount of liquid on top of the column is negligibly small. Therefore, at equilibrium in each barometer

$$P_{atm} = P_{Hg} + P_{vapor}$$

In Figure 12.16a, $P_{vapor} = 0$; therefore, $P_{atm} = P_{Hg} = 760$ torr. In Figures 12.16b, 12.16c, and 12.16d, $P_{Hg} = 736$ mm, 695 mm, and 215 mm, respectively. Therefore, at 25 °C the vapor pressure of water is 24 torr, that of ethyl alcohol is 65 torr, and that of diethyl ether is 545 torr.

Factors that affect the vapor pressure

There are two principal factors that determine the magnitude of the vapor pressure. One is the nature of the attractive forces in the liquid itself, and the other is the temperature. Both of these influence the rate at which molecules evaporate. In liquids where the intermolecular attractions are strong, only molecules that have very large kinetic energies will be able to escape, and the rate of evaporation will be low. On the other hand, if the intermolecular attractions are weak, a large fraction of the molecules will have sufficient kinetic energy to leave and the rate of evaporation will be large. When the rate of evaporation is small, only a small concentration of vapor molecules will be needed in order to have the rate of condensation equal to the rate of evaporation, but if the rate of evaporation is large, a large concentration of vapor molecules will have to exist to reach equilibrium. This means that liquids that have large intermolecular attractions will have low vapor pressures, and vice versa.

If we examine the three liquids whose vapor pressure measurements were described above, we see that water has the lowest vapor pressure and diethyl ether has the largest. This means that the attractive forces are largest in water and smallest in ether. This is reasonable, as we can see if we examine the structures of these molecules.

Organic molecules with the structure shown below are called alcohols.

$$-\overset{\displaystyle |}{\underset{\displaystyle |}{C}}-O-H$$

water   ethyl alcohol   diethyl ether

Both water and the alcohol have O—H groups and can form hydrogen bonds, which are strong attractions. Water with its two O—H groups can form more hydrogen bonds, so its attractions are stronger than those in the alcohol. Ether molecules, on the other hand, lack the ability to form hydrogen bonds and can only interact with each other through much weaker dipole–dipole and London forces.

The effect of temperature on the vapor pressure can be seen in Figure 12.17, which contains a graph of vapor pressure versus temperature for ether, alcohol, and water. We see that as the temperature is increased, the vapor pressure rises. The reason for this is easily understood by recalling that raising the temperature increases the rate of evaporation. As you know by now, this leads to an increase in the vapor pressure.

Vapor pressure–temperature curves like those in Figure 12.17 do not continue

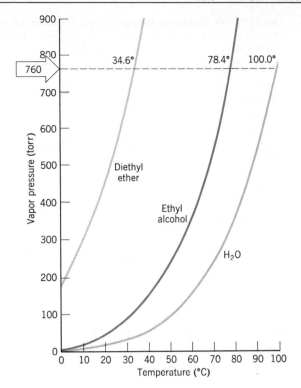

Figure 12.17. *Vapor pressure curves for three liquids. The vapor pressure increases with increasing temperature.*

indefinitely. Consider, for example, what happens as a sealed container half full of liquid is heated. At first there is a distinct boundary or surface between the more dense liquid and the less dense vapor (Figure 12.18a). As the temperature is raised, more liquid evaporates and more molecules accumulate in the vapor. Since the number of molecules per cubic centimeter is increasing, the density of the vapor is also increasing. At the same time, the liquid becomes less dense. It expands just as the liquid mercury in a thermometer does. As the temperature rises, therefore, the increasing density of the vapor gradually approaches the decreasing density of the liquid. Eventually a temperature is reached at which they become identical: Both have the same number of molecules per cubic centimeter and there is really no difference between them. If we were watching the liquid–vapor boundary in the sealed container, we would see that it suddenly disappears when the two phases become the same (Figure 12.18b).

The highest temperature at which a distinct liquid phase exists is called the **critical temperature,** symbolized T_c. The vapor pressure at the critical temperature is called the **critical pressure,** P_c. Vapor pressure–temperature curves terminate at T_c and P_c and

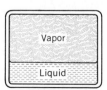

More dense liquid at the bottom can be detected by the interface between the phases

(a)

Densities of ''liquid'' and ''vapor'' have become the same — there is only one phase

(b)

Figure 12.18. (a) *When a liquid is below its critical temperature, there is a distinct liquid phase, which we can identify because the liquid is denser than the vapor.* (b) *Above the critical temperature the density is everywhere the same, so there are no separate liquid and vapor phases.*

Table 12.2. Some critical temperatures and pressures

Compound	T_c (°C)	P_c (atm)
Methane (CH_4)	−82.1	45.8
Ethane (C_2H_6)	32.2	48.2
Benzene (C_6H_6)	288.9	48.6
Ammonia	132.5	112.5
Carbon dioxide	31	72.9
Water	374.1	217.7
Helium	−267.8	2.3

Above its critical temperature, a substance behaves like a gas.

the point at the end of the curve is the **critical point.** Critical temperatures and pressures of some substances are given in Table 12.2.

Each substance has its own characteristic values for T_c and P_c that are controlled by the strengths of the intermolecular attractions. When these attractions are very weak, as with helium, the molecules must be slowed down a great deal before they will stick together when they collide. This means that the gas must be cooled to a low temperature for it to condense, so helium has a very low critical temperature.

Factors that do not affect the vapor pressure

We have seen that the vapor pressure of a liquid depends on the nature of the liquid and its temperature. What happens to the vapor pressure if the volume of the vapor is changed? Suppose that a liquid–vapor equilibrium has been established in the apparatus in Figure 12.19a. The arrows indicate that evaporation and condensation are occurring at equal rates. Now suppose that the piston is pulled up, as shown in Figure 12.19b. This expansion will cause a sudden drop in the pressure of the vapor, which means that fewer collisions will be occurring each second with each square centimeter of the walls. Each square centimeter of the liquid's surface will now be receiving fewer returning vapor molecules, but we've done nothing to affect the rate of evaporation. As a result, evaporation will be occurring faster than condensation. More liquid will therefore evaporate until the concentration of molecules in the vapor is sufficient to make the rate of condensation equal to the rate of evaporation once again. At that point, equilibrium will be reestablished and, provided the temperature hasn't changed, the vapor pressure will have returned to its previous value. At this newly established equilibrium the larger volume of gas is now occupied by more molecules. The vapor pressure will be the same as before the volume change occurred, but the volume of the liquid will be slightly smaller.

Figure 12.19. *The effect of volume changes on the vapor pressure. (a) Equilibrium between liquid and vapor. (b) The equilibrium is upset. The rate of evaporation is greater than the rate of condensation. (c) The equilibrium is also upset. The rate of concensation is greater than the rate of evaporation.*

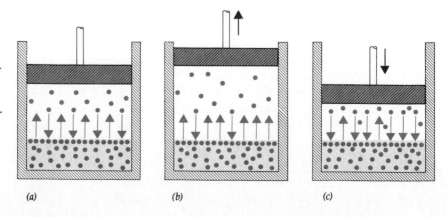

(a)   (b)   (c)

Decreasing the volume of the vapor by lowering the piston (Figure 12.19c) will also disturb the equilibrium. Increasing the pressure of the vapor will cause an increase in the number of collisions each second with each square centimeter of the walls. This, in turn, will lead to an increase in the rate of condensation, but will have essentially no effect on the rate of evaporation. The rate at which the molecules leave the vapor will consequently be greater than the rate at which molecules leave the liquid. This imbalance in rates causes the pressure exerted by the vapor to decrease and the volume of the liquid to increase. Eventually, the rate of condensation will decrease to a point where it exactly equals the rate of evaporation, reestablishing equilibrium. At this new equilibrium the smaller vapor volume, caused by the movement of the piston, will be occupied by fewer gaseous molecules. The vapor pressure will have returned to its initial value, and the volume of the liquid will have increased slightly.

From this discussion we see that the *vapor pressure of a liquid is independent of the volume of the container, provided that there is some liquid present so that an equilibrium can be established.*

Because the equilibrium vapor pressure is independent of the volume of the vapor, it is possible to condense a gas simply by decreasing its volume, provided the temperature of the gas is below its critical temperature. For example, suppose a sample of water vapor had a pressure of 500 torr at 100 °C. If this gas were compressed, it would obey Boyle's law and its pressure would rise until it reached 760 torr, the equilibrium vapor pressure of water at 100 °C. Further compression would then cause some of the water vapor to condense to a liquid, because only in this way could the pressure be prevented from rising above 760 torr.

If the temperature of the gas is above its critical temperature, then the gas must be cooled before it can be condensed. For example, methane cannot be liquefied at *room temperature* by increasing its pressure because room temperature (approximately 25 °C) is above methane's critical temperature. At room temperature, methane can exist in only one phase—as a gas. However, if methane is cooled first to −82.1 °C or lower, compression of the gas will eventually cause some to condense. At −82.1 °C, for instance, some liquid will begin to form when the pressure on the methane reaches 45.8 atm. As the volume continues to be decreased, more and more liquid will be formed and the pressure will remain constant at 45.8 atm, because that is the vapor pressure of methane at this temperature.

Vapor pressures of solids

In Section 12.3 you learned that solids are able to evaporate just like liquids; the process is called sublimation. In a crystalline solid the molecules vibrate about their equilibrium positions and continually undergo collisions with their nearest neighbors. This gives rise to a distribution of kinetic energies in the solid, just as it does in a liquid or gas. A small fraction of the molecules at the surface of a solid possess large kinetic energies, large enough to be able to overcome the attractive forces within the solid and break away from the surface, entering the gaseous phase above. That's how sublimation occurs.

When sublimation takes place in a closed container, more and more molecules enter the gaseous state and the pressure exerted by the vapor increases. Meanwhile, the slower-moving gaseous molecules that are colliding with the surface of the solid become trapped and return to the solid state. In time, the rate at which the molecules leave will exactly equal the rate at which they return, and a dynamic equilibrium will be established. The pressure exerted by a vapor in equilibrium with its solid is known as the **equilibrium vapor pressure of the solid.** Just as in liquids, the vapor pressure of a solid depends on the ease with which the molecules are able to enter the gaseous state. For example, the attractive forces are stronger in ionic solids than molecular

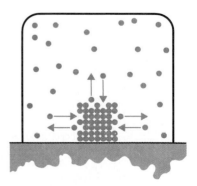

In a solid–vapor equilibrium, sublimation and condensation occur at equal rates.

Freeze-drying is a way of preventing foods from spoiling; thereafter they need no refrigeration. Freeze-dried foods such as these are often carried by campers.

solids and, as expected, we find that the vapor pressures of ionic solids are generally very much lower than those of molecular solids.

A practical application of sublimation that you've probably heard of is freeze drying. Freeze-dried instant coffee, for example, is manufactured by first freezing a batch of brewed coffee and then removing the ice component by vacuum. The vacuum creates an atmosphere of diminished pressure in which ice readily sublimes. Removing the water in this way preserves the delicate heat-sensitive molecules that give coffee its taste, so the quality of the product is improved. Solid foods, even entire meals, are also freeze-dried, both to protect their flavor as well as prevent spoilage, since bacteria that might cause harm cannot grow and multiply in the absence of moisture. Because bacteria are no threat, freeze-dried foods can be stored without refrigeration, and they are easily reconstituted by adding water to them. Campers often carry these products because of their convenience.

12.6 LE CHATELIER'S PRINCIPLE

The dynamic equilibrium between a liquid and its vapor can be represented by the equation

$$\text{liquid} \rightleftharpoons \text{vapor} \qquad (12.1)$$

Here the double arrows mean that the rates of evaporation and condensation are equal. If we in any way disturb this system so that the equilibrium is upset, a change occurs that tends to bring the system back to equilibrium. For example, we've seen that if we suddenly increase the volume of the vapor, the pressure drops and equilibrium is upset. In response, more of the liquid evaporates until the pressure is restored to its equilibrium value. In Equation (12.1), this corresponds to a change from left to right (i.e., liquid → vapor) and results in a new position of equilibrium in which there is less liquid and more vapor. In a sense, the position of equilibrium has shifted to the right in response to the disturbance imposed on it.

The way an equilibrium system behaves when it is disturbed can be predicted in a very simple fashion by applying a principle proposed in 1888 by the French chemist, Henry Le Châtelier (1850–1936). **Le Châtelier's principle** states that *when a system in a state of dynamic equilibrium is disturbed by some outside influence that upsets the equilibrium, the system responds by undergoing a change in a direction that reduces the disturbance and, if possible, brings the system back to equilibrium.*

To see how this works, let's analyze the effect that a volume increase has on a liquid–vapor equilibrium. When the piston of the apparatus in Figure 12.19 is pulled upward, the volume of the vapor is increased and the pressure drops. Since the pressure is no longer equal to the equilibrium vapor pressure, the system is no longer at equilibrium. How can this system counteract the disturbance and bring the pressure back to its equilibrium value? The answer, of course, is to produce more vapor, which will exert more pressure.[2] In other words, some additional liquid will evaporate and produce more vapor. We can also say that this volume increase shifts the position of equilibrium in Equation 12.1 to the right.

The effect of a temperature change on the liquid–vapor equilibrium can also be predicted with Le Châtelier's principle. This is simple if we include the energy change as part of the reaction. Since the conversion of liquid to vapor is endothermic, heat is absorbed during the change and can be included as a reactant in the equation.

$$\text{heat} + \text{liquid} \rightleftharpoons \text{vapor}$$

---

[2] If the volume of the vapor is increased sufficiently, all the liquid will evaporate and equilibrium cannot be reestablished. This will occur, for example, if the piston in Figure 12.19 is removed entirely so that the liquid is open to the atmosphere.

Now, suppose the temperature of a liquid–vapor equilibrium system is increased. This is accomplished by adding heat (with a Bunsen burner, for example). The system can counteract the buildup of heat energy by changing some liquid to vapor —a change that absorbs heat. The increased amount of vapor will exert more pressure, so our conclusion is that the vapor pressure will increase with increasing temperature (which, of course, it does). In summary, Le Châtelier's principle predicts that a temperature increase will shift the position of equilibrium in the direction of an endothermic change. It also is not difficult to see that a temperature decrease will shift an equilibrium in the direction of an exothermic change.

This analysis by Le Châtelier's principle is certainly much less tedious than our earlier analysis of the same phenomenon.

12.7 BOILING POINTS AND FREEZING POINTS

A glance at the photograph in Figure 12.20 tells you that the water in the pot is boiling. Large bubbles are forming within the liquid and rising to the surface. When a bubble is formed, the liquid that originally occupied this space is pushed aside and the level of the liquid in the container is forced to rise against the downward pressure exerted by the atmosphere. In other words, it is the pressure exerted by the vapor inside the bubble that pushes the surface of the liquid up against the opposing atmospheric pressure. This can occur only when the vapor pressure of the liquid becomes equal to the prevailing atmospheric pressure. If it were less, the atmospheric pressure would cause the bubble to collapse. The temperature at which the liquid boils, its **boiling point,** is therefore the temperature at which its vapor pressure equals the atmospheric pressure.

Figure 12.20. *Just a casual glance at the water in this pot tells you that it is boiling. Why?*

As long as bubbles are forming within the liquid — that is, as long as the liquid is boiling — the vapor pressure of the liquid is equal to the atmospheric pressure. Since the vapor pressure remains constant, the temperature of the boiling liquid also stays the same. An increase in the rate at which heat is supplied to the boiling liquid simply causes bubbles to form more rapidly. The liquid boils away more quickly, but the temperature does not increase.

It should be obvious from this discussion that the boiling point of a liquid depends on the prevailing atmospheric pressure: The larger the atmospheric pressure is, the higher the temperature must be to give a matching vapor pressure. The boiling point of a liquid at 1 atm (760 torr) is referred to as its **normal boiling point.** For water, the normal boiling point is 100 °C. At higher pressures its boiling point is greater; at lower pressures (for example, on a mountaintop) its boiling point is less. Boiling points given in reference tables are always normal boiling points, unless otherwise stated.

The normal boiling points of water, ethyl alcohol, and diethyl ether can be read from the graph in Figure 12.17.

The constant temperature maintained by a boiling liquid is relied on when we use water to cook foods. While water boils, its temperature remains constant. As long as water surrounds the food, we know that the food won't burn. A pressure cooker also takes advantage of the fact that the boiling point changes with pressure. These cookers are time-savers because they allow foods to be prepared at a much faster rate than possible in an open pot. The lid on a pressure cooker forms a tight seal on the pot and is equipped with a pressure-relief valve to prevent the pot from exploding. The heat supplied by the stove causes more and more liquid water to evaporate and the pressure inside the kettle increases until steam begins to exit from the relief valve. The pressure inside the cooker at this point is approximately 2 atm and the water boils at a temperature of about 120 °C, so foods cook faster.

The boiling point is another property that gives an indirect estimate of how strong the attractive forces are between the molecules in a liquid. Liquids whose molecules attract each other strongly tend to have high boiling points, and vice versa. In Table 12.1, for example, we see that among the hydrocarbons the boiling points rise along with the heats of vaporization.

Figure 12.21. *Boiling points of the hydrogen compounds of elements of Groups IVA, VA, VIA, and VIIA of the periodic table.*

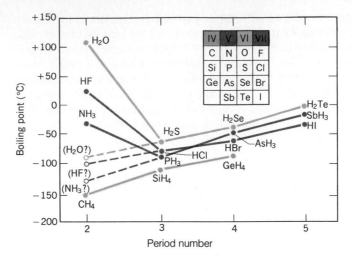

The dependence of boiling point on intermolecular attractive forces is also seen in Figure 12.21, in which the boiling points of some hydrogen compounds of the elements in Groups IVA, VA, VIA, and VIIA are compared. Let's look at the compounds of Group IVA first because they form a nearly ideal pattern. We see from the figure that as the size of the atoms increases going down the group (from CH_4 to GeH_4), the boiling points increase, corresponding to the increasing strengths of the London forces. Similar trends are observed among the hydrogen compounds of the elements in the other groups if we begin with the element in period 3. However, H_2O, NH_3, and HF have boiling points that are much higher than those we would expect if only London forces existed between their molecules, which we can estimate by extrapolation from the trends among the heavier elements in their respective groups (the dotted lines in Figure 12.21). Once again, we see the effects of hydrogen bonding, which exists among H_2O, NH_3, and HF, but not among the other hydrogen compounds of the nonmetallic elements.

You may have noticed that water has a higher boiling point than hydrogen fluoride, even though HF is more polar than H_2O. The reason seems to be the number of hydrogen bonds that they can form. In HF, each molecule is hydrogen bonded to two others, whereas in water, each molecule can be hydrogen bonded to four others (Figure 12.22). Therefore, even though HF forms somewhat stronger hydrogen bonds than water, the total strength of four hydrogen bonds to a water molecule exceeds the total strength of two hydrogen bonds to an HF molecule.

The hydrogen bonding in NH_3 is much weaker than in H_2O or HF because of the

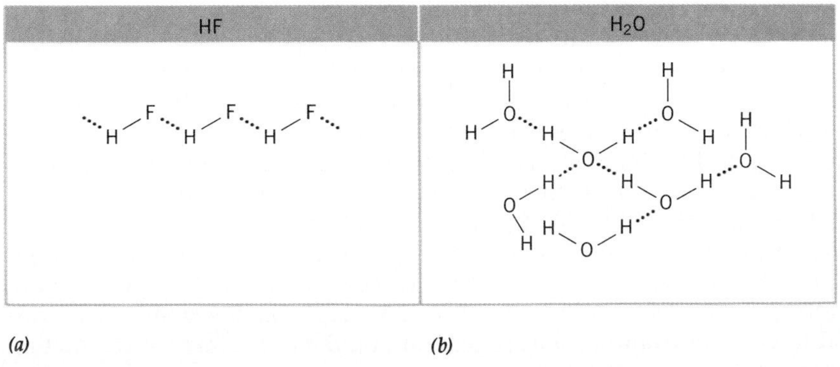

Figure 12.22. *Hydrogen bonding in HF and H_2O. (a) Each HF molecule has only one hydrogen atom that can hydrogen-bond to another HF molecule. (b) Each H_2O has two hydrogen atoms that can hydrogen-bond to other H_2O molecules.*

HF	H_2O

(a)   *(b)*

considerably lower electronegativity of nitrogen. In addition, the nitrogen in NH_3 has only one lone pair to which a hydrogen bond can be formed, so each molecule, on the average, can be held by only two hydrogen bonds: one *from* another NH_3 molecule and one *to* another NH_3 molecule. Altogether, the total strength of the hydrogen bonding in NH_3 is so small that NH_3 has a lower boiling point than either HF or H_2O.

Freezing point

Anyone who has ever made a tray of ice cubes in a refrigerator realizes that liquids freeze if heat is removed from them. You also know that ice cubes melt when they absorb heat. For any substance at a given pressure there is a characteristic tempera-ture at which the liquid and solid can coexist in equilibrium. This is called either the **freezing point** or **melting point,** depending on whether you imagine approaching it from a high or low temperature. At the freezing point (melting point) the rate at which particles leave the solid and enter the liquid is the same as the rate at which particles leave the liquid and join the solid. If heat is added, some solid melts and more liquid is formed, but the temperature stays the same as long as both phases are present. Similarly, if some heat is removed, some liquid freezes and more solid forms — again without a temperature change.

As with evaporation and condensation, there are energy changes associated with freezing and melting. The **molar heat of crystallization,** ΔH_{cryst}, *is the amount of energy that must be removed from one mole of a liquid to convert it to a solid at the same temperature.* For melting, which is also called *fusion,* there is the **molar heat of fusion,** ΔH_{fus}, *the energy needed to melt one mole of a solid.* Following the definition given for the molar heat of vaporization, we can express these as

$$\Delta H_{cryst} = H_{solid} - H_{liquid}$$

$$\Delta H_{fus} = H_{liquid} - H_{solid}$$

We see that the numerical values of ΔH_{cryst} and ΔH_{fus} must be identical; one is simply the negative of the other.

The size of the molar heat of fusion (or crystallization) is a measure of the difference in the strengths of the attractive forces between the liquid and solid, and it is always much smaller than the molar heat of vaporization, as shown in Table 12.3. When a solid melts, there are relatively small changes in the distances between the molecules. Therefore, only small changes in potential energy are involved. When a liquid is converted to a gas, however, the intermolecular distances increase tremen-dously and large energy changes occur. This means that the amount of energy (ΔH_{fus}) required to cause the molecules of a solid to overcome their attractive forces and form a liquid is small compared to the energy (ΔH_{vap}) required for liquid mole-cules to move apart, forming a gas.

At the freezing point (melting point), there is an equilibrium between solid and liquid.

liquid $\rightleftharpoons$ solid + heat

The normal freezing point of a substance can be defined as the freezing point at a pressure of 1 atm. However, the freezing point varies imperceptibly over pressure ranges ordinarily en-countered in the laboratory.

The actual values of H_{solid} and H_{liquid} cannot be measured. Only their difference can be obtained experimentally.

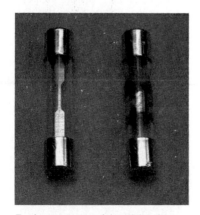

Fusion means melting. The thin metal band in an electrical fuse protects a circuit by melting if too much current is passed through it. On the right is a fuse that has done its job.

Table 12.3 Heats of fusion and vaporization

Substance	ΔH_{fus} (kJ/mol)	ΔH_{vap} (kJ/mol)
Water	5.98	40.6
Benzene	9.92	30.7
Chloroform	12.4	31.9
Diethyl ether	6.86	26.0
Ethanol	7.61	38.6

EXAMPLE 12.2. USING THE HEAT OF FUSION

PROBLEM: Calculate the energy, in kilojoules, necessary to melt 1.00 g of ice.

SOLUTION: As before, we have a unit conversion requiring ΔH_{fus}. The problem reduces to

$$1.00 \text{ g } H_2O \Longleftrightarrow (?) \text{ kJ}$$

From Table 12.3, $\Delta H_{fus} = 5.98$ kJ/mol for H_2O. Therefore,

$$1.00 \text{ g } H_2O \times \left(\frac{1 \text{ mol } H_2O}{18.0 \text{ g } H_2O}\right) \times \left(\frac{5.98 \text{ kJ}}{1 \text{ mol } H_2O}\right) \Longleftrightarrow 0.332 \text{ kJ}$$

The energy needed is 0.332 kJ.

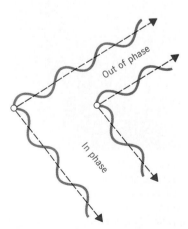

Figure 12.23. *Snowflakes are crystals of ice. Like other crystals, they have regular features that reflect the ordered arrangement of the particles within them.*

12.8 CRYSTALLINE SOLIDS

When most substances freeze, or when they are created in a precipitation reaction, they form crystals that have highly regular, symmetrical shapes. You have probably seen photographs of snowflakes similar to the one in Figure 12.23. Notice how symmetrical the ice crystal is and how it has a characteristic hexagonal form.

The highly regular surface features of a crystal are a reflection of an orderly repeating pattern of atoms, molecules, or ions that exist within it. This order has made possible detailed analyses of the structures of solids and has led to much of our knowledge of the shapes of molecules and the sizes of atoms and ions.

X-ray diffraction

In 1912 a German physicist named Max von Laue pointed out that a crystal could serve as a three-dimensional diffraction grating if the wavelength of the incident radiation was of the same order of magnitude as the distance between particles in the solid. This condition is fulfilled by X rays, which have wavelengths of approximately 0.1 nm (100 pm or 1 Å).

When a crystal is bathed in X rays, each atom of the crystal within the path of an X ray absorbs some of its energy and then reemits it in all directions. Thus, each atom is a source of secondary wavelets, and the X rays are said to be scattered by the atoms. These secondary wavelets from the different sources interfere with each other, either reinforcing or cancelling each other. In certain directions the waves emanating from nearly all the atoms in any orderly array are in phase—that is, the peaks and troughs of the waves coincide as shown in Figure 12.24—and intense beams of X rays are observed in these directions. In all other directions the waves from various atoms are out of phase and cancel each other, so no intensity is detected.

Two English scientists, William Bragg and his son Lawrence, treated the diffraction of X rays as if the process were reflection. In Bragg's treatment, the X rays that penetrate a crystal are thought of as being reflected by successive layers of particles within the substance (Figure 12.25). We can see from this diagram that beams reflected from deeper layers must travel farther to reach the detector. For there to be any intensity at the detector these waves have to be in phase with those reflected from the upper layers, which must mean that the extra distance traveled by the more penetrating beam has to be some integral multiple of the wavelength of the X rays.

Figure 12.24. *X rays that are scattered by atoms in a crystal are in phase in some directions and out of phase in other directions.*

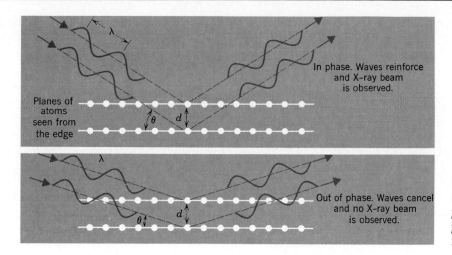

In phase. Waves reinforce and X-ray beam is observed.

Planes of atoms seen from the edge

Out of phase. Waves cancel and no X-ray beam is observed.

Figure 12.25. *Bragg showed that X rays reflected from different planes of atoms in a crystal are in phase only when the equation* 2d sin θ = nλ *is fulfilled.*

Bragg showed that in order to observe any intensity in the emerging X rays, a relatively simple relationship had to be fulfilled. This relationship, known as the **Bragg equation,** is

$$2d \sin \theta = n\lambda \qquad (12.2)$$

where d is the spacing between the successive layers that are reflecting the X rays, θ is the angle at which the X rays enter and leave the particular set of layers, λ is the wavelength of the X rays, and n is an integer (that is, $n = 1, 2,$ or 3, etc.). The Bragg equation serves as the basis for the study of crystalline structure by X-ray diffraction.

In practice, X rays of known wavelength are directed at a crystal and the angles at which they are reflected are recorded — for example, on a piece of photographic film (Figure 12.26). By measuring the angles at which the X rays are reflected, it is a simple matter to calculate the distances between planes of atoms within a crystal, as illustrated in Example 12.3. If, in addition, the intensities of the reflected X rays are measured, a crystallographer may be able to deduce, through a rather complex procedure, the actual positions of atoms within the solid. In this way the molecular structures of many substances have been found. In recent years X-ray diffraction has become a powerful tool in biochemistry by which the structures of even very complex molecules have been investigated. In fact, Rosalind Franklin's X-ray data led

In the older scientific literature, X-ray wavelengths and atomic dimensions are given in angstroms. Today, the SI units nanometers or picometers are preferred.

$$1 \text{ Å} = 0.1 \text{ nm}$$
$$1 \text{ Å} = 100 \text{ pm}$$

Dorothy Hodgkin won the Nobel prize in 1964 for her X-ray structure determination of vitamin B_{12}.

Photographic plate

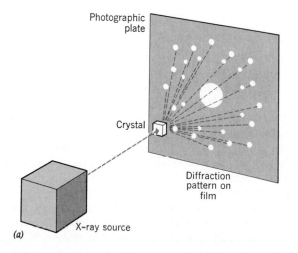

Crystal

Diffraction pattern on film

X-ray source

(a)

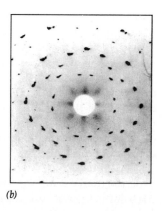

(b)

Figure 12.26. (a) *The production and recording of an X-ray diffraction pattern.* (b) *The X-ray diffraction pattern for sodium chloride.*

James Watson, Francis Crick, and Maurice Wilkins to their deduction of the double-helix structure of DNA — a feat that won Watson, Crick, and Wilkins the Nobel prize in 1962.

EXAMPLE 12.3. WORKING WITH BRAGG'S LAW

PROBLEM: X rays of wavelength 154 pm (0.154 nm) strike a crystal and are observed to be reflected at an angle of 22.5°. Assuming that $n = 1$, calculate the spacing in picometers between the planes of atoms that are responsible for this reflection.

SOLUTION: We wish to calculate d. Solving Equation 12.2 for d, we have

$$d = \frac{n\lambda}{2 \sin \theta}$$

From the data, $n = 1$, $\lambda = 154$ pm, $\theta = 22.5°$. Substituting, we get

$$d = \frac{(1)(154 \text{ pm})}{2 \sin(22.5)}$$

$$= \frac{154 \text{ pm}}{2(0.383)}$$

$$= 201 \text{ pm}$$

12.9 LATTICES AND CRYSTAL STRUCTURES

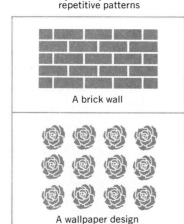

Symmetry in some repetitive patterns

A brick wall

A wallpaper design

Packing of atoms in a crystal

Any repetitive pattern has a symmetrical aspect about it, whether it be the stacking of bricks in a wall, a wallpaper design, or the orderly packing of particles in a crystal. For example, certain repeating distances between the elements of the pattern can be easily recognized, and the lines along which the elements of the pattern are repeated are at certain angles to each other.

To avoid having to deal with the details of a repeating structure and concentrate on its symmetrical features, it is convenient to describe the structure simply in terms of a set of points that have the same repeat distances as the structure, arranged along lines oriented at the same angles. This kind of pattern of points is called a **lattice,** and when applied to the description of the packing of particles in a solid, we often use the term **crystal lattice.**

In a crystal the number of particles is enormous. If you could imagine being in the center of even the tiniest crystal, you would find that the particles go on as far as you could see in every direction. Describing the positions of all these particles or their lattice points is impossible and, fortunately, unnecessary. All we need to do is to describe the basic repeating unit of the lattice, which we call a unit cell. To see this, let's begin in two dimensions.

Figure 12.27a illustrates a two-dimensional lattice. It is a *square lattice* because the lattice points lie at the corners of squares. The repeating unit of this lattice, its **unit cell,** is the square drawn in white. If we began with this unit cell, we could produce the entire lattice simply by moving it repeatedly left and right, and up and down by distances equal to its edge length. In this sense, all the properties of a lattice are contained in the properties of its unit cell.

An important fact about lattices is that the same *kind* of lattice can be used to describe many different designs. For example, if we decided to place a diamond at each lattice point, we could create a wallpaper design like that shown in Figure

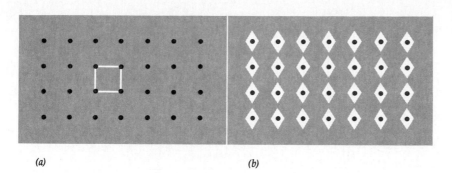

Figure 12.27. (a) *A portion of a square lattice.* (b) *A design based on a square lattice.*

(a)   (b)

12.27b. A different wallpaper pattern could be created by placing a rose at each lattice point, or by changing the lengths of the edges of the unit cell.

The extension of the lattice concept to crystals in three dimensions is quite straightforward. In Figure 12.28 we see an example of a simple cubic lattice in which a unit cell is shaded in color. By associating a particular chemical environment with each lattice point, we can arrive at a chemical structure, and by varying the chemical environment about each point, we can create an infinite number of chemical structures, all based on the same lattice.

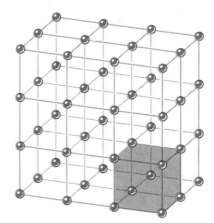

Figure 12.28. *A simple cubic lattice. The unit cell is shaded in color.*

The unit cells of all three-dimensional lattices are similar in that they all have eight corners, as shown in Figure 12.29. Unit cells differ by the lengths of their edges (a, b, and c) and the angles opposite them (α, β, and γ). In 1848 Auguste Bravais showed that *only* 14 different kinds of lattices are possible. These can be divided among seven basic crystal systems, whose properties are described in Table 12.4.

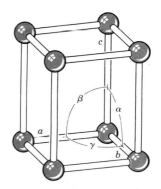

Figure 12.29. *A three-dimensional unit cell has lattice points at its eight corners. The edges* a, b, *and* c *intersect at angles* α, β, *and* γ.

Table 12.4. Properties of the unit cells of the seven crystal systems

System	Edge Lengths	Angles
Cubic	$a = b = c$	$\alpha = \beta = \gamma = 90°$
Tetragonal	$a = b \neq c$	$\alpha = \beta = \gamma = 90°$
Orthorhombic	$a \neq b \neq c$	$\alpha = \beta = \gamma = 90°$
Monoclinic	$a \neq b \neq c$	$\alpha = \beta = 90° \neq \gamma$
Triclinic	$a \neq b \neq c$	$\alpha \neq \beta \neq \gamma$
Rhombohedral	$a = b = c$	$\alpha = \beta = \gamma \neq 90°$
Hexagonal	$a = b \neq c$	$\alpha = \beta = 90°; \gamma = 120°$

Figure 12.30. *The three unit cells belonging to the cubic system. (a) Simple cubic. (b) Body-centered cubic. (c) Face-centered cubic.*

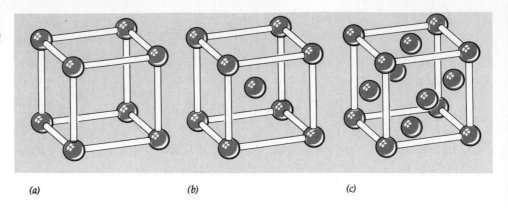

(a)   *(b)*   *(c)*

What this means is that the crystals of all the millions of different chemical compounds ever discovered can be described by just this small set of lattices. It is this fact that has made a detailed understanding of the solid state possible.

Three common kinds of lattices are characterized by the cubic unit cells shown in Figure 12.30. The simplest is called a **simple cubic unit cell** or a **primitive cubic unit cell.** It has lattice points only at the corners. When oxygen is frozen, it has a primitive cubic lattice with this kind of unit cell.

A **body-centered cubic unit cell** has lattice points at the corners, plus a lattice point in the center of the cell. Some common metals—chromium, iron, and tungsten, for example—form crystals with a body-centered cubic lattice.

A **face-centered cubic unit cell** has lattice points at the eight corners and one in the center of each of its six faces. This produces a very common kind of lattice that is found in crystals of such metals as nickel, copper, silver, gold, and aluminum.

Figure 12.31 illustrates the unit cells of copper and gold, sliced out of crystals of these metals. Notice that only parts of atoms are located at the corners and the face centers of the unit cells. The rest of these atoms are located in the unit cells that surround the ones we've selected. One of the things to notice here is how similar the structures are; they have the same kind of lattice. Only the lengths of the unit cell edges are different because of the differences in the sizes of copper and gold atoms.

Cubic lattices are not only characteristic of many elements. Some important and familiar compounds also have cubic lattices. For example, one of our most common compounds, sodium chloride, forms crystals having a face-centered cubic lattice. A portion of a NaCl crystal is shown in Figure 12.32 along with its unit cell. Notice how

Figure 12.31. *Copper and gold crystallize in a face-centered cubic lattice. Their atoms are arranged the same way, but the edges of their unit cells differ in length because gold atoms are larger than copper atoms.*

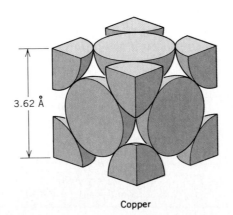

3.62 Å

Copper

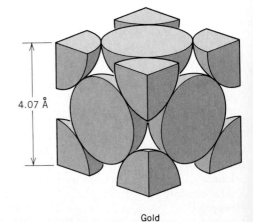

4.07 Å

Gold

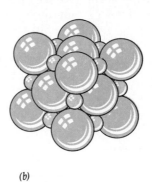

(a)

(b)

Figure 12.32. *The crystal structure of sodium chloride (rock salt). (a) A portion of a NaCl crystal showing the packing of chloride ions (green) and sodium ions (gray). (b) The arrangement of ions in the face-centered cubic unit cell of NaCl. The unit cell could also have been chosen with Na$^+$ at the lattice sites and the Cl$^-$ in between.*

the chloride ions are located at positions corresponding to the lattice points, with the sodium ions squeezed in between. This kind of packing of cations and anions is called the *rock salt structure.* Similar structures are found for many of the other alkali halides as well—for instance, KCl and LiCl.

Among the factors that determine the kind of lattice and structure an ionic compound can form are the relative sizes of the ions and the ratio of the numbers of anions to cations in the crystal. The question of size is rather complex, so we won't discuss it further, but the importance of the anion–cation ratio is not very difficult to see.

Since a crystal is composed of a large number of unit cells, whatever the anion-to-cation ratio is in the crystal as a whole, it must also be the same in the unit cell. Let's count the number of chloride and sodium ions in the unit cell of NaCl to show that their ratio is one-to-one. In doing this, however, we have to be careful, because ions at the corners, along the edges, and in the face centers are shared with one or more other unit cells.

Crystals of ordinary table salt are cubic because NaCl crystallizes with a cubic lattice.

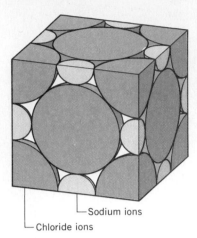

Sodium ions
Chloride ions

Figure 12.33. *The faces of the unit cell of NaCl slice through the ions at the corners, edges, and centers of the faces. Only parts of these ions are within this unit cell.*

This does not mean that all substances with a 1-to-1 anion–cation ratio must crystallize with the rock salt structure.

An ion at the corner of the unit cell is shared with seven others. In Figure 12.33 we see that only one-eighth of such an ion is in a given unit cell. We also see that an ion along an edge, which is shared among four unit cells, has only one-fourth of itself in a given unit cell. An ion in the center of a face contributes half to a given unit cell, because it is shared between two of them. In addition to these, there is one Na^+ ion that can't be seen in Figure 12.33. It is located in the center of the unit cell, as shown in Figure 12.34.

Now we can count the ions. For the chloride ions, we see that they are at the eight corners and in the centers of the six faces.

$$8 \text{ corners} \times \tfrac{1}{8}Cl^- \text{ per corner} = 1Cl^-$$
$$6 \text{ faces} \times \tfrac{1}{2}Cl^- \text{ per face} \quad = \underline{3Cl^-}$$
$$\text{Total} \quad 4Cl^-$$

Thus, altogether there are four chloride ions contained within the unit cell. For the sodium ions, we have one along each of the 12 edges of the cube, which each contribute one-quarter to the unit cell, plus one in the center that is entirely within the unit cell.

$$12 \text{ edges} \times \tfrac{1}{4}Na^+ \text{ per edge} = 3Na^+$$
$$1Na^+ \text{ in the center} \quad = \underline{1Na^+}$$
$$\text{Total} \quad 4Na^+$$

The number of sodium ions in the unit cell is also four. This means, therefore, that the Na^+ and Cl^- ions are in a one-to-one ratio, which is necessary for the crystal to be electrically neutral.

Any substance that crystallizes with the rock salt structure *must* have a one-to-one anion–cation ratio. Sodium chloride and the other alkali halides have formulas that satisfy this condition, and many of them form crystals having this structure. Calcium oxide, CaO, also has the rock salt structure, which is allowed because of its formula. However, $CaCl_2$ or Al_2O_3 could not possibly form crystals with the rock salt structure because their anion–cation ratios forbid it. Thus, we see that the formula of a compound places certain restrictions on the kinds of crystal structures it can and cannot have.

Na^+ion in the center of the unit cell

Figure 12.34. *An exploded view of the NaCl unit cell showing the Na^+ ion in the center.*

EXAMPLE 12.4. CALCULATING THE NUMBER OF IONS PER UNIT CELL

PROBLEM: Sodium forms crystals having a body-centered cubic lattice. How many sodium atoms are contained within one unit cell?

SOLUTION: The unit cell has sodium atoms at the eight corners, each of which is shared among a total of eight cells, plus one sodium atom in the center.

$$8 \text{ corners} \times \tfrac{1}{8}\text{Na per corner} = 1\text{Na}$$
$$1\text{Na in the center} \qquad\quad\; = \underline{1\text{Na}}$$
$$\text{Total} = 2\text{Na}$$

Each unit cell has two sodium atoms within it.

Atomic and ionic radii

The study of the structures of crystals has yielded much useful information. The applications to the determination of molecular structures — even those as complex as DNA — have already been mentioned, although they are far too complex to discuss in any detail here. We can see an illustration of the usefulness of studying crystals, however, by examining how information about unit cells can be used to calculate the radii of atoms and ions.

It was noted in the last section that copper is a metal that crystallizes with a face-centered cubic lattice. X-ray diffraction measurements show that the unit cell has an edge length of 362 pm (3.62 Å) (this is length AC in Figure 12.35). Copper atoms are in contact along the line joining points A and B (the face diagonal). This distance corresponds to four times the radius, r, of a copper atom. From geometry we know that

$$\overline{AB} = \sqrt{2}(\overline{AC}) = \sqrt{2}(362 \text{ pm})$$
$$= 512 \text{ pm}$$

Therefore,

$$4r = 512 \text{ pm}$$
$$r = 128 \text{ pm}$$

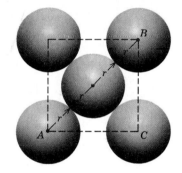

Figure 12.35. *Copper atoms in the face of a unit cell.*

The radius of a copper atom in metallic copper is therefore 128 pm (1.28 Å or 0.128 nm). In a similar fashion we can also use the results of X-ray diffraction to determine the radii of ions in ionic crystals.

12.10 CRYSTAL TYPES AND PROPERTIES

We have seen that there is only a limited number of ways of arranging particles in a crystalline solid. The particular arrangements, as well as the physical properties of the solid, are determined by the types of particles present at the lattice points and the nature of the attractive forces between them. As a result, we can divide crystals into distinct types that are each characterized by certain kinds of properties.

Molecular crystals

In molecular crystals, either molecules or individual atoms occupy lattice sites. The attractive forces between them are of the type described in Section 12.2 and are much weaker than the covalent bonds that exist within individual molecules. London forces are present in crystals of nonpolar substances such as Ar, O_2, naphthalene (moth crystals), and CO_2 (dry ice). In crystals of polar molecules such as SO_2, there are also dipole–dipole attractions, and in solids such as ice (H_2O), NH_3, and HF, the molecules are held in place primarily by hydrogen bonding. Since these are relatively weak forces (compared to covalent or ionic attractions), molecular crystals tend to have small lattice energies and are easily deformed; we say that they are soft. Also, relatively little thermal energy is required to overcome these attractions, and molecular solids generally tend to have low melting points.

Molecular crystals are poor conductors of electricity because the electrons are bound to individual molecules and are not able to move freely through the solid.

Ionic crystals

In an ionic crystal, such as NaCl, there are ions located at lattice sites and the binding between them is mainly electrostatic (which is essentially nondirectional). As a result, the kind of lattice that is formed is determined mostly by the relative sizes of the ions and their charges. When the crystal forms, the ions arrange themselves to maximize attractions and minimize repulsions.

Because electrostatic forces are strong, ionic crystals have large lattice energies. They are often hard and are characterized by relatively high melting points. They are also very brittle. When struck, they tend to shatter, because as planes of ions slip by one another, they pass from a condition of mutual attraction to one of mutual repulsion. You may recall that this was described in Section 4.7. Figure 4.14 on page 103 illustrates the principle.

In the solid state, ionic compounds are poor conductors of electricity because the ions are held rigidly in place. When melted, however, the ions are free to move about and ionic substances become good conductors.

Covalent crystals

In a covalent crystal there is a network of covalent bonds between the atoms that extends throughout the entire solid. An example of such a substance is diamond, whose structure is shown in Figure 12.36. Diamond is a form of elemental carbon in which each atom is covalently bonded to four nearest neighbors. Other common examples are carborundum (silicon carbide, SiC) and quartz (silicon dioxide, SiO_2, commonly recognized as a major constituent of many sands).

Because of the interlocking framework of covalent bonds, covalent crystals have very high melting points and are usually extremely hard. Diamond, of course, is the hardest substance known and is used in grinding and cutting tools. Silicon carbide is like diamond, except that half of the carbon atoms in the structure have been replaced by silicon atoms. It too is very hard and is used as an abrasive in sandpaper, as well as in other grinding and cutting applications.

Covalent crystals are poor conductors of electricity because the electrons in the solid are localized in the covalent bonds and are not free to move through the crystal.

Figure 12.36. *The unit cell of diamond. Notice how the carbon atoms inside the unit cell are each covalently bonded to four other carbon atoms. The carbon atoms at the corners are each similarly bonded.*

Metallic crystals

The simplest picture of a metallic crystal has positive ions (nuclei plus core electrons) situated at lattice points, with the valence electrons belonging to the crystal as a whole instead of to any single atom (Figure 12.37). The solid is held together by the electrostatic attraction between the lattice of positive ions and this sort of "sea of electrons." These electrons can move freely, so we find metals to be good conductors

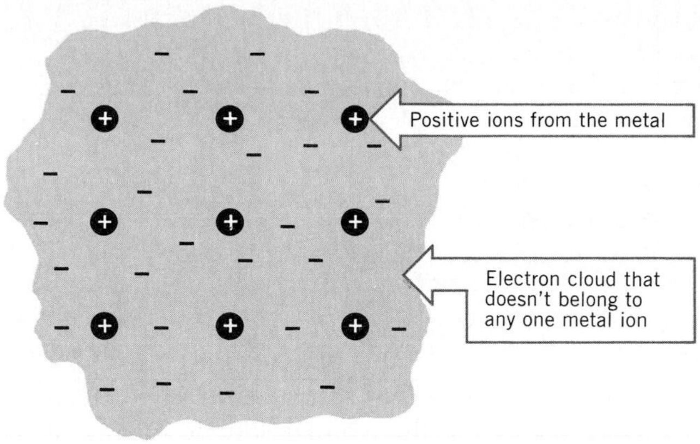

Figure 12.37. *The "electron sea" model of a metallic crystal.*

Positive ions from the metal

Electron cloud that doesn't belong to any one metal ion

of electricity. Since the melting points and hardness of metals vary over wide ranges, there must also be, at least in some cases, some degree of covalent bonding between atoms in the solid.

Table 12.5 summarizes the properties of these different kinds of solids.

Table 12.5. Types of solids

	Molecular	**Ionic**	**Covalent**	**Metallic**
Chemical units at lattice sites	Molecules or atoms	Positive and negative ions	Atoms	Positive ions
Forces holding the solid together	London forces, dipole–dipole, hydrogen bonds	Electrostatic attraction between + and − ions	Covalent bonds	Electrostatic attraction between + ions and electron "sea"
Some properties	Soft, generally low melting, nonconductors	Hard, brittle, high melting, nonconductors (but conduct when melted)	Very hard, high melting, nonconductors	Hard to soft, low to high melting, high luster, good conductors
Some examples	CO_2 (dry ice) H_2O (ice) $C_{12}H_{22}O_{11}$ (sugar) I_2	NaCl (salt) $CaCO_3$ (limestone; chalk) $MgSO_4$ (in Epsom salt)	SiC (carborundum) C (diamond) WC (tungsten carbide, used in cutting tools)	Na, Fe, Cu, Hg

12.11 HEATING AND COOLING CURVES; CHANGES OF STATE

When heat is added to a solid, initially at some temperature below its melting point, the temperature begins to rise. Once the melting point is reached, the temperature levels off — it stays constant until all the solid has melted. When more heat is added, the liquid becomes warmer until it reaches its boiling point. As heat is supplied to the boiling liquid, the temperature stays constant again until the liquid has all been converted to a gas. Then the temperature can rise further as still more heat is added. Figure 12.38, which illustrates this graphically for one mole of a typical substance, is called a **heating curve.**

Figure 12.38. *A typical heating curve for one mole of a substance. T_f is the freezing point (or melting point) and T_b is the boiling point.*

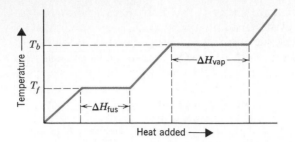

In those portions of the heating curve where the temperature is rising, heat that is added is increasing the average kinetic energy of the molecules. But what is happening during melting and boiling—those portions of the curve where the temperature stays the same? Because the temperature stays constant, the average kinetic energy also remains the same. This means that the heat energy being added must be increasing the potential energies of the molecules.

If the K.E. doesn't change, then the P.E. must change if energy is added.

When most solids melt, there is a slight expansion in volume. This means that the particles must be getting slightly further apart. Since there are attractive forces between the particles, energy must be supplied to separate them.[3] This energy is the heat of fusion. Similarly, when a liquid vaporizes, the molecules go from the closely packed liquid state to their widely spaced distribution in a vapor. This also requires an energy input to overcome the attractive forces—the heat of vaporization. In both melting and vaporization it is the distance between molecules that attract each other that is changing, and as we learned in Chapter 6, this involves changes in potential energy. Thus, ΔH_{fus} and ΔH_{vap} are energy changes that affect the potential energies of particles.

A cooling curve, like that in Figure 12.39a, is essentially the opposite of a heating curve. During condensation and freezing, the temperature stays constant while the potential energies of the particles decrease as they come together.

Some liquids do not follow a smooth transition into the solid state, but instead give rise to a cooling curve such as that shown in Figure 12.39b. As the temperature of the liquid drops, it eventually reaches point A, the expected freezing point of the substance. The molecules, however, may not be oriented properly to fit into the

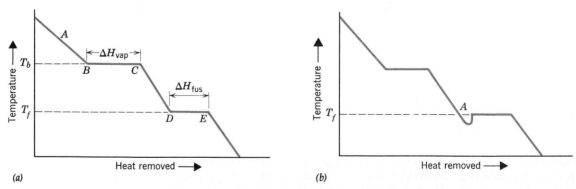

Figure 12.39. (a) *A typical cooling curve for one mole of a substance.* (b) *Supercooling. As the liquid is cooled, its temperature drops below the freezing point. After a short time freezing begins, and the temperature rises to the freezing point and stays there until all the liquid has frozen.*

[3] A most important exception to this, of course, is ice, in which the solid is less dense (more expanded) than the liquid. Here energy supplied to the solid disrupts some of the hydrogen bonding that exists in the solid ice, and the open structure of ice collapses to give a denser liquid.

crystalline lattice, and random motion continues as heat is further withdrawn from the liquid. As a result, the temperature of the liquid drops below its expected freezing point and the liquid is said to be **supercooled.** Once a small number of molecules have achieved the correct pattern, a tiny crystal is formed that serves as a seed on which additional molecules may rapidly accumulate. Potential energy is suddenly released as this crystal quickly grows, and the energy that is evolved increases the average kinetic energy of the molecules in the liquid and solid. Therefore, the temperature of the system rises again until it returns to the freezing point, after which the substance behaves normally. Further removal of heat eventually leads to the complete conversion of the liquid to a solid.

Some substances, such as glass, rubber, and many plastics, never do achieve a crystalline state when their liquids solidify on cooling. These compounds consist of long chainlike molecules that intertwine in the liquid. As they are cooled, their molecules move so slowly that they never do find the proper orientation to form a crystalline solid, and an **amorphous solid** results instead. The term amorphous comes from the Greek word meaning "without shape," and if you've ever examined pieces of broken glass, you will recall that the surfaces are not uniform and flat like those of a crystal. Instead, they curve in unpredictable ways. This is because amorphous "solids" are actually supercooled liquids and lack the internal order found within crystals.

A property of supercooled liquids is that they continue to flow, although very slowly, even at room temperature. For example, very old glass exhibits greater crystallinity, when examined by X-ray diffraction, then freshly formed glass does, showing that molecules are slowly finding their way into a crystalline lattice. A supercooled liquid familiar to many children is Silly Putty. In many ways it behaves like a solid, particularly when forced to flow rapidly (for example, it breaks when pulled apart suddenly). In other ways it flows like a liquid. Supercooled liquids such as glass, Silly Putty, and plastics in general do not have sharp, well-defined melting points, but instead gradually soften when heated.

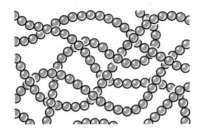

In an amorphous solid, long chainlike molecules become tangled and cannot crystallize.

12.12 PHASE DIAGRAMS

The vapor pressure of a solid, like that of a liquid, increases with increasing temperature. (The reasons are the same and are easily analyzed by applying Le Châtelier's principle.) This increase in vapor pressure with temperature continues until eventually the solid melts, as illustrated in Figure 12.40 for water. On this graph we have plotted two equilibrium lines. The lower one is the vapor pressure curve for solid

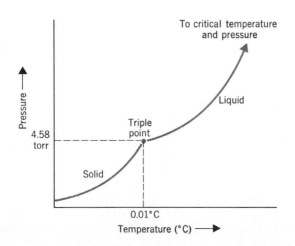

Figure 12.40. *The vapor pressure curves for solid and liquid water intersect at the triple point.*

Figure 12.41. *The phase diagram for water, somewhat distorted to emphasize certain features.* T_f *is the normal freezing point and* T_b *is the normal boiling point. The slope of the solid–liquid equilibrium line is exaggerated. The actual slope is much less to the left; it takes approximately 130 atm to lower the melting point of ice by 1 °C.*

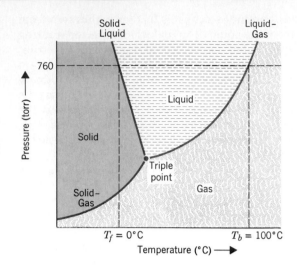

It appears that at atmospheric pressure, helium remains liquid all the way to absolute zero.

water; the upper one is the vapor pressure curve for the liquid. These two curves intersect at a temperature and pressure called the **triple point.** At the triple point there is an equilibrium between all three states: solid, liquid, and gas. In other words, at this unique temperature and pressure, all three states exist simultaneously in equilibrium with one another.

Every pure substance except helium has a triple point, and its temperature and pressure are determined by the nature of the attractive forces between the particles. For example, the triple point of water, which has strong intermolecular attractions, occurs at a temperature of 0.01 °C and a pressure of 4.58 torr. For carbon dioxide, however, the triple point occurs at −57 °C and a pressure of 5.2 atm.

There is still another equilibrium that can be represented on the same graph. This line corresponds to the combinations of temperatures and pressures that must be maintained to achieve a solid–liquid equilibrium. At a pressure of 1 atm, the melting point of water is 0 °C; therefore, the solid–liquid equilibrium line passes through both the triple point and the normal melting point, as shown in Figure 12.41. This graph is called a **phase diagram** because it allows us to pinpoint temperatures and pressures at which the various phases exist, as well as those conditions under which equilibrium can occur. For instance, at a pressure of 1 atm, water exists as a solid at all temperatures below 0 °C, and in fact the region bounded by the solid–liquid equilibrium line and the solid–gas equilibrium line corresponds to all the temperatures and pressures at which water exists as a solid. Similarly, in the region bounded by the solid–liquid and liquid–gas equilibrium lines, the substance can exist only as a liquid, while to the right of both the solid–gas and liquid–gas lines the substance must be a gas.

Table 12.6 lists some randomly chosen temperatures and pressures and the physical states of water that we can predict from its phase diagram. You might verify these predictions to illustrate how to use a phase diagram.

Table 12.6. Physical states of water at randomly chosen temperatures and pressures

Temperature (°C)	Pressure (atm)	State
25	1.0	Liquid
0	2.0	Liquid
0	0.5	Solid
100	0.5	Gas

EXAMPLE 12.5. INTERPRETING A PHASE DIAGRAM

PROBLEM: What phase of water would exist at a temperature of 5 °C and a pressure of 4 torr?

SOLUTION: To answer this question, we have to locate 5 °C and 4 torr on the phase diagram. Here it is best to examine Figure 12.40, which gives a more detailed view of the region around the triple point. Notice that 4 torr is below the pressure corresponding to the triple point, and that 5 °C lies to the right of the triple point. This places this point in the "gas" region of the phase diagram, so at this temperature and pressure, water is a gas.

To gain a further insight into the meaning of a phase diagram, let us follow the changes that take place as we move along a line of constant pressure, say 1 atm, by varying the temperature. In Figure 12.42, point *A* lies in the region of the diagram where a sample of the substance would exist entirely as a solid, as shown in Figure 12.43*a*. When the temperature rises to point *B* in Figure 12.42, the solid begins to melt and an equilibrium between the solid and liquid can occur (Figure 12.43*b*). At a still higher temperature, point *C*, all the solid will have been converted to a liquid (Figure 12.43*c*); and when the liquid–gas line is encountered at point *D* in Figure 12.42, vapor may at last begin to form and an equilibrium can exist (Figure 12.43*d*). Finally at a sufficiently high temperature, such as point *E*, all the water will exist in the gaseous state (Figure 12.43*e*).

We could also proceed with a similar analysis in which the temperature is held constant and the pressure is permitted to change. For example, at point *F* in Figure 12.42, the water would exist entirely as a gas (Figure 12.44*a*). At a higher pressure, point *G* in Figure 12.42, a solid–vapor equilibrium would exist (Figure 12.44*b*), and above that pressure, at point *H*, all the water would be converted to a solid (Figure 12.44*c*). As the pressure is increased further, we encounter the solid–liquid line at point *B* in Figure 12.42, where we again have an equilibrium as represented by Figure 12.44*d*. At still higher pressures, the water will melt so that at point *I* all the water is present in the liquid state (Figure 12.44*e*).

In the phase diagram for water, we see that the solid–liquid equilibrium line slants to the left. This is a direct consequence of the fact that liquid water at 0 °C has a higher density than the solid. Le Châtelier's principle requires that an increase in pressure on a system at equilibrium lead to the production of the denser phase; that

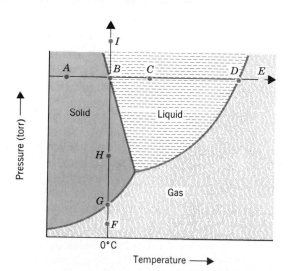

Figure 12.42. *The phase diagram for water, not drawn to scale.*

Figure 12.43. *Increasing the temperature of water at a constant pressure of 760 torr. Temperatures correspond to points A to E in Figure 12.42.*

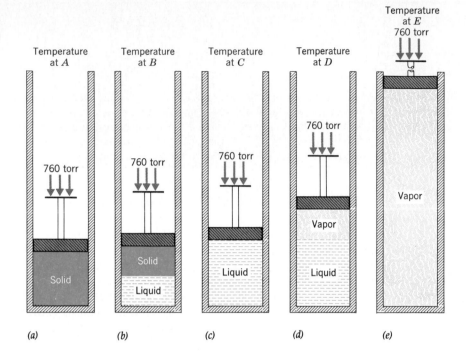

Figure 12.44. *Raising the pressure on water at a constant temperature of 0 °C. Pressures correspond to points in Figure 12.42.*

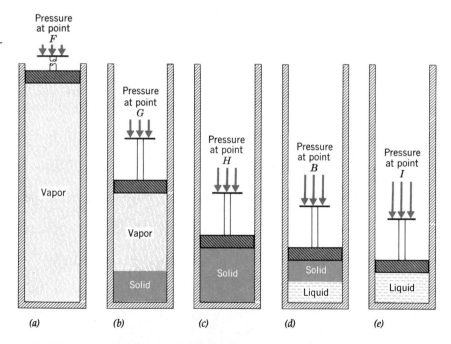

is, a rise in pressure favors the packing together of molecules—quite a reasonable expectation. This means that if we have solid and liquid water at equilibrium and we increase the pressure while holding the temperature at 0 °C, we should produce the higher-density, liquid phase.[4] On the phase diagram, a rise in pressure at constant

---

[4] In fact, it had long been thought that this melting of water that occurs at high pressure is responsible for our ability to skate on ice. It was believed that the high pressure produced by the skater's weight concentrated on the sharp edge of a blade causes the ice just beneath the blade to melt, producing a thin film of liquid water that serves as a lubricant and allows the skate to slide smoothly on the ice. Current feeling, however, is that this film of water is most probably the result of melting caused by friction between the moving skate blade and the ice.

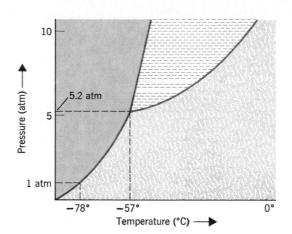

Figure 12.45. *The phase diagram for carbon dioxide.*

temperature amounts to moving upward along a vertical line. We can move from the solid–liquid equilibrium line upward into a region of all liquid only if the solid–liquid line leans to the left.

Water is quite an unusual substance. For nearly all other compounds the solid phase is more dense than the liquid and for these substances the solid–liquid line slants to the right, as is shown in the phase diagram for CO_2 that appears in Figure 12.45. An interesting feature of this phase diagram is that the entire liquid range lies above a pressure of 1 atm; therefore, it is impossible to form liquid CO_2 at atmospheric pressure. Instead, as the gas is cooled, the solid–gas equilibrium line is encountered at -78 °C and the vapor is converted directly to the solid. This also explains why dry ice sublimes rather than melts at ordinary pressures.

At room temperature CO_2 exists as a liquid at the high pressure inside a CO_2 fire extinguisher.

REVIEW QUESTIONS AND PROBLEMS

Problems whose numbers are in blue have their answers in Appendix D at the back of the book. The more difficult problems are marked with asterisks.

Intermolecular Attractions

12.1 Why are intermolecular attractions so strong in liquids and solids, but so weak in gases?

12.2 Why are there no "liquid laws" comparable to the gas laws?

12.3 What are dipole–dipole attractions? Why are they weaker in a gas than in a liquid or solid?

12.4 What are hydrogen bonds? Why are they most important in compounds containing N—H, O—H, and F—H bonds?

12.5 What are London forces?

12.6 What type (or types) of intermolecular attractive forces are found in the following?
(a) HCl   (e) NO
(b) Ar   (f) CO_2
(c) CH_4 (g) H_2S
(d) HF   (h) SO_2

12.7 How is the structure of ice affected by hydrogen bonding?

12.8 If you were asked to compare the strengths of the intermolecular attractive forces in liquid *A* with those in liquid *B*, what types of data would you collect?

12.9 How should the strengths of London forces vary from helium to argon in Group 0?

12.10 What is polarizability? How is it related to the strengths of intermolecular attractions?

12.11 What evidence is there for hydrogen bonding in H_2O, HF, and NH_3?

General Properties of the States of Matter

12.12 State how the following physical properties differ for the three states of matter—solid, liquid, and gas: (a) density, (b) rate of diffusion, (c) compressibility, (d) ability to flow.

12.13 Which of the properties in Question 12.12 are determined primarily by how tightly the molecules are packed?

12.14 Explain why the rate of diffusion in a liquid is less than in a gas.

12.15 At room temperature, diffusion in solids is virtually nonexistent. Why is this so? Why does solid-state diffusion occur at very high temperatures?

12.16 What is surface tension? Why do liquids tend to form *spherical* droplets?

12.17 As a child, you probably had the experience of filling a glass above its rim with water. Why doesn't the water simply overflow?

12.18 Why are soap bubbles spherical?

Evaporation and Sublimation

12.19 Sketch the kinetic energy distribution for a liquid at two temperatures. Indicate the minimum K.E. required for molecules to escape the liquid. On the basis of this diagram, explain why liquids evaporate faster at higher temperatures.

12.20 Why does evaporation lead to a lowering of temperature?

12.21 Clothes dry more rapidly on a dry day than on a humid day. They also dry more rapidly if there is a breeze blowing than if the air is still. Why?

12.22 At high altitudes, ice and snow gradually disappear without melting. Why?

12.23 Why does warm water evaporate more quickly than cold water?

12.24 Solid iodine vaporizes without melting if warmed gently. What is this process called?

12.25 Salt, NaCl, is harvested from seawater by evaporation of water from large shallow ponds. What is the purpose of spreading the water over such large areas?

12.26 At a given temperature, methyl alcohol, the fuel in canned heat, evaporates much more rapidly than propylene glycol, a food additive. Which substance has the weaker intermolecular attractions?

12.27 What *three* factors determine the rate of evaporation of a liquid?

Heats of Vaporization

12.28 What is the formal definition of the molar heat of vaporization? Why is it unimportant that we can't measure the total energy of either a liquid or its vapor?

12.29 Suppose that two substances, *X* and *Y*, have heats of vaporization equal to 37.6 and 27.2 kJ/mol, respectively. Which compound would you expect to have the higher boiling point? Which compound would be less likely to exhibit hydrogen bonding?

12.30 Among the hydrocarbons CH_4 through $C_{10}H_{22}$, both ΔH_{vap} and boiling points increase with increasing molecular mass, even though the molecules are composed of the same kinds of atoms. How can this be explained?

12.31 How would you expect the heats of vaporization to vary among the compounds PH_3, AsH_3, SbH_3?

12.32 How would you expect the heats of vaporization to vary among the compounds H_2S, H_2Se, H_2Te?

12.33 What is a reasonable explanation for the fact that ΔH_{vap} is larger for H_2O than HF?

12.34 The molar heat of vaporization of water is greater than that of any of the components of gasoline. Yet, if gasoline is spilled on your hand, it gives a greater cooling effect than if water is spilled. Why?

12.35 What is the source of energy in a thunderstorm?

12.36 Steam at 100 °C produces a much more severe burn than an equivalent amount of liquid water at 100 °C. Why?

12.37 *Trouton's rule* states that the ratio of the heat of vaporization to the boiling point (in kelvins) is approximately a constant. Verify this for the hydrocarbons CH_4 through $C_{10}H_{22}$ in Table 12.1. What conclusions can you draw concerning the relationship between ΔH_{vap} and boiling point?

12.38 Calculate the heat in kilojoules necessary to convert 55.0 g of ethanol (ethyl alcohol, C_2H_5OH) from liquid to vapor. $\Delta H_{vap} = 38.6$ kJ/mol.

12.39 How many kilojoules of energy are necessary to melt 35.0 g of benzene (C_6H_6)? $\Delta H_{fus} = 9.92$ kJ/mol.

12.40 A 14.5-g sample of liquid mercury required 4.29 kJ to completely convert it to vapor at the same temperature. What is ΔH_{vap} of Hg in kJ/mol? What is it in kcal/mol?

* 12.41 A 150-lb (68.2-kg) skater skids to a halt from a speed of 10.0 mi/hr. If we assume that all of his energy appears as frictional heat transferred to ice at 0 °C, how many grams of ice will be melted?

* 12.42 A student (with very slow reflexes) holds his hand in a stream of steam at 100 °C until exactly 1.00 g of water has condensed. If this water then cools to 40 °C, how many joules have been absorbed by the student's hand?

* 12.43 A cube of solid benzene (C_6H_6) at its melting point weighing 10.0 g is introduced into 50.0 g of H_2O at 30.0 °C. Given that ΔH_{fus} for C_6H_6 is 9.92 kJ/mol, to what temperature will the water have cooled by the time all of the benzene has melted?

* 12.44 A 50.0-g ice cube at 0.00 °C is added to 10.0 g of steam at 100.0 °C. What will be the final temperature of the 60.0 g of water?

Vapor Pressure

12.45 What is meant by the term *equilibrium vapor pressure*?

12.46 What are the two principal factors that affect the observed vapor pressure of a liquid?

12.47 At room temperature, the vapor pressure of acetone (in nail polish remover) is approximately 220 torr, whereas the vapor pressure of ethyl alcohol is approximately 60 torr at this same temperature. Which of these substances possesses the stronger intermolecular attractions?

12.48 Why does the vapor pressure of a liquid increase with increasing temperature?

12.49 Why is the vapor pressure a measure of the strengths of intermolecular attractive forces in a liquid?

12.50 Explain why decreasing the volume of a container does not alter the vapor pressure of a liquid.

12.51 The vapor pressure of a liquid does not depend on the liquid's surface area. Explain why this is so.

12.52 A glass of cola, or other beverage, with ice in it often becomes wet on the outside. Explain why.

12.53 At 0 °F there is less water in 1 L of air at 100% humidity (which means the air is saturated with water vapor) than at 75 °F when the humidity is 100%. Why is this so?

12.54 When warm, moist air is forced to rise over mountains, clouds form and rain frequently falls. On the basis of the concepts in this chapter, explain why this happens.

12.55 Define the terms critical temperature and critical pressure.

12.56 What happens to the boundary between liquid and vapor at temperatures above the critical temperature?

12.57 To what temperature must helium be cooled before it can be condensed to a liquid?

12.58 If you shake a CO_2 fire extinguisher on a cool day, you can feel a liquid sloshing around inside. On a hot summer day, when the temperature is around 90 °F, there is no sloshing when the same extinguisher is shaken. Why?

12.59 Why does blowing gently across the surface of a hot cup of coffee help to cool it?

12.60 At room temperature, solid iodine, I_2, has a much higher vapor pressure than sodium chloride. Why?

12.61 How does the vapor pressure of a solid vary with temperature?

12.62 If a vacuum pump is used to draw a vacuum on a frozen sample of brewed coffee, the water evaporates. What is this commercial process called?

Le Châtelier's Principle

12.63 State Le Châtelier's principle.

12.64 Using Le Châtelier's principle, predict the effect of a change in temperature on the following equilibria:
(a) solid + heat $\rightleftharpoons$ liquid
(b) liquid + heat $\rightleftharpoons$ vapor

12.65 How will an increase in pressure affect the equilibrium, solid $\rightleftharpoons$ vapor?

Boiling Point and Freezing Point

12.66 Define boiling point and normal boiling point.

12.67 From Figure 12.17, estimate the boiling point of water at a pressure of 500 torr.

12.68 At the top of Mount Everest in the Himalayas, which is 29,000 ft above sea level, the atmospheric pressure is approximately 270 torr. At what Celsius temperature would water boil at this altitude?

12.69 Explain why compounds with strong intermolecular attractive forces have higher boiling points than compounds with weak intermolecular attractive forces.

12.70 What is the enthalpy change called that is associated with the change, solid $\rightarrow$ liquid? What is the enthalpy change called for the change, liquid $\rightarrow$ solid? How are these two enthalpy changes related?

12.71 What is the name that we give to the temperature at which there is an equilibrium between solid and liquid?

12.72 What is the difference between a substance's melting point and freezing point?

12.73 As used in this chapter, what does *fusion* mean?

12.74 Explain why for any given substance, ΔH_{fus} is smaller than ΔH_{vap}.

Crystals and X-Ray Diffraction

12.75 What external features do crystals exhibit?

12.76 What is an amorphous solid? How does it differ from a crystalline solid?

12.77 What is a lattice? What is a unit cell? Why can one kind of lattice be used to describe many different chemical structures?

12.78 What is the Bragg equation? What do the symbols in the equation stand for?

12.79 What quantities determine the kind of lattice to which a particular unit cell belongs?

12.80 Sketch and name the three types of cubic lattices.

12.81 Describe the rock salt structure. What kind of lattice does it belong to? How many formula units are there per unit cell? Could a salt like K_2S crystallize in the rock salt structure? Explain your answer.

12.82 Calculate the angles at which X rays of wavelength 229 pm will be observed to be reflected from crystal planes spaced (a) 1000 pm apart and (b) 200 pm apart. Assume that $n = 1$.

12.83 Calculate the interplanar spacings (in picometers) that correspond to reflections at $\theta = 20.0°$, $27.4°$, and $35.8°$ by X rays of wavelength 0.141 nm. Assume that $n = 1$.

12.84 From the following list of angles, determine the angles at which X rays of wavelength 141 pm, diffracted from planes of atoms 200 pm apart, are in phase: $\theta = 17.3°$, $20.5°$, $44.8°$, and $55.3°$.

12.85 Chromium, used to protect and beautify other metals, crystallizes in a body-centered cubic structure in which the Cr atoms are in contact along the body diagonal of the unit cell. The edge of the unit cell is 288.4 pm. Calculate the atomic radius of a Cr atom in picometers.

12.86 Gold crystallizes with a face-centered cubic lattice. The length of the unit cell edge is 407.86 pm. What is the atomic radius of a gold atom in picometers?

12.87 Aluminum crystallizes in a face-centered cubic structure. If the Al atom has an atomic radius of 143 pm, what is the length of the unit cell edge in Al in picometers?

12.88 CsCl forms a simple cubic lattice in which there are Cs$^+$ ions at the corners of the unit cell and a Cl$^-$ ion in the center of the cell. The cation–anion contact occurs along the body diagonal of the unit cell. The length of the unit cell edge is 412.3 pm. The Cl$^-$ ion has a radius of 181 pm. What is the radius of the Cs$^+$ ion in picometers?

12.89 RbCl has the rock salt structure shown in Figure 12.32. The unit cell edge length is 658 pm. Cations and anions are in contact along the edges. The ionic radius of the chloride ion is 181 pm. Calculate the ionic radius of the Rb$^+$ ion in picometers. What is its radius in angstroms?

12.90 Silver has an atomic radius of 144 pm. What would be the density of Ag in g/cm^3 if it were to crystallize in the following structures: (a) simple cubic, (b) body-centered cubic, (c) face-centered cubic? The actual density of Ag is 10.6 g/cm^3. Which of these corresponds to the correct crystal structure for Ag?

*12.91 Calculate the amount of vacant (unoccupied) space (in pm^3) in a primitive cubic, a body-centered cubic, and a face-centered cubic packing of identical spheres of diameter 100 pm.

12.92 LiBr has the rock salt structure in which Br$^-$ ions, centered at lattice points, are in contact. Calculate the ionic radii of Br$^-$ and Li$^+$ in picometers if the unit cell edge is 550 pm. Why is the accepted value for the ionic radius of Li$^+$ (60 pm) smaller than the value that you just computed?

12.93 CsCl crystallizes with a cubic unit cell of edge length 412.3 pm. The density of CsCl is 3.99 g/cm^3. Show that the unit cell cannot be face-centered or body-centered.

12.94 Metallic sodium crystallizes with a body-centered lattice. The element has a density of 0.97 g/cm^3. What is the length of the edge of the unit cell in Na expressed in nanometers?

12.95 Calcium fluoride crystallizes with a cubic lattice. The unit cell has an edge length of 546.26 pm. The density

of CaF$_2$ is 3.180 g/cm^3. How many formula units of CaF$_2$ must there be per unit cell?

12.96 NaCl (which has the rock salt structure) has a density of 2.165 g/cm^3. The ionic radius of Cl$^-$ is 181 pm. What is the ionic radius of Na$^+$ in picometers?

Crystal Types

12.97 Identify the kinds of chemical units associated with each of the following kinds of crystals: molecular, metallic, covalent, ionic. Describe the properties of each of them. What kinds of attractive forces exist between the chemical units in these crystals?

12.98 Indicate which type of crystal (ionic, covalent, etc.) each of the following would form upon solidification: (a) O$_2$ (b) H$_2$S (c) Pt (d) KCl (e) Ge (f) Al$_2$(SO$_4$)$_3$ (g) Ne.

12.99 Indicate which type of crystal (ionic, covalent, etc.) each of the following would form upon solidification: (a) Br$_2$ (b) LiF (c) MgO (d) Cr (e) SiO$_2$ (f) PH$_3$ (g) NaOH.

12.100 SnCl$_4$ is a colorless liquid having a boiling point of 114 °C and a melting point of -33 °C. SnCl$_2$, on the other hand, is a white solid that melts at 246 °C. What type of solid (ionic, covalent, etc.) is most likely formed when SnCl$_4$ solidifies?

12.101 Elemental boron is extremely hard (nearly as hard as diamond) and has a melting point of 2300 °C. It is a poor conductor of electricity at room temperature. What kind of solid would you expect for boron based on these properties?

12.102 Paraffin (wax) is low-melting, soft, and a nonconductor in both the solid and liquid states. What kind of solid is expected for paraffin?

12.103 OsO$_4$ has a melting point of 39.5 °C and is a nonconductor of electricity in the molten state. It boils at 130 °C. What kind of solid is expected for OsO$_4$?

12.104 CaCO$_3$ (calcite) is brittle. It decomposes, before it melts, at a temperature of about 900 °C. What kind of solid is likely for calcite?

Changes of State: Heating and Cooling Curves

12.105 Aluminum has a melting point of 660 °C and a boiling point of 1800 °C. Its $\Delta H_{fus} = 10.7$ kJ/mol and $\Delta H_{vap} = 225$ kJ/mol. Sketch and label a heating curve for aluminum.

12.106 When a supercooled liquid begins to freeze, its temperature rises. Why doesn't the temperature ever rise above the melting point of the substance?

12.107 Why doesn't glass have a sharp melting point?

12.108 Is it possible to have only *liquid* water in a container at 32 °F (0 °C)?

12.109 Explain, on a molecular level, why the temperature remains constant as heat is added to vaporize a liquid at its boiling point.

Phase Diagrams

12.110 At a pressure of 760 torr a new compound was found to melt at 25 °C and boil at 95 °C. The triple point of the substance was determined to occur at a pressure of 150 torr and a temperature of 20 °C. Sketch the phase diagram for this substance. Label, on your drawing, the solid, liquid, and vapor regions as well as the solid–liquid, liquid–vapor, and solid–vapor equilibrium lines.

12.111 On the basis of the phase diagram in Question 12.110, describe the changes you would observe if, at a constant temperature of 22 °C, the pressure on a sample of the compound is gradually increased from 10 to 1000 torr. What would be observed if the same process were to occur at a constant temperature of 10 °C?

12.112 Sketch the heating curve you would expect to find when one mole of the compound described in Question 12.110 is heated at a constant rate under a constant pressure of 1.00 atm. On your drawing, indicate the melting point and boiling point of the substance. Also, label the intervals that correspond to ΔH_{fus} and ΔH_{vap}.

12.113 What can we conclude about the relative densities of the liquid and solid phases of the compound in Question 12.110?

12.114 With the aid of the phase diagram in Figure 12.45, predict the physical state of carbon dioxide under each of the following conditions of temperature and pressure:

Temperature (°C)	Pressure (atm)
−80	1.0
−60	1.0
−56	10.0
−56	2.0
−65	5.0
−40	10.0

12.115 Iodine, I_2, sublimes without melting when heated in an open container at atmospheric pressure. What can be said about the triple point of I_2?

12.116 Use Le Châtelier's principle to predict how variations in pressure will affect the melting point of (a) water and (b) carbon dioxide.

Additional Exercises

12.117 Arrange the substances whose structures are shown below in their expected order of increasing (a) vapor pressure, (b) boiling point at 1 atm, (c) heat of vaporization, (d) surface tension, (e) rate of evaporation at 25 °C (if liquid at this temperature), (f) heat of sublimation, (g) critical temperature.

(1) acetone

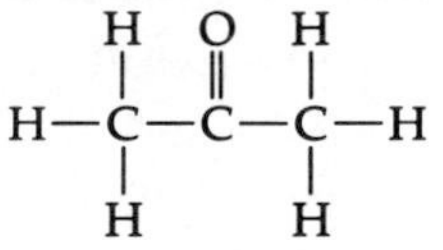

(2) isopropyl alcohol

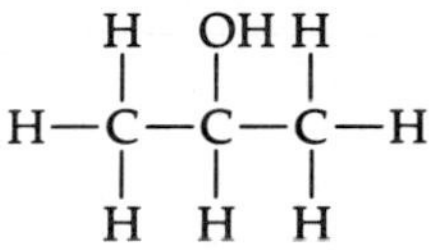

(3) methyl ethyl ether

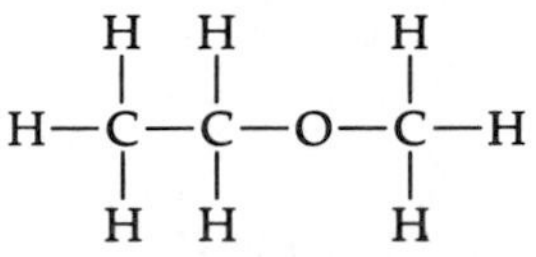

(4) butane

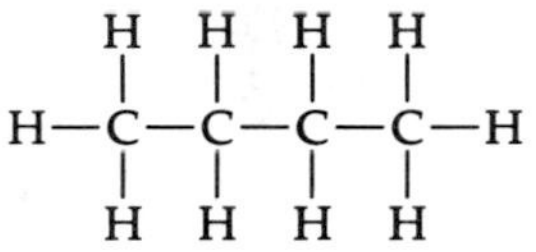

(5) propylene glycol

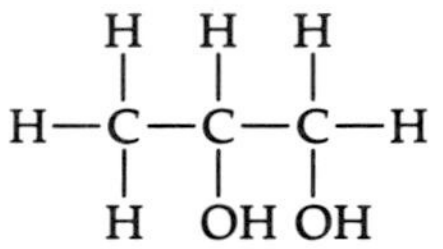

*12.118 Calculate the number of joules required to change 1.00 mol of ice at −20 °C to steam at 120 °C under a constant pressure of 1.00 atm. The specific heat of ice is 2.05 J g⁻¹ °C⁻¹ and the specific heat of steam is 2.01 J g⁻¹ °C⁻¹. Obtain the rest of the necessary data from tables in this chapter and from appropriate locations in preceding chapters, if necessary.

*12.119 Refer to the diagram below to derive the Bragg equation. Remember that the extra distance traveled by the more penetrating beam must be an integral multiple of the wavelength to have constructive interference.

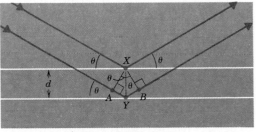

$\measuredangle XAY = \measuredangle XBY = 90°$

*12.120 Chromium crystallizes with a body-centered lattice. Its density is 7.19 g/cm³ and the unit cell edge is 288.4 pm. Use these data to compute Avogadro's number.

13 PHYSICAL PROPERTIES OF COLLOIDS AND SOLUTIONS

In the previous two chapters we discussed the physical properties associated with the three states of matter. For the most part, however, these discussions applied to pure substances, and it is only rarely that we encounter pure materials, either in our daily activities or in the laboratory. Usually, the chemicals we meet are found in mixtures of various kinds.

The physical properties of mixtures are often quite different than those of their isolated, pure components. It is not uncommon to take advantage of this for a variety of practical purposes. For example, steel is a mixture of iron and other elements, such as carbon and other metals. By combining these ingredients in controlled proportions, the finished product can have properties such as hardness and strength that differ substantially from those of iron itself.

In this chapter we study the physical properties of mixtures. We begin by examining the various kind of mixtures that can occur and what effect the sizes of the particles have on the properties of a mixture. One common and special kind of mixture is a solution, and we have already discussed the importance of solutions as a medium for carrying out chemical reactions. In this chapter we will spend much of our time studying another aspect of solutions: the effect that a solute has on a solution's physical properties. Many of these phenomena have very useful laboratory applications, such as the determination of molecular masses. They also have been applied to many practical problems, such as the refining of crude oil and the desalination of seawater.

13.1 KINDS OF MIXTURES: SUSPENSIONS, COLLOIDS, AND SOLUTIONS

In Chapter 1 you learned that any particular sample of matter can be classified as either a pure substance or a mixture. Pure substances include the elements and all the compounds that are formed from them. They are characterized by their constant composition. All samples of pure water, for example, are composed of the two elements hydrogen and oxygen in a ratio of one gram of hydrogen for each eight grams of oxygen. What is special about mixtures is their variable composition. There seems to be no end to the variety and complexity of mixtures because they can be composed of any number of components in any proportions by mass.

One of the principal features that differentiates one kind of mixture from another is the sizes of the particles. Accordingly, mixtures are divided into three general categories: suspensions, colloids, and solutions.

Suspensions

In a suspension, relatively large particles of at least one component are distributed throughout another. Examples are fine sand suspended in water, or snow being blown about through the air, or a precipitate in a reaction mixture. In all these instances the sizes of the suspended particles are large enough to be seen, either with the naked eye or through a microscope. Furthermore, if not continually agitated, the particles of a suspension will settle under the influence of gravity, although the rate at which they settle depends on their size. Coarse sand will settle rapidly in water, but fine mud will settle at a considerably slower rate.

In the laboratory we often find it necessary to separate suspended precipitates from a reaction mixture. One method is filtration. The mixture containing the suspended material is passed through a filter as described on page 123. Sometimes we also take advantage of the tendency of a suspension to settle under the influence of gravity, but we help the process along by using a centrifuge (Figure 13.1). In a centrifuge, a mixture is spun rapidly, and the centrifugal force thus produced behaves as a very powerful artificial gravity that drives the precipitate to the bottom of the container.

The physical properties of suspensions, such as the freezing point or vapor pressure of a suspension of a solid in a liquid, are little affected by the suspended particles. Thus, muddy water freezes at 0 °C, just as pure water does. The suspended particles are too large, and their number is too small compared to the number of molecules of water in the mixture to have any measurable effect.

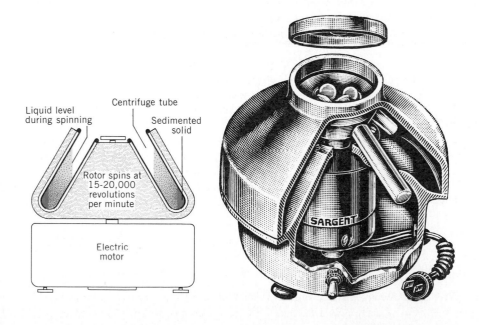

Figure 13.1. *Cutaway views of typical laboratory centrifuges. A suspended solid is driven toward the bottom of the centrifuge tubes by the centrifugal force created when the rotor spins at high speed.*

Solutions

Compared to suspensions, solutions stand at the opposite end of the particle size "spectrum." In a solution, all the particles—those of both the solute and the solvent—are of the dimensions of individual small molecules or ions. These particles are distributed uniformly throughout each other to yield a single homogeneous phase.

Because of the intimate way that the particles of a solute distribute themselves among those of a solvent, the physical properties of a solution often differ quite a bit from those of the solvent alone. In fact, most of this chapter will focus on the effects that a solute has on the physical properties of a solution. Before we turn to that, however, there is a third kind of mixture to discuss, which has some unique properties.

Colloids

Colloid is derived from the Greek *kolla*, meaning glue. Old-time glues were colloidal dispersions in water.

Colloids, also called **colloidal dispersions** or **colloidal suspensions,** are mixtures that are intermediate between true solutions and suspensions. An example is homogenized milk, which consists of very tiny drops of butterfat dispersed in an aqueous phase that also contains casein (a protein) and a few other ingredients. In a colloid such as milk, the solutelike particles are larger than those in a solution, but smaller than the floating particles in a suspension. Because of the way the sizes of colloidal particles compare to the dimensions of the particles of the medium in which they are distributed, we don't use the terms solute and solvent. Instead, we refer to the **dispersed phase** and the **dispersing medium.**

The particles in a colloid are generally too small to be removed by filtration through filter paper. The particles pass through the pores of the filter.

Typically, colloidal particles range in size from about 1 to 1000 nm. Usually, they consist of collections of many molecules or ions, although many of the large molecules in living systems, such as proteins, fall into this size range as well. Even though the particles are larger than those in a true solution, they are still small enough so that constant collisions with the surrounding medium keep them suspended for long periods. One of the general properties of colloids, therefore, is their stability toward separating under the influence of gravity. In fact, some colloids appear to be stable indefinitely. As with suspensions, the relative number of colloid particles in a mixture is small compared to the number of particles of the dispersing medium. Because of this, most of the physical properties of colloids differ very little from those of the dispersing medium.

Milk freezes at very near 0 °C, just as pure water does.

Table 13.1 describes the various combinations of phases that can be combined

Table 13.1. Types of colloidal dispersions

Dispersing Medium	Dispersed Phase	Colloid Type	Examples
Solid	Solid	Solid sol	Pearls, opals
Solid	Liquid	Solid emulsion	Cheese, butter
Solid	Gas	Solid foam	Pumice, marshmallow
Liquid	Solid	Sol, gel	Starch in water, Jello, paint
Liquid	Liquid	Emulsion	Milk, mayonnaise
Liquid	Gas	Foam	Whipped cream, shaving cream
Gas	Solid	Solid aerosols	Smoke, dust
Gas	Liquid	Liquid aerosols	Clouds, mist, fog

Note: A gas in a gas always produces a solution.

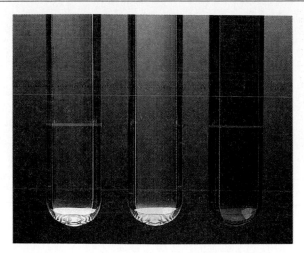

Figure 13.2. *The Tyndal effect. A thin red laser beam passes through two colloidal dispersions and one true solution. On the left is a colloidal dispersion of starch and on the right is a colloidal dispersion of Fe_2O_3. The tube in the center contains a solution of sodium chromate. Although all three appear transparent, the Tyndall effect shows us that the first and third are colloids.*

to give colloidal dispersions. As you can see from the examples given, we come across colloids on a daily basis. You might also notice that all the combinations are possible, except for a colloid formed by a gas dispersed in a gas. Since all gases mix uniformly at the molecular level, gases only form solutions with each other.

The particles in a colloid are too small to be visible with either the naked eye or an ordinary microscope. Nevertheless, they do have an influence on visible light; the particle sizes are just right to cause light to be scattered at large angles. When the concentration of colloid particles is large, this scattering can make the mixture opaque: Light is not allowed to pass through. Milk is an example. Incoming light is scattered by the particles and absorbed, so it never has a chance to exit. When less concentrated, the colloidal dispersion can appear cloudy, and if dilute enough, it can even appear transparent. A dilute colloidal dispersion of starch in water, for example, appears to be as transparent as a solution.

We can tell the difference between colloids and true solutions by observing a beam of light from the side as it passes through them. This is illustrated in Figure 13.2, which shows a laser beam passing through two colloidal dispersions and a true solution. Its path is visible through the colloids because the light is scattered to the side, a phenomenon known as the **Tyndall effect.** Solutions do not exhibit the Tyndall effect because the solute particles are too small to scatter light.

The larger the particles, the larger the angle at which they scatter light.

Stability of colloidal dispersions

In order for a colloid to be stable, its particles must be prevented from sticking to each other when they collide. If they do stick, the particles will grow in size and eventually separate from the mixture. For emulsions (liquids dispersed in liquids), stability is achieved by the action of an **emulsifying agent.** Two common examples of emulsions are milk and mayonnaise. Both consist of an oil dispersed in an aqueous phase. As you know, oil and water "don't mix," and if you shake a mixture of them, afterward they tend to separate quickly into two distinct phases. In mayonnaise, this is prevented by the addition of egg yolk, which forms a protective layer around the tiny drops of vegetable oil as the mixture is whipped. Casein in milk serves a similar purpose by preventing the coalescing of the tiny droplets of butterfat.

Colloids of solids in liquids (sols) are often stabilized by the adsorption of ions onto the surfaces of the colloid particles. (**Adsorption** is a process whereby something sticks to the surface of something else.) For example, the beautiful red sol shown in Figure 13.2 is formed if a solution of $FeCl_3$ is added slowly to boiling water.

Sunlight piercing a cloud cover is scattered by the Tyndall effect, producing a spectacular sight.

Many consumer products, including salad dressings, contain emulsifiers to keep them homogeneous.

Figure 13.3. *Stabilization of an* $Fe_2O_3 \cdot xH_2O$ *sol by adsorption of* Fe^{3+} *ions on the surfaces of the colloidal particles. Because the particles carry charges of the same sign, they repel one another and do not collide and stick together.*

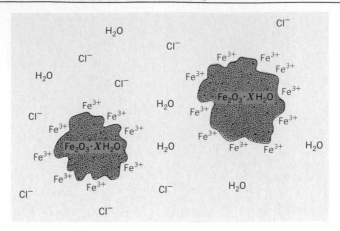

In a chemical reaction the hydrated iron(III) ions lose water and hydrogen ions and form a *hydrated oxide*, $Fe_2O_3 \cdot xH_2O$, which contains a variable amount, x, of water of hydration. The equation for the change can be written

$$2Fe^{3+}(aq) + (x + 3)H_2O \longrightarrow Fe_2O_3 \cdot xH_2O(sol) + 6H^+(aq)$$

As the particles of the sol begin to grow, they adsorb Fe^{3+} ions on their surfaces, which makes them positively charged, as illustrated in Figure 13.3. Because each of the oxide particles acquires the *same* electrical charge, they repel each other. As a result, they no longer collide, so they stop growing. By the time this has happened, they have reached colloidal size.

In aerosols such as smoke, the colloidal particles also pick up electrical charges, but these tend to be from static electricity. Nevertheless, the effect is the same. Because their electrical charges are of the same sign, they repel each other and don't stick together when they collide.

Destabilizing colloids

Colloids can be made unstable by countering those things that stabilize them. When this happens, the particles can come together and grow, and this causes them to separate, or coagulate. Sometimes this coagulation happens by accident, and other times we deliberately destabilize a colloid. For example, the first step in making cheese is curdling the milk.

Sols such as the one formed by the hydrated iron(III) oxide can be coagulated by adding an electrolyte capable of neutralizing the charges on the surfaces of their particles. The addition of a solution containing phosphate ion, for example, will coagulate the sol just mentioned. The negatively charged PO_4^{3-} ions gather around the positively charged Fe^{3+} ions on the surface of the colloidal particles. This effectively neutralizes the charges on the particles and allows them to collide and grow, which ultimately leads to their precipitation. Colloidal clays carried by rivers are precipitated by this same kind of action when they meet the salt water of the sea. River deltas like that at the mouth of the Mississippi have been formed, in part, in this way.

Aerosols consisting of solids dispersed in the air are also separated by neutralizing their electrical charges. Smoke and dust can be removed from the air by passing the mixture over an electrically charged wire grid that carries a charge opposite in sign to that carried by the colloidal particles. The particles are attracted to the grid where their charges are neutralized, which allows them to precipitate.

The first step in making cheese is the addition of rennet to milk, which destabilized the colloidal dispersion and causes the milk to curdle.

13.2 TYPES OF SOLUTIONS

The most common type of solution we encounter consists of a solute dissolved in a liquid, so most of our attention will be directed toward solutions of this kind. Liquid solutions can be prepared by dissolving a solid in a liquid (for example, NaCl in water), a liquid in a liquid (for example, ethylene glycol in water — antifreeze solution), or a gas in a liquid (for example, carbonated beverages, which contain dissolved carbon dioxide).

In addition to liquid solutions, it is possible to have solutions of gases, such as the atmosphere that surrounds the earth, as well as solid solutions that are formed when a substance is dissolved in a solid. The properties of gaseous solutions were discussed in Section 11.5 under the heading "Dalton's Law of Partial Pressures," and nothing more need be said about them here. Solid solutions, of which many **alloys** (mixtures of metals) are examples, are of two types. **Substitutional solid solutions** exist in which atoms, molecules, or ions of one substance take the place of particles of another substance in a crystalline lattice, as shown in Figure 13.4*a*. Zinc sulfide and cadmium sulfide form such mixtures in which cadmium ions randomly replace zinc ions in the ZnS lattice. Another example is brass, a substitutional solid solution of copper and zinc.

Interstitial solid solutions constitute the other type and are formed by placing

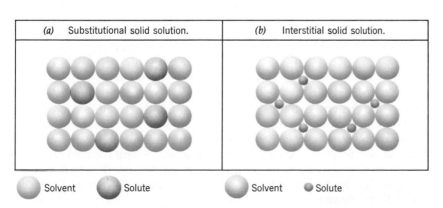

(a) Substitutional solid solution.	(b) Interstitial solid solution.

Solvent   Solute   Solvent   Solute

Figure 13.4. *Two types of solutions.* (a) *A substitutional solid solution in which particles of the solute replace particles in the host lattice (the solvent).* (b) *An interstitial solid solution in which the solute particles fit in spaces between particles of the host lattice.*

A tungsten carbide cutting tool removes metal chips rapidly from an object being machined on a lathe.

$$\text{molarity} = \frac{\text{moles of solute}}{\text{L of solution}}$$

$$\text{normality} = \frac{\text{equivalents of solute}}{\text{L of solution}}$$

atoms of one kind into voids, or interstices, that exist between atoms in the host lattice. This is illustrated in Figure 13.4b. Tungsten carbide, WC, an extremely hard substance that has found many uses in cutting tools for machining steels, is an example of an interstitial solid solution. The tungsten atoms are arranged in a face-centered cubic pattern with carbon atoms in *octahedral holes*—spaces within the crystal where the carbon atoms are surrounded by six tungsten atoms at the vertices of an octahedron.

13.3 CONCENTRATION UNITS

The physical properties of a solution are determined by the relative proportions, or concentrations, of the solution's various components. We have already discussed several ways that concentration can be expressed. For example, you learned about molarity and normality, which are useful concentration units for dealing with problems involving the stoichiometry of reactions that take place in solutions. Molarity and normality were created just for this purpose. In a similar way, it has been found that certain other concentration units are most convenient for dealing with the physical properties of solutions. An important point to remember about all concentration units is that they represent *ratios*. The way to remember them and to keep them straight in your mind is to learn the units associated with numerator and denominator.

Mole fraction and mole percent

You encountered mole fraction as a way of expressing the concentration of a mixture in Chapter 11 in our discussion of Dalton's law of partial pressures. A **mole fraction** *is defined as the ratio of the number of moles of a particular component to the total number of moles of everything in the solution.* Normally, we use the symbol X to stand for mole fraction, so the mole fraction of some component A in a solution is expressed as X_A.

$$X_A = \frac{n_A}{n_A + n_B + n_C + \cdots}$$

where n_A, n_B, etc. are the numbers of moles of the various components of the solution. For example, a solution composed of 2.0 mol water and 3.0 mol ethanol (C_2H_5OH) has a mole fraction of water, X_{H_2O}, given by

$$X_{H_2O} = \frac{2.0 \text{ mol } H_2O}{2.0 \text{ mol } H_2O + 3.0 \text{ mol } C_2H_5OH} = \frac{2.0 \text{ mol}}{5.0 \text{ mol}}$$
$$= 0.40$$

Similarly, the mole fraction of ethanol in the mixture is

We can also calculate the mole fraction of ethanol as

$$X_{C_2H_5OH} = 1 - X_{H_2O}$$

$$X_{C_2H_5OH} = \frac{3.0 \text{ mol}}{5.0 \text{ mol}} = 0.60$$

We see that the sum of all the mole fractions is equal to one, as of course it must be.

Another frequently used term is **mole percent** (abbreviated **mol %**), which is simply equal to 100 × mole fraction. Thus, the mixture above is composed of 40 mol % water and 60 mol % ethanol. Often, it is convenient to think of mole percent as specifying the number of moles of the component per 100 moles of solution. For example, the 60 mol % ethanol solution contains 60 mol of C_2H_5OH for every 100 mol of solution.

Weight fraction and weight percent

The **weight fraction** of a particular component of a solution, $w_{component}$, is the ratio of the number of grams of that component to the total number of grams of solution.

$$w_{component} = \frac{\text{grams of component}}{\text{total grams of solution}}$$

Thus, a solution composed of 12.5 g of water and 37.5 g of ethanol has a weight fraction of water, w_{H_2O}, that is given by

$$w_{H_2O} = \frac{12.5 \text{ g } H_2O}{12.5 \text{ g } H_2O + 37.5 \text{ g } C_2H_5OH} = \frac{12.5 \text{ g}}{50.0 \text{ g}}$$
$$= 0.250$$

Notice that weight fraction, like mole fraction, is without units. We can find the weight fraction of ethanol by a similar calculation, or we can realize that the sum of all the weight fractions must be equal to one. The weight fraction of ethanol is therefore $1 - 0.250 = 0.750$.

Weight percent is just weight fraction multiplied by 100. Often, it is convenient to think of it as the number of grams of solute per 100 g of solution. Thus, a solution that has a weight fraction of 0.750 ethanol has a weight percent equal to 75.0% ethanol. To emphasize that it is *weight* percent, this would often be written as "75.0% w/w ethanol." It tells us that for each 100.0 g of solution, we will find 75.0 g of ethanol.

> Even though these quantities are expressed in terms of masses, *weight fraction* and *weight percent* are the terms most commonly used by chemists.

> Parts per million (ppm) equals weight fraction $\times 10^6$. For an aqueous solution, it is essentially equal to milligrams of solute per liter of solution.

Molality

Molality is defined as the number of moles of solute per kilogram of solvent. It is thus a ratio of moles of solute to the mass of solvent expressed in kilograms.

$$\text{molality} = \frac{\text{moles of solute}}{\text{kilograms of solvent}}$$

A 1.00 molal solution (written 1.00 m) therefore contains 1.00 mol of solute for every 1.00 kg of the solvent.

It is very important not to confuse molality with molarity. Their spellings are almost the same, but they mean quite different things. To see this difference, let's consider how we would prepare typical 1.00 M and 1.00 m solutions using sucrose, $C_{12}H_{22}O_{11}$, as the solute and water as the solvent.

To prepare the 1.00 M solution, we place exactly 1.00 mol of sucrose (342 g) into a volumetric flask that is calibrated to contain precisely 1.00 L when filled to the line etched around its neck (Figure 13.5). Water is added while the mixture is stirred to dissolve the solute, until the flask is filled to the mark. At this point we have exactly 1.00 mol of sugar in a total volume of 1.00 L of solution, and the concentration is 1.00 mol/L, or 1.00 molar (1.00 M).

To prepare the 1.00 m solution, we place 1.00 mol of sucrose into a flask or beaker and *add to it* 1000 g of water. Since the density of water is nearly 1 g/mL, we are adding very nearly 1 L of water to the 342 g (1 mol) of solute. However, the final volume of this 1.00 m solution is somewhat larger than 1 L—it is actually 1110 mL—because part of the volume of the final solution is taken up by molecules of the sucrose. The molarity of this 1 m solution is 1.00 mol/1.110 L = 0.901 M. Because the mole of solute is contained in a larger volume in this 1 m solution, 1-mL portion will contain a smaller amount of solute than a 1-mL portion of the 1 M solution of sucrose.

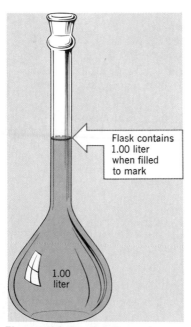

Flask contains 1.00 liter when filled to mark

1.00 liter

Figure 13.5. *A volumetric flask designed to hold exactly 1.00 L when filled to the mark etched around the neck.*

Figure 13.6. *The difference between a 1.00 molar and a 1.00 molal solution when the solvent is CCl₄.*

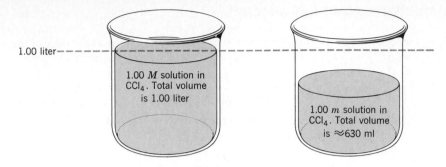

1.00 liter

1.00 *M* solution in CCl₄. Total volume is 1.00 liter

1.00 *m* solution in CCl₄. Total volume is ≈630 ml

The difference between molarity and molality becomes even greater if we choose a solvent whose density is far from 1 g/mL. For instance, in Figure 13.6 we see a 1 *M* solution of a solute in carbon tetrachloride.[1] It contains 1 mol of solute in a total volume of 1 L; however, because CCl₄ has a density of 1.59 g/mL (considerably greater than water), a 1 *m* solution would contain the 1 mol of solute in a volume of only about 630 mL. In other words, it takes much less than 1 L of CCl₄ to weigh 1000 g. The molarity of this 1 *m* solution is actually about 1.6 *M*.

In dilute aqueous solutions, molarity and molality are nearly equal. Can you explain why?

Conversions among concentration units

Mole fraction, mole percent, weight fraction, weight percent, and molality can all be easily converted from one to the other. In other words, if you are given the concentration of a solution expressed in one of these units, you can convert it to one of the other expressions of concentration. All that is needed is the molecular masses of the components.

The first step in performing these conversions is to disassemble the given concentration. We take it apart to obtain information about the amount of solute and either the amount of solvent or the total amount of the solution. For example, if you are told that a solution is 75.0% w/w C₂H₅OH in water, you know that this means

$$\frac{75.0 \text{ g C}_2\text{H}_5\text{OH}}{100.0 \text{ g solution}}$$

The given concentration gives us these two quantities: 75.0 g C₂H₅OH and 100.0 g solution. Next, we use these quantities to obtain the information we need to calculate the concentration in the desired units. For instance, if we wanted mole fraction, which is the ratio of moles of C₂H₅OH to total moles, we would change 75.0 g of C₂H₅OH to moles, subtract 75.0 g C₂H₅OH from 100.0 g of solution to get the number of grams of water in the solution, change grams of water to moles, add all the moles together to get total moles, and finally calculate the mole fraction of C₂H₅OH.

To help organize the conversion process, we will set up a table like the following in each of the example problems below.

	Grams	Moles
solute	A	D
solvent	B	E
Total	C	F

[1] Carbon tetrachloride is a very toxic solvent that should always be handled with care. It is absorbed through the skin and is a cumulative poison.

The method will be to obtain two entries in the table from the given concentration. Then use formula masses as necessary to obtain any other entries that we need to calculate the desired concentration. For example, given 75.0% w/w C_2H_5OH in water, we get values corresponding to A and C. To calculate mole fraction, we need D and F. We know that $A + B = C$, so we can get B. Changing from grams to moles just involves dividing grams by the formula mass, so we change A to D and B to E. Then we add D and E to get F and we have everything we need to obtain our answer. The table completely filled in for this is

	Grams	Moles
C_2H_5OH	75.0	1.63
H_2O	25.0	1.39
Total	100.0	3.02

The mole fraction of C_2H_5OH is therefore

$$X_{ethanol} = \frac{1.63 \text{ mol}}{3.02 \text{ mol}} = 0.540$$

The quantities printed in red come from the given concentration, 75% w/w C_2H_5OH.

EXAMPLE 13.1. CONVERTING FROM WEIGHT FRACTION TO MOLALITY

PROBLEM: A certain aqueous solution of Epsom salts contains a weight fraction of magnesium sulfate, $MgSO_4$, equal to 0.200. What is the molality of the $MgSO_4$ in this solution?

SOLUTION: The first step is to take the given concentration unit apart. A weight fraction of 0.200 means there is 0.200 g of solute per 1.000 g of solution. This gives us two entries (those in red) in the table below. To calculate molality, we need moles of solute, which we get by changing grams of $MgSO_4$ to moles, and we need kilograms of solvent (water). The latter we get by just converting grams of water to kilograms. The formula mass of $MgSO_4$ is 120.4. Therefore, the table with the required values is

	Grams	Moles
$MgSO_4$	0.200	1.66×10^{-3}
H_2O	0.800	—
Total	1.000	—

The moles of $MgSO_4$ were obtained by the calculation

$$0.200 \text{ g MgSO}_4 \times \left(\frac{1 \text{ mol MgSO}_4}{120.4 \text{ g MgSO}_4} \right) = 1.66 \times 10^{-3} \text{ mol MgSO}_4$$

Notice that we haven't bothered with some of the entries because we don't need them. The molality of the solution is

$$\text{molality} = \frac{\text{mol solute}}{\text{kg solvent}}$$
$$= \frac{1.66 \times 10^{-3} \text{ mol MgSO}_4}{8.00 \times 10^{-4} \text{ kg H}_2O}$$
$$= 2.08 \text{ } m \text{ MgSO}_4$$

To solve problems like these, you *must* know the definitions of the various concentration units.

$0.800 \text{ g H}_2O = 8.00 \times 10^{-4} \text{ kg H}_2O$

EXAMPLE 13.2. CONCENTRATION CONVERSIONS STARTING WITH MOLE FRACTION

PROBLEM: Benzene (C_6H_6) and chloroform ($CHCl_3$) are solvents that have proven to be highly toxic. They are mutually soluble in each other. In a certain solution of benzene and chloroform, the mole fraction of C_6H_6 is 0.450. What is the weight percent of C_6H_6 in this mixture?

SOLUTION: As in the preceding problem, the first step is to take the given concentration apart so that we can get two entries in our table. Since the mole fraction of C_6H_6 is 0.450, there must be 0.450 mol of C_6H_6 for every 1.000 mol of solution (total). This gives us the two entries in red in our table. To get weight fraction, we need the total mass of the solution, so we need the masses of both C_6H_6 and $CHCl_3$, which we get from the number of moles by using their formula masses.

$$0.450 \text{ mol } C_6H_6 \times \left(\frac{78.1 \text{ g } C_6H_6}{1 \text{ mol } C_6H_6} \right) = 35.1 \text{ g } C_6H_6$$

$$0.550 \text{ mol } CHCl_3 \times \left(\frac{119.4 \text{ g } CHCl_3}{1 \text{ mol } CHCl_3} \right) = 65.7 \text{ g } CHCl_3$$

The table with the necessary entries is therefore

	Grams	Moles
C_6H_6	35.1	0.450
$CHCl_3$	65.7	0.550
Total	100.8	1.000

The percent of C_6H_6 is found as

$$\%C_6H_6 = \frac{\text{weight } C_6H_6}{\text{weight of mixture}} \times 100\%$$

$$= \frac{35.1 \text{ g}}{100.8 \text{ g}} \times 100\% = 34.8\%$$

EXAMPLE 13.3. CONCENTRATION CONVERSIONS STARTING FROM MOLALITY

PROBLEM: Calcium chloride has a strong tendency to absorb moisture from the air and dissolve in this moisture to produce an aqueous solution. It is sold in hardware stores under a variety of trade names as a dehumidifying agent for use in drying out damp basements and other places with high humidity. Suppose that a certain solution of $CaCl_2$ has a concentration of 4.57 m. What are the mole fractions of $CaCl_2$ and water in the solution?

SOLUTION: A concentration of 4.57 m $CaCl_2$ means the following:

$$\frac{4.57 \text{ mol } CaCl_2}{1.00 \text{ kg } H_2O}$$

The numerator gives us one entry in the table, and the denominator gives us another. Notice that in the table we have changed 1.00 kg to 1000 g. To calculate mole fractions, we need the number of moles of water so we can calculate the total moles of solution.

$$1000 \text{ g H}_2\text{O} \times \left(\frac{1 \text{ mol H}_2\text{O}}{18.0 \text{ g H}_2\text{O}}\right) = 55.6 \text{ mol H}_2\text{O}$$

The table with the required entries is

	Grams	Moles
$CaCl_2$	—	4.57
H_2O	1000	55.6
Total	—	60.2

Notice we've left some entries blank again because we don't need them to calculate the desired concentration. (Why do more work than necessary?)

Now we can compute the mole fractions.

$$X_{CaCl_2} = \frac{4.57 \text{ mol CaCl}_2}{4.57 \text{ mol CaCl}_2 + 55.6 \text{ mol H}_2\text{O}}$$

$$= \frac{4.57 \text{ mol}}{60.2 \text{ mol}} = 0.0759$$

The simplest way to calculate the mole fraction of water is

$$X_{H_2O} = 1.0000 - 0.0759$$
$$= 0.9241$$

This should be rounded to three significant figures to give $X_{H_2O} = 0.924$.

To perform conversions among mole fraction, weight fraction, and molality, all we need are the formula masses of the solvent and solute. To convert between any of these concentration units and molarity, we need one additional piece of information, the density of the solution.

EXAMPLE 13.4. CONCENTRATION CONVERSIONS INVOLVING MOLARITY

PROBLEM: The painful sting of ant bites is caused by formic acid injected under the skin by the ant. Calculate the weight percent of formic acid ($HCHO_2$) in a solution that is 1.099 M $HCHO_2$. The density of the solution is 1.0115 g/mL.

SOLUTION: Molarity gives the ratio of moles of solute to the total *volume* of the solution.

$$1.099 \ M \ HCHO_2 \quad \text{means} \quad \frac{1.099 \text{ mol HCHO}_2}{1000 \text{ mL soln}}$$

The numerator here gives us one of the entries we need in the conversion table, but the denominator doesn't have the right units to fit. This is where we use the density, which provides us with a conversion factor relating the volume of the solution to its mass. It lets us convert the denominator from "1000 mL of soln" to "grams of solution."

$$\text{density} = \frac{1.0115 \text{ g soln}}{1.0000 \text{ mL soln}}$$

If each milliliter has a mass of 1.0115 g, then 1000 mL must have a mass of 1011.5 g, and that value goes into our table as the total mass of the solution. Now we have our two initial entries (those in red), and we can proceed to calculate the others we need.

To calculate the percentage of $HCHO_2$ in the solution, we need the mass of $HCHO_2$. Its formula mass is 46.03; therefore,

$$1.099 \text{ mol HCHO}_2 \times \left(\frac{46.03 \text{ g HCHO}_2}{1 \text{ mol HCHO}_2} \right) = 50.59 \text{ g HCHO}_2$$

The table with its necessary values is

	Grams	Moles
$HCHO_2$	50.59	1.099
H_2O	—	—
Total	1011.5	—

The total mass is obtained from the total volume by using the density as a conversion factor.

Now we can calculate the percentage of solute in the solution.

$$\% \; HCHO_2 = \frac{50.59 \text{ g}}{1011.5 \text{ g}} \times 100\%$$
$$= 5.001\% \text{ w/w HCHO}_2$$

As you have probably realized by now, in order to solve problems like those in the preceding examples, it is absolutely essential that you know the definitions of the various concentration units. If you have acquired this knowledge and proceed systematically, these conversions are not difficult.

13.4 ENERGY AND DISORDER IN THE FORMATION OF LIQUID SOLUTIONS

Experience has taught us that substances differ widely in their solubilities in different solvents. For example, as indicated earlier we know that oil and water "don't mix" and that they form separate phases because they are mutually insoluble. Alcohol and water, on the other hand, are soluble in each other in all proportions. Similarly, sugar is soluble in water, but it won't dissolve in gasoline. What is it that accounts for these differences? What determines whether some substance will dissolve in one solvent or another? To answer these questions, we have to look closely at what the formation of a solution involves.

The role of probability and disorder in the formation of a solution

In our discussion of gases, we said that the volume of a gas is the volume of its container, whether or not other gases are present. Without explicitly saying so, we described a fact about gases. Whenever two or more gases are in the same container, they mix completely to give a homogeneous solution. There is never any question about the solubility of gases in each other; they always mix completely in whatever proportions might be combined. The reason is easy to understand if we examine the formation of such a solution in the apparatus depicted in Figure 13.7.

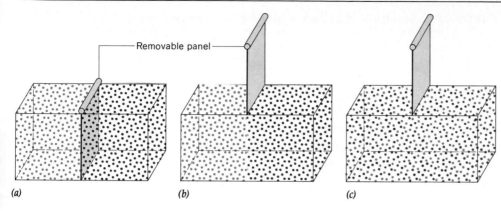

Figure 13.7. *The formation of a solution of two gases. (a) The gases are in two compartments, separated by a removable panel. (b) At the moment the panel is removed, all of one gas is on the left and all of the other is on the right. Because of the random motions of the molecules, it is a situation that won't last long. (c) After a while the gases have formed a homogeneous solution.*

In Figure 13.7a we see two gases in different compartments separated by a removable panel. When this panel is slid away as shown in Figure 13.7b, you know what will happen. Because of their random molecular motions, the gases will begin to mix, and after a while they will have formed a uniform solution, as illustrated in Figure 13.7c. You also probably realize that once formed, this solution will never separate spontaneously to give the condition that existed at the moment the panel was removed, with all of one gas on one side of the container and all of the other gas on the opposite side. That would be like the air in a room spontaneously separating, with all the oxygen going to one end of the room and all the nitrogen going to the other end. It just doesn't happen. But why not?

The reason for this is one of probability. It is just extremely improbable that the gases will not mix when they are first exposed to each other when the panel is slid back, and it is equally unlikely that, by random motions, the gas molecules will become unmixed once the solution has been formed. What we see here is nature's inherent tendency to move toward states of higher probability and greater disorder. For the mixing of gases, this is the principal factor that controls what happens.

The importance of intermolecular attractions in the formation of liquid solutions

Between molecules in a gas there are hardly any attractive forces, and in the mixing of gases the only factor to consider is nature's drive toward increased randomness. This same drive also exists in the formation of solutions of liquids in liquids and solids in liquids. It is, in fact, one of the main reasons such solutions form, but here the drive toward disorder is tempered by the effects of intermolecular attractive forces.

The tendency for systems to become disordered is the principal driving force for the formation of any solution.

Consider, for example, the formation of solutions between water and ethyl alcohol (ethanol), C_2H_5OH. These two liquids are completely soluble in each other in all proportions and are said to be **miscible.** As you learned in Chapter 12, alcohol molecules attract one another by the same kinds of hydrogen bonds that exist between water molecules, so a water molecule in alcohol is held almost the same as it is in pure water. In other words, taking a water molecule out of water and placing it in alcohol involves only relatively small changes in the attractive forces and can occur with little difficulty. As a result, nature's tendency toward randomness has no trouble causing these substances to mix freely.

Now consider the mixing of water and hexane, C_6H_{14}. Water molecules attract one another by strong hydrogen bonds, but they can attract the nonpolar hexane molecules only by London forces. Separating water molecules from each other and distributing them among hexane molecules requires that the hydrogen bonds be-

Table 13.2. Solubilities of some alcohols in water

Substance	Formula	Solubility (mol of solute/100 g H_2O)	
Methanol	CH_3OH	∞	
Ethanol	C_2H_5OH	∞	Completely miscible
Propanol	C_3H_7OH	∞	
Butanol	C_4H_9OH	0.12	
Pentanol	$C_5H_{11}OH$	0.031	
Hexanol	$C_6H_{13}OH$	0.0059	
Heptanol	$C_7H_{15}OH$	0.0015	

In mixtures of C_6H_{14} and H_2O, the tendency toward disorder alone can't overcome the strong attractions between the water molecules.

tween water molecules be overcome by the weak attractions offered by the hexane molecules. Even nature's drive toward disorder can't overcome this hurdle, and water isn't soluble in hexane. Neither can hexane molecules squeeze between water molecules whose attractions for one another effectively exclude the much less polar C_6H_{14} molecules. These two liquids are said to be **immiscible.**

If two nonpolar substances are mixed, such as C_6H_{14} and CCl_4, we again find that the attractions a molecule feels for others of like kind are of about the same strength as they feel for each other. Therefore, a C_6H_{14} molecule can leave its own kind and work its way between CCl_4 molecules in a solution with little difficulty. Once again, there is no resistance to the formation of a solution, so the tendency toward mixing proceeds readily and once again we find complete miscibility.

A rule becomes apparent from these analyses. When the intermolecular attractions are similar for a pair of substances (e.g., C_2H_5OH and H_2O), they tend to be mutually soluble. But if they differ greatly in the kinds and strengths of their intermolecular attractions (e.g., C_6H_{14} and H_2O), they tend to be mutually insoluble. Sometimes this is summed up in the expression "like dissolves like," with the term "like" referring to the similarity in the strengths of intermolecular attractions.

If A and B are partially miscible, a mixture of the two will consist of two phases, one with a little A in B and the other with a little B in A.

Between the extremes of complete miscibility and immiscibility we find many substances that are partially soluble (partially miscible) in each other. For example, Table 13.2 lists the solubilities of a series of alcohols expressed in moles of alcohol required to give a saturated solution in 100 g of H_2O. Notice that as the alcohol molecules become larger, and the OH group becomes a smaller part of the entire molecule, the solubilities decrease. Here we see that as their size increases, the alcohols become more and more like nonpolar hydrocarbons and less and less like water, and their solubilities simply reflect this fact.

Solutions of solids in liquids

For these kinds of solutions, the attractive forces between the solute particles play an even more important role than in the formation of solutions between two liquids. In a solid the molecules or ions are arranged in a very regular pattern and the attractive forces are at a maximum. In order for the solution to form, the attractions between solute and solvent particles must be sufficient to allow nature's drive toward disorder to work its way.

In molecular crystals that are held together by London forces, such as I_2, the attractive forces between the molecules are easily overcome and these substances are soluble in nonpolar solvents such as CCl_4. Molecular iodine is relatively insoluble in water, however, for the same reason that hexane is insoluble. The water molecules attract each other too strongly to be forced apart by I_2 molecules to which they are only weakly attracted.

When a solid is composed of polar molecules, we find it to be soluble in polar

Molecular I_2 gives the CCl_4 layer at the bottom of this test tube a purple color. The water layer above is colorless because I_2 is virtually insoluble in H_2O.

solvents. Sugar molecules, for instance, have many OH groups in their molecular structure. They hydrogen-bond to water molecules very well, so it is easy for them to be pulled from their crystals into the solvent. On the other hand, sugar isn't soluble in nonpolar solvents such as gasoline (a mixture of mostly nonpolar hydrocarbons) because the attractions between sugar molecules in the solid are too strong to be overcome by the attractions to the solvent.

In ionic solids the attractive forces are especially strong, so it takes an extremely polar solvent such as water to cause them to dissolve. Even moderately polar solvents such as methyl alcohol or ethyl alcohol are not up to the job, and salts such as NaCl are nearly insoluble in them, but soluble in water.

When an ionic substance dissolves in water, the ions adjacent to one another in the solid become separated and are surrounded by water molecules. In Chapter 5 we represented this dissociation of the solute by an equation such as

$$NaCl(s) \longrightarrow Na^+(aq) + Cl^-(aq)$$

In Figure 13.8 we take a closer look at what occurs during this process. In the immediate vicinity of a positive ion, the surrounding water molecules are oriented so that the negative ends of their dipoles point in the direction of the positive charge. The water molecules surrounding a negative ion have their positive ends directed at the ion. An ion enclosed within this "cage" of water molecules, such as the Al^{3+} ion shown below, is said to be **hydrated** and, in general, when a solute particle becomes surrounded by molecules of a solvent, we say that it is **solvated**; hydration is a special case of the more general phenomenon of solvation.

The layer of oriented water molecules that surrounds an ion helps to neutralize the ion's charge and serves to keep ions of opposite charge from attracting each other strongly over large distances within the solution. In a sense, the solvent insulates the ions from each other. Nonpolar solvents do not dissolve ionic compounds because they can neither tear an ionic lattice apart nor do they offer any shielding for the ions. In a nonpolar solvent, ions quickly congregate and separate from the solution as the solid.

Glucose, which is a sugar, has many polar OH groups in its molecular structure.

This orientation of water molecules may extend through several layers.

Water molecules

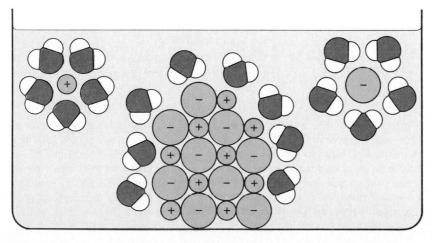

Figure 13.8. *Polar water molecules orient themselves so that their positive ends point at negative ions and their negative ends point at positive ions. In the solution the ions become completely surrounded by these oriented water molecules, and we say the ions are hydrated.*

Soaps and detergents

One of the more practical applications of the solubility relationships described above is the use of soaps and detergents to remove dirt and oil from fabrics. Long before anyone knew any modern chemistry, people had learned to make soap by reacting animal fats with aqueous basic solutions. The reaction of a base with fats produces anions of substances called fatty acids. An example is the stearate ion, $C_{18}H_{35}O_2^-$, which has the structure

$$CH_3—CH_2—CH_2—CH_2—CH_2—CH_2—CH_2—CH_2—CH_2—CH_2—CH_2—CH_2—CH_2—CH_2—CH_2—CH_2—CH_2—\overset{\overset{\displaystyle O}{\|}}{C}—O^-$$

hydrocarbon tail ⟵⟶ anionic head

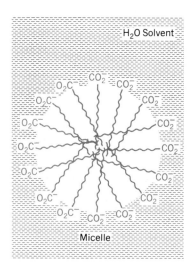

Figure 13.9. *The formation of micelles in soap solutions. The hydrophobic tails of the anions of fatty acids intermingle while the hydrophilic, negatively charged heads are turned outward to face the solvent water.*

Notice that one end of this particle consists of a long hydrocarbon chain and the other consists of the charge-carrying $—CO_2^-$ unit. The hydrocarbon end of the stearate ion resembles other hydrocarbons that we have seen, such as the molecules in gasoline or motor oil. As you know, these do not tend to be soluble in water, but they do dissolve in each other. We describe this behavior by saying that hydrocarbon molecules are **hydrophobic,** which strictly translated means "water fearing." Hydrophobic substances avoid water, and this is exactly what happens in a solution that contains soap. The hydrophobic tails of many fatty acid anions intermingle with each other to form an oillike globule, with their electrically charged **hydrophilic** ("water loving") heads pointing outward toward the solvent, as illustrated in Figure 13.9. These collections of fatty acid anions, which are colloidal in size, are called **micelles.**

The ability of soaps to remove oil and grease from fabrics is also traced to the principle of "like dissolves like." When soap contacts an oil stain on a cloth fiber, the hydrophobic tails of the anions dissolve in the oil. The oil is gradually separated from the fiber and encapsulated in micelles that trap bits of oil inside. This emulsifies the oil and holds it in suspension so it can be carried away with the wash water (Figure 13.10).

Synthetic detergents are very similar to the salts of fatty acids found in soap, except they are manufactured chemically from materials other than animal fats.

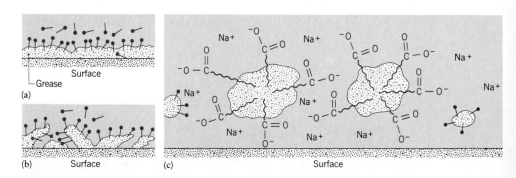

Figure 13.10. *The action of soap on grease that is attached to a surface such as a cloth fiber. (a) The hydrocarbon tails of the soap anions dissolve in the grease. (b) The grease spot gradually breaks up and becomes pincushioned by the soap anions. (c) Small bits of grease are held in colloidal suspension by the soap. The anionic heads keep the grease from coalescing because the particles carry the same electrical charge. (From J. R. Holum,* Fundamentals of Organic and Biological Chemistry, *2nd edition, 1982, John Wiley & Sons, New York, p. 459. Copyright © 1982, by John Wiley & Sons. Used with permission.)*

Examples include salts called sodium alkylbenzenesulfonates, which have the general structure

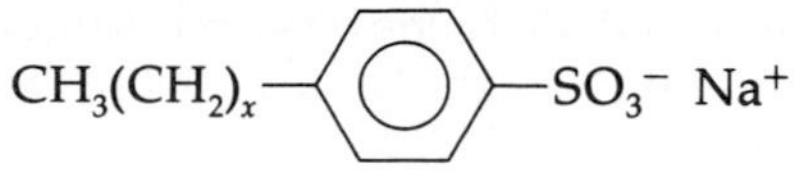

$$CH_3(CH_2)_x\text{---}\langle\bigcirc\rangle\text{---}SO_3^- \ Na^+$$

Their advantage over natural soaps is that they work in *hard water*, which contains ions such as Ca^{2+} that form a precipitate with the anions found in soap. This removes the anions and thereby reduces the effectiveness of the soap solution. The anions of synthetic detergents don't precipitate in the presence of Ca^{2+}, however, so their cleansing action is not affected by hard water.

13.5 HEATS OF SOLUTION

The solution process nearly always occurs with either an absorption or release of energy. For example, when potassium iodide is dissolved in water, the mixture becomes cool, indicating that for potassium iodide the solution process is endothermic. On the other hand, when lithium chloride is added to water, the mixture becomes warm, signifying that the solution process in this case evolves heat and is therefore exothermic. *The amount of heat that is absorbed or released when a substance enters solution is called the heat of solution and is given the symbol, ΔH_{soln}. It represents the difference between the energy possessed by the solution after it has been formed and the energy that the components of the solution had before they were mixed;* that is,

The heat of solution is an enthalpy change, just like the heat of vaporization and the heat of fusion.

$$\Delta H_{soln} = H_{soln} - H_{components}$$

Neither H_{soln} nor $H_{components}$ can actually be measured, but their difference, ΔH_{soln}, can be. When energy is evolved during the solution process, the resulting solution possesses less energy than the components from which it was prepared, so the difference represented by ΔH_{soln} is a negative number. Conversely, an endothermic solution process would have a positive ΔH_{soln}. Heats of solution for some typical ionic solids in water are shown in Table 13.3.

Table 13.3. Heats of solution in water

Substance	Heat of Solution[a] (kJ/mol of solute)
KCl	17.2
KBr	19.9
KI	20.3
LiCl	−37.0
LiI	−59.0
LiNO$_3$	−1.3
AlCl$_3$	−321
Al$_2$(SO$_4$)$_3 \cdot$ 6H$_2$O	−230
NH$_4$Cl	16
NH$_4$NO$_3$	26

[a] At "infinite" dilution. The heat of solution depends, to an extent, on the concentration of the solution produced. A negative sign signifies an exothermic process.

The magnitude of the heat of solution can provide us with information about the relative forces of attraction between the various particles that make up a solution. To analyze the factors that contribute to the absorption or evolution of energy, let us imagine that we could create the solution in a stepwise fashion.

Solutions of liquids in liquids

When one liquid dissolves in another, we can imagine that the molecules of the solvent are caused to move apart so as to allow room for the solute molecules. Similarly, for the solute to enter solution, its molecules must also become separated so that they can take their places in the mixture. Since there are attractive forces between molecules in both the solvent and solute, the process of separating their molecules requires an input of energy—that is, work must be done on both the solute and solvent to separate their molecules from one another. Finally, as the solute and solvent, in their expanded states, are brought together, energy is released because of the attractions that exist between the solute and solvent molecules. This sequence of steps we have just described is illustrated in Figure 13.11.

In some substances, such as benzene and carbon tetrachloride, the intermolecular attractive forces are of very nearly the same magnitude; therefore, these compounds form solutions with virtually no evolution or absorption of heat. Solutions in which the solute–solute, solute–solvent, and solvent–solvent interactions are all the same are called **ideal solutions.** The enthalpy changes that occur along the series of steps that we have devised to arrive at the solution are shown graphically in Figure 13.12. We see that for an ideal solution the energy released in the final step is the same as that absorbed in the first two; thus, the net change is zero.

When the molecules of the solute and solvent attract each other more strongly than they do molecules of their own kind, more energy can be released as the expanded solute and solvent are brought together than was required to separate them in the first place (Figure 13.13a). Under these circumstances the overall solution process can evolve heat and be exothermic. Heat evolves when acetone, an important solvent (nail polish remover), and water are mixed.

When the solute–solvent attractive forces are weaker than those between pure solute or pure solvent, the formation of a solution requires a net input of energy.

Although this three-step process is purely hypothetical, it allows us to analyze the factors that contribute to ΔH_{soln}.

$\Delta H_{\text{soln}} = 0$ for an ideal solution.

$$
\begin{array}{c}
\text{O} \\
\parallel \\
\text{CH}_3\text{—C—CH}_3
\end{array}
$$
acetone

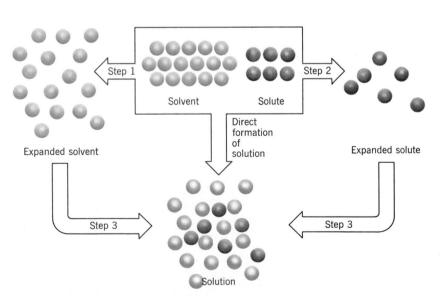

Figure 13.11. *Analysis of the formation of a liquid solution.*

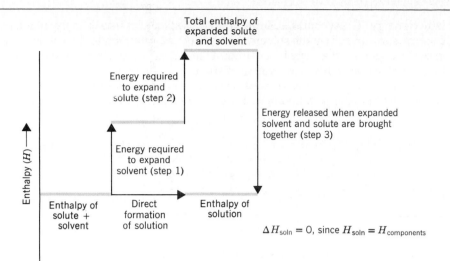

Total enthalpy of
expanded solute
and solvent

Energy required
to expand
solute (step 2)

Energy released when expanded
solvent and solute are brought
together (step 3)

Energy required
to expand
solvent (step 1)

Enthalpy of
solute +
solvent

Direct
formation
of solution

Enthalpy of
solution

$\Delta H_{soln} = 0$, since $H_{soln} = H_{components}$

Figure 13.12. *An enthalpy diagram for the formation of an ideal solution. The actual solution process follows the direct path, although the net enthalpy change is the same.*

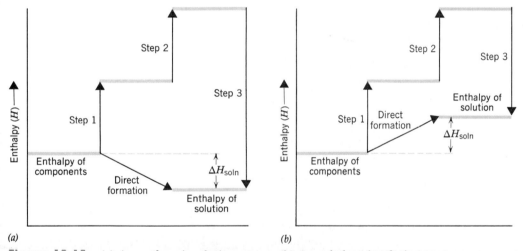

(a)

(b)

Figure 13.13. (a) *An exothermic solution process.* (b) *An endothermic solution process.*

More energy is absorbed in separating the molecules in the first two steps than is recovered when the solute and solvent are brought together in the third (Figure 13.13*b*). The solution becomes cool as it is formed, signifying that an endothermic change has occurred. There is cooling, for example, when ethanol and a hydrocarbon solvent such as hexane are mixed. The nonpolar hexane molecules come between the hydrogen-bonded ethanol molecules, effectively destroying the hydrogen bonds. This absorbs energy and the solution becomes cool.

Solutions of solids in liquids

The analysis here is done a little differently. We imagine the solution being formed in two steps. The first is the vaporization of the solid to give the isolated gaseous solute particles. This step requires an increase in potential energy corresponding to the

This instant cold pack contains plastic pouches of ammonium nitrate, NH₄NO₃, and water. When the pouches are broken and the contents of the package are mixed, the salt dissolves and produces an amazing cooling effect.

lattice energy. The second step brings the solute particles into the solvent where they become surrounded by the solvent. Since they are attracted to the solvent, this step lowers the potential energy by an amount called the **solvation energy** for a solvent in general, or the **hydration energy** if the solvent is water.

Consider as an example the formation of a solution of potassium iodide in water. The first step is the vaporization of the solid to give its gaseous ions.

$$KI(s) \longrightarrow K^+(g) + I^-(g) \tag{13.1}$$

The second step brings the ions into water where they become hydrated.

$$K^+(g) + I^-(g) \xrightarrow{\text{H}_2\text{O}} K^+(aq) + I^-(aq) \tag{13.2}$$

If we add Equations 13.1 and 13.2, we get the net equation for the formation of the solution

$$KI(s) \xrightarrow{\text{H}_2\text{O}} K^+(aq) + I^-(aq) \tag{13.3}$$

The enthalpy diagram that depicts both the individual steps and the net change is shown in Figure 13.14.

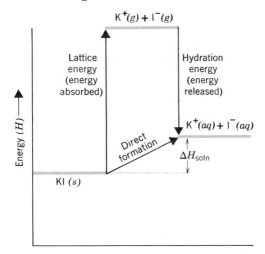

Figure 13.14. *Enthalpy changes that occur when KI dissolves in water.*

Remember that the solution is actually formed by the direct path in Figure 13.14. The alternative path is just hypothetical, but it helps us understand the energy change that occurs along the direct route.

For many salts, such as potassium iodide, the lattice energy is larger than the hydration energy of the ions, so the formation of the solution is endothermic. For other salts, such as LiCl, the hydration energy is larger than the lattice energy and formation of the solution is exothermic.

13.6 SOLUBILITY AND TEMPERATURE

At any given temperature, a saturated solution in contact with undissolved solute represents yet another example of a dynamic equilibrium. Figure 13.15, for example, illustrates such an equilibrium for a solid in contact with its saturated solution, and we see that solute particles are passing into the solution at the same rate as they are crystallizing on the solid solute. Similar situations also exist for other types of solutions in which there is some limit on the amount of solute that is able to dissolve.

Because we have an equilibrium system, we can use Le Châtelier's principle to analyze how a disturbance will affect the position of equilibrium. Such a disturbance is a temperature change, and as you learned in the last chapter, a rise in temperature tends to shift the position of equilibrium in a direction that absorbs heat.

If dissolving more solute is endothermic, as illustrated by the equation

$$\text{heat} + \text{solute} + \text{solution(1)} \rightleftharpoons \text{solution(2)}$$

Figure 13.15. *A dynamic equilibrium exists between a solid and its saturated solution. The solid is dissolving and crystallizing at the same rate.*

[in which solution(2) is more concentrated than solution(1)], then increasing the temperature increases the solubility. In other words, the equilibrium is shifted to the right by an increase in temperature. For most solids and liquids dissolved in liquid solvents, this is what happens; their solubilities increase with increasing temperature.

For gases, the formation of a solution in a liquid is almost always exothermic. Since the molecules in a gas are already separated from each other, the dominant effect is the solvation of the molecules of the gas as they enter the solution. This is nearly always exothermic, so the equilibrium can be represented as

$$\text{gas} + \text{solution(1)} \rightleftharpoons \text{solution(2)} + \text{heat}$$

For this equilibrium an increase in temperature drives gas from the solution because a shift to the left is endothermic. Therefore, gases almost always become less soluble in liquids as the temperature is increased, which is something you probably already knew. You store opened bottles of soda in the refrigerator so that they remain cold and don't lose their carbonation too rapidly. And perhaps you may have noticed that when you heat water in a pot, small bubbles appear on the surface of the pot before the water begins to boil. This is air being driven out of solution as the temperature of the water rises. Temperature also affects the amount of dissolved oxygen in streams, lakes, and rivers. In the summer the amount of O_2 in the water decreases. If the water becomes warm enough, some species of fish will die through lack of oxygen.

The solubilities of most inorganic salts in water increase with increasing temperature. This is a useful generalization to remember.

For some solvents, dissolving a gas in a liquid can be endothermic, because the solvation energy is so small that it doesn't compensate for the energy needed to separate the solvent molecules from each other.

Fractional crystallization

Figure 13.16 illustrates graphically the way in which solubility changes with temperature for a variety of typical solids in water. From these solubility curves it is evident that the variation of solubility with temperature is quite different for different substances. For some substances, such as NH_4NO_3, the solubility changes very rapidly with temperature, but for others the change is more gradual. These differences in behavior provide the basis for a useful laboratory technique called **fractional crystallization,** which is often used to separate impurities from the products of a chemical reaction.

In this technique the impure product is first dissolved in a small amount of hot solvent — generally one in which the desired product is less soluble than the impurities. As the hot solution is allowed to cool, the pure product separates from the mixture, leaving the impurities behind. Finally, the crystals of the product are fil-

Figure 13.16. *The variation of solubility as a function of temperature for some typical solids in water.*

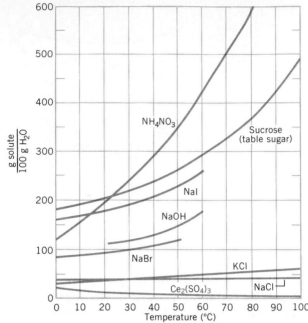

Rock candy is formed by the crystallization of sugar from a saturated solution that is slowly cooled.

tered from the cool solution and dried. The amount of pure product that can be recovered in this fashion depends on the concentrations of the impurities and their solubilities relative to that of the desired material.

13.7 THE EFFECT OF PRESSURE ON SOLUBILITY

In general, pressure has very little effect on the solubility of liquids or solids in liquid solvents. The solubility of gases, however, always increases with increasing pressure. Carbonated beverages, for example, are bottled under pressure to ensure a high concentration of CO_2, and once the bottle has been opened, the beverage quickly loses its carbonation unless it is recapped. The same phenomenon is responsible for decompression sickness, also known as the bends. When a deep-sea diver or a tunnel worker comes to the surface too quickly, nitrogen and oxygen that have dissolved in the blood at high pressure are suddenly released in the form of bubbles in the blood vessels. This is very painful and in extreme cases can even cause death.

Astronauts have to worry about their space suits tearing for the same reason.

The effect of pressure on the solubility of a gas is not difficult to understand. Let's imagine that a liquid is saturated with a gaseous solute, and that this solution is in contact with the gas at some particular pressure. Once again we have a dynamic equilibrium whereby molecules of the solute are entering the vapor phase at the same rate at which molecules from the gas are entering the solution, as shown in Figure 13.17a. As we might expect, the rate at which molecules go into solution depends on the number of collisions per second that they experience with the surface of the liquid. Similarly, the rate at which the solute molecules leave the solution depends on their concentration. If we suddenly increase the pressure of the gas, we pack the molecules close together and increase the rate at which they dissolve (Figure 13.17b). As a result, the dissolved gas becomes more concentrated until equilibrium is reestablished (Figure 13.17c).

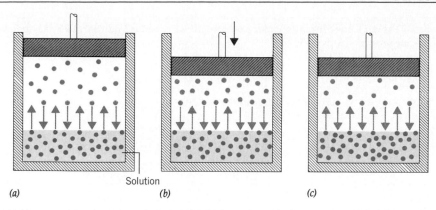

Figure 13.17. *The effect of pressure on the solubility of a gas in a liquid. (a) An equilibrium exists between the gas and its solution. (b) Decreasing the volume increases the pressure of the gas and the equilibrium is upset. (c) Equilibrium is restored when more gas dissolves.*

Solution

(a)   (b)   (c)

The effect that pressure has on the solubility of a gas is even easier to explain in terms of Le Châtelier's principle. We can represent the equilibrium as

$$gas + solution(1) \rightleftharpoons solution(2)$$

If we decrease the volume of the gas above the solution, the pressure increases, upsetting the equilibrium. The system can bring the pressure down again by shifting the position of equilibrium to the right. This decreases the amount of gas and therefore lowers the pressure. At the same time, we see that the concentration of the gas in the solution increases.

Henry's law

The effect of pressure on the solubility of a gas is given quantitatively by **Henry's law,** which states that *the solubility of a gas in a liquid solution, C_g, is directly proportional to the pressure of the gas above the solution.* In equation form this is

$$C_g = k_g p_g \qquad (13.4)$$

where k_g is a proportionality constant called the **Henry's law constant.** This relationship allows us to compute the solubility of a gas at some particular pressure, provided that we know its solubility at some other pressure, as shown in Example 13.5. Actually, Henry's law is accurate only for relatively low concentrations and pressures, and for gases that do not react significantly with the solvent.

The value of k_g depends on the temperature and, for a given temperature, is different for different gases.

EXAMPLE 13.5. USING HENRY'S LAW TO CALCULATE THE SOLUBILITY OF A GAS

PROBLEM: At 25 °C, oxygen gas collected over water at a *total* pressure of 1.00 atm is soluble to the extent of 0.0393 g/L. What would its solubility be if its partial pressure over water was 800 torr?

SOLUTION: The data given us permit the calculation of the Henry's law constant if we know the partial pressure of oxygen above the solution. The total pressure is the sum of the partial pressures of the H_2O vapor and oxygen,

$$P_{total} = p_{H_2O} + p_{O_2}$$

From Table 11.1 we find the vapor pressure of water to be 23.8 torr at 25 °C; therefore, the partial pressure of oxygen is

$$p_{O_2} = P_{total} - p_{H_2O}$$
$$= 760 \text{ torr} - 24 \text{ torr} = 736 \text{ torr}$$

The Henry's law constant is obtained as the ratio

$$k_{O_2} = \frac{C_{O_2}}{p_{O_2}}$$

$$= \frac{0.0393 \text{ g/L}}{736 \text{ torr}} = 5.34 \times 10^{-5} \frac{\text{g}}{\text{L torr}}.$$

Now we can use Henry's law to determine that at a partial pressure of 800 torr the solubility of oxygen is

$$C_{O_2} = \left(5.34 \times 10^{-5} \frac{\text{g}}{\text{L torr}}\right)(800 \text{ torr})$$

$$= 0.0427 \frac{\text{g}}{\text{L}}$$

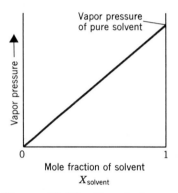

Figure 13.18. *If a solution obeyed Raoult's law for all concentrations, its vapor pressure would vary linearly from zero to the vapor pressure of the pure solvent.*

Raoult's law is easy to use if you remember the definitions of the various terms.

13.8 VAPOR PRESSURES OF SOLUTIONS

The formation of a solution has very little effect on the *chemical* properties of its components. Sodium, for instance, reacts with the water in an aqueous solution to yield exactly the same products as when it reacts with distilled water. The physical properties of substances, however, are often dramatically altered when they become part of a solution. The same water that can freeze and crack the block of an automobile engine at 0 °F will remain liquid if it is mixed with ethylene glycol, an antifreeze.

The vapor pressure of a solution is one physical property that is affected by the presence of a solute. If a *nonvolatile solute* (one that has no tendency itself to escape from a solution) is dissolved in a liquid solvent, the solvent's vapor pressure is lowered. If we exclude solutes that can dissociate in the solvent, such as electrolytes in water, the equilibrium pressure exerted by the solvent vapor, which we call the **vapor pressure of the solution,** ($P_{solution}$), is directly proportional to the mole fraction of the solvent in the solution (Figure 13.18). This relationship is called **Raoult's law** and in equation form is

$$P_{solution} = X_{solvent} P^0_{solvent} \tag{13.5}$$

where $X_{solvent}$ is the mole fraction of the solvent in the solution and $P^0_{solvent}$ is the vapor pressure of the pure solvent. For example, a solution that contains 95 mol % water and 5 mol % of a nonvolatile solute such as sugar has $X_{H_2O} = 0.95$. At a temperature where the vapor pressure of pure water is 100 torr, the vapor pressure of the solution will be

$$P_{solution} = 0.95 \ (100 \text{ torr})$$
$$= 95 \text{ torr}$$

It isn't difficult to see the origin of Raoult's law from a molecular view. Suppose we prepare a solution of a nonvolatile solute in which the mole fraction of the *solvent* is 0.6. In this solution only 60% of the molecules are the solvent, so at the surface there are only 60% as many solvent molecules as in the pure solvent, as illustrated in Figure 13.19. The other 40% are the solute molecules. This means that for a given surface area, only 60% as many solvent molecules can evaporate from the solution as from the pure solvent, so the vapor pressure of the solution will be reduced to 60% of that of the pure solvent. And this is exactly what Raoult's law says.

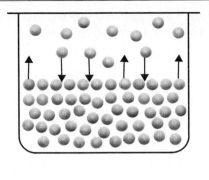 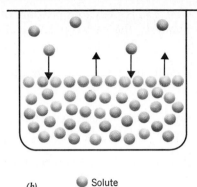

Figure 13.19. *A molecular view of Raoult's law. (a) The liquid–vapor equilibrium for the pure solvent. (b) When a nonvolatile solute is present, the rate of evaporation of the solvent is less and the vapor pressure is lower.*

(a)   *(b)*   ● Solute

EXAMPLE 13.6. APPLYING RAOULT'S LAW

PROBLEM: 10.0 g of the paraffin $C_{20}H_{42}$, a nonvolatile solute, was dissolved in 50.0 g of benzene, C_6H_6. At 53 °C, the vapor pressure of pure benzene is 300 torr. What is the vapor pressure of the solution at this tempurature?

SOLUTION: To calculate the vapor pressure of the solution, we need the mole fraction of the *solvent*. The first thing to do therefore is to calculate the number of moles of each component. The formula mass of $C_{20}H_{42}$ is 282, and the formula mass of C_6H_6 is 78.1.

$$10.0 \text{ g } C_{20}H_{42} \times \left(\frac{1 \text{ mol } C_{20}H_{42}}{282 \text{ g } C_{20}H_{42}} \right) = 0.035 \text{ mol } C_{20}H_{42}$$

$$50.0 \text{ g } C_6H_6 \times \left(\frac{1 \text{ mol } C_6H_6}{78.1 \text{ g } C_6H_6} \right) = 0.640 \text{ mol } C_6H_6$$

The mole fraction of benzene in the mixture is therefore

$$X_{C_6H_6} = \frac{0.640 \text{ mol}}{0.675 \text{ mol}} = 0.948$$

Raoult's law states that

$$P_{\text{solution}} = X_{\text{solvent}} P^0_{\text{solvent}}$$

Substituting values gives

$$P_{\text{solution}} = 0.948 \, (300 \text{ torr})$$
$$= 284 \text{ torr}$$

Solutions that contain more than one volatile component

In many solutions, such as benzene and carbon tetrachloride, for example, both solute and solvent have appreciable tendencies to undergo evaporation. In this case, the vapor will contain both solute and solvent molecules, and the vapor pressure of the solution will be the sum of the partial pressures exerted by each component. If we follow the same line of reasoning as above, we conclude that the partial pressure of any component above such a mixture is also given by Raoult's law. Thus, the partial pressure of component A, p_A, is given by

$$p_A = X_A P^0_A \qquad (13.6)$$

where P_A^0 is the vapor pressure of pure A and X_A is its mole fraction in the solution. Similarly, the partial pressure of a second component, p_B, is given as

$$p_B = X_B P_B^0 \tag{13.7}$$

Finally, the total vapor pressure of a mixture of A and B is given by Dalton's law,

$$P_T = p_A + p_B \tag{13.8}$$

Substituting Equations 13.6 and 13.7 into Equation 13.8 gives

$$P_T = X_A P_A^0 + X_B P_B^0$$

Example 13.7 illustrates how we can apply these relationships to a problem.

EXAMPLE 13.7. CALCULATING THE VAPOR PRESSURE OF A SOLUTION OF TWO VOLATILE COMPONENTS

PROBLEM: A mixture was prepared that contained 50.0 g of carbon tetrachloride, CCl_4, and 50.0 g of chloroform, $CHCl_3$. At 50 °C, the vapor pressure of pure CCl_4 is 317 torr and that of pure $CHCl_3$ is 526 torr. What is the vapor pressure of the mixture at 50 °C?

SOLUTION: To solve this problem, we must use Raoult's law to calculate the partial pressures of each component above the solution. Then the total vapor pressure is simply the sum of the partial pressures. We begin, therefore, by calculating the mole fractions of each of the components.

$$50.0 \text{ g } CCl_4 \times \left(\frac{1 \text{ mol } CCl_4}{153.8 \text{ g } CCl_4} \right) = 0.325 \text{ mol } CCl_4$$

$$50.0 \text{ g } CHCl_3 \times \left(\frac{1 \text{ mol } CHCl_3}{119.4 \text{ g } CHCl_3} \right) = 0.419 \text{ mol } CHCl_3$$

The mole fractions are therefore

$$X_{CCl_4} = \frac{0.325 \text{ mol}}{0.744 \text{ mol}} = 0.437$$

$$X_{CHCl_3} = \frac{0.419 \text{ mol}}{0.744 \text{ mol}} = 0.563$$

Now we can calculate the partial pressures of each component above the solution. For the CCl_4 we have

$$\begin{aligned} p_{CCl_4} &= X_{CCl_4} P_{CCl_4}^0 \\ &= (0.437) \times (317 \text{ torr}) \\ &= 139 \text{ torr} \end{aligned}$$

and for $CHCl_3$

$$\begin{aligned} p_{CHCl_3} &= X_{CHCl_3} P_{CHCl_3}^0 \\ &= (0.563) \times (526 \text{ torr}) \\ &= 296 \text{ torr} \end{aligned}$$

The total vapor pressure is the sum of these.

$$\begin{aligned} P_{soln} &= p_{CCl_4} + p_{CHCl_3} \\ &= 139 \text{ torr} + 296 \text{ torr} \\ &= 435 \text{ torr} \end{aligned}$$

The vapor pressure of the solution is 435 torr.

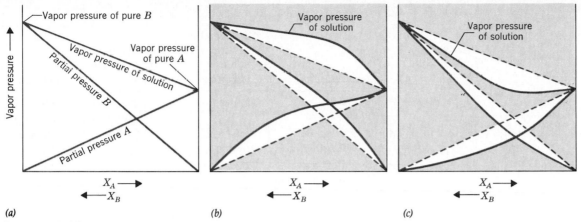

Figure 13.20.   *The vapor pressure of a two-component system as a function of composition.*
(a) An ideal solution. (b) A solution that shows positive deviations from Raoult's law. (c) A
solution that shows negative deviations from Raoult's law.

Ideal and nonideal solutions

Figure 13.20*a* illustrates how the vapor pressures of two hypothetical volatile substances, *A* and *B*, vary for a two-component solution that obeys Raoult's law. The
upper straight line is just the sum of the partial pressures of each of the components.
If a pair of substances followed this behavior perfectly, they would be said to form
ideal solutions, and an example of two substances that come quite close are benzene
and carbon tetrachloride.

Actually, very few mixtures really obey Raoult's law very closely over wide
ranges of composition. Usually, the measured vapor pressure of the solution is either
larger or smaller than Raoult's law would predict. If the vapor pressure is larger, the
solution is said to exhibit **positive deviations** from Raoult's law; if the vapor pressure is lower than predicted, the solution shows **negative deviations.** These deviations are illustrated in Figures 13.20*b* and 13.20*c*. Notice that we relate these deviations in the total pressure of the mixture to deviations in the partial pressures of each
of the components.

As we saw in our discussion of heats of solution, the origin of nonideal behavior
lies in the relative strengths of the attractions between molecules of the solute and
solvent. When the attractive forces between the solute and solvent molecules are
weaker than those between solute molecules or between solvent molecules, neither
the solute nor solvent particles are held as tightly in the solution as they are in the
pure substances. The escaping tendency of each is therefore greater in the solution
than in the solute or solvent alone. As a result, the partial pressures of both of them
over the solution are greater than predicted by Raoult's law. Therefore, the solution
has a larger vapor pressure than expected and exhibits a positive deviation from
Raoult's law.

Just the opposite effect is produced when the solute–solvent attractions are
larger than the solute–solute and solvent–solvent attractions. In this case, each
substance is held more tightly in the presence of each other than in their pure liquids.
As a result, their partial pressures over the solution are less than Raoult's law would
predict and their solutions show negative deviations.

The same factors that cause solutions to deviate from Raoult's law also cause
solutions to be formed with the absorption or release of heat. In other words, there is
a correlation between the heat of solution and deviations from Raoult's law. These
are summarized, with examples, in Table 13.4.

Raoult's law is generally obeyed
only for very dilute solutions.

Table 13.4. Summary of solution properties

Relative Attractive Forces	ΔH_{soln}	Temperature Change When Solution Is Formed	Deviations from Raoult's Law	Example
$A - A, B - B = A - B$	Zero	None	None (ideal solution)	Benzene–chloroform
$A - A, B - B < A - B$	Negative (exothermic)	Increase	Negative	Acetone–water
$A - A, B - B > A - B$	Positive (endothermic)	Decrease	Positive	Ethanol–hexane

13.9 FRACTIONAL DISTILLATION

In a simple distillation process — one that could be used to separate sodium chloride and water, for example — a volatile solvent is vaporized from a solution and subsequently condensed to provide a pure liquid (see Figure 13.21). If the process is continued, eventually all the solvent will be removed and only the solid solute will remain.

The separation of mixtures of volatile liquids into their components presents

Figure 13.21. *A simple distillation apparatus.*

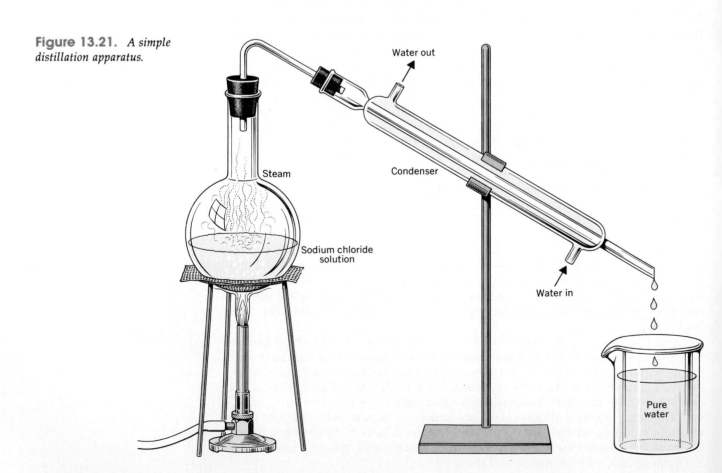

Water out

Steam

Condenser

Sodium chloride solution

Water in

Pure water

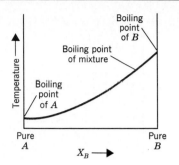

Figure 13.22. *The boiling point curve for mixtures of* A *and* B.

more of a problem. A technique that can frequently be used successfully to accomplish this task is called **fractional distillation.**

Let's suppose that we had a mixture of two volatile liquids, A and B, that form an ideal solution. This mixture will boil when the sum of the partial pressures of A and B equals the prevailing atmospheric pressure; that is, when

$$P_{atm} = p_A + p_B$$

The boiling points of various mixtures of A and B will increase gradually from that of the more volatile component (let us say, A) to that of the less volatile one, B, as shown in Figure 13.22.

Suppose, now, that when 1.00 mol of A is mixed with 2.00 mol of B, the resulting mixture boils (at 1 atm) at a temperature at which the vapor pressure of *pure A* is 1140 torr and that of *pure B* is 570 torr. Under these conditions the partial pressure of A is

$$p_A = X_A P_A^0$$

$$= \left(\frac{1.00 \text{ mol } A}{1.00 \text{ mol } A + 2.00 \text{ mol } B} \right) \times (1140 \text{ torr})$$

$$= \left(\frac{1.00 \text{ mol}}{3.00 \text{ mol}} \right) \times (1140 \text{ torr})$$

$$= (0.333)(1140 \text{ torr}) = 380 \text{ torr}$$

Similarly, the partial pressure of B would be

$$p_B = \left(\frac{2.00 \text{ mol}}{3.00 \text{ mol}} \right) \times (570 \text{ torr})$$

$$= 380 \text{ torr}$$

The sum of p_A and p_B is 760 torr as, of course, it must be if the solution is to boil.

What can we say about the composition of the vapor? In Section 11.7, under our discussion of Dalton's law of partial pressures, it was stated that the partial pressure of a gas in a mixture is equal to its mole fraction multiplied by the total pressure *exerted by the gas.* We can write this as

$$p_A = X_{A(vapor)} P_{T(vapor)}$$

where $X_{A(vapor)}$ is the mole fraction of A in the vapor and $P_{T(vapor)}$ is the total vapor pressure. In the vapor over our solution, the partial pressure of each gas is 380 torr and the total pressure is 760 torr. This means that the mole fraction of both A and B *in the vapor* must be 0.500. In the liquid the mole fraction of A was only 0.333, so the vapor contains a greater proportion of the more volatile component (A) than the solution. In fact, any time we boil a mixture of these two substances, the vapor will be richer than the solution in the more volatile compound.

Note that this equation applies to the *vapor.*

When a liquid solution made from two volatile liquids is boiled, the vapor always is richer in the more volatile component.

Figure 13.23. *The boiling point diagram for a two-component system. A solution of composition X_1 boils at a temperature T_1 and gives a vapor with a composition X_2. When this vapor is condensed and then reheated, it boils at a temperature T_2 and gives a vapor with a composition X_3.*

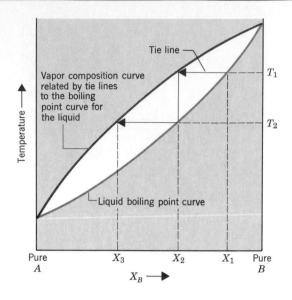

On our boiling point diagram we can indicate the composition of the vapor by the upper curve drawn in Figure 13.23. Here, points corresponding to the compositions of liquid and vapor in equilibrium can be obtained by drawing a horizontal line, called a **tie line,** between the curve for the liquid and that for the vapor. When the composition of the mixture is X_1, it boils at a temperature T_1 to provide a vapor that has a composition X_2. If this vapor is condensed and then reheated, it will boil at a temperature T_2 and give a vapor whose composition is X_3. Repetition of this process will produce fractions ever richer in A. This procedure is called **fractional distillation** and is useful not only in the laboratory, where it is employed for purifying the products of chemical reactions, but also industrially. For instance, the petroleum industry uses fractional distillation to separate crude oil into its various components, which include gasoline, kerosene, oils, and paraffin.

Solutions that exhibit large deviations from Raoult's law

There are some solutions that exhibit very large deviations from ideality; as a result, they cannot be totally separated into their components even by fractional distillation. Ethanol (grain alcohol) and water form such mixtures. Solutions of these two substances have such large positive deviations from Raoult's law that there is a maximum in the vapor pressure curve and hence a minimum in the boiling point diagram as shown in Figure 13.24. A solution with such a minimum boiling point is called a **minimum-boiling azeotrope.** Fractional distillation of solutions lying on either side of this azeotropic composition is capable of separating them into, at best, one pure component plus a solution having the minimum boiling point. As any moonshiner will agree, ethyl alcohol–water mixtures (obtained by fermentation of sugars, for example) are rich in water. Fractional distillation is able to concentrate the alcohol to, at best, the azeotropic composition of approximately 95% by volume of ethyl alcohol. Once this composition has been achieved, the liquid and vapor have the same composition, and no additional fractionation takes place.

There are also solutions that show large negative deviations from ideality and therefore have a minimum in their vapor pressure curves. This leads to a maximum on the boiling point diagram and hence to a **maximum-boiling azeotrope.** Hydrochloric acid, for instance, forms a maximum-boiling azeotrope having the approximate composition, 20% HCl and 80% H_2O by weight, with a boiling point of 109 °C.

Towers such as these in this oil refinery are fractional distillation columns used to separate the components of crude oil into gasoline, kerosene, diesel oil, and so on.

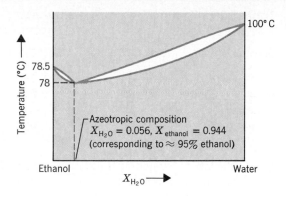

Figure 13.24. *The boiling point diagram for water–ethanol mixtures (not drawn to scale)*

13.10 FREEZING POINT DEPRESSION AND BOILING POINT ELEVATION

In Section 13.8, you learned that a nonvolatile solute lowers the vapor pressure of a solution. This phenomenon also influences other physical properties—especially, the freezing point and boiling point.

Figure 13.25 illustrates the phase diagram for water. As you know, we can use this diagram to read off the normal boiling point and the normal freezing point; the normal boiling point is the temperature at which the vapor pressure of the liquid equals 1 atm, and the normal freezing point is the temperature at which the solid–liquid equilibrium line crosses the 1 atm pressure line.

In Figure 13.25, we have also plotted the vapor pressure curve for a solution that contains a nonvolatile solute. Notice that at any particular temperature, the vapor pressure of the solution is lower than that of the pure solvent. Also notice that the vapor pressure of the solution reaches 1 atm at a higher temperature than does the vapor pressure of the pure solvent. In other words, the boiling point of the solution is higher than that of the solvent itself. The amount that the boiling point is raised is indicated on the diagram by ΔT_b, and we call this increase the *boiling point elevation*.

Examining Figure 13.25, we also find that the solution has a new triple point that occurs at the intersection of the vapor pressure curve for the solution and the solid vapor pressure curve for the pure solvent. Generally, the solute particles don't

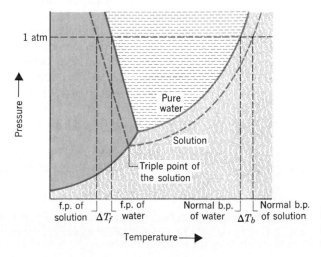

Figure 13.25. *The effect of a nonvolatile solute on the phase diagram for water.*

Table 13.5. K_b and K_f for some solvents

Solvent	Boiling Point (°C)	K_b (°C/m)	Melting Point (°C)	K_f (°C/m)
Water	100.0	0.51	0.0	1.86
Benzene	80.1	2.53	5.5	5.12
Camphor	—	—	179	39.7
Acetic acid	118.2	2.93	17	3.90

The solid vapor-pressure curve for the solution is the same as the solid vapor-pressure curve for the solvent.

fit into the lattice formed by the solvent when it freezes, so the solid that forms is the pure solvent. As a result, there is no separate solid vapor pressure curve for the solution. As you learned in Chapter 12, the solid–liquid equilibrium line (which defines the freezing point as a function of pressure) rises from the triple point. Since the new triple point for the solution lies to the left of that for the pure solvent, the freezing point of the solution is lower than the freezing point of the solvent. The amount that the freezing point is lowered (the *freezing point depression*) is shown on the diagram as ΔT_f.

In summary, the presence of a solute increases the liquid range of the solution by both raising the boiling point and lowering the freezing point. One of the most common practical applications of this phenomenon is the use of an antifreeze solution in the radiator of an automobile. The solute is usually ethylene glycol, $C_2H_4(OH)_2$, which is completely miscible with water and has a very low vapor pressure itself. It is a nonvolatile solute. When dissolved in water, it both lowers the freezing point and raises the boiling point. In the winter it protects a car by preventing the liquid in the radiator from freezing, as water would if it were used instead. In hot summer weather, the antifreeze solution also protects the radiator from boiling over as easily as it would if it were filled with pure water.

For dilute solutions it has been found that the extent to which the boiling point is raised, and the freezing point is lowered, depends on the *molality* of the solute in the solution,

$$\Delta T_b = K_b m \tag{13.9}$$

and

$$\Delta T_f = K_f m \tag{13.10}$$

where K_b and K_f are referred to as the **molal boiling point elevation constant** and the **molal freezing point depression constant,** respectively. The magnitudes of K_b and K_f are characteristics of each solvent. Table 13.5 contains a list of some typical solvents and their values of K_b and K_f.

If we know the concentration of the solution, in moles of solute per kilogram of solvent, the relationships expressed in Equations 13.9 and 13.10 permit us to calculate the extent to which the boiling point and freezing point are changed. For example, a solution containing 1 mol of sugar in 1000 g of water will have its freezing point lowered by 1.86 °C and its boiling point raised by 0.51 °C. The solution will therefore freeze at -1.86 °C and boil at 100.51° when the atmospheric pressure is 1 atm. Example 13.8 provides another sample calculation.

EXAMPLE 13.8. CALCULATING THE FREEZING AND BOILING POINTS OF A SOLUTION

PROBLEM: What are the freezing point and boiling point of a solution containing 6.50 g of ethylene glycol ($C_2H_6O_2$), commonly used as an automotive antifreeze, in 200 g of water?

SOLUTION: To determine ΔT_f and ΔT_b we must know the molality of the solution—the ratio of moles of solute to kilograms of solvent. The number of moles of $C_2H_6O_2$ is

$$6.50 \ g \ C_2H_6O_2 \times \left(\frac{1 \ mol \ C_2H_6O_2}{62.1 \ g \ C_2H_6O_2} \right) = 0.105 \ mol \ C_2H_6O_2$$

The number of kilograms of solvent is

$$200 \ g \ H_2O \times \left(\frac{1 \ kg}{1000 \ g} \right) = 0.200 \ kg \ H_2O$$

The molality is therefore

$$\frac{0.105 \ mol \ C_2H_6O_2}{0.200 \ kg \ H_2O} = 0.525 \ m \ C_2H_6O_2$$

For H_2O, $K_f = 1.86$ °C/m and $K_b = 0.51$ °C/m. Therefore, the changes in the freezing point and boiling point are

$$\Delta T_f = \left(1.86 \ \frac{°C}{m} \right) \times (0.525 \ m) = 0.976 \ °C$$

$$\Delta T_b = \left(0.51 \ \frac{°C}{m} \right) \times (0.525 \ m) = 0.27 \ °C$$

The freezing and boiling points of the solution are then -0.976 °C and 100.27 °C. We see that solutions considerably more concentrated than this (approximately 3%) are necessary to protect an automobile's cooling system in frigid weather. See Problem 13.92.

Determining molecular masses from freezing point depression and boiling point elevation

In describing the effects that a solute has on lowering the vapor pressure of a solution, increasing its boiling point, and lowering its freezing point, the only restriction that we gave was that the solute be nonvolatile. The actual chemical makeup of the solute does not matter. Only the concentration of the solute particles is important. Because of this, we can use these phenomena to measure molecular masses of substances, as illustrated in the following example.

Properties of solutions that depend only on the relative numbers of solute and solvent particles, such as vapor pressure, boiling point, and freezing point, are known as **colligative properties.**

EXAMPLE 13.9. DETERMINING THE MOLECULAR MASS OF A SOLUTE FROM FREEZING POINT DEPRESSION MEASUREMENTS

PROBLEM: A 5.50-g sample of a newly synthesized compound was dissolved in exactly 250.0 g of benzene. It was found that the freezing point of the solution was 1.20 °C below that of pure benzene. For this solvent, $K_f = 5.12$ °C/m. What is the molecular mass of the compound?

ANALYSIS: As you learned in Chapter 11, to obtain the molecular mass of a substance experimentally, two quantities have to be determined. One is the mass of a sample of the substance and the other is the number of moles in the sample. Once this information is known, the ratio of grams to moles gives the molecular mass.

The data in this problem give us the mass of a sample, so all we need to do is figure out how many moles this corresponds to. But where do we begin?

From the freezing point depression, ΔT_f, and the freezing point depression constant, we can calculate the molality of the solute in the solution by using Equation 13.10. This gives us the number of moles of the solute per kilogram of solvent. Multiplying this ratio by the actual number of kilograms of solvent in the solution that was prepared will give the number of moles of solute in the solution, which is just what we are looking for. Finally, the molecular mass is just the ratio of grams of solute to moles of solute.

SOLUTION: First, let's solve Equation 13.10 for the molality

$$m = \frac{\Delta T_f}{K_f}$$

Substituting the values for the freezing point depression and K_f, we have

$$\text{molality} = \frac{1.02\ °C}{5.12\ °C/m}$$
$$= 0.199\ m$$

This translates to mean that the solution we have prepared contains 0.199 mol of solute for every 1.00 kg of benzene. In the actual solution we have only 0.2500 kg of benzene, so the number of moles of solute in the solution must be

$$\left(\frac{0.199\ \text{mol solute}}{1.00\ \text{kg benzene}}\right) \times 0.2500\ \text{kg benzene} = 0.0498\ \text{mol solute}$$

To get the formula mass of the solute, we take the ratio of the number of grams of solute to the number of moles of solute.

$$\frac{5.50\ \text{g solute}}{0.0498\ \text{mol solute}} = 110\ \text{g/mol}$$

If there are 110 g per mole for this substance, its molecular mass must be 110.

13.11 OSMOTIC PRESSURE

Osmosis is a process whereby a solvent passes from a dilute solution into a more concentrated one by moving through a thin film that selectively permits the passage of the solvent, but restricts the passage of the solute. Such films are called **semipermeable membranes,** and typical examples include certain types of parchment paper and some gelatinlike inorganic substances. A similar phenomenon called **dialysis,** which occurs at cell walls in plants and animals, allows the passage of water, small ions, and small molecules but restricts passage of large molecules such as proteins. Osmosis is the limiting case of dialysis.

In the process of osmosis there is a drive toward equalization of concentrations between the two solutions in contact with one another across the membrane. The rate of passage of solvent molecules through the membrane into the more concentrated solution is greater than their rate of passage in the opposite direction, presumably because at the surface of the membrane the solvent concentration is greater in the more dilute solution (Figure 13.26). We observe a similar effect if two solutions, with unequal concentrations of a nonvolatile solute, are placed in a sealed enclosure, as shown in Figure 13.27. The rate of evaporation from the dilute solution is greater than that from the concentrated solution, but the rate of return to each is the same (both solutions are in contact with the same gas phase). As a result, neither solution is in equilibrium with the vapor. In the dilute solution molecules are evaporating faster than they are condensing, but in the concentrated solution the reverse occurs. Consequently, there is a gradual net transfer of solvent from the dilute solution into the more concentrated one until they both achieve the same concentration.

If we perform an osmosis experiment using the apparatus shown in Figure

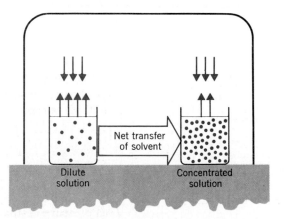

A patient uses a portable artificial kidney machine. It cleanses the blood of impurities by dialysis, a process whereby small molecules such as urea are able to pass through a semipermeable membrane into another fluid, which is ultimately discarded. The membrane restricts the flow of large molecules so that they stay in the blood.

13.28, in which we have a solution inside the bulb of the *osmometer* and the pure solvent outside, the net passage of the solvent into the solution through the semipermeable membrane will increase the volume and cause the height of the liquid in the capillary to rise. The column of liquid in the capillary that rises above the surface of the solvent exerts a pressure just like the mercury in a barometer does. This pressure tends to increase the rate of passage of solvent from the solution into the pure solvent, so the net rate of osmosis slows as the liquid level climbs higher and higher. This continues until eventually osmosis stops. The pressure on the solution at this

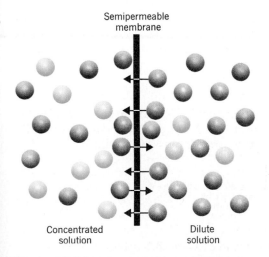

Figure 13.26. *Osmosis. Solvent molecules pass more rapidly from the more dilute solution into the more concentrated solution.*

Figure 13.27. *Because the two solutions have unequal vapor pressures, there is a gradual net transfer of solvent from the more dilute solution to the more concentrated one.*

Figure 13.28. *An apparatus for the measurement of osmotic pressure. When equilibrium is reached, the height of the liquid in the capillary is proportional to the osmotic pressure of the solution.*

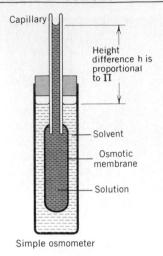

Simple osmometer

Π is the capital Greek letter pi.

point, which is proportional to the height of the liquid in the capillary, h, is called the **osmotic pressure** of the *solution* and is symbolized by the Greek letter Π.

For solutions that are dilute, it can be shown that the osmotic pressure is proportional to the molarity (M) of the solute, and that the proportionality constant is RT, where R is the gas constant and T is the absolute temperature.

$R = 0.0821$ L atm mol^{-1} K^{-1}

$$\Pi = MRT \tag{13.11}$$

Another form of this equation can be obtained if we realize that the molarity of the solute is obtained as the ratio of the number of moles of solute (n) to the volume of the solution (V).

$$M = \frac{n}{V} \tag{13.12}$$

Substituting this into Equation 13.11 gives

$$\Pi = \frac{n}{V} RT$$

or

$$\Pi V = nRT \tag{13.13}$$

Jacobus Hendricus van't Hoff, a Dutch chemist, received the first Nobel prize in chemistry in 1901.

It is interesting to note the similarity between Equation 13.13 (called the van't Hoff equation) and the ideal gas law

$$PV = nRT$$

The magnitude of the osmotic pressure, even in very dilute solutions, is quite large. For instance, with a concentration of 0.010 mol of solute particles per liter (0.010 M) at room temperature (298 K), the osmotic pressure would be

$$\Pi = MRT$$
$$= \left(\frac{0.010 \text{ mol}}{1 \text{ L}} \right) \times \left(\frac{0.0821 \text{ L atm}}{\text{mol K}} \right) \times 298 \text{ K}$$
$$= 0.24 \text{ atm}$$

This pressure is sufficient to support a column of water 8.1 ft high!

Because the osmotic pressure that can be developed between solutions of only slightly different concentrations is so great, it is very important that fluids added to the body intravenously not alter significantly the osmotic pressure of the blood. If the blood fluids become too dilute, the osmotic pressure that develops within the blood cells can cause them to rupture. On the other hand, if the fluids are too concentrated, water will diffuse out of the cells and they will no longer function properly. For this reason care is taken to use solutions with the same osmotic pressure as the solution within the cells. Solutions that have the same osmotic pressure are called **isotonic** solutions.

The large differences in pressure developed between solutions of very similar concentrations provide us with a method of measuring the very large molecular masses of polymers (both of synthetic and biological origin). Freezing point lowering and boiling point elevation just won't work in these cases. For instance, a solution containing even as much as 15 g of a solute whose molecular mass is 30,000 in 1000 g of water produces a solution with a concentration of only 0.0005 m. The freezing point depression of this solution is approximately 0.0009 °C, which is virtually undetectable and unmeasurable.

$$\frac{15 \text{ g solute}}{1.00 \text{ kg H}_2\text{O}} \times$$
$$\frac{1 \text{ mol}}{3 \times 10^4 \text{ g}} = 0.0005 \ m$$

A solution containing only 15 g of this solute per 1000 g of water at 25 °C, however, would have an easily measured osmotic pressure. The concentration of the solution is 5.0×10^{-4} m. Because the solvent is water and the solution is so dilute, the molality and molarity will be nearly the same, so we can express the concentration as 5.0×10^{-4} M with very little error. Now we can calculate the osmotic pressure.

$$\Pi = \left(\frac{5.0 \times 10^{-4} \text{ mol}}{L}\right) \times \left(\frac{0.0821 \ L \text{ atm}}{\text{mol K}}\right) \times 298 \text{ K}$$

$$= 1.2 \times 10^{-2} \text{ atm}$$

This corresponds to a pressure of 9.1 torr, or 9.1 mm Hg. If we assume the solution has a density of 1 g/mL, this pressure is able to support a column of the solution that is 12.3 cm high. Heights such as this are easily measured with both precision and accuracy, so we see that osmotic pressure measurements are a useful tool for determining the molecular masses of large molecules.

Reverse osmosis and water purification

The osmosis process can actually be reversed if a pressure larger than the osmotic pressure is applied to the solution. This phenomenon has found uses in water purification, especially the desalination of sea water. An apparatus for this process is shown schematically in Figure 13.29. If no pressure were applied to the salt water solution, osmosis would transfer water into the solution and gradually dilute it, as

Desalination is the removal of salt from seawater.

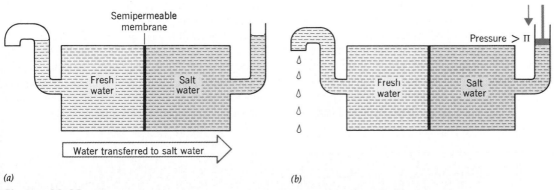

(a)   *(b)*

Figure 13.29. *Desalination by reverse osmosis. (a) If no pressure is applied to the salt solution, osmosis occurs in the normal direction. (b) Reverse osmosis occurs when a pressure larger than the osmotic pressure is applied to the solution.*

Portable hand-operated reverse osmosis devices are available for use in life rafts. The one shown in use here is capable of producing enough fresh water from seawater each day to keep 25 survivors alive.

A large-scale reverse osmosis unit has been used to provide water to the U.S. Naval Base at Guantanamo Bay, Cuba since Castro's takeover.

illustrated in Figure 13.29*a*. When the applied pressure is greater than the osmotic pressure, the osmosis is driven in the opposite direction and pure water is "squeezed" out of the sea water (Figure 13.29*b*). The pressures required for this are quite high and one type of membrane able to withstand the stress is a film of cellulose acetate placed over a suitable supporting structure. Cellulose acetate is permeable to water but impermeable to the ions and impurities in sea water. Desalination plants have been constructed that can produce as much as 3,000,000 gallons of fresh water daily.

13.12 SOLUTIONS OF ELECTROLYTES

For simplicity, we have limited our discussion thus far to solutions that do not contain an electrolyte. The reason for this is that the vapor pressure lowering, freezing point depression, boiling point elevation, and osmotic pressure depend on the *number* of particles present in the solution. One mole of a nonelectrolyte, such as sugar, when placed in water yields 1 mol of particles, so a solution labeled 1 *m* sucrose would have a freezing point that is 1.86 °C below that of pure water. However, a solution containing 1 mol of an electrolyte such as NaCl contains 2 mol of particles — 1 mol of Na^+ ions and 1 mol of Cl^- ions. As a result, a solution labeled 1 *m* NaCl actually contains 2 mol of particles per 1000 g of water and theoretically should have a freezing point depression of 2×1.86 °C = 3.72 °C. Similarly, a 1 *m* solution of $CaCl_2$, which contains 3 mol of ions per 1000 g of water, should have a freezing point depression three times as great as a 1 *m* solution of sucrose. (Actually, these predictions for NaCl and $CaCl_2$ are not entirely accurate, but they are close.) Similar predictions also hold fairly well for the other colligative properties.

EXAMPLE 13.10. CALCULATING THE FREEZING POINT DEPRESSION OF A SOLUTION THAT CONTAINS AN ELECTROLYTE

PROBLEM: What is the expected freezing point of a 0.150 *m* aqueous solution of $Al_2(SO_4)_3$?

SOLUTION: Of course, the first thing we have to recognize here is that aluminum sulfate is a strong electrolyte and we can assume it to be fully dissociated in aqueous solution. The equation for this dissociation is

$$Al_2(SO_4)_3 \longrightarrow 2Al^{3+} + 3SO_4^{2-}$$

so we see that 1 mol of the salt yields 5 mol of ions. Therefore, a solution labeled 0.150 m $Al_2(SO_4)_3$ doesn't contain 0.150 mol of particles per kilogram of water; it actually contains five times 0.150 mol, or 0.750 mol of particles. In other words, the solution is effectively 0.750 m in terms of its effect on the freezing point depression.

For water, K_f = 1.86 °C/m, so the expected freezing point depression for the solution is

$$\Delta T_f = 0.750 \ m \times 1.86 \ °C/m$$
$$= 1.40 \ °C$$

Since water freezes at 0 °C, the solution should freeze at -1.40 °C.

EXAMPLE 13.11. CALCULATING THE VAPOR PRESSURE OF A SOLUTION OF AN ELECTROLYTE

PROBLEM: A solution of calcium chloride was prepared by dissolving 25.0 g of $CaCl_2$ in exactly 500 g of H_2O. What is the expected vapor pressure of this solution at 80 °C? At 80 °C, water has a vapor pressure of 355 torr. What would the vapor pressure of the solution be if $CaCl_2$ were not an electrolyte?

SOLUTION: To solve the problem we must use Raoult's law.

$$P_{solution} = X_{solvent}P^0_{solvent}$$

so we need to calculate the mole fraction of water. For the $CaCl_2$,

$$25.0 \ g \ CaCl_2 \times \left(\frac{1 \ mol \ CaCl_2}{111 \ g \ CaCl_2} \right) = 0.225 \ mol \ CaCl_2$$

and for water

$$500 \ g \ H_2O \times \left(\frac{1 \ mol \ H_2O}{18.0 \ g \ H_2O} \right) = 27.8 \ mol \ H_2O$$

Before we compute X_{H_2O}, we have to realize that the 0.255 mol $CaCl_2$ will produce three times as many moles of ions, that is, 0.675 mol ions. Therefore,

$$X_{H_2O} = \frac{27.8 \ mol \ H_2O}{27.8 \ mol \ H_2O + 0.675 \ mol \ ions} = 0.975$$

The vapor pressure of the solution should therefore be

$$P_{solution} = 0.975 \times 355 \ torr$$
$$= 346 \ torr$$

If $CaCl_2$ were not an electrolyte, the mole fraction of water would have been

$$X_{H_2O} = \frac{27.8 \ mol \ H_2O}{27.8 \ mol \ H_2O + 0.225 \ mol \ CaCl_2} = 0.993$$

and

$$P_{solution} = 0.993 \times 355 \ torr$$
$$= 352 \ torr$$

Interionic attractions

Earlier in this section we mentioned that the actual freezing points of solutions of electrolytes such as NaCl and $CaCl_2$ are not exactly the same as those calculated if we

Table 13.6. Values of the van't Hoff factor at various concentrations

Salt	Concentration (mol/kg of H_2O)			*i* Factor If Completely Dissociated
	0.1	0.01	0.001	
NaCl	1.87	1.94	1.97	2.00
KCl	1.85	1.94	1.98	2.00
K_2SO_4	2.32	2.70	2.84	3.00
$MgSO_4$	1.21	1.53	1.82	2.00

assume complete dissociation. This happens because the ions in a solution are not really completely independent particles. Although the solvent shields them from each other's charge, this shielding is not perfect, and it becomes worse as the solution becomes more concentrated—that is, as the average distance between the ions becomes smaller. As a result, a concentrated solution behaves as if it has fewer ions than a dilute solution, and the ions' effectiveness at altering the properties of the solution (boiling point, freezing point, osmotic pressure) diminishes as the concentration of the solute becomes larger. Thus, ionic compounds behave as though they are less fully dissociated in concentrated solutions than when they are dilute.

Van't Hoff discovered the effects of interionic attractions.

Quantitatively, the degree to which an electrolyte behaves as though it were dissociated can be expressed by the **van't Hoff factor,** *i*. This quantity can be defined as the ratio of the observed freezing point depression produced by a solution to the freezing point that the solution would exhibit if the solute were a nonelectrolyte.

$$i = \frac{(\Delta T_f) \text{ measured}}{(\Delta T_f) \text{ calculated as nonelectrolyte}}$$

Table 13.6 has values of the *i* factor for several strong electrolytes. For NaCl, KCl, and $MgSO_4$, *i* approaches 2 as the solution becomes more dilute. For K_2SO_4, *i* approaches 3, as we expect.

It is interesting to compare the effects that the charges on the ions have on the interionic attractions. For NaCl, the value of *i* changes by about 5%, going from 0.1 *m* to 0.001 *m*. For K_2SO_4, which has a doubly charged SO_4^{2-} ion, *i* changes by about 22% for the same dilution. When there are two doubly charged ions in $MgSO_4$, the *i* factor changes by about 50% for the same dilution. These observations are not surprising, because as the charges on the cations and anions become larger, so should their attractions for each other. This leads to a lower degree of independence of the ions as their charges increase.

REVIEW QUESTIONS AND PROBLEMS

Problems whose numbers are in blue have their answers in Appendix D at the back of the book. The more difficult problems are marked with asterisks.

Mixtures; Suspensions and Colloids

13.1 What differentiates a mixture from a pure substance?

13.2 What range of particle sizes are found in suspensions? Give two examples of suspensions.

13.3 What range of particle sizes are found in colloids? What are these dimensions expressed in (a) millimeters and (b) inches?

13.4 How can a suspension of a solid in a liquid be separated? (Give two methods.)

13.5 How does a centrifuge work?

13.6 Describe the Tyndall effect. What kind of mixture displays the Tyndall effect? Why?

13.7 For the following colloids, identify the nature of (1) the dispersing phase, (2) the dispersed phase, and (3) the kind of colloid:

(a) Styrofoam
(b) cream
(c) lard
(d) jelly
(e) liquid rubber cement

13.8 What is an emulsifying agent?

13.9 Suppose you wished to determine whether a clear, colorless mixture was a solution or a colloid. What simple test could you perform?

13.10 What effect does electrical charge have on stabilizing a colloidal dispersion? How can a colloid that is stabilized in this way be coagulated?

13.11 When $AgNO_3$ is first added to a solution of NaCl, the mixture has a milky appearance that persists until enough Ag^+ has been added to react with all the Cl^-. At that point the precipitate coagulates and settles to the bottom of the container. Based on what you've learned in Section 13.1, explain these observations.

13.12 What is the one combination of phases for the dispersing agent and the dispersed phase that is incapable of yielding a colloidal dispersion? Why?

Types of Solutions

13.13 What kinds of solutions are possible?

13.14 How do the sizes of the particles in a solution compare to those in a colloidal dispersion or a suspension?

13.15 What is a substitutional solid solution? Give an example.

13.16 What is an interstitial solid solution? Give an example.

13.17 Suppose a solid solution is formed between two substances, one whose particles are very large and the other whose particles are very small. Which type of solid solution is this likely to be?

Concentration Units

13.18 State the definitions of the following concentration units: mole fraction, mole percent, weight fraction, weight percent, molarity, molality.

13.19 What feature do all concentration units have in common?

13.20 Calculate the mole fraction, weight fraction, weight percent, and molality of glycerin in a solution prepared by dissolving 45.0 g of glycerin, $C_3H_5(OH)_3$, in 100.0 g of H_2O.

13.21 A mixture is prepared from 45.0 g of benzene (C_6H_6) and 80.0 g of toluene (C_7H_8). Calculate (a) the weight percent of each component, (b) the mole fraction of each component, (c) the molality of the solution if toluene is taken to be the solvent.

13.22 A solution containing 121.8 g of $Zn(NO_3)_2$ per liter has a density of 1.107 g/mL. Calculate (a) the weight per-cent of $Zn(NO_3)_2$ in the solution, (b) the molality of the solution, (c) the mole fraction of $Zn(NO_3)_2$, (d) the molarity of the solution.

13.23 What are the mole fraction, molality, and weight percent of $CuCl_2$ in a solution prepared by dissolving 0.30 mol of $CuCl_2$ in 40.0 mol of H_2O?

13.24 A sample of drinking water was found to be severely contaminated with chloroform, $CHCl_3$, a known carcinogen. The level of contamination was 12.4 ppm (by weight).
(a) Express this in percent by weight.
(b) What is the molarity of the $CHCl_3$ in the water?

13.25 A solution of isopropyl alcohol (rubbing alcohol), C_3H_7OH, in water has a mole fraction of alcohol equal to 0.250. What is the weight percent alcohol and the molality of the alcohol in the solution?

13.26 The solubility of baking soda, $NaHCO_3$, in water at 20 °C is 9.6 g/100 g of H_2O. What is the mole fraction of $NaHCO_3$ in a saturated solution? What is the molality of the solution?

13.27 A saturated solution of NaCl at 30 °C has a molality of 6.25 m. What is the mole fraction and weight fraction of NaCl in the solution?

13.28 A solution of sodium carbonate was prepared containing 14.0% Na_2CO_3 by weight. What is the mole fraction and molality of Na_2CO_3 in this solution?

13.29 The World Health Organization's drinking water standards specify that the maximum permissible concentration of magnesium ion in drinking water is 150 mg/L. What does this correspond to in terms of (a) molarity and (b) molality (assume the density of the water is 1.00 g/mL)?

13.30 Concentrated sulfuric acid is 96.0% H_2SO_4 by weight. What is the molality of the H_2SO_4? What are the mole fractions of H_2SO_4 and water?

13.31 The molality of an aqueous solution of ammonium nitrate was 2.25 m. What was the weight percent of ammonium nitrate in the solution and what were the mole fractions of ammonium nitrate and water?

13.32 A solution of benzene, C_6H_6, dissolved in chloroform, $CHCl_3$, had a mole fraction of C_6H_6 equal to 0.240.
(a) What was the mole percent $CHCl_3$ in the solution?
(b) What was the molality of the C_6H_6 in chloroform?
(c) If we take benzene to be the solvent, what was the molality of the $CHCl_3$ in the C_6H_6?
(d) What were the weight percents of $CHCl_3$ and C_6H_6 in the solution?

13.33 An antifreeze solution is prepared from 222.6 g of ethylene glycol, $C_2H_4(OH)_2$, and 200.0 g of water. Its density is 1.072 g/mL. Calculate the molality and molarity of the solution.

13.34 A 4.03 M solution of ethylene glycol, $C_2H_4(OH)_2$, has a density of 1.045 g/mL. Calculate the weight percent $C_2H_4(OH)_2$, mole fraction of $C_2H_4(OH)_2$ and the molality of the solution.

13.35 Below is a list of the most abundant ions in seawater.

Ion	Molality
Chloride	0.566
Sodium	0.486
Magnesium	0.055
Sulfate	0.029
Calcium	0.011
Potassium	0.011
Bicarbonate	0.002

Calculate the mass, in grams, of each component contained in 3.78 L (1.00 gal) of seawater with a density of 1.024 g/mL. What is the total mass of ions in this sample?

*13.36 Suppose that you wish to prepare a solution containing 10.0% Na_2CO_3 by weight. The bottle of chemical that you have lists the contents as $Na_2CO_3 \cdot 10H_2O$. How many grams of the hydrate would be needed to prepare 50.0 g of the 10.0% Na_2CO_3 solution?

The Solution Process

13.37 Viewed on a molecular level, why is the distribution of gas molecules illustrated in Figure 13.7b not maintained indefinitely? (In other words, why do the gas molecules mix spontaneously?)

13.38 What role does the tendency toward disorder play in the formation of an alcohol–water solution?

13.39 When NaCl is added to a beaker containing liquid hexane, C_6H_{14}, none of the salt dissolves. Why doesn't nature's tendency toward disorder cause them to form a solution?

13.40 The solubilities in water of alcohols with the general formula $C_nH_{(2n+1)}OH$ decrease as the value of n increases. Why?

13.41 Molecular H_2 and O_2 are relatively insoluble in water, but ammonia is quite soluble. Why?

13.42 Which gas, NH_3 or PH_3, would you expect to be more soluble in water? Why?

13.43 As applied to solubility, what is the significance, on a molecular level, of the phrase "like dissolves like"?

13.44 Frequently a liquid that is soluble in water can be made less soluble (and can therefore be made to separate as a distinct phase) by addition of salt to the solution. How can this be explained?

13.45 Small amounts of water in the fuel tank of an automobile can cause severe difficulties in engine performance. The problem can be overcome by adding "dry gas" to the fuel. The "dry gas" consists mostly of methyl alcohol, CH_3OH, which allows the water to dissolve in the gasoline. Can you explain how the dry gas accomplishes this?

13.46 What is meant when we say an ion is hydrated? What is the meaning of the term solvation?

13.47 What role does water play in dissolving an ionic compound?

13.48 What is a micelle? Why does soap form micelles? How does a soap or detergent remove a grease spot from clothing? • •

13.49 What advantages do synthetic detergents have over soap?

Heats of Solution

13.50 How is the heat of solution defined? What happens to the temperature of a mixture as a solution is formed if the ΔH_{soln} is negative?

13.51 The combined attractions between ethanol molecules in pure ethanol and CCl_4 molecules in pure CCl_4 are larger than the attractions between ethanol and CCl_4 molecules. When ethanol dissolves in CCl_4, is ΔH_{soln} positive or negative?

13.52 Acetone has the structure shown in the margin on page 426. Why are acetone molecules attracted to water molecules more strongly than they are attracted to other acetone molecules?

13.53 Ethanol, C_2H_5OH, and hexane, C_6H_{14}, form solutions in which more energy is expended overall separating the solvent and solute molecules from each other than is recovered when the two components are brought together. Why?

13.54 Compare the definitions of an ideal gas and an ideal solution.

13.55 Discuss the relationship between lattice energy and hydration energy in determining the magnitude and sign of the heat of solution of a solid in a liquid.

13.56 Why is ΔH_{soln} for gases nearly always negative?

13.57 The heat of solution for $AlCl_3$ is -321 kJ/mol. What is the probable reason for this highly exothermic ΔH_{soln}?

13.58 For NaCl, $\Delta H_{soln} = +4.94$ kJ/mol. For this salt, which is larger: the lattice energy or the hydration energies of the ions?

13.59 Use the data in Table 13.3 to calculate the amount of heat liberated by dissolving 10.0 g of $AlCl_3$ in 1.00 L water.

13.60 How many calories are absorbed by dissolving 115 g of NH_4NO_3 in 100 mL of H_2O? Use the data in Table 13.3.

The Effect of Temperature on Solubility

13.61 The solution process for KI in water is endothermic (ΔH_{soln} is positive). Would you expect KI to become more or less soluble as the temperature is increased? Explain this on the basis of Le Châtelier's principle.

13.62 Describe qualitatively the procedure called fractional crystallization.

13.63 Why do gases nearly always become less soluble in liquids as the temperature is raised?

The Effect of Pressure on Solubility

13.64 Why do gases become more soluble in liquids as the pressure is increased?

13.65 Why do pressure changes have very little effect on the solubility of a solid in a liquid?

13.66 The partial pressure of ethane over a saturated solution containing 6.56×10^{-2} g of ethane is 751 torr. What is its partial pressure when the saturated solution contains 5.00×10^{-2} g of ethane?

13.67 The Henry's law constant for a gas dissolved in water was found to be 6.50×10^{-5} g/L torr at 25 °C. In an experiment the gas was collected over water and its concentration was found to be 0.0478 g/L. What was the *total* pressure of gas above the solution?

13.68 Methane (natural gas) has a solubility in water at 25 °C of 2.09×10^{-4} g/L at a pressure of 0.968 atm. If we assume Henry's law is valid, what would be the concentration of methane (in grams per liter) in ground water deep below the earth's surface at a pressure of 1000 atm and a temperature of 25 °C?

* 13.69 Air contains approximately 20% O_2 by volume. The Henry's law constant for O_2 at 25 °C is 5.34×10^{-5} g/L torr. Calculate the mass, in grams, of O_2 per liter of water in a stream that has a temperature of 25 °C if the atmospheric pressure is 760 torr. (Assume equilibrium with the atmosphere.)

Vapor Pressures of Solutions: Raoult's Law

13.70 Explain, on a molecular level, why the vapor pressure of the solvent is expected to be directly proportional to its mole fraction in the solution (Raoult's law).

13.71 What are meant by positive and negative deviations from Raoult's law?

13.72 How is the sign of ΔH_{soln} related to positive and negative deviations from Raoult's law?

13.73 The vapor pressure of benzene (C_6H_6) at 25 °C is 93.4 torr. What will be the vapor pressure in torr at 25 °C of a solution prepared by dissolving 56.4 g of the nonvolatile solute, $C_{20}H_{42}$, in 1000 g of benzene?

13.74 The vapor pressure of pure methyl alcohol at 30 °C is 160 torr. What mole fraction of glycerol (a nonvolatile nondissociating solute) would be required to lower the vapor pressure to 130 torr?

13.75 Heptane (C_7H_{16}) has a vapor pressure of 791 torr at 100 °C. At this same temperature, octane (C_8H_{18}) has a vapor pressure of 352 torr. What will be the vapor pressure, in torr, of a mixture of 25.0 g of heptane and 35.0 g of octane? Assume ideal solution behavior.

13.76 At 25 °C the vapor pressures of benzene (C_6H_6) and toluene (C_7H_8) are 93.4 and 26.9 torr, respectively. At what applied pressure, in torr, will a solution prepared from 60.0 g of benzene and 40.0 g of toluene boil at 25 °C?

13.77 The vapor pressure of a mixture containing 400 g of carbon tetrachloride and 43.3 g of an unknown substance is 137 torr at 30 °C. The vapor pressure of pure carbon tetrachloride at 30 °C is 143 torr, while that of the pure unknown is 85 torr. What is the approximate molecular mass of the unknown?

13.78 A solution containing 8.3 g of a nonvolatile nondissociating substance dissolved in 1 mol of chloroform, $CHCl_3$, has a vapor pressure of 511 torr. The vapor pressure of pure $CHCl_3$ at the same temperature is 526 torr. Calculate (a) the mole fraction of the solute, (b) the number of moles of solute, (c) the molecular mass of the solute.

13.79 To what Celsius temperature would a solution containing 150 g of glycerol, $C_3H_5(OH)_3$, in 100 g of H_2O have to be heated to have a vapor pressure of 91.1 torr? Assume ideal solution behavior, $C_3H_5(OH)_3$ to be a nonvolatile solute, and refer to Table 11.2.

Boiling Point Diagram and Fractional Distillation

13.80 Describe, qualitatively, the procedure called fractional distillation.

13.81 If we refer to Figure 13.23, approximately how many times must boiling, followed by condensation of the resulting vapor, be repeated in order to obtain a portion of liquid having a mole fraction of A of at least 0.80 if the original mole fraction of A was 0.20?

13.82 Benzene has a boiling point of 80.1 °C; carbon tetrachloride boils at 76.8 °C. Sketch a boiling point diagram for benzene–carbon tetrachloride mixtures. Assume ideal solution behavior.

13.83 Water and butyl alcohol form an azeotrope that boils at 92.4 °C at 760 torr. At this same pressure butyl alcohol boils at 117.8 °C. The composition of the azeotrope is 28.4 mol % butyl alcohol and 71.6 mol % water. Sketch the boiling point diagram for butyl alcohol–water mixtures. Do these substances show positive or negative deviations from ideality?

Boiling Point Elevation and Freezing Point Depression

13.84 What is a *colligative property*?

13.85 We've described how the addition of a nonvolatile solute to a solvent reduces the escaping tendency of the solvent from the solution, and in Section 13.10 we saw that this leads to a boiling-point elevation. On a molecular level, account for the fact that the presence of a solute also reduces the tendency of the solvent to escape from the liquid onto the solid. Explain why a lower temperature must be achieved to establish equilibrium between the solid solvent and the solution than between the pure solid and liquid solvent.

13.86 What will be the freezing point and boiling point of an aqueous solution containing 55.0 g of glycerol, $C_3H_5(OH)_3$, dissolved in 250 g of water? Glycerol is a nonvolatile, undissociated solute.

13.87 What is the molecular mass and molecular formula of a nondissociating compound whose empirical formula is C_4H_2N if 3.84 g of the compound in 500 g of benzene gives a freezing point depression of 0.307 °C?

13.88 A solution containing 16.9 g of a nondissociating substance in 250 g of water has a freezing point of −0.774 °C. The substance is composed of 57.2% C, 4.77% H, and 38.1% O. What is the molecular formula of the compound?

13.89 How many grams of glucose, $C_6H_{12}O_6$ (a nondissociating solute), are required to lower the freezing point of 150 g of H_2O by 0.750 °C? What will be the boiling point of this solution in °C?

13.90 An aqueous solution freezes at −2.47 °C. What is its boiling point in °C?

*13.91 Calculate the freezing point in °C of a 0.100 m aqueous solution of a weak electrolyte that is 7.5% dissociated.

*13.92 The cooling system of an automobile usually contains a solution of antifreeze prepared by mixing equal volumes of ethylene glycol, $C_2H_4(OH)_2$, and water. The density of ethylene glycol is 1.113 g/mL. Calculate the freezing point of this mixture. The label of the antifreeze container states that this mixture will protect your engine to a temperature of −34 °F. How does your computed freezing point compare with this?

Osmotic Pressure

13.93 What is the difference between osmosis and dialysis?

13.94 What is a *semipermeable membrane*?

13.95 What are *isotonic solutions*? Why must the solute concentration be carefully controlled in intravenous feeding?

13.96 Calculate the osmotic pressure, in torr, of an aqueous solution containing 5.0 g of sucrose, $C_{12}H_{22}O_{11}$, per liter at 25 °C.

13.97 A solution of 0.400 g of a polypeptide in 1.00 L of an aqueous solution has an osmotic pressure at 27 °C of

3.74 torr. What is the approximate molecular mass of this polymer?

Solutions of Electrolytes

13.98 If we assume complete dissociation, what would the expected freezing point of a 0.10 m solution of $MgSO_4$ be?

13.99 What would be the expected freezing point, in °C, of 0.10 m $CaCl_2$?

13.100 If the measured freezing point depression of a solution is *less* than that calculated for a solute, what does this suggest?

13.101 What would be the osmotic pressure, in torr, of a 0.010 M aqueous solution of the electrolyte NaCl at 25 °C? (Assume 100% dissociation of NaCl in water.)

13.102 What is the osmotic pressure, in atm, of seawater at 25 °C? What is the minimum pressure, in atm, needed to desalinate seawater by reverse osmosis? (See the data in Problem 13.35.)

Interionic Attractions

13.103 What would be the actual freezing point, in °C, of a 0.10 m $MgSO_4$ solution if we take into account interionic attractions?

13.104 From the data in Table 13.6, which 1 : 1 electrolyte (one positive ion to one negative ion) appears to be *least* fully dissociated in concentrated solutions? How does this agree (or disagree) with what might be predicted based on the charges on the ions involved?

13.105 If we assume equal concentrations in water, which of the following salts would you expect to appear to be least fully dissociated: KCl, $NiCl_2$, or $Al_2(SO_4)_3$?

13.106 If we assume complete dissociation, what i factor would you expect for each of the salts in Question 13.105?

Additional Exercises

13.107 A 0.130 m solution of Hg_2Cl_2 has a freezing point of −0.72 °C. Explain why this is consistent with the mercury(I) ion having the formula Hg_2^{2+} rather than Hg^+.

13.108 What is the vapor pressure at 25 °C of a solution of a nonvolatile solute in benzene if the solution freezes at 2.63 °C? The vapor pressure of pure benzene at 25 °C is 93.4 torr and pure benzene freezes at 5.50 °C.

13.109 A solution consists of 10.0% w/w ethylene glycol, $C_2H_4(OH)_2$, in water. What are the freezing and boiling points of this solution in °C? What is the vapor pressure of this solution at 100 °C?

FACTORS THAT CONTROL THE OUTCOME OF REACTIONS

14 CHEMICAL THERMODYNAMICS

By this time you have learned quite a few facts about the chemical and physical properties of substances. You've learned about the kinds of compounds that many of the elements form, about the physical properties of the states of matter, and about the properties of solutions. In studying these topics, you've learned that not all changes are possible. For example, you learned that certain metals such as zinc dissolve in HCl, but others such as copper do not. In the last chapter you learned that certain substances are mutually soluble, such as water and ethanol, but others are mutually insoluble.

A question that has always fascinated those who have studied these phenomena is "What is it that determines whether or not a chemical or physical change is possible?" In other words, are there any criteria that we can apply to chemical and physical changes to determine which can happen and which can't? The answer, as you may have guessed, is yes, and these criteria come to us through the study of a subject called thermodynamics.

Thermo implies heat; *dynamics* implies movement or change.

Thermodynamics is basically concerned with the energy changes that accompany chemical and physical processes. If this sounds familiar, it is no accident; you've already studied one aspect of thermodynamics — heats of reaction — in Chapter 6. We will extend this study a little further in this chapter, and you will see that a knowledge of heats of reaction allows us to learn about the strengths of chemical bonds between atoms. You will also see that the energy change is *one of two* factors determining whether or not an event can happen.

Finally, before we move on to the study of thermodynamics, it should be mentioned that two conditions must be fulfilled in order for us to actually observe a chemical or physical change. One of these, which is addressed by thermodynamics, is the question of whether or not the change is possible. The other factor is the speed at which the change occurs. Even if thermodynamics predicts that a change is possible, it must occur at a reasonably rapid pace, otherwise, we might never live long enough to see it happen. For example, we will see that thermodynamics predicts that H_2 and O_2 should react spontaneously to form water. However, if we place a mixture of these gases in a container at room temperature, nothing appears to happen. In fact, the mixture appears to be stable indefinitely (unless someone strikes a match). This is because at room temperature the reaction of H_2 and O_2 is so slow that, even though their reaction is spontaneous, it takes nearly forever to occur.

14.1 SOME COMMON TERMS

We defined most of the special terms that you will need to know for thermodynamics in Chapter 6. These include the concepts of **system** and **surroundings.** You should know what is meant by the **state** of a system and you should know what it means for a quantity to be a **state function.** You should also know the difference between an **isothermal change** and an **adiabatic change.** It has probably been some time since you've used these terms, and if you are not sure of their meanings, you would be wise to return to Section 6.4 beginning on page 176 to review them. It is important that they be part of your vocabulary as you follow the discussions in this chapter.

14.2 CONSERVATION OF ENERGY AND THE FIRST LAW OF THERMODYNAMICS

Thermodynamics is organized around a set of fundamental laws of nature that relate to energy and energy transfers between a system and its surroundings. The laws are specified by number; thus, there is the first law of thermodynamics, the second law, and so forth.

The first law of thermodynamics is essentially a restatement of the law of conservation of energy. Recall that this law tells us that *in chemical and physical changes, energy cannot be created or destroyed; it can only be changed from one kind to another.* Thermodynamics, however, says this in a somewhat more elegant way. The **first law of thermodynamics** states that *if a system undergoes some series of changes that ultimately brings it back to its original state, the net energy change is zero.* In effect, this is telling us that energy is a state function.

Normally, we are not very interested in processes that bring us back to where we started. Instead, we are concerned with how the energy changes when a system is transformed from one state to another. Thermodynamics therefore defines a quantity, called the **internal energy, E,** that is used in describing these changes when they take place in either chemical or physical systems. The internal energy is a state function that corresponds to the total energy of the system. It is equal to the sum of all the energies that the system possesses as a consequence of the kinetic energy of its atoms, ions, or molecules, plus all the potential energy that arises from the binding forces between the particles making up the system. A change in the internal energy, ΔE, is defined as

$$\Delta E = E_{final} - E_{initial}$$

or, for a chemical system,

$$\Delta E = E_{products} - E_{reactants}$$

The first law says that if E_{final} and $E_{initial}$ are the same, then $\Delta E = 0$.

Notice that we have used the same convention here (final minus initial) as in our previous discussions of enthalpy changes. Here, too, we cannot actually measure or calculate E_{final} or $E_{initial}$. The reason, which was discussed in Chapter 6, is that we have no way of knowing exactly how fast something is moving (so we can't determine its kinetic energy), and we have no way of taking into account all the attractive and repulsive forces felt by the particles of a system (so we can't determine its total potential energy). But, as in earlier discussions, this is really no terrible crisis, because we are interested only in energy changes, and these we can measure.

When a system changes from one state to another, there are two ways for it to exchange energy with its surroundings. One is for it to gain or lose heat energy. If the system absorbs heat, its energy rises, and if it loses heat, its energy drops. The second way for the system to exchange energy with its surroundings is to do work or have work done on it. If the system does work, its energy drops (just as you have less energy after doing work). On the other hand, if work is done on the system, its energy rises. For example, when you (the surroundings) do work on a watch (the system) by winding its spring, the energy of the watch increases. The energy bookkeeping for both heat and work is taken care of by the equation

$$\Delta E = q + w \tag{14.1}$$

where q is defined as the *heat absorbed by the system from the surroundings* when the system undergoes a change, and w is the *work done on the system by the surroundings*. Thus, if a system absorbs heat, its energy increases. That's pretty obvious. The energy of the system also increases if work is done on it. For example, suppose the system is a spring. If you compress the spring, you do work on it and in the process you give the spring more energy than it had before.

Since Equation 14.1 deals with the transfer of amounts of energy, it is important that we establish certain sign conventions to avoid confusion in our bookkeeping. Heat *added to* a system and work *done on* a system are both defined as positive quantities. On the other hand, if heat is lost by the system, its value is given a negative sign, and if it is the system that does the work on the surroundings, then the amount of work involved is assigned a negative sign. Let's look at some examples.

Suppose a change occurs in which the system absorbs 50 J of heat and does 30 J of work. For this change, $q = +50$ J and $w = -30$ J. The change in the internal energy of the system is

$$\Delta E_{\text{system}} = (+50 \text{ J}) + (-30 \text{ J})$$
$$= +20 \text{ J}$$

Thus, the system undergoes a net increase of 20 J.

We could also focus our attention on the change in the surroundings. If the system absorbs 50 J of heat, then the surroundings must lose this much heat, so *for the surroundings*, $q = -50$ J. Similarly, the surroundings have 30 J of work done on it, so for the surroundings, $w = +30$ J. Therefore,

$$\Delta E_{\text{surroundings}} = (-50 \text{ J}) + (+30 \text{ J})$$
$$= -20 \text{ J}$$

Notice that

$$\Delta E_{\text{system}} + \Delta E_{\text{surroundings}} = 0$$

which, in a sense, is another statement of the law of conservation of energy. The total amount of energy gained by the system is *exactly equal* to the total amount of energy lost by the surroundings. Nothing is lost and nothing is gained.

In summary

q positive ($q > 0$): Heat is added to the system.

w positive ($w > 0$): Work is done on the system; energy is added.

q negative ($q < 0$): Heat is evolved by (removed from) the system.

w negative ($w < 0$): Work is done by the system; energy is removed.

(When q and w are positive, energy enters the system; when they are negative, energy leaves the system.)

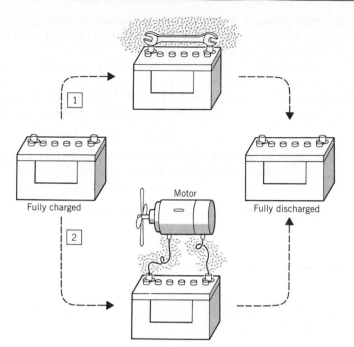

Figure 14.1. *The complete discharge of a battery gives the same net energy change, ΔE, regardless of the path, but how much of the energy appears as work and how much as heat depends on how the discharge is carried out. Along path 1, all the energy is released as heat and no work is done. Along path 2, some of the available energy is harnessed as work and the rest appears as heat.*

q and *w* are not state functions

When a system changes from one state to another, the value of ΔE for the change is independent of how the change takes place because E is a state function. The same is not true, however, for q and w. These quantities are not state functions and their values depend very much on the path that's followed. As an illustration, let's consider the discharge of an ordinary automobile battery, in which a redox reaction provides the source of electricity. If we take one state to be the fully charged battery and another to be the dead battery, we know that regardless of how we go from the first state to the second, ΔE will be the same. But how about q and w?

Suppose we decide to discharge the battery by placing a heavy wrench across the terminals, as shown in Figure 14.1. If you've every done this, even by accident, you know that sparks fly and the wrench gets very hot. (In fact, so much heat is generated that the wrench may melt and the battery may explode!) When we discharge the battery along this path, no work is done ($w = 0$) and the entire energy change appears as heat. On the other hand, if we connect the terminals to a motor, part of the net energy change now can be harnessed as work, so less of it appears as heat. Thus, depending on how we draw energy from the battery, q and w can vary even though their sum, ΔE, is the same.

Work in physical and chemical systems

Electrical work is just one kind of work a system can do, or have done on it. Another important kind of work is related to the expansion or compression of a system that feels an opposing pressure. For example, when gasoline burns in the cylinder of an automobile engine, the reaction produces very hot gases that exert a very high pressure. This pressure pushes a piston, which in turn is attached to machinery that ultimately turns the wheels and makes the car move. In the process the hot gases lose energy because of the work they've done. In an air compressor, just the opposite happens. A motor drives a piston that does work on air in a cylinder, compressing it to a high pressure. This gives energy to the gas, which can later do work by expanding again. Thus, compression and expansion provide another way for work to add or remove energy from a system.

The amount of work associated with an expansion or compression is related to the pressure exerted on the system as well as the volume change that the system undergoes. We can see this by first realizing that work is accomplished by moving an opposing force by some distance. Expressed mathematically,

$$work = force \times distance$$

Now, suppose we have a gas in a cylinder that's fitted with a piston, as illustrated in Figure 14.2, and that the gas undergoes an expansion by pushing the piston upward. The piston exerts a pressure on the gas that comes from a force F spread over the area of the piston A.

$$P = \frac{F}{A}$$

The volume of the gas in the cylinder is equal to its cross-sectional area, A, multiplied by the height of the column of gas, h.

$$V = Ah$$

When the gas expands and pushes back the piston, A remains the same but h changes. The volume change is therefore

$$\begin{aligned} \Delta V &= V_f - V_i \\ &= Ah_f - Ah_i \\ &= A(h_f - h_i) = A(\Delta h) \end{aligned}$$

The product of pressure times volume change is

$$P\,\Delta V = \frac{F}{A}\,A(\Delta h) = F\,\Delta h$$

We see that $P\,\Delta V$ is equivalent to force (F) times distance (Δh) and is therefore equal to work. In fact, $P\,\Delta V$ is the amount of work that the system does on the surroundings when it expands, so we can say

$$w_{\text{surroundings}} = P\,\Delta V$$

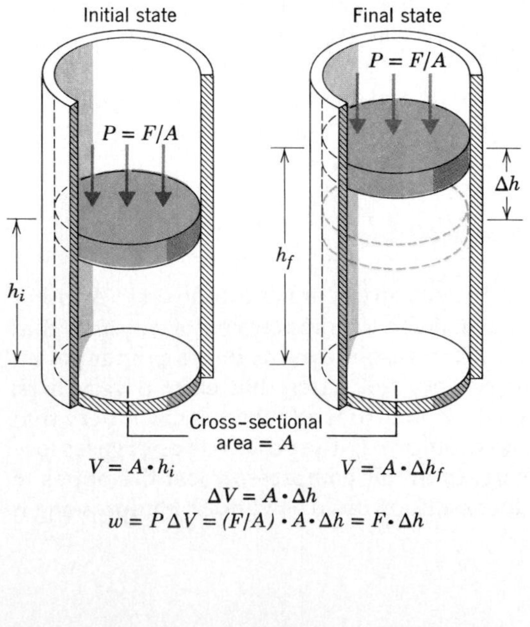

In the SI, pressure is expressed in pascals (newtons per square meter)

$$1\ Pa = 1\ N/m^2$$

and volume is expressed in cubic meters, m^3. Therefore, the units of $P\,\Delta V$ are

$$Pa\ m^3 = \frac{N}{m^2} \cdot m^3 = N \cdot m$$

but $1\ N \cdot m = 1\ J$. Therefore,

$$1\ Pa\ m^3 = 1\ J$$

Figure 14.2. *Expansion work equal to P ΔV is done by a gas when it expands against an opposing pressure.*

Initial state   Final state

$P = F/A$

$P = F/A$

h_i h_f Δh

Cross-sectional area = A

$$V = A \cdot h_i \qquad V = A \cdot \Delta h_f$$
$$\Delta V = A \cdot \Delta h$$
$$w = P\Delta V = (F/A) \cdot A \cdot \Delta h = F \cdot \Delta h$$

But if the surroundings gain energy by having work done on them, the system must lose an equivalent amount of work, so *for the system,*

$$w_{system} = -P \, \Delta V \qquad \qquad (14.2)$$

This expansion work is done by any system when it expands against an external pressure imposed by the surroundings; the system doesn't have to be a gas. Conversely, when a system contracts under the influence of some external pressure, work is done on it.

If we express pressure in units of atmospheres and the volume change in units of liters, then $P \, \Delta V$ has the units *liter* $\times$ *atmosphere* (L atm). We can convert this to the more familiar energy units of joules or calories by these relationships:

$$1 \text{ L atm} = 101.3 \text{ J}$$

$$1 \text{ L atm} = 24.2 \text{ cal}$$

Expansion work and the first law of thermodynamics

We can use what we've learned about expansion work to demonstrate quantitatively that q and w are not state functions. To do this, let's consider the isothermal expansion of a gas (for simplicity, an ideal gas). This could be accomplished by placing a cylinder of the gas in a large vat of water held at constant temperature by a thermostat.

If the temperature is constant during the expansion, then the average kinetic energy of the molecules stays the same. Furthermore, since there are no attractive forces between the molecules of an ideal gas, there is no potential energy. This means that when an ideal gas expands or contracts at constant temperature, there is no net change in its total energy, so $\Delta E = 0$. Thus,

Since the change in K.E. and P.E. are both zero, $\Delta E = 0$.

$$\Delta E = q + w = 0$$

and

$$q = -w$$

What this tells us is that if the gas does some work on the surroundings by expanding, it must absorb an equal amount of heat from the surroundings in order for ΔE to be zero.

This really only holds for an ideal gas. For a real gas ΔE is small, but not equal to zero.

If you've ever lifted a heavy box, you know that the amount of work you have to do depends on how heavy the box is — that is, on how hard the box pushes back. So it is with the expansion of a gas. In Figure 14.3a, we see a compressed gas with a

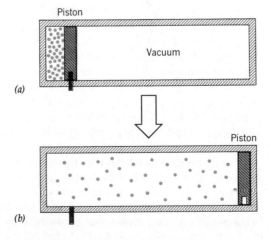

(a)

(b)

Figure 14.3. *Isothermal expansion of an ideal gas against an opposing pressure equal to zero. (a) Before expansion, P — 10.0 atm and V = 1.00 L. (b) After expansion P = 1.00 atm and V = 10.0 L.*

pressure of 10.0 atm and a volume of 1.00 L pushing against a piston held by a pin through the cylinder wall. On the other side of the piston is a vacuum, and the apparatus is designed so that when the piston is moved to the other end, the gas can occupy a volume of 10.0 L.

When the pin is pulled out, the compressed gas will drive the piston all the way to the opposite end of the cylinder (Figure 14.3b). The gas expands to a volume of 10.0 L and, by applying the combined gas law, we can calculate that its pressure drops to 1.00 atm. However, the gas doesn't actually do any work in this expansion because the piston doesn't push back; there is no opposing pressure during the expansion, so $P \Delta V = 0(10.0 \text{ L} - 1.0 \text{ L}) = 0$. This means that $w = 0$ and since $\Delta E = 0$ and $q = -w$, $q = 0$.

Now let's consider the expansion illustrated in Figure 14.4. Here we begin with the same gas at the same temperature as before and at the same pressure and volume ($P = 10.0$ atm and $V = 1.00$ L). When the gas is allowed to push back the piston against the opposing pressure of 1.00 atm, it expands until its pressure drops to 1 atm, which means that its volume increases to 10.0 L. Thus, in this experiment the gas changes to the same volume and pressure as in the previous expansion. As far as the gas is concerned, then, the initial and final states are the same in both expansions, and $\Delta E = 0$ for both. However, in the expansion pictured in Figure 14.4, q and w are *not* zero because the pressure opposing the expansion is no longer zero. For the system,

$$w = -P \Delta V$$

The volume change is $\Delta V = V_f - V_i = 10.0 \text{ L} - 1.0 \text{ L} = +9.0 \text{ L}$. Therefore,

$$w = -(1.00 \text{ atm}) (+9.0 \text{ L}) = -9.0 \text{ L atm}$$

opposing pressure

This is -912 J or -218 cal.

Since w equals -9.0 L atm, q must equal $+9.0$ L atm.

In summary, both of the experiments that we've described have the same initial and final states and the same value for ΔE. However, for the first, $w = 0$ and $q = 0$, and for the second $w = -9.0$ L atm and $q = +9.0$ L atm. Thus, we have shown that q and w are not state functions, even though their sum is.

At constant temperature,

$$P_i V_i = P_f V_f$$

$$(10 \text{ atm})(1 \text{ L}) = (1 \text{ atm})(10 \text{ L})$$

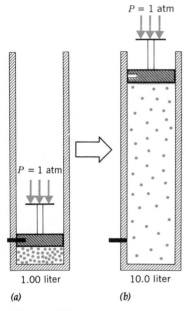

$P = 1$ atm

$P = 1$ atm

1.00 liter   10.0 liter

(a)   (b)

Figure 14.4. *Isothermal expansion of an ideal gas against a constant opposing pressure of 1 atm. (a) Initial state: P = 10.0 atm, V = 1.00 L. (b) Final state: P = 1.00 atm, V = 10.0 L.*

Reversible processes and maximum work

One of the main reasons we carry out chemical reactions in our modern society is to provide energy for the production of work. But we have seen in the previous discussions that the amount of work that is accomplished by a system depends on how a change is carried out. For example, if we discharge a battery quickly by offering essentially no resistance to the flow of electricity, no work is done and all the energy appears as heat. As we increase the resistance (by attaching a motor across the terminals, for example), we slow down the rate at which energy is removed, but now some of the available energy accomplishes work. As you might imagine, the limit of this would be the point where the energy is withdrawn infinitely slowly by having the battery just barely able to overcome the opposing resistance, in which case the amount of work done is a maximum.

Any change in which the opposing "force" is virtually balanced by the driving force is called a **reversible process,** because just the smallest increase in the opposing force will reverse the direction of the change. An example in a physical system is the piston–cylinder apparatus illustrated in Figure 14.5. As the water evaporates one molecule at a time, the pressure opposing the expansion is very gradually decreased and the gas expands very gradually, with the internal pressure always nearly matching the external pressure. This is essentially a reversible process because if just one water molecule were to condense, the pressure would increase just slightly and the direction of the change would be reversed.

In the expansion in Figure 14.5, the opposing pressure is always kept as high as it can be while still permitting the expansion to take place. Because the amount of work done by the gas depends on the external pressure, this process will produce more work than any other expansion that we could carry out between the same two states. Similarly, the discharge of the battery against a resistance just barely less than the force that drives the electricity through the wires also produces the maximum work. In fact, it is a general rule that *the maximum work available from any change is obtained if the change occurs by a reversible process.* Unfortunately, like the infinitely slow evaporation of the water one molecule at a time from the apparatus in Figure 14.5, reversible processes take forever to occur. All real spontaneous changes do not take place in a "reversible" manner, and the amount of work that we can obtain from them is always less than the theoretical maximum.

14.3 HEATS OF REACTION

In Chapter 6 we described the measurement of heats of reaction, and you may recall that we made a distinction between the heat of reaction measured under conditions of constant volume and that measured under conditions of constant pressure. Let's look at how thermodynamics views these quantities, so we can see why they are different.

One of the devices used for measuring heats of reaction is a bomb calorimeter, which was described on page 175. As we mentioned in our discussion, this apparatus measures the heat of reaction at constant volume because during the reaction, no change in volume takes place. Since $\Delta V = 0$, $P\,\Delta V = 0$ and $w = 0$; no expansion work is done by either the system or its surroundings. Therefore, the heat of reaction equals ΔE.

$$\Delta E = q + w = q + 0$$

Labeling the heat with the subscript v to denote constant volume conditions, we obtain

$$\Delta E = q_v$$

Because most of the reactions that are of interest to us take place at constant pressure instead of constant volume, thermodynamicists invented the quantity we call enthalpy. The enthalpy, H, of a system is defined as

$$H = E + PV \tag{14.3}$$

and for a change at constant pressure,

$$\Delta H = \Delta E + P\,\Delta V \tag{14.4}$$

If only expansion work is possible for the system, then

$$\Delta E = q - P\,\Delta V$$

and Equation 14.4 becomes

$$\Delta H = (q - P\,\Delta V) + P\,\Delta V$$

which reduces to

$$\Delta H = q_p$$

Thus, as we said in Chapter 6, the enthalpy change for a reaction equals the heat of reaction measured at constant pressure, q_p.

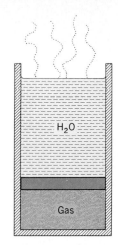

Figure 14.5. *A reversible expansion of a gas. As water molecules gradually evaporate one at a time, the pressure decreases and the gas expands. It is a reversible process because if one H_2O molecule were to condense, the direction of the change would be reversed.*

A bomb calorimeter is normally used to measure ΔE for exothermic reactions.

The difference between ΔE and ΔH

If you look at Equation 14.4, you will see that ΔE and ΔH differ by the term $P \, \Delta V$. For reactions occurring in open containers, this is the amount of work done by or on the system when it expands or contracts under the opposing pressure of the atmosphere. For example, suppose a certain reaction is exothermic and therefore liberates energy when it occurs. If the reaction takes place at constant volume, no work can be done and all the energy can appear as heat, which is the heat of reaction at constant volume, ΔE. However, if the reaction occurs at constant pressure instead, its volume can change. If the reaction produces gases, their expansion at constant pressure pushes back the atmosphere and the system does some work. This means it uses some of the energy of the reaction to do work, so less is available to come off as heat. Under these conditions the heat of reaction at constant pressure (ΔH) is a little bit less than the heat of reaction at constant volume. In a sense, we pay a penalty equal to $P \, \Delta V$ to carry out the reaction at constant pressure. On the other hand, if the system contracts during the reaction, we actually get a bonus; a little more heat is given off because the $P \, \Delta V$ work done by the surroundings adds some energy to the system.

The difference between ΔE and ΔH is never very large, even when gases are produced or consumed in a reaction.

14.4 BOND ENERGIES AND HEATS OF REACTIONS

In a chemical reaction, potential energy changes occur when bonds are broken and new ones are formed, and they are reflected in the amount of energy absorbed or given off when the reaction occurs. It seems reasonable, therefore, that we should be able to relate heats of reaction to the potential energy changes associated with bond forming and bond breaking. Strictly speaking, we should use ΔE for this; however, since $P \, \Delta V$ contributions to ΔH are relatively small for chemical reactions, we can use ΔH in place of ΔE and still expect to obtain quite reasonable results. Therefore, we shall use the terms bond energy and bond enthalpy interchangeably.

If we are dealing with a molecule, the **bond energy** is defined as the amount of energy needed to break a chemical bond to produce electrically neutral fragments. Thus, for H_2, the energy absorbed in the following reaction is the bond energy.

$$H_2(g) \longrightarrow 2H(g)$$

For a complex molecule, the energy needed to break all the bonds and reduce the gaseous molecule entirely to neutral gaseous atoms is called the **atomization energy**, which we will represent as ΔH_{atom}. Its value is the sum of all the bond energies in the molecule. Thus, for methane, CH_4, the atomization energy is the energy absorbed in the process

$$CH_4(g) \longrightarrow C(g) + 4H(g)$$

Simple diatomic molecules such as H_2, Cl_2, and HCl have only one bond so the bond energy and atomization energy are the same. For these simple species, the bond energies can be obtained from the spectra produced when the molecules absorb or emit light. For more complex molecules, such as CH_4, however, we must use an indirect method that makes use of measured heats of formation.

The heat of formation of methane has been determined to be -74.9 kJ/mol. This corresponds to the enthalpy change ΔH_f° for the reaction

$$C(s, \text{graphite}) + 2H_2(g) \longrightarrow CH_4(g)$$

We can envision an alternative path to take us from the free elements to the compound that follows a series of successive reactions

Table 14.1. Heats of formation of gaseous atoms from the elements in their standard states

Atom	ΔH_f° per mole of atoms (kJ/mol)
H	218
Li	161
Be	327
B	555
C	715
N	473
O	249
F	79.1
Na	108
Si	454
Cl	121
Br	112
I	107

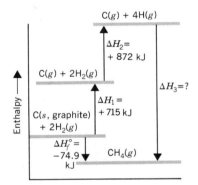

$$(1) \qquad C(s, graphite) \longrightarrow C(g) \qquad \Delta H_1$$

$$(2) \qquad 2H_2(g) \longrightarrow 4H(g) \qquad \Delta H_2$$

$$(3) \qquad C(g) + 4H(g) \longrightarrow CH_4(g) \qquad \Delta H_3$$

The sum of these three reactions will give us the desired overall reaction, so the sum of their ΔH's must equal the ΔH of the overall reaction (that is, the ΔH_f° for CH_4).

$$\Delta H_f^\circ = \Delta H_1 + \Delta H_2 + \Delta H_3 \qquad (14.5)$$

Steps 1 and 2 involve the formation of gaseous atoms from their elements, and their heats have been measured. They are found in Table 14.1 along with the heats of formation of gaseous atoms of some other typical elements. We've already noted that ΔH_f° for CH_4 is known, so the only unknown quantity is ΔH_3, which is the negative of the atomization energy (negative because step 3 involves the *formation* of chemical bonds).

$$\Delta H_3 = -\Delta H_{atom} \text{ for } CH_4$$

Substituting and solving for the atomization energy in Equation 14.5 give

$$\Delta H_{atom} = \Delta H_1 + \Delta H_2 - \Delta H_f^\circ$$

From the data in Table 14.1, we see that $\Delta H_1 = +715$ kJ (the heat of formation of gaseous carbon atoms). Similarly, $\Delta H_2 = 4(+218$ kJ), which is four times the heat of formation of 1 mol of gaseous H atoms. Substituting these values and the heat of formation of CH_4 into the equation above yields

$$\Delta H_{atom} = (+715kJ) + (+872 \text{ kJ}) - (-74.9 \text{ kJ})$$

$$= +1662 \text{ kJ}$$

This quantity is the total amount of energy needed to break all four C—H bonds in 1 mol of CH_4. Division by 4 provides us with an *average* bond energy of 415 kJ per mol of C—H bonds. This value, along with some other average bond energies, appears in Table 14.2.

Table 14.2. Average bond energies

Bond	Bond Energy (kJ/mol)	Bond	Bond Energy (kJ/mol)
H—C	415	C=O	724
H—O	463	C—N	292
H—N	391	C=N	619
H—F	563	C≡N	879
H—Cl	432	C—C	348
H—Br	366	C=C	607
H—I	299	C≡C	833
C—O	356		

EXAMPLE 14.1. CALCULATING BOND ENERGIES FROM THERMODYNAMIC DATA

PROBLEM: The heat of formation of ethylene, C_2H_4, is $+51.9$ kJ/mol. The structure of the molecule is

$$\begin{array}{ccc} H & & H \\ \diagdown & & \diagup \\ & C=C & \\ \diagup & & \diagdown \\ H & & H \end{array}$$

Assuming the C—H bond energy to be 415 kJ/mol, calculate the C=C bond energy.

ANALYSIS: We begin here by recognizing that the atomization energy for C_2H_4 corresponds to the reaction

$$C_2H_4(g) \longrightarrow 2C(g) + 4H(g)$$

This reaction involves the breaking of all four C—H bonds as well as the C=C bond, so ΔH_{atom} equals four times the C—H bond energy plus the C=C bond energy. Therefore, if we can calculate ΔH_{atom}, we can multiply the C—H bond energy by four and subtract it from ΔH_{atom} to get the C=C bond energy.

In the problem, we are given ΔH_f° for C_2H_4, which corresponds to the reaction

$$2C(s) + 2H_2(g) \longrightarrow C_2H_4(g) \tag{14.6}$$

The question we need to answer, then, is how can we incorporate the atomization energy in an alternative sequence of reactions that add up to give the equation for the heat of formation of C_2H_4? We know we can find the heats of formation of the gaseous atoms from the elements in Table 14.1, so we should be able to see a path that takes us from the reactants in Equation 14.6 to the gaseous elements; then the reverse of the atomization energy will take us to the products in Equation 14.6, as illustrated in the figure in the margin. According to Hess's law, the sum of the ΔH values along the upper path must equal the ΔH along the lower path, which is ΔH_f°.

$$\Delta H_1 + \Delta H_2 + \Delta H_3 = \Delta H_f^\circ$$

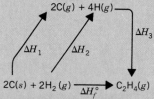

Alternative paths for the formation of C_2H_4 from its elements in their standard states.

SOLUTION: Now that we've established the general approach, let's begin to substitute some values. First, ΔH_1 equals twice the heat of formation of a mole of carbon atoms, which we get from Table 14.1.

$$\Delta H_1 = 2 \text{ mol} \times (+715 \text{ kJ/mol}) = +1430 \text{ kJ}$$

Second, ΔH_2 equals four times the heat of formation of a mole of hydrogen atoms.

$$\Delta H_2 = 4 \text{ mol} \times (+218 \text{ kJ/mol}) = +872 \text{ kJ}$$

And finally, ΔH_3 equals the negative of the atomization energy

$$\Delta H_3 = -\Delta H_{atom}$$

Therefore, combining these gives

$$1430 \text{ kJ} + 872 \text{ kJ} + (-\Delta H_{atom}) = +51.9 \text{ kJ}$$

Solving for ΔH_{atom} yields

$$\Delta H_{atom} = 1430 \text{ kJ} + 872 \text{ kJ} - 51.9 \text{ kJ}$$
$$= 2250 \text{ kJ}$$

This is the energy required to break four moles of C—H bonds plus one mole of C=C bonds.

$$\Delta H_{atom} = 4\Delta H_{C-H} + \Delta H_{C=C}$$

Substituting and solving for $\Delta H_{C=C}$, we get

$$\Delta H_{C=C} = \Delta H_{atom} - 4\Delta H_{C-H}$$
$$= 2250 \text{ kJ} - 4(415 \text{ kJ})$$
$$= 590 \text{ kJ per mol of C=C bonds}$$

It's worth noting that this value is within 3% of the accepted average C=C bond energy of 607 kJ/mol.

> For a particular kind of bond, the bond energy actually varies somewhat from compound to compound.

The average bond energies given in Table 14.2 can be used to estimate heats of formation of compounds, as illustrated in Example 14.2. The very fact that we *can* do this, however, suggests an important property of chemical bonds. It tells us that a bond between two atoms has very nearly the same strength in one molecule as it does in another. For example, it means that nearly all C—H bonds are pretty much alike, whether in a small molecule such as CH_4 or in a large complex one such as $C_{42}H_{86}$. The same applies to other bonds as well. This phenomenon has had a great impact on the theories concerning chemical bonding.

EXAMPLE 14.2. USING BOND ENERGIES TO ESTIMATE ΔH_f° FOR A COMPOUND

PROBLEM: Use the data in Tables 14.1 and 14.2 to compute the heat of formation of liquid ethyl alcohol in kilojoules per mole. This compound has a heat of vaporization, $\Delta H_{vap}^\circ = 39$ kJ/mol and the structural formula

$$\begin{array}{ccc} & H & H \\ & | & | \\ H- & C-C & -O-H \\ & | & | \\ & H & H \end{array}$$

ANALYSIS: We are being asked to determine ΔH° for the reaction

$$2C(s) + 3H_2(g) + \tfrac{1}{2}O_2(g) \longrightarrow C_2H_5OH(l)$$

Let's construct an alternative path from the reactants to the products that uses the information given in the problem. For identification purposes, we will label the steps A, B, etc.

From the data in Table 14.1, we can obtain the heats of formation of the gaseous atoms of each of the elements. Let's call this step A. Next, for step B, we can use the bond energies in Table 14.2 to calculate the negative of the atomization energy, which corresponds to the ΔH° for the reaction

$$2C(g) + 6H(g) + O(g) \longrightarrow C_2H_5OH(g)$$

This would give us the heat of formation of gaseous C_2H_5OH, but we want the $\Delta H°$ for the formation of the liquid. We can get this by including as step C the ΔH for the condensation of the vapor to the liquid

$$C_2H_5OH(g) \longrightarrow C_2H_5OH(l)$$

The enthalpy change for this is the negative of the heat of vaporization.

All of this together gives us the alternative path shown in Figure 14.6, which is what we need to solve the problem.

Figure 14.6. *Alternative paths for the formation of liquid ethyl alcohol, $C_2H_5OH(l)$.*

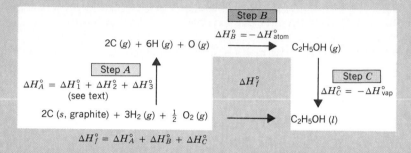

$$\Delta H_f° = \Delta H_A° + \Delta H_B° + \Delta H_C°$$

SOLUTION: Now that we have an alternate path along which all the enthalpy changes are known, we simply combine the ΔH values to obtain $\Delta H_f°$.

For step A, we use the data in Table 14.1 to calculate $\Delta H_A°$.

$$2C(s, \text{graphite}) \longrightarrow 2C(g) \qquad \begin{aligned} \Delta H_1° &= 2\Delta H_{fC(g)}° \\ &= 2(+715 \text{ kJ}) = +1430 \text{ kJ} \end{aligned}$$

$$3H_2(g) \longrightarrow 6H(g) \qquad \begin{aligned} \Delta H_2° &= 6\Delta H_{fH(g)}° \\ &= 6(+218 \text{ kJ}) = +1308 \text{ kJ} \end{aligned}$$

$$\tfrac{1}{2}O_2(g) \longrightarrow O(g) \qquad \Delta H_3° = \Delta H_{fO(g)}° = +249 \text{ kJ}$$

The total energy needed to give gaseous atoms, $\Delta H_A° = \Delta H_1° + \Delta H_2° + \Delta H_3° = +2987$ kJ.

Next, we compute the energy liberated when these atoms combine to form 1 mol of gaseous C_2H_5OH (step B). This is the negative of the atomization energy of C_2H_5OH, which involves five C—H bonds, one C—C bond, one C—O bond, and one O—H bond. The energies are obtained from Table 14.2.

5(C—H)	5(415) kJ
1(C—C)	348 kJ
1(C—O)	356 kJ
1(O—H)	463 kJ
$\Delta H_{\text{atom}}°$ for $C_2H_5OH(g) =$	3242 kJ

Therefore, for step B, $\Delta H_B° = -3242$ kJ.

Finally, energy is liberated when $C_2H_5OH(g)$ is condensed to a liquid (step C). The $\Delta H°$ for this process is the negative of $\Delta H_{\text{vap}}°$. Therefore, $\Delta H_C° = -39$ kJ.

Now we can compute $\Delta H_f°$ for $C_2H_5OH(l)$ by adding the $\Delta H°$ values of each step in the alternative path.

$$\begin{aligned} \Delta H_f° &= \Delta H_A° + \Delta H_B° + \Delta H_C° \\ &= +2987 \text{ kJ} + (-3242 \text{ kJ}) + (-39 \text{ kJ}) \\ &= -294 \text{ kJ} \end{aligned}$$

Since the computation was performed for 1 mol, we can write

$$\Delta H_f° = -294 \text{ kJ/mol}$$

We can compare this to the value reported in Table 6.1.

$$\Delta H_f° = -278 \text{ kJ/mol}$$

We see that the agreement (within about 6%) is not really too bad, considering that we have assumed that each kind of bond has the same energy in *all* compounds.

14.5 ENERGY, ENTROPY, AND THE SPONTANEITY OF CHEMICAL AND PHYSICAL CHANGES

A **spontaneous change** is one that takes place without outside assistance. Once begun, such a change continues on its own. An example is the reaction between hydrogen and oxygen. Once the mixture is ignited, the reaction to form water continues until all of one of the reactants is consumed. A nonspontaneous change, on the other hand, must be continually helped along or it will cease. For instance, water can be decomposed to hydrogen and oxygen by passing an electric current through an aqueous solution that contains an electrolyte such as H_2SO_4. However, this change is not spontaneous and the decomposition only continues as long as the electric current is supplied.

At the beginning of this chapter we said that thermodynamics is able to tell us whether or not a change can happen on its own — in other words, whether or not a change is spontaneous. Before we can see this, however, we must first consider the factors that tend to promote spontaneity.

Energy changes and spontaneity

Everyone has had the experience of placing a ball on an incline and watching it roll downhill. It is also common knowledge that rivers cascade over waterfalls and fuels burn in air when ignited. Each of these events is spontaneous and each is accompanied by a decrease in potential energy. For the ball and the water, their potential energy decreases along with their altitude. For the fuel and oxygen, their exothermic reaction is accompanied by a decrease in the potential energy of their atoms as the reaction takes place.

In general, there is a tendency in nature for the potential energy to be at a minimum, and any change that tends to lower the potential energy also tends to occur spontaneously. We have used the word "tend" here because the energy change is not the only factor to consider. Nevertheless, it is one factor that points the way toward spontaneity.

In thermodynamics, the quantity that is related to the lowering of the potential energy for changes that take place at constant pressure is ΔH, and for exothermic

The lowering of the potential energy of the water as it cascades downhill helps make the process spontaneous, much to the delight of those in the raft.

changes, $\Delta H < 0$. Therefore, changes that have negative enthalpy changes, such as the combustion of a fuel, have a tendency to occur spontaneously.

The tendency toward disorder

The second factor involved in determining spontaneity is nature's tendency toward disorder. You first learned of this in Chapter 13 when we discussed the reasons that solutions form. For example, Figure 13.7 was used to illustrate what happens when two gases are mixed in the same container, and you saw that the random motions of the molecules of the gases cause them to form a solution spontaneously (although not necessarily instantaneously). This tendency toward disorder is a general phenomenon in nature and is not restricted just to the formation of solutions. (Consider, for example, the tendency that your room has to become disordered quite spontaneously!) But what is it in general that promotes disorder?

To find an answer, let's imagine opening a new deck of playing cards. When you take them out of the box, you find that they are sorted according to suit and within each suit they are arranged in order. Now, suppose you toss the cards into the air, let them fall to the floor, and sweep them up into a pile. Do you expect to find them in the same order as before? The answer, of course, is no. Their order will be jumbled, with some face up and others face down, and the reason is simply one of probability. There is an enormous number of ways that the cards can be arranged, and the original sorted version is just one of them. When the cards are thrown in the air, all the arrangements become possible, so the probability that the cards will come down the same way they went up is extremely small. It is so much more likely that they will fall in some other order, and that's what they do.

If cards are tossed in the air, they become disordered spontaneously.

The cards are highly ordered

After they've been tossed they are disordered

The same laws of chance that determine the fate of playing cards, and on which the casinos in Las Vegas and Atlantic City depend, also influence the fates of molecules and atoms in chemical and physical systems. Consider, for example, a collection of molecules that have no attractions for each other. Imagine placing these molecules into a container in the weightless environment of a spacecraft, so there aren't even gravitational attractions to be concerned with. If we were to look into this container, which of the distributions of particles shown in Figure 14.7 would be the most likely and which the least likely? Intuitively, we choose the gas as the distribution of greatest probability because the particles are randomly spread throughout the container. Less likely is the arrangement of particles corresponding to a liquid, with all of them crowded into one end. And the least likely is the arrangement found in a solid, with the particles neatly stacked in a crystalline pattern. If left purely to chance, both the solid and liquid particle distributions would tend to change spontaneously to give the more probable distribution corresponding to the gas.

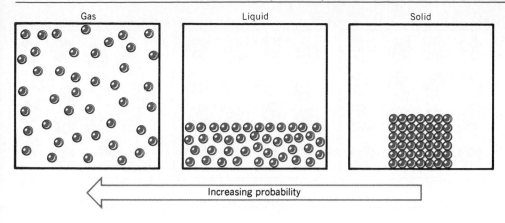

Figure 14.7. *In the absence of attractive forces, the most likely distribution of molecules in a container corresponds to that of a gas. Less likely is the distribution in a liquid, which places all the molecules at one end of the container. Least likely is the highly ordered stacking of molecules that we find in a solid.*

Entropy

The thermodynamic quantity related to probability is called **entropy** and is given the symbol S. The larger the entropy of a system, the greater is its statistical probability. Since systems tend to change spontaneously in the direction of greater probability, they tend to change spontaneously in the direction of greater entropy. In other words, if for a given change there is an increase in the entropy, the change has a tendency to occur spontaneously. If we define an entropy change as $\Delta S = S_{final} - S_{initial}$, then when ΔS is positive ($\Delta S > 0$), spontaneity is favored.

The entropy of a system depends on a number of factors. One is the physical state of the system. For example, within a given container, a solid has the least disorder and the lowest entropy. The liquid state is more disordered, but all the molecules still stay at one end, so its entropy is somewhat higher. The gas, which completely fills the container, has a much higher entropy.

$$S_{solid} < S_{liquid} \ll S_{gas}$$

For a gas, the entropy increases with volume — the larger the volume, the larger the entropy. In fact, in thermodynamic terms, this is why gases expand spontaneously to fill a vacuum. Consider the container in Figure 14.8*a*, which shows a gas in a compartment separated from a vacuum by a panel that can be removed. The moment the panel is withdrawn, the gas exists in a very improbable state, with all its molecules at one end of the now larger container. By filling the entire container, the gas achieves a more probable particle distribution and therefore a larger entropy.

The entropy of a system also increases with increasing temperature, as we see in Figure 14.9. When a substance is a solid at 0 K (-273 °C), the nuclei of all the atoms are located exactly at their lattice points, which are marked by the black dots. There is no molecular motion and the system is perfectly ordered. The entropy is a

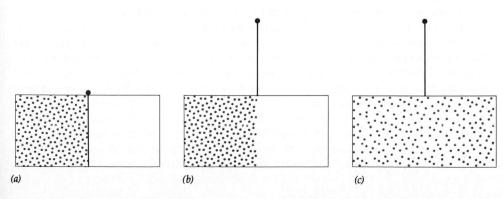

(a)   *(b)*   *(c)*

Figure 14.8. *The expansion of a gas into a vacuum. (a) A gas in a container separated from a vacuum by a removable wall. (b) At the moment the wall is removed, the gas finds itself in an improbable particle distribution within the now larger container. (c) The gas expands spontaneously to a more probable distribution of its molecules.*

Figure 14.9. *Increasing the temperature increases the entropy of a substance.* (a) *At absolute zero all the atoms are at their equilibrium lattice positions (the black dots) and the disorder is a minimum.* (b) *In this stop-action view of the solid at a temperature above absolute zero, the particles are slightly displaced from their equilibrium positions because of their random molecular motions.* (c) *At a still higher temperature the particles have more kinetic energy and their motions are more violent, which causes a greater degree of disorder.*

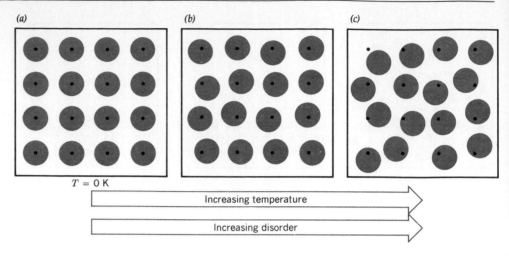

minimum. If we increase the temperature a little (Figure 14.9*b*), the molecules begin to move and jiggle about their equilibrium positions in the lattice. At any particular instant, like the one shown frozen in Figure 14.9*b*, the atoms are slightly displaced from their lattice points, and the system is somewhat disordered. In other words, the system's entropy increased when the temperature increased. At a still higher temperature (Figure 14.9*c*), molecular motion is more violent and the displacements from the lattice points are greater. This produces even more disorder and therefore an even larger entropy.

Predicting the sign of entropy changes for chemical reactions

For chemical reactions, it is sometimes (but not always) possible to predict the sign of ΔS. For example, consider the reaction

$$2NaHCO_3(s) \longrightarrow Na_2CO_3(s) + CO_2(g) + H_2O(g)$$

which occurs when sodium bicarbonate (baking soda) is heated. The reactant is a solid, and as we know, solids have low entropies. Among the products there is one solid, but there are also two gases, which have much larger entropies than any of the solids. Therefore, the products represent a condition of higher entropy than the reactants, so ΔS is positive.

There is another factor that helps make the reaction above have a positive ΔS, and that is the change in the complexity of the particles involved in the reaction. To see the importance of this, consider the following reaction:

$$H_2(g) \longrightarrow 2H(g)$$

The reactant is a gas composed of pairs of hydrogen atoms; the product is a gas composed of single atoms. To get a feel for the relative probabilities of these particle distributions, let's compare the H atoms to marbles. If you were to toss marbles into a box from across a room, which is a more likely way for them to land, as individual marbles scattered across the bottom of the box, or arranged in groups of two? The answer, of course, is that single marbles are more probable than pairs, and in the reaction just given, the single atoms correspond to a more probable distribution of particles than H_2 molecules. Two moles of H atoms therefore have a larger entropy than one mole of H_2 molecules. In fact, the general rule is that for a given number of atoms, the more complex the molecules in which they are found, the *lower* the entropy (volume and temperature being the same). In the decomposition of $NaHCO_3$ described above, the atoms among the products are assembled into less

 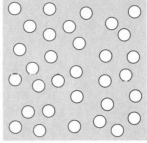

Hydrogen atoms grouped in pairs are a less likely (lower entropy) particle distribution than single atoms.

Lower entropy   Higher entropy

complex groups than among the reactants, so this is another factor that leads to a higher entropy for the products.

EXAMPLE 14.3. DETERMINING THE SIGN OF THE ENTROPY CHANGE QUALITATIVELY

PROBLEM: Ammonia is made by the reaction of hydrogen with nitrogen in the reaction

$$3H_2(g) + N_2(g) \longrightarrow 2NH_3(g)$$

What should be the sign of ΔS for this reaction?

SOLUTION: On the left side of the equation we can count 4 moles of gas (3 mol H_2 plus 1 mol N_2). On the right side of the equation there are only 2 moles of gas. This means that as the products are formed, there are fewer molecules in the system, which must also mean that the atoms are combining to form more complex particles. More complexity means more order and therefore lower entropy. As a result, we can conclude that there is a decrease in entropy associated with the formation of NH_3 in this reaction, and ΔS is negative.

14.6 THE SECOND LAW OF THERMODYNAMICS

The **second law of thermodynamics** deals with the criteria for spontaneity, and it applies to changes of all kinds. One statement of the second law is that *in any spontaneous change, there is always an increase in the entropy of the universe.* This has a very profound meaning not only in chemistry but also in the way our world works. Whenever we see something happening, we know that the entropy of the universe is increasing. Even if some change has been accompanied by a lowering of the entropy, such as the creation of a brick wall from a pile of bricks, we can be sure that something else has occurred that has produced an even larger entropy increase, such as the activities of a bricklayer.

The second law of thermodynamics also gives us some important news about pollution and our attempts to clean it up. When some chemical spreads through the environment, there is a very large increase in the entropy. To remove the pollutant requires that this entropy change be reversed, but the second law tells us that any attempt to do this must lead to some even larger increase in the entropy somewhere else in the universe. Often, this means that attempts to remove one pollutant causes an even more widespread distribution of other pollutants, such as the exhausts and spilled fuel and oil from vehicles used in the cleanup.

In general, the spread of pollutants in the environment is virtually unstoppable once they are released because of the enormous entropy increase that occurs. The only practical way to limit pollution is to not let it begin.

Figure 14.10. *The Hotel Madison in Boston was demolished in seconds by explosives placed strategically on its ground floor. Once begun, this spontaneous collapse was sure to happen because it was accompanied by both a decrease in potential energy and an increase in entropy.*

Free energy and spontaneity

The two factors that control the spontaneity of events, the energy change and the entropy change, often work in concert. For example, the collapse of the building shown in Figure 14.10 is bound to happen when the columns that support it are blown away. The fall of the bricks and other materials is accompanied by a large lowering of the potential energy and the resulting pile of debris has a much larger entropy than the ordered structure that the building represents.

There are other instances, however, in which there is a conflict between ΔH and ΔS. An example is the melting of a solid such as ice described in Figure 14.11. The change from solid to liquid is favored by the entropy change, which is positive, but not by the enthalpy change, ΔH_{fus}, which is also positive. In these cases the temperature plays an important role, and as you know, when the temperature is high, melting is favored, but when it is low, just the opposite occurs.

Figure 14.11. *The melting of a solid is accompanied by an enthalpy increase and an entropy increase, so both ΔH and ΔS are positive quantities.*

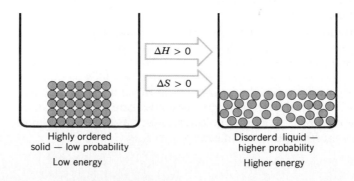

$\Delta H > 0$

$\Delta S > 0$

Highly ordered
solid — low probability

Low energy

Disorderd liquid —
higher probability

Higher energy

The second law of thermodynamics provides a way of bringing the enthalpy change and the entropy change together into a single thermodynamic quantity, G, called the **Gibbs free energy.** This is defined as

$$G = H - TS$$

For a change at constant T and P, we have

$$\Delta G = \Delta H - T\,\Delta S \tag{14.7}$$

According to the second law, any spontaneous change that takes place in a system must lead to a lowering of the free energy, which means that ΔG must be negative ($\Delta G < 0$). Thus, the Gibbs free energy change, ΔG, gives us a composite of the two factors that contribute to spontaneity, ΔH and ΔS.

For changes in which ΔH is negative and ΔS is positive, the algebraic sign of ΔG must be negative regardless of the temperature. (The temperature is in kelvins, so it must be a positive number; negative absolute temperatures are impossible.) Therefore, just in terms of the signs of the quantities,

$$\Delta G = \underset{\Delta H}{(-)} - [\underset{T}{(+)}\underset{\Delta S}{(+)}] \longrightarrow (-)$$

When ΔG is negative, the change is spontaneous, so thermodynamics is telling us that changes such as the collapse of the building mentioned earlier must occur, regardless of the temperature.

Conversely, of course, if ΔH is positive and ΔS is negative, then ΔG will be positive at all temperatures, and the change will never be spontaneous. That's why it is impossible for a building to form all by itself from a pile of rubble.

In situations in which ΔH and ΔS have the same sign, the temperature plays the determining role in controlling spontaneity. For example, if ΔH and ΔS are both positive, we have the following situation:

$$\Delta G = \underset{\Delta H}{(+)} - [\underset{T}{(+)}\underset{\Delta S}{(+)}]$$

The only way that ΔG can be negative is if the second term is larger than the first, and the size of the second term is determined by the temperature. When T is large, the $T\,\Delta S$ term can be larger than the ΔH term, and ΔG will be negative. This is what happens in the melting of a solid such as ice, in which both ΔH and ΔS are positive. Ice melts spontaneously if T is large (larger than 0 °C, or 273 K), but not if T is small.

When ΔH and ΔS are both negative, the change is spontaneous only at low temperature. In this case,

$$\Delta G = \underset{\Delta H}{(-)} - [\underset{T}{(+)}\underset{\Delta S}{(-)}]$$

Only if the second term is small will ΔG be negative, so such changes are only spontaneous at low temperature.

The effects of the algebraic signs of ΔH and ΔS and the effect of temperature on spontaneity can be summarized as follows:

ΔH	ΔS	Outcome
(−)	(+)	Spontaneous at all temperatures
(+)	(−)	Nonspontaneous regardless of temperature
(+)	(+)	Spontaneous only at high temperature
(−)	(−)	Spontaneous only at low temperature

An example of a thermodynamic analysis of a real-world problem

We mentioned earlier the implications of the second law of thermodynamics regarding pollution and the inherent problems that we face in cleaning up the environment. This was a very simple demonstration of how thermodynamics helps us understand which changes are possible and which are not. Sometimes such thermodynamic analyses can be approached quantitatively to solve problems, and a classic example is in the conversion of graphite to diamond.

People had been fascinated by the possibility of changing graphite to diamond ever since 1797, when it was first discovered that they are just two different forms of elemental carbon. Over the years, many experiments were devised in an attempt to accomplish this conversion, but none were successful. Finally, in 1938, a careful thermodynamic analysis of the problem was performed, the results of which are summarized in Figure 14.12.

In this figure, we have curves that show how the ratio $\Delta G/T$ for the reaction

$$C(s, \text{graphite}) \longrightarrow C(s, \text{diamond})$$

varies with the absolute temperature for various pressures.

For this reaction to occur, ΔG must be negative and so must the ratio of $\Delta G/T$. In other words, the change is spontaneous only for $\Delta G/T$ values that lie in the shaded area below the $\Delta G/T = 0$ line.

Figure 14.12. *Thermodynamics of graphite to diamond conversion. (From* Chemical and Engineering News, *April 5, 1971, p. 51, used by permission.)*

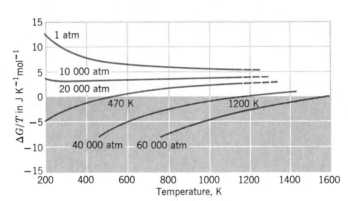

It is easy to see from this diagram why experiments performed at 1 atm were doomed to failure; at this pressure $\Delta G/T$ is never negative. However, we can also see that the higher the pressure, the easier it becomes to carry out the conversion. This fact guided the search for suitable materials and conditions needed for success, and on February 16, 1955, the General Electric Company announced that their scientists had accomplished the conversion in their laboratories in Schenectady, New York. They had used pressures in excess of 100,000 atm and temperatures above 2800 °C. The diamonds were of industrial quality, and before long G.E. was manufacturing over 1 million carats of diamonds per year.

14.7 FREE ENERGY AND USEFUL WORK

As we noted earlier, one of the principal applications of chemical reactions is in the generation of energy in the form of useful work. Often these reactions involve combustion, and the heat produced might yield the hot gases in the cylinder of an engine, or it might be used to boil water for steam, which is then used to drive other machinery. Other reactions are used in batteries to produce electrical energy. In each

A turbine and generator of a modern steam-powered electric generating plant. Although efficient by engineering standards, the thermodynamic efficiency is only about 35 to 40%.

case we are interested in how much work we can obtain from the energy of the reaction, and thermodynamics provides the answer.

The quantity G is called the *free energy* because ΔG is the *maximum* amount of energy released in a process occurring at constant temperature and pressure that is "free," meaning available, to perform useful work. Thus, the driving force for chemical change, which is the lowering of the free energy, is what is harnassed when we make a change do work for us.

The "free" in free energy does not mean *without cost.* This is the energy that is free in the sense of not being tied up for some other purpose.

Earlier, you learned something about *maximum* work — that it is the amount of work that is obtained from a reversible process. Therefore, ΔG gives us a goal to reach for. It tells us how much work we could get if we were able to carry out a change reversibly. Because all *real* changes do not follow reversible paths, the amount of work we actually obtain is always less than the amount specified by ΔG. Nevertheless, knowing ΔG allows us to assess the efficiency of a system. For example, living systems are able to convert only about 40% of the free energy available in the oxidation of glucose to other forms of stored chemical energy (for example, ATP). What happens to the other 60%? It appears as heat, which is why the body needs an effective temperature control system. Mechanical systems such as automobile engines are even less efficient, and they too must be cooled as they operate.

ATP is adenosine triphospate, the substance that the body uses to store energy. It provides the energy for muscle contractions.

14.8 STANDARD ENTROPIES AND FREE ENERGIES

You were almost introduced to the third law in Section 14.5 when we described the effect of temperature on entropy. In Figure 14.9a on page 470, all the particles of the solid are shown to be perfectly ordered by being at their equilibrium lattice positions at absolute zero, and you were told that at this temperature the entropy is a minimum. The **third law of thermodynamics** goes one step further by stating that *at absolute zero, the entropy of any pure crystalline substance is zero.*

Because the zero point on the entropy scale is known, it is possible to actually measure the absolute amount of entropy a substance has in its standard state (25 °C and 1 atm) by adding up entropy-change increments from 0 to 298 K. Table 14.3 contains a number of these standard entropies. In particular, notice that the standard entropies of the elements are *not* zero, as was the case for their standard heats of formation. This is because we are dealing with absolute amounts of entropy here, not changes.

Additional values of standard entropies are located in Appendix C.

Table 14.3. Absolute entropies of some substances at 25 °C and 1 atm

Substance	$S°$ (J/mol K)	Substance	$S°$ (J/mol K)	Substance	$S°$ (J/mol K)	Substance	$S°$ (J/mol K)
$Al(s)$	28.3	$HCN(g)$	201.8	$CuSO_4 \cdot 5H_2O(s)$	300.4	$N_2O_5(g)$	356
$AlCl_3(s)$	110.7	$CH_4(g)$	186.2	$F_2(g)$	202.7	$HNO_3(l)$	155.6
$Al_2O_3(s)$	51.0	$C_2H_2(g)$	201	$HF(g)$	173.5	$O_2(g)$	205.0
$Al_2(SO_4)_3(s)$	239	$C_2H_4(g)$	220	$H_2(g)$	130.6	$O_3(g)$	238.8
$As(s)$	35.1	$C_2H_6(g)$	230	$H_2O(l)$	70.0	$P(s, white)$	41.09
$AsH_3(g)$	223	$C_3H_8(g)$	269.9	$H_2O(g)$	188.7	$P_4O_{10}(s)$	228.9
$As_4O_6(s)$	214	$C_4H_{10}(g)$	310.2	$H_2O_2(l)$	109.6	$H_3PO_4(s)$	110.5
$As_2O_5(s)$	105	$C_6H_6(l)$	173.3	$I_2(s)$	116.1	$K(s)$	64.18
$Ba(s)$	66.9	$CH_3OH(l)$	126.8	$I_2(g)$	260.7	$KCl(s)$	82.6
$BaCO_3(s)$	112	$C_2H_5OH(l)$	160.7	$HI(g)$	206	$Si(s)$	19
$BaCl_2(s)$	125	$HCHO_2(g)$ (formic acid)	251	$Fe(s)$	27.3	$SiH_4(g)$	205
$BaO(s)$	70.4	$HC_2H_3O_2(l)$ (acetic acid)	159.8	$Fe_2O_3(s)$	87.4	$SiO_2(s, alpha)$	41.8
$BaSO_4(s)$	132			$Fe_3O_4(s)$	146.4	$Na(s)$	51.0
$Br_2(l)$	152.2	$HCHO(g)$ (formaldehyde)	218.8	$Pb(s)$	64.8	$NaF(s)$	51.5
$Br_2(g)$	245.4	$CH_3CHO(g)$ (acetaldehyde)	250	$PbO(s, yellow)$	68.7	$NaCl(s)$	72.8
$HBr(g)$	198.5			$PbO_2(s)$	68.6	$NaBr(s)$	83.7
$Ca(s)$	41.4	$(CH_3)_2CO(l)$ (acetone)	200.4	$Pb(OH)_2(s)$	88	$NaI(s)$	91.2
$CaCO_3(s)$	92.9	$C_6H_5CO_2H(s)$ (benzoic acid)	167.6	$PbSO_4(s)$	149	$NaHCO_3(s)$	155
$CaCl_2(s)$	104.6			$Li(s)$	29.1	$Na_2CO_3(s)$	136
$CaO(s)$	39.8	$CO(NH_2)_2(s)$ (urea)	104.6	$LiCl(s)$	59.33	$Na_2O_2(s)$	94.6
$Ca(OH)_2(s)$	76.1			$Mg(s)$	32.5	$NaOH(s)$	64.5
$Ca_3(PO_4)_2(s)$	241	$Cl_2(g)$	223.0	$MgCl_2(s)$	89.5	$Na_2SO_4(s)$	149.6
$CaSO_3 \cdot 2H_2O(s)$	184	$HCl(g)$	186.7	$MgCl_2 \cdot 2H_2O(s)$	180	$S(s, rhombic)$	31.8
$CaSO_4(s)$	107	$HCl(aq)$	56.5	$Mg(OH)_2(s)$	63.1	$SO_2(g)$	248
$CaSO_4 \cdot \frac{1}{2}H_2O(s)$	131	$Cr(s)$	23.8	$Mn(s)$	32.0	$SO_3(g)$	256
$CaSO_4 \cdot 2H_2O(s)$	194	$Cr_2O_3(s)$	81.2	$MnSO_4(s)$	112	$H_2SO_4(l)$	157
$C(s, graphite)$	5.69	$K_2Cr_2O_7(s)$	292.5	$KMnO_4(s)$	171.7	$Sn(s, white)$	51.6
$C(s, diamond)$	2.4	$Cu(s)$	33.15	$N_2(g)$	191.5	$SnCl_4(l)$	258.6
$CCl_4(l)$	214.4	$CuCl_2(s)$	119	$NH_3(g)$	192.5	$SnO_2(s)$	52.3
$CO(g)$	197.9	$CuO(s)$	42.6	$N_2H_4(l)$	121.2	$Zn(s)$	41.6
$CO_2(g)$	213.6	$Cu_2S(s)$	121	$NH_4Cl(s)$	94.6	$ZnO(s)$	43.6
$CO_2(aq)$	117.5	$CuS(s)$	66.5	$NO(g)$	210.6	$ZnSO_4(s)$	120
$H_2CO_3(aq)$	187.4	$CuSO_4(s)$	109	$NO_2(g)$	240.5		
$CS_2(l)$	151.3			$N_2O(g)$	220.0		
$CS_2(g)$	237.7			$N_2O_4(g)$	304.3		

Standard entropies can be used to calculate $\Delta S°$ for reactions by a Hess's law type of calculation. In general,

$$\Delta S° = (\text{Sum of } S° \text{ of products}) - (\text{Sum of } S° \text{ of reactants}) \qquad (14.8)$$

From standard heats of formation and standard entropies we can also calculate **standard free energies of formation, $\Delta G_f°$**. For example, consider the formation of CO_2 from the elements, with all reactants and products in their standard states

$$C(s, \text{graphite}) + O_2(g) \longrightarrow CO_2(g)$$

Table 6.1 gives us the standard enthalpy of formation of $CO_2(g)$, ΔH_f°, as -394 kJ/mol. From the data in Table 14.3 we can calculate ΔS_f°.

Notice that we have to calculate ΔS_f°; they're not tabulated.

$$\Delta S_f^\circ = S_{CO_2}^\circ - (S_C^\circ + S_{O_2}^\circ)$$

$$= [213.6 - (5.69 + 205.0)]\ J/mol\ K = +2.9\ J/mol\ K$$

We can then obtain ΔG_f° as

$$\Delta G_f^\circ = \Delta H_f^\circ - T\Delta S_f^\circ$$

At 25 °C (298 K), then,

$$\Delta G_f^\circ = -394\ kJ/mol - (298\ K)(2.9\ J/mol\ K)$$

$$= -394\ kJ/mol - 864\ J/mol$$

Converting entirely to kilojoules per mole gives

$$\Delta G_f^\circ = (-394 - 0.864)\ kJ/mol$$

$$= -395\ kJ/mol \quad \text{(rounded)}$$

This and other standard free energies of formation are given in Table 14.4.

In Chapter 6 you learned that ΔH° for a reaction can be computed from standard heats of formation. The same rules also apply for the calculation of ΔG° using standard free energies of formation; that is,

Appendix C contains a more complete table of standard free energies of formation.

$$\Delta G^\circ = (\text{Sum of } \Delta G_f^\circ \text{ of products}) - (\text{Sum of } \Delta G_f^\circ \text{ of reactants}) \qquad \textbf{(14.9)}$$

EXAMPLE 14.4. CALCULATING THE STANDARD ENTROPY CHANGE FROM TABULATED VALUES OF S°

PROBLEM: Calculate the standard entropy change (in units of joules per kelvin) for the reaction

$$2NaHCO_3(s) \longrightarrow Na_2CO_3(s) + CO_2(g) + H_2O(g)$$

SOLUTION: We use Equation 14.8 and the data in Table 14.3.

$$\Delta S_{reaction}^\circ = [S_{Na_2CO_3(s)}^\circ + S_{CO_2(g)}^\circ + S_{H_2O(g)}^\circ] - [2S_{NaHCO_3(s)}^\circ]$$

$$= [1\ mol\ (136\ J/mol\ K) + 1\ mol\ (213.6\ J/mol\ K) + 1\ mol\ (188.7\ J/mol\ K)] - [2\ mol\ (155\ J/mol\ K)]$$

$$= (538\ J/K) - (310\ J/K)$$

$$= +228\ J/K$$

It's interesting to note that this reaction also has a positive ΔH° (Example 6.6), so it falls into that category in which both ΔH and ΔS are positive. The decomposition of $NaHCO_3$ is nonspontaneous at low temperature but becomes spontaneous at high temperature, and that is why it can be used as a fire extinguisher.

Many dry-chemical fire extinguishers contain $NaHCO_3$ whose decomposition products help smother a fire.

EXAMPLE 14.5. CALCULATING ΔG° FOR A REACTION FROM TABULATED VALUES OF ΔG_f°

PROBLEM: Silane, SiH_4, is the silicon analog of the main constituent in natural gas — methane, CH_4. Like methane, silane burns in air. The product is silica, a solid quite unlike carbon dioxide.

$$SiH_4(g) + 2O_2(g) \longrightarrow SiO_2(s) + 2H_2O(g)$$
$$\text{silicon dioxide}$$
$$\text{(silica)}$$

Silica, SiO_2, is the major component in ordinary sand.

Calculate $\Delta G°$ for this reaction in kilojoules.

SOLUTION: We have to apply Equation 14.9, using the data in Table 14.4.

$$\Delta G° = [\Delta G°_{f\,SiO_2(s)} + 2\Delta G°_{f\,H_2O(g)}] - [\Delta G°_{f\,SiH_4(g)} + 2\Delta G°_{f\,O_2(g)}]$$

(As with $\Delta H°_f$, we take the standard free energy of formation of a free element to be zero.) Therefore,

$$\Delta G° = \left[1\ \text{mol} \left(\frac{-856\ \text{kJ}}{\text{mol}} \right) + 2\ \text{mol} \left(\frac{-228\ \text{kJ}}{\text{mol}} \right) \right] - \left[1\ \text{mol} \left(\frac{+52.3\ \text{kJ}}{\text{mol}} \right) + 0 \right]$$

$$= -1364\ \text{kJ}$$

Table 14.4. Standard free energies of formation of some substances at 25 °C and 1 atm

Substance	$\Delta G°_f$ (kJ/mol)	Substance	$\Delta G°_f$ (kJ/mol)	Substance	$\Delta G°_f$ (kJ/mol)	Substance	$\Delta G°_f$ (kJ/mol)
$Al(s)$	0	$HCN(g)$	+124.7	$CuSO_4 \cdot 5H_2O(s)$	-1879.7	$N_2O_5(g)$	+115
$AlCl_3(s)$	-629	$CH_4(g)$	-50.6	$F_2(g)$	0	$HNO_3(l)$	-79.9
$Al_2O_3(s)$	-1577	$C_2H_2(g)$	+209	$HF(g)$	-273	$O_2(g)$	0
$Al_2(SO_4)_3(s)$	-3100	$C_2H_4(g)$	+68.2	$H_2(g)$	0	$O_3(g)$	+163
$As(s)$	0	$C_2H_6(g)$	-33	$H_2O(l)$	-237	$P(s, \text{white})$	0
$AsH_3(g)$	+68.9	$C_3H_8(g)$	-23	$H_2O(g)$	-228	$P_4O_{10}(s)$	-2698
$As_4O_6(s)$	-1153	$C_4H_{10}(g)$	-17.0	$H_2O_2(l)$	-120.3	$H_3PO_4(s)$	-1119
$As_2O_5(s)$	-782	$C_6H_6(l)$	+124.3	$I_2(s)$	0	$K(s)$	0
$Ba(s)$	0	$CH_3OH(l)$	-166	$I_2(g)$	+19.3	$KCl(s)$	-409.1
$BaCO_3(s)$	-1139	$C_2H_5OH(l)$	-175	$HI(g)$	+1.30	$Si(s)$	0
$BaCl_2(s)$	-810.8	$HCHO_2(g)$ (formic acid)	335	$Fe(s)$	0	$SiH_4(g)$	+52.3
$BaO(s)$	-525.1	$HC_2H_3O_2(l)$ (acetic acid)	-392	$Fe_2O_3(s)$	-741.0	$SiO_2(s, \text{alpha})$	-856
$BaSO_4(s)$	-1353	$HCHO(g)$ (formaldehyde)	-102.5	$Fe_3O_4(s)$	-1015.4	$Na(s)$	0
$Br_2(l)$	0			$Pb(s)$	0	$NaF(s)$	-545
$Br_2(g)$	+3.11	$CH_3CHO(g)$ (acetaldehyde)	-129	$PbO(s, \text{yellow})$	-187.9	$NaCl(s)$	-384
$HBr(g)$	-53.1	$(CH_3)_2CO(l)$ (acetone)	-155.4	$PbO_2(s)$	-219	$NaBr(s)$	-349
$Ca(s)$	0			$Pb(OH)_2(s)$	-420.9	$NaI(s)$	-286
$CaCO_3(s)$	-1129	$C_6H_5CO_2H(s)$ (benzoic acid)	-245.3	$PbSO_4(s)$	-811.3	$NaHCO_3(s)$	-852
$CaCl_2(s)$	-748.1			$Li(s)$	0	$Na_2CO_3(s)$	-1048
$CaO(s)$	-604.2	$CO(NH_2)_2(s)$ (urea)	-197.3	$LiCl(s)$	-384.4	$Na_2O_2(s)$	-447.7
$Ca(OH)_2(s)$	-896.6			$Mg(s)$	0	$NaOH(s)$	-379.5
$Ca_3(PO_4)_2(s)$	-3852	$Cl_2(g)$	0	$MgCl_2(s)$	-592.5	$Na_2SO_4(s)$	-1270.2
$CaSO_4(s)$	-1320	$HCl(g)$	-95.4	$MgCl_2 \cdot 2H_2O(s)$	-1118	$S(s, \text{rhombic})$	0
$CaSO_3 \cdot 2H_2O(s)$	-1555	$HCl(aq)$	-131.2	$Mg(OH)_2(s)$	-833.9	$SO_2(g)$	-300
$CaSO_4 \cdot \frac{1}{2}H_2O(s)$	-1435	$Cr(s)$	0	$Mn(s)$	0	$SO_3(g)$	-370
$CaSO_4 \cdot 2H_2O(s)$	-1796	$Cr_2O_3(s)$	-1059	$MnSO_4(s)$	-956	$H_2SO_4(l)$	-689.9
$C(s, \text{graphite})$	0	$K_2Cr_2O_7(s)$	-1864	$KMnO_4(s)$	-737.6	$Sn(s, \text{white})$	0
$C(s, \text{diamond})$	+2.9	$Cu(s)$	0	$N_2(g)$	0	$SnCl_4(l)$	-440.2
$CCl_4(l)$	-65.3	$CuCl_2(s)$	-131	$NH_3(g)$	-17	$SnO_2(s)$	-519.6
$CO(g)$	-137	$CuO(s)$	-127	$N_2H_4(l)$	+149.4	$Zn(s)$	0
$CO_2(g)$	-395	$Cu_2S(s)$	-86.2	$NH_4Cl(s)$	-202.9	$ZnO(s)$	-318.3
$CO_2(aq)$	-386.02	$CuS(s)$	-53.6	$NO(g)$	+86.8	$ZnSO_4(s)$	-874.5
$H_2CO_3(aq)$	-623.16	$CuSO_4(s)$	-661.8	$NO_2(g)$	+51.9		
$CS_2(l)$	+65.3			$N_2O(g)$	+104		
$CS_2(g)$	+67.2			$N_2O_4(g)$	+97.9		

14.9 FREE ENERGY AND EQUILIBRIUM

Earlier we said that ΔG determines the maximum amount of energy that is available to perform useful work as a system passes from one state to another. As a reaction proceeds, its capacity to perform work, as measured by G, diminishes until finally, at equilibrium, the system is no longer able to supply additional work. This means that both reactants and products possess the same free energy, and therefore, $\Delta G = 0$. We see then that the value of ΔG for a particular change determines where the system stands with respect to equilibrium. When ΔG is negative—meaning that the free energy of the system is decreasing—the reaction is spontaneous and proceeds in the forward direction toward a state of equilibrium. When ΔG is zero, the system is in a state of dynamic equilibrium, and when ΔG is positive, the reaction is really spontaneous in the reverse direction.

At this point, it should be reemphasized that although ΔG may predict that a particular process is spontaneous, nothing is implied about how rapid the change will be.

Figure 14.13 illustrates graphically the free energy changes in a typical chemical reaction. Notice that the free energy curve has a minimum that lies below the free energies of *both* the reactants and the products. Therefore, if we begin with pure reactants, we proceed in the direction of the products because this gives a lowering of the free energy: ΔG is negative. Also notice that if we begin with pure products, we proceed in the direction of the reactants: This lowers the free energy, too. Therefore, whether we begin with reactants or products, we *always* head in the direction of the minimum on the free energy curve. When the reaction gets there, equilibrium is reached ($\Delta G = 0$). No change in the composition can occur because going toward either the reactants or products involves going uphill on the free energy curve, and that is nonspontaneous.

The $\Delta G°$ for the reaction is also shown in Figure 14.13. It is simply the difference between the free energies of the pure products and the pure reactants. In other words, for a given reaction, $\Delta G°$ is a constant whose sign and magnitude depend on $G°_{products}$ and $G°_{reactants}$, and in this case $\Delta G°$ is positive because $G°_{products}$ is larger than $G°_{reactants}$. But the sign of $\Delta G°$ doesn't tell us about spontaneity directly; we've just seen that for the system described in Figure 14.13, a chemical change would take place spontaneously whether we start with the pure reactants or pure products. To

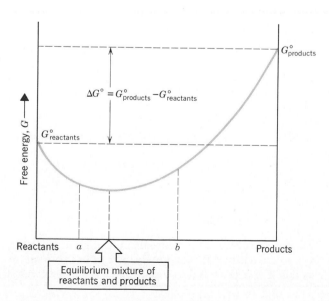

Figure 14.13. *The variation in the free energy of a homogeneous chemical system as the reaction proceeds from pure reactants on the left to pure products on the right. The minimum on the curve marks the position of equilibrium. If the system has a composition corresponding to point a, the reaction is spontaneous in the forward direction. If the composition corresponds to point b, the reaction is spontaneous in the reverse direction.*

Figure 14.14. *The position of equilibrium depends on the value of ∆G° for the reaction. (a) When ∆G° is positive, the position of equilibrium favors the reactants. (b) When ∆G° is zero, the position of equilibrium is intermediate between reactants and products. (c) When ∆G° is negative, the position of equilibrium favors the products.*

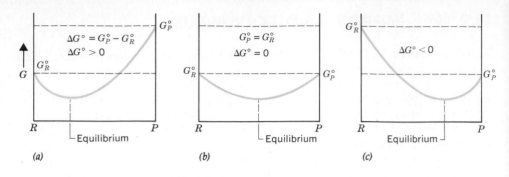

It is whether *G* is increasing or decreasing that determines whether a reaction is spontaneous in one direction or the other.

know about spontaneity, we need to know what the composition of the system is relative to the minimum on the free energy curve. For example, if the composition of the system corresponded to that at point *a* in Figure 14.23, going in the direction of the products is spontaneous because it moves the system "downhill" toward the minimum. On the other hand, if the composition corresponded to point *b*, the reverse reaction would occur spontaneously, because this leads to a lowering of the free energy.

At this point, you may ask, "Why bother with ∆G° at all?" The answer can be seen in Figure 14.14. In Figure 14.14*a* we see a reaction for which ∆G° is positive. In Figure 14.14*b*, the reaction has ∆G° = 0, and in Figure 14.14*c*, the reaction has a negative ∆G°. Notice how the location of the minimum on the free energy curve varies with ∆G°. When ∆G° is positive, the minimum is near the reactants; hardly any reaction has occurred by the time equilibrium is reached. When ∆G° equals zero, equilibrium is reached about halfway between reactants and products, and when ∆G° is negative, the reaction goes almost to completion by the time equilibrium is reached.

Now we can see the usefulness of ∆G° calculations. The value of ∆G° allows us to predict the ultimate outcome of a reaction if it is allowed to proceed to equilibrium. When ∆G° is positive by about 20 kJ or so, so little product will be present at equilibrium that we might conclude that nothing has happened in the reaction mixture. In other words, the reaction will appear to be nonspontaneous. On the other hand, when ∆G° is negative by about 20 kJ, the reaction will appear to have gone to completion. It is in this sense that ∆G° predicts the spontaneity of a reaction.

EXAMPLE 14.6. USING ∆G° AS A GUIDE TO SPONTANEITY

PROBLEM: Is the reaction $2SO_2(g) + O_2(g) \rightarrow 2SO_3(g)$ expected to be spontaneous at 25 °C and 1 atm (i.e., will it proceed far toward completion by the time equilibrium is reached)?

SOLUTION: Let's calculate ∆G° for the reaction using the data in Table 14.4.

$$\Delta G° = 2\,\Delta G°_{f\ SO_3(g)} - [2\,\Delta G°_{f\ SO_2(g)} + \Delta G°_{f\ O_2(g)}]$$

$$= 2\ \text{mol}\left(\frac{-370\ \text{kJ}}{\text{mol}}\right) - \left[2\ \text{mol}\left(\frac{-300\ \text{kJ}}{\text{mol}}\right) + 0.0\right] = (-740\ \text{kJ}) - (-600\ \text{kJ})$$

$$= -140\ \text{kJ}$$

This very negative ∆G° indicates that the position of equilibrium is very largely in favor of SO_3, so the reaction should appear to be quite spontaneous. Actually, this is true but the reaction is very slow in the absence of agents called catalysts.

REVIEW QUESTIONS AND PROBLEMS

Problems whose numbers are in blue have their answers in Appendix D at the back of the book. The more difficult problems are marked with asterisks.

Terminology of Thermodynamics

14.1 What does the word *thermodynamics* imply?

14.2 Define the following: system, surroundings, isothermal change, adiabatic change, and state function.

14.3 What are some different kinds of work that a system can perform on its surroundings?

14.4 What is meant by the term spontaneous change?

14.5 What is a reversible process?

14.6 From an experimental point of view, how can a change be carried out isothermally? How can a change be carried out adiabatically?

First Law of Thermodynamics

14.7 State the first law of thermodynamics.

14.8 On a molecular basis, why is E a state function?

14.9 What is a perpetual motion machine? Why are such devices impossible according to the first law of thermodynamics?

14.10 Why is it not possible to measure or calculate E for a system?

14.11 A real gas usually cools slightly if it is allowed to expand adiabatically into an evacuated container.
(a) What should happen to the temperature of an ideal gas when it is treated in the same way?
(b) What conclusions can you draw about ΔE, q, and w for the *isothermal* expansion of a real gas?

14.12 Why is ΔE equal to zero for an isothermal expansion of an ideal gas?

14.13 Suppose that a steel spring is compressed, tied so it cannot expand, and then dissolved in hydrochloric acid. What happens to the potential energy stored in the spring?

14.14 What advantages do the SI units of pressure and volume offer in calculations of pressure–volume work?

14.15 Why are we usually concerned more about changes in enthalpy than changes in internal energy?

14.16 A gas, initially under a pressure of 15.0 atm and having a volume of 10.0 L, is permitted to expand isothermally in two steps. In the first step the external pressure is held constant at 7.50 atm, and in the second step the external pressure is maintained at 1.00 atm. What are q and w for each step? What is the net change in the internal energy, expressed in L atm, for both the system and its surroundings? Assume ideal gas behavior.

14.17 A gas having an initial volume of 50.0 m³ at an initial pressure of 200 kPa is allowed to expand against a constant opposing pressure of 100 kPa. Calculate the work done by the gas in kilojoules. If the gas is ideal and the expansion is isothermal, what is q for the gas in kilojoules?

14.18 If 500 mL of a gas is compressed to 250 mL under a constant external pressure of 3.00 atm, and if the gas also absorbs 12.6 kJ, what are the values of q, w, and ΔE for the gas expressed in kilojoules? What is the value of ΔE for the surroundings?

14.19 In a particular change, a system absorbs 35 J of heat and has 40 J of work done on it.
(a) What are q, w, and ΔE for the system?
(b) What are q, w, and ΔE for the surroundings?

14.20 For reactions involving just liquids and solids, the values of ΔE and ΔH are nearly identical. Why?

14.21 Consider the following exothermic reactions. For which of them is ΔE larger (i.e., more exothermic) than ΔH?
(a) $CaCO_3(s) + 2HCl(aq) \longrightarrow$
$$CaCl_2(aq) + H_2O(l) + CO_2(g)$$
(b) $NH_3(g) + HCl(g) \longrightarrow NH_4Cl(s)$
(c) $NH_3(g) + H^+(aq) \longrightarrow NH_4^+(aq) + H_2O(l)$
(d) $2C_6H_6(l) + 15O_2(g) \longrightarrow 12CO_2(g) + 6H_2O(l)$

Bond Energies

14.22 What is meant by atomization energy? Write an equation representing the process for which ΔH is the atomization energy of H_2O. Indicate the physical state (gas, liquid, or solid) for each substance in the equation.

14.23 Why don't values of ΔH_f° computed from tabulated bond energies agree precisely with ΔH_f° measured experimentally?

14.24 Use the data in Tables 14.1 and 14.2 to calculate ΔH_f° in kilojoules for acetylene, $C_2H_2(g)$. The structure of the molecule is $H-C\equiv C-H$.

*14.25 Benzene is often written as a resonance hybrid of two equivalent structures

The ΔH_f° for gaseous benzene has been determined from its heat of combustion to be $+82.8$ kJ/mol

$$6C(s) + 3H_2(g) \longrightarrow C_6H_6(g) \qquad \Delta H_f^\circ = +82.8 \text{ kJ/mol}$$

Use the data in Tables 14.1 and 14.2 to calculate ΔH_f° in kJ/mol. How does your calculated value compare with the experimental value? The difference between the calculated and experimental values is called the resonance energy. What might you conclude about the stability of species that exist as a composite of two or more resonance structures?

14.26 Use data in Tables 14.1 and 14.2 to calculate ΔH_f° (in kJ/mol) for gaseous propylene, the substance used to make the plastic polypropylene. The structure of propylene is

$$
\begin{array}{ccccc}
 & & \text{H} & & \\
 & & | & & \\
\text{H} & - \text{C} & - \text{C} & = \text{C} & - \text{H} \\
 & | & | & | & \\
 & \text{H} & \text{H} & \text{H} &
\end{array}
$$

14.27 Use the average bond energies in Table 14.2 and the data in Table 14.1 to compute the standard heat of formation of $C_3H_8(g)$ in kJ/mol. Its structure is

$$
\begin{array}{ccccc}
 & \text{H} & \text{H} & \text{H} & \\
 & | & | & | & \\
\text{H} - & \text{C} & - \text{C} & - \text{C} & - \text{H} \\
 & | & | & | & \\
 & \text{H} & \text{H} & \text{H} &
\end{array}
$$

How well does your computed value compare with that reported in Table 6.1 on page 184.

Entropy, Energy, and Spontaneity

14.28 How does a spontaneous change differ from a nonspontaneous change?

14.29 How is the energy change in a reaction related to its likelihood of being spontaneous?

14.30 In what way is statistical probability related to spontaneity? How is entropy related to statistical probability?

14.31 If a spontaneous chemical reaction is endothermic, what can be said about ΔS for the reaction?

14.32 What are the units of ΔS?

14.33 Which of the following changes will produce an increase in the entropy of a system?
(a) an increase in temperature
(b) formation of gaseous products from solid reactants
(c) formation of a precipitate in a liquid solution
(d) an increase in the volume of the system
(e) the formation of a solid from a gas

14.34 What is the sign of the entropy change for each of the following processes?
(a) A solute crystallizes from a solution.
(b) Water evaporates.
(c) A deck of playing cards is shuffled.
(d) A card player is dealt 13 spades.
(e) Solid AgCl precipitates from a solution of $AgNO_3$ and NaCl.
(f) $^{235}_{92}U$ is extracted from a mixture of $^{235}_{92}U$ and $^{238}_{92}U$.

(g) A sample of H_2 expands from 500 to 1500 mL at a constant temperature.
(h) A sample of iron is cooled from 100 °C to -30 °C.

14.35 Predict the sign of the entropy change for the following reactions:
(a) $HCl(g) \rightarrow H(g) + Cl(g)$
(b) $2H_2(g) + O_2(g) \rightarrow 2H_2O(g)$
(c) $H_2(g) + I_2(s) \rightarrow 2HI(g)$
(d) $CaO(s) + 2NH_4Cl(s) \rightarrow 2NH_3(g) + CaCl_2(s)$
(e) $AgNO_3(aq) + NaCl(aq) \rightarrow AgCl(s) + NaNO_3(aq)$

14.36 Predict the sign of the entropy change for the following reactions:
(a) $Cl_2(g) + F_2(g) \rightarrow 2ClF(g)$
(b) $Na_2CO_3(aq) + 2HCl(aq) \rightarrow$
$$2NaCl(aq) + H_2O(l) + CO_2(g)$$
(c) $CO_2(g) + CaO(s) \rightarrow CaCO_3(s)$
(d) $4NH_3(g) + 3O_2(g) \rightarrow 2N_2(g) + 6H_2O(g)$
(e) $NH_3(g) + HCl(g) \rightarrow NH_4Cl(s)$

Second Law of Thermodynamics

14.37 State the second law of thermodynamics.

14.38 What two criteria must be met in order for a process to be spontaneous, regardless of the temperature?

14.39 Why is it so difficult to remove pollutants from the environment?

14.40 ΔH and ΔS are nearly independent of temperature. Why is this not true for ΔG?

14.41 If we refer to Figure 14.12, what is the minimum pressure necessary for the conversion of graphite to diamond at a temperature of 200 K? Is it theoretically possible to change graphite to diamond at 1 atm?

Third Law of Thermodynamics

14.42 State the third law of thermodynamics.

14.43 Explain why the entropy of a pure substance is zero at 0 K. Would the entropy of an alloy such as brass be zero at 0 K? Explain your answer.

14.44 Which of the following reactions is accompanied by the greatest entropy change? Use data in Table 14.3.
(a) $SO_2(g) + \frac{1}{2}O_2(g) \rightarrow SO_3(g)$
(b) $CO(g) + \frac{1}{2}O_2(g) \rightarrow CO_2(g)$

14.45 Use the data in Table 14.3 to calculate ΔS_f° for the following:
(a) $CCl_4(l)$ (c) $PbSO_4(s)$ (e) $NH_3(g)$
(b) $Mg(OH)_2(s)$ (d) $NaHCO_3(s)$

14.46 Why is it possible to tabulate values of S°, but not values of H°?

14.47 Use the data in Table 14.3 and Appendix C to calculate ΔS° in joules/kelvin for the following reactions:
(a) $2Al(s) + Fe_2O_3(s) \rightarrow Al_2O_3(s) + 2Fe(s)$
(b) $SiH_4(g) + 2O_2(g) \rightarrow SiO_2(s) + 2H_2O(g)$
(c) $CaO(s) + SO_3(g) \rightarrow CaSO_4(s)$

(d) $CuO(s) + H_2(g) \rightarrow Cu(s) + H_2O(g)$

(e) $C_2H_4(g) + H_2(g) \rightarrow C_2H_6(g)$

14.48 Use the data in Table 14.3 and Appendix C to calculate $\Delta S°$ in joules/kelvin for the following reactions:

(a) $C_2H_2(g) + H_2(g) \rightarrow C_2H_4(g)$

(b) $SO_3(g) + H_2O(l) \rightarrow H_2SO_4(l)$

(c) $Mg(OH)_2(s) + 2HCl(g) \rightarrow MgCl_2 \cdot 2H_2O(s)$

(d) $CO_2(g) + H_2(g) \rightarrow CO(g) + H_2O(g)$

(e) $10N_2O(g) + C_3H_8(g) \rightarrow$
$$10N_2(g) + 3CO_2(g) + 4H_2O(g)$$

Standard Free-Energy Changes

14.49 Calculate $\Delta G°$ in kilojoules for each of the reactions in Exercise 14.47.

14.50 Calculate $\Delta G°$ in kilojoules for each of the reactions in Exercise 14.48.

14.51 The standard free energy of formation of glucose is $\Delta G_f° = -910.2$ kJ/mol. Calculate $\Delta G°$ for the reaction

$$C_6H_{12}O_6(s) + 6O_2(g) \longrightarrow 6CO_2(g) + 6H_2O(l)$$

14.52 Calculate $\Delta G°$ in kilojoules for the following reactions:

(a) $Pb(s) + PbO_2(s) + 2H_2SO_4(l) \rightarrow 2PbSO_4(s) + 2H_2O(l)$

(b) $CH_4(g) + 4Cl_2(g) \rightarrow CCl_4(l) + 4HCl(g)$

(c) $13N_2O(g) + C_4H_{10}(g) \rightarrow 13N_2(g) + 4CO_2(g) + 5H_2O(g)$

14.53 Given the following reactions and their values of $\Delta G°$,

$$CaCO_3(s) \longrightarrow CaO(s) + CO_2(g) \qquad \Delta G° = +130 \text{ kJ}$$

$$3CaO(s) + 2H_3PO_4(l) \longrightarrow$$
$$Ca_3(PO_4)_2(s) + 3H_2O(l) \qquad \Delta G° = -512 \text{ kJ}$$

calculate $\Delta G°$ for the reaction

$$3CaCO_3(s) + 2H_3PO_4(l) \longrightarrow$$
$$Ca_3(PO_4)_2(s) + 3CO_2(g) + 3H_2O(l)$$

14.54 Given the following reactions and their values of $\Delta G°$,

$$2CO(g) + O_2(g) \longrightarrow 2CO_2(g) \qquad \Delta G° = -516 \text{ kJ}$$

$$4MnO(s) + O_2(g) \longrightarrow 2Mn_2O_3(s) \qquad \Delta G° = -312 \text{ kJ}$$

calculate $\Delta G°$ for the reaction

$$Mn_2O_3(s) + CO(g) \longrightarrow 2MnO(s) + CO_2(g)$$

Free Energy and Useful Work

14.55 In terms of converting energy into useful work, what advantage and disadvantage does a reversible process offer?

14.56 What is the maximum amount of useful work, expressed in kilojoules, that could be obtained at 25 °C and 1 atm by the oxidation of 1.00 mol of propane, C_3H_8, according to the equation

$$C_3H_8(g) + 5O_2(g) \longrightarrow 3CO_2(g) + 4H_2O(g)$$

Why is it that we always obtain less than this maximum amount of work in any real process that uses propane as a fuel?

Free Energy and Equilibrium

14.57 Sketch a free-energy diagram for a reaction that has a negative value of $\Delta G°$. In this reaction, does the position of equilibrium favor the reactants or products?

14.58 Explain why $\Delta G = 0$ for a system that is in a state of equilibrium.

14.59 The heat of fusion of water at 0 °C is 6.02 kJ/mol; its heat of vaporization is 40.7 kJ/mol at 100 °C. What are ΔS for the melting and boiling of 1 mol of water? Can you explain why ΔS_{vap} is greater than $\Delta S_{melting}$?

14.60 From the data in Tables 6.1 and 14.3, calculate the boiling point of liquid bromine in °C [i.e., the temperature at which $Br_2(l)$ and $Br_2(g)$ can coexist in equilibrium with each other when $p_{Br_2} = 1$ atm].

14.61 Why is it possible for a chemical reaction to occur spontaneously even though $\Delta G°$ for the reaction is positive?

14.62 Why can $\Delta G°$ be used to predict whether or not a reaction can be observed?

14.63 What relationship is there between $\Delta G°$ and the speed with which the products of a reaction are formed?

14.64 What two factors ultimately determine whether we will be able to observe the formation of products in a chemical reaction?

14.65 Describe the relationship between $\Delta G°$ and the position of equilibrium in a chemical reaction.

14.66 Which of the following reactions could *potentially* serve as a practical method for the preparation of NO_2? (*Note:* The equations are not balanced.)

(a) $N_2(g) + O_2(g) \rightarrow NO_2(g)$

(b) $HNO_3(l) + Ag(s) \rightarrow AgNO_3(s) + NO_2(g) + H_2O(l)$

(c) $NH_3(g) + O_2(g) \rightarrow NO_2(g) + NO(g) + H_2O(g)$

(d) $CuO(s) + NO(g) \rightarrow NO_2(g) + Cu(s)$

(e) $NO(g) + O_2(g) \rightarrow NO_2(g)$

(f) $H_2O(g) + N_2O(g) \rightarrow NH_3(g) + NO_2(g)$

Additional Exercises

14.67 What is the maximum amount of useful work, expressed in kilojoules, that could be obtained from the combustion of 4.00 L of $C_4H_{10}(g)$ measured at 25 °C and 1 atm?

14.68 If the thermodynamic efficiency of a gasoline engine is 35.3%, how many grams of isooctane, $C_8H_{18}(l)$, would have to be burned to give $CO_2(g)$ and $H_2O(l)$ in order to deliver 2.00×10^4 kJ of useful work? For $C_8H_{18}(l)$, $\Delta G_f° = +12.8$ kJ/mol. Assume all the products are returned to 25 °C and 1 atm.

15 CHEMICAL EQUILIBRIUM IN GASEOUS SYSTEMS

When a chemical reaction takes place spontaneously, the concentrations of the reactants and products change and the free energy of the system gradually decreases. As you learned in the preceding chapter, eventually the free energy reaches a minimum and the system comes to a state of equilibrium. If we follow the concentrations of the reactants and products while this is happening, we find that they gradually approach steady values, as illustrated in Figure 15.1. During this time the rate at which the reactants form the products approaches the rate at which the products form the reactants. When equilibrium is finally reached, both the forward and reverse reactions are occurring at equal rates and the concentrations no longer change with time. It is the continuation of both the forward and reverse reactions without a change in concentrations, you recall, that identifies this as a *dynamic equilibrium.*

All chemical systems tend toward equilibrium, so the study of the equilibrium state naturally is one of the central topics in chemistry. In this chapter we will explore the quantitative relationships that can be used to describe the equilibrium state, and we will see how the principles of thermodynamics studied in the preceding chapter are able to tell us about the composition of a system at equilibrium. We will focus primarily on gaseous equilibria in this chapter, but in the following two chapters we will extend the concepts learned here to equilibria in aqueous solutions.

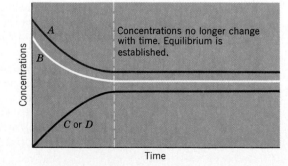

Figure 15.1. *As the reaction* A + B → C + D *approaches equilibrium, the concentrations approach steady values. Once equilibrium is reached, the concentrations no longer change with time.*

15.1 THE EQUILIBRIUM LAW FOR A CHEMICAL REACTION

When we set up an experiment to study a chemical reaction, we are usually free to choose almost any values that we wish for the initial concentrations of the reactants. However, once the reaction has begun, the stoichiometry of the reaction takes over and determines the relative amounts of the reactants that are consumed as well as the amounts of the products that are formed. When equilibrium is finally reached, there will be a mixture of the reactants and products and each substance will have its own equilibrium concentration. The values of these concentrations are no longer arbitrary, however. They are determined partly by the initial concentrations that were chosen and partly by the stoichiometry, but all the equilibrium concentrations are related by an expression called the *equilibrium law* or *equilibrium expression* for the reaction.

As an example, let's consider the reaction between H_2 and N_2 to form NH_3.

$$3H_2(g) + N_2(g) \rightleftharpoons 2NH_3(g)$$

This is one of our most important equilibria because it is used to capture nitrogen from the atmosphere in a form that can be used to make fertilizers and many other chemicals.

Table 15.1 shows the results of five experiments dealing with this equilibrium taking place in a closed container at 500 °C. We introduce here a symbolic notation that we will use throughout the remainder of this book: The concentration of a substance in units of moles per liter (i.e., molar concentration) is expressed by enclosing the chemical formula of the substance within square brackets, [   ]. Thus, $[NH_3]$ means the molar concentration of NH_3, and $[H_2]$ means the molar concentration of H_2.

In examining Table 15.1, we see that in each experiment the initial molar concentrations of H_2 and N_2 are different. We also see that when equilibrium has been reached in each experiment, we have entirely different sets of equilibrium molar concentrations. What is very interesting, however, is that if we use the equilibrium concentrations to compute the fraction indicated in the rightmost column of the table, we obtain essentially the same value each time. In other words, the value of this fraction is a constant for this reaction at this temperature when it is computed from the equilibrium concentrations. This constant is called the **equilibrium constant** and is given the symbol K_c. The subscript c indicates that K_c is calculated with molar concentrations. Therefore, for this reaction we can write the equation

$$\frac{[NH_3]^2}{[H_2]^3[N_2]} = K_c = 6.00 \times 10^{-2} \quad \text{(at equilibrium)}$$

Table 15.1. Initial and equilibrium concentrations, in moles per liter, at 500 °C and the value of the mass action expression for the reaction $N_2(g) + 3H_2(g) \rightleftharpoons 2NH_3(g)$

Initial Concentrations			Equilibrium Concentrations			
$[H_2]$	$[N_2]$	$[NH_3]$	$[H_2]$	$[N_2]$	$[NH_3]$	$\dfrac{[NH_3]^2}{[H_2]^3[N_2]}$
0.168	0.756	0	0.150	0.750	1.23×10^{-2}	5.98×10^{-2}
0.630	1.04	0	0.500	1.00	8.66×10^{-2}	6.00×10^{-2}
1.97	1.36	0	1.35	1.15	4.12×10^{-1}	6.00×10^{-2}
4.33	2.48	0	2.43	1.85	1.27	6.08×10^{-2}
2.03	0.938	0	1.47	0.750	3.76×10^{-1}	5.93×10^{-2}
					Average	6.00×10^{-2}

The equilibrium law is sometimes called the *equilibrium condition* because it is a condition that must be fulfilled in order for the reaction to be at equilibrium.

In general, the equilibrium constant for a reaction varies with temperature.

a, *b*, *c*, and *d* are the coefficients of *A*, *B*, *C*, and *D*.

This relationship is called the **equilibrium law** or **equilibrium expression** for the reaction, and as long as we study the reaction at 500 °C, we will always find that the concentrations of the reactants and products satisfy the law when the system is at equilibrium. In fact, this same relationship holds at any temperature with the exception that the value of the equilibrium constant will be different for different temperatures.

A useful fact about the equilibrium law is that we can predict what it will be for a given reaction just from the stoichiometry of the overall equation. Notice that the *exponents* on the concentrations of NH_3, H_2, and N_2 in the equilibrium law are exactly the same as the *coefficients* of these gases in the balanced equation. In fact, for a general equation

$$aA + bB \rightleftharpoons cC + dD$$

the equilibrium law will be

$$\frac{[C]^c[D]^d}{[A]^a[B]^b} = K_c \qquad (15.1)$$

where the quantities in square brackets indicate equilibrium molar concentrations of the reactants *A* and *B* and the products *C* and *D*.

For historical reasons, which we will not discuss, the fraction on the left in Equation 15.1 is called the **mass action expression,** and, of course, at equilibrium the mass action expression has a value equal to the equilibrium constant. But this fraction can have other values when the system is not at equilibrium, and to be able to discuss these situations, we define a term called the **reaction quotient, Q_c,** which is the numerical value of the mass action expression. It follows that at equilibrium the reaction quotient equals K_c.

$$\frac{[C]^c[D]^d}{[A]^a[B]^b} = Q_c$$

At equilibrium, $Q_c = K_c$

For example, if we choose any of the experiments in Table 15.1 and substitute the *initial* concentrations into the mass action expression,

$$\frac{[NH_3]^2}{[H_2]^3[N_2]}$$

For reasons that we will not discuss further, equilibrium constants are usually given without units.

we find that $Q_c = 0$. Using equilibrium concentrations, however, we obtain $Q_c = 6.00 \times 10^{-2}$, which is the value of K_c.

EXAMPLE 15.1. WRITING THE EQUILIBRIUM LAW FOR A CHEMICAL REACTION

PROBLEM: Write the equilibrium law using molar concentrations for the reaction

$$2PCl_3(g) + O_2(g) \rightleftharpoons 2POCl_3(g)$$

SOLUTION: We construct the mass action expression by raising the molar concentrations of each substance to a power equal to its coefficient in the equation. Following the usual convention, the product appears in the numerator and the reactants in the denominator. This gives

$$\frac{[POCl_3]^2}{[PCl_3]^2[O_2]} = K_c$$

For reactions involving gases, the partial pressures of the reactants and products are proportional to their molar concentrations. The equilibrium expressions for these reactions can therefore be written using partial pressures instead of concentrations. For example, the equilibrium law for the reaction between $N_2(g)$ and $H_2(g)$ can also be expressed as

$$\frac{p^2_{NH_3}}{p_{N_2}p^3_{H_2}} = K_P$$

We will use the symbol K_P to denote equilibrium constants derived from partial pressures and K_c to indicate equilibrium constants having molar concentrations in the mass action expression. In general, K_c and K_P are not numerically equal. We will discuss this further in Section 15.4.

We have written the mass action expression with the concentrations (or partial pressures) of the products in the numerator and those of the reactants in the denominator. Since this fraction is equal to a constant at equilibrium, its reciprocal must also be a constant. Thus,

$$\frac{[NH_3]^2}{[N_2][H_2]^3} = K_c \qquad \frac{p^2_{NH_3}}{p_{N_2}p^3_{H_2}} = K_P$$

and

$$\frac{[N_2][H_2]^3}{[NH_3]^2} = \frac{1}{K_c} = K'_c \qquad \frac{p_{N_2}p^3_{H_2}}{p^2_{NH_3}} = \frac{1}{K_P} = K'_P$$

Either form is a valid description of the equilibrium state. However, chemists have chosen, somewhat arbitrarily, always to write the equilibrium expression with the concentrations or partial pressures of the products appearing in the numerator. This allows us then to tabulate equilibrium constants without always having to state explicitly the form of the mass action expression. It is only necessary to specify the chemical equation and whether we are dealing with K_c or K_P.

$$molarity = \frac{moles}{liters} = \frac{n}{V}$$

For an ideal gas,

$$\frac{n}{V} = \frac{P}{RT}$$

The mass action expression that is equal to K can always be constructed from the chemical equation.

Manipulating chemical equilibria and their equilibrium laws

In Chapter 6 you learned that there are times when it is advantageous to be able to manipulate thermochemical equations by adding them, reversing their directions, and multiplying their coefficients by factors, and you also learned how these operations affect the values of ΔH. Similarly, there are times when it is useful to perform similar operations on chemical equilibria, but the rules we follow for dealing with their equilibrium expressions are different.

The same rules apply whether we use concentrations or partial pressures in the equilibrium expressions.

Adding Chemical Equilibria Chemical equilibria can be combined by addition of their equations to give the equation for an overall equilibrium reaction. When this is done, the equilibrium constant corresponding to the overall equation is the *product* of the equilibrium constants for the equations that were combined. For example, consider the addition of the following two equations:

(1) $\qquad 2NO(g) + O_2(g) \rightleftharpoons 2NO_2(g) \qquad K_{c1} = \dfrac{[NO_2]^2}{[NO]^2[O_2]}$

(2) $\qquad N_2(g) + O_2(g) \rightleftharpoons 2NO(g) \qquad K_{c2} = \dfrac{[NO]^2}{[N_2][O_2]}$

(3) $\qquad N_2(g) + 2O_2(g) \rightleftharpoons 2NO_2(g) \qquad K_{c3} = \dfrac{[NO_2]^2}{[N_2][O_2]^2}$

As shown below, if we multiply the mass action expression for K_{c1} by that for K_{c2}, we obtain the mass action expression for K_{c3}. Therefore, the value of K_{c3} is the product of K_{c1} and K_{c2}.

$$K_{c1} \times K_{c2} = \frac{[NO_2]^2}{\cancel{[NO]^2}[O_2]} \times \frac{\cancel{[NO]^2}}{[N_2][O_2]} = \frac{[NO_2]^2}{[N_2][O_2]^2} = K_{c3}$$

Therefore, $K_{c3} = K_{c1} \times K_{c2}$.

Reversing the Direction of an Equilibrium Equation When we reverse the direction of an equilibrium equation, the new equilibrium constant is the reciprocal of the original. For example, when we reverse the equilibrium equation

$$2NO(g) + O_2(g) \rightleftharpoons 2NO_2(g) \qquad\qquad K_c = \frac{[NO_2]^2}{[NO]^2[O_2]}$$

we get the equation

$$2NO_2(g) \rightleftharpoons 2NO(g) + O_2(g) \qquad K_c' = \frac{[NO]^2[O_2]}{[NO_2]^2}$$

Notice that the mass action expression for the second reaction is the reciprocal of that for the first, so $K_c' = 1/K_c$.

Multiplying the Coefficients of an Equation by Some Factor When we multiply the coefficients of an equilibrium equation by a factor, we raise the equilibrium constant to a power equal to that factor. For example, suppose we multiply the coefficients in the following equation by two.

$$2NO(g) + O_2(g) \rightleftharpoons 2NO_2(g) \qquad K_c = \frac{[NO_2]^2}{[NO]^2[O_2]}$$

This gives

$$4NO(g) + 2O_2(g) \rightleftharpoons 4NO_2(g) \qquad K_c' = \frac{[NO_2]^4}{[NO]^4[O_2]^2}$$

The mass action expression for the second reaction is the square of the mass action expression for the first.

$$\frac{[NO_2]^4}{[NO]^4[O_2]^2} = \left(\frac{[NO_2]^2}{[NO]^2[O_2]}\right)^2$$

Therefore $K_c' = K_c^2$

15.2 WHAT EQUILIBRIUM CONSTANTS TELL US

The equilibrium constant is a quantity that must be calculated from experimental data. One method involves the use of standard free energies of reaction to calculate a *thermodynamic equilibrium constant* and is discussed in Section 15.3. Another method involves the direct measurement of equilibrium concentrations that can then be substituted into the mass action expression to obtain a numerical value for K. We will look at a sample calculation of this type in Section 15.7.

Knowing the value of an equilibrium constant is useful because it allows us to perform computations that relate the concentrations of reactants and products in an equilibrium system. But even *without* doing calculations, the magnitude of K provides us with useful qualitative information about the extent to which a reaction proceeds toward completion. For example, consider the simple reaction

$$A \rightleftharpoons B$$

for which we write

$$\frac{[B]}{[A]} = K_c$$

Suppose that $K_c = 10$. This means that at equilibrium

$$\frac{[B]}{[A]} = 10 = \frac{10}{1}$$

which tells us that at equilibrium the concentration of B must be ten times larger than that of A. In other words, the position of equilibrium lies in favor of the product, B. On the other hand, if $K_c = 0.1$, then

$$\frac{[B]}{[A]} = 0.1 = \frac{1}{10}$$

In this case the equilibrium concentration of A would have to be ten times larger than the concentration of B at equilibrium, and the position of equilibrium would lie in favor of the reactant, A. It is a general rule that when K is large, the position of equilibrium lies far to the right. Conversely, when K is small, only relatively small amounts of the products are present in the system at equilibrium.

The same generalization applies for K_c and K_P.

Let's look at two examples of real chemical reactions: first, the reaction of hydrogen with chlorine,

$$H_2(g) + Cl_2(g) \rightleftharpoons 2HCl(g)$$

for which $K_c = 4.4 \times 10^{32}$ at 25 °C. This very large value of K tells us that at equilibrium the reaction will have proceeded far toward completion. If 1 mol each of H_2 and Cl_2 are combined, very little H_2 and Cl_2 will remain unreacted at equilibrium. By way of contrast, the decomposition of water vapor at room temperature (25 °C),

$$2H_2O(g) \rightleftharpoons 2H_2(g) + O_2(g)$$

$$K_c = \frac{[HCl]^2}{[H_2][Cl_2]} = 4.4 \times 10^{32}$$

$$K_c = \frac{[H_2]^2[O_2]}{[H_2O]^2} = 1.1 \times 10^{-81}$$

has $K_c = 1.1 \times 10^{-81}$. Examining this value of K_c, we can conclude that the decomposition takes place to only a very small degree, because in order to have such a very small value of K_c, the concentrations of the products (which appear in the numerator of the mass action expression) must be very small.

When we use the value of K to compare the positions of equilibrium for two reactions, we do have to be careful. If the values of K differ by an enormous amount, as do the values of K_c for the reactions described in the preceding paragraph, there is no question about which one proceeds farthest toward completion. Usually, however, such comparisons can only be done safely if the equations for the reactions being compared have the same stoichiometry (i.e., the same number of reactants and products with the same coefficients).

15.3 THERMODYNAMICS AND CHEMICAL EQUILIBRIUM

In Section 14.9 we saw, qualitatively, that there is a relationship between $\Delta G°$ for a reaction and the position of equilibrium. In addition, the direction in which a reaction proceeds toward equilibrium is determined by where the system lies with respect to the free-energy minimum. The reaction proceeds spontaneously only in a direction that gives rise to a decrease in free energy — that is, when ΔG is negative.

All of this is summed up quantitatively by the equation (which we will not attempt to justify)

$$\Delta G = \Delta G° + RT \ln Q \tag{15.2}$$

The symbol "ln" means we are to take the *natural logarithm* of Q (the reaction

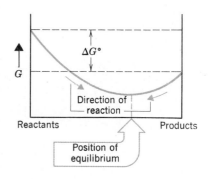

quotient). Natural logarithms occur frequently in the sciences, and if you haven't encountered them before, you should be sure to study Appendix A at the back of the book. In applying this equation, we have to be careful about the units that we use for R. In earlier gas law calculations we have used $R = 0.0821$ L atm mol^{-1} K^{-1}. You have also learned that the product L $\times$ atm is an energy unit, and that it can be related to the more familiar energy units of joules and calories by appropriate conversion factors. Applying these gives R with the following units that we will find useful in calculations later on.

$$R = 8.314 \text{ J mol}^{-1} \text{ K}^{-1}$$
$$= 1.987 \text{ cal mol}^{-1} \text{ K}^{-1}$$

Natural logarithms and common logarithms (logarithms to the base 10) are related by the equation
$$\ln x = 2.303 \log x$$

so Equation 15.2 can also be written

$$\Delta G = \Delta G° + 2.303 \, RT \log Q$$

However, if you are using a scientific calculator to work arithmetic, you will find it easier to work directly with natural logs.

As noted above, the symbol Q in Equation 15.2 stands for the reaction quotient. For reactions that involve gases, Q is obtained from partial pressures (not concentrations). For reactions in solution, however, molar concentrations are used.[1]

Equation 15.2 tells us how ΔG varies with temperature and the relative proportions of the reactants and products. For example, for the reaction

$$2NO_2(g) \rightleftharpoons N_2O_4(g) \tag{15.3}$$

Equation 15.2 takes the form

$$\Delta G = \Delta G° + RT \ln \left(\frac{p_{N_2O_4}}{p_{NO_2}^2} \right) \tag{15.4}$$

which we can use to determine where the system stands with respect to equilibrium. If, for a particular system, we substitute the partial pressures of these gases into this equation and obtain a negative value for ΔG, we know that the reaction will proceed spontaneously in the forward direction. Conversely, if the value of ΔG turns out to be positive, the forward reaction is not spontaneous, but the reverse reaction is.

EXAMPLE 15.2. DETERMINING WHERE A REACTION STANDS WITH RESPECT TO EQUILIBRIUM

PROBLEM: For the equilibrium between NO_2 and N_2O_4 described in Equation 15.3, determine the direction the reaction will proceed in order to get to equilibrium if the temperature of the system is 25 °C (298 K), the partial pressure of NO_2 is 0.200 atm, and the partial pressure of N_2O_4 is 0.100 atm. For this reaction, $\Delta G° = -5.40$ kJ/mol of N_2O_4.

SOLUTION: To solve the problem, we simply substitute the partial pressures of the gases into Equation 15.4 along with the appropriate values for $\Delta G°$, R, and T.

Since $\Delta G°$ has energy units of kilojoules, we will use $R = 8.314$ J mol^{-1} K^{-1} and convert kilojoules to joules so that the energy units can cancel. This gives $\Delta G° = -5400$ J/mol. We also have to express the temperature in kelvins, so $T = 298$ K.

---

[1] Actually, to make Equation 15.2 fit exactly, "effective pressures" or "effective concentrations" must be used in Q. These are called **activities**. Fortunately, at low pressures in gaseous reactions and low concentrations in solutions, the use of actual pressures and concentrations leads to only small errors.

Substituting all these quantities into Equation 15.4 gives

$$\Delta G = (-5400 \text{ J mol}^{-1}) + (8.314 \text{ J mol}^{-1} \text{ K}^{-1})(298 \text{ K}) \ln\left[\frac{0.100}{(0.200)^2}\right]$$

$$= (-5400 \text{ J mol}^{-1}) + (8.314 \text{ J mol}^{-1} \text{ K}^{-1})(298 \text{ K}) \ln(2.50)$$

$$= (-5400 \text{ J mol}^{-1}) + (8.314 \text{ J mol}^{-1} \text{ K}^{-1})(298 \text{ K})(0.916)$$

$$= (-5400 \text{ J mol}^{-1}) + (2270 \text{ J mol}^{-1})$$

$$= -3130 \text{ J mol}^{-1}$$

Since the value of ΔG is negative, the reaction will proceed spontaneously in the forward direction.

Equation 15.2 can be used to derive an extremely valuable relationship between the value of $\Delta G°$ and the equilibrium constant. In Chapter 14 you learned that when a system reaches equilibrium, the products have the same free energy as the reactants and ΔG is equal to zero. If we apply Equation 15.2 to an equilibrium system, therefore, we set $\Delta G = 0$ and the reaction quotient equal to the equilibrium constant ($Q = K$ at equilibrium). This gives

$$0 = \Delta G° + RT \ln K$$

or, solving for $\Delta G°$, we obtain

$$\Delta G° = -RT \ln K \tag{15.5}$$

For reactions involving gases, the K is K_p, and for reactions in solution, $K = K_c$.

For reactions involving gases

$$\Delta G° = -RT \ln K_p$$

For reactions in solution

$$\Delta G° = -RT \ln K_c$$

Equilibrium constants computed from these relationships are sometimes called **thermodynamic equilibrium constants**.

EXAMPLE 15.3. COMPUTING A THERMODYNAMIC EQUILIBRIUM CONSTANT

PROBLEM: What is the thermodynamic equilibrium constant for the reaction

$$2SO_2(g) + O_2(g) \rightleftharpoons 2SO_3(g)$$

SOLUTION: From the data in Table 14.4, we can obtain the standard free energies of formation of SO_3 and SO_2

$$\Delta G°_{f\,SO_3} = -370 \text{ kJ/mol}$$

$$\Delta G°_{f\,SO_2} = -300 \text{ kJ/mol}$$

By definition, $\Delta G°_{f\,O_2} = 0.0$ kJ/mol. Using these data, we can compute $\Delta G°$ for the reaction

$$\Delta G° = 2 \text{ mol} \times \left(-370 \frac{\text{kJ}}{\text{mol}}\right) - 2 \text{ mol} \times \left(-300 \frac{\text{kJ}}{\text{mol}}\right)$$

$$= -140 \text{ kJ}$$

Next, we solve Equation 15.5 for $\ln K_P$ (we are dealing with a gaseous reaction):

$$\ln K_P = \frac{-\Delta G°}{RT}$$

Be sure $\Delta G°$ and R have exactly the same energy units.

We must express $\Delta G°$ in joules ($\Delta G° = -140,000$ J), T in kelvins (298 K), and use $R = 8.314$ J/mol K). Substituting numerical values gives

$$\ln K_P = \frac{-(-140,000)}{(8.314)(298)}$$
$$= 56.5$$

$K_P = e^{56.5} = 3 \times 10^{24}$

The antilogarithm is taken by raising e to the 56.5 power, which gives[2]

$$K_P = 3 \times 10^{24}$$

Thermodynamic data can also be used to calculate equilibrium constants at temperatures other than 25 °C. This is shown in Example 15.4.

EXAMPLE 15.4. CALCULATING K_P AT A TEMPERATURE OTHER THAN 25 °C FROM THERMODYNAMIC DATA

PROBLEM: For the reaction, $2NO_2(g) \rightleftharpoons N_2O_4(g)$, $\Delta H°_{298K} = -56.8$ kJ and $\Delta S°_{298K} = -175$ J/K. Calculate K_P at 100 °C.

ANALYSIS: To calculate K_P from Equation 15.5, we must have a numerical value for the equivalent of $\Delta G°$, but at 100 °C (373 K) instead of 25 °C. Let's call this quantity $\Delta G°_{373}$. In Chapter 14, you learned that

We have been writing $\Delta G°_{298}$ simply as $\Delta G°$. When no temperature is specified, it is assumed to be 298 K.

$$\Delta G° = \Delta H° - (298 \text{ K}) \Delta S°$$

And for some temperature T other than 298 K, we can write

$$\Delta G°_T = \Delta H°_T - T \Delta S°_T$$

It happens that ΔH and ΔS vary only slightly with temperature, so for the purposes of many calculations we can assume temperature independence and write $\Delta H°_{298} = \Delta H°_T$ and $\Delta S°_{298} = \Delta S°_T$. Thus,

$$\Delta G°_T = \Delta H°_{298} - T \Delta S°_{298}$$

SOLUTION: At this point, all we need to do is substitute values. Therefore, at 100 °C (373 K),

$$\Delta G°_{373} = 56,900 \text{ J} - (373 \text{ K})(-175 \text{ J/K})$$
$$= +8380 \text{ J} \quad \text{(rounded to three significant figures)}$$

Solving Equation 15.5 for $\ln K_P$ gives us

$$\ln K_P = \frac{-\Delta G°}{RT}$$

Substituting numerical values, using $R = 8.314$ J mol^{-1} K^{-1} and $T = 373$ K, gives

$$\ln K_P = \frac{-8380}{(8.314)(373)}$$
$$= -2.70$$

[2] When working with logarithms, the following rule for significant figures applies. *The number of significant figures in a number equals the number of decimal places that should appear in the logarithm of the number. When taking the antilogarithm, the number of decimal places in the log value equals the number of significant figures that should appear in the antilogarithm.* In this calculation the logarithm of K_P has one digit after the decimal place, so only one significant figure appears in the answer.

Taking the antilogarithm yields[3]

$$K_P = 6.7 \times 10^{-2}$$

$$K_p = e^{-2.70} = 6.7 \times 10^{-2}$$

The measurement of equilibrium constants also provides a very convenient method for obtaining thermodynamic data. This is illustrated in the next example.

EXAMPLE 15.5. USING K_p TO OBTAIN THERMODYNAMIC DATA

PROBLEM: At 25 °C it was found that $K_p = 7.13$ for the reaction

$$2NO_2(g) \rightleftharpoons N_2O_4(g)$$

What is $\Delta G°$ for this reaction in kilojoules?

SOLUTION: We can calculate $\Delta G°$ by substituting appropriate values into Equation 15.5,

$$\Delta G° = -RT \ln K_p$$

Since we wish $\Delta G°$ expressed in kilojoules, we must use $R = 8.314$ J/mol K. As usual, T is the absolute temperature ($T = 298$ K in this example). Substituting numerical values, we get

$$\Delta G° = -(8.314)(298) \ln (7.13)$$

$$= -4870 \text{ J}   \text{(to three significant figures)}$$

The value of $\Delta G°$ is in joules because R is in joules. To convert to kilojoules, simply divide by 1000:

$$\Delta G° = -4.87 \text{ kJ}$$

15.4 THE RELATIONSHIP BETWEEN K_P and K_c

It was stated earlier that for reactions involving gases, K_P and K_c are not necessarily equal. For the general equation,

$$aA + bB \rightleftharpoons eE + fF$$

$$K_P = \frac{p_E^e p_F^f}{p_A^a p_B^b}$$

and

$$K_c = \frac{[E]^e[F]^f}{[A]^a[B]^b}$$

Molar concentration, you recall, has the units *moles per liter*. Using the symbols n for moles and V for liters and assuming ideal gas behavior, we can use the ideal gas law

$$PV = nRT$$

to obtain the concentration of a gas, X, in a mixture as

$$[X] = \frac{n_X}{V} = \frac{p_X}{RT}$$

where p_X is its partial pressure. From this it follows that

$$p_X = [X]RT$$

[3] In this calculation, the logarithm of K_p has two digits after the decimal place, so when we take the antilogarithm, two significant figures are reported in the answer.

Substituting this relationship into the expression for K_P, we have

$$K_P = \frac{p_E^e p_F^f}{p_A^a p_B^b} = \frac{[E]^e(RT)^e[F]^f(RT)^f}{[A]^a(RT)^a[B]^b(RT)^b}$$

This can be rearranged to give

$$K_P = \frac{[E]^e[F]^f}{[A]^a[B]^b}(RT)^{(e+f)-(a+b)}$$

or

$$K_P = K_c(RT)^{\Delta n_g} \tag{15.6}$$

When $\Delta n_g = 0$, $K_P = K_c$.

where Δn_g is the change in the number of moles of **gas** when going from reactants to products.

$\Delta n_g = $ (number of moles of gaseous products) $-$ (number of moles of gaseous reactants)

Thus, K_P and K_c are related in a very simple fashion for reactions between ideal gases, a relationship that also holds adequately for many real gases.

EXAMPLE 15.6. CONVERTING FROM K_P TO K_c

PROBLEM: In Example 15.3 we determined the value of K_P for the reaction of SO_2 with O_2 to produce SO_3. What is K_c for this equilibrium at 25 °C?

SOLUTION: Solving Equation 15.6 for K_c, we obtain

$$K_c = \frac{K_P}{(RT)^{\Delta n_g}} = K_P(RT)^{-\Delta n_g}$$

The chemical equation we are dealing with is

$$2SO_2(g) + O_2(g) \rightleftharpoons 2SO_3(g)$$

To calculate Δn_g, we interpret the coefficients as moles—there are two moles of gaseous products and three moles of gaseous reactants. Therefore,

$$\Delta n_g = (2 - 3)$$
$$= -1$$

In Example 15.3 we found $K_P = 3 \times 10^{24}$. From the equilibrium expression

$$\frac{p_{SO_3}^2}{p_{SO_2}^2 p_{O_2}} = K_P$$

We don't normally include units with K, but in this case it helps us choose the correct value for R.

K_P has the units atm^{-1} if the partial pressures are expressed in atm. As a result, we must use $R = 0.0821$ L atm mol^{-1} K^{-1} to obtain the proper units for K_c (why?). Thus,

$$K_c = (3 \times 10^{24} \text{ atm}^{-1})[(0.0821 \text{ L atm mol}^{-1} \text{ K}^{-1})(298 \text{ K})]^{-(-1)}$$

Therefore,

$$K_c = 7 \times 10^{25} \text{ K mol}^{-1}$$

15.5 HETEROGENEOUS EQUILIBRIA

Up to now our discussion has focused on homogeneous reactions—reactions in which all the reactants and products are in the same phase. Heterogeneous reactions, of which there are many examples, also eventually arrive at a state of equilibrium. A typical reaction that we might consider is the decomposition of solid $NaHCO_3$ to produce solid Na_2CO_3, gaseous CO_2, and gaseous H_2O.[4]

$$2NaHCO_3(s) \rightleftharpoons Na_2CO_3(s) + CO_2(g) + H_2O(g)$$

We can write the equilibrium law for this reaction as

$$\frac{[Na_2CO_3(s)][CO_2(g)][H_2O(g)]}{[NaHCO_3(s)]^2} = K_c' \tag{15.7}$$

For reasons that will be apparent shortly, we have temporarily indicated the equilibrium constant as K_c'.

In this reaction we have an equilibrium involving the two gases, CO_2 and H_2O, and the two pure solid phases, $NaHCO_3$ and Na_2CO_3. Any pure solid substance such as $NaHCO_3$ has a density that is the same for all samples taken of it, regardless of their size. In addition, this density is unaffected by the nature of any chemical reaction that the substance is undergoing. This means that even during a chemical reaction the amount of $NaHCO_3$ per unit volume of the pure solid is always the same. In other words, the concentration of $NaHCO_3$ in pure solid $NaHCO_3$ is a constant. We cannot alter the number of moles per liter of $NaHCO_3$ in the pure solid, nor can we change the concentration of Na_2CO_3 in pure solid Na_2CO_3. Therefore, the concentrations of these two substances in the equilibrium expression take on constant values that can be incorporated into the equilibrium constant. Rearranging Equation 15.7 to place all the constants on the same side gives

$$[CO_2(g)][H_2O(g)] = K_c' \underbrace{\frac{[NaHCO_3(s)]^2}{[Na_2CO_3(s)]}}_{K_c}$$

or

$$[CO_2(g)][H_2O(g)] = K_c \tag{15.8}$$

Thus, we find that for heterogeneous reactions, *the equilibrium constant expression does not include the concentrations of pure solids.* Similarly, in reactions in which a reactant or product occurs as a pure liquid phase, the concentration of that substance in the pure liquid is also constant. As a result, *the concentrations of pure liquid phases also do not appear in an equilibrium constant expression.*[5] These simplifications apply *only* when we are dealing with *pure* condensed phases. When substances occur in liquid or solid solutions, their concentrations are variable and their concentration terms in the mass action expression therefore cannot be incorporated into K.

If we wish to work with K_P rather than K_c, we need to take into account only the substances present in the gas phase. For the decomposition of $NaHCO_3$, therefore, we have

$$K_P = p_{CO_2(g)} p_{H_2O(g)}$$

As noted in the last section, if we know K_c, we can evaluate K_P as

$$K_P = K_c(RT)^{\Delta n_g} \tag{15.9}$$

where, for this reaction, $\Delta n_g = +2$.

We learned that this reaction makes $NaHCO_3$ a good fire extinguisher.

[4] This reaction is used in the commercial production of Na_2CO_3, which ranked eleventh among industrial chemicals produced in 1988. The total output of the chemical amounted to approximately 17 billion pounds. Sodium carbonate is used in the manufacture of glass, detergents, and many other important products.

[5] Thermodynamics handles this question in a slightly more elegant way by defining the activity of a pure solid or liquid as numerically equal to one. This simply makes terms involving pure solids or liquids disappear from the mass action expression. For instance, substituting values of 1 for $[NaHCO_3(s)]$ and $[Na_2CO_3(s)]$ in Equation 15.7 gives Equation 15.8 directly.

EXAMPLE 15.7. WRITING EQUILIBRIUM EXPRESSIONS FOR HETEROGENEOUS REACTIONS

PROBLEM: What are the values of K_P and K_c for the reaction

$$H_2O(l) \rightleftharpoons H_2O(g)$$

at 25 °C, given that the vapor pressure of water at 25 °C equals 23.8 torr?

SOLUTION: Since liquid water is a pure liquid phase, we can write

$$K_P = p_{H_2O(g)}$$

and

$$K_c = [H_2O(g)]$$

(a)   If we express the vapor pressure of water in atmospheres,

$$p_{H_2O} = 23.8 \text{ torr} \times \left(\frac{1 \text{ atm}}{760 \text{ torr}}\right) = 0.0313 \text{ atm}$$

Therefore,

$$K_P = p_{H_2O} = 3.13 \times 10^{-2} \text{ atm}$$

Note that this equilibrium expression states that the partial pressure of water must be a constant when the liquid and vapor are in equilibrium.

(b)   We can evaluate K_c as

$$K_c = K_P(RT)^{-\Delta n_g}$$

For this reaction, $\Delta n_g = 1$; therefore,

$$K_c = K_P(RT)^{-1} = \frac{K_P}{RT}$$

$$= \frac{3.13 \times 10^{-2} \text{ atm}}{(0.0821 \text{ L atm/mol K})(298 \text{ K})}$$

or

$$K_c = 1.28 \times 10^{-3} \frac{\text{mol}}{\text{L}}$$

Solving Equation 15.6 for K_c gives this equation.

15.6 LE CHÂTELIER'S PRINCIPLE AND CHEMICAL EQUILIBRIA

The equilibrium expression, in the form of either K_P or K_c, can be used to perform numerical computations of various kinds dealing with equilibrium systems. This is discussed in the next section. Often, however, it is desirable simply to be able to predict how some disturbance imposed on a system from outside will influence the position of equilibrium. For instance, we may wish to predict, in a qualitative way, the conditions that favor the greatest production of products. Should we run our reaction at high or low temperature? Should the pressure on the system be high or low? These are questions we would like to answer quickly without having to perform tedious computations. We can do this by applying Le Châtelier's principle, which was introduced in Chapter 12.

Le Chatelier's principle states that when a system in dynamic equilibrium is subjected to a disturbance that upsets the equilibrium, the system changes in a way to reduce the disturbance and, if possible, return to equilibrium.

Let's examine some of the ways that a chemical equilibrium can be upset and how the system is able to respond by changing its position of equilibrium (i.e., the relative proportions of reactants and products).

Changes in the concentration of a reactant or product

A chemical equilibrium can be upset by adding or removing one of the reactants or products. For example, consider the equilibrium

$$H_2(g) + I_2(g) \rightleftharpoons 2HI(g)$$

$$\frac{[HI]^2}{[H_2][I_2]} = K_c$$

If we add H_2 to a reaction mixture that is at equilibrium, the H_2 concentration increases, which causes the denominator of the mass action expression to become larger. This means that the reaction quotient becomes smaller than K, which tells us that equilibrium no longer exists.

Le Chatelier's principle lets us predict what will happen. The disturbance is the increased amount of H_2 in the system. The system can reduce the disturbance by eliminating some of the added H_2. This is exactly what happens; some of the additional H_2 reacts with some of the I_2 to form more HI. As a result, the position of equilibrium shifts to the right. When equilibrium is finally reestablished, there will be a larger concentration of HI than before and there will be a smaller concentration of I_2, as illustrated in Figure 15.2. Notice that after equilibrium has been reestablished, the concentration of H_2 is larger than in the original reaction mixture. The system is never able to completely overcome the effect of the addition or removal of a reactant or product, and all the final concentrations are different from the original ones.

By applying Le Chatelier's principle, we can also predict the effect that removing a reactant or product will have on the position of equilibrium. For example, if we could somehow remove some I_2 from the reaction mixture, the system will adjust by having some HI decompose to replace it. In this case the position of equilibrium shifts to the left.

In general, for the addition or removal of a substance involved in a chemical equilibrium:

1. The position of equilibrium shifts in a direction away from a substance that has been added.

2. The position of equilibrium shifts in the direction of a substance that has been removed.

In a chemical equation, the position of equilibrium shifts away from a substance that's been added or toward a substance that's removed.

Now we can use Le Châtelier's principle to predict what must be done to drive a reaction far toward completion: We can either add a large excess of one of the

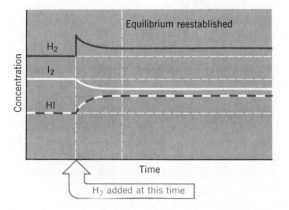

Figure 15.2. *Addition of H_2 to the equilibrium $H_2(g) + I_2(g) \rightleftharpoons 2HI(g)$ causes a sudden jump in the H_2 concentration. After equilibrium is reestablished, the concentration of HI has increased, the concentration of I_2 has decreased, and the final H_2 concentration is a little higher than before the H_2 was added.*

reactants or remove the products as they are formed. Recall that removing the products serves as the driving force for ionic reactions (Chapter 5) in which a product is a precipitate, a gas, or a weak electrolyte. The creation of these products removes ions from solution and therefore forces the reaction to proceed toward completion.

Closer to home, Le Châtelier's principle can help us understand the origin of tooth decay. Tooth enamel consists of an insoluble substance called hydroxyapatite, $Ca_5(PO_4)_3OH$. The dissolving of this substance from the teeth is called demineralization, and its formation is called remineralization. In the mouth there is an equilibrium

$$Ca_5(PO_4)_3OH(s) \underset{\text{remineralization}}{\overset{\text{demineralization}}{\rightleftharpoons}} 5Ca^{2+}(aq) + 3PO_4^{3-}(aq) + OH^-(aq)$$

which is established even with healthy teeth. However, when sugar is absorbed on teeth and ferments, H^+ is produced and upsets the equilibrium by combining with OH^- to form water and with PO_4^{3-} to form HPO_4^{2-}. Removing OH^- and PO_4^{3-} causes more of the $Ca_5(PO_4)_3OH$ to dissolve, resulting in tooth decay.

The effect of temperature on the position of equilibrium

Up to now we have been careful to imply that the equilibrium constant for a reaction has a fixed numerical value only as long as the temperature remains constant. Temperature, as well as the concentrations of reactants and products, affects the position of equilibrium. However, the temperature, unlike the concentrations of reactants and products, affects the value of the equilibrium constant itself.

Let's look at the exothermic reaction for the formation of N_2O_4 from NO_2. The equation for the reaction can be written

For this reaction, $\Delta H = -58.8$ kJ.

$$2NO_2(g) \rightleftharpoons N_2O_4(g) + 58.8 \text{ kJ}$$

where the heat of reaction has been indicated as a product. If we have a system of these two gases in equilibrium and wish to lower its temperature, we remove heat from the system by placing the reaction mixture in a cold bath. Le Châtelier's principle tells us that when we remove this heat, the system will undergo a change that attempts to replace the heat. Since the formation of N_2O_4 is exothermic, the position of equilibrium is shifted to the right, and when the equilibrium is reestablished, there will be a higher concentration of N_2O_4 and a lower concentration of NO_2. Figure 15.3 illustrates this phenomenon in a visual way. Nitrogen dioxide, NO_2, is a reddish-brown gas that is one of the pollutants in smog. On the other hand, N_2O_4 is colorless. In Figure 15.3, we see that at room temperature the mixture of gases has a distinctly deep reddish-brown color. However, when immersed in a

In a sense, we think of heat as a reactant or product in the equation.

Figure 15.3. (a) *The glass tube contains a mixture of NO_2 and N_2O_4 at room temperature. The mixture has a deep reddish-brown color because the NO_2 concentration is high. (b) When the mixture is cooled, the position of equilibrium is shifted in the direction of the exothermic reaction, which converts the colored NO_2 into colorless N_2O_4. Because the concentration of NO_2 has decreased, the color is less intense.*

(a)

(b)

cooling bath, the color becomes noticeably lighter because the position of equilibrium has been shifted in the direction of the colorless N_2O_4.

The effect of temperature on the position of equilibrium can be summarized as follows:

A decrease in temperature causes the position of equilibrium to shift in the direction of an exothermic change, while an increase in temperature shifts an equilibrium in the direction of an endothermic change.

In our discussion of the $NO_2 \rightleftharpoons N_2O_4$ equilibrium, we have seen that a temperature change causes the concentrations of the reactant and product to change as equilibrium is restored. The fact that this occurs without the addition or removal of any of the chemicals and without a change in volume means that the temperature change has altered the equilibrium constant.

When equilibrium is reached at the lower temperature for this reaction, the NO_2 concentration has decreased and the N_2O_4 concentration has increased. This means that at the lower temperature, the reaction quotient

$$Q_c = \frac{[N_2O_4]}{[NO_2]^2}$$

is larger, and since the system is at equilibrium again at the lower temperature, the value of the equilibrium constant is also larger. From this, we can draw another conclusion. *For an exothermic reaction, the equilibrium constant becomes larger as the temperature is decreased. Just the opposite happens, of course, when the reaction is endothermic.*

The effects of pressure and volume changes on the position of equilibrium

At constant temperature, a change in the volume of a system also causes a change in pressure and vice versa. We would expect, therefore, that an increase in the external pressure on a system would favor a change that leads to a smaller volume. Since liquids and solids are virtually incompressible, reactions in which only these phases are involved can't respond to changes in the external pressure, so their equilibria are not affected appreciably by changes in pressure. However, reactions in which gases are involved are another matter, because gases are very compressible.

Let's choose as an example the reaction of N_2 and H_2 to form NH_3.

Changing the external pressure on a chemical system containing only liquids and solids has virtually no effect on the position of equilibrium.

$$N_2(g) + 3H_2(g) \rightleftharpoons 2NH_3(g)$$

If we have this system at equilibrium and suddenly increase the pressure by decreasing the volume of the container, the system will want to respond in a direction that will bring the pressure back down again. But how can this be brought about?

We know that the pressure of a gas is caused by the collisions of the molecules with the walls of the container, and that at a given temperature, the larger the number of molecules, the greater the pressure. In this reaction the number of gas molecules decreases as the product is formed; four gaseous reactant molecules produce two gaseous product molecules. This means that the pressure in the system can be decreased if the position of equilibrium shifts to the right. This analysis gives us another generalization:

Decreasing the volume of a mixture of gases that are in chemical equilibrium shifts the position of equilibrium in the direction of the fewest number of molecules of gas.

Finally, note that when there are the same number of molecules of gaseous

$$K_p = \frac{p_{HI}^2}{p_{H_2} p_{I_2}}$$

Halving the volume doubles the partial pressure of each of the gases. This causes both the numerator and denominator to increase by a factor of four, but their ratio remains equal to K_p.

reactants and products on both sides of the equation, as in the reaction between H_2 and I_2,

$$H_2(g) + I_2(g) \rightleftharpoons 2HI(g)$$

pressure changes brought about by volume changes will not influence the amounts of the various substances present in the reaction mixture at equilibrium. This is because there is no way for the system to counteract pressure changes placed on it.

Addition of an inert gas

If an inert (nonreacting) gas is introduced into a reaction vessel containing other gases at equilibrium, it will cause an increase in the total pressure within the container. This kind of pressure increase, however, will not affect the position of equilibrium because it will not alter the partial pressures or the concentrations of any of the substances already present.

The effect of a catalyst on the position of equilibrium

In Chapter 19, we will examine substances called *catalysts* that increase the speeds of chemical reactions without themselves being used up. Biological enzymes are examples. Catalysts alter the path of a reaction, but because the initial and final states are the same (i.e., the reactants and products are the same overall), catalysts have no effect on the free energy change for the reaction. Since it is the free energy change that determines the equilibrium constant, catalysis cannot change K and cannot influence the position of equilibrium.

15.7 EQUILIBRIUM CALCULATIONS

This section is intended to illustrate the types of computations that one might perform either to evaluate an equilibrium constant from measured concentrations or to use the equilibrium constant to calculate the concentrations of the reactants and products in a particular equilibrium mixture. First let's see how we might evaluate K in a typical experiment.

EXAMPLE 15.8. CALCULATING K_c FROM CONCENTRATION DATA AT EQUILIBRIUM

PROBLEM: As you learned earlier, the brown gas NO_2 and the colorless gas N_2O_4 exist in equilibrium.

$$2NO_2 \rightleftharpoons N_2O_4$$

In an experiment, 0.625 mol of N_2O_4 was introduced into a 5.00-L vessel at a certain temperature. The N_2O_4 gradually decomposed until it reached equilibrium with NO_2. At equilibrium, the concentration of N_2O_4 was 0.075 M. What is K_c for this reaction at this temperature?

ANALYSIS: The first thing to do for any equilibrium problem is to write the balanced chemical equation and then use it to construct the equilibrium law. In this problem, the equation is given to us, so we proceed to the equilibrium law, which is

$$\frac{[N_2O_4]}{[NO_2]^2} = K_c$$

This problem asks us to calculate the value of K_c, which means we need to have numerical values for all the *equilibrium concentrations* that go into the mass action expression.

In the problem, we are given just one of the equilibrium concentrations ($[N_2O_4] = 0.0750\ M$), so we have to find a way of figuring out the other one. To help us in this and other equilibrium problems, we are going to set up a *concentration table* like that shown below.

In the first column of the table, we enter the initial concentrations. These are quantities that the experimentalist has control over, and we get them from the information stated in the problem. It is important to remember that these must be *molar concentrations,* not moles. Therefore, for N_2O_4, we have to divide the number of moles given (0.625 mol) by the total volume of the reaction vessel (5.00 L) to get the initial molar concentration of N_2O_4 (0.125 M). Notice that we have entered a value of 0.000 M for the initial concentration of NO_2. This wasn't stated explicitly in the problem, but since we weren't told that any NO_2 was put in the reaction vessel, we have to assume none was. We have entered these initial concentrations in red because they are part of the initial data given in the problem.

The center column describes how the concentrations change from their initial values to the values present at equilibrium. Because this change occurs as a result of the chemical reaction, the various entries must be in the same ratio as the coefficients in the balanced chemical equation. We will use positive changes to represent increases in concentration and negative values to indicate decreases. In a moment we will describe how the values in this center column were obtained.

The rightmost column has the values that we need to substitute in the mass action expression. One of them is given and this is entered in red. The question, now, is how do we get the others?

Remember, the equilibrium law is only satisfied by equilibrium concentrations.

When the question asks you to calculate K_c, you can be sure there is enough data for you to obtain numerical values for all of the equilibrium concentrations.

Concentration table

	Initial Concentrations	Change	Equilibrium Concentrations
N_2O_4	0.125 M		0.075 M
NO_2	0.000 M		

Since we know the initial and equilibrium concentrations of N_2O_4, their difference is the change in concentration. Once we know this change, we can use the stoichiometry of the equation to calculate the change in the NO_2 concentration. Finally, we can apply this change to calculate the equilibrium concentration of NO_2. After all the equilibrium concentrations are known, we use them in the equilibrium law to calculate K_c.

SOLUTION: The concentration of N_2O_4 drops from 0.125 to 0.075 M, so the change in the N_2O_4 concentration is

$$0.125\ M - 0.075\ M = 0.050\ M$$

This value is entered in the change column with a negative sign (see below) to indicate a decrease in concentration.

The change in the NO_2 concentration is obtained from the change for the N_2O_4 by a simple stoichiometry calculation using the coefficients in the balanced chemical equation. To get to equilibrium in this reaction, 0.050 mol/L of N_2O_4 decomposes, so the number of moles of NO_2 that are formed per liter is

$$0.050\ \text{mol } N_2O_4 \times \frac{2\ \text{mol } NO_2}{1\ \text{mol } N_2O_4} = 0.10\ \text{mol } NO_2$$

Since the $[NO_2]$ is increasing, this value goes into the table with a positive sign. Finally, we obtain the equilibrium NO_2 concentration by adding the change to the initial concentration

$$[NO_2]_{\text{equilibrium}} = 0.000\ M + 0.10\ M$$

$$= 0.10\ M$$

This gives the completed concentration table below.

	Initial Concentrations	Change	Equilibrium Concentrations
N_2O_4	0.125 M	-0.050 M	0.075 M
NO_2	0.000 M	$+0.10$ M	0.10 M

Now we can substitute the equilibrium concentrations into the mass action expression to calculate K_c.

$$\frac{(0.075)}{(0.10)^2} = K_c$$

and, finally,

$$K_c = 7.5$$

Calculating equilibrium concentrations when the equilibrium constant is known

Knowledge of the equilibrium constant for a reaction allows us to calculate the concentrations or partial pressures of the substances present in a reaction mixture at equilibrium. The ease with which these computations can be carried out depends on the complexity of the mass action expression, the concentrations of the various species in the reaction mixture, and the magnitude of the equilibrium constant. We will look only at some of the more simple examples of problems of this type. The following sample problems, however, illustrate the type of reasoning employed in these computations, as well as some of the concepts that have been presented up to this point.

EXAMPLE 15.9. USING K_P TO CALCULATE AN EQUILIBRIUM PRESSURE

PROBLEM: At 25 °C, $K_P = 7.13$ atm^{-1} for the reaction

$$2NO_2(g) \rightleftharpoons N_2O_4(g)$$

The temperature here is different from that in the experiment described in Example 15.8.

At equilibrium the partial pressure of NO_2 in a container is 0.15 atm. What is the partial pressure of N_2O_4 in the mixture?

SOLUTION: The first step in the solution of any equilibrium problem is to write down the equilibrium expression. For K_P, we have

$$K_P = \frac{p_{N_2O_4}}{p_{NO_2}^2} = 7.13 \text{ atm}^{-1}$$

We are given the equilibrium partial pressure of $NO_2(p_{NO_2} = 0.15$ atm). There is only one unknown quantity, $p_{N_2O_4}$. Substituting, we get

$$\frac{p_{N_2O_4}}{(0.15 \text{ atm})^2} = 7.13 \text{ atm}^{-1}$$

$$p_{N_2O_4} = 7.13 \text{ atm}^{-1}(0.15 \text{ atm})^2$$

$$= 0.16 \text{ atm}$$

The partial pressure of N_2O_4 at equilibrium is 0.16 atm.

EXAMPLE 15.10. USING K_c TO CALCULATE AN EQUILIBRIUM CONCENTRATION

PROBLEM: At a temperature of 500 °C, the equilibrium constant, K_c, for the nitrogen fixation reaction for the production of ammonia

$$3H_2(g) + N_2(g) \rightleftharpoons 2NH_3(g)$$

has a value of 6.0×10^{-2}. If, in a particular reaction vessel at this temperature, there are 0.250 mol/L of H_2 and 0.0500 mol/L of NH_3 present at equilibrium, what is the concentration of N_2?

SOLUTION: Let's first write the equilibrium constant expression. For this reaction we have

$$K_c = \frac{[NH_3]^2}{[H_2]^3[N_2]} = 6.0 \times 10^{-2}$$

We wish to calculate the concentration of N_2. This can be accomplished if we know the values of the equilibrium concentrations of both NH_3 and H_2 and, in this problem, these are given to us.

$$\left.\begin{array}{l} [NH_3] = 0.0500\ M \\ [H_2] = 0.250\ M \end{array}\right\} \text{ at equilibrium}$$

In this problem we don't need the concentration table because we're given K_c and all but one equilibrium concentration.

Substituting these numerical values into the mass action expression gives us

$$\frac{(0.0500)^2}{(0.250)^3[N_2]} = 6.0 \times 10^{-2}$$

If we solve for $[N_2]$, we get

$$[N_2] = \frac{(0.0500)^2}{(0.250)^3(6.0 \times 10^{-2})}$$

$$= 2.7\ M$$

The equilibrium concentration of N_2 is thus 2.7 mol/L.

EXAMPLE 15.11. CALCULATING EQUILIBRIUM CONCENTRATIONS FROM K_c AND INITIAL CONCENTRATIONS

PROBLEM: At 440 °C the equilibrium constant K_c for the reaction

$$H_2(g) + I_2(g) \rightleftharpoons 2HI(g)$$

has a value of 49.5. If 0.200 mol of H_2 and 0.200 mol of I_2 are placed into a 10.0-L vessel and permitted to react at this temperature, what will be the concentration of each substance at equilibrium?

ANALYSIS: The first step is to write the equilibrium law.

$$\frac{[HI]^2}{[H_2][I_2]} = K_c = 49.5$$

In this problem, we are given information from which we can calculate the initial concentrations of the reactants and product, and we are given the value of the equilibrium constant. Although we don't know what the equilibrium concentrations are, we do know that the equilibrium law provides the relationship among them. What we will do, therefore, is obtain algebraic expressions for the equilibrium concentrations, as illustrated below, and then use them in the equilibrium law to obtain an algebraic equation that can be solved to give the final concentrations that we seek.

SOLUTION: The initial concentrations of the reactants and products are as follows:

$$[H_2]_{initial} = \frac{0.200 \text{ mol}}{10.0 \text{ L}} = 0.0200 \ M$$

$$[I_2]_{initial} = \frac{0.200 \text{ mol}}{10.0 \text{ L}} = 0.0200 \ M$$

$$[HI]_{initial} = 0.0 \ M$$

If any HI had been present initially, its concentration would have been given in the statement of the problem.

These values are entered in red in the concentration table below.

We don't know what the final (equilibrium) concentrations are and we don't know how much the concentrations change. However, since there is no HI present initially, we know that its concentration must increase. (It can't be smaller than 0.0 M). If the HI concentration is increasing, then the reaction is proceeding to the right to get to equilibrium and the concentrations of H_2 and I_2 must decrease. But how large are these changes? If we knew, we could calculate the equilibrium concentrations that we seek. Therefore, we will express these changes in algebraic terms. Let's let the change in the concentration of H_2 be equal to x. Then the change in the I_2 concentration must be the same, because H_2 and I_2 have the same coefficients in the equation. (Remember, the *changes* are controlled by the stoichiometry of the reaction.) The change in the HI concentration must be twice as large as that for either H_2 or I_2 because of its coefficient in the equation. With their appropriate signs, then

The coefficients of x *must* be in the same ratio as the coefficients in the balanced chemical equation.

$$\text{Change in } [H_2] = -x$$

$$\text{Change in } [I_2] = -x$$

$$\text{Change in } [HI] = +2x$$

These are the quantities entered in the center column of the concentration table. The expressions for the equilibrium concentrations are obtained by algebraically adding the change to the initial value.

	Initial Concentrations	Change	Equilibrium Concentrations
H_2	0.0200 M	$-x$	$(0.0200 - x) \ M$
I_2	0.0200 M	$-x$	$(0.0200 - x) \ M$
HI	0.0 M	$+2x$	$0.0 + 2x = 2x \ M$

If the sign for the HI change is positive, then the signs for both the H_2 and I_2 changes must be negative.

Now that we have expressions for the equilibrium concentrations, we substitute them into the mass action expression of the equilibrium law. This gives an algebraic equation that we can then solve (we hope) for x.

$$\frac{(2x)^2}{(0.0200 - x)(0.0200 - x)} = 49.5$$

or

$$\frac{(2x)^2}{(0.0200 - x)^2} = 49.5$$

In this particular problem, the algebra is relatively simple because the left side of the equation is a perfect square. Therefore, taking the square root of both sides of the equation gives

$$\frac{2x}{0.0200 - x} = 7.04$$

Solving for x, we obtain

$$2x = 7.04(0.0200 - x) = 0.141 - 7.04x$$

$$2x + 7.04x = 0.141$$

$$9.04x = 0.141$$

$$x = 0.0156$$

Finally, the equilibrium concentrations are

$$[H_2] = 0.0200 - 0.0156 = 0.0044 \ M$$

$$[I_2] = 0.0200 - 0.0156 = 0.0044 \ M$$

$$[HI] = 2(0.0156) = 0.0312 \ M$$

In this last problem we employed some relatively simple algebra to help us arrive at the solution. Let us look at another example of this type.

EXAMPLE 15.12. CALCULATING EQUILIBRIUM CONCENTRATIONS FROM K_c AND INITIAL CONCENTRATIONS

PROBLEM: A 10.0-L vessel is filled with 0.40 mol of HI at 440 °C. What will be the concentration of H_2, I_2, and HI at equilibrium?

ANALYSIS: In this problem we are concerned about the same equilibrium as in the previous example. The chemical equation is

$$H_2(g) + I_2(g) \rightleftharpoons 2HI(g)$$

and the equilibrium expression is

$$\frac{[HI]^2}{[H_2][I_2]} = 49.5$$

Initially, there is no H_2 or I_2 in the container, so their concentrations are given values of zero in the concentration table. The concentration of HI is 0.40 mol/10.0 L = 0.040 M.

Next we need quantities in the change column. These are our unknowns, and we will use the symbol x once again. Since the magnitudes of the changes must be in the same ratio as the coefficients of the balanced equation, we can let the coefficients of x be equal to the coefficients in the chemical equation. Thus, the coefficients of x for the changes in the concentrations of H_2 and I_2 are each 1, and the coefficient of x for the change in the HI concentration is 2.

Next we have to decide whether the changes will be positive or negative. Since the initial concentrations of H_2 and I_2 are zero, they must increase. (They can't decrease if they are already zero.) Their changes are therefore positive. This means that the change for the HI must be negative, because if the H_2 and I_2 increase, they must be formed from the HI.

	Initial Concentrations	Change	Equilibrium Concentrations
H_2	0.0 M	$+x$	$0.0 + x = x \ M$
I_2	0.0 M	$+x$	$0.0 + x = x \ M$
HI	0.040 M	$-2x$	$(0.040 - 2x) \ M$

If the initial concentration of a reactant or product is zero, the change *must* be positive. The equilibrium concentration can't be negative.

SOLUTION: As in the preceding problem, we substitute the quantities that are in the "Equilibrium Concentrations" column into the mass action expression of the equilibrium law to obtain an equation that we can then solve for x.

$$\frac{(0.040 - 2x)^2}{(x)(x)} = 49.5$$

or

$$\frac{(0.040 - 2x)^2}{(x)^2} = 49.5$$

Once again, we are able to take the square root of both sides of the equation. This gives

$$\frac{0.040 - 2x}{x} = 7.04$$

Solving for x, we get

$$0.040 - 2x = (x)(7.04)$$

$$x = 0.0044$$

We now calculate the equilibrium concentrations to be

$$[H_2] = x = 0.0044 \ M$$

$$[I_2] = x = 0.0044 \ M$$

$$[HI] = 0.040 - 2x = 0.031 \ M$$

Observe that we have obtained essentially the same answers in both Examples 15.11 and 15.12. If all the H_2 and I_2 in Example 15.11 had completely reacted, it would have produced 0.40 mol of HI—the same amount of HI that we began with in Example 15.12. We find, therefore, that the same equilibrium composition can be approached from either direction.

Solving equilibrium problems when K_c is very small

In the last two examples the solution of the algebra was simple because we were able to take the square root of both sides of the equation. You can't always expect the algebra to work out so easily, however, and in some cases it can really prove to be quite a challenge. Fortunately, though, when the equilibrium constant is either extremely large or extremely small, it is frequently possible to greatly reduce the difficulty of even rather complex algebra by making some simple approximations. Example 15.13 illustrates the kinds of simplifying approximations that we will find useful in Chapter 16.

EXAMPLE 15.13. CALCULATING EQUILIBRIUM CONCENTRATIONS FROM K_c AND INITIAL CONCENTRATIONS WHEN K_c IS VERY SMALL

PROBLEM: The equilibrium constant, K_c, for the decomposition of gaseous water at 500 °C has a value of 6.0×10^{-28}. If 2.0 mol of H_2O is placed into a 5.0-L container, what will be the equilibrium concentrations of the three gases, H_2, O_2, and H_2O, at 500 °C?

ANALYSIS: The equation for the reaction is

$$2H_2O(g) \rightleftharpoons 2H_2(g) + O_2(g)$$

Therefore, we can write

$$\frac{[H_2]^2[O_2]}{[H_2O]^2} = 6.0 \times 10^{-28}$$

The initial H_2O concentration is 2.0 mol/5.0 L = 0.40 M. As before, we let the coefficients of x be the same as the coefficients in the balanced chemical equation. Constructing our table, we get

	Initial Concentrations	Change	Equilibrium Concentrations
H_2O	0.40 M	$-2x$	$(0.40 - 2x)\ M$
H_2	0.0 M	$+2x$	$2x\ M$
O_2	0.0 M	$+x$	$x\ M$

The coefficients of x can be the same as the coefficients in the balanced chemical equation for the equilibrium.

Substituting equilibrium quantities into the mass action expression gives

$$\frac{(2x)^2(x)}{(0.40 - 2x)^2} = 6.0 \times 10^{-28}$$

Unless we can somehow simplify this equation, we have a real mess on our hands. Fortunately, in this case the problem can be made easy to solve.

From the magnitude of the equilibrium constant, 6.0×10^{-28}, we know that the position of equilibrium lies almost all the way toward the reactants. In other words, we know from the size of K_c that hardly any of the H_2O will decompose. If this is true, then the equilibrium concentration of H_2O will be very close to its initial value, and this means that the quantity $(0.40 - 2x)$ will be very nearly equal to 0.40. (Stated differently, the quantity $2x$ is expected to be so small that it is negligible compared to 0.40). To simplify the algebra in this problem, therefore, we will make the assumption that $(0.40 - 2x)\ M \approx 0.40\ M$. Therefore, the values that we will use in the mass action expression are $[H_2] = 2x$, $[O_2] = x$, and $[H_2O] = (0.40 - 2x) \approx 0.40$.

We can only neglect a small x if it is *added to* or *subtracted from* some other value that's much larger.

Substituting these into the mass action expression gives

$$\frac{(2x)^2(x)}{(0.40)^2} = 6.0 \times 10^{-28}$$

from which we get

$$\frac{4x^3}{0.16} = 6.0 \times 10^{-28}$$

SOLUTION: This problem is now simple to solve. First we solve for x^3

$$x^3 = \frac{0.16}{4}(6.0 \times 10^{-28}) = 2.4 \times 10^{-29}$$

At this point x can be obtained by extracting the cube root. Many hand-held calculators can perform this operation directly or by raising 2.4×10^{-29} to the $\frac{1}{3}$ power.

$$x = \sqrt[3]{2.4 \times 10^{-29}} = (2.4 \times 10^{-29})^{1/3} = 2.9 \times 10^{-10}$$

Having obtained a value for x, we should check to see whether our assumption was valid, and we see that 5.8×10^{-10} is indeed much smaller than 0.40. Having satisfied ourselves of this, we use the calculated value of x to calculate the equilibrium concentrations.

$0.40 - 0.00000000058 = 0.39999999942$. When rounded, this gives 0.40.

$$[H_2] = 2x = 5.8 \times 10^{-10}\ M$$

$$[O_2] = x = 2.9 \times 10^{-10}\ M$$

$$[H_2] = 0.40 - (5.8 \times 10^{-10}) = 0.40\ M$$

In working out a problem of this sort, look for any assumption that will make the algebra easier to handle. If the assumption you make is invalid, you will discover this when you check the assumption after obtaining a value for x. Sometimes no assumption of the kind we made will be valid, and then some other method of solving the equation for x will have to be sought. You will learn one such method in the next chapter.

REVIEW QUESTIONS AND PROBLEMS

Problems whose numbers are in blue have their answers in Appendix D at the back of the book. The more difficult problems are marked with asterisks.

Equilibrium Expressions

15.1 What is meant by a *dynamic equilibrium*?

15.2 Write the mass action expression in terms of molar concentration for each of the following reactions:
(a) $N_2(g) + O_2(g) \rightleftharpoons 2NO(g)$
(b) $2NO(g) + O_2(g) \rightleftharpoons 2NO_2(g)$
(c) $2H_2(g) + S_2(g) \rightleftharpoons 2H_2S(g)$
(d) $2N_2O_5(g) \rightleftharpoons 4NO_2(g) + O_2(g)$
(e) $P_4O_{10}(g) + 6PCl_5(g) \rightleftharpoons 10POCl_3(g)$

15.3 Give the mass action expressions for the reactions in Question 15.2 in terms of partial pressures.

15.4 Write equilibrium expressions for K_P and K_c for each of the following reactions:
(a) $CO(g) + 2H_2(g) \rightleftharpoons CH_3OH(g)$
(b) $CO(g) + H_2O(g) \rightleftharpoons CO_2(g) + H_2(g)$
(c) $PCl_3(g) + Cl_2(g) \rightleftharpoons PCl_5(g)$
(d) $2NO_2(g) + 4H_2(g) \rightleftharpoons N_2(g) + 4H_2O(g)$
(e) $2H_2S(g) + 3O_2(g) \rightleftharpoons 2H_2O(g) + 2SO_2(g)$

15.5 Why do we always write the concentrations (or partial pressures) of the products in the numerator and those of the reactants in the denominator in the mass action expression?

15.6 Show that the following equilibrium data, obtained for the reaction

$$PCl_5(g) \rightleftharpoons PCl_3(g) + Cl_2(g)$$

demonstrate the constancy of the mass action expression for a system at equilibrium. What is K_c for this reaction?

Experiment	[PCl$_5$]	[PCl$_3$]	[Cl$_2$]
1	0.0023	0.23	0.055
2	0.010	0.15	0.37
3	0.085	0.99	0.47
4	1.00	3.66	1.50

15.7 Write equilibrium expressions for the following two reactions:
(a) $H_2(g) + Cl_2(g) \rightleftharpoons 2HCl(g)$
(b) $\frac{1}{2}H_2(g) + \frac{1}{2}Cl_2(g) \rightleftharpoons HCl(g)$

How would the magnitude of the K for reaction (a) compare with that for reaction (b)?

15.8 At 25 °C, the equilibrium constant $K_c = 1.4 \times 10^7$ for the reaction $CO(g) + 2H_2(g) \rightleftharpoons CH_3OH(g)$. What is the value of K_c at this temperature for the following reactions?
(a) $CH_3OH(g) \rightleftharpoons CO(g) + 2H_2(g)$
(b) $2CO(g) + 4H_2(g) \rightleftharpoons 2CH_3OH(g)$

15.9 Consider the following equilibria and their equilibrium constants:

$$2H_2(g) + O_2(g) \rightleftharpoons 2H_2O(g) \qquad K_p = 8.6 \times 10^{79}$$

$$N_2(g) + 3H_2(g) \rightleftharpoons 2NH_3(g) \qquad K_p = 9.1 \times 10^5$$

$$4NH_3(g) + 5O_2(g) \rightleftharpoons$$
$$4NO(g) + 6H_2O(g) \qquad K_p = 9 \times 10^{172}$$

What is the equilibrium constant K_p for the reaction

$$N_2(g) + O_2(g) \rightleftharpoons 2NO(g)$$

15.10 Consider the following equilibria and their equilibrium constants:

$$BrF_3(g) + F_2(g) \rightleftharpoons BrF_5(g) \qquad K_c = 8.6 \times 10^{35}$$

$$ClF(g) + F_2(g) \rightleftharpoons ClF_3(g) \qquad K_c = 7.8 \times 10^{12}$$

$$BrF(g) + F_2(g) \rightleftharpoons BrF_3(g) \qquad K_c = 7.3 \times 10^{27}$$

Determine the equilibrium constant K_c for the reaction

$$BrF(g) + 2ClF_3(g) \rightleftharpoons 2ClF(g) + BrF_5(g)$$

Significance of K

15.11 What general information can be gathered by observing the magnitude of the equilibrium constant?

15.12 Arrange the following reactions in order of their increasing tendency to proceed toward completion:
(a) $4NH_3(g) + 3O_2(g) \rightleftharpoons$
$$2N_2(g) + 6H_2O(g) \qquad K = 1 \times 10^{228}$$
(b) $N_2(g) + O_2(g) \rightleftharpoons 2NO(g) \qquad K = 5 \times 10^{-31}$
(c) $2HF(g) \rightleftharpoons H_2(g) + F_2(g) \qquad K = 1 \times 10^{-13}$
(d) $2NOCl(g) \rightleftharpoons 2NO(g) + Cl_2(g) \qquad K = 4.7 \times 10^{-4}$

15.13 Which of the following reactions proceeds farthest toward completion if allowed to come to equilibrium?

(a) $BrF_3(g) + F_2(g) \rightleftharpoons BrF_5(g)$ $K_c = 8.6 \times 10^{35}$
(b) $ClF(g) + F_2(g) \rightleftharpoons ClF_3(g)$ $K_c = 7.8 \times 10^{12}$
(c) $BrF(g) + F_2(g) \rightleftharpoons BrF_3(g)$ $K_c = 7.3 \times 10^{27}$

15.14 Which of the following reactions proceeds farthest toward completion if allowed to come to equilibrium?
(a) $2H_2(g) + O_2(g) \rightleftharpoons 2H_2O(g)$ $K_p = 8.6 \times 10^{79}$
(b) $N_2(g) + 3H_2(g) \rightleftharpoons 2NH_3(g)$ $K_p = 9.1 \times 10^5$
(c) $4NH_3(g) + 5O_2(g) \rightleftharpoons$
     $4NO(g) + 6H_2O(g)$ $K_p = 9 \times 10^{172}$

Thermodynamics and Equilibrium

15.15 What value would $\Delta G°$ have for a reaction if $K = 1$?

15.16 For reactions between gases, what kind of equilibrium constant is calculated from $\Delta G°$?

15.17 Using Equations 14.7 (p. 473) and 15.5, show that a straight line should be obtained if $\log K_P$ is plotted against $1/T$ (that is, $\log K_P$ along the vertical axis, $1/T$ along the horizontal axis). What does the slope of this line give? What does the value of $\log K_P$ at $1/T = 0$ (the y intercept of the line) give?

15.18 Use the data in Table 14.4 (p. 478) to calculate K_P at 25 °C for the reaction

$$2HCl(g) + F_2(g) \rightleftharpoons 2HF(g) + Cl_2(g)$$

15.19 An air pollutant produced by burning high-sulfur fuels is sulfur dioxide. In smog, which contains appreciable amounts of NO_2, the sulfur dioxide can be oxidized to sulfur trioxide, which forms H_2SO_4 when it reacts with moisture. The reaction is

$$SO_2(g) + NO_2(g) \rightleftharpoons NO(g) + SO_3(g)$$

Use the data in Table 14.4 (p. 478) to calculate K_P for this reaction at 25 °C.

15.20 The following thermodynamic data apply at 25 °C:

Substance	$\Delta G_f°$ (kJ/mol)
$NiSO_4 \cdot 6H_2O(s)$	−2222
$NiSO_4(s)$	−773.6
$H_2O(g)$	−228

(a) What is $\Delta G°$ (in kJ) for the reaction

$$NiSO_4 \cdot 6H_2O(s) \rightleftharpoons NiSO_4(s) + 6H_2O(g)$$

(b) What is K_P for this reaction?

15.21 At 700 K, $\Delta G_{700K}° = -13.5$ kJ for the reaction $CO(g) + 2H_2(g) \rightleftharpoons CH_3OH(g)$. Calculate the value of K_P for the reaction at 700 K.

15.22 The equilibrium constant, K_P, for the reaction $COCl_2(g) \rightleftharpoons CO(g) + Cl_2(g)$ has a value of 4.56×10^{-2} atm at 395 °C. What is the value of $\Delta G_{668K}°$ (in kilojoules) for this reaction?

15.23 At 527 °C the reaction

$$CO(g) + H_2O(g) \rightleftharpoons CO_2(g) + H_2(g)$$

has $K_P = 5.10$. What is $\Delta G_{800K}°$ for this reaction expressed in kilojoules?

15.24 Use the data in Tables 6.1 (p. 184) and 14.3 (p. 476) to compute $\Delta G_{773K}°$ (in kilojoules) and K_P at 500 °C for the reaction

$$2HCl(g) \rightleftharpoons H_2(g) + Cl_2(g)$$

Assume that $\Delta H°$ and $\Delta S°$ are independent of temperature.

15.25 Use the data in Tables 6.1 (p. 184) and 14.3 (p. 476) to compute the temperature at which $K_P = 1$ for the reaction $C_2H_4(g) + H_2(g) \rightleftharpoons C_2H_6(g)$. Assume $\Delta H°$ and $\Delta S°$ are independent of temperature.

15.26 Methyl alcohol, CH_3OH, is a fuel that can be made from carbon monoxide (produced by burning coal) and hydrogen. The equilibrium is

$$CO(g) + 2H_2(g) \rightleftharpoons CH_3OH(g)$$

At 427 °C (700 K) a mixture of CO, H_2, and CH_3OH having the following partial pressures was prepared: $p_{CO} = 2 \times 10^{-3}$ atm, $p_{H_2} = 1 \times 10^{-2}$atm, $p_{CH_3OH} = 3 \times 10^{-6}$ atm. For this reaction, $\Delta G_{700K}° = -13.5$ kJ. Use Equation 15.2 to determine whether this system is at equilibrium. If not, will the reaction proceed spontaneously to the left or the right?

Relationship Between K_P and K_c

15.27 For which of the reactions in Questions 15.2 and 15.4 would $K_P = K_c$?

15.28 The reaction

$$CO(g) + H_2O(g) \rightleftharpoons CO_2(g) + H_2(g)$$

is used industrially as a source of hydrogen. The value of K_c for this reaction at 500 °C is 4.05. What is its value of K_P at this temperature?

15.29 The reaction

$$CH_4(g) + H_2O(g) \rightleftharpoons CO(g) + 3H_2(g)$$

is also a source of hydrogen. At 1500 °C, its value of $K_c = 5.67$. What is its value of K_P at this temperature?

15.30 At 100 °C, $K_P = 6.5 \times 10^{-2}$ for the reaction

$$2NO_2(g) \rightleftharpoons N_2O_4(g)$$

What is the value of K_c at this temperature?

Heterogeneous Equilibria

15.31 On the basis of the equilibrium

$$H_2O(l) \rightleftharpoons H_2O(g)$$

explain why the vapor pressure of water depends only on temperature and not the amount of liquid water in equilibrium with its vapor.

15.32 Why is it *not* necessary to include the concentrations of pure liquid or solid phases in the equilibrium constant expression?

15.33 Write equilibrium expressions in terms of K_c for each of the following reactions:
(a) $CaCO_3(s) \rightleftharpoons CaO(s) + CO_2(g)$
(b) $Ni(s) + 4CO(g) \rightleftharpoons Ni(CO)_4(g)$
(c) $5CO(g) + I_2O_5(s) \rightleftharpoons I_2(g) + 5CO_2(g)$
(d) $Ca(HCO_3)_2(aq) \rightleftharpoons CaCO_3(s) + H_2O(l) + CO_2(g)$
(e) $AgCl(s) \rightleftharpoons Ag^+(aq) + Cl^-(aq)$

Le Châtelier's Principle

15.34 Consider the equilibrium $PCl_3(g) + Cl_2(g) \rightleftharpoons PCl_5(g)$. How would the following affect the position of equilibrium?
(a) addition of PCl_3
(b) removal of Cl_2
(c) removal of PCl_5
(d) decrease in the volume of the container
(e) addition of He without a change in volume

15.35 Which, if any, of the changes in Exercise 15.34 above will change the value of the equilibrium constant for the reaction?

15.36 Indicate how each of the following changes affects the amount of H_2 in the system below, for which $\Delta H_{reaction} = +41$ kJ.
$$H_2(g) + CO_2(g) \rightleftharpoons H_2O(g) + CO(g)$$
(a) addition of CO_2
(b) addition of H_2O
(c) addition of a catalyst
(d) increase in temperature
(e) decrease in the volume of the container

15.37 How will each of the changes in Exercise 15.36 affect the equilibrium constant?

15.38 Consider the equilibrium
$$2N_2O(g) + O_2(g) \rightleftharpoons 4NO(g)$$
How will the amount of NO at equilibrium be affected by
(a) adding N_2O?
(b) removing O_2?
(c) increasing the volume of the container?
(d) adding a catalyst?

15.39 For the reaction
$$4NH_3(g) + 3O_2(g) \rightleftharpoons 2N_2(g) + 6H_2O(l)$$
how will the amount of NH_3 at equilibrium be affected by
(a) adding O_2 to the system?
(b) adding N_2 to the system?
(c) removing H_2O from the system?
(d) decreasing the volume of the container?

15.40 Sketch a graph to show how the concentrations of H_2, N_2, and NH_3 would change with time after N_2 had been added to a mixture of these gases initially at equilibrium.

15.41 In the equilibrium
$$CaCO_3(s) + heat \rightleftharpoons CaO(s) + CO_2(g)$$
how will the amount of $CaCO_3(s)$ change if
(a) $CaO(s)$ is added?
(b) $CO_2(g)$ is added?
(c) the volume of the container is increased?
(d) the temperature is lowered?

Equilibrium Calculations

15.42 Referring to Exercise 15.6, calculate K_c for the reaction
$$PCl_3(g) + Cl_2(g) \rightleftharpoons PCl_5(g)$$

15.43 What is the value of K_P for the reaction
$$PCl_5(g) \rightleftharpoons PCl_3(g) + Cl_2(g)$$
Refer to the data in Exercise 15.6 ($T = 298$ K).

15.44 At 460 °C, $K_c = 85.0$ for the reaction
$$SO_2(g) + NO_2(g) \rightleftharpoons NO(g) + SO_3(g)$$
A mixture of these gases has the following concentrations of the reactants and products: $[SO_2] = 0.040\ M$, $[NO_2] = 0.50\ M$, $[NO] = 0.30\ M$, $[SO_3] = 0.020\ M$. Is this system at equilibrium? If not, in which direction must the reaction proceed to reach equilibrium?

15.45 For the reaction $PCl_5(g) \rightleftharpoons PCl_3(g) + Cl_2(g)$, $K_c = 33.3$ at 760 °C. In a container at equilibrium there are 1.29×10^{-3} mol/L of PCl_5 and 1.87×10^{-1} mol/L of Cl_2. Calculate the equilibrium molar concentration of PCl_3 in the vessel.

15.46 At 25 °C, in a mixture of N_2O_4 and NO_2 in equilibrium at a total pressure of 0.844 atm, the partial pressure of N_2O_4 is 0.563 atm. Calculate for the reaction
$$N_2O_4(g) \rightleftharpoons 2NO_2(g)$$
(a) K_P, (b) K_c, (c) ΔG°_{298K} in kJ.

15.47 At a certain temperature the following equilibrium concentrations were found for the reactants and products in the reaction
$$2HI(g) \rightleftharpoons H_2(g) + I_2(g)$$
$$[H_2] = 1.0 \times 10^{-3}\ M$$
$$[I_2] = 2.5 \times 10^{-2}\ M$$
$$[HI] = 2.2 \times 10^{-2}\ M$$
What is the value of K_c for this reaction?

15.48 In a particular experiment the following partial pressures were determined for the reaction at equilibrium
$$2NO(g) + Cl_2(g) \rightleftharpoons 2NOCl(g)$$
$$p_{NO} = 0.65\ atm \quad p_{Cl_2} = 0.18\ atm \quad p_{NOCl} = 0.15\ atm$$
What is K_P for this reaction at the temperature at which the experiment was performed?

15.49 At 25 °C, 0.0560 mol of O_2 and 0.020 mol N_2O were

placed in a 1.00-L vessel and allowed to react according to the equation $2N_2O(g) + 3O_2(g) \rightleftharpoons 4NO_2(g)$. When the system reached equilibrium, the concentration of the NO_2 was found to be 0.020 mol/L.
(a) What were the equilibrium molar concentrations of N_2O and O_2?
(b) What is the value of K_c for this reaction at 25 °C?

15.50 At 460 °C, the reaction

$$SO_2(g) + NO_2(g) \rightleftharpoons NO(g) + SO_3(g)$$

has $K_c = 85.0$. What will the equilibrium molar concentrations of the four gases be if a mixture of SO_2 and NO_2 is prepared in which they each have an initial concentration of 0.0500 M?

15.51 For the reaction

$$H_2(g) + CO_2(g) \rightleftharpoons CO(g) + H_2O(g)$$

$K_c = 0.771$ at 750 °C. If 1.00 mol of H_2 and 1.00 mol of CO_2 are placed into a 5.00-L container and permitted to react, what will the concentrations of all four gases be at equilibrium?

15.52 Suppose a mixture of SO_2, NO_2, NO, and SO_3 having the initial concentrations $[SO_2] = 0.0100 \, M$, $[NO_2] = 0.0200 \, M$, $[NO] = 0.0100 \, M$, and $[SO_3] = 0.0150 \, M$ is prepared at 460 °C. At this temperature the reaction

$$SO_2(g) + NO_2(g) \rightleftharpoons NO(g) + SO_3(g)$$

has $K_c = 85.0$. What will be the molar concentrations of the four gases at equilibrium?

15.53 The reaction $2CO_2 \rightleftharpoons 2CO + O_2$ has $K_c = 6.4 \times 10^{-7}$ at 2000 °C. If 1.0×10^{-3} mol of CO_2 is placed into a 1.0-L vessel at this temperature,
(a) What will be the equilibrium concentrations of CO and O_2?
(b) What fraction of the CO_2 will have decomposed?

15.54 At 100 °C the equilibrium constant, K_c, for the reaction

$$CO(g) + Cl_2(g) \rightleftharpoons COCl_2(g)$$

has a value of 4.6×10^9. If 0.20 mol of $COCl_2$ is placed into a 10.0-L flask at 100 °C, what will be the molar concentrations of all species at equilibrium?

15.55 Sodium bicarbonate (baking soda) has many useful properties, among them the ability to serve as a fire extinguisher. When heated it decomposes to produce CO_2, which smothers the fire.

$$2NaHCO_3(s) \rightleftharpoons Na_2CO_3(s) + CO_2(g) + H_2O(g)$$

At 125 °C the value of K_P for this reaction is 0.25 atm². What are the partial pressures of $CO_2(g)$ and $H_2O(g)$ in this system at equilibrim? Can you explain why $NaHCO_3$ is used in baking?

15.56 In a 10.0-L mixture of H_2, I_2, and HI at equilibrium at 425 °C, there are 0.100 mol of H_2, 0.100 mol of I_2, and 0.740 mol of HI. If 0.50 mol of HI is now added to this system, what will be the molar concentrations of H_2, I_2, and HI once equilibrium has been reestablished?

* 15.57 At a certain temperature, $K_c = 7.5$ for the reaction

$$2NO_2(g) \rightleftharpoons N_2O_4(g)$$

If 2.0 mol of NO_2 is placed in a 2.0-L container and permitted to react, what will be the concentrations of NO_2 and N_2O_4 at equilibrium? What will be the equilibrium concentrations if the size of the container is doubled? Does this conform to what you would expect from Le Châtelier's principle?

Additional Exercises

* 15.58 In Exercise 15.54 it was stated that at 100 °C the value of K_c for the reaction $CO(g) + Cl_2(g) \rightleftharpoons COCl_2(g)$ is 4.6×10^9. Suppose that 0.15 mol of CO and 0.30 mol of Cl_2 are placed into a 1.0-L vessel and allowed to react. What will the molar concentration of each of the gases in the system be at equilibrium? (*Hint:* First assume 100% reaction; then work backward toward equilibrium.)

15.59 At 750 °C, $K_c = 0.771$ for the reaction

$$H_2(g) + CO_2(g) \rightleftharpoons CO(g) + H_2O(g)$$

If 0.100 mol of H_2 and 0.200 mol of CO_2 are placed in a 5.00-L reaction vessel and allowed to come to equilibrium, what will be the molar concentrations of each of the gases in the mixture?

15.60 At 25 °C, $K_c = 8.6 \times 10^{35}$ for the reaction

$$BrF_3(g) + F_2(g) \rightleftharpoons BrF_5(g)$$

(a) If 0.200 mol of BrF_3 and 0.200 mol of F_2 are placed in a 4.00-L vessel and allowed to come to equilibrium, what will the molar concentrations of each of the gases in the reaction mixture be?
(b) In a certain reaction vessel, the concentration of BrF_3 was set initially to 0.100 M and the concentration of F_2 was set initially to 0.200 M. When allowed to come to equilibrium, what will the molar concentrations of all the gases in the reaction mixture be?

* 15.61 The production of NO by reaction of N_2 and O_2 in an automobile engine is an important source of nitrogen oxide pollution. At 1000 °C the reaction $N_2(g) + O_2(g) \rightleftharpoons 2NO(g)$ has $K_P = 4.8 \times 10^{-7}$. Suppose that the partial pressures of N_2 and O_2 in the cylinder of an engine after the gasoline vapor has been ignited are $p_{N_2} = 33.6$ atm and $p_{O_2} = 4.0$ atm. Assume that the temperature of the mixture is 1000 °C. Calculate the partial pressure (in atm) of NO in the mixture if the system has time to reach equilibrium.

* 15.62 If it is assumed that the reactants and products in the preceding question are unable to react further when the exhaust gases are suddenly cooled as they exit the engine, calculate the partial pressure of the NO (in atm) when the partial pressure of N_2 has dropped to 0.80 atm and the temperature has dropped to 150 °C.

16

ACID-BASE EQUILIBRIA IN AQUEOUS SOLUTIONS

Acids and bases are among the most common substances found in aqueous solutions. You no doubt have already encountered some of them in the lab—acids such as HCl and H_2SO_4 and bases such as NaOH and NH_3. Actually, though, you've been around acids and bases your whole life, because many of the molecules in living systems fall into this category, as do many of the other substances that surround us in the environment.

Some acids and bases are strong, meaning they are virtually 100% dissociated into their respective ions in aqueous solution. Most, however, are weak and exist in solution in a dynamic equilibrium with their ions. Our goal in this chapter is to study these equilibria quantitatively by applying the principles learned in Chapter 15. We begin with the most important acid–base equilibrium of all, the reaction of water with itself to form hydronium ion and hydroxide ion.

16.1 THE IONIZATION OF WATER

In Chapter 5 we commented that water itself is a very weak electrolyte because of the reaction

$$H_2O + H_2O \rightleftharpoons H_3O^+(aq) + OH^-(aq)$$

This kind of reaction, in which two molecules of the solvent react with each other to form ions, is called **autoionization.** This is a very important equilibrium because it is present in any aqueous solution, regardless of what other reactions may also be taking place. Since the autoionization is an equilibrium, we can write an equilibrium expression for it. Following the approach of the last chapter gives

$$K_c = \frac{[H_3O^+][OH^-]}{[H_2O][H_2O]}$$

The molar concentration of water, which appears in the denominator of this expression, is very nearly constant ($\approx 55.6 \, M$) in both pure water and dilute aqueous solutions. Therefore, $[H_2O]^2$ can be included with the equilibrium constant, K_c, on the left side of the equation. This gives

$$K_c \cdot [H_2O]^2 = [H_3O^+][OH^-]$$

The left side of this expression is the product of two constants which, of course, must also equal a constant. This combined constant is written as

$$K_w = K_c[H_2O]^2$$

The equilibrium law therefore becomes

$$K_w = [H_3O^+][OH^-]$$

Because $[H_3O^+][OH^-]$ is the product of ion concentrations, K_w is called the **ion product constant** for water, or frequently simply the **ionization constant** or **dissociation constant** of water. At 25 °C, $K_w = 1.0 \times 10^{-14}$, and this is one equilibrium constant that you should be sure to memorize.

The equation for the autoionization of water is often simplified by omitting the water molecule that picks up the H^+ and abbreviating the hydronium ion as H^+. The chemical equation becomes

$$H_2O \rightleftharpoons H^+ + OH^-$$

[In this and most future chemical equations in this chapter we will omit (*aq*) after the formulas of the molecules and ions in the solution. These species are of course hydrated, but the equations are easier to work with if they are not cluttered by this notation.] When we use this simplified equation, the expression for the ionization constant for water is

$$K_w = [H^+][OH^-] \tag{16.1}$$

It is very important to remember that *in any aqueous solution the relationship expressed in Equation 16.1 must always be satisfied, regardless of any other equilibria that may also exist in the solution.*

Equation 16.1 can be used to calculate the molar concentrations of both the H^+ and OH^- ions in pure water. From the stoichiometry of the dissociation, we see that for each 1 mol of H^+ formed, 1 mol of OH^- is also produced. This means that at equilibrium, $[H^+] = [OH^-]$. If we let x equal the hydrogen ion concentration, then

$$x = [H^+] = [OH^-]$$

Substituting into Equation 16.1 gives

$$K_w = x \cdot x = x^2$$

or, because $K_w = 1.0 \times 10^{-14}$,

$$x^2 = 1.0 \times 10^{-14}$$

Taking the square root yields

$$x = 1.0 \times 10^{-7}$$

which means that the concentrations of hydrogen ion and hydroxide ion in pure water are

$$[H^+] = [OH^-] = 1.0 \times 10^{-7} \, M \quad \text{(at 25 °C)}$$

Whenever the hydrogen ion concentration equals the hydroxide ion concentration, as it does in pure water, the solution is said to be *neutral*. An acid is a substance that makes the H^+ concentration greater than the OH^- concentration; conversely, a base makes the OH^- concentration greater than the H^+ concentration. However, remember that there is always *some* OH^- present in an acidic solution, just as there is

---

The density of H_2O is 1.00 g/mL, so 1.00 L has a mass of 1.00×10^3 g. Therefore,

$$[H_2O] = \frac{1.00 \times 10^3 \text{ g } H_2O}{1.00 \text{ L}}$$

$$\times \frac{1 \text{ mol } H_2O}{18.0 \text{ g } H_2O} = 55.6 \text{ mol } H_2O/L$$

The mass action expression $[H_3O^+][OH^-]$ is called the *ion product* for water.

K_w varies with temperature. At 37 °C (body temperature)

$$K_w = 2.42 \times 10^{-14}$$

always *some* H^+ present even if the solution is basic. At all times, Equation 16.1 is obeyed if the solution is at equilibrium.

16.2 SOLUTIONS OF STRONG ACIDS AND BASES

Since we will be dealing with acids and bases throughout this chapter, it is very important that you begin by recognizing those that are strong. We discussed these in Chapter 5, and Table 5.2 contains a list of acids and bases with the strong ones marked by asterisks.

The most common strong monoprotic acids are HCl and HNO_3. The most important thing to remember about them is that they are essentially 100% dissociated. When we write an equation for their reaction with water, therefore, we *do not* represent it as an equilibrium. Thus, for HNO_3 we have

$$HNO_3 + H_2O \longrightarrow H_3O^+ + NO_3^-$$

or, omitting the water and using H^+ in place of H_3O^+, we obtain

$$HNO_3 \longrightarrow H^+ + NO_3^-$$

In a solution of a strong monoprotic acid, the amount of H^+ contributed to the solution by the acid is determined by the specified concentration of the acid. Thus, if a solution is labeled 0.020 M HNO_3, it contributes 0.020 M H^+ and 0.020 M NO_3^-.

The strong bases in water are the soluble metal hydroxides. These include solutions of the hydroxides of the metals of Group IA, such as NaOH and KOH, as well as dilute solutions of the hydroxides of the Group IIA elements from calcium down to barium. These are ionic compounds that are also completely dissociated in solution, so we *never* write equilibria for their dissolving in water. Thus, typically we have

Metal hydroxides of low solubility are also strong electrolytes, but we don't tend to think of them as strong bases because they don't produce a high concentration of OH^- when placed in water.

$$NaOH(s) \xrightarrow{H_2O} Na^+ + OH^-$$

$$Ca(OH)_2(s) \xrightarrow{H_2O} Ca^{2+} + 2OH^-$$

Notice that for each mole of $Ca(OH)_2$ that dissolves, 2 mol of OH^- are added to the solution.

In an aqueous solution of an acid we will often want to know what the H^+ concentration is. In these cases it is almost always safe to assume that essentially all the H^+ in the solution comes from the dissolved acid. In other words, it is usually safe to assume that the dissociation of water contributes a negligible amount of H^+ to the solution. This is because the presence of H^+ from an acid (for example, HCl) shifts the equilibrium

$$H_2O \rightleftharpoons H^+ + OH^-$$

to the left. Therefore, the amount of water that is dissociated in a solution of an acid is even less than in pure water, which means that the H^+ coming from the dissociation of water is less than 10^{-7} M. Similarly, the OH^- concentration in a solution of a base can be calculated just from the concentration of the solute. The OH^- contributed by the dissociation of water is negligible. Examples 16.1 and 16.2 illustrate this point.

EXAMPLE 16.1. CALCULATING CONCENTRATIONS IN A SOLUTION OF A STRONG ACID

PROBLEM: (a) What is the OH^- concentration in a 0.0010 M HCl solution? (b) What is the H^+ concentration derived from the dissociation of the solvent?

SOLUTION: (a)   At equilibrium we must have

$$[H^+][OH^-] = 1.0 \times 10^{-14}$$

HCl is a strong acid and is essentially 100% dissociated.

$$HCl \longrightarrow H^+ + Cl^-$$

Therefore, 0.0010 mol of HCl per liter gives 0.0010 mol of H^+ per liter. The total hydrogen ion concentration, then, is 0.0010 M *plus* the amount contributed by the dissociation of water. Let's assume for the moment that this contribution is negligible, as suggested above, and that it can be ignored. This gives

$$[H^+]_{total} = 0.0010\ M + [H^+]_{(from\ H_2O)} \approx 0.0010\ M$$

Solving for the hydroxide ion concentration in Equation 16.1, we get

$$[OH^-] = \frac{1.0 \times 10^{-14}}{[H^+]}$$
$$= \frac{1.0 \times 10^{-14}}{0.0010} = 1.0 \times 10^{-11}\ M$$

(b)   The hydroxide ion in part (a) comes entirely from the dissociation of water. Therefore, the concentration of H^+ derived from H_2O must *also* be $1.0 \times 10^{-11}\ M$, as can be seen from the stoichiometry of the ionization reaction. Note that this value ($1.0 \times 10^{-11}\ M$) is indeed negligible compared to the H^+ concentration produced by the HCl ($1.0 \times 10^{-3}\ M$), so the assumption made in part (a) was valid.

EXAMPLE 16.2. CALCULATING THE CONCENTRATIONS IN A SOLUTION OF A STRONG BASE

PROBLEM: What are the concentrations of H^+ and OH^- in a 0.0040 M solution of $Ca(OH)_2$? What part of the $[OH^-]$ in the solution comes from the ionization of water?

SOLUTION: Because $Ca(OH)_2$ is a strong base, we know that each mole of $Ca(OH)_2$ that dissolves contributes 2 mol of OH^- to the solution. Therefore, from 0.0040 mol per liter of $Ca(OH)_2$, we obtain 0.0080 mol per liter of OH^-. In addition to this, there is the small contribution to the OH^- concentration made by the ionization of the solvent, which we will assume is negligible.

$$[OH^-]_{total} = 0.0080\ M + [OH^-]_{(from\ H_2O)} \approx 0.0080\ M$$

To obtain the $[H^+]$, we use Equation 16.1.

$$[H^+] = \frac{1.0 \times 10^{-14}}{[OH^-]}$$
$$= \frac{1.0 \times 10^{-14}}{0.0080} = 1.2 \times 10^{-12}\ M$$

Because H^+ and OH^- are formed in equal numbers in the dissociation of water, the value of $[H^+]$ above is the same as the value of $[OH^-]_{(from\ H_2O)}$. Therefore,

$$[OH^-]_{(from\ H_2O)} = 1.2 \times 10^{-12}\ M$$

Notice that this value is negligibly small compared to the OH^- contribution made by the solute, $Ca(OH)_2$, and that the assumption made above was indeed valid.

The approximations that we made in these last two examples work quite well almost all the time. Only if the acid or base is extremely dilute (or if we are working with very weak acids or bases), do we need to worry about the contribution of H^+ or OH^- from the solvent. You can use the following as a rule of thumb:

> If the $[H^+]$ from a dissolved acid or the $[OH^-]$ from a dissolved base is larger than 5×10^{-7} M, it is safe to ignore the contribution of that species from the dissociation of water.

16.3 THE pH CONCEPT

Hydrogen ion and hydroxide ion enter into many equilibria in addition to the dissociation of water, so it is frequently necessary to specify their concentrations in aqueous solutions. These concentrations may range from relatively high values to very small ones (for example, 10 M to 10^{-14} M), and a logarithmic notation[1] has been devised to simplify the expression of these quantities. In general, for some quantity X, the quantity pX is defined as

Notice that logs to the base 10 (common logs) are used here, *not* natural logs.

$$pX = \log \frac{1}{X} = -\log X \qquad (16.2)$$

For example, if we wish to specify the hydrogen ion concentration in a solution, we speak of **pH.** This is defined as

$$pH = \log \frac{1}{[H^+]} = -\log[H^+] \qquad (16.3)$$

In a solution where the hydrogen ion concentration is 10^{-3} M, we therefore have

$$pH = -\log(10^{-3}) = -(-3)$$

or

$$pH = 3$$

Similarly, if the hydrogen ion concentration is 10^{-8} M, the pH of the solution is 8.

Following the same approach for the hydroxide ion concentration, we can define the **pOH** of a solution as

$$pOH = -\log[OH^-] \qquad (16.4)$$

Just as the H^+ and OH^- ion concentrations in a solution are related to each other, so also are the pH and pOH. From the equilibrium expression for the dissociation of water (Equation 16.1), we obtain the following by taking the logarithm of both sides:

$$\log K_w = \log[H^+] + \log[OH^-]$$

Multiplying through by -1 gives

$$(-\log K_w) = (-\log[H^+]) + (-\log[OH^-])$$

By definition, $-\log K_w = pK_w$. Therefore,

$$pK_w = pH + pOH$$

$$pK_w = -\log K_w$$
$$= -\log (1.0 \times 10^{-14})$$
$$= 14.00$$

Since $K_w = 1.0 \times 10^{-14}$, $pK_w = 14.00$. This gives the useful relationship

$$pH + pOH = 14.00 \qquad (16.5)$$

In a neutral solution, $[H^+] = [OH^-] = 1.0 \times 10^{-7}$ M, and pH = pOH = 7.00, so that in a neutral solution we say that the pH = 7.00. In an acidic solution the hydrogen ion concentration is greater than 1.0×10^{-7} M (for example, 10^{-3} M) and the pH is less than 7.00. By the same token, in basic solutions the $[H^+]$ is less than 1.0×10^{-7} M (for example, 10^{-10} M) and the pH is greater than 7.00. This is summarized as follows:

---

[1] A discussion of the use of logarithms can be found in Appendix A.

	$[H^+]$	$[OH^-]$	pH	pOH
Acidic solution	$>(1.0 \times 10^{-7})$	$<(1.0 \times 10^{-7})$	<7.00	>7.00
Neutral solution	1.0×10^{-7}	1.0×10^{-7}	7.00	7.00
Basic solution	$<(1.0 \times 10^{-7})$	$>(1.0 \times 10^{-7})$	>7.00	<7.00

These relationships apply at 25 °C. Only at this temperature does $K_w = 1.0 \times 10^{-14}$.

Many common substances are either acidic or basic, and their degree of acidity or basicity is conveniently expressed in terms of pH (Figure 16.1). Notice that substances having a pH less than 7 — that is, those that are acidic — have characteristically sour tastes. Lemon juice contains citric acid, for example, and vinegar contains acetic acid. On the other hand, basic substances such as milk of magnesia, a suspension of $Mg(OH)_2$ in water, have a bitter taste. Although sour and bitter taste is the body's way of judging the acidity of foods, *never* taste chemicals in the lab; many of them are poisonous and will definitely ruin your day!

Never taste any chemical in the laboratory!

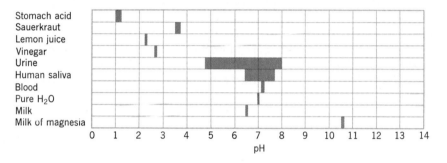

Stomach acid
Sauerkraut
Lemon juice
Vinegar
Urine
Human saliva
Blood
Pure H_2O
Milk
Milk of magnesia

pH: 0 1 2 3 4 5 6 7 8 9 10 11 12 13 14

Figure 16.1. *The pH ranges found for some common substances.*

Measuring pH; acid–base indicators and pH meters

One of the earliest ways of judging the acidity or basicity of a solution was by the use of substances called **acid–base indicators.** These are organic compounds whose color depends on the pH of the solution in which they're dissolved. A typical example is litmus, which is pink in an acidic solution and blue in a basic solution. In the lab, you probably will have occasion to use litmus paper — strips of absorbent paper impregnated with the litmus dye. It comes in two varieties, blue and red, and is used to test whether a solution is acidic or basic. For example, to test whether a solution is basic, you place a drop of the solution on red litmus paper, which will turn blue if it is basic. Similarly, blue litmus paper is used to determine whether a solution is acidic. If you touch a drop of an acidic solution to blue litmus paper, the dye turns pink.

Other pH test papers are available that contain mixtures of indicator dyes and can actually be used to estimate the approximate value of the pH, as illustrated in Figure 16.2.

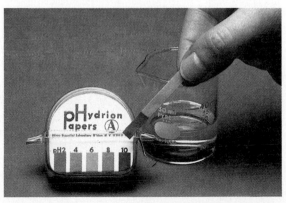

Figure 16.2. *The pH of a solution can be estimated using test papers impregnated with acid–base indicators. Here the color of the Hydrion® strip has changed to purple after being touched with a drop of the solution in the beaker. According to the color code, the solution is basic with a pH of approximately 10.*

Figure 16.3. *A pH meter. The instrument is first calibrated by dipping the pair of electrodes (one sensitive to the H^+ concentration and the other a reference electrode) into a solution of known pH, and then adjusting the scale appropriately. Afterward, the pH of any other solution is obtained simply by dipping the electrodes into the solution and reading the pH from the scale.*

It is quite common today to find instruments called **pH meters** in labs of all kinds. These are electronic devices that enable the pH of a solution to be measured with a high degree of precision and accuracy. A typical pH meter is shown in Figure 16.3.

pH and pOH calculations

Equations 16.3 through 16.5 allow us to calculate the pH or pOH of a solution from concentrations of H^+ or OH^-, or to calculate the concentrations of these ions if the pH or pOH is known.

The next three example problems illustrate the kinds of calculations that you may encounter. We will assume that you are using a scientific electronic calculator, from which you can obtain logarithms and antilogarithms at the press of a button or two. With the availability of these electronic marvels, pH calculations are so routine that you should be able to perform them almost without effort, and in future problems dealing with hydrogen ion concentrations we will assume you have mastered this topic. If you do not have a scientific calculator, don't be discouraged, however. A description of the use of log tables in obtaining pH from $[H^+]$, and vice versa, is contained in Appendix A, where you will find the pH calculations in Examples 16.3 through 16.5 worked out using values taken from a table of common logarithms.

EXAMPLE 16.3. CALCULATING THE pH OF A SOLUTION OF A STRONG ACID

PROBLEM: What is the pH of a 0.0020 M solution of HCl?

ANALYSIS: The key to solving this particular problem is realizing that HCl is a strong acid, which means that it is completely dissociated in the solution.

SOLUTION: The complete dissociation of HCl gives 1 mol of H^+ and 1 mol of Cl^- for every 1 mol of HCl that dissociates. Since the solution contains 0.0020 mol of HCl per liter, it contains 0.0020 mol of H^+ per liter and therefore $[H^+] = 0.0020$ M. (The solution also contains 0.0020 mol of Cl^- per liter, but we don't need that to calculate the pH.)

The pH is defined as

$$pH = -\log[H^+]$$

Substituting yields

$$pH = -\log(2.0 \times 10^{-3})$$

Performing this operation gives

$$pH = 2.70$$

The rules for significant figures when using logarithms were described on page 72.

Notice that the number of digits after the decimal is equal to the number of significant figures in the hydrogen ion concentration.

EXAMPLE 16.4. CALCULATING THE pH OF A SOLUTION OF A STRONG BASE

PROBLEM: What is the pH of a 5.0×10^{-4} M solution of NaOH at 25 °C?

SOLUTION: Metal hydroxides are strong bases, which means they are 100% dissociated.

Therefore, in this solution the OH^- concentration is equal to 5.0×10^{-4} M. To calculate the pH, we can proceed in either of two ways.

1. We can calculate $[H^+]$ from K_w and the known value of $[OH^-]$, and then proceed as in the previous example.

2. We can use the $[OH^-]$ to calculate the pOH and then subtract this value from 14 to obtain the pH.

METHOD 1 We know that

$$K_w = [H^+][OH^-]$$

Therefore,

$$[H^+] = \frac{1.0 \times 10^{-14}}{5.0 \times 10^{-4}}$$

$$= 2.0 \times 10^{-11} M$$

From this we caculate the pH

$$pH = -\log(2.0 \times 10^{-11})$$

$$= 10.70$$

METHOD 2 By definition

$$pOH = -\log[OH^-]$$

In this problem

$$[OH^-] = 5.0 \times 10^{-4} M$$

Therefore,

$$pOH = -\log(5.0 \times 10^{-4})$$

$$= 3.30$$

The pH is therefore

$$pH = 14.00 - 3.30$$

Remember, pH + pOH = 14.00.

$$= 10.70$$

EXAMPLE 16.5. CALCULATING $[H^+]$ AND $[OH^-]$ FROM pH

PROBLEM: A sample of orange juice was found to have a pH of 3.80. What were the H^+ and OH^- concentrations in the juice?

SOLUTION: The pH is defined as

$$pH = -\log[H^+]$$

The antilogarithm of this gives the following relationship:

$$[H^+] = 10^{-pH}$$

For the sample of orange juice, therefore,

$$[H^+] = 10^{-3.80}$$

$$= 1.6 \times 10^{-4}$$

(Check the direction booklet that came with your calculator to be sure you can take the antilogarithm of -3.80. Practice using your calculator until you can obtain the answer given above.) Once we have obtained $[H^+]$, we can calculate $[OH^-]$ from K_w.

$$[OH^-] = \frac{1.0 \times 10^{-14}}{1.6 \times 10^{-4}}$$

$$= 6.3 \times 10^{-11} M$$

Alternatively, we could first calculate pOH from pH and then take the antilogarithm.

$$pOH = 14.00 - pH$$
$$= 14.00 - 3.80 = 10.20$$
$$[OH^-] = 10^{-pOH}$$
$$= 10^{-10.20}$$
$$= 6.3 \times 10^{-11} \ M$$

16.4 CONJUGATE ACID–BASE SYSTEMS IN AQUEOUS SOLUTIONS

In Section 10.5 we discussed the Brønsted–Lowry approach to acids and bases. Recall that in this system, an acid is defined as a proton (H^+) donor and a base is defined as a proton acceptor. Although this system doesn't rely on the presence of a solvent, it is nevertheless very powerful in helping us analyze acid–base equilibria in aqueous solutions.

The most important equilibrium in aqueous solutions is the ionization of water, which was described at the beginning of this chapter. If we look at this equilibrium in Brønsted–Lowry terms and consider one water molecule the acid and the other the base, we have

$$
\overbrace{H_2O + H_2O \rightleftharpoons H_3O^+ + OH^-}^{\text{conjugate pair}}
$$

acid   base   acid   base

conjugate pair

Let's focus on the water molecule that's behaving as an acid in this reaction (the one printed in red). Notice that when it reacts, it forms its conjugate base, OH^-, and a hydronium ion. In general, this is how acids behave in water; they each transfer an H^+ ion to a molecule of H_2O and form their conjugate bases. Some examples are

Cl^- is the conjugate base of HCl.

$C_2H_3O_2^-$ is the conjugate base of $HC_2H_3O_2$.

$$HCl + H_2O \longrightarrow H_3O^+ + Cl^-$$
$$HC_2H_3O_2 + H_2O \rightleftharpoons H_3O^+ + C_2H_3O_2^-$$

The first reaction is not written as an equilibrium, you recall, because HCl is a strong acid; it is effectively 100% dissociated in water.

Acids are not always neutral molecules. Some anions, such as the bisulfate ion, HSO_4^-, are also acids. This anion, which is formed in the first step in the ionization of H_2SO_4, reacts with water as follows.

$$HSO_4^- + H_2O \rightleftharpoons H_3O^+ + SO_4^{2-}$$

Acids can even be cations, such as the ammonium ion, NH_4^+, which undergoes this reaction.

$$NH_4^+ + H_2O \rightleftharpoons H_3O^+ + NH_3$$

Equilibrium constant expressions for acids in water

For strong acids, such as HCl, we never bother to write an equilibrium law because the dissociation is taken to be complete. Acetic acid, however, is a weak acid and its reaction with water is an equilibrium. We can write an equilibrium law for the reaction following the same principles as before. For the reaction

$$HC_2H_3O_2 + H_2O \rightleftharpoons H_3O^+ + C_2H_3O_2^-$$

we write

$$K_c = \frac{[H_3O^+][C_2H_3O_2^-]}{[H_2O][HC_2H_3O_2]}$$

As we noted earlier, the concentration of water in dilute aqueous solutions is essentially a constant, and as such it can be incorporated into the equilibrium constant to give a new constant that we will call K_a. The subscript a is a reminder that the equilibrium constant is for the substance behaving as an acid and the equilibrium constant is called the **ionization constant** or **dissociation constant** for the acid. (Using the subscript a may seem unnecessary now, but later when we've encountered a variety of equilibrium constants, it is very helpful to know what kind of reaction the equilibrium constant is for.)

$$K_c \times [H_2O] = K_a = \frac{[H_3O^+][C_2H_3O_2^-]}{[HC_2H_3O_2]}$$

We can also simplify this somewhat by using H^+ in place of H_3O^+ and writing the simplified chemical equation

$$HC_2H_3O_2 \rightleftharpoons H^+ + C_2H_3O_2^-$$

from which we obtain the simplified equilibrium expression

$$K_a = \frac{[H^+][C_2H_3O_2^-]}{[HC_2H_3O_2]}$$

We will usually work with these simplified equations and equilibrium laws.

We can write similar equilibrium laws for acids such as HSO_4^- and NH_4^+, as shown below. Referring to the chemical equation above for the reaction of the bisulfate ion with water, we have

$$K_a = \frac{[H_3O^+][SO_4^{2-}]}{[HSO_4^-]}$$

The simplified equation for the reaction is

$$HSO_4^- \rightleftharpoons H^+ + SO_4^{2-}$$

for which we write the equilibrium expression

$$K_a = \frac{[H^+][SO_4^{2-}]}{[HSO_4^-]}$$

Similarly, for the ammonium ion, we have

$$K_a = \frac{[H_3O^+][NH_3]}{[NH_4^+]}$$

The simplified equation for the equilibrium and the corresponding equilibrium law are

$$NH_4^+ \rightleftharpoons H^+ + NH_3$$

$$K_a = \frac{[H^+][NH_3]}{[NH_4^+]}$$

By now, you may have noticed the pattern. In water, all acids react in essentially the same way, regardless of whether they are neutral molecules, anions, or cations. If we represent an acid by the general formula HA, then it reacts with water according to the equation

$$HA + H_2O \rightleftharpoons H_3O^+ + A^- \qquad (16.6)$$

and its equilibrium law is

$$K_a = \frac{[H_3O^+][A^-]}{[HA]} \qquad (16.7)$$

Usually, it is easier to write everything in the abbreviated simpler form.

$$HA \rightleftharpoons H^+ + A^- \tag{16.8}$$

$$K_a = \frac{[H^+][A^-]}{[HA]} \tag{16.9}$$

For solving equilibrium problems dealing with acids in solution, it is very important to be able to write the appropriate chemical equation and to construct the equilibrium expression. Usually, the latter isn't a problem, once you've gotten the correct equation. Therefore, it is imperative that you learn and understand how to write equations such as 16.6 and 16.7 or 16.8 and 16.9.

Equilibrium constant expressions for bases in water

As with acids, bases also undergo the same kinds of reactions with water. For instance, in the ionization of water,

conjugate pair

$$H_2O + H_2O \rightleftharpoons H_3O^+ + OH^-$$

acid   base   acid   base

conjugate pair

notice that the water molecule that functions as a base (the one printed in blue) accepts a proton from the other water molecule and becomes its conjugate acid. This is how bases in general react in water. An example is ammonia (the conjugate *base* of NH_4^+).

$$NH_3 + H_2O \rightleftharpoons NH_4^+ + OH^-$$

The equilibrium law for this reaction is

$$K_c = \frac{[NH_4^+][OH^-]}{[H_2O][NH_3]}$$

Once again, the concentration of water appearing in the denominator is essentially a constant that we can incorporate into the equilibrium constant K_c to give a new constant that we will call K_b.

$$K_c \times [H_2O] = K_b = \frac{[NH_4^+][OH^-]}{[NH_3]}$$

This time we can't write a simplified equation by leaving the water out because we need it to balance the oxygen in the hydroxide ion. However, we will be writing the equilibrium law in the form just described for K_b.

Bases don't have to be neutral molecules either. Many anions are bases and produce hydroxide ion in water just as ammonia does. For example, acetate ion, $C_2H_3O_2^-$, which is the conjugate *base* of acetic acid, reacts with water as follows.

$$C_2H_3O_2^- + H_2O \rightleftharpoons HC_2H_3O_2 + OH^-$$

We write the equilibrium expression as

$$K_b = \frac{[HC_2H_3O_2][OH^-]}{[C_2H_3O_2^-]}$$

If the base is an anion, then the reaction is

$$B^- + H_2O \rightleftharpoons BH + OH^-$$

or even

$$B^{2-} + H_2O \rightleftharpoons HB^- + OH^-$$

In general, cations have little tendency to behave as bases because they resist acquiring H^+ ions (like charges repel).

There is a general equation that we can write for the reaction of a base with water. If we use the symbol B to stand for a base, then

$$B + H_2O \rightleftharpoons HB^+ + OH^- \tag{16.10}$$

and the equilibrium law is written as

$$K_b = \frac{[HB^+][OH^-]}{[B]} \tag{16.11}$$

Be sure you learn Equations 16.10 and 16.11 so you can apply them to equilibrium problems that involve weak bases in water.

The relationship between K_a and K_b for an acid–base conjugate pair

A very simple, yet interesting relationship exists between the values of K_a and K_b for the members of an acid–base conjugate pair, which is

$$K_a \times K_b = K_w \tag{16.12}$$

That this is true can be easily seen if we consider the pair NH_4^+ and NH_3. For the ammonium ion the expression corresponding to K_a is

$$K_a = \frac{[H^+][NH_3]}{[NH_4^+]}$$

and for ammonia the expression for K_b is

$$K_b = \frac{[NH_4^+][OH^-]}{[NH_3]}$$

Therefore, the product of K_a and K_b is

$$K_a \times K_b = \frac{[H^+][\cancel{NH_3}]}{[\cancel{NH_4^+}]} \times \frac{[\cancel{NH_4^+}][OH^-]}{[\cancel{NH_3}]}$$

$$K_a \times K_b = [H^+][OH^-]$$

$$K_a \times K_b = K_w$$

Be sure you know Equation 16.12, because normally either K_a or K_b, but not both, will be found in tables. Thus, it is common to find listed the value of K_b for NH_3, but rarely will you find a table that gives the value of K_a for NH_4^+. This is no problem, however, because if you know K for one of the members of the pair, you can easily calculate the K for the other.

The relative strengths of the members of conjugate acid–base pairs

You learned in the last chapter that the larger the value of K for a reaction, the farther the reaction proceeds toward completion if it is allowed to come to equilibrium. Let's apply this to the ionization of an acid.

$$HA \rightleftharpoons H^+ + A^-$$

The stronger the acid, the greater its degree of ionization, and the farther the reaction proceeds in the forward direction. What this means is that the larger the value of K_a for the acid, the stronger the acid is. This allows us to compare the strengths of acids according to their values of K_a.

Similarly, we can compare the strengths of bases by comparing their values of K_b. For the base B, which reacts according to the equation

$$B + H_2O \rightleftharpoons HB^+ + OH^-$$

the larger the value of K_b, the stronger the base.

Figure 16.4. *The relative strengths of the members of acid–base conjugate pairs. The stronger the acid, the weaker its conjugate base.*

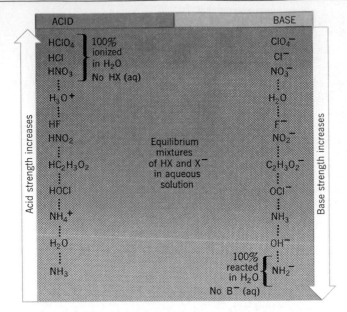

For an acid–base conjugate pair, there is an inverse relationship between the strength of the acid and the strength of its conjugate base: *The stronger an acid is, the weaker is its conjugate base.* This should be obvious from the relationship given in Equation 16.12. The larger K_a is, the smaller K_b must be in order for their product to remain equal to K_w.

For strong acids, the extent of reaction is essentially 100% and the value of K_a is *very large.* This also means that the K_b for the corresponding conjugate base is *very small.* Thus, HCl is a very strong acid, and Cl^- is a very weak base. There are some very strong bases, too. An example is the oxide ion, which reacts with water as follows.

$$\overset{\displaystyle\text{conjugate pair}}{\underset{\text{base}\qquad\qquad\qquad\text{acid}}{O^{2-} + H_2O \rightleftharpoons OH^- + OH^-}}$$

Although we've written the reaction as an equilibrium, it proceeds to completion 100%. Oxide ion is therefore a very strong base, and its conjugate acid, the hydroxide ion, is a *very weak* acid.

Many of the acids and bases that we will discuss later in this chapter have strengths that lie between *very strong* and *very weak.* We could say that they are moderately weak, which means that their conjugates are also moderately weak. Thus, acetic acid ($HC_2H_3O_2$) is a moderately weak acid (although we usually leave out the adjective "moderately") with $K_a = 1.8 \times 10^{-5}$. The conjugate base, $C_2H_3O_2^-$, is also moderately weak with $K_b = 5.6 \times 10^{-10}$. Similarly, NH_3 is a moderately weak base (but we usually leave out "moderately"), and its conjugate acid, NH_4^+, is a moderately weak acid. Figure 16.4 illustrates how the relative acid and base strengths of typical acids and bases vary relative to each other.

Substances that we have earlier called weak acids and bases are the ones that we are referring to here, for comparative purposes, as moderately weak.

16.5 EQUILIBRIA INVOLVING WEAK MOLECULAR ACIDS AND BASES

In the preceding section you learned how to write the chemical equations and equilibrium laws of the dissociation of weak acids and bases. In this section we

Table 16.1. Ionization constants for some weak acids and bases

Weak Acid	Ionization	K_a	pK_a
Dichloroacetic acid	$HC_2HO_2Cl_2 \rightleftharpoons H^+ + C_2HO_2Cl_2^-$	5.0×10^{-2}	1.30
Chloroacetic acid	$HC_2H_2O_2Cl \rightleftharpoons H^+ + C_2H_2O_2Cl^-$	1.4×10^{-3}	2.85
Hydrofluoric acid	$HF \rightleftharpoons H^+ + F^-$	6.5×10^{-4}	3.19
Nitrous acid	$HNO_2 \rightleftharpoons H^+ + NO_2^-$	4.5×10^{-4}	3.35
Formic acid	$HCHO_2 \rightleftharpoons H^+ + CHO_2^-$	1.8×10^{-4}	3.74
Lactic acid	$HC_3H_5O_3 \rightleftharpoons H^+ + C_3H_5O_3^-$	1.38×10^{-4}	3.86
Benzoic acid	$HC_7H_5O_2 \rightleftharpoons H^+ + C_7H_5O_2^-$	6.5×10^{-5}	4.19
Acetic acid	$HC_2H_3O_2 \rightleftharpoons H^+ + C_2H_3O_2^-$	1.8×10^{-5}	4.74
Butyric acid	$HC_4H_7O_2 \rightleftharpoons H^+ + C_4H_7O_2^-$	1.5×10^{-5}	4.82
Nicotinic acid	$HC_6H_4NO_2 \rightleftharpoons H^+ + C_6H_4NO_2^-$	1.4×10^{-5}	4.85
Propionic acid	$HC_3H_5O_2 \rightleftharpoons H^+ + C_3H_5O_2^-$	1.4×10^{-5}	4.85
Barbituric acid	$HC_4H_3N_2O_3 \rightleftharpoons H^+ + C_4H_3N_2O_3^-$	1.0×10^{-5}	5.00
Veronal (diethylbarbituric acid)	$HC_8H_{11}N_2O_3 \rightleftharpoons H^+ + C_8H_{11}N_2O_3^-$	3.7×10^{-8}	7.43
Hypochlorous acid	$HOCl \rightleftharpoons H^+ + OCl^-$	3.1×10^{-8}	7.51
Hydrocyanic acid	$HCN \rightleftharpoons H^+ + CN^-$	4.9×10^{-10}	9.31

Weak Base	Ionization	K_b	pK_b
Diethylamine	$(C_2H_5)_2NH + H_2O \rightleftharpoons (C_2H_5)_2NH_2^+ + OH^-$	9.6×10^{-4}	3.02
Methylamine	$CH_3NH_2 + H_2O \rightleftharpoons CH_3NH_3^+ + OH^-$	3.7×10^{-4}	3.43
Ammonia	$NH_3 + H_2O \rightleftharpoons NH_4^+ + OH^-$	1.8×10^{-5}	4.74
Hydrazine	$N_2H_4 + H_2O \rightleftharpoons N_2H_5^+ + OH^-$	1.7×10^{-6}	5.77
Hydroxylamine	$NH_2OH + H_2O \rightleftharpoons NH_3OH^+ + OH^-$	1.1×10^{-8}	7.97
Pyridine	$C_5H_5N + H_2O \rightleftharpoons C_5H_5NH^+ + OH^-$	1.7×10^{-9}	8.77
Aniline	$C_6H_5NH_2 + H_2O \rightleftharpoons C_6H_5NH_3^+ + OH^-$	3.8×10^{-10}	9.42

examine the kinds of calculations that are involved when we deal with solutions that contain molecular acids and bases, such as acetic acid and ammonia.

The extent to which a weak acid or base undergoes ionization, as well as the value of the ionization constant, must be determined experimentally. (One way of doing this is to measure the pH of a solution prepared by dissolving a known quantity of the weak acid or base in a given volume of solution, as illustrated in Examples 16.6 and 16.7.) Ionization constants of a number of weak acids and bases are listed in Table 16.1. Notice that their values are quite small, ranging from 10^{-2} to 10^{-10}. Recalling that numbers this small can be simplified by applying the logarithmic notation of Equation 16.2, we can write for any K_a,

A more extensive table of ionization constants of acids and bases is located in Appendix C.

$$pK_a = -\log K_a$$

and for any K_b,

$$pK_b = -\log K_b$$

For example, the pK_a of acetic acid is

$$pK_a = -\log K_a = -\log(1.8 \times 10^{-5})$$

$$= 4.74$$

and for pyridine, a bad-smelling liquid,

$$pK_b = -\log(1.7 \times 10^{-9})$$

$$= 8.77$$

From our discussion in Section 15.2, we know that the smaller the value of an equilibrium constant, the smaller the extent of reaction. Therefore, the smaller the value of K_a or K_b, the smaller the extent of ionization and the weaker the acid or base. Relative strengths of acids and bases can also be indicated by their pK_a's and pK_b's. In this case the smaller the value of pK_a or pK_b, the *stronger* is the acid or base. Let's compare, for example, the pK_a's for acetic, chloroacetic, and dichloroacetic acid. Their pK_a's are

$$HC_2H_3O_2 \qquad pK_a = 4.74$$
$$HC_2H_2ClO_2 \qquad pK_a = 2.85$$
$$HC_2HCl_2O_2 \qquad pK_a = 1.30$$

The order of increasing acidity is, therefore,

$$\text{acetic} < \text{chloroacetic} < \text{dichloroacetic acid}$$

Let's look now at some examples showing how these equilibrium constants can be calculated.

EXAMPLE 16.6. USING THE pH OF A SOLUTION OF AN ACID TO CALCULATE K_a

PROBLEM: A student prepared a 0.10 M acetic acid solution and measured its pH to be 2.88. Calculate K_a for acetic acid.

ANALYSIS: As in any equilibrium problem, the first step is to write the chemical equation for the equilibrium and its equilibrium law. For acetic acid, we have

$$HC_2H_3O_2 \rightleftharpoons H^+ + C_2H_3O_2^-$$

and

$$K_a = \frac{[H^+][C_2H_3O_2^-]}{[HC_2H_3O_2]}$$

To evaluate K_a, we need to have all the equilibrium concentrations to substitute into the mass action expression. They are obtained from the information given in the problem as follows.

SOLUTION: We've been given the pH, from which we can obtain the H^+ concentration.

$$pH = 2.88$$

$$[H^+] = 1.3 \times 10^{-3}\ M$$

$[H^+] = 10^{-2.88} = 1.3 \times 10^{-3}\ M$

We will assume that the measurement of pH always yields the equilibrium H^+ concentration.

Equilibria involving the ionization of acids and bases are established rapidly, so this must be the equilibrium value for $[H^+]$. We saw earlier that in a solution of an acid, it is usually a good approximation to assume that all the H^+ comes from the acid itself, so we can take this as the amount of H^+ per liter that comes from the acetic acid. From the stoichiometry of the dissociation, we see that H^+ and $C_2H_3O_2^-$ are formed in a 1-to-1 ratio in the solution, so for each mole of H^+ that is formed, we also get 1 mol of $C_2H_3O_2^-$. In this solution, therefore, the concentrations of the ions must be equal.

$$[H^+] = [C_2H_3O_2^-] = 1.3 \times 10^{-3}\ M$$

The concentration of undissociated $HC_2H_3O_2$ at equilibrium is equal to the original concentration, 0.10 M, *minus* the number of moles per liter of acetic acid that have dissociated. At equilibrium, then, we have

	Equilibrium Concentrations
H^+	$1.3 \times 10^{-3}\ M$
$C_2H_3O_2^-$	$1.3 \times 10^{-3}\ M$
$HC_2H_3O_2$	$1.0 \times 10^{-1} - 0.013 \times 10^{-1} = 1.0 \times 10^{-1}\ M$

Note that when we compute the acetic acid concentration *rounded to the proper number of significant figures*, the amount that has dissociated is negligible compared to the amount initially present. Thus,

$$0.10\ M - 0.0013\ M = (0.0987\ M) = 0.10\ M \quad \text{(rounded)}$$

Substituting the equilibrium concentrations into the expression for K_a, we have

$$K_a = \frac{(1.3 \times 10^{-3})(1.3 \times 10^{-3})}{(1.0 \times 10^{-1})}$$

$$= 1.7 \times 10^{-5}$$

This value of K_a differs from that given in Table 16.1 because of round-off.

Another way of specifying the extent of the dissociation of a weak acid or base is by giving its **percent dissociation,** which is defined as

$$\text{percent dissociation} = \frac{(\text{mol/L of acid or base dissociated})}{(\text{mol/L of acid or base available})} \times 100\%$$

This could also be called **percent ionization.** We often use the terms dissociation and ionization interchangeably.

For the acetic acid in the preceding example, the mol/L dissociated equals the mol/L of H^+ formed, which is $1.3 \times 10^{-3}\ M$. The mol/L available is the original concentration, $0.10\ M$. Therefore,

$$\text{percent dissociation} = \frac{1.3 \times 10^{-3}\ M}{0.10\ M} \times 100\% = 1.3\%$$

EXAMPLE 16.7. USING THE PERCENT DISSOCIATION TO CALCULATE AN EQUILIBRIUM CONSTANT

PROBLEM: A $0.010\ M\ NH_3$ solution was prepared and it was determined that the NH_3 had undergone 4.2% ionization. Calculate the K_b for NH_3.

SOLUTION: Ammonia ionizes in water according to the reaction

$$NH_3 + H_2O \rightleftharpoons NH_4^+ + OH^-$$

for which we write

$$K_b = \frac{[NH_4^+][OH^-]}{[NH_3]}$$

From the stoichiometry of the ionization we see that at equilibrium

$$[NH_4^+] = [OH^-]$$

Since the $0.010\ M$ solution undergoes 4.2% ionization, the number of moles per liter of these ions present at equilibrium is

$$[NH_4^+] = [OH^-] = 0.042 \times 0.010\ M = 4.2 \times 10^{-4}\ M$$

The amount ionized per liter is 4.2% of 0.010 mol/L.

The number of moles per liter of NH_3 at equilibrium would be

$$[NH_3] = 1.0 \times 10^{-2} - 0.042 \times 10^{-2} = 0.958 \times 10^{-2}\ M$$

When this is rounded off to the appropriate number of significant figures, we have $[NH_3] = 1.0 \times 10^{-2}\ M$ (once again, the amount lost by ionization is negligible). The equilibrium concentrations are shown in the margin.

When these concentrations are substituted into the equation for K_b, we have

$$K_b = \frac{(4.2 \times 10^{-4})(4.2 \times 10^{-4})}{(1.0 \times 10^{-2})}$$

or

$$K_b = 1.8 \times 10^{-5}$$

	Equilibrium Concentrations
NH_4^+	$4.2 \times 10^{-4}\ M$
OH^-	$4.2 \times 10^{-4}\ M$
NH_3	$1.0 \times 10^{-2}\ M$

In Examples 16.6 and 16.7, we computed K from a knowledge of equilibrium concentrations. We can also use our knowledge of K to calculate the concentrations in an equilibrium mixture. Let's look at some examples.

EXAMPLE 16.8. USING K_a TO CALCULATE THE CONCENTRATIONS OF THE SPECIES IN A SOLUTION OF A WEAK ACID

PROBLEM: What are the concentrations of all the species present in a 0.50 M solution of acetic acid. For $HC_2H_3O_2$, $K_a = 1.8 \times 10^{-5}$.

ANALYSIS: In this problem we know the value of K_a, so the unknown quantities are in the mass action expression. We tackled similar problems in Chapter 15, where we used a concentration table to organize the information, and that's how we will organize our problem solution here. In fact, our approach is very much the same as in Chapter 15.

To start, let's imagine that we could form the solution by dissolving $HC_2H_3O_2$ in water without any of it dissociating. This allows us to determine the initial concentrations based on the information given in the statement of the problem. Next, we imagine the reaction proceeding to equilibrium. This will cause the concentrations to change, and the magnitudes of these changes, which are our unknown quantities, are related to each other by the coefficients in the balanced equation for the equilibrium. The equilibrium quantities in the rightmost column of the table are obtained by adding the "changes" algebraically to the initial concentrations. Finally, the equilibrium quantities are substituted into the mass action expression to give an algebraic equation that we solve to get the values to substitute back into the change column so we can calculate the equilibrium concentrations.

SOLUTION: As always, we begin with the chemical equation and the equilibrium law.

$$HC_2H_3O_2 \rightleftharpoons H^+ + C_2H_3O_2^-$$

$$K_a = \frac{[H^+][C_2H_3O_2^-]}{[HC_2H_3O_2]} = 1.8 \times 10^{-5}$$

We are told that the solution contains 0.50 M $HC_2H_3O_2$, which we take as the initial concentration of acetic acid. The only solute in the solution is the acid; there is no other source of $C_2H_3O_2^-$, so we assign the initial $[C_2H_3O_2^-]$ a value of zero. We also expect that we can neglect any contributions of the solvent toward $[H^+]$, and we therefore take the initial concentration of H^+ to be zero.

In reaching equilibrium, some of the $HC_2H_3O_2$ will dissociate, so its concentration will decrease. We don't know by how much, so let's say that the change in the $HC_2H_3O_2$ concentration is x. Since all the coefficients in the equation are equal to 1, the concentrations of H^+ and $C_2H_3O_2^-$ will increase by x. This gives us the quantities in the center column of the table. The quantities corresponding to the equilibrium concentrations are obtained by adding the "changes" to the initial concentrations algebraically.

> Of course, before the $HC_2H_3O_2$ ionizes, there is *some* H^+ present because of the dissociation of water, but we include in the "Initial Concentrations" column only those contributions made by solutes.

> The $HC_2H_3O_2$ concentration at equilibrium equals its initial concentration minus the number of moles per liter that ionized.

	Initial Molar Concentrations	Change	Equilibrium Molar Concentrations
H^+	0.0	$+x$	x
$C_2H_3O_2^-$	0.0	$+x$	x
$HC_2H_3O_2$	0.50	$-x$	$0.50 - x$

Substituting these values into the equilibrium expression gives us

$$\frac{(x)(x)}{(0.50 - x)} = 1.8 \times 10^{-5}$$

Without simplification, this expression leads to a quadratic equation that can be solved using the quadratic formula. However, in Example 15.13 we saw that it is sometimes possible to make simplifying assumptions that greatly reduce the effort required to obtain solutions to problems of this type. Because K is small, very little $HC_2H_3O_2$ will have actually undergone dissociation, so x will be small. Let us assume that x will be negligible compared to 0.50; that is

$$0.50 - x \approx 0.50$$

Our equation then becomes

$$\frac{x^2}{0.50} = 1.8 \times 10^{-5}$$

or

$$x = 3.0 \times 10^{-3}$$

If we look back on our assumption, we see that x is in fact small compared to 0.50 and that, when *rounded to the proper number of significant figures,*

$$0.50 - 0.0030 = 0.50 \quad \text{(rounded)}$$

Therefore, the equilibrium concentrations of the species involved in the dissociation of the acid are those given in the table in the margin. Since the question asks for *all* concentrations, we must also calculate the concentration of OH^-, which comes from the dissociation of water. Here we use K_w.

$$[OH^-] = \frac{K_w}{[H^+]}$$
$$= \frac{1.0 \times 10^{-14}}{3.0 \times 10^{-3}}$$
$$= 3.3 \times 10^{-12} \, M$$

	Equilibrium Concentrations (M)
H^+	3.0×10^{-3}
$C_2H_3O_2^-$	3.0×10^{-3}
$HC_2H_3O_2$	0.50

In the last example the only source of H^+ and $C_2H_3O_2^-$ was from the dissociation of the weak acid. Example 16.9 shows how we would handle a problem dealing with a solution for which there are two sources of one of the ions.

EXAMPLE 16.9. CALCULATING THE pH OF A SOLUTION THAT CONTAINS A WEAK ACID AND A STRONG ACID

PROBLEM: What is the pH of a solution that contains 0.10 M HCl and 0.10 M $HC_2H_3O_2$? For acetic acid, $K_a = 1.8 \times 10^{-5}$.

ANALYSIS: The first thing to realize here is that one of the solutes—the HCl—is a strong electrolyte. Failure to recognize this fact will prevent you from solving the problem. Next, we need a strategy.

When the solution contains a solute that is a strong electrolyte and that contributes one of the ions involved in the equilibrium, the best approach is to imagine that the solution is formed by dissolving the weak electrolyte in a solution that *already* contains the other solute. Thus, we imagine in this case that we dissolve 0.10 mol of $HC_2H_3O_2$ in 1 L of a 0.10 M solution of HCl. As before, it is also helpful to imagine that the $HC_2H_3O_2$ dissolves initially without dissociating; then we allow the reaction to take the system to equilibrium.

SOLUTION: The 0.10 M solution of HCl contains 0.10 mol/L of H^+, because HCl is a strong acid. This is the concentration of H^+ before the $HC_2H_3O_2$ is added. Then we place

0.10 M $HC_2H_3O_2$ into it, but we imagine that none of it dissociates. This gives us the initial concentrations of H^+ and $HC_2H_3O_2$.

$$[H^+]_{initial} = 0.10\ M$$

$$[HC_2H_3O_2]_{initial} = 0.10\ M$$

Now we have to consider the ionization of the acetic acid. The chemical equilibrium and equilibrium law are the same as in the preceding problem.

$$HC_2H_3O_2 \rightleftharpoons H^+ + C_2H_3O_2^-$$

$$K_a = \frac{[H^+][C_2H_3O_2^-]}{[HC_2H_3O_2]} = 1.8 \times 10^{-5}$$

Since there is no $C_2H_3O_2^-$ present initially, some of it must form by the dissociation of the $HC_2H_3O_2$. Therefore, we let the concentration of $HC_2H_3O_2$ decrease by x and the concentrations of H^+ and $C_2H_3O_2^-$ increase by x, and then add the initial concentrations and the changes to get the equilibrium quantities.

	Initial Molar Concentrations	Change	Equilibrium Molar Concentrations
H^+	0.10	$+x$	$0.10 + x \approx 0.10$
$C_2H_3O_2^-$	0	$+x$	x
$HC_2H_3O_2$	0.10	$-x$	$0.10 - x \approx 0.10$

We expect x to be small, so we've assumed $0.10 \pm x \approx 0.10$. Substituting into the K_a expression, we get

$$1.8 \times 10^{-5} = K_a = \frac{[H^+][C_2H_3O_2^-]}{[HC_2H_3O_2]}$$

$$1.8 \times 10^{-5} = \frac{(0.10)(x)}{(0.10)}$$

$$x = 1.8 \times 10^{-5}$$

We have learned here that when a solution contains both a strong acid and a weak acid, it is the concentration of the strong acid that determines the pH of the solution.

We see that x is indeed small compared to 0.10, so in the solution $[H^+] = 0.10\ M$. This gives a pH of 1.00.

In the preceding examples, solving for x was made easy because the extent of dissociation of the acid was very small. Sometimes, however, this assumption fails, as illustrated in the next example.

EXAMPLE 16.10. CALCULATING THE pH OF A SOLUTION OF A WEAK BASE

PROBLEM: What is the pH of a 0.010 M solution of the weak base diethylamine, $(C_2H_5)_2NH$, for which $K_b = 9.6 \times 10^{-4}$?

SOLUTION: Our approach is essentially the same as in the preceding problem. The first step is to write the equation for the equilibrium and the equilibrium law. Since the solute is a base, the equilibrium is

You should know by now how bases react with water.

$$(C_2H_5)_2NH + H_2O \rightleftharpoons (C_2H_5)_2NH_2^+ + OH^-$$

Therefore,

$$K_b = \frac{[(C_2H_5)_2NH_2^+][OH^-]}{[(C_2H_5)_2NH]} = 9.6 \times 10^{-4}$$

We take the $(C_2H_5)_2NH$ concentration specified in the problem as its initial concentration and the concentration of $(C_2H_5)_2NH_2^+$ to be zero. As usual, we assume the solvent makes no contribution to $[OH^-]$. Therefore, we take the initial OH^- concentration to be zero, too. This gives us the entries for the initial concentration column in the concentration table. Next, we realize that some of the diethylamine must react, so we let its concentration decrease by x, and we increase the concentrations of $(C_2H_5)_2NH_2^+$ and OH^- by x. Adding the "changes" to the initial concentrations gives the equilibrium quantities.

	Initial Molar Concentrations	Change	Equilibrium Molar Concentrations
$(C_2H_5)_2NH_2^+$	0.0	x	x
OH^-	0.0	x	x
$(C_2H_5)_2NH$	0.010	$-x$	$0.010 - x$

Substituting these into the mass action expression gives

$$\frac{(x)(x)}{0.010 - x} = 9.6 \times 10^{-4}$$

Now let's see what happens if we assume that x is negligibly small compared to 0.010 M. This gives the equation

$$\frac{(x)(x)}{0.010} = 9.6 \times 10^{-4}$$

$$x^2 = 9.6 \times 10^{-6}$$

$$x = 3.1 \times 10^{-3}$$

If we aren't careful, we might accept this as the answer. However, if we check our assumption, we find that x is *not* negligible compared to 0.010. In decimal form, $x = 0.0031$, so $0.010 - 0.0031 = 0.007$ (rounded to the correct number of significant figures). Certainly 0.007 is not the same as 0.010. In other words, in this problem, it was *not* safe to assume that the initial concentration of $(C_2H_5)_2NH$ is the same as its equilibrium value.

0.0031 M is about 30% of the initial value of 0.010 M, so it is certainly not negligible.

The next question, of course, is what do we do now? There are actually two choices. One is to expand the equation into a quadratic equation and apply the quadratic formula to it.

$$\textit{The quadratic formula:} \quad x = \frac{-b \pm \sqrt{b^2 - 4ac}}{2a}$$

Although tedious, the solution is relatively straightforward. It gives two values for x because of the "$\pm$" sign before the radical. These turn out to be

$$x = 2.7 \times 10^{-3}$$

$$x = -3.6 \times 10^{-3}$$

We can discard the second value because the concentrations of $(C_2H_5)_2NH_2^+$ and OH^- can't be negative and also because $0.010 - (-3.6 \times 10^{-3}) = 0.014$ (rounded), which is impossible. Therefore, we are left with $x = [OH^-] = 2.7 \times 10^{-3}$ M. This gives a pOH of 2.57 and a pH of 11.43.

With $(C_2H_5)_2NH$ as the only solute, its equilibrium concentration cannot *possibly be larger than its initial concentration.*

Successive Approximations If you work through the solution of this problem with the quadratic formula (and you should try it, just to appreciate the following discussion), you will see how tiresome it can be. An alternative that is just as accurate and much faster, especially with a calculator, is the method of successive approximations. In this method we sort of sneak up on the answer.

Our original assumption (later found to be incorrect) was that

$$0.010 - x = 0.010$$

and we obtained a value of $x = 0.0031$. Let's call this value our *first approximation of x*. In the method of successive approximations, we substitute the first approximation into our equation and solve it again to get a second better approximation. Thus,

$$\frac{x^2}{0.010 - x} = \frac{x^2}{0.010 - 0.0031} = \frac{x^2}{0.0069} = 9.6 \times 10^{-4}$$

Substitute the first approximation, 0.0031

Solving this for x gives $x = 0.00257 = 0.0026$ (rounded). Notice that this *second approximation* is much closer to the answer we get by the quadratic equation. Let's try a third approximation. This time we substitute our second approximation into the algebraic equation and solve for x again.

$$\frac{x^2}{0.010 - 0.0026} = \frac{x^2}{0.0074} = 9.6 \times 10^{-4}$$

Solving for x gives $x = 0.00266$ that rounds to 0.0027, which is the same value obtained by using the quadratic formula.

If you study this method, you will see that it is really easy to apply. Set up the algebra and simplify, solve for x, and then substitute this into the unsimplified equation at the place where you subtract x from the initial concentration. Now, solve for x again, then substitute the new x into the unsimplified equation, and repeat the process, over and over. As you do this, you will find that each approximation differs from the preceding one by less and less. You finally stop when the new approximation no longer differs from the preceding one by a significant amount. As you see, this usually happens quickly, and the method is much faster and less tedious than using the quadratic formula.

This method must be used in more complex situations for which the quadratic formula doesn't apply.

In most of the equilibrium problems you will encounter, you will not find the kind of difficulty we experienced in the preceding example. In fact, when you begin the problem, try the approximation that the initial concentration of the acid or base is essentially the same as its equilibrium value. Usually, x will be so small that the value of an initial concentration won't change when x is added to it or subtracted from it, provided the sum or difference is rounded to the correct number of significant figures. Even when this condition is not quite fulfilled, we can sometimes avoid having to work the problem through the "long way." This is because there are built-in inaccuracies in the calculations caused by our use of molar concentrations instead of quantities called *activities*, which are "effective concentrations." We won't get into this any further, but it does allow us some leeway. The general rule of thumb that you can follow is

If the value of x that you obtain by making the approximation

(Initial concentration) $\pm x \approx$ (Initial concentration)

is less than about 5% of the initial concentration, you are safe in assuming that the approximation is valid.

16.6 BUFFERS: THE CONTROL OF pH

If a solution contains *both* a weak acid and a weak base, it has the ability to absorb small additions of either a strong acid or strong base with very little change in pH. When a small amount of a strong acid is added, its H_3O^+ is neutralized by the weak base, and when a small amount of a strong base is added, its OH^- is neutralized by the weak acid. Such solutions are said to be **buffered** and the mixture of solutes that produces the effect is called a **buffer.**

Buffers usually consist of a weak acid and its conjugate base. An example is acetic acid and acetate ion, and a solution buffered with this pair is prepared by dissolving $HC_2H_3O_2$ and an acetate salt such as $NaC_2H_3O_2$ in some appropriate amount of water. In this case, the buffer has a pH lower than 7 because the acid, $HC_2H_3O_2$, is stronger than the base, $C_2H_3O_2^-$. (For $HC_2H_3O_2$, $K_a = 1.8 \times 10^{-5}$ and for $C_2H_3O_2^-$, $K_b = 5.6 \times 10^{-10}$.) If a small amount of a strong acid is added to this buffer, its H_3O^+ can react with the acetate ion,

$$H_3O^+ + C_2H_3O_2^- \longrightarrow HC_2H_3O_2 + H_2O$$

or, more simply,

$$H^+ + C_2H_3O_2^- \longrightarrow HC_2H_3O_2$$

Acetic acid is a weak acid and acetate ion is its weak conjugate base.

Notice that this reaction removes H^+ and changes the base into its conjugate acid. A similar reaction occurs if a strong base is added. The OH^- supplied by the strong base reacts with the acetic acid and converts it to its conjugate base.

$$HC_2H_3O_2 + OH^- \longrightarrow C_2H_3O_2^- + H_2O$$

Buffers with a pH higher than 7 can be prepared by using a base that is stronger than its conjugate acid. A common basic buffer is formed by mixing ammonia with an ammonium salt such as NH_4Cl and contains the conjugate acid–base pair NH_4^+ and NH_3. If a strong acid is added to the buffer, it reacts as follows,

$$H_3O^+ + NH_3 \longrightarrow NH_4^+ + H_2O$$

$$H^+ + NH_3 \longrightarrow NH_4^+ \text{ (simplified)}$$

Ammonia is a weak base and ammonium ion is its weak conjugate acid.

and if a strong base is added, the reaction is

$$OH^- + NH_4^+ \longrightarrow NH_3 + H_2O$$

Buffer calculations

One of the kinds of calculation we have to be able to do is to calculate the pH of a buffer solution. This is illustrated in the next example.

EXAMPLE 16.11. CALCULATING THE pH OF A BUFFER SOLUTION

PROBLEM: What is the pH of a buffer solution prepared by dissolving 0.10 mol of $NaC_2H_3O_2$ and 0.20 mol of $HC_2H_3O_2$ in enough water to give 1.00 L of solution?

SOLUTION: There is only one equilibrium here that we must be concerned with

$$HC_2H_3O_2 \rightleftharpoons H^+ + C_2H_3O_2^-$$

$$\frac{[H^+][C_2H_3O_2^-]}{[HC_2H_3O_2]} = 1.8 \times 10^{-5}$$

When $NaC_2H_3O_2$ dissolves, it is completely dissociated. It is important to remember that virtually all salts are 100% dissociated in solution. Therefore, 0.10 mol/L of $NaC_2H_3O_2$ gives 0.10 mol/L of Na^+ and 0.10 mol/L of $C_2H_3O_2^-$. We are interested only in the $C_2H_3O_2^-$; the Na^+ is simply a *spectator ion* and we can ignore it. The initial concentration of the $C_2H_3O_2^-$ is therefore 0.10 M. As usual, we take the initial concentration of the weak acid to be the value specified in the problem; in this case $[HC_2H_3O_2]_{initial} = 0.20$ M. We also ignore the contribu-

It is very important to remember that salts are strong electrolytes.

tion that the solvent makes to $[H^+]$, so we set this value to zero. These are the values that go into the first column of the concentration table.

Next, we enter the changes in the center column. Since no H^+ is present, some $HC_2H_3O_2$ must ionize; so let's allow x to equal the number of moles per liter of $HC_2H_3O_2$ that dissociates to give H^+ and $C_2H_3O_2^-$. This will increase $[H^+]$ and $[C_2H_3O_2^-]$ by x and decrease $[HC_2H_3O_2]$ by x. The equilibrium concentrations are then found in the last column for our table.

	Initial Molar Concentrations	Change	Final Molar Concentrations
H^+	0.0	$+x$	x
$C_2H_3O_2^-$	0.10	$+x$	$0.10 + x \approx 0.10$
$HC_2H_3O_2$	0.20	$-x$	$0.20 - x \approx 0.20$

As before, we look at K_a and see that x will probably be small. We will therefore assume that $0.10 + x \approx 0.10$ and $0.20 - x \approx 0.20$. Substituting into the expression for K_a gives

$$\frac{(x)(0.10)}{(0.20)} = 1.8 \times 10^{-5}$$

$$x = 3.6 \times 10^{-5}$$

Now let's check our assumption. If 3.6×10^{-5} is added to 0.10 and the result rounded to the correct number of significant figures, the sum is 0.10. If 3.6×10^{-5} is subtracted from 0.20, the difference is 0.20 when rounded correctly. This means that we were correct in our assumption that x was small compared to 0.10 and 0.20. The equilibrium concentrations are therefore

$$[H^+] = 3.6 \times 10^{-5}\ M$$

$$[C_2H_3O_2^-] = 0.10\ M$$

$$[HC_2H_3O_2] = 0.20\ M$$

Since we are interested in the pH, we use the H^+ concentration.

$$pH = -\log[H^+] = -\log(3.6 \times 10^{-5})$$
$$= 4.44$$

If we look back at the preceding example, we see that the H^+ concentration in the buffer solution, and therefore its pH, is controlled by relative concentrations of the weak acid and its conjugate base, the anion. Let's look at this a little more closely. Continuing to use acetic acid as an example, we have

$$K_a = \frac{[H^+][C_2H_3O_2^-]}{[HC_2H_3O_2]}$$

Solving for $[H^+]$ gives

For the basic buffer composed of NH_3 and NH_4^+, we would use the K_a for NH_4^+. Alternatively, we could use the equation

$$[OH^-] = K_b \left(\frac{[NH_3]}{[NH_4^+]} \right)$$

to obtain $[OH^-]$ and then calculate the pH.

$$[H^+] = K_a \left(\frac{[HC_2H_3O_2]}{[C_2H_3O_2^-]} \right) \qquad (16.13)$$

To calculate $[H^+]$, and then pH, we must know K_a for the weak acid (from Table 16.1) as well as the ratio of the concentrations of the weak acid and its anion.

Since salts completely dissociate in aqueous solution, the number of moles of the anion in the solution from this source is determined by the formula of the salt and the number of moles of salt dissolved. Thus, a 1.0 M $NaC_2H_3O_2$ solution contains 1.0 mol/L of $C_2H_3O_2^-$, while a 1.0 M $Ca(C_2H_3O_2)_2$ solution contains 2.0 mol/L of $C_2H_3O_2^-$. In the buffer there is also an additional amount of acetate ion that comes from the dissociation of $HC_2H_3O_2$. The amount of H^+ and $C_2H_3O_2^-$ stemming from this source is very small even in solutions containing only acetic acid, and this small amount is reduced even further in the buffer because of the presence of the large concentration of $C_2H_3O_2^-$ from the salt.[2] The total anion concentration in the buffer is essentially determined by the salt concentration alone, since the contribution from the dissociation of the weak acid is negligible. As in our previous calculations on weak acids, the concentration of $HC_2H_3O_2$ will not be reduced appreciably by its dissociation and, in a mixture of 1.0 M $NaC_2H_3O_2$ and 1.0 M $HC_2H_3O_2$, for example, the concentrations of both the molecular acid and the anion are 1.0 M. The H^+ concentration in such a buffer could be found from Equation 16.13 using $K_a = 1.8 \times 10^{-5}$ (Table 16.1).

> In buffer calculations, it is safe to assume that the initial concentrations of the acid and its conjugate base are the same as the equilibrium values. This applies to both acidic and basic buffers.

$$[H^+] = (1.8 \times 10^{-5})\frac{1.0}{1.0}$$

$$= 1.8 \times 10^{-5} M$$

The pH of this solution is 4.74.

Notice that when the concentrations of the acid and anion are the same in a buffer, the H^+ concentration in that solution is equal to the K_a of the weak acid, and the $pH = pK_a$. Thus, if a buffer was prepared by mixing 0.1 mol of formic acid ($K_a = 1.8 \times 10^{-4}$ from Table 16.1) and 0.1 mol of sodium formate into a liter of solution, the resulting H^+ concentration would be

$$[H^+] = 1.8 \times 10^{-4} M$$

and

$$pH = 3.74$$

We can also use Equation 16.13 to calculate the concentrations of acid and salt that would be needed to achieve a certain pH buffer.[3] This is illustrated in Example 16.12.

EXAMPLE 16.12. PREPARING A BUFFER WITH A SPECIFIED pH

PROBLEM: What ratio of acetic acid to sodium acetate concentration is needed to form a buffer whose pH is 5.70?

SOLUTION: To solve this problem we need to rearrange Equation 16.13 to solve for the ratio of concentrations. Thus,

$$\frac{[HC_2H_3O_2]}{[C_2H_3O_2^-]} = \frac{[H^+]}{K_a}$$

The H^+ concentration when the pH is 5.70 is

$$[H^+] = 2.0 \times 10^{-6} M$$

---

[2] This can easily be seen by applying Le Châtelier's principle to the dissociation of $HC_2H_3O_2$. The presence of acetate ion from a salt causes the dissociation equilibrium of the acid to be shifted to the left and actually suppresses the dissociation.

[3] If you are taking a biology or biochemistry course, you will probably encounter an equation called the Henderson–Hasselbalch equation, which is derived essentially from Equation 16.13.

$$pH = pK_a + \log\frac{[anion\ A^-]}{[acid\ HA]}$$

A similar equation that would apply to a basic buffer is

$$pOH = pK_b + \log\frac{[cation\ BH^+]}{[base\ B]}$$

Therefore,

$$\frac{[HC_2H_3O_2]}{[C_2H_3O_2^-]} = \frac{2.0 \times 10^{-6}}{1.8 \times 10^{-5}} = \frac{2.0 \times 10^{-6}}{18 \times 10^{-6}}$$

or

$$\frac{[HC_2H_3O_2]}{[C_2H_3O_2^-]} = \frac{1}{9}$$

As long as this ratio is maintained, the pH of an acetic acid–sodium acetate buffer is 5.70. For example, if 0.2 mol of $HC_2H_3O_2$ and 1.8 mol of $NaC_2H_3O_2$ are dissolved in 1 L of solution, the pH is 5.70. This same pH will result if 0.1 mol of $HC_2H_3O_2$ and 9.9 mol of $NaC_2H_3O_2$ are dissolved.

The larger the concentrations of the components of a buffer, the more effective it is at resisting changes in pH.

We have seen that by adjusting the ratio of concentrations of the weak acid to that of its conjugate base, we can adjust the pH of the buffer solution. For example, a buffer composed of 0.010 mol of $HC_2H_3O_2$ and 1.0 mol of $NaC_2H_3O_2$ would have a pH of 6.74. However, when as little as 0.010 mol of base is added to this buffer, all of the acetic acid is neutralized and a large change in the pH of the buffer results. Therefore, *the most effective pH range for any buffer is at or near the pH where the acid and salt concentrations are equal (that is, pK_a)*. Also, in order to be most effective, the numbers of moles of weak acid and base used to prepare the buffer must be considerably greater than the numbers of moles of acid or base that may later be added to the buffer.

Let us now see how effective a buffer is at holding the pH nearly constant. Suppose we have an acetic acid–acetate buffer in which the concentrations of $HC_2H_3O_2$ and $C_2H_3O_2^-$ are each 1.00 M. We saw earlier that its pH will be 4.74 and $[H^+] = 1.8 \times 10^{-5}$ M. What happens to the pH if we add, say, 0.20 mol of HCl to a liter of this buffer?

Any H^+ from a strong acid converts the conjugate weak base in the buffer to its corresponding conjugate acid.

When a strong acid such as HCl is added, its H^+ reacts with acetate ion.

$$H^+ + C_2H_3O_2^- \longrightarrow HC_2H_3O_2.$$

This decreases the $C_2H_3O_2^-$ concentration and increases the $HC_2H_3O_2$ concentration. Below are the number of moles per liter of all species before and after the addition.

We are assuming a negligible change in volume.

Before	After
$[H^+] = 1.8 \times 10^{-5}$ M	$[H^+] = ?$
$[C_2H_3O_2^-] = 1.00$ M	$[C_2H_3O_2^-] = 1.00 - 0.20$ $M = 0.80$ M
$[HC_2H_3O_2] = 1.00$ M	$[HC_2H_3O_2] = 1.00 + 0.20$ $M = 1.20$ M

Substituting the concentrations in the column on the right into Equation 16.13 gives

$$[H^+] = (1.8 \times 10^{-5}) \times \left(\frac{1.20}{0.80}\right)$$

$$= 2.7 \times 10^{-5} M$$

The pH of the buffer after the 0.20 mol of H^+ is added is 4.57 — a change of only 0.17 pH units from its initial pH of 4.74.

Suppose now that we were to add 0.20 mol of H^+ to 1 L of a solution of HCl whose pH = 4.74 (that is, a 1.8×10^{-5} M HCl solution). Since Cl^- has virtually no tendency to react with H^+, the final H^+ concentration will be 0.20 M (0.20 + 1.8 × $10^{-5} = 0.20$), and the pH of the solution will be 0.70. The change in pH in this case is

4.04 pH units, as opposed to a change of only 0.17 pH units when the same quantity of H^+ is added to the buffer.

Additions of strong base are also absorbed by the buffer. When 0.20 mol of OH^- is added to the 1 L of our original buffer, it is neutralized according to the reaction

$$OH^- + HC_2H_3O_2 \longrightarrow H_2O + C_2H_3O_2^-$$

The number of moles per liter before the addition are the same as above, but the number of moles per liter after 0.20 mol of OH^- is added would be

Any OH^- from a strong base converts the conjugate weak acid in the buffer to its corresponding conjugate base.

Concentration after Addition of 0.20 mol of OH^-
$[H^+] = ?$
$[C_2H_3O_2^-] = 1.00 + 0.20 = 1.20\ M$
$[HC_2H_3O_2] = 1.00 - 0.20 = 0.80\ M$

Substituting these values into Equation 16.13, we find that the H^+ concentration after the addition is $1.2 \times 10^{-5}\ M$ and the resulting pH is 4.92 — once again, a small change of 0.18 pH units. Finally, note that the addition of an acid to the buffer lowers the pH, while the addition of a base raises the pH. Although the change in pH is small, the *direction* of the change is as expected.

EXAMPLE 16.13. CALCULATING THE pH OF A BUFFER AND THE CHANGE IN pH WHEN A STRONG ACID IS ADDED

PROBLEM: A buffer was prepared by mixing exactly 200 mL of a 0.60 M NH_3 solution and 300 mL of a 0.30 M NH_4Cl solution. (a) What is the pH of this buffer, if we assume a final volume of 500 mL? (b) What will be the pH after 0.020 mol of HCl is added? For NH_3, $K_b = 1.8 \times 10^{-5}$.

ANALYSIS: The first thing to realize here is that two solutions are being mixed and that one solution dilutes the other. Therefore, we need to calculate the concentrations of the NH_3 and NH_4^+ in the final mixture. We can do this by first calculating the number of moles of each of these substances in their respective solutions (these are quantities that don't change when the two solutions are mixed) and then dividing the numbers of moles by the total volume of the mixture to obtain molar concentrations.

(a) The equilibrium involves the ionization of ammonia, a weak base. Therefore, we want to write the chemical equation for this reaction and set up the equilibrium law. Then we can substitute the concentrations of NH_3 and NH_4^+ into the mass action expression and solve for $[OH^-]$. Once we have this, it is easy to calculate the pH.

(b) There are two main points that we must realize to solve this part of the problem. The first is that HCl is a strong acid, so we are really adding 0.020 mol of H^+ to the buffer. The second point is that the H^+ reacts with NH_3 and changes it to NH_4^+. We use this fact to adjust the concentrations of the base and its conjugate acid in the final solution.

SOLUTION: The number of moles of NH_3 added to the mixture is

$$0.60\ \frac{\text{mol}}{\text{L}} \times 0.200\ \text{L} = 0.12\ \text{mol}$$

and the number of moles of NH_4^+ added is

$$0.30\ \frac{\text{mol}}{\text{L}} \times 0.300\ \text{L} = 0.090\ \text{mol}$$

Therefore, the concentrations of these in the 500 mL are

$$[NH_3] = \frac{0.12 \text{ mol}}{0.500 \text{ L}} = 0.24 \text{ } M$$

$$[NH_4^+] = \frac{0.090 \text{ mol}}{0.500 \text{ L}} = 0.18 \text{ } M$$

(a) The OH⁻ concentration for this buffer is found by using the K_b for NH_3.

$$NH_3 + H_2O \rightleftharpoons NH_4^+ + OH^-$$

$$K_b = \frac{[NH_4^+][OH^-]}{[NH_3]} \quad \text{from salt conc.}$$

Rearranging and solving for [OH⁻], we have

$$[OH^-] = K_b \frac{[NH_3]}{[NH_4^+]}$$

Substituting the values of K_b and the concentrations of NH_3 and NH_4^+ into this equation gives

$$[OH^-] = (1.8 \times 10^{-5}) \times \left(\frac{0.24}{0.18}\right)$$

$$= 2.4 \times 10^{-5} \text{ } M$$

$$pOH = 4.62$$

$$pH = 9.38 \quad \text{— buffer only}$$

(b) The neutralization reaction for H⁺ in this buffer is

$$H^+ + NH_3 \longrightarrow NH_4^+$$

We are adding 0.020 mol of H⁺ to 500 mL, or 0.040 mol of H⁺ per liter. Therefore, the concentrations before and after the addition of the acid are

Before	After
$[OH^-] = 2.4 \times 10^{-5} \text{ } M$	$[OH^-] = ?$
$[NH_3] = 0.24 \text{ } M$	$[NH_3] = 0.24 - 0.040 \text{ } M = 0.20 \text{ } M$
$[NH_4^+] = 0.18 \text{ } M$	$[NH_4^+] = 0.18 + 0.040 \text{ } M = 0.22 \text{ } M$

The new OH⁻ concentration is

$$[OH^-] = (1.8 \times 10^{-5}) \times \left(\frac{0.20}{0.22}\right)$$

$$= 1.6 \times 10^{-5} \text{ } M$$

$$pOH = 4.80$$

The pH is therefore 9.20, a decrease of 0.18 pH units.

Notice that we calculate the amount of H⁺ added per liter.

Adding a strong acid to a solution will always lower the pH.

Buffers find many important applications. Living systems employ buffers to maintain nearly constant pH so that biochemical reactions can follow their correct paths. For example, blood contains, among other things, a H_2CO_3/HCO_3^- buffer system that helps maintain the pH at 7.4.

In the laboratory many inorganic and organic chemical reactions are performed in buffered solutions to minimize any adverse effects caused by acids or bases that might be consumed or produced during reaction.

Sodium bicarbonate (baking soda) is added to a swimming pool because it acts as a buffer and controls the pH of the water in the pool.

16.7 ACID–BASE EQUILIBRIA IN SALT SOLUTIONS

In Section 16.4 you learned that ions can be acids and bases, just like neutral molecules can. In fact, we've been using this concept continually in describing acetate ion, $C_2H_3O_2^-$, as the conjugate base of acetic acid and ammonium ion, NH_4^+, as the conjugate acid of ammonia. Therefore, if we dissolve an acetate salt in water, the solution contains a weak base, and if we dissolve an ammonium salt in water, the solution contains a weak acid. With this in mind, it's not surprising to find that these solutions tend not to have a neutral pH.

You have already learned how to deal with equilibria involving weak acids and bases, so there doesn't seem to be much new here. However, there is one complication that we have to consider if we want to calculate the pH of a salt solution. Salts are composed of *two* ions, so each could potentially affect the pH of a solution. To deal with this, we need to examine each of the ions to determine how each will behave in a solution.

Cations

Cations of salts are either metal ions or the conjugate acids of molecular bases, like NH_4^+. If they are conjugate acids, then we certainly expect them to affect the pH of a solution. For example, if a salt contains the hydrazinium ion, $N_2H_5^+$, which is the conjugate acid of hydrazine, N_2H_4, we expect it to react with water as follows.

$$N_2H_5^+ + H_2O \rightleftharpoons N_2H_4 + H_3O^+$$

The K_a for $N_2H_5^+$ can be calculated from the K_b for N_2H_4, as you learned earlier, by the relationship

$$K_a \times K_b = K_w$$

Metal ions with large positive charges also tend to be weak acids. In fact, we discussed the acidity of the $Al(H_2O)_6^{3+}$ ion in an earlier chapter. For our purposes here, however, we will not examine their equilibria quantitatively. Metal ions with positive charges of $1+$ or $2+$, on the other hand, tend to have little influence on the acidity or basicity of an aqueous solution. In solutions of sodium salts, for example, the Na^+ ion is effectively a bystander. In other words, we don't have to consider any acid–base equilibria involving the Na^+ ion. Nor would we worry about the Ca^{2+} ion affecting the pH of a solution of a calcium salt.

Anions

For now, we will only consider anions of monoprotic acids.

In general, anions can be viewed as the conjugate bases of molecular acids. Thus, Cl^- is the conjugate base of HCl, NO_3^- is the conjugate base of HNO_3, and $C_2H_3O_2^-$ is the conjugate base of $HC_2H_3O_2$. To judge the effects that these anions have on the pH of a solution, we need to recall the discussion in Section 16.4 about the relative strengths of the members of an acid–base conjugate pair.

The stronger an acid is, the weaker is its conjugate base, and vice versa.

This means that if we have a *very strong* acid, its conjugate base will be *very weak*. Acids such as HCl and HNO_3 are indeed very strong and are 100% dissociated in solution. Their anions are likewise extremely weak bases — so weak that they do not react to any appreciable degree with water. Thus, we can say that the reaction

$$Cl^- + H_2O \rightleftharpoons HCl + OH^-$$

occurs to a negligible extent and that the Cl^- ion is also a bystander in acid–base equilibria. This is very useful to know, because if you are dealing with a salt of a strong acid, you don't have to be concerned about any equilibria involving the anion.

We also noted earlier that moderately weak acids have conjugate bases that are also moderately weak. Acetic acid and acetate ion fall into this category as do the other acids and their conjugate bases listed in Table 16.1. Their K_a and K_b are also related by the expression $K_a \times K_b = K_w$.

EXAMPLE 16.14. CALCULATING THE pH OF A SALT SOLUTION

PROBLEM: Calculate the pH of (a) a 0.10 M solution of potassium nitrite, KNO_2 and (b) a 0.10 M solution of sodium chloride.

ANALYSIS: When the only solute is a salt, the first step is to determine which of the ions, if any, are involved in acid–base equilibria. In part (a), the salt contains the ions K^+ and NO_2^-. From our preceding discussion, we expect that the K^+ ion will have no effect on the pH. (It will be a bystander, just like Na^+.) The NO_2^- ion, on the other hand, is the conjugate base of HNO_2, and since this is not on the list of strong acids, we expect HNO_2 to be weak. And that is what it is; in Table 16.1 we find its K_a listed to be 4.5×10^{-4}. Because HNO_2 is a weak acid, we expect that the NO_2^- ion is a weak base, so the pH of the solution will be determined by the reaction

$$NO_2^- + H_2O \rightleftharpoons HNO_2 + OH^-$$

For part (b), the salt is NaCl, which gives Na^+ and Cl^- ions in solution. We know that the Na^+ ion has no effect on the pH and we know that the Cl^- ion comes from the very strong acid HCl and aso has no effect. Neither of these ions reacts with water, so we expect a solution of NaCl to be neutral. If made with pure distilled water, the pH of the solution should be 7.

SOLUTION: We really only have to work to get the answer to part (a), because our analysis of the problem took care of part (b).

The equation for the equilibrium, arrived at above, gives us the equilibrium law

$$K_b = \frac{[HNO_2][OH^-]}{[NO_2^-]}$$

The value of K_b is not given to us and it isn't tabulated either. However, we know we can calculate K_b from K_a, which equals

$$K_b = \frac{K_w}{K_a}$$

$$= \frac{1.0 \times 10^{-14}}{4.5 \times 10^{-4}}$$

$$= 2.2 \times 10^{-11}$$

Because K_b is so small, we know that the position of equilibrium lies far to the left in favor of NO_2^-. Therefore, if we let x be the number of moles per liter of NO_2^- that react, we can set up our concentration table and expect that x will be small compared to 0.10.

	Initial Molar Concentrations	Change	Equilibrium Molar Concentrations
NO_2^-	0.10	$-x$	$0.10 - x \approx 0.10$
HNO_2	0.0	$+x$	x
OH^-	0.0	$+x$	x

Substituting values into the equilibrium expression gives

$$K_b = \frac{(x)(x)}{0.10} = 2.2 \times 10^{-11}$$

$$x^2 = 2.2 \times 10^{-12}$$

$$x = 1.5 \times 10^{-6}$$

Thus,

$$[OH^-] = 1.5 \times 10^{-6} \, M$$

$$pOH = 5.82$$

and therefore,

$$pH = 8.18$$

1.5×10^{-6} is negligible compared to 0.10.

Notice that the pH indicates that the solution is basic.

EXAMPLE 16.15. CALCULATING THE pH OF A SALT SOLUTION

PROBLEM: What is the pH of a 0.10 M solution of N_2H_5Cl?

SOLUTION: First we have to realize that this is a salt that gives the ions $N_2H_5^+$ and Cl^- in solution. We also must recognize that these ions come from a strong acid (HCl) and a weak base (N_2H_4). This means that only the cation, $N_2H_5^+$, affects the pH of the solution. Since $N_2H_5^+$ is the conjugate acid of N_2H_4, its reaction in water is

$$N_2H_5^+ \rightleftharpoons H^+ + N_2H_4$$

for which we can write

$$K_a = \frac{[H^+][N_2H_4]}{[N_2H_5^+]}$$

The only strong bases you have to worry about are metal hydroxides. Since N_2H_4 isn't a metal hydroxide, it must be a *weak base*.

Searching Table 16.1, we find K_b for N_2H_4, but not K_a for $N_2H_5^+$, so we have to calculate K_a. For N_2H_4, $K_b = 1.7 \times 10^{-6}$; therefore,

$$K_a = \frac{K_w}{K_b}$$

$$K_a = \frac{1.0 \times 10^{-14}}{1.7 \times 10^{-6}} = 5.9 \times 10^{-9}$$

Once again, because K_a is so small, we expect very little reaction to occur as the system approaches equilibrium. If we let x equal the number of moles of $N_2H_5^+$ that react, then our concentration table becomes

	Initial Molar Concentrations	Change	Equilibrium Molar Concentrations
$N_2H_5^+$	0.10	$-x$	$0.10 - x \approx 0.10$
H^+	0.0	$+x$	x
N_2H_4	0.0	$+x$	x

Substituting equilibrium quantities into the expression for K_a gives

$$\frac{(x)(x)}{0.10} = 5.9 \times 10^{-9}$$

$$x^2 = 5.9 \times 10^{-10}$$

$$x = 2.4 \times 10^{-5}$$

Therefore,

$$[H^+] = 2.4 \times 10^{-5} \, M$$

and

$$pH = 4.62$$

Notice that the solution is acidic, which is what is expected.

2.4×10^{-5} is negligible compared to 0.10.

In the two preceding examples, we find aqueous solutions of salts that are not neutral. In one we find a pH higher than 7 and in the other a pH lower than 7. These acid–base properties are caused by one of the ions of the salt reacting with water, and a general term that means "reaction with water" is **hydrolysis.** Therefore, the acid or base behavior of salts such as these is sometimes also called hydrolysis, and the salt is said to *hydrolyze* when placed in water.

Solutions of salts of a weak acid and a weak base

The kinds of solutions that fall under this heading contain two ions that are both able to affect the pH. An example is ammonium acetate, $NH_4C_2H_3O_2$. The pH of such a salt solution is governed by the relative extent of the reaction of each ion (i.e., by their relative acid and base strengths). If the K_a of the acid in the solution is larger than the K_b of the base, then the acid is stronger than the base and the solution will be acidic. If the situation is reversed ($K_b > K_a$), then the base is stronger than the acid and the solution should be basic. And if K_a and K_b are equal, then both ions react to the same extent and the solution should be neutral. Let's examine some examples.

Consider the salt NH_4CN, which yields NH_4^+ and CN^- in solution. The ammonium ion, NH_4^+, is the conjugate acid of NH_3; cyanide ion, CN^-, is the conjugate base of HCN. Both NH_3 and HCN are weak, so we can expect that their conjugates will react with water to affect the pH. The two equilibria that we have to contend with are therefore

$$NH_4^+ + H_2O \rightleftharpoons H_3O^+ + NH_3$$

and

$$CN^- + H_2O \rightleftharpoons HCN + OH^-$$

The K_a for NH_4^+ is calculated from K_b for NH_3 and the K_b for CN^- is calculated from K_a for HCN.

For NH$_4^+$ $$K_a = \frac{1.0 \times 10^{-14}}{1.8 \times 10^{-5}} = 5.6 \times 10^{-10}$$

For CN$^-$ $$K_b = \frac{1.0 \times 10^{-14}}{4.9 \times 10^{-10}} = 2.0 \times 10^{-5}$$

These equilibrium constants tell us that CN^- as a base is stronger than NH_4^+ as an acid. Therefore, more OH^- will be produced than H^+, and the solution will be basic.

Following similar reasoning, we would conclude that a solution of $NH_4C_2H_3O_2$ is neutral. Ammonium ion is the conjugate acid of ammonia, whose value of K_b is 1.8×10^{-5}. Acetate ion is the conjugate base of acetic acid, whose value of K_a is also 1.8×10^{-5}. This means that the K_a for NH_4^+ will be equal to the K_b for $C_2H_3O_2^-$, and both hydrolysis reactions will proceed to the same degree. This will give equal amounts of H^+ and OH^-, so the solution will have to be neutral.

EXAMPLE 16.16. DETERMINING HOW THE SALT OF A WEAK ACID AND A WEAK BASE AFFECTS THE pH OF A SOLUTION

PROBLEM: Will a solution of ammonium formate, NH_4CHO_2, be acidic, neutral, or basic?

SOLUTION: The ions of this salt are NH_4^+ and CHO_2^-. Ammonium ion is the conjugate acid of NH_3, and we've calculated its K_a to be 5.6×10^{-10}. The formate ion, CHO_2^-, is the conjugate base of formic acid, $HCHO_2$, whose K_a is 1.8×10^{-4}. The K_b for CHO_2^- is therefore

$$K_b = \frac{1.0 \times 10^{-14}}{1.8 \times 10^{-4}} = 5.6 \times 10^{-11}$$

In this case, the K_a for NH_4^+ is slightly larger than the K_b for CHO_2^-, so the solution will be slightly acidic.

16.8 IONIZATION OF POLYPROTIC ACIDS

Acids containing more than one atom of hydrogen that can be lost on dissociation are known as polyprotic acids. Some examples are H_2SO_4 and H_2CO_3, both of which contain two ionizable hydrogens, and H_3PO_4, which contains three. These acids lose their hydrogens one at a time. Thus, we write two steps for the dissociation of sulfuric acid, each with a corresponding equation for K_a,

$$H_2SO_4 \rightleftharpoons H^+ + HSO_4^- \qquad K_{a_1} = \frac{[H^+][HSO_4^-]}{[H_2SO_4]}$$

$$HSO_4^- \rightleftharpoons H^+ + SO_4^{2-} \qquad K_{a_2} = \frac{[H^+][SO_4^{2-}]}{[HSO_4^-]}$$

Table 16.2. Stepwise dissociation of some polyprotic acids at 25 °C

Acid	Stepwise Dissociation	Dissociation Constant for Each Step	pK_a
Phosphoric	$H_3PO_4 \rightleftharpoons H^+ + H_2PO_4^-$	$K_{a_1} = 7.5 \times 10^{-3}$	2.13
	$H_2PO_4^- \rightleftharpoons H^+ + HPO_4^{2-}$	$K_{a_2} = 6.2 \times 10^{-8}$	7.21
	$HPO_4^{2-} \rightleftharpoons H^+ + PO_4^{3-}$	$K_{a_3} = 2.2 \times 10^{-12}$	11.66
Phosphorous	$H_3PO_3 \rightleftharpoons H^+ + H_2PO_3^-$	$K_{a_1} = 3 \times 10^{-2}$	1.5
	$H_2PO_3^- \rightleftharpoons H^+ + HPO_3^{2-}$	$K_{a_2} = 1.6 \times 10^{-7}$	6.80
Sulfuric	$H_2SO_4 \rightleftharpoons H^+ + HSO_4^-$	$K_{a_1} = $ very large	<0
	$HSO_4^- \rightleftharpoons H^+ + SO_4^{2-}$	$K_{a_2} = 1.2 \times 10^{-2}$	1.92
Sulfurous	$H_2SO_3 \rightleftharpoons H^+ + HSO_3^-$	$K_{a_1} = 1.5 \times 10^{-2}$	1.82
	$HSO_3^- \rightleftharpoons H^+ + SO_3^{2-}$	$K_{a_2} = 1.0 \times 10^{-7}$	7.00
Carbonic	$H_2CO_3 \rightleftharpoons H^+ + HCO_3^-$	$K_{a_1} = 4.3 \times 10^{-7}$	6.37
	$HCO_3^- \rightleftharpoons H^+ + CO_3^{2-}$	$K_{a_2} = 5.6 \times 10^{-11}$	10.26
Ascorbic (vitamin C)	$H_2C_6H_6O_6 \rightleftharpoons H^+ + HC_6H_6O_6^-$	$K_{a_1} = 7.9 \times 10^{-5}$	4.10
	$HC_6H_6O_6^- \rightleftharpoons H^+ + C_6H_6O_6^{2-}$	$K_{a_2} = 1.6 \times 10^{-12}$	11.79

and for H_2CO_3,

$$H_2CO_3 \rightleftharpoons H^+ + HCO_3^- \qquad K_{a_1} = \frac{[H^+][HCO_3^-]}{[H_2CO_3]} \qquad (16.14)$$

$$HCO_3^- \rightleftharpoons H^+ + CO_3^{2-} \qquad K_{a_2} = \frac{[H^+][CO_3^{2-}]}{[HCO_3^-]} \qquad (16.15)$$

The three steps in the dissociation of H_3PO_4 are

$$H_3PO_4 \rightleftharpoons H^+ + H_2PO_4^- \qquad K_{a_1} = \frac{[H^+][H_2PO_4^-]}{[H_3PO_4]}$$

$$H_2PO_4^- \rightleftharpoons H^+ + HPO_4^{2-} \qquad K_{a_2} = \frac{[H^+][HPO_4^{2-}]}{[H_2PO_4^-]}$$

$$HPO_4^{2-} \rightleftharpoons H^+ + PO_4^{3-} \qquad K_{a_3} = \frac{[H^+][PO_4^{3-}]}{[HPO_4^{2-}]}$$

Table 16.2 gives some polyprotic acids and their stepwise dissociation constants. We see from this table that the first dissociation goes essentially to completion for sulfuric acid, while the second occurs only to a relatively limited degree. Because of the virtual completion of its first dissociation, sulfuric acid is considered a strong acid. We also see that the first dissociation step of each of these acids occurs with the largest value of K_a, and that each successive step occurs with an ever-decreasing value of K_a. This trend in K_a is reasonable when we consider that it should be easiest to remove an H^+ ion from an uncharged species, and that it should become progressively more difficult to do so as the negative charge on the ion increases.

Because the equilibria involving polyprotic acids are more complex than those of monoprotic acids, equilibrium calculations are somewhat more complicated. The next example shows, however, that a number of approximations can be made that simplify the approach to these kinds of problems.

EXAMPLE 16.17. CALCULATING THE CONCENTRATIONS OF SPECIES IN A SOLUTION OF A WEAK DIPROTIC ACID

PROBLEM: Carbonic acid, H_2CO_3, is a weak diprotic acid formed by the reaction of carbon dioxide with water, and its presence in natural waters is one reason that water seeping through limestone deposits carves out enormous caverns. For this acid, $K_{a_1} = 4.3 \times 10^{-7}$ and $K_{a_2} = 5.6 \times 10^{-11}$. What are the equilibrium concentrations in a 0.10 M solution of carbonic acid?

SOLUTION: The equilibria involved in the ionization of carbonic acid were given in Equations 16.14 and 16.15.

Because K_{a_1} is *so much larger* than K_{a_2}, we can safely assume that nearly all the hydrogen ion in the solution is derived from the first step of the dissociation. In addition, only very little of the HCO_3^- formed in the first step will undergo further dissociation. On the basis of this we can calculate the H^+ and HCO_3^- concentrations using the expression for K_{a_1} alone.

$$K_{a_1} = \frac{[H^+][HCO_3^-]}{[H_2CO_3]}$$

If we let x equal the number of moles per liter of H_2CO_3 that dissociate, we obtain, from the stoichiometry of the first step, x mol/L of H^+ and x mol/L of HCO_3^-. At equilibrium, there will be $(0.10 - x)$ mol/L of H_2CO_3 remaining.

	Initial Molar Concentrations	Change	Equilibrium Molar Concentrations
H^+	0.0	$+x$	x
HCO_3^-	0.0	$+x$	x
H_2CO_3	0.10	$-x$	$0.10 - x \approx 0.10$

Note that as before, because K_{a_1} is very small, we may assume that x will be negligible compared to 0.10 and write

$$[H_2CO_3] = 0.10 - x \approx 0.10\ M$$

Substituting these equilibrium quantities into the expression for K_{a_1}, we have

$$\frac{(x)(x)}{0.10} = 4.3 \times 10^{-7}$$

$$x^2 = 4.3 \times 10^{-8}$$

$$x = 2.1 \times 10^{-4}$$

Therefore, the equilibrium concentrations from this first dissociation are

$$[H^+] = 2.1 \times 10^{-4}\ M$$

$$[HCO_3^-] = 2.1 \times 10^{-4}\ M$$

$$[H_2CO_3] = 0.10 - 2.1 \times 10^{-4} = 0.10\ M$$

Notice that our approximation in the H_2CO_3 concentration is valid, because 2.1×10^{-4} is in fact negligible compared to 0.10.

By employing K_{a_2}, we can now calculate the equilibrium concentration of CO_3^{2-}

$$HCO_3^- \rightleftharpoons H^+ + CO_3^{2-}$$

and

$$K_{a_2} = \frac{[H^+][CO_3^{2-}]}{[HCO_3^-]}$$

Suppose we let y equal the number of moles per liter of HCO_3^- that dissociates. Then, from the stoichiometry of this second dissociation step, the additional number of moles per liter of H^+ and the number of moles per liter of CO_3^{2-} produced would also be y. Thus, the total hydrogen ion concentration from both the first and second dissociations will be $[H^+] = (2.1 \times 10^{-4} + y)$, and the concentration of HCO_3^- that remains at equilibrium will be $(2.1 \times 10^{-4} - y)$. For this second dissociation,

	Initial Concentrations	Change	Equilibrium Concentrations
H^+	2.1×10^{-4}	$+y$	$2.1 \times 10^{-4} + y$
CO_3^{2-}	0.0	$+y$	y
HCO_3^-	2.1×10^{-4}	$-y$	$2.1 \times 10^{-4} - y$

These quantities may be simplified by recognizing that because K_{a_2} is so very small, the amount of HCO_3^- that will dissociate will also be very small. We can therefore make the assumption that y will be negligible compared to 2.1×10^{-4}. Our equilibrium concentrations then become

$$[H^+] = 2.1 \times 10^{-4} + y \approx 2.1 \times 10^{-4}\ M$$

$$[CO_3^{2-}] = y$$

$$[HCO_3^-] = 2.1 \times 10^{-4} - y \approx 2.1 \times 10^{-4}\ M$$

Substituting, we obtain

$$K_{a_2} = \frac{(2.1 \times 10^{-4})(y)}{(2.1 \times 10^{-4})} = 5.6 \times 10^{-11}$$

$$y = 5.6 \times 10^{-11}$$

Note that the value of y is very much smaller than the value of x obtained for the first step in the dissociation.

Therefore,

$$[CO_3^{2-}] = 5.6 \times 10^{-11}\ M$$

In summary, the concentrations of all solute species present at equilibrium in a $0.10\ M$ H_2CO_3 solution are

$$[H^+] = 2.1 \times 10^{-4}\ M$$

$$[HCO_3^-] = 2.1 \times 10^{-4}\ M$$

$$[CO_3^{2-}] = 5.6 \times 10^{-11}\ M$$

$$[H_2CO_3] = 0.10\ M$$

Reviewing this last example, we see that in any solution containing H_2CO_3 as the only solute, the concentration of the carbonate ion will be equal to K_{a_2}. In fact, for any polyprotic acid in which $K_{a_2} \ll K_{a_1}$, the concentration of the anion formed in the second dissociation will always equal K_{a_2}, provided, of course, that the acid is the only solute. For example, K_{a_2} for H_3PO_4 has a value of 6.2×10^{-8}, and in a solution containing only H_3PO_4 and H_2O, the concentration of HPO_4^{2-} is $6.2 \times 10^{-8}\ M$.

HPO_4^{2-} is the anion formed in the second step of the dissociation of H_3PO_4.

16.9 SOLUTIONS OF SALTS OF POLYPROTIC ACIDS

Like the anions of molecular monoprotic acids, the anions of polyprotic acids are weak bases and salts of polyprotic acids therefore produce solutions that are more or less basic. An example is sodium carbonate, Na_2CO_3. The carbonate ion is a weak base that reacts with water in two steps to form ultimately carbonic acid, H_2CO_3.

Each step has its own equilibrium constant.

Step 1. $CO_3^{2-} + H_2O \rightleftharpoons HCO_3^- + OH^-$

$$K_{b_1} = \frac{[HCO_3^-][OH^-]}{[CO_3^{2-}]}$$

Step 2. $HCO_3^- + H_2O \rightleftharpoons H_2CO_3 + OH^-$

$$K_{b_2} = \frac{[H_2CO_3][OH^-]}{[HCO_3^-]}$$

The first step involves the reaction of CO_3^{2-}, which is the conjugate base of HCO_3^-. This means that in computing K_{b_1}, we must use the K_a for HCO_3^-. This K_a is actually the second ionization constant for H_2CO_3, K_{a_2}, which is found in Table 16.2. Similarly, to calculate K_{b_2}, we must use the K_a for H_2CO_3, which is K_{a_1}. Thus,

$$K_{b_1} = \frac{K_w}{K_{a_2}}$$

$$= \frac{1.0 \times 10^{-14}}{5.6 \times 10^{-11}} = 1.8 \times 10^{-4}$$

and

$$K_{b_2} = \frac{K_w}{K_{a_1}}$$

$$= \frac{1.0 \times 10^{-14}}{4.3 \times 10^{-7}} = 2.3 \times 10^{-8}$$

Notice that the value of K_{b_2} is much smaller than that of K_{b_1}. The relative magnitudes of these two equilibrium constants indicate that the second step occurs to a negligible extent compared to the first, so virtually all the OH^- in the solution comes from the first step. Therefore, if we are interested only in the pH of the solution, we need consider only the first step and the value of K_{b_1}.

EXAMPLE 16.18. CALCULATING THE pH OF A SOLUTION OF A SALT OF A POLYPROTIC ACID

PROBLEM: What is the pH of a 0.50 M solution of Na_2CO_3?

SOLUTION: As explained above, we need consider only the first step in the stepwise reaction of carbonate ion with water. The equilibrium and the equilibrium law are

$$CO_3^{2-} + H_2O \rightleftharpoons HCO_3^- + OH^-$$

$$K_{b_1} = \frac{[HCO_3^-][OH^-]}{[CO_3^{2-}]} = 1.8 \times 10^{-4}$$

The initial concentration of CO_3^{2-} is 0.50 M. We let the number of moles per liter that reacts equal x. This gives

	Initial Molar Concentrations	Change	Equilibrium Molar Concentrations
HCO_3^-	0.0	x	x
OH^-	0.0	x	x
CO_3^{2-}	0.50	$-x$	$0.50 - x$

As usual, because K_{b_1} is small, we will assume that x is small compared to the initial concentration of CO_3^{2-}.

$$0.50 - x \approx 0.50$$

Substituting into the mass action expression gives

$$\frac{(x)(x)}{0.50} = 1.8 \times 10^{-4}$$

Solving for x gives

$$x = 9.5 \times 10^{-3}$$

Next, we check the validity of our assumption. This time, it is just barely valid.

$$0.50 - 0.0095 = 0.49 \quad \text{(rounded)}$$

The rounded value, 0.49, differs from the initial concentration by 0.01, which amounts to 2% of the initial concentration.

$$\frac{0.01}{0.50} \times 100\% = 2\%$$

Earlier we noted that as long as the difference is less than 5%, we can take the assumption as valid.

At this point, we have $x = 9.5 \times 10^{-3}\ M = [OH^-]$. From this we can calculate the pOH to be 2.02 and the pH to be 11.98. This is quite basic, and as a result, sodium carbonate is used in various detergents in place of phosphates to make them caustic, meaning basic.

16.10 ACID–BASE TITRATIONS

In Chapter 5 we saw that a titration is a useful way of determining the concentrations of solutions of acids and bases, provided that the equivalence point can be detected. The equivalence point, you should remember, occurs when equal numbers of equivalents of acid and base have been combined. In this section we will see how the pH of a solution changes during the course of typical acid–base titrations and what the pH is at the equivalence point.

Titration of a strong acid with a strong base

The titration of 25.00 mL of 0.10 M HCl with 0.10 M NaOH is a typical example of the titration of a strong acid with a strong base. We can mathematically determine the pH throughout the titration by calculating the H^+ concentration present in the flask each time a quantity of NaOH is added to the HCl. For example, the number of moles of H^+ present in the 25 mL of a 0.10 M HCl solution is

$$\left(\frac{0.10\ \text{mol}}{1000\ \text{mL}}\right) \times 25\ \text{mL} = 2.5 \times 10^{-3}\ \text{mol of }H^+$$

When 10 mL of the 0.10 M NaOH is added, we in fact have added

$$\left(\frac{0.10\ \text{mol}}{1000\ \text{mL}}\right) \times 10\ \text{mL} = 1.0 \times 10^{-3}\ \text{mol of }OH^-$$

The neutralization reaction

$$H^+ + OH^- \longrightarrow H_2O$$

Table 16.3. Titration of 25 mL of 0.10 *M* HCl with a 0.10 *M* NaOH solution

Volume of HCl (mL)	Volume of NaOH (mL)	Total Volume (mL)	Moles of H^+	Moles of OH^-	Molarity of Ion in Excess	pH
25.00	0.00	25.00	2.5×10^{-3}	0	0.10 (H^+)	1.00
25.00	10.00	35.00	2.5×10^{-3}	1.0×10^{-3}	4.3×10^{-2} (H^+)	1.37
25.00	24.99	49.99	2.5×10^{-3}	2.499×10^{-3}	2.0×10^{-5} (H^+)	4.70
25.00	25.00	50.00	2.5×10^{-3}	2.50×10^{-3}	0	7.00
25.00	25.01	50.01	2.5×10^{-3}	2.501×10^{-3}	2.0×10^{-5} (OH^-)	9.30
25.00	26.00	51.00	2.5×10^{-3}	2.60×10^{-3}	2.0×10^{-3} (OH^-)	11.30
25.00	50.00	75.00	2.5×10^{-3}	5.0×10^{-3}	3.3×10^{-2} (OH^-)	12.52

occurs, and the amount of H^+ remaining is

$$(2.5 \times 10^{-3}) - (1.0 \times 10^{-3}) = 1.5 \times 10^{-3} \text{ mol of } H^+$$

The molar concentration of H^+ is now

$$[H^+] = \frac{1.5 \times 10^{-3}\,\text{mol}}{0.035\,\text{L}} = 4.3 \times 10^{-2}\,M$$

The total volume

and the pH is calculated to be 1.37. The concentrations of H^+ after further additions of NaOH have occurred are summarized in Table 16.3.

Our calculations show that the pH increases slowly at first, then rises rapidly near the equivalence point, and finally levels off gradually after the equivalence point is reached.

If a graph is drawn of pH versus the volume of base added, we obtain the plot shown in Figure 16.5. The equivalence point occurs, in this case, at a pH of 7. At the equivalence point the solution is neutral because neither of the ions of the salt that is left in solution (NaCl) undergoes hydrolysis.

Titration using a weak acid and a strong base

In an acid–base titration in which one substance is strong and the other weak, the solution is not neutral at the equivalence point because one of the ions of the salt is able to function as a weak acid or base. For example, consider the titration of 25.0 mL of 0.10 *M* $HC_2H_3O_2$ with 0.10 *M* NaOH. Before any base is added, the only solute is acetic acid. The pH of the solution is calculated as shown in Section 16.5 — for 0.10 *M* $HC_2H_3O_2$, the pH = 2.89.

When we begin to add NaOH, acetic acid molecules are converted to acetate ions.

$$HC_2H_3O_2 + OH^- \longrightarrow H_2O + C_2H_3O_2^-$$

Since the solution then contains both $HC_2H_3O_2$ and $C_2H_3O_2^-$, it is a buffer, and we've also learned how to calculate the pH for this kind of mixture. For instance, when 10.0 mL of 0.10 *M* NaOH has been added, 1.0×10^{-3} mol of OH^- has been supplied. This neutralizes 1.0×10^{-3} mol of $HC_2H_3O_2$ and converts it to 1.0×10^{-3} mol of $C_2H_3O_2^-$. The original 25.0 mL of 0.10 *M* $HC_2H_3O_2$ contained 2.5×10^{-3} mol $HC_2H_3O_2$, so the amount that is left is

$$(2.5 \times 10^{-3} \text{ mol}) - (1.0 \times 10^{-3} \text{ mol}) = 1.5 \times 10^{-3} \text{ mol } HC_2H_3O_2$$

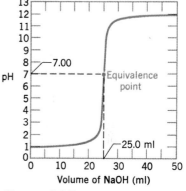

Figure 16.5. *Titration of 0.10 M HCl with 0.10 M NaOH.*

The concentrations of acetic acid and acetate ion in the total volume of 35.0 mL are therefore

$$[HC_2H_3O_2] = \frac{1.5 \times 10^{-3} \text{ mol}}{0.0350 \text{ L}} = 4.3 \times 10^{-2} \text{ } M$$

$$[C_2H_3O_2^-] = \frac{1.0 \times 10^{-3} \text{ mol}}{0.0350 \text{ L}} = 2.9 \times 10^{-2} \text{ } M$$

If we solve the K_a expression for acetic acid for the H^+ concentration and substitute these values for $[HC_2H_3O_2]$ and $[C_2H_3O_2^-]$, we obtain

$$[H^+] = K_a \times \frac{[HC_2H_3O_2]}{[C_2H_3O_2^-]}$$

$$= 1.8 \times 10^{-5} \left(\frac{4.3 \times 10^{-2}}{2.9 \times 10^{-2}} \right)$$

$$= 2.7 \times 10^{-5} \text{ } M$$

Therefore,

$$pH = 4.57$$

From the time of the first addition of base until the equivalence point is reached, the solution contains both acetic acid and acetate ion, and the pH may be computed in this fashion.

When a total of 25.0 mL of NaOH is added, all the acetic acid is "neutralized," and we have produced 2.5×10^{-3} mol of $NaC_2H_3O_2$ in 50.0 mL of solution. The resulting 0.050 M $NaC_2H_3O_2$ solution is not neutral, however, because it contains the anion of a weak acid. We have seen that for this anion the equilibrium is

For the $NaC_2H_3O_2$,

$$\frac{2.5 \times 10^{-3} \text{ mol}}{0.050 \text{ L}} = 0.050 \text{ } M$$

$$C_2H_3O_2^- + H_2O \rightleftharpoons HC_2H_3O_2 + OH^-$$

and we know that

$$K_b = \frac{[HC_2H_3O_2][OH^-]}{[C_2H_3O_2^-]} = \frac{K_w}{K_a} = 5.6 \times 10^{-10}$$

We can calculate the OH^- concentration in this solution as we did previously by letting x equal the number of moles per liter of $C_2H_3O_2^-$ that reacts. This allows us to construct our table.

	Initial Molar Concentrations	Change	Equilibrium Molar Concentrations
$HC_2H_3O_2$	0.0	$+x$	x
OH^-	0.0	$+x$	x
$C_2H_3O_2^-$	0.050	$-x$	$0.050 - x \approx 0.050$

Substituting into the K_b expression gives

$$K_b = \frac{(x)(x)}{0.050} = 5.6 \times 10^{-10}$$

$$x^2 = 2.8 \times 10^{-11}$$

$$x = 5.3 \times 10^{-6}$$

This means that $[OH^-] = 5.3 \times 10^{-6} \text{ } M$, from which we obtain

$$pOH = 5.28$$

and finally,

$$pH = 8.72$$

Table 16.4. Titration of 25.0 mL of 0.10 M $HC_2H_3O_2$ with 0.10 M NaOH

Milliliters of Base Added	Molar Concentration of Species in Parentheses	pH
0.0	1.3×10^{-3} (H^+)	2.89
10.0	2.7×10^{-5} (H^+)	4.57
24.99	7.2×10^{-9} (H^+)	8.14
25.0	5.3×10^{-6} (OH^-)	8.72
25.01	2.0×10^{-5} (OH^-)	9.30
26.0	2.0×10^{-3} (OH^-)	11.30

Thus, the pH at which the equivalence point occurs is greater than 7. We find that this is true for any weak acid–strong base titration.

Thus far we have discussed only the first half of the titration (see Table 16.4). What takes place beyond the equivalence point? As soon as all the weak acid has been neutralized, any further addition of NaOH suppresses the reaction of the anion and the pH is then solely dependent on the concentration of OH^- coming from the added NaOH. Thus, we generate the last half of Table 16.4 in the same manner as we did Table 16.3 in the HCl–NaOH titration.

A graph of these data is shown in Figure 16.6, where we have plotted pH versus volume of base added. From both Table 16.4 and Figure 16.6, we can see that the change in pH near the equivalence point is not as drastic as in the case of the HCl–NaOH titration. This less rapid change near the equivalence point becomes even more pronounced for weaker acids such as HCN.

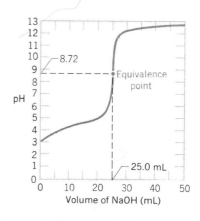

Figure 16.6. *Titration of 25.0 mL of 0.10 M acetic acid with 0.10 M sodium hydroxide.*

Titration of a weak base with a strong acid

When a weak base is titrated with a strong acid, the titration curve that is generated is very similar in shape to that obtained by reaction of a weak acid with a strong base. During the initial addition of acid, the solution contains unreacted weak base as well as its conjugate acid. It therefore constitutes a buffer. At the equivalence point the solution contains the salt of the weak base, and the pH of the mixture is determined by the hydrolysis of the conjugate acid (the cation). Finally, beyond the equivalence point the pH of the solution is controlled by the excess hydrogen ion from the strong acid. The shape of the titration curve for such a titration is shown in Figure 16.7 for the titration of 25.0 mL of 0.10 M NH_3 with 0.10 M HCl. We can show that the pH at the equivalence point is less than 7 by considering the reaction of the NH_4^+ ion produced in the reaction.

Recall that the K_a for NH_4^+ is written as

$$K_a = \frac{[H^+][NH_3]}{[NH_4^+]} = \frac{K_w}{K_b} = 5.6 \times 10^{-10}$$

All the NH_3 is "neutralized" in this titration when exactly 25.0 mL of the 0.10 M HCl (2.5×10^{-3} mol HCl) has been added. At this point, the concentration of NH_4^+ is

$$\frac{2.5 \times 10^{-3} \text{ mol}}{0.0500 \text{ L}} = 5.0 \times 10^{-2} M$$

If we let x equal the number of moles per liter of NH_4^+ that undergo reaction, then

$$[H_3O^+] = x$$

$$[NH_3] = x$$

$$[NH_4^+] = 5.0 \times 10^{-2} - x \approx 5.0 \times 10^{-2} M$$

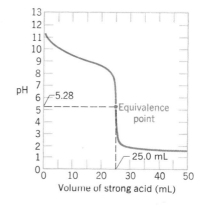

Figure 16.7. *Titration of 25.0 mL of 0.10 M NH$_3$ with 0.10 M HCl.*

Substituting these concentrations into the equation for K_a gives

$$K_a = \frac{(x)(x)}{5.0 \times 10^{-2}} = 5.6 \times 10^{-10}$$

$$x^2 = 2.8 \times 10^{-11}$$

$$x = 5.3 \times 10^{-6}$$

$$[H^+] = 5.3 \times 10^{-6}\ M$$

and

$$pH = 5.28$$

The pH at the equivalence point of this titration is less than 7, which is typical for all weak base–strong acid titrations.

Phenolphthalein is often used as an indicator in the titration of a strong acid with a strong base or the titration of a weak acid with a strong base.

16.11 ACID–BASE INDICATORS

Earlier we mentioned acid–base indicators and their role in determining whether or not a solution is acidic or basic. One of the principal uses of acid–base indicators is in the detection of the equivalence point in acid–base titrations.

Indicators are usually weak organic acids or bases that change color over a range of pH values. Not all indicators change color at the same pH, however. The choice of indicator for a particular titration depends on the pH at which the equivalence point is expected to occur. A list of some common indicators, with their color changes and the pH ranges over which the color changes are observed, is found in Table 16.5. Let us examine briefly how these indicators work.

If we denote an indicator by the general formula HIn, we have the dissociation reaction

$$HIn \rightleftharpoons H^+ + In^-$$

Applying Le Châtelier's principle to this equilibrium, we see that in an acidic solution (excess H^+) the predominant species is HIn. On the other hand, in basic solutions the equilibrium is shifted to the right and the predominant species is In^-. Therefore, HIn is said to be the acid form and In^- the basic form of the indicator. The

Some indicators, like thymol blue, have two color changes over separate pH ranges.

Table 16.5. Some common indicators

Indicator	Color Change	pH Range in Which Color Change Occurs
Thymol blue	Red to yellow	1.2–2.8
Bromophenol blue	Yellow to blue	3.0–4.6
Congo red	Blue to red	3.0–5.0
Methyl orange	Red to yellow	3.2–4.4
Bromocresol green	Yellow to blue	3.8–5.4
Methyl red	Red to yellow	4.8–6.0
Bromocresol purple	Yellow to purple	5.2–6.8
Bromothymol blue	Yellow to blue	6.0–7.6
Cresol red	Yellow to red	7.0–8.8
Thymol blue	Yellow to blue	8.0–9.6
Phenolphthalein	Colorless to pink	8.2–10.0
Alizarin yellow	Yellow to red	10.1–12.0

pH 8.2   pH 10.0

Phenolphthalein

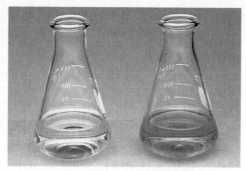

pH 6.0   pH 7.6

Bromothymol blue

pH 3.2   pH 4.4

Methyl orange

pH 9.4   pH 10.6

Thymolphthalein

The colors of some common acid–base indicators.

ability of HIn to function as an indicator is based on the difference in color between acid and basic forms. For example, with litmus the acid form (HIn) is pink while the basic form (In$^-$) is blue.

The dissociation constant, K_a, for an indicator is

$$K_a = \frac{[\text{H}^+][\text{In}^-]}{[\text{HIn}]}$$

Let's solve this for the ratio [In$^-$]/[HIn].

$$\frac{[\text{In}^-]}{[\text{HIn}]} = \frac{K_a}{[\text{H}^+]}$$

We have seen that as we pass through the equivalence point, the pH changes very rapidly. For example, in the NaOH–HCl titration described earlier—Table 16.3 and Figure 16.5—the pH changed from 4.7 to 9.3 with the addition of only 0.02 mL of base, which is only about one-half drop of solution. This pH change corresponds to a change in [H$^+$] from 2×10^{-5} M to 5×10^{-10} M. How does this affect the [In$^-$]-to-[HIn] ratio?

Suppose that we were using an indicator whose $K_a = 1 \times 10^{-7}$. Then, before the equivalence point,

$$\frac{[\text{In}^-]}{[\text{HIn}]} = \frac{1 \times 10^{-7}}{2 \times 10^{-5}} = \frac{1}{200}$$

This tells us that there is 200 times as much HIn as In$^-$, so the color observed is that resulting from HIn.

After the equivalence point,

$$\frac{[\text{In}^-]}{[\text{HIn}]} = \frac{1 \times 10^{-7}}{5 \times 10^{-10}} = \frac{200}{1}$$

Visually, we observe the color of the species that is present in largest concentration.

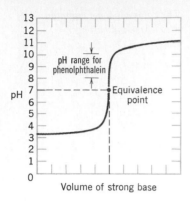

Figure 16.8. *Titration curve for the titration of a strong acid with a strong base.*

Now there is 200 times as much In⁻ as HIn, and the color that we see is due to In⁻. Thus, as we pass through the equivalence point, there is a sudden change in the relative amounts of the acid and basic forms of the indicator, which we notice as a change in color.

If the indicator changes color at the equivalence point, the end point in the titration — that point at which we observe the color change — occurs at the same pH as the equivalence point. Often, however, we find ourselves using an indicator whose color change takes place at a pH slightly different from that of the equivalence point. This is shown in Figure 16.8 for phenolphthalein. When the color change occurs, we have actually gone slightly past the equivalence point.

In choosing an indicator, we wish to have it change color very close to the equivalence point. Phenolphthalein, for example, would be a poor choice of indicator for the titration depicted in Figure 16.7, because its color change would occur long before the equivalence point. We would find that we had stopped adding acid before the equivalence point had been reached, thereby defeating the purpose of using an indicator. A better choice would be an indicator such as methyl red, where the center of the color change range occurs very near the pH at the equivalence point.

REVIEW QUESTIONS AND PROBLEMS

Problems whose numbers are in blue have their answers in Appendix D at the back of the book. The more difficult problems are marked with asterisks.

Ionization of Water and Solutions of Strong Acids and Bases

16.1 Write the equation for the autoionization of water. Why is this equilibrium important?

16.2 Which are the common strong monoprotic acids? Which are the strong bases?

16.3 Write chemical equations to show what happens to HCl and KOH when they are dissolved in water.

16.4 Why can we almost always ignore the H^+ contributed by the dissociation of water when we calculate the H^+ concentration in solutions containing an acid? Under what conditions would we have to consider the H^+ from the dissociation of H_2O, even though the solute was an acid?

16.5 Calculate the H^+ and OH^- concentrations (in mol/L) and the pH of the following solutions of acids and bases:
(a) 0.0010 M HCl
(b) 0.125 M HNO₃
(c) 0.0031 M NaOH
(d) 0.012 M Ba(OH)₂
(e) 2.1 × 10⁻⁴ M HClO₄
(f) 1.3 × 10⁻⁵ M HCl
(g) 8.4 × 10⁻³ M NaOH
(h) 4.8 × 10⁻² M KOH

16.6 At 37 °C (body temperature), $K_w = 2.42 \times 10^{-14}$. What are the molar concentrations of H^+ and OH^- in a neutral aqueous solution at 37 °C?

pH and pOH

16.7 How is pH defined? How is pOH defined? Why does pH + pOH = 14?

16.8 Identify the following as representing acidic, basic, or neutral solutions:
(a) pH = 3.54
(b) pH = 8.25
(c) pOH = 7.00
(d) pOH = 10.43
(e) pOH = 2.25

16.9 Arrange the solutions in Problem 16.8 in order of increasing acidity.

16.10 What is the pH of a neutral solution at 37 °C? (See Problem 16.6.)

16.11 Calculate the H^+ and OH^- concentrations (in mol/L) in a solution having a pH equal to
(a) 1.30
(b) 5.73
(c) 4.00
(d) 7.80
(e) 10.94
(f) 12.61

16.12 What is the pOH of each solution in Problem 16.11?

16.13 Calculate the pH and pOH of each of the solutions in Exercise 16.5.

Conjugate Acid–Base Relationships

16.14 Write chemical equations for the Brønsted–Lowry equilibria involved in the reactions of the following acids with water: (a) HOBr, (b) HCN, (c) (CH₃)₃NH⁺, (d) C₅H₅NH⁺, (e) HCO₃⁻.

16.15 Write chemical equations for the Brønsted–Lowry equilibria involved in the reactions of the following bases with water: (a) C₅H₅N, (b) CO₃²⁻, (c) H₂PO₄⁻, (d) NO₂⁻, (e) C₆H₅NH₂.

16.16 For the members of an acid–base conjugate pair, what is the relationship between K_a and K_b in water?

16.17 From Figure 16.4, place the following reactions in order of increasing tendency to proceed toward completion:
(a) $H_2O + NH_3 \rightleftharpoons NH_4^+ + OH^-$
(b) $HClO_4 + NH_2^- \rightleftharpoons ClO_4^- + NH_3$
(c) $H_2O + NO_2^- \rightleftharpoons HNO_2 + OH^-$
(d) $NH_3 + Cl^- \rightleftharpoons NH_2^- + HCl$

16.18 Use Figure 16.4 to place the following reactions in order of increasing tendency to proceed toward completion:
(a) $OCl^- + HCl \rightleftharpoons HOCl + Cl^-$
(b) $HF + C_2H_3O_2^- \rightleftharpoons HC_2H_3O_2 + F^-$
(c) $NH_4^+ + ClO_4^- \rightleftharpoons NH_3 + HClO_4$
(d) $HNO_2 + F^- \rightleftharpoons HF + NO_2^-$

16.19 Hydrogen sulfide is a stronger acid than phosphine, PH_3. What may we conclude about the strengths of their conjugate bases, HS^- and PH_2^-?

16.20 Hydrogen cyanide, HCN, is a weaker acid in water than nitrous acid, HNO_2. What can we say about the relative strengths of CN^- and NO_2^- as bases?

16.21 Ammonia is a stronger base in water than hydrazine, N_2H_4. If we were to use these two substances as solvents for a very weak acid, in which of them would the acid be more fully ionized?

16.22 Given the following equilibria and equilibrium constants:
(a) $HOCl + H_2O \rightleftharpoons H_3O^+ + OCl^-$ $K_a = 3.2 \times 10^{-8}$
(b) $NH_4^+ + H_2O \rightleftharpoons H_3O^+ + NH_3$ $K_a = 5.6 \times 10^{-10}$
(c) $HC_2H_3O_2 + H_2O \rightleftharpoons H_3O^+ + C_2H_3O_2^-$ $K_a = 1.8 \times 10^{-5}$
(d) $H_2CO_3 + H_2O \rightleftharpoons H_3O^+ + HCO_3^-$ $K_a = 4.2 \times 10^{-7}$
(e) $HSO_4^- + H_2O \rightleftharpoons H_3O^+ + SO_4^{2-}$ $K_a = 1.3 \times 10^{-2}$

Arrange the conjugate bases in order of *increasing* strength.

Solutions of Molecular Acids and Bases

16.23 A weak acid has an equilibrium constant $K_a = 3.8 \times 10^{-9}$. What is the pK_a of the acid?

16.24 A base has $pK_b = 3.84$. What is K_b of the base?

16.25 Refer to Table 16.1 and write the appropriate equilibrium constant expressions for the ionization of:
(a) benzoic acid   (d) Veronal
(b) hydrazine   (e) pyridine
(c) formic acid

16.26 What is the H^+ concentration (in mol/L) in each of the following solutions:
(a) 0.30 M HNO_2 (d) 0.10 M butyric acid
(b) 1.00 M HF   (e) 0.050 M barbituric acid
(c) 0.025 M HCN

16.27 Calculate the OH^- concentration (in mol/L) in the following solutions:
(a) 0.15 M NH_3 (d) 0.35 M hydroxylamine
(b) 0.20 M N_2H_4 (e) 0.010 M pyridine
(c) 0.80 M CH_3NH_2

16.28 What is the OH^- concentration (in mol/L) in each solution in Problem 16.26?

16.29 What is the pH of each solution in Exercise 16.27?

16.30 A 0.25 M solution of a monoprotic weak acid was observed to have a pH = 1.35. What is K_a for this acid?

16.31 A 0.10 M solution of a weak monoprotic acid was found to have a pH = 5.37. What is K_a for the acid?

16.32 A weak base was found to give a solution with pH = 8.75 when its concentration was 0.10 M. What is K_b for the base?

16.33 What is the percent ionization of the acid in each of the following six solutions:
(a) 1.0 M formic acid   (d) 0.35 M nicotinic acid
(b) 0.010 M propionic acid (e) 0.50 M HOCl
(c) 0.025 M HCN   (f) 0.25 M HNO_3

16.34 Calculate the percent ionization of each of the following acetic acid solutions. What conclusions can you draw? Can you explain on a molecular level why you obtain these results?
(a) 1.00 M $HC_2H_3O_2$
(b) 0.10 M $HC_2H_3O_2$
(c) 0.010 M $HC_2H_3O_2$

16.35 The pH of a 0.012 M solution of a weak base, B, was experimentally determined to be 11.40. Calculate K_b for the base.

16.36 How many grams of HCl gas would have to be dissolved in 500 mL of 1.0 M $NaC_2H_3O_2$ to give a solution having a pH = 4.74?

16.37 When 10 mg of sodium barbituate is swallowed, what fraction is converted to barbituric acid if the pH of the stomach is 1.0 and there are 250 mL of fluid in the stomach?

16.38 Nicotinic acid is another name for the important vitamin, niacin. What is the pH of a 0.010 M solution of nicotinic acid?

16.39 What is the molarity of a solution of acetic acid whose pH is 2.50?

16.40 What molar concentration of hydrazine, N_2H_4, yields a solution whose pH = 10.64?

16.41 A 0.010 M solution of a weak acid, HA, is found to have a pH of 4.55. What is the value of K_a for this acid?

*16.42 Calculate the H^+ concentration (in mol/L) in 0.0010 M $HC_2H_3O_2$.

*16.43 Calculate the molar concentrations of all species in 0.010 M formic acid solution.

Buffers

16.44 What is a buffer? Explain how the following solutes function as buffers:
(a) $NaCHO_2$ and $HCHO_2$ (c) $NH_4C_2H_3O_2$
(b) C_5H_5N and C_5H_5NHCl (d) $NaHCO_3$

16.45 Would a solution that contains a mixture of NaCl and HCl be an effective buffer? Explain.

16.46 Among the buffer systems in blood is the pair HPO_4^{2-}

and $H_2PO_4^-$. Using appropriate equations, show how this system is able to respond to the addition of (a) a strong acid and (b) a strong base.

16.47 Calculate the hydrogen ion concentration in mol/L for each of the following solutions of an acid or base and its salt:
(a) $0.25\ M\ HC_2H_3O_2$, $0.15\ M\ NaC_2H_3O_2$
(b) $0.50\ M\ HCHO_2$, $0.50\ M\ NaCHO_2$
(c) $0.30\ M\ HNO_2$, $0.40\ M\ NaNO_2$
(d) $0.25\ M\ NH_3$, $0.15\ M\ NH_4Cl$
(e) $0.30\ M\ N_2H_4$, $0.50\ M\ N_2H_5NO_3$

16.48 What ratio of molar concentrations of lactic acid to sodium lactate is required to give a solution having a pH = 4.25?

16.49 Calculate the pH of each of the following buffers prepared by placing, in 1.0 L of solution,
(a) 0.10 mol of NH_3 and 0.10 mol of NH_4Cl
(b) 0.20 mol of $HC_2H_3O_2$ and 0.40 mol of $NaC_2H_3O_2$
(c) 0.15 mol of N_2H_4 and 0.10 mol of N_2H_5Cl
(d) 0.20 mol of HCl and 0.30 mol of NaCl

16.50 How many grams of $NaC_2H_3O_2$ must be added to 1.00 mol of $HC_2H_3O_2$ in order to prepare 1.00 L of a buffer whose pH equals 5.15?

16.51 What must the ratio of molar concentrations of NH_3 to NH_4^+ be to have a buffer with a pH of 10.00?

*16.52 How many moles of HCl must be added to 1.0 L of a mixture containing $0.010\ M\ HC_2H_3O_2$ and $0.010\ M$ $NaC_2H_3O_2$ in order to give a solution whose pH = 3.00?

16.53 Calculate the pH change produced by adding 0.10 mol of solid NaOH to each of the following buffers:
(a) 500 mL of $1.00\ M\ HC_2H_3O_2$ and $1.00\ M\ NaC_2H_3O_2$
(b) 500 mL of $0.50\ M\ HC_2H_3O_2$ and $0.50\ M\ NaC_2H_3O_2$
(c) 500 mL of $0.30\ M\ HC_2H_3O_2$ and $0.70\ M\ NaC_2H_3O_2$
(d) 500 mL of $0.20\ M\ HC_2H_3O_2$ and $0.80\ M\ NaC_2H_3O_2$
(e) 500 mL of $0.10\ M\ HC_2H_3O_2$ and $0.90\ M\ NaC_2H_3O_2$

16.54 How much will the pH change if 0.10 mol of HCl is added to 1.0 L of a formic acid–sodium formate buffer containing 0.45 mol of $HCHO_2$ and 0.55 mol of $NaCHO_2$?

16.55 How much will the pH change if 0.20 mol of NaOH is added to the original buffer in Exercise 16.54?

16.56 What happens to the pH of a buffer solution if it is diluted with water? Illustrate your answer using the buffer in part (a) of Exercise 16.49.

Acid–Base Properties of Salt Solutions

16.57 Without performing any computations, predict whether the following solutions will be acidic, basic, or neutral:
(a) KCl   (c) $NaC_4H_7O_2$
(b) NH_4NO_3 (d) $C_6H_5NH_3NO_3$

16.58 If the concentration of each solute in Problem 16.57 was $0.10\ M$, which solution would be most basic? Which would be most acidic?

16.59 Determine the pH of each of the following salt solutions:
(a) $1.0 \times 10^{-3}\ M\ NaC_2H_3O_2$ (d) $0.10\ M\ NaCN$
(b) $0.125\ M\ NH_4Cl$ (e) $0.20\ M\ NH_3OHCl$
(c) $0.10\ M\ KCHO_2$

16.60 What percent of $C_5H_5NH^+$ reacts in a $0.10\ M$ solution of pyridinium chloride, C_5H_5NHCl? For pyridine, C_5H_5N, the value of $K_b = 1.7 \times 10^{-9}$.

16.61 A $0.10\ M$ solution of the sodium salt of a weak monoprotic acid has a pH of 9.35. What is the K_a of the weak acid?

16.62 Liquid chlorine bleach is really nothing more than a dilute solution of NaOCl, usually about 5% NaOCl by weight. A particular sample of bleach was found to contain 0.67 mol/L NaOCl. Calculate the pH of the solution.

16.63 Veronal, a barbiturate drug, is generally administered as its sodium salt. What is the pH of a solution of $NaC_8H_{11}N_2O_3$ that contains 10 mg of the drug in 250 mL of solution? For veronal, $HC_8H_{11}N_2O_3$, the value of $K_a = 3.7 \times 10^{-8}$.

16.64 What would be the molar concentration of barbituric acid in a $0.0010\ M$ solution of sodium barbiturate?

16.65 Sodium benzoate, $NaC_7H_5O_2$, is used as a preservative in foods. Calculate the pH of a $0.30\ M$ solution of this compound. For $HC_7H_5O_2$, $K_a = 6.5 \times 10^{-5}$.

16.66 What is the pH of a $0.020\ M$ solution of sodium benzoate, $NaC_7H_5O_2$?

*16.67 What would be the pH of a $0.0010\ M$ solution of potassium cyanide (a deadly poison)?

Polyprotic Acids

16.68 Write appropriate mass action expressions for K_{a_1} and K_{a_2} for ascorbic acid (vitamin C).

16.69 Citric acid, which is present in many fruits and vegetables, has the formula, $H_3C_6H_5O_7$. It is a triprotic acid. Write the three equilibria for the dissociation of the acid and the appropriate equilibrium constant expression for each step.

16.70 Selenious acid, H_2SeO_3, has $K_{a_1} = 3 \times 10^{-3}$ and $K_{a_2} = 5 \times 10^{-8}$. What is the pH of a $0.50\ M$ solution of H_2SeO_3? What are the equilibrium molar concentrations of H_2SeO_3, $HSeO_3^-$, and SeO_3^{2-}?

16.71 Calculate the pH obtained by dissolving a 500-mg tablet of vitamin C in 250 mL (approximately 8 oz) of H_2O.

16.72 Calculate the molar concentrations of all the species in a $0.050\ M$ solution of the diprotic acid, vitamin C.

16.73 In the stomach the fluids have a pH ≈ 1.0 produced by the strong acid, HCl. What fraction of the vitamin C in a

500-mg tablet will be dissociated if the volume of fluid in the stomach is 200 mL?

16.74 What is the HCO_3^- concentration (in mol/L) in a 0.10 M solution of H_2CO_3 whose pH = 3.00? What is the CO_3^{2-} concentration in this solution?

* 16.75 A sample of arterial blood was found to contain 2.6 $\times$ 10^{-2} mol of dissolved CO_2 per liter. The pH of the sample was 7.43. If it is assumed that in solution the CO_2 forms H_2CO_3, what is the HCO_3^- concentration (in mol/L) in this blood sample?

Solutions of Salts of Polyprotic Acids

16.76 What is the pH of a 0.20 M solution of sodium ascorbate?

16.77 What is the pH of a 0.25 M solution of sodium sulfite?

16.78 Write the chemical equations for the equilibria that take place in a solution of sodium carbonate. Calculate the molar concentrations of all the species that exist in a 0.20 M solution of Na_2CO_3. What is the pH of the solution?

Acid–Base Titrations

16.79 Is it possible to have a pH other than 7 at the equivalence point in an acid–base titration?

16.80 In a titration, the equivalence point and the end point are often not exactly the same. Justify this statement.

16.81 A 15.0-mL portion of a solution of 0.0200 M HNO_3 is titrated with 0.0100 M KOH.
(a) What will the pH be at the equivalence point?
(b) How many milliliters of the base will be required to reach the equivalence point?
(c) What will the pH be when 10.0 mL of the KOH solution has been added?
(d) What will the pH be when 35.0 mL of the KOH solution has been added?

16.82 What would the pH be at the equivalence point if 25.0 mL of 0.010 M barbituric acid is titrated with 0.020 M NaOH?

* 16.83 When 50.0 mL of 0.200 M HF is titrated with 0.100 M NaOH, what is the pH
(a) after 5.0 mL of base has been added?
(b) when half of the HF has been neutralized?
(c) at the equivalence point?

16.84 Plot a curve showing the pH of a solution of 100 mL of 0.10 M butyric acid that is gradually neutralized by the addition of solid NaOH. Do this by calculating the pH after addition of 0.0, 0.0010, 0.0050, 0.0090, 0.010, and 0.011 mol of NaOH. Assume no change in volume. What is the pH at the equivalence point? Which indicator in Table 16.5 could be used for this titration?

16.85 Determine the shape of the titration curve for the titration of 50.0 mL of 0.10 M HCl with 0.10 M NaOH.

* 16.86 Determine the shape of the titration curve when 100 mL of 0.20 M H_2CO_3 is titrated with 0.10 M NaOH. Determine the pH at each equivalence point.

Indicators

16.87 Explain how an indicator works. Why do we want to use as little of the indicator as possible when we perform a titration?

16.88 What indicators might be acceptable for the titration depicted in Figure 16.6? Why would we not wish to use congo red as an indicator?

16.89 Would congo red be an acceptable indicator for the titration depicted in Figure 16.8? Explain your answer.

16.90 Using the data in Tables 16.1 and 16.5, choose an indicator that is suitable for the titration of
(a) hydrocyanic acid with sodium hydroxide
(b) aniline with hydrochloric acid

16.91 An indicator, HIn, has an ionization constant, K_a, equal to 1 $\times$ 10^{-5}. If the molecular form of the indicator is yellow and the In^- ion green, what is the color of a solution containing this indicator when its pH is 7.0?

Additional Exercises

16.92 A solution is prepared by dissolving 25.0 g of lactic acid, $HC_3H_5O_3$, and 38.0 g of potassium lactate in 500 mL of solution.
(a) What is the pH of this solution?
(b) By how much will the pH change if 50.0 mL of 0.20 M HCl is added?
(c) By how much will the pH change if 50.0 mL of 1.0 M KOH is added?

* 16.93 Calculate the pH of 0.50 M $NaHCO_3$. How much will the pH change if 0.05 mol/L of HCl is added?

* 16.94 How many milliliters of 6.0 M HCl are required to be added to 100 mL of 0.10 M $NaC_2H_3O_2$ to give a solution having a pH = 4.25?

* 16.95 Calculate the pH of 0.10 M NH_4NO_2.

* 16.96 A dilute solution of hydrochloric acid was labeled 1.0 $\times$ 10^{-8} M HCl. Is this solution acidic or basic? What is its pH?

16.97 Calculate the pH of a 0.20 M solution of iodic acid. For HIO_3, $K_a = 1.7 \times 10^{-1}$.

16.98 Calculate the pH of a 0.60 M solution of trichloroacetic acid, $HC_2O_2Cl_3$. For this acid, $K_a = 2.2 \times 10^{-1}$. (Solve this problem by using the quadratic formula and by applying the method of successive approximations.)

16.99 Trisodium phosphate, Na_3PO_4, is used as a cleaning agent. Calculate the pH of a 1.0 M solution of trisodium phosphate. What molar concentration of lye (NaOH) would be necessary to give a solution of the same pH? (Solve this problem by the method of successive approximations.)

17

SOLUBILITY AND COMPLEX ION EQUILIBRIA

In the preceding chapter, we studied ionic equilibria involving acids and bases. These are not, however, the only equilibria that can take place in aqueous solutions. In this chapter we turn our attention to the equilibria of salts that have a low solubility in water — those that we described as being "insoluble" in our discussions of metathesis reactions in Chapter 5. We will also examine equilibria associated with the formation of complex ions in solution, and we will see how complex ion formation can affect the solubilities of salts.

17.1 SOLUBILITY PRODUCT

In Chapter 5 you were presented with a list of solubility rules that described certain salts as soluble and others as quite insoluble. Even the most insoluble salts, however, dissolve in water to at least some degree, and their saturated solutions constitute dynamic equilibria that can be studied by the same principles that we applied to acid–base equilibria in the last chapter.

Nearly all salts are completely dissociated in water. There are some exceptions, such as $HgCl_2$ and $CdSO_4$, but they are rare. For simplicity, therefore, our discussions will not include them, and we will assume that in a saturated solution an equilibrium exists between the solid salt and its dissolved ions. For example, in a saturated solution of silver chloride we have the equilibrium

$$AgCl(s) \rightleftharpoons Ag^+(aq) + Cl^-(aq)$$

It is safe to assume that there is no molecular AgCl in the solution.

for which we can write

$$K_c = \frac{[Ag^+][Cl^-]}{[AgCl(s)]}$$

In Section 15.5 we saw that the *concentration* of a pure solid is independent of the amount of solid present. In other words, the concentration of the solid is a constant and can therefore be included with the constant K_c, so that

$$K_c[AgCl(s)] = K_{sp} = [Ag^+][Cl^-]$$

The equilibrium constant K_c multiplied by the concentration of solid AgCl yields yet another equilibrium constant that is called the **solubility product constant, K_{sp}**. The name comes from the nature of the mass action expression, which is a product of ion concentrations raised to appropriate powers (in this case, they are each 1). The mass action expression itself is called the **ion product** for the salt, and when a saturated solution exists, the ion product is equal to K_{sp}.

In general, the K_{sp} expression is easily obtained from the equation for the solubility equilibrium. For example, for silver acetate, $AgC_2H_3O_2$, the equilibrium is

$$AgC_2H_3O_2(s) \rightleftharpoons Ag^+(aq) + C_2H_3O_2^-(aq)$$

The equilibrium expression is therefore

$$K_{sp} = [Ag^+][C_2H_3O_2^-]$$

In the case of an insoluble solid such as $Mg(OH)_2$, the coefficients in the equilibrium are not all equal to one

$$Mg(OH)_2(s) \rightleftharpoons Mg^{2+}(aq) + 2OH^-(aq)$$

The K_{sp} for $Mg(OH)_2$ is then given by

$$K_{sp} = [Mg^{2+}][OH^-]^2$$

Thus, the solubility product constant is equal to the product of the molar concentrations of the ions in a saturated solution, each raised to a power equal to its coefficient in the balanced equation. A list of some ionic solids and their K_{sp}'s at temperatures ranging between 18 and 25 °C is given in Table 17.1.

Creamy milk of magnesia is a suspension of magnesium hydroxide in water.

A more extensive table of solubility product constants is given in Appendix C.

Table 17.1. Solubility product constants

Anion	Compound	K_{sp}	Anion	Compound	K_{sp}
Fluorides	MgF_2	7.3×10^{-9}	Hydroxides	$Mg(OH)_2$	7.1×10^{-12}
	CaF_2	1.7×10^{-10}		$Ca(OH)_2$	6.5×10^{-6}
	BaF_2	1.7×10^{-6}		$Fe(OH)_2$	2×10^{-15}
	PbF_2	3.2×10^{-8}		$Fe(OH)_3$	1.1×10^{-36}
Chlorides	$AgCl$	1.7×10^{-10}		$Al(OH)_3$	2×10^{-33}
	$PbCl_2$	1.6×10^{-5}		$Sn(OH)_2$	5×10^{-26}
	Hg_2Cl_2	2×10^{-18}		$Mn(OH)_2$	1.2×10^{-11}
	$AuCl_3$	3.2×10^{-25}		$Ni(OH)_2$	1.6×10^{-14}
Bromides	$AgBr$	5×10^{-15}		$Cu(OH)_2$	4.8×10^{-20}
	$PbBr_2$	2.1×10^{-6}		$Zn(OH)_2$	4.5×10^{-19}
Iodides	AgI	8.5×10^{-17}	Sulfates	$CaSO_4$	2×10^{-4}
	PbI_2	1.4×10^{-8}		$SrSO_4$	3.2×10^{-7}
Carbonates	$MgCO_3$	3.5×10^{-8}		$BaSO_4$	1.5×10^{-9}
	$CaCO_3$	9×10^{-9}		$PbSO_4$	6.3×10^{-7}
	$SrCO_3$	9.3×10^{-10}		Ag_2SO_4	1.5×10^{-5}
	$BaCO_3$	8.9×10^{-9}	Chromates	$CaCrO_4$	1.0×10^{-4}
	$PbCO_3$	7.4×10^{-14}		$BaCrO_4$	2.4×10^{-10}
Oxalates	CaC_2O_4	2.3×10^{-9}		Ag_2CrO_4	1.9×10^{-12}
	MgC_2O_4	8.6×10^{-5}		$PbCrO_4$	1.8×10^{-14}
	BaC_2O_4	1.2×10^{-7}	Other anions	$AgC_2H_3O_2$	2.3×10^{-3}
	FeC_2O_4	2.1×10^{-7}		$AgCN$	1.6×10^{-14}
	PbC_2O_4	2.7×10^{-11}		$Pb(IO_3)_2$	2.6×10^{-13}

Calculations involving K_{sp} can be divided into three categories:

1. Calculating K_{sp} from solubility data.
2. Calculating solubility from K_{sp}.
3. Problems dealing with precipitation.

We will begin (as you might expect) with the first type.

EXAMPLE 17.1. CALCULATING K_{sp} FROM SOLUBILITY DATA

PROBLEM: It was experimentally determined that at 25 °C the solubility of $BaSO_4$ in water is 0.0091 g/L. What is the value of K_{sp} for barium sulfate?

ANALYSIS: As in earlier calculations of the same type, when we wish to calculate the equilibrium constant from equilibrium data, we must obtain values corresponding to the equilibrium concentrations. These concentrations must be in moles per liter, so our first job will be to calculate a quantity called the **molar solubility,** the solubility of the compound expressed as the number of moles of solute per liter. We will then be able to use this to determine the number of moles of each of the ions per liter, which correspond to the ion concentrations. Once we have these, we substitute them into the ion product to calculate K_{sp}.

SOLUTION: From the solubility we can calculate the number of moles of $BaSO_4$ that are dissolved in 1 L of solution.

$$\left(0.0091 \frac{g}{L}\right) \times \left(\frac{1 \text{ mol}}{233 \text{ g}}\right) = 3.9 \times 10^{-5} \frac{\text{mol}}{L}$$

The solubility equilibrium for $BaSO_4$ is

$$BaSO_4(s) \rightleftharpoons Ba^{2+}(aq) + SO_4^{2-}(aq)$$

so that for every mole of $BaSO_4$ that dissolves, 1 mol of Ba^{2+} and 1 mol of SO_4^{2-} are produced. Therefore, the molar concentrations of Ba^{2+} and SO_4^{2-} in this saturated solution at 25 °C are

$$[Ba^{2+}] = 3.9 \times 10^{-5} \, M$$
$$[SO_4^{2-}] = 3.9 \times 10^{-5} \, M$$

and the K_{sp} is

$$K_{sp} = [Ba^{2+}][SO_4^{2-}]$$
$$= (3.9 \times 10^{-5})(3.9 \times 10^{-5})$$
$$= 1.5 \times 10^{-9}$$

EXAMPLE 17.2. CALCULATING K_{sp} FROM SOLUBILITY DATA

PROBLEM: The molar solubility in water of lead iodate, $Pb(IO_3)_2$, is 4.0×10^{-5} mol/L at 25 °C. What is the value of K_{sp} for this salt?

SOLUTION: First we write the chemical equation and the K_{sp} expression.

$$Pb(IO_3)_2(s) \rightleftharpoons Pb^{2+}(aq) + 2IO_3^-(aq)$$
$$K_{sp} = [Pb^{2+}][IO_3^-]^2$$

When the $Pb(IO_3)_2$ dissolves, we get 1 mol of Pb^{2+} and 2 mol of IO_3^- for each mole of $Pb(IO_3)_2$. Therefore, when 4.0×10^{-5} mol of $Pb(IO_3)_2$ is dissolved in 1 L, we obtain

Remember, the molar solubility is the number of moles of solute per liter of the saturated solution.

$$[Pb^{2+}] = 4.0 \times 10^{-5} \, M$$

$$[IO_3^-] = 2(4.0 \times 10^{-5}) = 8.0 \times 10^{-5} \, M$$

These quantities are now substituted into the K_{sp} expression.

$$K_{sp} = (4.0 \times 10^{-5})(8.0 \times 10^{-5})^2$$
$$= 2.6 \times 10^{-13}$$

Let us now look at how we can determine solubility from a known value of K_{sp}, the second type of problem on our list.

EXAMPLE 17.3. CALCULATING MOLAR SOLUBILITY FROM K_{sp}

PROBLEM: What is the molar solubility of AgCl in water at 25 °C? For AgCl, $K_{sp} = 1.7 \times 10^{-10}$.

ANALYSIS: In working problems of this type, we again set up a concentration table. As we do this, it is helpful to formulate a mental picture of how the final solution is formed. We will imagine that the solid is added to the liquid solvent (pure water, in this case), and once there, that it dissolves to reach equilibrium.

The values in the "initial concentrations" column are the ion concentrations that are already in the solvent. Since the solvent is pure water in this problem, there are no Ag^+ or Cl^- ions present to start with, so the "initial" entries for these ions are both zero. (In later problems, this will not necessarily be the case, however.)

The changes in concentrations are caused by the AgCl dissolving. To obtain these values, we will call the molar solubility of the solute x. Then we use the stoichiometry of the equilibrium to determine the coefficients of x, as described below.

SOLUTION: First, let's write the equilibrium equation

$$AgCl(s) \rightleftharpoons Ag^+(aq) + Cl^-(aq)$$

for which

$$K_{sp} = [Ag^+][Cl^-]$$

As noted above, there are no solutes already in the water that give any Ag^+ or Cl^-, so the concentrations of these ions in the solvent are both zero. These values are entered in the "Initial Molar Concentrations" column.

For each mole of AgCl that dissolves, we obtain 1 mol of Ag^+ and 1 mol of Cl^-. If we let x be the molar solubility, which is the number of moles of AgCl that dissolves per liter, then when the salt dissolves, the concentrations of each of the ions will increase by x mol/L. Both entries in the "Change" column are therefore $+x$. Finally, the values in the "equilibrium concentrations" column are obtained by adding the change to the initial concentration, a rather trivial operation in this particular instance.

	Initial Molar Concentrations	Change	Equilibrium Molar Concentrations
Ag^+	0.0	$+x$	x
Cl^-	0.0	$+x$	x

Substituting the equilibrium quantities and the value of K_{sp} into the K_{sp} expression, we obtain

$$K_{sp} = (x)(x) = 1.7 \times 10^{-10}$$

$$x^2 = 1.7 \times 10^{-10}$$

$$x = 1.3 \times 10^{-5}$$

Therefore, the molar solubility of AgCl in water is 1.3×10^{-5} M.

EXAMPLE 17.4. CALCULATING MOLAR SOLUBILITY FROM K_{sp}

PROBLEM: What are the molar concentrations of Ag^+ and CrO_4^{2-} in a saturated solution of Ag_2CrO_4 at 25 °C? For Ag_2CrO_4, $K_{sp} = 1.9 \times 10^{-12}$.

SOLUTION: Ag_2CrO_4 dissolves in water according to the equation

$$Ag_2CrO_4(s) \rightleftharpoons 2Ag^+(aq) + CrO_4^{2-}(aq)$$

and the K_{sp} expression is

$$K_{sp} = [Ag^+]^2[CrO_4^{2-}]$$

Once again, neither of the ions involved in the equilibrium is present in the solvent before the salt is added, so the initial concentrations are zero. Next, we let x be the molar solubility. Since two Ag^+ and one CrO_4^{2-} are produced for each Ag_2CrO_4 that dissolves, then when x mol of Ag_2CrO_4 dissolves per liter, the Ag^+ concentration increases by $2x$ and the CrO_4^{2-} concentration increases by x. Finally, we add the change to the initial concentration to obtain the equilibrium concentration in the last column.

	Initial Molar Concentrations	Change	Equilibrium Molar Concentrations
Ag^+	0.0	$+2x$	$2x$
CrO_4^{2-}	0.0	$+x$	x

With x defined as the molar solubility, the coefficients of x in the change column are the coefficients of the ions in the chemical equation for the equilibrium.

Remember, $(2x)^2 = 4x^2$

Substituting the value of K_{sp} and solving for x, we have

$$K_{sp} = (2x)^2(x) = 1.9 \times 10^{-12}$$

$$(4x^2)x = 4x^3 = 1.9 \times 10^{-12}$$

$$x^3 = 4.8 \times 10^{-13}$$

and

$$x = 7.8 \times 10^{-5}$$

Therefore,

$$[Ag^+] = 2(7.8 \times 10^{-5}) = 1.6 \times 10^{-4} \; M$$

$$[CrO_4^{2-}] = 7.8 \times 10^{-5} \; M$$

Determining when a precipitate will form in a solution

You should recall from earlier discussions that a saturated solution is one in which the undissolved solute is in dynamic equilibrium with the solution. This is precisely the situation to which we apply K_{sp}. In other words, a saturated solution exists *only* when the **ion product,** *the product of the concentrations of the dissolved ions each raised to its proper power* is exactly equal to K_{sp}. When the ion product is less than K_{sp}, the solution is unsaturated, because more salt would have to dissolve in order to raise the concentrations to the point at which the ion product equals K_{sp}. On the other hand, when the ion product exceeds K_{sp}, a supersaturated solution exists because some of the salt would have to precipitate to lower the concentrations until the ion product is equal to K_{sp} once again.

In a solution, a precipitate will be formed only when the mixture is supersaturated. Therefore, we can use the value of the ion product in a solution to tell us whether or not precipitation will occur. In summary, we find that

Unsaturated: Ion product $< K_{sp}$ $\Big\}$ No precipitate will form.
Saturated: Ion product $= K_{sp}$
Supersaturated: Ion product $> K_{sp}$ Precipitation will occur.

EXAMPLE 17.5. DETERMINING WHETHER A PRECIPITATE WILL FORM IN A SOLUTION

PROBLEM: Will a precipitate of $PbCl_2$ form in a solution having a $Pb(NO_3)_2$ concentration of 0.010 M and an HCl concentration of 0.010 M? For $PbCl_2$, $K_{sp} = 1.6 \times 10^{-5}$.

SOLUTION: For lead chloride, we write the following equation

$$PbCl_2(s) \rightleftharpoons Pb^{2+}(aq) + 2Cl^-(aq)$$

so the ion product that we will use in our test for precipitation is $[Pb^{2+}][Cl^-]^2$. In a 0.010 M $Pb(NO_3)_2$ solution, $[Pb^{2+}] = 0.010$ M and in 0.010 M HCl, $[Cl^-] = 0.010$ M. Using these values we now compute the ion product.

We can ignore the H^+ and NO_3^- because they are simply spectator ions here.

$$[Pb^{2+}][Cl^-]^2 = (0.010)(0.010)^2 = 1.0 \times 10^{-6}$$

Since this is less than the value of K_{sp} (1.6×10^{-5}), we conclude that no precipitate will form.

EXAMPLE 17.6. DETERMINING WHETHER A PRECIPITATE WILL FORM WHEN TWO SOLUTIONS ARE MIXED

PROBLEM: Will a precipitate of $PbSO_4$ form when exactly 100 mL of a 0.0030 M $Pb(NO_3)_2$ solution is mixed with exactly 400 mL of 0.040 M Na_2SO_4? For $PbSO_4$, $K_{sp} = 6.3 \times 10^{-7}$.

ANALYSIS: For lead sulfate, the ion product that we must examine is

$$[Pb^{2+}][SO_4^{2-}]$$

but this time we have to take into account that one solution dilutes the other when they are mixed. We approach this by imagining that the solutions can be combined before any reaction can take place, and then we look at the value of the ion product in the final mixture.

SOLUTION: The first step is to calculate the concentrations of Pb^{2+} and SO_4^{2-} in our total volume of 500 mL. The original 100 mL of the 0.0030 M $Pb(NO_3)_2$ solution contains

$$(0.100 \ L) \times \left(0.0030 \ \frac{mol}{L}\right) = 0.00030 \ \text{mol of } Pb^{2+}$$

(This solution also contains 0.00060 mol of NO_3^-, but this species is unimportant in this calculation — it is a spectator ion.)

The 400 mL of Na_2SO_4 contains

$$(0.400 \ L) \times \left(0.040 \ \frac{mol}{L}\right) = 0.016 \ \text{mol of } SO_4^{2-}$$

If necessary, review dilution calculations in Section 3.6 on page 72.

(This solution also contains 0.032 mol of Na^+, but it also is unimportant in this problem.)

The molar concentration of the Pb^{2+} in the 500 mL is then

$$[Pb^{2+}] = \frac{0.00030 \text{ mol}}{0.500 \text{ L}} = 0.00060\ M = 6.0 \times 10^{-4}\ M$$

and the molar concentration of SO_4^{2-} in the 500 mL is

$$[SO_4^{2-}] = \frac{0.016 \text{ mol}}{0.500 \text{ L}} = 0.032\ M = 3.2 \times 10^{-2}\ M$$

The ion product in the final solution is therefore

$$[Pb^{2+}][SO_4^{2-}] = (6.0 \times 10^{-4})(3.2 \times 10^{-2}) = 1.9 \times 10^{-5}$$

When we compare the value of the ion product to the value of K_{sp} for $PbSO_4$ ($K_{sp} = 6.3 \times 10^{-7}$), we find that the ion product is *larger* than K_{sp} and therefore a precipitate will form.

17.2 THE COMMON ION EFFECT AND SOLUBILITY

When a salt is dissolved in a solution that already contains one of its ions, its solubility is less than in pure water. Silver chloride, for example, is less soluble in a solution of NaCl than it is in pure water. In this case both solutes have an ion in common: chloride ion. The reduction in the solubility in the presence of a **common ion** is called the **common ion effect**.

The effect of a common ion on solubility is really just an example of Le Châtelier's principle in action. Suppose that solid silver chloride is placed into pure water and allowed to come to equilibrium with its ions in solution.

$$AgCl(s) \rightleftharpoons Ag^+(aq) + Cl^-(aq)$$

If a soluble chloride salt such as NaCl is now added to this solution, the chloride ion concentration will increase and drive this equilibrium to the left, thereby causing some AgCl to precipitate. In other words, AgCl is less soluble in aqueous NaCl than in pure water. Let's look at a few examples.

EXAMPLE 17.7. CALCULATING THE SOLUBILITY OF A SALT IN A SOLUTION THAT CONTAINS A COMMON ION

PROBLEM: What is the molar solubility of AgCl in a 0.010 M solution of NaCl? For AgCl, $K_{sp} = 1.7 \times 10^{-10}$.

ANALYSIS: Once again, we examine the solvent *before* the addition of the AgCl to see what ions are already present. In this case the solution contains 0.010 M NaCl, so the initial $[Cl^-] = 0.010\ M$. Then we imagine adding the AgCl. Some of it dissolves, increasing the Cl^- concentration slightly, and also producing some Ag^+ in the solution.

NaCl is soluble and completely dissociated. Don't attempt to write an equilibrium equation for it.

SOLUTION: For silver chloride we have

$$K_{sp} = [Ag^+][Cl^-] = 1.7 \times 10^{-10}$$

Before any AgCl dissolves, the initial Cl^- concentration is 0.010 M. We can ignore the Na^+ because it is not involved in the equilibrium. We now let x equal the number of moles per liter of AgCl that dissolve. This increases both $[Cl^-]$ and $[Ag^+]$ by x. Thus,

	Initial Molar Concentrations	Change	Equilibrium Molar Concentrations
Ag^+	0.0	$+x$	x
Cl^-	0.010	$+x$	$0.010 + x \approx 0.010$

If very little AgCl dissolves, the equilibrium Cl^- concentration will be very close to the initial Cl^- concentration, and that is what we are assuming.

Note that we have assumed that we can neglect x in computing the equilibrium Cl^- concentration. We make this assumption because the value of K_{sp} is very small, and we expect that the amount of AgCl that will dissolve will be very small. By doing this, we greatly simplify the algebra. Substituting the equilibrium concentrations into the expression for K_{sp} gives

$$(x)(0.010) = 1.7 \times 10^{-10}$$

or

$$x = 1.7 \times 10^{-8} M$$

Thus, the molar solubility of AgCl is $1.7 \times 10^{-8} M$. We might compare this to the molar solubility of AgCl in pure water, which we found to be $1.3 \times 10^{-5} M$ in Example 17.3. The solubility of AgCl is indeed much less in a solution containing a common ion.

Notice that 1.7×10^{-8} is negligible compared to 0.010, which justifies our approximation.

EXAMPLE 17.8. CALCULATING THE SOLUBILITY OF A COMPOUND IN A SOLUTION THAT CONTAINS A COMMON ION

PROBLEM: What is the molar solubility of $Mg(OH)_2$ in 0.10 M NaOH? For $Mg(OH)_2$, $K_{sp} = 7.1 \times 10^{-12}$.

ANALYSIS: The solvent is 0.10 M NaOH, which contains 0.10 M OH^-. Into this we place the $Mg(OH)_2$. When some of it dissolves, the OH^- concentration increases slightly and some Mg^{2+} also enters the solution. The relative amounts of Mg^{2+} and OH^- added per liter are determined by the formula of the $Mg(OH)_2$; for each mole of Mg^{2+}, we obtain 2 mol of OH^-. Therefore, the increase in the concentration of OH^- is twice the increase in the Mg^{2+} concentration. This is the thinking we need to do to obtain the entries in the concentration table, and then we proceed as usual.

SOLUTION: The K_{sp} expression for $Mg(OH)_2$ is

$$K_{sp} = [Mg^{2+}][OH^-]^2$$

As noted in the analysis, the solution contains 0.10 M NaOH even before the addition of the magnesium hydroxide. Since NaOH is a strong electrolyte, the initial OH^- concentration is 0.10 M. This value is entered in the first column. There are no sources of Mg^{2+} in the solvent, so this initial value is set to zero.

When the $Mg(OH)_2$ dissolves, the concentrations change. If x mol/L of $Mg(OH)_2$ dissolves, then the $[Mg^{2+}]$ increases by x and $[OH^-]$ increases by $2x$. These are entered in the center column, and the equilibrium concentrations are obtained by adding the quantities in the first two columns (as usual).

	Initial Molar Concentrations	Change	Equilibrium Molar Concentrations
Mg^{2+}	0.0	$+x$	x
OH^-	0.10	$+2x$	$0.10 + 2x \approx 0.10$

Again, we simplify the algebra by assuming that $2x$ is negligible compared to 0.10. Substituting the equilibrium concentrations and the K_{sp} for $Mg(OH)_2$ into the solubility product expression gives

$$(x)(0.10)^2 = 7.1 \times 10^{-12}$$

or

$$x = \frac{7.1 \times 10^{-12}}{(0.10)^2}$$

$$= 7.1 \times 10^{-10}$$

Note that $2x$, which equals 1.4×10^{-9}, is negligible compared to 0.10.

Thus, 7.1×10^{-10} mol/L of $Mg(OH)_2$ dissolves in a $0.10\ M$ solution of NaOH.

EXAMPLE 17.9. CALCULATING THE SOLUBILITY OF A SALT IN A SOLUTION THAT CONTAINS A COMMON ION

PROBLEM: What is the molar solubility of PbI_2 in $0.10\ M$ $Pb(NO_3)_2$ solution? For lead iodide, $K_{sp} = 1.4 \times 10^{-8}$.

SOLUTION: The K_{sp} expression is

$$K_{sp} = [Pb^{2+}][I^-]^2$$

To construct the concentration table, we begin by asking, "What are the initial concentrations of Pb^{2+} and I^-?" The solution initially contains $0.10\ M$ $Pb(NO_3)_2$, so the initial Pb^{2+} concentration is $0.10\ M$. No iodide is present initially, so the initial I^- concentration is $0.0\ M$. These values go in the first column.

Next, we let x be the molar solubility of PbI_2. When x mol/L of PbI_2 dissolves, $[Pb^{2+}]$ increases by x and $[I^-]$ increases by $2x$. These quantities are entered in the change column. Then the first and second columns are added to give the equilibrium concentrations.

	Initial Molar Concentrations	Change	Equilibrium Molar Concentrations
Pb^{2+}	0.10	$+x$	$0.10 + x \approx 0.10$
I^-	0.0	$+2x$	$2x$

After we make our usual simplification (expecting the solubility to be very small), the equilibrium quantities are substituted into the K_{sp} expression.

$$K_{sp} = (0.10)(2x)^2 = 1.4 \times 10^{-8}$$

$$4x^2 = 1.4 \times 10^{-7}$$

$$x^2 = 3.5 \times 10^{-8}$$

The simplification was justified because 1.9×10^{-4} is negligible compared to 0.10.

$$x = 1.9 \times 10^{-4}$$

The molar solubility of PbI_2 in $0.10\ M$ $Pb(NO_3)_2$ is $1.9 \times 10^{-4}\ M$.

17.3 SEPARATION OF IONS BY SELECTIVE PRECIPITATION

In a chemical analysis, often we are satisfied just to know what substances are present in a sample. Since no numbers are involved, determining the composition in this way is called **qualitative analysis.** One type of qualitative analysis involves determining which metal ions are present in a solution and is accomplished by performing a series chemical tests that are able to confirm either the presence or absence of each of the ions. Usually, to be effective, these tests require that the ions be separated from one another so that one ion doesn't interfere with the test for another.

From the solubility rules presented in Chapter 5, we know that it is possible to separate certain ions from each other when they are present together in a solution. For instance, the addition of chloride ion to a solution containing both Mg^{2+} and Ag^+ yields a precipitate of AgCl but not $MgCl_2$. If enough Cl^- is added, virtually all the silver ion is precipitated as AgCl, leaving the Mg^{2+} still in solution. Filtration separates the solid from the solution, so we have effectively separated the silver from the magnesium.

The separation just described is accomplished because one chloride salt is insoluble (AgCl) and the other ($MgCl_2$) is soluble. But even when both products are "insoluble," it is still frequently possible to achieve some degree of separation. Consider, for example, the salts $CaSO_4$ and $BaSO_4$. Although both have very low solubilities, as evidenced by their respective K_{sp}'s, we can compute that $CaSO_4$ is about 400 times more soluble, on a mole basis, than $BaSO_4$. As a result, if we had a solution containing equal concentrations of Ca^{2+} and Ba^{2+}, we would find that as the SO_4^{2-} concentration is increased slowly, the less soluble $BaSO_4$ would precipitate first. Conceivably, one could separate Ca^{2+} from Ba^{2+} by appropriately adjusting the SO_4^{2-} concentration so that the Ca^{2+} would remain in solution while nearly all the Ba^{2+} would be removed as $BaSO_4$.

Accomplishing a separation like that described above requires precise control of the concentration of the precipitating agent (e.g., sulfate ion). If any of the equilibria involved in the reactions relate in any way to hydrogen ion or hydroxide ion, this can usually be done by controlling the pH of the solution.

One of the simplest examples is the separation of ions according to the solubilities of their hydroxides. Consider, for example, the hydroxides of calcium and magnesium.

$$Ca(OH)_2 \qquad K_{sp} = 6.5 \times 10^{-6}$$

$$Mg(OH)_2 \qquad K_{sp} = 7.1 \times 10^{-12}$$

> For salts with the same general formula, it is safe to say that the salt with the smaller value for K_{sp} will have the lower solubility in water.

Suppose we had a solution in which the concentrations of Ca^{2+} and Mg^{2+} were both 0.10 M. As we increase the OH^- concentration, we expect the less soluble $Mg(OH)_2$ to begin to precipitate first, and if we are careful, we should be able to adjust the $[OH^-]$ so that *only* the $Mg(OH)_2$ precipitates. What range of $[OH^-]$ values will accomplish this? Let's calculate the $[OH^-]$ that would give a saturated solution for each of these hydroxides.

For calcium,

$$K_{sp} = [Ca^{2+}][OH^-]^2 = 6.5 \times 10^{-6}$$

Substituting 0.10 M for $[Ca^{2+}]$ and solving for $[OH^-]$ gives

$$[OH^-]^2 = \frac{6.5 \times 10^{-6}}{0.10} = 6.5 \times 10^{-5}$$

$$[OH^-] = 8.1 \times 10^{-3}\ M$$

This corresponds to a solution with a pH = 11.91. If the pH is higher than this, meaning a hydroxide ion concentration larger than 8.1×10^{-3} M, then $Ca(OH)_2$ will precipitate. If we want to avoid the precipitation of $Ca(OH)_2$, then this is an upper limit on the pH and the hydroxide ion concentration.

For magnesium,

$$K_{sp} = [Mg^{2+}][OH^-]^2 = 7.1 \times 10^{-12}$$

Substituting 0.10 M for $[Mg^{2+}]$ and solving for $[OH^-]$ yields

$$[OH^-]^2 = \frac{7.1 \times 10^{-12}}{0.10} = 7.1 \times 10^{-11}$$

$$[OH^-] = 8.4 \times 10^{-6}\ M$$

This corresponds to a pH of 8.92. For this solution a pH higher than 8.92 or an OH^- concentration greater than 8.4×10^{-6} M, will cause a precipitate of $Mg(OH)_2$ to form.

Looking back at the results of these calculations, we realize that if we keep the pH higher than 8.92 but not higher than 11.91 (i.e., make the $[OH^-]$ larger than 8.4×10^{-6} M, but not larger than 8.1×10^{-3} M), we will obtain a precipitate of $Mg(OH)_2$ but not a precipitate of $Ca(OH)_2$, and we will achieve separation of the ions. The maximum separation will be obtained if we keep the $[OH^-]$ just below 8.1×10^{-3} M, because this will suppress the solubility of the $Mg(OH)_2$ to a maximum degree by the common ion effect. Controlling the pH is the job of a buffer, so to separate Ca^{2+} from Mg^{2+}, we would adjust the pH of the solution to some value between 8.92 and 11.91 with some appropriate buffer, perhaps with an ammonia–ammonium ion buffer.

Selective precipitation of salts

In selectively precipitating metal hydroxides by the control of pH, we are directly controlling the concentration of the precipitating ion, OH^-. For other salts, the control of pH influences the availability of the precipitating ion *indirectly*. An example is the separation of metal carbonates according to their solubilities. Consider these:

$$MgCO_3 \qquad K_{sp} = 3.5 \times 10^{-8}$$

$$SrCO_3 \qquad K_{sp} = 9.3 \times 10^{-10}$$

For the less soluble $SrCO_3$ to precipitate selectively, we need to control the CO_3^{2-} concentration. This can be done through the control of the pH because hydrogen ion is a member of the equilibria formed by H_2CO_3, HCO_3^-, and CO_3^{2-}.

$$H_2CO_3 \rightleftharpoons H^+ + HCO_3^- \qquad K_{a_1} = 4.3 \times 10^{-7}$$

$$HCO_3^- \rightleftharpoons H^+ + CO_3^{2-} \qquad K_{a_2} = 5.6 \times 10^{-11}$$

By applying Le Châtelier's principle, we can see how control is possible. If the H^+ concentration is increased, both equilibria are shifted to the left, and this causes the CO_3^{2-} concentration to decrease. Conversely, decreasing the H^+ concentration causes the reactions to shift to the right, and this increases the CO_3^{2-} concentration. Thus, by varying the pH, we indirectly alter the concentration of the CO_3^{2-} ion.

In dealing with this situation quantitatively, it is often useful to combine the two equilibria above to give an overall equation relating the concentrations of carbonic acid, hydrogen ion, and carbonate ion. Adding the two equations gives the following:

$$\begin{array}{r} H_2CO_3 \rightleftharpoons H^+ + HCO_3^- \\ HCO_3^- \rightleftharpoons H^+ + CO_3^{2-} \\ \hline H_2CO_3 \rightleftharpoons 2H^+ + CO_3^{2-} \end{array} \qquad (17.1)$$

In Chapter 15, you learned that when we add two equilibria to obtain a third, the K for the third is the product of the K's of the two equilibria that were combined. The equilibrium constant for Equation 17.1, which we will call K_a, is therefore

It is safe to use Equation 17.2 *only* if two of the three concentrations are known.

$$K_a = K_{a_1} \times K_{a_2} = \frac{[H^+]^2[CO_3^{2-}]}{[H_2CO_3]} \qquad (17.2)$$

$$= (4.3 \times 10^{-7}) \times (5.6 \times 10^{-11})$$

$$= 2.4 \times 10^{-17}$$

Example 17.10 illustrates how this could be applied.

EXAMPLE 17.10. SEPARATING IONS BY SELECTIVE PRECIPITATION

PROBLEM: A solution contains $0.10\ M\ Mg^{2+}$ and $0.10\ M\ Sr^{2+}$. The concentration of H_2CO_3 is adjusted to a value of $0.050\ M$. What range of pH values would permit the separation of these metal ions as their carbonates?

SOLUTION: As in our illustration of the separation of metal hydroxides, let's first determine the limits on the carbonate ion concentration. The K_{sp} expressions for the salts are

$$MgCO_3 \qquad K_{sp} = [Mg^{2+}][CO_3^{2-}] = 3.5 \times 10^{-8}$$

$$SrCO_3 \qquad K_{sp} = [Sr^{2+}][CO_3^{2-}] = 9.3 \times 10^{-10}$$

For the more soluble magnesium carbonate, the $[CO_3^{2-}]$ needed to give a saturated solution is

$$[CO_3^{2-}] = \frac{K_{sp}}{[Mg^{2+}]} = \frac{3.5 \times 10^{-8}}{0.10}$$

$$= 3.5 \times 10^{-7}\ M$$

We must not let the carbonate concentration become larger than this, otherwise, the $SrCO_3$ precipitate will be contaminated with $MgCO_3$.

For strontium carbonate, the $[CO_3^{2-}]$ needed for saturation is

$$[CO_3^{2-}] = \frac{K_{sp}}{[Sr^{2+}]} = \frac{9.3 \times 10^{-10}}{0.10}$$

$$= 9.3 \times 10^{-9}\ M$$

The carbonate concentration must be larger than this to obtain a precipitate of $SrCO_3$.

Now let's determine the H^+ concentrations that will give us the limiting CO_3^{2-} concentrations. If we solve Equation 17.2 for the square of the H^+ concentration, we get

$$[H^+]^2 = K_a \times \frac{[H_2CO_3]}{[CO_3^{2-}]}$$

The $[H_2CO_3]$ concentration was specified as $0.050\ M$ and we determined the value of K_a to be 2.4×10^{-17}. Therefore, for the limiting $[CO_3^{2-}]$ for magnesium carbonate, we have

$$[H^+]^2 = 2.4 \times 10^{-17} \times \frac{0.050}{3.5 \times 10^{-7}}$$

$$= 3.4 \times 10^{-12}$$

Taking the square root, we get

$$[H^+] = 1.9 \times 10^{-6}\ M$$

which corresponds to a pH of 5.72. To prevent the precipitation of $MgCO_3$, the pH cannot be higher than 5.72.

For the limiting $[CO_3^{2-}]$ for the strontium carbonate, we have

$$[H^+]^2 = 2.4 \times 10^{-17} \times \frac{0.050}{9.3 \times 10^{-9}}$$

$$= 1.3 \times 10^{-10}$$

Therefore,

$$[H^+] = 1.1 \times 10^{-5}\ M$$

which corresponds to a pH of 4.96. If the pH is larger than 4.96, $SrCO_3$ will precipitate.

We can now summarize our findings. If we keep the pH higher than 4.96, but not higher than 5.72, $SrCO_3$ will precipitate but $MgCO_3$ will not.

In qualitative analysis, pH control is used to separate the sulfides of metal ions. The sulfide ions are formed from hydrogen sulfide, H_2S, which is added to a solution containing the metal ions. The chemistry here is a bit more complex, as are the equilibrium expressions, because H_2S is not really a diprotic acid in water. Precipitation actually appears to occur by reaction of the HS^- ion with the metal ion. For example, with Cu^{2+} ion, the reaction is

$$Cu^{2+}(aq) + HS^-(aq) \rightleftharpoons CuS(s) + H^+(aq) \qquad (17.3)$$

Because there is the equilibrium

$$H_2S(aq) \rightleftharpoons H^+(aq) + HS^-(aq)$$

the pH of the solution controls the amount of HS^- available, so different metal ions can be separated according to the values of the equilibrium constants for reactions analogous to that in Equation 17.3. The principles are the same, but because the equilibria are more complex, we will not explore them further.

17.4 COMPLEX ION EQUILIBRIA

Complex ions were discussed briefly in Chapter 10. Recall that they are formed when neutral molecules or anions become bonded to a metal ion. Examples are the complexes formed by copper(II) ion with ammonia and cyanide ion: $Cu(NH_3)_4{}^{2+}$ and $Cu(CN)_4{}^{2-}$. Similar complexes are formed by silver ion, but with a different number of ligands attached to the metal ion. For silver, the complexes are $Ag(NH_3)_2{}^+$ and $Ag(CN)_2{}^-$.

Complexes of any given metal ion vary in their stabilities, as shown by their relative tendencies to dissociate. Such dissociation reactions are equilibria. For example, the overall reaction for the dissociation of the $Ag(NH_3)_2{}^+$ complex is

The formation of a complex actually occurs in a series of steps, each of which is an equilibrium, but we will study only the equilibrium constant for the net overall reaction.

$$Ag(NH_3)_2{}^+(aq) \rightleftharpoons Ag^+(aq) + 2NH_3(aq)$$

The equilibrium constant for this reaction is called an **instability constant,** K_{inst}. This is because the larger the value of K_{inst}, the less stable the complex is, as reflected by its tendency to dissociate. For the $Ag(NH_3)_2{}^+$ ion the equilibrium expression is

$$K_{inst} = \frac{[Ag^+][NH_3]^2}{[Ag(NH_3)_2{}^+]}$$

The value of the instability constant for this complex has been found to be 6.0×10^{-8}. We can see by the size of this constant that this particular complex is quite stable and will readily form whenever Ag^+ and NH_3 are added to the same solution. Some examples of complex ions and their instability constants are given in Table 17.2.

An alternative way of writing the equilibrium for a complex ion is as an equation representing its formation. For example,

$$Ag^+(aq) + 2NH_3(aq) \rightleftharpoons Ag(NH_3)_2{}^+(aq)$$

The equilibrium expression, of course, is simply the reciprocal of the K_{inst} expression. In this case the equilibrium constant (which equals the reciprocal of K_{inst}) is called a **formation constant,** K_{form}, or **stability constant.** These are also given in Table 17.2.

$$K_{form} = \frac{[Ag(NH_3)_2{}^+]}{[Ag^+][NH_3]^2}$$

$$K_{form} = \frac{1}{K_{inst}}$$

Table 17.2. Instability constants and formation constants for some complexes at 25 °C

Complex Ion	K_{inst}	K_{form}
AlF_6^{3-}	1.5×10^{-20}	6.7×10^{19}
$Cd(CN)_4^{2-}$	1.3×10^{-17}	7.7×10^{16}
$Co(NH_3)_6^{2+}$	1.3×10^{-5}	7.7×10^4
$Co(NH_3)_6^{3+}$	2.0×10^{-34}	5.0×10^{33}
$Cu(NH_3)_4^{2+}$	2.1×10^{-13}	4.8×10^{12}
$Cu(CN)_2^{-}$	1.0×10^{-16}	1.0×10^{16}
$Fe(CN)_6^{4-}$	1.0×10^{-35}	1.0×10^{35}
$Fe(CN)_6^{3-}$	1.1×10^{-42}	9.1×10^{41}
$Ni(NH_3)_4^{2+}$	1.1×10^{-8}	9.1×10^7
$Ni(NH_3)_6^{2+}$	2.0×10^{-9}	5.0×10^0
$Ag(NH_3)_2^{+}$	6.0×10^{-8}	1.7×10^7
$Ag(CN)_2^{-}$	1.9×10^{-19}	5.3×10^{18}
$Zn(OH)_4^{2-}$	3.6×10^{-16}	2.8×10^{15}

In the chemical literature, the equilibrium constants for complex ions are sometimes tabulated as instability constants and at other times as formation constants or stability constants. You should know the difference between them.

17.5 COMPLEX IONS AND SOLUBILITY

Suppose ammonia is added to a saturated solution of silver chloride in which the solid AgCl is in equilibrium with its ions.

$$AgCl(s) \rightleftharpoons Ag^+(aq) + Cl^-(aq) \tag{17.4}$$

Because NH_3 forms a complex ion with the Ag^+, a second equilibrium is established.

$$Ag^+(aq) + 2NH_3(aq) \rightleftharpoons Ag(NH_3)_2^+(aq) \tag{17.5}$$

As the concentration of NH_3 is increased, the equilibrium in Equation 17.5 is shifted to the right, which removes Ag^+ ions from the solution. However, Ag^+ is also involved in the equilibrium in Equation 17.4, and as Ag^+ ions are removed by the second reaction, AgCl dissolves in an attempt to replace them; Equation 17.4 is also shifted to the right. The result, as we see, is that silver chloride is more soluble in a solution that contains ammonia than it is in pure water.

The phenomenon that we just examined for AgCl and NH_3 also applies in a general way. *When a complex is formed by the metal ion of an insoluble salt, the solubility of the salt is increased.* This has some useful and practical applications. For example, in Chapter 10 you learned that EDTA is added to shampoos to prevent ions such as Ca^{2+} from interfering with the action of the soap. It does this by forming a complex with the Ca^{2+} ions, which prevents the precipitation of the calcium salt of the soap. In terms of the discussion just presented, we can say that the ability of Ca^{2+} to form a complex with EDTA increases the solubility of the calcium–soap precipitate to the extent that the appropriate ion product does not exceed the K_{sp} for the calcium–soap compound, and precipitation is prevented.

Another application of this phenomenon is in the processing of photographic film. The light-sensitive emulsion on the film contains silver bromide, AgBr, which is quite insoluble in water. When the film is exposed to light, some of the AgBr crystals are sensitized and become especially easy to reduce to metallic silver, which is accomplished by a chemical called a "developer." After development, however, unexposed grains of AgBr must be removed from the film and this is done by treating the film with a solution called the "fixer," which contains sodium thiosulfate, $Na_2S_2O_3$. Thiosulfate ion forms a very stable complex with silver ion, so in the presence of this ion the following pair of simultaneous equilibria is established:

$$AgBr(s) \rightleftharpoons Ag^+(aq) + Br^-(aq)$$

$$Ag^+(aq) + 2S_2O_3^{2-}(aq) \rightleftharpoons Ag(S_2O_3)_2^{3-}(aq)$$

As the $S_2O_3^{2-}$ forms the complex with the silver ion, the second equilibrium is shifted to the right. This removes Ag^+ from the solution and causes the first equilibrium to be also shifted to the right, and the AgBr dissolves.

The effect of complex formation on solubility can be treated quantitatively, provided the appropriate K_{sp} and K_{form} values are known. For example, the effect of NH_3 on the solubility of AgCl can be analyzed by combining Equations 17.4 and 17.5 by addition.

$$\begin{array}{r} AgCl(s) \rightleftharpoons Ag^+ + Cl^- \\ Ag^+ + 2NH_3 \rightleftharpoons Ag(NH_3)_2^+ \\ \hline AgCl(s) + 2NH_3 \rightleftharpoons Ag(NH_3)_2^+ + Cl^- \end{array}$$

The equilibrium constant for this overall reaction is

$$K_c = \frac{[Ag(NH_3)_2^+][Cl^-]}{[NH_3]^2}$$

We can obtain this same expression by multiplying the K_{sp} of AgCl by the K_{form} of the complex ion. Thus,

$$K_{sp} \times K_{form} = [Ag^+][Cl^-] \times \frac{[Ag(NH_3)_2^+]}{[Ag^+][NH_3]^2} = K_c$$

Recall that when we add equilibria, we multiply their equilibrium constants.

Therefore, with a knowledge of K_{sp} of the salt, K_{form} of the complex ion (or K_{inst}), and the concentration of NH_3, it is possible to calculate the concentrations of Ag^+ and Cl^- present at equilibrium and thus determine the solubility of AgCl in NH_3, as shown by the next example.

A *complexing agent* is a substance that can form a complex with a metal ion (e.g., NH_3).

EXAMPLE 17.11. CALCULATING THE SOLUBILITY OF A SALT IN A SOLUTION THAT CONTAINS A COMPLEXING AGENT

PROBLEM: What is the molar solubility of AgCl in 1.0 M NH_3 at 25 °C?

SOLUTION: As we have seen, the overall equilibrium equation for this problem is

$$AgCl(s) + 2NH_3(aq) \rightleftharpoons Ag(NH_3)_2^+(aq) + Cl^-(aq)$$

for which we write

$$K_c = \frac{[Ag(NH_3)_2^+][Cl^-]}{[NH_3]^2}$$

where

$$K_c = K_{sp} \times K_{form} = (1.7 \times 10^{-10}) \times (1.7 \times 10^7) = 2.9 \times 10^{-3}$$

If we let x equal the number of moles per liter of AgCl that dissolves, then we have the following initial and equilibrium concentrations.

	Initial Molar Concentrations	Change	Equilibrium Molar Concentrations
NH_3	1.0	$-2x$	$(1.0 - 2x)$
$Ag(NH_3)_2^+$	0.0	$+x$	x
Cl^-	0.0	$+x$	x

Before AgCl begins to dissolve, $[NH_3] = 1.0\ M$ and there is no $Ag(NH_3)_2^+$ or Cl^- in the solution. This is how we obtain the initial concentrations.

Substituting the concentrations at equilibrium into the K_c equation, we have

$$K_c = \frac{(x)(x)}{(1.0 - 2x)^2} = \frac{x^2}{(1.0 - 2x)^2} = 2.9 \times 10^{-3}$$

Since the left side of the equation is a perfect square, we need make no approximations. Simply taking the square root of both sides gives

$$\frac{x}{1.0 - 2x} = 5.4 \times 10^{-2}$$

from which we obtain

$$x = 0.049$$

Therefore, we find that 0.049 mol of AgCl will dissolve in 1.00 L of 1.0 M NH_3.

In Example 17.11 we assumed that when the AgCl dissolves in the ammonia solution, essentially all the Ag^+ becomes complexed by NH_3. In other words, we said that the chloride ion concentration was equal to the concentration of $Ag(NH_3)_2^+$. Note that this assumption is valid only if K_{form} is very large indicating that the complex is very stable.

EXAMPLE 17.12. DISSOLVING AN INSOLUBLE COMPOUND BY FORMING A COMPLEX ION

PROBLEM: How many moles of solid NaOH must be added to 1.00 L of H_2O in order to dissolve 0.10 mol of $Zn(OH)_2$ according to the reaction,

$$Zn(OH)_2(s) + 2OH^- \rightleftharpoons Zn(OH)_4^{2-}$$

$Zn(OH)_2$ is amphoteric and dissolves in base as well as acid.

SOLUTION: The two equilibria involved in this system are

$$Zn(OH)_2(s) \rightleftharpoons Zn^{2+} + 2OH^- \qquad K_{sp} = 4.5 \times 10^{-17}$$

$$Zn^{2+} + 4OH^- \rightleftharpoons Zn(OH)_4^{2-} \qquad K_{form} = 2.8 \times 10^{15}$$

As before, the overall reaction can be written as the sum of these two equilibria

$$Zn(OH)_2(s) + 2OH^- \rightleftharpoons Zn(OH)_4^{2-}$$

for which

$$K_c = K_{sp} \times K_{form} = (4.5 \times 10^{-17})(2.8 \times 10^{15})$$
$$= 1.3 \times 10^{-1}$$

Therefore,

$$\frac{[Zn(OH)_4^{2-}]}{[OH^-]^2} = 1.3 \times 10^{-1}$$

In this problem we know that 0.10 mol of zinc goes into solution where it is present as either free Zn^{2+} or $Zn(OH)_4{}^{2-}$. Because K_{form} is so very large, essentially all the zinc will be present as the complex ion; therefore, we can write

$$[Zn(OH)_4{}^{2-}] = 0.10\ M$$

Substituting this into the equilibrium expression gives

$$1.3 \times 10^{-1} = \frac{0.10}{[OH^-]^2}$$

Therefore,

$$[OH^-]^2 = \frac{0.10}{1.3 \times 10^{-1}} = 7.7 \times 10^{-1}$$

and

$$[OH^-] = 0.88\ M$$

This corresponds to the equilibrium concentration of free OH^-. In this solution, however, we also have 0.10 mol of $Zn(OH)_4{}^{2-}$, which contains an additional 0.40 mol of OH^-, 0.20 mol of which was contained in the original 0.10 mol of $Zn(OH)_2$ that dissolved. Therefore, the total number of moles of NaOH that must be *added* to the water is $0.88 + 0.20 = 1.08$ mol.

REVIEW QUESTIONS AND PROBLEMS

Problems whose numbers are in blue have their answers in Appendix D at the back of the book. The more difficult problems are marked by asterisks.

Solubility Product, General

17.1 Why is the concentration of the solid left out of the equilibrium expression for the solubility of a salt?

17.2 Write the K_{sp} expression for each of the following substances:
(a) Ag_2SO_3
(b) CaF_2
(c) $Fe(OH)_3$
(d) MgC_2O_4
(e) $AuCl_3$
(f) $BaCO_3$

17.3 Write the K_{sp} expression for these salts:
(a) PbF_2
(b) Ag_3PO_4
(c) $Fe_3(PO_4)_2$
(d) Li_2CO_3
(e) $Ca(IO_3)_2$
(f) $Ag_2Cr_2O_7$

Calculating K_{sp} from Solubility

17.4 The molar solubility of CuCl in water is 1.0×10^{-3} mol/L. What is its value of K_{sp}?

17.5 The molar solubility of $PbCO_3$ in water is 1.8×10^{-7} mol/L. What is K_{sp} for $PbCO_3$?

17.6 The solubility of barium oxalate, BaC_2O_4, in water is 0.0781 g/L. Calculate K_{sp} for BaC_2O_4.

17.7 The molar solubility of $CaCrO_4$ in water is 1.0×10^{-2} mol/L. What is K_{sp} for $CaCrO_4$?

17.8 The molar solubility of lead iodide, PbI_2, in water is 1.5×10^{-3} mol/L. Calculate its K_{sp}.

17.9 A student determined that 0.0981 g of PbF_2 was dissolved in 200 mL of saturated PbF_2 solution. What is K_{sp} for PbF_2?

17.10 The solubility of MgF_2 in water is 7.6×10^{-2} g/L. Calculate K_{sp} for this salt.

17.11 The solubility of Ag_2SO_4 in water is 4.99 g/L. What is K_{sp} for Ag_2SO_4?

17.12 The pH of a saturated solution of $Ni(OH)_2$ is 9.50. Calculate K_{sp} for $Ni(OH)_2$.

*__17.13__ A 500-mL portion of 0.0020 M $Na_2C_2O_4$ (sodium oxalate) solution is able to dissolve 0.47 g of MgC_2O_4. What is K_{sp} for MgC_2O_4?

Calculating Solubility from K_{sp}

17.14 Using the data in Table 17.1, calculate the molar solubility in water of each of the following:
(a) $AuCl_3$
(b) $Fe(OH)_2$
(c) $BaSO_4$
(d) Hg_2Cl_2 (which yields $Hg_2{}^{2+}$ and $2Cl^-$)
(e) CaF_2
(f) MgC_2O_4

17.15 Milk of magnesia is a suspension of solid $Mg(OH)_2$ in water. Calculate the pH of the aqueous phase, assuming that it is composed of pure water saturated with $Mg(OH)_2(s)$.

17.16 How many grams of $CaSO_4$ will dissolve in 600 mL of water?

*17.17 Plaster is composed of $CaSO_4$. Suppose that there is a leak above a ceiling through which water is seeping at the rate of 2.0 L/day. If the plaster in the ceiling is 1.50 cm thick, how long will it take to dissolve a circular hole 1 cm in diameter? Assume that the density of the plaster is 0.97 g/mL.

*17.18 Calculate the molar solubility of $Al(OH)_3$ in water.

K_{sp} and Precipitation

17.19 What condition must be met to have a precipitate form in a solution?

17.20 Would a precipitate form in the following solutions?
(a) 5.0×10^{-2} mol of $AgNO_3$ and 1.0×10^{-3} mol of $NaC_2H_3O_2$ dissolved in 1.0 L of solution
(b) 1.0×10^{-2} mol of $Ba(NO_3)_2$ and 2.0×10^{-2} mol of NaF dissolved in 1.0 L of solution
(c) 500 mL of 1.4×10^{-2} M $CaCl_2$ and 250 mL of 0.25 M Na_2SO_4 mixed to give a final volume of 750 mL

17.21 What is the minimum pH necessary to cause a precipitate of $Fe(OH)_2$ to form in a 0.010 M $FeCl_2$ solution?

*17.22 A solution is prepared by mixing 100 mL of 0.20 M $AgNO_3$ with 100 mL of 0.10 M HCl. What are the molar concentrations of all species present in the solution when equilibrium is reached?

17.23 Will a precipitate form in a solution containing
(a) 0.025 M $CaCl_2$ and 0.0050 M Na_2CO_3
(b) 0.010 M $Pb(NO_3)_2$ and 0.030 M $CaCl_2$
(c) 1.5×10^{-3} M $FeCl_2$ and 2.2×10^{-3} M $Na_2C_2O_4$

The Common Ion Effect

17.24 What is the common ion effect?

17.25 What is the molar solubility of $CaCO_3$ in 0.50 M Na_2CO_3?

17.26 What is the molar solubility of AgCl in 0.020 M $AlCl_3$? Assume that $AlCl_3$ gives Al^{3+} and Cl^- in solution.

17.27 What is the molar solubility of $PbCl_2$ in 0.020 M $AlCl_3$? Assume that $AlCl_3$ gives Al^{3+} and Cl^- in solution.

17.28 How many moles of Ag_2CrO_4 will dissolve in 1.0 L of 0.10 M $AgNO_3$?

17.29 How many moles of Ag_2CrO_4 will dissolve in 1.0 L of 0.10 M Na_2CrO_4?

17.30 What is the molar solubility of CaF_2 in 0.010 M NaF?

17.31 How many grams of NaF must be added to 1.00 L of solution to reduce the molar solubility of BaF_2 to 6.8×10^{-4} mol/L?

17.32 What is the molar solubility of $Fe(OH)_2$ in a buffer that has a pH of 9.50?

17.33 What is the molar solubility of $Ca(OH)_2$ in (a) 0.10 M $CaCl_2$ and (b) 0.10 M NaOH?

*17.34 Suppose that 25.0 mL of 0.10 M HCl is added to 1.000 L of saturated $Mg(OH)_2$ in contact with more than enough $Mg(OH)_2(s)$ to react with all the HCl. After reaction has ceased, what will the molar concentration of Mg^{2+} be? What will the pH of the solution be?

*17.35 In an experiment 2.20 g of NaOH(s) is added to 250 mL of 0.10 M $FeCl_2$ solution. What weight of $Fe(OH)_2$ will be formed? What will the molar concentration of Fe^{2+} be in the final solution?

*17.36 Suppose that 1.75 g of NaOH(s) is added to 250 mL of 0.10 M $NiCl_2$ solution. What mass, in grams, of $Ni(OH)_2$ will be formed? What will be the pH of the final solution? For $Ni(OH)_2$, $K_{sp} = 1.6 \times 10^{-14}$.

*17.37 Solid $Mn(OH)_2$ is added to a solution of 0.100 M $FeCl_2$. After reaction, what will be the molar concentrations of Mn^{2+} and Fe^{2+} in the solution? What will be the pH of the solution?

Selective Precipitation

17.38 What will precipitate first when $Na_2CrO_4(s)$ is gradually added to a solution containing 0.010 M Pb^{2+} and 0.010 M Ba^{2+}? What will be the molar concentration of the ion that precipitates first when the other ion just begins to form a precipitate?

17.39 Magnesium oxalate, MgC_2O_4, has $K_{sp} = 8.6 \times 10^{-5}$ and calcium oxalate, CaC_2O_4, has $K_{sp} = 2.3 \times 10^{-9}$. What pH must be maintained to achieve *maximum* separation of Ca^{2+} from Mg^{2+} if both have a concentration of 0.10 M and the concentration of oxalic acid, $H_2C_2O_4$, is maintained at 0.10 M? For $H_2C_2O_4$, $K_{a_1} = 6.5 \times 10^{-2}$ and $K_{a_2} = 6.1 \times 10^{-5}$.

17.40 What range of pH values would permit the selective precipitation of Cu^{2+} as $Cu(OH)_2$ from a solution that contains 0.10 M Mn^{2+} and 0.10 M Cu^{2+}? For $Mn(OH)_2$, $K_{sp} = 1.2 \times 10^{-11}$ and for $Cu(OH)_2$, $K_{sp} = 4.8 \times 10^{-20}$.

17.41 The H_2CO_3 concentration in a solution that contains 0.050 M Ca^{2+} and 0.050 M Pb^{2+} is maintained at a value of 0.050 M by adjusting the partial pressure of CO_2 over the reaction mixture. To what pH must the solution be adjusted in order to achieve maximum separation of these ions by selectively precipitating $PbCO_3$?

Complex Ion Equilibria

17.42 Write equilibrium expressions corresponding to K_{form} for the complex ions
(a) $AgCl_2^-$ (b) $Ag(S_2O_3)_2^{3-}$ (c) $Zn(NH_3)_4^{2+}$

17.43 Write equilibrium expressions corresponding to K_{inst} for the complex ions
(a) $Fe(CN)_6^{4-}$ (b) $CuCl_4^{2-}$ (c) $Ni(NH_3)_6^{2+}$

17.44 The overall formation constant for $Ag(CN)_2^-$ equals 5.5×10^{18}, and the K_{sp} for AgCN equals 1.6×10^{-14}. Calculate K_c for the reaction

$$AgCN(s) + CN^-(aq) \rightleftharpoons Ag(CN)_2^-(aq)$$

Complex Ions and Solubility

17.45 Silver forms a relatively stable complex ion, AgI_2^-. When a solution containing this ion is diluted with water, AgI precipitates. Explain why this happens in terms of the equilibria that are involved.

17.46 Use the data in Tables 17.1 and 17.2 to determine the molar solubility of AgI in 0.010 M KCN solution.

17.47 The molar solubility of $Zn(OH)_2$ in 1.0 M NH_3 is 5.7×10^{-3} mol/L. Determine the value of the instability constant of the complex ion, $Zn(NH_3)_4^{2+}$. Ignore the reaction, $NH_3 + H_2O \rightleftharpoons NH_4^+ + OH^-$.

17.48 Calculate the molar solubility of $Cu(OH)_2$ in 2.0 M NH_3. (For simplicity, ignore the reaction of NH_3 as a base.)

17.49 Calculate the molar solubility of AgI in 1.0 M NaI solution. For AgI_2^-, $K_{form} = 1.0 \times 10^{11}$.

Additional Exercises

17.50 On the basis of Le Châtelier's principle, explain why the addition of solid NH_4Cl to a beaker containing solid $Mg(OH)_2$ in contact with water causes the $Mg(OH)_2$ to dissolve.

* 17.51 Will a precipitate form in a solution made by dissolving 1.0 mol of $AgNO_3$ and 1.0 mol $HC_2H_3O_2$ in 1.0 L of solution?

* 17.52 How many moles of solid NH_4Cl must be added to 1.0 L of water in order to dissolve 0.10 mol of solid $Mg(OH)_2$? *Hint:* Consider the simultaneous equilibria

$$Mg(OH)_2 \rightleftharpoons Mg^{2+} + 2OH^-$$

$$NH_3 + H_2O \rightleftharpoons NH_4^+ + OH^-$$

* 17.53 How many grams of solid sodium acetate must be added to 200 mL of a solution containing 0.200 M $AgNO_3$ and 0.10 M nitric acid to cause silver acetate to begin to precipitate. For $HC_2H_3O_2$, $K_a = 1.8 \times 10^{-5}$ and for $AgC_2H_3O_2$, $K_{sp} = 2.3 \times 10^{-3}$.

* 17.54 How many grams of solid potassium fluoride must be added to 200 mL of a solution that contains 0.20 M $AgNO_3$ and 0.10 M acetic acid to cause silver acetate to begin to precipitate? For HF, $K_a = 6.5 \times 10^{-4}$; for $HC_2H_3O_2$, $K_a = 1.8 \times 10^{-5}$; for $AgC_2H_3O_2$, $K_{sp} = 2.3 \times 10^{-3}$.

* 17.55 What is the molar solubility of $Mg(OH)_2$ in 0.10 M NH_3 solution? Remember that NH_3 is a weak base.

* 17.56 If 100 mL of 2.0 M NH_3 is added to 400 mL of a solution containing 0.10 M Mn^{2+} and 0.10 M Sn^{2+}, what minimum number of grams of NH_4Cl would have to be added to the mixture to prevent $Mn(OH)_2$ from precipitating? Assume that virtually all the tin is precipitated as $Sn(OH)_2$ by the reaction

$$Sn^{2+}(aq) + 2NH_3(aq) + 2H_2O \longrightarrow Sn(OH)_2(s) + 2NH_4^+(aq)$$

* 17.57 What is the molar concentration of Cu^{2+} ion in a solution prepared by mixing 0.50 mol of NH_3 and 0.050 mol of $CuSO_4$ in 1.00 L of solution. For NH_3, $K_b = 1.8 \times 10^{-5}$; for $Cu(OH)_2$, $K_{sp} = 4.8 \times 10^{-20}$; and for $Cu(NH_3)_4^{2+}$, $K_{form} = 4.8 \times 10^{12}$.

ELECTROCHEMISTRY

Electrochemistry is the study of the relationships that exist between chemical reactions and the flow of electricity. Included here are electrolysis reactions, in which nonspontaneous changes are forced to occur by the passage of electricity through chemical systems. Also included are spontaneous oxidation–reduction reactions (redox reactions) that are able to supply electricity. For convenience, we describe all these changes under the general heading of **electrochemical reactions.**

Electrochemical reactions are of great practical importance in the study of chemistry and also in our day-to-day activities. In the chemistry lab, they allow us to gain information about the energy changes involved in chemical reactions and help in the analysis of chemical systems. This has proven to be so useful that other sciences (biology, for example) have adapted electrochemical techniques to the study of systems of particular interest to them.

The impact of electrochemical reactions on modern society is found nearly everywhere. Critically important industrial chemicals such as aluminum, chlorine, and sodium hydroxide, without which modern society could not function, are manufactured exclusively by electrolysis reactions. And all of the tiny (and not so tiny) portable sources of electricity that we commonly call "batteries" are able to function solely because electrochemical reactions within them are able to provide electrical energy.

18.1 METALLIC AND ELECTROLYTIC CONDUCTION

Before we can begin to understand how electrochemical systems function, we first must examine in some detail the way electrical conduction takes place. This process is different for metals and for chemical systems that consist of molten salts and aqueous solutions.

For a substance to be classified as a conductor of electricity, it must be able to allow electrical charges within it to be moved from one place to another. From our earlier discussion of solids, you learned that metals are electrical conductors because electrons within the metallic lattice are able to move relatively easily. Thus, an electron injected into one end of a wire is able to move through the solid and eventually emerge at the other end. This kind of electrical conduction, which occurs by the transport of *electrons,* is called **metallic conduction.**

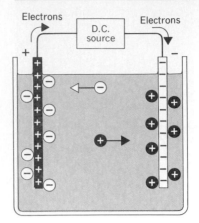

Figure 18.1. *Oxidation–reduction reactions at the electrodes remove ions from the layers of ions that surround them. As these ions are replaced by others, there is a gradual net flow of ions within the bulk of the liquid, with positive ions moving in the direction of the negative electrode and negative ions moving in the direction of the positive electrode.*

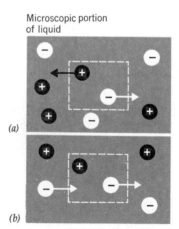

Figure 18.2. *During electrolytic conduction, electrical neutrality tends to be maintained by the flow of ions, even in the smallest portion of the liquid. Here we see two ways for this to happen. (a) When a negative ion leaves, a positive ion can also leave. (b) When a negative ion leaves, another negative ion can move in to take its place.*

You also learned earlier that molten ionic compounds and solutions of substances we call electrolytes are able to conduct electricity. The apparatus illustrated in Figure 4.12 on page 102 is able to demonstrate this phenomenon. When the two electrodes are connected to a source of electricity and dipped into a molten salt such as NaCl (Figure 4.13, page 102) or a solution of an ionic compound (Figure 5.2, page 119), the light bulb glows brightly, indicating that electricity is flowing. But neither molten NaCl nor an aqueous solution of NaCl contains free mobile electrons, so how are they able to conduct electricity?

The answer is seen if we examine in detail what takes place in the solution and at the electrodes of the conductivity apparatus. When the electricity is provided by a battery or other direct current (D.C.) source, the electrodes acquire electrical charges as shown in Figure 18.1. Each electrode attracts to itself ions of opposite sign that are in the liquid. Thus, positive ions are pulled toward the negative electrode, and negative ions are pulled toward the positive electrode, so each electrode becomes surrounded by a layer of ions of opposite charge.

For electrical conduction to take place, chemical reactions — redox reactions — must occur at the electrodes. At the positive electrode, electrons are pulled from negative ions, which therefore become oxidized. Each time a negative ion is oxidized, it is replaced by another from the surrounding liquid, so there is a gradual flow of negative ions from the liquid to the positive electrode.

The electrons that are removed from the negative ions are pumped through the external circuit by the direct-current source and are delivered to the negative electrode where they are gained by positive ions. This causes the positive ions to be reduced. As this happens, other positive ions in the surrounding liquid move in to take their place, so there is a gradual flow of positive ions from the liquid to the negative electrode.

The overall effect is that as these redox reactions are taking place, there is a flow of electrons through the wires of the external circuit but a flow of ions through the liquid, a flow that corresponds to the migration of positive ions toward the negative electrode and the migration of negative ions toward the positive electrode. This flow of ions through the liquid is called **electrolytic conduction.**

In electrolytic conduction, the migration of the ions takes place primarily because a large buildup of either positive or negative charges is extremely unstable. Therefore, throughout the liquid there is a strong tendency to maintain electrical neutrality, which is accomplished by the flow of ions. Even in the most microscopic portion of the liquid, if a negative ion leaves, either a positive ion also leaves or another negative ion moves in to take its place, as illustrated in Figure 18.2.

18.2 ELECTROLYSIS

The chemical reactions that take place at the electrodes during electrolytic conduction constitute **electrolysis,** and an apparatus designed specifically to carry out such reactions is called an **electrolysis cell** or an **electrolytic cell.** When a substance is subjected to electrolysis, it is said to be **electrolyzed.**

An example of electrolysis occurs when molten (liquid) NaCl is electrolyzed (Figure 18.3). In this system, Na^+ ions migrate toward the negative electrode where they are reduced to Na atoms. The Cl^- ions migrate toward the positive electrode where they become oxidized to Cl atoms, which combine to form Cl_2 molecules. The reactions that take place at the electrodes are

Positive electrode	$2Cl^-(l) \longrightarrow Cl_2(g) + 2e^-$	oxidation
Negative electrode	$Na^+(l) + e^- \longrightarrow Na(l)$	reduction

You might recognize these as half-reactions of the type that you learned to construct

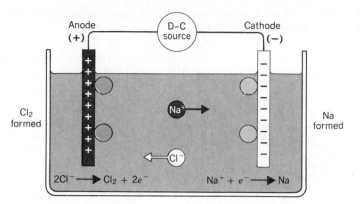

Figure 18.3. *Electrolysis of molten NaCl.*

in Chapter 5. In fact, the methods that you learned to obtain balanced half-reactions in Chapter 5 give the electrode reactions that we will discuss in this chapter.

In electrochemistry we always assign the terms **cathode** and **anode** according to the chemical reaction that is taking place at the electrode. *Reduction always takes place at the cathode and oxidation always takes place at the anode.* Thus, in the electrolysis reactions above, we label the negative electrode the cathode and the positive electrode the anode.

The net chemical change that takes place in an electrolytic cell is called the **cell reaction.** It is obtained by adding together the anode and cathode half-reactions in such a way that the same number of electrons are gained and lost. This is the same procedure we used in the ion electron method of balancing oxidation-reduction reactions in Chapter 5. Thus, in this case, we must multiply the reduction half-reaction by 2 to obtain

$$2Cl^-(l) \longrightarrow Cl_2(g) + 2e^-$$
$$2Na^+(l) + 2e^- \longrightarrow 2Na(l)$$
$$\overline{2Na^+(l) + 2Cl^-(l) \longrightarrow Cl_2(g) + 2Na(l)}$$

For electrolysis, the negative electrode is the cathode *and the positive electrode is the* anode.

In any redox reaction, the total electron gain must equal the total electron loss.

Therefore, in this electrolytic cell sodium is formed at the cathode and chlorine gas is produced at the anode.

The electrolysis of aqueous solutions of electrolytes is somewhat more complex because of the ability of water to be oxidized as well as reduced. The oxidation reaction for water is

$$2H_2O(l) \longrightarrow O_2(g) + 4H^+(aq) + 4e^- \tag{18.1}$$

and the reduction reaction takes the form

$$2H_2O(l) + 2e^- \longrightarrow H_2(g) + 2OH^-(aq) \tag{18.2}$$

In acidic solutions another reaction that may take place is the reduction of H^+, which is

$$2H^+(aq) + 2e^- \longrightarrow H_2(g) \tag{18.3}$$

Reaction 18.3 is not, however, a major reaction in most dilute aqueous solutions that we will consider.

When electrolysis is carried out on an aqueous solution, there is the possible oxidation and reduction of water as well as the possible oxidation and reduction of the solute. In other words, at each electrode there are competing reactions. At the anode there is the possible oxidation of water or of the anion, and at the cathode there is the possible reduction of water or the cation. What is actually observed in any particular electrolytic cell is determined, therefore, by the relative tendencies of the competing reactions to occur. This, in turn, is controlled both by the thermodynamic ease of reaction and by the nature of the materials used to construct the electrodes. As a result, it is not an easy matter to predict what will actually take place

Figure 18.4. *The electrolysis of aqueous sodium bromide solution.*

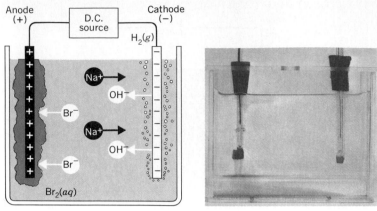

Cathode: $2 H_2O + 2e^- \longrightarrow H_2(g) + 2 OH^-(aq)$
Anode: $2 Br^-(aq) \longrightarrow Br_2(aq) + 2 e^-$

when electrolysis is carried out without actually carrying out the experiment. As you will see below, however, what we learn from one experiment can often be used to predict the outcome of others.

Electrolysis of aqueous NaBr

In the electrolysis of aqueous NaBr, the following two anode reactions (oxidation reactions) are possible:

(1) $\qquad\qquad\qquad 2Br^-(aq) \longrightarrow Br_2(aq) + 2e^-$

(2) $\qquad\qquad\qquad 2H_2O(l) \longrightarrow O_2(g) + 4H^+(aq) + 4e^-$

The following two cathode reactions (reduction reactions) are also possible:

(3) $\qquad\qquad\qquad Na^+(aq) + e^- \longrightarrow Na(s)$

(4) $\qquad\qquad\qquad 2H_2O(l) + 2e^- \longrightarrow H_2(g) + 2OH^-(aq)$

We can, of course, experimentally determine the outcome of the electrolysis simply by examining the products formed at the electrodes. When we actually carry out the reaction, we find that H_2 bubbles are produced at the cathode and the solution around the anode takes on a red color due to the formation of Br_2 (Figure 18.4). Therefore, during the electrolysis of NaBr solutions, the two electrode half-reactions and the overall cell reaction must be

$$
\begin{array}{ll}
2Br^-(aq) \longrightarrow Br_2(aq) + 2e^- & \text{(Anode)} \\
\underline{2H_2O(l) + 2e^- \longrightarrow H_2(g) + 2OH^-(aq)} & \text{(Cathode)} \\
2Br^-(aq) + 2H_2O(l) \longrightarrow Br_2(aq) + H_2(g) + 2OH^-(aq) & \text{(Cell reaction)}
\end{array}
$$

The results of this experiment tell us that in an electrolysis reaction, H_2O has a greater tendency to be reduced than Na^+ and that Br^- has a greater tendency to be oxidized than H_2O.

Electrolysis of aqueous CuSO₄

As in our last example, there are two possible oxidation and reduction reactions for the electrolysis of $CuSO_4$. These are

Oxidation

(1) $\qquad\qquad\qquad 2SO_4{}^{2-}(aq) \longrightarrow S_2O_8{}^{2-}(aq) + 2e^-$

(2) $$2H_2O(l) \longrightarrow O_2(g) + 4H^+(aq) + 4e^-$$

and

Reduction

(3) $$Cu^{2+}(aq) + 2e^- \longrightarrow Cu(s)$$

(4) $$2H_2O(l) + 2e^- \longrightarrow H_2(g) + 2OH^-(aq)$$

During this electrolysis we find experimentally that oxygen gas bubbles form at the anode and a coating of copper metal is deposited on the cathode. Therefore, we would write for the electrolysis of aqueous $CuSO_4$

$$
\begin{array}{ll}
2H_2O(l) \longrightarrow O_2(g) + 4H^+(aq) + 4e^- & \text{(Anode)} \\
\underline{2Cu^{2+}(aq) + 4e^- \longrightarrow 2\ Cu(s)} & \text{(Cathode)} \\
2H_2O(l) + 2Cu^{2+}(aq) \longrightarrow O_2(g) + 4H^+(aq) + 2Cu(s) & \text{(Cell reaction)}
\end{array}
$$

Notice that we multiplied the equation for the reduction of Cu^{2+} by 2 so that we have equal numbers of electrons lost and gained.

Because of the products that are observed, we can conclude that in the electrolysis of $CuSO_4$, Cu^{2+} has a greater tendency to be reduced than water and water has a greater tendency to be oxidized than SO_4^{2-}.

Electrolysis of aqueous CuBr₂

Let's see if we can apply what we have learned from the preceding two experiments to the electrolysis of aqueous $CuBr_2$. In an aqueous solution, this salt dissociates to give Cu^{2+} and Br^- ions. In the preceding experiment we saw that Cu^{2+} has a greater tendency to be reduced than H_2O, so at the cathode, we expect to observe the reduction of Cu^{2+} to Cu. We also saw that Br^- has a greater tendency to be oxidized than H_2O, so at the anode we expect to observe the oxidation of Br^- to Br_2.

$$
\begin{array}{ll}
2Br^-(aq) \longrightarrow Br_2(aq) + 2e^- & \text{(Anode)} \\
\underline{Cu^{2+}(aq) + 2e^- \longrightarrow Cu(s)} & \text{(Cathode)} \\
Cu^{2+}(aq) + 2Br^-(aq) \longrightarrow Cu(s) + Br_2(aq) & \text{(Cell reaction)}
\end{array}
$$

This is, in fact, exactly what is observed experimentally, as we see in Figure 18.5.

Electrolysis of aqueous Na₂SO₄

Once again we call on what we have learned previously. We know that water has a greater tendency to be oxidized than SO_4^{2-}, so we expect to observe the formation of O_2 at the anode. We also know that water has a greater tendency to be reduced than Na^+, so at the cathode we expect to observe the formation of H_2. In this cell, therefore, water is both oxidized and reduced, giving us

$$
\begin{array}{ll}
2H_2O(l) \longrightarrow O_2(g) + 4H^+(aq) + 4e^- & \text{(Anode)} \\
\underline{4H_2O(l) + 4e^- \longrightarrow 2H_2(g) + 4OH^-(aq)} & \text{(Cathode)} \\
6H_2O(l) \longrightarrow O_2(g) + 2H_2(g) + 4OH^-(aq) + 4H^+(aq) & \text{(Cell reaction)}
\end{array}
$$

Notice that we had to multiply the reduction reaction by 2 to have the same number of electrons as in the oxidation reaction.

During the electrolysis, H^+ ions are formed at the anode and OH^- ions are formed at the cathode. The solution in the immediate vicinity of the anode therefore becomes acidic, and around the cathode the solution becomes basic. We can verify this by performing the electrolysis in the presence of an acid–base indicator, as shown in Figure 18.6. We can also see from this figure that if the solution is stirred afterward, the H^+ and OH^- ions react to give water, and the solution becomes

Figure 18.5. *Electrolysis of aqueous copper(II) bromide. At the anode on the left, Br^- is oxidized to Br_2, which gives an orange color to the solution. On the right, Cu^{2+} is reduced to Cu, which appears as a black deposit on the cathode.*

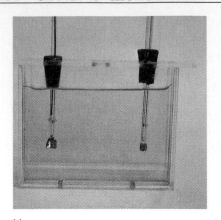
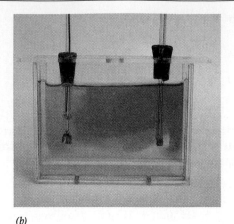

(a) *(b)* *(c)*

Figure 18.6. *Electrolysis of a solution of sodium sulfate, Na_2SO_4, in the presence of an indicator. (a) Initially, the solution is neutral and the indicator gives it a yellow color. (b) During electrolysis, H^+ is formed along with O_2 at the anode and causes the solution there to become acidic. This changes the indicator solution around the anode to pink. Around the cathode, the indicator takes on a bluish-violet color because the solution is made basic by OH^- which is formed along with H_2. (c) After electrolysis, the solution is stirred. The H^+ and OH^- neutralize each other, which restores the original yellow color.*

neutral again. Using the amounts of these ions in the equation above, we obtain the reaction

$$4OH^- + 4H^+ \longrightarrow 4H_2O$$

Thus, the net overall reaction for the electrolysis of a stirred aqueous Na_2SO_4 solution is

$$2H_2O(l) \longrightarrow O_2(g) + 2H_2(g)$$

which is simply the reaction for the electrolysis of H_2O. Sodium sulfate does not participate in this electrolysis in the sense that it is not consumed at the electrodes. Yet we would find experimentally that it or some other similar salt is needed if electrolysis of water is to occur. What, then, is the role of the Na_2SO_4? The Na_2SO_4 is needed to maintain electrical neutrality (Figure 18.7). During the oxidation of H_2O, H^+ ions are produced in the immediate vicinity of the anode. A negative ion must also be present in that region to neutralize the positive charges. This is fulfilled by SO_4^{2-} ions. Likewise, at the cathode, where OH^- ions are produced, there must be positive ions present to neutralize the charges on the OH^- ions and thereby keep the solution electrically neutral.

Figure 18.7. *Electrolysis of aqueous Na_2SO_4. The sodium ions and sulfate ions are needed to counter the charges of the ions formed at the electrodes and thereby maintain electrical neutrality.*

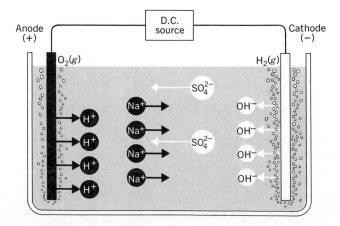

18.3 PRACTICAL APPLICATIONS OF ELECTROLYSIS

Each day our lives are touched either directly or indirectly by the products of electrolysis reactions. For example, the drinking water in most places is treated with chlorine to kill bacteria, and chlorine is used to manufacture many chemicals, from pesticides that protect crops to plastics such as poly(vinyl chloride), usually just called vinyl. Yet elemental chlorine does not occur free in nature. It must be extracted from its compounds, and this is done most economically by electrolysis. In this section we will study how chlorine and some other commercially important substances are made.

Electrolysis of molten sodium chloride

The chemistry of this process was described in our introduction to the discussion of electrolysis reactions on page 578. The products, sodium and chlorine, are both commercially important.

Sodium and chlorine are very reactive chemicals. Therefore, when they are produced from NaCl, they must be kept apart; otherwise, they will react and reform sodium chloride. The Downs cell, illustrated in Figure 18.8, accomplishes this. The chlorine, of course, comes off as a gas. Because of the temperature at which the cell operates, the sodium is formed as a liquid and also easily removed. This allows the cell to operate continuously as fresh sodium chloride is added and the products are taken away.

Molten means melted.

Cathode:

$$Na^+ + e^- \rightleftharpoons Na(l)$$

Anode:

$$2Cl^- \rightleftharpoons Cl_2(g) + 2e^-$$

Melting points:
NaCl 801 °C
Na   98 °C

Electrolysis of brine (NaCl solution)

In the electrolysis of a sodium chloride solution there is a competition at the anode between the oxidation of chloride ion and the oxidation of water.

(1) $$2Cl^-(aq) \longrightarrow Cl_2(g) + 2e^-$$

(2) $$2H_2O(l) \longrightarrow O_2(g) + 4H^+(aq) + 4e^-$$

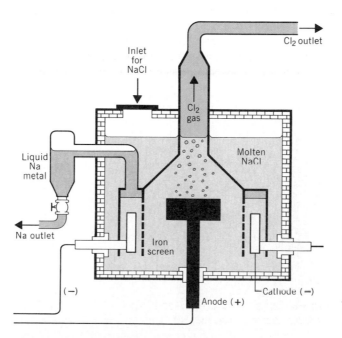

Figure 18.8. *A cross section of the Downs cell used for the electrolysis of molten sodium chloride. The cell is constructed to keep the metallic sodium away from the gaseous chlorine and prevent the two from reacting with each other.*

As the NaCl becomes more dilute, reaction (2) begins to compete and some O_2 is formed as well.

When a concentrated salt solution **(brine)** is used, the first reaction is the one that is observed. At the cathode, the reaction is the reduction of water (you've learned that H_2O is reduced more readily than Na^+).

$$2H_2O(l) + 2e^- \longrightarrow H_2(g) + 2OH^-(aq)$$

When the anode and cathode reactions are combined and the Na^+ ion is included (Na^+ is actually a spectator ion), the overall reaction is

$$2Na^+ + 2Cl^- + 2H_2O \longrightarrow Cl_2(g) + H_2(g) + \underbrace{2Na^+ + 2OH^-}_{NaOH(aq)}$$

In recent years, industrial production of NaOH by this reaction has amounted to over 22 billion pounds annually.

If we examine this reaction, we can appreciate its commercial importance. All the products—chlorine, hydrogen, and sodium hydroxide—are important industrial chemicals. Chlorine is used to make plastics such as poly(vinyl chloride) (PVC) and to purify drinking water. Hydrogen is used to make ammonia and to manufacture hydrogenated vegetable oils such as Crisco®. And sodium hydroxide is used in huge quantities to neutralize acids in various chemical processes, process pulp and paper, purify aluminum ores, and in the manufacture of textiles and the refining of petroleum.

This is only a small sampling of the uses of Cl_2, H_2, and NaOH.

The electrolysis of NaCl can be carried out in an apparatus such as that illustrated in Figure 18.9, but doing so always leads to some contamination of the NaOH by unreacted NaCl and by hypochlorite ion, which is formed in a reaction between OH^- and the chlorine produced at the anode. (This reaction is discussed further later.) To prevent the formation of hypochlorite and avoid dangerously explosive mixtures of H_2 and Cl_2, the products formed at the cathode must be prevented from coming into contact with those formed at the anode. There have been several approaches to accomplishing this, and most of the NaOH made today is produced by an apparatus called a **diaphragm cell.** There are several variations in the design of this cell, but the drawing in Figure 18.10 illustrates its essential features.

The cell consists of a compartment into which a NaCl solution is fed slowly. Graphite anodes dip into this solution. The cylindrical wall of the compartment consists of a porous asbestos paper layer (the diaphragm), which is supported by a steel wire mesh that serves as the cathode. When the cell operates, the NaCl solution seeps slowly through the asbestos and comes into contact with the steel cathode where hydrogen is evolved and hydroxide ion is formed. This solution, now containing NaOH, drips to the bottom of the container that holds the cell and is removed for purification. Meanwhile, the anode reaction within the compartment generates Cl_2 that bubbles through the electrolyte and is removed at the top.

Figure 18.9. *Electrolysis of aqueous sodium chloride (brine) solutions.*

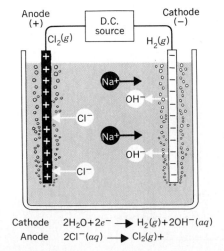

Cathode $2H_2O + 2e^- \longrightarrow H_2(g) + 2OH^-(aq)$

Anode $2Cl^-(aq) \longrightarrow Cl_2(g) +$

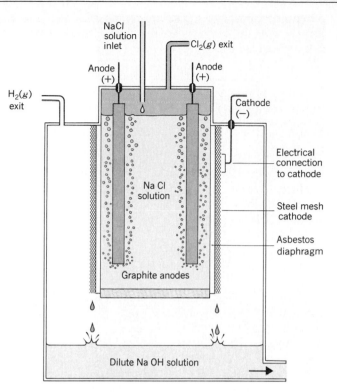

Figure 18.10. *A diaphragm cell for the production of NaOH by the electrolysis of aqueous NaCl solutions. The diagram shows a cross section of a cylindrical cell in which the NaCl solution is surrounded by an asbestos diaphragm held in place by the steel wire mesh cathode.*

The NaOH produced in the diaphragm cell is always contaminated by a little unreacted NaCl. If very pure NaOH is needed, a mercury cell, illustrated in Figure 18.11, can be used. In this cell, sodium ion is actually the substance that is reduced, and the free metal dissolves in the liquid mercury as it is formed. The solution of sodium in mercury — it's called sodium amalgam — is pumped to a separate vessel where the metallic sodium at the surface of the mercury can react with water. This reaction liberates hydrogen and leaves pure NaOH in solution.

$$2Na \text{ (in Hg)} + 2H_2O \longrightarrow 2Na^+ + 2OH^- + H_2(g)$$

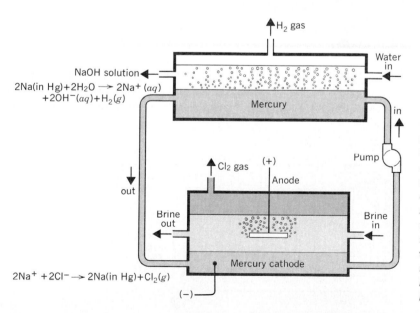

Figure 18.11. *Electrolysis of aqueous sodium chloride using a mercury cell. Chlorine gas is evolved at the anode where chloride ions are oxidized. At the cathode sodium ions are reduced to sodium atoms that dissolve in the mercury. The mercury is pumped to a tank where it comes in contact with water. Sodium atoms react there with water to liberate hydrogen gas and produce sodium hydroxide.*

A row of electrolytic cells in an aluminum production plant in Ghana (a country on the west coast of Africa). In the foreground, a crane is about to lift a large ladle filled with molten aluminum.

The mercury cell has the disadvantage of posing a serious threat of mercury water pollution and so must be carefully monitored.

If the electrolysis of brine is carried out in a vigorously stirred solution, the OH^- produced at the cathode reacts with the Cl_2 formed at the anode. The reaction is

$$Cl_2 + 2OH^- \longrightarrow Cl^- + OCl^- + H_2O$$

Continued electrolysis therefore gradually converts nearly all the chloride ion to hypochlorite ion, OCl^-, and the sodium chloride solution is changed to a solution of sodium hypochlorite. When diluted to about 5 to 6% by weight, this is is sold as liquid laundry bleach (e.g., Clorox®).

Aluminum

For many years aluminum was an exotic metal, available only in small amounts and at great expense, because it could only be prepared by the chemical reduction of its compounds by more reactive metals such as sodium. Attempts at preparing it by

Figure 18.12. *Production of aluminum by the Hall process.*

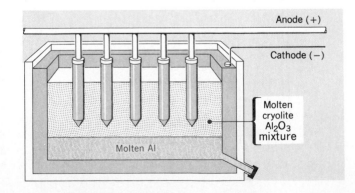

electrolysis had been unsuccessful because when an aqueous solution of an aluminum salt is electrolyzed, water is more readily reduced than Al^{3+}. This means that H_2 is formed at the cathode instead of Al. Electrolysis of molten $AlCl_3$ was unsuccessful for two reasons: The liquid is not particularly ionic and the compound vaporizes easily at high temperatures. Electrolysis of the oxide, Al_2O_3, was also impractical because of its high melting point (about 2000 °C).

In 1886 a 22-year old graduate of Oberlin College named Charles Hall invented a process that permitted the electrolysis of aluminum oxide. He worked with a mixture of Al_2O_3 and a mineral called cryolite, Na_3AlF_6. Addition of cryolite to the aluminum oxide, he found, reduced the melting point to about 1000 °C for the mixture, which was low enough to make electrolysis of the molten mixture feasible. A diagram of the electrolysis cell is shown in Figure 18.12. The vessel holding the molten mixture is made of iron lined with carbon and serves as the cathode. Carbon rods that serve as anodes are inserted into the molten mixture from above. As the redox reactions proceed, pure aluminum is produced at the cathode and sinks to the bottom of the vessel. The reactions at the electrodes are

$$3O^{2-}(l) \longrightarrow \tfrac{3}{2}O_2(g) + 6e^- \quad \text{(Anode)}$$
$$2Al^{3+}(l) + 6e^- \longrightarrow 2Al(l) \quad \text{(Cathode)}$$
$$\overline{2Al^{3+}(l) + 3O^{2-}(l) \longrightarrow \tfrac{3}{2}O_2(g) + 2Al(l)}$$

At the high temperature of the cell, the O_2 attacks the carbon electrodes, and they must be replaced periodically.

Today, cryolite has been replaced by a synthetic electrolyte composed of a mixture of NaF, CaF_2, and AlF_3. This mixture permits operation at still lower temperatures and is less dense than the cryolite used by Hall. This lower density of the electrolyte mix permits easier separation of the molten aluminum.

This electrolytic method for producing Al from Al_2O_3 is now called the *Hall process*.

Magnesium

Magnesium, another structural metal that is important because of its light weight, occurs to an appreciable extent in seawater. Magnesium ions are precipitated from seawater as the hydroxide and the $Mg(OH)_2$ is then converted to the chloride by treatment with hydrochloric acid. After evaporation of the water, the $MgCl_2$ is melted and electrolyzed. Magnesium is produced at the cathode and chlorine is evolved at the anode. The overall net reaction is simply

$$MgCl_2(l) \longrightarrow Mg(l) + Cl_2(g)$$

Mg^{2+} is the third most abundant ion in seawater.

Copper

An interesting application of electrolysis is the refining, or purification, of copper metal. When first separated from its ore, copper metal is about 99% pure, with iron, zinc, silver, gold, and platinum as major impurities. In the refining process the impure copper is used as the anode in an electrolytic cell that contains aqueous copper sulfate as the electrolyte. The cathode of the cell is constructed of high-purity copper (Figure 18.13).

When electrolysis is carried out, the voltage across the cell is adjusted so that only copper and other more active metals, such as iron or zinc, are able to be oxidized and dissolve at the anode. The silver, gold, and platinum do not dissolve, and simply fall off and settle to the bottom of the electrolysis cell. At the cathode only the most easily reduced species, Cu^{2+}, is caused to pick up electrons; hence, only copper is deposited.

The net result of the operation of this cell is that copper is transferred from the anode to the cathode while the Fe and Zn impurities remain in solution as Fe^{2+} and Zn^{2+}. Afterward the silver, gold, and platinum sludge is removed from the apparatus and sold for enough money to nearly pay for the cost of the electricity required in the electrolysis. As a result, the purification of copper (about 99.96% pure) is relatively

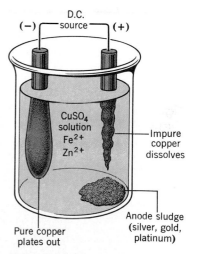

Figure 18.13. *Purification of copper by electrolysis.*

Copper cathodes, 99.96% pure, are pulled from the electrolytic refining tanks at Kennecott's Utah copper refinery. It takes about 28 days for the impure copper anodes to dissolve and deposit pure copper on the cathodes.

inexpensive. Nevertheless, the total production cost of copper is still considerable, because it includes the mining of the crude ore and its initial purification.

Electroplating

We have just seen how copper can be deposited on the surface of a copper cathode by electrolysis. This same kind of deposition would occur regardless of the metal used for the cathode, so if an iron cathode were used, it too would be coated with a layer of metallic copper. Covering one metal by another by electrolysis is called **electroplating,** and it is a common commercial process. For example, metallic objects such as automobile bumpers are electroplated with chromium to provide protection against corrosion and to provide an attractive appearance. Silver plating of objects such as knives and forks is also common, as is gold plating of jewelry.

18.4 QUANTITATIVE ASPECTS OF ELECTROLYSIS

Michael Faraday was the first to describe quantitatively the relationships that exist during electrolysis between the amount of electricity used and the amount of chemical change produced. In modern terms, we can say that the amount of chemical change is proportional to the number of moles of electrons that are exchanged during the oxidation–reduction reactions that occur, and this information can be obtained from the balanced half-reactions. For example, in the reduction of silver ion to metallic silver,

$$Ag^+(aq) + e^- \longrightarrow Ag(s)$$

1 mol of silver is produced for every 1 mol of electrons supplied by the cathode.

To use this kind of information, we need to have some way of relating moles of electrons to measurable properties of electricity. We can do this by expressing the charge on 1 mol of electrons in units of coulombs (C), the SI unit for electrical charge. It is found experimentally that

$$1 \text{ mol } e^- \Longleftrightarrow 96{,}494 \text{ C}$$

Normally, we will use this to just three significant figures and express the relationship as

$$1 \text{ mol } e^- \Longleftrightarrow 96{,}500 \text{ C} \quad \text{(to 3 significant figures)}$$

One mole of electrons is known as a *faraday* ($\mathscr{F}$) in honor of Michael Faraday (1791–1867).

$$1 \, \mathscr{F} = 96{,}500 \text{ C}$$

The relationship between moles of electrons and coulombs is important because a coulomb is the amount of electrical charge that passes a given point in an electrical circuit when a current of 1 ampere (1 A) flows for 1 second (1 s). Thus,

$$1 \text{ coulomb} = 1 \text{ ampere} \times 1 \text{ second}$$

$$1 \text{ C} = 1 \text{ A} \cdot \text{s}$$

This means that by measuring the amount of current that passes through an electrolytic cell and by measuring the length of time that it flows, we can calculate the number of coulombs that pass through the cell. From this we can then calculate the number of moles of electrons and proceed from there with stoichiometric calculations, as illustrated below.

EXAMPLE 18.1. QUANTITATIVE PROBLEMS ON ELECTROLYSIS

PROBLEM: In an electrolytic cell like that in Figure 18.1, how many grams of Cu will be deposited from a solution of $CuSO_4$ by a current of 1.5 A flowing for 2.0 hr?

ANALYSIS: The first step in solving a problem of this kind is to write the chemical equation for the reaction. In this instance, it is the reduction of Cu^{2+} to Cu.

$$Cu^{2+}(aq) + 2e^- \longrightarrow Cu(s)$$

From the equation, we see that 1 mol of copper is produced when 2 mol of e^- are gained from the cathode.

$$1 \text{ mol Cu} \Longleftrightarrow 2 \text{ mol } e^-$$

We can calculate the number of moles of electrons by first obtaining the product of amperes and seconds, which is coulombs, and then changing coulombs to mol e^-.

SOLUTION: To calculate coulombs, we multiply amperes by time in seconds. We are given hours, so we must also use the relationship

$$1 \text{ hr} = 3600 \text{ s}$$

Then,

$$1.5 \text{ A} \times 2.0 \text{ hr} \times \left(\frac{3600 \text{ s}}{1 \text{ hr}}\right) = 11,000 \text{ A} \cdot \text{s (rounded)}$$

$$11,000 \text{ A s} \times \left(\frac{1 \text{ C}}{1 \text{ A s}}\right) = 11,000 \text{ C}$$

$$11,000 \text{ C} \times \left(\frac{1 \text{ mol } e^-}{96,500 \text{ C}}\right) \Longleftrightarrow 0.11 \text{ mol } e^-$$

Next, we calculate the number of moles of Cu formed and then change this to grams of Cu.

$$0.11 \text{ mol } e^- \times \left(\frac{1 \text{ mol Cu}}{2 \text{ mol } e^-}\right) \times \left(\frac{63.55 \text{ g Cu}}{1 \text{ mol Cu}}\right) \Longleftrightarrow 3.5 \text{ g Cu}$$

EXAMPLE 18.2. QUANTITATIVE PROBLEMS ON ELECTROLYSIS

PROBLEM: How many hours would it take to produce 25.0 g of Cr from a solution of $CrCl_3$ by a current of 2.75 A?

SOLUTION: Again, we begin by writing the equation for the reaction. It also is a reduction,

$$Cr^{3+}(aq) + 3e^- \longrightarrow Cr(s)$$

from which we can say that

$$1 \text{ mol Cr} \Longleftrightarrow 3 \text{ mol } e^-$$

First we calculate the number of moles of electrons required.

$$25.0 \text{ g Cr} \times \left(\frac{1 \text{ mol Cr}}{52.0 \text{ g Cr}}\right) \times \left(\frac{3 \text{ mol } e^-}{1 \text{ mol Cr}}\right) \Longleftrightarrow 1.44 \text{ mol } e^-$$

Next we calculate the number of coulombs required.

$$1.44 \text{ mol } e^- \times \left(\frac{96,500 \text{ C}}{1 \text{ mol } e^-}\right) \Longleftrightarrow 139,000 \text{ C}$$

Since 1 C is equal to 1 A · s,

$$139,000 \text{ C} \times \left(\frac{1 \text{ A} \cdot \text{s}}{1 \text{ C}}\right) \times \left(\frac{1}{2.75 \text{ A}}\right) \Longleftrightarrow 50,500 \text{ s}$$

If we convert this to hours, we get

$$50,500 \cancel{s} \times \left(\frac{1 \text{ hr}}{3600 \cancel{s}}\right) = 14.0 \text{ hr}$$

18.5 GALVANIC CELLS

In an electrolytic cell, a nonspontaneous chemical change is forced to occur by an electrical voltage that is placed across a pair of electrodes that dip into the system. We now turn to the opposite situation, in which a spontaneous redox reaction is used to provide a voltage and an electron flow through some electrical circuit. Chemical systems set up to function in this way are called either **galvanic cells** or **voltaic cells.**

An example of a spontaneous redox reaction that you have seen before occurs when metallic zinc is dipped into a solution of copper sulfate, as shown in Figure 10.2 on page 301. Over a period of time, a dark brown, spongy layer of metallic copper forms on the zinc. The blue color of the $CuSO_4$ also fades and if the solution is analyzed, it is found to contain Zn^{2+} formed by the dissolving zinc strip. The reaction that has taken place here is

$$Cu^{2+}(aq) + Zn(s) \longrightarrow Cu(s) + Zn^{2+}(aq)$$

This can be divided into a pair of redox half-reactions:

$$Cu^{2+}(aq) + 2e^- \longrightarrow Cu(s)$$

$$Zn(s) \longrightarrow Zn^{2+}(aq) + 2e^-$$

We see from these reactions that the Cu^{2+} ions are spontaneously removed from the solution and replaced by the colorless Zn^{2+} ions. Thus, the blue color of the solution gradually disappears as more and more Zn^{2+} ions are formed.

As long as these reactions take place at the surface of the zinc, no useful flow of electrons can be obtained. The reaction, being exothermic, simply generates heat. We can, however, take advantage of the spontaneous electron transfer by making the oxidation and reduction half-reactions occur in separate compartments of a galvanic cell, as shown in Figure 18.14. Each compartment is called a **half-cell,** and

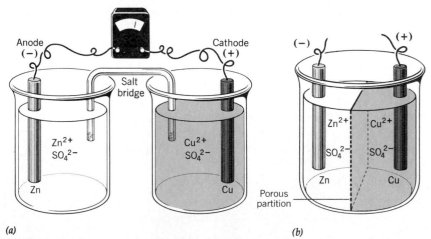

(a)   *(b)*

Figure 18.14. *A galvanic cell that uses the reaction*

$$Zn(s) + Cu^{2+}(aq) \longrightarrow Zn^{2+}(aq) + Cu(s)$$

(a) *The salt bridge serves the same purpose as* **(b)** *the porous partition; they allow electrical neutrality to be maintained as the reactions occur at the electrodes.*

when they are properly connected, the electrons produced by the oxidation of the zinc must travel through the wire and into the electrode in the $CuSO_4$ solution. The electrons are then picked up by the Cu^{2+} ions and reduction takes place. The electrons flowing through the external wire constitute an electrical current, and the galvanic cell therefore serves as a source of electricity.

Although the zinc and copper have to be separated to obtain a useful flow of electrons, complete isolation of the two species would lead to an electrical imbalance at the electrodes, and the electron flow would soon cease. To see why, let's imagine that the two half-cells are completely isolated from each other and the oxidation–reduction reactions still continue to take place. On the left side of this hypothetical setup, Zn^{2+} ions entering the solution would give the solution an overall positive charge. This would prevent additional Zn^{2+} from entering. On the right, as Cu^{2+} ions leave the solution, the SO_4^{2-} ions left behind would give the solution a negative charge and the electrode would become positively charged. This would cause the electrode to repel Cu^{2+} ions and prevent their further removal from the solution.

From the preceding discussion we see that a flow of current, accompanied by a continuous chemical activity, can only occur if the solution around each electrode is kept electrically neutral. For this to happen, however, ions must flow into or out of the cell compartments. In the zinc half-cell, for example, Zn^{2+} must leave the electrode compartment or anions must enter it. Similarly, in the copper half-cell, cations must enter to balance the charge of the SO_4^{2-}, or SO_4^{2-} ions must leave.

Although ions must be able to diffuse from one cell compartment to another, the two solutions cannot be allowed to mix freely; otherwise, Cu^{2+} would react directly at the Zn electrode and electrons would not be forced to flow through the external circuit.

The salt bridge in Figure 18.14*a* and the porous partition in Figure 18.14*b* allow for the slow mixing of the ions in the two solutions. A **salt bridge** is usually a tube filled with an electrolyte such as KNO_3 or KCl in gelatin. Cations from the salt bridge can move into one compartment to compensate for the excess negative charge, while the anions from the salt bridge diffuse into the other compartment to neutralize the excess positive charge. The porous partition in Figure 18.14*b* serves the same purpose as the salt bridge. With either the salt bridge or the porous partition in place, there is a continuous electron flow through the external wire and an ion flow through the solution as a result of the spontaneous oxidation–reduction reactions taking place at the electrodes.

The signs of the electrodes in galvanic cells

Earlier we defined the anode in electrochemistry as the electrode at which oxidation takes place and the cathode as the one at which reduction occurs. This applies regardless of whether the cell is an electrolytic cell or a galvanic cell. In the galvanic cell just described, oxidation takes place in the zinc compartment, so the zinc bar would be the anode and the copper electrode would be the cathode. As zinc ions leave the solid anode and enter the solution, electrons are left behind and the zinc electrode acquires a negative charge. At the copper cathode, Cu^{2+} ions become attached to the electrode and seek electrons to become reduced. This gives the copper electrode a positive charge. Thus we see that in a galvanic cell, the anode is negative and the cathode is positive, quite the opposite of what we found to be true in electrolytic cells.[1]

In a galvanic cell, the positive electrode is the cathode and the negative electrode is the anode.

[1] This labeling of electrodes in galvanic cells is, however, consistent with electrolytic cells when we consider the movement of the ions within the solution. The Zn^{2+} ions produced at the anode and the SO_4^{2-} ions freed at the cathode must mingle with each other if electrical neutrality is to prevail in the solution. To accomplish this, some of the Zn^{2+} ions must move toward the cathode and some of the SO_4^{2-} ions must move toward the anode. Thus, we have cations moving toward the cathode and anions moving toward the anode, which is precisely the same situation as in electrolytic cells.

Cell potentials

The electric current obtained from a galvanic cell is a result of electrons being pushed or forced to flow from the negative electrode, through an external wire, to the positive electrode. The force with which these electrons move through the wire is called the **electromotive force, or emf,** and is measured in **volts** (V). Actually, the volt is a measure of the energy that is capable of being extracted from the flowing electric charge. If the emf is 1 V, the passage of 1 coulomb is able to accomplish 1 joule of work.

$$1 \text{ volt} = \frac{1 \text{ joule}}{\text{coulomb}}$$

$$1 \text{ V} = 1 \text{ J/C} \tag{18.4}$$

The emf produced by a galvanic cell is called the **cell potential,** E_{cell}. This emf depends on the concentrations of the ions in the cell, the temperature, and the partial pressures of any gases that might be involved in the cell reactions. When all ion concentrations are 1 M, all partial pressures of gases are 1 atm, and the temperature of the cell is 25 °C, the emf is called the **standard cell potential,**[2] designated as E_{cell}°.

Notice that the standard conditions here are the same as those used in defining standard thermodynamic quantities.

To measure the cell potential accurately, we must take care not to draw current from the cell. This is because some of the cell's voltage is required to overcome the cell's own internal resistance when current is drawn. The remaining voltage that can be measured under these conditions is less than the maximum.

One device that can be used to measure the emf of a cell is called a **potentiometer.** In this instrument the potential generated by the cell is balanced by an opposing potential from within the potentiometer. When the two opposing potentials are equal, no current flows and the cell potential is equal to the opposing emf, which can be read directly from the instrument. The voltage that is measured in this way is the maximum emf of the cell. Today modern advances in electronics have led to a variety of other devices that are able to measure the emf of a cell quickly and simply, without drawing significant amounts of current.

Cell diagrams

To give someone a complete description of a galvanic cell, the kinds of information we would want to specify are (1) the nature of the electrode materials, (2) the nature of the solutions in contact with the electrodes (including the concentrations of ions in solution), (3) which of the half-cells is the anode and which is the cathode, and (4) the reactants and products in each of the half cells. To provide this information in a compact way, electrochemists have developed a standard notation called a **cell diagram.**

In constructing a cell diagram, we specify the anode half-cell first, followed by the cathode half-cell. A pair of vertical bars is used to represent the salt bridge between them. Within a given half-cell, the reactants are specified first, followed by the products. A single vertical bar is used to represent a phase boundary, as between a solution and a solid electrode. For substances in solution, molar concentrations are indicated in parentheses after the formulas.

As an example, let's construct the cell diagram for the zinc–copper cell dis-

---

[2] In footnote 1 in Chapter 15, it was mentioned that activities (effective concentrations and pressures) should be used in the mass action expression when computing ΔG. This applies also to the effect of concentration on E. The standard potential is obtained when all species are at unit activity. Only a small error is introduced, however, by using actual concentrations and pressures when solutions are relatively dilute.

cussed earlier. In this cell, zinc is the anode and undergoes oxidation to give Zn^{2+}. If the half-cell contains 1.00 M Zn^{2+}, its diagram is

$$Zn(s)|Zn^{2+}(1.00\ M)$$

Copper is the cathode, and in the half-cell Cu^{2+} is reduced to give $Cu(s)$. If we take the concentration of Cu^{2+} to be 1.00 M, this half-cell is diagrammed as

$$Cu^{2+}(1.00\ M)|Cu(s)$$

The galvanic cell is made by joining these two half-cells by a salt bridge (indicated by double bars, ||), which gives

$$Zn(s)|Zn^{2+}(1.00\ M)||Cu^{2+}(1.00\ M)|Cu(s)$$

This is the complete cell diagram.

18.6 REDUCTION POTENTIALS

A very important and useful concept can be developed if we attempt to answer the question, "What is the origin of the cell potential?" To answer this question, we will use the Zn/Cu cell described earlier. In this cell we have a solution containing Zn^{2+} ions around one electrode and a solution containing Cu^{2+} ions around the other. Each of these ions has a certain tendency to acquire electrons from its respective electrode and become reduced. In other words, each reduction half-reaction

$$Zn^{2+}(aq) + 2e^- \longrightarrow Zn(s)$$

and

$$Cu^{2+}(aq) + 2e^- \longrightarrow Cu(s)$$

has a certain intrinsic tendency to proceed from left to right that we can describe by its **reduction potential.** The larger the reduction potential for any half-reaction, the greater its tendency to undergo reduction.

When the cell reaction takes place, what we are actually observing is a kind of "tug-of-war." Each of the species in solution attempts to pull electrons from its electrode so as to become reduced. The species with the greater tendency to acquire electrons (that is, the one with the larger reduction potential) wins the tug-of-war and does undergo reduction. The loser, on the other hand, must supply the electrons to the winner, so the loser is oxidized. In the zinc–copper cell, therefore, copper must have a larger reduction potential than zinc, because copper is reduced.

The potential that we measure for a cell arises from the *difference* in the tendencies of the two ions to become reduced and is equal to the reduction potential for the substance that actually undergoes reduction minus the reduction potential for the substance that is forced to undergo oxidation. In terms of standard reduction potentials,

$$E^{\circ}_{cell} = E^{\circ}_{substance\ reduced} - E^{\circ}_{substance\ oxidized} \tag{18.5}$$

In the zinc–copper cell, therefore,

$$E^{\circ}_{cell} = E^{\circ}_{Cu^{2+}|Cu} - E^{\circ}_{Zn^{2+}|Zn}$$

where $E^{\circ}_{Cu^{2+}|Cu}$ is the standard reduction potential for copper and $E^{\circ}_{Zn^{2+}|Zn}$ is the standard reduction potential for zinc. Since $E^{\circ}_{Cu^{2+}|Cu}$ is larger than $E^{\circ}_{Zn^{2+}|Zn}$, E°_{cell} is positive.

Experimentally, it is only possible to measure overall cell potentials. This means that we are only capable of obtaining *differences* between the reduction potentials for any two half-reactions. Then how can we obtain the reduction potential for any specific half-reaction? Clearly, if the cell potential and the E° for one of the half-

Figure 18.15. (a) *The hydrogen electrode.* (b) *The hydrogen electrode used in a galvanic cell.*

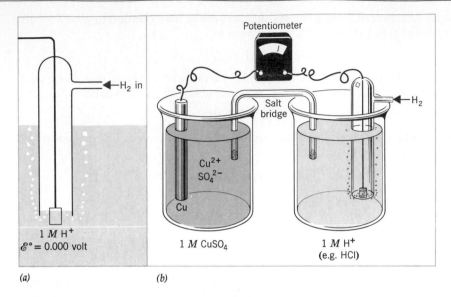

reactions were known, the $E°$ for the other half-reaction could be calculated. What has been done, therefore, is to choose a half-reaction arbitrarily and assign to it a standard reduction potential of zero volts. All other half-reactions can then be compared to this standard and a set of relative values of $E°$ obtained.

The electrode chosen to be the standard is called the **hydrogen electrode,** shown in Figure 18.15a. It consists of a platinum wire encased in a glass sleeve through which hydrogen gas is passing at a pressure of 1 atm. The platinum wire is attached to a platinum foil that is coated with a black velvety layer of finely divided platinum, which serves as a catalyst for the reaction

$$2H^+(aq) + 2e^- \rightleftharpoons H_2(g)$$

This assembly is then immersed in an acidic solution whose hydrogen ion concentration is $1.00\ M$.

By definition, the standard hydrogen electrode is assigned a reduction potential $E°_{H^+|H_2}$, of exactly 0.000 V. Any substance that is more easily reduced than H^+ has a positive value of $E°$, and any substance that is more difficult to reduce has a negative $E°$.

When the hydrogen electrode is paired with another half-cell in a galvanic cell, it can undergo either oxidation or reduction, depending on the reduction potential of the other half-cell. For instance, if the reduction potential of the species in the other half-cell is greater than that for the hydrogen electrode—that is, if its $E°$ is positive—the hydrogen electrode is forced to undergo oxidation. The corresponding half-reaction for the oxidation of the hydrogen electrode is

$$H_2(g) \longrightarrow 2H^+(aq) + 2e^-$$

On the other hand, if the reduction potential of the other half-reaction is less than 0.000 V, this species has a negative reduction potential and the hydrogen electrode undergoes reduction.

$$2H^+(aq) + 2e^- \longrightarrow H_2(g)$$

This causes the other species to become oxidized.

To illustrate how the hydrogen electrode is used, let's examine the galvanic cell in Figure 18.15b. When we connect a potentiometer to this cell to measure its potential, proper readings can be obtained only if we connect the terminal labeled (+) to the positive electrode and the terminal labeled (−) to the negative electrode.

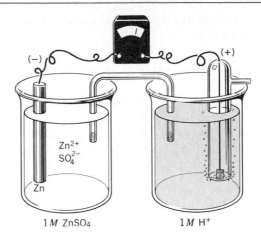

Figure 18.16. *A galvanic cell that can be used to determine the standard reduction potential of Zn^{2+}.*

$1 M$ ZnSO₄ $1 M$ H⁺

That's the way these instruments are designed. In this case, we find that to obtain a proper reading the (+) terminal must be connected to the copper electrode and the (−) terminal to the hydrogen electrode. From what we learned earlier, this tells us that the copper electrode is the cathode and the hydrogen electrode is the anode. The spontaneous half-reactions in this cell are therefore

$$Cu^{2+}(aq) + 2e^- \longrightarrow Cu(s) \qquad \text{(Cathode)}$$

$$H_2(g) \longrightarrow 2H^+(aq) + 2e^- \quad \text{(Anode)}$$

Since copper is reduced, when we apply Equation 18.5, we must write

$$E^\circ_{cell} = E^\circ_{Cu^{2+}|Cu} - E^\circ_{H^+|H_2}$$

The measured cell potential (as read from the potentiometer) is 0.34 V. Therefore,

$$0.34 \text{ V} = E^\circ_{Cu^{2+}|Cu} - 0.000 \text{ V}$$

or

$$E^\circ_{Cu^{2+}|Cu} = +0.34 \text{ V}$$

Now let's look at the cell in Figure 18.16. Here we have a zinc half-cell paired with the hydrogen electrode. To obtain a reading with the potentiometer we find, by experiment, that the positive terminal has to be connected to the hydrogen electrode, so this time it is the cathode. The spontaneous half-reactions in this cell are therefore

$$2H^+(aq) + 2e^- \longrightarrow H_2(g) \qquad \text{(Cathode)}$$

$$Zn(s) \longrightarrow Zn^{2+}(aq) + 2e^- \quad \text{(Anode)}$$

The measured value of the cell potential, as read from the potentiometer, is 0.76 V. Using Equation 18.5 again, we obtain

$$E^\circ_{cell} = \underbrace{E^\circ_{H^+|H_2}}_{\substack{\text{substance} \\ \text{reduced}}} - \underbrace{E^\circ_{Zn^{2+}|Zn}}_{\substack{\text{substance} \\ \text{oxidized}}}$$

Substituting yields

$$0.76 \text{ V} = 0.000 \text{ V} - E^\circ_{Zn^{2+}|Zn}$$

Solving for $E^\circ_{Zn^{2+}|Zn}$ gives

$$E^\circ_{Zn^{2+}|Zn} = -0.76 \text{ V}$$

The cathode carries the positive charge in a galvanic cell.

A *measured* cell potential is always a positive number.

If the hydrogen electrode is the cathode, reduction must be taking place in this half-cell and oxidation must be taking place in the zinc half-cell.

E° is negative for species that are more difficult to reduce than H⁺.

The negative sign for $E^{\circ}_{Zn^{2+}|Zn}$ reflects the fact that Zn^{2+} is more difficult to reduce than H^+.

Using reduction potentials

With a knowledge of the reduction potentials for the zinc and copper electrodes, we can now *predict* the cell potential and the spontaneous cell reaction for the Zn/Cu cell. This can be done even if we had no previous knowledge of the species undergoing oxidation or reduction.

First, simply by examining the reduction potentials,

$$Cu^{2+}(aq) + 2e^- \rightleftharpoons Cu(s) \qquad E^{\circ}_{Cu^{2+}|Cu} = +0.34 \text{ V}$$

$$Zn^{2+}(aq) + 2e^- \rightleftharpoons Zn(s) \qquad E^{\circ}_{Zn^{2+}|Zn} = -0.76 \text{ V}$$

Remember, the substance more easily reduced *is* the substance actually reduced in the spontaneous reaction.

we know immediately that Cu^{2+} is more easily reduced than Zn^{2+}. This is because Cu^{2+} has the higher (more positive) reduction potential. The cell reaction must therefore be

$$Zn(s) + Cu^{2+}(aq) \longrightarrow Cu(s) + Zn^{2+}(aq)$$

To calculate E°_{cell} we use Equation 18.5 and subtract $E^{\circ}_{Zn^{2+}|Zn}$ (the E° of the substance oxidized) from $E^{\circ}_{Cu^{2+}|Cu}$ (the E° of the substance reduced).

$$E^{\circ}_{cell} = E^{\circ}_{Cu^{2+}|Cu} - E^{\circ}_{Zn^{2+}|Zn}$$

$$E^{\circ}_{cell} = +0.34 \text{ V} - (-0.76 \text{ V})$$

$$E^{\circ}_{cell} = +1.10 \text{ V}$$

This is precisely the value that we observe experimentally when we measure the potential of the cell. Also notice that the calculated E°_{cell} is positive, which it will always be for the spontaneous cell reaction.

The double arrows in the half-reactions in Table 18.1 signify that the reactions are reversible, not that they are equilibria.

We are now in a position to determine the reduction potentials for many different half-reactions, because all we have to do is construct galvanic cells in which the reduction potential of one half-cell is known, relative to the hydrogen electrode. Some standard reduction potentials, E°, determined in this manner are given in Table 18.1. In this table the reduction potential of the hydrogen electrode is placed in the middle, with the species more difficult to reduce than hydrogen listed below it and those more easily reduced placed above it. Such a table of reduction potentials serves many useful purposes.

1. From a table of reduction potentials we can, at a glance, pick out substances that are good oxidizing agents and those that are good reducing agents. Any species that appears on the left of the double arrow serves as an oxidizing agent if it undergoes reduction during the course of a chemical reaction. Since the substances at the top left side of the table are more easily reduced than those at the bottom, their ability to serve as oxidizing agents decreases as we proceed down the table. Thus, we could conclude, from their positions in this table, that H^+ is a better oxidizing agent than Zn^{2+}, and that F_2 is a better oxidizing agent than Cl_2. In brief, *good oxidizing agents are those species on the left of the double arrow at the top of the table.*

Notice that the order of increasing ease of oxidation of the metals (top to bottom) in Table 18.1 is the same as the order (bottom to top) in the activity series in Table 10.1 on p. 302.

Each of the half reactions listed in Table 18.1 is reversible. We saw, for example, that H_2 is oxidized to H^+ when placed in a cell with copper, and that H^+ is reduced to H_2 when placed against zinc. When the reactions in Table 18.1 are forced to proceed from right to left — that is, when they are caused to be the oxidation step in an overall reaction — then the species appearing at the right in Table 18.1 are functioning as reducing agents by being oxidized. *All the substances appearing on the right side of the reactions in Table 18.1 could*

Table 18.1. Standard reduction potentials at 25 °C

$E°$ (volts)	Half-Reaction
2.87	$F_2 + 2e^- \rightleftharpoons 2F^-$
2.00	$S_2O_8^{2-} + 2e^- \rightleftharpoons 2SO_4^{2-}$
1.78	$H_2O_2 + 2H^+ + 2e^- \rightleftharpoons 2H_2O$
1.69	$PbO_2 + SO_4^{2-} + 4H^+ + 2e^- \rightleftharpoons PbSO_4 + 2H_2O$
1.49	$8H^+ + MnO_4^- + 5e^- \rightleftharpoons Mn^{2+} + 4H_2O$
1.47	$2ClO_3^- + 12H^+ + 10e^- \rightleftharpoons Cl_2 + 6H_2O$
1.36	$Cl_2(g) + 2e^- \rightleftharpoons 2Cl^-$
1.33	$Cr_2O_7^{2-} + 14H^+ + 6e^- \rightleftharpoons 2Cr^{3+} + 7H_2O$
1.28	$MnO_2 + 4H^+ + 2e^- \rightleftharpoons Mn^{2+} + 2H_2O$
1.23	$O_2 + 4H^+ + 4e^- \rightleftharpoons 2H_2O$
1.09	$Br_2(aq) + 2e^- \rightleftharpoons 2Br^-$
0.80	$Ag^+ + e^- \rightleftharpoons Ag$
0.77	$Fe^{3+} + e^- \rightleftharpoons Fe^{2+}$
0.54	$I_2(aq) + 2e^- \rightleftharpoons 2I^-$
0.52	$Cu^+ + e^- \rightleftharpoons Cu$
0.34	$Cu^{2+} + 2e^- \rightleftharpoons Cu$
0.27	$Hg_2Cl_2 + 2e^- \rightleftharpoons 2Hg + 2Cl^-$
0.22	$AgCl + e^- \rightleftharpoons Ag + Cl^-$
0.00	$2H^+ + 2e^- \rightleftharpoons H_2$
−0.04	$Fe^{3+} + 3e^- \rightleftharpoons Fe$
−0.13	$Pb^{2+} + 2e^- \rightleftharpoons Pb$
−0.14	$Sn^{2+} + 2e^- \rightleftharpoons Sn$
−0.25	$Ni^{2+} + 2e^- \rightleftharpoons Ni$
−0.36	$PbSO_4 + 2e^- \rightleftharpoons Pb + SO_4^{2-}$
−0.44	$Fe^{2+} + 2e^- \rightleftharpoons Fe$
−0.74	$Cr^{3+} + 3e^- \rightleftharpoons Cr$
−0.76	$Zn^{2+} + 2e^- \rightleftharpoons Zn$
−0.83	$2H_2O + 2e^- \rightleftharpoons H_2 + 2OH^-$
−1.03	$Mn^{2+} + 2e^- \rightleftharpoons Mn$
−1.67	$Al^{3+} + 3e^- \rightleftharpoons Al$
−2.38	$Mg^{2+} + 2e^- \rightleftharpoons Mg$
−2.71	$Na^+ + e^- \rightleftharpoons Na$
−2.76	$Ca^{2+} + 2e^- \rightleftharpoons Ca$
−2.90	$Ba^{2+} + 2e^- \rightleftharpoons Ba$
−2.92	$K^+ + e^- \rightleftharpoons K$
−3.05	$Li^+ + e^- \rightleftharpoons Li$

behave as reducing agents; those species at the bottom right of the table, such as Li, are the best and those at the top right, such as F⁻, are the poorest.

2. Using Table 18.1, we can find rather quickly which combinations of reactants lead to spontaneous oxidation–reduction reactions (when the concentrations of the reactants and products are 1 M and the partial pressures of any gases involved are 1 atm). We can also determine whether or not a given reaction, as written, will proceed spontaneously in the forward direction.

Consider, for example, a mixture consisting of pieces of solid zinc and solid chromium in contact with a solution containing 1 M Zn^{2+} and 1 M Cr^{3+}. What reaction will occur in this mixture? To answer this question we look at the following half-reactions found in Table 18.1.

$$Cr^{3+}(aq) + 3e^- \rightleftharpoons Cr(s) \qquad E^\circ_{Cr^{3+}|Cr} = -0.74 \text{ V}$$

$$Zn^{2+}(aq) + 2e^- \rightleftharpoons Zn(s) \qquad E^\circ_{Zn^{2+}|Zn} = -0.76 \text{ V}$$

-0.74 V is more positive than -0.76 V.

The reduction potentials tell us that Cr^{3+} is more easily reduced than Zn^{2+}, so in this mixture reduction of Cr^{3+} will occur and the zinc half-reaction will be forced to occur as oxidation. To obtain the cell reaction we combine the half-reactions in a way that makes the total number of electrons gained equal the total number lost.

$$2 \times [Cr^{3+}(aq) + 3e^- \longrightarrow Cr(s)] \qquad \text{(Reduction)}$$
$$\underline{3 \times [Zn(s) \longrightarrow Zn^{2+}(aq) + 2e^-]} \qquad \text{(Oxidation)}$$
$$2Cr^{3+}(aq) + 3Zn(s) \longrightarrow 2Cr(s) + 3Zn^{2+}(aq) \quad \text{(Cell reaction)}$$

To calculate the E°_{cell} for this reaction, we use Equation 18.5.

$$E^\circ_{cell} = E^\circ_{\text{substance reduced}} - E^\circ_{\text{substance oxidized}}$$
$$= E^\circ_{Cr^{3+}|Cr} - E^\circ_{Zn^{2+}|Zn}$$
$$= (-0.74 \text{ V}) - (-0.76 \text{ V})$$
$$= +0.02 \text{ V}$$

Notice that E°_{cell} is positive, as it must be for a spontaneous change. Also notice that even though the half-reactions are multiplied by factors before they are combined to give the net cell reaction, the reduction potentials are not. They are simply subtracted one from the other. This is because reduction potentials are intensive quantities and are therefore independent of the number of moles of reactants and products involved.

We have also said that we can determine whether a reaction, as written, will occur spontaneously. Let's consider the possible reaction

$$Fe^{2+}(aq) + Ni(s) \longrightarrow Fe(s) + Ni^{2+}(aq)$$

If the reaction is spontaneous, the calculated E°_{cell} will be positive, as it was above. On the other hand, if the reaction is not spontaneous in the direction written, the calculated E°_{cell} will be negative.

For the reaction we are considering, the first step is to divide the overall equation into two half-reactions.

$$Fe^{2+}(aq) + 2e^- \longrightarrow Fe(s)$$

$$Ni(s) \longrightarrow Ni^{2+}(aq) + 2e^-$$

We see that the first equation is a reduction and we can find its reduction potential, $E^\circ_{Fe^{2+}|Fe} = -0.44$ V, from Table 18.1. The second equation is an oxidation. If we were to rewrite it as a reduction, we would also be able to find its reduction potential, $E^\circ_{Ni^{2+}|Ni} = -0.25$ V. Since in our overall equation iron(II) is reduced and nickel is oxidized, when we substitute into Equation 18.5, we get

$$E^\circ_{cell} = E^\circ_{Fe^{2+}|Fe} - E^\circ_{Ni^{2+}|Ni}$$
$$= -0.44 - (-0.25)$$
$$= -0.19 \text{ V}$$

Because E°_{cell} is computed to be negative, we know the reaction of $Fe^{2+}(aq)$ with $Ni(s)$ is *not* spontaneous. In fact, it is the reverse reaction that is spontaneous.

EXAMPLE 18.3. PREDICTING THE emf OF A CELL FROM STANDARD REDUCTION POTENTIALS

PROBLEM: What will be the spontaneous reaction between the following set of half-reactions? What is the value of $E°_{cell}$?

(1) $$Cr^{3+}(aq) + 3e^- \rightleftharpoons Cr(s)$$

(2) $$MnO_2(s) + 4H^+(aq) + 2e^- \rightleftharpoons Mn^{2+}(aq) + 2H_2O(l)$$

SOLUTION: From Table 18.1, reaction (1) has $E° = -0.74$ V and reaction (2) has $E° = +1.28$ V. Since the reduction potential of reaction (2) is larger (more positive) than that of reaction (1), reaction (2) will occur as reduction. Reaction (1) must be reversed and written as oxidation. To obtain the net overall reaction we multiply by appropriate coefficients so that electrons cancel.

$$3[MnO_2(s) + 4H^+(aq) + 2e^- \longrightarrow Mn^{2+}(aq) + 2H_2O(l)]$$
$$2[Cr(s) \longrightarrow Cr^{3+}(aq) + 3e^-]$$
$$\overline{2Cr(s) + 3MnO_2(s) + 12H^+(aq) \longrightarrow 2Cr^{3+}(aq) + 3Mn^{2+}(aq) + 6H_2O(l)}$$

To calculate $E°_{cell}$ we apply Equation 18.5 and subtract reduction potentials to obtain a positive value.

$$E°_{cell} = +1.28 - (-0.74)$$
$$= +2.02 \text{ V}$$

3. In the process of combining half-reactions from Table 18.1, we see that some of the reactants in the spontaneous reaction appear on the left side of one half-reaction, while the rest of the reactants are found on the right side of another half-reaction. Among the reactants in Example 18.3, for instance, we have $MnO_2 + 4H^+$ from the left side of one half-reaction and $Cr(s)$ from the right side of the other. The order of these reactions in Table 18.1 is

$$MnO_2(s) + 4H^+(aq) + 2e^- \rightleftharpoons Mn^{2+}(aq) + 2H_2O(l) \qquad E° = 1.28 \text{ V}$$
$$Cr^{3+}(aq) + 3e^- \rightleftharpoons Cr(s) \qquad E° = -0.74 \text{ V}$$

and we see that the reactants in the overall spontaneous reaction are those substances related by the diagonal line (colored arrow) running from upper left to lower right. As a general statement, *we can say that when comparing reactants and products having unit concentrations, any species on the left of a given half-reaction will react spontaneously with a substance that is found on the right of a half-reaction located below it in Table 18.1.* We could use this rule of thumb, for example, to tell us that Br_2 will react spontaneously with I^- to produce Br^- and I_2, while Br_2 will *not* react spontaneously with Cl^-. Our rule, therefore, permits us to determine the course of a reaction without having to worry about subtracting electrode potentials in the proper sequence.

4. A point worth noting is that a collection of half-reactions, such as that found in Table 18.1, enables us to predict the outcome of many chemical reactions when we know only a relatively few half-reactions and their corresponding reduction potentials. From the 36 half-reactions listed in Table 18.1, for example, we can predict the results of 630 different chemical reactions! A table of this type, therefore, provides us with a very compact way of storing chemical information.

The reduction potentials of hundreds of half-reactions have been measured.

5. With a knowledge of the standard reduction potentials listed in Table 18.1,

we account for the course of electrolysis reactions. For example, we know from experiment that we can produce copper by electrolyzing an aqueous solution containing Cu^{2+}, but that we cannot obtain aluminum in this same fashion. From Table 18.1 we see that the reduction potential of copper is $+0.34$ V and for H_2O it is -0.83 V. Thus, copper ion is more readily reduced than H_2O and will plate out on the electrode according to the half-reaction

$$Cu^{2+}(aq) + 2e^- \longrightarrow Cu(s)$$

In the case of aluminum, however, we find that the reduction potential for Al^{3+} is -1.66 V, which makes it more difficult to reduce than water. This means that when an aqueous solution containing Al^{3+} ions is electrolyzed, the H_2O will preferentially be reduced.

Complicating factors that we haven't discussed make it difficult to predict, accurately, based on emf data, what will or will not be formed at an electrode during electrolysis.

18.7 SPONTANEITY OF OXIDATION–REDUCTION REACTIONS

It was pointed out in Section 14.6 that the thermodynamic criterion for spontaneity of a chemical reaction taking place at constant temperature and pressure is that the change in free energy, ΔG, has to be a negative quantity. In Section 14.7 we saw that ΔG also represents the maximum amount of useful work obtainable from a chemical reaction. The relationship between ΔG and maximum useful work (w_{max}) for any system takes the form

$$\Delta G = -w_{max}$$

But what is w_{max} for an electrochemical cell?

The work derived from an electrochemical cell might be compared to that obtained from a waterwheel, shown in Figure 18.17. The amount of work that can be obtained from this waterwheel depends on two factors: (1) the volume of water flowing over the blades of the wheel, and (2) the energy given to the wheel per unit volume of water as it drops to the lower level of the stream.

$$\text{work} = (\text{volume of water}) \times \left(\frac{\text{energy released}}{\text{unit volume}} \right)$$

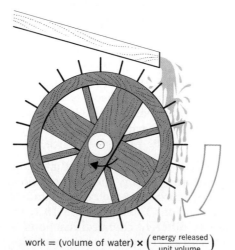

Figure 18.17. *The work obtained from a waterwheel depends on the volume of water flowing over the wheel and the amount of energy given to the wheel per unit volume.*

$$\text{work} = (\text{volume of water}) \times \left(\frac{\text{energy released}}{\text{unit volume}} \right)$$

Similarly, the work that can be done by an electrochemical cell is dependent on (1) the number of coulombs that flow and (2) the energy available per coulomb

$$\text{work} = (\text{number of coulombs}) \times \left(\frac{\text{energy available}}{\text{coulomb}} \right)$$

The number of coulombs that flow is equal to the number of moles of electrons exchanged in the redox reaction, n, multiplied by the number of coulombs per mole of electrons (96,500 C/mol e^-). As noted in the margin on page 588, this quantity is known as a faraday and is usually given the symbol $\mathscr{F}$, so the number of coulombs that flow can be represented as

$$\text{number of coulombs} = n\mathscr{F}$$

The value of n depends on the nature of the half-reactions taking place in the cell and can be derived once the specific reactions are known. For example, in the Zn/Cu cell there are two electrons involved in both of the half-reactions; therefore, n for this cell is 2.

The energy available per coulomb is simply the emf of the cell, because the volt is equal to the energy per coulomb (Equation 18.4)

$$\frac{\text{energy available}}{\text{coulomb}} = \text{emf}$$

When the emf is a maximum, the work derived from the cell is also a maximum. In Section 18.6 we saw that the maximum emf is the cell potential, E_{cell}.

Thus, the equation for maximum work for an electrochemical cell is

$$w_{\text{max}} = \quad n \quad \times \quad \mathscr{F} \quad \times \quad E$$
$$\updownarrow \qquad\qquad \updownarrow \qquad\qquad \updownarrow$$
$$\text{joules} = (\text{moles of electrons}) \times \left(\frac{\text{coulombs}}{\text{mole } e^-} \right) \times \left(\frac{\text{joules}}{\text{coulomb}} \right)$$

Since $\Delta G = -w_{\text{max}}$, then

$$\Delta G = -n\mathscr{F}E_{cell} \qquad\qquad (18.6)$$

When all species are at unit concentration as identified by the superscript $^\circ$ in E_{cell}°, then ΔG becomes the standard free energy change for the reaction, ΔG°. Thus, Equation 18.6 becomes

$$\Delta G^\circ = -n\mathscr{F}E_{cell}^\circ \qquad\qquad (18.7)$$

Equation 18.7 is valuable because it allows us to determine thermodynamic data from measurements of cell potentials. For example, for the zinc–copper cell, we have these half-reactions and net cell reaction.

$$\text{Zn}(s) \longrightarrow \text{Zn}^{2+}(aq) + 2e^-$$
$$\underline{\text{Cu}^{2+}(aq) + 2e^- \longrightarrow \text{Cu}(s)}$$
$$\text{Zn}(s) + \text{Cu}^{2+}(aq) \longrightarrow \text{Zn}^{2+}(aq) + \text{Cu}(s)$$

The n for this reaction is 2 (because two electrons are transferred), $\mathscr{F} = 96,500$ C/mol of electrons, and E_{cell}°, either derived from Table 18.1 or determined experimentally, is $+1.10$ V; therefore,

$$\Delta G^\circ = -2 \text{ mol } e^- \times \left(\frac{96,500 \text{ ¢}}{\text{mol } e^-} \right) \times \left(\frac{+1.10 \text{ J}}{\text{¢}} \right)$$
$$= -212,000 \text{ J}$$
$$= -212 \text{ kJ}$$

18.8 THERMODYNAMIC EQUILIBRIUM CONSTANTS FROM STANDARD CELL POTENTIALS

In Section 15.3 we saw that for reactions in solution

$$\Delta G° = -RT \ln K_c$$

Scientists who have studied electrochemistry have traditionally worked with common logarithms, so this equation is expressed as

$$\Delta G° = -2.303RT \log K_c \qquad (18.8)$$

Combining Equations 18.7 and 18.8 gives us

$$\Delta G° = -n\mathscr{F}E° = -2.303RT \log K_c$$

or simply

$$n\mathscr{F}E° = 2.303RT \log K_c$$

Solving for $E°$, we have

$$E° = \frac{2.303RT}{n\mathscr{F}} \log K_c \qquad (18.9)$$

If we choose to restrict ourselves to discussing reactions that take place at 25 °C (298 K), the quantity $2.303RT/\mathscr{F}$ becomes a constant.

$$\frac{2.303RT}{\mathscr{F}} = \frac{2.303(8.314 \text{ J/mol K})(298 \text{ K})}{96,500 \text{ C/mol}} = 0.0592 \text{ J/C}$$

Since I V = 1 J/C,

$$\frac{2.303RT}{\mathscr{F}} = 0.0592 \text{ V}$$

Thus, at 25 °C, Equation 18.9 becomes

$$E° = \frac{0.0592 \text{ V}}{n} \log K_c$$

Solving for $\log K_c$ gives

Standard cell potentials are an important source of equilibrium constants.

$$\log K_c = \frac{nE°}{0.0592 \text{ V}} \qquad (18.10)$$

We see, therefore, that from a knowledge of the standard cell potential, the equilibrium constant for the cell reaction can be calculated. For the Zn–Cu cell, we have

$$\log K_c = \frac{nE°}{0.0592 \text{ V}} = \frac{2(+1.10 \text{ V})}{0.0592 \text{ V}} = 37.2$$

Hence,

$$K_c = 10^{37.2}$$

$$= 2 \times 10^{37} \quad \text{(rounded)}$$

Recall the rules for significant figures for logarithms.

From the magnitude of this equilibrium constant, we could certainly say that the spontaneous Zn/Cu cell reaction will go very nearly to completion.

EXAMPLE 18.4. DETERMINING THE SPONTANEITY OF A REDOX REACTION AND CALCULATING ITS EQUILIBRIUM CONSTANT

PROBLEM: Using Table 18.1, determine whether the oxidation–reduction reaction

$$Sn(s) + Ni^{2+} \longrightarrow Sn^{2+} + Ni(s)$$

is spontaneous and calculate its equilibrium constant at 25 °C.

SOLUTION: The two half-reactions for this overall reaction are

$$Sn(s) \longrightarrow Sn^{2+}(aq) + 2e^- \qquad \text{(Oxidation)}$$

$$Ni^{2+}(aq) + 2e^- \longrightarrow Ni(s) \qquad \text{(Reduction)}$$

From Table 18.1 we find the reduction potential for the half-reaction involving Sn and Sn^{2+} to be $E^\circ_{Sn^{2+}|Sn} = -0.14$ V; for the nickel half-reaction $E^\circ_{Ni^{2+}|Ni} = -0.25$ V. Applying Equation 18.5, we get

$$E^\circ_{cell} = -0.25 - (-0.14)$$

$$= -0.11 \text{ V}$$

This means that under *standard conditions* (unit concentrations), the reaction in this direction is nonspontaneous. We can still calculate the equilibrium constant in the same fashion as outlined above. Since two electrons are transferred during the reaction,

$$\log K_c = \frac{2(-0.11 \text{ V})}{0.0592 \text{ V}} = -3.7$$

Taking the antilogarithm gives

$$K_c = 2 \times 10^{-4}$$

From the size of this equilibrium constant, we can say that this reaction will not occur to an appreciable extent in the forward direction.

18.9 THE EFFECTS OF CONCENTRATION ON CELL POTENTIALS: THE NERNST EQUATION

Until now, we have only discussed cells containing reactants at unit concentration. In the laboratory, however, we usually do not restrict ourselves to only this one set of conditions, and it is found that the cell emf, and in fact the direction of the cell reaction, can be controlled by the concentrations of the species taking part in the reaction. Let us examine this now from a quantitative point of view.

An equation that summarizes the way the free-energy change for a reaction varies with temperature and the concentrations of the reactants and products was given in Equation 15.2 on page 489.

$$\Delta G = \Delta G^\circ + RT \ln Q$$

where Q is the reaction quotient (the value of the mass action expression) for the reaction in question. Expressed in common logarithms, this equation becomes

$$\Delta G = \Delta G^\circ + 2.303RT \log Q$$

Equations 18.6 and 18.7 give the relationships between ΔG and E, and ΔG° and E°, respectively. Substituting these expressions into the preceding equation gives

$$-n\mathscr{F}E = -n\mathscr{F}E^\circ + 2.303RT \log Q$$

which can be rearranged to give

$$E = E^\circ - \frac{2.303RT}{n\mathscr{F}} \log Q \qquad\qquad (18.11)$$

This equation, developed by Walter Nernst in 1889, now bears his name and is called the **Nernst equation**.

It was Walter Nernst who discovered the third law of thermodynamics.

We've seen that at 25 °C, the value of $2.303RT/\mathscr{F}$ is 0.0592 V. Therefore, for cell reactions at 25 °C, the Nernst equation is

$$E = E° - \frac{0.0592 \text{ V}}{n} \log Q \qquad (18.12)$$

An interesting point to notice about this equation is that if all the ionic concentrations in the cell are 1 M and the partial pressures of any gases involved in calculating Q are 1 atm, then all the values in the mass action expression equal 1 and the value of Q equals 1. Since the log of 1 is zero, when the cell reaction takes place under standard conditions, $E = E°$ (which, of course, it must).

We can also derive Equation 18.10 from the Nernst equation. When a galvanic cell reaches equilibrium, it can no longer deliver any work and its potential becomes zero. (A dead battery is a galvanic cell that has reached equilibrium!) When this is true, $E_{cell} = 0$ and the reaction quotient is equal to K_c. Making these substitutions and rearranging the equation give Equation 18.10.

To write the Nernst equation for a galvanic cell, we need to have the balanced chemical equation. Consider, for example, the zinc–copper cell, for which the net cell reaction is

$$Zn(s) + Cu^{2+}(aq) \longrightarrow Zn^{2+}(aq) + Cu(s)$$

The Nernst equation as applied to this reaction is

$$E = E° - \frac{0.0592 \text{ V}}{n} \log \frac{[Zn^{2+}]}{[Cu^{2+}]}$$

Notice that we have omitted $Zn(s)$ and $Cu(s)$ from the mass action expression, which is the usual practice for heterogeneous reactions.[3] For this reaction, $n = 2$ and $E° = 1.10$ V, so the equation becomes

$$E = 1.10 \text{ V} - 0.0296 \text{ V} \log \frac{[Zn^{2+}]}{[Cu^{2+}]}$$

Using this equation, we can calculate the cell potential for any particular set of concentrations of Zn^{2+} and Cu^{2+} as illustrated in the following example.

EXAMPLE 18.5. USING THE NERNST EQUATION TO CALCULATE THE POTENTIAL OF A CELL UNDER NONSTANDARD CONDITIONS

PROBLEM: Calculate the emf of the Zn/Cu cell at 25 °C under the following conditions:

$$Zn(s)|Zn^{2+}(0.40 \text{ } M)||Cu^{2+}(0.020 \text{ } M)|Cu(s)$$

SOLUTION: The cell diagram tells us that the reaction is

$$Zn(s) + Cu^{2+}(0.020 \text{ } M) \longrightarrow Cu(s) + Zn^{2+}(0.40 \text{ } M)$$

In the cell diagram, the anode half-cell is given first, followed by the cathode half-cell. This tells us that Zn is oxidized and Cu^{2+} is reduced.

We have just seen that for this system the Nernst equation is

$$E = 1.10 \text{ V} - 0.0296 \text{ V} \log \frac{[Zn^{2+}]}{[Cu^{2+}]}$$

---

[3] The Nernst equation applies exactly only if we use activities. The activity of any pure solid or liquid is equal to 1. Errors introduced by using the concentrations of the ions instead of their activities are small, as mentioned before, provided that all the solutions are relatively dilute.

Substituting the concentrations of the Zn^{2+} and Cu^{2+} gives

$$E = 1.10 \text{ V} - 0.0296 \text{ V} \log \frac{(0.40)}{(0.020)}$$

$$= 1.10 \text{ V} - 0.0296 \text{ V} \log(20)$$

$$= 1.10 \text{ V} - 0.0296 \text{ V} (1.30)$$

$$= 1.10 \text{ V} - 0.0385 \text{ V}$$

$$= 1.06 \text{ V}$$

Thus, we see that under these conditions of concentration the voltage obtained from this cell is slightly less than that obtained at unit concentration.

18.10 APPLICATIONS OF THE NERNST EQUATION

One of the most common ways that the Nernst equation is used is in measuring and monitoring concentrations. To see how this can be done, suppose we connect a zinc half-cell in which the Zn^{2+} concentration is accurately known to a copper half-cell in which the copper ion concentration is unknown. We know the value of $E°$ for the cell ($E° = 1.10$ V) and the potential of the cell can be measured. Therefore, the only unknown quantity in the Nernst equation is the concentration of the Cu^{2+} ion in the copper half-cell, which can be obtained with little difficulty.

EXAMPLE 18.6. USING THE NERNST EQUATION TO OBTAIN AN UNKNOWN CONCENTRATION

PROBLEM: The apparatus shown in Figure 18.18 was used to measure the concentration of cadmium ion, Cd^{2+}, in an unknown solution. The galvanic cell is

$$Cd(s)|Cd^{2+}(?~M)\|Ag^+(1.000~M)|Ag(s)$$

for which the standard cell potential $E°_{cell}$ has been accurately measured to be 1.2022 V. The measured cell potential was 1.2871 V. What is the concentration of Cd^{2+} in the unknown solution?

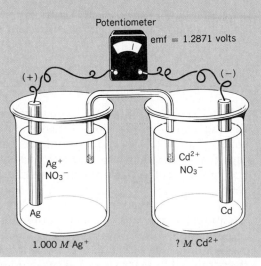

Figure 18.18. *An electro-chemical cell set up to determine the concentration of Cd^{2+} ion in a solution.*

ANALYSIS: The first step is to write the cell reaction and the Nernst equation, including the proper mass action expression. From the equation for the cell reaction, we can obtain the value of n (the number of electrons transferred). The values of E_{cell} and $E°_{cell}$ are known, and these can be substituted into the Nernst equation. Finally, we solve for the unknown Cd^{2+} concentration.

SOLUTION: The galvanic cell here is described by its cell diagram, which specifes the anode half-reaction first. Since oxidation occurs at the anode, the half-reaction for the cadmium is

$$Cd(s) \longrightarrow Cd^{2+}(aq) + 2e^-$$

For the silver, the half-reaction is

$$Ag^+(aq) + e^- \longrightarrow Ag(s)$$

When these are combined, the net cell reaction is

$$Cd(s) + 2Ag^+(aq) \longrightarrow Cd^{2+}(aq) + 2Ag(s)$$

Notice that the number of electrons transferred is 2. The Nernst equation for this cell is therefore

$$E_{cell} = E°_{cell} - \frac{0.0592 \text{ V}}{2} \log \frac{[Cd^{2+}]}{[Ag^+]^2}$$

Substituting the values for E_{cell} and $E°_{cell}$ gives

$$1.2871 \text{ V} = 1.2022 \text{ V} - 0.0296 \text{ V} \log \frac{[Cd^{2+}]}{[Ag^+]^2}$$

Let's solve for the log of the mass action expression.

$$\log \frac{[Cd^{2+}]}{[Ag^+]^2} = \frac{1.2871 \text{ V} - 1.2022 \text{ V}}{-0.0296 \text{ V}}$$

$$= -2.87$$

Taking the antilog gives the value of the mass action expression.

$$\frac{[Cd^{2+}]}{[Ag^+]^2} = 10^{-2.87} = 1.3 \times 10^{-3}$$

Solving for the cadmium ion concentration gives

$$[Cd^{2+}] = 1.3 \times 10^{-3} [Ag^+]^2$$

The silver ion concentration in its half-cell is 1.000 M, so

$$[Cd^{2+}] = 1.3 \times 10^{-3} M$$

This is the value of the unknown Cd^{2+} concentration.

In the preceding example, you have seen that a potential measurement permits the calculation of the molar concentration of an ion in a solution of unknown concentration. This unknown solution doesn't have to be in a beaker. It could be in a pipe that carries the waste water from a chemical plant. To monitor the concentration of Cd^{2+} in the waste stream, we need only make it part of the overall galvanic cell.

Potential measurements are very easy to make, and with today's electronic equipment, a galvanic cell such as this could be easily interfaced with a computer programmed to perform all the necessary calculations. A virtually continuous monitoring of the concentration is possible in a totally automated way. It is easy to see, therefore, why electrochemical measurements have become important tools for chemical analyses.

Determination of pH

This is a special application of the technique described above in which the species monitored is hydrogen ion. One way to accomplish this would be to connect a hydrogen electrode, dipping into a solution of unknown pH, with some other electrode whose potential is accurately known. For instance, we might use the Cu/H_2 cell discussed in Section 18.6. The cell reaction is

$$Cu^{2+}(aq) + H_2(g) \longrightarrow Cu(s) + 2H^+(aq)$$

and the corresponding form of the Nernst equation at 25 °C is

$$E = E° - \frac{0.0592 \text{ V}}{2} \log \frac{[H^+]^2}{[Cu^{2+}]p_{H_2}}$$

The symbol p_{H_2} designates the partial pressure of H_2.

If the concentration of Cu^{2+} is 1 M and the pressure of H_2 is 1 atm, this equation reduces to

$$E = E° - \frac{0.0592 \text{ V}}{2} \log[H^+]^2 \qquad (18.13)$$

EXAMPLE 18.7. USING THE NERNST EQUATION TO CALCULATE THE pH OF A SOLUTION

PROBLEM: A galvanic cell consisting of a Cu versus a hydrogen electrode was used to determine the pH of an unknown solution. The unknown was placed in the hydrogen electrode compartment and the pressure of the hydrogen gas was controlled at 1 atm. The concentration of Cu^{2+} was 1 M and the emf of the cell at 25 °C was determined to be +0.48 V. Calculate the pH of this unknown solution.

SOLUTION: The cell reaction for the Cu/H_2 cell is

$$Cu^{2+}(1 \text{ } M) + H_2(g)(1 \text{ atm}) \longrightarrow Cu(s) + 2H^+(? M)$$

for which we write the Nernst equation as

$$E = E° - \frac{0.0592}{n} \log \frac{[H^+]^2}{[Cu^{2+}]p_{H_2}}$$

Because $[Cu^{2+}] = 1$ M and $p_{H_2} = 1$ atm, and because $\log[H^+]^2 = 2 \log[H^+]$,

$$E = E° - \frac{(0.0592)(2)}{n} \log[H^+]$$

The $E°$ for this cell is +0.34 V; the value of n is 2. Substituting these values as well as the measured value of E into the equation gives

$$+0.48 \text{ V} = +0.34 \text{ V} - 0.0592 \text{ V} \log[H^+]$$

Let's rearrange this to solve for $-\log[H^+]$.

$$-\log[H^+] = pH = \frac{0.48 \text{ V} - 0.34 \text{ V}}{0.0592 \text{ V}} = 2.4$$

Notice that the unit volt cancels to give us a unitless value, which is what we want for pH.

Therefore, the pH of this solution is 2.4.

In Example 18.7, we saw that we could write Equation 18.13 as

$$E = E° - \frac{(0.0592 \text{ V})(\cancel{2})}{\cancel{2}} \log[H^+]$$

Figure 18.19. (a) *A photograph of a typical glass electrode.* (b) *A cutaway view of the glass electrode. In practice, it is always used in conjunction with another reference electrode, which is shown in cross section at the right.*

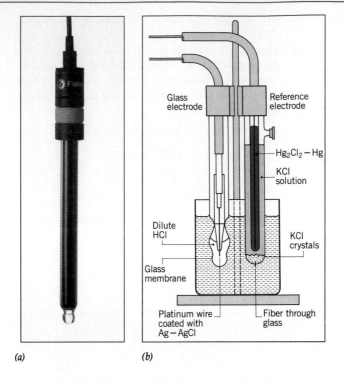

(a)   *(b)*

Let's rewrite this equation as

$$E = E° + 0.0592 \text{ V } (-\log[H^+])$$

By definition, pH $= -\log[H^+]$. Therefore, this equation becomes

$$E = E° + (0.0592 \text{ V}) \times (\text{pH})$$

$$\updownarrow \quad \updownarrow \qquad \updownarrow \qquad\quad \updownarrow$$

$$y = b + \quad m \qquad x \qquad\qquad (18.14)$$

As noted, Equation 18.14 has the same form as an equation for a straight line, which means that the value of *E* varies linearly with the pH of the solution. This fact forms the basis of operation of all modern pH meters, with one modification—the nature of the electrodes used to monitor the hydrogen ion concentration.

The glass electrode is an example of an *ion selective electrode,* an electrode that is sensitive to the concentration of just one ion in the solution.

Modern pH meters use a specialized electrode, called a **glass electrode,** to follow the hydrogen ion concentration. The basic construction of a glass electrode is illustrated in Figure 18.19. It consists of a silver wire covered with a layer of silver chloride that dips into a dilute solution of HCl. The HCl solution is enclosed within a very thin-walled glass membrane sealed into the end of a hollow glass tube. The emf of this electrode is sensitive to the difference between the concentrations of H^+ inside and outside. Since the H^+ concentration inside the electrode is constant, the electrode in effect becomes sensitive to the H^+ concentration in the solution in which it is dipped. The glass electrode is always used with another reference electrode to generate a potential difference that can be measured. The measurement of this potential difference, and translating it to pH, are the jobs of the pH meter.

18.11 PRACTICAL APPLICATIONS OF GALVANIC CELLS

The use of galvanic cells for the production of electricity has a long history. There is evidence, for example, that the Parthians (early inhabitants of the land now known as Iran) may have used primitive galvanic cells for gold-plating jewelry as early as 250 B.C. Today we enjoy the choice of a variety of galvanic cells, or batteries as they are better known, and they have become an intimate part of our modern society. The following discussion describes the chemistry of some of the more important and common types.

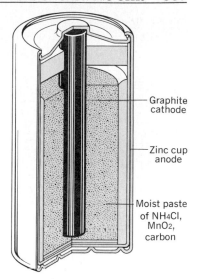

Graphite cathode

Zinc cup anode

Moist paste of NH4Cl, MnO2, carbon

Figure 18.20. *Cutaway view of a common dry cell.*

Zinc–carbon dry cell

This type of cell, technically known as a Leclanché cell, is used in flashlights, portable radios, toys, and the like. A cutaway diagram of a typical zinc–carbon dry cell is shown in Figure 18.20. It has an exterior layer of either cardboard or metal that serves only as a seal. Inside this outer shell is a zinc cup that serves as the anode. The zinc cup is filled with a moist paste consisting of ammonium chloride, manganese dioxide, and finely divided carbon. Immersed in this paste is a graphite rod, which serves as the cathode. The chemical reactions that take place when the circuit is completed are actually quite complex and, in fact, are not completely understood. The following, however, is perhaps a reasonable estimate of what occurs.

At the anode zinc is oxidized,

$$Zn(s) \longrightarrow Zn^{2+}(aq) + 2e^- \quad \text{(Anode)}$$

while at the carbon cathode the MnO_2/NH_4Cl mixture undergoes reduction to give a complex mixture of products. One of these reactions appears to be

$$2MnO_2(s) + 2NH_4^+(aq) + 2e^- \longrightarrow Mn_2O_3(s) + 2NH_3(aq) + H_2O(l) \quad \text{(Cathode)}$$

The ammonia produced at the cathode reacts with part of the Zn^{2+} formed at the anode to give the complex ion, $Zn(NH_3)_4^{2+}$. Because of the complex nature of the dry cell, no simple overall cell reaction can be written.

The common zinc–carbon dry cell suffers from the disadvantage that under heavy use it rather quickly appears to become "dead." Yet, if allowed to rest for a while, it appears to come back to life and can deliver additional current. When the cell delivers current, the reaction products can't diffuse away from electrodes very quickly. As they accumulate, it becomes more difficult for the electrode reactions to occur, and the voltage of the cell drops. After sitting idle for a while, however, these products diffuse away from the electrodes and the cell regains the ability to function.

Dry cells cannot be effectively recharged and, therefore, have a relatively short lifetime (as compared to the rechargeable lead storage and nickel–cadmium batteries, for example).

The physical size of a cell doesn't affect its voltage, but it does affect the amount of current that can be delivered. A small AA battery doesn't produce as much current as a larger D-size cell.

We say the cell has become polarized.

Alkaline battery

Another type of dry cell that uses zinc and manganese dioxide as reactants is the alkaline dry cell. Once again, zinc serves as the anode and manganese dioxide functions as the cathode. The electrolyte, however, contains potassium hydroxide and is therefore basic (alkaline). The zinc anode is also slightly porous, giving it a larger effective area. This allows the cell to deliver more current than the common zinc cell. As you've probably learned from TV commercials, these batteries are able to stand up better under heavy use and have a longer shelf life. The reactions in the alkaline battery are

$$Zn(s) + 2OH^-(aq) \longrightarrow Zn(OH)_2(s) + 2e^- \quad \text{(Anode)}$$

$$2MnO_2(s) + 2H_2O + 2e^- \longrightarrow 2MnO(OH)(s) + 2OH^-(aq) \quad \text{(Cathode)}$$

The cell produces an emf of about 1.5 V.

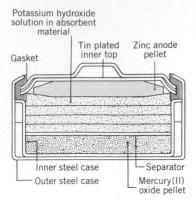

Figure 18.21. *Cross section of a mercury battery.*

Mercury battery

This was one of the first small batteries developed for commercial use. It consists of a zinc anode and a mercury(II) oxide (HgO) cathode in contact with a concentrated solution of potassium hydroxide (the electrolyte). The KOH solution is held in absorbent pads that separate the electrodes, as shown in Figure 18.21. The reactions that take place at the electrodes are

$$Zn(s) + 2OH^-(aq) \longrightarrow ZnO(s) + H_2O + 2e^- \quad \text{(Anode)}$$
$$HgO(s) + H_2O + 2e^- \longrightarrow Hg(l) + 2OH^-(aq) \quad \text{(Cathode)}$$

The net cell reaction is

$$Zn(s) + HgO(s) \longrightarrow ZnO(s) + Hg(l)$$

The mercury cell has a potential of about 1.35 V, which remains very nearly constant throughout its useful life. This is one of its principal advantages. (By comparison, the potential of an ordinary dry cell gradually decreases as it is discharged.)

Silver oxide battery

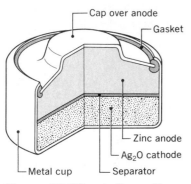

Figure 18.22. *A silver oxide battery.*

These tiny and rather expensive batteries (Figure 18.22) have become popular as power sources in electronic wristwatches, auto exposure cameras, and electronic calculators. The cathode reactant is silver oxide, Ag_2O, and the anode once again is zinc. The electrode reactions occur in a basic electrolyte.

$$Zn(s) + 2OH^-(aq) \longrightarrow Zn(OH)_2(s) + 2e^- \quad \text{(Anode)}$$
$$Ag_2O(s) + H_2O + 2e^- \longrightarrow 2Ag(s) + 2OH^-(aq) \quad \text{(Cathode)}$$

The emf of this battery is about 1.5 V.

Lead storage battery

The common automobile battery is a lead storage battery that usually delivers either 6 or 12 V, depending on the number of cells used in its construction. The inside of the battery consists of a number of galvanic cells connected to each other in series.

To increase the current output, each of the individual cells (Figure 18.23) contains a number of lead anodes connected together, plus a number of cathodes,

Figure 18.23. *A 12-volt lead storage battery, such as those used in most automobiles, consists of six cells like the one illustrated here. Notice that the anode and cathode each consist of several plates connected together. This allows the cell to produce the large currents necessary to start a car.*

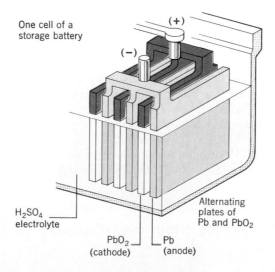

composed of PbO_2, also joined together. These electrodes are immersed in an electrolyte composed of dilute sulfuric acid (actually about 30% by weight in a fully charged cell). A single lead storage cell delivers 2 V, so that a 12-V battery contains six such cells connected in series.

When the external circuit is complete and the battery is in operation, the following oxidation–reduction reactions take place:

$$Pb(s) + SO_4^{2-}(aq) \longrightarrow PbSO_4(s) + 2e^- \qquad \text{(Anode)}$$

$$PbO_2(s) + 4H^+(aq) + SO_4^{2-}(aq) + 2e^- \longrightarrow PbSO_4(s) + 2H_2O \quad \text{(Cathode)}$$

and the overall reaction is

$$Pb(s) + PbO_2(s) + 4H^+(aq) + 2SO_4^{2-}(aq) \longrightarrow 2PbSO_4(s) + 2H_2O$$

These batteries have the advantage that the electrode reactions can be reversed by placing a voltage across the electrodes that is slightly larger than that which the battery can deliver. The recharging operation is performed in such a way that the negative external voltage is applied to the negative pole and the positive voltage to the positive pole. As this is done, the H_2SO_4 that is used up while the battery is in operation is restored. Recharging is accomplished by the generator or alternator of the car, or if the battery is really run down, with the aid of a battery charger.

A convenient method of estimating the degree to which the battery has been discharged is by checking the density (or specific gravity) of the electrolyte. If the battery is in a weakened state, the electrolyte will be mostly water — a product of the overall reaction — and will have a density somewhere near 1 g/mL. If, however, the battery is in good operating order, with a full charge, the density of the electrolyte will be somewhat higher than 1 g/mL (the density of concentrated sulfuric acid is 1.8 g/mL). The mechanic in a garage can perform this test with the aid of a **hydrometer,** a device having a float that sinks to a depth that is a function of the density of the liquid in which it is immersed.

The anodes in the most modern lead storage batteries are made of an alloy of lead and calcium that inhibits the electrolysis of water during recharging. Because water doesn't have to be added to these batteries periodically, they can be sealed to prevent leakage.

Nickel–cadmium cell

A storage cell that has acquired widespread use in recent years is the *nicad*, or nickel–cadmium battery. The anode in the cell is composed of cadmium, which undergoes oxidation in an alkaline (basic) electrolyte.

$$Cd(s) + 2OH^-(aq) \longrightarrow Cd(OH)_2(s) + 2e^- \quad \text{(Anode)}$$

The cathode is composed of NiO_2, which undergoes reduction.

$$NiO_2(s) + 2H_2O + 2e^- \longrightarrow Ni(OH)_2(s) + 2OH^-(aq) \quad \text{(Cathode)}$$

The net cell reaction during discharge is therefore

$$Cd(s) + NiO_2(s) + 2H_2O \longrightarrow Cd(OH)_2(s) + Ni(OH)_2(s)$$

The voltage of the cell is about 1.4 V, somewhat less than the dry cell.

The nicad battery has some appealing features. First, it has a longer life than a lead storage battery. Second, it can be packaged in a sealed unit, much like the common dry cell. These advantages have made the nicad the choice among manufacturers of such devices as rechargeable calculators and electronic flash units in photography.

Fuel cells

A fuel cell is an electrochemical cell that "burns" fuel under conditions that produce a potential between a pair of suitable electrodes. An example is the hydrogen–

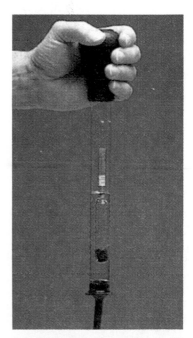

A battery hydrometer. Battery acid is drawn into the glass tube. The depth to which the float sinks is inversely proportional to the density of the sulfuric acid solution and the state of charge of the battery.

Figure 18.24. *A hydrogen–oxygen fuel cell.*

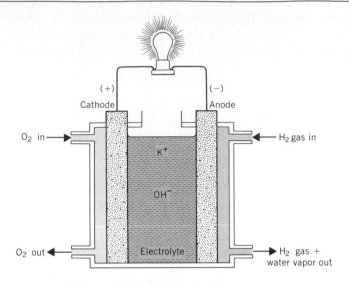

oxygen fuel cell illustrated in Figure 18.24. In this cell there are three compartments separated from one another by porous electrodes (either porous graphite or nickel have been used). The hydrogen gas is fed into one compartment and the oxygen gas into another. These gases diffuse through the electrodes and react with an electrolyte solution in the center compartment.

At the cathode the oxygen undergoes reduction, producing OH^- ions.

$$O_2(g) + 2H_2O(l) + 4e^- \longrightarrow 4OH^-(aq) \quad \text{(Cathode)}$$

At the anode, hydroxide ions react with H_2 gas by the equation

$$H_2(g) + 2OH^-(aq) \longrightarrow 2H_2O(l) + 2e^- \quad \text{(Anode)}$$

When these two half-reactions are properly combined, the net cell reaction is

$$2H_2(g) + O_2(g) \longrightarrow 2H_2O(l)$$

The fuel cell is operated at high temperature, so the water that is produced is removed as steam.

The H_2/O_2 fuel cell has been used successfully in space vehicles for years and has proven to be a reliable source of power. A secondary advantage is that the water produced as a product of the cell reaction can be condensed and used as drinking water for the astronauts.

Fuel cells offer some advantages over other sources of energy. Unlike the dry cell or storage battery, the cathode and anode materials can be continuously fed to the cell so that a constant and prolonged flow of electricity can be produced.

Another advantage of the fuel cell is its efficiency. The conventional method of generating electricity is to burn the fuel to generate heat. This is then used to make steam that drives turbines to generate electricity. The thermodynamic efficiency of this process is very low; only about 35 to 40% of the available energy is actually harnessed as electrical energy. The rest is given off as heat to the surroundings.

In a fuel cell, the redox reactions occur under much more nearly thermodynamically reversible conditions than in simple combustion, and that leads to a greatly improved thermodynamic efficiency, approaching 75%. Thus, for the oxidation of a given amount of fuel, a fuel cell produces nearly twice as much usable energy as combustion.

Despite these advantages, fuel cell applications are few at the present time because of their high cost of construction and the rather large physical size required to produce useful amounts of electrical energy.

The Space Shuttle uses H_2/O_2 fuel cells to provide electrical power.

REVIEW QUESTIONS AND PROBLEMS

Problems whose numbers are in blue have their answers in Appendix D at the back of the book. The more difficult problems are marked with asterisks.

General

18.1 Define *electrochemistry*. What is an electrochemical reaction?

18.2 How is electric charge carried in metallic conduction? How is electric charge transported in electrolytic conduction?

18.3 What must happen at the electrodes in order for electrolytic conduction to occur in an aqueous solution of an electrolyte?

18.4 Identify the kinds of chemical changes that take place at the cathode and at the anode (a) in an electrolytic cell and (b) in a galvanic cell.

Electrolysis

18.5 Sketch an electrolysis cell in which the net cell reaction is

$$MgCl_2(l) \longrightarrow Mg(l) + Cl_2(g)$$

(the electrolysis of molten $MgCl_2$).
(a) Identify the anode and cathode.
(b) Write the half-reactions that occur at each electrode.
(c) Indicate the direction of ion flow in the molten $MgCl_2$.
(d) Indicate the direction of electron flow in the external circuit.

18.6 Write equations for the half-reactions for the oxidation and reduction of water.

18.7 From the reactions discussed in Section 18.2, predict the products that you would obtain in the electrolysis of an aqueous solution of H_2SO_4.

18.8 What is the function of an electrolyte such as Na_2SO_4 or H_2SO_4 during the electrolysis of water? Why can't we carry out electrolysis on pure H_2O in the absence of an electrolyte?

18.9 The oxidation of I^- to I_2 occurs more easily than the oxidation of water, and the reduction of Ni^{2+} occurs more easily than the reduction of water. Write the anode and cathode half-reactions and the cell reaction for the electrolysis of aqueous NiI_2.

18.10 The reduction of Ni^{2+} to Ni occurs more easily than the reduction of water. Write the anode and cathode half-reactions and the cell reaction for the electrolysis of an aqueous solution of $NiSO_4$.

Practical Applications of Electrolysis

18.11 Write equations for the electrode reactions and the net reaction for (a) electrolysis of an unstirred brine solution, (b) electrolysis of a stirred brine solution.

18.12 What are the advantages and disadvantages of the diaphragm cell used to produce NaOH and Cl_2?

18.13 What are the advantages and disadvantages of the mercury electrolysis cell used to produce NaOH and Cl_2?

18.14 What function does the design of the Downs cell serve?

18.15 In the Hall process, why was cryolite mixed with the Al_2O_3 prior to its electrolysis to produce Al?

18.16 Why can Al not be produced by electrolysis of an aqueous solution containing a salt such as $Al_2(SO_4)_3$?

18.17 Write a series of chemical equations representing the reactions involved with the recovery of Mg from seawater.

18.18 Describe the electrolytic purification of metallic copper. Why is the process economically feasible?

18.19 What is electroplating? If an object is to be plated with nickel from a solution of $NiSO_4$, to which electrode (anode or cathode) would the object be connected? Why?

Quantitative Aspects of Electrolysis

18.20 What relationships connect moles of electrons to measurements of current and time in the laboratory?

18.21 What is a faraday?

18.22 How many moles of electrons are required to reduce 1 mol of each of the following to the indicated product?
(a) Cu^{2+} to Cu   (d) F_2 to $2F^-$
(b) Fe^{3+} to Fe^{2+} (e) NO_3^- to NH_3
(c) MnO_4^- to Mn^{2+}

18.23 Calculate the number of electrons that corresponds to 1 coulomb of charge.

18.24 How many moles of electrons are required to oxidize 1 mol of each of the following to give the indicated product?
(a) Cu^+ to Cu^{2+}
(b) Pb to PbO_2
(c) Cl_2 to $2ClO_3^-$
(d) O_2 to H_2O_2 (hydrogen peroxide)
(e) NH_3 to NO_3^-

18.25 How many moles of electrons are given by
(a) 8950 C
(b) a current of 1.5 A for 30 s
(c) a current of 14.7 A for 10 min

18.26 State how many minutes it would take to
(a) deliver 10,500 C using a current of 25.0 A
(b) deliver 0.65 mol e^- using a current of 15 A
(c) reduce 0.20 mol of Cu^{2+} to Cu using a current of 12 A

18.27 State how many minutes it would take to
(a) deliver 84,200 C using a current of 6.30 A
(b) deliver 1.25 mol e^- using a current of 8.40 A
(c) produce 0.500 mol of Al from molten Al_2O_3 (in cryolite) using a current of 18.3 A.

18.28 How many moles of electrons are required to produce the following?
(a) 10.0 mL of O_2 gas (at STP) from aqueous Na_2SO_4
(b) 10.0 g of Al from molten Al_2O_3 (in cryolite)
(c) 5.00 g of Na from molten NaCl
(d) 5.00 g of Mg from molten $MgCl_2$

18.29 How many grams of Na and Cl_2 would be produced if a current of 25 A was applied for 8.0 hr to the cell shown in Figure 18.3?

18.30 How many grams of O_2 and H_2 are produced in 1.0 hr when water is electrolyzed at a current of 0.50 A? What would be the volumes in liters, at STP, of O_2 and H_2?

18.31 How many grams of copper could be purified by a current of 115 A for 8.00 hr? Refer to Figure 18.13.

18.32 How many grams of silver could be plated out on a serving tray by electrolysis of a solution containing Ag in the +1 oxidation state for a period of 8.00 hr at a current of 8.46 A? What area would this cover, if we assume that the density of Ag is 10.5 g/cm³ and the thickness of the silver plate is 0.00100 in. (0.00254 cm)?

18.33 How many seconds would it take to deposit 21.4 g of Ag from a solution of $AgNO_3$ by a current of 10.0 A?

18.34 How many hours would it take to deposit 35.3 g of Cr from a solution of $CrCl_3$ at a current of 6.00 A?

18.35 How many minutes would it take to plate out 5.00 g of copper from a solution of $CuSO_4$ at a current of 5.00 A?

18.36 What current in amperes is required to deposit 0.225 g of Ni from a solution of $NiSO_4$ in 10.0 min?

18.37 What current in amperes is required to produce 1.33 g of Cl_2 from a solution of NaCl in 45.0 min?

18.38 What current in amperes is required to produce 50.0 mL of O_2 gas, measured at STP, by the electrolysis of H_2O for a period of 3.00 hr?

18.39 If there are no other competing reduction reactions, how many minutes will it take to remove all the Cr from 500 mL of 0.270 M $Cr_2(SO_4)_3$ by a constant current of 3.00 A?

18.40 In an experiment two electrolytic cells were connected so that the same current passed through both of them. One cell contained $CuSO_4$, the other an unknown salt. It was found that 1.25 g of copper was plated out during the same period of time that 3.42 g of the unknown metal was plated out.

(a) How many moles of electrons passed through these cells?
(b) If the oxidation state of the unknown metal ion in the solution was 2 +, what is the atomic mass of the unknown?

18.41 Two electrolytic cells were connected in series so that the same current passed through each of them. In an experiment, 0.125 mol of Cu was deposited from a solution of $CuSO_4$ in one of the cells. How many moles of Cr were deposited at the same time from a $Cr_2(SO_4)_3$ solution in the other?

*18.42 A current of 0.250 A is passed through 400 mL of a 2.00 M solution of NaCl for 35.0 min. What will be the pH of the solution after the current is turned off?

Galvanic Cells

18.43 In Section 18.5 we saw that electrical energy can be extracted from a Zn/Cu cell. If Cu^{2+} is brought into contact with metallic Zn, it is reduced to Cu while the Zn is oxidized, without the generation of electricity. In this case, what happens to the energy that is *not* being extracted as electrical energy?

18.44 What is the function of a salt bridge in a galvanic cell?

18.45 What are the signs of the anode and cathode in galvanic and electrolytic cells.

18.46 Sketch a galvanic cell in which the following net reaction occurs: $Ni^{2+}(aq) + Fe(s) \rightarrow Ni(s) + Fe^{2+}(aq)$.
(a) Label the cathode and anode.
(b) Indicate the charges on the electrodes.
(c) Indicate the direction of electron flow.
(d) Indicate the direction of the flow of cations and anions.
(e) If the concentrations of the ions are each 1 M, what is the potential of the cell?

18.47 How can one identify *experimentally* which electrode in a galvanic cell is the anode and which is the cathode?

Cell Diagrams

18.48 Write the cell diagrams that correspond to the following net cell reactions.
(a) $Ni^{2+}(1\ M) + Fe(s) \rightarrow Fe^{2+}(1\ M) + Ni(s)$
(b) $2Cr(s) + 3Sn^{2+}(1\ M) \rightarrow 3Sn(s) + 2Cr^{3+}(1\ M)$
(c) $Cu(s) + 2Ag^+(1\ M) \rightarrow Cu^{2+}(1\ M) + 2Ag(s)$

18.49 Write the chemical equations for the cell reactions that occur in the following galvanic cells. For each, identify the anode and the cathode.
(a) $Zn(s)|Zn^{2+}(1\ M)||Pb^{2+}(1\ M)|Pb(s)$
(b) $Al(s)|Al^{3+}(1\ M)||Ag^+(1\ M)|Ag(s)$
(c) $Mn(s)|Mn^{2+}(1\ M)||Fe^{3+}(1\ M)|Fe(s)$

Reduction Potentials

18.50 Sketch a hydrogen electrode. Under what conditions is its reduction potential defined as exactly 0.000 V?

18.51 A galvanic cell was set up with a zinc electrode dipping into 1.00 M $ZnSO_4$ as one half-cell and a gallium electrode dipping into 1.00 M $GaCl_3$ solution as the other. The emf of the cell was measured to be 0.23 V. The zinc electrode was found to be negatively charged.
   (a) Write the anode and cathode half-reactions that are taking place in this cell.
   (b) Write the cell reaction.
   (c) Calculate $E°_{Ga^{3+}|Ga}$ for $Ga^{3+} + 3e^- \rightleftharpoons Ga(s)$.
   (d) What would be the standard potential of a gallium–copper galvanic cell?

18.52 Which is the better oxidizing agent? (Refer to Table 18.1.)
   (a) Li^+ or Ca^{2+}
   (b) Cl_2 or F_2
   (c) H_2O or Al^{3+}
   (d) $S_2O_8^{2-}$ or Cl_2
   (e) Br_2 or H_2O

18.53 Which is the better oxidizing agent? (Refer to Table 18.1.)
   (a) Cl_2 or ClO_3^-
   (b) O_2 or $Cr_2O_7^{2-}$
   (c) MnO_4^- or $Cr_2O_7^{2-}$
   (d) PbO_2 or Hg_2Cl_2

18.54 Which is the better reducing agent? (Refer to Table 18.1.)
   (a) Ni or Fe
   (b) H_2 or Mg
   (c) Br^- or I^-
   (d) SO_4^{2-} or F^-
   (e) Sn or Mn

18.55 Which is the better reducing agent? (Refer to Table 18.1.)
   (a) Na or Cr
   (b) $PbSO_4$ or Cl_2
   (c) Ag or Cu
   (d) I^- or Sn
   (e) H_2 or H_2O

18.56 Calculate the cell potential for each of the galvanic cells described in Question 18.48.

18.57 Calculate the cell potential for each of the galvanic cells described in Question 18.49.

Cell Potentials and Spontaneity of Reactions

18.58 Without computing $E°$, determine what reactions will occur spontaneously among the following sets of reactants in aqueous solution:
   (a) $Al(s)$, $Ni(s)$, $NiSO_4(aq)$, $Al_2(SO_4)_3(aq)$
   (b) $PbO_2(s)$, $K_2Cr_2O_7(aq)$, $H_2SO_4(aq)$, $PbSO_4(s)$, $Cr_2(SO_4)_3(aq)$
   (c) $Ag(s)$, $AgNO_3(aq)$, $Pb(s)$, $Pb(NO_3)_2(aq)$
   (d) $MnO_2(s)$, $HCl(aq)$, $Cl_2(g)$, $MnCl_2(aq)$
   (e) $Mn(s)$, $HCl(aq)$, $MnCl_2(aq)$, $H_2(g)$

18.59 Without computing $E°$, determine whether the following reactions will occur spontaneously:
   (a) $2Fe^{3+} + Sn \rightarrow 2Fe^{2+} + Sn^{2+}$
   (b) $Cu + 2H^+ \rightarrow Cu^{2+} + H_2$
   (c) $3Mg^{2+} + 2Al \rightarrow 3Mg + 2Al^{3+}$
   (d) $Mn + Zn^{2+} \rightarrow Mn^{2+} + Zn$
   (e) $PbO_2 + SO_4^{2-} + 4H^+ + 2Hg + 2Cl^- \rightarrow$
   $$Hg_2Cl_2 + PbSO_4 + 2H_2O$$

18.60 Compute $E°$ and use its value to determine whether the following reactions will occur spontaneously:
   (a) $Ca^{2+} + Mg \rightarrow Ca + Mg^{2+}$
   (b) $Pb^{2+} + 2Cl^- \rightarrow Pb + Cl_2$
   (c) $2Cl^- + S_2O_8^{2-} \rightarrow Cl_2 + 2SO_4^{2-}$
   (d) $6Mn^{2+} + 5Cr_2O_7^{2-} + 22H^+ \rightarrow$
   $$6MnO_4^- + 10Cr^{3+} + 11H_2O$$
   (e) $O_2 + 4Cl^- + 4H^+ \rightarrow 2H_2O + 2Cl_2$

18.61 Given the following sets of half-reactions, write the net cell reaction and calculate $E°$ for the spontaneous changes that will occur:
   (a) $Hg_2Cl_2 + 2e^- \rightleftharpoons 2Hg + 2Cl^-$
   $PbSO_4 + 2e^- \rightleftharpoons Pb + SO_4^{2-}$
   (b) $AgCl + e^- \rightleftharpoons Ag + Cl^-$
   $Cu^{2+} + 2e^- \rightleftharpoons Cu$
   (c) $Mn^{2+} + 2e^- \rightleftharpoons Mn$
   $Cl_2(g) + 2e^- \rightleftharpoons 2Cl^-$
   (d) $Al^{3+} + 3e^- \rightleftharpoons Al$
   $Br_2(aq) + 2e^- \rightleftharpoons 2Br^-$

18.62 Determine the value of $E°$ for each of the spontaneous reactions in Problem 18.58.

18.63 Determine the value of $E°$ for each of the reactions as written from left to right in Problem 18.59.

Calculating Equilibrium Contants from $E°$

18.64 Calculate the equilibrium constants for the following cell reactions:
   (a) $Ni(s) + Sn^{2+}(aq) \rightleftharpoons Ni^{2+}(aq) + Sn(s)$
   (b) $Cl_2(aq) + 2Br^-(aq) \rightleftharpoons Br_2(aq) + 2Cl^-(aq)$
   (c) $Fe^{2+}(aq) + Ag^+(aq) \rightleftharpoons Ag(s) + Fe^{3+}(aq)$

18.65 Calculate the equilibrium constants for the reactions in Problem 18.59.

18.66 Calculate the equilibrium constants for the reactions in Problem 18.60.

Standard Cell Potentials and $\Delta G°$

18.67 Calculate $\Delta G°_{298}$ in kilojoules for each reaction in Problem 18.59.

18.68 Calculate $\Delta G°_{298}$ in kilojoules for each reaction in Problem 18.60.

Nernst Equation

18.69 Write the Nernst equation and calculate $E°$ and E for the following reactions:
   (a) $Cu^{2+}(0.1\ M) + Zn(s) \rightarrow Cu(s) + Zn^{2+}(1.0\ M)$
   (b) $Sn^{2+}(0.5\ M) + Ni(s) \rightarrow Sn(s) + Ni^{2+}(0.01\ M)$
   (c) $F_2(g, 1\ atm) + 2Li(s) \rightarrow 2Li^+(1\ M) + 2F^-(0.5\ M)$
   (d) $Zn(s) + 2H^+(0.01\ M) \rightarrow Zn^{2+}(1\ M) + H_2(1\ atm)$
   (e) $2H^+(1.0\ M) + Fe(s) \rightarrow H_2(1\ atm) + Fe^{2+}(0.2\ M)$

18.70 Calculate $E°$, E, and ΔG (in kilojoules) for the following cell reactions (not balanced):

(a) $Al(s) + Ni^{2+}(0.80\ M) \rightarrow Al^{3+}(0.020\ M) + Ni(s)$
(b) $Ni(s) + Sn^{2+}(1.10\ M) \rightarrow Sn(s) + Ni^{2+}(0.010\ M)$
(c) $Ag^+(0.050\ M) + Zn(s) \rightarrow Ag(s) + Zn^{2+}(0.010\ M)$

18.71 Calculate the cell potential for the following:

(a) $Sn(s) + Pb^{2+}(0.050\ M) \rightarrow Sn^{2+}(1.50\ M) + Pb(s)$
(b) $3Zn(s) + 2Cr^{3+}(0.010\ M) \rightarrow$
$$3Zn^{2+}(0.020\ M) + Cr(s)$$
(c) $PbO_2(s) + SO_4{}^{2-}(0.010\ M) + 4H^+(0.10\ M) +$
$Cu(s) \rightarrow PbSO_4(s) + 2H_2O + Cu^{2+}(0.0010\ M)$

18.72 What is the reduction potential of a half-cell composed of a copper wire dipping into $2 \times 10^{-4}\ M\ CuSO_4$? (The Nernst equation applies to individual half-cells, too.)

*18.73 Calculate the value of ΔG (in kilojoules) for a system containing the following species: $Mn^{2+}(0.10\ M)$, $Cr_2O_7{}^{2-}(0.010\ M)$, $MnO_4{}^-(0.0010\ M)$, $Cr^{3+}(0.0010\ M)$. The pH of the solution is 6.00. The reaction that you should consider is

$$6Mn^{2+}(aq) + 5Cr_2O_7{}^{2-}(aq) + 22H^+(aq) \rightleftharpoons$$
$$6MnO_4{}^-(aq) + 10Cr^{3+}(aq) + 11H_2O(l)$$

In which direction will this reaction proceed to get to equilibrium from the starting conditions given above?

18.74 The solubility product constant of AgBr is 5×10^{-15}. What will be the potential of a cell constructed using the H_2 electrode ($[H^+] = 1.0\ M$, $p_{H_2} = 1$ atm) versus a half-cell containing a silver wire coated with AgBr immersed in $0.010\ M$ HBr?

*18.75 A Ag/AgCl electrode dipping into 1 M HCl has a standard reduction potential of $+0.22$ V [$AgCl(s) + e^- \rightleftharpoons Ag(s) + Cl^-(aq)$]. A second Ag/AgCl electrode is dipped into a solution containing Cl^- at an unknown concentration. The cell generates a potential of 0.0435 V, with the electrode in the unknown solution serving as the anode. What is the molar concentration of Cl^- in the unknown?

*18.76 A galvanic cell was set up using silver as one electrode dipping into 200 mL of $0.100\ M\ AgNO_3$ and magnesium as the other electrode dipping into 250 mL of $0.100\ M\ Mg(NO_3)_2$ solution.

(a) What is the potential of the cell?
(b) Suppose that current was drawn from the cell for a period of time until 1.00 g of silver plated out on the silver electrode. What is the potential of the cell now?
(c) Suppose that the original magnesium electrode had only weighed 0.080 g (it consisted of 0.080 g of magnesium deposited on an inert platinum electrode). What would the potential of the cell be just before the last tiny bit of magnesium dissolved?

$E°$ and Work

*18.77 How many hours will a 25-watt light bulb burn if it is powered by a lead storage battery that has available 25.0 g of Pb that can react as an anode? Assume a constant voltage of 1.5 V. (1 watt = 1 J/s)

*18.78 What masses of H_2 and O_2 in grams would have to react each second in a fuel cell at $110°$ C to provide 1.0 kilowatt (kW) of power, if we assume a thermodynamic efficiency of 70%? (*Hint:* Use the data in Chapters 6 and 14 to compute $\Delta G°$ for the reaction $H_2(g) + \frac{1}{2}O_2(g) \rightarrow H_2O(g)$ at 110 °C. 1 watt = 1 J/s.)

*18.79 How much work, expressed in kilojoules, is able to be accomplished by a 5.00-min flow of electricity having a voltage of 110 V and a current of 1.00 A?

*18.80 If we assume that the typical electric generating plant has an efficiency of only about 30%, what volume (in liters) of fuel having an average formula of $C_{12}H_{26}$ must be burned, giving $H_2O(g)$ and $CO_2(g)$, to produce 1.0 kilowatt hour (kW·h) of electricity? Assume $\Delta H_f°$ of $C_{12}H_{26}(l) = 291$ kJ/mol and a density of 0.74 g/mL. 1 watt = J/s.

Using Cell Potentials to Measure Concentrations

18.81 The standard cell potential $E° = 0.6365$ V for the galvanic cell

$$\overset{\smile}{Z}n(s)|Zn^{2+}(0.500\ M)\|Pb^{2+}(?\ M)|Pb(s)$$

The potential of the cell was measured to be 0.4438 V. What was the molar concentration of Pb^{2+} in its half-cell?

18.82 The cell reaction

$$3Ag^+(aq) + Ga(s) \longrightarrow Ga^{3+}(aq) + 3Ag(s)$$

has $E° = 1.360$ V. A galvanic cell was prepared in which a gallium metal electrode was dipping into $0.800\ M\ Ga^{3+}$ and a silver electrode was dipping into a solution of unknown silver ion concentration. The potential of the cell was measured to be 1.122 V. What was the value of the silver ion concentration in moles per liter?

Practical Galvanic Cells

18.83 Describe the anode and cathode reactions in the ordinary dry cell (the Leclanché cell).

18.84 Why does a dead dry cell seem to come back to life if left idle for a while?

18.85 What are the anode, cathode, and net cell reactions in the alkaline zinc–manganese dioxide battery? What is the electrolyte?

18.86 What are the anode, cathode, and net cell reactions in the mercury battery?

18.87 What are the cathode, anode, and net cell reactions in the silver oxide battery?

18.88 What are the reactions that take place during the discharge of the lead storage battery? What reactions occur when this battery is being charged?

18.89 Write the anode, cathode, and overall cell reactions for the discharge of the nickel–cadmium battery.

18.90 What is a fuel cell? What advantages does a fuel cell offer over the lead storage battery?

18.91 What possible advantages do fuel cells offer over current electrical power plants?

Additional Exercises

* **18.92** An unstirred solution of 2.00 M NaCl was electrolyzed for a period of 25.0 min and then titrated with 0.250 M HCl. The titration required 15.5 mL of the acid. What was the average current in amperes during the electrolysis?

* **18.93** A student set up an electrolysis apparatus and passed a current of 1.22 A through a 3 M H_2SO_4 solution for 30.0 min. He collcted the H_2 evolved and found that it occupied a volume, over water at 27 °C, of 288 mL at a total pressure of 767 torr. Use these data to calculate the charge on the electron, expressed in coulombs.

* **18.94** What current would be required to deposit 1 m² of chrome plate having a thickness of 0.050 mm in 25 min from a solution containing H_2CrO_4? The density of Cr is 7.19 g/mL.

* **18.95** A hydrogen electrode is immersed in a 0.10 M solution of acetic acid. This electrode is connected to another consisting of an iron nail dipping into 0.10 M $FeCl_2$. What will the measured emf of this cell be? Assume $p_{H_2} = 1$ atm.

* **18.96** A cell was constructed using the standard hydrogen electrode ([H^+] = 1.0 M, $p_{H_2} = 1$ atm) in one compartment and a lead electrode in a 0.10 M K_2CrO_4 solution in contact with undissolved $PbCrO_4$. The potential of the cell was measured to be 0.51 V with the Pb electrode as the anode. Determine the K_{sp} of $PbCrO_4$ from these data.

18.97 Calculate the potential of the galvanic cell

$$Zn(s)|Zn^{2+}(0.100\ M)||Ag^+(0.0500\ M)|Ag(s)$$

at a temperature of 45 °C.

18.98 What is the value of the equilibrium constant for the reaction

$$Sn^{2+}(aq) + Pb(s) \rightleftharpoons Pb^{2+}(aq) + Sn(s)$$

at 50 °C?

CHEMICAL KINETICS: THE STUDY OF THE RATES OF CHEMICAL REACTIONS

In the previous four chapters, we studied the ultimate fate of chemical reactions—that is, what chemical systems will be like when they reach equilibrium. This kind of information is provided by thermodynamics, which tells us also whether specific chemical changes are spontaneous. But there is another factor that plays a major role in determining what happens in our universe, and this is the speed at which events take place.

If all spontaneous chemical reactions occurred instantaneously, our lives would be over in a flash and our universe would have come to equilibrium long ago. Fortunately for us, some reactions occur slowly and others at a more rapid pace, and our knowledge of this constantly affects the decisions we make in our daily lives. For example, an architect will decide on building materials based in part on their relative rates of reaction with oxygen and moisture. Thus, if a strong metal is needed in a severely corrosive environment, stainless steel will be chosen in preference to ordinary steel because it is oxidized much more slowly.

The study of rates of reaction is sometimes called *chemical dynamics.*

Chemical kinetics is the name we give to the study of the speeds, or rates of chemical reactions. One of our major goals is to learn about the factors that control how rapidly chemical changes occur. These can be divided into four main categories:

1. *The nature of the reactants and products.* All other factors being equal, some reactions are just naturally fast and others are naturally slow, depending on the chemical makeup of the molecules or ions involved.

2. *The concentrations of the reacting species.* For two molecules to react with each other, they must meet, and the probability that this will happen in a

homogeneous mixture increases as their concentrations increase. For heterogeneous reactions — those in which the reactants are in separate phases — the rate also depends on the area of contact between the phases. Since many small particles have a much larger area than one large particle of the same total mass, decreasing the particle size increases the reaction rate. Sometimes this can have devastating results, as we can see in Figure 19.1.

Chopping a log into small pieces of kindling with a large total surface area makes a campfire easier to start.

3. *The effect of temperature.* Nearly all chemical reactions take place faster when their temperatures are increased.

4. *The influence of outside agents called catalysts.* The rates of many reactions, including virtually all biochemical reactions, are affected by substances called catalysts that undergo no net chemical change during the course of the reaction.

Studying how these factors affect the rate of a reaction serves several purposes. For example, it allows us to adjust the conditions of a reaction system to obtain the products as quickly as possible. The importance of this in the commercial manufacture of chemicals is obvious. It also allows us to adjust conditions to make a reaction occur as slowly as possible. This is helpful, for instance, in controlling the growth of fungi and other microorganisms that spoil foods.

For chemists, one of the most significant benefits that comes from studying reaction rates is knowledge about the details of how chemical changes take place. We will see later in this chapter that a chemical reaction usually does not occur in one single step that involves the simultaneous collision of all the reactant molecules described in the balanced overall equation. Instead, the net change is the result of a sequence of simple reactions. This sequence is called the **mechanism** of the reaction, and studying reaction rates gives clues to what the mechanism is. In this way we gain insight into the fundamental reasons of why substances react the way they do.

Figure 19.1. *The remains of a grain elevator in New Orleans, Louisiana, after it exploded in December 1977, killing 35 people. Rapid combustion of very fine dust caused the explosive effects seen here.*

19.1 REACTION RATES AND THEIR MEASUREMENT

Before we examine in more detail the factors that influence rates of reaction, let's be sure that we know what is meant by rate. In general, the rate (or speed) of any chemical reaction can be expressed as the ratio of the change in the concentration of a reactant (or product) to a change in time. This is exactly analogous to giving the speed of an automobile as the change in position (that is, the distance traveled) divided by its time of travel. Here the speed might be given in miles per hour. With chemical reactions, the rate is usually expressed in moles per liter per second.

$$\text{speed of auto} = \text{rate of travel} = \frac{\text{change in position}}{\text{time}} = \frac{\text{miles}}{\text{hour}}$$

$$\text{rate of chemical reaction} = \frac{\text{change in concentration}}{\text{time}}$$

$$= \frac{\text{moles/liter}}{\text{second}} = \frac{\text{mol/L}}{\text{s}}$$

$$= \text{mol L}^{-1}\text{ s}^{-1}$$

To determine the rate of a given chemical reaction, we must measure how fast the concentration of a reactant or product changes. In practice, the species whose concentration is easiest to follow is determined at various time intervals. The simplest example is a reaction in which only one reactant undergoes a change to form a single product. An example is the conversion of cyclopropane to propylene.

cyclopropane   propylene

In general, the balanced equation for a reaction with this stoichiometry is

$$A \longrightarrow B \tag{19.1}$$

When the reaction is carried out, no product (B) is present initially and, as time goes on, the concentration of B increases with a corresponding decrease in the concentration of A (see Figure 19.2). Notice that the rate of this chemical reaction changes with time. For instance, near the start of the reaction the concentration of A is decreasing rapidly and the concentration of B is rising rapidly. Much later during the reaction, however, only small changes in concentration occur with time, and the rate is therefore much less. In general, this type of behavior is observed with nearly every chemical reaction: As the reactants are consumed, the rate of reaction gradually decreases.

In more complex reactions, the rates of formation of the various products and the rates of disappearance of the various reactants are not all equal, but are related by the coefficients in the overall balanced equation. For example, consider the reaction

$$N_2(g) + 3H_2(g) \longrightarrow 2NH_3(g)$$

We see that for every molecule of N_2 that reacts, three molecules of H_2 react. This means that the hydrogen is disappearing three times as fast as the nitrogen. The coefficients also tell us that two molecules of NH_3 are formed from each N_2, so the rate at which NH_3 is formed must be twice as fast as the rate at which N_2 disappears.

Measurement of reaction rates

An accurate, quantitative estimate of the rate of reaction at any given moment during the reaction can be obtained from the slope of the tangent to the concentration–time

Cyclopropane is used as a fast-acting anesthetic.

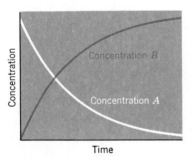

Figure 19.2. *Changes in the concentrations of the reactants and products as a function of time for the reaction* A → B.

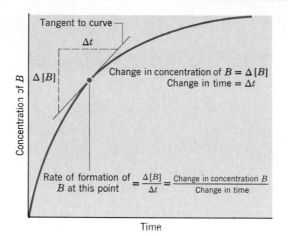

Figure 19.3. *Estimating the rate of reaction at a particular time during a reaction. The slope of the tangent to the curve gives the change in the concentration of* B *with time, which is the rate. The reaction is* A → B.

curve at that particular instant. This is shown in Figure 19.3. As in our earlier discussions, square brackets, [ ], are used here to denote concentration in moles per liter. From the tangent to the curve we can write

$$\frac{moles}{liters} = molar\ concentration$$

$$\text{rate} = \frac{\Delta[B]}{\Delta t} \qquad (19.2)$$

We can also express the rate of the above reaction in terms of the concentration of the reactant A, because its concentration is also changing with time. The rate measured in terms of the concentration of A would be

$$\text{rate} = \frac{-\Delta[A]}{\Delta t} \qquad (19.3)$$

The minus sign indicates that the concentration of A is decreasing with time. A minus sign is always used whenever reactants are employed to express the rate.

When measuring the rate of any chemical reaction, the concentration that is monitored and the technique that is used to measure the change depend on the nature of the reaction. For example, for gaseous reactions the pressure can be followed, provided that there is a change in the number of moles of gas as the reaction proceeds. On the other hand, if a colored reactant or product is involved, the intensity of color can be monitored during the reaction. Whatever method of analysis is employed, it must be fast, accurate, and in no way interfere with the normal course of the reaction being studied.

EXAMPLE 19.1. CALCULATING RELATIVE RATES OF REACTION

PROBLEM: Ammonia can be made to burn according to the reaction

$$4NH_3(g) + 5O_2(g) \longrightarrow 4NO(g) + 6H_2O(g)$$

Suppose that at a particular moment during the reaction the ammonia is reacting at the rate of 0.24 mol L^{-1} s^{-1}. (a) What is the rate at which oxygen is reacting? (b) What is the rate at which H_2O is being formed?

SOLUTION: The solution of this problem simply uses the coefficients of the equation to construct conversion factors relating the various numbers of moles of the reactants and products. We are given

$$\text{rate (for NH}_3) = \frac{-0.24 \text{ mol NH}_3}{\text{L s}}$$

The negative sign, you recall, means that the concentration is *decreasing* with time.

The coefficients in the equation allow us to construct the conversion factors

$$\frac{5 \text{ mol } O_2}{4 \text{ mol } NH_3} \quad \text{and} \quad \frac{6 \text{ mol } H_2O}{4 \text{ mol } NH_3}$$

We use these as follows.

(a)   For the rate at which oxygen is consumed

$$\text{rate (for } O_2) = \frac{-0.24 \text{ mol } NH_3}{\text{L s}} \times \left(\frac{5 \text{ mol } O_2}{4 \text{ mol } NH_3} \right)$$

$$= \frac{-0.30 \text{ mol } O_2}{\text{L s}}$$

(b)   For the rate at which water is formed

$$\text{rate (for } H_2O) = \frac{0.24 \text{ mol } NH_3}{\text{L s}} \times \left(\frac{6 \text{ mol } H_2O}{4 \text{ mol } NH_3} \right)$$

$$= \frac{0.36 \text{ mol } H_2O}{\text{L s}}$$

This rate is given with a positive sign because the concentration of H_2O is increasing.

19.2 RATE LAWS

In this section we begin to examine the factors that control the rate of reaction. Not all reactions take place at the same rate. Ionic reactions like those described in Chapter 5 are virtually instantaneous; the speed is determined by how rapidly we can mix the chemicals. Other reactions, such as the digestion of food, take place more slowly. These different rates exist primarily because of chemical differences among the reacting substances.

For any given reaction, one of the most important controlling influences is the concentrations of the reactants. Generally, if we follow a chemical reaction over a period of time, we find that its rate gradually decreases as the reactants are consumed. From this we conclude that the rate is related, in some way, to the concentrations of the reacting species. In fact, the rate is nearly always proportional to the concentrations of the reactants, each raised to some power. This means that for the general reaction

$$A \longrightarrow B$$

the rate can be written as

$$\text{rate} \propto [A]^x \tag{19.4}$$

where the exponent, x, is called the **order of the reaction.** When $x = 1$, we have a first-order reaction. An example is the decomposition of cyclopropane mentioned earlier

$$\text{rate} \propto [\text{cyclopropane}]^1$$

In some cases an exponent can even be negative, which means that increasing the concentration of that reactant decreases the reaction rate.

Second-order ($x = 2$), third-order ($x = 3$), and higher-order reactions are also possible, as are reactions in which x is a fraction. There are also examples of zero-order reactions, for which $x = 0$. For a zero-order reaction the rate is constant and does not depend on the concentrations of the reactants. An example is the decomposition of ammonia on a platinum or tungsten metal surface. The rate at which the ammonia decomposes is always the same, regardless of its concentration. Another example of a zero-order process is the elimination of ethyl alcohol by the body. Regardless of how much alcohol is present in the bloodstream, its rate of expulsion from the body is constant. Thus, the rate is independent of concentration.

A very important fact is that there is not necessarily any direct obvious relationship between the coefficients in the balanced chemical equation for a reaction and the order of the reaction. *The value of x can only be determined from experiment.*

If we consider a slightly more complex reaction, for example,

$$A + B \longrightarrow \text{products}$$

the rate usually depends on the concentrations of both A and B. Normally, increasing the concentration of either A or B will increase the reaction rate, and the rate is proportional to the product of the concentrations of A and B, each raised to some power.

$$\text{rate} \propto [A]^x[B]^y \qquad (19.5)$$

In this case, we say that the order of the reaction with respect to A is x and the order with respect to B is y. We can also describe the **overall order** of the reaction, which is the sum of the exponents on the concentration terms. In this example, the overall order is the sum $x + y$. Once again, x and y can have whole-number, fractional, negative or even zero values. When one of the exponents is zero, this simply means that the rate of the reaction does not depend on the concentration of that substance. For example, if the exponent y in Equation 18.5 were zero, the equation would become

$$\text{rate} \propto [A]^x[B]^0$$

Any quantity raised to the zero power is equal to 1, so the equation reduces to

$$\text{rate} \propto [A]^x(1)$$

or

$$\text{rate} \propto [A]^x$$

A real chemical example is the reaction

$$NO_2(g) + CO(g) \longrightarrow CO_2(g) + NO(g)$$

at temperatures below 225 °C, the relationship between concentration and rate is

$$\text{rate} \propto [NO_2]^2$$

The rate is independent of the CO concentration but depends on the *square* of the NO_2 concentration. We say that the reaction is second-order with respect to NO_2 and zero-order with respect to CO. Notice that there is no relationship between the coefficients and the exponent. As mentioned previously, the order of a reaction can only be determined experimentally.

The proportionality represented by Equation 19.5 can be converted to an equality by introducing a proportionality constant, k, which we call the **rate constant.** The resulting equation, termed the **rate law** for the reaction, is

$$\text{rate} = k[A]^x[B]^y$$

For example, the rate law for the reaction between ICl and H_2,

$$2ICl(g) + H_2(g) \longrightarrow I_2(g) + 2HCl(g)$$

at 230 °C has been found experimentally to be

$$\text{rate} = 0.163 \text{ L mol}^{-1} \text{ s}^{-1} [ICl][H_2]$$

This reaction therefore is *first-order* with respect to both ICl and H_2 (hence, *second-order*, overall) and has as its rate constant $k = 0.163$ L mol^{-1} s^{-1}. It should be noted that this value of k applies only at 230 °C for this particular reaction. Other reactions have other values of k, and as we will see later, k varies with the temperature.

Unlike with chemical equilibria, we *cannot* write the rate law just by using the coefficients in the balanced chemical equation. The exponents *must* be determined by experiment.

Reactions are categorized according to their orders of reaction because the mathematics for dealing with a rate law is the same for a given order.

The rate law lets us calculate the rate for any particular set of concentrations.

Determining the rate law

How can a rate law such as the one above be determined? One way is to perform a series of experiments in which the initial concentration of each reactant is systematically varied. Once again we can use as our example the simple reaction

$$A \longrightarrow B$$

The rate law for this reaction would take the form

$$\text{rate} = k[A]^x$$

If the reaction were first-order, the value of x would be 1, and the rate expression would then be

$$\text{rate} = k[A]$$

For a first-order reaction, increasing the concentration by a factor of 2 increases the rate by a factor of 2^1.

This means that the rate of the reaction varies directly with the concentration of A raised to the first power. As a result, if we were to double the concentration of A from one experiment to another, we would also find that the rate increases by a factor of 2. We conclude, therefore, that *when the reaction rate is doubled by doubling the concentration of a reactant, the order with respect to that react is 1.*

Suppose, now, that the rate law were, instead,

$$\text{rate} = k[A]^2$$

In this instance, a twofold increase in the concentration would cause a *fourfold* increase in rate. To see this, let's imagine that the initial rate was measured with the concentration of A equal to, say, a mol/L. This rate would be given by

$$\text{rate} = k(a)^2$$

Now, if the reaction were repeated with $[A] = 2a$, the rate would be

$$\text{rate} = k(2a)^2$$

or

$$\text{rate} = 4ka^2$$

For a second-order reaction, increasing the concentration by a factor of 3 increases the rate by a factor of 9, which is 3^2; the order of the reaction is 2.

which is four times the previous rate. Thus, *if the rate is increased by a factor of 4 when the concentration of a reactant is doubled, the reaction is second-order with respect to that reactant. Similarly, we predict that the rate of a third-order reaction would undergo an eightfold increase when the concentration is doubled ($2^3 = 8$).* From analyses like these, we can formulate a general rule.

> When the concentration of a reactant changes by some factor (i.e., the concentration is multiplied or divided by some number), the rate changes by this same factor raised to a power that is equal to the exponent in the rate law for that reactant.

The following examples illustrate how we can use these ideas to obtain the rate law for a reaction by varying the concentrations of reactants.

EXAMPLE 19.2. DETERMINING THE RATE LAW FOR A REACTION

PROBLEM: Below are some data collected in a series of experiments on the reaction of nitric oxide with bromine at 273 °C:

$$2NO(g) + Br_2(g) \longrightarrow 2NOBr(g)$$

Experiment	Initial Concentrations (mol/L)		Initial Rate of Formation of NOBr (mol L^{-1} s^{-1})
	NO	Br$_2$	
1	0.10	0.10	12
2	0.10	0.20	24
3	0.10	0.30	36
4	0.20	0.10	48
5	0.30	0.10	108

Determine the rate law for the reaction and compute the value of the rate constant.

ANALYSIS: The rate law for the reaction will have the form

$$\text{rate} = k[\text{NO}]^x[\text{Br}_2]^y$$

To determine each exponent, we will study how the rate changes when the concentration of one reactant varies while that of the other reactant stays the same. For instance, when the NO concentration is held constant, we can see how changes in the Br$_2$ concentration affect the rate and thereby determine what y must be. The value of x is determined in a similar way. With this strategy in mind, let's study the data.

SOLUTION: In experiments 1 to 3, the concentration of NO is constant and the concentration of Br$_2$ is varied. When the concentration of Br$_2$ is doubled (experiments 1 and 2), the rate is increased by a factor of 2; when it is tripled (experiments 1 and 3), the rate is increased by a factor of 3. The only way this could happen is if the concentration of Br$_2$ appears to the first power in the rate law. Therefore, $y = 1$.

Comparing experiments 1 and 4, we see that when the Br$_2$ concentration is held constant, the rate increases by a factor of 4 when the NO concentration is multiplied by 2. Similarly, raising the concentration of NO by a factor of 3 causes a ninefold increase in rate (experiments 1 and 5). This means that the exponent on the NO concentration in the rate law must be 2. Therefore, $x = 2$ and the rate law is

$$\text{rate} = k[\text{NO}]^2[\text{Br}_2]$$

The rate constant can be evaluated using the data from any of these experiments. Working with experiment 1, we have

$$12 \text{ mol L}^{-1} \text{ s}^{-1} = k(0.10 \text{ mol L}^{-1})^2(0.10 \text{ mol L}^{-1})$$

$$12 \text{ mol L}^{-1} \text{ s}^{-1} = k(0.0010 \text{ mol}^3 \text{ L}^{-3})$$

Solving for k, we get

$$k = \frac{12 \text{ mol L}^{-1} \text{ s}^{-1}}{1.0 \times 10^{-3} \text{ mol}^3 \text{ L}^{-3}} = 1.2 \times 10^4 \text{ L}^2 \text{ mol}^{-2} \text{ s}^{-1}$$

You might wish to verify for yourself that the same rate constant is obtained from the other data. (That's why it's called the rate *constant*.)

EXAMPLE 19.3. DETERMINING THE RATE LAW FOR A REACTION

PROBLEM: The following data were collected for the reaction of *t*-butyl bromide, $(CH_3)_3CBr$, with hydroxide ion at 55 °C.

$$(CH_3)_3CBr + OH^- \longrightarrow (CH_3)_3COH + Br^-$$

Experiment	Initial Concentrations (M)		Initial Rate of Formation of $(CH_3)_3COH$ (mol L^{-1} s^{-1})
	$(CH_3)_3CBr$	OH$^-$	
1	0.10	0.10	0.0010
2	0.20	0.10	0.0020
3	0.30	0.10	0.0030
4	0.10	0.20	0.0010
5	0.10	0.30	0.0010

What is the rate law and rate constant for this reaction?

SOLUTION: Based on the equation, we expect a rate law of the form

$$\text{rate} = k[(CH_3)_3CBr]^x[OH^-]^y$$

To obtain x and y, we follow the same approach as in the previous example.

Let's examine experiments 1, 2, and 3 first. In each of these the OH$^-$ concentration is the same. Doubling the $(CH_3)_3CBr$ concentration doubles the rate; tripling it triples the rate. The order with respect to $(CH_3)_3CBr$ must therefore be 1.

In Experiments 1, 4, and 5, the $(CH_3)_3CBr$ concentration is held constant while the OH$^-$ concentration is varied. Notice, however, that no matter what the OH$^-$ concentration is, the rate is the same. In effect, we find that when the [OH$^-$] is increased by a factor of 2 in experiments 1 and 4, the rate changes by a factor of $2^0 = 1$ (i.e., the rate in experiment 4 equals the rate in experiment 1 multiplied by a factor of 1). This means that the reaction is zero-order with respect to OH$^-$. Therefore,

If the rate is independent of the concentration of a reactant, the reactant doesn't appear in the rate law and its order is zero.

$$\text{rate} = k[(CH_3)_3CBr]^1[OH^-]^0$$

Since anything raised to the zero power is 1,

$$\text{rate} = k[(CH_3)_3CBr]^1 \cdot 1$$

When no exponent is written, it's assumed to be 1.

$$= k[(CH_3)_3CBr]$$

The final rate law contains only the concentration of $(CH_3)_3CBr$, because this is the only concentration that affects the rate. To solve for the rate constant we can use the results of any of the experiments. Using experiment 1 and substituting the rate and concentration into the rate law give

$$0.0010 \text{ mol L}^{-1} \text{ s}^{-1} = k(0.10 \text{ mol L}^{-1})$$

$$k = \frac{0.0010 \text{ mol L}^{-1} \text{ s}^{-1}}{0.10 \text{ mol L}^{-1}}$$

$$= 0.010 \text{ s}^{-1}$$

In Example 19.2, the exponents in the rate law just happen to be the same as the coefficients in the balanced equation. This is not true in Example 19.3. Please keep in mind that the *only* way we can find the exponents in the rate law for a chemical reaction is by experimentally measuring the way that the concentrations of the reactants affect the rate. It is also important to remember that since temperature is another factor that influences the rate, a given value of k applies only at *one* temperature (the temperature at which it was measured).

19.3 CONCENTRATION AND TIME: HALF-LIVES

The rate law for a reaction tells us how the rate of a reaction is related to the concentrations of the reactants. Applying calculus to the rate law, which we won't attempt to go through here, yields an expression relating the concentration to time. For example, a first-order rate law such as

$$\text{rate} = k[A]$$

gives the expression

$$\ln \frac{[A]_0}{[A]_t} = kt \tag{19.6}$$

Expressed in terms of common logarithms, Equation 19.6 is

$$2.303 \log \frac{[A]_0}{[A]_t} = kt$$

where $[A]_0$ is the initial concentration of A (at time t equal to zero), $[A]_t$ is the concentration at some time t after the beginning of the reaction, and the symbol "ln" tells us to take the natural logarithm of the ratio of $[A]_0$ divided by $[A]_t$.

Equation 19.6 or its counterpart in common logarithms is useful because it allows us to calculate, for a first-order reaction, the concentration of the reactant at any time during the course of the reaction, provided that we know the value of k. If we know $[A]_0$, $[A]_t$, and k, we can also compute t, the length of time that the reaction has progressed. We will see in Chapter 24 that this is useful in archaeological dating using radioactive isotopes.

EXAMPLE 19.4. CALCULATING THE CONCENTRATION OF A REACTANT AT SOME TIME AFTER THE START OF THE REACTION

PROBLEM: At 400 °C, the first-order conversion of cyclopropane into propylene has a rate constant of 1.16×10^{-6} s^{-1}. If the initial concentration of cyclopropane is 1.00×10^{-2} mol/L at 400 °C, what will its concentration be 24.0 hours after the reaction begins?

SOLUTION: To solve this problem we use Equation 19.6. Our data are

$$[\text{cyclopropane}]_0 = 1.00 \times 10^{-2} \text{ mol/L}$$

$$k = 1.16 \times 10^{-6} \text{ s}^{-1}$$

$$t = 24.0 \text{ hr}$$

The first step is to solve for the ratio of concentrations.

$$\ln \frac{[\text{cyclopropane}]_0}{[\text{cyclopropane}]_t} = (1.16 \times 10^{-6} \text{ s}^{-1})(24.0 \text{ hr})\left(\frac{3600 \text{ s}}{1 \text{ hr}}\right)$$

$$= 0.100$$

1 hr = 3600 s

To obtain the concentration ratio we must take the antilog. Recall that if $\ln x = a$, then $x = e^a$. Therefore,

$$\frac{[\text{cyclopropane}]_0}{[\text{cyclopropane}]_t} = e^{0.100}$$

$$= 1.11$$

Now we solve for $[\text{cyclopropane}]_t$.

$$[\text{cyclopropane}]_t = \frac{[\text{cyclopropane}]_0}{1.11}$$

$$= \frac{1.00 \times 10^{-2} \text{ mol/L}}{1.11}$$

$$= 9.01 \times 10^{-3} \text{ mol/L}$$

After 24 hours the concentration of cyclopropane will have dropped to 9.01×10^{-3} M.

The equation relating concentration to time is different for different reaction orders. For instance, for a second-order reaction having the rate law

$$\text{rate} = k[B]^2$$

the relationship is

$$\frac{1}{[B]_t} - \frac{1}{[B]_0} = kt \tag{19.7}$$

Even more complicated equations occur when the rate law is more complex, but we won't discuss them.

Half-lives

An important quantity, particularly for first-order reactions, is the **half-life, $t_{1/2}$,** which is defined as *the length of time required for the concentration of a reactant to decrease to half of its initial value.* At this point, $t = t_{1/2}$, and

$$[A]_t = \tfrac{1}{2}[A]_0$$

Substituting into Equation 19.6 gives

$$\ln \frac{[A]_0}{\tfrac{1}{2}[A]_0} = kt_{1/2}$$

Since ln 2 = 0.693, Equation 19.8 can also be written

$$t_{1/2} = \frac{0.693}{k}$$

$$\ln 2 = kt_{1/2}$$

Solving for $t_{1/2}$ gives

$$t_{1/2} = \frac{\ln 2}{k} \tag{19.8}$$

For a first-order reaction

Number of Half-Lives	Fraction of Reactant Remaining
0	1
1	1/2
2	1/4
3	1/8
4	1/16
5	1/32
⋮	⋮
n	$1/2^n$

Equation 19.8 tells us that for a first-order reaction, $t_{1/2}$ depends only on k: It is constant throughout the reaction. If the half-life for a particular first-order reaction is 1 hour, then during the first hour the concentration drops to half of its initial value. During the second hour the concentration is again cut in half, so after a total of 2 hours the concentration is $\tfrac{1}{4}$ of its initial value (Figure 19.4).

The half-life of a second-order reaction differs from that of a first-order process by being concentration-dependent. Following the same procedure as above, we find that a second-order reaction whose rate law is

$$\text{rate} = k[B]^2$$

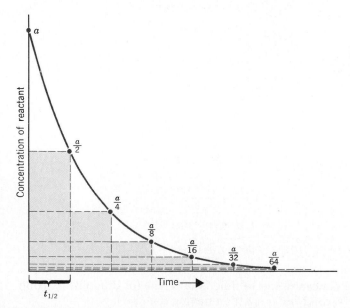

Figure 19.4. *A graph of concentration versus time for a first-order reaction. Over each half-life the concentration of the reactant is halved. Notice that the half-life of the reaction is a constant.*

has a half-life given by

$$t_{1/2} = \frac{1}{k[B]_0} \tag{19.9}$$

This means that if we cut the concentration of B in half, the value of $t_{1/2}$ will double. Therefore, during the course of the reaction, each successive half-life is twice as large as the preceding one. If a second-order reaction such as this starts with a half-life of 20 minutes, the concentration at the beginning of the second half-life is half of the initial value, so the second half-life will be twice as long. Similarly, the third half-life will be twice as long as the second, and so forth.

EXAMPLE 19.5. CALCULATING THE HALF-LIFE FOR A REACTION

PROBLEM: The decomposition of N_2O_5 dissolved in carbon tetrachloride is a first-order reaction. The chemical change is

$$2N_2O_5 \longrightarrow 4NO_2 + O_2$$

At 45 °C the reaction was begun with an initial N_2O_5 concentration of 1.00 mol/L. After 3.00 hours the N_2O_5 concentration had decreased to 1.21×10^{-3} mol/L. What is the half-life of N_2O_5 expressed in minutes at 45 °C?

SOLUTION: To obtain the half-life, we must know the value of k. Since the reaction is first-order, we can use Equation 19.6

$$\ln \left(\frac{1.00 \text{ mol/L}}{1.21 \times 10^{-3} \text{ mol/L}} \right) = k(3.00 \text{ hr}) \left(\frac{3600 \text{ s}}{1 \text{ hr}} \right)$$

$$\ln (826) = (1.08 \times 10^4 \text{ s})k$$

$$6.72 = (1.08 \times 10^4 \text{ s})k$$

Therefore

$$k = \frac{6.72}{1.08 \times 10^4 \text{ s}}$$

$$= 6.22 \times 10^{-4} \text{ s}^{-1}$$

Now we can calculate the half-life

$$t_{1/2} = \frac{\ln 2}{k}$$

$$= \frac{0.693}{6.22 \times 10^{-4} \text{ s}^{-1}}$$

$$= 1.11 \times 10^3 \text{ s}$$

This is 18.5 minutes.

19.4 COLLISION THEORY

For a chemical reaction to occur, the reacting molecules must collide with each other. This common-sense idea forms the basis of the **collision theory** of chemical kinetics. Basically, this theory states that the rate of a reaction is proportional to the number of collisions occurring each second between the reacting molecules:

$$\text{rate} \propto \frac{\text{number of collisions}}{\text{second}} \tag{19.10}$$

As we will see shortly, this permits us to explain the dependence of the reaction rate on the concentrations of the reactants. In Section 19.6 we will also see that only a fraction of the total number of collisions each second are effective at producing a net chemical change, and that this fraction depends both on the nature of the reactants and the temperature.

At this point, let's see how collision theory accounts for the way the rate of a reaction depends on the concentrations of the reactants. Suppose that we have a reaction that occurs by the collision of two molecules, such as

$$A + B \longrightarrow \text{products}$$

In this case we are assuming that we know precisely what occurs between A and B — that is, we assume for the sake of this discussion that the products are formed in **bimolecular** (two-molecule) **collisions** between A and B.

According to the theory, the rate of the reaction is proportional to the number of collisions each second between molecules of A and B. If the concentration of A were doubled, then the number of A–B collisions per second would also be doubled because there would be twice as many A molecules that can collide with B. This would increase the rate by a factor of 2. Similarly, if the concentration of B were doubled, there would be a twofold increase in the number of A–B collisions per second and the rate would also increase by a factor of 2. From our previous discussion we conclude that the order with respect to each reactant is 1, so the rate law for this bimolecular collision process is

$$\text{rate} = k[A][B]$$

Now let's look at what would happen for a reaction of the type

$$2A \longrightarrow \text{products}$$

in which the reaction occurs by the collision of two A molecules. In this instance, if we double the concentration of A, we double the number of collisions per second that each *single* A molecule makes with its neighbors, because we have doubled the number of neighbors. However, we have also doubled the number of A molecules that are colliding. The number of A–A collisions per second has therefore doubly doubled — that is, increased by a factor of 2 squared. Consequently, the rate law for this bimolecular reaction between identical molecules must be

We are doubling both the number of molecules colliding and the number of collisions that each of them makes.

$$\text{rate} = k[A]^2$$

What we find, then, is that *if* we know what collision process is involved in the formation of products, we can predict, on the basis of collision theory, what the rate law for that process will be. *The exponents in the rate law for a simple collision process are equal to the coefficients in the balanced equation for that collision process.*

At this point we might ask, why is it necessary to determine the rate law for a reaction experimentally? Why can't we simply use the coefficients of the balanced overall equation to deduce the rate law? The answer is that when we begin to study a reaction, we don't know what the collision processes are. Are A–B collisions important, or do A–A collisions determine the rate? We can't answer that question until we know the mechanism of the reaction, and that's our next topic.

19.5 REACTION MECHANISMS

The overall balanced equation for a reaction represents the net chemical change that occurs as the reaction proceeds to completion. This does not mean, however, that all the reactants must come together simultaneously to undergo a change that gives the products. In fact, the net change can (and usually does) actually represent the sum of a series of simpler reactions. These simpler reactions are referred to as **elementary processes.** The sequence of elementary processes that ultimately leads to the formation of the products is called the **reaction mechanism.** For example, it appears that the reaction

$$2NO + 2H_2 \longrightarrow 2H_2O + N_2$$

proceeds by the three-step mechanism

$$
\left.
\begin{aligned}
2NO &\longrightarrow N_2O_2 \\
N_2O_2 + H_2 &\longrightarrow N_2O + H_2O \\
N_2O + H_2 &\longrightarrow N_2 + H_2O
\end{aligned}
\right\}
\quad \text{These are elementary processes.}
$$

Notice that the sum of these three steps gives us the net overall balanced equation.

Reaction mechanisms, such as the one just described, are usually arrived at by bringing together both theory and experiment. For instance, suppose we wished to study the following hypothetical reaction in hopes of discovering the mechanism:

$$2A + B \longrightarrow C + D$$

We begin by first determining the rate law, perhaps by studying how the rate changes as we vary the concentrations of A and B as described earlier. Let us say it turned out to be

$$\text{rate} = k[A]^2[B]$$

Next, we attempt to propose a mechanism that, by the application of the principles of collision theory, gives us a predicted rate law that is the same as the one found by experiment.

Since we are beginners at proposing mechanisms, we might be tempted to propose a one-step mechanism in which two molecules of A and one of B come together simultaneously — that is, a three-body or *termolecular* collision. This process

$$2A + B \longrightarrow C + D$$

indeed leads to the rate law

$$\text{rate} = k[A]^2[B]$$

The fact that the predicted rate law matches the experimental one could mean our guess is correct. But we must now ask ourselves, is this a *reasonable* mechanism? Anyone who has ever played billiards or pool knows that it is a very rare event when three billiard balls collide simultaneously so that only one click is heard. Such three-body collisions are improbable. Yet, this is exactly what we are proposing for our mechanism. If the reaction relies on such collisions, it should be slow. And in fact, it has been found that reactions that must proceed by such termolecular collisions are very slow. Therefore, a third-order reaction such as this, if it is fairly rapid, is usually interpreted as taking place by way of a series of simple bimolecular collisions. (Back to the drawing board!)

Let's suppose that our chemical thinking next leads us to suggest the following sequence of reactions.

$$
\begin{aligned}
2A &\longrightarrow A_2 \\
A_2 + B &\longrightarrow C + D
\end{aligned}
$$

Here we have two steps in which we propose that some intermediate, A_2, is first formed by the collision of two molecules of A. In a second step, a reaction between A_2 and B produces the products C and D. Again, the sum of these elementary processes gives us our net overall change.

Both of these reactions are unlikely to occur at the same rate, so let's suppose that the first reaction is slow, and that once the intermediate, A_2, is formed, it reacts

A mechanism must make chemical sense and also be capable of yielding the experimentally measured rate law. Proposing such mechanisms requires a lot of practice and chemical knowledge, plus a bit of "chemical intuition."

rapidly with B in the second step to produce the products. If this is what happens, the rate at which the final products appear is actually determined by how fast A_2 is produced. This first step serves as a bottleneck in the reaction path. We refer to this slowest step as the **rate-determining step** in the reaction because it governs how rapidly the overall reaction takes place. Because the rate-determining step is an elementary process (in this instance, a bimolecular collision between two A molecules), we can predict by applying the collision theory that its rate law should be

$$\text{rate} = k[A]^2$$

If this is the rate law for the rate-determining step, it will also be the rate law for the overall reaction. However, this rate law *cannot* be the correct one because it is not the same as the one determined from experiment. This does not necessarily mean that our mechanism is wrong, but it does mean that the first step cannot be the rate-determining step. Let us see what we would expect to observe if the second step were slow, instead of the first. In this case the rate law would be

$$\text{rate} = k[A_2][B] \tag{19.11}$$

However, this rate law contains the concentration of the proposed intermediate (A_2) and the experimental rate law contains only the concentrations of reactants A and B. How can we express the concentration of A_2 in terms of A and B?

If the first step is fast and the second is slow, A_2 could not be a particularly stable species; otherwise, we would observe it being formed in the reaction mixture and there would be no question about the mechanism. Therefore, we propose that A_2 is very unstable and that once it has been formed, it can react in either of two ways. One way is that it can decompose almost immediately to form two molecules of A. The other is that it might combine with a molecule of B to give the products, C and D. Our mechanism, therefore, should include a reaction that allows A_2 to decompose.

$$A_2 \longrightarrow 2A$$

Our total mechanism is now

$$2A \longrightarrow A_2$$
$$A_2 \longrightarrow 2A$$
$$A_2 + B \longrightarrow C + D$$

If the rate at which the intermediate is formed from reactant A is equal to the rate at which A is formed from intermediate A_2, then these two reactions represent a state of dynamic equilibrium. We could therefore write our first two equations as an equilibrium, which would take the form

$$2A \rightleftharpoons A_2$$

We are now back to a two-step mechanism in which the first step is an equilibrium. Our mechanism is now

$$2A \rightleftharpoons A_2 \qquad \text{fast}$$
$$A_2 + B \longrightarrow C + D \quad \text{slow}$$

Since, in an equilibrium situation, the rate of the forward reaction (rate_f) is equal to the rate of the reverse reaction (rate_r),

$$\text{rate}_f = k_f[A]^2 = \text{rate}_r = k_r[A_2]$$

or simply

$$k_f[A]^2 = k_r[A_2]$$

Solving this equation for $[A_2]$, we have

$$[A_2] - \frac{k_f[A]^2}{k_r}$$

The reaction cannot proceed any faster than its slowest step, and anything that alters the rate of this step also alters the rate of the reaction.

Recall that in a dynamic equilibrium two opposing processes occur at equal rates.

In this mechanism, we are assuming that a very rapidly established equilibrium maintains a constant concentration of A_2 with which B can react. This is called a *steady-state* assumption.

We can now substitute this into Equation 19.11 and combine all the constants to give still another constant, let us say k'.

$$\text{rate} = k \underbrace{\frac{k_f}{k_r}}_{k'} [A]^2[B]$$

This gives the rate law

$$\text{rate} = k'[A]^2[B]$$

Now the rate law does agree with the one found by experiment, so the proposed mechanism is in agreement with experiment. This could mean (and we hope it does) that the mechanism is correct. However, a mechanism is, in essence, a theory. It is a sequence of steps that we propose to explain the chemistry and to provide a rate law that agrees with experiment. It frequently happens, though, that more than one mechanism can be written to satisfy both criteria, so we can never be certain we have truly discovered the actual path of the reaction. Nevertheless, studying the kinetics of a reaction gives clues to what the mechanism may be and allows us to discard many alternatives. After that, we can hope to gather further information that either supports (or proves wrong) our guess.

EXAMPLE 19.6. STUDYING THE MECHANISM OF A REACTION

PROBLEM: The decomposition of NO_2Cl is believed to involve the two-step mechanism

$$NO_2Cl \longrightarrow NO_2 + Cl$$

$$NO_2Cl + Cl \longrightarrow NO_2 + Cl_2$$

What would be the observed experimental rate law if the first step is slow and the second fast?

SOLUTION: If the first reaction is the slow step, it is also the rate-determing step. The rate law for the overall reaction should be the same as the rate law for the rate-determining step. Since only one molecule of NO_2Cl is involved, the rate law for the first reaction, as well as for the overall reaction, would be

$$\text{rate} = k[NO_2Cl]$$

19.6 EFFECTIVE COLLISIONS AND THE EFFECT OF TEMPERATURE ON REACTION RATE

If all the collisions that take place in a reaction vessel were effective in producing chemical change, all chemical reactions, including biochemical ones, would be over almost instantaneously. Since living creatures have finite life spans, it is clear that some factor (or factors) must intervene to decrease reaction rates to a reasonable level. Consider, for example, the decomposition of hydrogen iodide

$$2HI(g) \longrightarrow H_2(g) + I_2(g)$$

At a concentration of only 10^{-3} mol/L of HI there are approximately 3.5×10^{28} collisions per liter per second at 500 °C. This is equivalent to 5.8×10^4 moles of collisions per liter per second; if each of these collisions were effective, we would expect a rate of reaction of 5.8×10^4 mol L^{-1} s^{-1}. Actually, the rate under these conditions is only about 1.2×10^{-8} mol L^{-1}s^{-1}. This is smaller by a factor of approx-

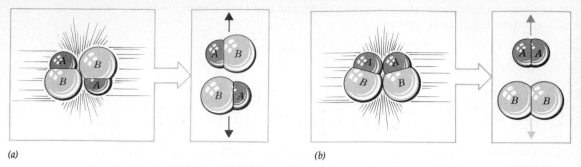

(a)   *(b)*

Figure 19.5. *Collisions between A–B molecules. (a) A collision that cannot produce any net chemical change, even when the A's and B's are exchanged. (b) A collision that can lead to a net reaction.*

imately 5×10^{12} than we would observe if all collisions led to reaction, which means that only one out of every five thousand billion collisions actually leads to the formation of the products! Clearly, not all encounters between HI molecules result in the production of H_2 and I_2. In fact, only a very small fraction of the total number of collisions are effective.

Similar observations are found for many other reactions, which suggests that there are common factors that determine whether or not a collision will yield a chemical change. One of these is the relative orientations of the molecules at the moment of impact. For example, consider the decomposition of hypothetical *AB* molecules whose collisions result in the formation of A_2 and B_2.

$$2AB \longrightarrow A_2 + B_2$$

In order for the products A_2 and B_2 to be produced, the two atoms of *A* and two atoms of *B* must approach each other very closely so that *A–A* and *B–B* bonds can be formed. Suppose, now, that two *A–B* molecules come together in a collision oriented as shown in Figure 19.5*a*. We certainly do not expect this collision to be effective in forming the products. However, a collision in which the *A–B* molecules are aligned as shown in Figure 19.5*b* can lead to the creation of *A–A* and *B–B* bonds and, hence, a net chemical change. Thus, the number of effective collisions—and, therefore, the rate of the reaction—is decreased by a factor that is a measure of the importance of the molecular orientations during collision.

A second factor that determines whether a collision is effective is the energy that the molecules possess at the moment they collide. To understand this, we have to examine molecular collisions in some detail.

A collision between two molecules is quite unlike a collision between two billiard balls. The electron cloud of a molecule has no sharp boundary; Its outer reaches are somewhat soft and "fuzzy." Therefore, when two molecules approach each other in a collision, the electron clouds experience a gradual increase in their mutual repulsions, and the molecules begin to slow down. As this happens, the kinetic energy of the molecules is gradually converted to potential energy— somewhat like compressing a spring. If the pair of colliding molecules has little kinetic energy to begin with—that is, if they are not moving very fast—they come to a stop before their electron clouds have penetrated each other very much, and then they fly apart again, chemically unchanged (Figure 19.6*a*).

When two fast-moving molecules collide, they have a lot of kinetic energy that can be converted to potential energy. This means that they are able to overcome substantial forces of repulsion between their electron clouds, and they approach each other quite closely. As illustrated in Figure 19.6*b*, the interpenetration of the electron clouds that occurs permits a reshuffling of electrons. Old bonds are broken as new ones form and a net chemical change takes place.

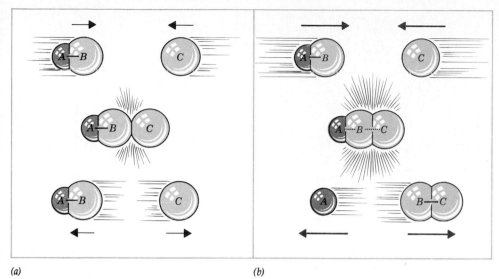

Figure 19.6. (a) *When two slow-moving molecules collide, their electron clouds cannot interpenetrate much and they just bounce off each other, chemically unchanged.* (b) *When fast-moving molecules collide, atoms approach each other much more closely as their electron clouds interpenetrate. This can lead to bond making and bond breaking. The net change here is*

$$AB + C \rightarrow A + BC.$$

(a)   (b)

From the preceding discussion we see that an effective collision, one that changes reactant molecules into product molecules, can occur only if the molecules collide with sufficient force. Expressed differently, there is a minimum kinetic energy that the molecules must possess jointly to overcome the repulsions between their electron clouds when they collide. This minimum kinetic energy, which is changed to potential energy at the moment of impact, is called the **activation energy, E_a.**

Transition state theory

The collision theory, discussed in Section 19.4, focuses attention primarily on the relationship between reaction rate and the *number* of collisions per second between reactant molecules. In this way, the dependence of rate on concentration was explained. The **transition state theory** is concerned with what actually happens *during* a collision. It follows the energy and orientation of the reactant molecules as they collide, and in doing so seeks explanations of why so few collisions out of the many that occur are actually effective.

The change in potential energy that takes place during the course of a reaction is shown in Figure 19.7. The horizontal axis is called the **reaction coordinate:** It follows the path of the reaction as molecules come together in a collision and product molecules emerge. In a sense, positions along the reaction coordinate represent the extent to which the reaction has progressed toward completion. On the left of this

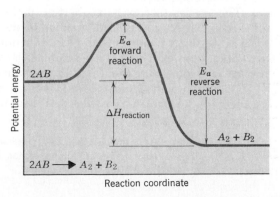

Figure 19.7. *Potential energy diagram for an exothermic reaction.*

Figure 19.8. *Potential energy diagram for an endothermic reaction.*

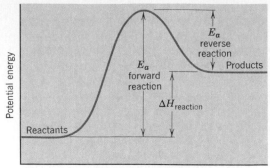

particular potential energy diagram we find two molecules of *AB*. As they approach each other, their potential energy increases to a maximum. As we continue toward the right along the reaction coordinate, the potential energy of the system decreases as the products, A_2 and B_2, move apart. When the A_2 and B_2 molecules are finally separated from each other, the total potential energy drops to essentially a constant value.

The activation energy for the decomposition of *AB* corresponds to the difference between the energy of the reactants and the maximum on the potential energy curve. Slow-moving molecules of *AB* do not possess sufficient energy to overcome this potential energy barrier, while fast-moving ones do.

In Figure 19.7 we have drawn the potential energy of the products so that it is lower than that of the reactants. The difference between them corresponds to the heat of reaction.[1] In this case, because the products are at a lower energy than the reactants, the reaction is exothermic. The energy released appears as an increase in the kinetic energy of the products; therefore, the temperature of the system rises as the reaction progresses.

In the reaction mixture there are also collisions between A_2 and B_2 molecules. Such collisions, if energetic enough, can reform *AB* molecules. In Figure 19.7 the activation energy for the reaction

$$A_2 + B_2 \longrightarrow 2AB$$

is indicated as the difference in energy between the products and the top of the potential energy hill. Since the forward reaction (the reaction from left to right) is exothermic, the reverse reaction is endothermic.

Figure 19.8 depicts the energy changes for a reaction that is endothermic in the forward direction. In this case the products are at a higher potential energy than the reactants. The net increase in potential energy that takes place as the products are formed occurs at the expense of kinetic energy. As a result, there is a net overall decrease in the average kinetic energy as the reaction proceeds and the reaction mixture becomes cool.

The species that exists at the top of the potential energy barrier during an effective collision corresponds neither to the reactants nor to the products but, instead, to some highly unstable combination of atoms that we speak of as the **activated complex.** This activated complex is said to exist in a **transition state** along the reaction coordinate (hence, the name transition state theory).

Transition state theory views chemical kinetics in terms of the energy and geometry of the activated complex that, once it has formed, can come apart to yield

In a collision, the total energy (K.E. plus P.E.) is constant. If the P.E. becomes smaller after a collision, the K.E. must become larger.

An endothermic reaction is one in which K.E. is converted to P.E. In this sense a system absorbs or stores energy.

---

[1] Strictly speaking, the difference between the potential energy of the reactants and the potential energy of the products corresponds to ΔE for the reaction, but we learned in Chapter 14 that ΔE and ΔH for a process are of very nearly the same magnitude.

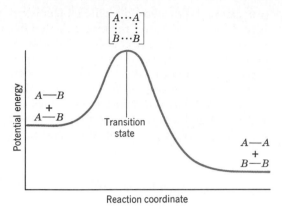

Figure 19.9. *Transition state theory and the potential energy diagram for the reaction described in Equation 19.12.*

the reactants again or go on to produce the products. For example, let us examine again the decomposition of hypothetical *AB* molecules to produce A_2 and B_2. The change that takes place along the reaction coordinate can be represented as

$$
\begin{matrix} A \\ | \\ B \end{matrix} + \begin{matrix} A \\ | \\ B \end{matrix} \rightleftharpoons \begin{bmatrix} A \cdots A \\ \vdots \quad \vdots \\ B \cdots B \end{bmatrix} \longrightarrow \begin{matrix} A-A \\ + \\ B-B \end{matrix} \tag{19.12}
$$

where we have used solid dashes to denote ordinary covalent bonds and dotted lines to symbolize the partially broken and partially formed bonds in the transition state, which is enclosed within brackets. Figure 19.9 illustrates this change as it occurs on the potential energy diagram for the reaction.

If the potential energy of the transition state is very high, then a great deal of energy must be available in a collision to form the activated complex. This results in a high activation energy and a slow reaction.

The effect of temperature on reaction rates

In nearly every instance, an increase in temperature causes an increase in the rate of reaction. In fact, for many reactions, the rate approximately doubles for every 10 °C rise in temperature. Why do reactions behave in such a predictable way?

In any chemical system there is a distribution of kinetic energies that is determined by the temperature. Some molecules are moving rapidly and have large kinetic energies, while others are moving more slowly and have smaller kinetic energies, but the average kinetic energy is controlled by the temperature. Figure 19.10 illustrates this kinetic energy distribution for two different temperatures.

If we assume that these kinetic energy distributions are for a system of reacting molecules, we can plot the activation energy along the kinetic energy axis. This is the

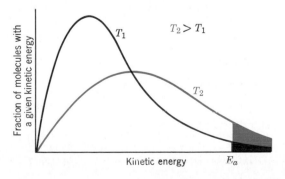

Figure 19.10. *The effect of temperature on the number of molecules that have kinetic energies larger than E_a. At temperature T_2 a greater fraction of the total number of molecules have sufficient kinetic energy to react, so the reaction is faster at the higher temperature.*

minimum energy needed by the molecules in order to react and is the same regardless of the temperature; any molecule that has this much energy or more will be able to react successfully in a collision. The *total* fraction of all the molecules with energies equal to or greater than E_a corresponds to the shaded area under each of the curves. When we compare this area for the two different temperatures, we see that it is larger at the higher temperature. This means that at the higher temperature, a larger fraction of the total number of molecules are able to react, so the reaction proceeds faster. Stated more simply, *as the temperature increases, the number of molecules with enough kinetic energy to react also increases, so the reaction proceeds more rapidly.*

19.7 MEASURING THE ACTIVATION ENERGY

The magnitude of the rate constant, which is the rate of the reaction when all the concentrations have a value of one, depends on a number of factors. In Section 19.6 we saw that it depends in part on the collision frequency. We have also learned that the rate of reaction is affected by the orientations of the molecules during a collision and the kinetic energy that they must have when they collide. These various factors are related quantitatively by the equation

$$k = Ae^{-E_a/RT} \qquad (19.13)$$

You have seen the name Arrhenius before in our discussion of acids and bases in Chapter 5. In 1903 Arrhenius received the third Nobel prize ever awarded in chemistry.

which is known as the **Arrhenius equation** after its discoverer, the Swedish chemist Svante Arrhenius. In the equation, e is the base of the natural logarithms, R is the gas constant, T is the absolute temperature and, of course, k is the rate constant and E_a is the activation energy. The factor A is a proportionality constant whose magnitude is related to the collision frequency and also to the importance of molecular orientations during a collision.

The Arrhenius equation provides us with a means of determining the value of the activation energy (as well as the factor A) from measurements of the rate constant at a minimum of two different temperatures. Taking the natural logarithm of Equation 19.13 gives

$$\ln k = \ln A - \frac{E_a}{RT} \qquad (19.14)$$

We can compare this equation to the equation for a straight line.

$$\ln k = \ln A - \frac{E_a}{R}\left(\frac{1}{T}\right)$$

$$\updownarrow \qquad \updownarrow \qquad \updownarrow \quad \updownarrow$$

$$y \;\;=\;\; b \;\;+\; m \;\; x$$

Thus, a plot of $\ln k$ versus $1/T$ gives a straight line whose slope m is equal to $-E_a/R$ and whose intercept b with the ordinate (the vertical axis) is $\ln A$ (Figure 19.11).

We can also obtain E_a by direct computation if we have k values determined at two different temperatures. The equation needed is derived from Equation 19.13. For a temperature T_1, Equation 19.13 becomes

$$k_1 = Ae^{-E_a/RT_1}$$

and for some other temperature, T_2,

$$k_2 = Ae^{-E_a/RT_2}$$

Dividing k_1 by k_2, we have

$$\frac{k_1}{k_2} = \frac{Ae^{-E_a/RT_1}}{Ae^{-E_a/RT_2}}$$

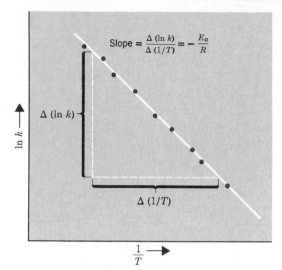

Figure 19.11. *The graphical determination of the activation energy, E_a. Points along the line represent the natural logarithms of rate constants experimentally measured at various temperatures. The straight line that best fits the data has a slope that can be used to calculate E_a.*

or

$$\frac{k_1}{k_2} = e^{(E_a/R)[(1/T_2)-(1/T_1)]}$$

Taking the natural logarithm of both sides, we get

$$\ln\left(\frac{k_1}{k_2}\right) = \frac{E_a}{R}\left(\frac{1}{T_2} - \frac{1}{T_1}\right) \qquad (19.16)$$

Equation 19.16 can be used to compute E_a if rate constants at two different temperatures are known. It can also be used to calculate the rate constant of some specific temperature if E_a and k at some other temperature are known. The value of R that is used in either case depends on the units of the energy. When E_a is in joules, we use $R = 8.314$ J mol^{-1} K^{-1}.

In terms of common logarithms, Equation 19.16 becomes

$$\log\left(\frac{k_1}{k_2}\right) = \frac{E_a}{2.303R}\left(\frac{1}{T_2} - \frac{1}{T_1}\right)$$

EXAMPLE 19.7. CALCULATING THE ACTIVATION ENERGY FOR A REACTION FROM RATE CONSTANTS AT TWO TEMPERATURES

PROBLEM: At 300 °C the rate constant for the reaction

$$\underset{\text{cyclopropane}}{H_2C\overset{\overset{\overset{H_2}{C}}{}}{\underline{\qquad}}CH_2} \longrightarrow \underset{\text{propylene}}{H_3C-\overset{\overset{H}{|}}{C}=CH_2}$$

is 2.41×10^{-10} s^{-1}. At 400 °C, k equals 1.16×10^{-6} s^{-1}. What are the values of E_a (in kilojoules per mole) and A for this reaction?

SOLUTION: We can obtain E_a by substituting values of k_1, k_2, T_1, and T_2 into Equation 19.16 and then solving for E_a. To avoid confusion, let's first tabulate the data.

	k	T
1	2.41×10^{-10} s^{-1}	$300 + 273 = 573$ K
2	1.16×10^{-6} s^{-1}	$400 + 273 = 673$ K

Notice that we have converted the temperatures to kelvins. Substituting these values into Equation 19.16 gives

$$\ln\left(\frac{2.41 \times 10^{-10} \text{ s}^{-1}}{1.16 \times 10^{-6} \text{ s}^{-1}}\right) = \frac{E_a}{8.314 \text{ J mol}^{-1} \text{ K}^{-1}}\left(\frac{1}{673 \text{ K}} - \frac{1}{573 \text{ K}}\right)$$

$$\ln(2.08 \times 10^{-4}) = \frac{E_a}{8.314 \text{ J mol}^{-1} \text{ K}^{-1}}(0.00149 \text{ K}^{-1} - 0.00175 \text{ K}^{-1})$$

$$-8.479 = E_a(-3.1 \times 10^{-5} \text{ J}^{-1} \text{ mol})$$

$$E_a = \frac{-8.479}{-3.1 \times 10^{-5} \text{ J}^{-1} \text{ mol}}$$

$$= 2.7 \times 10^5 \text{ J/mol}$$

$$= 270 \text{ kJ/mol}$$

The easiest way to compute A is to use Equation 19.13 and solve for A.

$$A = \frac{k}{e^{-E_a/RT}} = k\, e^{E_a/RT}$$

Using the first set of data and our calculated E_a, we get

$$A = (2.41 \times 10^{-10} \text{ s}^{-1})e^{(2.7 \times 10^5 \text{ J/mol})/(8.314 \text{ J mol}^{-1}\text{K}^{-1} \times 573 \text{ K})}$$

$$= (2.41 \times 10^{-10} \text{ s}^{-1}) \times (4.1 \times 10^{24})$$

$$= 9.9 \times 10^{14} \text{ s}^{-1}$$

Notice that A must have the same units as k.

19.8 CATALYSTS

A **catalyst** is a substance that increases the rate of a reaction without being consumed; after the reaction has ceased, it can be recovered from the reaction mixture chemically unchanged. The catalyst participates in the reaction by providing a lower-energy alternative mechanism for the production of the products. In Figure 19.12, note that the energy curve of the catalyzed reaction is drawn along a different reaction coordinate to emphasize that a different mechanism is involved. In addition, the energy barrier for the catalyzed path is lower than for the uncatalyzed reaction. This smaller activation energy means that in the reaction mixture there is a

Figure 19.12. *The effect of a catalyst on the path of a reaction. The catalyst changes the mechanism of the reaction by providing a different lower-energy path from the reactants to the products. Although the activation energy is different for the catalyzed mechanism, ΔH is the same for both because ΔH is a state function.*

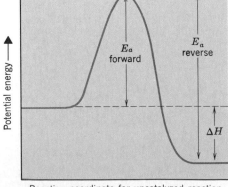
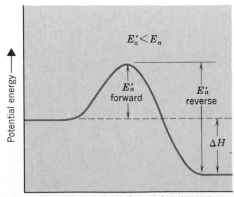

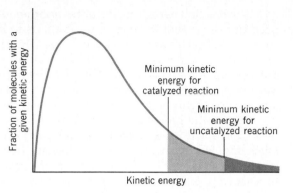

Figure 19.13. *More molecules possess the minimum kinetic energy needed for effective collisions when the catalyst is present.*

greater total fraction of molecules possessing sufficient kinetic energy to react (Figure 19.13). Therefore, in the presence of the catalyst there is an increased number of effective collisions. Of course, an increased number of effective collisions means a greater reaction rate.

Since a catalyst emerges from a reaction chemically unchanged, it does not appear either as a reactant or as a product in the overall balanced chemical equation. Instead, its presence is usually indicated by writing its name or formula over the arrow. For example, oxygen can be prepared by the thermal decomposition of potassium chlorate, $KClO_3$. In the absence of a catalyst the reaction is slow, and the $KClO_3$ must be heated to a high temperature to make it decompose at a reasonable rate. However, if a small amount of manganese dioxide (MnO_2) is added to the $KClO_3$, the decomposition proceeds smoothly at a relatively low temperature. Analysis of the reaction mixture after the evolution of oxygen has ceased reveals that all the MnO_2 added initially is still present, showing that MnO_2 has served as a catalyst. The equation for the catalyzed reaction is given as

$$2KClO_3 \xrightarrow{MnO_2} 2KCl + 3O_2$$

Homogeneous catalysts

Catalysts are broadly classified into two categories: homogeneous catalysts and heterogeneous catalysts. A **homogeneous catalyst** is in the same phase as the reactants (and usually the products as well). It participates in the reaction by combining chemically with the reactants to produce an intermediate that ultimately goes on to form the products and regenerate the catalyst. An example is the catalytic effect that HBr has on the thermal decomposition of t-butyl alcohol, $(CH_3)_3COH$, to give water and isobutylene, $(CH_3)_2C{=}CH_2$.

$$(CH_3)_3COH \longrightarrow (CH_3)_2C{=}CH_2 + H_2O$$

In the absence of a catalyst, this reaction is extremely slow even at elevated temperatures because it has a very large activation energy of 274 kJ/mol. In the presence of HBr, however, the reaction proceeds smoothly with an activation energy of just 127 kJ/mol, and it is believed that the catalyzed reaction proceeds by the attack of HBr on the alcohol,

$$(CH_3)_3COH + HBr \longrightarrow (CH_3)_3CBr + H_2O$$

followed by the rapid decomposition of the t-butyl bromide, $(CH_3)_3CBr$,

$$(CH_3)_3CBr \longrightarrow (CH_3)_2C{=}CH_2 + HBr$$

Notice that HBr combines with the reactant in one step of the mechanism and is formed again in a later step. The presence of HBr thus opens the door to a low-energy path to the products, even though the HBr is not part of the overall stoichiometry.

t-Butyl means tertiary butyl and stands for the group of atoms

$$CH_3{-}\underset{\underset{CH_3}{|}}{\overset{\overset{CH_3}{|}}{C}}{-}$$

The reaction described illustrates another important property of catalysts; they need not be present in large concentrations to influence dramatically the rate of a reaction. In the decomposition of $(CH_3)_3COH$, each molecule of HBr can be recycled over and over again as it first combines with the reactant and is then reformed in the second step of the mechanism. This phenomenon is particularly important in biological systems, for in them practically every reaction is catalyzed by very small amounts of highly specific biochemical catalysts called enzymes.

Because a catalyst is recycled over and over again, it can be very effective even if present in only trace amounts.

One of the problems with homogeneous catalysts is that after the reaction is over, they must be separated from the reaction mixture. This can sometimes pose problems. Furthermore, reactions that are homogeneously catalyzed usually are batch processes; the catalyst is added to the reaction mixture which then is allowed to react completely before the products and the catalyst are separated. These drawbacks are virtually eliminated by the use of heterogeneous catalysts.

Heterogeneous catalysts

A **heterogeneous catalyst** exists in a separate phase from the reactants and products and functions by providing a surface on which the reaction can proceed with a low activation energy. An example is the reaction of H_2 with O_2, which normally proceeds at an imperceptible rate at room temperature. However, in the presence of certain metals such as nickel, copper, or silver, the reaction is able to occur much more rapidly.

Adsorption means sticking to a surface.

Heterogeneous catalysts appear to function through a process whereby reactant molecules are adsorbed on a surface where the reaction then takes place. The high reactivity of hydrogen in the presence of certain metals, for example, is thought to occur by the adsorption of H_2 molecules onto the catalytic surface. On the surface of the metal the bonds between hydrogen atoms are apparently stretched or broken, as illustrated in Figure 19.14, so that the metal surface actually behaves as if it contained highly reactive hydrogen atoms.

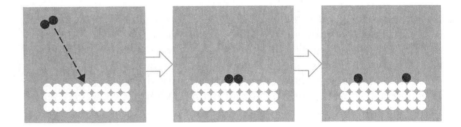

Figure 19.14. *Adsorption of hydrogen molecules on a metal surface is believed to yield hydrogen atoms, which are more reactive than the H_2 molecules themselves.*

Unless the reactant molecules can be adsorbed on the catalyst, no increase in reaction rate can occur. A substance whose presence during a reaction interferes with the adsorption process will therefore reduce the effectiveness of the catalyst and is thus called an **inhibitor.** These substances, by being strongly adsorbed on the catalytic surface, decrease the available space on which the reaction can occur. In some cases the catalyst eventually becomes useless and is said to be *poisoned.* The destruction of catalytic activity by poisoning is very important in biological systems.

Heterogeneous catalysts are preferred in industrial applications because the reactants can be continuously fed through the reaction vessel that contains the catalyst.

Heterogeneous catalysts have found many uses in the petroleum industry. One particularly interesting application is in the production of synthetic liquid fuels such as gasoline and diesel fuel by the reaction of hydrogen, H_2, with carbon monoxide, CO. (These chemicals themselves can be inexpensively produced from coal and natural gas by reactions that we won't discuss further at this time.) In the presence of an appropriate metal catalyst such as nickel or cobalt, the carbon monoxide reacts

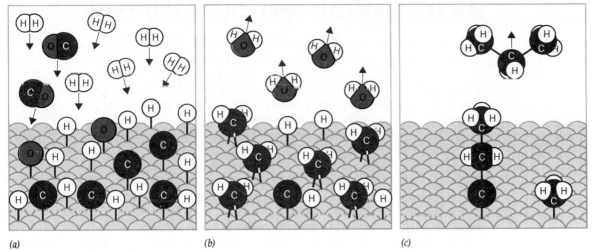

(a)   *(b)*   *(c)*

Figure 19.15. *Suggested mechanism for the reaction of CO and H_2 on a catalytic surface.*
(a) CO molecules become attached to the catalytic surface where they dissociate into C and O
atoms. Molecules of H_2 also become adsorbed. (b) Hydrogen and oxygen combine to form water
molecules that detach from the surface and leave. Atoms of H become attached to C atoms to
form CH_2 groups bound to the catalyst. (c) Additional H atoms bind to CH_2 groups to give CH_3
groups. These attach to other carbon units, and hydrocarbon chains begin to grow up from the
surface. The ultimate length of the chain depends on the catalyst.

with hydrogen to give water and a hydrocarbon. The mechanism of the reaction is believed to involve the kinds of changes illustrated in Figure 19.15.

Closer to home, we find heterogeneous catalysts used in the control of auto exhaust emissions. Gasoline-powered autos sold today in the United States are equipped with "catalytic converters." These converters employ a mixed metal oxide bed over which the exhaust gases pass after they are mixed with additional air (Figure 19.16). The catalyst quite effectively promotes the oxidation of CO and hydrocarbons to less harmful CO_2 and H_2O. The catalysts in newer catalytic converters also remove nitrogen oxide pollutants by promoting their decomposition to nitrogen and oxygen. Catalytic mufflers suffer from the disadvantage of being poisoned by lead. As a result, lead-free fuels must be employed in autos fitted with this type of antipollution device. Another disadvantage is that they also catalyze the oxidation of SO_2 to SO_3, which then reacts with water vapor to produce a mist of sulfuric acid. Since SO_2 is produced from the combustion of high-sulfur fuels, this problem is a serious one.

Small, portable flameless heaters used by campers have a catalyst that permits the fuel and oxygen to combine without producing a flame. They must be used with caution, however, because small amounts of carbon monoxide are sometimes produced.

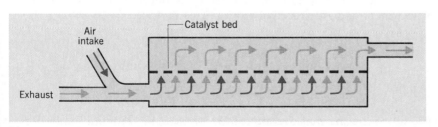

(a)   *(b)*

Figure 19.16. *(a) Hot exhaust gases mixed with air pass over the catalyst in a catalytic*
converter. The CO in the exhaust is oxidized to CO_2 and the nitrogen oxides are decomposed to
N_2 and O_2. (b) A cutaway view of a modern catalytic converter.

19.9 FREE RADICALS AND CHAIN REACTIONS

The reactions that we have discussed so far have been rather simple and straightforward, with uncomplicated rate laws. There are some reactions, however, that have very complex rate laws and whose elementary processes involve extremely reactive intermediate substances that possess unpaired electrons. As a group, these intermediates are called **free radicals** and might consist of single atoms or either neutral or electrically charged particles composed of a number of atoms. In chemical systems, molecules or ions can form free radicals either thermally (at high temperature) or by the absorption of photons of appropriate frequencies. At high temperatures, the violent vibrational motions within the molecules can cause bonds to break, whereas in the absorption of radiation, the photons supply the energy to break chemical bonds. In either case, the fragments that are formed have unpaired electrons. (Normally, the energy required to break chemical bonds is such that the photons must have frequencies placing them in the ultraviolet or X-ray region of the spectrum. This is the reason that exposure to X rays or other forms of high-energy radiation is dangerous.)

Free radicals have been blamed for a variety of ill effects in biological systems. These range from the effects of aging to the formation of cancers.

As you learned in earlier chapters, electrons become paired when atoms form chemical bonds, and it is the drive toward the pairing of electrons that is responsible in part for the formation of bonds between atoms. This same drive is what makes free radicals so reactive; the particles seek out other atoms or molecules with which they are able to form bonds and thus pair their electrons. In these reactions it often happens that a product molecule is formed along with *another* free radical, a process that can be repeated over and over. Therefore, once a free radical is formed through some initiation process, its reaction with other species is *self-propagating,* and the entire series of reactions that follows the production of the very reactive intermediate is called a **chain reaction.**

An oxygen molecule has two unpaired electrons, which is why O_2 is so reactive.

An example of a chain reaction is the mechanism that has been proposed to explain the rate law observed for the reaction between hydrogen and bromine. The overall reaction between these substances is

$$H_2 + Br_2 \longrightarrow 2HBr$$

If this reaction proceeded simply by a bimolecular collision between H_2 and Br_2, the expected rate law would be

$$rate = k[H_2][Br_2]$$

However, the actual rate law turns out to be

$$rate = k \frac{[H_2][Br_2]^{1/2}}{1 + [HBr]/k'[Br_2]}$$

which is very complex indeed. A mechanism that has been proposed to account for this rate law is the chain reaction shown below. A dot is used to represent the unpaired electron on a highly reactive atom or free radical.

1. $Br_2 \rightarrow 2Br\cdot$		Initiation
2. $Br\cdot + H_2 \rightarrow HBr + H\cdot$	⎫	
3. $H\cdot + Br_2 \rightarrow HBr + Br\cdot$	⎬	Propagation
4. $H\cdot + HBr \rightarrow H_2 + Br\cdot$		Inhibition
5. $2Br\cdot \rightarrow Br_2$		Termination

Reaction 1 is the thermal decomposition of diatomic bromine molecules to produce bromine atoms (the reactive intermediate). The overall reaction proceeds very slowly when the two reactants are mixed at room temperature. However, at high temperature reaction 1 takes place to an appreciable extent and rapidly sets off the remaining reactions. Step 1 is therefore called the *initiation step* because it begins the chain. In steps 2 and 3 the product HBr is formed as well as additional free atoms that serve to keep the reaction going. These steps are called *propagation steps* in the chain. Step 5, which leads only to the formation of a stable species, serves to end the chain and is called the *termination step*. Step 4 is referred to as an *inhibition step*, because its occurrence removes the product and thus decreases the overall rate of production of HBr. It is included in the mechanism because the presence of HBr decreases the reaction rate (note the appearance of [HBr] in the denominator of the rate law).

In general, chain reactions have very low activation energies once the initial free radical has been created, so they tend to be very rapid. In fact, many explosive reactions are chain reactions. Often, they contain reactions called *branching* steps in which more free radicals are produced than are consumed. Therefore, as they progress, the number of free radicals rises very rapidly and the reaction becomes so fast that it becomes explosive.

REVIEW QUESTIONS AND PROBLEMS

Problems whose numbers are in blue have their answers in Appendix D at the back of the book. The more difficult problems are marked with asterisks.

General Facts About Reaction Rates

19.1 What are the four factors that control the rates of chemical reactions?

19.2 What effect does particle size have on the rate of a heterogeneous reaction?

19.3 What criteria must be met by methods used to study the rate of a reaction?

19.4 Make a list of five reactions that occur in the world around you and compare their rates. Try to think of some that are fast and some that are slow.

19.5 Why do we say that one of the factors that influences reaction rate is the nature of the reactants?

19.6 Which reaction, or set of reaction conditions, should give the larger rate of formation of $H_2(g)$:
(a) Magnesium or iron in 1.00 M HCl?
(b) Zinc in 1.00 M HCl, or in 0.10 M HCl?
(c) 1.0 g of powdered zinc in 1.00 M HCl or a 1.0-g zinc bar in 1.00 M HCl?
(d) An iron nail in 1.00 M HCl at 25 °C, or in 1.00 M HCl at 40 °C?

Reaction Rate and Its Measurement

19.7 Define *reaction rate*. What are the units that are normally used?

19.8 For each of the following reactions, express the reaction rate in terms of the disappearance of the reactants and the appearance of the products. Predict the role of the coefficients in the balanced overall equation in determining the relative rates of disappearance of reactants and formation of products.
(a) $2H_2 + O_2 \rightarrow 2H_2O$
(b) $2NOCl \rightarrow 2NO + Cl_2$
(c) $NO + O_3 \rightarrow NO_2 + O_2$
(d) $H_2O_2 + H_2 \rightarrow 2H_2O$

19.9 Consider the reaction for the combustion of methane, CH_4,

$$CH_4(g) + 2O_2(g) \longrightarrow CO_2(g) + 2H_2O(g)$$

If the methane is burning at a rate of 0.16 mol $L^{-1} s^{-1}$, at what rates are CO_2 and H_2O being formed?

19.10 For the reaction

$$4NH_3(g) + 3O_2(g) \longrightarrow 2N_2(g) + 6H_2O(g)$$

it was found that at a particular instant N_2 was being formed at a rate of 0.68 mol $L^{-1} s^{-1}$.
(a) At what rate was water being formed?
(b) At what rate was NH_3 reacting?
(c) At what rate was O_2 being consumed?

19.11 The following data were collected for the reaction $2A \rightarrow 4B + C$

Time (min)	Concentration of A (mol/L)	Concentration of B (mol/L)
0	1.000	0.000
10	0.800	0.400
20	0.667	0.667
30	0.571	0.858
40	0.500	1.000
50	0.444	1.112

Make a graph of the concentrations of A and B versus time (concentrations along the vertical axis, time along the horizontal axis). Estimate the rate of disappearance of A and the rate of formation of B at $t = 25$ min and $t = 40$ min. Compare the rates of disappearance of A and formation of B. What would you expect the rate of formation of C to be at $t = 25$ min and $t = 40$ min?

Rate Laws

19.12 What is a rate law? What factors affect the value of the rate constant for a given reaction?

19.13 What is meant by the *order of a reaction*?

19.14 What are the units of the rate constant for (a) a first-order reaction, (b) a second-order reaction, and (c) a third-order reaction?

19.15 The rate at which CO is removed from the Earth's atmosphere by fungi in the soil is constant. What is the apparent order of this process?

19.16 What is the order with respect to each reactant and the overall order of the reactions described by the following rate laws:
(a) Rate $= k_1[A][B]$ (c) Rate $= k_3[G]^2[H]^2$
(b) Rate $= k_2[E]^2$

19.17 What would be the units of each of the rate constants in the preceding question if rate has the units *mol* $L^{-1} s^{-1}$?

19.18 When the concentration of a reactant is doubled, by what factor would the rate of reaction be changed if the order with respect to that reactant is (a) 1, (b) 2, (c) 3, (d) 4, (e) $\frac{1}{2}$, (f) -2?

19.19 Suppose that when the concentration of a reactant was doubled, the rate of reaction decreased by a factor of 2. What would be the exponent on the concentration term for that reactant in the rate law?

19.20 The rate constant for the reaction

$$2ICl + H_2 \longrightarrow I_2 + 2HCl$$

is 1.63×10^{-1} L mol^{-1} s^{-1}. The rate law is given by

$$\text{rate} = k[ICl][H_2]$$

What is the rate of the reaction for each of the sets of concentrations given below?

ICl Concentration (mol/L)	H$_2$ Concentration (mol/L)
0.25	0.25
0.25	0.50
0.50	0.50

19.21 The rate law for a reaction was found to be

$$\text{rate} = (2.35 \times 10^{-6} \text{ L}^2 \text{ mol}^{-2} \text{ s}^{-1})[A]^2[B]$$

What would the rate of reaction be if:
(a) The concentrations of A and B are 1.00 mol/L?
(b) $[A] = 0.250$ M, $[B] = 1.30$ M?

19.22 For the decomposition of dinitrogen pentoxide,

$$2N_2O_5 \longrightarrow 4NO_2 + O_2$$

the following data were collected.

N$_2$O$_5$ Concentration (mol/L)	Time (s)
5.00	0
3.52	500
2.48	1000
1.75	1500
1.23	2000
0.87	2500
0.61	3000

(a) Make a graph of the concentration of N$_2$O$_5$ versus time. Draw tangents to the curve at $t = 500$, 1000, and 1500 s. Determine the rate at these different reaction times.
(b) Determine the value of the rate constant at 500, 1000, and 1500 s, given the rate law,

$$\text{rate} = k[N_2O_5].$$

19.23 One of the reactions that can take place in polluted air is the reaction of nitrogen dioxide, NO$_2$, with ozone, O$_3$.

$$NO_2(g) + O_3(g) \longrightarrow NO_3(g) + O_2(g)$$

The following data were collected on this reaction at 25 °C:

Initial NO$_2$ Concentration (mol/L)	Initial O$_3$ Concentration (mol/L)	Initial Rate of Formation of O$_2(g)$ (mol L^{-1} s^{-1})
5.0×10^{-5}	1.0×10^{-5}	0.022
5.0×10^{-5}	2.0×10^{-5}	0.044
2.5×10^{-5}	2.0×10^{-5}	0.022

(a) What is the rate law for the reaction?
(b) What is the value of the rate constant (give the correct units)?

19.24 A certain reaction that follows the equation

$$2A(g) + B(g) \longrightarrow C(g) + 2D(g)$$

was studied at 25 °C. The following data were obtained.

Initial Concentration of A (mol/L)	Initial Concentration of B (mol/L)	Initial Rate of Formation of C (mol L^{-1} s^{-1})
0.010	0.010	1.20×10^{-3}
0.020	0.010	2.40×10^{-3}
0.030	0.010	3.60×10^{-3}
0.030	0.020	1.44×10^{-2}

(a) What is the rate law for the reaction?

(b) What is the rate constant at this temperature (including its units)?

(c) If $[A] = 0.020\ M$ and $[B] = 0.060\ M$, what will be the rate of formation of D?

19.25 At 27 °C, the reaction, $2NOCl \rightarrow 2NO + Cl_2$, is observed to exhibit the following dependence of rate on concentration.

Initial NOCl Concentration (mol/L)	Initial Rate of Formation of NO (mol L^{-1} s^{-1})
0.30	3.60×10^{-9}
0.60	1.44×10^{-8}
0.90	3.24×10^{-8}

(a) What is the rate law for the reaction?

(b) What is the rate constant (including its units)?

(c) By what factor would the rate increase if the initial concentration of NOCl is increased from 0.30 to 0.45 M?

19.26 The reaction of NO with Cl_2 follows the equation

$$2NO + Cl_2 \longrightarrow 2NOCl$$

The following data were collected.

Initial NO Concentration (mol/L)	Initial Cl$_2$ Concentration (mol/L)	Initial Rate of Formation of NOCl (mol L^{-1} s^{-1})
0.10	0.10	2.53×10^{-6}
0.10	0.20	5.06×10^{-6}
0.20	0.10	10.1×10^{-6}
0.30	0.10	22.8×10^{-6}

(a) What is the rate law for the reaction?

(b) What is the value of the rate constant (be sure to give the proper units)?

Concentration and Time; Half-Lives

19.27 Define *half-life of a reaction*.

19.28 How is the half-life of a first-order reaction affected by the concentrations of the reactants?

19.29 A reaction $2B(g) \rightarrow C(g)$ follows the rate law of rate $= k[B]^2$. Use what you have learned about half-lives for second-order reactions to construct a graph similar to Figure 19.4 for a second-order reaction. Take the initial concentration of B to be 1.00 M and follow the reaction through four half-lives. You can use any arbitrary time scale for the horizontal axis.

19.30 The rate constant for the first-order decomposition of $N_2O_5(g)$ at 100 °C is 1.46×10^{-1} s^{-1}.

(a) If the initial concentration of N_2O_5 in a reaction vessel is 4.50×10^{-3} mol/L, what will be the concentration 20.0 s after the decomposition begins?

(b) What is the half-life (in seconds) of N_2O_5 at 100 °C?

(c) If the initial concentration of N_2O_5 is $4.50 \times 10^{-3}\ M$, what will be the concentration be after three half-lives?

19.31 The decomposition of hydrogen iodide, HI, is second-order. At 500 °C, the half-life of HI is 2.11 min when the initial HI concentration is 0.10 M. What will be the half-life (in minutes) when the initial HI concentration is 0.010 M?

19.32 If we refer to Exercise 19.31, what is the rate constant for the decomposition of HI at 500 °C in the units, L mol^{-1} s^{-1}?

19.33 The decomposition of C_2H_5Cl has the following rate law: rate $= k[C_2H_5Cl]$. At 550 °C, $k = 3.2 \times 10^{-2}$ s^{-1}.

(a) What is the half-life of this reaction at 550 °C?

(b) If the concentration of C_2H_5Cl is 0.010 M after 1.00 min, what was the initial concentration of C_2H_5Cl?

Collision Theory and Reaction Mechanisms

19.34 How does collision theory account for the dependence of rate on the concentrations of the reactants?

19.35 Why can't we use the ideas of collision theory to predict in general the rate laws of chemical reactions? What must we know in order to predict the rate law?

19.36 What is a *reaction mechanism*?

19.37 Propane, C_3H_8, is a fuel used in many rural areas for cooking. It is known as LPG. It is unlikely that the combustion of propane occurs by a simple one-step mechanism,

$$C_3H_8(g) + 5O_2(g) \longrightarrow 3CO_2(g) + 4H_2O(g)$$

Explain why.

19.38 A mechanism for the reaction, $2NO + Br_2 \rightarrow 2NOBr$, has been suggested to be

Step 1	$NO + Br_2 \longrightarrow NOBr_2$
Step 2	$NOBr_2 + NO \longrightarrow 2NOBr$

(a) What would the rate law for the reaction be if the first step in this mechanism were slow and the second fast?

(b) What would the rate law be if the second step were slow, with the first reaction being a rapidly established dynamic equilibrium?

(c) Experimentally, the rate law has been found to be

$$\text{rate} = k[NO]^2[Br_2]$$

What can we conclude about the relative rates of steps 1 and 2?

(d) Why do we not prefer a simple, one-step mechanism

$$NO + NO + Br_2 \longrightarrow 2NOBr$$

(e) Can we, on the basis of the experimental rate law definitely exclude the mechanism in part (d)?

19.39 The reaction $NO_2(g) + CO(g) \rightarrow CO_2(g) + NO(g)$ appears to have the mechanism (at low temperature),

$$NO_2 + NO_2 \longrightarrow NO_3 + NO \qquad slow$$

$$NO_3 + CO \longrightarrow NO_2 + CO_2 \qquad fast$$

Explain why the reaction is zero-order with respect to CO.

19.40 The reaction of methyl bromide, CH_3Br, with OH^- appears to occur through a one-step mechanism involving collision of CH_3Br with OH^-,

$$CH_3Br + OH^- \longrightarrow CH_3OH + Br^-$$

The rate law for the reaction is found to be

$$rate = k[CH_3Br][OH^-]$$

In Example 19.3 we found that the rate law of the reaction of $(CH_3)_3CBr$ with OH^- is

$$rate = k[(CH_3)_3CBr]$$

Try to propose a mechanism that can account for the rate law for the reaction of $(CH_3)_3CBr$ with OH^-.

19.41 Suppose that the following sequence of reactions is proposed for a reaction:

Step 1 $\qquad\qquad 2A \longrightarrow A_2$

Step 2 $\qquad\qquad A_2 + B \longrightarrow C + 2D$

(a) What would be the overall net chemical reaction?
(b) What would the rate law be if step 1 is slow and step 2 is fast?
(c) What would the rate law be if step 2 is slow and step 1 is fast?

19.42 One of the reactions that occurs in polluted air in urban areas is $2NO_2(g) + O_3(g) \rightarrow N_2O_5(g) + O_2(g)$. It is believed that a species with the formula NO_3 is involved in the mechanism, and the observed rate law for the reaction is rate $= k[NO_2][O_3]$. Propose a mechanism for this reaction.

19.43 How do we know that not all collisions between reactant molecules lead to chemical change? What determines whether a particular collision will be effective?

19.44 How do the orientations of molecules influence whether a collision between them can be effective at producing chemical change?

19.45 One step in the mechanism for the decomposition of NO_2Cl into NO_2 and Cl_2 appears to be $NO_2Cl + Cl \rightarrow NO_2 + Cl_2$. Given the structure of NO_2Cl,

show how molecular orientation at the moment of a collision can be important in determining whether or not the products will be formed.

19.46 Nitric acid is one of the world's most important chemicals. One of its principal uses is in making fertilizers. Approximately 14 billion pounds $(1.4 \times 10^{10}$ lb) of HNO_3 are produced annually, mostly by oxidation of ammonia. This reaction gives nitric oxide, NO, which then reacts with oxygen to form nitrogen dioxide, NO_2. Nitrogen dioxide finally reacts with water to form nitric acid. The oxidation of NO to NO_2 follows the reaction

$$2NO(g) + O_2(g) \longrightarrow 2NO_2(g)$$

and has, as a rate law, rate $= k[NO]^2[O_2]$. Predict a possible mechanism for this reaction.

Effect of Temperature on Reaction Rate

19.47 Define the term activation energy.

19.48 Explain qualitatively, in terms of the kinetic theory, why an increase in temperature leads to an increase in reaction rate.

19.49 Insects, which are cold-blooded animals whose changes in body temperature tend to follow changes in the temperature of their environment, become quite sluggish in cool weather. On the basis of chemical kinetics, explain this phenomenon.

Transition State Theory

19.50 Draw a potential energy diagram for an endothermic reaction. Indicate on the drawing (a) the potential energy of the reactants, (b) the potential energy of the products, (c) the energies of activation for the forward and reverse reactions, (d) the heat of reaction.

19.51 What are meant by the terms transition state and activated complex? Where on the potential energy diagram for a reaction will we find the transition state?

Calculations Involving the Activation Energy

19.52 The rate constants for the reaction between ICl and H_2 (Exercise 19.20) at 230 and 240 °C have been found to be 0.163 and 0.348 L mol^{-1} s^{-1}, respectively. What are the values of E_a (in kilojoules per mole) and A for this reaction?

19.53 The rate constant for the reaction

$$CH_3I(g) + HI(g) \longrightarrow CH_4(g) + I_2(g)$$

at 200 °C is 1.32×10^{-2} L mol^{-1} s^{-1}. At 275 °C the rate constant is 1.64 L mol^{-1} s^{-1}. What is the activation energy (in kilojoules per mole) and the value of A (including its units)?

19.54 The activation energy for the decomposition of HI

$$2HI(g) \longrightarrow H_2(g) + I_2(g)$$

is 182 kJ/mol. The rate constant for the reaction at 700 °C is 1.57×10^{-3} L mol^{-1} s^{-1}. What is the value of the rate constant at 600 °C?

19.55 The activation energy for the reaction

$$HI(g) + CH_3I(g) \longrightarrow CH_4(g) + I_2(g)$$

is 138 kJ/mol. At 200 °C the rate constant has a value of 1.32×10^{-2} L mol^{-1} s^{-1}. What is the rate constant at 300 °C?

* **19.56** A chemist was able to determine that the rate of a particular reaction at 100 °C was four times faster than at 30 °C. The concentrations of the reactants were the same at both temperatures. Calculate the approximate energy of activation for the reaction in kJ/mol.

19.57 The decomposition of C_2H_5Cl is a first-order reaction having $k = 3.2 \times 10^{-2}$ s^{-1} at 550 °C and $k = 9.3 \times 10^{-2}$ s^{-1} at 575 °C. What is the activation energy, in kilojoules per mole, for this reaction?

19.58 The rate constant for the reaction

$$H_2(g) + I_2(g) \longrightarrow 2HI(g)$$

was measured at a series of temperatures. The data are given below.

Temperature (°C)	k (L mol^{-1} s^{-1})
283	1.2×10^{-4}
302	3.5×10^{-4}
355	6.8×10^{-3}
393	3.8×10^{-2}
430	1.7×10^{-1}

Graphically determine the value of E_a in kJ/mol for this reaction.

* **19.59** The development of a photographic image on film is a process controlled by the kinetics of the reduction of silver halide by a developer. The time required for development at a particular temperature is inversely proportional to the rate constant for the process. Below are published data on development times for Kodak's Tri-X film using Kodak D-76 developer. From these data, estimate the activation energy for the development process in kilojoules per mole.

Temperature (°C)	Time for Development (min)
18	10
20	9
21	8
22	7
24	6

Estimate the development time at 15 °C.

Catalysis

19.60 What is the difference between a homogeneous and a heterogeneous catalyst?

19.61 What is a heterogeneous catalyst? How does it function? What is an inhibitor?

19.62 Why is it unlawful for cars with catalytic converters to use leaded gasoline?

19.63 How does a catalyst play a part in lowering the activation energy for a reaction?

19.64 What effect does a catalyst have on
(a) the heat of reaction?
(b) the potential energy of the reactants?
(c) the transition state?

Chain Reactions

19.65 What is a free radical? How can free radicals be formed?

19.66 The C–C bond energy in C_2H_6 is 348 kJ/mol. Calculate the frequency and wavelength (in nanometers) of light that would be able to promote the reaction

$$C_2H_6(g) \longrightarrow 2CH_3\cdot(g)$$

19.67 Why are free radicals dangerous in living organisms?

19.68 Why are chain reactions often so fast?

19.69 The decomposition of acetaldehyde, CH_3CHO, follows the overall reaction

$$CH_3CHO \longrightarrow CH_4 + CO$$

with small amounts of H_2 and C_2H_6 also being produced. The reaction is thought to proceed by a chain reaction involving free radicals (note again that a free radical is indicated by using a dot to represent its unpaired electron). A proposed mechanism is
(1) $CH_3CHO \rightarrow CH_3\cdot + CHO\cdot$
(2) $2CH_3\cdot \rightarrow C_2H_6$
(3) $CHO\cdot \rightarrow H\cdot + CO$
(4) $H\cdot + CH_3CHO \rightarrow H_2 + CH_3CO\cdot$
(5) $CH_3\cdot + CH_3CHO \rightarrow CH_4 + CH_3CO\cdot$
(6) $CH_3CO\cdot \rightarrow CH_3\cdot + CO$

Identify (a) the initiation step, (b) the propagation step(s), and (c) the termination step(s).

Additional Exercises

19.70 Suppose a reaction has the rate law

$$rate = k[B]^x$$

in which the order of the reaction is unknown. Show that a graph of log(rate) versus log[B] should give a straight line that has a slope equal to the order of the reaction, x.

*** 19.71** The rate at which crickets chirp depends on the ambient temperature because they are cold-blooded insects whose body temperature follows the temperature of their surroundings. It has been found that the Celsius temperature can be estimated by counting the number of chirps in 8 seconds and then adding 4. In other words, $t(°C) = $ (number of chirps in 8 s) $+ 4$. From this information,

(a) Calculate the number of chirps in 8 seconds for temperatures of 20 °C, 25 °C, 30 °C, and 35 °C.

(b) The number of chirps per unit time is directly proportional to the rate constant for the biochemical reactions involved in the cricket's chirp. Based on this assumption, make a graph of ln(chirps in 8 s) versus $(1/T)$. Calculate the activation energy of the biochemical reaction involved.

(c) How many chirps would a cricket make in 8 seconds at a temperature of 120 °C?

*** 19.72** The cooking of an egg involves the denaturation of a protein called albumin, and the time required to achieve a particular degree of denaturation is inversely proportional to the rate constant for the process. This reaction has a high activation energy; $E_a = 418$ kJ/mol. Calculate how long it would take to cook a traditional three-minute egg on top of Mt. McKinley in Alaska on a day when the atmospheric pressure there is 355 torr.

19.73 Chlorofluorocarbons, CFCs, are compounds containing carbon, chlorine, and fluorine (e.g., CCl_2F_2). In the upper atmosphere they undergo reactions that ultimately lead to the destruction of ozone (O_3) that protects the earth's inhabitants from harmful ultraviolet radiation. This ozone depletion has been particularly severe over the South Pole during the Antarctic winter. Approximately 80% of the ozone loss is atrributed to the reaction sequence

$$ClO + ClO \longrightarrow Cl_2O_2$$
$$Cl_2O_2 + hv \longrightarrow Cl + ClOO$$
$$ClOO \longrightarrow Cl + O_2$$
$$Cl + O_3 \longrightarrow ClO + O_2$$
$$Cl + O_3 \longrightarrow ClO + O_2$$

What is the net chemical change for this sequence of reactions? Explain why a small amount of CFCs in the stratosphere might produce a large decrease in the ozone concentration.

*** 19.74** For a first-order reaction, a graph of ln[A] versus time (where A is a reactant) gives a straight line having a slope equal to $-k$. On the other hand, if the reaction is second-order with respect to A, a straight line is obtained when $1/[A]$ is plotted against time. In this case the slope of the line is equal to k. From this information, determine whether the reaction in Exercise 19.11 is first- or second-order. Calculate the rate constant for the reaction.

SURVEY OF
THE CHEMISTRY
OF THE ELEMENTS

20

METALS AND THEIR COMPOUNDS; THE REPRESENTATIVE METALS

At various times throughout this book we have described some of the physical and chemical properties of the elements. In Chapter 4, for example, we studied some of the physical and chemical properties of the metals and nonmetals, and in Chapter 10 we examined some of the redox reactions that typical metals and nonmetals undergo. Now that you have acquired a broader understanding of basic chemical principles—thermodynamics, kinetics, and equilibrium, for example—we again turn our attention toward the properties of the elements. Our goal will be not only to learn about some of their specific chemical properties, but also to see how intimately our daily lives are tied to chemistry.

In this chapter we begin by studying how metals are obtained from natural sources and how the properties of metal compounds are related to bonding and structure. Then we will discuss some of the specific chemical properties of the representative metals. These include the elements in Group IA (excluding hydrogen) and IIA, as well as the heavier members of Groups IIIA, IVA, and VA.

20.1 OCCURRENCE AND METALLURGY

Metallurgy is the science and technology of extracting metals from the earth and sea and bringing them to the point at which they can be put to practical use. It is a subject with a long and rich history, dating from the time when humans first fashioned tools and weapons from natural deposits of metals such as copper. Modern metallurgy touches our lives at every turn, from special alloys used in automobile engines to stainless steels used in surgical implants.

Sources of Metals

Most metals are obtained from land-based deposits that are called ores. An **ore** is a material that contains a particular constituent in a sufficiently high concentration that its extraction from the ore is *economically* worthwhile. Notice the stress on the economics of the process. Aluminum, for example, is present in the form of *aluminosilicates* in various kinds of rock, such as granite. However, there is no economical way of recovering the aluminum from these sources, so they are not considered aluminum ores. Instead, the ore of aluminum is a mineral called bauxite, in which the aluminum occurs as its oxide, Al_2O_3. As you learned in Chapter 18, aluminum can be extracted from its oxide by electrolysis in the Hall process.

Some metals, such as gold, silver, and (occasionally) copper, occur as free metals in nature. However, most metals exist in nature in compounds. Some are found in deposits of their carbonates or sulfates. Limestone, for example, is primarily $CaCO_3$, and gypsum is a hydrate of calcium sulfate. Many important metal ores contain oxides. Two examples are aluminum (Al_2O_3) and iron (Fe_2O_3 and Fe_3O_4). Sulfides are the primary sources of lead (as PbS) and copper (as Cu_2S).

The oceans also are a source of some metals, which exist there primarily as soluble halides and sulfates. The principal source of magnesium, for example, is sea water, even though Mg^{2+} is present at a concentration of only about 0.13%. On the ocean floor, large deposits of orange-sized metal-rich *manganese nodules* that are rich in manganese (~25%) and iron (~15%) are found in certain locations. There has been some interest expressed in attempting to recover them, although at this time no commercial mining operations are under way.

Once a metal ore has been obtained from either the earth or the sea, it must undergo a series of metallurgical processes that ultimately provide the finished metal for commercial use. Because of the wide variety of sources and the varying nature of the metal compounds in the ores, no single method can apply to the production of all metals. Nevertheless, it is possible to divide metallurgical processes into three principal categories.

1. *Concentration.* Ores that contain substantial amounts of impurities, such as rock, must often be treated to concentrate the metal-bearing constituent. Pretreatment of an ore is also carried out to convert some metal compounds into substances that can be more easily reduced.

2. *Reduction.* In their compounds, metals nearly always exist in positive oxidation states. Therefore, to obtain a metal from its ore it must be reduced. The particular procedure employed for a given metal depends on its ease of reduction to the free state.

3. *Refining.* Often, during reduction, substantial amounts of impurities become introduced into the metal. Refining is the process whereby these impurities are removed and the composition of the metal adjusted (alloys formed) to meet specific applications.

Let us take a brief look at each of these as they apply to some important metals.

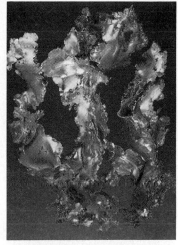

A large sample of naturally occurring gold discovered in California.

A large pile of manganese nodules on the ocean floor. Each is about the size of an orange.

Concentration

Not all ores have to be subjected to a pretreatment step prior to reduction, although most of them must. These pretreatment procedures involve the separation of the metal-bearing component of the ore from unwanted or interfering impurities. This is particularly important for low-grade ores in which the desired metal is present only in small amounts.

As expected, different methods are applied to different ores, depending on the

Many tourists try their hand at panning for gold when they visit Alaska.

An ***amalgam*** is a solution of a metal in mercury.

You are familiar with the reddish-brown color of Fe_2O_3 that is characteristic of rust.

specific properties of the impurities and the metal compounds. We can divide these procedures into two classes: *physical separations,* in which the chemical compositions of the constituents are not altered, and *chemical separations,* which use the chemical properties of the different substances in the ore.

As noted above, some metals, such as silver and gold, are found in deposits as the free element, and their recovery simply involves removing them from the rock and sand with which they are mixed. One of the earliest forms of physical separation was used by the "forty-niners" in panning for gold. A mixture of sand containing (it is hoped!) particles of metallic gold is placed in a shallow pan with water. The mixture is swirled about and the sand is washed over the rim, leaving the gold dust in the bottom of the pan. The success of this procedure is based on the fact that gold is about nine times as dense as the sand and gravel impurities. As a result, the lighter impurities are more easily washed away than the more dense metal.

Another way of removing metallic gold and silver from their ores is to treat the mixture with metallic mercury, a liquid in which silver and gold dissolve to form an alloy called an **amalgam.** The silver and gold are later recovered by distilling away the mercury, which is reclaimed and used again. You are probably familiar with silver and gold amalgams as the material used by dentists to fill teeth.

A physical separation technique that can be applied to the sulfide ores of zinc, copper, and lead is called **flotation.** In this process, illustrated in Figure 20.1, the ore is pulverized and added to large vats of a mixture of water and oil containing suitable additives. The finely powdered metal-bearing sulfide particles of the ore become coated by the oil while the unwanted material, called the **gangue,** is wetted by the water. A stream of air is then blown through the mixture, and the oil-covered mineral is carried to the surface by bubbles where it is trapped in a froth that can be removed to recover the metal compound. The gangue, on the other hand, simply settles to the bottom of the apparatus and is later discarded.

Iron, as you are probably aware, is society's most important metal, and its ores have traditionally been rich deposits of reddish-brown **hematite,** composed primarily of Fe_2O_3. Through decades of mining, hematite deposits in regions such as the Mesabi range of Minnesota have been nearly exhausted and other sources of iron ore have been sought. One such ore being mined today is *taconite.* Taconites have Fe_3O_4 as their iron-containing compound, and the better of them contain 30 to 40% iron by

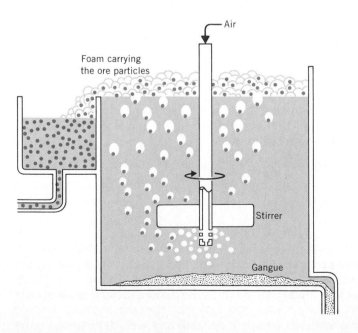

Figure 20.1. *The flotation process.*

weight. The compound Fe_3O_4 is called **magnetite,** and as its name suggests, it is magnetic. This provides the basis for the enrichment of taconite ores by a physical separation process. The rock-hard ore is first ground to give particles of very small size and then the Fe_3O_4 is pulled out with a powerful electromagnet.

Chemical methods of concentrating the metal-bearing component of an ore vary considerably because of the variety of chemical properties exhibited by the metals and their compounds. As mentioned earlier, aluminum, whose electrolytic reduction was described in Chapter 18, occurs in deposits of *bauxite,* a form of Al_2O_3. In this case the ore is concentrated by taking advantage of the amphoteric behavior of aluminum. The bauxite is treated with concentrated base, which dissolves the Al_2O_3 to produce aluminate ion, AlO_2^-.

$$Al_2O_3 + 2OH^- \longrightarrow 2AlO_2^- + H_2O$$

After the solution is removed from the gangue, it is neutralized with acid. This precipitates $Al(OH)_3$, which yields pure Al_2O_3 when heated.

$$2Al(OH)_3 \xrightarrow{\text{heat}} Al_2O_3 + 3H_2O$$

This purified aluminum oxide serves as the raw material in the Hall process discussed previously.

Another chemical pretreatment, often given to a sulfide ore, is called **roasting.** Here the ore is heated in air, converting the metal sulfide to an oxide that is more conveniently reduced.

$$2PbS + 3O_2 \longrightarrow 2PbO + 2SO_2$$

$$2ZnS + 3O_2 \longrightarrow 2ZnO + 2SO_2$$

The sulfur dioxide produced in the roasting process has been a severe source of air pollution, as described later in Chapter 23.

Reduction

Most metals, including those of both the representative and transition elements, are always found in nature in the combined state where they exist in positive oxidation states. Therefore, recovering them from their compounds involves a chemical reduction, and the nature of the reduction process depends on the ease with which the metal can be reduced.

Some metals are so easily reduced that many of their compounds can be decomposed just by heating them at relatively low temperatures. Priestley, for example, in his experiments on oxygen, produced metallic mercury and oxygen from mercuric oxide by simply heating it with sunlight focused on the HgO by means of a magnifying glass. In this case, HgO decomposes quite spontaneously at elevated temperatures according to the equation

$$2HgO(s) \longrightarrow 2Hg(g) + O_2(g)$$

The practicality of using a thermal decomposition reaction of this type is controlled by the thermodynamics of the reaction. To be of practical use, the reaction must proceed toward completion to a significant degree at a reasonable temperature. As you learned earlier, this requires that the value of ΔG_T° at that temperature be negative.

ΔG_T° is ΔG° at a temperature T.

The sign and magnitude of ΔG_T° depend on the signs and magnitudes of ΔH_T° and ΔS_T°.

$$\Delta G_T^\circ = \Delta H_T^\circ - T\,\Delta S_T^\circ$$

For decomposition, ΔS_T° is positive, so the term $T\,\Delta S_T^\circ$ when subtracted from ΔH_T° will

Remember, T is always positive.

tend to make ΔG_T° negative. Whether or not ΔG_T° is negative at a given temperature is therefore determined by the magnitude and sign of ΔH_T°.

Most metals have oxides, sulfides, and halides that have rather large negative heats of formation, so their "heats of decomposition" are correspondingly large positive quantities. For ΔG_T° to be negative therefore requires that the $T\,\Delta S_T^\circ$ term be very large, which means that the temperature must be very high. In effect, these substances are very thermally stable and resist decomposition at temperatures normally attainable in the laboratory or industrial processes.

Some metals, on the other hand, form compounds with either very small negative heats of formation or heats of formation that are positive. Examples are silver oxide (Ag_2O, $\Delta H_f^\circ = -30.5$ kJ/mol) and gold(III) oxide (Au_2O_3, $\Delta H_f^\circ = +80.8$ kJ/mol). For silver oxide, it can be estimated that the decomposition to Ag and O_2 should have an equilibrium constant equal to 1.0 if the temperature is raised to a mere 188 °C, so silver oxide decomposes relatively easily. The decomposition of gold(III) oxide has a negative ΔG° at all temperatures. The fact that it can be isolated at all at room temperature is because at this temperature the *rate* of decomposition is very slow.

Except in a few cases, thermal decomposition is not a practical way of extracting a metal from its compounds. Usually, the metal compound is allowed to react with some other substance that is a better reducing agent than the metal being sought. Titanium, for example, is obtained from $TiCl_4$ by reacting it with the more active metal magnesium. The $TiCl_4$ is obtained from the mineral rutile, a fairly pure source of TiO_2, by reaction with chlorine gas and carbon.

$$TiO_2 + 2C + 2Cl_2 \longrightarrow TiCl_4 + 2CO$$

The $TiCl_4$ is a volatile liquid (b.p. = 136 °C) and can be separated easily from less volatile impurities by distillation. The final reduction to the metal then follows the equation

$$TiCl_4 + 2Mg \longrightarrow Ti + 2MgCl_2$$

Active metals such as magnesium are very expensive reducing agents because they themselves are difficult or costly to prepare. Therefore, less expensive reducing agents are employed whenever possible. One such reducing agent is hydrogen, which can be used to liberate metals of moderate chemical activity from their compounds. For example, tin and lead oxides can be reduced by heating them in the presence of H_2.

$$SnO + H_2 \xrightarrow{\text{heat}} Sn + H_2O$$

$$PbO + H_2 \xrightarrow{\text{heat}} Pb + H_2O$$

Although hydrogen is sometimes used to reduce metal oxides, it is still a rather expensive reducing agent compared to the least expensive of all, carbon. Carbon is generally used in the form of **coke,** which is made from coal by heating it at high temperatures in the absence of air (Figure 20.2). This treatment drives off the volatile components of the coal (from which other important chemicals are derived) and leaves nearly pure carbon behind.

Tin and lead are normally prepared by heating their oxides with carbon.

$$2SnO + C \xrightarrow{\text{heat}} 2Sn + CO_2$$

$$2PbO + C \xrightarrow{\text{heat}} 2Pb + CO_2$$

Copper oxide can also be reduced with carbon.

$$2CuO + C \xrightarrow{\text{heat}} 2Cu + CO_2$$

The low thermal stabilities of compounds of silver and gold is one of the reasons these metals are found in the free state in nature.

Here is an example in which kinetics, not thermodynamics, controls what we are able to observe.

Titanium is used in making many aircraft parts. It is very strong, but only about 60% as dense as iron. This means that a part made of Ti will weigh only about 60% as much as the same size part made of steel.

Very reactive metals, such as sodium and magnesium, are produced by electrolysis, as described in Chapter 18.

Figure 20.2. *Coke is made in large ovens where coal is heated to drive off volatile materials. Here we see a fresh batch of white-hot coke being pushed from a coke oven. It will be transported to a blast furnace where it will be used to reduce iron ore.*

However, this is unnecessary for some copper sulfide ores if the conditions under which the ore is roasted are carefully controlled. For an ore that contains Cu_2S, roasting in air converts part of the Cu_2S to Cu_2O.

$$2Cu_2S + 3O_2 \xrightarrow{\text{heat}} 2Cu_2O + 2SO_2$$

At the appropriate time, the air supply is cut off and roasting is continued as the Cu_2O and the remaining Cu_2S react.

$$Cu_2S + 2Cu_2O \xrightarrow{\text{heat}} 6Cu + SO_2$$

Undoubtedly the most important chemical reduction brought about by carbon is that of iron oxide, either Fe_2O_3 or Fe_3O_4. This is accomplished in an apparatus called a **blast furnace,** developed in about 1300 A.D. and which in modern times has taken the form shown in Figure 20.3 on page 658.

The mixture of ingredients added to the blast furnace is called the **charge,** and its composition depends on the type of ore being reduced. For a hematite ore, which contains Fe_2O_3 as the iron-bearing compound, a typical charge consists of a mixture of limestone, coke, and the iron ore. The ore is normally composed of mostly Fe_2O_3 with impurities of sand (SiO_2, about 10%) and smaller amounts of compounds containing sulfur, phosphorus, aluminum, and manganese.

During the operation of the blast furnace, heated air is forced in at the bottom of the furnace where it reacts with carbon in a very exothermic reaction to produce carbon dioxide.

$$C + O_2 \longrightarrow CO_2 \qquad \Delta H = -394 \text{ kJ}$$

This hot blast of air gives the blast furnace its name.

The large amount of heat generated in this region of the furnace raises the temperature to nearly 1900 °C. As the hot gases rise, the CO_2 reacts with additional carbon in an endothermic reaction to form carbon monoxide, the active reducing agent in the furnace.

$$CO_2 + C \longrightarrow 2CO \qquad \Delta H = +173 \text{ kJ}$$

Figure 20.3. *A typical blast furnace for the reduction of iron ore.*

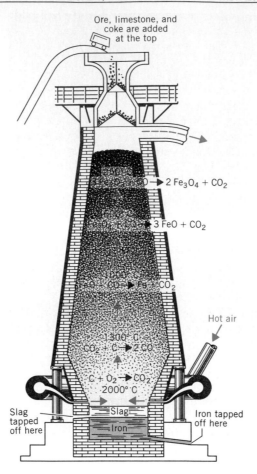

Ore, limestone, and coke are added at the top

$3Fe_2O_3 + CO \longrightarrow 2Fe_3O_4 + CO_2$

$Fe_3O_4 + CO \longrightarrow 3FeO + CO_2$

$FeO + CO \longrightarrow Fe + CO_2$

Hot air

$CO_2 + C \longrightarrow 2CO$
1300° C

$C + O_2 \longrightarrow CO_2$
2000° C

Slag tapped off here

Slag

Iron

Iron tapped off here

The reduction of the iron oxide takes place in a series of steps. Near the top of the furnace, Fe_2O_3 is reduced to Fe_3O_4.

$$3Fe_2O_3 + CO \longrightarrow 2Fe_3O_4 + CO_2$$

Farther down, in a hotter region of the furnace, this is reduced to FeO.

$$Fe_3O_4 + CO \longrightarrow 3FeO + CO_2$$

Finally, still farther down the FeO is reduced to the metal that, at these high temperatures, is a liquid and trickles down to form a pool of molten metal at the base of the tower.

$$FeO + CO \longrightarrow Fe + CO_2$$

Some blast furnaces are as tall as a 15-story building and produce up to 2400 tons of iron each day.

The function of the limestone in the furnace is to provide a basic medium with which acidic oxides, such as SiO_2 and P_4O_{10}, or amphoteric oxides such as Al_2O_3 can react. At elevated temperatures limestone, $CaCO_3$, decomposes to form lime (CaO) and CO_2 according to the equation

$$CaCO_3 \longrightarrow CaO + CO_2$$

The lime then reacts as follows:

$$CaO + SiO_2 \longrightarrow CaSiO_3$$

$$6CaO + P_4O_{10} \longrightarrow 2Ca_3(PO_4)_2$$

$$CaO + Al_2O_3 \longrightarrow Ca(AlO_2)_2$$

The products of these reactions have relatively low melting points and are liquids when they are formed. The mixture, called **slag,** also runs to the base of the furnace, where it floats atop the molten iron. As these two layers are formed, the charge in the furnace settles and additional limestone – coke – ore mixture is added at the top. In this way the blast furnace operates continuously, with fresh charge being added at the top and molten iron and slag being tapped off at the bottom. These furnaces are often run for months at a time before they are shut down for routine maintenance.

The liquid iron, when it is withdrawn from the blast furnace, is called **pig iron** and consists of about 95% Fe and approximately 4% carbon, with small amounts of silicon, manganese, phosphorus, and sulfur. This somewhat impure iron is very hard and can be poured into molds as **cast iron.** The slag that comes from the furnace can be used in making cement.

Refining

In the process of separating a metal from its ore, impurities are often introduced that impart undesirable properties to the final product. Therefore, it is generally necessary to purify the metal before it can be put to practical use. This purification process is called **refining.**

The specific procedure employed for refining a given metal depends on the chemical and physical properties of the metal as well as the properties of the impurities. As a result, there is no single method applicable to a very large number of different metals. We saw in Chapter 17 that copper can be economically refined electrolytically. This is possible, however, primarily because the silver and other precious metals recovered from the electrolytic cell offset the generally high cost of electricity.

An interesting process for refining nickel, called the **Mond process,** makes use of the relative ease of formation of nickel carbonyl — a compound formed between nickel and carbon monoxide.

$$Ni + 4CO \longrightarrow Ni(CO)_4$$
nickel carbonyl

Besides being easily formed, nickel carbonyl is also very volatile (and very poisonous). The impure nickel is therefore treated with CO at a moderately low temperature of 60 °C, where the $Ni(CO)_4$ that is formed exists as a gas. This is circulated to another portion of the apparatus, where it is heated to about 200 °C and decomposes to give pure nickel plus CO, which can be recycled through the process.[1]

The most important commercial refining processes involve the conversion of pig iron into steel. This requires the removal of impurities such as silicon, sulfur, and phosphorus and lowering the carbon content significantly from the approximately 4% introduced into the pig iron in the blast furnace.

Modern steel making began with the introduction of the **Bessemer converter** in England in 1856. A batch of molten pig iron from the blast furnace, weighing about 25 tons, is transferred to a tapered cylindrical vessel containing a refractory lining (Figure 20.4a). The composition of the lining is determined in part by the nature of the impurities in the iron. Since these impurities are usually silicon, phosphorus, and sulfur, whose oxides are acidic, a basic lining of dolomite (a $MgCO_3$, $CaCO_3$ mineral) is generally used. A blast of air (or oxygen) is blown through the metal from a set of small holes at the bottom of the vessel. The oxygen passing through the molten metal

A *refractory substance* is one that is difficult to melt.

---

[1] For a time it was believed that extremely toxic $Ni(CO)_4$ was responsible for the so-called Legionnaires' disease that killed a group of people attending an American Legion Convention in Philadelphia in 1976. Later, however, this idea was abandoned.

converts the silicon, phosphorus, and sulfur to oxides that then react with the lining to form a slag. The carbon in the pig iron is also oxidized to CO, so its concentration is reduced, too. The conversion of the pig iron to steel by this process is rapid, requiring about 15 minutes, and gives rise to a spectacular display of fire and showers of sparks. The reaction is difficult to control, however, and the quality of the steel produced in the Bessemer converter can be quite variable.

A somewhat newer method that virtually replaced the Bessemer process employs an **open-hearth furnace** (Figure 20.4b), a large, shallow hearth usually lined with a basic oxide refractory (for example, MgO, CaO). The furnace is charged with a mixture of pig iron, Fe_2O_3, scrap iron, and limestone. A mixture of burning gases and hot air is played over the surface of the charge to maintain it in a molten state while a series of chemical reactions take place. Impurities in the steel are oxidized by the Fe_2O_3 and air. Carbon dioxide, formed by oxidation of the carbon in the pig iron, bubbles out of the mixture, keeping it stirred, while the SiO_2 and other acidic oxides combine with CaO (from the limestone) and the refractory lining to form a slag. This entire process takes much longer than the Bessemer process, requiring 8 to 10 hours to complete. However, the quality of the steel is much more easily controlled because chemical analyses can be constantly carried out on samples of the mixture. The increased length of time required to process a batch of steel is also offset by the fact that much larger quantities (about 200 tons) can be handled at one time. In addition, other metals (e.g., cobalt, chromium, nickel, vanadium, and tungsten) can be added to the steel to form alloys with special properties prior to the steel being poured from the furnace. A typical stainless steel, for instance, is composed of approximately 72% iron, 19% chromium, and 9% nickel.

Modern methods of chemical analysis, making use of high-speed computers, have enabled a return to a modified form of the Bessemer process called the **basic oxygen process.** This newer procedure, which has largely replaced the open hearth

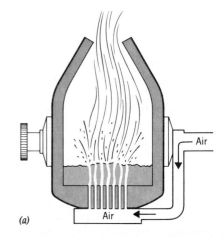

Figure 20.4. (a) *A Bessemer converter.* (b) *An open-hearth furnace.*

(a)

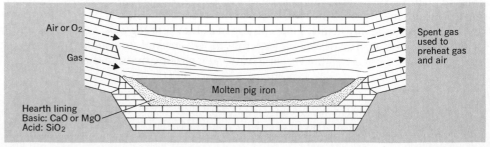

(b)

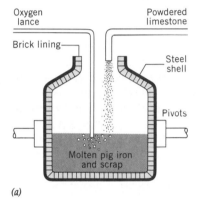

Figure 20.5. (a) *Cross section of a basic oxygen furnace used for the production of steel.* (b) *Molten steel is poured from a basic oxygen furnace.*

Oxygen lance

Powdered limestone

Brick lining

Steel shell

Pivots

Molten pig iron and scrap

(a)

(b)

furnace because of its speed, involves forcing a mixture of powdered $CaCO_3$ and oxygen gas into the molten pig iron (Figure 20.5). This rapidly burns away the impurities, which form a slag. The characteristic emission spectra of the elements in the steel permit rapid chemical analysis, and additives can be incorporated into the steel in the proper proportions to give a product with the desired properties. This process takes only about 20 to 25 minutes to complete, thereby yielding very substantial savings in time (and, of course, money) over the open-hearth process.

20.2 PERIODIC TRENDS IN METALLIC BEHAVIOR

In the earlier chapters we described some of the physical and chemical properties normally associated with metals, such as their tendencies to form ionic compounds with nonmetals and the basicity of their oxides. All metals are not equally metallic, however, especially in their chemical properties. For example, we commonly think of compounds formed from a metal and a halogen, such as chlorine or bromine, as being ionic. Beryllium chloride, however, does not fit this mold. When melted, $BeCl_2$ is a poor conductor of electricity, which suggests that molecules are present instead of ions. Similarly, aluminum forms molecular species such as Al_2Cl_6 and Al_2Br_6 in which the atoms are held together by covalent rather than ionic bonds.

Within the periodic table there are clear trends in the tendency of atoms to form covalent bonds. In Chapter 8 we learned that ionic bonds are preferred when a metal from the far left of the table combines with a nonmetal from the upper right corner. Covalent bonds are preferred, however, if two nonmetals combine. The deciding factor in each case is the difference in electronegativity between the bonded atoms. Metals have low electronegativities, which reflect their low ionization energies and

Metals with low electronegativities are said to be *electropositive*.

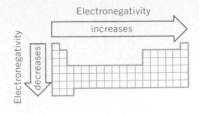

Electronegativity increases

Electronegativity decreases

For this discussion, refer to the periodic table on the inside front cover of the book.

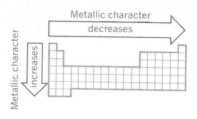

Metallic character decreases

Metallic character increases

$$O^{2-} + H_2O \longrightarrow 2OH^-$$

electron affinities. Nonmetals, on the other hand, have high electronegativities. An atom's electronegativity, therefore, is a measure of how metallic it is chemically.

You learned in Chapter 5 that electronegativity increases from left to right across a period. If we compare the bonds formed by a metal with a given nonmetal —chlorine, for example—they should become less ionic moving from left to right across the table. This is because the difference in electronegativity between the metal and nonmetal becomes progressively smaller. In period 2 we see this by comparing $LiCl$, $BeCl_2$, and BCl_3. Lithium chloride is distinctly ionic—it conducts electricity well in the molten state. As we've noted, $BeCl_2$ is a poor conductor when molten, which suggests that Be—Cl bonds are covalent. Boron trichloride is also covalent and exists as distinct molecules of BCl_3.

You also learned that electronegativity decreases going down a group. Metal–nonmetal compounds should therefore become more ionic as we descend a group, which they do. In Group IIA, for example, $BeCl_2$ is covalent, but $MgCl_2$ is ionic. In Group IIIA, BCl_3 is molecular and boron shows no evidence of forming B^{3+} ions. Aluminum chloride, on the other hand, appears to be ionic in the solid, although it exists in the vapor as molecules of Al_2Cl_6.

The elements of Group IVA, perhaps more than any others, illustrate the transition toward increasing metallic character as we descend a group. Carbon, at the head of the group, is nonmetallic in virtually every way.[2] Below carbon are the metalloids, silicon and germanium, and below these the metals, tin and lead.

Tin is unusual. At high temperatures it forms crystals that are metallic in appearance. At low temperatures (below 13.2 °C) these very gradually change to a nonmetallic, powdery form in which tin has a diamond-type lattice. Tin articles left for long periods in the cold therefore gradually crumble. At one time it was thought that some pest attacked the tin, and this disintegration of tin articles was called tin disease.

Lead, at the bottom of the group, demonstrates only metallic properties in the elemental state.

The trends in the metallic character of the elements that we have just seen — decreasing from left to right across a period and increasing from top to bottom in a group—can also be demonstrated by considering acid–base properties. Recall that metal oxides such as Na_2O are typically basic. They react with water to form hydroxides.

$$Na_2O + H_2O \longrightarrow 2NaOH$$

Nonmetal oxides are acidic. They form acids if they are able to react with water.

$$SO_3 + H_2O \longrightarrow H_2SO_4$$

The acid–base properties of the oxide of an element can therefore be used as an indicator of its metallic character.

If we look again at period 2, the trend in metallic character is clearly seen. Lithium oxide reacts with water to form LiOH. It also reacts with acids to neutralize them.

$$Li_2O(s) + 2HCl(aq) \longrightarrow 2LiCl(aq) + H_2O$$

Lithium hydroxide reacts with acids, but it does not react with bases. This purely basic behavior of Li_2O and LiOH identifies lithium as a distinctly metallic element.

When we examine beryllium oxide, we find quite a different state of affairs. Although BeO is rather inert (resistant to chemical attack), under appropriate conditions it will react with *either acids or bases*. The hydroxide, $Be(OH)_2$, is also able to react with either acids or bases. Substances that behave this way are said to be **amphoteric:** They show both acidic and basic characteristics. Because beryllium

---

[2] One form of carbon, graphite, does conduct electricity. This, however, is more a result of the bonding in graphite than evidence for metallic behavior. The structure of graphite and the mechanism of its electricial conductivity are discussed in Section 22.2.

compounds show some acidic characteristics, while the corresponding lithium compounds are purely basic, beryllium is less metallic than lithium. As we move farther to the right, we see that boron is even less metallic than beryllium. Its oxide shows only acidic properties.

The same trend toward decreasing metallic character is also seen in moving from left to right in period 3. The hydroxides of sodium and magnesium are only basic; they react with acids but not with other bases. Aluminum, on the other hand, is amphoteric, indicating that is is less metallic than either sodium or magnesium. The free element as well as its oxide and hydroxide reacts with both acids and bases.

$$2Al(s) + 6H^+(aq) \longrightarrow 2Al^{3+}(aq) + 3H_2(g)$$

$$2Al(s) + 2OH^-(aq) + 2H_2O \longrightarrow 2AlO_2^-(aq) + 3H_2(g)$$

The ion AlO_2^- is called aluminate ion.

The reaction of metallic aluminum with a base explains why you will find warnings on many oven cleaners not to use them on aluminum pots and pans. These oven cleaners contain lye (NaOH), which would quickly attack aluminum cookware. The popular drain cleaner Drano® also makes use of this reaction, as mentioned in Example 11.10. It consist of crystals of lye mixed with bits of metallic aluminum. When the Drano® is placed into a plugged drain, the reaction of the NaOH with the aluminum releases bubbles of H_2 that stir the mixture and help the cleaner dissolve any hair and grease that might be causing the stoppage.

20.3 IONIC–COVALENT CHARACTER OF METAL–NONMETAL BONDS

In the last section we saw that there is a gradual transition in the properties of the elements from metallic to nonmetallic as we move from left to right across a period. We also saw that the elements become more metallic going down a group. Although the changes in the covalence of metal–nonmetal compounds can be related to the variations in electronegativity, there is another way of looking at this that is useful in accounting for many properties.

Metal cations are almost always smaller than nonmetal anions. The positive charge on the cation is therefore concentrated in a rather small volume, while the negative charge on the anion is spread over a much larger volume. When a cation is placed near an anion, it can distort the electron cloud of the anion, pulling part of the electron density into the region between the two nuclei as shown in Figure 20.6. This causes the anion to become something of a dipole (besides being an anion) and we say that the electron cloud of the anion has been *polarized* by the cation. Since a covalent bond consists of electrons shared *between* nuclei, polarization of the anion results in the partial formation of a covalent bond. Furthermore, *the greater the degree of polarization, the greater the amount of covalent character of the bond.*

There are two major factors that contribute to a cation's ability to polarize a given anion. One of these is the number of positive charges on the cation. All other things being equal, a cation with a 2+ charge will distort the electron cloud of an

B_2O_3 reacts with water to give $B(OH)_3$, boric acid.

In this close-up photograph of the drain cleaner Drano®, we see tiny bits of aluminum metal surrounded by crystals of sodium hydroxide, which have been colored slightly blue with a dye. The aluminum reacts with the NaOH to liberate H_2 gas, which stirs the reaction mixture in the clogged drain.

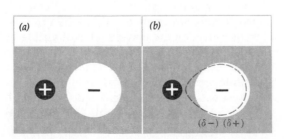

Figure 20.6. *Polarization of an anion by the positive charge on the cation. (a) Unpolarized anion and cation. (b) The electron cloud of the anion is distorted (polarized) by the cation.*

anion more than one with a 1 + charge. The second factor is the size of the cation. In a small cation, the positive charge is highly concentrated and has a strong effect on the anion. The same positive charge on a large cation will be more spread out and diffuse, so it won't distort the anion's electron cloud as well. A cation's effect on an anion is therefore directly proportional to its positive charge and inversely proportional to its size. We could also say that the effect of the cation is proportional to the *ratio* of its charge to its size. This ratio is called the **ionic potential,** ϕ,

$$\phi = \frac{q}{r}$$

where q is the cation's charge and r is its radius.

To see how this relates to the covalent character of metal–nonmetal bonds, let's look once more at the chlorides of the period 2 elements: $LiCl$, $BeCl_2$, and BCl_3. Suppose, for the moment, that these compounds could all be made ionic. The cations would then be Li^+, Be^{2+}, and B^{3+}, and each would be surrounded by one or more chloride ions. Suppose we now ask the question, "How would the ionic potentials of these cations vary?" Studying their electronic structures, we find that each has just a filled 1s subshell. All the valence electrons have been stripped away. As the nuclear charge increases from lithium ($Z = 3$) to boron ($Z = 5$), this 1s subshell is drawn more and more closely to the nucleus, so the size of the cations should decrease from Li^+ to B^{3+}. At the same time, the charge increases. The ionic potentials, therefore, increase very rapidly from Li^+ to B^{3+}, which means that their polarizing effects on the Cl^- ions surrounding them should also increase dramatically. This would cause a rapid increase in the amount of covalent bonding going from Li^+—Cl^- to B^{3+}—Cl^-, which, of course, is exactly what we've found. In $LiCl$ the bonding is ionic, in $BeCl_2$ it is largely covalent, and in BCl_3 the bonds are even more covalent.

> Simple ions with large positive charges are rare; their ionic potentials tend to be so large that they form covalent bonds to neighboring molecules or anions.

Going down a group, the charge on the cations stays the same (for example, Be^{2+}, Mg^{2+}, Ca^{2+}, etc.). The sizes of the cations increase, however, so their ionic potentials decrease. This means that if we compare $BeCl_2$ with $MgCl_2$, we would expect Mg^{2+} to have less of a polarizing effect on Cl^- than Be^{2+}. The result is that $MgCl_2$ should be more ionic than $BeCl_2$, which is also what we've seen before.

The usefulness of this approach toward discussing the covalent character of metal–nonmetal bonds is that it lets us easily understand why certain metal compounds are so clearly covalent. Consider, for instance, tin(IV) chloride, $SnCl_4$. This substance is a clear colorless liquid at room temperature. It boils at 114 °C, and its colorless cubic crystals are soft and melt at − 33 °C. These are properties we learned to identify with molecular substances, so the Sn—Cl bonds must be covalent. This is really not surprising in light of our previous discussions. If $SnCl_4$ were initially ionic, the cation would be Sn^{4+}. The large charge on this cation would give it a very large ionic potential and we would expect very extensive polarization of the chloride ions to occur. This would lead to quite covalent Sn—Cl bonds.

> Of course, we can also explain the bonding in $SnCl_4$ by valence bond theory.

20.4 COLORS OF METAL COMPOUNDS

Many inorganic compounds, such as common table salt or sodium bicarbonate, appear white, which means that when they are bathed in visible white light, their crystals reflect all the wavelengths of visible light equally well. You've only to look around, of course, to also appreciate that there is a huge number of very brightly colored compounds. Although many of these are organic substances, many also are inorganic compounds formed from a metal and a nonmetal.

A substance appears colored if it absorbs some wavelengths (or a band of wavelengths) from white light. The light that is reflected has these wavelengths

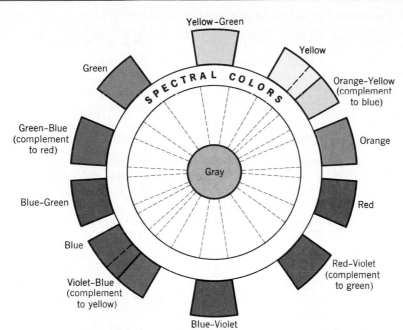

Figure 20.7. *A color wheel. Colors that are across from each other are called complementary colors. When a substance absorbs a particular color, light that is reflected or transmitted has the complementary color. Thus, something that absorbs red light appears green-blue and vice versa.*

missing and the color we perceive results from the wavelengths that remain.[3] Figure 20.7 illustrates a color wheel that we can use to anticipate the colors resulting when selected wavelength regions are absorbed. Across from each other are complementary colors. Green is the complementary color to red; yellow is the complementary color to violet. If a substance absorbs a particular color when bathed in white light, the reflected light is perceived as its complement. A substance that absorbs yellow light, for example, will appear violet. Chlorophyll, the substance that gives plants their green color, absorbs both red and blue light as shown by its absorption spectrum, Figure 20.8. The light that is reflected is yellow-green, and that's the color we see.

When a substance absorbs light of a particular color, the energy of the absorbed photon raises an electron from one energy level to another. Thus, the energy levels in substances affect the wavelengths (colors) of light absorbed, just as we saw in Chapter 7 that they affect the wavelengths of light emitted.

Typical ionic compounds such as NaCl or Na$_2$S don't absorb visible light. They have no electronic transitions with energies corresponding to visible wavelengths, so they appear white or colorless. The absorption that does take place occurs in the shorter-wavelength (higher-frequency) ultraviolet region of the spectrum. Here the energy that is absorbed is used to shift an electron from the anion to the cation.

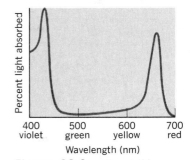

Figure 20.8. *The visible absorption spectrum of chlorophyll. The light that is not absorbed is in the green and yellow parts of the spectrum, which is why the leaves of plants appear green.*

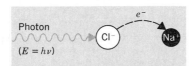

This is a **charge transfer process,** and the band of wavelengths absorbed from the ultraviolet rainbow is called a **charge transfer absorption band.**

Recall that the energy of a photon is $E = h\nu$.

[3] The perception of color is actually somewhat more complex than this because of the varying sensitivity of the human eye to various wavelengths. For example, the eye is much more sensitive to green than to red. If a compound reflects both of these colors with equal intensity, it wll appear greenish simply because the eye sees green better than it sees red.

Table 20.1. Colors of the silver halides

Compound	Color	Anion Radius (pm)
AgF	White	136
AgCl	White	181
AgBr	Cream	195
AgI	Yellow	216

The cream color of unexposed photographic film results from AgBr that is used as the light sensitive agent.

As the bond between a metal and a nonmetal becomes somewhat covalent (that is, as electron density shifts away from the anion *toward* the cation), less energy is required to produce the charge transfer. This means that light of lower frequency (longer wavelength) is required, and the absorption band shifts from the high-frequency ultraviolet region toward the violet and blue end of the visible spectrum. As a portion of the absorption band moves into the visible region, some violet light is absorbed and the compound takes on a pale yellow color. As the bond becomes even more covalent, the band shifts farther into the visible region and even more violet light is absorbed, so the compound appears even more intensely yellow. The depth or intensity of the color, therefore, is related to the degree of covalent character of the metal–nonmetal bonds.

Using color as a way to gauge the covalent character of bonds, we can now look at two other factors related to the ease with which a cation is able to polarize an anion. One is the size of the anion. For anions of the same charge, the ease of polarization increases with size. This occurs because as the size gets larger, the outer electrons are farther from the nucleus and are not held as tightly, so the cloud becomes more easily deformed. We can see this effect by looking at the colors of the silver halides (Table 20.1). Comparing AgCl, AgBr, and AgI, we see that as the anion becomes larger, the colors of the compounds become progressively deeper. This indicates a progressive increase in the covalent character of the Ag—X bonds as the halide ions become easier to polarize.[4]

We also find a similar relationship between covalent character and anion size if we compare metal oxides ($r_{O^{2-}} = 140$ pm) and sulfides ($r_{S^{2-}} = 184$ pm), as shown in Table 20.2. Both of these anions are colorless as evidenced by the fact that both Na_2O and Na_2S are colorless. However, with aluminum we find Al_2O_3 to be white while Al_2S_3 is yellow, suggesting a greater degree of ionic bonding in the oxide.

A second factor related to the ease of anion polarization is the charge on the anion. For anions of the same size, the ease of polarization increases as the negative charge becomes greater. For example, we can compare the colors of some com-

Table 20.2. Colors of some metal oxides and sulfides

Oxides	Color	Sulfides	Color
Al_2O_3	White	Al_2S_3	Yellow
Ga_2O_3	White	Ga_2S_3	Yellow
Sb_2O_3	White	Sb_2S_3	Yellow-red
Bi_2O_3	Yellow	Bi_2S_3	Brown-black
SnO_2	White	SnS_2	Yellow

[4] The salts NaF, NaCl, NaBr, and NaI, which are predominantly ionic, are colorless. This indicates that the halide ions themselves are colorless. The colors of the AgX compounds are therefore a reflection of covalent bonding.

Table 20.3. Colors of metal chlorides and sulfides

Chloride (r = 181 pm)		Sulfide (r = 184 pm)	
AgCl	White	Ag_2S	Black
CuCl	White	Cu_2S	Black
AuCl	Yellow	Au_2S	Brown-black
$CdCl_2$	White	CdS	Yellow
$HgCl_2$	White	HgS	Black or red, depending on crystal structure
$PbCl_2$	White	PbS	Black
$SnCl_2$	White	SnS	Black
$AlCl_3$	White	Al_2S_3	Yellow
$GaCl_3$	White	Ga_2S_3	Yellow
$BiCl_3$	White	Bi_2S_3	Brown-black

(a)

(b)

Figure 20.9. (a) *Solutions of Na_2S and $Cd(NO_3)_2$ are colorless because their ions are colorless.* (b) *When the solutions are combined, a precipitate of orange CdS is formed. The color of the CdS can range from orange to yellow, depending on the particle size. Cadmium sulfide is used as a pigment in artists' oil paints.*

pounds containing chloride (r = 181 pm) and sulfide (r = 184 pm) as shown in Table 20.3. In each case the sulfide has a deeper color than the corresponding chloride salt, indicating that the compounds containing the more highly charged sulfide ion are more covalent. We also see that, in general, most metal sulfides, except those of the alkali and alkaline earth metals, are deeply colored and possess quite a substantial degree of covalent bonding.

Some of the colored compounds described above are not at all uncommon. For instance, when silver tarnishes, it reacts with traces of H_2S in the air to give a dull film of black Ag_2S. This hydrogen sulfide, produced generally from decomposing organic matter, also darkens lead-based paint. The pigment, $Pb_3(OH)_2(CO_3)_2$, called *white lead*, reacts with H_2S to form black PbS.

The fact that many of these compounds possess rather striking colors has also been put to use throughout history. For example, the brilliant yellow or orange of CdS (Figure 20.9) and the vermilion red of HgS have led them to be employed as pigments for the oil paints used by artists. Not long ago, several sticks of black PbS that had been used as a type of mascara were recovered from an ancient Egyptian burial ground.

20.5 GROUP IA: THE ALKALI METALS

The elements of Group IA consist of hydrogen, lithium, sodium, potassium, rubidium, cesium, and francium. Except for hydrogen, all of them are very reactive metals. They are called alkali metals because their oxides are quite soluble in water and produce very basic (alkaline) solutions. For example, sodium oxide, the basic anhydride of sodium hydroxide, reacts as follows:

$$Na_2O(s) + H_2O(l) \longrightarrow 2Na^+(aq) + 2OH^-(aq)$$

The oxides of the other alkali metals react in a similar fashion.

Each of the alkali metals has only a single, rather loosely held electron in its outer shell. Loss of this electron gives an ion with a charge of 1+. This is the only oxidation state (other than zero, of course) that these elements exhibit, which leads

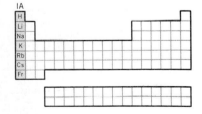

to some rather simple chemistry. In fact, the chemical and physical similarities among the metals of Group IA illustrate in a very striking way the empirical basis of the periodic table: Elements having similar properties are placed in the same vertical column.

Occurrence and preparation

The most abundant of the alkali metals are sodium and potassium. It is not surprising, therefore, that they are also the most biologically important. In animals, for instance, both sodium and potassium ions are needed, and the proper balance between them must be maintained in order for the organism to function properly. In plants, potassium is significantly more important than sodium.

Lithium, rubidium, and cesium are present in the earth's crust in much smaller amounts than sodium and potassium. They and their compounds are therefore more difficult to come by and, as a result, they are more expensive. Because of this, their practical applications are limited.

Francium is radioactive, and even its longest-lived isotope, ^{223}Fr, has a half-life of only 21 minutes. (It is produced in the radioactive decay of another radioactive isotope, ^{227}Ac.) Because of francium's short half-life, it is estimated that at any given time there is less than 30 g of it in the entire Earth's crust.

The alkali metals never occur free in nature. They always exist in compounds, which are found in both the Earth's crust as well as the ocean. For example, gigantic salt deposits (Figure 20.10a) are located in many places below the surface of the Earth — in Louisiana, New York, Michigan, Oklahoma, California, and Texas, to name just some. Where the climate is arid, salt deposits even occur on the surface (Figure 20.10b). Nearly all the compounds of the alkali metals are soluble in water,

Sodium is not required at all by plants, except for some salt marsh species.

Clays, mica, and silicate ores also contain alkali metal ions.

Molar Concentrations in Seawater	
Li$^+$	6×10^{-5}
Na$^+$	0.47
K$^+$	0.010
Rb$^+$	$\approx 10^{-6}$
Cs$^+$	$\approx 10^{-8}$

(a)

(b)

Figure 20.10. *Sodium chloride occurs in large deposits both below and above ground.* (a) *The interior of a salt mine located in Texas.* (b) *The Bonneville Salt Flats in Utah.*

Figure 20.11. *Salt crystals being harvested in Newark, California. The salt crystals were formed by the evaporation of seawater.*

and where rainfall is plentiful, they have been largely washed away into the oceans and salt lakes, or leached into subterranean waters. The recovery of the alkali metals therefore occurs from both land and water—from salt mines like that shown in Figure 20.10, by the evaporation of seawater in large ponds (Figure 20.11), and from brine wells.

The separation of the alkali metals from their compounds can be accomplished by the electrolytic reduction of a molten compound. Normally, halides are chosen because they have relatively low melting points (compared to oxides, for instance). An example is the electrolysis of molten sodium chloride using the Downs cell, which was described in Chapter 18.

One of the principal uses of sodium is the manufacture of tetraethyllead, a gasoline octane booster.

The Downs cell does not work well for the electrolysis of molten KCl, RbCl, and CsCl because at the temperatures required to melt these salts, the metals are very volatile. This makes their collection somewhat difficult. Instead, the molten chlorides are exposed to sodium vapor. The equilibrium

$$Na(g) + MCl(l) \rightleftharpoons NaCl(l) + M(g)$$

is shifted to the right because potassium, rubidium, and cesium are considerably more volatile than sodium. Sodium condenses and the other metal evaporates. What makes this reaction especially interesting is that it is driven to the right even though sodium is not as strong a reducing agent as potassium, rubidium, or cesium when they are all compared under identical conditions.

Physical properties

The alkali metals exhibit many of the typical properties that we've come to expect of metals: high luster and good thermal and electrical conductivity. Nevertheless, applications of the free metals rarely exploit these properties because the metals are so reactive. An exception has been the use of sodium in cooling certain nuclear reactors, which takes advantage of its low melting point, its relatively high boiling point, and its good thermal conductivity. These characteristics allow the metal to be easily melted and pumped through pipes that pass through the hot core of the reactor where the sodium quickly absorbs heat. The sodium is then pumped through

The operation of a nuclear reactor is described further in Chapter 24.

Table 20.4. Some physical properties of the alkali metals

Element	Ionization Energy (kJ/mol)	$E°$ (V)	M^+ Radius (pm)	Melting Point (°C)	Boiling Point (°C)
Lithium	520.1	−3.05	60	180.5	1326
Sodium	495.8	−2.71	95	97.8	883
Potassium	418.8	−2.92	133	63.7	756
Rubidium	402.9	−2.99	148	38.98	688
Cesium	375.6	−3.02	169	28.59	690

the pipes of a heat exchanger outside the reactor where the heat is transferred to water, which can be made into steam to generate electricity.

The softness and low melting points of the alkali metals (Table 20.4) reflect the existence in the metallic lattice of singly charged cations that attract the surrounding "electron sea" weakly. These cations are formed by the loss of the single s electrons from the outer shells of the atoms. Because these outer-shell electrons are only weakly held, the alkali metals also have low ionization energies, electron affinities, and electronegativities

An important physical property of the alkali metals is their emission spectra, which can be produced by passing an electric discharge through their vapors or introducing one of their salts into a Bunsen burner flame. Lithium salts, for example, impart a beautiful red color to a flame, sodium salts give a brilliant yellow color, whereas potassium salts produce a violet-colored flame. See Figure 20.12. These colors are intense enough to serve as useful qualitative tests called **flame tests** that can be used in analyzing mixtures of unknown composition. For example, if a drop of a solution of an unknown is placed into a flame and the yellow color is observed, sodium ions are in the unknown. If no yellow color is seen, then sodium is absent. The violet color of the potassium flame test is not as bright as the yellow sodium

Figure 20.12. *The colors given to a Bunsen burner flame by salts of* (a) *sodium,* (b) *potassium, and* (c) *lithium.*

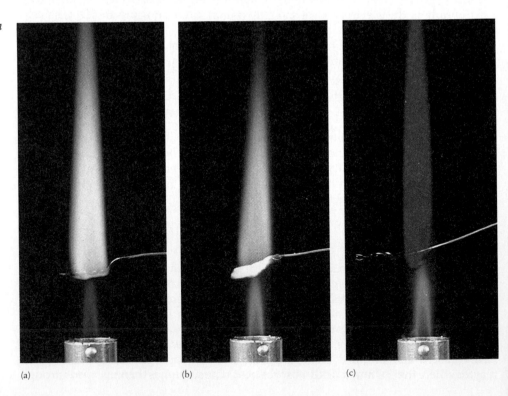

(a) (b) (c)

flame and is easily masked, even by traces of sodium in the unknown. Viewing the flame through blue glass, called *cobalt glass,* filters out the yellow and allows the violet potassium flame to be seen.

The brilliant yellow emission by sodium accounts for one of this element's growing commercial uses: in sodium vapor lamps (Figure 7.5 on p. 196). These are becoming more and more widely used throughout the country for street lighting because they are much more economical to operate than incandescent lamps. An incandescent lamp, such as an ordinary light bulb, gives off much of its energy in the form of invisible infrared radiation, so much of the electrical energy used to operate it is wasted. However, in a sodium vapor lamp, which is really nothing more than a gas discharge tube containing sodium vapor as the gas, most of the energy of the electrical discharge appears as visible yellow light ($\lambda = 589$ nm). This actually corresponds to a pair of closely spaced lines in the emission spectrum of sodium. See Figure 7.4 on p. 195.

Chemical properties and important compounds

The alkali metals are the most reactive metals. They are very powerful reducing agents and they are able to reduce water with the evolution of hydrogen. The general reaction is

$$2M(s) + 2H_2O(l) \longrightarrow 2M^+(aq) + 2OH^-(aq) + H_2(g)$$

The effectiveness of the alkali metals as reducing agents is reflected in the extremely negative reduction potentials of their ions (Table 20.4). These suggest that the half-reactions

$$M^+(aq) + e^- \longrightarrow M(s)$$

occur with great difficulty, and therefore that the oxidation half-reactions

$$M(s) \longrightarrow M^+(aq) + e^-$$

take place easily.

A close examination of the trends in the ionization energy and in $E°$ reveals an apparent contradiction, however. Note that the ionization energy (IE) decreases as we proceed down within the group, suggesting that it becomes progressively easier to strip an electron from the atom as we go from Li to Cs. In Chapter 7 we saw that this is, in fact, expected. We would also anticipate that the reduction potentials should become more negative as the IE becomes smaller since the elements should become more easily oxidized. This trend is indeed followed from Na downward; however, Li has an $E°$ that is more negative than Na (or any of the other alkali metals for that matter). Why is this so?

The ionization energy, remember, is a measure of the ease with which a *gaseous* atom loses electrons to produce a *gaseous* cation. The reduction potential, on the other hand, is concerned with the transfer of electrons between the *solid* metal and the corresponding cation in *aqueous solution,* where it is hydrated by the water molecules surrounding it. This latter process is more complex than simply removing an electron from the isolated metal atom. To understand the trends in the reduction potentials, we must break down the overall reaction into several steps. If we concentrate on the enthalpy changes involved in the reaction, we can construct the diagram in Figure 20.13. We see that the net enthalpy change is the sum of three energy terms. Two of these are endothermic — the sublimation energy, ΔH_{subl}, which is the energy needed to convert the solid into gaseous atoms, and the ionization energy, which we have already examined. The third quantity, called the hydration energy, is strongly exothermic. It corresponds to the energy *released* when the cation is placed into the solvent cage where it is surrounded by the water dipoles oriented in such a way that their negative ends are directed at the positive ion (Figure 20.14).

Figure 20.13. *Enthalpy diagram for the reaction*

$$M(s) \longrightarrow M^+(g) + e^-$$

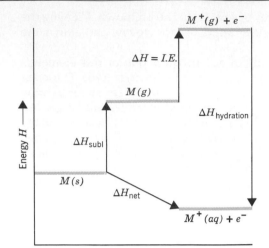

Figure 20.14. *Solvation of a cation by water molecules. The negative ends of the water dipoles surround the positive ion.*

Figure 20.15. *When an alkali metal dissolves in liquid ammonia, it forms a deep blue solution. The same color is produced by all the alkali metals. The color is believed to be from electrons that are solvated by ammonia molecules in the solution.*

Na_2O has a strong affinity for moisture and is an effective drying agent.

Among the alkali metals the sublimation energy remains approximately constant as we descend the group while the ionization energy decreases. To reach the peak on the energy diagram, we require the greatest amount of energy for Li and the least amount for Cs. However, because of its small size and its correspondingly high ionic potential, upon hydration Li^+ interacts much more strongly with the water dipoles than any of the other Group IA ions do. As a result, the hydration energy of Li^+ is unusually large, and much greater than for the other M^+ ions in the group. Therefore, the net overall enthalpy change is most exothermic for Li. This in turn causes Li to be more easily oxidized in an aqueous medium than any other alkali metal. In other words, the extraordinarily high hydration energy of the small Li^+ ion more than compensates for its relatively high ionization energy and causes Li to have an unexpectedly large negative $E°$.

Associated with the easy loss of electrons from the alkali metals is their interesting behavior in liquid ammonia. We have already seen that these metals are capable of reducing water to liberate H_2. Ammonia is not as easily reduced as water and, when placed into this solvent, alkali metals dissolve without reaction to form deep blue solutions (Figure 20.15). It is generally agreed that this color, which is identical for liquid ammonia solutions of all the alkali metals (as well as for Ca, Sr, and Ba from Group IIA), is a result of the presence of free electrons that have become solvated by ammonia molecules. Apparently, when the metal dissolves in NH_3, it loses its valence electron to become a cation. This electron becomes surrounded by NH_3 molecules arranged so that the positive ends of their dipoles are directed at the negatively charged electron, as shown in Figure 20.16, thereby stabilizing it through solvation. Solutions containing alkali metals in liquid ammonia are, as we would expect from the presence of readily available electrons, excellent reducing agents.

The strong tendency of the alkali metals to undergo oxidation permits them to react readily with most of the elemental nonmetals. The design of the Downs cell, you recall, is based on the need to keep Cl_2 and Na apart after they are formed by electrolysis. All the halogens (F_2, Cl_2, Br_2, and I_2) react with all the alkali metals to form the corresponding salts.

Among the most interesting reactions of the alkali metals is their behavior toward elemental oxygen. Only lithium burns in air to form the normal oxide, Li_2O.

$$4Li(s) + O_2(g) \longrightarrow 2Li_2O(s)$$

Sodium undergoes a similar reaction, but only if the supply of oxygen is limited. In the presence of excess oxygen sodium forms the pale yellow peroxide.

$$2Na(s) + O_2(g) \longrightarrow Na_2O_2(s)$$
$$\text{sodium peroxide}$$

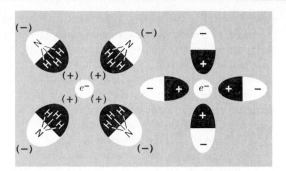

Figure 20.16. *The solvated electron in liquid ammonia.*

This solid contains the peroxide ion, O_2^{2-}, which itself is an effective oxidizing agent. Sodium peroxide is therefore used commercially as an oxidizing agent. It is also used as a bleaching agent, because the oxidation of intensely colored molecules often gives colorless reaction products. When dissolved in water, the peroxide ion hydrolyzes extensively

$$Na_2O_2(s) + 2H_2O(l) \longrightarrow 2Na^+(aq) + 2OH^-(aq) + H_2O_2(aq)$$

This causes solutions of Na_2O_2 to be very alkaline.

Potassium, rubidium, and cesium react with oxygen to form *superoxides.* For example,

$$K(s) + O_2(g) \longrightarrow KO_2(s)$$
$$\text{potassium}$$
$$\text{superoxide}$$

These compounds contain the paramagnetic superoxide ion, O_2^-. When the superoxides are dissolved in water, oxygen is evolved.

$$2MO_2(s) + 2H_2O(l) \longrightarrow 2M^+(aq) + 2OH^-(aq) + O_2(g) + H_2O_2(aq)$$

Because of this reaction, potassium superoxide has found applications in breathing equipment designed to recirculate the air of the user. When air is exhaled, it contains both moisture and carbon dioxide. This is circulated through a canister containing KO_2. Moisture reacts as above, and carbon dioxide is removed from the air and replaced with oxygen by the overall reaction

$$4KO_2(s) + 2CO_2(g) \longrightarrow 2K_2CO_3(s) + 3O_2(g)$$

This allows the user to continue to breathe without having to draw in possibly contaminated air from outside the apparatus.

Because of their vigorous reactions with both moisture and oxygen, the alkali metals must be stored under an inert (nonreactive) liquid—often an oil or kerosene. Use is made of their affinity for H_2O in drying solvents. Sodium, for instance, is often employed to effectively remove traces of moisture from laboratory solvents. Commercially, the alkali metals have been used as *getters* in the production of electronic vacuum tubes. Getters combine with the last traces of H_2O and O_2 that would otherwise interfere with the operation of the tube.

Besides combining with the halogens and oxygen, the alkali metals react directly with virtually all the other nonmetals as well. Lithium, however, is the only one that combines directly with gaseous nitrogen to form a *nitride.*

$$6Li(s) + N_2(g) \longrightarrow 2Li_3N(s)$$
$$\text{lithium nitride}$$

Most of the important compounds of the alkali metals are those of sodium and potassium. This is simply because among the Group IA metals these two elements have the largest abundance. Sodium chloride is by far the most readily available and easily recoverable of all the alkali metal compounds, which makes it an inexpensive raw material for preparing other sodium compounds. For this reason, sodium compounds are considerably cheaper than those of potassium or the other alkali metals. Since there are such close similarities among the chemical properties of alkali metal compounds of a given type, when given a choice industry will almost always select a sodium compound simply because of economics.

The primary raw material for the preparation of potassium compounds is potassium chloride. It is obtained from *sylvite*, a mineral form of KCl, and *carnallite*, $KCl \cdot MgCl_2 \cdot 6H_2O$. These are important as fertilizers because of the need of plants for both potassium and magnesium.

Over 24 billion pounds of Cl_2 and NaOH are made annually from NaCl by electrolysis.

Sodium hydroxide, made from salt by the electrolysis of brine, is industry's most important strong base. Its common name is *lye* or *caustic soda*. Around the home it is found in oven and drain cleaners. It is used industrially to make soap and detergents, pulp and paper, textiles, in removing sulfur from petroleum, and in the manufacture of myriad other chemicals. In the majority of its applications, NaOH is used to neutralize acids.

Potassium hydroxide, which is produced by electrolysis of KCl solution, is more expensive than NaOH, so its uses have been limited. It serves as the electrolyte in alkaline Zn/MnO_2 batteries, as we learned in Chapter 18.

The carbonates of the alkali metals constitute another class of important compounds. Once again, the sodium salts are used most widely because of their low cost. The principal source of sodium carbonate is *trona* ore, a mixture of sodium carbonate and sodium bicarbonate with the composition $Na_2CO_3 \cdot NaHCO_3 \cdot 2H_2O$. Large amounts of this ore are mined from deposits in Wyoming.

Sodium carbonate is also made chemically from salt and limestone ($CaCO_3$) by the **Solvay process,** which takes advantage of the fact that $NaHCO_3$ is less soluble in cold water than NaCl. The process begins by thermally decomposing limestone to give CaO and CO_2.

$$CaCO_3(s) \xrightarrow{\text{heat}} CaO(s) + CO_2(g)$$

The carbon dioxide, along with ammonia, is bubbled into a concentrated solution of salt. The ammonia partially neutralizes carbonic acid that is formed from the carbon dioxide.

$$CO_2(g) + H_2O(l) \longrightarrow H_2CO_3(aq)$$

$$H_2CO_3(aq) + NH_3(aq) \longrightarrow NH_4^+(aq) + HCO_3^-(aq)$$

As the concentration of HCO_3^- builds up, the less soluble $NaHCO_3$ begins to precipitate. The overall reaction can be written

$$Na^+(aq) + Cl^-(aq) + NH_4^+(aq) + HCO_3^-(aq) \longrightarrow NaHCO_3(s) + NH_4^+(aq) + Cl^-(aq)$$

After the solid $NaHCO_3$ is separated by filtration, it is heated to convert it to sodium carbonate.

$$2NaHCO_3(s) \xrightarrow{\text{heat}} Na_2CO_3(s) + H_2O(g) + CO_2(g)$$

Meanwhile, the solution containing NH_4Cl is treated with the calcium oxide left over from the decomposition of the limestone. This regenerates ammonia, which is recycled and used again. The reactions are

$$CaO + H_2O \longrightarrow Ca(OH)_2$$

$$Ca(OH)_2 + 2NH_4Cl \longrightarrow CaCl_2 + 2H_2O + 2NH_3(g)$$

The hydrate, $Na_2CO_3 \cdot 10H_2O$, is called washing soda. Its solutions are basic because of hydrolysis of the CO_3^{2-} ion.

If care is taken not to lose any ammonia, the process consumes NaCl and $CaCO_3$ and produces Na_2CO_3 and $CaCl_2$. The net overall change is

$$2NaCl + CaCO_3 \longrightarrow Na_2CO_3 + CaCl_2$$

Sodium carbonate is used in huge amounts by industry (about 17 billion pounds of it are produced each year). About half is used to manufacture glass; the rest is used to manufacture other chemicals, process pulp and paper, and make soap and detergents.

Sodium bicarbonate, the intermediate product in the Solvay process, is also a useful chemical. A solution of it is mildly basic and serves as a buffer because of the ability of the HCO_3^- ion to react with both acids and bases.

$$HCO_3^- + H^+ \longrightarrow H_2CO_3$$

$$HCO_3^- + OH^- \longrightarrow H_2O + CO_3^{2-}$$

The bicarbonate ion is simultaneously a weak Brønsted acid and a weak Brønsted base.

For this reason, sodium bicarbonate is recommended as an additive to swimming pools because it helps control the pH of the water when other chemicals are added to destroy bacteria.

The common name for sodium bicarbonate is *baking soda*. When added to dough, it decomposes during baking to release the gas carbon dioxide.

$$2NaHCO_3(s) \xrightarrow{heat} Na_2CO_3(s) + CO_2(g) + H_2O(g)$$

The CO_2 produces tiny bubbles throughout the dough and makes the baked product rise and become light and appealing.

The carbonates and bicarbonates of the other alkali metals have properties similar to the sodium compounds. Potassium carbonate, K_2CO_3, is called *potash* and is a major constituent of wood ashes. When needed in quantity, it is prepared by first reacting potassium hydroxide with carbon dioxide to give potassium bicarbonate.

$$KOH + CO_2 \longrightarrow KHCO_3$$

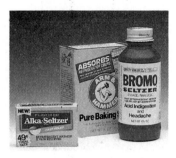

Some consumer products that contain sodium bicarbonate.

Thermal decomposition gives K_2CO_3.

Lithium carbonate, Li_2CO_3, has been found to be useful as a drug in the treatment of manic depression.

20.6 GROUP IIA: THE ALKALINE EARTH METALS

The elements of Group IIA consist of beryllium, magnesium, calcium, strontium, barium, and radium. All are very reactive metals, although not as reactive as the metals of Group IA. They are called *alkaline earth* metals because their oxides are basic, and because many of their compounds are of low solubility in water and are therefore found in mineral deposits in the Earth's crust.

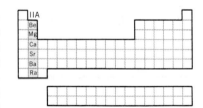

Each of the alkaline earth metals has two electrons in an *s* subshell that lies outside a noble gas core. Each exhibits only a 2+ oxidation state in its compounds and, with the exception of the compounds of beryllium and some of those of magnesium, their compounds are largely ionic. As in Group IA, there are strong similarities among all members of this group, but among the heavier ones, calcium through radium, the similarities are particularly close.

Occurrence and preparation

Calcium and magnesum are among the most abundant of the elements in the Earth's crust, ranking fifth and eighth in terms of percentage by weight. They are found in many locations in large mineral deposits of various compositions. Examples are gypsum ($CaSO_4 \cdot 2H_2O$), limestone ($CaCO_3$), dolomite ($CaCO_3 \cdot MgCO_3$), and carnallite ($MgCl_2 \cdot KCl \cdot H_2O$). In seawater, Ca^{2+} and Mg^{2+} are both major constituents among the dissolved ions. On a mole basis, Mg^{2+} is the third most abundant ion in

Calcium is the third most abundant metal in the Earth's crust.

In seawater, the concentration of Mg^{2+} is 0.056 M and that of Ca^{2+} is 0.011 M.

the sea, and Ca^{2+} places sixth. Calcium and magnesium are also the most biologically important alkaline earth metals. Calcium is found in the bones of animals, and shellfish extract Ca^{2+} from seawater to form their $CaCO_3$ shells. Magnesium is vital to plants where it is found at the center of the chlorophyll molecule, the substance that captures solar energy and begins the biological food chain.

The major source of beryllium is the mineral *beryl*, $Be_3Al_2(SiO_3)_6$. Sometimes beautiful large pure crystals of this substance are found and, when polished, become the gems emerald and aquamarine.

Radium was discovered by Pierre and Marie Curie.

Strontium and barium are recovered from deposits of their insoluble sulfates and carbonates. Radium, however, occurs principally as an impurity in *pitchblende*, a mineral from which uranium is extracted. Radium, which itself is radioactive, is formed as a product of the radioactive decay of heavier elements. For example, ^{226}Ra is radium's longest-lived isotope, with a half-life of 1600 years. It is one of a chain of isotopes produced from ^{238}U when it decays (see Chapter 24).

Of the metals in Group IIA, only beryllium and magnesium are produced and used in significant quantities. This is because they are the only alkaline earth metals that do not react rapidly with air and moisture at room temperature.

Beryllium is obtained by the electrolysis of molten beryllium chloride. However, sodium chloride must be added to the melt as an electrolyte because $BeCl_2$ is essentially covalent and therefore is a very poor electrical conductor. During the electrolysis, the less active metal, Be, is produced at the cathode and Cl_2 is evolved at the anode.

$$BeCl_2(l) \xrightarrow[\text{(NaCl)}]{\text{electrolysis}} Be(l) + Cl_2(g)$$

In recent times beryllium has found a variety of practical uses. It is a very lightweight, strong metal that has structural applications in missiles and spacecraft. It absorbs X rays less than any other metal that's stable toward air and is used as windows for X-ray tubes. Beryllium–copper alloy is fashioned into springs, electrical contacts, and welding rods. It is also made into tools — hammers, for example — that will not create sparks when used. This is particularly important when workers must labor in an explosive atmosphere. A major drawback of beryllium, however, is that its compounds are highly toxic, and stringent safety standards must be maintained when it or its alloys are machined.

Magnesum is extracted both from its land-based ores and the sea. On land its principal sources are its chloride, found in carnallite, $MgCl_2 \cdot KCl \cdot 6H_2O$, and its carbonate in dolomite, $CaCO_3 \cdot MgCO_3$. Extraction from dolomite involves first heating the ore strongly, a process called *calcining*, which decomposes the carbonates and forms the oxides.

$$CaCO_3 \cdot MgCO_3(s) \xrightarrow{\text{heat}} CaO \cdot MgO(s) + 2CO_2(g)$$

The mixed oxides are then treated with an excess of seawater, which contains appreciable amounts of dissolved Mg^{2+}. The water converts the oxides to the hydroxides.

$$CaO(s) + H_2O \longrightarrow Ca^{2+}(aq) + 2OH^-(aq)$$

$$MgO(s) + \ \ H_2O \longrightarrow Mg(OH)_2(s)$$

Calcium hydroxide, although not extremely soluble, is appreciably more soluble than $Mg(OH)_2$. As it dissolves, it makes the water basic, causing the Mg^{2+} that's in the seawater to precipitate as $Mg(OH)_2$. The combined precipitate of $Mg(OH)_2$ is then filtered and dissolved in hydrochloric acid to convert it to the chloride. This solution is evaporated and the solid $MgCl_2$ is melted and electrolyzed, giving magnesium and chlorine. The chlorine produced is made into HCl again and recycled.

In the absence of dolomite, the magnesium in seawater can be recovered by

Magnesium hydroxide settling ponds at Dow Chemical Company's plant in Freeport, Texas, where magnesium is recovered from seawater.

making the water basic with calcium oxide. This is obtained by calcining limestone, $CaCO_3$, or even sea shells, which are also composed of $CaCO_3$.

$$CaCO_3(s) \xrightarrow{heat} CaO(s) + CO_2(g)$$

$$CaO(s) + H_2O + Mg^{2+}(aq) \longrightarrow Ca^{2+}(aq) + Mg(OH)_2(s)$$

The precipitate of $Mg(OH)_2$ is treated as described previously.

Magnesium metal has a number of practical uses. Its low density (light weight) and moderate strength when alloyed with aluminum make it a useful structural metal. (Perhaps you or your family owns a magnesium alloy stepladder.) Presently magnesium is more expensive than aluminum, but its virtually inexhaustible supply in the sea is likely to make it comparatively less expensive as land-based supplies of aluminum ore are depleted.

Another application of magnesium — one you've surely seen — is in flashbulbs and signal flares. The combustion of magnesium

$$2 Mg(s) + O_2(g) \longrightarrow 2 MgO(s)$$

produces not only a great deal of heat, but also intense light. As noted in Chapter 10, a flashbulb (Figure 10.8 on p. 308) contains a fine magnesium wire in an atmosphere of pure oxygen. The flashbulb is fired by passing a small electrical current through the wire, which heats it and sets off the combustion reaction. Afterward, the interior of the bulb is coated with a thin deposit of the white powder MgO.

Calcium, strontium, and barium have very few commercial applications as free metals because of their reactivity toward oxygen and moisture. As a result, they are prepared only in small quantities, usually by electrolysis of their molten chlorides.

Barium is used as a *getter* to remove traces of oxygen in the manufacture of vacuum tubes.

Physical properties

Table 20.5 lists some physical properties of the alkaline earth metals. We see that in general they have higher melting points and higher densities than their neighbors in Group IA. This is not difficult to explain. The greater effective nuclear charges experienced by their outer electrons make the atoms of the alkaline earth metals smaller than those of the alkali metals alongside them in the periodic table, so more mass is packed into a smaller volume, which leads to higher densities. The 2+ charge on the cations on the metallic lattice of an alkaline earth metal causes them to be attracted more strongly to the "sea of electrons" and makes it more difficult to pull

Table 2.5. Some properties of the alkaline earth metals

| Element | Ionization Energy (kJ/mol) | | $E°$ (V)[a] | M^{2+} Radius (pm) | Melting Point (°C) |
	First	Second			
Beryllium	900	1757	−1.70	31[b]	1278
Magnesium	737.6	1450	−2.38	65	651
Calcium	589.5	1146	−2.76	99	843
Strontium	549	1064	−2.89	113	769
Barium	503	965	−2.90	135	725
Radium	509	979	−2.92	140	700

[a] For $M^{2+}(aq) + 2e^- \rightarrow M(s)$.
[b] Estimated.

them apart than the 1+ cations in the metallic lattice of an alkali metal. The Group IIA metals therefore have the higher melting points. The Group IIA metals are also harder than the Group IA metals for the same reason, and they have larger ionization energies.

Salts of the heavier alkaline earth metals, like those of the metals of Group IA, produce striking colors when introduced into a Bunsen burner flame. These colors serve as flame tests for them. Calcium salts, for example, give a brick-red color; strontium salts produce a crimson flame; and barium salts give a yellowish-green flame (Figure 20.17). Salts of these metals are often used to give spectacular colors to fireworks displays.

Chemical properties and important compounds

As we've already noted, the alkaline earth metals are all very reactive elements. They are easily oxidized and therefore serve as excellent reducing agents, as evidenced by

Figure 20.17. *The colors given to a Bunsen burner flame by salts of* (a) *calcium,* (b) *strontium, and* (c) *barium.*

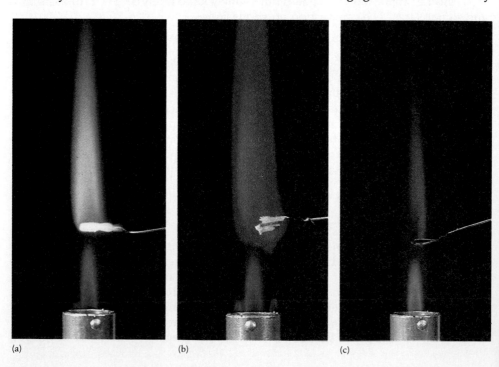

(a) (b) (c)

their very negative reduction potentials. The heavier of them, in fact, have reduction potentials comparable to those of the alkali metals. At first glance, this may appear somewhat surprising, because the sum of the first two ionization energies of an alkaline earth metal is considerably greater than just the first ionization energy of an alkali metal. In other words, removing two electrons from an isolated alkaline earth metal atom is much more difficult than removing a single electron from an isolated metal atom of Group IA. However, as we learned earlier in our discussion of lithium's unusually negative $E°$, the size of the reduction potential is controlled by more than simply the ionization energy. The hydration energy of the ion also plays a very important role.

The magnitude of the hydration energy of an ion depends both on its size and charge. A small ion can get closer to the water dipoles than a large ion can, so the smaller ion interacts more strongly with the solvent and its hydration energy is larger. A highly charged ion attracts the water dipoles more strongly than one of low charge, so the higher the ion's charge, the larger its hydration energy. Therefore, because the ions of the alkaline earth metals are both smaller *and* more highly charged than those of their neighbors in Group IA, their hydration energies are much larger. This serves to offset their much larger ionization energies. As a result, the alkaline earth metals have reduction potentials almost as negative as those of the alkali metals.

Ionic Radii (pm)			
Li$^+$	60	Be^{2+}	31^a
Na$^+$	95	Mg^{2+}	65
K$^+$	133	Ca^{2+}	99
Rb$^+$	148	Sr^{2+}	113
Cs$^+$	169	Ba^{2+}	135

^{a} Estimated.

As reducing agents, the Group IIA metals are all powerful enough to reduce water, at least in principle. However, beryllium and magnesium both form insoluble oxide coatings that protect them from attack by water. Beryllium, in particular, is quite resistant to oxidation, even by acids, because of its BeO coating. Magnesium is more reactive than beryllium. Even though it is not attacked by cold water, magnesium reacts slowly with boiling water and quite rapidly with steam to liberate hydrogen.

Their protective oxide coatings permit beryllium and magnesium to serve as useful structural metals.

The remaining alkaline earth metals, calcium through radium, form oxides that are at least moderately soluble in water, so their oxides are unable to protect them. As a result, they reduce water and liberate hydrogen. The general reaction is

$$M(s) + 2H_2O \longrightarrow M(OH)_2(aq) + H_2(g)$$

The vigor with which this reaction occurs increases from calcium to strontium to barium.

Like the alkali metals, the Group IIA elements react directly with most elemental nonmetals. For example, magnesium, we learned, reacts directly with oxygen to form MgO. It also is able to react with nitrogen to give magnesium nitride.

$$3Mg(s) + N_2(g) \longrightarrow Mg_3N_2(s)$$

With sulfur it gives MgS, and with the halogens it yields MgX_2 (X = halide ion). The other Group IIA metals react similarly.

An interesting phenomenon among the Group IIA elements is the variation in their metallic character. Although physically they all exhibit metallic characteristics, chemically we find that beryllium and, to some slight extent, magnesium exhibit a degree of nonmetallic character as well.

Beryllium and its oxide are amphoteric: They dissolve in both acids and bases. For example, in base they react as follows:

$$Be + 2H_2O + 2OH^- \longrightarrow Be(OH)_4^{2-} + H_2(g)$$

$$BeO + H_2O + 2OH^- \longrightarrow Be(OH)_4^{2-}$$

The remainder of the alkaline earth metals and their oxides are not amphoteric, so beryllium is chemically less metallic than the other members of its group.

Another way that beryllium is less metallic than the other Group IIA elements is in the degree of covalence of its compounds. There is no evidence that beryllium

The reaction of calcium with water liberates H_2 gas, which bubbles to the surface. Notice that the reaction of calcium with water is much less violent than the reaction of sodium with water shown on page 23.

The ion $Be(OH)_4^{2-}$ is called the beryllate ion.

exists as Be^{2+} in any of its compounds; they all show a significant degree of covalent character. Beryllium chloride, for example, is a poor conductor when melted, and we saw that an electrolyte, sodium chloride, has to be added to molten $BeCl_2$ so that it can be electrolyzed. In the solid state, $BeCl_2$ exists as a covalently linked chain of $BeCl_2$ units in which Be completes its octet by forming covalent bonds with chlorine atoms on adjacent $BeCl_2$ molecules.

gives

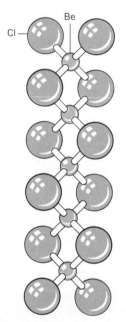

Figure 20.18. *The structure of* $(BeCl_2)_x$.

	K_{sp}
$CaSO_4$	2×10^{-4}
$SrSO_4$	3.2×10^{-7}
$BaSO_4$	1.5×10^{-9}

$$O^{2-} + H_2O \longrightarrow 2\ OH^-$$

Crushed limestone or dolomite is often spread on lawns and gardens to make the soil less acidic.

About 19 million tons of CaO are produced each year.

Since each Be atom has four separate electron pairs around it, the arrangement of the chlorines around the Be is tetrahedral, as shown in Figure 20.18.

The covalent character of bonds to beryllium can be related to the small size and high charge that a true Be^{2+} ion would have. We saw in Section 20.3 that this would give the Be^{2+} a very large ionic potential—sufficiently large to polarize other ions next to it and cause the bonds to become mostly covalent.

Although most magnesium compounds are ionic, magnesium does form a variety of covalently bonded compounds in which portions of organic molecules—molecules derived from methane (CH_4), ethane (C_2H_6), and other hydrocarbons—are bonded to magnesium. Examples are C_2H_5MgBr and $Mg(C_2H_5)_2$. These are called *organomagnesium compounds* and are important reagents in organic chemistry. The compounds of the rest of the alkaline earth metals are nearly all predominantly ionic.

The most important compounds of the alkaline earth metals are their carbonates, oxides, hydroxides, and sulfates. The carbonates are all quite insoluble in water, but there are trends in the solubilities of the oxides, hydroxides, and sulfates that influence their practical applications.

The solubilities of the oxides increase going down the group. BeO and MgO are insoluble, but CaO, SrO, and BaO react with water to form the corresponding hydroxides.

$$MO + H_2O \longrightarrow M(OH)_2 \quad (M = Ca, Sr, Ba)$$

The solubilities of the hydroxides also increase going down the group. Magnesium hydroxide is insoluble, calcium hydroxide is slightly soluble, and the hydroxides of strontium and barium are even more soluble. In contrast, the solubilities of the sulfates vary in the opposite direction: $BaSO_4$, $SrSO_4$, and $CaSO_4$ are insoluble, but their K_{sp}'s increase from Ba to Ca. Magnesium sulfate, however, is quite soluble.

We learned earlier that calcium and magnesium occur in large deposits of their carbonates, $CaCO_3$ (limestone and marble) and $MgCO_3 \cdot CaCO_3$ (dolomite). Calcium carbonate is a very common chemical. Seashells are composed almost entirely of $CaCO_3$, as is the chalk your teacher uses to write on the blackboard. Calcium carbonate is also used as a mild abrasive in toothpaste and household cleansers, and as an antacid. Limestone is an extremely important raw material in many industrial reactions. As we saw earlier, its thermal decomposition produces calcium oxide and carbon dioxide.

$$CaCO_3 \xrightarrow[900\ °C]{heat} CaO + CO_2$$

Some things that are composed of calcium carbonate.

Calcium oxide is called *lime,* or sometimes *quicklime,* and its annual production ranks second among industrial chemicals because it is an inexpensive, relatively strong base. When treated with water, a process called *slaking,* calcium hydroxide *(slaked lime)* is formed with considerable evolution of heat.

$$CaO + H_2O \longrightarrow Ca(OH)_2$$

Calcium oxide is an ingredient in Portland cement, and this reaction is among the first to occur when water is added to the cement.

Magnesium oxide can also be formed by decomposing its carbonate. However, unlike CaO, magnesium oxide is relatively unreactive toward water, especially if heated to very high temperatures. It is used as a component in refractory bricks, those used to line the interiors of high-temperature furnaces. It is also used in the manufacture of paper and medicinally as an antacid.

MgO melts at about 2800 °C.

Magnesium hydroxide, $Mg(OH)_2$, is much less soluble in water than $Ca(OH)_2$. An antacid and a laxative, it is the creamy white substance in milk of magnesia.

Important sulfates of the alkaline earth metals are those of magnesium, calcium, and barium. Magnesium sulfate in the form of its hydrate, $MgSO_4 \cdot 7H_2O$, is called *epsom salts* and is used to treat fabrics so that they readily accept dyes, to fireproof fabrics, as a fertilizer, and medicinally.

The mineral gypsum is $CaSO_4 \cdot 2H_2O$ and is used to make plaster and plasterboard, a building material commonly called sheet rock. When gypsum is partially dehydrated by heating, *plaster of Paris* is formed.

$$2CaSO_4 \cdot 2H_2O \xrightarrow{\text{heat}} (CaSO_4)_2 \cdot H_2O + 3H_2O$$
$$\text{gypsum} \qquad\qquad \text{plaster of Paris}$$

In plaster of Paris there is 1 mol of H_2O for each 2 mol of $CaSO_4$.

The formula for plaster of Paris is normally written $CaSO_4 \cdot \frac{1}{2}H_2O$. When water is added to $CaSO_4 \cdot \frac{1}{2}H_2O$, the crystals absorb it and reform the dihydrate. This is an exothermic reaction, and anyone who has had a broken limb set in a plaster cast has probably noticed how warm the cast became as it hardened.

Barium sulfate has a number of uses that are based on its whiteness and low solubility in water. It is used as a whitener in photographic papers and a filler in papers and polymeric fibers. Medicinally it is used for X-ray diagnosis of intestinal tract disorders because $BaSO_4$ is quite opaque to X rays. Even though barium salts are normally quite poisonous, $BaSO_4$ is so insoluble that a suspension of it can be swallowed without harm. This is because so little Ba^{2+} is in solution. As the suspension of $BaSO_4$ passes through a patient's intestines, its path can be followed by X-ray photos like that in Figure 20.19.

Figure 20.19. *An X-ray photograph of a patient's large intestine which has been filled with a suspension of BaSO₄. The BaSO₄ is opaque to X rays and allows the shape of the intestine to be seen.*

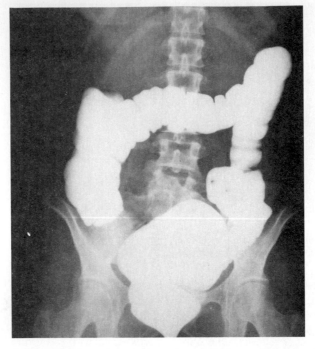

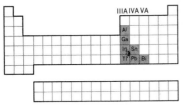

There are no genuine metals in Groups VIA or VIIA.

The pseudonoble gas core is $ns^2np^6nd^{10}$.

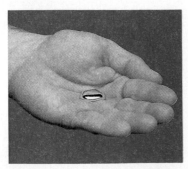

Gallium metal melts at 29.8 °C. Body temperature is 37 °C, which is high enough to cause gallium to melt in your hand. Gallium has the largest liquid range of any metal. It boils at 2403 °C.

20.7 METALS OF GROUPS IIIA, IVA, AND VA

In Section 20.2, you learned that the metallic character of the elements decreases from left to right in a period and increases from top to bottom in a group. The impact of these trends is especially apparent in Groups IIIA, IVA, and VA, where fewer metallic elements are found as the group number increases. In Group IIIA, for example, the metals are aluminum, gallium, indium, and thallium. In Group IVA, only tin and lead are metals, and in Group VA the only metal is bismuth. In general, the metals in these three groups tend to be less reactive and less metallic in their chemical behavior than the metals of Groups IA and IIA. For instance, many are amphoteric and many form covalent compounds.

Aluminum has three electrons in its valence shell outside a neon core ([Ne]$3s^23p^1$), and in its compounds aluminum always occurs in the +3 oxidation state. The metals in periods 4 (Ga), 5 (In, Sn), and 6 (Tl, Pb, Bi) occur after a row of transition elements and are called *post-transition metals.* They have pseudonoble gas cores beneath their valence shells, and their chemistry is characterized by the occurrence of two oxidation states. In each case, the lower oxidation state corresponds to the loss of the outer *p* electron(s) and the higher one corresponds to the further loss of the pair of *s* electrons. Thus, in Group IIIA the two oxidation states are +1 and +3; in Group IVA they are +2 and +4; and in Group VA they are +3 and +5.

Among the post-transition metals the relative stability of the lower oxidation state increases with increasing atomic number within a group. For instance, Ga^{3+} is more stable than Ga^+, while Tl^+ is more stable than Tl^{3+}. This trend in the relative stabilities of the high and low oxidation states persists in Groups IVA and VA and results, it appears, from a decreasing stability of *M—X* bonds with the increasing size of the metal atom. Since atoms become larger going down a group, it becomes increasingly more difficult to recover, by bond formation, the extra energy that must be expended to remove the pair of *s* electrons. This causes the lower oxidation state to become much more stable toward the bottom of a group.

Occurrence and preparation

Aluminum Of the metals in Group IIIA, aluminum is the only one of much practical importance and the only one whose chemistry we will examine in much detail. On a weight basis, aluminum is the third most abundant element in the Earth's crust. For example, it is found combined with silicon and oxygen in *aluminosilicates,* which occur in rock such as granite and various clays. Unfortunately, no practical method has yet been found to extract aluminum from these sources.

The Earth's crust is composed of about 8.8% aluminum by weight, which makes aluminum the most abundant *metal* in the crust.

The major ore of aluminum is *bauxite,* which contains its oxide Al_2O_3. However, before the metal can be obtained electrolytically by the Hall process, which was described in Chapter 18, the oxide must first be purified. The method of purification takes advantage of the amphoteric nature of aluminum and its compounds. The ore is first treated with a concentrated NaOH solution, which dissolves the Al_2O_3 and leaves most of the impurities behind.

$$Al_2O_3(s) + 2OH^- \longrightarrow 2AlO_2^- + H_2O$$
$$\text{aluminate ion}$$

Then the basic aluminum-containing solution is neutralized with acid, which precipitates the insoluble hydroxide.

$$AlO_2^- + H_3O^+ \longrightarrow Al(OH)_3(s)$$

After filtration the aluminum hydroxide is heated. This drives off water and gives the oxide again, which is now pure.

Purification of bauxite consumes about 1.2 billion pounds of NaOH each year.

$$2Al(OH)_3(s) \xrightarrow{\text{heat}} Al_2O_3(s) + 3H_2O(g)$$

The Al_2O_3 is then dissolved in molten cryolite and electrolyzed as described in Chapter 18.

Aluminum finds many uses in modern society. Large amounts of it serve as a structural metal in kitchen utensils, automobiles, aircraft, beverage cans, aluminum foil, and other consumer products. Aluminum is a good electrical conductor and is used in electrical wiring. Alloyed with magnesium, it is employed in structural applications, and an alloy called *alnico* — 50% Fe, 20% Ni, 20% Al, 10% Co — forms powerful magnets.

Tin and Lead Tin occurs as SnO_2 in an ore called *cassiterite.* To recover the metal, the ore is first heated strongly in air to drive off volatile oxides of some of its impurities — for example, arsenic and sulfur. Afterward, the SnO_2 is reduced with carbon.

$$SnO_2 + C \longrightarrow Sn + CO_2$$

The metal can be further purified in a manner similar to the electrolytic purification of copper. The impure tin is made the anode in an electrolytic cell and pure tin is made the cathode. Operation of the cell causes the impure tin to dissolve and pure tin to plate out on the cathode.

Elemental tin occurs in three different physical forms. This phenomenon, you recall, is termed allotropism, and the various forms of the element are called allotropes. The most common allotrope of tin is called *white tin* or *malleable tin* — you've seen it as the shiny coating on tin cans. Below 13.2 °C, the white form very gradually changes to a powdery, nonmetallic form called *gray tin.* A third form, called *brittle tin,* is obtained when white tin is heated. Its properties are obvious from its name.

The allotropes of tin differ in their crystal structures.

One of the principal uses of tin is as a protective coating over steel — the familiar tin can, for example. The coating is thin and usually applied electrolytically. It protects the steel simply by excluding air and moisture. However, once the coating is scratched and the steel below is exposed, corrosion occurs very rapidly. Iron is more easily oxidized than tin, so when both metals are exposed to moisture, a galvanic cell is established in which iron is the anode and is therefore oxidized in preference to the tin. Other applications of tin are in alloys such as bronze (copper and tin), solder (tin and lead) and pewter (about 90% tin, 7% antimony, and 3% copper).

Lead is found in nature as its sulfate, $PbSO_4$, its carbonate, $PbCO_3$, and its sulfide, PbS. The principal ore of lead is called *galena* and contains PbS. The metal is obtained from the ore by first heating it strongly in air, which converts it to the oxide.

$$2PbS + 3O_2 \longrightarrow 2PbO + 2SO_2$$

The oxide is then reduced with carbon.

$$2PbO + C \longrightarrow 2Pb + CO_2$$

The metallic lead that comes from this process contains impurities of silver, gold, and other metals. It also can be purified by electrolysis in the same manner as copper and tin. The silver and gold, of course, are recovered and help offset the cost of the electricity used in the purification process.

Metallic lead is used to make lead storage batteries and as a raw material in the manufacture of tetraethyl lead, $Pb(C_2H_5)_4$—the additive in leaded gasoline. It is also used to make compounds found in lead-based paints. Lead's ability to absorb high-energy radiation makes it a useful shielding material for X rays and high-energy radiation produced in nuclear reactors.

Bismuth Bismuth is found as its oxide, Bi_2O_3, and sulfide, Bi_2S_3. The sulfide must first be heated in air to convert it to the oxide before reduction.

$$2Bi_2S_3 + 9O_2 \longrightarrow 2Bi_2O_3 + 6SO_2$$

$$2Bi_2O_3 + 3C \longrightarrow 4Bi + 3CO_2$$

Bismuth is also obtained as a byproduct in the production of lead.

Bismuth is a hard, brittle, slightly reddish-colored metal. It is fairly resistant to corrosion and is one of only a few known substances that expand when they freeze. It is used to make alloys whose volumes stay very nearly constant when they solidify.

In general, alloys of bismuth have low melting points. One of them, called *Wood's metal,* is composed of 50% Bi, 25% Pb, and $12\frac{1}{2}$% each of Sn and Cd. It has a melting point of about 70 °C, well below the boiling point of water, and is used in the triggering mechanism of automatic sprinkler systems. Similar alloys are used in fuses designed to protect electrical circuits. When too much current is drawn through the alloy wire in the fuse, it becomes hot and quickly melts. This breaks the electrical circuit and prevents damage to the device that it's meant to protect.

Chemical properties and compounds

Aluminum Aluminum is a very reactive element, a fact that prevented the simple recovery of the metal from its compounds until the invention of the Hall process. It is a good reducing agent, and its reduction potential ($E° = -1.67$ V) is sufficiently negative that aluminum should react with water and liberate hydrogen. However, it forms a tough oxide coating that adheres to the metal and protects it from attack both by moisture and oxygen.[5] This allows aluminum to be a useful structural metal.

---

[5] A freshly exposed aluminum surface can be amalgamated—coated with a film of mercury in which the metallic aluminum dissolves. The Al_2O_3 that's formed on exposure to air does not adhere to the amalgamated surface, and the aluminum reacts easily with oxygen, corroding very rapidly. This very exothermic reaction can make the aluminum hot.

In fact, this oxide coating is sometimes deliberately made especially thick by having aluminum be the anode in an electrolytic cell. The product is called *anodized aluminum.*

Earlier, you learned that aluminum metal is amphoteric and is able to dissolve in both dilute acids and in base. Both reactions liberate hydrogen.

$$2Al(s) + 6H^+(aq) \longrightarrow 2Al^{3+}(aq) + 3H_2(g)$$

$$2Al(s) + 2OH^-(aq) + 2H_2O \longrightarrow 2AlO_2^-(aq) + 3H_2(g)$$

However, toward concentrated HNO_3 aluminum appears passive—that is, it doesn't react. The strong oxidizing power of HNO_3 creates an oxide coating that protects the metal from further attack.

Besides protecting the metal from oxidation, aluminum oxide is an important compound in its own right. It occurs in two principal forms, which are designated as $\gamma\text{-}Al_2O_3$ and $\alpha\text{-}Al_2O_3$. The $\gamma\text{-}Al_2O_3$ is formed by dehydrating aluminum hydroxide at a relatively low temperature.

$$2Al(OH)_3(s) \xrightarrow{<450\,°C} \gamma\text{-}Al_2O_3(s) + 3H_2O(g)$$

This form of the oxide is quite reactive, dissolving readily in both acids and bases.

If $\gamma\text{-}Al_2O_3$ is heated strongly, its crystal structure changes to that of the $\alpha\text{-}Al_2O_3$ form, which is called *corundum.* This substance has a very high melting point (about 2050 °C), is very hard and quite inert, especially toward acids. Corundum from deposits that are found in nature is used as an abrasive in sandpaper and in making refractory bricks that line furnace interiors.

A number of familiar gemstones are also composed of almost pure $\alpha\text{-}Al_2O_3$. For example, ruby consists of $\alpha\text{-}Al_2O_3$ with small amounts of dissolved Cr^{3+}. Other impurities in the $\alpha\text{-}Al_2O_3$ produce gems with other colors. Sapphire, for instance, contains traces of Fe^{2+} and Ti^{4+}.

Aluminum oxide has a very large exothermic heat of formation ($\Delta H_f^° = -1676$ kJ/mol). This allows aluminum to extract oxygen from other metal oxides with the simultaneous release of large amounts of heat—enough to melt the products of the reaction. For example, the reaction of Fe_2O_3 with aluminum

$$2Al(s) + Fe_2O_3(s) \longrightarrow Al_2O_3(l) + Fe(l)$$

produces temperatures approaching 3000 °C. This is called the **thermite reaction** and is used often to weld large masses of iron and steel. Thermite bombs have also been used by the military as incendiary devices because of the intense heat of the reaction.

The large exothermic heat of formation of Al_2O_3 has had another interesting application in recent times: It provides the thrust for the booster rockets that enable the space shuttle to take off. The solid propellant in these booster rockets is a mixture of powdered aluminum (the fuel) and ammonium perchlorate, NH_4ClO_4 (the oxidizer). They are mixed with a small amount of iron oxide catalyst, and the entire mixture is held together in a solid mass by an epoxy plastic. When the rocket is ignited, the aluminum is oxidized, and the formation of Al_2O_3 liberates large amounts of heat that cause the gases also formed to expand with great force. That is what lifts the rocket.

The anhydrous aluminum halides are interesting compounds because of their tendency to form *dimeric species,* molecules formed by the pairing of two AlX_3 units. This is especially so in the vapor and solutions of these compounds in nonpolar solvents, such as benzene and carbon tetrachloride. The pairing occurs in a way similar to the linking together of $BeCl_2$ molecules in solid $BeCl_2$—by the formation of a covalent bond from the halogen on one AlX_3 unit to the aluminum atom of another. For example, with $AlCl_3$ the species Al_2Cl_6 is formed.

The very exothermic thermite reaction in which Fe_2O_3 is reduced by Al yields liquid iron as the product. Here it is being used to weld steel rails together.

White clouds of aluminum oxide produced by burning aluminum powder billow from the solid booster rockets that lift the space shuttle Challenger from its launch pad at the Kennedy Space Center in Florida. The very exothermic reaction for the formation of Al_2O_3 provides the energy for the thrust.

The chemistries of beryllium and aluminum are similar in many ways. This has been attributed to their similar ionic potentials.

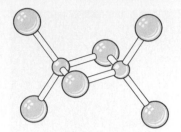

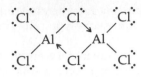

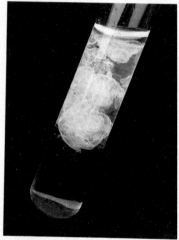

The structure of Al_2X_6. The small spheres are Al, the large spheres X.

gives

Addition of ammonia to a solution containing an aluminum salt yields a gelatinous precipitate of $Al(OH)_3(H_2O)_3$. The solid is usually called aluminum hydroxide, and the formula is often written without the water as $Al(OH)_3$.

$Al_2(SO_4)_3$ ranks about 37th in total production among industrial chemicals.

When aluminum chloride or other aluminum salts are dissolved in water, their solutions are acidic. In Chapter 10 we saw that this could be accounted for by considering how the Al^{3+} ion polarizes the water molecules that surround it in solution, making it easier for the hydrogens to be removed as H^+.

$$H_2O + Al(H_2O)_6^{3+} \longrightarrow Al(H_2O)_5(OH)^{2+} + H_3O^+$$

As base is added to a solution containing Al^{3+}, it neutralizes H_3O^+ and gradually strips protons from the water molecules that surround the Al^{3+} until the insoluble hydroxide is produced.

$$Al(H_2O)_6^{3+} + 3OH^- \longrightarrow Al(H_2O)_3(OH)_3(s) + 3HOH$$

Because of all the water contained within it, the aluminum hydroxide formed in this way is gelatinous (gelatinlike), rather than crystalline.

When base is added to the aluminum hydroxide precipitate, it dissolves, presumably because the OH^- extracts a proton from yet another water molecule that is attached to the Al^{3+}.

$$Al(H_2O)_3(OH)_3 + OH^- \longrightarrow Al(H_2O)_2(OH)_4^- + HOH$$

The species $Al(H_2O)_2(OH)_4^-$ is called the aluminate ion, although its precise composition is uncertain. Notice, for example, the equivalence

$$Al(H_2O)_2(OH)_4^- \quad \text{and} \quad AlO_2^- + 4H_2O$$

The reactions described above can be reversed by the addition of acid. For instance, when a basic solution containing aluminate ion is gradually neutralized, the hydroxide precipitates and then redissolves as more acid is added

$$Al(H_2O)_2(OH)_4^- + H_3O^+ \longrightarrow Al(H_2O)_3(OH)_3(s) + H_2O$$

$$Al(H_2O)_3(OH)_3(s) + H_3O^+ \longrightarrow Al(H_2O)_4(OH)_2^+ + H_2O$$

and ultimately, in solutions that are sufficiently acidic

$$Al(H_2O)_5OH^{2+} + H_3O^+ \longrightarrow Al(H_2O)_6^{3+} + H_2O$$

A compound of aluminum produced in large quantities, much of which is destined to be changed to the hydroxide, is aluminum sulfate. It is usually made from bauxite and sulfuric acid.

$$Al_2O_3(s) + 3H_2SO_4(aq) \longrightarrow Al_2(SO_4)_3(aq) + 3H_2O$$

When crystallized, it forms hydrates with as many as 18 water molecules [$Al_2(SO_4)_3 \cdot 18H_2O$].

The principal use of $Al_2(SO_4)_3$ is in the paper industry, where it is used to adjust acidity and treat paper to make it water-resistant. Aluminum sulfate is also employed in municipal water treatment. It is added to the water, which is then made basic by the addition of lime (CaO). This causes a gelatinous aluminum hydroxide precipitate to be formed that settles slowly to the bottom, taking fine sediment and bacteria with it.

When an aqueous mixture of aluminum sulfate and sodium sulfate is evaporated, crystals having the composition $NaAl(SO_4)_2 \cdot 12H_2O$ are formed. Similar crystals are also formed if $(NH_4)_2SO_4$ or K_2SO_4 is substituted for the Na_2SO_4 and if sulfates of Cr^{3+} or Fe^{3+} are substituted for $Al_2(SO_4)_3$. These crystals are called **alums.** They are characterized by the general formula $M^+M^{3+}(SO_4)_2 \cdot 12H_2O$ and are examples of **double salts.** If conditions are right, large well-formed octahedrally shaped crystals can be produced, as shown in Figure 20.20.

One use of sodium alum is in baking powder, in which it is combined with sodium bicarbonate. When added to moist dough, the acidity of the aluminum ion in water causes carbon dioxide to be released by reaction of hydronium ion with the bicarbonate ion

$$HCO_3^- + H_3O^+ \longrightarrow 2H_2O + CO_2(g)$$

As mentioned earlier, when the dough is baked, the small bubbles of CO_2 give the finished product a light and porous texture.

Tin, Lead, and Bismuth The chemistries of these elements are characterized by the existence of two oxidation states. Tin and lead, for example, form compounds in both the $+2$ and $+4$ states, while bismuth forms compounds in the $+3$ and $+5$ states. The relative stabilities of their higher and lower oxidation states are reflected in the way the metals react with various substances. For example, tin reacts with nonoxidizing acids such as HCl to form Sn^{2+}, but with oxidizing acids the $+4$ state is produced.

$$Sn + 2HCl \longrightarrow SnCl_2 + H_2$$
$$Sn + 4HNO_3 \longrightarrow SnO_2 + 4NO_2 + 2H_2O$$

Lead, however, forms only lead(II) compounds, even when strong oxidizing acids such as concentrated HNO_3 are used.

$$3Pb + 8HNO_3 \longrightarrow 3Pb(NO_3)_2 + 2NO + 4H_2O$$

This tells us that when tin and lead are compared, the higher oxidation state is relatively more stable for tin than lead. This is also revealed in their reaction products with chlorine. Tin combines with chlorine to form $SnCl_4$, but lead reacts to produce $PbCl_2$.

$$Sn + 2Cl_2 \longrightarrow SnCl_4$$
$$Pb + Cl_2 \longrightarrow PbCl_2$$

In the case of bismuth, the $+3$ state is much more stable than the $+5$ one. Direct combination of bismuth with oxygen or chlorine, for example, produces Bi_2O_3 and $BiCl_3$, respectively. The oxidation of Bi_2O_3 to Bi_2O_5 requires very severe oxidizing conditions, and the compound $BiCl_5$ does not exist at all because the $+5$ oxidation state of bismuth is such a powerful oxidizing agent—that is, it has such a strong tendency to be reduced—that it would oxidize Cl^- to Cl_2.

Besides dissolving in acids, tin and lead also dissolve in base with the evolution of hydrogen.

$$Sn + 2OH^- + 2H_2O \longrightarrow Sn(OH)_4^{2-} + H_2(g)$$
$$Pb + 2OH^- + 2H_2O \longrightarrow Pb(OH)_4^{2-} + H_2(g)$$

The *stannite ion*, $Sn(OH)_4^{2-}$, is a very powerful reducing agent and is easily oxidized. On the other hand, the *plumbite ion*, $Pb(OH)_4^{2-}$, is less easily oxidized, which again reveals that the $+2$ state is more stable for lead than for tin.

Compounds of tin, lead, and bismuth have a variety of practical uses. For

$M^+ = Na^+, K^+, NH_4^+$

$M^{3+} = Al^{3+}, Cr^{3+}, Fe^{3+}$

Figure 20.20. *An octahedrally shaped crystal of $KAl(SO_4)_2 \cdot 12H_2O$, potassium alum, shown in actual size.*

Tin and lead are both amphoteric.

The use of $[(C_4H_9)_3Sn]_2O$ in marine paints has been severely restricted because it is toxic to shellfish.

Red lead, Pb_3O_4.

Lead-based paints, of course, are toxic. Their use indoors has been banned because of this.

The BiO_3^- ion is such a powerful oxidizing agent that in acidic solution, yellow $NaBiO_3$ oxidizes the nearly colorless Mn^{2+} ion to violet MnO_4^- ion.

example, tin(II) fluoride, SnF_2, also called stannous fluoride, is the decay-inhibiting ingredient in some fluoride toothpastes. Some tin compounds are also useful fungicides. An example is $[(C_4H_9)_3Sn]_2O$, whose chemical name is bis(tributyltin) oxide. It is used in some antifouling paints that are applied to boat hulls to prevent marine growth, and in preparations that are applied to wood to prevent rotting.

Among the most useful compounds of lead are its oxides. Lead(II) oxide, PbO, a yellow powder that is also called litharge, is used in pottery glazes and in making fine lead crystal. If PbO is heated carefully in air, it can be oxidized to Pb_3O_4. This is a mixed oxide containing both lead(II) and lead(IV). Its common name is *red lead*, and it is used in corrosion-inhibiting paints that are applied to structural steel.

If solutions of plumbite ion are subjected to strong oxidizing conditions, PbO_2 can be prepared. For example,

$$Pb(OH)_4^{2-} + OCl^- \longrightarrow PbO_2 + H_2O + 2OH^- + Cl^-$$

Unlike SnO_2, PbO_2 is a strong oxidizing agent. Its most common use is as the cathode material in lead storage batteries. Recall that during the discharge of this battery the net reaction is

$$Pb + PbO_2 + 2H_2SO_4 \longrightarrow 2PbSO_4 + 2H_2O$$

As you are probably aware, in relatively recent times there has been a great deal of concern over the past use of lead-based paints, particularly on surfaces in living areas children may be exposed to. This is because lead compounds are very toxic. The pigment used in lead-based paints is a white basic carbonate, $Pb_3(OH)_2(CO_3)_2$, and its use in interior paints is now restricted. Although it provides excellent covering power as a pigment, it suffers from the disadvantage of being darkened on contact with H_2S produced in the environment by decaying vegetation. The darkening is caused by the production of black PbS. Another lead-based pigment that is used in artists' oil colors is lead chromate, $PbCrO_4$. This compound has a bright yellow color.

Bismuth compounds are utilized frequently in the cosmetic and pharmaceutical industry. In fact, these industries account for about 30% of the bismuth produced each year. For example, a substance known as bismuth subnitrate, produced by partial hydrolysis of $Bi(NO_3)_3$, is used medicinally as an antacid in the treatment of gastric ulcers. Its exact composition varies according to how it is prepared, but it can be approximately formulated as $BiO(NO_3)$.

The hydrolysis of bismuth(III) compounds is not unusual. When $BiCl_3$ is dissolved in water, it hydrolyzes, producing the bismuthyl ion, BiO^+. As the solution is diluted, an insoluble precipitate of BiOCl is formed that redissolves if the solution is made more acidic by the addition of hydrochloric acid.

The oxide of bismuth, Bi_2O_3, is formed when bismuth is heated in air. As mentioned earlier, under extreme oxidizing conditions this can be oxidized to give bismuth in the +5 oxidation state. Compounds containing bismuth(V), such as $NaBiO_3$ *(sodium bismuthate)*, are extremely powerful oxidizing agents as a result of bismuth's strong tendency to revert to the +3 oxidation state by acquiring electrons from other substances.

REVIEW QUESTIONS AND PROBLEMS

Problems whose numbers are in blue have their answers in Appendix D at the back of the book. The more difficult problems are marked with asterisks.

Metallurgy

20.1 How are most metals found in nature? What are some important sources of metals?

20.2 What is an *ore*?

20.3 What are the three steps involved in extracting a metal from its ore and making it ready for practical use?

20.4   Why is "panning" able to separate sand and mud from tiny particles of gold?

20.5   What is an amalgam? How is it used in the recovery of gold from gold ores?

20.6   Describe the process called flotation. What is meant by *roasting* as applied to metallurgy?

20.7   Write chemical equations for the purification of bauxite, Al_2O_3.

20.8   Why must metal compounds always be reduced to extract the metals from them?

20.9   What property must a metal compound possess to be easily decomposed thermally?

20.10 What is implied thermodynamically when we say that a particular compound is thermally stable?

20.11 Above what temperature would the decomposition of silver oxide, Ag_2O, into silver and gaseous O_2 be expected to have an equilibrium constant larger than 1.0? For Ag_2O at 25 °C, $\Delta H_f^\circ = -30.5$ kJ/mol and $\Delta S_f^\circ = -66.1$ J/mol K. Assume that $\Delta H°$ and $\Delta S°$ are independent of temperature.

20.12 At 25 °C, $\Delta H_f^\circ = +80.8$ kJ/mol for Au_2O_3. Also for Au_2O_3, $S° = 125$ J/mol K; for Au, $S° = 47.7$ J/mol K; and for O_2, $S° = 205$ J/mol K. Assuming that $\Delta H°$ and $\Delta S°$ are independent of temperature, show that the decomposition of Au_2O_3 should occur spontaneously at all temperatures.

20.13 From the data in Tables 6.1 and 14.3, calculate the temperature (in °C) above which the thermal decomposition of ZnO should become feasible. Assume that $\Delta H°$ and $\Delta S°$ are independent of temperature.

20.14 Given the data below, determine the temperature in °C at which $K_P = 1$ for the reaction

$$CuO(s) \longrightarrow Cu(s) + \tfrac{1}{2}O_2(g)$$

For $CuO(s)$, $\Delta H_f^\circ = -155$ kJ/mol. Absolute entropies: $CuO(s)$, 43.5 J/mol K; $Cu(s)$, 33.3 J/mol K; $O_2(g)$, 205.0 J/mol K. Assume $\Delta H°$ and $\Delta S°$ are independent of temperature.

20.15 Calculate K_P at 100, 500, and at 2000 °C for the reaction

$$MoO_3(s) \longrightarrow Mo(s) + \tfrac{3}{2}O_2(g)$$

given the following data:

	ΔH_f° (kJ/mol)	$S°$ (J/mol K)
$MoO_3(s)$	−754.4	78.16
$Mo(s)$	0.0	28.6
$O_2(g)$	0.0	205.0

Assume $\Delta H°$ and $\Delta S°$ are independent of temperature.

20.16 Write chemical equations for (a) the reduction of Fe_2O_3 in the blast furnace and (b) the production of slag from SiO_2 and $CaCO_3$.

20.17 Why is carbon a preferred reducing agent in commercial metallurgy?

20.18 Write equations showing the use of carbon and hydrogen as reducing agents in the extraction of a metal from one of its compounds.

20.19 Why isn't sodium produced commercially by reduction of its compounds with a reducing agent that is stronger than sodium?

20.20 Why are halide salts often used when preparing metals by electrolysis?

20.21 Write equations for the commercial electrolytic production of sodium and aluminum.

20.22 Why must pig iron be refined to be useful as a strong structural metal? Describe the Bessemer converter; the open hearth furnace; the basic oxygen process. Which of these is the principal method used today to make steel?

20.23 Describe the Mond process.

Trends in Metallic Behavior

20.24 How does the metallic character of the elements depend on electronegativity? What vertical and horizontal trends in metallic character exist in the periodic table? Illustrate these trends for the elements in the second period and in Group IVA.

20.25 In each pair below, choose the element expected to have the more metallic character.
(a) Li or Be   (d) Sn or P
(b) B or Al   (e) Ga or I
(c) Al or Cs

20.26 Which oxide should be more basic, Al_2O_3 or Ga_2O_3?

20.27 Which should be more acidic, MgO or Al_2O_3?

20.28 What is meant by amphoteric? Write chemical equations to illustrate the amphoteric behavior of beryllium and aluminum.

20.29 In what way are the metals in Groups IIIA, IVA, and VA less metallic than those in Groups IA and IIA?

Covalent Character of Metal Compounds

20.30 Define ionic potential. How is it related to the degree of covalent character in a metal–nonmetal bond?

20.31 Which should have the greater degree of covalent bonding?
(a) $GaCl_3$ or $GeCl_4$ (d) LiCl or Li_2S
(b) Bi_2O_3 or Bi_2O_5 (e) Na_2S or MgS
(c) PbO or PbS

20.32 Which should be more ionic?
(a) SnO or SnS   (b) $AlCl_3$ or $AlBr_3$

(c) $BeCl_2$ or BeF_2 (d) SnS or PbS

(e) SnS or SnS_2

Colors of Metal Compounds

20.33 What is responsible for the color in compounds such as SnS_2 or PbS?

20.34 Which compound would you expect to be more deeply colored?
(a) Ag_2O or Ag_2S (c) SnS or SnS_2
(b) CuCl or CuBr   (d) MgS or Al_2S_3

20.35 If a particular compound absorbs green light, what color will it appear to be?

Group IA Metals: General

20.36 Why are the Group IA elements called *alkali metals*? What oxidation states are observed for the Group IA metals?

20.37 Why do compounds of lithium, rubidium, and cesium have little commercial importance?

20.38 Why are sodium compounds more important, commercially, than compounds of the other alkali metals?

20.39 Given the following thermodynamic data, calculate the hydration energy for the Na^+ ion in kJ/mol:

$$\Delta H_f^\circ \text{ of } Na^+(aq) = -239.7 \text{ kJ/mol}$$

$$\Delta H_{subl}^\circ \text{ of } Na(s) = 108.7 \text{ kJ/mol}$$

$$\text{IE of } Na(g) = 493.7 \text{ kJ/mol}$$

20.40 Using the data for the atomization energy of Na (Problem 20.39) and the first and second ionization energies for Na in Table 7.4, compute the value of the hydration energy (in kJ/mol) required to produce a negative ΔH_f for $Na^{2+}(aq)$.

*20.41 The standard reduction potential of potassium is -2.92 V. ΔH_{hyd}° of K^+ in $1\,M$ aqueous solution is -759 kJ/mol. The sublimation energy of $K(s)$ is $+90.0$ kJ/mol and the ionization energy of $K(g)$ is $+418$ kJ/mol. Calculate ΔS° (in J/mol K) for the process

$$K(s) \longrightarrow K^+(1\,M) + e^-$$

20.42 Which alkali metals are most abundant? Which one is least abundant? Why?

20.43 Where do the alkali metals occur in nature?

20.44 Write a chemical equation to show how elemental potassium, rubidium, and cesium are usually made.

Group IA Metals: Physical and Chemical Properties

20.45 Describe two applications of metallic sodium related to its physical properties.

20.46 What colors are given to a Bunsen burner flame by compounds of (a) sodium, (b) lithium, and (c) potassium?

20.47 How can potassium ion in a mixture be detected if the mixture also contains sodium ion?

20.48 Write a chemical equation for the reaction of rubidium with water.

20.49 Why is the reduction potential of lithium more negative than the reduction potential of sodium?

20.50 What happens when an alkali metal is added to liquid ammonia? Why are these solutions such good reducing agents?

20.51 Write chemical equations to show the reaction (if any) of lithium and sodium with (a) bromine, (b) sulfur, and (c) nitrogen.

20.52 Write equations that illustrate the reactions of each of the alkali metals with oxygen (present in excess).

20.53 What compound of the alkali metals is used in a recirculating breathing apparatus? Write chemical equations to show how it functions.

20.54 Why are solutions of Na_2O_2 basic?

20.55 What is a *getter*?

20.56 What are other common names for sodium hydroxide? What are some of its uses?

20.57 What is trona ore? Give the chemical reactions involved in the Solvay process. What is the net reaction in the Solvay process?

20.58 How does sodium bicarbonate serve as a buffer? What is the common name for sodium bicarbonate? How can $NaHCO_3$ serve as a fire extinguisher?

20.59 What is potash? How is it made?

20.60 What alkali metal compound is used to treat manic depression?

Group IIA Metals: General

20.61 Why are the Group IIA elements called alkaline earth metals? How do their densities, hardness, melting points, and ionization energies compare to their neighbors in Group IA?

20.62 Where are calcium and magnesium found? What is the source of radium?

20.63 How is magnesium recovered from dolomite? How is magnesium recovered from sea water?

Group IIA Metals: Physical and Chemical Properties

20.64 Why do the alkaline earth metals have reduction potentials that are nearly as negative as the metals in Group IA?

20.65 Define *calcining*. What is lime? What happens when water is added to lime? Why is lime such an important industrial chemical?

20.66 What reaction takes place inside a flashbulb when it is fired?

20.67 What color is given to a Bunsen burner flame by compounds of (a) calcium, (b) strontium, and (c) barium?

20.68 What chemical fact is responsible for the structural uses of metallic beryllium and magnesium? Why are metallic calcium, strontium, and barium not used as structural metals?

20.69 Write chemical equations for the reaction of magnesium with elemental oxygen, sulfur, and nitrogen.

20.70 Write chemical equations that illustrate the amphoteric behavior of metallic beryllium.

20.71 What is the structure of solid $BeCl_2$? How does this support the statements made in earlier chapters that molecules of $BeCl_2$ have less than an octet in the valence shell of beryllium?

20.72 Why are beryllium compounds covalent? What are organomagnesium compounds?

20.73 Write an equation for the reaction of water with (a) calcium and (b) potassium.

20.74 How do the solubilities of the alkaline earth hydroxides vary from top to bottom in the group? How do the solubilities of the sulfates vary?

20.75 What are some uses of calcium carbonate?

20.76 What is milk of magnesia composed of?

20.77 What is gypsum? Write chemical equations showing how plaster of Paris is made and what happens when it combines with water during hardening.

20.78 What property of barium sulfate alows it to be used for X-ray diagnosis of intestinal tract disorders?

20.79 What is *Epsom salts?* What are some uses of magnesium oxide?

Metals in Group IIIA, IVA, and VA

20.80 What oxidation states are observed for the metals in Groups IIIA, IVA, and VA?

20.81 Which are the post-transition metals? Why do their lower oxidation states become more stable relative to the higher ones going down a group?

Aluminum

20.82 What is the ore of aluminum? Write chemical equations to show how it is purified.

20.83 What are some commercial uses of metallic aluminum?

20.84 Why doesn't aluminum corrode rapidly in the presence of air and moisture? How can it be made to corrode very quickly?

20.85 Write chemical equations showing the amphoteric behavior of metallic aluminum.

20.86 How do the properties of γ-Al_2O_3 and α-Al_2O_3 differ? What gems are composed primarily of aluminum oxide?

20.87 What is the thermite reaction?

20.88 What is the structure of dimeric aluminum chloride?

20.89 Why are solutions of aluminum salts acidic?

20.90 Write chemical equations for the gradual neutralization of $Al(H_2O)_6{}^{3+}$ and the dissolving in base of the gelatinous aluminum hydroxide precipitate.

20.91 What are two ways of writing the formula of the aluminate ion?

20.92 How is aluminum sulfate used in water treatment plants?

20.93 What is an alum? Give an example. Which alum is used in baking powders and how does it function?

20.94 When aluminum oxide dissolves in base, the aluminate ion, which can be written as $AlO_2{}^-$, is formed. Write an equation for this reaction.

Tin, Lead, and Bismuth

20.95 Write chemical equations showing how tin, lead, and bismuth are recovered from their ores.

20.96 Why does a tin can rust so rapidly if the tin coating is scratched, exposing the steel beneath it?

20.97 What are allotropes? Which of the representative metals exhibits allotropism?

20.98 What unusual property is possessed by metallic bismuth? What is Wood's metal? What are its properties and uses?

20.99 How do the elements tin and lead differ in their behavior toward nitric acid and chlorine? Illustrate using chemical equations.

20.100 Why doesn't $BiCl_5$ exist?

20.101 Write chemical equations for the reactions of metallic tin and lead with base.

20.102 What are the oxides of lead and what are some of their uses?

20.103 Why do lead-based paints gradually darken over a period of time?

20.104 What are some uses of bismuth compounds? What is the formula of the bismuthyl ion?

CHAPTER

21

THE TRANSITION METALS

In this chapter we conclude our discussion of the chemistry of the metallic elements by examining those that are found in the center of the periodic table. These are called the transition elements, or because they are all metals, the transition metals. As you learned in Chapter 7, they arise because of the gradual filling of the d or f subshells that lie one or two levels below the outer shell. Many chemists prefer to reserve the term *transition element* for an element whose atoms possess a partially filled d or f subshell. For completeness, however, we will also include in our discussion in this chapter elements that are found in Group IIB (zinc, cadmium, and mercury), even though their inner subshells are filled.

21.1 SOME GENERAL PROPERTIES

For discussion it is convenient to divide the transition elements into two categories: the **d-block elements** (or **main transition elements**) and the **inner transition elements.** In the condensed version of the periodic table on the inside front cover of the book, the d-block elements are located in the main body of the table between Groups IIA and IIIA. They consist of three rows of elements that are often called the *first, second, and third transition series.* The inner transition elements include the lanthanides and actinides and are the elements in the two long rows of 14 elements each that are placed just below the main body of the table. See also Figure 21.1.

Like the representative elements, most of the d-block transition elements possess certain vertical similarities in chemical and physical properties and are therefore divided into groups, designated as B groups. They begin with Group IIIB on the left and proceed through Group VIIB. Next is a set of nine elements collectively termed Group VIII, and finally, on the right, we find Groups IB and IIB. The group numbers were chosen to correspond to the highest positive oxidation state that their elements normally exhibit.

The division of the periodic table into A and B groups — for example, IIIA and IIIB — suggests that there may be certain parallels between the two, and to a limited degree this is true. The similarities, however, are restricted primarily to likenesses of composition, structure, and maximum positive oxidation state, instead of chemical reactivity. Some examples of these similarities are found in Table 21.1.

The Group VIII elements, which lie between Groups VIIB and IB, are classed

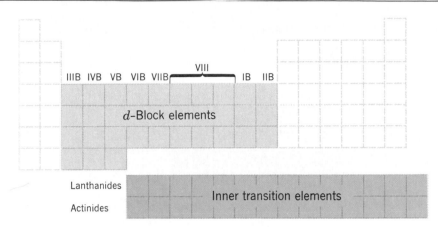

Figure 21.1. *The locations of the transition elements in the periodic table.*

differently from the other *d*-block elements because they have no counterparts among the representative elements. Within this group there are greater *horizontal similarities* than vertical ones, and the description of the behavior of these elements is usually organized on the basis of horizontal groups of three elements each, called **triads.** Each triad is named after the best-known element within it. Thus, we have the iron triad, the palladium triad, and the platinum triad.

As a class, the transition elements are all typical metals; they possess a characteristic metallic luster and are good conductors of heat and electricity. Silver has the highest electrical and thermal conductivity of any metal, followed closely by copper. Copper, of course, is used in vast amounts in electrical wiring. Silver is the preferred coating for mirrors because of its high reflectivity.

The chemical and physical properties of the transition elements cover a wide range and account for the large variety of uses to which they are applied. Some of these metals are very hard and strong and are used as structural metals, either in the pure state or as alloys. Iron is the prime example; steels with a variety of different properties are formed by incorporating iron with other transition elements such as chromium, cobalt, and nickel. Even copper, which is very soft when pure, can be made very strong by forming an alloy with beryllium. Beryllium–copper alloys are used in place of steel in nonsparking tools for use in explosive atmospheres and as high-quality springs in cameras and other precision instruments.

The melting points of the transition elements also vary over a wide range (Figure 21.2 and 21.3). Most are high-melting. Tungsten, with a melting point of approximately 3400 °C, is used for filaments in light bulbs. At the other extreme is mercury, which is a liquid at room temperature and is used in thermometers.

The chemical reactivity of the free elements varies greatly too. Most react directly with nonmetals such as oxygen and the halogens to produce the corre-

About one-quarter of the silver consumed by industry each year is used to make electronic equipment.

The filament in an electric light bulb is made of tungsten, which has the highest melting point of any metal.

Table 21.1. Some similarities of chemical composition between A and B group compounds

Group	Compounds	Group	Compounds
IVA	CCl_4, $SnCl_4$, CO_2	IVB	$TiCl_4$, TiO_2
VA	PO_4^{3-} $POCl_3$	VB	VO_4^{3-}, $VOCl_3$
VIA	SO_4^{2-}, $S_2O_7^{2-}$	VIB	CrO_4^{2-}, $Cr_2O_7^{2-}$
VIIA	ClO_4^-, Cl_2O_7	VIIB	MnO_4^-, Mn_2O_7
IA	NaCl	IB	CuCl, AgCl
IIA	$CaCl_2$	IIB	$ZnCl_2$

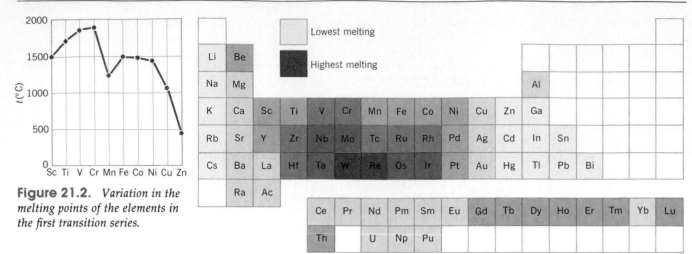

Figure 21.2. *Variation in the melting points of the elements in the first transition series.*

Figure 21.3. *Variations in the melting points of the metals according to their locations in the periodic table.*

sponding oxides and halides. In fact, some transition elements are so easily oxidized that they react with water to liberate hydrogen. This is true for scandium (Sc), yttrium (Y), lanthanum (La), and the lanthanide elements (atomic numbers 58 to 71). They have very negative reduction potentials and react according to the equation

$$2M + 6H_2O \longrightarrow 3H_2 + 2M(OH)_3$$

Some other transition elements, such as platinum and gold, are very resistant to oxidation. They are insoluble in both nonoxidizing acids such as HCl as well as in oxidizing acids such as HNO_3, although they do dissolve slowly in *aqua regia*.

Despite some rather marked differences in behavior, the transition elements have several characteristics in common with each other.

1. *Most have multiple oxidation states.* With only a few exceptions, the transition elements tend to exhibit more than one oxidation state.

2. *Many of their compounds are paramagnetic.* Because the transition elements tend to have partially completed *d* or *f* subshells in both the free state and in their compounds, the metal atoms and ions often possess unpaired electrons. These impart the property of paramagnetism.

3. *Many (if not most) of their compounds are colored.* The origin of the colors of complex ions of the transition elements will be discussed in Section 21.7.

4. *They have a strong tendency to form complex ions.* As a group, these elements form a huge number of complex ions of varying degrees of complexity. The last three sections of this chapter are devoted to discussions of their structures and bonding.

21.2 PERIODIC TRENDS IN OXIDATION STATES AND ATOMIC RADII

In Chapter 7 you learned that as we proceed across a period from left to right through the *d*-block elements, there is a gradual filling of the *d* subshell that lies just below the outer shell. In period 4, for example, this is the 3*d* subshell, and except for irregularities at chromium and copper (which were discussed in Chapter 7), each of the elements in the first transition series has two electrons in its outer 4*s* subshell plus some number of electrons in the $3d$. You also learned that when atoms lose electrons, they always come first from the shell with the largest value of the principal quantum number. This means that for the first transition series, the $4s$ subshell is emptied before electrons are removed from the $3d$. It should be no surprise, therefore, to find that one of the most common oxidation states observed among these elements, as well as the other transition elements, is $+2$.

As noted in the previous section, the transition elements commonly exhibit more than one oxidation state. In general, these states form because electrons are lost not only from the outer s subshell but from the underlying d subshell as well. Table 21.2 lists most of the oxidation states observed for the various d-block elements, with the most stable oxidation states printed in color. Don't be overwhelmed; you're not expected to memorize the contents of the table. We are just looking for trends in the stabilities of oxidation states, and two of them become apparent.

1. Elements at the left in a transition series tend to prefer higher oxidation states. As we move to the right, the lower oxidation states become increasingly more stable relative to the higher ones. (Remember, however, that the highest oxidation state is usually equal to the group number.

2. If we go down a group among the d-block elements, the higher oxidation states become progressively more stable relative to the lower ones.

Knowing these trends is useful because it helps us make comparisons of the strengths of oxidizing agents. For example, in the first row of transition elements (the first *transition series*), we see an increasing tendency toward a stable $+2$ oxidation state and fewer and fewer high oxidation states. Suppose we use this to compare the

The $+2$ oxidation state is common because many transition elements have a pair of s electrons that are the first to be lost.

Table 21.2. Oxidation states^a of the transition metals.

Group Number									
IIIB	IVB	VB	VIB	VIIB	VIII			IB	IIB
Sc	Ti	V	Cr	Mn	Fe	Co	Ni	Cu	Zn
+3	+2	+1	+2	+2	+2	+1	+2	+1	+2
	+3	+2	+3	+3	+3	+2	+3	+2	
	+4	+3	+6	+4	+4	+3			
		+4		+6	+6				
		+5		+7					
Y	Zr	Nb	Mo	Tc	Ru	Rh	Pd	Ag	Cd
+3	+2	+2	+2	+2	+2	+1	+2	+1	+2
	+3	+3	+3	+3	+3	+2	+3	+2	
	+4	+4	+4	+4	+4	+3	+4	+3	
		+5	+5	+5	+5	+4			
			+6	+6	+6	+5			
			+8	+7	+7	+6			
					+8				
La	Hf	Ta	W	Re	Os	Ir	Pt	Au	Hg
+3	+3	+2	+2	+3	+2	+1	+2	+1	+1
	+4	+3	+3	+4	+3	+2	+3	+3	+2
		+4	+4	+5	+4	+3	+4		
		+5	+5	+6	+5	+4	+5		
			+6	+7	+6	+5	+6		
					+8	+6			

^{a} The most stable oxidation states are in color.

Remember, an oxidizing agent becomes reduced in a redox reaction.

MnO_4^- is *permanganate* ion.
ReO_4^- is *perrhenate* ion.

relative stabilities of Fe^{2+} and Fe^{3+} with Ni^{2+} and Ni^{3+}. The horizontal trend suggests that Ni^{3+} has a greater tendency to become Ni^{2+} than Fe^{3+} has to become Fe^{2+}. In other words, Ni^{3+} has a greater tendency to gain an electron than Fe^{3+}. Since these species are functioning as oxidizing agents when they gain electrons, we conclude that Ni^{3+} is a stronger oxidizing agent than Fe^{3+}. In fact, this is what is found experimentally.

We can also make vertical comparisons. For example, suppose we consider the ions MnO_4^- and ReO_4^-. Since high oxidation states become more stable going down, ReO_4^- should be a weaker oxidizing agent than MnO_4^-. This is also found to be true experimentally.

EXAMPLE 21.1. PREDICTING RELATIVE STABILITIES OF OXIDATION STATES BASED ON AN ELEMENT'S POSITION IN THE PERIODIC TABLE

PROBLEM: Which would you expect to be a more powerful oxidizing agent, TiO_2 or MnO_2?

SOLUTION: First we locate Ti and Mn in the periodic table. Since the higher oxidation states become less stable relative to the lower ones going from left to right, Mn(IV) should be relatively less stable than Ti(IV) toward reduction; Mn(IV) should have a greater tendency to be reduced than Ti(IV). Therefore, MnO_2 should be the stronger oxidizing agent. (Actually, TiO_2 is a stable white pigment used in paint, and MnO_2 is the oxidizing agent in the common dry cell.)

In contrast to the wide range of chemical properties of the d-block elements, the lanthanides exhibit a remarkable sameness of properties. The $4f$ subshell, which is only partially filled for most of these elements, is buried beneath the outer $5d$ and $6s$ subshells and does not interact to an appreciable extent with the surrounding chemical environment. Consequently, the chemistry of the lanthanides, like that of lanthanum itself, is predominantly that of the $3+$ ion, and differences in behavior depend primarily on differences in ionic size.

The actinide elements exhibit a greater variation in oxidation numbers than the lanthanides (e.g., uranium forms compounds in the $+3$, $+4$, $+5$, and $+6$ oxidation states). An explanation sometimes given for this is that the $5f$ orbitals of the actinides project farther toward the outer parts of the atoms than the $4f$ orbitals of the lanthanides. As a result, the $5f$ orbitals are able to become involved to a greater degree in chemical bonding, so more complex chemistry is observed.

Atomic radii

We have seen before that many trends in properties can be correlated with variations in atomic and ionic radii. This is true among the transition elements as well as among the representative elements.

In Chapter 7 the horizontal and vertical trends in atomic size were discussed. Let's briefly review them by referring to Figure 21.4. Here we see that as we move from left to right across a given transition series, there is a gradual decrease in atomic radius. The change in size from one element to the next is not as large as among the representative elements because the electrons being added to the atom enter the d subshell just beneath the outer shell. They are fairly effective at shielding the outer shell from the nuclear charge, so only a small increase in effective nuclear charge is felt by the outer electrons and only a small decrease in size takes place.

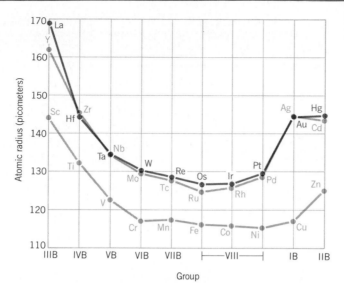

Figure 21.4. *Size variations among the transition elements.*

In Figure 21.4 we also see that there is a minimum in atomic radius that occurs near the center of each transition series. One explanation for this is that once the *d* subshell becomes half-filled, adding additional electrons causes the *d* orbitals to expand in size because of inter-electron repulsions. In turn, this causes the atoms as a whole to grow slightly larger.

The effects on atomic radius of the filling of the 4*f* subshell between atomic numbers 57 and 72, known as the lanthanide contraction, can also be seen in Figure 21.4. For the elements in Group IIIB (Sc, Y, and La), there is the gradual increase in atomic radius that's expected as we descend the group. However, for the elements in Group IVB, the radius increases from the first to the second transition series (i.e., from Ti to Zr), but there is virtually no change in size going from the second series to the third (i.e., from Zr to Hf). The reason, it appears, is because on going from La to Hf in period 6, there is an *extra* decrease in size that occurs because of the existence of the intervening 14 elements of the lanthanide series. In the crossing of the lanthanide series, 14 electrons are added to compensate for the 14 protons added to the nuclei. But because these 14 4*f* electrons are not completely effective at shielding the outer electrons from the nuclear charge, there is a size decrease that is ultimately reflected in Hf being almost the same size as Zr.

The lanthanide contraction has some profound effects on the physical and chemical properties of the period 6 elements that follow the lanthanide series. One property that is affected is density. Although the atoms are virtually the same size, their masses are larger by approximately 90 u. As a result, their densities are correspondingly much larger. Another physical property affected is ionization energy, which is exceptionally large for many of these elements because of the high effective nuclear charge felt by their outer electrons. This causes elements such as platinum and gold to resist the loss of electrons and therefore to resist oxidation. As a result, they are relatively inert to chemical attack.

Trends in the ease of oxidation of the metals were summarized in Figure 10.5 on page 304.

21.3 MAGNETIC PROPERTIES

In Chapter 7 we saw that the presence of unpaired electrons in an atom or molecule imparts the property called paramagnetism to the substance. The tiny electron magnets cause the atom or molecule as a whole to behave as a small magnet. When these are placed into a magnetic field, the microscopic magnets tend to align them-

Figure 21.5. *Domains in a ferromagnetic solid.*

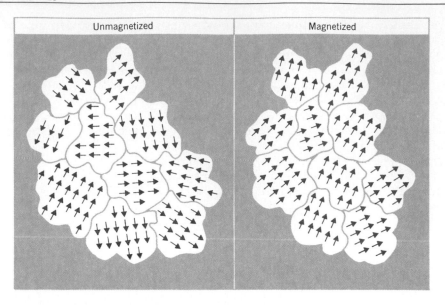

selves with and be attracted toward the field. However, thermal motion operates to randomize the orientations of the little magnets. The net result is that only a relatively small fraction of all of the tiny magnets are aligned with the field at any particular instant, so an ordinary paramagnetic substance is attracted only weakly to an external magnetic field.

Characteristically, the transition elements and their compounds possess a partially filled d or f subshell, so many of them exhibit the phenomenon of paramagnetism. The prediction of magnetic properties of transition metal compounds is therefore one requisite of a theory of bonding that is applicable to these substances. We will explore this a little further when we discuss complex ions later in this chapter.

Related to the property of paramagnetism is the phenomenon called **ferromagnetism,** observed for the three pure elements, iron, cobalt, and nickel. Ferromagnetic materials, like paramagnetic ones, are also attracted to a magnetic field; however, the magnitude of the interaction for a ferromagnetic substance is approximately a million times stronger than it is with paramagnetic materials. How does this occur?

The origin of ferromagnetism is the same as paramagnetism — that is, the existence of unpaired electrons in the ferromagnetic material. In these substances it is believed that regions exist, called **domains,** that contain very large numbers of paramagnetic atoms with their atomic magnets all lined up in the same direction, as illustrated in Figure 21.5. Ordinarily, these domains are randomly oriented in a ferromagnetic solid, so even though each domain behaves as a relatively large magnet, their combined effects cancel. When the ferromagnetic material is placed in a magnetic field, the domains tend to become aligned, much the same as the atomic magnets of a paramagnetic substance. In this case, however, each time one domain becomes aligned with the field, millions of tiny atomic magnets become aligned all at once. As a result, the interaction between the ferromagnetic solid and the magnetic field is very much larger than that experienced by paramagnetic substances.

When the magnetic field is removed from a paramagnetic substance, its atomic magnets very quickly become randomly oriented and no permanent magnetism is induced. For a ferromagnetic substance, however, the domains tend to remain in the orientation in which they found themselves when the external magnetic field was present. This alignment of domains in the absence of an external field causes the substance to possess a residual magnetism, and we say that it has become *permanently magnetized.* Any piece of iron (e.g., a pin) can be magnetized simply by stroking it with another permanent magnet.

A permanent magnet is not really permanent, because the magnetism may be destroyed either by heating the solid or by pounding it. Increased thermal motion causes the domains to become randomly oriented. Similarly, violent vibrational motions will twist and turn them and also make them become disoriented.

The phenomenon of ferromagnetism is associated only with the solid state. Iron, for example, is no longer ferromagnetic when it is melted. Instead it exhibits only paramagnetism. Melting of the solid thus appears to destroy the domains, and each individual atom in the liquid behaves more or less independently of the others nearby.

Even in the solid state not all elements containing unpaired electrons are ferromagnetic. Manganese, for example, possesses five unpaired electrons compared to only four for iron; yet pure iron is ferromagnetic but pure manganese is not. Apparently a requirement for ferromagnetism is for the spacings between paramagnetic ions to be just right so that they may lock onto each other to form a domain. Nonferromagnetic metals, in which the ions are too close together, can sometimes be made ferromagnetic by forming an alloy. This is the case with manganese; the addition of the proper amount of copper permits the Mn^{2+} ions in the metallic lattice to interact strongly and form domains, thereby producing a ferromagnetic alloy.

Many ferromagnetic alloys have been made. Alnico magnets, for example, contain iron, aluminum, nickel, copper, and cobalt.

21.4 PROPERTIES OF SOME IMPORTANT TRANSITION METALS

Many transition metals have useful applications that depend on both their chemical and physical properties. We find them almost anywhere we look — iron in the many steel products that are all around us, chromium on automobile bumpers, the zinc coating on galvanized steel, and the titanium dioxide pigment that is in nearly all paints. In this section we will examine the chemical and physical properties of some of the more important transition elements.

Titanium

Titanium is a strong, lightweight corrosion-resistant metal with a density only about 60% that of iron. These properties make it especially useful in the aircraft industry where it is used often in place of steel and aluminum. It is also used in aircraft jet engines because it doesn't lose its strength at high temperatures.

Lightweight in this case means low density.

Titanium is difficult to extract from its compounds because the metal reacts with carbon, oxygen, and even nitrogen at very high temperatures. As described in Chapter 20, the metal is made by first converting its oxide, TiO_2 (found in the ore *rutile*), to $TiCl_4$. This is accomplished by heating the oxide with carbon and chlorine.

$$TiO_2 + C + 2Cl_2 \longrightarrow TiCl_4 + CO_2$$

The low-boiling $TiCl_4$ is vaporized to separate it from impurities in the ore and then reduced using magnesium.

$$TiCl_4 + 2Mg \longrightarrow Ti + 2MgCl_2$$

Titanium tetrachloride consists of covalently bonded molecules of $TiCl_4$. It is a liquid at room temperature and boils at 136 °C. It is very sensitive to moisture and its vapor reacts with water vapor in the air to give a white smoke containing TiO_2. (See Figure 21.6.) The reaction is

$$TiCl_4(g) + 2H_2O(g) \longrightarrow TiO_2(s) + 4HCl(g)$$

The most important compound of titanium is its oxide, TiO_2, which is usually called titanium dioxide. It is a very white solid and is a common pigment in paint,

Figure 21.6. *Vapors of titanium tetrachloride, $TiCl_4$, react with moisture in the air to give a dense smoke of TiO_2.*

Figure 21.7. *Chrome alum,* $KCr(SO_4)_2 \cdot 12H_2O$, *contains the* $Cr(H_2O)_6^{3+}$ *ion both in the solid state and in aqueous solutions.*

Cr^{2+} is also called *chromous ion;* Cr^{3+} is called *chromic ion.*

H₂O

The $Cr(H_2O)_6^{3+}$ ion

where it possesses good hiding power. It is better than white lead, $Pb_3(OH)_2(CO_3)_2$, because it appears to be of low toxicity and because it doesn't darken on exposure to H_2S as lead-based pigments do. Titanium dioxide is also used as a brightener in the manufacture of paper.

Chromium

Chromium is a hard, brittle, lustrous metal that is very resistant to corrosion. That's why it is used as a protective coating over steel for automobile bumpers. Thin layers of chromium are also deposited by electroplating on brass or bronze objects for decorative purposes.

One of the principal uses of chromium is in stainless steel, a type of steel that is very resistant to corrosion. A typical stainless steel contains about 19% chromium, 9% nickel, and the rest iron. Unlike ordinary iron and steel, a high-quality stainless steel is not ferromagnetic. A small magnet is therefore a handy tool to check whether or not a metal object claimed to be made of stainless steel is, in fact, composed of this alloy.

The principal oxidation states of chromium are $+2$, $+3$, and $+6$. The $+2$ state, characterized by the blue Cr^{2+} ion in aqueous solutions, is very easily oxidized to the $+3$ state, which is the most stable oxidation state.

The most important compound of chromium(III) is the oxide, Cr_2O_3, the most stable green pigment known. It is used for coloring paints, roofing shingles, cements, and plaster. Chromium(III) ion forms many stable complex ions, and in aqueous solutions it actually exists as the violet complex ion, $Cr(H_2O)_6^{3+}$ (Figure 21.7). This ion is what gives many chromium(III) salts their violet color.

Like the $Al(H_2O)_6^{3+}$ ion, the $Cr(H_2O)_6^{3+}$ ion is slightly acidic. When base is added to a solution of the ion, protons are extracted from the water molecules attached to the chromium and a pale blue-violet precipitate is formed. The net reaction is

$$Cr(H_2O)_6^{3+}(aq) + 3OH^-(aq) \longrightarrow Cr(H_2O)_3(OH)_3(s) + 3H_2O$$

This precipitate redissolves in additional base *or* in acid, so chromium(III) is amphoteric (Figure 21.8).

In base

$$Cr(H_2O)_3(OH)_3(s) + OH^-(aq) \longrightarrow Cr(H_2O)_2(OH)_4^-(aq) + H_2O$$

In acid

$$Cr(H_2O)_3(OH)_3(s) + H^+(aq) \longrightarrow Cr(H_2O)_4(OH)_2^+(aq) + H_2O$$

$$Cr(H_2O)_4(OH)_2^+(aq) + H^+(aq) \longrightarrow Cr(H_2O)_5OH^{2+}(aq) + H_2O$$

$$Cr(H_2O)_5OH^{2+}(aq) + H^+(aq) \longrightarrow Cr(H_2O)_6^{3+}(aq) + H_2O$$

Figure 21.8. *On the left is a solution containing the* $Cr(H_2O)_6^{3+}$ *ion. In the center, a precipitate of* $Cr(H_2O)_3(OH)_3$ *has been formed by making the first solution slightly basic. On the right a green solution containing* $Cr(H_2O)_2(OH)_4^-$ *has been formed by dissolving the precipitate in excess strong base.*

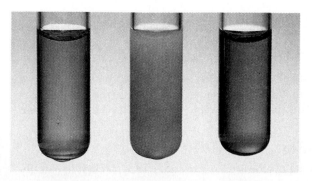

Figure 21.9. *Chromium(VI) oxide,* CrO_3.

Figure 21.10. *Dichromate ion,* $Cr_2O_7{}^{2-}$, *in the solution at the left has a reddish-orange color. The chromate ion,* $CrO_4{}^{2-}$, *in the solution at the right is yellow.*

In the $+6$ oxidation state, chromium forms the red oxide, CrO_3 (Figure 21.9). Chromium (VI) oxide is a strong oxidizing agent and the acid anhydride of *chromic acid,* H_2CrO_4. In very acidic solutions, H_2CrO_4 is the principal species. As the pH is raised, however, two other species are formed. One is the yellow *chromate ion,* $CrO_4{}^{2-}$, and the other is the red-orange *dichromate ion,* $Cr_2O_7{}^{2-}$. These are shown in Figure 21.10. Both are strong oxidizing agents and exist in equilibrium with each other.

In general, oxides of metals in high oxidation states tend to be acidic instead of basic.

$$2CrO_4{}^{2-} + 2H^+ \rightleftharpoons Cr_2O_7{}^{2-} + H_2O$$

Chromate ion
CrO_4^{2-}

Dichromate ion
$Cr_2O_7^{2-}$

By Le Châtelier's principle, we see that $Cr_2O_7{}^{2-}$ predominates at low pH and $CrO_4{}^{2-}$ is the major species at high pH.

Manganese

Manganese is much less corrosion-resistant than its neighbor chromium. It corrodes in moist air, and dilute acids dissolve it, much like iron. Manganese is used mostly in making alloys such as the ferromagnetic manganese–copper alloy mentioned in the last section.

The most stable oxidation state of manganese is $+2$. In solution and in many compounds it exists as the very pale pink $Mn(H_2O)_6{}^{2+}$ ion (Figure 21.11). An important compound of manganese in the $+4$ oxidation state is MnO_2, commonly called manganese dioxide. This is the substance, you recall, that undergoes reduction at the cathode in the ordinary dry cell (see Section 18.11).

When MnO_2 is added to molten KOH and oxidized with O_2 or KNO_3, the green manganate ion, $MnO_4{}^{2-}$, is formed. It is stable only in very alkaline (basic) solutions, and when they are acidified, the $MnO_4{}^{2-}$ **disproportionates:** Some of it is oxidized to $MnO_4{}^-$ (permanganate ion) while the rest is reduced to MnO_2.

In a *disproportionation reaction,* a portion of a reactant is oxidized while the rest is reduced. It is sometimes called an autooxidation–reduction reaction.

$$3MnO_4{}^{2-} + 4H^+(aq) \longrightarrow 2MnO_4{}^-(aq) + MnO_2(s) + 2H_2O$$

Solutions of $MnO_4{}^-$ ion have a deep violet color and are strong oxidizing agents.

Figure 21.11. *The* $Mn(H_2O)_6^{2+}$ *ion gives a pink color to* $MnCl_2 \cdot 6H_2O$ *both in the solid state and in its solutions.*

Solutions of iron(II) salts have a pale greenish-blue color because they contain the $Fe(H_2O)_6^{2+}$ *ion. Solutions that contain iron(III), like the one on the right, have a yellow-orange color from the reaction of* $Fe(H_2O)_6^{3+}$ *with water to give* $Fe(H_2O)_5(OH)^{2+}$ *and* H_3O^+.

Iron is the second most abundant metal and the fourth most abundant element in the Earth's crust.

An open-pit iron mine in the Mesabi Range in Minnesota. The red hematite (Fe_2O_3) ore in this region has been nearly depleted. Instead, mining operations today focus on processing a rock-hard ore called taconite, which contains magnetic Fe_3O_4.

When MnO_4^- is reduced in an acidic solution, the pale pink Mn^{2+} ion is formed.

$$MnO_4^-(aq) + 8H^+(aq) + 5e^- \longrightarrow Mn^{2+}(aq) + 4H_2O$$
deep violet $\qquad\qquad\qquad\qquad$ very pale pink

This makes MnO_4^- a useful analytical reagent for performing redox titrations. As the aqueous MnO_4^- is added from a buret to a solution of a reducing agent, reduction of the MnO_4^- takes place. This produces Mn^{2+} ion whose pale pink color is invisible at the concentrations used in the titration. Therefore, when the MnO_4^- is reduced, its violet color disappears and the solution appears colorless. When all the reducing agent has finally been consumed, the next drop of MnO_4^- solution gives an excess of this ion and that makes the solution appear pink, signaling the endpoint.

Reduction of MnO_4^- in neutral or somewhat basic solutions produces a dark brown precipitate of MnO_2.

$$MnO_4^-(aq) + 2H_2O + 3e^- \longrightarrow MnO_2(s) + 4OH^-(aq)$$

Because a precipitate is formed, MnO_4^- is not used for redox titrations in neutral or basic solutions. The dark brown color of the precipitate obscures the endpoint.

Iron

Iron is the most used of the transition metals because it is relatively abundant and easily extracted from its ores. In the pure state, iron is not very hard, but when small amounts of carbon and other metals are added to it, strong steel alloys are formed.

Iron is a fairly reactive metal. It forms compounds principally in two oxidation states: $+2$ and $+3$. Generally, iron(II) compounds are easily oxidized to give the corresponding iron(III) compounds. Three oxides of iron are formed: FeO, Fe_2O_3, and Fe_3O_4. Iron(II) oxide is difficult to prepare and disproportionates into Fe and Fe_2O_3 when heated.

$$3FeO(s) \longrightarrow Fe(s) + Fe_2O_3(s)$$

Iron(III) oxide is the principal component of many iron ores, and its hydrated form is produced when iron rusts (see below). The oxide Fe_3O_4, known as *magnetite*, con-

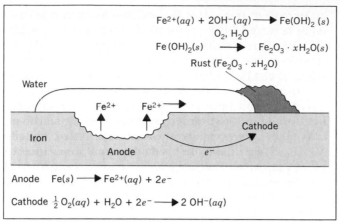

Anode $Fe(s) \longrightarrow Fe^{2+}(aq) + 2e^-$

Cathode $\frac{1}{2}O_2(aq) + H_2O + 2e^- \longrightarrow 2\,OH^-(aq)$

Figure 21.12. *The corrosion of iron. Iron dissolves at anodic sites and diffuses through the water as Fe^{2+} ions to cathodic sites where it precipitates as $Fe(OH)_2$. Iron(II) hydroxide is easily oxidized by O_2 in the presence of water to give the hydrated iron(III) oxide, $Fe_2O_3 \cdot xH_2O$, which is called rust.*

tains iron in two different oxidation states and can be formulated as $Fe^{II}Fe_2^{III}O_4$ (Roman numerals are used here to express the oxidation states of the iron atoms). As its name suggests, magnetite is magnetic. It is an important iron ore because a strong magnet can easily separate it from useless rock and because it is rich in iron.

Metallic iron dissolves in nonoxidizing acids such as HCl or H_2SO_4 with the release of hydrogen (see photo on page 299).

$$Fe(s) + 2H^+(aq) \longrightarrow Fe^{2+}(aq) + H_2(g)$$

One of the most important reactions of iron, at least from an economic point of view, is its corrosion in the presence of air and moisture. The mechanism for the reaction appears to be electrochemical, as shown in Figure 21.12. The iron—in contact with water—is oxidized to the +2 state.

$$Fe(s) \longrightarrow Fe^{2+}(aq) + 2e^-$$

The electrons released by the iron during this oxidation are transmitted to sites on the iron that are in contact with oxygen and moisture, where reduction of O_2 to hydroxide ions occurs.

Iron is acting as the anode in a galvanic cell. That's why it is oxidized.

$$\tfrac{1}{2}O_2(aq) + H_2O + 2e^- \longrightarrow 2\,OH^-(aq)$$

The iron(II) ions diffuse through the water and when they contact the OH^-, a precipitate of $Fe(OH)_2$ is formed, which is very easily oxidized to $Fe(OH)_3$ by oxygen.

$$4Fe(OH)_2(s) + O_2(aq) + 2H_2O \longrightarrow 4Fe(OH)_3(s)$$

Iron(III) hydroxide is easily dehydrated and is better represented as a hydrated oxide, $Fe_2O_3 \cdot xH_2O$, in which the proportion of water is somewhat variable. This hydrated oxide is **rust**.

This mechanism for the rusting of iron explains some interesting observations. First, rusting only occurs if *both* oxygen and water are present. Iron won't rust in dry

Notice that complete dehydration of $Fe(OH)_3$ would give Fe_2O_3:

$$2Fe(OH)_3 \longrightarrow Fe_2O_3 + 3H_2O$$

Rusting has taken place beneath the paint on this very badly corroded car.

Figure 21.13. *The deep blue color of Prussian blue is formed when a drop of a dilute solution containing Fe^{3+} is added to a dilute solution containing ferrocyanide ion, $Fe(CN)_6^{4-}$. After a few moments, the blue precipitate, $Fe_4[Fe(CN)_6]_3 \cdot 16H_2O$, settles to the bottom of the test tube.*

Blue glass is made using cobalt salts. This is the *cobalt glass* used in the flame test for potassium.

air or oxygen-free water. Second, the formation of rust often occurs at a site somewhat removed from the location where the iron is pitting. Have you noticed, for example, that a car rusts out *under* the paint around places where the paint has been scratched. The iron that dissolves migrates under the paint to the place of the scratch where it can be oxidized by O_2 to give the hydrated oxide.

Iron, like other transition elements, forms many complex ions. Among the most interesting are those with cyanide ion. These are commonly called ferrocyanide ion, $Fe(CN)_6^{4-}$, and ferricyanide ion, $Fe(CN)_6^{3-}$. Since the cyanide ion has a $1-$ charge, the ferrocyanide ion contains iron(II) and the ferricyanide ion contains iron(III). When a solution of Fe^{3+} is added to a solution of $Fe(CN)_6^{4-}$ ion, a deep blue precipitate called *Prussian blue* is formed, as shown in Figure 21.13. The formula of the compound, which contains both iron(II) and iron(III), is $Fe_4[Fe(CN)_6]_3 \cdot 16H_2O$. It is interesting that exactly the same compound is formed if a solution containing Fe^{2+} is added to a solution containing $Fe(CN)_6^{3-}$ ion.

Cobalt

Cobalt is an important metal because is it used in many alloys having special properties. *Stellite*, for example, is an alloy containing cobalt, chromium, and tungsten that retains its hardness even when very hot. This property allows it to be used to make cutting tools such as drill bits for high-speed machining of steel parts. Cobalt is also used in catalysts.

The principal oxidation states of cobalt are $+2$ and $+3$. In water, the $+2$ state is the most stable and the cobalt ion exists as the pink complex ion $Co(H_2O)_6^{2+}$. This ion is also found in salts of cobalt(II) and gives them a red color. One such salt, $Co(H_2O)_6Cl_2$, is particularly interesting. On the labels of bottles of this chemical that come from chemical suppliers, the formula is usually written as $CoCl_2 \cdot 6H_2O$, indicating it to be a hydrate. When this salt loses some of its water (by being heated, for example), it becomes blue as shown in Figure 21.14. It is thought that the loss of water produces a change in the nature of the complex as shown by the equilibrium

$$[Co(H_2O)_6]Cl_2(s) \rightleftharpoons [Co(H_2O)_4]Cl_2(s) + 2H_2O(g)$$
$$\text{pink} \qquad\qquad\qquad \text{blue}$$

The change is reversible, and if left in a moist atmosphere, the blue crystals regain their pink color. In fact, this color change from blue to pink is used to detect moisture.

When complex-forming agents other than water are present, cobalt is able to form stable complexes in the $+3$ oxidation state. Many such complexes have been studied.

(a)   *(b)*

Figure 21.14. *(a) Pink hydrated $CoCl_2 \cdot 6H_2O$, which actually contains $[Co(H_2O)_6]Cl_2$. (b) After heating and driving off some of the water, blue $[Co(H_2O)_4]Cl_2$ remains.*

Nickel

Nickel is very useful because it resists corrosion, and its alloys have desirable properties. Electroplating of steel parts with nickel gives them a thin protective coating. Nickel, along with chromium, is added to iron to produce stainless steel, and iron–nickel alloys are used for armor plating because they are impact-resistant. Even the familiar five-cent piece is a copper–nickel alloy.

The most stable oxidation state of nickel is $+2$. In water, the Ni^{2+} ion is green because it actually exists as the complex $Ni(H_2O)_6^{2+}$. This green ion imparts a green color to many nickel salts, as shown in Figure 10.15 on page 323. Nickel forms many other complexes as well. Another example is the blue complex formed when ammonia is added to an aqueous solution containing $Ni(H_2O)_6^{2+}$.

An important compound of nickel in the $+4$ oxidation state is NiO_2, the cathode material in the nickel–cadmium battery. When the battery discharges, the nickel is reduced to the $+2$ state.

$$NiO_2(s) + Cd(s) + 2H_2O \longrightarrow Ni(OH)_2(s) + Cd(OH)_2(s)$$

Nickel salts are added to glass to give it a green color.

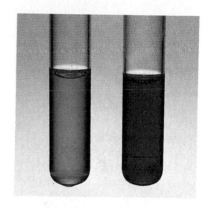

A solution of nickel(II) is green because it contains the $[Ni(H_2O)_6]^{2+}$ ion. The addition of NH_3 gives the deep blue complex $[Ni(NH_3)_6]^{2+}$.

Photofinishers recover and recycle large amounts of silver from the films and prints they process.

The coinage metals: copper, silver, and gold

The name "coinage metals" given to the Group IB elements is not surprising since these metals have been used for many centuries to make coins as well as jewelry. They are relatively unreactive and their reactivity decreases going down the group. As a result, gold is found free in nature, as are silver and copper, but silver and especially copper also are found in deposits of their compounds.

All the coinage metals have practical uses. As mentioned earlier, copper has the second highest electrical conductivity of any metal and is used extensively for electrical wiring. Silver is an even better conductor than copper, but its lower abundance and its value as a decorative metal make it too expensive to use generally in place of copper. Nevertheless, U.S. industries use over 175 million ounces of silver each year; about 24% is consumed by the electronics industry. Large amounts of silver are also used to make photographic film and paper. Even gold has practical applications. Low-voltage electrical contacts in computers are often gold-plated because of gold's resistance to corrosion. Even a thin film of corrosion on contacts made of other metals would seriously impede the flow of electricity in low-voltage circuits.

Copper, silver, and gold have more positive reduction potentials than hydrogen, which means that they cannot displace hydrogen from acids such as HCl or H_2SO_4 in which the strongest oxidizing agent is H^+. Copper and silver do dissolve in HNO_3, however, because it contains NO_3^-, which serves as the oxidizing agent. In dilute nitric acid, the reactions are

$$3Cu(s) + 8H^+(aq) + 2NO_3^-(aq) \longrightarrow 3Cu^{2+}(aq) + 2NO(g) + 4H_2O$$

$$3Ag(s) + 4H^+(aq) + NO_3^-(aq) \longrightarrow 3Ag^+(aq) + NO(g) + 2H_2O$$

Gold is much more difficult to oxidize than copper or silver, and even concentrated HNO_3 is not a strong enough oxidizing agent to do the job. However, aqua regia, a 3-to-1 mixture of concentrated HCl and HNO_3, does dissolve gold because the chloride ion stabilizes the Au^{3+} by forming a complex ion with it.

$$Au(s) + 6H^+(aq) + 3NO_3^-(aq) + 4Cl^-(aq) \longrightarrow AuCl_4^-(aq) + 3H_2O + 3NO_2(g)$$

Each of the Group IB metals forms compounds in the $+1$ oxidation state. Examples are CuCl, AgCl, and AuCl, all of which are insoluble in water. For copper and gold, the $+1$ oxidation states are not especially stable and tend to dispropor-

Figure 21.15. *A pellet of superconducting ceramic floats above a magnet at Argonne National Laboratory. The pellet had been previously cooled to 77 K by liquid nitrogen. The magnet induces an electric current in the superconducting pellet that generates a counteracting magnetic force. The pellet will remain levitated like this until it warms sufficiently to lose its superconducting properties.*

Liquid helium boils at 4.2 K, which is about −270 °C.

Liquid nitrogen boils at 77 K, or −196 °C. It is often used in refrigeration and costs about as much as milk or beer.

Many of the new superconducting materials contain lanthanide elements.

tionate. In aqueous solution, for example, copper(I) ion spontaneously gives metallic copper and copper(II).

$$2Cu^+(aq) \longrightarrow Cu(s) + Cu^{2+}(aq)$$

The most stable oxidation state of gold is +3, so gold(I) compounds tend to disproportionate to give the free metal and gold(III).

By far, the most stable oxidation state for silver is the +1 state. The most important compound of silver is silver nitrate, $AgNO_3$, made by dissolving silver in nitric acid. It serves as the starting material for nearly all other silver compounds — for instance, the halides AgCl, AgBr, and AgI that are used in photographic films and papers.

The ions of the Group IB metals form many complex ions. Copper salts are pale blue in aqueous solutions because of the pale blue ion $Cu(H_2O)_4^{2+}$. Addition of ammonia produces the deep blue $Cu(NH_3)_4^{2+}$ ion (see Figure 10.16 on page 324). The qualitative test for silver ion in a solution uses a similar complex between Ag^+ and NH_3. First the solution to be tested is made acidic with HCl, which precipitates white AgCl, as well as the chlorides of lead $(PbCl_2)$ and mercury(I) (Hg_2Cl_2), if any of these metal ions are in the solution. Then the precipitate is treated with aqueous NH_3, which causes any AgCl that might be present to dissolve. The ammonia does not dissolve $PbCl_2$ or Hg_2Cl_2, however, because lead and mercury do not form soluble complex ions with ammonia.

$$AgCl(s) + 2NH_3(aq) \rightleftharpoons Ag(NH_3)_2^+(aq) + Cl^-(aq)$$

The solution, which would contain $Ag(NH_3)_2^+$ and Cl^-, is then acidified. If any AgCl had dissolved when the aqueous NH_3 was added, it is now reprecipated because the equilibrium is shifted to the left as NH_3 is converted to NH_4^+ by the acid.

Superconductors Metals such as copper or silver, although excellent conductors of electricity, do offer some resistance to the flow of an electric current. This resistance causes some of the electrical energy to be lost as heat, so when electricity is transmitted over long distances, significant losses of power occur. However, if these metals are cooled to very low temperatures (near absolute zero), they lose all their electrical resistance and become **superconductors.** It would seem, therefore, that substantial savings in the cost of power transmission could be realized by using these metals in their superconducting states in which no loss of electrical energy would occur. Unfortunately, the cost of maintaining metals at such cold temperatures by bathing them in liquid helium far exceeds any economic benefits. As a result, many scientists had essentially given up any hope for widespread commercial use of the superconductivity phenomenon.

With this as background, it is easy to appreciate the enthusiasm expressed by scientists when it was reported in 1987 that a new class of compounds had been discovered that becomes superconducting at much higher temperatures than had ever been found for metals. These are temperatures that can easily be reached by cooling with liquid nitrogen. Amazingly, the new superconducting materials are not metals. They are, instead, ceramics made from mixtures of metal oxides that are baked at high temperatures. Almost all of them contain copper. Typical examples have compositions such as $Ba_2YCu_3O_7$, which becomes superconducting at 93 K, and $Tl_2Ca_2Ba_2Cu_3O_{10}$, which becomes superconducting at a temperature of 125 K.

One of the remarkable properties of a superconducting substance is that when it is placed in a magnetic field, electrical currents induced within it create an opposing magnetic field that causes the superconductor to be repelled by the magnet. In Figure 21.15 we see an example of this; the cold superconducting pellet is suspended in air above a strong magnet. It will stay suspended until it warms to a temperature at which it loses its superconductivity.

The remarkable properties of superconductors and the prospects of achieving them at reasonable temperatures, perhaps even room temperature, have rekindled a long-standing hope that these substances will form the basis for a new age of technology. Before this comes to pass, however, improvements in many areas are needed, and there are many technical problems to be solved.

Zinc, cadmium, and mercury

These metals are often not considered true transition elements because their d subshells are completed. Each has a valence shell consisting of only two electrons in an s orbital, so their highest oxidation state is $+2$. In fact, zinc and cadmium show only a $+2$ oxidation state (other than zero, of course), while mercury has oxidation states of $+2$ and $+1$.

Zinc and cadmium are both silvery, rather reactive metals. They dissolve readily in nonoxidizing acids such as HCl and H_2SO_4 with the evolution of hydrogen. In fact, the reaction of zinc with dilute H_2SO_4 is a common way of preparing hydrogen in the laboratory.

$$Zn(s) + H_2SO_4(aq) \longrightarrow ZnSO_4(aq) + H_2(g)$$

The reactivities of zinc and cadmium account for one of their principal uses as free metals—providing corrosion protection for iron and steel. Coating an iron or steel object with zinc is called **galvanizing.** The zinc protects the steel in two ways. First, it reacts with moisture and CO_2 to form a film of $Zn_2(OH)_2CO_3$ that prevents oxygen and moisture from coming into contact with and reacting with the zinc below or with the iron. But even if this zinc coating is scratched through to the iron, the zinc still protects the iron by electrolytic action. Zinc is more easily oxidized than iron, so when the metals are in contact, the zinc becomes the anode of a galvanic cell and iron becomes the cathode. Since oxidation always occurs at the anode, zinc is oxidized, but iron is protected by being the cathode. The phenomenon is called **cathodic protection.** Cadmium plating on steel functions in a similar way, but cadmium is used less often than zinc for a number of reasons. One is that cadmium is less abundant than zinc and is therefore more expensive. Another is that cadmium compounds are very toxic: They cause high blood pressure, heart disease, and can even lead to painful death.

Zinc and cadmium have uses other than as protective coatings over other

In recent years, pennies have been made of zinc with just a thin copper coating. Here we see a new penny that has been dipped into nitric acid, which stripped away the copper coating to expose the gray color of the zinc beneath.

The shiny zinc coating on a galvanized steel object such as a garbage pail becomes dull as the zinc reacts with air and moisture to form $Zn_2(OH)_2CO_3$.

Cadmium generally occurs in nature as an impurity in zinc ores.

If you look carefully, you can see the zinc wire exposed alongside the Alaskan pipeline. By being connected electrically to the pipeline, the zinc provides cathodic protection and prevents corrosion damage to the pipe. After a time, the zinc wire will have to be replaced.

metals. Zinc is the metal used as the anode in the common dry cell, and it is alloyed with copper to produce *brass* and with copper and tin to produce *bronze*. Cadmium is used to make rechargeable nickel–cadmium batteries.

Mercury is the only metal that is a liquid at ordinary room temperatures. It freezes at $-38.9\ °C$ and boils at $357\ °C$. This liquid range has led to its widespread use as the fluid in thermometers. As we learned earlier, mercury has the ability to dissolve many metals to form solutions called amalgams.

Cadmium and zinc have many similar chemical properties, which differ considerably from those of mercury. For instance, we have seen that both Zn and Cd dissolve in dilute acids with the evolution of hydrogen. Mercury, however, is considerably less reactive. It doesn't dissolve in acids such as HCl or H_2SO_4, although it does react with nitric acid.

$$3Hg(l) + 8H^+(aq) + 2NO_3^-(aq) \longrightarrow 3Hg^{2+}(aq) + 2NO(g) + 4H_2O$$

One important difference between zinc and cadmium is that zinc is amphoteric but cadmium is not. For example, zinc dissolves in base.

$$Zn(s) + 2OH^-(aq) + 2H_2O \longrightarrow Zn(OH)_4^{2-}(aq) + H_2(g)$$

Its hydroxide is also amphoteric. When base is added to Zn^{2+} in aqueous solution, a precipitate of $Zn(OH)_2$ is formed.

$$Zn^{2+}(aq) + 2OH^-(aq) \longrightarrow Zn(OH)_2(s)$$

The zinc hydroxide dissolves either in acid or additional base.

$$Zn(OH)_2(s) + 2H^+(aq) \longrightarrow Zn^{2+}(aq) + 2H_2O$$

$$Zn(OH)_2(s) + 2OH^-(aq) \longrightarrow Zn(OH)_4^{2-}(aq)$$

This chemical difference between zinc and cadmium explains why metals are sometimes cadmium-plated rather than zinc-plated. Cadmium is used if the metal is to be exposed to an alkaline environment, because the base will destroy a zinc coating but not a cadmium one.

Many zinc compounds have important commercial use. Zinc oxide, which forms when zinc is heated in air, is used as a paint pigment, in creams that are spread on the skin as sun screens, and in fast-setting dental cements. Zinc chloride is used in many different ways, including deodorants, embalming, and fireproofing lumber.

As mentioned earlier, mercury forms compounds in both the $+1$ and $+2$ oxidation states. In the $+1$ state, two mercury(I) ions are joined by a covalent bond to give Hg_2^{2+}. The existence of this ion is supported by equilibrium studies (see Exercise 21.69), X-ray determination of crystal structures, and the fact that mercury(I) compounds are diamagnetic. If there were a simple Hg^+ ion, it would have an odd number of electrons, so all the electrons couldn't be paired and Hg^+ would be paramagnetic. Pairing of the odd electron with an electron from another Hg^+ gives a covalent bond and a diamagnetic Hg_2^{2+} species.

Mercury compounds have quite different properties from those of zinc and cadmium. They are considerably less ionic, for example, and $HgCl_2$ in aqueous solution exists primarily (99%) as $HgCl_2$ molecules.

$$HgCl_2(aq) + H_2O \rightleftharpoons Hg(OH)Cl + H^+ + Cl^-$$

As you probably know, mercury and its compounds are very toxic. Mercury spills should be avoided in the laboratory because the vapor pressure of mercury (about 10^{-3} torr at room temperature) is sufficient to cause mercury poisoning if the vapors are breathed for long periods of time. Soluble mercury compounds, such as mercury(II) chloride, are especially poisonous because they can quickly provide sufficient mercury to the body to cause death. By contrast, mercury(I) chloride, Hg_2Cl_2, is

Not all metals dissolve in mercury. Iron is an example of one that doesn't.

$HgCl_2$ is a weak electrolyte.

One or two grams of $HgCl_2$ is a fatal dose.

very insoluble in water and at one time — before the discovery of penicillin — it was used medicinally as a treatment for syphilis. Its low solubility prevents the body from absorbing lethal doses of mercury. Mercury is a cumulative poison, however, so even small amounts absorbed over extended periods can lead to serious medical problems.

Because Hg_2Cl_2 is insoluble in water, the presence of Hg_2^{2+} ion in a solution can be tested by treating the solution with HCl, which causes the Hg_2Cl_2 to precipitate. However, since AgCl and $PbCl_2$ are also white and insoluble, and would also be formed if Ag^+ and Pb^{2+} were present, the precipitate must be tested further. Addition of aqueous ammonia, which is part of the test for silver, causes the Hg_2Cl_2 to disproportionate.

$$Hg_2Cl_2(s) + 2NH_3(aq) \longrightarrow Hg(NH_2)Cl(s) + Hg(l) + NH_4^+(aq) + Cl^-(aq)$$

The mixture of Hg and $Hg(NH_2)Cl$ appears black or gray because the black color of the finely divided mercury masks the white of the $Hg(NH_2)Cl$.

$Hg(NH_2)Cl$ is called *mercury(II) amido chloride.*

21.5 COMPLEXES OF TRANSITION METALS; COORDINATION NUMBER AND STRUCTURE

As we noted earlier, one of the properties that characterize the transition metals is their ability to form large numbers of stable complexes or complex ions. In fact, in our discussion of the chemistries of some of the transition metals we commented on the tendencies of the ions to exist in aqueous solutions as complex ions with water molecules serving as the ligands.

The formation of complexes from metal ions was described in Section 10.7. If necessary, you should review the kinds of substances that function as ligands as well as the rules that we follow in writing the formulas for complexes.

Recall that compounds that contain metal complexes are called *coordination compounds.*

One of the most interesting aspects of studying metal complexes is their structures. It is convenient to classify these structures according to the number of ligand atoms (donor atoms) that surround the metal. The term used to describe this is **coordination number (C.N.)**, which is defined as the number of ligand atoms that are bonded to a given metal ion. These atoms can be supplied by monodentate ligands (ligands capable of attaching themselves to a metal by only one atom), by polydentate ligands (ligands that can attach themselves to a metal ion by two or more atoms), or by both. Consider, for example, the complexes $[CoCl_6]^{3-}$, $[Co(en)_2Cl_2]^+$, and $[Co(en)_3]^{3+}$. The second and third contain the bidentate ligand ethylenediamine (abbreviated en). In each of these complexes the C.N. of the Co^{3+} ion is the same, because each has *six* donor atoms bonded to the cobalt.

Coordination numbers ranging from 2 to more than 8 are observed in various coordination compounds, with the C.N. in any given instance determined by the nature of the metal ion, its oxidation state, and to some extent, the ligands and the environment surrounding the complex. The most common coordination numbers are observed to be 2, 4, and 6. The basic structural types found for them are shown in Figure 21.16.

By far the most frequently occurring coordination number in transition metal complexes is 6, and the geometry that is observed in nearly all instances is octahedral. A simple two-dimensional way of representing the octahedral geometry was described in Chapter 9 on page 261 and is shown in Figure 21.17. The dashed rectangle represents the square plane that joins the upper and lower pyramids in the octahedron. The six solid lines connect the center of the metal cation to the coordinated ligand atoms. This arrangement is illustrated in Figure 21.17 for the complex ion, $[CoCl_6]^{3-}$.

Figure 21.16. *Structures commonly found for coordination numbers of 2, 4, and 6.*

C.N. = 2	Linear	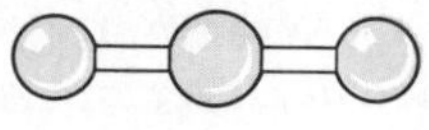
C.N. = 4	Tetrahedral	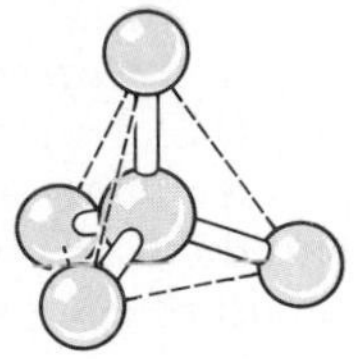
	Square planar	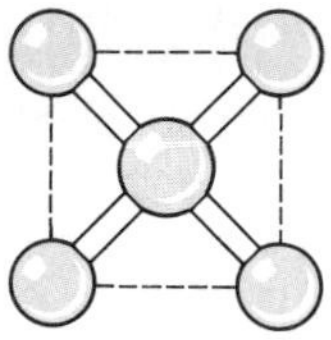
C.N. = 6	Octahedral	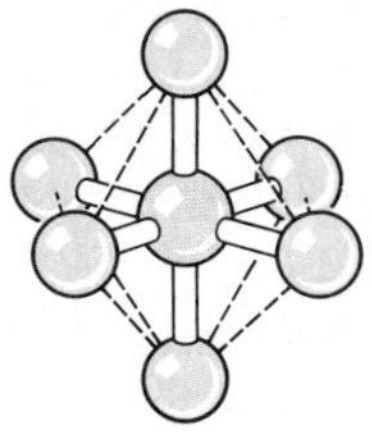

Figure 21.17. *Two-dimensional representation of octahedral coordination.*

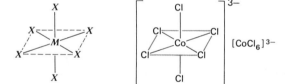

$[CoCl_6]^{3-}$

21.6 ISOMERISM AMONG METAL COMPLEXES

When two different compounds have the same molecular formula, but differ in the way that their atoms are arranged, they are said to be **isomers** of one another. For example, there are two compounds with the general formula

$$Cr(NH_3)_5SO_4Br$$

One of these we should formulate as

$$[Cr(NH_3)_5SO_4]Br$$

because it yields a precipitate of AgBr when treated in aqueous solution with $AgNO_3$ but does not give a precipitate of $BaSO_4$ when treated with $Ba(NO_3)_2$. This last observation means that the SO_4^{2-} is not free in the solution and, hence, must be bound to the chromium.

The second compound is written as

$$[Cr(NH_3)_5Br]SO_4$$

Figure 21.18. *Cis–trans isomers for complexes with the general formula* Ma_4b_2.

and produces $BaSO_4$ when treated with $Ba(NO_3)_2$. On the other hand, addition of $AgNO_3$ to a solution of the compound does not yield AgBr.

The two compounds just described have different chemical properties and are clearly different chemical substances, even though they are composed of the same number of the same kinds of atoms. This particular type of isomerism is not uncommon among coordination compounds and is called **ionization isomerism.**

Another type of isomerism that is very important is called **stereoisomerism,** and results when a given molecule or ion can exist in more than one structural form in which the same atoms are bonded to one another but find themselves oriented differently in space. To illustrate this, we will focus our attention on octahedral complexes because they represent the most common structural type.

The simplest form of stereoisomerism results when a complex has the general formula Ma_4b_2 in which a and b represent monodentate ligands. An example would be the ion $[Co(NH_3)_4Cl_2]^+$. This complex can exist in two different isomeric forms, called **geometrical isomers,** as shown in Figure 21.18. As you can see, in one of these isomers the two b ligands are located across from each other on opposite sides of the metal ion. Such an isomer is given the designation **trans** (Latin *trans* means across). The other isomer has the two b ligands adjacent to each other and is referred to as the **cis** isomer (L. *cis* means on the same side). Thus, the two isomers would be specified as

$$trans\text{-}[Co(NH_3)_4Cl_2]^+$$

and

$$cis\text{-}[Co(NH_3)_4Cl_2]^+$$

Because *cis* and *trans* isomers possess different structures, they are different chemical species, each with its own set of chemical and physical properties. While these properties may often be similar, the fact that they are not identical clearly tells us that the two structures represent truly different compounds.

Geometrical isomers also occur when the complex contains polydentate ligands. This is illustrated in Figure 21.19 for the complex $[Cr(en)_2Cl_2]^+$, which contains two bidentate ligands. In this figure the ethylenediamine ligands, $H_2N{-}CH_2{-}CH_2{-}NH_2$, are represented by

$$N\frown N$$

As you can see, the two CH_2 groups between the nitrogens are represented by the curved line and the hydrogens bonded to the nitrogens have been omitted. The purpose is to make it easier to view the arrangement of the bidentate ligands around the chromium(III) ion. This kind of simplification is often made when representing polydentate ligands in the structure of a complex.

An important thing to notice about the structures in Figure 21.19 is that the bidentate ligands span adjacent *(cis)* positions in the octahedron. This is virtually always the case in complexes of this type, and for complexes that contain two bidentate ligands and two monodentate ligands, two isomers, cis and trans are possible. In the cis isomer, the Cl^- ligands are alongside each other, but in the trans isomer, they are on opposite sides of the metal ion.

A second form of stereoisomerism is called **optical isomerism.** (As we will see,

To appreciate this simplification, you might attempt to draw the structures of the $[Cr(en)_2Cl_2]^+$ or $[Cr(en)_3]^{3+}$ ions using the full structural formula for ethylenediamine.

Bidentate ligands always span adjacent or *cis* positions in a complex.

Figure 21.19. *Cis–trans isomerism for [Cr(en)₂Cl₂]⁺.*

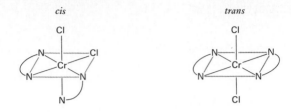

N⌒N represents the bidentate ethylenediamine ligand

optical isomers affect polarized light differently.) They bear the same structural relationship to each other as do your left and right hands — that is, they are *nonsuperimposable mirror images* of one another. To see what this means, try this simple experiment. Place your right hand in front of a mirror, with the palm toward the mirror, and hold your left hand alongside with the palm facing you (Figure 21.20). Notice that the image of your *right* hand in the mirror looks the same as your left hand. That is why we say that your left and right hands are mirror images of each other. But your left and right hands also have a nonsuperimposable aspect because, while similar in appearance, they do not match exactly when one is placed over the other, both with palms down; the thumbs point in different directions. This difference is perhaps seen even more clearly if you attempt to place your right hand into a left-hand glove; it doesn't fit properly. Thus, your left and right hands, and optical isomers too, cannot be superimposed on each other.

Molecules or ions that can have two structures related to each other in the same way that your left and right hands are related — that is, as nonsuperimposable mirror images — are said to be **chiral** (from the Greek *cheir*, meaning "hand"). An example of a chiral complex is the [Co(en)₃]³⁺ ion. The mirror image relationship between the two isomers is shown in Figure 21.21. A pair of isomers related in this manner are called **enantiomers.**

In the complex [Co(en)₃]³⁺ it is the arrangement of the metal–ligand rings (called *chelate rings*[1]) that gives rise to the chirality, or optical isomerism. In fact, any

If you make a model of a molecule and use its reflection as a guide in constructing its mirror image, you can test for superimposability.

Figure 21.20. *Illustration of nonsuperimposable mirror images.*

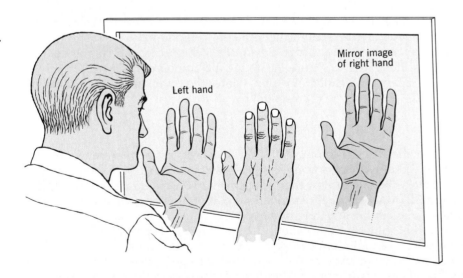

[1] The origin of chemical terminology is sometimes rather colorful. Complexes of this general type are often called **chelates,** from the Greek *chele*, meaning claw. The bidentate ligand in this case bites the metal with two claws (donor atoms) much like a crab.

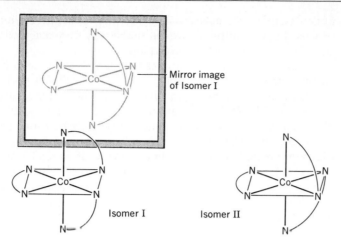

Figure 21.21. *The mirror image relationship between isomers of the chiral $[Co(en)_3]^{3+}$ ion. Isomer II is the mirror image of isomer I, but I and II are not identical. They are nonsuperimposable isomers.*

octahedral complex containing three bidentate ligands is chiral and can exist as two optical isomers.

Optical isomerism is also important for the cis form of complex ions containing two bidentate ligands and two monodentate ligands—for example, cis-$[Co(en)_2Cl_2]^+$ (Figure 21.22). The trans form of this complex is not chiral and does not exhibit optical isomerism, however, because it and its mirror image are identical. They are superimposable.

In general, the properties of optical isomers are identical except for the way in which they interact with outside influences that are able to distinguish between left- and right-handedness. The situation here is analogous to having a group of baseball players, some of whom are left-handed and some right-handed. Since they are all able to toss a baseball with equal ease, a baseball will not differentiate between left- and right-handed players. The same applies to a bat, since there are no left- or right-handed baseball bats. A fielder's glove, however, will fit only one hand. A glove designed to be worn on the left hand cannot be used by a player who catches a ball in the right hand. In this case, the glove differentiates between these two kinds of players because it also has a left- or right-handedness to it. In this same fashion,

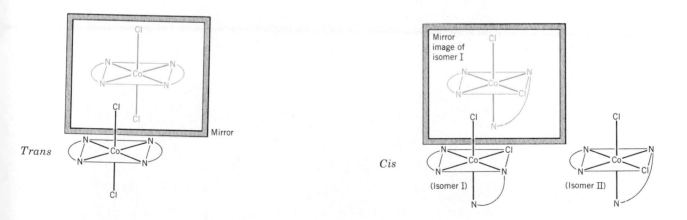

Figure 21.22. *Isomers of $[Co(en)_2Cl_2]^+$. The mirror image of the trans isomer is identical to the original, so the trans isomer is not chiral. The mirror image of the cis isomer is not superimposable on the original, so the cis isomer exists as two nonidentical optical isomers.*

optical isomers interact in an identical way with most chemical reagents and physical probes. They do differ, however, in the way they react toward other optically active chemicals and toward polarized light.

Light, in general, is composed of electromagnetic radiation that possesses both electric and magnetic components that behave like vectors. These vectors oscillate in a sinusoidal fashion perpendicular to the direction in which the light wave is traveling (Figure 21.23). If we examine the electric vectors, all different orientations are observed in an unpolarized beam. However, when such a beam is passed through a polarizing medium, only the vibrations in one plane remain. The result is called **plane-polarized light.** A unique feature of optical isomers is that when plane-polarized light is passed through them (or their solutions), the plane of polarization is rotated through some angle, θ, as shown in Figure 21.24. Substances that affect polarized light this way are said to be optically active. One enantiomer (optical isomer) causes the light to be rotated to the right — that is, clockwise when viewed down the axis of the oncoming light beam — and is said the be **dextrorotatory.** The other enantiomer causes the polarized beam to be rotated to the left and is described as **levorotatory.** The two isomers are therefore designated as d or l depending on the direction of rotation of the polarized light.

An equal mixture of two enantiomers tends to rotate polarized light to both the left and right simultaneously. These effects therefore cancel each other, and such a mixture shows no optical activity; it is said to be **racemic.** In almost all cases, when enantiomers are produced in a chemical reaction carried out in the laboratory, they are formed in equal numbers so that a racemic mixture results. One of the arts in chemistry is the separation of optical isomers from one another.

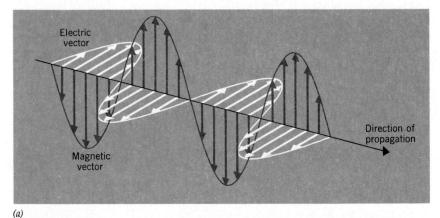

(a)

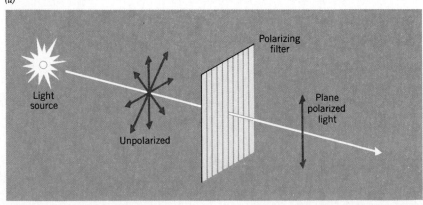

(b)

Figure 21.23. *Polarized light.* (a) *Electromagnetic radiation is composed of electric and magnetic vectors.* (b) *The orientations of the electric vectors in polarized and unpolarized light.*

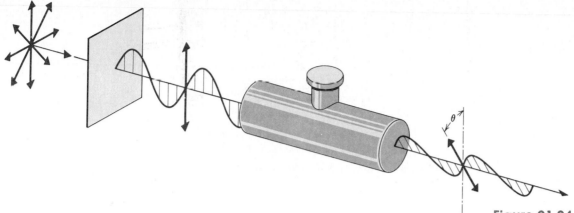

21.7 BONDING IN TRANSITION METAL COMPLEXES: CRYSTAL FIELD THEORY

In our earlier discussions of the formation of metal complexes, we viewed the metal as a Lewis acid and the ligands as Lewis bases. The bonds that form between the metal ion and the ligands would therefore be considered to arise from coordinate covalent bonding. This is a very simplistic picture, however, and it fails to explain many of the interesting properties of complex ions, such as their colors and certain of their magnetic properties.

A theory that has proven to be very useful in explaining these properties is called the **crystal field theory** (CFT). It was developed during the late 1930s by physicists who were studying metal ions trapped within crystalline lattices. Chemists saw similarities between the structures of metal complexes and the symmetrical arrangements of ions in crystals and discovered that the crystal field theory worked well for complexes, too. Over the years the crystal field theory has evolved and taken on many of the trappings of quantum theory, but for simplicity we will look at the theory in an elementary way.

In its simplest form, crystal field theory ignores covalent bonding! Instead, it focuses on the way the electric charge on the ligands affects the energies of the d orbitals of the metal ion. In general, the ligands in a complex are either anions or they are polar molecules. When they are polar molecules, the negative ends of the ligand dipoles point in the direction of the metal cation. Let's examine how these ligands affect the d orbitals. One of the simplest complex ions that we can consider for this purpose is the $[Ti(H_2O)_6]^{3+}$ cation, consisting of a Ti^{3+} ion surrounded octahedrally by six water molecules. Titanium(III) has a single $3d$ electron

$$Ti^{3+} \quad \uparrow \;\underline{\quad}\;\underline{\quad}\;\underline{\quad}\;\underline{\quad} \qquad \underline{\quad} \qquad \underline{\quad}\;\underline{\quad}\;\underline{\quad}$$

$$\qquad\qquad\quad 3d \qquad\qquad\quad 4s \qquad\quad 4p$$

Which of the five $3d$ orbitals will this electron prefer to occupy? Before we can answer this question, we must examine the shapes and directional properties of d orbitals.

In Chapter 7 we discussed the shapes and directional properties of the d orbitals. That description is repeated here in Figure 21.25. Notice that four of the d orbitals, labeled d_{xy}, d_{xz}, d_{yz}, and $d_{x^2-y^2}$, have the same shape and are composed of four lobes each. The fifth, the d_{z^2}, consists of two large lobes directed along the positive and negative z axis plus a donut of charge in the xy plane. For our purposes here, it is important to notice that two of these d orbitals have lobes that are pointed *along* the

The labels for the d orbitals have their origins in the mathematics of quantum mechanics.

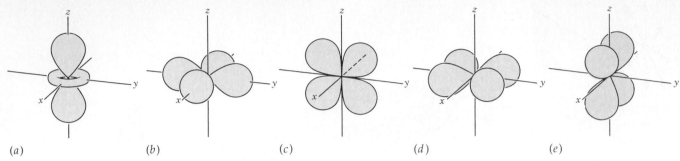

Figure 21.25. *Directional properties of the* d *orbitals.* (a) d_{z^2}, (b) $d_{x^2-y^2}$, (c) d_{yz}, (d) d_{xy}, (e) d_{xz}.

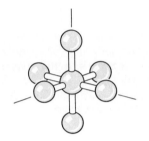

Figure 21.26. *An octahedral arrangement of ligands around a metal ion is produced by placing the ligands along the* x, y, *and* z *axes.*

coordinate axes—the $d_{x^2-y^2}$ and d_{z^2} orbitals. The other three—the d_{xy}, d_{xz}, and d_{yz}—have lobes that point *between* the axes at 45° angles to them.

Now let's consider constructing an octahedral complex by placing the six ligands along the *xyz* axes as shown in Figure 21.26. Since the ligands are negatively charged, or have the negative ends of their dipoles pointing at the metal ion, when the complex is formed, electrons in the *d* orbitals will feel an electrostatic repulsion and their energies will be raised. What is especially important, however, is that an electron in a $d_{x^2-y^2}$ or d_{z^2} orbital will be repelled *more* than an electron in one of the d_{xy}, d_{xz}, or d_{yz} orbitals because the $d_{x^2-y^2}$ and d_{z^2} orbitals point directly at the ligands (Figure 21.27). As a result, the energies of the $d_{x^2-y^2}$ and d_{z^2} are raised more than the energies of the d_{xy}, d_{xz}, and d_{yz} orbitals, as shown in Figure 21.28. This splits the *d* subshell into two energy levels. The lower one consists of the d_{xy}, d_{xz}, and d_{yz} orbitals and the higher one has the $d_{x^2-y^2}$ and d_{z^2} orbitals. For reasons beyond the scope of this book, in an octahedral complex the lower level is labeled t_{2g} and the upper level is labeled e_g. The energy difference between the t_{2g} and e_g levels is called the **crystal field splitting** and is usually indicated by the symbol Δ.

Returning to the $[Ti(H_2O)_6]^{3+}$ ion, we see that its single *d* electron will have the lowest energy if it occupies one of the orbitals of the t_{2g} level. We can also understand what happens when the complex absorbs light (Figure 21.29). If the energy of the light that strikes the complex is equal to Δ, the light can be absorbed and the electron raised from the t_{2g} level to the e_g level. The energy of this absorbed light, of course, depends on the magnitude of Δ and, as you learned in Chapter 7, the energy of a light wave, E, is also related to its frequency, v.

$$E = hv \quad (h \text{ is Planck's constant})$$

For most complexes, the magnitude of Δ is such that the frequencies of light that are absorbed lie in the visible portion of the spectrum. Since the color of light is related to its frequency, the color of the complex depends on the frequencies absorbed when white light is reflected from it or passes through it. In other words, we see the complement of the color of the light that is absorbed. For example, the magnitude of Δ for the $[Ti(H_2O)_6]^{3+}$ ion corresponds to the energy of light in the yellow-green portion of the spectrum. Therefore, when white light passes through a solution of this complex, yellow-green light is absorbed and the light that emerges appears violet.

For a given metal ion, different ligands have different effects on the magnitude of the splitting of the *d* orbitals—that is, on Δ. For example, in Figure 21.30 we see six different complexes of Co^{3+}, each with a different set of ligands. The different colors tell us that different amounts of energy are needed to raise an electron from the lower to the higher energy level, and these differences in energy must reflect

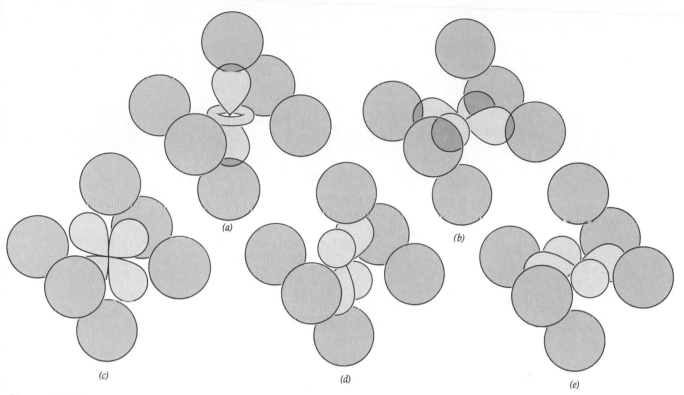

Figure 21.27. *Interaction of the ligands with the* d *orbitals of the metal.* (a) d_{z^2}, (b) $d_{x^2-y^2}$, (c) d_{yz}, (d) d_{xz}, (e) d_{xy}.

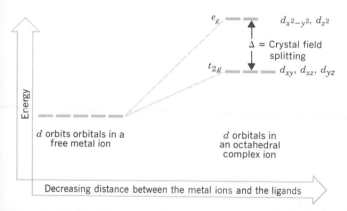

Figure 21.28. *Repulsions caused by the ligands along the* xyz *axes raise the energy of the* $d_{x^2-y^2}$, *and* d_{z^2} *orbitals more than the energies of the* d_{xy}, d_{xz}, *and* d_{yz} *orbitals.*

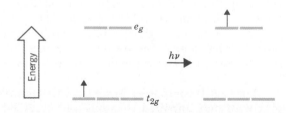

Figure 21.29. *Absorption of light* (hν) *by the* $Ti(H_2O)_6^{3+}$ *ion promotes the electron from the* t_{2g} *to the* e_g *level.*

Figure 21.30. *Each of these brightly colored compounds contains a complex ion of Co^{3+}. The variety of colors arise because of the different ligands that are bonded to the metal ion.*

differences in the abilities of the various ligands to split the energies of the d-orbital energy levels.

By examining the absorption spectra of various complexes like those in Figure 21.30, we can arrange the ligands in the order of their ability to produce a large Δ. This series is called the **spectrochemical series,** which in an abbreviated form is

$$I^- < Br^- < Cl^- < F^- < OH^- < H_2O < NH_3 < en < NO_2^- < CN^-$$

Thus, I^- is poorest at splitting the energies of the t_{2g} and e_g levels, and CN^- is best. What is particularly interesting is that this same series applies for essentially any metal in any oxidation state. However, although the order is usually the same, the actual magnitude of Δ for a given complex in a given geometry depends on the ligand, the metal, and its oxidation state.

As with nearly any generalization we might make in attempting to describe chemical properties, there are exceptions. This is true here with the order of ligands in the spectrochemical series, because in some instances the relative positions of neighboring ligands in the series are reversed. With cobalt(III), for example, Cl^- appears to produce a greater crystal field splitting than F^-. Nevertheless, the spectrochemical series often serves as a useful guide in understanding, and sometimes even predicting, the properties of complexes. For instance, we have just seen how CFT accounts for the colors of complexes. We can explain their magnetic properties as well. Consider, for example, the cobalt(III) complexes of F^- and Cl^-. The metal ion here contains six d electrons and in a weak crystal field — one that produces a small Δ — they will be unpaired as much as possible, as shown in Figure 21.31a, to give a complex with four unpaired electrons. This is what occurs with F^- in the $[CoF_6]^{3-}$ ion.

The $[CoF_6]^{3-}$ ion is paramagnetic.

When the ligand produces a large crystal field splitting, we have the possibility of pairing all the d electrons in the t_{2g} level (Figure 21.31b) to produce a diamagnetic complex. This will occur if the magnitude of Δ is greater than the energy needed to pair the electrons in a given orbital. In other words, when the pairing energy (let's call it P) is less than Δ, more energy is required to place the electrons in the e_g orbital than to pair them and place them in the t_{2g} level. This happens when the ligand is Cl^-.

For Co(III) complexes (or for that matter, any d^6 system), a paramagnetic complex with four unpaired electrons will occur whenever $\Delta < P$; diamagnetic com-

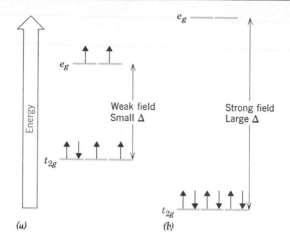

Figure 21.31. *The distribution of the electrons among the* t_{2g} *and* e_g *levels of cobalt(III) in (a) weak and (b) strong crystal fields.*

plexes will be formed when $\Delta > P$. In general, we speak of these two possibilities as **high-spin** complexes (minimum pairing of electrons) and **low-spin** complexes (maximum pairing of electrons from the e_g into the t_{2g}).

The possibility of both low-spin and high-spin complexes exists when the central metal ion contains four, five, six, or seven d electrons. The electron configurations of the t_{2g} and e_g levels in these species are left to you as a problem (Exercise 21.92). For d^1, d^2, and d^3 systems the electrons will naturally prefer the three low-energy t_{2g} orbitals because no pairing is required; therefore, only one type of electron configuration will be found for them. Likewise, with a d^8 or d^9 configuration, six electrons will be forced to occupy the t_{2g} (thereby filling it) and the e_g level will contain either two or three electrons, respectively. Once again we see that d^8 and d^9 ions will each have only one type of electron configuration in an octahedral complex.

Sometimes crystal field theory can also explain the relative stabilities of the oxidation states of metals. For example, we noted earlier that chromium(II) ion is a very good reducing agent in aqueous solutions. This means that it is easily oxidized to give chromium(III) ion. The reason for this is easily understood if we examine the populations of the t_{2g} and e_g levels in the complexes $Cr(H_2O)_6^{2+}$ and $Cr(H_2O)_6^{3+}$, which is the form in which we would expect to find these ions in aqueous solution. These are illustrated in Figure 21.32.

First, notice that the magnitude of Δ is larger for the chromium(III) species. This is always the case; the larger the charge on a metal ion, the larger the value of Δ for a given ligand. The reason is that the more highly charged ion is smaller and the ligands can approach the d orbitals more closely. This causes greater repulsions along the x, y, and z axes, which leads to a larger crystal field splitting.

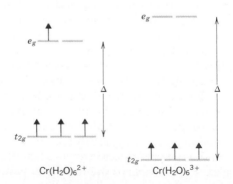

Figure 21.32. *The distributions of electrons among the* d *orbitals of chromium in the* $Cr(H_2O)_6^{2+}$ *and* $Cr(H_2O)_6^{3+}$ *ions. Oxidation of the chromium(II) ion removes a high-energy electron and lowers the energies of the electrons that remain.*

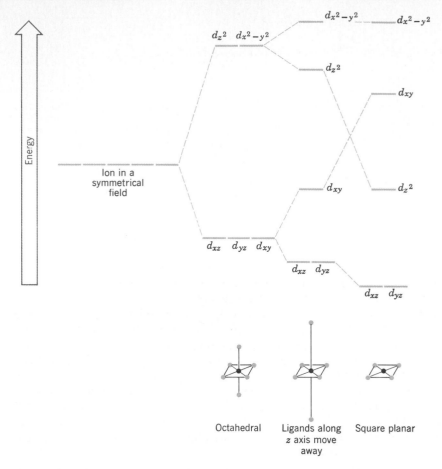

Figure 21.33. *The splitting pattern of the d orbitals changes as the geometry of the complex changes.*

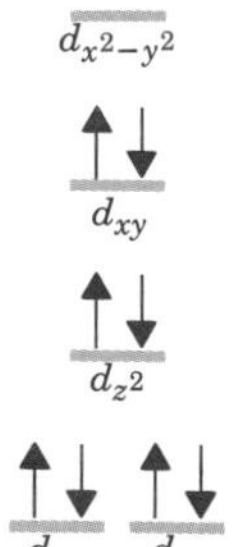

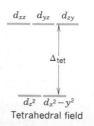

Figure 21.34. *Distribution of the electrons among the d orbitals of nickel in the diamagnetic $[Ni(CN)_4]^{2-}$ ion.*

Second, notice that the electron that is removed from the chromium(II) ion is in the higher-energy e_g level. Oxidizing the chromium(II) ion therefore leads to the loss of a high-energy electron from the e_g level and a lowering of the energies of the remaining electrons in the chromium(III) ion. Both of these factors favor oxidation of $Cr(H_2O)_6{}^{2+}$ and help make this ion a very good reducing agent.

Crystal field theory can be extended to other geometries besides octahedral; the difference is that other splitting patterns are observed. For example, a square planar complex can be thought of as being derived from an octahedral complex by removing the ligands that lie along the z axis. As shown in Figure 21.33, when this occurs, the energies of the d_{z^2}, d_{xz}, and d_{yz} orbitals decrease because an electron placed into them experiences less repulsion than in an octahedral complex. Also, by removing the ligands along the z axis, those along the x and y axis can move in slightly and therefore the energies of the $d_{x^2-y^2}$ and d_{xy} orbitals rise somewhat. In the diamagnetic $[Ni(CN)_4]^{2-}$ ion, the energy separation between the d_{xy} and $d_{x^2-y^2}$ is large enough so that the eight d electrons of the Ni^{2+} ion can exist as four pairs (Figure 21.34).

Finally, in tetrahedral complexes the splitting pattern of the d orbitals is that shown in Figure 21.35. Notice that the order of the energy levels is exactly opposite that found in octahedral complexes. The magnitude of Δ is also considerably smaller (actually, $\Delta_{tet} \approx \frac{4}{9}\Delta_{oct}$ for the same ligands and metal ion). The small Δ observed for tetrahedral complexes is always less than the pairing energy, and tetrahedral complexes are always of the high-spin variety.

Figure 21.35. *Splitting pattern for a tetrahedral field, $\Delta_{tet} \approx \frac{4}{9}\Delta_{oct}$.*

REVIEW QUESTIONS AND PROBLEMS

Problems whose numbers are in blue have their answers in Appendix D at the back of the book. The more difficult questions are marked by asterisks.

General Properties of Transition Metals

21.1 In general, what distinguishes a transition element from a representative element?

21.2 Which elements are called *inner transition elements?*

21.3 What similarities exist between elements in the A and B groups?

21.4 Why do the Group VIII elements consist of *three* columns?

21.5 Which elements are in the iron triad?

21.6 Make a list of the d-block elements you are familiar with and give as many applications of each as you can.

21.7 What are the four general properties we normally associate with the transition metals?

21.8 What oxoanion of sulfur is analogous to CrO_4^{2-}?

21.9 What manganese compound is analogous to $KClO_4$?

Electron Configurations and Oxidation States

21.10 Write the electron configurations of the members of the first transition series.

21.11 Why do many transition metals exhibit a $+2$ oxidation state in their compounds?

21.12 Why do many of the transition metals exhibit multiple oxidation states?

21.13 How do the relative stabilities of high and low oxidation states vary within the d-block elements?

21.14 Which is probably the better oxidizing agent, CrO_4^{2-} or WO_4^{2-}?

21.15 Which would you expect to be the stronger oxidizing agent, Cr^{3+} or Ni^{3+}?

21.16 Which would you expect to be more easily oxidized, Cr^{2+} or Fe^{2+}?

21.17 Which would you expect to be a better oxidizing agent, Cu^{3+} or Au^{3+}?

Atomic and Ionic Radii

21.18 What is the lanthanide contraction? Why does it occur?

21.19 Why are the chemical properties of the lanthanide elements so similar?

21.20 In crossing a transition series from left to right, why do the atomic radii pass through a minimum?

21.21 The chemistries of Zr and Hf are very similar; the elements are always found together in nature and are difficult to separate from one another. What explanation can be offered to account for the similar properties of Zr and Hf compounds?

21.22 The density of molybdenum (Mo) is 10.2. Calculate the density of tungsten and compare your value to the experimental value of 19.3 g/cm³. The atomic radii of both Mo and W are 139 pm.

21.23 In general, there are only small changes in atomic size crossing a row of transition elements. Among the representative elements, the size changes across a period are much larger. Why?

Magnetism

21.24 Which transition elements are ferromagnetic in the pure state?

21.25 Compare paramagnetism and ferromagnetism. Why can a ferromagnetic material become permanently magnetized?

21.26 What happens to the ferromagnetism of iron when the metal is melted? Why?

Properties of Transition Metals and Their Compounds

21.27 Which d-block elements react with water with the evolution of hydrogen? Give a chemical equation for the reaction.

21.28 CrO_3 is the anhydride of $CrO_2(OH)_2$, which was written as H_2CrO_4 in the text because it is acidic. On the other hand, Cr_2O_3 is the anhydride (at least in a formal sense) of $Cr(OH)_3$, which exhibits both basic and acidic properties. On the basis of what you learned in Chapter 10 about the factors that affect the acidity of X—O—H bonds, explain why $CrO_2(OH)_2$ is more acidic than $Cr(OH)_3$.

21.29 On the basis of your answer to Question 21.28, explain why the oxides of metals in high oxidation states tend to be acidic anhydrides instead of basic anhydrides.

21.30 Why is TiO_2 a better paint pigment than white lead? Why is titanium metal useful in the manufacture of aircraft?

21.31 $TiCl_4$ is a low-boiling molecular substance. Why isn't $TiCl_4$ ionic?

21.32 Write the equation for the reaction of $TiCl_4$ with water.

21.33 Why is chromium used to coat other metals such as steel? What are the most importnt oxidation states of chromium?

21.34 What is stainless steel? How does it differ from ordinary steel?

21.35 Why is chromium(III) ion in aqueous solution acidic?

21.36 Write chemical equations showing the amphoteric behavior of chromium(III) hydroxide.

*21.37 When solutions containing chromate ion (CrO_4^{2-}) are acidified, dichromate ion ($Cr_2O_7^{2-}$) is produced. Use structural formulas to indicate how this polymerization occurs.

21.38 What are the principal oxidation states of manganese? Which one is most stable with respect to redox in aqueous solution?

21.39 Why is $KMnO_4$ a useful titrant for redox reactions in acidic solutions? Why is it not used for reactions in neutral or basic solutions?

21.40 Write an equation for the reaction of a strong acid, such as HCl, with manganese.

21.41 Write an equation for the disproportionation of manganate ion in an acidic solution.

21.42 Why does iron have so many practical uses?

21.43 What are the oxides of iron? Which one is magnetic?

21.44 What is the apparent mechanism for the rusting of iron in moist air?

21.45 What compound is formed when a solution that contains Fe^{3+} is added to a solution that contains $Fe(CN)_6^{4-}$?

21.46 Write a chemical equation for the reaction of hydrochloric acid with iron.

21.47 Why is cobalt an important metal? What are the principal oxidation states of cobalt?

21.48 Why is nickel such a useful metal?

21.49 Why are aqueous solutions of many nickel salts green? What practical application is there for the compound, NiO_2?

21.50 How does the ease of oxidation of the coinage metals vary? How is this related to how they are found in nature?

21.51 Give a practical application for each of the following metals: copper, silver, and gold.

21.52 What oxidation states are observed for each of the coinage metals?

21.53 Why can't hydrochloric acid be used to dissolve the coinage metals? Which of them react with nitric acid? Give chemical equations.

21.54 Write a chemical equation for the reaction of gold with *aqua regia*.

21.55 Which compounds of silver are used in photography?

21.56 Describe how a solution can be tested for the presence of Ag^+. Give all the important chemical equations.

21.57 What happens when concentrated aqueous ammonia is added to a solution containing copper(II) ion?

21.58 What is a superconductor? Why are the new superconducting ceramics generating so much interest among scientists?

21.59 What simple test can you perform to determine whether a substance is in a superconducting condition?

21.60 What reactions occur, if any, when dilute sulfuric acid is added to (a) zinc, (b) cadmium, (c) mercury?

21.61 What is *cathodic protection*?

21.62 What is galvanizing? How does it protect steel?

21.63 Why is cadmium sometimes used in place of zinc as a protective coating over steel? Why is cadmium not used more often as a protective coating over other metals?

21.64 What are two common alloys that contain zinc?

21.65 What are some common uses for zinc oxide?

21.66 Aqueous solutions of $HgCl_2$ are poor conductors of electricity. Why? What equilibrium accounts for the small degree of conductivity that is observed?

21.67 Describe how a solution can be tested for the presence of Hg_2^{2+}.

21.68 Salts of which metals are added to glass to give it (a) a blue color and (b) a green color?

21.69 Saturated solutions of mercurous chloride were prepared by adding the solid to solutions having various chloride ion concentrations. The total concentration of mercury in each solution was then determined. Use the concepts that you learned having to do with equilibrium and solubility product to show that the data below are only consistent with mercury(I) having the formula Hg_2^{2+} [and hence mercury(I) chloride being Hg_2Cl_2], and not with Hg^+ (and therefore HgCl).

Chloride Ion Concentration	Moles of Mercury per Liter
1.0 M	2.2×10^{-18}
0.5 M	8.8×10^{-18}
0.2 M	5.5×10^{-17}
0.1 M	2.2×10^{-16}

21.70 What are the colors of the following ions?
(a) $Ni(H_2O)_6^{2+}$ (d) $Cu(H_2O)_4^{2+}$
(b) $Ni(NH_3)_6^{2+}$ (e) $Cu(NH_3)_4^{2+}$
(c) $Co(H_2O)_6^{2+}$ (f) $Mn(H_2O)_6^{2+}$

*21.71 On the basis of what you have learned in this chapter, suggest a series of reactions that would allow you to detect the presence of small amounts of cyanide in an aqueous solution.

Coordination Compounds

21.72 Define ligand, coordination compound, monodentate ligand, polydentate ligand, chelate, and coordination number.

21.73 Sketch the structures of the following ligands and circle the atoms that each ligand uses when it becomes attached to a metal ion:
(a) oxalate ion
(b) ethylenediamine
(c) ethylenediaminetetraacetate ion

21.74 What are the full names of these ligands: (a) en and (b) EDTA?

21.75 What structures are commonly found for coordination numbers 2, 4, and 6? Sketch them.

Structure and Isomerism

21.76 Sketch the structure of an octahedral EDTA complex.

21.77 Nitrilotriacetic acid, NTA (structure shown below), was used by detergent manufacturers for a while in place of phosphates because it is biodegradable and does not promote the growth of algae. However, it was found to increase the solubility of some heavy metals that are poisonous to a variety of life forms, and its use has been discontinued. Can you suggest, using appropriate chemical equations and structural formulas, how NTA dissolves metal compounds?

$$CH_2-\overset{\overset{\displaystyle O}{\|}}{C}-O^{(-)}$$

$$:N-CH_2-\overset{\overset{\displaystyle O}{\|}}{C}-O^{(-)}$$

$$CH_2-\overset{\overset{\displaystyle O}{\|}}{C}-O^{(-)}$$

nitrilotriacetate ion

21.78 What is meant by *isomer*? What are stereoisomers?

21.79 Sketch the isomers of $[Co(NH_3)_2Cl_4]^-$. Identify cis and trans isomers. How many isomers are there for the complex, $[Co(NH_3)_3Cl_3]$? Sketch them.

21.80 What condition must be met for a molecule or ion to be *chiral*?

21.81 Why are chiral substances said to be optically active?

21.82 Sketch the isomers of $[Cr(en)_2Cl_2]^+$. Identify cis and trans isomers and indicate any isomers that exhibit optical isomerism.

21.83 Draw the two optical isomers of $[Co(EDTA)]^-$.

21.84 What are enantiomers? What is meant by *racemic*?

Bonding: Crystal Field Theory

21.85 Predict the number of unpaired electrons in (a) $[Cr(H_2O)_6]^{2+}$ and (b) $[Cr(CN)_6]^{4-}$.

21.86 Sketch on appropriate coordinate axes the shapes of the five *d* orbitals.

21.87 Diagram the crystal field splitting of the *d* orbitals and indicate the electron population of each energy level in the paramagnetic complex $[Mn(H_2O)_6]^{3+}$. Label the energy levels.

21.88 How does crystal field theory account for the colors of complex ions?

21.89 What relationship exists between Δ (the crystal field splitting) and the pairing energy in determining whether a given complex will be paramagnetic or diamagnetic?

21.90 What are meant by high-spin and low-spin complexes? How do these compare with inner orbital and outer orbital complexes in the valence bond theory?

21.91 Sketch the CFT splitting patterns of the *d* orbitals for (a) square planar complexes and (b) tetrahedral complexes.

21.92 Using the CFT splitting pattern for an octahedral complex, indicate the high-spin and low-spin distribution of electrons among the t_{2g} and e_g levels for the configurations: (a) d^4, (b) d^5, (c) d^6, (d) d^7.

21.93 Use crystal field theory to predict the number of unpaired electrons in each of the following:
(a) $[VCl_6]^{3-}$ (d) $[Co(CN)_6]^{3-}$
(b) $[Ni(NH_3)_6]^{2+}$ (e) $[CrCl_6]^{3-}$
(c) $[Fe(CN)_6]^{3-}$

21.94 The $[Co(NO_2)_6]^{4-}$ ion has a single unpaired electron. Sketch the CFT energy-level diagram for this ion and indicate the electron populations in the t_{2g} and e_g levels.

21.95 On the basis of your answer to Question 21.94, should the $[Co(NO_2)_6]^{4-}$ ion be easy or difficult to oxidize to $[Co(NO_2)_6]^{3-}$? Justify your answer.

21.96 Which ion in Question 21.95 would absorb visible light of the shorter wavelength? Explain your answer.

22

HYDROGEN, OXYGEN, NITROGEN, AND CARBON, AND AN INTRODUCTION TO ORGANIC CHEMISTRY

In Chapters 20 and 21 we discussed the chemical and physical properties of metals. We now turn our attention, in this chapter and the next, to the remainder of the representative elements: the nonmetals and metalloids. In many ways, the chemistry of these nonmetallic elements is more interesting than that of the metals because of the large variety of compounds that they form. Not only do they combine with metals to form substances that are often predominantly ionic, but they also combine with each other by way of covalent bonds to form molecules and polyatomic ions that range from simple clusters of atoms to gigantic molecules, such as the DNA that controls heredity and guides the chemical functions of our cells.

In this chapter we begin by studying the properties of hydrogen, oxygen, nitrogen, and carbon (with a brief introduction to organic chemistry). Although our discussions will not be biased toward biology, these four nonmetals are uniformly important to all life as we know it, and much of our interest in their chemical properties is tied to how they and their compounds influence our lives and affect the quality of our environment.

22.1 HYDROGEN

Hydrogen, first recognized as an element by the English chemist, Henry Cavendish (1731–1810), is the most abundant of all the elements in the universe—

approximately 93% if counted by atoms. It is the principal element in the solar atmosphere, and it is the nuclear fuel that stars consume in their generation of energy. Here on Earth, hydrogen is much less abundant—about 3% by atoms, or about 0.14% by mass—presumably because the Earth's gravity was insufficient to hold onto most of the hydrogen that was present when the planet was formed.

How stars produce energy is discussed further in Chapter 24.

Virtually all the Earth's hydrogen exists in the combined state. Its reactivity, especially toward oxygen, is too great to allow it to be present as the free element in the atmosphere except in trace amounts. Water, of course, is two-thirds hydrogen on an atom basis (about 11% by weight) and the oceans represent a huge storehouse of this element. Hydrogen is also a principal element in all organic material. This includes living things, both animal and vegetable, as well as their fossils—petroleum and natural gas.

In its elemental state hydrogen exists as diatomic molecules of H_2. Its boiling point is −253 °C and its freezing point (melting point) is −259 °C. At room temperature, of course, hydrogen is a gas, and its very small molecular mass makes it the least dense gas of all (it is only half as dense as helium). Hydrogen therefore has great lifting power in balloons, but its extreme flammability poses a threat of disaster, as shown in Figure 22.1.

The official name for H_2 is dihydrogen.

Only helium has a lower boiling point than hydrogen.

$$H\ 1s^1$$

In the periodic table, hydrogen is placed in Group IA because its valence shell has the same electron configuration as the other members of this group. Here any similarities cease, however. Hydrogen has no electrons below its valence shell, and for this reason its chemistry does not resemble that of the Group IA metals at all. In fact, hydrogen really does not fit well into any group within the periodic table.

Isotopes of hydrogen

Many elements occur in nature as mixtures of isotopes, but hydrogen is the only one whose isotopes have their own names. The nucleus of ordinary hydrogen, ^1_1H, consists of a single proton. It is the most abundant of hydrogen's three isotopes and on rare occasions is called **protium.**

Figure 22.1. *The fierce reaction of hydrogen with oxygen (from the air) led to the fiery destruction of the* Hindenburg, *a German airship filled with hydrogen, in Lakehurst, New Jersey, in 1937. Thirty-six people died in the disaster.*

Deuterium is called heavy hydrogen.

Only 1 of every 10 million atoms of naturally occurring hydrogen is ^3_1H.

A *mass spectrometer* is an instrument that is able to determine very precisely the masses of ions formed from molecules when they are bombarded by electrons. It is a valuable analytic tool in chemistry and biochemistry.

Similar reactions can occur with other hydrocarbons.

Combustion of water gas:

$$2CO + O_2 \longrightarrow 2CO_2$$
$$2H_2 + O_2 \longrightarrow 2H_2O$$

Deuterium, ^2_1H, has a nucleus consisting of one proton and one neutron and is often given the symbol D. For example, the formula for deuterium oxide, also known as **heavy water,** is usually written D_2O. Only about 1 of every 5000 atoms of naturally occurring hydrogen is deuterium.

The third isotope of hydrogen, ^3_1H, is called **tritium** (symbol T). It is radioactive, and because of its relatively short half-life of 12.3 years, it is found in only very minute amounts in naturally occurring hydrogen. However, it can be made in nuclear reactions — for example, by bombarding lithium or boron with neutrons. In fact, tritium is a by-product of the operation of nuclear power plants, where it is produced by a variety of nuclear reactions including those involving lithium and boron additives in the reactor cooling system.

Chemically, the isotopes of hydrogen are identical except for small differences in the rates at which they react. This is both a blessing and a curse. One beneficial use of hydrogen's isotopes is in the study of reaction mechanisms. For example, a compound can be labeled by replacing one or more of the hydrogens in its structure with atoms of deuterium. After this compound is allowed to react, the locations of the labels in the product molecules can be determined using a mass spectrometer, and the information gathered in this way can help a chemist deduce the mechanism of the reaction.

A danger also exists in the chemical similarities of isotopes. The tritium isotope can easily be incorporated into biological molecules because it behaves chemically just like ordinary hydrogen, and the radiation that it would give off within an organism could cause many problems, including cancer and other radiation-related maladies.

Preparation and uses

Hydrogen is an important industrial chemical. Its pricipal source is natural gas, from which it is extracted by reactions with steam at high temperature in the presence of a catalyst.

$$CH_4(g) + H_2O(g) \xrightarrow[\text{catalyst}]{\text{heat}} CO(g) + 3H_2(g)$$
methane
(natural gas)

$$CO(g) + H_2O(g) \xrightarrow[\text{catalyst}]{\text{heat}} CO_2(g) + H_2(g)$$

The hydrogen and carbon dioxide can be easily separated from each other by bubbling the gas mixture through water, in which CO_2 is fairly soluble and H_2 virtually insoluble.

Hydrogen can also be extracted from water by allowing steam to react with carbon (from coal, for instance) at temperatures of about 1000 °C.

$$C(s) + H_2O(g) \xrightarrow{1000\ °C} CO(g) + H_2(g)$$

The mixture of CO and H_2 is called *water gas* and the reaction is referred to as the *water gas reaction.* It is used by industry because it changes a solid fuel (coal), which is awkward to handle in large quantities, into a gaseous combustible mixture that is easily piped to where it is needed.

Another way of obtaining hydrogen from water is by electrolysis. In Chapter 18 we saw that the electrolysis of brine is used to produce huge quantities of caustic soda (sodium hydroxide). A second product of this electrolysis is hydrogen, which also becomes part of the industrial supply of this element. The overall cell reaction is

$$2NaCl(aq) + 2H_2O \longrightarrow 2NaOH(aq) + Cl_2(g) + H_2(g)$$

If hydrogen is needed in small quantities in the laboratory, it can be conveniently made by reacting an active metal with an acid. Any metal that is below hydrogen in Table 18.1 is a better reducing agent than hydrogen and, in principle, should cause H^+ to be reduced. However, those metals having reduction potentials near zero (lead and tin, for example) react with acids very slowly. At the other extreme, those metals near the bottom of Table 18.1 are such active reducing agents that they react very vigorously with water and almost explosively with an acid. Metals having intermediate reduction potentials react smoothly with acids and serve as practical sources of hydrogen. A metal frequently used for this purpose is zinc. With dilute sulfuric acid it reacts as follows:

$$H_2SO_4(aq) + Zn(s) \longrightarrow ZnSO_4(aq) + H_2(g)$$

Figure 22.2 shows the type of apparatus that can be used to prepare H_2 in the laboratory. The gas is collected by allowing it to displace water from the inverted bottle, which has its mouth below the surface of the water in the tray. This method of collection works because H_2 has a very low solubility in water.

Zinc metal readily dissolves in sulfuric acid with the generation of hydrogen gas.

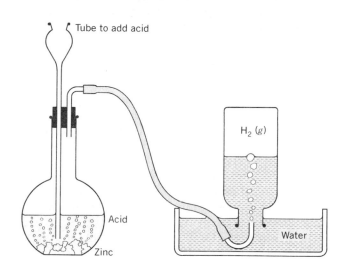

Figure 22.2. *The laboratory preparation of hydrogen by the reaction of a metal such as zinc with an acid such as H_2SO_4.*

The greatest single use of hydrogen is in the production of ammonia from nitrogen.

$$3H_2(g) + N_2(g) \longrightarrow 2NH_3(g)$$

This reaction, which consumes approximately two-thirds of the annual world production of hydrogen, is discussed further in Section 22.3.

Hydrogen is also used in large quantities in the manufacture of methanol (also called methyl alcohol or wood alcohol[1]). The reaction combines carbon monoxide and hydrogen at high pressure and temperature over a catalyst.

$$CO + 2H_2 \xrightarrow[\substack{200-300 \text{ atm} \\ \text{catalyst}}]{300-400 \text{ °C}} CH_3OH$$

This reaction is important because it provides a simple route by which coal can be converted to a liquid fuel. First the coal can be converted to a mixture of CO and H_2 by the water gas reaction. Then CO and H_2 can be combined to produce methanol, which is itself a potentially useful fuel. What makes matters even more interesting is

By the use of appropriate catalysts, CO and H_2 can be made to form other alcohols and hydrocarbons. Industrially, mixtures of CO and H_2 are also known as **synthesis gas.**

[1] At one time methanol was obtained as one of the products formed by heating wood in the absence of air, a process called *destructive distillation.* Although methanol is still called wood alcohol in some places, very little of it is currently produced by this older method.

A methanol-to-gasoline conversion plant is in commercial operation in New Zealand.

a catalytic process discovered by Mobil Oil Corporation that is able to change CH_3OH into high-octane gasoline.[2] At the present time, however, in the United States gasoline produced from methanol is more expensive than that obtained directly from petroleum.

Still another commercial use of hydrogen is the **hydrogenation** of vegetable oils, in which hydrogen is added chemically to carbon–carbon double bonds. For example, ethylene can be converted to ethane by hydrogenation.

Because organic molecules having double bonds possess the abiity to take on additional hydrogen, they are said to be *unsaturated*. On the other hand, molecules with only carbon–carbon single bonds cannot react with more hydrogen, so they are termed *saturated*. Hydrogenation therefore converts unsaturated vegetable oils into saturated fats.

Compounds of hydrogen

Hydrogen is found in more compounds than any other element. Virtually all organic compounds contain hydrogen: They are either hydrocarbons or are formed from hydrocarbons by replacing some of their hydrogen atoms with other elements. Hydrogen is found in both binary acids and oxoacids, and hydrogen even combines directly with active metals.

In some hydrides of the transition metals, hydrogen atoms or molecules become lodged in spaces (interstices) between metal atoms. These are called **interstitial hydrides** and are more like solutions of hydrogen in a metal.

Binary compounds with hydrogen are called **hydrides,** and most fall into two principal categories: ionic or saltlike hydrides and covalent hydrides. The ionic hydrides are formed from hydrogen and an active metal. Because of hydrogen's rather high electronegativity, it combines directly with the active metals in Groups IA and IIA by acquiring an electron and forming the **hydride ion,** H^-, thereby completing its valence shell.

$$2Li(l) + H_2(g) \xrightarrow{\text{heat}} 2LiH(s)$$

$$Ca(s) + H_2(g) \xrightarrow{\text{heat}} CaH_2(s)$$

These compounds are very sensitive toward moisture because of the basicity of the hydride ion. In water their overall reactions are as follows,

$$LiH(s) + H_2O \longrightarrow LiOH(aq) + H_2(g)$$

$$CaH_2(s) + 2H_2O \longrightarrow Ca(OH)_2(aq) + 2H_2(g)$$

but the reaction is really one between H^- and H_2O.

This can be viewed as either an acid–base reaction or a redox reaction.

$$H^- + H_2O \longrightarrow H_2 + OH^-$$

Hydrogen, of course, can also complete its $1s$ subshell by electron sharing. You no doubt recall drawing many Lewis structures showing hydrogen covalently bonded to another atom. Hydrogen most often binds to only one other atom by way of electron sharing, because it seeks only one additional electron. As a result, the structural chemistry of the simple hydrides is rather straightforward and, perhaps, even somewhat mundane. With the halogens it forms compounds with the general

---

[2] Direct conversion of coal to liquid hydrocarbon fuels by catalytic reactions of carbon in the form of coal dust with hydrogen has also been under study, and some success has been achieved.

Table 22.1. Catenation among nonmetal hydrides

Group IVA					
	CH_4	SiH_4	GeH_4	SnH_4	
	C_2H_6	Si_2H_6	Ge_2H_6	Sn_2H_6	
	C_3H_8	Si_3H_8	Ge_3H_8		
	$\vdots$	$\vdots$			
	C_nH_{2n+2}	Si_6H_{14}			
	+				
	many others				
Group VA	NH_3	PH_3	AsH_3	SbH_3	BiH_3
	N_2H_4	P_2H_4			
Group VIA	H_2O	H_2S	H_2Se	H_2Te	H_2Po
	H_2O_2	H_2S_2			
		H_2S_n ($n = 1 - 18$)			

formula HX (e.g., HF, HCl). Similarly, we find, quite expectedly, that the Group VIA elements form molecules of general formula H_2X (e.g., H_2O, H_2S), the elements of Group VA form H_3X (usually written XH_3, for example, NH_3, PH_3), and those in Group IVA form H_4X (or XH_4, for example, CH_4, SiH_4). The geometries of these molecules are all readily predicted by the VSEPR theory discussed in Section 9.2.

In addition to these simple hydrides, which contain a single atom of the non-metal, there are others that contain hydrogen and two or more atoms of a given nonmetal. Some examples are given in Table 22.1. All these compounds are charac-terized by nonmetal atoms of the same element linked directly to one another, a phenomenon called **catenation.** Thus, hydrogen peroxide, H_2O_2, has the Lewis structure

Similarly, there are others such as

hydrazine   disilane   propane

The ability of nonmetals to form compounds in which they bond to other like atoms varies greatly. You will notice, for example, that in Group VIA only oxygen and sulfur form such compounds. In Group VA we find that both nitrogen and phosphorus catenate, but the chain length seems to be limited to two atoms for the hydrides. When we proceed to Group IVA, all the elements, down to and including tin, exhibit this property and here we find chains containing three, four, and even more atoms. We also see that the tendency toward catenation generally decreases downward in a group, as evidenced by the trend toward shorter chains demon-strated by the heavier elements in Group IVA, Ge and Sn.

Of all the elements, carbon has the greatest capacity to form bonds simulta-neously to other elements and to itself. In fact, the broad area of organic chemistry is concerned entirely with hydrocarbons and compounds that are derived from them by substituting other elements for hydrogen. As discussed later in this chapter, organic compounds are compounds in which the molecular framework consists primarily of carbon–carbon chains. The unique ability of carbon to form such

We will see later that catenation is not restricted to nonmetal hydrides.

diverse compounds containing these long stable carbon chains is undoubtedly the reason why life has evolved around the element carbon instead of around another element such as silicon.

Preparation of nonmetal hydrides

Nonmetal hydrides are formed as products of many different chemical reactions; however, we will consider only two general methods of preparation here. One of these is the direct combination of the elements, as illustrated, for example, by the reaction of hydrogen with either chlorine,

$$H_2 + Cl_2 \longrightarrow 2HCl$$

or with oxygen,

$$2H_2 + O_2 \longrightarrow 2H_2O$$

However, this method is not applicable to all the hydrides, as we can see by examining some of their thermodynamic properties shown in Table 22.2. Here we see that only the hydrides of the more active nonmetals possess negative free energies of formation. Those lying below the heavy red line in the table have positive free energies of formation and from a practical standpoint cannot be prepared directly from the free elements. Instead an indirect procedure must be employed.

The rates of reaction toward hydrogen vary substantially among the nonmetals. In period 2, for instance, fluorine reacts immediately with hydrogen when they come in contact. On the other hand, H_2 and O_2 mixtures are stable virtually indefinitely, unless the reaction is initiated in some way—for example, by applying heat or introducing a catalyst.

Nitrogen is even less reactive than oxygen, not only toward hydrogen but toward nearly all other chemical reagents as well. Presumably this is because of the high stability of the N_2 molecule that arises as a consequence of its strong triple bond (the bond energy of N_2 is 946 kJ/mol, compared to 502 and 159 kJ/mol for O_2 and F_2, respectively).

The second method of preparation of nonmetal hydrides involves the addition

Table 22.2. Standard free energies and enthalpies of formation of nonmetal hydrides

	XH_n ΔG_f°(kJ/mol) ΔH_f°(kJ/mol)			
BH$_3$ Not stable, simplest hydride is B$_2$H$_6$	CH$_4$ -50.6 -74.9	NH$_3$ -17 -46.0	H$_2$O -228 -242	HF -273 -271
	SiH$_4$ $+52.3$ $+33$	PH$_3$ $+12.9$ $+5.4$	H$_2$S -33.6 -20.6	HCl -95.4 -92.5
	GeH$_4$ $+117$ $+90.4$	AsH$_3$ $+68.9$ $+66.4$	H$_2$Se $+62.3$ $+76$	HBr -53.1 -36
		SbH$_3$ $+148$ $+145$	H$_2$Te $+138$ $+154$	HI $+1.3$ $+26$

of protons, from a Brønsted acid, to the conjugate base of a nonmetal hydride, a reaction that we might depict as

$$X^{n-} + nHA \longrightarrow H_nX + nA^-$$

where X^{n-} is the conjugate base of the hydride H_nX, and HA is the Brønsted acid. Let's look at some examples.

The hydrogen halides are commonly prepared in the laboratory by treating a halide salt with a nonvolatile acid such as sulfuric or phosphoric acid.

$$NaCl(s) + H_2SO_4(l) \longrightarrow HCl(g) + NaHSO_4(s)$$

$$NaCl(s) + H_3PO_4(l) \xrightarrow{heat} HCl(g) + NaH_2PO_4(s)$$

In these examples HCl is removed as a gas, which causes the reaction to proceed to completion.

With the heavier halogens—bromine and iodine—sulfuric acid cannot be used because when concentrated, it is a sufficiently strong oxidizing agent to oxidize the halide ion to the free halogen. For example, when treated with concentrated H_2SO_4, I^- reacts as follows:

$$2I^- + HSO_4^- + 3H^+ \longrightarrow I_2 + SO_2 + 2H_2O$$

Phosphoric acid, being a much weaker oxidizing agent than H_2SO_4, simply supplies protons to I^-, and HI can therefore be produced in a reaction analogous to the production of HCl above, that is,

$$NaI(s) + H_3PO_4(l) \longrightarrow HI(g) + NaH_2PO_4(s)$$

The reaction is slow and the reaction mixture must be warmed to expel the HI.

As we proceed from right to left across a period—for example, from fluorine toward carbon—we have seen that the acid strengths of the H_nX compounds decrease. Thus, HF is a stronger acid than H_2O, which, in turn, is stronger than NH_3, and so forth. This means that the strengths of their corresponding conjugate bases *increase* from right to left ($C^{4-} > N^{3-} > O^{2-} > F^-$). As a result, the strength of the Brønsted acid required to react with the anion of the nonmetal to produce the hydride decreases. For example, the production of HF, whose conjugate base, F^-, is relatively weak, requires a strong acid such as H_2SO_4. Oxide ion, on the other hand, is a much stronger base than F^- and when treated with even a relatively weak source of protons—for example, acetic acid—the oxide ion gobbles them up to produce water.

$$O^{2-} + 2HC_2H_3O_2 \longrightarrow H_2O + 2C_2H_3O_2^-$$

This is a reaction we have seen before in Chapter 5.

Nitride ion, N^{3-}, is expected to be even a stronger base than O^{2-}. Therefore, it is not surprising to find that Mg_3N_2 reacts with an acid as weak as H_2O to produce NH_3 in a reaction that we can interpret as a hydrolysis of the N^{3-} ion.

$$Mg_3N_2 + 6H_2O \longrightarrow 3Mg(OH)_2 + 2NH_3$$

Metal carbides, which can be prepared by heating an active metal with carbon, also react with water in the same fashion. Aluminum carbide, for instance, which contains C^{4-} ions, hydrolyzes according to the reaction

$$Al_4C_3 + 12H_2O \longrightarrow 4Al(OH)_3 + 3CH_4$$

This general method of preparation also extends to the third, fourth, and fifth periods, with the same trends in the strength of the Brønsted acid required to liberate the hydride. In period 3 we have the following anions.

Group	IV	V	VI	VII
Anion	Si^{4-}	P^{3-}	S^{2-}	Cl^-

We again expect the anions to become increasingly basic as we move from right to left (from Cl^- to Si^{4-}); therefore, the strength of the Brønsted acid needed to protonate the anion decreases. To form HCl from NaCl, a strong acid is required. Sulfide ion, on the other hand, is sufficiently basic to be extensively hydrolyzed in aqueous solution, and solutions containing a soluble sulfide such as Na_2S always are very basic and have a strong odor of H_2S because of the reaction

$$S^{2-} + 2H_2O \longrightarrow H_2S + 2OH^-$$

The ability of S^{2-} to pick up protons also explains why many insoluble metal sulfides dissolve in acid with the evolution of H_2S.

Phosphides, like sulfides, also hydrolyze on contact with water. However, because the P^{3-} ion is more basic than the S^{2-} ion, the hydrolysis proceeds essentially to completion. Thus, aluminum phosphide, AlP, reacts with water to produce phosphine, PH_3.

PH_3 = phosphine

$$AlP + 3H_2O \longrightarrow Al(OH)_3 + PH_3$$

Moving left to Group IVA, we again find that a hydrolysis reaction serves to prepare silicon hydrides. A metal silicide such as Mg_2Si (which can be formed by simply heating Mg and Si together) reacts with water to generate a mixture of silanes; SiH_4, Si_3H_8, and so on, up to Si_6H_{14}.

SiH_4 = silane

The heavier nonmetals behave in much the same fashion as those above them. Thus, H_2Se and H_2Te, like H_2S, can be prepared by adding an acid to the metal selenide or telluride. Arsine, AsH_3, like phosphine, PH_3, is made by the hydrolysis of a metal arsenide such as Na_3As or AlAs, and the germanes, GeH_4, Ge_2H_6, and Ge_3H_8, are produced by the action of dilute HCl on Mg_2Ge.

AsH_3 = arsine

GeH_4 = germane

A hydrogen economy

As the world's supplies of petroleum dwindle and become increasingly more expensive, the search for alternative fuels will grow more urgent. Coal, and its conversion to synthetic petroleumlike products, has been mentioned as a means of providing substitutes, as we've noted earlier. Another interesting alternative is suggested by proponents of the large-scale use of hydrogen as a fuel, around which virtually our entire energy economy could be built.

Hydrogen has some very attractive features as a fuel. Its reaction with oxygen

$$2H_2(g) + O_2(g) \longrightarrow 2H_2O(l) \qquad \Delta H° = -572 \text{ kJ}$$

is highly exothermic and produces no pollutants. The gas could be either burned according to this reaction to produce heat or it could be used as a fuel in hydrogen–oxygen fuel cells to generate electricity directly. Since hydrogen is a gas, it could be pumped through the already existing network of pipelines that crisscross the nation carrying natural gas. Hydrogen is also available in virtually infinite amounts from water.

Despite all these advantages, however, there are problems. One of them is that hydrogen is not a very convenient portable fuel for use in an automobile. High-pressure tanks of gaseous hydrogen are heavy and can carry only relatively small amounts of this fuel. Other approaches are being tried, including combining hydrogen with certain metals to form metal hydrides that can be decomposed when the hydrogen is needed, but many technical difficulties still exist.

It makes no sense, of course, to burn petroleum to make electricity to produce hydrogen by electrolysis!

The main problem with hydrogen as a fuel is that it is not a primary energy source like petroleum and natural gas, which already exist in a state ready for use. Hydrogen can only be obtained by first investing energy to extract it from water, so hydrogen will become a viable fuel only if some inexpensive method can be found to produce it. A number of approaches to this problem have been proposed that make use of solar or nuclear energy.

Solar energy can be harnessed in several ways. One is to employ large arrays of solar cells to generate electricity that can be used in the electrolysis of water. Another is to focus solar energy by mirrors into solar furnaces that would heat water vapor to very high temperatures where it would decompose into H_2 and O_2. The mixture of gases would then be rapidly cooled before they could recombine to form H_2O. Nuclear reactors could also be used either to generate the electricity needed to decompose water by electrolysis, or to produce the high temperatures necessary to crack water into H_2 and O_2.

Advances have been made in harnessing solar energy directly to split water into H_2 and O_2. Using special catalysts, scientists in France have been able to achieve the efficient decomposition of water under visible and ultraviolet illumination. If this process can be made industrially practicable, a convenient means of converting solar energy directly to a useful form of stored chemical energy will be available.

> It has also been shown that solar energy can be harnessed to change a mixture of CH_4 and CO_2 into CO and H_2. Conversion back to CH_4 and CO_2 is exothermic, so the solar energy that was originally absorbed can be recovered as heat when it is needed.

22.2 OXYGEN

Very few people are unaware of the importance of oxygen to our existence. Breathing oxygen keeps us alive, and fuels couldn't be burned without it. Oxygen is literally everywhere—not only as a free element in the atmosphere, but also in all living creatures as well as in most of the substances that surround them. Water, which covers much of the Earth, is one-third oxygen on an atom basis, but about 89% oxygen by mass. In the Earth's crust, which is composed mostly of silicate minerals, oxygen is the *most* abundant element—46.6% by mass, 62.6% if counted by atoms, and an amazing 93.8% if measured by volume!

> The atmosphere is 20.9% oxygen by volume.

In its ground state, the element oxygen has the electron configuration $1s^2 2s^2 2p^4$ and therefore needs only two electrons to complete its octet. As you know, it is able to accomplish this either by acquiring electrons or by electron sharing. In the free state, oxygen exists as diatomic molecules, O_2. These molecules each contain two unpaired electrons, so O_2 is paramagnetic. When cooled, oxygen forms a pale blue paramagnetic liquid that boils at $-183\ ^\circ C$ (see Figure 22.3). Oxygen freezes at $-219\ ^\circ C$, giving a pale blue solid.

> The official IUPAC name for O_2 is dioxygen, but it is rarely used.

(a)

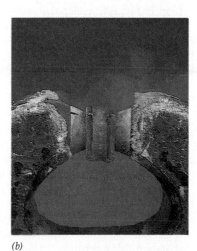
(b)

Figure 22.3. (a) *Liquid oxygen is a pale blue liquid that boils at* $-183\ ^\circ C$. *(b) Two unpaired electrons in each* O_2 *molecule cause molecular oxygen to be paramagnetic. Here we see liquid oxygen held between the poles of a powerful magnet by magnetic forces.*

Preparation and uses

The most obvious source of oxygen when it is needed in large quantities is the atmosphere, and the commercial preparation of oxygen involves separating it from

Figure 22.4. *The laboratory preparation of oxygen by the thermal decomposition of potassium chloride, KClO₃, using manganese dioxide as a catalyst. Care must be taken not to heat the rubber stopper or allow the hot KClO₃ to come into contact with it.*

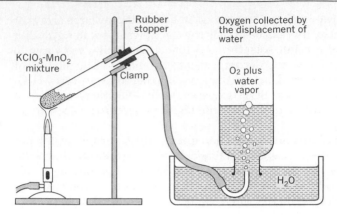

Liquid oxygen has been used as an oxidizer in rocket engines. Along with liquid hydrogen, it is used to power the space shuttle.

Because it is so freely available, O_2 is normally produced at the plant site where it will be used.

liquefied air. Nitrogen, which has a lower boiling point than O_2, is removed from the liquid air by allowing it to boil off, thereby leaving behind liquid oxygen contaminated with small amounts of N_2 and argon (another component of air). Warming the liquid, of course, converts the oxygen to a gas. Most of the oxygen prepared this way (about 85%) is used in the steel industry and metal fabrication. Some is also used to make chemical intermediates in the manufacture of plastics, in waste water treatment, for life support in hospitals, and in bleaching pulp and paper.

In the laboratory, small amounts of pure oxygen can be prepared in a number of ways. Joseph Priestley (1733–1804), credited with the discovery of oxygen, obtained the gas by thermally decomposing mercury(II) oxide. Usually, however, oxygen is made in the laboratory either by electrolysis of water or by thermally decomposing potassium chloride, $KClO_3$, using manganese dioxide as a catalyst.

$$2KClO_3(s) \xrightarrow[\text{heat}]{\text{MnO}_2} 2KCl(s) + 3O_2(g)$$

Like hydrogen, oxygen can be collected by the displacement of water because of its relatively low solubility (Figure 22.4).

In nature, oxygen is generated by green plants during photosynthesis. The chemical equation for the conversion of CO_2 and H_2O into glucose and O_2 is described in Section 22.4. Large forests and the plankton in the sea are responsible for maintaining the balance of oxygen in the Earth's atmosphere. As you may be aware, environmentalists are concerned that the clearing of the Amazon rain forests in South America, which has been under way in recent years, may severely upset this balance.

Ozone

When an electric discharge is passed through molecular oxygen, a second allotrope of the element is formed that is called **ozone,** O_3.

$$3O_2(g) \longrightarrow 2O_3(g) \qquad \Delta H = +284 \text{ kJ}$$

Its pungent odor can sometimes be detected after a severe thunderstorm or near electrical machinery. In low concentrations it is poisonous (a few parts per million in air can be dangerous). In Chapter 9 we saw that the structure of ozone can be represented by the resonance formulas

As we would expect from the number of groups of electrons around the central oxygen atom, the molecule is nonlinear.

When ozone is formed, a large amount of energy must be supplied, and an equally large amount of energy, of course, is released when the ozone decomposes. Therefore, since highly exothermic reactions tend to be quite spontaneous, ozone decomposes very easily. Ozone is also a very powerful oxidizing agent—considerably more powerful than ordinary O_2. This has both advantages and disadvantages. On the positive side, the oxidizing power of ozone shows promise as an alternative to chlorine in the treatment of municipal drinking water supplies. It has been found that Cl_2 in drinking water is able to form chlorine compounds with some of the organic substances that are also present in small amounts. Some of these products, such as chloroform, $CHCl_3$, have been shown to be carcinogenic, and the long-range toxic effects of others are open to question. All these problems are avoided if bacteria in the water are killed by O_3 instead of Cl_2. However, because little residual O_3 remains after treatment, any bacteria that get into the water after the treatment process are not destroyed.

Recall that carcinogenic means cancer-causing.

On the negative side, the powerful oxidizing ability of ozone can cause extensive damage to plants and articles made of natural rubber. As we will see in the next section, ozone is one of the major constituents of smog, so in locations that experience severe smog episodes, damage caused by ozone can be particularly troublesome.

In the Earth's upper atmosphere, at altitudes ranging from about 9 to 15 miles, ozone is formed in appreciable amounts from O_2 by absorption of ultraviolet radiation from the sun. The light energy first splits oxygen molecules into oxygen atoms.

The ozone concentration in the upper atmosphere approaches 27% by mass. However, because the air is so thin there, this is 27% of very little of anything.

$$O_2 \xrightarrow{hv} 2O$$

Reaction of oxygen atoms with oxygen molecules produces O_3.

$$O_2 + O \longrightarrow O_3$$

Ozone also absorbs ultraviolet light, especially at wavelengths that prove harmful to living organisms. This causes the O_3 to decompose and form O_2 again. The absorption of UV radiation by ozone converts the energy of the UV light into heat and also protects the inhabitants of our planet from the radiation's harmful effects.

You are probably aware of the controversy that has arisen in recent years concerning the effects of certain human activity on this ozone shield. For a time there was worry that high-flying supersonic airliners such as the Concorde would be emitting nitrogen oxide pollutants in quantities that would have a significant effect on the ozone concentration. Even small amounts of NO could have damaging effects because of such reactions as

$$NO + O_3 \longrightarrow O_2 + NO_2$$
$$NO_2 + O \longrightarrow O_2 + NO$$

Thus, molecules of NO remove not only O_3, but also oxygen atoms needed to reform the O_3. Since the product of the second reaction is the reactant in the first, the cycle can be repeated many times by each NO molecule. Despite these early concerns, it appears that the Cocorde, as presently operated, has not caused any noticeable change in the stratospheric ozone concentration.

The release into the atmosphere of compounds known as chlorofluorocarbons (CFC) poses another danger to the ozone shield. These compounds are commonly called Freons; an example is the Freon-12 that is used in the air-conditioning systems of automobiles. CFCs have been used in refrigerating systems of all kinds and as propellants in aerosol cans.

Under ordinary circumstances CFCs are unreactive chemicals. When they diffuse into the upper atmosphere, however, they can absorb ultraviolet light, which

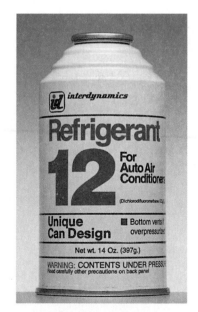

Freon-12, CCl_2F_2, is the refrigerant used in the air conditioners in automobiles. Containers of it can be purchased in any automotive store.

ruptures carbon–chlorine bonds to give chlorine atoms. For example,

$$CFCl_3 \xrightarrow{hv} CFCl_2 + Cl$$

Although both products are reactive, it is the chlorine atoms that have been implicated in the destruction of ozone. Among a number of possible reaction sequences, an important one seems to be the following:

$$
\begin{aligned}
2Cl + 2O_3 &\longrightarrow 2ClO + 2O_2 \\
2ClO + M &\longrightarrow Cl_2O_2 + M \\
Cl_2O_2 + hv &\longrightarrow Cl + ClOO \\
ClOO + M &\longrightarrow Cl + O_2 + M
\end{aligned}
$$

$$\text{Net:} \quad 2O_3 \longrightarrow 3O_2$$

In this reaction scheme, M stands for another molecule that is unchanged in the reaction. It can either promote the decomposition of a molecule or can absorb some of the energy of a collision so the product can stay intact.

Destruction of the ozone layer is known to be particularly severe over the polar caps during winter when ice particles in the atmosphere can catalyze the reactions. However, global losses of ozone have also been found. This has led 24 nations, under the auspices of the United Nations Environment Program, to sign a pact known as the Montreal protocol, which will ultimately cut the consumption of CFCs. Many scientists, however, feel that the agreement does not move swiftly enough.

The two Cl atoms consumed in the first step are regenerated in the last two steps. They can then go on to react again with ozone. In this way, trace amounts of Cl atoms can do a lot of damage to the ozone shield.

Destruction of stratospheric ozone could lead to increased incidents of skin cancer and a rise in the Earth's average temperature.

Compounds of oxygen

Oxygen forms compounds with every element except helium, neon, and argon. These compounds are usually called **oxides** and are generally of two types: ionic or covalent. *Ionic oxides* are formed with many metals and can usually be made by direct combination of the elements, for example,

$$4Li(s) + O_2(g) \longrightarrow 2Li_2O(s)$$

$$2Ca(s) + O_2(g) \longrightarrow 2CaO(s)$$

An exception, for example, is oxygen difluoride, OF_2.

Metal oxides are basic and, if soluble, give solutions containing hydroxide ion caused by the virtual complete hydrolysis of O^{2-}.

$$Li_2O(s) + H_2O(l) \longrightarrow 2LiOH(aq) \quad \text{(Molecular equation)}$$

$$O^{2-} + H_2O(l) \xrightarrow{100\%} 2OH^-(aq) \quad \text{(Net ionic equation)}$$

Remember, metal oxides are basic anhydrides.

Even insoluble metal oxides are basic because they neutralize acids. Iron(III) oxide, for example, dissolves in acids such as HCl or H_2SO_4.

$$Fe_2O_3(s) + 6H^+(aq) \longrightarrow 2Fe^{3+}(aq) + 3H_2O$$

This reaction, in fact, is often used to remove rust (Fe_2O_3) from iron or steel prior to being given a protective coating of zinc or tin. The acid treatment is called **pickling.**

Some oxides of metals show acidic properties as well as basic ones and are said to be amphoteric. Examples that we have discussed earlier are the oxides of beryllium and aluminum, which dissolve in both acids and bases. For instance,

$$Al_2O_3 + 6H^+ \longrightarrow 2Al^{3+} + 3H_2O$$

$$Al_2O_3 + 2OH^- \longrightarrow 2AlO_2^- + H_2O$$

Covalent oxides are generally formed by the nonmetals. In them, oxygen completes its octet by electron sharing, normally by forming either two single bonds (as in H_2O) or one double bond (as in CO_2). Exceptions are CO and NO, which contain triple bonds. Table 22.3 contains a list of many oxides of the nonmetallic elements.

Table 22.3. Typical oxides of the nonmetallic elements

Group III	B_2O_3			
Group IV	CO	SiO_2	GeO_2	
	CO_2			
Group V	N_2O	P_4O_6	As_4O_6	Sb_4O_6
	NO	P_4O_{10}	$As_2O_5{}^a$	$Sb_2O_5{}^a$
	N_2O_3			
	NO_2; (N_2O_4)			
	N_2O_5			
Group VI	O_2	SO_2	SeO_2	TeO_2
	O_3	SO_3	SeO_3	TeO_3
Group VII	OF_2	Cl_2O	Br_2O	I_2O_5
	O_2F_2	ClO_2	BrO_2	I_2O_7
		Cl_2O_7		

^{a} Molecular structure unknown.

There are several ways that oxides of the nonmetallic elements can be made. Except for the halogens and the noble gases, most oxides can be prepared simply by the direct union of the elements, as typified by the reactions

$$S + O_2 \longrightarrow SO_2$$

$$C + O_2 \longrightarrow CO_2 \qquad \text{(Excess oxygen)}$$

$$2C + O_2 \longrightarrow 2CO \qquad \text{(Limited supply of oxygen)}$$

$$2H_2 + O_2 \longrightarrow 2H_2O$$

Not all oxides can be prepared effectively in this manner, however. For example, in Table 22.4 we see that many oxides of nitrogen have positive free energies of formation and, therefore, from what we know of thermodynamics, they cannot be synthesized in high yields directly from the elements. In these instances, and in others too, indirect methods of preparation are employed.

The indirect procedures are, expectedly, many in number. However, a few generalizations can be made. In some cases an oxide can be prepared from a lower oxide by further reaction with oxygen. The synthesis of SO_3, for example, consists of the catalytic oxidation of SO_2,

By *lower oxide* we mean an oxide of the element in a lower oxidation state.

$$2SO_2 + O_2 \longrightarrow 2SO_3$$

In Chapter 19 it was mentioned that this reaction is promoted in catalytic mufflers originally designed to speed up the conversion of CO to CO_2.

$$2CO + O_2 \longrightarrow 2CO_2$$

Table 22.4. Thermodynamic properties of some nitrogen oxides

Oxide	ΔH_f° (kJ/mol)	ΔG_f° (kJ/mol)
$N_2O(g)$	$+81.5$	$+104$
$NO(g)$	90.4	86.8
$NO_2(g)$	34	51.9
$N_2O_4(g)$	9.2	97.9
$N_2O_5(g)$	11	115

The combustion of CO, you recall, is an important industrial reaction because CO is often used as a fuel.

Another technique that serves to produce oxides is the combustion of nonmetal hydrides. Methane, the chief constituent of natural gas, and other hydrocarbons burn to produce CO_2 and H_2O when an excess of O_2 is present.

$$CH_4 + 2O_2 \longrightarrow CO_2 + 2H_2O$$
$$\text{methane}$$

$$2C_8H_{18} + 25O_2 \longrightarrow 16CO_2 + 18H_2O$$
$$\text{octane}$$
$$\text{(in gasoline)}$$

When insufficient O_2 is available, as in an automobile engine, CO may be produced instead of CO_2.

A reaction of this general type that is of great commercial importance is the oxidation of ammonia. In this case a platinum catalyst is used and the reaction is

$$4NH_3 + 5O_2 \xrightarrow{\text{Pt}} 4NO + 6H_2O$$

The NO formed in this reaction is readily oxidized further to produce NO_2,

$$2NO + O_2 \longrightarrow 2NO_2$$

These reactions are essential to the production of nitric acid and nitrates, which are used in manufacturing many chemicals, from explosives to fertilizers. We will discuss the synthesis of HNO_3 in the next section.

Nonmetal oxides are also formed in oxidation–reduction reactions that do not involve molecular oxygen. For instance, when nitric acid serves as an oxidizing agent, the nitrate ion is reduced and, depending on conditions, nitrogen in various oxidation states may be produced. When concentrated nitric acid is used, the reduction product is frequently NO_2. Dilute solutions of HNO_3 often yield NO as the reduction product.

$$4HNO_3 + Cu \longrightarrow Cu(NO_3)_2 + 2NO_2 + 2H_2O \qquad \text{(Concentrated)}$$

$$8HNO_3 + 3Cu \longrightarrow 3Cu(NO_3)_2 + 2NO + 4H_2O \qquad \text{(Dilute)}$$

Similarly, hot concentrated sulfuric acid is a fairly potent oxidizing agent, and the reduction product is usually SO_2; for example,

$$Cu + 2H_2SO_4 \longrightarrow CuSO_4 + SO_2 + 2H_2O$$

In Chapter 5 we saw that nonmetal oxides are acidic anhydrides; when they react with water, they produce oxoacids. For example,

$$SO_2 + H_2O \longrightarrow H_2SO_3$$

Neutralization gives the corresponding oxoanion. Thus, sulfurous acid (H_2SO_3) gives sulfite ion (SO_3^{2-}). Table 22.5 provides a summary of the simple oxoacids and oxoanions of the nonmetallic elements. As indicated in the table, some of the oxoacids cannot actually be isolated, although their corresponding anions can be.

Other compounds of oxygen

Peroxides and Superoxides In addition to forming normal oxides with metals (i.e., compounds that contain the O^{2-} ion), oxygen combines with the more active alkali metals to form peroxides and superoxides, which contain the O_2^{2-} and O_2^- ions, respectively. Sodium forms the yellow peroxide.

$$2Na(s) + O_2(g) \longrightarrow Na_2O_2(s)$$
$$\text{sodium peroxide}$$

In the absence of a catalyst, combustion of ammonia in air gives N_2 and H_2O.

$$4NH_3(g) + 3O_2(g) \longrightarrow$$
$$2N_2(g) + 6H_2O(g)$$

Peroxides and superoxides of the alkali metals were discussed on page 673.

Table 22.5. Simple oxoacids and oxoanions of the nonmetallic elements

Group IIIA	H_3BO_3 (no simple borates)			
Group IVA	H_2CO_3 (CO_3^{2-})	$H_4SiO_4{}^a$ $(SiO_4{}^{4-})$	$H_4GeO_4{}^a$ $(GeO_4{}^{4-})$	
Group VA	HNO_2 $(NO_2{}^-)$	H_3PO_3 $(HPO_3{}^{2-})$		
	HNO_3 $(NO_3{}^-)$	H_3PO_4 $(PO_4{}^{3-})$	H_3AsO_4 $(AsO_4{}^{3-})$	
Group VIA		$H_2SO_3{}^a$ $(SO_3{}^{2-})$	H_2SeO_3 $(SeO_3{}^{2-})$	H_2TeO_3 $(TeO_3{}^{2-})$
		H_2SO_4 $(SO_4{}^{2-})$	H_2SeO_4 $(SeO_4{}^{2-})$	$Te(OH)_6$ $(TeO(OH)_5{}^-)$
Group VIIA	HOF	HOCl (OCl^-)	HOBr (OBr^-)	HOI (OI^-)
		$HClO_2$ $(ClO_2{}^-)$	$HBrO_2$ $(BrO_2{}^-)$	HIO_3 $(IO_3{}^-)$
		$HClO_3$ $(ClO_3{}^-)$	$HBrO_3$ $(BrO_3{}^-)$	H_5IO_6 $(H_2IO_6{}^{3-})$
		$HClO_4$ $(ClO_4{}^-)$	$HBrO_4$ $(BrO_4{}^-)$	HIO_4 $(IO_4{}^-)$

^{a} Not observed to exist.

Potassium, rubidium, and cesium form yellow to orange superoxides.

$$K(s) + O_2 \longrightarrow KO_2(s)$$
potassium
superoxide

Peroxides and superoxides are powerful oxidizing agents. In Figure 22.5, a drop of water is placed on a mixture of Na_2O_2 and sugar. The H_2O reacts with the $O_2{}^{2-}$ ion

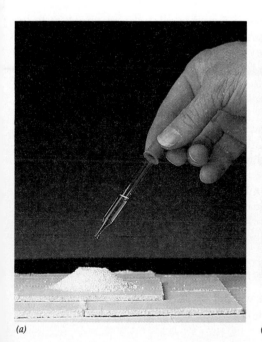

(a) (b)

Figure 22.5. (a) *A drop of water is about to be added to a mixture of sodium peroxide, Na_2O_2, and ordinary table sugar. (b) The hydrolysis of the peroxide ion is so exothermic that it causes the mixture to ignite. Sodium peroxide is a powerful oxidizing agent and causes the sugar to burn.*

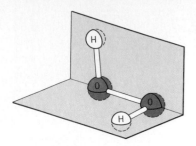

Figure 22.6. *The structure of the hydrogen peroxide molecule.*

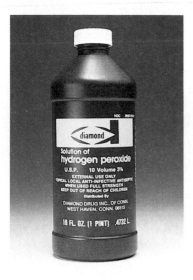

Hydrogen peroxide solutions such as this can be purchased in a pharmacy.

The nitrogen in organic matter represents a small fraction of the Earth's total.

to yield H_2O_2. This is a very exothermic reaction and causes the rest of the mixture to ignite. The sugar burns fiercely as it is oxidized by the peroxide.

Hydrogen Peroxide As just noted, when Na_2O_2 is placed in water, the peroxide ion hydrolyzes to form hydrogen peroxide, H_2O_2.

$$Na_2O_2(s) + 2H_2O(l) \xrightarrow{\text{cold}} 2NaOH(aq) + H_2O_2(aq)$$

Hydrogen peroxide is a molecular substance having an oxygen–oxygen bond.

$$\overset{\overset{\displaystyle H}{|}}{:\!\ddot{O}}-\overset{\overset{\displaystyle |}{\ddot{O}:}}{\underset{\displaystyle H}{|}}$$

The structure of the molecule is shown in Figure 22.6. In water, H_2O_2 is a weak acid.

$$H_2O_2 \rightleftharpoons H^+ + HO_2^- \qquad K_a = 1.5 \times 10^{-12}$$

Pure H_2O_2 is a colorless liquid with a boiling point of 150 °C, but in its pure state it is an extremely hazardous substance to work with. Its decomposition to water and oxygen can occur explosively.

$$2H_2O_2(l) \longrightarrow 2H_2O(l) + O_2(g)$$

This decomposition is promoted by heat or by traces of many heavy metal ions.

Hydrogen peroxide is usually purchased as a solution in water. Solutions of 3% H_2O_2 by weight can be purchased in drug stores for use as an antiseptic. Its oxidizing power destroys bacteria, while blood catalyzes its decomposition. The fizzing action caused by escaping O_2 dislodges dirt and other foreign matter from a wound. More concentrated solutions of H_2O_2 are effective oxidizing agents and are used as bleaches for hair and industrially in the bleaching of cotton fabrics.

22.3 NITROGEN

Nitrogen is another of the principal elements found in all living creatures. It is an essential constituent of amino acids, the primary building blocks of proteins, and is the key element in a precise lock-and-key fit between molecules that cells use to decipher the genetic code incorporated in the DNA residing in their nuclei. Because of the importance of nitrogen to the growth of plants and animals, it is not surprising that nitrogen compounds, particularly fertilizers, rank very high in total annual commercial production.

The element nitrogen has as its ground state electron configuration, $1s^2 2s^2 2p^3$. It achieves a completed octet either by gaining electrons to form ionic nitrides that contain the N^{3-} ion or by covalent bonding. In many of its compounds, nitrogen forms three bonds.

Very little nitrogen is present in a combined state in the Earth's crust. Instead, most occurs in the atmosphere as diatomic molecules of N_2 (officially called dinitrogen). Molecular nitrogen is a colorless, odorless, and quite unreactive gas which boils at -196 °C and freezes at -210 °C. Its lack of reactivity is attributed to the strength of the nitrogen–nitrogen triple bond,

$$:N\equiv N:$$

The bond energy of N_2 (946 kJ/mol) is very high, so the molecule is very stable. For example, at 25 °C (room temperature), N_2 reacts directly only with lithium to form

Li_3N, although the nitrogen-fixing bacteria in the root nodules of clover, beans, peas, and certain other plants are also able to convert N_2 into usable nitrogen compounds. The high bond energy of N_2 also causes many nitrogen compounds to have positive enthalpies and free energies of formation.

The large bond energy of N_2 acts as an activation energy barrier toward reaction.

Preparation and uses of elemental nitrogen

The Earth's atmosphere is composed of about 78% N_2 by volume and, as with oxygen, the commercial source of N_2 is from the liquefaction of air. As liquid air boils, the lower-boiling nitrogen escapes and is collected. Generally, commercial nitrogen from this source consists of about 99% N_2, with a small amount of argon and traces of oxygen.

In the laboratory, nitrogen can be made chemically by warming a solution that contains an ammonium salt (such as NH_4Cl) and a nitrite salt (such as $NaNO_2$). The net ionic reaction corresponds to the decomposition of NH_4NO_2.

$$NH_4^+(aq) + NO_2^-(aq) \xrightarrow{\text{warm}} N_2(g) + 2H_2O(l)$$

Solid NH_4NO_2 explodes if warmed and forms the same products.

Commercially, the leading use of nitrogen is in the manufacture of ammonia, which is described later in this section. Because of nitrogen's low reactivity, it is also used in large quantities as an inert gaseous blanket to exclude oxygen during the manufacture of chemicals, the fabrication of metals, and in the production of electronic devices. Large amounts of liquid nitrogen are employed in the food industry where its low temperature (-196 °C) is used to rapidly freeze foods.[3]

Gaseous N_2 is often used in chemical apparatus in the laboratory to provide an inert atmosphere.

Ionic compounds of nitrogen

Nitrides Elemental nitrogen combines directly with some of the very reactive metals on the left side of the periodic table. Earlier it was mentioned that lithium reacts with N_2 at room temperature to form an ionic nitride.

$$6Li(s) + N_2(g) \longrightarrow 2Li_3N$$

Similar nitrides are formed at higher temperatures by magnesium, calcium, strontium, and barium and have the general formula M_3N_2. When placed into water, the nitrides immediately hydrolyze, liberating ammonia.

$$Li_3N(s) + 3H_2O(l) \longrightarrow 3LiOH(aq) + NH_3(g)$$

$$Mg_3N_2(s) + 6H_2O(l) \longrightarrow 3Mg(OH)_2(s) + 2NH_3(g)$$

The N^{3-} ion is a very strong Brønsted–Lowry base.

Covalent compounds of nitrogen

Nitrogen forms covalent compounds with many nonmetals. The most important are those with hydrogen and oxygen, and among them we find nitrogen in every oxidation state from -3 to $+5$, as shown in Table 22.6. Actually, the oxidation numbers have no real physical significance for these substances—for example, the nitrogen in NH_3 certainly doesn't carry a charge of $3-$. Nevertheless, the oxidation numbers are useful in balancing redox equations and organizing the discussion of the chemistry of nitrogen.

[3] Businesses have been established that, for a fee, offer to freeze your body in liquid nitrogen after you've died and maintain it in a frozen state until (hopefully) a cure is found for whatever killed you. No one has yet been returned to life, however, and most scientists question whether the cell damage caused by rapid freezing can ever be reversed in humans.

Table 22.6. Oxidation states of nitrogen

Oxidation State	Example	Preparative Reaction
-3	NH_3 (ammonia)	$N_2 + 3H_2 \xrightarrow[\text{pressure}]{\text{heat}} \mathbf{2NH_3}$
-2	N_2H_4 (hydrazine)	$2NH_3 + NaOCl \longrightarrow \mathbf{N_2H_4} + NaCl + H_2O$
-1	NH_2OH (hydroxylamine)	$NaNO_2 + NaHSO_3 + SO_2 + 2H_2O \longrightarrow$ $2NaHSO_4 + \mathbf{NH_2OH}$
0	N_2 (dinitrogen)	$NH_4NO_2 \xrightarrow{\text{heat}} \mathbf{N_2} + 2H_2O$
$+1$	N_2O (nitrous oxide)	$NH_4NO_3 \xrightarrow{\text{heat}} \mathbf{N_2O} + 2H_2O$
$+2$	NO (nitric oxide)	$4NH_3 + 5O_2 \longrightarrow \mathbf{4NO} + 6H_2O$
$+3$	N_2O_3 (dinitrogen trioxide)	$NO + NO_2 \xrightarrow{-30\ °C} \mathbf{N_2O_3}$
$+4$	NO_2 (nitrogen dioxide) N_2O_4 (dinitrogen tetroxide)	$2NO + O_2 \longrightarrow \mathbf{2NO_2} \rightleftharpoons N_2O_4$
$+5$	HNO_3 (nitric acid)	$3NO_2 + H_2O \longrightarrow \mathbf{2HNO_3} + NO$

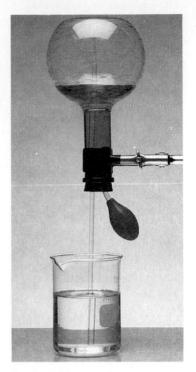

Figure 22.7. *An ammonia fountain. Water with a trace of phenolphthalein (an acid–base indicator that is pink in a basic solution and colorless in an acidic or neutral solution) is being sucked into the inverted flask. Initially the flask contained only gaseous ammonia. Then a little water was squirted into the flask by the squeeze bulb. Ammonia is so soluble in water that enough of it dissolved to create a partial vacuum in the flask. As the water–phenolphthalein mixture is drawn into the flask from the beaker, the indicator becomes pink because aqueous ammonia solutions are slightly basic.*

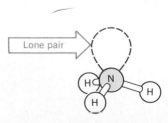

The pyramidal ammonia molecule.

Ammonia(−3 Oxidation State) Ammonia (NH_3) is by far the most important compound of nitrogen. It serves as a route to virtually all other nitrogen compounds and is itself a useful fertilizer. Ammonia is prepared industrially by the **Haber process,** which combines hydrogen and nitrogen on a catalytic iron surface.

$$N_2(g) + 3H_2(g) \xrightarrow[\text{catalyst}]{\text{iron}} 2NH_3(g) \qquad \Delta H° = -92 \text{ kJ}$$

The reaction is run at pressures of several hundred atmospheres, which favors the production of NH_3. It is also run at temperatures between 400 and 500 °C in order to cause the reaction to proceed at a reasonable rate, even though in principle the position of equilibrium is less favorable than at lower temperatures.

Ammonia is a colorless gas with a powerful irritating odor that most people would recognize immediately. It has a boiling point of −33.35 °C and freezes at −77.7 °C. As a liquid, it has many solvent properties similar to those of water.

The Lewis structure of ammonia is

$$H - \overset{\displaystyle ..}{N} - H$$
$$|$$
$$H$$

and, as we would expect from VSEPR theory, NH_3 is a pyramidal molecule. It is quite polar and is extremely soluble in water as illustrated in Figure 22.7. A saturated solution at room temperature contains about 28% w/w of NH_3 and is about 15 molar. Its high solubility in water is the result of its ability to form hydrogen bonds with water molecules.

Aqueous solutions of ammonia are basic, as you recall from Chapter 16. The reaction between NH_3 and H_2O is

$$NH_3 + H_2O \rightleftharpoons NH_4^+ + OH^- \qquad K_b = 1.8 \times 10^{-5}$$

Bottles of concentrated ammonia purchased from chemical supply companies are almost always labeled "ammonium hydroxide"; however, there is no evidence that the species NH_4OH actually exists. These solutions simply consist of molecules of NH_3 dissolved in water (along with small amounts of NH_4^+ and OH^- produced by the ionization of NH_3).

Ammonia can serve as either a proton donor or a proton acceptor, depending on conditions, and therefore forms two types of salts. When acting as a proton donor, NH_3 forms *amide salts* such as $NaNH_2$ that contain the NH_2^- ion. These can only be made in nonaqueous media, because NH_2^- is a very strong base and is completely hydrolyzed in water to give NH_3.

$$NH_2^- + H_2O \xrightarrow{100\%} NH_3 + OH^-$$
amide ion

As a proton acceptor, ammonia forms *ammonium salts*—for example, NH_4Cl.

$$NH_3 + H_3O^+ \longrightarrow NH_4^+ + H_2O \qquad K_c = 1.8 \times 10^9$$

As we learned in Chapter 16, NH_4^+ also is a weak acid, so solutions of NH_4^+ tend to be slightly acidic (providing the anion present is that of a strong acid).

$$NH_4^+ + H_2O \rightleftharpoons NH_3 + H_3O \qquad K_a = 5.5 \times 10^{-10}$$

In the laboratory, ammonia can be prepared by reacting an ammonium salt with a strong base such as oxide ion or hydroxide ion.

$$CaO + 2NH_4Cl \longrightarrow 2NH_3 + CaCl_2 + H_2O$$

$$NaOH + NH_4Cl \longrightarrow NH_3 + NaCl + H_2O$$

These strong bases displace the weaker base, NH_3, from the NH_4^+ ion. The reaction of ammonium ion with a base serves as part of the qualitative test for ammonium ion in a mixture. If a sample of the mixture is warmed with base, an odor of ammonia serves to confirm the presence of NH_4^+ in the mixture. Alternatively, the NH_3 can be detected by the basic reaction of its vapors with moist litmus paper (pink → blue).

One of the principal reactions of ammonia is its catalytic oxidation to nitric oxide in the **Ostwald process,** shown in Figure 22.8.

$$4NH_3(g) + 5O_2(g) \xrightarrow[\substack{\text{catalyst} \\ 750-900\ °C}]{Pt} 4NO(g) + 6H_2O(g)$$

The NO is quickly oxidized to NO_2 in the presence of excess oxygen,

$$2NO(g) + O_2(g) \longrightarrow 2NO_2(g)$$

and when the NO_2 is dissolved in water, it **disproportionates**—that is, it enters into a redox reaction with itself so that some of it is oxidized while the rest is reduced.

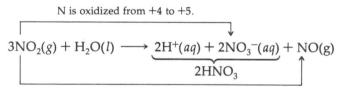

N is oxidized from +4 to +5.

$$3NO_2(g) + H_2O(l) \longrightarrow \underbrace{2H^+(aq) + 2NO_3^-(aq)}_{2HNO_3} + NO(g)$$

N is reduced from +4 to +2.

The commercial application of this sequence of reactions accounts for the major source of nitric acid and nitrates used in the manufacture of explosives, fertilizers, plastics, and many other useful substances. In fact, the development of this process in Germany by Wilhelm Ostwald, accompanied by the successful preparation of NH_3 from N_2 and H_2 by Haber, is said to have prolonged World War I, since the Allied blockade of Germany was unable to halt the German manufacture of munitions that had depended, prior to these processes, on the importation of nitrates from other countries.

Hydrazine (−2 Oxidation State) Hydrazine has the formula N_2H_4 and can be considered the nitrogen analog of hydrogen peroxide (although hydrazine's reactions are vastly different than those of H_2O_2). Its Lewis structure is

$$\overset{\displaystyle H \quad\ H}{\underset{\displaystyle \ \ \ \bullet\bullet\ \ \ \bullet\bullet}{H-N-N-H}}$$

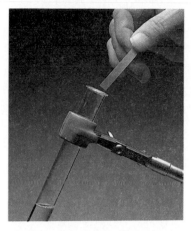

When a base is added to a solution of an ammonium salt, NH_3 is generated. Some of the ammonia escapes from the solution and can be detected by noting how it changes the color of moist litmus paper from red to blue.

Figure 22.8. *In this illustration of the Ostwald process, ammonia escaping from a concentrated aqueous solution reacts with oxygen on a catalytic platinum surface, which glows from the very exothermic heat of reaction. The colorless NO formed in the reaction is further oxidized by O_2 to give the reddish-brown gas NO_2.*

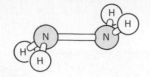

Figure 22.9. *The structure of hydrazine,* N_2H_4.

Traces of Cu^{2+} are common in water that passes through the copper water pipes found in many homes.

and its molecular geometry is shown in Figure 22.9. The staggered arrangement of the hydrogens probably reflects the tendency of the lone pairs of the nitrogens to be as far apart as possible so that the repulsions between them are at a minimum.

Hydrazine is prepared by the reaction of ammonia with hypochlorite ion in aqueous solutions.

$$2NH_3 + NaOCl \longrightarrow N_2H_4 + NaCl + H_2O$$

Hydrazine is a violent poison, and this reaction is one of the reasons that warnings are given about the dangers of mixing household cleaning agents—specifically household ammonia and liquid bleach, which contains NaOCl. The danger of such a mixture is somewhat lessened fortunately, because an intermediate in the reaction (NH_2Cl) has a tendency to react with N_2H_4, yielding harmless NH_4Cl, especially if traces of Cu^{2+} are present.

Pure hydrazine is a liquid that freezes at 2 °C and boils at 114 °C. It has a strong affinity for water and its solutions are weakly basic because of the reaction

$$N_2H_4 + H_2O \rightleftharpoons N_2H_5^+ + OH^- \qquad K_b = 1.7 \times 10^{-6}$$
$$\text{hydrazinium ion}$$

A major industrial use of hydrazine is as an oxygen scavenger in high-temperature boilers.

In basic solutions, hydrazine is a powerful reducing agent. For example, it reacts with iodine to liberate nitrogen.

$$N_2H_4 + 2I_2 + 4OH^- \longrightarrow N_2 + 4H_2O + 4I^-$$

The combustion of hydrazine is very exothermic.

$$N_2H_4(l) + O_2(g) \longrightarrow N_2(g) + 2H_2O(g) \qquad \Delta H° = -534 \text{ kJ}$$

Because of this, hydrazine as well as some compounds derived from it have been used as rocket fuels.

Monomethylhydrazine, one of the fuels used by the space shuttle.

Hydroxylamine (−1 Oxidation State) Hydroxylamine can be thought of as an ammonia molecule in which one hydrogen has been replaced by an —O—H group.

$$H-\overset{\overset{\displaystyle ..}{}}{N}-OH$$
$$\overset{|}{H}$$

It can be made by reduction of nitrites with SO_2.

$$NaNO_2 + NaHSO_3 + SO_2 + 2H_2O \longrightarrow 2NaHSO_4 + NH_2OH$$

Pure hydroxylamine is a white solid that melts at 33 °C, but it decomposes very easily. In water it is a weak base,

$$NH_2OH + H_2O \rightleftharpoons NH_3OH^+ + OH^- \qquad K_b = 1.1 \times 10^{-8}$$

NH_3OH^+ is called the *hydroxylammonium ion.*

and forms salts such as $[NH_3OH]Cl$ and $[NH_3OH]_2SO_4$. The salts are stable and are used as mild reducing agents in photography. The sulfate salt is used for removing hair from animal hides.

Nitrous Oxide (+1 Oxidation State) Nitrous oxide (N_2O) is made by decomposing molten ammonium nitrate, NH_4NO_3.

$$NH_4NO_3(l) \xrightarrow{\text{heat}} N_2O(g) + 2H_2O(g)$$

Although the reaction proceeds smoothly under most circumstances, it has the capability of occurring explosively. In fact, in 1947 a cargo ship filled with NH_4NO_3 exploded after burning quietly for four days in the harbor of Texas City, Texas. The blast destroyed a major part of the city and caused great loss of life. Ammonium nitrate can also be detonated by other explosives and one of its uses is as a high explosive.

The structure of N_2O can be described by the resonance formulas

$$:\ddot{N}\!=\!N\!=\!\ddot{O}: \longleftrightarrow :N\!\equiv\!N\!-\!\ddot{\underset{\cdot\cdot}{O}}:$$

As the VSEPR theory would predict, it is a linear molecule.

Thermodynamically, N_2O is unstable with respect to decomposition to the elements. This is because its heat of formation and free energy of formation are both positive ($\Delta H_f^\circ = +81.5$ kJ/mol, $\Delta G_f^\circ = +104$ kJ/mol). Nitrous oxide is stable at room temperature only because its *rate* of decomposition is extremely slow, but at elevated temperatures it decomposes easily to N_2 and O_2 with the *release* of heat. Oxygen, of course, supports combustion, so burning a fuel with N_2O releases more heat than with just O_2 because of the added heat released by the decomposition of the N_2O. This is the reason that race car drivers often use N_2O as an oxidizer for their fuel — it gives their engines extra power.

Nitrous oxide is also used as an anesthetic. It produces a mild intoxicating effect and is commonly called laughing gas. Another application is as a propellant in aerosol cans.

N_2O was the first anesthetic used for surgery.

Nitrogen Oxide (+2 Oxidation State) and Nitrogen Dioxide (+4 Oxidation State)

Nitrogen oxide (also called nitric oxide), NO, and nitrogen dioxide, NO_2, are the two most important oxides of nitrogen. They are intermediates in the conversion of ammonia to nitric acid, and both play a major role in the formation of a type of air pollution called photochemical smog. In this context, they are generally referred to together as NO_x.

The path from ammonia to nitric acid was discussed earlier in this section. Oxidation of ammonia on a platinum catalyst gives NO, which is rapidly oxidized to NO_2 by excess oxygen. Dissolving the NO_2 in water gives nitric acid plus NO (which is oxidized again to NO_2). This can be represented schematically as

$$NH_3(g) \xrightarrow[\text{catalyst}]{O_2} NO(g) \xrightarrow{O_2} NO_2(g) \xrightarrow{H_2O} HNO_3(aq) + NO(g)$$

Nitrogen oxide is a fairly reactive, colorless gas. It has an odd number of electrons, so they can't all be paired. NO is therefore paramagnetic. *Nitrogen dioxide* is a toxic, reddish-brown gas. Its structure is given by the resonance formulas

and in the gas phase there is an equilibrium between NO_2 and its dimer, dinitrogen tetroxide, N_2O_4. (A *dimer* is a molecule formed by joining two simpler ones.)

$$\underset{\text{reddish-brown}}{2NO_2(g)} \rightleftharpoons \underset{\text{colorless}}{N_2O_4(g)} + 57 \text{ kJ}$$

The formation of N_2O_4 from NO_2 can be explained by the tendency of the unpaired electron in NO_2, which spends most of its time on the nitrogen, to become paired with another electron from a neighboring NO_2 molecule.

NO_2 is paramagnetic, but N_2O_4 is not.

(a) A glass tube filled with gaseous NO_2. Notice its reddish-brown color. (b) After the tube has been cooled, the color of the NO_2 is much less intense because it has formed more of the colorless N_2O_4. Lowering the temperature has shifted the equilibrium,

$$2NO_2(g) \rightleftharpoons N_2O_4(g) + heat$$

to the right.

 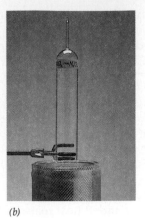

(a)   (b)

Dinitrogen tetroxide is a low-boiling liquid (bp = 21 °C) that is deep brown at its boiling point, caused by some NO_2 produced by its dissociation (pure N_2O_4 is colorless). It is a good oxidizing agent and has been used as an oxidizer in liquid-fueled rockets. In fact, the space shuttle burns monomethylhydrazine (see page 744) with N_2O_4 in the rocket engine that it uses to drop out of orbit on its return to Earth.

Photochemical Smog Within the last several decades, the growing number of automobiles in urban areas has produced a kind of air pollution never before experienced by civilization. It is characterized by a reddish-brown haze that contains substances irritating to the eyes, nose, and lungs, and that cause extensive damage to vegetation and rubber products not containing antioxidants. This haze has come to be known as **photochemical smog** (or often just smog) and has been traced to abnormally high levels of ozone and oxides of nitrogen in the atmosphere.

Nitrogen oxides, along with sulfur oxides, have been implicated as the source of acid rain. This problem is discussed further in the next chapter.

A typical smog episode begins in the early morning when urban rush-hour traffic spews out the primary pollutant, nitrogen oxide. This is formed in the gasoline engine during combustion by the direct combination of N_2 and O_2, which are present in the air drawn into the engine to burn the fuel. Even though the reaction

$$N_2(g) + O_2(g) \rightleftharpoons 2NO(g)$$

has an extremely small equilibrium constant at ordinary temperatures, the K increases with increasing temperature, so at the high temperature inside the engine small amounts of NO are produced. When the exhaust gases leave the engine, they cool so rapidly that the NO doesn't have an opportunity to decompose back to N_2 and O_2. As a result, it is released into the atmosphere. In Figure 22.10 we see that during the early hours, the NO concentration rises.

Figure 22.10. *The concentrations of pollutants during a typical photochemical smog episode.*

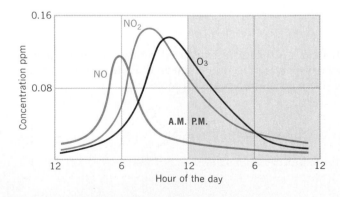

The next step in the sequence of reactions is the oxidation of NO to NO_2. This occurs slowly as the morning wears on, and in Figure 22.10 we see that the NO concentration starts to drop as the NO_2 concentration rises. The presence of the NO_2 in smog is what gives this form of air pollution its characteristic reddish-brown color, as shown in the photograph below.

$$2NO + O_2 \longrightarrow 2NO_2$$

As the sun climbs higher in the sky, the NO_2 begins to undergo a photochemical reaction—a reaction brought on by the absorption of light. Here ultraviolet light from the sun's rays causes NO_2 to decompose.

$$NO_2 \xrightarrow{hv} NO + O$$

hv is the energy absorbed from a photon of frequency *v*.

The oxygen atoms produced in this reaction combine with oxygen molecules and ozone is formed

$$O_2 + O \longrightarrow O_3$$

In Figure 22.10 we see that the ozone concentration peaks about noon as the NO_2 concentration begins to decline.

As you learned in Section 22.2, ozone is an extremely reactive substance. It reacts with NO to form NO_2, and this reaction helps moderate the rate of buildup of O_3.

$$O_3 + NO \longrightarrow NO_2 + O_2$$

Ozone also attacks other substances and is particularly hard on vegetation because it reacts with chlorophyll. It is especially reactive toward molecules that contain carbon–carbon double bonds. Until antipollution devices were installed on automobiles, considerable amounts of hydrocarbons (including those with double bonds) were released into the atmosphere by the evaporation of gasoline and as unburned hydrocarbons in auto exhausts. These substances react with ozone and nitrogen oxides in a series of complex reactions to give final products with the formula

$$R-\overset{\overset{\displaystyle :O:}{\|}}{C}-\overset{\cdot\cdot}{\underset{\cdot\cdot}{O}}-\overset{\cdot\cdot}{\underset{\cdot\cdot}{O}}-NO_2$$

where R is a portion of a hydrocarbon molecule—for example, $-CH_3$, $-C_2H_5$, and so on. They are collectively called *peroxyacyl nitrates,* or PAN for short. Molecules of PAN are oxidizing agents, a property that is typical of compounds having an

$$CH_3-\overset{\overset{\displaystyle :O:}{\|}}{C}-\overset{\cdot\cdot}{\underset{\cdot\cdot}{O}}-\overset{\cdot\cdot}{\underset{\cdot\cdot}{O}}-NO_2$$
peroxyacetyl nitrate

The reddish-brown color of smog, seen here hanging over New York City, is caused chiefly by nitrogen dioxide, NO_2.

oxygen–oxygen bond (e.g., H_2O_2), and contribute to the oxidizing nature of photochemical smog. Molecules of PAN are also eye irritants, even at concentration levels of parts per billion. This is one of the reasons that smog is so unpleasant.

Finally, as late afternoon approaches, the O_3 concentration begins to fall off as it reacts with airborne hydrocarbons and other substances, and the smog attack gradually subsides — until daybreak the next day.

Dinitrogen Trioxide, Nitrous Acid, and Nitrites (the +3 Oxidation State) Dinitrogen trioxide, N_2O_3, is produced by condensing an equimolar mixture of NO and NO_2 at very low temperatures (below -30 °C).

$$NO + NO_2 \longrightarrow N_2O_3$$

It is a blue liquid in which there are molecules having the structure

$$\overset{..}{:}\!O\!\!:\!\!\overset{N-N}{\underset{\overset{|}{\overset{..}{:O:}}}{}}\overset{..}{:}\!O\!\!:$$

At least in a formal sense, N_2O_3 is the acid anhydride of nitrous acid, HNO_2. In fact, if an equimolar mixture of NO and NO_2 is bubbled into water, nitrous acid is produced.

$$N_2O_3 + H_2O \longrightarrow 2HNO_2$$

Nitrous acid is a weak acid ($K_a = 4.5 \times 10^{-4}$) and is only stable in solution and in the gas phase. It cannot be isolated in pure liquid form because it decomposes by a disproportionation reaction.

$$3HNO_2 \longrightarrow HNO_3 + H_2O + 2NO$$

When HNO_2 is needed for a chemical reaction, it is usually generated in aqueous solution by adding a strong acid, such as HCl, to a salt of HNO_2 such as $NaNO_2$ (sodium nitrite). Hydrogen ion and nitrite ion combine in the net reaction

$$H^+ + NO_2^- \longrightarrow HNO_2$$

The formation of the weak electrolyte HNO_2 drives this reaction to the right.

Nitrous acid is able to function as either an oxidizing agent or a reducing agent, depending on the ease of oxidation or reduction of the other reactant. For example, it is oxidized to HNO_3 by permanganate ion,

$$H^+ + 5HNO_2 + 2MnO_4^- \longrightarrow 5NO_3^- + 2Mn^{2+} + 3H_2O$$

Here HNO_2 is oxidized and is a reducing agent.

but, in the presence of iodide ion and excess H^+, HNO_2 is reduced.

$$2H^+ + 2HNO_2 + 2I^- \longrightarrow I_2 + 2NO + 2H_2O$$

Here HNO_2 is reduced and is an oxidizing agent.

Neutralization of nitrous acid gives nitrite ion, although nitrite salts are normally made by reducing a metal nitrate. For example,

$$NaNO_3 + Pb \xrightarrow{\text{heat}} NaNO_2 + PbO$$

The nitrite ion has an angular structure that is predicted by VSEPR theory from either of its resonance formulas

$$\left[\overset{..}{:}\!O\overset{\overset{\displaystyle\overset{..}{N}}{\diagup\;\diagdown}}{}O\overset{..}{:}\right]^- \longleftrightarrow \left[\overset{..}{:}\!O\overset{\overset{\displaystyle\overset{..}{N}}{\diagup\;\diagdown}}{}O\overset{..}{:}\right]^-$$

A major but controversial use of nitrites is in preserving meats such as ham, frankfurters, bologna, and bacon. The nitrite serves two functions. The most important one is that it inhibits the growth of bacteria, especially *Clostridium botulinum*, which produces the very poisonous botulinus toxin that causes fatal food poisoning. The second is that it preserves the red color of the meat and thereby maintains the food's appetizing appearance.

The controversy over the use of nitrites in cured meat products stems from the effect that nitrous acid has on organic compounds called amines. The reaction of HNO_2 with some amines produces compounds called nitrosoamines, which have been shown to cause cancer, mutations, and birth defects in experimental animals. Meat and body fluids contain many amines, and it is feared by some scientists that when the NO_2^- contacts the acidic condition in the stomach, some of the HNO_2 that is formed might react with amines that are there and thereby give nitrosoamines. Although no direct evidence has been found linking nitrites in meat to cancer in either animals or humans, the FDA has set limits on the maximum allowable NO_2^- concentrations in these food products.

Some scientists feel that the risk of food poisoning from meats untreated with nitrites is greater than the risk of cancer from meats that are treated.

Dinitrogen Pentoxide, Nitric Acid, and Nitrates (the +5 Oxidation state)

Dinitrogen pentoxide, N_2O_5, exists in molecular form in the vapor. Its structure is

$$\overset{..}{\underset{..}{O}}\diagdown \underset{\overset{..}{\underset{..}{O}}}{N}-\overset{..}{\underset{..}{O}}-\underset{\overset{..}{\underset{..}{O}}}{N}\diagup \overset{..}{\underset{..}{O}}$$

In the solid state, N_2O_5 dissociates into ions and exists as $NO_2^+NO_3^-$. When allowed to react with water, N_2O_5 gives nitric acid, HNO_3.

$$N_2O_5 + H_2O \longrightarrow 2HNO_3$$

N_2O_5 is the acidic anhydride of HNO_3.

This reaction can be reversed, and N_2O_5 is produced by removing the components of water from a pair of HNO_3 molecules. This requires a very powerful dehydrating agent such as P_4O_{10}. (The reaction of P_4O_{10} with water is discussed in the next chapter.)

Nitric acid is one of the world's most vital chemicals, and huge amounts of it are produced each year. Much of it is made into ammonium nitrate to be used as a nitrogen fertilizer, because plants can utilize both NH_4^+ and NO_3^- ions. The manufacture of explosives such as TNT and nitroglycerine also require nitric acid, and nitrates are used in curing meats along with nitrites.

About 8.5 million tons of HNO_3 are made annually.

The commercial production of nitric acid by dissolving NO_2 in water was described earlier in our discussion of the Ostwald process. In the laboratory, nitric acid can be made by heating a mixture of KNO_3 and concentrated sulfuric acid.

$$KNO_3(s) + H_2SO_4(l) \xrightarrow{heat} KHSO_4(s) + HNO_3(g)$$

The nitric acid vapors condense to a liquid when they are cooled.

Pure nitric acid is a colorless liquid that tends to decompose above 0 °C into NO_2, H_2O, and O_2. The concentrated laboratory reagent (Figure 22.11) is about 70% HNO_3 and is often slightly yellow in color from the presence of small amounts of NO_2 formed by a photochemical decomposition.

$$4HNO_3 \xrightarrow{hv} 4NO_2 + O_2 + 2H_2O$$
$$\text{colorless} \qquad \begin{array}{c}\text{(red-brown)}\\\text{(appears yellow}\\\text{when dilute)}\end{array}$$

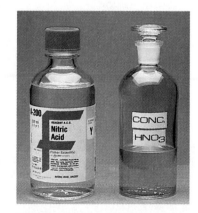

Figure 22.11. *On the left, a fresh bottle of concentrated HNO_3. On the right, an aged sample. The reddish-brown color of the aged sample is from NO_2 produced by the gradual decomposition of the HNO_3.*

Nitric acid is a strong acid, and the presence of nitrate ion in its solutions makes it an especially powerful oxidizing agent. It is therefore able to dissolve many metals that fail to dissolve in an acid such as HCl, which contains H^+ as its strongest (and only) oxidizing agent. For example, copper dissolves in both concentrated and dilute HNO_3, but the reduction products differ. These reactions were described previously.

$$Cu + 2NO_3^- + 4H^+ \longrightarrow Cu^{2+} + 2NO_2 + 2H_2O \qquad \text{(Concentrated)}$$

$$3Cu + 2NO_3^- + 8H^+ \longrightarrow 3Cu^{2+} + 2NO + 4H_2O \qquad \text{(Dilute)}$$

With stronger reducing agents and more dilute solutions, the nitrogen can be reduced all the way to the -3 oxidation state. For example,

$$4Zn + NO_3^- + 10H^+ \longrightarrow 4Zn^{2+} + NH_4^+ + 3H_2O$$

A mixture of one part by volume of concentrated HNO_3 and three parts by volume of concentrated HCl is called **aqua regia.** As discussed earlier, this mixture is able to dissolve the noble metals such as gold and platinum that won't dissolve in concentrated HNO_3 alone. The chloride ion in aqua regia forms complex ions with the ions of these metals and that helps draw them into solution. For instance, gold dissolves as follows:

$$4H^+(aq) + 4Cl^-(aq) + NO_3^-(aq) + Au(s) \longrightarrow AuCl_4^-(aq) + NO(g) + 2H_2O$$

22.4 CARBON

The element carbon is found in every living thing; all life as we know it is based on carbon-containing compounds. It is this fact that is responsible for the term *organic* that is used to describe hydrocarbons, and those compounds that come from hydrocarbons by substituting other atoms for some of the hydrogen atoms in their molecules.

Carbon is found in period 2 at the head of Group IVA. Its atoms each have four valence electrons that they tend to share with other atoms in the formation of four covalent bonds. Although in most of its compounds carbon forms four bonds, we will see some exceptions in this section.

Carbon occurs in nature as both the free element and in compounds. Coal contains elemental carbon, and when it is heated strongly in the absence of air, volatile substances are driven off and the material that remains, called **coke,** is almost all carbon. Diamonds are, for all practical purposes, pure carbon. In the combined state, we find carbon in all living things and in fossil fuels such as methane, CH_4, and petroleum. Carbon also occurs in large amounts in carbonates such as limestone.

The free element

At room temperature and pressure, graphite is the most thermodynamically stable form of carbon.

In Chapter 9 we saw that carbon can exist in two different allotropic forms. One of them is graphite, a soft black slippery solid that is a reasonably good conductor of electricity. The other is diamond, an extremely hard, nonconducting substance well known for its gem-quality crystals. Their different properties can be traced to the differences in the way the carbon atoms are bonded to each other.

Graphite's loosely stacked layer structure (Figure 9.26 on page 292) explains both its slippery feel and electrical conductivity. Because the layers are only weakly attracted to each other by London forces, they can slide over each other easily. Gas molecules trapped between the layers also help considerably by acting as molecular "ball bearings." The electrical conductivity of graphite is also a consequence of the solid's structure. Within each layer there is extensive delocalization of the pi-electron cloud. An electron forced into one end of a layer can produce a flow of charge through the pi cloud and cause an electron to exit from the other end. (Interestingly, it has been found by studying single crystals of graphite that the electrical conductivity is very good in a direction parallel to the layers, but very poor in a direction perpendicular to the layers.) Because of these properties, graphite is used commercially as a lubricant and to construct electrodes (for example, in ordinary dry cells).

In diamond, each carbon atom is surrounded tetrahedrally by four others. The interlocking network of covalent bonds that results gives crystals that are actually single molecules. As you learned earlier, this makes diamond extremely hard; in fact, it is the hardest substance known.

The hardness of diamond is one of its best-known properties and it accounts for many industrial uses of this allotrope. Diamond has other less well-known properties, however, that are also very desirable. For example, it has the greatest thermal conductivity of any solid, more than four times that of copper or silver. It also has unusual optical and acoustical properties. In fact, the properties of diamond are so valuable that synthetic methods for making diamond from other carbon-containing substances have been developed.

The synthesis of diamond under high pressure and at high temperatures was described earlier in Chapter 14. This method has been refined so that it now is capable of routinely producing diamonds of gem quality which are about 8 mm in their largest dimension. They are not used for jewelry, however, but for their thermal properties instead. The diamonds are cut with lasers into thin chips for use as heat sinks for microelectronic devices. (A heat sink is something used to conduct heat away from a device that generates enough heat to damage itself.)

One of the most exciting advances in materials science in recent years has been the development of techniques that permit diamond coatings to be deposited on other materials. Mixtures of CH_4 and H_2 at temperatures approaching 1000 °C are passed through a microwave discharge, which cracks the CH_4 molecules into carbon and hydrogen atoms. The carbon atoms then condense on a nearby surface, and under the right conditions they form a thin film of diamond rather than graphite.

The prospects for diamond coating technology are very bright. For example, thin diamond films on cutting tools greatly increase their resistance to wear. Japanese Victor Corporation (JVC) is currently marketing high-fidelity tweeters (high-frequency loudspeakers) that use a diamond-coated diaphragm. Such speakers are said to yield a sound reproduction quality never before achieved.

Graphite and diamond represent quite pure forms of elemental carbon. Less pure, predominantly graphitic forms of carbon are also known. One of these, charcoal, is formed by heating wood strongly in the absence of air. Charcoal has a particularly open structure with an enormous amount of surface area for a given mass of carbon. *Activated charcoal,* a finely pulverized form having a surface area of about 1000 m^2 per gram, has many commercial uses. Its large surface area permits small amounts of it to adsorb large numbers of molecules, a property that allows it to remove molecules from the air that have offensive odors and to remove toxic impurities from water. In fact, several municipalities located in areas polluted by chemical spills have had to install activated carbon filtration systems to make contaminated well water pure enough to drink.

When a hydrocarbon is burned in a very limited supply of oxygen, the hydrogen combines with the oxygen to form water, and finely divided carbon, called *carbon black,* remains.

$$CH_4(g) + O_2(g) \longrightarrow C(s) + 2H_2O(g)$$

Carbon black is used as a pigment in black inks, and large amounts of it are used in making automobile tires.

Synthetic diamonds. The diameters of the larger stones are about 8 mm. Their yellow color is due to small amounts of nitrogen incorporated into the crystals. When cut and polished they become the thin diamond wafer heat sinks used in the electronics industry. The small wafers shown here have dimensions of $3 \times 3 \times 1$ mm and cost approximately $150 each.

Nearly all the carbon black made in the U.S. is now made from waste oils.

Oxides and oxoacids of carbon

Carbon forms two principal oxides: carbon monoxide and carbon dioxide. Carbon monoxide has the Lewis structure

$$:C{\equiv}O:$$

A slurry of granular activated carbon (activated charcoal) is pumped into a water treatment facility where it will be used to remove offensive odors and toxic impurities from municipal drinking water.

and is one of a rather small number of species in which carbon has only three bonds. Carbon monoxide is a nearly nonpolar substance with a low melting point and boiling point (mp = −199 °C, bp = −192 °C). It is unreactive toward water and has a low solubility. It is also flammable and burns with a hot flame, which is why the mixture of CO and H_2 from the water gas reaction (Section 22.1) is a useful industrial fuel.

One way that carbon monoxide can be formed is by burning carbon or a hydrocarbon in a limited supply of oxygen. These are exactly the conditions that exist in the fire in a charcoal barbecue or gasoline engine, and both produce some carbon monoxide. As you are probably aware, carbon monoxide is quite toxic. It binds to hemoglobin in the blood, thereby preventing it from carrying oxygen — and that is why it is dangerous to be in an enclosed space with either a charcoal fire or a running automobile engine.

Carbon monoxide is particularly dangerous because it is both colorless and odorless.

In the laboratory, small amounts of carbon monoxide can be made by treating formic acid, $HCHO_2$, with concentrated sulfuric acid. Sulfuric acid has a strong affinity for water and actually extracts the components of water (two hydrogens and an oxygen) from a formic acid molecule.

$$HCHO_2(l) \xrightarrow{H_2SO_4} H_2O(l) + CO(g)$$

Industrially, carbon monoxide is proving to be an extremely important chemical. More and more interest is being shown in reactions like those discussed in the last section in which CO and H_2 are combined catalytically to form hydrocarbons, a replacement for petroleum. We saw in the last section that CO can be made by reacting steam with white-hot carbon in the form of coke.

$$C(s) + H_2O(g) \xrightarrow{1000\ °C} CO(g) + H_2(g)$$

At high temperatures, carbon monoxide is an effective reducing agent. For example,

$$Fe_2O_3(s) + 3CO(g) \xrightarrow{heat} 2Fe(s) + 3CO_2(g)$$

This overall reaction is used to extract iron from its ore and was discussed in greater detail in Chapter 20.

An interesting chemical property of carbon monoxide is its ability to form covalent compounds with transition metals in which the metal is in a low (or zero) oxidation state. These are called **metal carbonyls.** An example is $Ni(CO)_4$, nickel carbonyl, which is formed by simply warming metallic nickel in the presence of CO. The compound is quite volatile and very toxic.

$$Ni(s) + 4CO(g) \longrightarrow Ni(CO)_4(g)$$

Some other examples of metal carbonyls are $Cr(CO)_6$, $Mn_2(CO)_{10}$, $Fe(CO)_5$, and $Co_2(CO)_8$.

The second major oxide of carbon is carbon dioxide.

$$\overset{\cdot\cdot}{O}=C=\overset{\cdot\cdot}{O}$$

As predicted by VSEPR theory, CO_2 is a linear molecule. It is also nonpolar, because of its symmetrical structure. The triple point of CO_2 is above 1 atm, and when CO_2 is cooled at atmospheric pressure, it condenses to a solid rather than to a liquid. Solid carbon dioxide is called **dry ice** and its temperature is -78 °C, the temperature at which CO_2 sublimes at 1 atm. Because gaseous CO_2 is denser ("heavier") than air and because it doesn't support combustion, CO_2 is used to smother fires, especially those that are difficult to extinguish using water.

Carbon dioxide can be formed in a number of ways. Complete combustion of carbon, a hydrocarbon, or carbon monoxide gives CO_2.

$$C + O_2 \longrightarrow CO_2$$

$$CH_4 + 2O_2 \longrightarrow CO_2 + 2H_2O$$

$$2CO + O_2 \longrightarrow 2CO_2$$

Industrially, it is often made by the thermal decomposition of limestone.

$$CaCO_3(s) \xrightarrow{\text{heat}} CaO(s) + CO_2(g)$$
$$\text{limestone}$$

Carbon dioxide is also formed naturally by decomposing organic matter.

In the laboratory, CO_2 can be conveniently prepared by the reaction of an acid with a carbonate — for example, calcium carbonate.

$$CaCO_3(s) + 2HCl(aq) \longrightarrow CaCl_2(aq) + H_2O(l) + CO_2(g)$$

This reaction, you may recall, is used in the Mond process for purifying nickel.

Dry ice, which is solid CO_2. At atmospheric pressure, its temperature is -78 °C.

Similar reactions with bicarbonates (e.g., $NaHCO_3$) were mentioned in Chapter 20.

Carbon dioxide is consumed in large quantities by industry. You learned in Chapter 20 that it is used to make sodium carbonate by the Solvay process. Major uses also include refrigeration (including dry ice) and beverage carbonation. The process of carbonation consumes about 35% of a total annual production of about 2.5 million tons of CO_2.

In nature, green plants consume carbon dioxide in photosynthesis and produce glucose, $C_6H_{12}O_6$, a sugar from which they manufacture starch, cellulose, and other chemicals. In the process, oxygen is released to the atmosphere.

$$6CO_2(g) + 6H_2O(l) \xrightarrow[\text{chlorophyll}]{\text{light}} C_6H_{12}O_6(aq) + 6O_2(g)$$

Carbon dioxide is moderately soluble in water, and its solutions are slightly acidic due to the formation of carbonic acid.

$$CO_2(aq) + H_2O(l) \rightleftharpoons H_2CO_3(aq)$$
$$\text{carbonic acid}$$

Carbonic acid, as we've seen earlier, is a weak diprotic acid that ionizes in two steps. Neutralization with a base is able to give two types of salts: carbonates, formed by complete neutralization, and bicarbonates (hydrogen carbonates), formed by partial neutralization.

$$NaOH + H_2CO_3 \longrightarrow NaHCO_3 + H_2O$$
$$2NaOH + H_2CO_3 \longrightarrow Na_2CO_3 + 2H_2O$$

Bicarbonates such as $NaHCO_3$, you recall, are easily decomposed by heating to give carbonates ($2NaHCO_3 \rightarrow Na_2CO_3 + H_2O + CO_2$).

Carbon dioxide in the atmosphere has played a key role in the earth's environment by absorbing infrared radiation that would otherwise be radiated into space. In a very real sense, it has made our world function like a greenhouse and has made the planet habitable. However, in recent times scientists have begun to be concerned about this greenhouse effect. Since the beginning of the industrial revolution, the combustion of fuels has poured huge amounts of CO_2 into the atmosphere. One of the predicted results of this accumulation is a gradual global warming. This expected warming has led in turn to dire warnings of melting polar ice caps, rising oceans, and drastic climatic changes in the years ahead. However, whatever changes are occurring, they are taking place very slowly and only recently have some climate modelers claimed to have detected the first signs of global warming.

In Chapter 5 it was pointed out that when groundwater containing dissolved CO_2 trickles slowly through large deposits of limestone, it gradually dissolves the $CaCO_3$ and forms huge limestone caverns. The equilibrium equations are

$$H_2O + CO_2(g) \rightleftharpoons H_2CO_3(aq)$$
$$H_2CO_3(aq) + CaCO_3(s) \rightleftharpoons Ca(HCO_3)_2(aq)$$

When a drop of water containing a dilute solution of $Ca(HCO_3)_2$ collects on the ceiling of a limestone cave, it gradually evaporates. This causes the first of these equilibria to be shifted to the left as H_2CO_3 decomposes and gaseous CO_2 is released. The loss of H_2CO_3 also affects the second reaction. That equilibrium is also shifted to the left, causing $CaCO_3$ to precipitate. Over a period of many years, drop after drop evaporates and a stalactite composed of calcium carbonate is slowly formed. Drops that fall to the floor evaporate, too, and give rise to stalagmites. Limestone caves are not the only places to observe the growth of stalactites. Calcium carbonate is one of the substances formed during the curing of concrete, and when water seeps through the concrete of a highway overpass, stalactites looking a bit like whiskers form and hang from the overpass above the roadway.

CO_2 has replaced chlorofluoro-carbons as a propellant in many aerosol products.

The oceans represent a huge reservoir of dissolved CO_2.

It has been estimated that if it were not for traces of CO_2 in the atmosphere, the average global temperature would be lower by about 35 °C.

Photograph of stalagtites formed on the ceiling of a highway overpass.

Because of reactions like those just described, the groundwater in many parts of the country contains dissolved calcium as Ca^{2+}, as well as other ions such as Mg^{2+} and Fe^{3+}. If their concentrations are sufficiently large, these ions can interfere with the action of ordinary soap by forming a precipitate with it. The water is then described as **hard water,** and the ions causing the problem are called **hardness ions.** Hardness ions can be removed from water in a number of ways. If the solution also contains bicarbonate ion, as it would if the ions in it came from the dissolving of limestone, heating the water will drive off CO_2 and cause a precipitate to form. For example,

Soaps and detergents were discussed in Chapter 13.

$$Ca(HCO_3)_2(aq) \xrightarrow{\text{heat}} CaCO_3(s) + H_2O(l) + CO_2(g)$$

This type of hard water causes serious problems because the precipitate (called *boiler scale*) can clog hot water pipes and make it more difficult to heat water in a boiler (see Figure 22.12).

If the hard water doesn't contain HCO_3^-, the ions can still be precipitated by adding *washing soda*, a hydrate of sodium carbonate, $Na_2CO_3 \cdot 10H_2O$. The carbonate ion precipitates the hardness ions

$$Ca^{2+}(aq) + CO_3^{2-}(aq) \longrightarrow CaCO_3(s)$$

and, by removing them, allows the soap to do its job.

Other inorganic compounds of carbon

Carbides Binary compounds formed between carbon and a metal or metalloid are generally referred to as carbides. They are of three types. In *covalent carbides,* carbon is bonded covalently to the other element. An example is silicon carbide, SiC, known

Figure 22.12. *Boiler scale clogs a pipe.*

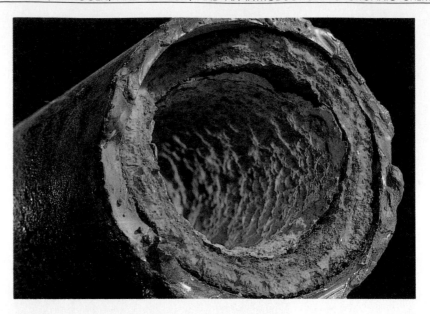

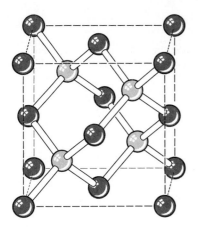

Figure 22.13. *The crystal structure of carborundum. Gray spheres represent silicon. In diamond or pure silicon (which has the same crystal structure as diamond), all the atoms would be the same.*

commercially as **carborundum.** It is made by heating silicon dioxide (sand) and carbon to very high temperatures.

$$SiO_2(s) + 3C(s) \xrightarrow{\text{heat}} SiC(s) + 2CO(g)$$

This solid has the same structure as diamond, except that silicon atoms alternate with carbon atoms (Figure 22.13). Like diamond, silicon carbide is very hard and is used as an abrasive—for example, in sandpaper.

Ionic carbides (also called *saltlike carbides*) are formed from carbon and active metals and contain more or less discrete ions of carbon in their solids. An example mentioned earlier (page 731) is Al_4C_3, which contains the anion C^{4-}. When placed into water, it hydrolyzes and methane is formed. A particularly important saltlike carbide is calcium carbide, CaC_2, which contains the ion, $[:C\equiv C:]^{2-}$, called acetylide ion. It is made by reacting calcium oxide (lime) with carbon at high temperatures.

$$CaO(s) + 3C(s) \longrightarrow CaC_2(s) + CO(g)$$

Calcium carbide reacts with water to form acetylene, the gas used in welding torches.

$$CaC_2(s) + 2H_2O(l) \longrightarrow C_2H_2(g) + Ca(OH)_2(s)$$
$$\text{acetylene}$$

The third type of carbide is called an *interstitial carbide.* In these substances, carbon atoms fit into voids (empty spaces called *interstices*) between the metal atoms in a crystal. An example is tungsten carbide, WC, which is extremely hard and brittle and is used to make cutting tools used to machine other metals.

Cyanides Hydrogen cyanide, HCN, can be made by a number of reactions, but one that is especially important commercially is the reaction of ammonia with methane. It is carried out at high temperatures over a platinum catalyst.

$$NH_3(g) + CH_4(g) \xrightarrow[\text{Pt}]{1200\ °C} HCN(g) + 3H_2(g)$$

CN⁻ forms many stable complex ions.

Hydrogen cyanide is an intermediate in the synthesis of a number of well-known plastics, including nylon. It is also a very potent, fast-acting poison that both deactivates critical enzymes in the body and binds irreversibly to iron in hemoglobin in the blood. In water, HCN is a weak acid and salts such as NaCN can be formed by neutralization. Like HCN, NaCN is a deadly poison. It is used in certain electroplating baths and to extract gold from its ores.

Carbon Disulfide Carbon reacts directly with sulfur to form carbon disulfide, CS_2.

$$C + 2S \xrightarrow[\text{temp}]{\text{high}} CS_2$$

$$:\overset{..}{S}=C=\overset{..}{S}:$$

carbon disulfide

It is a useful solvent, but it is extremely flammable: Boiling water is hot enough to ignite it! When it burns, CO_2 and SO_2 are formed. Besides being a solvent, carbon disulfide is a useful chemical reagent. It is used commercially to manufacture carbon tetrachloride.

$$CS_2 + 3Cl_2 \longrightarrow CCl_4 + S_2Cl_2$$

carbon
tetrachloride

22.5 FUNDAMENTALS OF ORGANIC CHEMISTRY

Organic chemistry is the study of the compounds of carbon and hydrogen (**hydrocarbons**) and of other compounds that can be considered to be derived from hydrocarbons by substituting various groups of atoms in place of hydrogen. These substances are possible because carbon has the ability to form stable covalent bonds not only to other carbon atoms but to most of the other nonmetals as well. In the process, chains and rings of carbon atoms are formed, linking the carbon atoms to one another and to other atoms such as hydrogen, oxygen, nitrogen, sulfur, and phosphorus. The outcome is a multitude of molecular structures of varying degrees of complexity and chemical reactivity. They range from methane (CH_4), the simplest organic compound, to complex biochemical substances such as proteins and the DNA in our cells that controls heredity. In fact, the name "organic chemistry" derives from the early belief that such compounds could only be synthesized by living organisms (a belief we now know to be untrue).

Just from the preceding paragraph, it should be obvious that we can hardly do more here than introduce you to some of the kinds of organic compounds that exist, and this is our goal. If your field of specialization requires greater knowledge, you will no doubt take a full-year course in the subject.

Hydrocarbons

The definition of organic chemistry given above reflects the way chemists have chosen to organize the subject. In broadest terms, organic compounds are viewed as being derived from basic sets of hydrocarbon molecules that consist of chains or rings of carbon atoms of varying size. The three simplest sets of hydrocarbons are called the **alkanes, alkenes,** and **alkynes.**

The Alkanes The alkanes are hydrocarbons in which there are only single bonds between carbon atoms. An example is ethane, C_2H_6, which has the structure

$$
\begin{array}{c}
\ \ \ \ \text{H}\ \ \text{H} \\
\ \ \ \ |\ \ \ \ | \\
\text{H}-\text{C}-\text{C}-\text{H} \qquad \text{ethane} \\
\ \ \ \ |\ \ \ \ | \\
\ \ \ \ \text{H}\ \ \text{H}
\end{array}
$$

As a general rule, carbon forms four covalent bonds to other atoms.

Other examples are propane (C_3H_8), butane (C_4H_{10}), pentane (C_5H_{12}), and hexane (C_6H_{14}). See Table 22.7. Notice that on going from one formula to the next, the

Table 22.7. The first ten members of the straight-chain alkanes

Formula	Name	Boiling Point (°C) at 1 atm
CH_4	Methane	−161
C_2H_6	Ethane	−89
C_3H_8	Propane	−44
C_4H_{10}	Butane	−0.5
C_5H_{12}	Pentane	36
C_6H_{14}	Hexane	68
C_7H_{16}	Heptane	98
C_8H_{18}	Octane	125
C_9H_{20}	Nonane	151
$C_{10}H_{22}$	Decane	174

numbers of atoms increase by one carbon and two hydrogens, which corresponds to a CH_2 group inserted in the chain.

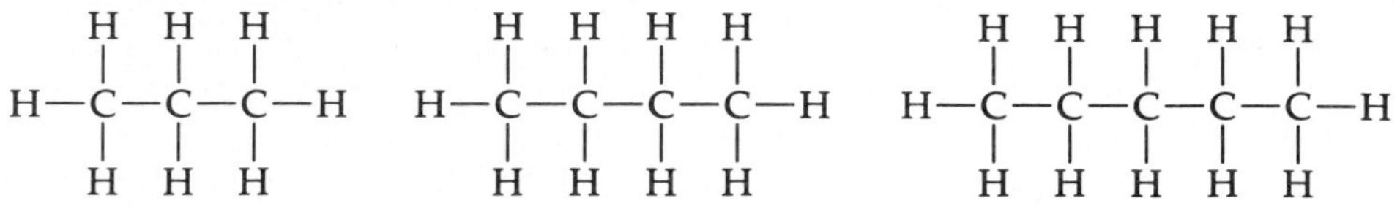

propane   butane   pentane

These structures can be written in a more compact form as follows:

$$CH_3CH_2CH_3 \qquad CH_3CH_2CH_2CH_3 \qquad CH_3CH_2CH_2CH_2CH_3$$

propane   butane   pentane

It is taken for granted that the various CH_3 and CH_2 groups are attached to each other by carbon–carbon bonds and that the hydrogens in each group are bonded to the carbon.

The alkanes can be represented by the general formula C_nH_{2n+2}, in which n is the number of carbon atoms in the molecule. If $n = 1$, the formula is CH_4, so methane is considered to be the first member of the alkanes even though it has no carbon–carbon bonds.

One of the most interesting aspects of organic chemistry is the existence of different compounds that share the same molecular formula. The following two compounds, for example, have the formula C_4H_{10}.

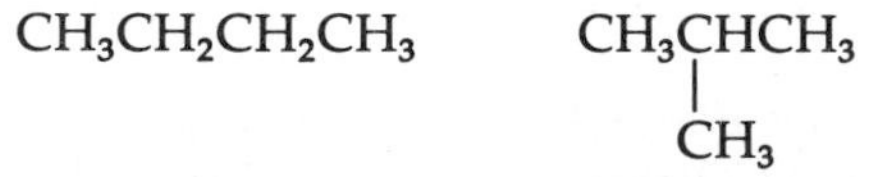

$$CH_3CH_2CH_2CH_3 \qquad \begin{array}{c} CH_3CHCH_3 \\ | \\ CH_3 \end{array}$$

butane
bp = −0.5 °C
$d = 0.601$ g/cm³

methylpropane
bp = −11.6 °C
$d = 0.549$ g/cm³

That they are different substances is evidenced by their distinctly different physical properties. Substances such as these are said to be **isomers** of each other, and the phenomenon of the existence of isomers is called **isomerism.**

The number of isomers with a particular molecular formula depends on the

Many of the organic compounds in petroleum are alkanes and have practical uses as fuels.

methane (CH_4): natural gas

propane (C_3H_8): LPG gas; a fuel for gas barbecues

butane (C_4H_{10}): disposble cigarette lighters

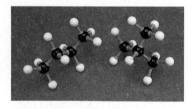

The isomers of C_4H_{10}. Butane is on the left and methylpropane (also called isobutane) is on the right.

bp = boiling point

d = density

Isomers of metal complexes were discussed in Chapter 21.

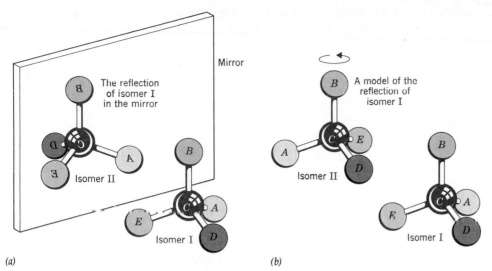

Figure 22.14. *Optical isomers of organic compounds. (a) Isomer II is a reflection of isomer I in the mirror; that is, it is the mirror image of isomer I. (b) When isomer II is rotated around the B—C bond so that atoms D in both structures are in the same relative position, the positions of A and E do not match. Therefore, isomers I and II are not superimposable and are not identical. The conclusion is that there are two optical isomers of the compound CABDE.*

number of alternative ways of arranging the atoms. For the four-carbon alkane, there are only two isomers, but for C_8H_{18} there are 18 and for $C_{10}H_{22}$ there are 75. Thus, as the number of carbons increases, the number of isomers increases even faster. For example, the compound $C_{20}H_{42}$ has 366,319 possible isomers (although only a handful have actually been made).

The isomers just described are called **structural isomers** because they differ in the way the atoms are grouped. Another kind of isomerism found for organic compounds is optical isomerism, which we discussed in Chapter 21 for metal complexes. Recall that two molecules are chiral and optical isomers of each other if one is the nonsuperimposable mirror image of the other. For organic compounds, this isomerism occurs when a carbon atom is bonded to four different groups. An example is

$$CH_3CH_2\underset{\underset{\displaystyle CH_3}{|}}{C}HCH_2CH_2CH_3$$

Notice that the carbon printed in red is bonded to the H, CH_3, C_2H_5, and C_3H_7 groups. Figure 22.14 illustrates how this leads to optical isomerism.

The naming of organic compounds is described in some detail later in this section. Nevertheless, we can use the names of the two isomers of C_4H_{10} to illustrate the way we view one compound as being derived from another. The compound "methylpropane" is considered to be a **derivative** of the compound propane (C_3H_8) obtained by substituting a *methyl* group, CH_3—, in place of a hydrogen.

$$CH_3 \diagdown \qquad \longrightarrow H$$
$$CH_3—CH—CH_3 \longrightarrow CH_3—\underset{\underset{\displaystyle CH_3}{|}}{C}H—CH_3$$

The CH_3 group is an example of an **alkyl group**, a group of atoms formed by removing one hydrogen from an alkane molecule.

Butane is said to be a *straight-chain* hydrocarbon because its structure can be written with all the carbon atoms in a straight line. Methylpropane, on the other hand, is said to be a *branched-chain* hydrocarbon because the carbon chain splits, or branches. In addition to these possibilities, there are also *cyclic* hydrocarbons in

A chiral carbon atom is often called an *asymmetric carbon atom.*

Methylpropane is also called *isobutane,* meaning an isomer of ordinary or *normal-butane* (*n*-butane).

This is only a hypothetical "reaction" written to illustrate the concept of one compound being derived from another. Methylpropane would not actually be made this way.

which the carbon atoms are arranged in rings. An example is cyclohexane, whose structure can be written in several ways, as illustrated below.

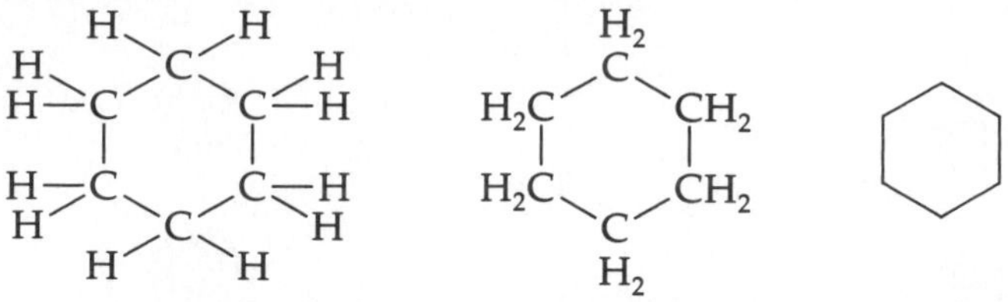

Various ways of representing the structure of cyclohexane, C_6H_{12}

In the drawing on the right, it is assumed that there is a carbon at each apex of the hexagon. The straight lines represent carbon–carbon bonds, and it is further assumed that there are enough hydrogen atoms bonded to each carbon to give a total of four bonds. (That would require 2H atoms per apex.)

In the alkanes, carbon atoms form four single covalent bonds by using sp^3 hybrid orbitals. This gives a tetrahedral arrangement of other atoms around each carbon. Because of this, even straight-chain hydrocarbons do not have linear carbon backbones, as illustrated in the margin. ("Straight chain" simply refers to how we can write the carbon chain on paper.)

The Alkenes and Alkynes An alkene is a hydrocarbon in which there is a carbon–carbon double bond. In an alkyne, there is a carbon–carbon triple bond. This gives the general formula C_nH_{2n} for an alkene and C_nH_{2n-2} for an alkyne. Examples are ethene (C_2H_4) and ethyne (C_2H_2).

The formation of acetylene by the reaction of CaC_2 with water was described earlier.

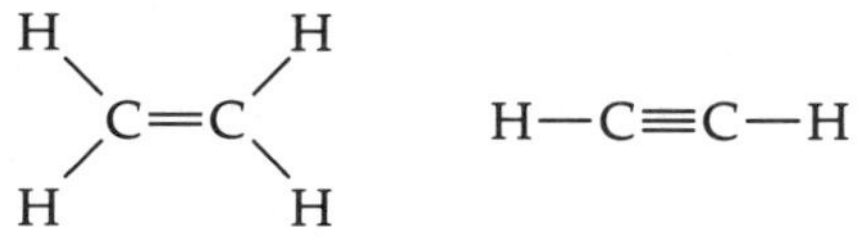

ethene ethylene

ethyne acetylene

Their common names appear below their preferred chemical names.

The bonding in these molecules was described earlier in Section 9.6. One of the significant features of the bonding among the alkenes is the restricted rotation of one part of the molecule relative to the other around the double bond. Rotation around the bond axis misaligns the p orbitals that give rise to the π bond, so rotation ruptures a bond. The π-bond energy therefore serves as an energy barrier (in effect, an activation energy) that must be overcome to have rotation, and at ordinary temperatures, the atoms in the molecule do not have enough kinetic energy to overcome this barrier. This makes it possible to isolate isomers such as

trans-2-butene *cis*-2-butene

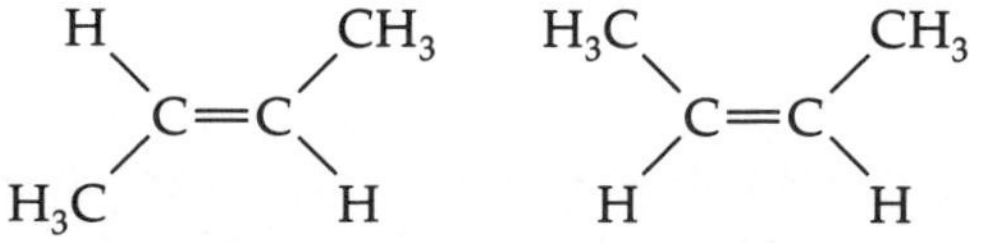

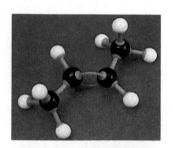

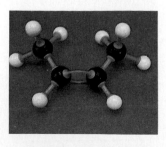

trans-2-*Butene and* cis-2-*butene*

These are called **geometric isomers.** They have the same molecular formula as well as the same basic arrangements of their atoms and bonds. They differ only in the directions in which the CH_3 groups are pointing. By contrast, similar isomers of butane are not possible because rotation around single bonds is not restricted.

Cyclic structures containing carbon–carbon double bonds are also possible, as shown on the next page. Similar-sized ring structures with triple bonds are not observed, however, because the $C—C\equiv C—C$ sequence of atoms is linear and can't fit into the structure of a small ring.

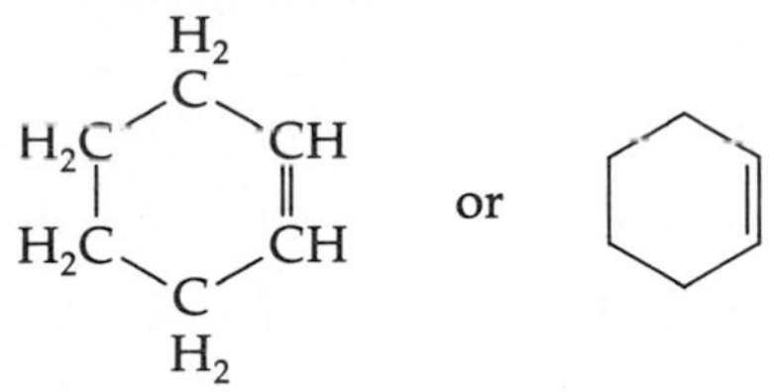

cyclohexene

Aromatic hydrocarbons

In Section 9.7 we described the bonding in benzene, a cyclic hydrocarbon with the formula C_6H_6. This molecule consists of a hexagonal ring of carbon atoms as shown below.

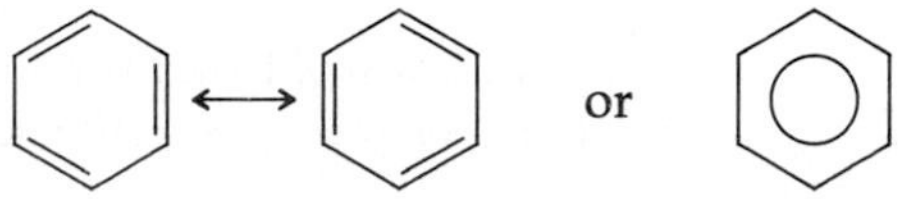

The pair of resonance structures on the left is usually represented by the single structure on the right, in which the delocalized π-electron cloud is shown by the circle within the hexagon. Some compounds that contain benzene rings have pleasant odors, which is the origin of the name of this class of substances. Examples are vanillin and oil of wintergreen, which contain atoms other than C and H.

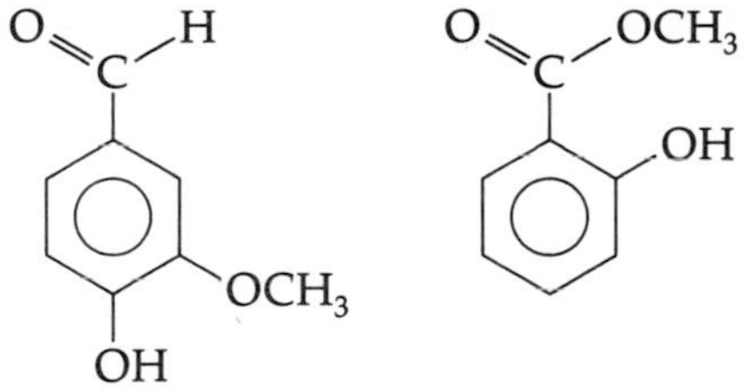

vanillin   oil of wintergreen

Not all compounds that contain the benzene ring have aromatic odors, however. An example is aspirin.

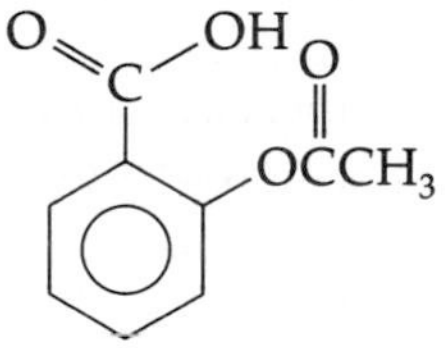

aspirin

The benzene ring has some special properties, most notably that it is more stable than would be predicted based on the alternating double bond structure of either of its resonance forms. As we will see, this causes the ring to behave chemically in ways that are different from the alkenes and alkynes.

Reactions of hydrocarbons

The alkanes are relatively resistant to attack by chemicals other than those that completely disrupt the molecular framework (e.g., oxygen, with which they react to form CO_2 and H_2O). At high temperature, in the presence of an appropriate catalyst, they can be made to lose hydrogen in a process known as *cracking*. Ethane, obtained from petroleum, is cracked to give ethene (ethylene), which is an important raw material in the organic chemicals industry.

$$CH_3-CH_3 \longrightarrow CH_2{=}CH_2 + H_2$$

The alkanes are also attacked by Cl_2, with which they react by **substitution.**

$$CH_3-CH_3 + Cl_2 \longrightarrow CH_3-CH_2Cl + HCl$$

Substitution reactions of this kind are difficult to control and yield products formed by further substitution of H by Cl (e.g., $C_2H_4Cl_2$, $C_2H_3Cl_3$, etc.).

Compared to the alkanes, the alkenes and alkynes are much more reactive. The π bond is a relatively weak bond and its electron density projects out from the bond axis where it can be easily attacked by electron-seeking reagents. In such reactions, the π bond is destroyed and additional atoms become incorporated into the structure by forming bonds to the carbon atoms. Because of this, these are called **addition reactions.**

One of the simplest of these is the addition of hydrogen to a double bond, which is accomplished at high temperature under pressure with the use of a platinum or nickel catalyst. For example,

Similar hydrogenation reactions occur with the alkynes. For example, addition of H_2 to acetylene yields ethylene first, followed by further hydrogenation to give ethane.

As noted earlier, the ability of alkenes and alkynes (as well as other compounds with carbon–carbon double and triple bonds) to add atoms to give more complex molecules explains why they are called **unsaturated** compounds—unsaturated in the sense that they accept additional atoms in chemical reactions. The alkanes and other organic compounds having only carbon–carbon single bonds, however, cannot do this and are said to be **saturated.**

Hydrogenation reactions are important in the food industry in which unsaturated liquid vegetable oils are hydrogenated to give saturated fats, which are usually solids at room temperature.

Under appropriate conditions, other reagents add to π bonds just as H_2 does. Two other examples are

The partial hydrogenation of a liquid vegetable oil gives a solid similar to animal fat.

Although it may appear that the benzene ring in aromatic compounds contains double bonds, the ring does not easily undergo addition reactions. In almost every case, reactions involving the benzene ring are substitution reactions and the delocalized π-electron system of the ring remains intact. The reason for this appears to be the extra stability provided by delocalization of the π electrons, which would be destroyed if the ring were to undergo addition.

IUPAC rules for naming hydrocarbons

The same international organization that gave us the rules discussed in Chapter 4 for naming inorganic compounds (the IUPAC) has also established a set of rules for naming organic compounds. In fact, the names we've given to various compounds earlier in this section are based on these rules, which are as follows.

1. **The longest unbroken chain of carbon atoms in any molecule serves to define the parent name for any hydrocarbon or its derivative. For the alkenes and alkynes, the parent chain *must* include the double or triple bond.** The number of carbon atoms in the chain is specified by an appropriate prefix such as *meth-* for one carbon, *eth-* for two carbons, and so forth. The prefixes for chains having from one to ten carbon atoms are as follows:

meth	1C	hex	6C
eth	2C	hept	7C
prop	3C	oct	8C
but	4C	non	9C
pent	5C	dec	10C

The names of the first ten alkanes in Table 22.7 illustrate how this rule applies to saturated hydrocarbons. Notice that for the alkanes, the prefix is followed by the suffix *-ane* to indicate that the compounds contain only carbon–carbon single bonds.

For branched compounds, we also seek the longest carbon chain. Thus, each of the following is named as a derivative of pentane, because in each case the longest carbon chain contains five carbon atoms. (We will refer to these compounds later.)

$$
\begin{array}{ll}
& \quad\;\; CH_3 \\
& \quad\;\; | \\
(1) \;\; CH_3\!-\!C\!-\!CH_2\!-\!CH_2\!-\!CH_3 \\
& \quad\;\; | \\
& \quad\;\; CH_3
\end{array}
\qquad
\begin{array}{l}
\quad\quad\quad CH_3 \;\; CH_3 \\
\quad\quad\quad | \quad\;\; | \\
(2) \;\; CH_3\!-\!CH_2\!-\!CH\!-\!CH\!-\!CH_3
\end{array}
$$

$$(3) \quad CH_3-\overset{\overset{\displaystyle CH_3}{|}}{\underset{\underset{\displaystyle CH_3}{|}}{C}}-CH_2-\overset{\overset{\displaystyle CH_3}{|}}{\underset{\underset{\displaystyle CH_3}{|}}{C}}-CH_3$$

In the following molecule, the longest chain has six carbons, but since the double bond *must* be included in the parent chain, it is also named as a derivative of a five-carbon hydrocarbon. This time, however, the suffix is *-ene*, so the compound is a derivative of pentene.

$$\begin{array}{c} CH_3-CH-CH_3 \\ | \\ CH_3-CH-CH_2-C=CH_2 \\ | \\ CH_3 \end{array}$$

Notice that in the molecule above, the longest carbon chain is not written in a straight line. This means that in a complex structure, you have to search out the longest chain. For example, the compound below has an eight-carbon chain indicated by the atoms printed in red, so it would be named as a derivative of octene (note the double bond).

$$\begin{array}{c} CH_3 \\ | \\ CH_3-CH-CH-CH_2-CH_2-CH_3 \\ | \\ CH_2 \\ | \\ CH_3-C-CH=CH_2 \\ | \\ CH_3 \end{array}$$

2. **Branched hydrocarbons are named as derivatives of straight-chain compounds in which one or more hydrogen atoms are replaced by hydrocarbon fragments.** An example of this is the compound methylpropane described earlier.

$$\begin{array}{c} CH_3 \\ | \\ CH_3-CH-CH_3 \\ \text{methylpropane} \end{array}$$

As noted earlier, hydrocarbon fragments that are derived from the alkanes by removal of a hydrogen are called alkyl groups. The name of the alkyl group is obtained by dropping the *-ane* ending and adding *-yl*. Some common alkyl groups and their structures are

$$\begin{array}{cccc} \overset{\displaystyle H}{\underset{\displaystyle H}{H-\overset{|}{\underset{|}{C}}-}} & \overset{\displaystyle H \;\; H}{\underset{\displaystyle H \;\; H}{H-\overset{|}{\underset{|}{C}}-\overset{|}{\underset{|}{C}}-}} & \overset{\displaystyle H \;\; H \;\; H}{\underset{\displaystyle H \;\; H \;\; H}{H-\overset{|}{\underset{|}{C}}-\overset{|}{\underset{|}{C}}-\overset{|}{\underset{|}{C}}-}} & \overset{\displaystyle H \;\;\;\;\; H}{\underset{\displaystyle H \;\; H \;\; H}{H-\overset{|}{\underset{|}{C}}-\overset{|}{\underset{|}{C}}-\overset{|}{\underset{|}{C}}-H}} \\ & & & | \\ CH_3- & CH_3CH_2- & CH_3CH_2CH_2- & CH_3CHCH_3 \\ \text{methyl} & \text{ethyl} & \text{propyl} & \text{isopropyl} \end{array}$$

3. **The locations of branches along the parent chain are indicated by numbers that identify the carbon atoms to which the branches are attached.** In alkanes the numbering starts from whichever end gives the lower

Methyl group

Ethyl group

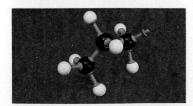

Propyl group

Isopropyl group

number to the first branching group. When multiple bonds are present, the numbering begins from the end nearer the first double or triple bond. The number identifying the position of an alkyl group immediately precedes its name, and the number is separated from the word by a hyphen. An example is

$$CH_3{-}CH_2{-}CH_2{-}\underset{\underset{\displaystyle CH_3}{|}}{CH}{-}CH_2{-}CH_3$$

3-methylhexane

This compound is chiral, as you learned earlier.

4. **If more than one of a given alkyl group is present, the prefixes di-, tri-, tetra-, and so on are used to indicate their number. It is also necessary to specify the location number of each one along the parent chain.** Some examples are provided by the compounds that were assigned numbers in our discussion of rule 1.

Compound 1:   2,2-dimethylpentane
Compound 2:   2,3-dimethylpentane
Compound 3:   2,2,4,4-tetramethylpentane

5. **When there are different alkyl groups among the branches, they are specified in alphabetical order according to the name of the group.** Thus, ethyl would be specified before methyl. An example is

$$\underset{\underset{\underset{\underset{\displaystyle CH_3}{|}}{\displaystyle CH_2}}{|}}{CH_3}{-}\underset{\underset{\displaystyle CH_3}{|}}{CH}{-}\underset{\underset{\underset{\displaystyle CH_2}{|}}{|}}{CH}{-}\underset{\underset{\displaystyle CH_3}{|}}{\overset{\overset{\displaystyle CH}{|}}{\overset{H_3C\diagdown\quad\diagup CH_3}{}}}{C}{-}CH_2{-}CH_3$$

3-ethyl-4-isopropyl-2,4-dimethylhexane

6. **The locations of double and triple bonds are indicated by numbering the bonds.** The numbering begins at the end of the parent chain that gives the lower number to the first multiple bond. Thus, we have

$CH_3{-}CH{=}CH{-}CH_3$ 2-butene

$CH_3{-}CH{=}CH{-}CH_2{-}CH{=}CH_2$ 1,4-hexadiene

$CH_3{-}\underset{\underset{\displaystyle CH_3}{|}}{CH}{-}CH{=}CH{-}CH_3$ 4-methyl-2-pentene

Notice the way we've indicated the presence of two double bonds in the second compound.

7. **A cyclic hydrocarbon is named by adding the prefix *cyclo-* to the name of the corresponding open-chain hydrocarbon.** The locations of alkyl groups that are attached to the ring are indicated by number. Some examples are

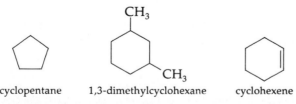

cyclopentane   1,3-dimethylcyclohexane   cyclohexene

Nomenclature of the aromatic hydrocarbons is more complex, so we will not attempt to name them.

Common names

Quite a large number of organic compounds were known before the IUPAC system was developed and had already been named according to other systems of nomenclature. Simply from common usage, some of these names carry over into current usage. For example, methylpropane, which we have mentioned on two occasions before, is commonly known as isobutane. Common names are also used for naming large complex molecules whose names according to the IUPAC system are long, complex, and cumbersome. Thus, it is much easier to remember the name "cholesterol" for the following molecule rather than attempting to name the molecule according to the IUPAC rules. (In fact, we would have to discuss the nomenclature rules to a much greater extent to be able to apply them here.)

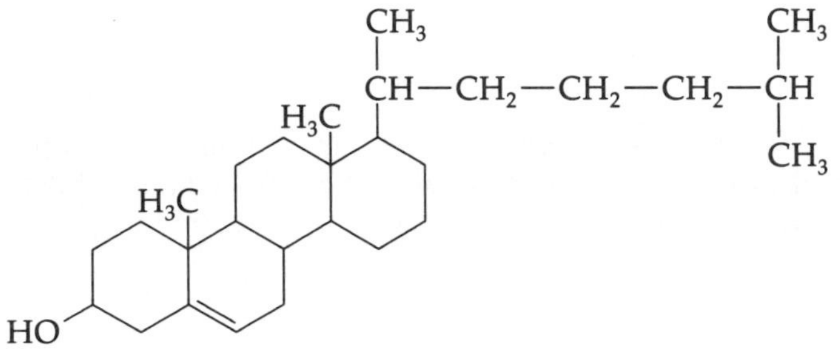

cholesterol

22.6 ORGANIC COMPOUNDS CONTAINING OXYGEN AND NITROGEN

A phenomenon that has greatly simplified the study of organic chemistry is the existence of groups of atoms, called **functional groups,** that impart particular chemical and physical properties to organic compounds that contain them. An example of a functional group we have already looked at is the double bond. Molecules that contain double bonds, be they large molecules or small, all tend to undergo the same kinds of chemical reactions because chemical attack is always at the same group of atoms, $\overset{\diagdown}{}\overset{\diagup}{}C{=}C\overset{\diagup}{}\overset{\diagdown}{}$.

Most functional groups contain atoms other than carbon and hydrogen, and in this section we will look briefly at some that contain the elements oxygen and nitrogen. They are listed in summary form in Table 22.8.

Alcohols

Alcohols are molecules that contain an OH group attached to a carbon atom, which in turn is attached to three other groups by single bonds.

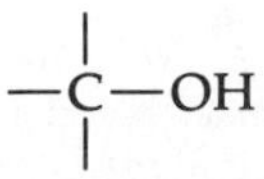

$$-\overset{\displaystyle |}{\underset{\displaystyle |}{C}}-OH$$

alcohol system

Table 22.8 Some functional groups in organic compounds

Functional Group	Compound Class	Example	Name
$-\overset{\mid}{C}=\overset{\mid}{C}-$	Alkenes	$CH_2\!=\!CH_2$	Ethene
$-C\equiv C-$	Alkynes	$HC\equiv CH$	Ethyne
F, Cl, Br, I	Halides	CH_3Cl	Chloromethane
$-OH$	Alcohols	CH_3OH	Methanol
$-C\overset{O}{\underset{H}{\diagdown}}$	Aldehydes	$CH_3-\overset{O}{\overset{\|}{C}}-H$ or CH_3CHO	Ethanal (acetaldehyde)
$-C\overset{O}{\diagdown}$	Ketones	$CH_3-\overset{O}{\overset{\|}{C}}-CH_3$	Propanone (acetone)
$-C\overset{O}{\underset{OH}{\diagdown}}$	Carboxylic acids	$CH_3-\overset{O}{\overset{\|}{C}}-OH$	Ethanoic acid (acetic acid)
$-\overset{\mid}{\underset{\mid}{C}}-N\overset{H}{\underset{H}{\diagdown}}$	Amines	CH_3NH_2	Methylamine
$-\overset{\mid}{\underset{\|}{\underset{O}{C}}}-N\overset{H}{\underset{H}{\diagdown}}$	Amides	$CH_3-\overset{O}{\overset{\|}{C}}-NH_2$	Ethanamide (acetamide)
$-\overset{\mid}{\underset{\mid}{C}}-O-\overset{\mid}{\underset{\mid}{C}}-$	Ethers	CH_3-O-CH_3	Dimethyl ether
$-\overset{\mid}{\underset{\mid}{C}}-\overset{O}{\overset{\|}{C}}-O-\overset{\mid}{\underset{\mid}{C}}-$	Esters	$CH_3-\overset{O}{\overset{\|}{C}}-O-CH_3$	Methyl acetate

Some common examples of alcohols are the following; their IUPAC names are written below their common names.

$$CH_3OH \qquad CH_3CH_2OH \qquad CH_3CH_2CH_2OH \qquad CH_3\overset{OH}{\overset{\mid}{C}H}CH_3$$

methyl alcohol   ethyl alcohol   propyl alcohol   isopropyl alcohol

methanol   ethanol   1-propanol   2-propanol

Methyl alcohol is the fuel used in "canned heat" products such as Sterno. It is an important industrial chemical. Ethyl alcohol is the alcohol found in alcoholic beverages. It is also mixed with gasoline to give a product called gasohol. Isopropyl alcohol is the alcohol in "rubbing alcohol."

Some alcohols have more than one OH group. Some common ones are

$$\underset{OH\quad OH}{CH_2-CH_2} \qquad \underset{OH\quad OH}{CH_3-CH-CH_2} \qquad \underset{OH\quad OH\quad OH}{CH_2-CH-CH_2}$$

ethylene glycol   propylene glycol   glycerol

Ethylene glycol is used as an antifreeze in automobiles. It is toxic, however, and cannot be used to protect water systems from freezing on boats and RVs. Often, antifreeze mixtures for these systems contain propylene glycol or glycerol, both of which are nontoxic. Propylene glycol is also used in various food products to improve texture.

The OH group in alcohols allows them to form hydrogen bonds to each other. As a result, these molecules have much higher boiling points than hydrocarbons with the same number of atoms. (We compare molecules with approximately the same number of atoms because the strengths of the London forces would be approximately the same for both of them. For example, ethyl alcohol boils at 78.5 °C, but propane, which has about the same number of atoms, boils at -44 °C. Ethylene glycol, with its two OH groups, can form more hydrogen bonds than ethyl alcohol and boils at 198 °C.

Alcohols can also hydrogen bond to water molecules, and the lower-molecular-weight alcohols such as CH_3OH and CH_3CH_2OH are water-soluble. When the hydrocarbon chain becomes longer than about five carbons, however, the solubility becomes quite low. Molecules such as ethylene glycol, propylene glycol, and glycerol are completely miscible with water, as you might expect based on the number of OH groups they contain.

R is an alkyl group.

Ethers

Ethers are molecules in which there are two hydrocarbon groups attached to the same oxygen.

$$R—O—R' \quad \text{R and R' stand for hydrocarbon groups.}$$
ether

$$CH_3—O—CH_3 \qquad CH_3—O—CH_2CH_3 \qquad CH_3CH_2—O—CH_2CH_3$$
dimethyl ether   methyl ethyl ether   diethyl ether

Diethyl ether was once the most widely used anesthetic for surgery.

Ethers have much lower boiling points than alcohols because they lack the ability to form hydrogen bonds. For example, dimethyl ether and ethyl alcohol have the same molecular formula, C_2H_6O, and are therefore isomers of each other. The alcohol boils at 78.5 °C, but the ether boils at -23 °C. Because ethers are relatively nonpolar substances, they also have a low solubility in water.

Aldehydes and ketones

Aldehydes and ketones have something in common; they both contain a carbon–oxygen double bond, $C=O$. This group of atoms is called a **carbonyl group.**

$$R(H)—C=O \qquad\qquad R—C—R'$$
aldehyde   ketone

Notice, however, that in an aldehyde there is at least one hydrogen attached to the carbon of the carbonyl group. This feature is absent from ketones and the difference has a major impact on the chemical properties of these two classes of compounds.

Examples of aldehydes are

$$H—C=O \qquad\qquad CH_3—C=O$$
formaldehyde   acetaldehyde
methanal   ethanal

Formaldehyde, of course, is used in preserving biological specimens and for embalming. It is also used in the manufacture of some plastics.

Some important ketones are

$$CH_3-\overset{\overset{\displaystyle O}{\|}}{C}-CH_3 \qquad CH_3-\overset{\overset{\displaystyle O}{\|}}{C}-CH_2CH_3$$

acetone
propanone

methyl ethyl ketone
butanone

Acetone is the solvent in nail polish remover, and methyl ethyl ketone is a useful industrial solvent.

Ketones are polar molecules because of the presence of the polar carbonyl group. When the molecular mass is low (i.e., when the hydrocarbon portions of the molecules are small), ketones tend to be water-soluble. Acetone, for example, is completely miscible with water.

Carboxylic acids

Carboxylic acids are substances that contain the carboxyl group

$$-\overset{\overset{\displaystyle O}{\|}}{C}-OH$$

carboxyl group

The name is a contraction for *carb*onyl + hydr*oxyl*. The carboxyl group is often written as either $-COOH$ or $-CO_2H$.

The hydrogen of the OH group is much more acidic in the carboxyl group than in the alcohols because of the presence of carbonyl group. The electronegative oxygen of the carbonyl group draws electron density from the carbon, which in turn further polarizes the already-polar $O-H$ bond. This makes the carboxyl group a reasonably good proton donor, and in aqueous solutions compounds that contain this group tend to be weak Brønsted acids. For example, acetic acid (which we've discussed many times before) reacts as follows:

$$CH_3-\overset{\overset{\displaystyle O}{\|}}{C}-OH + H_2O \rightleftharpoons CH_3-\overset{\overset{\displaystyle O}{\|}}{C}-O^- + H_3O^+$$

acetic acid

acetate ion

Other often-encountered carboxylic acids include formic acid, which you should recall is what gives ant bites their painful sting, and butyric acid, which is responsible for the bad odor of rancid butter.

$$H-\overset{\overset{\displaystyle O}{\|}}{C}-OH \qquad CH_3CH_2CH_2-\overset{\overset{\displaystyle O}{\|}}{C}-OH$$

formic acid
methanoic acid

butyric acid
butanoic acid

Oxidation reactions

Alcohols, aldehydes, ketones, and carboxylic acids are related to one another through oxidation–reduction reactions. The oxidation reactions can be summarized as follows, using the symbol [O] to stand for some arbitrary oxidizing agent:

$$
\begin{array}{c}
\underset{|}{\text{OH}} \\
\text{R}-\overset{|}{\underset{|}{\text{C}}}-\text{H} \\
\text{H}
\end{array}
\xrightarrow{\text{[O]}}
\text{R}-\overset{\text{O}}{\overset{\|}{\text{C}}}-\text{H}
\xrightarrow{\text{[O]}}
\text{R}-\overset{\text{O}}{\overset{\|}{\text{C}}}-\text{OH}
$$

$$
\begin{array}{c}
\underset{|}{\text{OH}} \\
\text{R}-\overset{|}{\underset{|}{\text{C}}}-\text{H} \\
\text{R}
\end{array}
\xrightarrow{\text{[O]}}
\text{R}-\overset{\text{O}}{\overset{\|}{\text{C}}}-\text{R}
\xrightarrow{\text{[O]}}
\text{CO}_2 + \text{H}_2\text{O}
$$

$$
\begin{array}{c}
\underset{|}{\text{OH}} \\
\text{R}-\overset{|}{\underset{|}{\text{C}}}-\text{R} \\
\text{R}
\end{array}
\xrightarrow{\text{[O]}}
\text{CO}_2 + \text{H}_2\text{O}
$$

If at least one hydrogen is attached to the carbon that is being oxidized, oxidation can proceed under relatively mild conditions. In the first reaction, the product of the oxidation of RCH$_2$OH is an aldehyde. But aldehydes are very easily oxidized, so this compound is oxidized further to give the carboxylic acid. In the second reaction, only one hydrogen is attached to the carbon that has the OH group. Mild oxidation of R$_2$CHOH gives the ketone, but this compound resists further oxidation. Mild oxidizing agents have no effect, but under more strenuous conditions the carbon skeleton is disrupted, ultimately giving CO$_2$ and H$_2$O. When there are no hydrogens attached to the carbon with the OH group, the alcohol resists oxidation. Under severe conditions oxidation once again breaks up the carbon skeleton and ultimately gives CO$_2$ and H$_2$O.

When wine changes to vinegar, the ethanol in the wine is oxidized to acetic acid.

Esters

Esters are formed by the reaction of alcohols and carboxylic acids. They contain the functional group

$$
\text{R}-\overset{\text{O}}{\overset{\|}{\text{C}}}-\text{O}-\text{R}'
$$

For example, acetic acid reacts with ethyl alcohol in the presence of H$^+$ to give the compound ethyl acetate.

$$
\underset{\text{acetic acid}}{\text{CH}_3-\overset{\text{O}}{\overset{\|}{\text{C}}}-\text{O}-\text{H}} + \underset{\text{ethyl alcohol}}{\text{H}-\text{O}-\text{CH}_2\text{CH}_3}
\underset{}{\overset{\text{H}^+}{\rightleftharpoons}}
\underset{\text{ethyl acetate}}{\text{CH}_3-\overset{\text{O}}{\overset{\|}{\text{C}}}-\text{O}-\text{CH}_2\text{CH}_3} + \text{H}_2\text{O}
$$

The reaction is an equilibrium, and a large concentration of the alcohol drives the equilibrium to the right by Le Châtelier's principle.

Esters tend to have very pleasant odors, and many of the fragrances associated with fruits are due to the presence of these compounds. Some examples are

$$
\underset{\text{isopentyl acetate (banana fragrance)}}{\text{CH}_3-\overset{\text{O}}{\overset{\|}{\text{C}}}-\text{O}-\text{CH}_2\text{CH}_2\overset{\overset{\text{CH}_3}{|}}{\text{CH}}\text{CH}_3}
\qquad
\underset{\text{ethyl butyrate (pineapple fragrance)}}{\text{CH}_3\text{CH}_2\text{CH}_2-\overset{\text{O}}{\overset{\|}{\text{C}}}-\text{O}-\text{CH}_2\text{CH}_3}
$$

$$
\underset{\text{octyl acetate (orange fragrance)}}{\text{CH}_3-\overset{\text{O}}{\overset{\|}{\text{C}}}-\text{O}-\text{CH}_2\text{CH}_2\text{CH}_2\text{CH}_2\text{CH}_2\text{CH}_2\text{CH}_2\text{CH}_3}
$$

The reaction by which esters are formed can be reversed if a large concentration of water is present. This causes the reaction shown for the formation of ethyl acetate to be shifted to the left. Esters are also decomposed by base, which helps the reaction along by neutralizing the carboxylic acid as it is formed. The process is called **saponification.**

$$CH_3-\overset{\overset{\textstyle O}{\|}}{C}-O-CH_2CH_3 + OH^- \longrightarrow CH_3-\overset{\overset{\textstyle O}{\|}}{C}-O^- + H-O-CH_2CH_3$$

This reation is used to change animal fats to soap. Fats are esters of glycerol with **fatty acids,** acids that have a long hydrocarbon chain attached to the carboxyl group. An example is stearic acid, $C_{17}H_{35}CO_2H$.

$$CH_3CH_2CH_2CH_2CH_2CH_2CH_2CH_2CH_2CH_2CH_2CH_2CH_2CH_2CH_2CH_2CH_2-\overset{\overset{\textstyle O}{\|}}{C}-OII$$
stearic acid

A typical saponification reaction is

$$
\begin{array}{l}
CH_2-O-\overset{\overset{\textstyle O}{\|}}{C}-C_{17}H_{35} \\[2mm]
| \\[1mm]
CH-O-\overset{\overset{\textstyle O}{\|}}{C}-C_{17}H_{35} + 3OH^- \longrightarrow \\[2mm]
| \\[1mm]
CH_2-O-\overset{\overset{\textstyle O}{\|}}{C}-C_{17}H_{35}
\end{array}
\qquad
\begin{array}{l}
CH_2-OH \\[2mm]
| \\[1mm]
CH-OH + 3\ ^-O-\overset{\overset{\textstyle O}{\|}}{C}-C_{17}H_{35} \\[2mm]
| \\[1mm]
CH_2-OH
\end{array}
$$

The way the $C_{17}H_{35}CO_2^-$ anions function in soaps was described in Chapter 13.

Nitrogen-containing functional groups

Amines could be considered to be organic derivatives of ammonia in which one or more hydrogens of the NH_3 molecule are replaced by hydrocarbon groups. Examples are

$$
\begin{array}{ccc}
H & H & CH_3 \\
| & | & | \\
H-N-CH_3 & CH_3-N-CH_3 & CH_3-N-CH_3 \\
\cdot\cdot & \cdot\cdot & \cdot\cdot \\
\text{methylamine} & \text{dimethylamine} & \text{trimethylamine}
\end{array}
$$

Because of the lone pair of electrons on the nitrogen, these compounds are basic in the same way that ammonia is. Amines generally have unpleasant odors; those with low molecular masses are responsible for the odor of overripe fish.

Amino acids are compounds that contain both an amine group and a carboxyl group. Especially important in biological systems are α-amino acids in which a NH_2 group is attached to the carbon adjacent to the carboxyl group. For example,

In water for example, CH_3NH_2 reacts as follows:

$$CH_3NH_2(aq) + H_2O \rightleftharpoons$$
$$CH_3NH_3^+(aq) + OH^-(aq)$$

$$
\begin{array}{ccc}
NH_2-CH-\overset{\overset{\textstyle O}{\|}}{C}-OH & NH_2-CH-\overset{\overset{\textstyle O}{\|}}{C}-OH & NH_2-CH-\overset{\overset{\textstyle O}{\|}}{C}-OH \\
| & | & | \\
R & H & CH_3 \\
\text{general formula for} & \text{glycine} & \text{alanine} \\
\text{an } \alpha\text{-amino acid} & &
\end{array}
$$

Actually, because the NH_2 group is basic and the CO_2H group is acidic, these compounds tend to exist in a *dipolar form* as illustrated below.

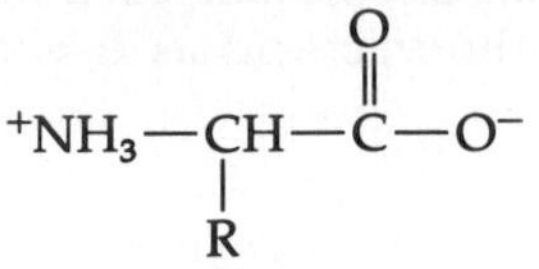

$$^+NH_3-CH-\overset{\overset{\displaystyle O}{\|}}{C}-O^-$$

$$\underset{R}{|}$$

dipolar form of
the amino acid

Amides

Amides are similar to esters except that a carbonyl group links to an amine group instead of an alcohol. The amide system is

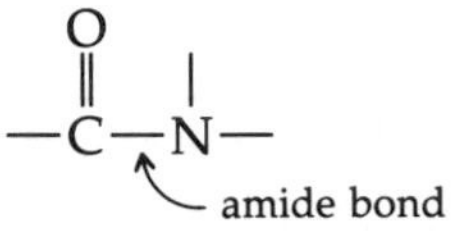

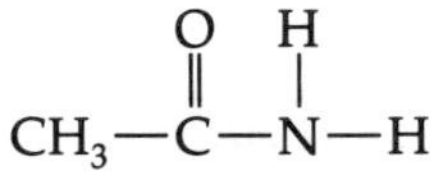

$$-\overset{\overset{\displaystyle O}{\|}}{C}-\overset{}{N}-$$
amide bond

$$CH_3-\overset{\overset{\displaystyle O}{\|}}{C}-\overset{\overset{\displaystyle H}{|}}{N}-H$$
acetamide
ethanamide

In biological systems amino acids become joined by amide bonds to give complex structures known as polypeptides. These are the principal constituents of proteins.

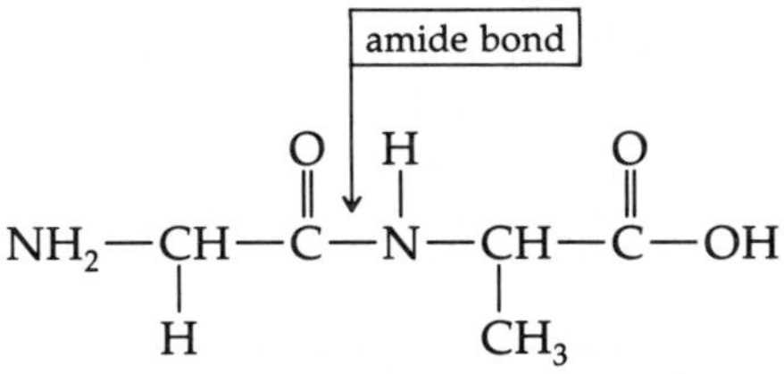

amide bond

$$NH_2-\underset{\underset{H}{|}}{CH}-\overset{\overset{\displaystyle O}{\|}}{C}-\overset{\overset{\displaystyle H}{|}}{N}-\underset{\underset{CH_3}{|}}{CH}-\overset{\overset{\displaystyle O}{\|}}{C}-OH$$

glycine   alanine

A dipeptide formed by
linking glycine and alanine

REVIEW QUESTIONS AND PROBLEMS

Problems whose numbers are in blue have their answers in Appendix D at the back of the book.

Hydrogen: General

22.1   What is the correct IUPAC name for H_2?

22.2   Which is the most abundant element in the universe?

22.3   What is the probable reason that the abundance of hydrogen on Earth is less than it is in the rest of the universe?

22.4   Why is there little free hydrogen in the Earth's atmosphere?

22.5   What advantage does hydrogen have over helium for use in lighter-than-air balloons? What is a disadvantage?

22.6   Why are the properties of hydrogen so different from the properties of the other elements in Group IA? Why is hydrogen placed in Group IA?

22.7   Give the names and symbols of the three isotopes of hydrogen. Which one is radioactive? Why does it potentially pose an environmental problem?

22.8   What are the advantages and disadvantages of hydrogen as a fuel?

Hydrogen: Preparation and Uses

22.9   Give chemical reactions for the preparation of hydrogen from water and (a) methane (natural gas); (b) coal.

22.10   Write chemical equations for the combustion of water gas.

22.11   In what way is the commercial production of caustic soda related to the commercial production of hydrogen?

22.12 What is the greatest industrial use for hydrogen? Give the appropriate chemical equation.

22.13 Write a chemical equation for the laboratory preparation of hydrogen. Sketch the apparatus you would use.

22.14 How is hydrogen used in the synthesis of methanol? What importance is this commercially?

22.15 Use Lewis structures to illustrate the hydrogenation of acetylene, C_2H_2. What effect does hydrogenation have on vegetable oils?

Hydrogen Compounds

22.16 What general name is given to binary compounds of hydrogen? What would be the name for the compound NaH? How could it be formed? What is its reaction with water?

22.17 How many covalent bonds does hydrogen normally form? Why?

22.18 Predict the molecular structures of (a) H_2S; (b) PH_3; (c) SiH_4.

22.19 Explain how the reaction $H^- + H_2O \rightarrow H_2 + OH^-$ can be interpreted as both an acid–base reaction and a redox reaction.

22.20 Define *catenation*. How does the tendency to catenate vary within a group in the periodic table?

22.21 Why can't all the nonmetal hydrides be made in high yield by direct combination of the elements? Which nonmetal hydrides can be made by this method?

22.22 How do the basicities of the conjugate bases of the nonmetal hydrides vary within periods and groups in the periodic table? Give two examples.

Oxygen: General

22.23 What are the commercial sources of oxygen and nitrogen?

22.24 What is the correct IUPAC name for O_2?

22.25 How is oxygen conveniently prepared in the laboratory?

22.26 What is the largest use for O_2 industrially?

22.27 Give the structure for ozone. How can it be made in the laboratory? What advantage does ozone have over Cl_2 for the purification of drinking water?

22.28 How is ozone formed in the Earth's upper atmosphere? How does it protect the Earth's inhabitants from ultraviolet radiation from the sun? How could nitrogen oxide pollutants affect the Earth's ozone layer (give equations)? How could Freons affect the ozone layer (give equations)?

Oxygen Compunds

22.29 Write a chemical equation for the hydrolysis of oxide ion.

22.30 What does "pickling" mean with respect to the treatment of iron and steel objects?

22.31 Write equations that show the amphoteric behavior of a metal oxide.

22.32 Give three ways that the oxide, CO_2, can be made. Illustrate each with a chemical equation.

22.33 Why can N_2O not be synthesized from its elements?

22.34 Write an equation for the reaction of ammonia with oxygen in the absence of a catalyst. What nitrogen-containing product is formed if a platinum catalyst is present?

22.35 Write chemical equations for the reaction of copper with (a) concentrated HNO_3; (b) dilute HNO_3.

22.36 Give equations showing reactions of oxygen to form (a) a metal peroxide; (b) a metal superoxide.

22.37 What is the structure of hydrogen peroxide?

22.38 Write an equation for the decomposition of hydrogen peroxide.

Nitrogen: General

22.39 Why is dinitrogen so unreactive?

22.40 Give three commercial uses of N_2.

22.41 What are *nitrogen-fixing bacteria*?

22.42 How can small amounts of N_2 be made in the laboratory?

22.43 Which is the only element N_2 reacts with at room temperature?

22.44 Write the equation for the reaction of Mg_3N_2 with water.

Ammonia

22.45 Why is ammonia so soluble in water?

22.46 Give the formula for the principal nitrogen-containing species in a solution of *ammonium hydroxide*.

22.47 How can ammonia be made in the laboratory? Could it be collected by displacement of water in the same manner as hydrogen? Explain.

22.48 How could you determine whether a particular unknown solid was an ammonium salt?

Nitric Acid

22.49 Write equations for the chemical reactions in the Ostwald process.

22.50 Why does concentrated HNO_3 usually have a pale yellow color?

22.51 How could pure HNO_3 be prepared in the laboratory?

22.52 Give three commercial uses of nitric acid and nitrates.

22.53 What nitrogen-containing product is formed when zinc reacts with HNO_3?

22.54 Why does nitric acid dissolve metals such as silver and copper, which are unaffected by hydrochloric acid?

22.55 What is aqua regia? How does it function in dissolving noble metals such as gold?

Other Compounds and Reactions of Nitrogen

22.56 Give an example of a nitrogen compound in which the oxidation number of nitrogen is (a) -3, (b) -1, (c) $+1$, (d) $+3$, and (e) $+5$.

22.57 Give the chemical equation and reaction conditions in the Haber process. Why are those particular conditions chosen?

22.58 Give the formula for potassium amide. Why can potassium amide not be made in an aqueous solution?

22.59 What is a disproportionation reaction?

22.60 What influence did the Haber and Ostwald processes have on the waging of World War I?

22.61 Give the Lewis structure of hydrazine.

22.62 How is hydrazine prepared? Why has hydrazine been used as a rocket fuel?

22.63 Give the Lewis structure of hydroxylamine. Compare the basicities of ammonia, hydrazine, and hydroxylamine.

22.64 Why is it potentially dangerous to mix household ammonia with liquid laundry bleach?

22.65 Give a chemical equation for the preparation of nitrous oxide. What are the resonance structures for nitrous oxide?

22.66 Look up the standard free energies of formation of N_2O, NO, and NO_2. What accounts for the apparent stability of these oxides of nitrogen?

22.67 Give the Lewis structures for NO_2 and N_2O_4. Write the equilibrium that exists between these two molecules.

22.68 What is the structure of N_2O_3? Under what conditions can it be formed? Of which acid is N_2O_3 the acid anhydride?

22.69 Write a chemical equation for the decomposition of nitrous acid.

22.70 Write Lewis structures for NO_2 and NO_2^-. On the basis of the principles of VSEPR theory, which one should have the larger bond angle?

22.71 Write equations that show nitrous acid functioning as (a) a reducing agent; (b) an oxidizing agent.

22.72 How is $NaNO_2$ normally made? What are the functions of $NaNO_2$ in meat products such as bologna? What are the pros and cons concerning its use in such products?

22.73 Give the Lewis structure for N_2O_5 as it exists in the vapor. How does N_2O_5 exist in the solid state?

22.74 How can N_2O_5 be made?

22.75 Write a chemical equation for the reaction of N_2O_5 with water.

22.76 For octane, $\Delta H_f^\circ = -255.1$ kJ/mol. Use this and the data in Table 6.1 to calculate the standard heats of reaction for the following:

$$C_8H_{18}(l) + 12\tfrac{1}{2}O_2(g) \longrightarrow 8CO_2(g) + 9H_2O(g)$$
$$C_8H_{18}(l) + 25N_2O(g) \longrightarrow 8CO_2(g) + 9H_2O(g) + 25N_2(g)$$

Smog

22.77 Describe the series of chemical reactions responsible for photochemical smog.

22.78 What substance causes the reddish-brown haze of photochemical smog? What is PAN?

Carbon: General

22.79 How is coke made?

22.80 How do the structures of diamond and graphite differ?

22.81 What are two uses of graphite?

22.82 What are three properties of diamond that make it useful industrially?

22.83 What is activated charcoal? What property does it possess that makes it commercially useful?

22.84 How is carbon black made? What are two of its uses?

Oxides and Oxoacids of Carbon

22.85 Give the structures of carbon monoxide and carbon dioxide.

22.86 How can carbon monoxide be prepared in the laboratory?

22.87 Write a chemical equation showing the reducing properties of carbon monoxide toward Fe_2O_3.

22.88 What is a metal carbonyl? Give an example.

22.89 How is CO_2 usually made industrially? How can it be made conveniently in the laboratory?

22.90 Give three major industrial uses of CO_2.

22.91 How do green plants use CO_2?

22.92 Why are environmentalists concerned about the clearing of the rain forests in the Amazon basin in South America?

22.93 How is carbon dioxide involved in the "greenhouse effect?" Why is CO_2 implicated in global warming?

22.94 Give the equations for the dissolving of limestone in water by dissolved carbon dioxide.

22.95 How are stalagmites and stalactites formed in limestone caves?

22.96 What is hard water? How can hardness ions be removed from hard water?

22.97 Can you suggest a chemical method for removing boiler scale from the insides of pipes and water boilers?

Other Inorganic Compounds of Carbon

22.98 What is carborundum? How is its structure related to that of diamond? How is carborundum made?

22.99 Write an equation for the reaction of aluminum carbide with water.

22.100 How is acetylene made?

22.101 What is an interstitial carbide? Give an example.

22.102 How is hydrogen cyanide synthesized? How does it function as a poison?

22.103 Give the structure of carbon disulfide. What makes CS_2 a hazardous solvent with which to work?

Organic Chemistry

22.104 What is an *organic* compound?

22.105 What is the difference between a saturated and an unsaturated hydrocarbon?

22.106 The straight-chain alkanes are nonpolar molecules. How do we explain the fact that their boiling points increase from CH_3 to $C_{10}H_{22}$?

22.107 What would be the molecular formulas for (a) an alkane having 30 carbon atoms, (b) an alkene having 27 carbon atoms, (c) an alkyne having 33 carbon atoms?

22.108 What would be the molecular formula for a straight-chain hydrocarbon having 17 carbon atoms and (a) all C—C single bonds, (b) one C—C double bond, (c) one C—C triple bond, (d) three C—C double bonds, and (e) two C—C triple bonds?

Isomerism

22.109 Draw the remaining isomers of the compound

$$CH_3-\underset{\underset{CH_3}{|}}{C}=\underset{\underset{CH_3}{|}}{C}-CH_3$$

22.110 How does optical isomerism arise in organic compounds? Sketch the optical isomers of

$$Br-\underset{\underset{Cl}{|}}{\overset{\overset{H}{|}}{C}}-I$$

22.111 Sketch the cis and trans isomers of 2-pentene.

22.112 Draw the structural formulas for and name the nine isomers of heptane. Which of these isomers would give rise to optical isomerism?

22.113 Draw all possible open chain isomers of hexene (those that do not contain rings of carbon atoms) and show geometric isomers wherever possible.

22.114 Show by sketches that the molecule

$$CH_3-\underset{\underset{Cl}{|}}{\overset{\overset{F}{|}}{C}}-H$$

is chiral.

22.115 Show by sketches that the compound CH_3CHCl_2 is not chiral.

Structure and Bonding

22.116 What geometry do we expect for the molecules described in Question 22.110?

22.117 The molecule C_2Cl_4 is planar. Why?

22.118 The carbon atoms in 2-butyne lie in a straight line while those in butane do not. Explain why this is so.

Nomenclature

22.119 Name the following compounds:

(a)
$$\begin{array}{c} H_3C \\ \diagdown \\ H_3C \diagup \end{array} CH-CH_2-\underset{\underset{CH_3}{|}}{CH}-CH_2-CH_3$$

(b)
$$CH_3-\underset{\underset{\underset{\underset{CH_3}{|}}{CH_2}}{|}}{CH}-CH_2-\underset{\underset{CH_3}{|}}{CH}-CH_2-CH_3$$

(c)
$$CH_3-CH_2-\underset{\underset{CH_3}{|}}{CH}-CH_2-\underset{\underset{CH_2-CH_2-CH_3}{|}}{CH}-CH_2-CH_3$$

(d)
$$CH_3-CH_2-CH=CH-\underset{\underset{CH_2-CH_3}{|}}{CH}-CH_3$$

(e)
$$CH_3-CH_2-\underset{\underset{CH_3}{|}}{CH}-CH_2-\underset{\underset{CH_3}{|}}{CH}-CH_3$$

22.120 Name the following compounds:

(a)
$$CH_3-\underset{\underset{CH_3-C=CH-CH_3}{|}}{\overset{\overset{CH_2-CH_3}{|}}{C}}=C-CH_2-CH_3$$

(b)
$$CH_3-CH_2-C\equiv C-\underset{\underset{CH_2-CH_3}{|}}{CH}-CH_3$$

(c)

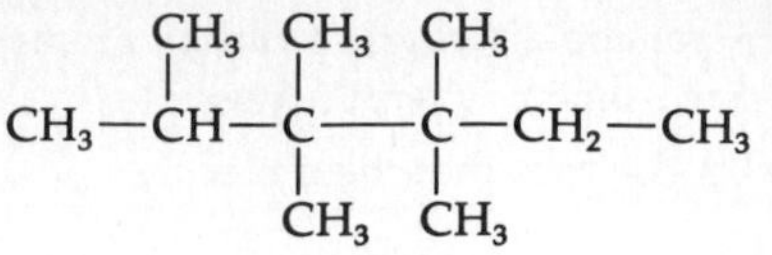

$$CH_3-CH-\underset{\underset{CH_3}{|}}{\overset{\overset{CH_3}{|}}{C}}-\underset{\underset{CH_3}{|}}{\overset{\overset{CH_3}{|}}{C}}-CH_2-CH_3$$

(d)

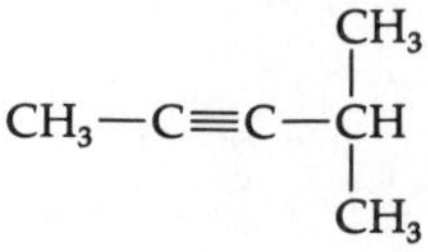

$$CH_3-C\equiv C-\underset{\underset{CH_3}{|}}{\overset{\overset{CH_3}{|}}{CH}}$$

(e)

$$\underset{CH_3-CH_2}{\overset{H_3C}{\diagdown}}C=C\underset{\diagdown CH_3}{\overset{\diagup CH=CH}{\diagdown}}CH_2-CH_3$$

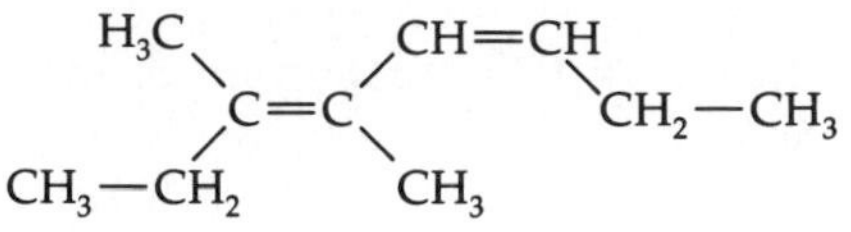

22.121 Write structural formulas for the following:
    (a) 2-methylpentane
    (b) 2,3-dimethylbutane
    (c) 2,3-dimethyl-2-butene
    (d) 1,3,5-octatriene
    (e) 3,3,4-trimethyl-1-pentyne

22.122 What is the proper IUPAC name for 2,4-diethyl-3,3,4-trimethylpentane?

Functional Groups

22.123 What is a functional group? Give the structural formula for
    (a) an aldehyde   (e) an alcohol
    (b) a ketone   (f) an ester
    (c) a carboxylic acid (g) an ether
    (d) an amine

22.124 Vanillin (page 761) contains three types of functional groups. What are they?

22.125 Draw the structures of the esters formed from

    (a) acetic acid and 2-propanol
    (b) acetic acid and 1-pentanol
        ($CH_3CH_2CH_2CH_2CH_2OH$)
    (c) benzoic acid and methanol
    (d) formic acid (methanoic acid) and methanol

22.126 Diethylamine gives a basic solution in water. Why? Write the equation for the reaction.

22.127 What kinds of compounds often have fruity odors?

22.128 The odor of slightly spoiled fish is characteristic of which kind of compound?

22.129 What type of reaction is characteristic of the alkenes and alkynes? What type of reaction is characteristic of the alkanes?

22.130 What products (if any) are obtained by the *mild* oxidation, using $K_2Cr_2O_7$ in acid solution, of each of the following:

    (a) CH_3-CH_2-OH

    (b) 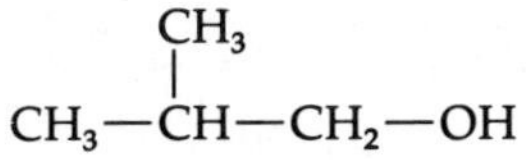
$$CH_3-\underset{\underset{CH_3}{|}}{CH}-CH_2-OH$$

    (c) 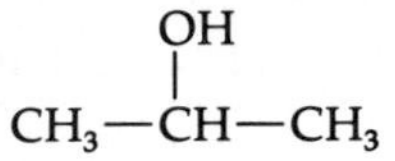
$$CH_3-\underset{\underset{OH}{|}}{CH}-CH_3$$

    (d) CH_3-CH_2-CHO

    (e) 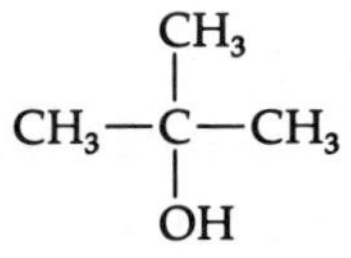
$$CH_3-\underset{\underset{OH}{|}}{\overset{\overset{CH_3}{|}}{C}}-CH_3$$

    (f) CH_3-COOH

    (g) 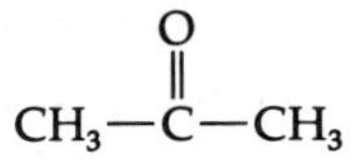
$$CH_3-\overset{\overset{\displaystyle O}{\|}}{C}-CH_3$$

22.131 Write equations for the saponification of

    (a) 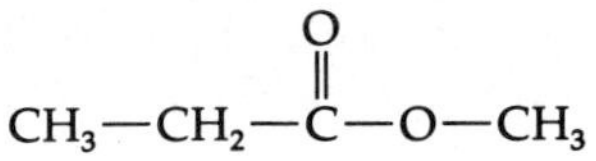
$$CH_3-CH_2-\overset{\overset{\displaystyle O}{\|}}{C}-O-CH_3$$

    (b) 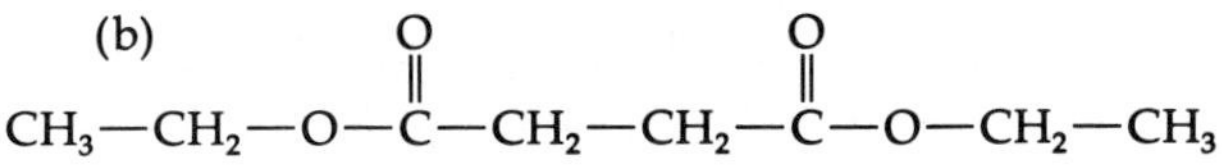
$$CH_3-CH_2-O-\overset{\overset{\displaystyle O}{\|}}{C}-CH_2-CH_2-\overset{\overset{\displaystyle O}{\|}}{C}-O-CH_2-CH_3$$

22.132 What is an alpha amino acid? Describe how amino acids interact to form a peptide bond.

PHOSPHORUS, SULFUR, THE HALOGENS, THE NOBLE GASES, AND SILICON, AND AN INTRODUCTION TO POLYMER CHEMISTRY

The nonmetallic elements discussed in this chapter cover a broad range of chemical reactivity and commercial importance. They range from fluorine, the most reactive element, to helium, the least reactive. Here we find sulfuric and phosphoric acids— chemicals that consistently rank among the top 10 in total tonnage manufactured annually. At the other extreme we have the compounds of the heavier noble gases, substances for which no practical uses have been found. (In fact, the discovery that some of the noble gases form compounds was a bombshell in the chemical community just three decades ago.)

We also see in this chapter a range of biological importance. Compounds of phosphorus and sulfur occur in all living organisms, but many compounds that contain the halogens are deadly poisons. On the other hand, living things appear to be indifferent toward the noble gases.

Finally, we conclude the chapter with a discussion of polymers. Our focus here is on synthetic materials of the type that have come to play such an important role in modern civilization.

23.1 PHOSPHORUS

The Earth's crust contains about 0.1% phosphorus by weight.

Phosphorus is a relatively abundant element, ranking twelfth in the Earth as a whole. It is always found in the combined state, normally in *phosphate rock*, which consists primarily of minerals that contain calcium phosphate, $Ca_3(PO_4)_2$. In living systems, phosphorus serves a number of critical functions. Phosphate units play a role in the structure of DNA, which directs the chemistry of our cells and transmits genetic information from one generation to another. Phosphate units are also present in phospholipids—substances that make up cell membranes—and phosphorus-oxygen bonds store the energy that we derive from the metabolism of foods.

About 85% of the phosphate rock mined in the United States now comes from Florida and North Carolina.

Because phosphorus is so important to biological systems of all kinds, it is not surprising that most industrial applications of phosphorus compounds, especially phosphates, are ultimately related in one way or another to providing sufficient phosphorus for the growth of agricultural products. For this reason also, deposits of phosphate rock are an important national resource. Until recently, some rich U.S. deposits of phosphate rock appeared on the verge of depletion during the 1990s. However, discoveries of huge phosphate deposits covering hundreds of square miles below the ocean floor within 60 miles of North Carolina have relieved fears of imminent shortages.

The free element

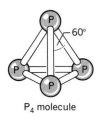

P₄ molecule

Phosphorus atoms in their ground state have the electron configuration [Ne] $3s^2 3p^3$ and therefore require three electrons to complete an octet. In the free element this is accomplished by linking each phosphorus atom to three others by single covalent bonds. As we learned in Chapter 9, this gives rise to three principal allotropes.

About 0.05 g of white phosphorus is a fatal dose.

White phosphorus (Figure 23.1) consists of individual tetrahedral P_4 molecules. It is a waxy solid that melts at 44.1 °C and boils at 280 °C. This allotrope is extremely reactive, especially toward oxygen, and must be stored under water. When exposed to O_2, white phosphorus tends to ignite spontaneously, and contact with the skin produces painful slow-healing burns. White phosphorus is also very toxic, and long-term exposure to even low levels of P_4 vapor produces a gradual deterioration and softening of the bones of the jaw.

White phosphorus was first prepared (unintentionally) in 1669 by an alchemist named Hennig Brand, who heated a mixture of dried urine and sand and condensed the vapors that were given off by passing them through water. The new element was named phosphorus, from the Greek *phosphoros* = light bringer, because the dry solid in a sealed bottle glows in the dark. The glow is actually caused by a slow oxidation of the phosphorus surface by residual oxygen in the container.

Modern methods of producing white phosphorus differ only slightly from Brand's. A mixture of phosphate rock, $Ca_3(PO_4)_2$, silica, SiO_2, and coke is heated to

Figure 23.1. *White phosphorus (left) is a yellowish-white waxy solid that is extremely reactive toward oxygen. Red phosphorus (right) is much less reactive than the white allotrope.*

about 1300 °C in an **electric furnace,** a furnace in which the contents are heated to very high temperatures by passing an electric current through them. The reaction is

$$2Ca_3(PO_4)_2(s) + 6SiO_2(s) + 10C(s) \longrightarrow 6CaSiO_3(l) + 10CO(g) + P_4(g)$$

The CO and P_4 vapor is passed through water, which causes the P_4 to condense. Most of the phosphorus produced in this way is ultimately used to make phosphoric acid.

Red phosphorus (also shown in Figure 23.1) is formed when white phosphorus is heated or exposed to ultraviolet light. It is amorphous and as we learned earlier, probably consists of P_4 tetrahedra joined at their corners. It is also relatively non-poisonous and is considerably less reactive than white phosphorus. Red phosphorus is used to make incendiary devices (bombs, fireworks, flares, etc.) and, when mixed with fine sand, is used on the striking surfaces of matchbooks.

Black phosphorus is the least reactive form of phosphorus. It consists of layers of phosphorus atoms. Within each layer the phosphorus atoms are covalently bonded to each other, but the attractions between layers are the result of much weaker London forces. This gives it a flaky, graphitelike appearance.

Compounds of phosphorus

Phosphorus combines with most of the nonmetals as well as with active metals. With the metals of Groups IA and IIA, for example, binary *phosphides* are formed that are predominantly ionic and contain the *phosphide ion,* P^{3-}. In water, this ion hydrolyzes to give *phosphine*, PH_3, a very bad-smelling poisonous gas.

$$Na_3P(s) + 3H_2O(l) \longrightarrow 3NaOH(aq) + PH_3(g)$$

The most important compounds of phosphorus are formed with nonmetals, particularly oxygen and the halogens.

Oxides of Phosphorus When any of the allotropes of phosphorus are burned in an excess supply of oxygen, white, powdery phosphorus(V) oxide, P_4O_{10}, is formed. For example,

$$P_4(s) + 5O_2(g) \longrightarrow P_4O_{10}(s)$$

This oxide is often called *phosphorus pentoxide* because its empirical formula, P_2O_5, was recognized long before its molecular formula and structure were discovered. The structure of P_4O_{10} is shown in Figure 23.2. Notice that it is related to the basic P_4 tetrahedron by insertion of an oxygen bridge between phosphorus atoms along each edge of the tetrahedron, plus an additional oxygen at each vertex.

The most striking property of P_4O_{10} is its strong affinity for water. When placed into water, it reacts vigorously and exothermically to form phosphoric acid, H_3PO_4.

$$P_4O_{10}(s) + 6H_2O(l) \longrightarrow 4H_3PO_4(l)$$

This oxide is such a strong dehydrating agent that it is even able to extract the components of water — two hydrogens and an oxygen — from other molecules. For example, in Chapter 22 we saw that P_4O_{10} is able to remove hydrogen and oxygen from HNO_3 molecules to form N_2O_5.

$$P_4O_{10} + 12HNO_3 \longrightarrow 4H_3PO_4 + 6N_2O_5$$

In the laboratory, P_4O_{10} is often used as a **desiccant,** an agent that removes moisture from air or other gases. When the moist air or gas mixture comes in contact with the P_4O_{10}, the water vapor reacts to form H_3PO_4, thereby leaving the gas virtually moisture-free.

Brand's mixture contained phosphates, carbon, and silica, too.

White phosphorus under water burns spontaneously when O_2 is bubbled into the mixture at a temperature above 70 °C.

The IUPAC name of P_4O_{10} is *tetraphosphorus decaoxide.*

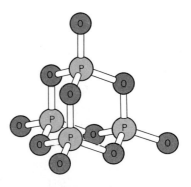

Figure 23.2. *The structure of P_4O_{10}.*

The IUPAC name of P_4O_6 is *tetraphosphorus hexaoxide.*

Note the spelling: Phosphorus is the name of the element; phosphorous implies the lower oxidation state of phosphorus in its oxoacids.

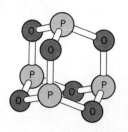

Figure 23.3. *The structure of P_4O_6.*

About 85% of the elemental phosphorus made each year is used to manufacture pure H_3PO_4.

If white phosphorus is burned in a limited supply of oxygen, not all of it is oxidized to P_4O_{10}. About half is oxidized to phosphorus(III) oxide, P_4O_6 (often called *phosphorus trioxide* because its empirical formula is P_2O_3). The structure of P_4O_6, shown in Figure 23.3, is similar to that of P_4O_{10}, except that the oxygen at each apex of the tetrahedron is missing in P_4O_6.

Phosphorus(III) oxide is a crystalline solid that melts at 23.8 °C and boils at 175 °C. Its vapors are quite poisonous. When stirred with cold water, P_4O_6 reacts to form phosphorous acid, H_3PO_3.

$$P_4O_6(s) + 6H_2O(l) \longrightarrow 4H_3PO_3(aq)$$

Phosphoric Acid and Phosphates The most important oxoacid of phosphorus, by far, is **phosphoric acid**, H_3PO_4 (also called **orthophosphoric acid**). It is produced in huge amounts, about 10 million tons annually, for use in the manufacture of fertilizers such as ammonium phosphate, as a food additive, and for the production of detergents.

Most phosphoric acid is made directly from phosphate rock by reaction with sulfuric acid.

$$Ca_3(PO_4)_2(s) + 3H_2SO_4(aq) + 6H_2O \longrightarrow 3CaSO_4 \cdot 2H_2O(s) + 2H_3PO_4(aq)$$

After reaction, the insoluble calcium sulfate is separated from the mixture by filtration and the phosphoric acid is then concentrated by evaporating water from the solution. This gives a concentrated solution that is about 85% H_3PO_4 by weight. It is known as *syrupy phosphoric acid* because of its viscous nature.

A purer form of phosphoric acid is obtained from elemental phosphorus by burning the element to give P_4O_{10} and then dissolving the oxide in water. As we saw earlier, P_4O_{10} reacts with water to give H_3PO_4. The phosphoric acid produced in this way is used to make detergents and in food preparations. For example, H_3PO_4 is added to many carbonated beverages to give them tartness. The acid is also used to make the acid salt $Ca(H_2PO_4)_2$, which is used in baking powders.

Pure phosphoric acid is a solid at room temperature; its clear, colorless crystals melt at 42.4 °C. In water, H_3PO_4 is a weak triprotic acid and is a very poor oxidizing agent. Its structure is

$$H-\overset{\displaystyle ..}{\underset{\displaystyle ..}{O}}-\overset{\displaystyle \overset{..}{O}..}{\underset{\displaystyle \underset{..}{O}..}{P}}-\overset{\displaystyle ..}{\underset{\displaystyle ..}{O}}-H$$

$$\overset{|}{H}$$

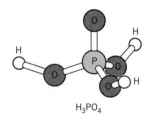

H_3PO_4

Neutralization of H_3PO_4 with a base produces three different ions, depending on the amount of base that is added.

$$H_3PO_4 \xrightarrow{OH^-} \underset{\substack{\left(\text{dihydrogen} \atop \text{phosphate ion}\right) \\ + \\ H_2O}}{H_2PO_4^-} \xrightarrow{OH^-} \underset{\substack{\left(\text{hydrogen} \atop \text{phosphate ion}\right) \\ + \\ H_2O}}{HPO_4^{2-}} \xrightarrow{OH^-} \underset{\substack{\left(\text{phosphate ion} \atop \text{[orthophosphate ion]}\right) \\ + \\ H_2O}}{PO_4^{3-}}$$

It is possible to isolate salts of each of the three anions.

Salts of phosphoric acid have a variety of applications, and one of the most common is in fertilizers. Phosphate rock itself can be pulverized and used as a phosphate fertilizer, but because $Ca_3(PO_4)_2$ has a very low solubility, it is able to deliver phosphate only slowly and in small amounts. However, treatment of

$Ca_3(PO_4)_2$ with dilute sulfuric acid gives a fertilizer known as *superphosphate,* a mixture of $CaSO_4$ and $Ca(H_2PO_4)_2$.

$$Ca_3(PO_4)_2 + 2H_2SO_4 + 4H_2O \longrightarrow \underbrace{2CaSO_4 \cdot 2H_2O + Ca(H_2PO_4)_2}_{superphosphate}$$

Because $Ca(H_2PO_4)_2$ is water-soluble, this mixture is a more effective fertilizer than phosphate rock — hence, the name superphosphate.

The anions of phosphoric acid are also found in the blood where they serve as one of the buffer systems. This system consists of the ions $H_2PO_4^-$ and HPO_4^{2-}. If base is added to the blood, it is able to react with $H_2PO_4^-$.

$$H_2PO_4^- + OH^- \longrightarrow HPO_4^{2-} + H_2O$$

and if acid is added, it reacts with HPO_4^{2-}.

$$HPO_4^{2-} + H^+ \longrightarrow H_2PO_4^-$$

These reactions help prevent large changes in the pH of the blood.

Trisodium phosphate, Na_3PO_4, often called TSP, is an effective water softener and cleansing agent. In hard water it reacts with hardness ions such as Ca^{2+}, Mg^{2+}, and Fe^{3+} to form precipitates or complex ions, which removes the ions from solution so they can't interfere with the action of the soap. Solutions of Na_3PO_4 are basic because of the hydrolysis of the PO_4^{3-} ion.

$$PO_4^{3-} + H_2O \rightleftharpoons HPO_4^{2-} + OH^-$$

A 1.0 M Na_3PO_4 solution has a pH of about 12.8. This is very basic, so you should always wear rubber gloves when working with concentrated TSP solutions.

Polymeric Phosphoric Acids and Their Anions Orthophosphoric acid is the parent of a host of more complex acids and anions that contain more than one phosphorus atom. They are formed by eliminating the components of water from —OH groups on neighboring acid molecules and linking together adjacent PO_4 tetrahedra by the mutual sharing of an oxygen atom. For example, when H_3PO_4 is heated to 250 °C, **pyrophosphoric acid,** $H_4P_2O_7$, is formed.

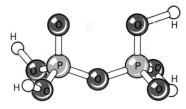

Pyrophosphoric acid

$H_4P_2O_7$ is also called diphosphoric acid.

oxygen bridge

pyrophosphoric acid

Pyrophosphoric acid is a colorless, glassy solid that is very soluble in water. It forms salts such as $Na_4P_2O_7$ and $Na_2H_2P_2O_7$. Both the acid and its salts are very slowly hydrolyzed to phosphoric acid or its anions. The reaction is slow enough, however, so that $Na_4P_2O_7$ is used as the phosphate ingredient in many liquid detergents.

If orthophosphoric acid is heated above 400 °C, extensive polymerization occurs by elimination of water, as shown below. The product is called **metaphosphoric acid.** Its formula is $(HPO_3)_n$ where n is a large number.

Hydrolysis adds water to the P—O—P bridge to regenerate the pair of P—OH units.

$$\cdots O-\overset{\overset{\textstyle O}{\|}}{\underset{\underset{\textstyle OH}{|}}{P}}-O-\overset{\overset{\textstyle O}{\|}}{\underset{\underset{\textstyle OH}{|}}{P}}\left(-O-\overset{\overset{\textstyle O}{\|}}{\underset{\underset{\textstyle OH}{|}}{P}}-\right)O-\overset{\overset{\textstyle O}{\|}}{\underset{\underset{\textstyle OH}{|}}{P}}-O-\overset{\overset{\textstyle O}{\|}}{\underset{\underset{\textstyle OH}{|}}{P}}-O-\cdots$$

HPO_3 repeating unit that
occurs n times in $(HPO_3)_n$

The sodium salt of metaphosphoric acid, which contains the $(PO_3^-)_n$ ion, is normally made by heating sodium dihydrogen phosphate.

$$n\,NaH_2PO_4 \xrightarrow{\text{heat}} (NaPO_3)_n + n\,H_2O$$

Polyphosphates of intermediate chain length can also be made. For example,

$$NaH_2PO_4 + 2Na_2HPO_4 \xrightarrow{\text{heat}} Na_5P_3O_{10} + 2H_2O$$
sodium
tripolyphosphate

Sodium tripolyphosphate is used in many solid detergent mixtures and contains the ion

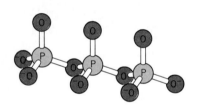

Tripolyphosphate ion, $P_3O_{10}{}^{5-}$

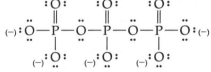

In many parts of the United States the use of phosphate-based detergents has been banned or severely restricted because of the effects that high phosphate concentrations have on lakes. In general, the rate of algae growth in lakes is determined by the nutrient present in the most limited amount. Normally, this limiting nutrient is phosphate, and when soluble phosphates from fertilizers and detergents become washed into lakes, episodes of rapid growth of algae—*algae blooms*—can occur. When the algae die and settle to the bottom, they begin to decompose. This causes the waters to be depleted of oxygen, so other marine organisms—fish, for example—begin to die. These processes speed the natural aging or *eutrophication* of the lake and gradually reduce the oxygen level to the point where no aquatic life can survive in the waters.

Phosphorous Acid   We learned earlier that **phosphorous acid,** H_3PO_3, is formed when phosphorus(III) oxide is dissolved in water.

$$P_4O_6(s) + 6H_2O(l) \longrightarrow 4H_3PO_3(aq)$$

Despite the way its formula is written, H_3PO_3 is only a diprotic acid. Its structure is

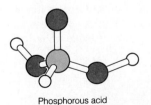

Phosphorous acid

$$H-\overset{..}{\underset{..}{O}}-\overset{\overset{\textstyle H}{|}}{\underset{\underset{\textstyle\|}{}}{P}}-\overset{..}{\underset{..}{O}}-H$$
$$\underset{\textstyle :\overset{..}{\underset{..}{O}}:}{}$$

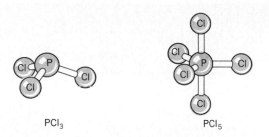

PCl_3

PCl_5

Only the O—H bonds break to give H^+; the hydrogen attached directly to the phosphorus is not acidic. This means that only two kinds of salts can be prepared — for example, NaH_2PO_3 and Na_2HPO_3. Phosphorous acid and phosphites are reasonably good reducing agents. For example, they reduce silver compounds to metallic silver.

Halogen Compounds Phosphorus forms two kinds of binary compounds with the halogens: the trihalides, PX_3 (X = F, Cl, Br, and I) and the pentahalides, PX_5 (X = F, Cl, and Br). The compound PI_5 is not known, presumably because iodine atoms are so large that five of them simply cannot be packed around a phosphorus atom.

The most important halogen compounds of phosphorus are the chlorides, PCl_3 and PCl_5. Their structures are shown in Figure 23.4. Phosphorus trichloride is made by reacting molten phosphorus with chlorine. If an excess of chlorine is present, the pentachloride can also be formed.

$$P_4(l) + 6Cl_2(g) \longrightarrow 4PCl_3(g)$$

$$PCl_3(g) + Cl_2(g) \rightleftharpoons PCl_5(g)$$

The second reaction is an equilibrium, and PCl_5 decomposes if heated.

Phosphorus trichloride is a volatile liquid that boils at 76 °C. It is used as a starting material for the preparation of many other phosphorus compounds. When exposed to water, PCl_3 hydrolyzes to give phosphorous acid.

$$PCl_3 + 3H_2O \longrightarrow H_3PO_3 + 3HCl$$

In fact, this is the method usually used to make phosphorous acid.

Phosphorus trichloride reacts with oxygen to give **phosphoryl chloride,** $POCl_3$ (also called **phosphorus oxychloride**). About half of the PCl_3 manufactured each year is oxidized to $POCl_3$ and much of that is used to make compounds that are employed as flame retardants.

Phosphorus pentachloride in the vapor or liquid has the trigonal bipyramidal structure shown in Figure 23.4. In the solid, however, PCl_5 appears to exist as an ionic compound composed of tetrahedral PCl_4^+ ions and octahedral PCl_6^- ions (i.e., in the solid, PCl_5 is really $PCl_4^+PCl_6^-$). One of the principal uses of PCl_5 is in the manufacture of $POCl_3$ by the reaction

$$P_4O_{10} + 6PCl_5 \longrightarrow 10POCl_3$$

Many phosphorus-containing pesticides are made using PCl_3.

23.2 SULFUR

The element sulfur is widely distributed in nature, although it is only about half as abundant as phosphorus. It occurs as the free element, and long before sulfur was recognized as an element, it was known as *brimstone*, meaning stone that burns. In the combined state it is found in mineral sulfides (those of lead and copper, for example), in sulfates such as gypsum and Epsom salts, and as sulfate ion in the ocean. Sulfur compounds also occur as impurities in natural gas, petroleum, and coal.

Brimstone is mentioned in the Bible.

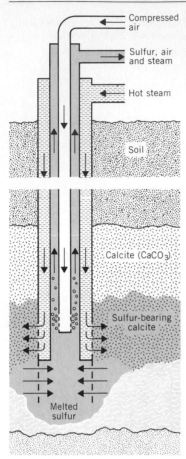

Figure 23.5. *Sulfur is extracted from deposits of the free element deep below the earth's surface by the Frasch process.*

Figure 23.6. *Liquid sulfur is pumped into large ponds to cool and solidify after being extracted from deep below the surface of the earth by the Frasch process. In the background are enormous piles of sulfur.*

Somewhat more than half of the sulfur used by industry each year is mined from large underground deposits of the free element, chiefly in Texas and Louisiana. A rather clever method to accomplish this was invented in 1890 by an American engineer, Herman Frasch, and has come to bear his name. It involves pumping superheated water under pressure into the sulfur deposit, which causes the sulfur to melt (Figure 23.5). The hot sulfur–water mixture is then foamed to the surface with compressed air and sprayed into huge piles to solidify and dry (Figure 23.6).

The rest of the industrial supply of sulfur is recovered during the purification of natural gas and petroleum, which contain sulfur compounds such as H_2S that would create pollution problems if burned along with these fuels.

The free element

Sulfur is also found in Spain, Mexico, Japan, and Italy.

Sulfur atoms in their ground state have the electron configuration [Ne] $3s^2 3p^4$. As we saw in Chapter 9, these combine with each other by forming two covalent bonds to separate sulfur atoms, and eight-member rings are formed. The most stable allotrope of sulfur at room temperature is known as rhombic sulfur in which the S_8 rings are stacked in a way that gives a rhombic crystal structure. Crystals of yellow rhombic sulfur are shown in Figure 23.7. A second allotrope of sulfur is known as monoclinic sulfur in which S_8 rings are stacked in a monoclinic crystal structure. Rhombic sulfur melts at about 112 °C, and when heated above 120° and then allowed to cool slowly, needlelike crystals of the monoclinic allotrope (mp 119 °C) are formed (Figure 23.8).

The behavior of elemental sulfur as it is heated from its melting point to its boiling point is quite interesting. When it first melts, a straw-colored, relatively nonviscous liquid is formed. As this is heated, it gradually darkens and becomes very viscous, like molasses. At still higher temperatures it thins out again and the dark red liquid finally boils at 445 °C. These changes are shown in Figure 23.9.

The changes in the sulfur as it is heated are explained as follows. When the sulfur first melts, the liquid is composed of S_8 rings in the usual jumbled arrangement

Figure 23.7. *Yellow crystals of naturally occurring rhombic sulfur are seen here surrounded by crystals of quartz.*

Figure 23.8. *Needlelike crystals of monoclinic sulfur.*

characteristic of liquids in general. As the temperature is raised, thermal energy increases the vibrational motion of the sulfur atoms in the rings, and sulfur–sulfur bonds begin to break. This gives chains of sulfur atoms that have an unpaired electron at each end.

When an end sulfur atom of one chain encounters another from a different chain, a covalent bond is formed and an S_{16} chain is produced. Coupling can continue and

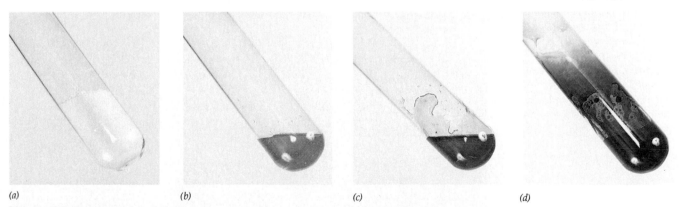

(a)   *(b)*   *(c)*   *(d)*

Figure 23.9. *Changes that take place when sulfur is heated. (a) Solid sulfur prior to being melted. (b) Liquid sulfur just above its melting point. (c) Liquid sulfur becomes dark and very viscous as its temperature is raised. (d) At its boiling point, the sulfur has become relatively nonviscous and has a dark red color.*

long chains of S_{24}, S_{32}, S_{40}, and so on are formed that intertwine and cause the liquid to become very viscous. At still higher temperatures the more violent motions of the sulfur atoms cause the long chains to start to break into smaller fragments, and the liquid becomes relatively nonviscous again.

It is interesting to note that if the thickened liquid sulfur is cooled rapidly—for example, by being poured into cold water—the sulfur atoms do not have a chance to rearrange into S_8 rings. As a result, a supercooled liquid called *amorphous sulfur* is produced that has many of the elastic properties of rubber. When allowed to stand, the S_x chains in the amorphous sulfur gradually revert to the thermodynamically more stable S_8 rings of the rhombic form.

Another name for amorphous sulfur is plastic sulfur. *See Figure 23.21 on p. 811.*

Compounds of sulfur

Sulfur Dioxide and Sulfurous Acid   Elemental sulfur is easily ignited and burns with a blue flame to give **sulfur dioxide,** SO_2, a colorless gas with a choking, irritating odor. If you've ever gotten a whiff of the fumes from an igniting match, you've smelled SO_2—sulfur is one of the substances used in matches. Sulfur dioxide is a nonlinear molecule and therefore is polar. This gives it a relatively high boiling point (-10 °C) and allows it to be easily liquefied under pressure. For this reason, SO_2 has been used as a gas in refrigeration systems.

$$S(s) + O_2(g) \longrightarrow SO_2(g)$$

Sulfur dioxide is also produced when sulfur compounds are burned and when metal sulfides are heated in air, an important step in the extraction of metals such as lead and copper from their ores. The SO_2 formed in metallurgical processes is now recovered, but the SO_2 released into the atmosphere by combustion of high-sulfur fuels, both petroleum-based and coal, presents a major pollution problem in some areas.

When sulfur dioxide dissolves in water, it forms hydrates that can be represented by the formula $SO_2 \cdot xH_2O$. These solutions are slightly acidic and have been attributed to the formation of sulfurous acid, H_2SO_3. However, no evidence of the existence of this molecular species exists. Nevertheless, it is convenient to *think* of solutions of SO_2 as containing H_2SO_3 as we have done in earlier chapters. Traditionally, the ionization of this acid has been represented by the equation

$$H_2SO_3(aq) \rightleftharpoons H^+(aq) + HSO_3^-(aq) \qquad K_{a_1} = 1.5 \times 10^{-2}$$

Heating metal sulfides in air to convert them to their oxides is called smelting and is a common metallurgical process, but the SO_2 that is released must not be allowed to escape into the atmosphere. Here we see the effects of a longtime smelting operation at Copper Hill, Tennessee where SO_2 was released into the atmosphere.

The second ionization is that of the HSO_3^- ion, which really does exist in solution.

$$HSO_3^-(aq) \rightleftharpoons H^+(aq) + SO_3^{2-}(aq) \qquad K_{a_2} = 1.0 \times 10^{-7}$$

Complete neutralization of aqueous sulfur dioxide yields sulfite ion, SO_3^{2-}, and salts of this ion can be isolated. However, true salts of bisulfite ion are not obtained. Although this ion exists in solution, when water is evaporated, the solid that is obtained contains the disulfite ion (also called the metabisulfite ion), $S_2O_5^{2-}$. It forms by loss of water from a pair of HSO_3^- ions.

$$2HSO_3^- \longrightarrow S_2O_5^{2-} + H_2O$$

Nevertheless, dissolving a solid such as $Na_2S_2O_5$ in water yields a solution that does contain the HSO_3^- ion because the reaction above is reversible.

In the laboratory sulfur dioxide can be obtained easily by acidifying a sulfite salt. For example, the following is the molecular equation for the reaction of Na_2SO_3 with the strong acid, H_2SO_4:

$$Na_2SO_3(s) + H_2SO_4(aq) \longrightarrow Na_2SO_4(aq) + SO_2(g) + H_2O$$

Sodium sulfite is the most important sulfite salt. Large amounts of it are made each year, mostly for the manufacture of paper pulp. Some is also used in food products and in making wine to prevent unwanted bacterial growth. Since some people are extremely allergic to sulfites, the U.S. government now requires that all products containing sulfites list them as an ingredient.

Sulfurous acid and sulfites are both rather good reducing agents. On exposure to air they are slowly oxidized to sulfuric acid and sulfates.

Acid Rain   The emissions of sulfur dioxide by industries and power plants that burn high- and moderate-sulfur fuels (especially coal) have produced a major environmental problem affecting many regions that lie down wind from industrialized areas. As mentioned, the combustion of sulfur and sulfur compounds gives SO_2 as the major sulfur-containing product. This SO_2 is carried by the wind and is slowly oxidized to sulfur trioxide, SO_3. When it rains, both of these sulfur oxides are washed from the atmosphere as they dissolve in the rain drops. Once dissolved, the oxides react to form dilute solutions of acids, sulfurous acid from the dissolved SO_2 and sulfuric acid from the dissolved SO_3. The rain is made even more acidic by reactions of nitrogen oxides. In the last chapter you learned that NO_2 dissolves in water where is disproportionates to give a solution of nitric acid.

As you have probably learned from the news media, this acidified rain has been called *acid rain*, and it is not unusual to find precipitation in northwestern Europe and in large areas of North America with a pH between 4 and 4.5. In cities this can cause structural damage to vehicles, buildings, and statues (Figure 23.10), and in the countryside it has damaged trees and caused lakes to become too acidic to support fish (Figure 23.11).

Rain as acidic as vinegar fell during a rainstorm in Pitlochry, Scotland, in April 1974.

One method of removing SO_2 from the exhaust gases of industrial furnaces is to allow it to pass over moist limestone. It is absorbed by the moisture and reacts as follows:

$$CaCO_3(s) + SO_2(g) \longrightarrow CaSO_3(s) + CO_2(g)$$

This is essentially the same reaction that occurs when acid rain falls on limestone and marble building materials.

Sulfur Trioxide and Sulfuric Acid   At room temperature sulfur trioxide is a solid consisting of SO_3 units linked together in various complex chain structures. The solid is easily vaporized and in the gas phase SO_3 exists as discrete planar triangular molecules.

Sulfur trioxide is made by the oxidation of sulfur dioxide with oxygen. As noted

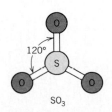

SO_3

Figure 23.10. *How polluted air and acid rain cause decay is seen in this statue made of Baumberg sandstone at the Herten Castle in Westphalia, West Germany. On the left is its appearance in 1908 after 206 years of exposure to the atmosphere. On the right is its appearance 60 years later after exposure to air pollution produced by European heavy industry.*

earlier, this is a highly favorable reaction thermodynamically, as can be seen by its large negative $\Delta G°$.

$$2SO_2(g) + O_2(g) \longrightarrow 2SO_3(g) \qquad \Delta G° = -740 \text{ kJ}$$

The reaction is slow in the absence of a catalyst, which explains the apparent stability of SO_2 in air. However, in the presence of an appropriate catalyst the oxidation of SO_2 occurs quickly. As mentioned in Chapter 19, catalysts used in automobile catalytic converters accelerate this reaction and the SO_3 produced combines with moisture in the exhaust gases to form a mist of H_2SO_4.

$$SO_3(g) + H_2O(g) \longrightarrow H_2SO_4(l)$$

The oxidation of SO_2 to SO_3 and its subsequent conversion to sulfuric acid are, by far, industry's most important chemical reactions. The amount of sulfuric acid

Figure 23.11. (a) *The effects of acid rain on fir trees in a forest in Bavaria, Germany.* (b) *In Boksjo, Sweden, "lime" (actually pulverized limestone, $CaCO_3$) is pumped into a lake to neutralize acidity caused by acid rain. The reaction is*

$$CaCO_3(s) + 2H^+(aq) \longrightarrow Ca^{2+}(aq) + CO_2(g) + H_2O$$

(a)   (b)

produced annually — approximately 42 million tons — is more than *twice* that of any other single chemical. About 60% of it is used to treat phosphate rock in the production of phosphate fertilizers and in making ammonium sulfate (also a fertilizer). Sulfuric acid is also used in the refining of petroleum, by the steel industry, in lead storage batteries, and in chemical reactions involved in the manufacture of paints, explosives, plastics, drugs, and many other chemicals.

Most sulfuric acid is made by the **contact process.** The raw material is sulfur, which is first burned in air to form sulfur dioxide. The SO_2 and additional oxygen (in air) are then passed over a vanadium pentoxide (V_2O_5) catalyst, which oxidizes the SO_2 to SO_3. Next, the SO_3 is absorbed into concentrated H_2SO_4, with which it reacts to form **disulfuric acid** (also called **pyrosulfuric acid**), $H_2S_2O_7$.

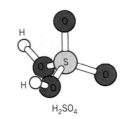

$$H_2SO_4(l) + SO_3(g) \longrightarrow H_2S_2O_7(l)$$

The H_2SO_4 is used to trap the SO_3 because it is more effective than water. Finally, the $H_2S_2O_7$ is diluted with water, which converts it to sulfuric acid.

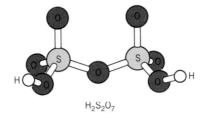

$$H_2S_2O_7(l) + H_2O(l) \longrightarrow 2H_2SO_4(l)$$

Pure H_2SO_4 is a dense, colorless, viscous liquid which tends to decompose into SO_3 and H_2O when heated strongly. The concentrated H_2SO_4 found in bottles on laboratory shelves is 96% H_2SO_4 by weight (18 M). Dilution of the concentrated acid with water is *very* exothermic because of the reaction

$$H_2SO_4(\text{conc.}) + H_2O(l) \longrightarrow H_3O^+(aq) + HSO_4^-(aq)$$

Care should always be taken to add concentrated H_2SO_4 *to the water* — never the other way around. If water is added to the concentrated acid, the heat generated causes the water to boil and the expanding steam spatters the neighborhood with H_2SO_4.

The strong affinity that H_2SO_4 has for water makes the concentrated acid an effective dehydrating agent. For example, if concentrated H_2SO_4 is poured onto sugar, $C_{12}H_{22}O_{11}$, it extracts the components of water and leaves a blackened, charred mass behind (Figure 23.12).

$$C_{12}H_{22}O_{11} \xrightarrow{\text{conc. } H_2SO_4} 12C + 11H_2O$$

Sulfuric acid is a strong diprotic acid. The first step in its dissociation is complete

(a)   (b)

Figure 23.12. *The effect of concentrated sulfuric acid on sugar. (a) The sugar before the acid is added. (b) A short time after adding the H_2SO_4. The concentrated acid has extracted the components of H_2O from the sugar molecules, leaving a foamy mass of carbon behind.*

$$HSO_4^- \rightleftharpoons H^+ + SO_4^{2-}$$

$$K_a = 1.2 \times 10^{-2}$$

A 0.10 M solution of NaHSO$_4$ has a pH of 1.4.

in aqueous solutions, but the second step only occurs to about 10% in a 1 M solution. This gives a 1.0 M solution of H_2SO_4 an H_3O^+ concentration of approximately 1.1 M.

By controlling the amount of base during the neutralization of H_2SO_4, two series of salts can be formed: sulfates such as Na_2SO_4 and hydrogen sulfates or bisulfates such as $NaHSO_4$. Solutions of bisulfates are quite acidic because the HSO_4^- ion is a relatively strong acid.

Concentrated sulfuric acid is a mild oxidizing agent when cold, but a fairly strong oxidizing agent when hot. For example, cold concentrated H_2SO_4 oxidizes iodide ion to iodine and, to some extent, bromide ion to bromine.

$$2X^- + 3H_2SO_4 \longrightarrow X_2 + SO_2 + 2H_2O + 2HSO_4^- \qquad (X = I \text{ or } Br)$$

Cold H_2SO_4 doesn't attack metallic copper, but the hot concentrated acid does. The reaction, which causes the sulfur to be reduced from the $+6$ state to the $+4$ state, is

$$Cu + 2H_2SO_4 \xrightarrow{\text{heat}} CuSO_4 + SO_2 + 2H_2O$$

A stronger reducing agent such as zinc reduces the sulfuric acid even further to free sulfur (oxidation number = zero) or even hydrogen sulfide (oxidation number of sulfur = -2).

Other Compounds of Sulfur Sulfur combines with most metals and nonmetals to form a large variety of different compounds. We will look only briefly at some of the more important types.

Metal sulfides tend to be less ionic than their oxides.

Binary compounds of metals with sulfur are sulfides and contain the sulfide ion, S^{2-}. They can often be prepared by direct reaction of the metal with sulfur. For example, zinc and sulfur react vigorously to produce zinc sulfide.

$$Zn(s) + S(s) \longrightarrow ZnS(s)$$

Since most metal sulfides are insoluble in water, they can also be formed by precipitation reactions using hydrogen sulfide, H_2S.

Hydrogen sulfide itself is a poisonous gas with an odor of rotten eggs. It is more poisonous than carbon monoxide, but its bad odor allows it to be detected at low concentrations. In nature, hydrogen sulfide is released into the air during volcanic eruptions and by the decomposition of organic matter in the absence of air — that's why rotten eggs smell of H_2S. Airborne, H_2S is responsible for the gradual tarnishing of silver and the darkening of lead-based paints. In each case a black metal sulfide (Ag_2S and PbS) is formed.

In the laboratory H_2S can be prepared by reacting a metal sulfide with a strong nonoxidizing acid — for example, HCl.

The S^{2-} ion is a strong Brønsted–Lowry base.

$$FeS(s) + 2H^+(aq) \longrightarrow Fe^{2+}(aq) + H_2S(g)$$

Another method, commonly used when one wants to generate the H_2S in an aqueous solution, is the hydrolysis of an organic compound called thioacetamide.

$$\underset{\text{thioacetamide}}{CH_3{-}\overset{\overset{\displaystyle S}{\|}}{C}{-}NH_2(aq)} + 2H_2O \longrightarrow H_2S(aq) + NH_4^+(aq) + \left(\underset{\text{acetate ion}}{CH_3\overset{\overset{\displaystyle O}{\|}}{C}{-}O}\right)^- (aq)$$

The advantage of this reaction is that it avoids the release of significant amounts of toxic H_2S into the atmosphere.

The prefix *thio-* in a chemical name implies that sulfur replaces oxygen in a compound. For example,

$$\underset{\text{acetamide}}{CH_3{-}\overset{\overset{\displaystyle O}{\|}}{C}{-}NH_2} \qquad \underset{\text{thioacetamide}}{CH_3{-}\overset{\overset{\displaystyle S}{\|}}{C}{-}NH_2}$$

An important oxoanion that fits this nomenclature pattern is the thiosulfate ion, $S_2O_3{}^{2-}$. It is formed by boiling a solution containing sulfite ion and sulfur.

$$S(s) + SO_3{}^{2-}(aq) \longrightarrow S_2O_3{}^{2-}(aq)$$

The Lewis structure for the $S_2O_3{}^{2-}$ ion is

$$\left[\begin{array}{c} :\!\ddot{O}\!: \\ \| \\ :\!\ddot{S}\!-\!S\!-\!\ddot{O}\!: \\ \| \\ :\!\ddot{O}\!: \end{array} \right]^{2-}$$

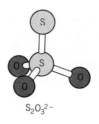

$S_2O_3{}^{2-}$

Notice that it can be viewed as a sulfate ion in which one oxygen is replaced by sulfur — hence, *thio*sulfate.

Thiosulfate ion forms quite stable complex ions with metal ions. The one with silver is particularly important in photography. Silver bromide is the usual light-sensitive substance found in photographic films and paper. After an image is developed, the photosensitive emulsion still contains unexposed silver bromide that must be removed to prevent the picture from darkening gradually when viewed in the light. Solutions of sodium thiosulfate, known as *hypo* to photographers, are used to remove the unexposed silver bromide. The equilibria involved are

$$AgBr(s) \rightleftharpoons Ag^+(aq) + Br^-(aq)$$

$$Ag^+(aq) + 2S_2O_3{}^{2-}(aq) \rightleftharpoons Ag(S_2O_3)_2{}^{3-}(aq)$$

Le Châtelier's principle explains how the thiosulfate works. As $S_2O_3{}^{2-}$ is added, the second equilibrium shifts to the right and the Ag^+ concentration drops. This causes the AgBr to dissolve in an attempt to restore the Ag^+ concentration to its former value. If sufficient thiosulfate is present, equilibrium is never reestablished and all the AgBr dissolves. Afterward, the $Ag(S_2O_3)_2{}^{3-}$ complex ion and residual thiosulfate are washed from the film or paper, leaving only the desired image.

Thiosulfate ion is also a good reducing agent. It reacts with strong oxidizing agents such as chlorine to give sulfate ion.

$$4Cl_2(g) + S_2O_3{}^{2-} + 5H_2O \longrightarrow 8Cl^- + 2SO_4{}^{2-} + 10H^+$$

Thiosulfate ion is often used in titrations of the weaker oxidizing agent iodine. Its reaction with iodine is

$$\underset{\substack{\text{thiosulfate}\\\text{ion}}}{2S_2O_3{}^{2-}} + I_2 \longrightarrow \underset{\substack{\text{tetrathionate}\\\text{ion}}}{S_4O_6{}^{2-}} + 2I^-$$

$$\left[\begin{array}{c} :\!\ddot{O}\!: \qquad\qquad :\!\ddot{O}\!: \\ \| \qquad\qquad\quad \| \\ :\!\ddot{O}\!-\!\ddot{S}\!-\!\ddot{S}\!-\!\ddot{S}\!-\!\ddot{S}\!-\!\ddot{O}\!: \\ \| \qquad\qquad\quad \| \\ :\!\ddot{O}\!: \qquad\qquad :\!\ddot{O}\!: \end{array} \right]^{2-}$$

tetrathionate ion

23.3 THE HALOGENS

Until now we've discussed the nonmetallic elements one at a time because the variations in their properties are rather substantial. For example, although oxygen and sulfur are the the same group and similar in some respects, they also have many differences, and the same applies to nitrogen and phosphorus. Among the halogens — the Group VIIA elements — many close similarities and trends in properties exist that are easy to follow, so we will discuss them as a group.

The halogens — fluorine, chlorine, bromine, and iodine — are never found free in nature because of their high reactivity. Until humans began their study of chemistry and began to produce the free halogens, these elements occurred almost exclusively in inorganic salts. In fact, the name halogen comes from the Greek *halos*, meaning salt. Today, halogens have been incorporated into many useful organic compounds, ranging from Teflon and poly(vinyl chloride) plastics to Freon refriger-

Astatine, At, is also a halogen, but it is rare and radioactive. We will not include it in our discussions.

ants and aerosol propellants to insecticides such as DDT. In recent times it has been found that some of these halogenated organic compounds [DDT, chloroform, and polychlorinated biphenyls (PCB's), for example] can have very harmful effects on the environment and its inhabitants—us!

Cryolite, you recall, was used by Hall in his process to recover aluminum from Al_2O_3 by electrolysis.

The principal sources of the halogens are their salts. Fluorine is found in the earth's crust principally in deposits of fluorspar, CaF_2, cryolite, Na_3AlF_6, and fluoroapatite, $Ca_5(PO_4)_3F$. The main source of chlorine is NaCl, which, as we learned earlier, is recovered from the sea and large underground deposits formed, presumably, by evaporation of ancient seas. Bromine and iodine are also found in the sea, but their concentrations as Br^- and I^- are *much* less than that of Cl^-. (Seawater contains these concentrations of the halide ions: $0.53\ M\ Cl^-$, $8.1 \times 10^{-4}\ M\ Br^-$, and $5 \times 10^{-7}\ M\ I^-$). Bromine and iodine are also found in the waters of brine (salt) wells, and iodine is obtained from sodium iodate, $NaIO_3$, that is recovered from deposits of saltpeter, $NaNO_3$, imported from Chile.

Properties of the free elements

Halogens atoms each have seven valence electrons and therefore need only one more to achieve a noble gas configuration. As a result, in the elemental state they form singly bonded diatomic molecules, X_2.

$X = F$, Cl, Br, or I.

Some physical properties of the halogens are given in Table 23.1. We see that as their molecules become larger, the melting points and boiling points increase, corresponding to an increase in the strengths of the London forces.

At room temperature fluorine is a pale yellow gas. It is the most reactive of all the elements because of the low F—F bond energy; that is, little energy is needed to split the molecule into very reactive fluorine atoms. This low bond energy is believed to be caused by electron–electron repulsions between the small, compact, electron-rich valence shells of the bonded fluorine atoms.

Chlorine is a pale yellow-green gas at room temperature. Chlorine molecules have a slightly larger bond energy than F_2 molecules, and chlorine atoms are less electronegative than fluorine. As a result, Cl_2 is somewhat less reactive than F_2.

At room temperature, bromine is a volatile, nonviscous red liquid, and iodine forms dark metallic-looking crystals that easily sublime to give a purple vapor. Chemical reactivity decreases from chlorine to bromine to iodine, which parallels the decrease in their electronegativities.

Because of the high electronegativities of the halogens compared to other elements, they tend to gain electrons from other substances and thereby serve as oxidizing agents. The ability of the halogens to serve as oxidizing agents decreases going down the group. As a result, we find that a given halogen is able to oxidize the

Gaseous chlorine.

Table 23.1. Some physical properties of the halogens

Halogen	Melting Point (°C)	Boiling Point (°C)	Bond Energy (kJ/mol)	Electronegativity
Fluorine (F_2)	−233	−188	157	4.1
Chlorine (Cl_2)	−103	−34.6	242	2.9
Bromine (Br_2)	−7.2	58.8	193	2.8
Iodine (I_2)	113.5	184.4	150	2.2

(a) *Bromine is a volatile red liquid.*
(b) *Iodine is a volatile purple-black solid.*

(a)   (b)

anions of the halogens below it in Group VIIA. Thus, F_2 will oxidize Cl^-, Br^-, and I^-, while Cl_2 will oxidize only Br^- and I^- but not F^-. This is illustrated by the following typical reactions:

Fluorine

$$F_2 + \begin{Bmatrix} 2NaCl \\ 2NaBr \\ 2NaI \end{Bmatrix} \longrightarrow 2NaF + \begin{Bmatrix} Cl_2 \\ Br_2 \\ I_2 \end{Bmatrix}$$

These reactions were discussed in Chapter 10. See the photograph on page 306.

Chlorine

$$Cl_2 + NaF \longrightarrow \text{No reaction}$$

$$Cl_2 + \begin{Bmatrix} 2NaBr \\ 2NaI \end{Bmatrix} \longrightarrow 2NaCl + \begin{Bmatrix} Br_2 \\ I_2 \end{Bmatrix}$$

Bromine

$$Br_2 + \begin{Bmatrix} NaF \\ NaCl \end{Bmatrix} \longrightarrow \text{No reaction}$$

$$Br_2 + 2NaI \longrightarrow 2NaBr + I_2$$

Iodine

$$I_2 + \begin{Bmatrix} NaF \\ NaCl \\ NaBr \end{Bmatrix} \longrightarrow \text{No reaction}$$

Preparation of the free elements

Fluorine Elemental fluorine is such an active oxidizing agent that it can only be made by electrolysis. The raw material is hydrogen fluoride, which is dissolved in molten KF. Electrolysis produces fluorine gas at the anode and hydrogen gas at the cathode. The KF in the mixture serves as an electrolyte because pure hydrogen fluoride is molecular and therefore is nonconducting.

$$2HF \xrightarrow[\text{KF}]{\text{electrolysis}} H_2(g) + F_2(g)$$

Fluorine is used to make a variety of fluorine-containing organic compounds. Examples are Teflon and the Freons. As you learned in Chapter 22, Freons are chlorofluorocarbons that are used as refrigerants and aerosol propellants, and they may have harmful effects on the Earth's ozone layer as they diffuse into the stratosphere.

Chlorine The production of chlorine by electrolysis both of molten NaCl and aqueous NaCl (brine) has been described previously. The reactions are

Industrially, chlorine ranks about eighth in total annual tons produced.

$$2NaCl(l) \xrightarrow{\text{electrolysis}} 2Na(l) + Cl_2(g)$$

$$2Na^+(aq) + 2Cl^-(aq) + 2H_2O \xrightarrow{\text{electrolysis}} 2Na^+(aq) + 2OH^-(aq) + Cl_2(g) + H_2(g)$$

Over 12 million tons of chlorine are produced each year. Its major use has been in the manufacture of chemical intermediates — chemicals that are used to make other chemicals. Chlorine is also used to treat municipal drinking water, make solvents and plastics such as poly(vinyl chloride) (vinyl plastics), and manufacture pesticides.

In the laboratory chlorine can be made by oxidation of chloride ion in an acidic solution by a strong oxidizing agent such as manganese dioxide, MnO_2, or potassium permanganate, $KMnO_4$. The simplest method is to react concentrated hydrochloric acid with MnO_2.

$$MnO_2(s) + 2Cl^-(aq) + 4H^+(aq) \longrightarrow Mn^{2+}(aq) + Cl_2(g) + 2H_2O$$

Bromine Although bromide ion occurs in a low concentration in seawater, bromine can be recovered from it by taking advantage of the ease of oxidation of Br^- by chlorine and of the volatility of Br_2. First, chlorine is dissolved in the water and oxidizes bromide ion to bromine.

$$2Br^-(aq) + Cl_2(aq) \longrightarrow Br_2(aq) + 2Cl^-(aq)$$

Air is then blown through the water and the volatile bromine and residual unreacted chlorine are flushed out. Cooling the air causes the Br_2 to condense to a liquid. Today, most bromine is extracted from brines obtained from wells in Arkansas and Michigan. These contain bromide ion in concentrations that are 50 to 60 times greater than in seawater.

About half the bromine produced each year has been used to make ethylene dibromide, $C_2H_4Br_2$, which is an additive in leaded gasoline. Its purpose is to prevent deposits of lead compounds from forming inside the engine. During combustion lead bromide, $PbBr_2$, is produced, which is volatile at temperatures that exist in the cylinders and therefore leaves the engine as a gas in the exhaust. Leaded gasoline should not be used in cars equipped with catalytic converters because the lead compounds would be absorbed on the catalyst surface and destroy its catalytic activity. Bromine is also used to make silver bromide, the principal ingredient in the light-sensitive emulsions on photographic film and paper.

In the laboratory, Br_2 can be made by oxidation of a bromide salt by MnO_2 in an acidic solution (e.g., a solution containing H_2SO_4).

$$MnO_2(s) + 2Br^-(aq) + 4H^+(aq) \longrightarrow Mn^{2+}(aq) + Br_2(l) + 2H_2O$$

It can also be made by oxidation of bromide ion by chlorine.

$$Cl_2 + 2Br^- \longrightarrow Br_2 + 2Cl^-$$

Iodine Iodide ion is present in very low concentrations in seawater, but seaweed extracts it from the water and concentrates it. Commercial quantities of iodine are recovered from the ashes of burned seaweed, in which the iodide concentrations approach 1%. The I^- is oxidized to I_2 using chlorine or other oxidizing agents.

Another commercial source of iodine is Chilean saltpeter, which contains sodium iodate, $NaIO_3$. The iodate is reduced to free iodine using bisulfite ion as the reducing agent.

$$2IO_3^-(aq) + 5HSO_3^-(aq) \longrightarrow I_2(s) + 5SO_4^{2-}(aq) + H_2O + 3H^+(aq)$$

The recovery of iodine is expensive and applications of the free element are limited. It is used to make various medicinal products (tincture of iodine, for example) and silver iodide, which is also employed in photographic film.

Compounds of the halogens

Binary Halides of Metals Halogen atoms easily gain an electron to form the halide ions—fluoride, chloride, bromide, and iodide—and their compounds with metals are quite common. Most metal halides are ionic as long as the metal is in a low oxidation state. When the metal is in a high oxidation state, polarization of the anion often produces covalently bonded species. For example, in the vapor, aluminum chloride exists as Al_2Cl_6 molecules. Tin(IV) chloride ($SnCl_4$) and titanium(IV) chloride ($TiCl_4$) are both covalent liquids at room temperature.

Hydrogen Halides The binary hydrogen compounds of the halogens have the general formula, HX. They can be prepared by direct combination of the elements

$$H_2 + X_2 \longrightarrow 2HX$$

but the vigor of the reaction varies substantially from fluorine to iodine.

Fluorine reacts violently with hydrogen as soon as the two gases are mixed. Chlorine and hydrogen react at a much slower rate, however, provided their mixtures are not heated or exposed to ultraviolet light. Light or heat splits Cl_2 molecules into Cl atoms and initiates a chain reaction that is explosively fast. Chain mechanisms are also involved in the reaction of H_2 with Br_2 and I_2, but the reactions are less vigorous than with chlorine.

The hydrogen halides can also be made from their binary salts by reaction with a nonvolatile acid. For example, hydrogen fluoride is prepared by reacting CaF_2 with sulfuric acid.

$$CaF_2(s) + H_2SO_4(l) \longrightarrow CaSO_4(s) + 2HF(g)$$

Similarly, HCl is evolved if concentrated H_2SO_4 is added to NaCl.

$$NaCl(s) + H_2SO_4(l) \longrightarrow HCl(g) + NaHSO_4(s)$$

Additional HCl can be produced by adding more salt to the $NaHSO_4$ and heating the mixture.

$$NaCl(s) + NaHSO_4(s) \xrightarrow{heat} Na_2SO_4(s) + HCl(g)$$

Concentrated sulfuric acid is too powerful an oxidizing agent, even when cold, to be used to generate HBr and HI from their salts. Oxidation of the halide ion to free Br_2 or I_2 occurs. Phosphoric acid, a very poor oxidizing agent, can be used in place of sulfuric acid, but the mixtures must be warmed to expel the hydrogen bromide or hydrogen iodide.

Hydrogen fluoride has a substantially higher boiling point than the other hydrogen halides because of extensive hydrogen bonding that produces long staggered chains of HF molecules in the liquid.

(Dots indicate hydrogen bonds.)

Water solutions of the hydrogen halides are the *hydrohalic acids*—hydrofluoric acid, hydrochloric acid, hydrobromic acid, and hydroiodic acid. In water, HF is a weak acid, whereas the others are all 100% ionized. Despite being a weak acid,

HCl is sometimes made commercially by the reaction

$$H_2 + Cl_2 \longrightarrow 2HCl$$

The chain mechanism for the reaction of H_2 with Br_2 was discussed in Chapter 19.

	Boiling Points (°C)
HF	+19.7
HCl	−85.1
HBr	−66.8
HI	−35.4

Even though HF is a weak acid, it causes very severe skin burns.

hydrofluoric acid attacks glass and sand. In this case the reaction produces silicon tetrafluoride, SiF_4, a volatile substance that can escape as a gas.

$$SiO_2(s) + 4HF(aq) \longrightarrow SiF_4(g) + 2H_2O(l)$$
(in sand and glass)

Concentrated solutions of HF contain the HF_2^- ion in which a hydrogen ion is shared between two fluoride ions.

$$[F \cdots H \cdots F]^-$$

Solutions of hydrochloric acid can be purchased in hardware stores under the name muriatic acid.

Hydrochloric acid is the most important of the hydrohalic acids because it is relatively inexpensive. It is used to remove rust from iron and steel before the metal is coated with zinc (galvanizing) and in the manufacture of many other chemicals. Hydrochloric acid is also produced in the stomach and helps us digest our foods.

Oxoacids and Oxoanions The halogens form four kinds of oxoacids and anions (Table 23.2). Those of chlorine are the most familiar and the most important.

As we learned earlier, the strengths of these acids increase from HOCl, which is a weak acid, to $HClO_4$, which is an extremely powerful acid. Recall that the explanation given is that the electron-withdrawing effect of the lone oxygens leads to an increased polarization of the O—H bond as the number of lone oxygens becomes larger and the anions that are formed become better at distributing the negative charge over their atoms.

Hypochlorous acid is formed by a disproportionation reaction (a reaction in which the same chemical is both oxidized and reduced) when chlorine is dissolved in cold water.

Similar reactions of Br_2 and I_2 occur to much lesser extents.

$$Cl_2(aq) + H_2O \rightleftharpoons HOCl(aq) + H^+(aq) + Cl^-(aq)$$

Approximately 30% of the dissolved chlorine exists in the form of HOCl and Cl^-.

Table 23.2. Oxoacids and oxoanions of the halogens

Fluorine	Chlorine	Bromine	Iodine
HOF	HOCl	HOBr	HOI
	(OCl^-)	(OBr^-)	(OI^-)
	$HClO_2$	$HBrO_2$	HIO_3
	(ClO_2^-)	(BrO_2^-)	(IO_3^-)
	$HClO_3$	$HBrO_3$	H_5IO_6
	(ClO_3^-)	(BrO_3^-)	$(H_2IO_6^{3-})$
	$HClO_4$	$HBrO_4$	HIO_4
	(ClO_4^-)	(BrO_4^-)	(IO_4^-)

HOCl	hypochlorous acid	$H-\overset{..}{\underset{..}{O}}-\overset{..}{\underset{..}{Cl}}:$
OCl^-	hypochlorite ion	
$HClO_2$ (or HOClO)	chlorous acid	$H-\overset{..}{\underset{..}{O}}-\overset{..}{\underset{..}{Cl}}=\overset{..}{\underset{..}{O}}$
ClO_2^-	chlorite ion	
$HClO_3$ (or $HOClO_2$)	chloric acid	$H-\overset{..}{\underset{..}{O}}-\overset{:\overset{..}{O}:}{\overset{\|}{\underset{..}{Cl}}}=\overset{..}{\underset{..}{O}}$
ClO_3^-	chlorate ion	
$HClO_4$ (or $HOClO_3$)	perchloric acid	$H-\overset{..}{\underset{..}{O}}-\overset{:\overset{..}{O}:}{\underset{:\underset{..}{O}:}{\overset{\|}{\underset{\|}{Cl}}}}=\overset{..}{\underset{..}{O}}$
ClO_4^-	perchlorate ion	

When chlorine is dissolved in base, the equilibrium is shifted to the right because the HOCl is neutralized. The net reaction is

$$Cl_2 + 2OH^- \longrightarrow OCl^- + Cl^- + H_2O$$

Electrolysis of a stirred NaCl solution, you recall, produces Cl_2, which reacts with the OH^- formed at the cathode. In this way the aqueous NaCl is gradually changed to aqueous NaOCl, which is diluted and sold as liquid laundry bleach (e.g., Clorox®).

Reaction of Cl_2 with lime, CaO, produces a calcium salt that is sold as a solid laundry bleach and in antimildew preparations.

$$CaO(s) + Cl_2(g) \longrightarrow CaCl(OCl)(s)$$

A similar compound, calcium hypochlorite, formed by the neutralization of hypochlorous acid with calcium oxide, is sold as an algicide for backyard swimming pools. It goes by the commercial name HTH, meaning "high test hypochlorite."

$$CaO(s) + 2HOCl(aq) \longrightarrow Ca(OCl)_2(aq) + H_2O$$

A chlorine bleach actually contains OCl^- ion which destroys dyes by oxidizing them to colorless products.

Hypochlorous acid and its salts are powerful oxidizing agents. This is why hypochlorites are used as bleaches; the oxidation of colored compounds often produces colorless oxidation products. This same strong oxidizing ability also kills bacteria and fungi.

The hypohalites (OCl^-, OBr^-, OI^-) all tend to disproportionate to form the corresponding halides (X^-) and halates (ClO_3^-, BrO_3^-, IO_3^-).

$$3OX^- \longrightarrow XO_3^- + 2X^- \quad (X = Cl, Br, I)$$

The equilibrium constants for this reaction are large for all three of these halogens; however, the *rates* of disproportionation differ greatly. The reaction of OI^- is very rapid at all temperatures, while OBr^- reacts moderately fast at room temperature. (Solutions of OBr^- ion can be prepared only if they are kept cold.) At room temperature the disproportionation of OCl^- is very slow, so its solutions can be stored for reasonable periods, which explains why they can be sold as liquid bleaches. This is an interesting example of stability being determined by the slow rate of reaction rather than thermodynamics.

HOCl is too unstable to be isolated in pure form.

Chlorous acid and the chlorite ion have been among the lesser important compounds of chlorine. Reaction of chlorine dioxide with base produces ClO_2^- ion.

$$2ClO_2(g) + 2OH^-(aq) \longrightarrow ClO_2^-(aq) + ClO_3^-(aq) + H_2O$$

Chlorous acid is made from its barium salt by a metathesis reaction using H_2SO_4.

$$2H^+ + SO_4^{2-} + Ba(ClO_2)_2(s) \longrightarrow BaSO_4(s) + 2HClO_2(aq)$$

The reaction proceeds because of the very low solubility of $BaSO_4$. Like HOCl, $HClO_2$ is unstable and decomposes when attempts are made to isolate it in pure form.

A disinfectant is being marketed that uses the catalytic decomposition of $HClO_2$ into ClO_2 and Cl^-. It is used for disinfecting the air-conditioning systems on automobiles and it has been shown to destroy the AIDS virus on surfaces.

The halate ions, XO_3^-, are obtained when Cl_2, Br_2, and I_2 are dissolved in hot basic solutions.

$$3X_2 + 6OH^-(aq) \longrightarrow 5X^-(aq) + XO_3^-(aq) + 3H_2O$$

Chloric acid, like chlorous acid, can be made in solution from its barium salt.

$$H_2SO_4(aq) + Ba(ClO_3)_2(aq) \longrightarrow BaSO_4(s) + 2HClO_3(aq)$$

or

Iodic acid, HIO_3, is the only halic acid that can be isolated in the pure state.

$$2H^+(aq) + SO_4^{2-}(aq) + Ba^{2+}(aq) + 2ClO_3^-(aq) \longrightarrow BaSO_4(s) + 2H^+(aq) + 2ClO_3^-(aq)$$

Chloric acid is a very powerful oxidizing agent. It is a strong acid, and it too cannot be isolated in pure form.

Perchlorates are commonly prepared by electrolytic oxidation of chlorates.

Disproportionation of $KClO_3$ at moderate temperatures in the absence of catalysts also produces $KClO_4$.

In the presence of a catalyst (MnO_2) or at high temperatures, $KClO_3$ decomposes to KCl and O_2.

$$4KClO_3 \longrightarrow 3KClO_4 + KCl$$

Perchloric acid itself can be obtained pure, but it is unstable and tends to explode. In concentrated solutions it is a very strong oxidizing agent, although its dilute solutions, for unknown reasons, have very little oxidizing power.

Other halogen compounds of the nonmetals

The halogens form compounds with almost all the nonmetals, and a list of some of them is given in Table 23.3. One of the interesting things to notice here is that the sizes of the atoms influence to a degree the complexity of the compounds that are formed. For example, among the interhalogen compounds (compounds formed between two different halogens), iodine is able to bond to seven fluorine atoms to form IF_7, but not to seven chlorine atoms. Presumably, the larger size of chlorine makes it impossible to squeeze seven of them around a single iodine. Also notice that bromine doesn't form BrF_7, probably because bromine is smaller than iodine and can't fit seven fluorines around it.

Most of the nonmetal halides react vigorously with water to yield the hydrogen halide and the corresponding oxoacid of the other nonmetal. Examples of these reactions are

$$PCl_5 + 4H_2O \longrightarrow H_3PO_4 + 5HCl$$

$$SiCl_4 + 2H_2O \longrightarrow SiO_2 + 4HCl$$

$$SF_4 + 2H_2O \longrightarrow SO_2 + 4HF$$

These reactions occur very rapidly and proceed to completion with the evolution of considerable amounts of heat. In fact, it is quite common for many halogen compounds of the elements below period 2 to react very rapidly in this same way. For example, the tin(IV) and lead(IV) chlorides, which are covalent, also hydrolyze in

Table 23.3. Some halogen compounds of the nonmetals

Group IIIA	BX_3 ($X = F, Cl, Br, I$) BF_4^-			
Group IVA	CX_4 ($X = F, Cl, Br, I$)	SiF_4 SiF_6^{2-} $SiCl_4$	GeF_4 GeF_6^{2-} $GeCl_4$	
Group VA	NX_3 ($X = F, Cl, Br, I$) N_2F_4	PX_3 ($X = F, Cl, Br, I$) PF_5 PCl_5 PBr_5	AsF_3 AsF_5	SbF_3 SbF_5
Group VIA	OF_2 (O_2F_2) OCl_2 OBr_2	SF_2 SCl_2 S_2F_2 S_2Cl_2 SF_4 SCl_4 SF_6	SeF_4 SeF_6 $SeCl_2$ $SeCl_4$ $SeBr_4$	TeF_4 TeF_6 $TeCl_4$ $TeBr_4$ TeI_4
Group VIIA	ICl IBr BrF $BrCl$ ClF	ClF_3 BrF_3 ICl_3 IF_3	ClF_5 BrF_5 IF_5	IF_7

this manner with the formation of a mixture of species including complexes of Sn^{4+} with the chloride ion.

There are also some nonmetal halogen compounds that are quite *unreactive* toward water, for example, CCl_4, SF_6, and NF_3. In these cases it is unfavorable kinetics, instead of thermodynamics, that prevents the hydrolysis from taking place.

Consider, for example, the potential hydrolysis reactions of CCl_4 and $SiCl_4$. Thermodynamics implies that both should proceed very nearly to completion.

$$SiCl_4(l) + 2H_2O(l) \longrightarrow SiO_2(s) + 4HCl(aq) \qquad \Delta G° = -282 \text{ kJ}$$

$$CCl_4(l) + 2H_2O(l) \longrightarrow CO_2(g) + 4HCl(aq) \qquad \Delta G° = -377 \text{ kJ}$$

In fact, we see from the values of $\Delta G°$ that the hydrolysis of CCl_4 is even more "spontaneous" than $SiCl_4$. Kinetically, however, the hydrolysis of CCl_4 is essentially prohibited. This is attributed to the absence of a low-energy path for the hydrolysis of CCl_4. Attack by water on the carbon atom of CCl_4 is prevented by the crowding of the Cl atoms. In $SiCl_4$, on the other hand, the larger Si atom provides a greater opportunity for attack and, in addition, the presence of low-energy $3d$ orbitals in the valence shell of the Si atom permits a temporary bonding of the water molecule to the Si atom prior to the expulsion of a molecule of HCl. The mechanism of this hydrolysis is believed to be

Repetition of this process eventually yields $Si(OH)_4$ (orthosilicic acid), which loses water spontaneously to give a hydrated SiO_2.

$$Si(OH)_4 \longrightarrow SiO_2 + 2H_2O$$

The stability of SF_6 and NF_3 toward hydrolysis can also be attributed to the absence of a low-energy reaction mechanism. Like CCl_4, SF_6 should also hydrolyze spontaneously; the value of $\Delta G°$ for the reaction is tremendous.

$$SF_6(g) + 4H_2O(l) \longrightarrow H_2SO_4(aq) + 6HF(g) \qquad \Delta G° = -423 \text{ kJ}$$

However, the crowding of the fluorine atoms around the sulfur atom apparently prevents attack by water (even up to 500 °C), as well as most other reagents. This crowding is absent with SF_4, and hydrolysis by water is instantaneous.

The resistance of NF_3 toward attack by water cannot be attributed to interference by the fluorine atoms as in SF_6, since the NF_3 molecule is pyramidal, with the nitrogen atom quite openly exposed to an attacking water molecule. We might compare NF_3 with NCl_3, which *does* hydrolyze (if it doesn't explode first — NCl_3 is extremely unstable). The mechanism for this reaction appears to involve the initial formation of a hydrogen bond from water to the lone pair on the nitrogen atom, followed by expulsion of hypochlorous acid.

SiO_2 is a complex solid. It is described in Section 23.5.

The ultimate products of the hydrolysis are ammonia and HOCl. This mechanism is not favorable for NF_3 because of its very low basicity. In this case the highly electronegative fluorine atoms draw electron density from the nitrogen atom. As a result, it has been suggested that the lone pair of electrons on nitrogen in NF_3 may not be available to serve as a point of attachment for the H_2O molecule that, as we see above, is a necessary step in the mechanism proposed for the hydrolysis of NCl_3.

23.4 NOBLE GAS COMPOUNDS

In our discussion of the nonmetals we have not mentioned compounds of the noble gases. These are rather unusual substances because, on the basis of the electronic structure of the noble gases, we would perhaps not have predicted their existence. In fact, until 1962 most chemists firmly believed that these elements were totally incapable of forming compounds (other than several **clathrates,** in which the noble gas atoms are trapped in cagelike sites within a crystalline lattice). For this reason chemists had referred to them as the *inert gases.* Today they are spoken of as the noble gases in recognition of the fact that although some do react, they nevertheless possess a very low degree of reactivity.

The first real chemistry of the noble gases was discovered in 1962 by Neil Bartlett at the University of British Columbia. He had found that molecular oxygen, O_2, reacts with PtF_6 to form an orange-red compound, O_2PtF_6, containing the ion, O_2^+. Since the ionization energies of O_2 and Xe are nearly the same (1210 and 1170 kJ/mol, respectively), he reasoned that Xe should react in the same way O_2 does. When he reacted Xe with PtF_6, he isolated a yellow compound, containing Xe, that was believed to be $XePtF_6$.

After the initial report by Bartlett, it was not long before chemists at Argonne National Laboratory found that Xe also reacts directly with fluorine at elevated temperatures. This reaction yields a series of fluorides: XeF_2, XeF_4, and XeF_6. Other reactions and compounds were soon discovered and a partial list of the known Xe compounds is given in Table 23.4. The oxides and oxofluorides result from the hydrolysis of the fluorides

$$XeF_6 + 3H_2O \longrightarrow XeO_3 + 6HF$$

$$XeF_6 + H_2O \longrightarrow XeOF_4 + 2HF$$

Some of these compounds are quite unstable and tend to decompose. This is particularly true for the oxides XeO_3 and XeO_4, which explode (XeO_3 has $\Delta H_f^\circ =$

Table 23.4. Some compounds of xenon

	Melting Point (°C)	Physical Form
Fluorides		
XeF_2	140	Colorless crystals
XeF_4	114	Colorless crystals
XeF_6	47.7	Colorless crystals
Oxides		
XeO_3	Explodes	Colorless crystals
XeO_4	Explodes	Colorless gas
Salts		
$XePtF_6$[a]	—	Red-orange crystals
$CsXeF_7$	Decomp > 50 °C	Colorless solid
Cs_2XeF_8	Decomp > 400 °C	Yellow solid

[a] Since shown to be more complex; $Xe(PtF_6)_x$, where x lies between 1 and 2.

$+400$ kJ/mol). Others, on the other hand, appear quite stable. For example, Cs_2XeF_8 does not decompose even when heated to 400 °C, and the fluorides have moderately high melting points, suggesting a modest degree of thermal stability.

The structure and bonding in these compounds are quite interesting. Since Xe has four pairs of electrons in its valence shell, corresponding to a completed $5s$ and $5p$ subshell, unpairing of electrons and expansion of the octet must occur to provide unpaired electrons for bonding. Let's consider the XeF_2 molecule.

The electronic structure of Xe can be represented as

Xe ⇅   ⇅ ⇅ ⇅   __ __ __ __ __
   $5s$ $5p$ $5d$

In order to form XeF_2, one electron must be promoted to the $5d$ subshell, followed by hybrid orbital formation. The smallest hybrid set that will accommodate all the electrons is sp^3d.

Xe ⇅   ⇅ ⇅ ↑   ↑ __ __ __ __
   $5s$ $5p$ $5d$

gives

Xe ⇅ ⇅ ⇅ ↑ ↑   __ __ __ __ __
      sp^3d unhybridized $5d$

The two unpaired electrons can now be used in bonding to fluorine to give

Xe (in XeF_2) ⇅ ⇅ ⇅ ⇅ ⇅   __ __ __ __ __
                   sp^3d unhybridized $5d$

(Colored arrows represent fluorine electrons.)

In Chapter 9 we saw that the sp^3d hybrids point to the vertices of a trigonal bipyramid. In terms of the valence shell electron-pair repulsion theory, these five electron pairs will also be situated in this fashion, and from our rules (presented on p. 268), we expect that the three lone pairs will locate themselves in the triangular plane with the fluorine atoms above and below (Figure 23.13). The XeF_2 molecule is therefore linear.

Since the initial discovery of noble gas compounds by Bartlett, three noble gases have been demonstrated to form compounds, Rn, Xe, and Kr. The lighter ones — helium, neon, and argon — do not appear able to form chemical compounds because of their much higher ionization energies.

Bartlett's work on Xe has taught chemists an important lesson. So firmly convinced were they that the noble gases were totally unreactive that after some initial experiments attempting to react Xe with fluorine had failed in the 1930s, no further efforts were made to explore the possibility that they were not inert. It is interesting that the noble gas compounds obtained do not present any particular problem in bonding theory. In fact, many of the interhalogen compounds that had already been found (for example, BrF_5, ICl_4^-, IF_7) have the same number of electrons in the valence shell of the central atom as do the noble gas compounds. The same concepts that we applied to other compounds in which the octet was exceeded, therefore, work with the noble gas compounds, too. The failure by chemists to recognize the possibility that the noble gases might react reflects a blind spot in their thinking that was probably founded in an overzealous acceptance of the stability and inertness of the ns^2np^6 octet of electrons.

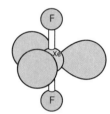

Figure 23.13. *The molecular structure of XeF_2.*

23.5 SILICON

The second most abundant element in the Earth's crust is silicon. It constitutes approximately 27.7% of the crust, by mass, where it occurs in rocks of various kinds

Figure 23.14. *Zone refining.*

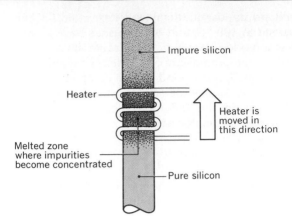

in the form of silicates, silicon--oxygen compounds. The free element is normally obtained by reduction of silicon dioxide (found in many kinds of sand) using carbon in an electric furnace. The reaction is

$$SiO_2(s) + 2C(s) \xrightarrow{\text{heat}} Si(s) + 2CO(g)$$

Silicon is a dark, metallic looking solid that melts at 1410 °C. In Chapter 4 you learned that it is a semiconductor, a fact that has made possible the fantastic world of miniature electronic devices such as pocket calculators and microcomputers. The silicon used in these devices must be extremely pure. It is usually made from a tetrahalide such as silicon tetrachloride ($SiCl_4$) by high-temperature reduction with hydrogen.

$$SiCl_4(g) + 2H_2(g) \longrightarrow Si(s) + 4HCl(g)$$

The silicon is then further purified by an interesting process called **zone refining** (Figure 23.14). This method is based on the fact that when a solution freezes, the crystal lattice of the solid formed does not accommodate the impurities very well, so the impurities tend to become concentrated in the remaining solution. In zone refining, a bar of silicon is placed in a device that melts a thin cross-sectional wafer of the solid near one end. The heat source is then gradually moved toward the other end of the bar, and as it moves, the melted zone follows along. Behind this molten section, the silicon solidifies in a very pure crystalline state, while the impurities collect in the molten band. In this way the impurities are brought to one end of the

Silica (left) and silicon (right).

bar. After the procedure is repeated several times, the impure end is cut off and discarded. The rest of the bar consists of very pure silicon with impurity levels ranging from a few parts per million to as little as a few parts per billion.

Chemical properties

Silicon is found in Group IVA of the periodic table, beneath carbon. Its atoms in their ground state have the electron configuration [Ne] $3s^2 3p^2$. Unlike carbon, silicon has little tendency to form π bonds by overlap of p atomic orbitals ($p\pi - p\pi$ bonds). Instead, silicon prefers to form single bonds. In the free element, each silicon atom completes its octet by bonding to four separate silicon atoms located at the corners of a tetrahedron, and, as we saw in Chapter 9, elemental silicon crystallizes in a diamond-type lattice.

Silicon is not a very reactive element at room temperature. It is unaffected by acids, but it does dissolve in hot basic solutions (NaOH, for example) to give hydrogen plus a mixture of silicates.

$$Si(s) + 4OH^-(aq) \longrightarrow SiO_4^{4-}(aq) + 2H_2(g)$$
(plus other silicates)

At high temperatures silicon also combines with halogens to produce the tetrahalides (for example, $SiCl_4$) and with hydrogen to form hydrides called silanes (SiH_4, Si_2H_6, etc.). The hydrides are not very stable, and the ability of silicon atoms to link to each other in these compounds is limited. The longest chain to be observed is in Si_6H_{14}, which tends to decompose to SiH_4.

Compounds with silicon-oxygen bonds

Silicon has a strong affinity for oxygen because it forms very stable silicon–oxygen bonds. In fact, all the naturally occurring silicon compounds are silicates in which the basic structural unit is the SiO_4 tetrahedron. Although some are rather complex, the structures of these naturally occurring silicates can be understood by considering them as polymers of more basic silicate units.

The simplest of the silicates contain the *orthosilicate ion*, SiO_4^{4-}, which is the anion of orthosilicic acid.

SiO₄ tetrahedron

orthosilicic acid
(cannot be isolated)

orthosilicate ion

An example of a substance containing the SiO_4^{4-} ion is the gem *zircon*, which consists of crystals of $ZrSiO_4$.

In Section 23.1 we saw that phosphoric acid units are able to be polymerized by the removal of the components of water from a pair of —OH groups on neighboring molecules and the formation of an oxygen bridge between the phosphorus atoms. Among the silicates this kind of polymerization also occurs, and as the pH of a solution of sodium silicate, Na_4SiO_4, is lowered, the SiO_4 units become joined by similar oxygen bridges. For example, the pyrosilicate ion, $Si_2O_7^{6-}$, can be considered to be formed as follows:

Notice that in the various silicates discussed in this section each nonbridging oxygen carries a negative charge.

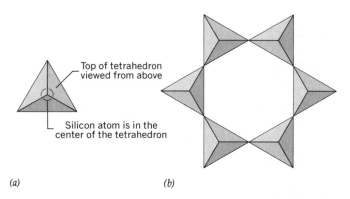

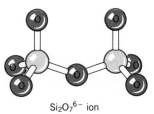

Si₂O₇⁶⁻ ion

Figure 23.15.   *The structure of the pyrosilicate anion, Si₂O₇⁶⁻. Each nonbridging oxygen carries a negative charge.*

In this anion a pair of SiO₄ tetrahedra share a corner in common (Figure 23.15). In nature, the Si₂O₇⁶⁻ ion is found in the mineral Sc₂Si₂O₇, called thortveitite.

When one SiO₄ tetrahedron shares *two* of its corners with other SiO₄ tetrahedra, rings and long-chain structures are able to be formed. For example, the polymeric anion, Si₆O₁₈¹²⁻, shown in Figure 23.16, is present in beryl, Be₃Al₂(Si₆O₁₈), which forms crystals of gem quality that are known as emeralds.

Linking the SiO₄ tetrahedra in long chains gives huge polymeric anions having the empirical formula SiO₃²⁻.

Figure 23.16.   (a) *A representation of the SiO₄ tetrahedron as viewed from above.* (b) *The structure of the Si₆O₁₈¹²⁻ ion showing the SiO₄ tetrahedra linked by oxygen bridges at their corners.*

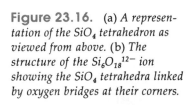

Top of tetrahedron viewed from above

Silicon atom is in the center of the tetrahedron

(a)   (b)

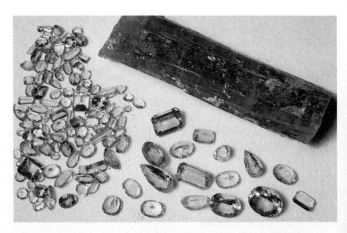

Naturally occurring emerald (left) and aquamarine (right). When cut and polished, they give beautiful gems. Both are forms of beryl, a silicate with the formula Be₃Al₂(SiO₃)₆.

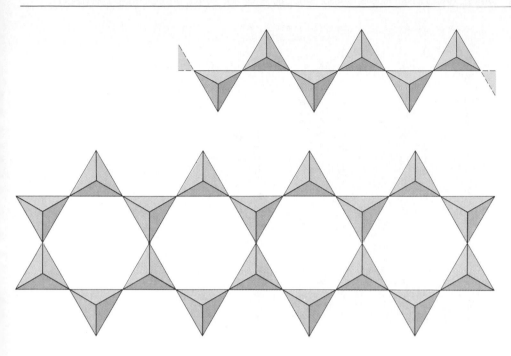

Figure 23.17. *The staggered arrangement of SiO$_4$ tetrahedra in a linear (SiO$_3{}^{2-}$)$_x$ metasilicate chain.*

Figure 23.18. *Linear double chain of the (Si$_4$O$_{11}{}^{6-}$)$_x$ anion found in asbestos.*

repeating
SiO$_3{}^{2-}$ unit

The staggered structure of an (SiO$_3$)$_x{}^{2x-}$ chain is shown in Figure 23.17. As you might expect, minerals such as MgSiO$_3$ and LiAl(SiO$_3$)$_2$, which contain these "linear" SiO$_3{}^{2-}$ chains, have a fibrous appearance. In addition to the simple strands of SiO$_4$ tetrahedra, double strands, formed by the sharing of three corners by *every other* SiO$_4$ unit, are also found. This time, an infinite *double* chain results, a small segment of which is illustrated in Figure 23.18. Once again, each unshared oxygen atom carries a negative charge, and the repeating unit along the chain is Si$_4$O$_{11}{}^{6-}$. This anion is found in asbestos, [Ca$_2$Mg$_5$(Si$_4$O$_{11}$)$_2$(OH)$_2$]$_x$, and, as we might expect, asbestos has a fiberlike nature because of the presence of the long (Si$_4$O$_{11}{}^{6-}$)$_x$ chains that line up more or less parallel to one another.

Asbestos has many useful properties. Its fibrous nature makes it excellent as a reinforcing filler in brake and clutch linings. Because it is fireproof, it has been spun and woven into fabrics used in fireproof curtains and ironing board covers. It has been used in insulation and sealers in cars, trucks, and planes. Once again, however, nature takes as well as gives. Asbestos fibers breathed into the lungs have been implicated in many cases of lung cancer. There is also evidence that if consumed in food, it can produce stomach cancer. Because of these problems, the uses of asbestos have been sharply curtailed.

Still another type of silicate is formed if *each* SiO$_4$ tetrahedron shares three of its corners so that each Si is attached to three others by oxygen bridges. When this occurs, a planar sheet of SiO$_4$ units results. A portion of one of these is shown in Figure 23.19 with the repeating unit, Si$_2$O$_5{}^{2-}$, outlined by the rectangle.

Asbestos fibers.

Figure 23.19. *Planar sheet silicate formed by sharing three corners of every SiO_4 tetrahedron.*

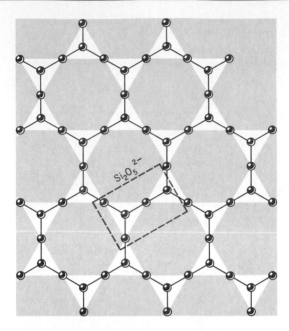

A number of different minerals are known to contain these $(Si_2O_5{}^{2-})_x$ sheets. They differ in the way the silicate layers are stacked, and the nature of the cations and other anions that are also present in the structure: However, all have certain similarities to each other. Some examples are talc (used in bath powder) and soapstone, in which the $(Si_2O_5{}^{2-})_x$ sheets are packed together with cations in such a way that there is a minimum of attractive forces between successive layers. These layers therefore slide over each other easily and both of these minerals feel slippery.

In other related minerals there is substitution of another element for Si. In mica, for instance every fourth Si atom is replaced by an Al^{3+} ion. The properties of mica are therefore different from talc; however, the layer structure is still apparent. A mica-type material you may be familiar with is *vermiculite,* used in place of soil in propagating house plants and as a cushioning filler in packaging items for shipment. This solid flakes into thin flat layers characteristic of mica.

When Si finally shares all four of its oxygen atoms with other Si atoms, a

Mica is a mineral that cleaves into thin flat sheets, reflecting the layer structure within the crystals.

three-dimensional framework is produced that has the empirical formula SiO_2. Silicon dioxide, also called silica, sometimes occurs in large beautiful crystals of quartz (Figure 23.20), although most people are more familiar with quartz in a finer state of subdivision, sand.

23.6 ORGANIC AND INORGANIC POLYMERS

We conclude our discussion of the chemistry of the elements by examining some of the special properties of substances that we call polymers. **Polymers** are very large molecules (often called **macromolecules**) formed by linking together a large number of smaller molecules called **monomers.** The resulting overall structure consists of a repetition of relatively small molecular fragments usually referred to as *repeating units.* For example, one of the most familiar polymers is polyethylene, which is made by linking together a very large number of ethylene molecules. In the process, the double bond in ethylene opens so the electrons can be used to bind neighboring C_2H_4 units. This gives a long chain of carbon atoms, each attached to two hydrogens.

Figure 23.20. *Quartz crystals.*

$$n \begin{matrix} H & H \\ | & | \\ C = C \\ | & | \\ H & H \end{matrix} \longrightarrow \cdots - \begin{matrix} H & H & H & H & H & H \\ | & | & | & | & | & | \\ C - C - C - C - C - C \\ | & | & | & | & | & | \\ H & H & H & H & H & H \end{matrix} - \cdots \quad \text{or} \quad \left(\begin{matrix} H & H \\ | & | \\ C - C \\ | & | \\ H & H \end{matrix} \right)_n$$

ethylene
(the monomer)

polyethylene
(the polymer)

$-CH_2-CH_2-$ is considered to be the repeating unit because it comes from the monomer, ethylene.

Chemically, polyethylene behaves much like other hydrocarbons with carbon–carbon single bonds. In fact, the chemical properties of polymers in general reflect the chemical properties of the various functional groups that they contain. The rates of reaction may be affected by molecular size, but not the basic chemical properties of the units that make up a polymer chain. What sets polymers apart, therefore, is not their chemical behavior but rather their physical properties, which result primarily from the enormous sizes of their molecules.

Recall that a functional group is a group of atoms in a molecule that impart characteristic chemical and physical properties.

Large molecules have many places along their structures where they can be attracted to neighbors, and the relative strengths of these attractions affect properties such as melting point, resistance to breaking when stretched, and solubility in solvents. The ability to chemically bind together polymer chains by *cross-linking* also affects these properties, and the tendencies of certain polymer chains to coil provide them with the property we call elasticity.

The resistance to breaking when stretched is called **tensile strength.** Cross-linking creates chemical bonds between one polymer strand and another.

Polymers can be classified into two broad categories, **addition polymers** and **condensation polymers,** according to the way the polymer is formed from its monomer units.

Addition polymers

Addition polymers are formed by simply adding monomer units together. Polyethylene, described earlier, is an example. Such reactions involve the opening of a double bond and the pairing of electrons between the monomer units to create the link. Generally, this requires the presence of a substance called an *initiator* to get the process started. In the polymerization of ethylene, for example, the initiator is an organic peroxide (a molecule with an O—O bond) that forms a free radical containing an unpaired electron. This free radical attaches itself to an ethylene molecule, causing the double bond to open and yielding a very reactive species that goes on to attack another ethylene molecule. Each addition step lengthens the chain by two

An initiator is somewhat like a catalyst, except that it can't be recovered unchanged after the reaction. It actually becomes part of the polymer.

carbon atoms and yields a reactive species that can go on to attack another C_2H_4 molecule. The result is a chain reaction that continues until the hydrocarbon chain is finally capped by another free radical. This is indicated below, where we have used the symbol $R\cdot$ to stand for the free radical formed by the initiator.

$$R\cdot + CH_2{=}CH_2 \longrightarrow R{-}CH_2{-}CH_2\cdot$$
$$R{-}CH_2{-}CH_2\cdot + CH_2{=}CH_2 \longrightarrow R{-}CH_2{-}CH_2{-}CH_2{-}CH_2\cdot$$
$$\vdots$$
$$R{-}(\!-CH_2{-}CH_2-\!)_n{-}CH_2{-}CH_2\cdot + R\cdot \longrightarrow$$
$$R{-}(\!-CH_2{-}CH_2-\!)_n{-}CH_2{-}CH_2{-}R$$

> Two hydrocarbon free radicals can also link by electron pairing, which terminates the chain reaction.

Because the chain is so long, the initiator is an insignificant part of the whole molecule and the formula is specified as simply $(C_2H_4)_n$.

Polyethylene is not the only important polymer made by addition polymerization. Some others whose names you will probably recognize include the following.

> The polyethylene chain length is approximately 40,000 repeating units with a total molecular mass of about 1,000,000.

Polypropylene   Polypropylene, which is similar to polyethylene, is formed by polymerizing propylene,

$$\underset{\underset{\displaystyle CH_3}{|}}{CH_2{=}CH}$$

Appropriate catalysts will control the orientations of the methyl groups relative to one another along the chain, and a polymer stronger than polyethylene can be formed.

Poly(vinyl chloride), Polyacrylonitrile, and Polystyrene   The $CH_2{=}CH{-}$ group that we find in propylene is called a *vinyl group*, and it forms the basis for a number of other polymers formed by addition polymerization. If the atom attached to the CH of the vinyl group is a chlorine, the monomer is called vinyl chloride; if a $-CN$

Many common objects, such as this Buick dashboard, are made of poly(vinyl chloride).

group is attached, the monomer is acrylonitrile, and if a benzene ring is attached, the monomer is styrene.

$$CH_2{=}CH \quad CH_2{=}CH \quad CH_2{=}CH$$
$$\mid \qquad\qquad \mid$$
$$Cl \qquad\qquad CN$$

vinyl chloride   acrylonitrile   styrene

Their polymerization yields poly(vinyl chloride), polyacrylonitrile, and polystyrene.

Poly(vinyl chloride), known better by the initials PVC, is a hard, brittle solid used to make such items as plastic pipes and phonograph records. Adding esters to the PVC softens it so that it can be used to make imitation leather, plastic shower curtains, garden hose, and other familiar "vinyl plastic" items.

Polyacrylonitrile is sometimes called *Orlon* and can be dissolved in a solvent from which it can be spun into fibers for use in carpets and cloth.

Polystyrene is the plastic from which model airplanes and other such objects are made.

Teflon The polymer teflon is made by polymerization of tetrafluoroethylene, $CF_2{=}CF_2$.

$$n\ CF_2{=}CF_2 \longrightarrow {+}CF_2{-}CF_2{\tfrac{}{}}_n$$

As you are probably aware, other substances do not adhere to teflon very well and it is used on cooking surfaces to make cleanup easy. Teflon is also very resistant to chemical attack and has a slippery surface, so it is used in greaseless bearings.

Other Addition Polymers Two other familiar addition polymers are poly(methyl methacrylate), marketed under the names *Lucite* and *Plexiglas,* and the polymer used to make Saran Wrap.

A teflon-coated frying pan.

$$n\ CH_2{=}\underset{\underset{\underset{O-CH_3}{|}}{\overset{\overset{C=O}{|}}{C}}}{\overset{\overset{CH_3}{|}}{}} \longrightarrow \left(-CH_2-\underset{\underset{\underset{O-CH_3}{|}}{\overset{\overset{C=O}{|}}{C}}}{\overset{\overset{CH_3}{|}}{}} \right)_n$$

methyl methacrylate   poly(methyl methacrylate)

$$x\ CH_2{=}\underset{\underset{Cl}{|}}{\overset{\overset{Cl}{|}}{C}} + y\ CH_2{=}\underset{\underset{Cl}{|}}{\overset{}{CH}} \longrightarrow \left[\left(-CH_2-\underset{\underset{Cl}{|}}{\overset{\overset{Cl}{|}}{C}} \right)_x \left(-CH_2-\underset{\underset{Cl}{|}}{\overset{\overset{H}{|}}{C}} \right)_y \right]$$

vinylidene   vinyl   Saran
chloride   chloride

In Saran the monomer units do not necessarily alternate regularly along the chain.

The Saran polymer, because it is made from two different monomers, is an example of a *copolymer.*

Condensation polymers

The complex phosphates and silicates discussed earlier in this chapter are polymeric species, but they are not the kinds of substances that most people think of when the word polymer is used.

In the formation of a condensation polymer, a small molecule such as water is eliminated from the functional groups of two monomer units as they become joined. One of the best known is nylon, a copolymer of a dicarboxylic acid (an organic acid with two —CO_2H groups) and a diamine (an organic molecule that contains two —NH_2 groups). There is a whole family of nylons, but the most common is formed from butanedicarboxylic acid and 1,6-hexanediamine. The reaction between the monomers can be diagrammed as follows:

$$\text{etc.} -\overset{\overset{\displaystyle H}{|}}{N} \boxed{H} \quad \overset{H_2O}{HO} \boxed{-\overset{\overset{\displaystyle O}{\|}}{C}(CH_2)_4 \overset{\overset{\displaystyle O}{\|}}{C}} \boxed{OH} \quad \overset{H_2O}{H} \boxed{-\overset{\overset{\displaystyle H}{|}}{N}(CH_2)_6 \overset{\overset{\displaystyle H}{|}}{N}} \boxed{H} \quad \overset{H_2O}{HO} \boxed{-\overset{\overset{\displaystyle O}{\|}}{C}(CH_2)_4 \overset{\overset{\displaystyle O}{\|}}{C}} \boxed{OH} \quad \overset{H_2O}{H} \boxed{-\overset{\overset{\displaystyle H}{|}}{N}} \text{etc.}$$

$$\downarrow$$

$$\text{etc.} -\underset{\text{amide group}}{\overset{\overset{\displaystyle H}{|}\;\overset{\displaystyle O}{\|}}{N-C}}-(CH_2)_4 -\underset{\text{amide group}}{\overset{\overset{\displaystyle O}{\|}\;\overset{\displaystyle H}{|}}{C-N}}-(CH_2)_6 -\underset{\text{amide group}}{\overset{\overset{\displaystyle H}{|}\;\overset{\displaystyle O}{\|}}{N-C}}-(CH_2)_4 -\underset{\text{amide group}}{\overset{\overset{\displaystyle O}{\|}\;\overset{\displaystyle H}{|}}{C-N}}- \text{etc.}$$

amide bond

A strand of Nylon 6,6 is pulled from a thin film of the polymer that is formed at the interface between a solution of adipoyl chloride, $ClOC(CH_2)_4COCl$, in a nonpolar solvent and hexamethylenediamine, $H_2N(CH_2)_6NH_2$, dissolved in water. The amide bonds are formed as the components of HCl are split out from between the monomers. This is the same polymer as described in the text, but the reaction for its formation is slightly different.

The bond joining the monomer units is an "amide bond," the same kind of bond that joins amino acid units in polypeptides and proteins, as discussed later.

Nylons are strong, tough plastics because the polymer chains attract each other strongly by hydrogen bonding between N—H groups in one chain and carbonyl groups in neighboring chains.

Dacron, a typical polyester, is formed in a similar condensation reaction with the elimination of methanol.

$$n\; CH_3-O-\overset{\overset{\displaystyle O}{\|}}{C}-\bigcirc\!\!\!\!\bigcirc-\overset{\overset{\displaystyle O}{\|}}{C}-O-CH_3 + n\; HO-CH_2-CH_2-OH \longrightarrow$$

methyl terephthalate   ethylene glycol

$$\left(\!\!\begin{array}{c}\overset{\overset{\displaystyle O}{\|}}{C}-\bigcirc\!\!\!\!\bigcirc-\overset{\overset{\displaystyle O}{\|}}{C}-O-CH_2-CH_2-O\end{array}\!\!\right)_{\!\!n} + n\; CH_3OH$$

Dacron

Inorganic polymers

The polymers we have discussed so far are organic polymers and are the kind most people are familiar with. Most have essentially a carbon "backbone" (the carbon chain in polyethylene, for example).

The organic polymers we have mentioned suffer from certain drawbacks. All except teflon are flammable, for example, and those that contain nitrogen can give off poisonous hydrogen cyanide when they burn. In addition, organic polymers usually become brittle and inflexible at low temperatures, and tend to deteriorate at high temperatures. Solvents often make organic polymers swell up, and only a few

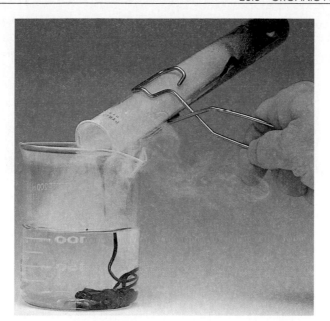

Figure 23.21. *Molten sulfur being poured into cold water yields plastic sulfur. It has many of the properties of rubber.*

organic polymers are truly compatible with living tissue and can be used for medical implants.

Many of the troubles that plague organic polymers can be traced to the nature of their molecular backbone. With inorganic polymers, whose molecular backbones consist of atoms other than carbon, these problems are often minimized, and some of them have become indispensable in modern technology.

In attempting to create useful inorganic polymers, chemists have met obstacles that are quite different from those encountered while making organic materials. Methods that work in the synthesis of organic polymers often fail, so entirely new approaches have had to be found. Furthermore, inorganic polymers are often unstable, or react with moisture and are destroyed. For example, one inorganic polymer that has been known for many years is plastic sulfur, formed when hot molten sulfur is poured into cold water (Figure 23.21). As described earlier in this chapter (page 786), when sulfur is heated, its eight-membered rings open and long polymeric chains of sulfur atoms form as linear S_8 units become linked. When cooled rapidly, these long polymeric chains remain intact and produce a rubberlike material. But plastic sulfur has never found any practical uses because with time the sulfur chains break down and reform S_8 rings of ordinary sulfur, and the solid becomes brittle.

Silicone polymers

Among the synthetic inorganic polymers that have received the most widespread uses are the silicones. These polymers contain an alternating silicon–oxygen backbone and have organic groups attached to the backbone at each silicon atom. Usually, these are methyl groups, —CH_3, which we can think of as being derived from methane, CH_4, by the removal of one hydrogen atom. For example, the structure of a typical linear silicone polymer is illustrated below.

$$CH_3\!-\!\overset{\displaystyle CH_3}{\underset{\displaystyle CH_3}{Si}}\!-\!O\!-\!\cdots\!-\!\overset{\displaystyle CH_3}{\underset{\displaystyle CH_3}{Si}}\!-\!O\!-\!\overset{\displaystyle CH_3}{\underset{\displaystyle CH_3}{Si}}\!-\!O\!-\!\overset{\displaystyle CH_3}{\underset{\displaystyle CH_3}{Si}}\!-\!O\!-\!\overset{\displaystyle CH_3}{\underset{\displaystyle CH_3}{Si}}\!-\!O\!-\!\overset{\displaystyle CH_3}{\underset{\displaystyle CH_3}{Si}}\!-\!O\!-\!\overset{\displaystyle CH_3}{\underset{\displaystyle CH_3}{Si}}\!-\!O\!-\!\cdots\!-\!\overset{\displaystyle CH_3}{\underset{\displaystyle CH_3}{Si}}\!-\!CH_3$$

A portion of a silicone polymer chain

Figure 23.22. *Water forms beads on a fabric treated with a silicone polymer.*

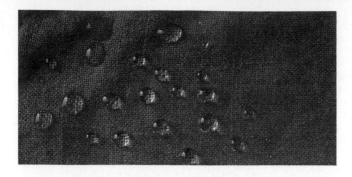

Silicone polymers are formed by hydrolysis of compounds such as $(CH_3)_2SiCl_2$ and $(CH_3)_3SiCl$. As shown below, this gives Si–OH bonds that subsequently expel the components of water and form Si–O–Si linkages. Thus, the first step in the polymerization of $(CH_3)_2SiCl_2$ is

$$
\underset{\underset{\displaystyle CH_3}{|}}{\overset{\overset{\displaystyle CH_3}{|}}{Cl-Si-Cl}} + 2H_2O \longrightarrow \underset{\underset{\displaystyle CH_3}{|}}{\overset{\overset{\displaystyle CH_3}{|}}{HO-Si-OH}} + 2HCl
$$

which is followed by the expulsion of two hydrogens and an oxygen from between pairs of silicon atoms to give the polymer.

$$
CH_3-\underset{\underset{\displaystyle CH_3}{|}}{\overset{\overset{\displaystyle CH_3}{|}}{Si}}-\boxed{OH} \quad H O-\underset{\underset{\displaystyle CH_3}{|}}{\overset{\overset{\displaystyle CH_3}{|}}{Si}}-\boxed{OH} \quad H O-\underset{\underset{\displaystyle CH_3}{|}}{\overset{\overset{\displaystyle CH_3}{|}}{Si}}-\boxed{OH} \quad H O-\underset{\underset{\displaystyle CH_3}{|}}{\overset{\overset{\displaystyle CH_3}{|}}{Si}}-\boxed{OH} \quad H O-\underset{\underset{\displaystyle CH_3}{|}}{\overset{\overset{\displaystyle CH_3}{|}}{Si}}-\boxed{OH} \cdots
$$

(This terminates a polymer chain.)   (These form the backbone.)

$$
CH_3-\underset{\underset{\displaystyle CH_3}{|}}{\overset{\overset{\displaystyle CH_3}{|}}{Si}}-O-\underset{\underset{\displaystyle CH_3}{|}}{\overset{\overset{\displaystyle CH_3}{|}}{Si}}-O-\underset{\underset{\displaystyle CH_3}{|}}{\overset{\overset{\displaystyle CH_3}{|}}{Si}}-O-\underset{\underset{\displaystyle CH_3}{|}}{\overset{\overset{\displaystyle CH_3}{|}}{Si}}-O-\underset{\underset{\displaystyle CH_3}{|}}{\overset{\overset{\displaystyle CH_3}{|}}{Si}}-O-\underset{\underset{\displaystyle CH_3}{|}}{\overset{\overset{\displaystyle CH_3}{|}}{Si}}-O- \cdots
$$

Depending on the chain length and the degree to which chains are cross-linked to each other, the silicones are oils, greases, or rubbery solids. Silicone oils repel water and water-based products. For this reason, they have been used in polishes for furniture and cars, and as waterproofing treatments for fabrics and leather (Figure 23.22). Silicone greases are used as permanent lubricants in clocks and ball bearings. Silicone resins are used in some paints because of their resistance to peeling and high temperatures. Silicone rubber has excellent electrical insulating properties and is used to coat electrical wire and in electric motors. These rubbery materials also retain their flexibility over wide temperature ranges and can be used in place of natural rubber for both high- and low-temperature applications. Silicones are even found in medical preparations. The polymer with the structure illustrated above is known medicinally as *simethicone* and is a common antigas agent found in antacid products.

Silicones have many medical applications. For example, silicone oils are used to lubricate the skin of burn victims.

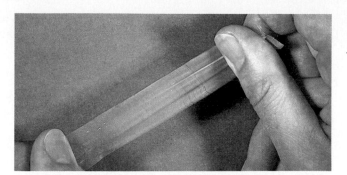

Figure 23.23. *A sample of a clear polyphosphazene polymer film.*

Phosphazenes

More recently, another breed of inorganic polymers has emerged with its own set of characteristic and useful properties. In these substances, called **polyphosphazenes,** the polymer backbone consists of alternating phosphorus and nitrogen atoms. They are prepared by polymerizing a ring-shaped molecule, $(PNCl_2)_3$, and then substituting various groups of atoms for the chlorines in the polymer. The general structure of the polyphosphazenes is

$$\cdots-\overset{\overset{\displaystyle X}{|}}{\underset{\underset{\displaystyle X}{|}}{P}}=N-\overset{\overset{\displaystyle X}{|}}{\underset{\underset{\displaystyle X}{|}}{P}}=N-\overset{\overset{\displaystyle X}{|}}{\underset{\underset{\displaystyle X}{|}}{P}}=N-\overset{\overset{\displaystyle X}{|}}{\underset{\underset{\displaystyle X}{|}}{P}}=N-\overset{\overset{\displaystyle X}{|}}{\underset{\underset{\displaystyle X}{|}}{P}}=N-\overset{\overset{\displaystyle X}{|}}{\underset{\underset{\displaystyle X}{|}}{P}}=N-\overset{\overset{\displaystyle X}{|}}{\underset{\underset{\displaystyle X}{|}}{P}}=N-\overset{\overset{\displaystyle X}{|}}{\underset{\underset{\displaystyle X}{|}}{P}}=N-\overset{\overset{\displaystyle X}{|}}{\underset{\underset{\displaystyle X}{|}}{P}}=N-\cdots$$

where X stands for any of a large number of different groups of atoms that can be attached to the phosphorus atoms. The large variety of possible side groups has led to the synthesis of many different polymers with a wide range of properties. The phosphazenes are generally nonflammable, and some are used as flame retardants for other polymers. Some phosphazenes are glasses, while others are elastomers that retain their elasticity to low temperatures (in one case, as low as $-90\ °C$). When X contains a nonpolar chain of atoms, as in the group $-O-CH_2-CF_3$, the polymer is water-insoluble. In fact, this side group produces flexible polymer films (Figure 23.23) that are more strongly water-repellent than the silicones. They also do not react with living tissue, which makes this polymer an excellent candidate for tubes used to replace blood vessels. On the other hand, when a polar group such as $-NHCH_3$ is attached to the phosphorus, a water-soluble polymer is obtained.

One of the most interesting potential applications of phosphazene polymers is in chemotherapy. Most chemotherapeutic agents are small molecules that can easily diffuse through cell membranes into regions of the body where they aren't wanted and where they can do more harm than good. The results are unwanted side effects. It is hoped that these chemicals can be immobilized by attaching them to phosphazene polymer chains, which are too large to diffuse through cell membranes, and thereby minimize such problems. Some progress in this direction has already been made. It has been demonstrated that certain biologically active molecules can be incorporated into these polymers without losing their activity.

Electrically conducting polymers

The earliest and one of the most interesting of these polymers is formed when tetrasulfur tetranitride, S_4N_4, is heated. The ring-shaped molecule initially breaks down into a smaller ring, S_2N_2, which then polymerizes to give crystals of poly(sulfur nitride), $(SN)_x$, a shiny metallic solid. The crystals contain long zigzag chains of alternating sulfur and nitrogen atoms.

What is amazing about these crystals is that they show metallic conduction along the chains. In fact, they even become superconductors if cooled to near absolute zero (that is, they lose all resistance to electrical conduction). As you might expect, these properties have generated great interest because of their potential practical importance, and research on this and other electrically conducting polymers is actively being pursued.

Biopolymers

Many (but certainly not all) of the compounds in living organisms fit the definition of polymers. Examples are the polypeptides that are found in proteins, the polysaccharides found in starch and cellulose, and the DNA found in the nuclei of cells and that controls heredity and the synthesis of proteins. Because of limited space, we surely can't describe all the details that are known about these substances; we leave that for other courses that you might take. We can, however, mention a few features of them.

α-Amino acids and the formation of peptide bonds were discussed in Chapter 22.

Polypeptides are formed by the linking together of α-amino acids with peptide bonds, as shown by the following.

R_1, R_2, and so forth are the different side groups that differentiate one amino acid from another.

Notice that the peptide bonds are very similar to the bonds that link monomer units in nylon. In living systems, the structures of polypeptides are made up of combinations of only about 20 different amino acids, but because their chains are so long, the number of combinations is virtually infinite. Polypeptide chains combine in various ways to form muscle tissue, hair, tendons, hemoglobin, and virtually all the enzymes that catalyze biochemical reactions.

Proteins have formula masses that range from about 6000 to 1,000,000.

Polysaccharides are polymers of sugar molecules. One of the most important is glucose, which exists in two cyclic forms as illustrated in Figure 23.24. Notice in particular the orientations of the H and OH groups marked by color. Polymerization of these yields two slightly different polysaccharides. The polymerization of α-D-glucose yields starch, whereas the polymerization of β-D-glucose yields cellulose. See Figure 23.25. Although the structural differences between starch and cellulose are subtle, they are nevertheless profound. For example, our digestive systems have enzymes that are able to decompose starch into its simpler units, which we can then use to generate energy by metabolism. However, we cannot digest cellulose, even though it is a polymer of essentially the same sugar. (Termites, however, contain microorganisms that produce an enzyme to digest cellulose, which is a major component of wood.)

Ribose is found in another nucleic acid known as RNA (ribonucleic acid).

DNA stands for **deoxyribonucleic acid.** It is a polymer of three different substances. One is a sugar called deoxyribose, shown along with the sugar ribose in Figure 23.26. At carbon 1 is attached one or another of molecules that are called *nitrogenous bases.* These substances are shown in Figure 23.27 and the way they are bonded to deoxyribose is illustrated in Figure 23.28. Also shown in Figure 23.28 is

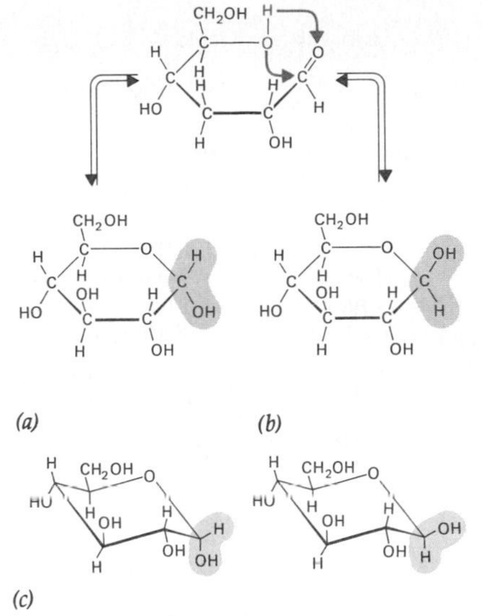

Figure 23.24. *Cyclic structures for glucose* (a) *α-D-glucose.* (b) *β-D-glucose.* (c) *Puckered ring. Note orientation of H and OH on the rightmost carbon.*

(a)   (b)

(c)

the involvement of the third substance in the polymer, phosphoric acid. The three primary units, deoxyribose, a nitrogenous base, and the phosphoric acid unit, when linked as shown in Figure 23.28 form the basic building blocks (called nucleotides) of DNA. The linking of nucleotides produces a long strand of DNA as illustrated in Figure 23.29. These long strands then intertwine and become interlocked by hydrogen bonding between bases on adjacent strands. This yields the now-famous double helix (Figure 23.30) discovered by Watson and Crick, for which they received the Nobel prize in 1962.

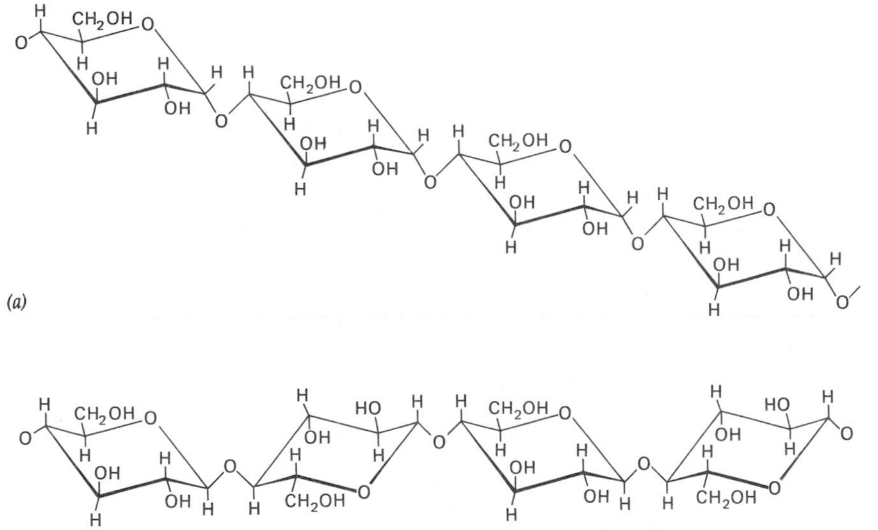

Figure 23.25. *Structures of the polysaccharides, starch and cellulose.* (a) *Amylose (starch).* (b) *Cellulose.*

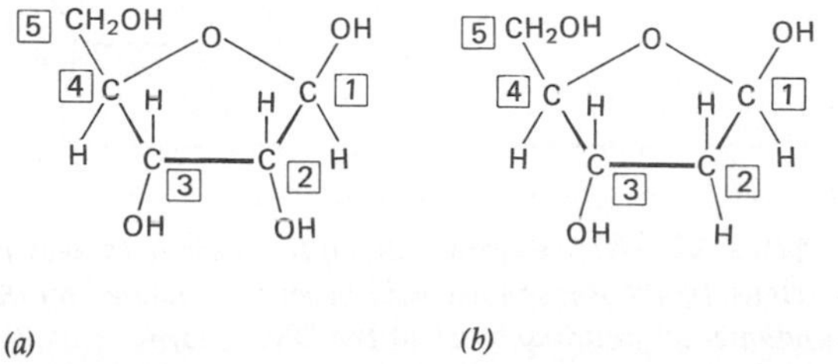

(a)

(b)

Figure 23.26. *Two sugars involved in the formation of nucleic acids.* (a) *Ribose.* (b) *Deoxyribose.*

(a)   (b)

Figure 23.27. *Nitrogenous bases found in DNA (deoxyribonucleic acid) and RNA (ribonucleic acid). They are called bases because of the nitrogen atoms that are able to be electron pair donors (Lewis bases).*

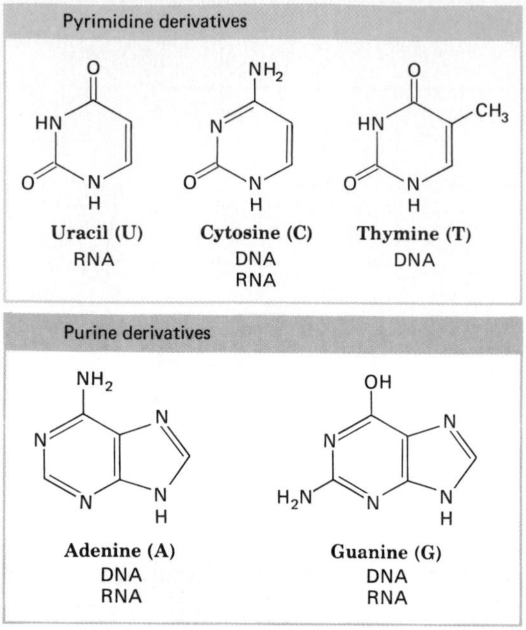

(a)

(b)

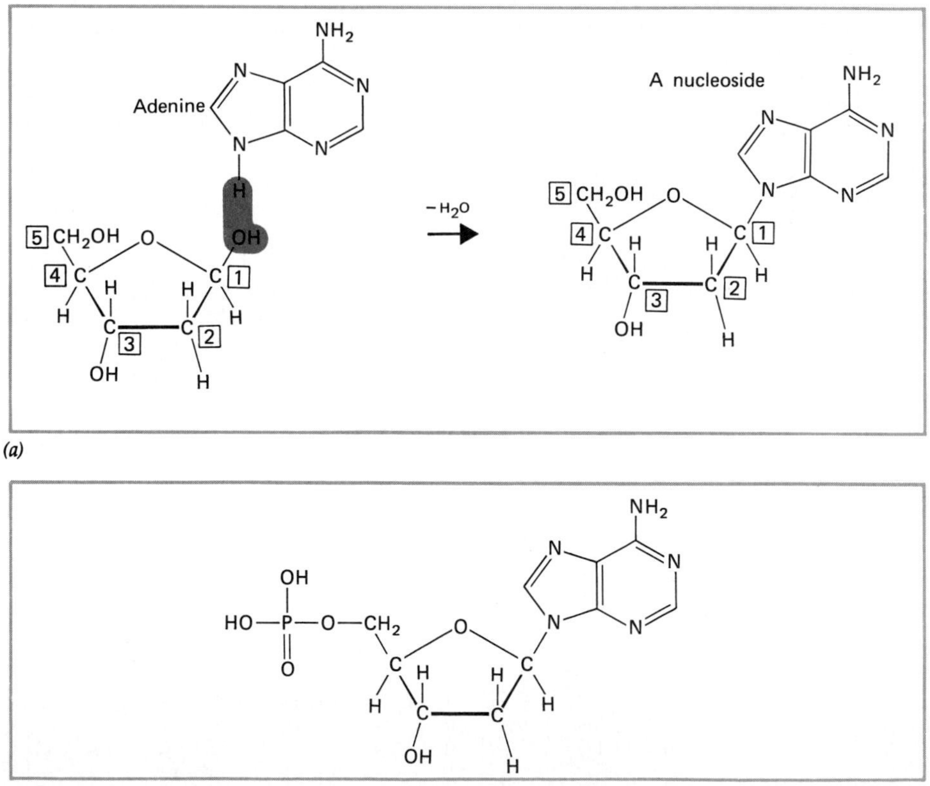

(a)

(b)

Figure 23.28. (a) *Bases attach themselves to deoxyribose by elimination of H_2O to yield a nucleoside.* (b) *Phosphoric acid becomes attached by the same process to yield a nucleotide, the fundamental building block of the DNA chain.*

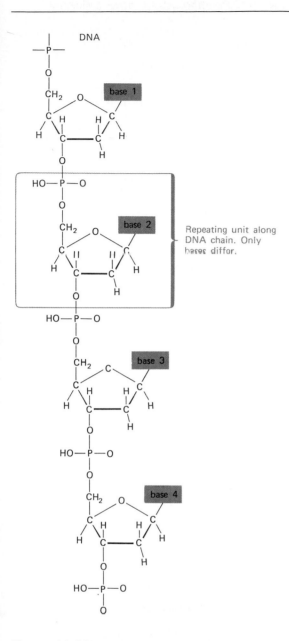

Figure 23.29. *Polymerization of nucleotides gives nucleic acids. Notice that the phosphate unit serves as the bridge between the sugar units in the chain.*

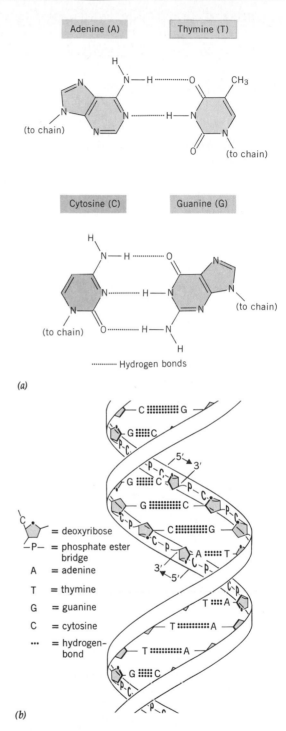

Figure 23.30. (a) *The bases are able to bind to each other by hydrogen bonding, but they fit together only one way. Adenine binds to thymine, and cytosine binds to guanine. (b) Base pairing as described in part a links DNA strands and holds them in a double helix. A cell's genetic information is stored in the sequence of bases along the DNA strands of the double helix.*

REVIEW QUESTIONS AND PROBLEMS

Problems whose numbers are in blue have their answers in Appendix D at the back of the book.

Phosphorus: General

23.1 What is the formula for the phosphorus-containing compound in phosphate rock?

23.2 What kinds of roles does phosphorus play in living systems?

23.3 Give the electron configuration of phosphorus.

23.4 What is the structure of the molecules found in white phosphorus? Why is white phosphorus so reactive?

23.5 How is white phosphorus made? Give a chemical equation.

23.6 What is an *electric furnace*?

23.7 Describe red phosphorus and black phosphorus. How do their reactivities compare to white phosphorus?

Phosphorus Oxides

23.8 Write chemical equations for the reaction of white phosphorus with oxygen (a) when the O_2 is present in excess and (b) when there is a limited amount of O_2.

23.9 Sketch the structures of the molecules found in (a) phosphorus(V) oxide and (b) phosphorus(III) oxide.

23.10 Write equations for the reaction of water with (a) phosphorus(V) oxide and (b) phosphorus(III) oxide.

23.11 What is a *desiccant*? Why is P_4O_{10} a good desiccant?

Phosphoric Acid and Phosphates

23.12 Give a chemical equation showing how phosphoric acid is made directly from phosphate rock.

23.13 What is *syrupy phosphoric acid*?

23.14 How is the phosphoric acid used in food products manufactured?

23.15 Give three commercial uses of phosphoric acid.

23.16 Give the formulas and names of three salts that can be formed from $Mg(OH)_2$ and phosphoric acid.

23.17 What is orthophosphoric acid?

23.18 What ions make up the phosphate buffer in the blood? How do they serve to help maintain a nearly constant blood pH?

23.19 What are some uses of TSP? Give a chemical equation to explan why its solutions are basic.

23.20 Give Lewis structures for H_3PO_4 and H_3PO_3.

23.21 What is superphosphate fertilizer? Give a chemical equation showing how it is made.

23.22 Why is phosphate rock itself a poor phosphate fertilizer?

23.23 What is the name of (a) Na_2HPO_4 and (b) NaH_2PO_4?

Phosphorus: Other Compounds

23.24 Write a chemical equation for the reaction of sodium with phosphorus. What happens when the product of this reaction is placed in water?

23.25 Use Lewis structures to illustrate how pyrophosphoric acid is formed from phosphoric acid.

23.26 What is the basic structural unit in the polymeric phosphoric acids and polyphosphates?

23.27 What is the empirical formula for the metaphosphate ion? Give its Lewis structure and show how it can be considered to be formed from phosphoric acid.

23.28 Give the structure of the tripolyphosphate ion. What uses does it have?

23.29 What mole ratio of NaH_2PO_4 and Na_2HPO_4 should be chosen to obtain a linear polyphosphate containing five phosphorus atoms?

23.30 Explain why phosphate pollution can be very harmful to lakes.

23.31 What is the name of the acid, H_3PO_3? What salts can be formed by reacting it with $Mg(OH)_2$?

23.32 How do H_3PO_4 and H_3PO_3 compare as oxidizing and/ or reducing agents?

23.33 Sketch the structures of PCl_3 and PCl_5. How does PCl_5 exist in the solid state? What kinds of hybrid orbitals does phosphorus use in PCl_3 and PCl_5?

Sulfur: General

23.34 Give the electron configuration of sulfur.

23.35 What is *brimstone*? What does this name mean?

23.36 How does the element sulfur occur in nature?

23.37 What are the two allotropic forms of sulfur? How do they differ? Outline the physical changes that take place when sulfur is gradually heated to its boiling point and relate them to the structural changes that occur.

23.38 What is plastic sulfur?

23.39 Describe the Frasch process for mining sulfur.

Sulfur: Oxides and Oxoacids

23.40 Which air pollutant is produced by the combustion of high-sulfur fuels?

23.41 Write chemical equations for the production of H_2SO_4 using O_2, S, and H_2O as starting materials.

23.42 Why is SO_2 stable toward oxidation to SO_3 in the air?

23.43 How can SO_2 be conveniently prepared in the laboratory?

23.44 Write chemical equations for the reaction of water with (a) sulfur dioxide and (b) sulfur trioxide.

23.45 Describe *acid rain*—how it is formed and what damage it does.

23.46 How do the acidities of H_2SO_3 and H_2SO_4 compare? How can the differences be explained?

23.47 List four uses for sulfuric acid.

23.48 Draw the resonance structures for and describe the shapes of SO_2 and SO_3.

23.49 What is the proper way to prepare dilute H_2SO_4 from concentrated H_2SO_4?

23.50 Even though H_2SO_4 isn't fully dissociated into H^+ and SO_4^{2-}, it is considered a strong acid. Why?

23.51 Write an equation showing the dehydrating action of concentrated H_2SO_4 on the sugar glucose, $C_6H_{12}O_6$.

23.52 Name the following two salts: (a) Na_2SO_3 and (b) $NaHSO_4$.

Sulfur: Other Compounds

23.53 What is the structure of pyrosulfuric acid? (*Hint:* What is the structure of pyrophosphoric acid?)

23.54 The cyanate ion has the Lewis structure

$$\left[:N\equiv C-\overset{\cdot\cdot}{\underset{\cdot\cdot}{O}}:\right]^-$$

What would be the structure of the thiocyanate ion?

23.55 Write an equation for the generation of H_2S by the hydrolysis of thioacetamide.

23.56 Why are hydrogen sulfide fumes to be avoided in the laboratory?

23.57 What is the structure of the thiosulfate ion? Write a chemical equation showing how thiosulfate ion is made.

23.58 Explain the function of sodium thiosulfate, $Na_2S_2O_3$, in photography.

23.59 Write balanced chemical equations for the oxidation of $S_2O_3^{2-}$ by (a) Cl_2 and (b) I_2.

23.60 What products, if any, are formed in the following reactions?
(a) cold concentrated H_2SO_4 and NaI
(b) hot concentrated H_2SO_4 and copper
(c) hot concentrated H_2SO_4 and zinc
(d) cold dilute H_2SO_4 and zinc
(e) cold dilute H_2SO_4 and copper

Halogens: General

23.61 Write the electron configuration of each of the halogens.

23.62 What is the origin of the name *halogen*?

23.63 How are the halogens normally found in nature?

23.64 What are the principal sources of each of the halogens?

23.65 How is fluorine prepared?

23.66 Describe how you could prepare chlorine in the laboratory. How is chlorine made industrially? What are three commercial uses for chlorine?

23.67 Describe the physical characteristics of the halogens.

23.68 Complete and balance the following equations. If no reaction occurs, write N.R.
(a) $Cl_2 + KI \rightarrow$ (c) $I_2 + NaCl \rightarrow$
(b) $F_2 + KBr \rightarrow$ (d) $Br_2 + NaI \rightarrow$

23.69 Why is fluorine so reactive?

23.70 How is bromine obtained industrially? Give two principal uses of Br_2. How can Br_2 be made in the laboratory? (Give a balanced equation.)

23.71 What are the commercial sources of iodine?

Halogens: Binary Compounds

23.72 How do the reactivities of the halogens toward hydrogen compare?

23.73 How is HF made?

23.74 How would you prepare HCl in the laboratory? What are two commercial uses of hydrochloric acid?

23.75 How would you prepare HBr and HI in the laboratory?

23.76 Why is glass attacked by hydrofluoric acid?

23.77 Compare the boiling points of the hydrogen halides. How do HF molecules exist in liquid HF?

23.78 HF is a weak acid, yet it is very dangerous to work with. Why?

Halogens: Oxoacids and Oxoanions

23.79 Name these compounds.
(a) HOBr   (e) HIO_4
(b) NaOCl   (f) $HBrO_3$
(c) $KBrO_3$ (g) $NaIO_3$
(d) $Mg(ClO_4)_2$ (h) $KClO_2$

23.80 What is a disproportionation reaction?

23.81 Write a chemical equation for the reaction that takes place when Cl_2 is dissolved in
(a) cold water   (b) cold NaOH solution

23.82 CaCl(OCl) is a solid water-soluble bleach. What reaction would occur if acid were added to this solid?

23.83 Compare the stabilities of the ions OCl^-, OBr^-, and OI^- in cold aqueous solutions. What accounts for their relative stabilities?

23.84 What reaction occurs if $KClO_3$ is heated at moderate temperatures? What happens if MnO_2 is present while the $KClO_3$ is heated?

23.85 What is the chemical formula for HTH, which is sold as a swimming pool chlorinating agent?

Halogens: Other Compounds

23.86 Use VSEPR theory to predict the molecular shapes of
(a) SF_2 (e) BrF_3
(b) PBr_5 (f) $TeBr_4$
(c) SiF_6^{2-} (g) BrF_5
(d) AsF_3 (h) $GeCl_4$

23.87 What is the likely reason that IF_7 can be made but ClF_7 cannot?

23.88 Suggest two reasons why oxygen doesn't form OF_4 even though sulfur forms SF_4.

23.89 Predict what products would be formed in the reaction between $GeCl_4$ and H_2O.

23.90 Why does $SiCl_4$ hydrolyze but CCl_4 does not?

23.91 Compare the effectiveness of NH_3 and NF_3 as Lewis bases.

Noble Gas Compounds

23.92 What is a *clathrate*?

23.93 Predict the molecular structures of XeF_4 and XeF_2 using the VSEPR theory.

23.94 Why were chemists so surprised when it was found that the noble gases were not inert?

Silicon and Silicates

23.95 How is elemental silicon prepared? Give an equation.

23.96 Describe *zone refining*.

23.97 Why doesn't silicon form graphitelike crystals?

23.98 Carbon forms CO_2, which is a gas, whereas silicon forms SiO_2, which is a complex solid. Why doesn't Si form an SiO_2 molecule with the same structure as CO_2?

23.99 What is the structure of the orthosilicate ion?

23.100 Use Lewis structures to describe the linking together of SiO_4 tetrahedra in the formation of the pyrosilicate ion.

23.101 Sketch the structure of the cyclic $Si_6O_{18}^{12-}$ anion. Which mineral contains this anion?

23.102 Sketch a portion of the anion found in asbestos. What is the repeating unit in the structure?

23.103 Sketch a portion of the polymeric metasilicate anion, $(SiO_3^{2-})_x$.

23.104 What kind of structure is formed when each SiO_4 tetrahedron shares three of its corners with neighboring SiO_4 tetrahedra? What are some common minerals containing this structure?

23.105 What is the empirical formula for quartz?

Polymers

23.106 What is a polymer?

23.107 What controls the chemical properties of polymers? What principally controls their physical properties?

23.108 How do addition polymers and condensation polymers differ in the way they are formed?

23.109 How does addition polymerization take place? What is the role of the initiator?

23.110 Sketch a portion of a polypropylene polymer chain. Include at least three repeating units.

23.111 What three properties of teflon make it an especially useful polymer?

23.112 Sketch a portion of a polyacrylonitrile polymer chain. Include at least three repeating units. What is another name for this polymer?

23.113 What is an amide bond? How is it related to a peptide bond? In which synthetic polymer do we find amide bonds?

23.114 Describe the structure of the Dacron polymer?

23.115 Sketch a portion of the backbone of a polyphosphazene polymer. What influence does the polarity of the groups attached to this backbone have on the physical properties of the polymer?

23.116 What is a use for poly(methyl methacrylate)? What is another name for this polymer?

23.117 What structure is found along the backbone of the silicone polymers?

23.118 What compound would be formed by hydrolysis of

$$CH_3-\underset{\underset{\displaystyle CH_3}{|}}{\overset{\overset{\displaystyle CH_3}{|}}{Si}}-Cl$$

23.119 Describe the structure of simethicone.

23.120 Show that the addition of $SiCl_4$ or CH_3SiCl_3 to the reaction mixture in the formation of a silicone polymer will lead to branching and cross-linking of the polymer chain.

23.121 Why is there so much interest in polymers like poly(sulfur nitride)?

23.122 What substances serve as monomers in the formation of a polypeptide? Why are so many polypeptides possible?

23.123 What is a polysaccharide?

23.124 What structural difference is there between starch and cellulose? If each is completely hydrolyzed, what sugar is obtained?

23.125 What three components are brought together in a nucleotide? How are nucleotides linked to give a strand of DNA?

23.126 How is hydrogen bonding important in determining the structure of the double helix of DNA?

NUCLEAR REACTIONS AND THEIR ROLE IN CHEMISTRY

CHAPTER

24 NUCLEAR CHEMISTRY

In our discussions of chemistry until now we have paid little attention to the nuclei of atoms, other than to note that the number of positive charges on the nucleus determines the number of electrons that a neutral atom must have and that the mass of the nucleus essentially determines the atomic mass. There are nuclear phenomena, however, that have applications in chemistry and one of the goals of this chapter is to explore some of them. Equally important in this nuclear age is the production of nuclear energy for both peaceful and defense applications. No matter how you feel about this topic politically, it is one that we all must face, now and in the future. Therefore, in this chapter we will also describe the origin of the tremendous amounts of energy available from nuclear reactions and attempts, both presently successful and promising for the future, at harnessing this energy for peaceful purposes.

24.1 SPONTANEOUS RADIOACTIVE DECAY

Natural radioactivity was discovered, quite by accident, by a French physicist named Antoine Henri Becquerel (1852–1908). Becquerel found that when uranium salts were left in contact with photographic plates, the plates were darkened in the same manner as when they were exposed to X rays. He reasoned correctly that uranium spontaneously emits a penetrating form of radiation that was responsible for blackening the photographic plates. Two colleagues of Becquerel, Pierre and Marie Curie, were successful in isolating two other radioactive elements from uranium ore — polonium (Po) and radium (Ra) — and found that they were even more intensely radioactive. For their discoveries, Becquerel and the two Curies were awarded the 1903 Nobel prize for physics.

We now know that the radiation emitted from radioactive substances is of three main types: **alpha (α) particles, beta (β) particles,** and **gamma (γ) rays.** Experimentally it is found that the alpha particle carries a positive charge, the beta particle a negative charge, and the gamma rays carry no charge. The actual charges and masses of these and other particles we will be concerned with are listed in Table 24.1.

When a substance spontaneously emits either an alpha or a beta particle, there is a change in the charge on the nucleus, and therefore a change in atomic number. For example, nuclei of the most abundant isotope of uranium, $^{238}_{92}U$, spontaneously emit alpha particles. This removes two units of charge and four units of mass from the nucleus and gives the isotope $^{234}_{90}Th$. Thus, the emission of an alpha particle changes a uranium atom into a thorium atom — we say that $^{238}_{92}U$ *decays* to $^{234}_{90}Th$.

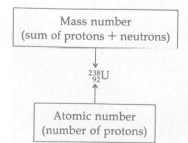

Mass number
(sum of protons + neutrons)

$^{238}_{92}U$

Atomic number
(number of protons)

Table 24.1. Principal types of radiation emitted by radioactive nuclei

Radiation	Approximate Mass (u)	Charge	Symbol	Type
Alpha	4	2+	^4_2He	Particle
Beta	0	1−	$^0_{-1}e$	Particle
Gamma	0	0	γ	Electromagnetic radiation
Neutron	1	0	$^1_0 n$	Particle
Proton	1	1+	$^1_1 p\ (^1_1\text{H})$	Particle
Positron	0	1+	$^0_1 e$	Particle

The changes that occur during a nuclear reaction such as the decay of $^{238}_{92}\text{U}$ can be represented by a *nuclear equation*. For example, the nuclear reaction just described is represented by the nuclear equation

$$^{238}_{92}\text{U} \longrightarrow \underset{\substack{\text{alpha}\\\text{particle}}}{^4_2\text{He}} + {}^{234}_{90}\text{Th}$$

Notice that for a nuclear equation to be balanced, both the total charge and the total mass must be balanced.

The thorium produced in the decay of $^{238}_{92}\text{U}$ is itself radioactive and decays by beta emission. The nuclear equation for the change is

$$^{234}_{90}\text{Th} \longrightarrow \underset{\substack{\text{beta}\\\text{particle}}}{^0_{-1}e} + {}^{234}_{91}\text{Pa}$$

Thus, beta emission increases the atomic number by one unit, but has (essentially) no effect on the mass. The overall result is that a neutron is changed to a proton.

Gamma radiation is really nothing more than a very energetic form of electromagnetic radiation. Its emission from a nucleus doesn't change the charge or mass number, so gamma radiation is often omitted from nuclear equations.

In discussions of nuclear reactions and radioactive decay, certain terms are used frequently, so you should become familiar with them. The most common is **nuclide,** a general term used when referring to the nucleus of a particular isotope. Radioactive nuclei are called **radionuclides** and the atoms having these nuclei are called **radio-isotopes.** In a radioactive decay, the isotope that decays is often called the **parent isotope** and the isotope formed is referred to as the **daughter isotope.** Thus, in the decay of uranium-238, the nuclide $^{238}_{92}\text{U}$ is the parent and $^{234}_{90}\text{Th}$ is the daughter.

> The algebraic sum of the subscripts must be the same on both sides, and the algebraic sum of the superscripts must match.

> Gamma rays are emitted by nearly all radioactive substances.

EXAMPLE 24.1. IDENTIFYING PRODUCTS IN A NUCLEAR EQUATION

PROBLEM: (a) ^{234}U decays by the emission of an alpha particle. Write the nuclear equation for the process and identify the daughter isotope.

(b) ^{214}Pb decays to ^{214}Bi. Write a nuclear equation for the change and identify the type of radiation emmited by the lead nucleus.

ANALYSIS: In each of these, we will use the fact that the mass and charge must balance. This means that for each nuclear equation, the sum of the mass numbers on both sides must be the same. The same requirement applies to the sum of the atomic numbers on each side.

We will use this to determine the mass number and atomic number for the unknown quantity and then use that information to identify it.

SOLUTION: (a) Let's let X represent the unknown nuclide and x and y its atomic number and mass number, respectively. The symbol for the nuclide is therefore $^y_x X$. This is a daughter nuclide, so we can write the nuclear equation as

$$^{234}_{92}U \longrightarrow {}^4_2He + {}^y_x X$$

To be balanced, the following must be true:

$$234 = 4 + y$$

$$92 = 2 + x$$

Therefore, y must equal 230 and x must equal 90. The nuclide is $^{230}_{90} X$. The element with an atomic number of 90 is thorium, so the daughter isotope is $^{230}_{90}Th$.

(b) Let's proceed as before, but this time we let the symbol X represent the kind of radiation. Keeping in mind the requirements for a balanced nuclear equation, we arrive at the following:

$$^{214}_{82}Pb \longrightarrow {}^{214}_{83}Bi + {}^0_{-1} X$$

From Table 24.1 we see that $_{-1}^0 X$ corresponds to $_{-1}^0e$; therefore, the decay occurs by beta emission.

Radioactive decay series

We mentioned earlier that ^{238}U decays to ^{234}Th, which is also radioactive and decays to ^{234}Pa. This isotope is also unstable and decays to ^{234}U (how?), which is also radioactive, and so on. This decay sequence continues until a stable (nonradioactive) isotope of an element is formed. The entire scheme, in which one isotope decays to another and that to another, and so on, is called a **radioactive series** or **decay series.** ^{238}U, for example, decays by some 14 steps to stable ^{206}Pb, as shown in Figure 24.1.

24.2 KINETICS OF RADIOACTIVE DECAY

In Chapter 19 we saw that the rate of a chemical reaction is given by the change in the concentration of a reactant per unit time. For a radioactive substance, its concen-

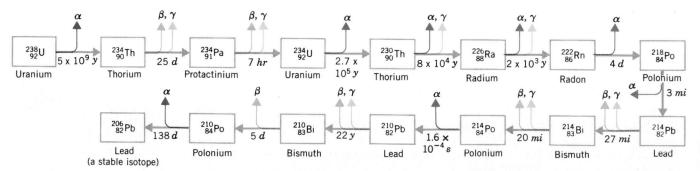

Figure 24.1. *The uranium-238 radioactive decay series. The time given beneath each arrow is the half-life period of the preceding isotope (y = year, m = month, d = day, mi = minute, s = second). (From J. E. Brady and J. R. Holum,* Fundamentals of Chemistry, *3rd edition, 1988, John Wiley & Sons, New York, p. 965. Copyright © 1988, by John Wiley & Sons. Used with permission.)*

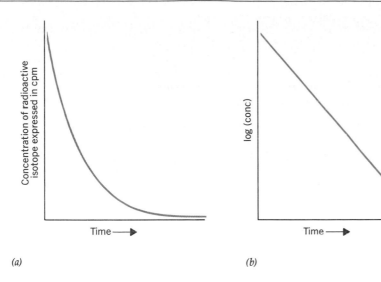

(a)   (b)

tration is directly proportional to the number of particles or rays that it emits per unit time. These emissions can be detected and counted by a suitable device such as the Geiger–Müller counter, which we will describe later. Experimentally, the concentration of the radioactive isotope is proportional to the number of counts per second (cps) or per minute (cpm) recorded by the counter.

Each *count* is a record of a nuclear event that is recorded by some measuring device that counts the individual radiations during the radioactive decay process.

If, for a particular radioactive isotope, we construct a graph in which the number of counts per minute is plotted against time, we obtain a curve similar to that shown in Figure 24.2a. If the log of the cpm is graphed versus time, a straight line is obtained as shown in Figure 24.2b. Both of these plots are indicative of a first-order process and, as it turns out, all radioactive decay reactions obey first-order kinetics.

In Chapter 19, we saw that for a first-order process the concentration of a reactant (for example, a radioactive isotope) at any time t is given by the equation

$$\ln \frac{[A]_0}{[A]_t} = kt \qquad (24.1)$$

We could also use the equation

$$\log \frac{[A]_0}{[A]_t} = \frac{kt}{2.303}$$

In this equation $[A]_0$ is the initial concentration of the isotope (concentration at time zero) and $[A]_t$ is its concentration at any time t. Equation 24.1 can be used to calculate rate constants, as shown in the following example.

EXAMPLE 24.2. CALCULATING THE RATE CONSTANT FOR A RADIOACTIVE DECAY

PROBLEM: A scientist determined that a sample contained 10.0 μg of radioactive ^{222}Rn. After exactly one week, the radon had decayed and the sample now contained only 2.82 μg of ^{222}Rn. On the basis of this information, calculate the rate constant for the decay of this radon isotope in units of day^{-1}.

SOLUTION: Equation 24.1 is written in terms of molar concentrations. However, because we are taking a ratio of concentrations, we can express these quantities in terms of anything that is proportional to the concentration. This is so because the proportionality constant, whatever it may be, will cancel from the numerator and denominator of the ratio.

Since we are dealing with the same sample, we can work with the number of micrograms directly in place of concentrations. Therefore, we will take $[A]_0$ to be 10.0 μg, $[A]_t$ to be

2.82 μg, and t to be 7 days. Now, let's solve Equation 24.1 for the rate constant.

$$k = \frac{1}{t} \ln \frac{[A]_0}{[A]_t}$$

Substituting the values given in the problem yields

$$k = \left(\frac{1}{7.000 \text{ day}}\right) \ln \left(\frac{10.0}{2.82}\right)$$

$$= 0.1810 \text{ day}^{-1}$$

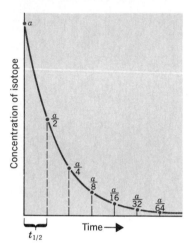

Figure 24.3. *A graph of a first-order radioactive decay illustrating the concept of half-life. The initial concentration is* a, *which drops to* $\frac{a}{2}$ *after one half-life, to* $\frac{a}{4}$ *after the second half-life, and so forth.*

In Chapter 19 you learned that the *half-life*, $t_{1/2}$, of a reactant in a first-order process is independent of the initial concentration of the reactant. In other words, the time required for half of a radioactive isotope to decay is the same regardless of the amount of the isotope that is present. Thus, if we were to begin with 10.0 g of a particular isotope, after one half-life only 5.0 g would remain. In a time equal to another half-life, this would decay to 2.5 g, and so on. This is represented graphically in Figure 24.3. In this figure we see in general how an isotope decays through many half-lives. We can also show that the half-life of any radioactive substance is inversely proportional to the rate constant for its decay. The relationship is

$$t_{1/2} = \frac{\ln 2}{k} = \frac{0.693}{k} \tag{24.2}$$

From this equation we see that if the rate constant for decay is large, its half-life will be small and vice versa. In addition, we see that if either $t_{1/2}$ or k is known, the other can be calculated.

EXAMPLE 24.3. USING THE RATE CONSTANT FOR RADIOACTIVE DECAY

PROBLEM: The rate constant for the decay of ^{222}Rn is 0.1810 day^{-1}.

(a) What is the half-life for ^{222}Rn expressed in days?

(b) What fraction of a sample of ^{222}Rn decays in a period of exactly one week?

ANALYSIS: For part (a), we simply use Equation 24.2 to calculate the rate constant. For part (b), we will use Equation 24.1. However, we must be careful here. This equation relates the initial amount of the isotope to the amount present after a time t. The fraction $[A]_t/[A]_0$ is therefore the fraction that remains after time t. To obtain the fraction that has decayed, we must subtract the value of $[A]_t/[A]_0$ from 1.

SOLUTION: (a) Substituting the value of k into Equation 24.2, we obtain

$$t_{1/2} = \frac{\ln 2}{0.1810 \text{ day}^{-1}}$$

which gives

$$t_{1/2} = 3.829 \text{ days}$$

(b)   Equation 24.1 gives us the value of $[A]_0/[A]_t$.

$$\ln \frac{[A]_0}{[A]_t} = kt$$

Substituting in $k = 0.1810$ days^{-1} and $t = 7.000$ days, we have

$$\ln \frac{[A]_0}{[A]_t} = (0.1810 \text{ days}^{-1})(7.000 \text{ days})$$

or

$$\ln \frac{[A]_0}{[A]_t} = 1.267$$

Taking the antilog gives

$$\frac{[A]_0}{[A]_t} = 3.55$$

The reciprocal, therefore, is

$$\frac{[A]_t}{[A]_0} = 0.282$$

This is the fraction left after one week, so the fraction that decayed is simply $1 - 0.282$, or 0.718.

24.3 MEASUREMENT OF RADIOACTIVITY

To study nuclear changes, we need to measure radiation and express it quantitatively. One of the earliest devices used to detect radioactive emissions was the Geiger–Müller counter, shown schematically in Figure 24.4. In this device alpha or beta particles pass through the thin window on the left and enter a chamber filled with low-pressure argon. As they pass through the argon, they knock electrons off the gaseous atoms, leaving behind a trail of positive argon ions and electrons. The presence of these charged particles causes the gas to suddenly become conducting, and this allows a discharge—a spark—to jump between the electrodes. Sensitive electronics (not shown in the figure) detect this momentary small burst of current and register it on some sort of meter or other counting device.

Newer methods of counting radioactive emissions often make use of **scintillation counters.** In these devices the radiation is absorbed and causes the emission of a photon of visible light. For example, if beta particles are being studied, a zinc sulfide phosphor can be used. It is this production of visible light that is called *scintillation*. Once the light is generated, it can be detected by photo-optical methods and the

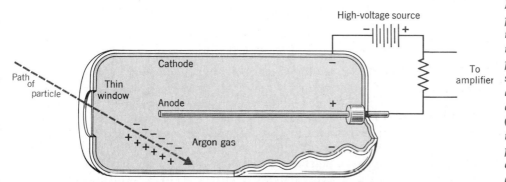

Figure 24.4. *The Geiger–Müller counter. An alpha or beta particle enters the Geiger tube through the thin window shown at the left of the apparatus. As the particle passes through the gas inside the tube, it ionizes argon atoms along its path. These ions cause an electrical breakdown (discharge) between the wire and the wall of the tube, thereby producing a current pulse. This current pulse is readily amplified and counted electronically.*

Figure 24.5. *A typical scintillation probe used to measure radioactive emissions. Pulses of light produced in the phosphor by the radiation are first amplified by a photomultiplier and then detected and counted electronically.*

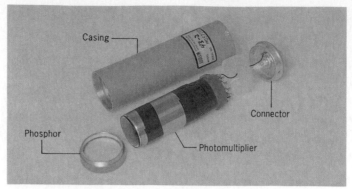

signal produced in these light detectors is then amplified. Figure 24.5 shows a typical scintillation probe.

In working with radioactive substances, it is desirable (and usually necessary) to be concerned with the amount of radiation emitted by the material. We express this by referring to its *activity,* which is the number of nuclear disintegrations or other nuclear changes that occur per second in the sample under study. The SI unit of activity is called the **becquerel (Bq),** named after Henri Becquerel, the discoverer of radioactivity. A becquerel is defined as one nuclear disintegration (or other nuclear change that produces radiation) per second.

$$1 \text{ Bq} = 1 \text{ disintegration/s}$$

> The curie and the becquerel are measures of the degree of radioactivity of a sample. They describe the amount of radiation that *comes from* a sample.

An older non-SI unit that is still widely used is the **curie (Ci),** which is the number of disintegrations per second in one gram of radium. (In other words, the activity in Ci is an activity relative to radium as a standard.) In one gram of radium there are 3.7×10^{10} disintegrations per second, so the relationship between the curie and the becquerel is

$$1 \text{ Ci} = 3.7 \times 10^{10} \text{ Bq}$$

Another measure of activity that is sometimes convenient to use is **specific activity,** which is the number of disintegrations per second per gram of sample. In SI units it would be Bq/g.

We are often interested in knowing not only the number of emissions from a radioactive source, but also the amount of radiation absorbed by some object in the path of the radiation. We refer to this as the **absorbed dose** or simply the **dose.** In the SI, absorbed dose is expressed in terms of the amount of energy (in joules) absorbed per kilogram of absorbing material. The unit is called the **gray (Gy)** and is named after British radiologist Harold Gray.

$$1 \text{ Gy} = 1 \text{ J/kg}$$

> The rad and the gray describe the amount of radiation *received by* a sample.

In the United States, the most common unit of absorbed dose is the **rad,** which is an acronym for **r**adiation **a**bsorbed **d**ose. It corresponds to the absorption of 10^{-5} J per gram of absorbing material. In terms of the gray, it is 0.01 Gy. Thus,

$$1 \text{ rad} = 10^{-5} \text{ J/g}$$
$$1 \text{ Gy} = 100 \text{ rad}$$

The gray and the rad do not differentiate between the effectiveness of various forms of radiation at causing damage to animal tissue, and that is certainly an important aspect of our concern about the amount of radiation that we are exposed to. Alpha, beta, and gamma radiation are absorbed to different degrees by tissue and therefore penetrate our bodies to different depths. Table 24.2 gives the depth of penetration of these forms of radiation in some familiar materials. From this table we can see that alpha radiation is far less dangerous to us than gamma radiation. Whatever damage is done by alpha radiation occurs at the very surface of our skin and only in rare cases

Table 24.2. Penetrating ability (approximate) of various kinds of radiation

Radiation Type	Approximate Depth of Penetration[a]		
	Dry Air	Animal Tissue	Lead
Alpha	4 cm	0.05 mm	0
Beta	6 to 300 cm	0.06 to 4 mm	0.005 to 0.3 mm
Gamma (data for reduction of intensity by 10%)	400 m	50 cm	30 mm

[a] Depth of penetration depends also on the initial energy of the radiation.

does this extend below our outer layer of dead skin cells. Gamma radiation, however, passes through this protective layer of skin and can cause damage to cells within our inner organs. This damage amounts to the breaking of chemical bonds and the production of free radicals (these were discussed in Chapter 19).

A unit that has been defined to take into account the *kind* of radiation that is absorbed by animal tissue is the **rem,** which stands for radiation equivalent for man. To obtain a dose in rem, the dose in rad is multiplied by an appropriate factor for the kind of radiation absorbed.

24.4 APPLICATIONS OF NUCLEAR REACTIONS TO CHEMISTRY

One of the most interesting applications of radioactive decay is in the dating of ancient objects of either natural origin, such as ancient rocks, or of human origin, such as objects made by prehistoric people. In these applications use is made of the known and constant rate of decay of radioactive isotopes. For example, the age of rocks containing uranium can be determined by measuring the ratio of ^{238}U to ^{206}Pb. Remember that ^{206}Pb is the stable isotope to which ^{238}U eventually decays. Using a $^{238}U/^{206}Pb$ ratio of 1 to 0 as corresponding to the ratio at zero time and a $^{238}U/^{206}Pb$ ratio of 1 to 1 as corresponding to the ratio after one half-life—that is, 4.5×10^9 years—we can approximate the age of these rocks. The oldest rocks that have been found on earth have an age of 3.9×10^9 years according to this method.

Rocks that do not contain uranium may be dated using a potassium–argon method. This makes use of the reaction

$$^{40}_{19}K + {}^{0}_{-1}e \longrightarrow {}^{40}_{18}Ar \qquad t_{1/2} = 1.3 \times 10^9 \text{ years}$$

The electron in this case comes from the $1s$ orbital of the potassium and is captured by the unstable $^{40}_{19}K$ nucleus. In the dating procedure the ratio of ^{40}K to ^{40}Ar is measured, and the age of the rock is determined in the same manner as the uranium dating just described.

The age of materials that were once living, such as bones and wood, can be estimated quite accurately by measuring their ratio of ^{14}C to ^{12}C. Carbon-14 is radioactive and is constantly being produced in the upper atmosphere by the bombardment of cosmic neutrons upon $^{14}_{7}N$, which is present there in large amounts. The equation for this reaction is

^{14}C dating is reliable provided the object isn't more than 70,000 years old.

$$^{14}_{7}N + {}^{1}_{0}n \longrightarrow {}^{14}_{6}C + {}^{1}_{1}p$$

The carbon-14 thus produced immediately begins to decay.

$$^{14}_{6}C \longrightarrow {}^{14}_{7}N + {}^{0}_{-1}e \qquad t_{1/2} = 5770 \text{ years}$$

Because ^{14}C is being both formed in the atmosphere and removed by its decay, a constant concentration is maintained. We say that a *steady-state* concentration is

The steady-state concentration of ^{14}C is 15 cpm per gram of carbon. This is an activity of 0.25 Bq/g.

achieved. This ^{14}C eventually becomes incorporated in carbon dioxide in the atmosphere where it can be taken in by plants through the process of photosynthesis. The intake of ^{14}C into animals is by the consumption of such plants or by the consumption of plant-eating animals.

While they are alive, plants and animals consume and excrete carbon so that they also maintain a steady-state concentration of ^{14}C and are thus in equilibrium with their surroundings. Once they die, however, the ^{14}C that they possess is not replaced as the organisms decay, so the ^{14}C concentration begins to decrease. The half-life of the ^{14}C is 5770 years; therefore, if we find that the carbon-14 concentration in an object that had once been living has dropped to half its initial value, we could conclude that the object is 5770 years old.

Atmospheric testing of nuclear weapons has made it impossible for our current history to be dated this way by future archaeologists.

EXAMPLE 24.4. USING THE KINETICS OF RADIOACTIVE DECAY FOR ARCHAEOLOGICAL DATING

PROBLEM: A piece of charcoal from the ruins of a settlement in Japan was found to have a ^{14}C/^{12}C ratio that was 0.617 times that found in living organisms. How old is this piece of charcoal?

SOLUTION: The answer to this problem can be obtained once again by using Equations 24.2 and 24.1. From Equation 24.2 we can obtain the k for the decay of ^{14}C.

$$k = \frac{0.693}{5770 \text{ years}} = 1.20 \times 10^{-4} \text{ year}^{-1}$$

Next, we need values to substitute for $[A]_0$ and $[A]_t$. Fortunately, we don't need actual values for them. If we *arbitrarily* take the ^{14}C/^{12}C ratio in a living organism to be equal to 1,000, then according to the data in the problem, the charcoal has a ^{14}C/^{12}C ratio of $0.617 \times 1.000 = 0.617$. Taking the ratio of these numbers is the *same* as taking the ratio of the concentrations. Therefore, substituting into Equation (24.1), we have

$$\ln \frac{[A]_0}{[A]_t} = kt$$

$$\ln \left(\frac{1.000}{0.617} \right) = \ln 1.62 = 0.482 = (1.20 \times 10^{-4} \text{ year}^{-1})t$$

Solving for t gives

$$t = 4020 \text{ years}$$

Thus, the charcoal is 4020 years old. It would be assumed that the Japanese settlement is of the same age.

Chemical applications

Ordinarily, nuclear changes such as those involved in radioactive decay have very little direct effect on chemical reactions, although in some cases the high-energy radiation emitted by radioactive nuclei can influence the products of reaction. This radiation is generally capable of disrupting chemical bonds. If the cleavage of a chemical bond occurs in DNA, for example, mutations of the DNA strand can be brought about. Mutational changes can also take place by reactions between DNA and the products of other cleavage reactions. For instance, free radicals are produced by splitting an H—O bond in water. This takes place through a series of steps, the first of which is the absorption of radiation by the water molecule and the subsequent ejection of an electron

$$H_2O + h\nu \longrightarrow H_2O^+ + e^-$$

$\underbrace{\hphantom{H_2O + h\nu}}$
energy from
radiation

The final products of the reaction are a hydrogen atom, H· (the dot indicates the unpaired electron in the radical) and a *hydroxyl radical,* ·OH. The net overall change is

$$H_2O + h\nu \longrightarrow H· + ·OH$$

Free radicals are extremely reactive because of the presence of their unpaired electrons, and their interactions with DNA can cause mutations or otherwise disrupt the replication of the DNA strands.

It is only in rare cases that the chemist makes direct use of the energy emitted in nuclear transformations. Most chemical applications of radioactive nuclides stem from their ease of identification and detection, even when they are present in very small amounts. Hence, radioactive isotopes are usually employed in **tracer studies.** They are added in very small amounts and used to follow, or trace, the course of a chemical reaction. The range of applications of these tracer techniques is limited only by the imagination and ingenuity of the experimenter. Let's take a brief look at some examples that demonstrate the scope of these applications.

Analytical chemistry

There are many examples of analytical uses for radioactive isotopes. One of these techniques, called **isotope dilution,** can be used when it is impossible to separate a desired substance completely from a mixture. In this case, a small measured amount of the substance containing a known quantity of a radioactive isotope is *added* to the mixture. After making sure that complete mixing has occurred, a small amount of the *pure* desired substance is separated from the mixture. This sample will contain some of the added radioactive isotope, and from the proportion of the labeled isotope present in the sample, the total quantity of the substance in the original mixture can be computed.

Consider, for instance, a mixture of salts of similar solubilities, such as a mixture of KNO_3 and $NaCl$. By fractional crystallization only a portion of the KNO_3 can be separated from the mixture. As a result, we cannot determine, in a simple fashion, how much of this salt is in the mixture.

Suppose, now, that 1.0 g of KNO_3 containing a small amount of radioactive ^{40}K is added to the salt mixture and then some KNO_3 (now containing K from the original mixture as well as from the added tagged KNO_3) is separated by fractional crystallization. If the specific activity of this KNO_3—that is, the number of counts per minute per gram (cpm/g)—has dropped to 1% of the specific activity of the added KNO_3, then we know that only 1% of the added solid has been recovered in our KNO_3 sample and the other 99% of the KNO_3 must have been present in the original mixture. In other words, after we had added the 1 g of labeled KNO_3 there was a 99-to-1 ratio of unlabeled to labeled salt. Therefore, the original mixture must have contained 99 g of KNO_3.

Isotope dilution methods are also used when the volume of a liquid in an irregular container must be measured. A small known volume of radioactive material is added and after mixing is complete, the extent of dilution allows one to calculate backward to find the initial liquid volume. This method has been used to measure blood volumes in living animals and the volumes of underground reservoirs of water.

Another technique applicable to analytical chemistry is called **neutron activation analysis.** When nonradioactive isotopes are bombarded by neutrons, heavy isotopes of these elements can be formed. The product of this reaction may be unstable and therefore radioactive. Even if another nonradioactive nucleus is produced, however, the absorption of these neutrons generally gives nuclei that are excited and emit gamma radiation in much the same way that an excited atom emits light when it returns to the ground state.

$$\ce{^{A}_{Z}X} + \ce{^{1}_{0}}n \longrightarrow \ce{^{A+1}_{Z}X^*} \qquad \text{(the asterisk indicates an excited nucleus)}$$

$$\ce{^{A+1}_{Z}X^*} \longrightarrow \ce{^{A+1}_{Z}X} + h\nu \qquad (h\nu = \gamma \text{ photon})$$

Since each element has its own characteristic gamma-emission spectrum, an analysis of the energies of the gamma emissions from the activated sample allows its elemental composition to be determined. In addition, from the intensities of the emitted gamma radiation, the concentration of each element can be computed.

This technique has some very useful advantages. First, it is nondestructive. Since the number of nuclei that must be activated to perform the analysis is small, most of the sample is unaffected. Second, as implied in the preceding sentence, the method is very sensitive and is, therefore, well suited to the analysis of trace amounts of impurities. In some cases, concentrations of the order of 10^{-9} % can be measured.

Descriptive chemistry

Many elements having atomic numbers greater than $Z = 83$ (bismuth) have short half-lives and, therefore, are not observed to occur naturally. Instead, they must be synthesized in particle accelerators. As a result, only extremely small quantities of these elements have ever been prepared. A question then arises: How can we study their chemistry if we cannot even obtain enough to be able to see them?

To arrive at the solution to this problem, let us consider the element astatine. Astatine was first produced in a device called a cyclotron by the reaction.

$$\ce{^{209}_{83}Bi} + \ce{^{4}_{2}He} \longrightarrow \ce{^{211}_{85}At} + 2\ce{^{1}_{0}}n$$

in which the ^{211}At produced has a half-life of only about 7.5 hr. The most stable isotope, ^{210}At, has a half-life of only 8.3 hr, so large quantities of the element cannot be accumulated.

Since astatine occurs in Group VIIA, we expect the element to be similar in some of its properties to iodine. To verify this, we add the astatine as a tracer in reactions involving iodine, and follow the fate of the At as the iodine undergoes reactions. If in a given reaction the At occurs in the products along with the iodine, we conclude that, in this reaction, At behaves just as I does. We have discovered something about the chemical behavior of an element that we cannot even see. For instance, it is observed that, like iodine, elemental astatine is rather volatile, since it is carried with the iodine when I_2 is sublimed. In solution At^- is carried from solution along with I^- upon the addition of Ag^+. Thus, we conclude that AgAt is insoluble just as is AgI.

How would you study the chemical properties of francium, Fr (Z = 87)?

Reaction mechanisms

In Chapter 19 we saw that a study of the effect of the concentrations of the reactants on the rate of a chemical reaction can often give some insight into the mechanism of the reaction. Such studies, however, seldom answer all the questions that we might ask about the reaction mechanism. Consider, for example, the reaction of an alcohol and an organic acid to produce an ester and water

$$C_2H_5-OH + HO-\overset{\displaystyle O}{\overset{\|}{C}}-CH_3 \longrightarrow C_2H_5-O-\overset{\displaystyle O}{\overset{\|}{C}}-CH_3 + H_2O$$

alcohol   acid   ester

In the formation of the ester molecule, two hydrogen atoms and one oxygen atom are removed from the alcohol and acid to become a molecule of water. There seems little doubt about the origin of the two hydrogen atoms; however, there is a question about which one of the $-OH$ oxygen atoms is removed and finds its way into the H_2O molecule.

This question can be resolved by carrying out the reaction with a labeled oxygen (for example, ^{18}O) incorporated into the OH group of either the alcohol or the acid. For instance, if the alcohol is labeled with ^{18}O, it is found that all the labeled oxygen becomes part of the ester. On the other hand, if the acid contains ^{18}O in the OH group, all the labeled oxygen ends up in the water with none in the ester. It is clear, therefore, that the reaction involves the removal of the OH from the acid and the H from the alcohol.

Reaction using labeled alcohol

$$C_2H_5-O-H + H-O-\overset{\displaystyle O}{\overset{\|}{C}}-CH_3 \longrightarrow C_2H_5-O-\overset{\displaystyle O}{\overset{\|}{C}}-CH_3 + H_2O$$

The symbols printed in red represent the oxygen-18 atoms.

Reaction using labeled acid

$$C_2H_5-O-H + H-O-\overset{\displaystyle O}{\overset{\|}{C}}-CH_3 \longrightarrow C_2H_5-O-\overset{\displaystyle O}{\overset{\|}{C}}-CH_3 + H_2O$$

Many similar experiments using tagged atoms have been employed to aid in the elucidation of a large number of reaction mechanisms, including biological and biochemical processes. For instance, labeled water can be added to the root system of a plant, and its progression into the stem — and ultimately into the leaves — can be traced. Experiments using ^{14}C-labeled CO_2 have been used to follow the course of carbon in photosynthesis in plants. In this case plants are exposed to CO_2 and, at various intervals, are killed and their cellular components separated to determine which compounds have had ^{14}C built into them. In this way the sequence of reactions in photosynthesis can be unraveled.

24.5 NUCLEAR STABILITY

Experimentally it is observed that all the elements with atomic numbers greater than 83 (bismuth) are radioactive and possess no known stable isotopes. On the other hand, all the lighter elements, with the exception of technetium ($Z = 43$) and promethium ($Z = 61$), have one or more stable, nonradioactive isotopes. In addition, radioactive isotopes undergo nuclear transformations that lead ultimately to stable nuclei. Sometimes this is accomplished by a simple one-step process, while in other cases a series of nuclear reactions occur before a stable isotope is reached. A question that naturally arises from these observations is, what factors give rise to stable or unstable nuclei?

There are some interesting facts about nuclear stability that emerge if we study the numbers of protons and neutrons that are found in stable nuclei. For example, if a graph is made of the numbers of neutrons (vertical axis) versus the numbers of protons (horizontal axis) that are observed for stable nuclei, we find that all the

Figure 24.6. *Stable nuclei fall within a narrow band when they are placed on a graph of number of neutrons versus number of protons. This band is known as the band of stability.*

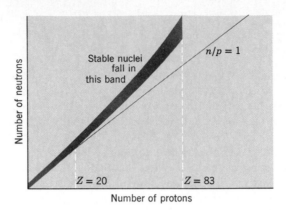

Number of neutrons

Stable nuclei fall in this band

$n/p = 1$

$Z = 20$ $Z = 83$

Number of protons

stable nuclei fall in a relatively narrow band, as shown in Figure 24.6. This band has been called a **band of stability.** In Figure 24.6, we see that at low atomic numbers stable nuclei possess approximately equal numbers of protons and neutrons. Above about $Z = 20$, however, the number of neutrons always exceeds the number of protons, and the neutron-to-proton ratio gradually increases to about 1.5 at the upper end of the band of stability. Apparently, as the number of protons in the nucleus increases, there must be more and more neutrons present to help overcome the strong repulsive forces between the protons. It also seems that there is an upper limit to the number of protons that can exist in a stable nucleus, with that number being reached at bismuth.

Nuclei that lie outside the band of stability are unstable and decay in a manner that tends to give them a stable neutron-to-proton (n/p) ratio. On this basis, we can understand why certain nuclei undergo the types of radioactive decay that they do. For instance, a nucleus that lies above the band of stability must either lose neutrons or gain protons to achieve stability. Thus, we can understand why elements such as ^{14}C (which lies above the band) decay by β-emission, because this process converts a neutron into a proton ($^1_1 p$).

$$^1_0 n \longrightarrow {}^1_1 p + {}^{\ 0}_{-1} e$$

For ^{14}C we have

$$^{14}_{6} C \longrightarrow {}^{14}_{7} N + {}^{\ 0}_{-1} e$$

Another way that an element located above the band can achieve a stable n/p ratio is by emitting a neutron, although this particular mode of decay is rare. An example is the decay of ^{137}I.

$$^{137}_{53} I \longrightarrow {}^{136}_{53} I + {}^1_0 n$$

Elements located *below* the band of stability must increase their n/p ratio to achieve stability. This is accomplished generally in either of two ways. One involves the emission of a positron, a particle having the same mass as the electron but with a unit positive charge. The positron is symbolized as $^0_1 e$. The ejection of a positron by an unstable nucleus converts a proton into a neutron

$$^1_1 p \longrightarrow {}^1_0 n + {}^0_1 e$$

An example is the decay of ^{11}C.

$$^{11}_{6} C \longrightarrow {}^{11}_{5} B + {}^0_1 e$$

The second mode of decay that results in an increased n/p ratio is called **electron capture.** In this case the unstable nucleus captures an electron, usually from its own $1s$ orbital. Since the captured electron most often originates in the K shell, the process is also called **K-capture.** The addition of this electron to the nucleus transforms a proton into a neutron.

$$^1_1p + ^{\ 0}_{-1}e \longrightarrow ^1_0n$$

Two examples of decay by K-capture are

$$^7_4\text{Be} + ^{\ 0}_{-1}e \xrightarrow{\ \ K\text{-capture}\ \ } ^7_3\text{Li}$$

and

$$^{40}_{19}\text{K} + ^{\ 0}_{-1}e \xrightarrow{\ \ K\text{-capture}\ \ } ^{40}_{18}\text{Ar}$$

The vacancy created in the $1s$ subshell as a result of K-capture is only temporary, and electrons from higher energy levels quickly drop to fill the $1s$ orbital. Since electrons are falling from higher energy levels to lower ones, energy is emitted in the form of electromagnetic radiation (light) — in this instance, in the X-ray region of the spectrum.

The K-shell has $n = 1$.

Elements having atomic numbers higher than 83 — that is, those beyond the end of the band of stability — cannot find their way to a stable n/p ratio by any of the decay modes we just discussed. In these cases the unstable nuclei must lose both protons *and* neutrons. As a result, their decay usually involves emission of alpha particles, since each alpha emission removes two protons and two neutrons simultaneously. Earlier, for example, we saw this type of decay process for uranium

$$^{238}_{92}\text{U} \longrightarrow ^4_2\text{He} + ^{234}_{90}\text{Th}$$

Another type of nuclear transformation that is available to the heavy elements is **fission,** in which a heavy nucleus splits into several much lighter fragments, many of which may also lie outside the band of stability and hence may be radioactive. The smaller nuclei that are produced by fission, if they are unstable, are able to undergo the simpler types of decay in order to produce a stable nucleus. We will take a closer look at nuclear fission in Section 24.8.

It is also observed that nuclei with even numbers of protons and neutrons are apparently more stable than those containing an odd number of these particles. For example, there are 157 stable isotopes in which there are even numbers of both protons and neutrons, 52 isotopes having an even number of protons and an odd number of neutrons, and 50 with an even number of neutrons but an odd number of protons. By contrast, there are only 5 stable nuclides in which there are odd numbers of both protons and neutrons.

Protons	Even	Even	Odd	Odd
Neutrons	Even	Odd	Even	Odd
Stable nuclei	157	52	50	5

This phenomenon suggests that in stable nuclei, protons and neutrons each tend to be paired, in much the same way that electrons become paired in the outer region of an atom. Apparently, extra stability, as evidenced by the number of stable nuclides, results when pairing takes place with both protons and neutrons. On the other hand, when pairing cannot occur, as must be true when the numbers of protons and neutrons are both odd, very few stable isotopes occur (most isotopes having odd numbers of both protons and neutrons are radioactive).

A final observation on nuclear stability is that nuclei that contain certain specific numbers of protons and neutrons possess a degree of extra stability. These so-called *magic numbers* for protons and neutrons are 2, 8, 20, 28, 50, and 82, with an additional magic number of 126 for neutrons. When nuclei contain a magic number of both protons and neutrons, they are said to be *doubly magic* and are extremely stable. Examples are ^4_2He, $^{16}_8\text{O}$, $^{40}_{20}\text{Ca}$, and $^{208}_{82}\text{Pb}$.

The occurrence of these magic numbers suggests a shell structure for the nucleus somewhat akin to the shell structure exhibited by electrons. For example, we have seen that very stable (unreactive) electron configurations occur when an atom contains magic numbers of 2, 8, 18, 36, or 54 electrons, corresponding to the noble gases, He through Kr. In the nucleus, it seems that nuclear shells of either protons or neutrons become completed when the nuclear magic numbers are reached and that a particularly stable nucleus occurs whenever there is a completed shell of either neutrons or protons. Exceptionally stable nuclei result when the nucleus contains filled shells of protons and neutrons simultaneously.

24.6 NUCLEAR TRANSFORMATIONS

A **nuclear transformation** is a nuclear reaction in which a bombarding particle is absorbed and causes the absorbing nucleus to change into a nucleus of another element. For example, when $^{14}_7\text{N}$ is bombarded with alpha particles, the isotope of oxygen, $^{17}_8\text{O}$, is produced. The equation for this reaction is written as follows:

$$^{14}_7\text{N} + {}^4_2\text{He} \longrightarrow {}^{18}_9\text{F}^* \longrightarrow {}^{17}_8\text{O} + {}^1_1\text{H}$$

When a nucleus absorbs a bombarding particle, the product is called a *compound nucleus* and is indicated by an asterisk.

The asterisk indicates that the isotope $^{18}_9\text{F}$ is a very unstable intermediate that rapidly decays to the products.

Other light elements can be similarly transformed by bombardment with alpha particles. For example, $^{27}_{13}\text{Al}$ can be transformed into $^{30}_{15}\text{P}$ by the nuclear reaction

$$^{27}_{13}\text{Al} + {}^4_2\text{He} \longrightarrow {}^{30}_{15}\text{P} + {}^1_0n$$

Reactions of this type, in which an alpha particle is used for bombardment and a neutron is one of the products, is known as an *alpha, neutron reaction*, symbolized by (α,n). A shorthand notation for the reaction is $^{27}_{13}\text{Al}(\alpha,n)^{30}_{15}\text{P}$.

Since 1933, when these discoveries were made, many isotopes have been produced by bombardment reactions, some in which particles other than alpha particles have been used. One of the main problems with such experiments is that a positively charged nucleus is being bombarded with positively charged particles. Heavy elements, with their very highly charged positive nuclei, repel particles with positive charges like the alpha particle. One way to circumvent this problem is to use neutrons as the bombarding particles.

Neutrons, which have no charge, are not repelled by the nucleus, so they are excellent for bombardment reactions. The supply of neutrons for these reactions can be obtained from either of two sources: a transformation reaction in which neutrons are produced or a nuclear reactor in which fission reactions (which will be examined in Section 24.8) occur at a controlled rate. Examples of neutron-producing reactions are the $^{27}_{13}\text{Al}(\alpha,n)^{30}_{15}\text{P}$ reaction mentioned above and

$$^9_4\text{Be} + {}^4_2\text{He} \longrightarrow {}^{12}_6\text{C} + {}^1_0n$$

in which the beryllium isotope undergoes an α,n reaction.

A second way that nuclear transformations can be brought about is through the use of particle accelerators. Particle accelerators, such as the cyclotron illustrated in Figure 24.7, speed up electrically charged particles to extremely high velocities and then direct them at target nuclei. At these speeds, positive particles are able to overcome the coulombic repulsion of a nucleus and collide with it. Accelerators have been used in producing many of the *transuranium elements* — that is, elements with atomic numbers larger than 92. In these accelerators, positive ions of such isotopes as

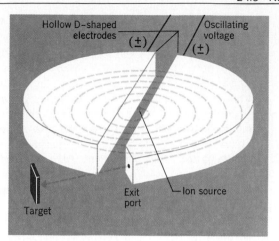

Figure 24.7. *Diagram of a cyclotron, one kind of particle accelerator.*

$^2_1$H (deuterium), $^{12}_6$C, $^{13}_6$C, $^{16}_8$O, $^{14}_7$N, $^{10}_6$B, as well as $^4_2$He have been used in producing new, artificial elements. A list of some of these bombardment reactions and the elements they produce is shown in Table 24.3. Many of these elements are extremely short-lived and, as a result, only a few atoms, especially of the high-atomic-numbered isotopes, have ever been formed.

Extension of the periodic table

The heaviest naturally occurring element is uranium and, as we saw in the previous section, elements beyond $Z = 92$ are all prepared artificially by bombarding lighter nuclei with protons, alpha particles, and the positive ions of some of the second-period elements. The discovery of these new elements quite expectedly prompted chemists to begin to think about a whole host of new elements with new and interesting properties to be studied. However, it soon became apparent that as the atomic number of the artificial element became higher, its half-life became shorter, so the prospects for stable elements of very high atomic numbers became dim.

Calculations by many nuclear physicists, based on the nuclear shell model, now suggest that a closed nuclear shell for protons exists at $Z = 114$ and that one for neutrons occurs at 184. As a result, chemists have once again begun to speculate about the possibilities of new stable elements.

Table 24.3. Some elements produced in particle accelerators

$^{238}_{92}$U + $^4_2$He $\longrightarrow$ $^{239}_{94}$Pu + 3 ^1_0n

$^{239}_{94}$Pu + $^4_2$He $\longrightarrow$ $^{240}_{95}$Am + $^1_1$H + 2 ^1_0n

$^{239}_{94}$Pu + $^4_2$He $\longrightarrow$ $^{242}_{96}$Cm + ^1_0n

$^{244}_{96}$Cm + $^4_2$He $\longrightarrow$ $^{245}_{97}$Bk + $^1_1$H + 2 ^1_0n

$^{238}_{92}$U + $^{12}_6$C $\longrightarrow$ $^{246}_{98}$Cf + 4 ^1_0n

$^{238}_{92}$U + $^{14}_7$N $\longrightarrow$ $^{247}_{99}$Es + 5 ^1_0n

$^{238}_{92}$U + $^{16}_8$O $\longrightarrow$ $^{249}_{100}$Fm + 5 ^1_0n

$^{253}_{99}$Es + $^4_2$He $\longrightarrow$ $^{256}_{101}$Md + ^1_0n

$^{246}_{96}$Cm + $^{13}_6$C $\longrightarrow$ $^{254}_{102}$No + 5 ^1_0n

$^{252}_{98}$Cf + $^{10}_5$B $\longrightarrow$ $^{257}_{103}$Lw + 5 ^1_0n

$^{249}_{98}$Cf + $^{12}_6$C $\longrightarrow$ $^{257}_{104}$Unq + 4 ^1_0n

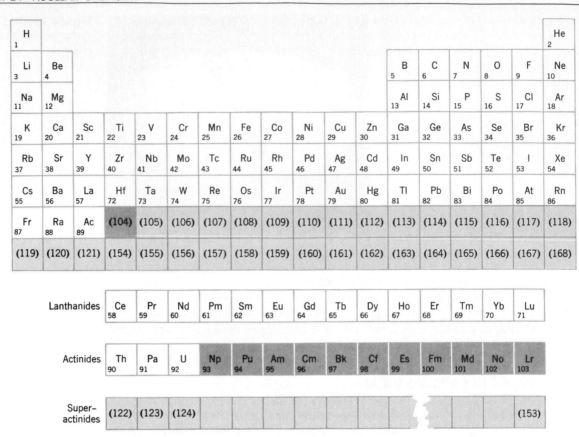

Figure 24.8. *Extended periodic table. From G. Seaborg,* Journal of Chemical Education, *Vol. 46, p. 626, October 1969, used by permission.*

One proposed extension of the periodic table to include these heavy and super-heavy elements is shown in Figure 24.8. Recall that the actinide series, which occurs as the result of the filling of the 5f subshell, ends at lawrencium, $Z = 103$. The next element, 104, therefore lies under hafnium if we follow the scheme for the filling of subshells developed in Chapter 7. Elements 104 to 112, therefore, would correspond to the filling of the 6d subshell. Next we have six elements in the p-block (113 to 118), which would have their 7p subshell gradually filled. Elements 119 and 120, in period 8, correspond to the completion of the 8s subshell, and following element 121 a sequence of 32 inner transition elements occurs, from $Z = 122$ to 153. These *super-actinides* would be accounted for by the completion of first the 6f subshell (14 elements) followed by the filling of a 5g subshell (a g subshell would contain nine orbitals that can accommodate 18 electrons; therefore, we would have 18 more elements, 136 to 153). After the superactinides we would fill the 7d subshell (elements 154 to 162) and then the 8p subshell (163 to 168).

In the search for new elements it is expected that nuclides that differ much from $Z = 114$ would be extremely unstable and decompose by fission with very short half-lives. However, in the vicinity of $Z = 114$ it has been suggested that fission should not occur and the half-lives of these elements with respect to alpha and beta decay should be long enough so that it should be possible to detect them and perhaps even investigate their chemical properties. There is even the possibility that these superheavy eements will not be radioactive at all.

The relative stabilities of nuclides containing differing numbers of protons and neutrons are dramatized in Figure 24.9. Here the stable nuclei of the band of stability are shown as a long peninsula extending out into a sea of instability. Stable nuclei correspond to points above sea level, whereas submerged regions constitute unstable nuclei. Notice that nuclei with magic numbers of either protons or neutrons are shown as higher, more stable ridges, while doubly magic nuclei are shown as mountains of stability.

The superheavy elements with approximately a magic number of 114 protons and either 184 or 196 neutrons are depicted as an island of stability separated from the peninsula by a region of high nuclear instability. As a result, in order to reach this island we cannot bombard stable (or relatively stable) nuclei with light particles such as $_{2}^{4}He$, because this simply places the product nuclei into the sea of instability where they decompose rapidly before any additional mass and charge can be added. Consequently, the jump to the island must be made in one step. At the present time, this presents substantial problems, because bombarding nuclei must contain a n/p ratio of at least 1.6 ($184/114 = 1.61$). However, light nuclei such as $_{18}^{40}Ar$, while possessing sufficient protons to give the desired atomic number by a reaction such as

$$_{96}^{248}Cm + {}_{18}^{40}Ar \longrightarrow {}_{114}^{284}X + 4{}_{0}^{1}n$$

do not contain enough neutrons to place the product isotope within the island of stability. For example, $_{114}^{284}X$ contains only $284 - 114 = 170$ neutrons, 14 less than the 184 that we would want to achieve a doubly magic nucleus. Research today is directed at obtaining suitable target and projectile nuclei that will give not only $Z = 114$ but also a sufficient number of neutrons to place the product nucleus within the bounds of the predicted island of stability.

While physicists continue their search for ways to synthesize these superheavy elements, other scientists are looking for evidence of their current or past existence in the universe. In 1975 evidence was discovered by Dr. Edward Anders at the University of Chicago's Fermi Institute that suggested the one-time existence of element 114 (or perhaps 115 or 113) in a meteorite that fell in Mexico in 1969. Other scientists

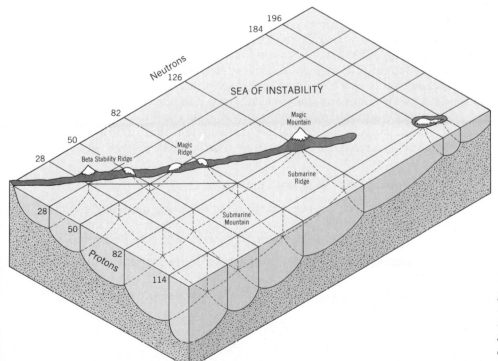

Figure 24.9. *Known and predicted regions of nuclear stability surrounded by a sea of instability. Adapted from G. Seaborg, Journal of Chemical Education, Vol. 46, p. 626, October 1969, used by permission.*

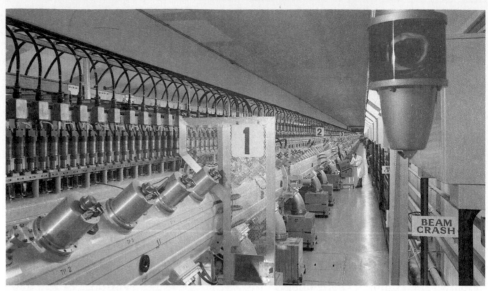

The linear accelerator at Brookhaven National Laboratory on Long Island, New York. Protons can be accelerated in this device to just under the speed of light before they strike their target.

at Oak Ridge National Laboratories have found crystals that may once have contained elements 116 and 126.

Naming elements with large atomic numbers

It has been historical practice to allow the discoverer of an element to assign the element's name. In recent times this had led to some controversy, because elements with very high atomic numbers are so unstable that only minute quantities of them (sometimes only one or two atoms) are prepared before scientists claim credit for their discovery. This has led on occasion to questions of the reliability of the data and whether the sought-after element had in fact been made. For example, both American and Soviet scientists claim credit for discovering element 104. The Americans named it rutherfordium and the Soviets named it kurchatovium.

Because of this problem, the International Union of Pure and Applied Chemistry (IUPAC) has made the official recommendation that until a new element's discovery has been proven, a systematic nomenclature be applied. The rules of this system are as follows:

1. All elements will end in the letters *-ium.*

2. The name will be constructed from the following numerical roots:

   0 = nil   1 = un   2 = bi   3 = tri   4 = quad
   5 = pent   6 = hex   7 = sept   8 = oct   9 = enn

3. The symbol will consist of three letters derived from the numerical roots above.

The following illustrates how this works for element 104:

$$104$$

un   nil   quad   -ium

The name for the element is *unnilquadium* and its symbol is Unq.

EXAMPLE 24.5. NAMING SUPERHEAVY ELEMENTS

PROBLEM: What would be the systematic name and symbol for element 114?

SOLUTION: The numerical root for 1 is un, so it has to be repeated; the root for 4 is quad. Therefore, the name is ununquadium, and its symbol is Uuq.

24.7 NUCLEAR BINDING ENERGY

Nuclei are composed of protons and neutrons. Naturally we would expect that if we added up the mass of all the protons that go into forming a nucleus, and then added in the mass of all the neutrons, we would obtain the mass of the nucleus. Actually, however, the mass of the nucleus is always somewhat *less* than the total mass of the individual protons and neutrons. How can this be?

Within the nucleus the nuclear particles are bound together by very strong forces, the nature of which is not understood very well. Nevertheless, enormous amounts of energy have to be supplied to separate the nucleus into its component protons and neutrons. It follows that if we were to form a nucleus, this same large amount of energy would be released.

Einstein showed that mass and energy are related by his now famous equation, $E = mc^2$, where c is the speed of light. The energy liberated when the nucleus is formed comes at the expense of some of the mass of the nucleons (the individual particles such as protons and neutrons that are found in the nucleus). In other words, some mass is converted to the energy that is liberated as the nucleus is formed. Consequently, the final mass of the nucleus is less than we might have expected.

Nucleons are the individual particles found in the nucleus.

The energy needed to decompose the nucleus (or the energy released when it is formed) is called the **binding energy.** The difference between the actual mass of a nucleus and the sum of the masses of its individual protons and neutrons is termed the **mass defect.** Let's look at an example.

A $^4_2$He atom is composed of two protons, two neutrons, and two electrons. These individual particles have the following masses:

$$p \quad 1.007277 \text{ u}$$
$$n \quad 1.008665 \text{ u}$$
$$e^- \quad 0.0005486 \text{ u}$$

In the SI, the symbol for the atomic mass unit is u.

The calculated mass of a $^4_2$He atom is therefore

$$(2 \times 1.007277 \text{ u}) + (2 \times 1.008665 \text{ u}) + (2 \times 0.0005486 \text{ u}) = 4.032981 \text{ u}$$

$$\text{calculated mass } ^4_2\text{He} = 4.032981 \text{ u}$$

The actual mass of $^4_2$He, as measured with a mass spectrometer, is 4.002603 u. The mass defect is the difference between the computed and measured mass; that is,

$$\text{mass defect} = 4.032981 \text{ u} - 4.002603 \text{ u} = 0.030378 \text{ u}$$

Thus, when a helium nucleus is formed from two protons and two neutrons, 0.030378 u of mass is converted to energy and released. How much energy does this represent?

Suppose that we were to form 1 mol of He atoms. The total mass lost would then be 0.030378 g or 3.0378×10^{-5} kg. We can use Einstein's equation to calculate the energy equivalent. Using $c = 2.9979 \times 10^8$ m/s, we have

If the mass defect *per atom* is 0.030378 u, the mass defect *per mole* is 0.030378 g.

$$E = (3.0378 \times 10^{-5} \text{ kg}) \times (2.9979 \times 10^8 \text{ m/s})^2$$

$$= 2.7302 \times 10^{12} \text{ kg m}^2/\text{s}^2$$

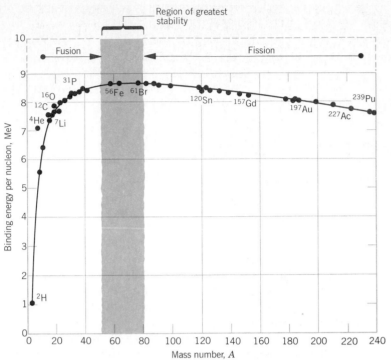

Figure 24.10. *Average nuclear binding energy per nucleon for atoms of different mass number.*

The energy liberated in forming just 4 g of helium from its nucleons is enough to raise the temperature of 230,000 cubic feet of water from 0 °C to 100 °C!

In the SI, 1 joule = 1 kilogram meter^2/second^2. Therefore, for 1 mol of He,

$$E = 2.73 \times 10^{12} \text{ J/mol}$$

$$= 2.73 \times 10^9 \text{ kJ/mol}$$

For comparison, combustion of 1 mol of CH_4 liberates only 8.9×10^2 kJ. The binding energy, therefore, represents a huge amount of energy.

Let us now look at the average binding energy per nucleon that occurs for various atoms. This is usually expressed in energy units of *MeV per nucleon*. Nuclear physicists generally deal in energy units of MeV. One MeV is 1 million **electron volts,** where the electron volt is the kinetic energy that an electron would acquire if it were accelerated from one electrode to another across a potential difference of 1 volt.

$$1 \text{ MeV} = 10^6 \text{ eV}$$

From Einstein's equation, the energy equivalence of 1 u can be calculated. Expressed in MeV, this is

$$1 \text{ u} \Longleftrightarrow 931 \text{ MeV}$$

For $_2^4$He the binding energy is therefore

$$0.030378 \text{ u} \times \left(\frac{931 \text{ MeV}}{1 \text{ u}} \right) = 28.3 \text{ MeV}$$

Since $_2^4$He is composed of four nucleons (2*p*, 2*n*),

1 MeV per molecule is equivalent to 9.65×10^7 kJ/mol.

$$\text{average binding energy per nucleon} = \frac{28.3 \text{ MeV}}{4 \text{ nucleons}} = 7.07 \text{ MeV/nucleon}$$

The average binding energy per nucleon varies, of course, for different atoms. Figure 24.10 is a plot of this average binding energy versus mass number.

24.8 FISSION, FUSION, AND NUCLEAR ENERGY

In 1939 two German chemists, Otto Hahn and Fritz Strassman, discovered that when uranium was bombarded with relatively slow-moving neutrons referred to as *thermal neutrons* (neutrons with kinetic energies that put them in thermal equilibrium with their surroundings), the unexpected products were the isotopes ^{139}Ba and ^{94}Kr. It was soon shown by physicists Lise Meitner and Otto Frisch that it was the nuclei of the isotope ^{235}U that absorb these slow neutrons. The unstable ^{236}U then split, yielding the two lighter fragments and additional neutrons.

$$^{235}_{92}U + ^{1}_{0}n \longrightarrow ^{236}_{92}U^* \longrightarrow ^{139}_{56}Ba + ^{94}_{36}Kr + 3\,^{1}_{0}n$$

The process that had been discovered is called **nuclear fission,** in which an atomic nucleus is split into two roughly equal parts. It was also found that a great deal of energy is emitted as well.

The origin of the enormous amounts of energy in nuclear fission can be seen by examining Figure 24.10. For the very heavy elements such as uranium, the binding energy per nucleon is less than for elements with intermediate atomic masses. Therefore, when a uranium-235 nucleus splits, the new nuclei have a greater binding energy per nucleon and therefore a greater mass defect. As a result, when the fission products are formed, there is actually a decrease in the total mass of the system. The mass that is lost has been changed to an equivalent amount of energy according to Einstein's equation $E = mc^2$. As you saw in the previous section, even a very tiny amount of mass yields a tremendous amount of energy when this conversion takes place. In fact, the fission of 1 kg of ^{235}U yields an amount of energy that is equivalent to the burning of 3000 tons of coal or 13,200 barrels of oil.

Uranium-235 is an example of a **fissile** isotope, meaning that it is capable of undergoing nuclear fission. Its natural abundance is only about 0.72%. Unfortunately, the much more abundant isotope, ^{238}U, is not fissile, although it can be changed into two other synthetic fissile isotopes, ^{239}Pu and ^{233}U in nuclear reactors.

As you can see in the equation given earlier, the fission of ^{235}U produces more neutrons than it consumes. These additional neutrons, if they are slowed down and become thermal neutrons, can go on to trigger other fission processes, which produce even more neutrons that can trigger more fissions, and so on, as illustrated in Figure 24.11. The result is a nuclear chain reaction whose rate increases very rapidly. For this to happen, however, requires a certain minimum amount of the fissile isotope. If too little is present, the neutrons that are formed escape before they are able to produce this rapidly increasing chain reaction. The minimum amount of the fissile isotope required to sustain the chain reaction is called the **critical mass.** In an atomic bomb, two subcritical masses of uranium-235 or plutonium-239 are brought together very rapidly to yield a critical mass and a nuclear explosion.

As you probably know, the United States was at war during the early 1940s when the discoveries of nuclear fission were being investigated. Albert Einstein alerted President Roosevelt to the military implications of the phenomenon. Roosevelt responded by establishing the Manhattan Project, which led to the two bombs that were dropped on Hiroshima and Nagasaki, and ultimately to the production of controlled nuclear fission for peaceful applications.

The neutrons that are given off in the fission reaction are fast neutrons and cannot trigger another fission event unless they slow down.

When less than the critical mass of a fissile isotope is present, a nuclear explosion cannot take place.

President Truman made the decision to use the atomic bomb in the hope that it would soon end the war and thereby save lives, overall.

Nuclear reactors

In a nuclear reactor, fission takes place at a controlled rate, slow enough to avoid a nuclear chain reaction but fast enough to produce useful amounts of energy. The reaction is controlled in two ways. First, the concentration of the fissile isotope is relatively low—about 2 to 4%. For a nuclear explosion to occur, essentially pure

Figure 24.11. *A nuclear chain reaction for* ^{235}U.

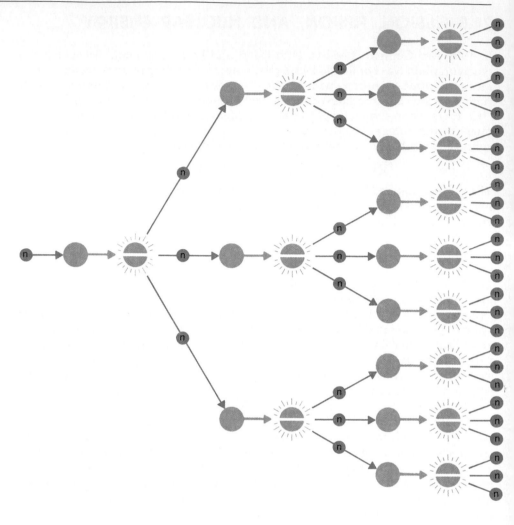

The Haddam Neck nuclear reactor located in Haddam Neck, Connecticut. The domed building serves as the containment vessel for the nuclear reactor.

^{235} U or ^{239} Pu is needed. The rate of reaction is also governed by the use of control rods which can be inserted into the reactor to absorb neutrons and thereby slow the reaction. These rods are positioned to permit the reaction to sustain itself without getting out of hand.

For nuclear fission to be sustained, the fast neutrons produced in the fission reaction must be slowed so they can initiate additional fission events within the reactor core (where the fuel elements are located). To slow these neutrons, the core of the reactor has a *moderator.* Often this is water, although graphite is also used. In fact, the reactor that experienced the nuclear accident at Chernobyl, Russia, was a graphite-moderated reactor.

Virtually all civilian nuclear plants today operate on the same overall principles. The heat generated in the fission reaction is transferred to a coolant that is circulated through the reactor. Most nuclear plants use a pressurized water system for this coolant and the pressure keeps the water in the liquid state even above its normal boiling point of 100 °C. In a *pressurized water reactor,* as this is called, this primary coolant then transfers heat through a heat exchanger to a secondary loop in which steam is produced to drive an electrical turbine to generate electricity. This is illustrated in Figure 24.12. (In some reactors, known as *boiling water reactors,* there is only one coolant loop in which water circulating through the loop is changed directly to steam to drive the turbines.)

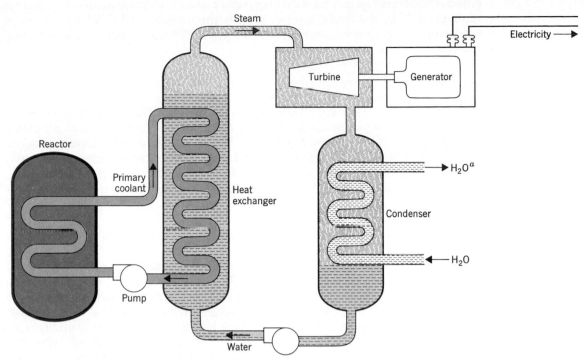

Figure 24.12. *The application of nuclear fission to the production of electricity. This is a schematic diagram of a typical pressurized water reactor.*

Nuclear Emergencies As you may know, one of the most feared nuclear emergencies occurs when there is a loss of coolant in the primary coolant loop of a nuclear reactor. The core temperature rises very rapidly and can cause any water remaining in the loop to change to steam. This can produce a steam explosion, which is believed to be what happened in the nuclear accident at the power plant in Chernobyl. The blast tore the top off the reactor and sent a shower of radioactive fuel and waste products into the air. The radioactive plume rose to 3200 feet and drifted downwind. The heat in the reactor core also ignited the graphite moderator, which burned and distilled many other volatile radioactive isotopes into the air.

In the United States, nuclear reactors are built with much more robust containment vessels than the reactor at Chernobyl. They are designed with many safety features, such as emergency cooling systems that can flood the reactor with water in the event of a loss of primary coolant. In the accident that occurred at Three Mile Island in Pennsylvania in 1979, the backup safety systems worked, although enough heat was generated in the core to yield some melting of the uranium oxide fuel elements.

Today, engineers are working on safer designs for nuclear reactors. One is a reactor that uses a metal alloy fuel instead of an oxide and that uses liquid sodium in its primary coolant loop. The liquid sodium has a much higher boiling point than pressurized water and is able to conduct heat away from the core more efficiently. It also seems that the metal fuel elements are inherently safe because as their temperature increases, the atoms become farther apart and the chance for them to absorb neutrons decreases. This slows the nuclear reaction and therefore slows the production of heat. In fact, in one test the flow of coolant was deliberately stopped. Instead of having a core meltdown, the core initially increased in temperature and then gradually cooled on its own to near its normal operating temperature. Other inherently safe gas-cooled reactors that refuse to "run away" are also under study.

As of April 1989, there were 414 nuclear plants operating worldwide in 26 countries and generating a total of 298,000 megawatts of electricity. This represents 16% of the world's total generating capacity.

Breeder Reactors As mentioned earlier, only a small fraction of naturally occurring uranium is ^{235}U, the fissile isotope. Most is the nonfissile ^{238}U. When a nuclear reactor operates, however, another fissile isotope, ^{239}Pu, is formed by the following reaction:

$$^{238}_{92}U + ^1_0n \longrightarrow ^{239}_{92}U^* \longrightarrow ^{239}_{94}Pu + 2_{-1}^{\;0}e$$

The formation of plutonium from uranium-238 in nuclear reactors therefore provides a source for another nuclear fuel.

If a mixture of ^{239}Pu and ^{238}U is used as a fuel in a reactor, it would seem that it would be able, in effect, to use the ^{238}U as a fuel. The fission of plutonium generates enough neutrons to change enough ^{238}U to ^{239}Pu that an excess of the latter isotope is formed. Thus, if properly designed, the reactor produces more nuclear fuel than it consumes and because of this it is called a **breeder reactor.** At the present time, however, the cost of constructing these reactors is too high to make them economically viable.

Nuclear Reactors and the Environment One of the problems that nuclear reactors have is the production of nuclear wastes, which must be processed to remove radioactive by-products of fission. Some of these waste products, such as ^{137}Cs and ^{90}Sr, are particularly harmful to humans. Because they form $1+$ charged cations, cesium ions can go places that sodium ions go in the body. Similarly, Sr^{2+} ions can replace Ca^{2+} ions in bone. These and other radioactive isotopes must be kept out of the environment, and some of them have long half-lives, so the storage problem is a serious one that critics of nuclear power believe will never be solved. However, proponents of nuclear energy point to the growing problem of atmospheric pollution and the greenhouse effect that accompanies the burning of fossil fuels such as coal and oil. They believe waste storage problems can be solved and that to prevent other problems in the environment we must rely on nuclear energy more heavily in the future.

Nuclear fusion

Nuclear fusion is the opposite of nuclear fission. In fusion, two isotopes, usually very light ones, come together to form a heavier one. When this occurs, a very large amount of energy is released. For example, the fusion reaction that takes place in the explosion of a hydrogen bomb is

$$^2_1H + ^3_1H \longrightarrow ^4_2He + ^1_0n + energy$$

This is also one of several fusion reactions that take place in the sun and which together are responsible for the energy that is radiated by the sun.

The origin of the energy released in nuclear fusion can also be seen by examining Figure 24.10. As we noted earlier, nuclei of intermediate atomic mass have larger binding energies per nucleon than light nuclei. Therefore, when light nuclei such as ^2_1H combine to form heavier ones, there is an increase in the binding energy per nucleon. As in fission, energy equal to the change in binding energy is released. However, in Figure 24.10 we see that the relative change in binding energy is much greater when light nuclei are fused than when heavy nuclei break apart. Thus, for reactions that involve the same number of nuclei, the total amount of energy given off by fusion is considerably larger than that given off by fission.

Although fusion reactions release more energy than fission reactions, they do not occur easily. Positive nuclei repel each other, and this nuclear repulsion must be overcome to bring the nuclei together so that fusion can take place. The nuclei must therefore be given an enormous amount of kinetic energy, which translated means

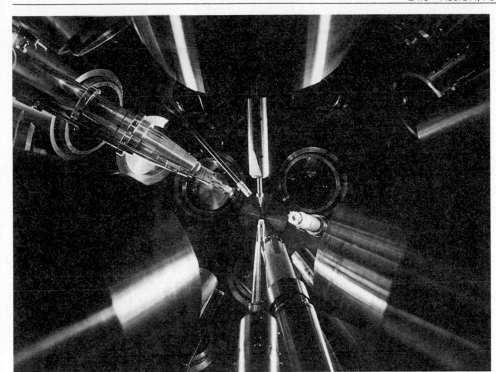

Several lasers focus on a tiny pellet of nuclear fuel in this apparatus being used to study nuclear fusion by laser implosion.

that they must be heated to a very high temperature. In fact, it is estimated that a temperature of approximately 40 million °C is needed. At these high temperatures, electrons are stripped from atoms and the entire "reaction mixture" becomes a high-temperature mixture of charged particles that is called a **plasma.**

In a fusion bomb (hydrogen bomb), the temperature required to initiate fusion is supplied by a fission bomb. In effect, an atomic bomb serves as a sort of nuclear match. Once fusion begins, the heat liberated is sufficient to cause other fusion events to occur and a chain reaction (explosion) occurs.

In controlled fusion applications, such as the generation of electrical power, the use of an atomic bomb to initiate fusion is, to say the least, unacceptable. Currently, much research work is centered on the use of high-energy lasers to initiate the fusion process. The fuel is encased in a tiny pellet that is injected into the intersecting paths of multiple laser beams (Figure 24.13). The heat of the laser beams causes the outer shell of the pellet to explode outward, which generates an equal force inward, causing the fuel to implode or be compressed to the point at which fusion takes place.

Another approach is to attempt to confine a plasma long enough to permit fusion to take place. There is no substance able to withstand the high temperatures involved, but because the plasma consists of charged particles, it can be contained in an appropriately shaped magnetic field which serves as a sort of "magnetic bottle." Figure 24.14 illustrates an apparatus called a *tokamak* (which is a Russian term for a magnetic containment device) that confines the plasma to a donut shape.

Unfortunately, neither laser implosion techniques nor plasma confinement has yet led to a condition in which more energy is released by fusion than is put in initially. If the problems that have been encountered can ever be overcome, nuclear fusion has some clear advantages over nuclear fission as a potential source of commercial power. First, as mentioned previously, the amount of energy liberated in a fusion reaction is much greater than in fission. A second advantage is that the

Figure 24.13. *Fusion by laser implosion. The energy of the neutrons given off in the fusion reaction is absorbed by the lithium blanket and carried to a heat exchanger that converts water into steam to generate electricity.*

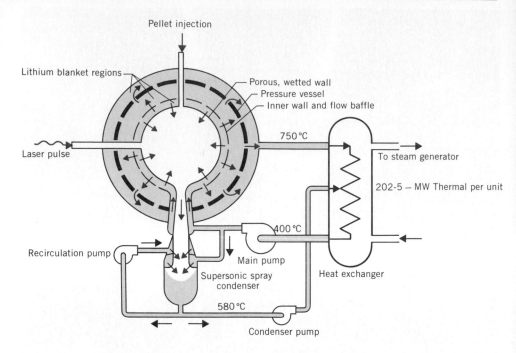

supply of fuel for nuclear fusion is virtually inexhaustible. One sequence of reactions being studied uses deuterium as the primary fuel.

$$\mathrm{^{2}_{1}H + ^{2}_{1}H \longrightarrow ^{1}_{1}H + ^{3}_{1}H}$$

$$\mathrm{^{2}_{1}H + ^{3}_{1}H \longrightarrow ^{4}_{2}He + ^{1}_{0}n + energy}$$

Deuterium atoms account for about 1% of the hydrogen atoms found in natural samples of water. Even if only a small fraction of all the deuterium present in the oceans could be separated and recovered, the energy that it could supply by fusion reactions such as those above is hundreds of thousands of times larger than all the energy contained in all the fossil fuels in the world! Still another advantage is that fusion reactions are relatively clean, in the sense that the products are generally not radioactive, so the waste disposal problem posed by fission reactors is avoided. In addition, the dangers of a runaway reactor are very small with fusion because all the

Even though the products of the fusion reaction are not themselves radioactive, the fusion reactor itself would become quite radioactive because of all the neutrons it would absorb.

Figure 24.14. *Fusion by magnetic confinement of a plasma. This is an artist's conception of a full-scale Ormak reactor nearing the end of its assembly. In the foreground are two partially assembled sectors of the donut-shaped magnetic confinement apparatus. During operation, the plasma will circulate through the circular tunnel under the influence of powerful magnetic forces.*

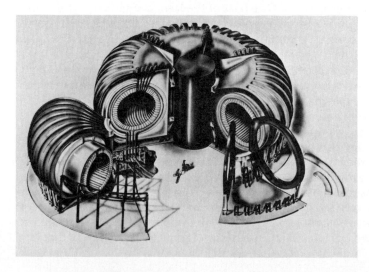

nuclear fuel will not be present at the same time as it is in a fission reactor. Instead, it will be added a bit at a time as it is needed.

Cold Fusion On March 23, 1989, an electrochemist at the University of Utah, B. Stanley Pons, and his British collaborator, Martin Fleischmann, astounded the scientific community with their announcement that they had observed nuclear fusion at room temperature in an electrolytic cell. In their apparatus they had electrolyzed deuterium oxide, ^2H_2O, using palladium electrodes to give deuterium gas and oxygen gas. They reported that in their experiment more heat had been generated than could be explained by the amount of electrical energy supplied to the cell. After considering all the possibilities, they concluded that nuclear fusion had occurred between deuterium atoms within the palladium electrode.

Palladium is known to absorb hydrogen, including the deuterium isotope. It was suggested that within the metallic lattice of palladium, deuterium atoms produced in the electrolysis reaction were forced to be in close proximity and that conditions became just right for fusion to occur. However, whether or not Pons and Fleischmann had actually observed nuclear fusion rather than some other phenomenon has been the subject of much skepticism and debate. There is conflicting evidence by other scientists who have attempted to repeat the experiments in their laboratories, and even theoreticians are of mixed opinion whether or not fusion at room temperature is possible.

Perhaps it is fitting to end this book on a note of hope. If room temperature fusion (or *cold fusion*, as it has come to be called) is possible, and if it takes place with little release of neutrons (as it must if the early experiments are correct), the discovery by Pons and Fleischmann could well change the direction of the world production of energy for the future. As this book goes to press, the jury is still out. But perhaps virtually infinite supplies of clean nuclear energy may be at our fingertips. Experiments in the near future will let us know.

The symbol for deuterium is D; deuterium oxide (heavy water) also can be represented by the formula D_2O.

Critics have claimed that fusion could not have taken place in the palladium electrode because few neutrons have been detected. Neutrons are a product of the fusion reaction sequence described on page 848.

It has been suggested that the principal fusion reaction within the palladium electrode is

$$^2_1H + ^2_1H \longrightarrow ^4_2He + energy$$

This would account for the negligible production of neutrons reported by those attempting to reproduce the work.

REVIEW QUESTIONS AND PROBLEMS

Problems whose numbers are in blue have their answers in Appendix D at the back of the book. The more difficult problems are marked with asterisks.

Radioactive Decay

24.1 What are the three main types of radiation emitted by radioactive nuclei? What are their properties?

24.2 What differences and similarities exist between beta particles and positrons?

24.3 Complete and balance the following nuclear equations:
(a) $^{81}_{36}Kr + ^0_{-1}e \rightarrow ?$
(d) $^{104}_{48}Cd \rightarrow ^{104}_{47}Ag + ?$
(b) $^{104}_{47}Ag \rightarrow ^0_1e + ?$
(e) $? + ^0_{-1}e \rightarrow ^{54}_{24}Cr$
(c) $^{73}_{31}Ga \rightarrow ^0_{-1}e + ?$

24.4 Complete and balance the following nuclear equations:
(a) $^{47}_{20}Ca \rightarrow ^{47}_{21}Sc + ?$
(b) $^{55}_{27}Co \rightarrow ^{55}_{26}Fe + ?$
(c) $^{220}_{86}Rn \rightarrow ^{216}_{84}Po + ?$
(d) $^{54}_{26}Fe + ^1_0n \rightarrow ^1_1H + ?$
(e) $^{46}_{20}Ca + ^1_0n \rightarrow ?$

24.5 Complete and balance the following nuclear equations:
(a) $^{135}_{53}I \rightarrow ^{135}_{54}Xe + ?$
(b) $^{245}_{97}Bk \rightarrow ^4_2He + ?$
(c) $^{238}_{92}U + ^{12}_6C \rightarrow ^{246}_{98}Cf + ?$
(d) $^{96}_{42}Mo + ^2_1H \rightarrow ^1_0n + ?$
(e) $^{20}_8O \rightarrow ^{20}_9F + ?$

24.6 Complete and balance the following nuclear equations:
(a) $^{35}_{17}Cl + ^1_0n \rightarrow ^{35}_{16}S + ?$
(b) $^{40}_{19}K \rightarrow ^0_{-1}e + ?$
(c) $^{98}_{42}Mo + ^1_0n \rightarrow ^0_{-1}e + ?$
(d) $^{229}_{90}Th \rightarrow ^4_2He + ?$
(e) $^{184}_{80}Hg \rightarrow ^{184}_{79}Au + ?$

24.7 Write balanced equations for the nuclear decay reactions below.
(a) alpha emission by $^{11}_5B$
(b) beta emission by $^{98}_{38}Sr$
(c) neutron absorption by $^{107}_{47}Ag$

(d) neutron emission by $^{88}_{35}Br$

(e) electron absorption by $^{116}_{51}Sb$

(f) positron emission by $^{70}_{33}As$

(g) proton emission by $^{41}_{19}K$

24.8 What is a radioactive decay series? For ^{238}U, why does it stop at ^{206}Pb?

24.9 What is a parent isotope? What is a daughter isotope?

24.10 In the nucleus of $^{41}_{19}K$, how many nucleons are there?

Kinetics of Radioactive Decay

24.11 Show that Equation 24.1 reduces to Equation 24.2 if we take $[A]_t = \frac{1}{2}[A]_0$.

24.12 Cobalt-60 has a half-life of 5.26 years. If 1.00 g of ^{60}Co was allowed to decay, how many grams would be present after (a) one half-life, (b) three half-lives, (c) five half-lives?

24.13 Selenium-75 has a half-life of 120.0 days. If we began with 8.00 g of ^{75}Se, how many grams would remain after (a) 240 days, (b) 480 days, (c) 960 days?

24.14 The rate constant for the decay of ^{45}Ca is 4.23×10^{-3} day^{-1}. What is the half-life of ^{45}Ca expressed in days?

24.15 The rate constant for the decay of ^{36}Cl is 2.30×10^{-6} year^{-1}; what is the half-life of ^{36}Cl (expressed in years)?

24.16 The half-life of ^{51}Cr is 27.72 days. What is the rate constant for decay of ^{51}Cr in units of second^{-1}?

24.17 The half-life for the decay of ^{109}Cd is 470 days. What is the value for the rate constant for this decay in units of day^{-1}?

* 24.18 The following data were obtained for the decay of ^{47}Ca:

Time (hr)	cpm
0.0	4720
8.0	4485
12.0	4372
24.0	4050
48.0	3475
72.0	2983
96.0	2560

Determine (a) the rate constant for the decay and (b) the half-life of ^{47}Ca.

Measurement of Radioactivity

24.19 How does a Geiger–Müller counter work? How does a scintillation counter work?

24.20 Define the following units: (a) becquerel, (b) curie, (c) specific activity, (d) gray, (e) rad.

24.21 What is the difference between a rad and a rem?

24.22 Suppose a hospital owns a radioactive source weighing 150 g. It has an activity of 1.24 Ci.

(a) What is its activity in Bq?

(b) What is the specific activity of the sample?

24.23 The isotope ^{145}Pr decays by emission of beta particles with an energy of 1.80 MeV each. Suppose a person swallowed, by accident, 1.0 mg of Pr having a specific activity of 140 Bq/g. What would be the absorbed dose from this isotope in Gy and rad over a period of 10 minutes? (1 MeV = 1.60×10^{-13} J) Assume all the beta particles are absorbed by the person's body.

24.24 Which kind of radiation is most dangerous to humans?

(a) alpha or beta

(b) beta or gamma

Applications of Radioactivity

24.25 A sample of rock was found to contain 2.07×10^{-5} mol of ^{40}K and 1.15×10^{-5} mol of ^{40}Ar. If we assume that all of the ^{40}Ar came from the decay of ^{40}K, what is the age of the rock in years ($t_{1/2} = 1.3 \times 10^9$ years for ^{40}K)?

24.26 The ^{14}C content of an ancient piece of wood was found to be one-eighth of that in living trees. How many years old is this piece of wood ($t_{1/2} = 5770$ years for ^{14}C)?

24.27 Dinitrogen trioxide, N_2O_3, is largely dissociated into NO and NO_2 in the gas phase where there exists the equilibrium, $N_2O_3 \rightleftharpoons NO + NO_2$. In an effort to determine the structure of N_2O_3, a mixture of NO and *NO_2 was prepared containing isotopically labeled N in the NO_2. After a period of time the mixture was analyzed and found to contain substantial amounts of both *NO and *NO_2. Explain how this is consistent with the structure for N_2O_3 being ONONO.

24.28 The reaction $(CH_3)_2Hg + HgI_2 \rightarrow 2CH_3HgI$ is believed to occur through a transition state with the structure

If this is so, what should be observed if CH_3HgI and *HgI_2 are mixed? Explain your answer.

24.29 *Racemization* is a chemical reaction in which one optical isomer of a compound is converted into its mirror image. One possible mechanism for the racemization of octahedral complex ions containing three bidentate ligands involves the temporary loss of one of the ligands

$$d\text{-}[M(AA)_3] \rightleftharpoons \begin{bmatrix} M(AA)_2 \\ + \\ AA \end{bmatrix} \rightleftharpoons l\text{-}[M(AA)_3]$$

This can be pictured as shown in Figure 24.15. Can you suggest a simple experiment, making use of radioisotopes, that would be able to confirm whether or not this mechanism is operative in the racemization of the $[Co(C_2O_4)_3]^{3-}$ ion?

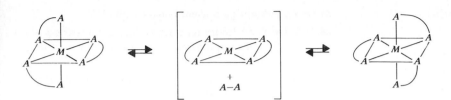

Figure 24.15. *A possible mechanism for the racemization of an octahedral* [M(AA)$_3$] *complex (AA = bidentate ligand).*

* **24.30** A large, complex piece of apparatus has built into it a cooling system containing an unknown volume of cooling liquid. It is desired to measure the volume of the coolant without draining the lines. To the coolant was added 10.0 mL of methanol labeled with ^{14}C and having a specific activity of 580 cpm per gram. The coolant was permitted to circulate to assure complete mixing before a sample was withdrawn that was found to have a specific activity of 29 cpm per gram. Calculate the volume of coolant in the system in milliliters. The density of methanol is 0.792 g/mL and the density of the coolant is 0.884 g/mL.

* **24.31** A complex ion of chromium(III) with oxalate ion was prepared from ^{51}Cr-labeled $K_2Cr_2O_7$, having a specific activity of 843 cpm per gram, and ^{14}C-labeled oxalic acid, $H_2C_2O_4$, having a specific activity of 345 cpm per gram. Chromium-51 decays by electron capture with the emission of a gamma ray, whereas ^{14}C is a pure beta emitter. Because of the characteristics of the beta and gamma detectors, each of these isotopes may be counted independently. A sample of the complex ion was observed to give a gamma count of 165 cpm and a beta count of 83 cpm. From these data, determine the number of oxalate ions bound to each Cr(III) in the complex ion. (*Hint:* For the starting materials, calculate the cpm per mole of Cr and oxalate, respectively.)

Nuclear Stability

24.32 What is the significance of the *band of stability?* What decay processes are likely to occur for nuclides that have n/p ratios that place them above the band of stability?

24.33 Elements with atomic numbers greater than 83 generally decay by either alpha emission or fission. Why are the other forms of decay less likely for these nuclides?

24.34 What is a *magic number?* What magic numbers occur for protons? For neutrons? What magic numbers do we observe for orbital electrons?

24.35 In the absence of any specific information about their actual stability, rank the following nuclides in their expected order of decreasing stability:
$^4_2$He $\quad$ $^{39}_{20}$Ca $\quad$ $^{10}_5$B $\quad$ $^{71}_{32}$Ge $\quad$ $^{58}_{28}$Ni

24.36 What would you anticipate for the order of increasing nuclear stability for the following nuclides?
$^3_2$He $\quad$ $^{40}_{20}$Ca $\quad$ $^{116}_{50}$Sn $\quad$ $^{13}_6$C $\quad$ $^{192}_{77}$Ir

24.37 The element $^{34}_{17}$Cl emits gamma radiation with energies

of 0.14, 1.15, 2.27, 3.22, and 4.80 MeV. How does this observation support the nuclear shell theory?

24.38 Technetium and promethium do not possess any stable isotopes. In light of the discussion in Section 24.5, comment on this observation.

Nuclear Transformation

24.39 What is a nuclear transformation?

24.40 How does a cyclotron work?

24.41 What are the transuranium elements? Where do they occur naturally?

24.42 Write nuclear equations for the following processes:
(a) $^{27}_{13}$Al$(\alpha,n)^{30}_{15}$P
(b) $^{209}_{83}$Bi$(d,n)^{210}_{84}$Po; $(d =$ deuteron, $^2_1$H$)$
(c) $^{15}_7$N$(p,\alpha)^{12}_6$C
(d) $^{12}_6$C$(p,\gamma)^{13}_7$N
(e) $^{14}_7$N$(\alpha,p)^{17}_8$O

24.43 Write nuclear equations for the following:
(a) $^{242}_{96}$Cm$(\alpha,n)^{245}_{98}$Cf
(b) $^{108}_{48}$Cd$(n,\gamma)^{109}_{48}$Cd
(c) $^{14}_7$N$(n,p)^{14}_6$C
(d) $^{27}_{13}$Al$(d,\alpha)^{25}_{12}$Mg
(e) $^{249}_{98}$Cf$(^{18}_8$O$,4n)^{263}_{106}$Xe

Extension of the Periodic Table

24.44 What chemical and physical properties would you predict for element number 114? If any of this element were formed at the time the universe came into being, where among earthly minerals would be a likely place to search for it?

24.45 If element 116 is found, what would be the expected formula of (a) its sodium salt, (b) its simple hydride, (c) its oxide? Would 116 be a metal or a nonmetal?

24.46 What would be the probable mass numbers of the most stable isotopes of element 114?

24.47 Explain the difficulties that must be overcome if a stable element with $Z = 114$ is to be made by nuclear bombardment.

24.48 What are the IUPAC names and symbols for these elements?
(a) 115 $\quad$ (b) 127

24.49 Specify the atomic numbers of
(a) unquadoctium $\quad$ (b) unbipentium $\quad$ (c) Unt

Nuclear Binding Energy and Nuclear Energy

24.50 What is the origin of the mass defect for a given nucleus?

24.51 What is the binding energy of $_{26}^{56}$Fe expressed in MeV (atomic mass, 55.9349 u)? Where is this on the binding energy versus mass number curve? Why don't we have to worry about an enemy that claims they have developed the iron bomb?

24.52 Calculate the energy (in kilojoules) liberated in the fusion reaction to produce one mole of helium from deuterium.

$$_1^2H + _1^2H \longrightarrow _2^4He$$

The accurate atomic masses are

$$_1^2H = 2.014102 \text{ u}$$

$$_2^4He = 4.002603 \text{ u}$$

24.53 When an electron and a positron (a positively charged electron) encounter each other, they destroy each other with the production of energy. How much energy, in joules, results from such an encounter? The rest mass of the electron is 9.1096×10^{-28} g.

* 24.54 Calculate the binding energy in kJ/mol and in MeV for the following isotopes, given the following masses: $p = 1.007277$ u, $n = 1.008665$ u, $e = 5.4859 \times 10^{-4}$ u.

Isotope	Actual Atomic Mass (u)
$_3^7$Li	7.01600
$_7^{14}$N	14.003074
$_9^{19}$F	18.99840

24.55 Define nuclear fission. Why is it possible to have a nuclear explosion with ^{235}U?

24.56 Why doesn't a nuclear reactor explode?

24.57 What is a breeder reactor? Write the nuclear equation for the reaction that generates the nuclear fuel that is produced.

24.58 In a nuclear reactor, what are control rods? What is the function of a moderator?

24.59 What is meant by the term critical mass?

24.60 What are thermal neutrons? What is their role in nuclear fission? Write the nuclear equation for the fission of ^{235}U.

24.61 Which is the only naturally occurring fissile isotope?

24.62 Define nuclear fusion. How does the energy obtained in fusion reactions compare with that obtained in fission?

24.63 Why is it so difficult to build a fusion reactor for the controlled production of nuclear energy?

24.64 What is a plasma? Why is it difficult to contain?

24.65 Give two reasons why nuclear fusion would be better than nuclear fission as a source of energy for our society.

* 24.66 Compare the energies liberated in the following fusion reactions:
(a) $2_1^2H \rightarrow _2^4He$
(b) $2_6^{12}C \rightarrow _{12}^{24}Mg$

Which reaction produces more energy per mole of product? If each were equally feasible from an engineering standpoint for producing energy by controlled fusion, which would be preferable? Actual atomic masses are $_1^2H = 2.014102$, $_2^4He = 4.002603$, $_{12}^{24}Mg = 23.98504$.

* 24.67 How many gallons of gasoline, C_8H_{18}, having a density of 0.703 g/mL, would have to be burned [to give $H_2O(l) + CO_2(g)$] to produce the same amount of energy as in the production of 1 mol of $_2^4$He by the fusion reaction described in Problem 24.52? 1 gal = 3.79 L. ΔH_f^0 of $C_8H_{18}(l)$ is -208.4 kJ/mol.

MATHEMATICS FOR GENERAL CHEMISTRY

For many students, solving numerical problems is often the most difficult part of any chemistry course. In this appendix we review some of the mathematical concepts that you will find useful in your study of chemistry.

A.1 THE FACTOR-LABEL METHOD OF PROBLEM SOLVING

Even after learning the principles of chemistry, students sometimes have difficulty in correctly setting up the arithmetic to give the proper numerical answer to a problem. The "factor-label" method uses the units associated with numbers as a guide in working out the arithmetic. The method is based on the fact that the units associated with numbers undergo the same kinds of mathematical operations of multiplication and division as the numbers themselves. For example, if two numbers are multiplied to give some desired quantity, the units belonging to those numbers are also multiplied together. As an illustration, suppose we wished to determine the area of a rectangular carpet whose dimensions had been measured to be 4.0 m (the length) and 3.0 m (the width). Area is calculated as length × width, so the area is

$$\text{area} = 4.0 \text{ m} \times 3.0 \text{ m} = 12 \text{ m}^2$$

We've obtained the numerical value of the area by multiplying the numbers, and we've obtained the units of the answer by multiplying the units for the length and the width (i.e., meter × meter = meter^2, or m × m = m^2).

An especially useful application of the properties of units is based on the mathematical operation of division: *Units that are the same in the numerator and denominator of a fraction cancel.* For instance, if the units cm appeared in numerator and denominator, they would cancel.

$$\frac{3 \text{ cm}}{2 \text{ cm}} = \frac{3}{2}$$

We use this in solving numerical problems by constructing **conversion factors** from relationships between units. A conversion factor is a fraction that we use to convert a given quantity to a desired quantity by multiplication.

(given quantity) × (conversion factor) = (desired quantity)

A conversion factor has the property of changing the *units* of the *given quantity* to the appropriate *units* of the *desired quantity*. It is easiest to see how this happens if we examine a specific example. Suppose we wished to convert a measured length of 4.0 yd (4.0 yards) into the equivalent length measured in feet. To do this, we must use the known relationship between yards and feet,

1 yd = 3 ft

This can be used to construct *two* conversion factors, which are

$$\frac{1\ yd}{3\ ft} \quad \text{and} \quad \frac{3\ ft}{1\ yd}$$

Each of these has a numerical equivalence of 1. For instance, since 3 ft equals 1 yd, we can substitute 1 yd wherever we find 3 ft. The first factor becomes

$$\frac{1\ yd}{3\ ft} = \frac{1\ \cancel{yd}}{1\ \cancel{yd}} = \frac{1}{1} = 1$$

Of course, if we multiply anything by 1, we don't change its size. Therefore, if we multiply any quantity by a conversion factor, its magnitude doesn't change. All we do is change the units.

To convert 4.0 yd to ft, we must multiply by one of the conversion factors that we've constructed. To make the choice, we simply examine the units. We know that the unit "yd" must not be in the answer, so we choose the factor that allows us to cancel yd. This is the second one.

$$4.0\ \cancel{yd} \times \frac{3\ ft}{1\ \cancel{yd}} = 12\ ft$$

Notice that we have used the units to guide us in the arithmetic. The units tell us that to make the conversion, we must multiply 4.0 yd by 3 to obtain the length in ft. Of course, you probably knew that, so the factor-label method just prolonged a simple problem. But in solving chemistry problems, the solution is not always so obvious. Furthermore, the units can often help you detect mistakes, because if you do the *wrong arithmetic,* you get the *wrong units.* For example, suppose we had used the first conversion factor by mistake.

$$4.0\ yd \times \frac{1\ yd}{3\ ft} = \frac{4.0\ yd^2}{3\ ft}$$

The units are yd^2/ft (square yards per foot). These are not the units we want, so we know that we made a mistake. In solving chemistry problems, it is very helpful to know when you've made a mistake.

It should be obvious that the key to using the factor-label method is choosing *correct* relationships between the units. For instance, if you didn't know how many feet equal 1 yd and decided to use the relationship

7 ft = 1 yd

you would certainly obtain the wrong answer, no matter how clever you are at arithmetic or applying the factor-label method. You will have many opportunities to use the factor-label method throughout your chemistry course. *Pay special attention to the correctness of the relationships that you use to make your conversion factors.*

Before concluding this section, we would like to point out that in many places throughout the book, we use the concept of an *equivalence* between quantities, instead of an equality. For example, if you have a job that pays 5 dollars per hour, there is an equivalence between the time you work and the dollars you earn: Each hour that you work is equivalent to 5 dollars pay. We express this in a mathematical way by using the symbol $\Leftrightarrow$ to mean "is equivalent to." Thus,

$$1 \text{ hr} \Leftrightarrow 5 \text{ dollars}$$

Equivalencies such as this are also used to make conversion factors. In this case, we can make these two factors,

$$\frac{1 \text{ hr}}{5 \text{ dollars}} \quad \text{and} \quad \frac{5 \text{ dollars}}{1 \text{ hr}}$$

If you worked for 12 hours, you could use the second factor to calculate your earnings.

$$12 \text{ hr} \times \frac{5 \text{ dollars}}{1 \text{ hr}} = 60 \text{ dollars}$$

We can't really say that one hour *equals* five dollars, so we express the relationship as an equivalence.

A.2 EXPONENTIAL NOTATION (SCIENTIFIC NOTATION)

Quite often in science it is necessary to deal with numbers that are very large, such as Avogadro's number,

$$602,300,000,000,000,000,000,000$$

or numbers that are very small, such as the mass of a single molecule of water,

$$0.000\ 000\ 000\ 000\ 000\ 000\ 000\ 03 \text{ g}$$

These numbers are very cumbersome and difficult to work with without making mistakes in arithmetic computations. To aid us in handling these large and small numbers, we use a system called **exponential notation** or **scientific notation.** In this system a number is expressed as a decimal part multiplied by 10 raised to an appropriate power. Thus,

$$200 = 2 \times 10 \times 10 = 2 \times 10^2$$

$$205,000 = 2.05 \times 100,000 = 2.05 \times (10 \times 10 \times 10 \times 10 \times 10)$$
$$= 2.05 \times 10^5$$

To determine the exponent on the 10, we can simply count the number of places the decimal must be moved to produce the number that precedes the 10 when the number is expressed in the scientific notation

$$\underset{5 \text{ places}}{\underleftarrow{205\ \ 000}}\ \circ\ = 2.05 \times 10^5$$

Note that the exponent on the 10 is positive when the decimal is moved to the left. When it is moved to the right, the exponent is negative.

$$0\ \circ\ \underset{7 \text{ places}}{\underrightarrow{000000315}} = 3.15 \times 10^{-7}$$

Most students today perform their computations using an electronic calculator, and virtually every "scientific calculator" is designed to handle calculations involving numbers expressed in scientific notation. As simple as these calculators are to manipulate, many students make mistakes in entering numbers in scientific notation. If you plan to use one of these calculators, take it out now and turn it on, so we can review the procedure.

4 means the key labeled 4, . means the key with the decimal point, and so forth.

Most scientific calculators have a key labeled EE or EXP. These are the keys that are used to activate the scientific notation function of the calculator. When you press this key, you should say to yourself, "times ten to the . . . " To enter the number 4.5×10^3, you use the key sequence

$$\boxed{4} \quad \boxed{.} \quad \boxed{5} \quad \boxed{\text{EE}} \text{ (or } \boxed{\text{EXP}} \text{)} \quad \boxed{3}$$

Try it on your calculator. As you press the keys, say to yourself, "four point five *times ten to the* third." On some calculators, the display will show a small 10 with the exponent 3; on other calculators, there will be a space after the 4.5, which is followed by 03.

To enter a negative exponent, you *do not* use the subtraction key, which appears as $\boxed{-}$. Instead, after entering the decimal part of the number and the EE (or EXP) key, press the change-sign key, which is usually labeled $\boxed{+/-}$. For example, to enter the number 4.5×10^{-3}, the key sequence is

$$\boxed{4} \quad \boxed{.} \quad \boxed{5} \quad \boxed{\text{EE}} \quad \boxed{+/-} \quad \boxed{3}$$

Once again, try it on your calculator. Notice that the exponent appears with a negative sign on the display.

Once you know how to enter numbers expressed in scientific notation, arithmetic operations are performed on them just as on other kinds of numbers. Refer to the directions that accompany your calculator and read Section A.5 of this appendix for additional help, if you need it.

Even though you may have learned to work with numbers in scientific notation using your calculator, someday the batteries may fail and you might find it necessary to do arithmetic the "old-fashioned" way. Therefore, it is a good idea to know the following rules for arithmetic.

Multiplication

In multiplication, the decimal portions of the numbers are multiplied and the exponents on the 10 are *added* algebraically.

Try these on your calculator to be sure you obtain the correct answers.

$$(2.0 \times 10^4) \times (3.0 \times 10^3) = (2.0 \times 3.0) \times 10^{(4+3)} = 6.0 \times 10^7$$
$$(4.0 \times 10^8) \times (-2.0 \times 10^{-5}) = [4.0 \times (-2.0)] \times 10^{[8+(-5)]} = -8.0 \times 10^3$$

Division

The decimal portions are divided, and the exponent on 10 in the denominator is *subtracted algebraically* from the exponent on 10 in the numerator.

$$\frac{8.0 \times 10^7}{4.0 \times 10^3} = \left(\frac{8.0}{4.0}\right) \times 10^{(7-3)} = 2.0 \times 10^4$$

$$\frac{6.0 \times 10^5}{2.0 \times 10^{-3}} = \left(\frac{6.0}{2.0}\right) \times 10^{[5-(-3)]} = 3.0 \times 10^8$$

$$\frac{9.0 \times 10^{-4}}{3.0 \times 10^{-6}} = \left(\frac{9.0}{3.0}\right) \times 10^{[-4-(-6)]} = 3.0 \times 10^2$$

You have probably noticed that the usual practice is to express a number with the decimal point located between the first and second digit. There are, of course, other ways that these numbers can be written that are all equivalent, and you will undoubtedly find occasions when it is convenient to use a number in other than its standard form. An example of a few equivalent expressions of the same number is

$$3.15 \times 10^{-7} = 315 \times 10^{-9} = 0.0315 \times 10^{-5}$$

Notice that in converting from one to another, one part of the number is increased while the other is decreased. For instance, to change 8.25×10^6 to 825×10^4, multiply *and* divide by 100 (or 10^2).

$$8.25 \times 10^6 \left(\frac{100}{100}\right) = (8.25 \times 100) \times \left(\frac{10^6}{10^2}\right) = 825 \times 10^4$$

Addition and subtraction

For carrying out addition and subtraction, each quantity must first be written with the same power of 10. Then addition or subtraction is performed on the decimal parts; the power of 10 remains the same. For example,

$$(2.17 \times 10^5) + (3.0 \times 10^4) = ?$$

If we express both numbers with the same power of 10, we have

$$\begin{array}{ccc}
2.17 \times 10^5 & \text{or} & 21.7 \times 10^4 \\
+0.30 \times 10^5 & & +3.0 \times 10^4 \\
\hline
2.47 \times 10^5 & & 24.7 \times 10^4
\end{array}$$

Taking a root

To extract a root (e.g., the square root), we make the exponent on the 10 divisible by the desired root. For instance, to take the square root of 3.7×10^7, we first change the number so that the power of 10 is divisible by 2. Then we take the square root of the decimal part and divide the exponent by 2.

$$\sqrt{3.7 \times 10^7} = \sqrt{37 \times 10^6} = \sqrt{37} \times 10^3 = 6.1 \times 10^3$$

A.3 LOGARITHMS

A logarithm is an exponent. **Common logarithms** are exponents to which 10 must be raised to give a specified number. They are also called base-10 logarithms. For instance, the $\log(100) = 2$ because $10^2 = 100$. Similarly, $\log(1000) = \log(10^3) = 3$.

Since logarithms are exponents, when we perform mathematical operations the same rules that apply to exponents also apply to logarithms. Thus, we have

Multiplication $\begin{cases} \text{add exponents} \\ \text{add logarithms} \end{cases}$

Division $\begin{cases} \text{subtract exponents} \\ \text{subtract logarithms} \end{cases}$

Your scientific calculator will handle logarithms with ease. See Section A.5 of this appendix.

For example,

$$\boxed{10^3 \times 10^4} = 10^{3+4} = \boxed{10^7}$$

$$\log(10^3 \times 10^4) = \log(10^3) + \log(10^4) = 3 + 4 = 7 = \log(10^7)$$

Similarly, for division

$$\boxed{\frac{10^8}{10^6}} = 10^{8-6} = \boxed{10^2}$$

$$\log\left(\frac{10^8}{10^6}\right) = \log(10^8) - \log(10^6) = 8 - 6 = 2 = \log(10^2)$$

For decimal numbers between 1 and 10, their logarithms lie between 0 and 1, since

$$\log(1) = 0 \quad (1 = 10^0)$$
$$\log(10) = 1 \quad (10 = 10^1)$$

For example, log 2 = 0.3010 or

$$10^{0.3010} = 2$$

The common logarithm of 2 and other numbers between 1 and 10 can be obtained from the table of logarithms at the end of this appendix.

To use this table to find the logarithm of a number, we use the extreme left column to locate the first two digits of the number, and the top horizontal row to locate the third digit. The value in the table corresponding to these is the logarithm of our number. For example, if we want log(4.61), we would locate 46 in the left column and proceed to the right until we were in the column headed by 1. The answer is

$$\log(4.61) = 0.6637$$

Notice that a decimal point is placed in front of the logarithm obtained from the table.

This table is extremely easy to use as long as our numbers are expressed in this fashion, that is, as a decimal number between 1 and 10. If the number whose logarithm we seek does not appear this way, we can first express the number in exponential notation and then take its logarithm. For example, what is log(728)?

$$\log(728) = \log(7.28 \times 10^2)$$
$$\log(7.28 \times 10^2) = \log(7.28) + \log(10^2)$$
$$\log(7.28) = 0.8621 \quad \text{(from table)}$$
$$+\log(10^2) = \underline{2.0000}$$
$$2.8621$$

Therefore,

$$\log(728) = 2.8621$$

What would be the value of log(0.00583)? Once again we first express the number in exponential notation,

$$\log(0.00583) = \log(5.83 \times 10^{-3})$$
$$\log(5.83 \times 10^{-3}) = \log(5.83) + \log(10^{-3})$$
$$\log(5.83) = +0.7657$$
$$+\log(10^{-3)} = \underline{-3.0000}$$

Numbers in the log table are positive quantities.

Adding these algebraically, we get

$$\log(0.00583) = -2.2343$$

Sometimes it is necessary to obtain the number whose logarithm is known. This is called taking the **antilogarithm.** The procedure is simply the reverse of that given above. For example, suppose that we wish to find the number whose logarithm is 3.253,

$$\log x = 3.253$$

First, we divide the number into two parts, an integer and a positive decimal.

$$3.253 = 3 + 0.253 = 0.253 + 3$$
$$\log x = (0.253 + 3)$$

We locate 0.253 in the body of the log table and find that it is the log of 1.79; we also know that 3 is the log of 10^3. Therefore,

$$\log x = \log(1.79) + \log(10^3) = \log(1.79 \times 10^3)$$
$$x = 1.79 \times 10^3$$

If the logarithm of the number is negative, the procedure for taking the anti-logarithm is just slightly different. For example, suppose we have the problem

$$\log x = -8.475$$

Once again, we divide the logarithm into two parts: a *positive decimal* and a *negative integer*.

$$-8.475 = (+0.525) + (-9)$$

Therefore,

$$\log x = (+0.525) + (-9)$$

Next, we locate 0.525 in the body of the log table and find that is in the log of 3.35. The −9 becomes the exponent on the 10, so our answer is

$$x = 3.35 \times 10^{-9}$$

What you've learned about logarithms here can be used in working problems dealing with pH in Chapter 16. The pH of an aqueous solution is defined as

$$pH = -\log[H^+]$$

where $[H^+]$ stands for the molar concentration of hydrogen ion in the solution. As promised in the text (page 518), here are the solutions to the logarithm calculations in Examples 16.3, 16.4, and 16.5.

EXAMPLE 16.3

SOLUTION: We are given that

$$[H^+] = 2.0 \times 10^{-3}$$

Therefore,

$$\begin{aligned}
pH &= -\log(2.0 \times 10^{-3}) \\
&= -(\log 2.0 + \log 10^{-3}) \\
&= -[0.30 + (-3)] \\
&= -(-2.70) \\
&= 2.70
\end{aligned}$$

EXAMPLE 16.4

SOLUTION: *Method 1*

$$[H^+] = 2.0 \times 10^{-11}$$
$$\begin{aligned}
pH &= -\log(2.0 \times 10^{-11}) \\
&= -(\log 2.0 + \log 10^{-11}) \\
&= -[0.30 + (-11)] \\
&= -(-10.70) \\
&= 10.70
\end{aligned}$$

Method 2

$$[OH^-] = 5.0 \times 10^{-4}$$

$$pOH = -\log[OH^-]$$

$$= -\log(5.0 \times 10^{-4})$$

$$= -[0.70 + (-4)]$$

$$= 3.30$$

Then

$$pH = 14.00 - pOH$$

$$= 10.70$$

EXAMPLE 16.5

SOLUTION: In this case, pH = 3.80 and we must compute $[H^+]$

$$pH = -\log[H^+] = 3.80$$

Therefore,

$$\log[H^+] = -3.80$$

To take the antilog, we have to express -3.80 as the sum of a negative whole number and a positive decimal.

$$\log[H^+] = (0.20) + (-4.00)$$

The antilog of 0.20, obtained from the log table, is 1.6; the antilog of -4 is 10^{-4}. Therefore,

$$\log[H^+] = \log(1.6) + \log(10^{-4})$$

$$= \log(1.6 \times 10^{-4})$$

This gives

$$[H^+] = 1.6 \times 10^{-4}$$

Also, the calculated pOH = 10.20. Therefore,

$$-\log[OH^-] = 10.20$$

$$\log[OH^-] = -10.20$$

$$= (0.80) + (-11)$$

$$= \log(6.3) + \log(10^{-11})$$

$$= \log(6.3 \times 10^{-11})$$

Hence,

$$[OH^-] = 6.3 \times 10^{-11}$$

Natural logarithms

A system of logarithms encountered frequently in the sciences, known as natural logarithms, has as its base $e = 2.71828. \ldots$ In other words, natural logarithms are exponents to which e must be raised to give a number. The relationship between common logs and natural logs is seen below.

$$\log_{10}(10) = 1 \qquad \text{or} \qquad 10^1 = 10$$

$$\log_e(10) = 2.303 \qquad \text{or} \qquad e^{2.303} = 10$$

With common logarithms we usually omit the base and write log 10 = 1. With natural logarithms the base e is omitted, and they are written

$$\ln 10 = 2.303$$

The conversion from base e to base 10 logarithm is accomplished by the equation

$$\ln x = 2.303 \log x$$

A.4 THE QUADRATIC EQUATION

When an equation can be written in the form

$$ax^2 + bx + c = 0$$

in which the coefficients a, b, and c are known, two values (called roots) of the variable x can be obtained by substituting the values of a, b, and c into the expression

$$x = \frac{-b \pm \sqrt{b^2 - 4ac}}{2a}$$

For example, given the equation

$$x^2 - 5x + 4 = 0$$

what is the value of x? In this equation $a = 1$, $b = -5$, and $c = 4$. Thus,

$$x = \frac{-(-5) \pm \sqrt{(-5)^2 - 4(1)(4)}}{2(1)} = \frac{5 \pm \sqrt{25 - 16}}{2}$$

$$= \frac{5 \pm \sqrt{9}}{2} = \frac{5 \pm 3}{2}$$

Therefore,

$$x = \frac{2}{2} = 1 \quad \text{and} \quad x = \frac{8}{2} = 4$$

Both values of x are mathematically correct. Usually when a quadratic equation is encountered in a chemical problem, only one of the roots has any real significance. Generally, the other root will be clearly meaningless—for instance, a negative concentration, which is impossible (you can't have a smaller amount of matter than no matter at all!)

A.5 ELECTRONIC CALCULATORS

Today, much of the tiresome work of arithmetic is relieved by the use of small hand-held electronic calculators. For example, we have already described how they can be useful for calculations that involve scientific notation. To obtain the most out of these electronic marvels, you should be aware of some simple mathematical relationships, and those that are most useful to you in chemistry are discussed below. Since operational procedures differ on various calculators, you will have to refer to the direction booklet that accompanies your calculator for specific instructions about how to apply these relationships.

When using your calculator, be sure to observe the rules for significant figures discussed in Section 1.4.

Successive multiplication and division

One of the most common kinds of computations you will encounter, especially if you apply the factor-label method, involves evaluating expressions that have several numbers in the numerator and several in the denominator. An example might be something like this:

$$\frac{14 \times 92 \times 32}{73 \times 43 \times 51} = ?$$

There are several ways that this expression can be evaluated and they all give the same result. For example, you could evaluate the numerator and write down the answer; then evaluate the denominator and write down its value; and then finally enter the value of the numerator and divide it by the value of the denominator. This will certainly give the correct answer, but it involves a lot of unnecessary writing and the error-prone reentering of numbers. The expression can be evaluated in one step without writing down any intermediate values as follows:

1. Enter the first value in the numerator (in this case, 14).
2. Each time you enter another value from the numerator, perform multiplication.
3. Each time you enter a value from the denominator, perform a division.

To evaluate the expression above, a suitable sequence of operations would be

$$14 \times 92 \times 32 \div 73 \div 43 \div 51 =$$

Actually, the order in which the operations are performed doesn't matter, so an equally suitable sequence is

$$14 \times 92 \div 73 \div 43 \times 32 \div 52 =$$

Obviously, there are other sequences, too. Try some of them, just for practice.

Logarithms and antilogarithms

We've seen that a logarithm is an exponent and that there are two systems of logarithms generally encountered, base e and base 10. If your calculator possesses logarithm capabilities, you will probably find a key labeled LN for base e (natural) logarithms and a key labeled LOG for base 10 (common) logarithms. Generally, if a number is entered and the LN key depressed, the display will show the natural log of that number. The common log will appear if you press the LOG key.

Useful relationships among logarithms are

$$10^{\log X} = X$$

$$e^{\ln X} = X$$

For example, log 2 = 0.3010, ln 2 = 0.6931.

$$10^{\log 2} = 10^{0.3010} = 2$$

$$e^{\ln 2} = e^{0.6931} = 2$$

Some calculators don't have keys labeled e^x and 10^x. They usually do antilogarithms by combining an inverse key, $\boxed{\text{INV}}$, with the appropriate logarithm key. For example, e^x is accomplished by the sequence $\boxed{\text{INV}}$ $\boxed{\text{LN}}$.

These provide a means to obtain the antilogarithms. If you have the natural log of a number, enter it and depress the e^x key. If you have the common log, enter it and press the 10^x key. In each case you will obtain the antilogarithm.

If your calculator does not have a 10^x key, but does have an x^y (or y^x) key, enter 10 and raise it to an exponent that corresponds to the common log. The result will be the antilogarithm.

Exponents and roots

Most calculators have X^2 and $\sqrt{X}$ keys, and these operations are simple. For higher powers and roots you can use either of two methods.

1. **Using the x^y key.** To compute, $X = a^b$, enter a and raise it to the power b. To compute $X = \sqrt[b]{a}$, enter a and raise it to the power, $1/b$. For example,

$$X = 2^3 = 8$$

$$X = \sqrt[3]{2} = 2^{1/3} = 2^{0.333...3} = 1.25992$$

2. **Using logarithms.** To compute $X = a^b$ with natural logarithms, we use the relationship that

$$\ln X = \ln a^b = b \ln a$$

Therefore,

$$X = e^{\ln X}$$
$$= e^{b \ln a}$$

Let's suppose that we wished to compute 3^5. On a typical calculator we would perform this computation in the following sequence.

1. Take ln 3. $\ln 3 = 1.098612$

2. Multiply ln 3 by 5. $5 \ln 3 = 5.493061$

3. Raise e to this exponent. $e^{5 \ln 3} = 243$

$$3^5 = 243$$

These calculations, of course, can also be done using common logarithms, in which case

$$X = 10^{b \log a}$$

To compute a root, $X = \sqrt[b]{a}$, we find that

$$X = e^{(\ln a)/b}$$

For example, suppose that we wished to calculate $\sqrt[5]{12}$. The sequence of operations is

1. Take ln 12. $\ln 12 = 2.484907$

2. Divide ln 12 by 5. $\dfrac{(\ln 12)}{5} = 0.496981$

3. Raise e to this exponent. $e^{(\ln 12)/5} = 1.643752$

$$\sqrt[5]{12} = 1.643752$$

COMMON LOGARITHMS

	0	1	2	3	4	5	6	7	8	9
10	0000	0043	0086	0128	0170	0212	0253	0294	0334	0374
11	0414	0453	0492	0531	0569	0607	0645	0682	0719	0755
12	0792	0828	0864	0899	0934	0969	1004	1038	1072	1106
13	1139	1173	1206	1239	1271	1303	1335	1367	1399	1430
14	1461	1492	1523	1553	1584	1614	1644	1673	1703	1732
15	1761	1790	1818	1847	1875	1903	1931	1959	1987	2014
16	2041	2068	2095	2122	2148	2175	2201	2227	2253	2279
17	2304	2330	2355	2380	2405	2430	2455	2480	2504	2529
18	2553	2577	2601	2625	2648	2672	2695	2718	2742	2765
19	2788	2810	2833	2856	2878	2900	2923	2945	2967	2989
20	3010	3032	3054	3075	3096	3118	3139	3160	3181	3201
21	3222	3243	3263	3284	3304	3324	3345	3365	3385	3404
22	3424	3444	3464	3483	3502	3522	3541	3560	3579	3598
23	3617	3636	3655	3674	3692	3711	3729	3747	3766	3784
24	3802	3820	3838	3856	3874	3892	3909	3927	3945	3962
25	3979	3997	4014	4031	4048	4065	4082	4099	4116	4133
26	4150	4166	4183	4200	4216	4232	4249	4265	4281	4298
27	4314	4330	4346	4362	4378	4393	4409	4425	4440	4456
28	4472	4487	4502	4518	4533	4548	4564	4579	4594	4609
29	4624	4639	4654	4669	4683	4698	4713	4728	4742	4757
30	4771	4786	4800	4814	4829	4843	4857	4871	4886	4900
31	4914	4928	4942	4955	4969	4983	4997	5011	5024	5038
32	5051	5065	5079	5092	5105	5119	5132	5145	5159	5172
33	5185	5198	5211	5224	5237	5250	5263	5276	5289	5302
34	5315	5328	5340	5353	5366	5378	5391	5403	5416	5428
35	5441	5453	5465	5478	5490	5502	5514	5527	5539	5551
36	5563	5575	5587	5599	5611	5623	5635	5647	5658	5670
37	5682	5694	5705	5717	5729	5740	5752	5763	5775	5786
38	5798	5809	5821	5832	5843	5855	5866	5877	5888	5899
39	5911	5922	5933	5944	5955	5966	5977	5988	5999	6010
40	6021	6031	6042	6053	6064	6075	6085	6096	6107	6117
41	6128	6138	6149	6160	6170	6180	6191	6201	6212	6222
42	6232	6243	6253	6263	6274	6284	6294	6304	6314	6325
43	6335	6345	6355	6365	6375	6385	6395	6405	6415	6425
44	6435	6444	6454	6464	6474	6484	6493	6503	6513	6522
45	6532	6542	6551	6561	6571	6580	6590	6599	6609	6618
46	6628	6637	6646	6656	6665	6675	6684	6693	6702	6712
47	6721	6730	6739	6749	6758	6767	6776	6785	6794	6803
48	6812	6821	6830	6839	6848	6857	6866	6875	6884	6893
49	6902	6911	6920	6928	6937	6946	6955	6964	6972	6981
50	6990	6998	7007	7016	7024	7033	7042	7050	7059	7067
51	7076	7084	7093	7101	7110	7118	7126	7135	7143	7152
52	7160	7168	7177	7185	7193	7202	7210	7218	7226	7235
53	7243	7251	7259	7267	7275	7284	7292	7300	7308	7316
54	7324	7332	7340	7348	7356	7364	7372	7380	7388	7396

COMMON LOGARITHMS — Continued

	0	1	2	3	4	5	6	7	8	9
55	7404	7412	7419	7427	7435	7443	7451	7459	7466	7474
56	7482	7490	7497	7505	7513	7520	7528	7536	7543	7551
57	7559	7566	7574	7582	7589	7597	7604	7612	7619	7627
58	7634	7642	7649	7657	7664	7672	7679	7686	7694	7701
59	7709	7716	7723	7731	7738	7745	7752	7760	7767	7774
60	7782	7789	7796	7803	7810	7818	7825	7832	7839	7846
61	7853	7860	7868	7875	7882	7889	7896	7903	7910	7917
62	7924	7931	7938	7945	7952	7959	7966	7973	7980	7987
63	7993	8000	8007	8014	8021	8028	8035	8041	8048	8055
64	8062	8069	8075	8082	8089	8096	8102	8109	8116	8122
65	8129	8136	8142	8149	8156	8162	8169	8176	8182	8189
66	8195	8202	8209	8215	8222	8228	8235	8241	8248	8254
67	8261	8267	8274	8280	8287	8293	8299	8306	8312	8319
68	8325	8331	8338	8344	8351	8357	8363	8370	8376	8382
69	8388	8395	8401	8407	8414	8420	8426	8432	8439	8445
70	8451	8457	8463	8470	8476	8482	8488	8494	8500	8506
71	8513	8519	8525	8531	8537	8543	8549	8555	8561	8567
72	8573	8579	8585	8591	8597	8603	8609	8615	8621	8627
73	8633	8639	8645	8651	8657	8663	8669	8675	8681	8686
74	8692	8698	8704	8710	8716	8722	8727	8733	8739	8745
75	8751	8756	8762	8768	8774	8779	8785	8791	8797	8802
76	8808	8814	8820	8825	8831	8837	8842	8848	8854	8859
77	8865	8871	8876	8882	8887	8893	8899	8904	8910	8915
78	8921	8927	8932	8938	8943	8949	8954	8960	8965	8971
79	8976	8982	8987	8993	8998	9004	9009	9015	9020	9025
80	9031	9036	9042	9047	9053	9058	9063	9069	9074	9079
81	9085	9090	9096	9101	9106	9112	9117	9122	9128	9133
82	9138	9143	9149	9154	9159	9165	9170	9175	9180	9186
83	9191	9196	9201	9206	9212	9217	9222	9227	9232	9238
84	9243	9248	9253	9258	9263	9269	9274	9279	9284	9289
85	9294	9299	9304	9309	9315	9320	9325	9330	9335	9340
86	9345	9350	9355	9360	9365	9370	9375	9380	9385	9390
87	9395	9400	9405	9410	9415	9420	9425	9430	9435	9440
88	9445	9450	9455	9460	9465	9469	9474	9479	9484	9489
89	9494	9499	9504	9509	9513	9518	9523	9528	9533	9538
90	9542	9547	9552	9557	9562	9566	9571	9576	9581	9586
91	9590	9595	9600	9605	9609	9614	9619	9624	9628	9633
92	9638	9643	9647	9652	9657	9661	9666	9671	9675	9680
93	9685	9689	9694	9699	9703	9708	9713	9717	9722	9727
94	9731	9736	9741	9745	9750	9754	9759	9763	9768	9773
95	9777	9782	9786	9791	9795	9800	9805	9809	9814	9818
96	9823	9827	9832	9836	9841	9845	9850	9854	9859	9863
97	9868	9872	9877	9881	9886	9890	9894	9899	9903	9908
98	9912	9917	9921	9926	9930	9934	9939	9943	9948	9952
99	9956	9961	9965	9969	9974	9978	9983	9987	9991	9996

B MOLECULAR ORBITAL THEORY

The valence bond theory, discussed in Sections 9.4 through 9.7, views covalent bonding from a rather different perspective than molecular orbital theory (abbreviated MO theory). In its simplest form, the MO theory is less interested in *how* a bond is formed than in what the bonding is like in the finished molecule.

Molecular orbital theory is an extension of the theory that we applied to the explanation of the electronic structures of atoms. According to that theory, an atom consists of a single positive nucleus surrounded by a set of atomic orbitals. The ground state electronic structure of a particular atom is derived by feeding the appropriate number of electrons into the set of atomic orbitals in such a way that (1) no more than two electrons populate a single orbital, (2) each electron is placed into the lowest energy orbital available, and (3) electrons are spread out as much as possible, with unpaired spins, over orbitals of the same energy.

Molecular orbital theory proceeds in much this same way. According to this theory, a molecule contains a certain arrangement of atomic nuclei, and spread out over these nuclei is a set of **molecular orbitals.** The electronic structure of the molecule is obtained by feeding the appropriate number of electrons into these molecular orbitals following the same rules that apply to the filling of atomic orbitals.

No one is quite sure what shapes molecular orbitals have in any particular molecule or ion. What *appears* to be an approximately correct picture is obtained by combining the atomic orbitals that reside on the nuclei composing the molecule. These combinations are achieved by considering the constructive interference and destructive interference of the electron waves of the atoms in the molecule. This is shown in Figure B.1 for the 1s orbitals on two identical nuclei. Notice that when the amplitudes of the two waves are *added*, the resulting molecular orbital has a shape that concentrates electron density between the two nuclei. Electrons placed in such a molecular orbital tend to hold the nuclei together and stabilize a molecule. For that reason this orbital is called a **bonding molecular orbital.** Because the electron density in the orbital is centered along the line joining the atomic nuclei, it is also a σ-type of orbital. Since it is derived in this case from two 1s atomic orbitals, we refer to it as the σ_{1s} **molecular orbital.**

An atom is one positive center (a nucleus) surrounded by a set of atomic orbitals; a molecule is a cluster of positive centers surrounded by a set of molecular orbitals.

Notice in Figures B.1 and B.2 that the orbitals have been assigned algebraic signs. These are the signs of their wave functions. When orbitals (or portions of orbitals) with the same sign overlap, there is constructive interference. When they are of opposite sign, there is destructive interference.

You will also observe in Figure B.1 that a second molecular orbital is obtained by the destructive interference of the electron waves. In this instance a molecular orbital is produced that places the maximum electron density outside the region between the two nuclei. If the electrons of a molecule are placed into this molecular orbital, they do not help to cement the nuclei together. In fact, the attractions between the electrons and the nuclei are less than the mutual repulsions of the electrons in this kind of molecular orbital, and this produces a net increase in energy rather than a decrease. Consequently, electrons placed into this molecular orbital lead to a destabilization of the molecule and, as a result, the orbital is said to be **antibonding.** This antibonding orbital also has its greatest electron density along the line that passes through the two nuclei and is thus a σ-type orbital. Its antibonding character is denoted by an asterisk superscript; thus it is called the σ_{1s}^* **molecular orbital.** As you might expect, we can also draw similar pictures for the combination of any pair of s orbitals; therefore, in a diatomic molecule we have $\sigma_{2s}, \sigma_{2s}^*, \sigma_{3s}, \sigma_{2s}^*, \ldots$, molecular orbitals.

In a molecule, the p orbitals are also capable of interacting to produce bonding and antibonding molecular orbitals, as illustrated in Figure B.2. Here we have arbitrarily chosen to denote the internuclear axis as the z axis of our coordinate system, so the p orbitals that point toward one another correspond to p_z orbitals. Again we find that one combination or orbitals gives a bonding molecular orbital, with electron density placed between the two nuclei, whereas the second combination places most of the electron density outside the region between the nuclei. The p_z orbitals, like s orbitals, form σ-type molecular orbitals, and for $2p_z$ orbitals they would be labeled σ_{2p_z} and $\sigma_{2p_z}^*$.

Having chosen the z axis at the internuclear axis, we find that the p_x and p_y orbitals on the two nuclei of our molecule are forced to overlap in a sideways fashion to produce two sets of π and π^* molecular orbitals (Figure B.2). Keep in mind that the π_{p_x} and $\pi_{p_x}^*$ orbitals are the same as the π_{p_y} and $\pi_{p_y}^*$ orbitals, respectively, except that they are situated at 90° to each other when viewed down the molecular axis.

For a diatomic molecule, we have now examined the *shapes* of the molecular orbitals that can be considered to arise as a consequence of the overlap of atomic orbitals. To discuss the electronic structure of a diatomic molecule, however, we must know their relative energies. Once this has been established, we can then

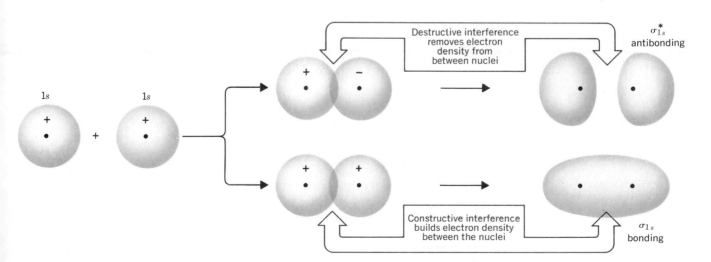

Figure B.1. *The combination of atomic* 1s *orbitals to give bonding and antibonding molecular orbitals.*

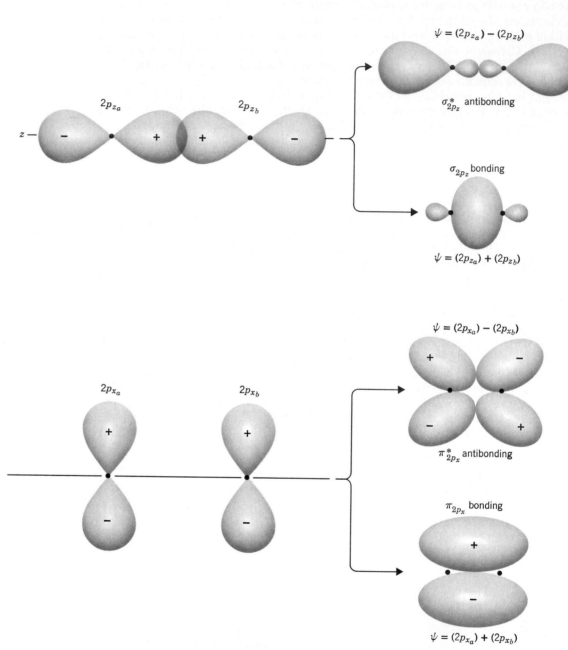

Figure B.2. *Formation of moleculear orbitals from atomic p orbitals.*

proceed with filling the orbitals with electrons, following the rules mentioned earlier.

Let us first consider the σ_{1s} and σ_{1s}^* orbitals. Electrons placed into the bonding orbital lead to stable bond formation and, therefore, an energy lower than that of two separate atoms. On the other hand, electrons placed into the antibonding orbital lead to a destabilization of the molecule and thus to a state higher in energy than that

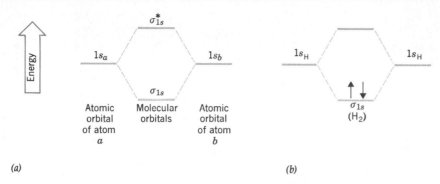

Figure B.3. (a) *The energies of the bonding and antibonding σ_{1s} molecular orbitals. (b) Bonding in H_2.*

of the atoms from which the molecule is formed. We can represent this schematically as shown in Figure B.3a, where the energies of the atomic orbitals of the separate atoms appear on either side of the energy-level diagram and the energies of the molecular orbitals appear in the center.

Using this simple diagram we can examine the bonding in the H_2 molecule. There are two electrons in H_2 that we place in the lowest-energy molecular orbital, the σ_{1s} (Figure B.3b). The electron distribution in H_2 is therefore that described by the shape of the σ_{1s} orbital. Notice that this picture is the same as that developed in the valence bond view of H_2. This should not be too surprising since both theories are attempting to describe the same molecular species.

Before moving on, let us also see why the molecule He_2 does *not* exist. The species He_2 would have four electrons, two of which would be placed into the σ_{1s} orbital. The other two would be forced to occupy the σ_{1s}^* orbital. The pair of electrons in the antibonding orbital would cancel out the stabilizing influence of the bonding pair. As a result, the **net bond order,** which we can define as

$$\left(\begin{array}{c}\text{Net bond}\\\text{order}\end{array}\right) = \frac{(\text{no. of } e^- \text{ in bonding MOs}) - (\text{no. of } e^- \text{ in antibonding MOs})}{2}$$

has a value of zero for He_2. Since the bond order in He_2 is zero, He_2 is not a stable molecule and is not observed to exist under normal conditions.

For diatomic molecules of second period elements, we really only need to consider molecular orbitals derived from the interaction of the $2s$ and $2p$ orbitals. The $1s$ orbitals are essentially buried beneath the valence shell orbitals and are therefore not involved to any appreciable extent in the bonding in these species. The energy-level diagram for the molecular orbitals created from the $2s$ and $2p$ orbitals is shown in Figure B.4a.[1] Let's see how this energy-level diagram can be used to account for the bonding in the molecules N_2, O_2, and F_2.

Nitrogen is in Group VA and, therefore, each nitrogen atom contributes five electrons to the N_2 molecule from its valence shell. This means that we must place ten electrons into our set of molecular orbitals. As shown in Figure B.4b, two electrons enter the σ_{2s}, two go into the σ_{2s}^*, two more into the σ_{2p_z} and, finally, two into each of the bonding π orbitals, π_{2p_x} and π_{2p_y}. As before, the two σ_{2s}^* antibonding electrons cancel the effect of the σ_{2s} bonding electrons, leaving us with a net total of six bonding electrons (two each in the σ_{2p_z}, π_{2p_x}, and π_{2p_y} orbitals). If, as usual, we take two electrons to represent a "bond," we find that N_2 is held together by a triple bond composed of one σ and two π bonds. As with H_2, we arrive at the same resultant description of the bonding in N_2 with both the valence bond and the molecular orbital theories.

The real mark of success for molecular orbital theory is seen in its description of the O_2 molecule. This species is found experimentally to be paramagnetic with two

[1] The relative energies of the σ_{2p_z} and the π_{2p_x}, π_{2p_y} set actually shift about somewhat as we cross the second period. This diagram will give qualitatively correct results for any species you will encounter here.

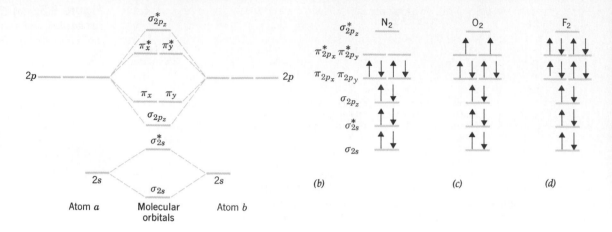

Figure B.4. (a) *The energies of molecular orbitals formed from atomic orbitals having* $n = 2$ *in diatomic molecules.* (b–d) *Molecular orbital electron configurations of* N_2, O_2, *and* F_2.

A dot structure such as

$$\cdot \ddot{O} \!:\! \ddot{O} \cdot$$

has the correct number of unpaired electrons, but it is unsatisfactory because there is only an O—O single bond.

unpaired electrons. In addition, its bond length and bond energy suggest that there is a double bond between the two oxygen atoms. An attempt to derive a valance bond picture for O_2, however, gives us

$$\ddot{O} \!::\! \ddot{O}$$

where, to satisfy the octet rule and give a double bond, all of the electrons appear in pairs.

The molecular orbital description of O_2 is seen in Figure B.4c. The first 10 of the 12 valence electrons populate all the same molecular orbitals as in N_2. The final two electrons must then be placed in the $\pi^*_{2p_x}$ and $\pi^*_{2p_y}$ antibonding orbitals. However, because these two orbitals are of the same energy, the electrons spread themselves out with their spins in the same direction. These two antibonding π electrons cancel the effects of two of the π-bonding electrons, so in the final analysis we see that O_2 is held together by a *net* double bond (one σ and one net π bond). Also notice that the molecule is predicted to have two unpaired electrons, in precise agreement with experimental evidence.

Finally, with F_2 (which contains two more electrons than O_2), we find that the two π^* antibonding orbitals are filled (Figure B.4d). This leaves one net single bond, and once again the valence bond and molecular orbital theories give the same result.

The success of the molecular orbital theory is not restricted merely to diatomic molecules. In more complex molecules and polyatomic ions, however, the energy-level diagrams are much more difficult to predict, so we will not attempt to extend the theory any further in this direction.

Delocalized molecular orbitals

A useful concept in molecular orbital theory is the idea that molecular orbitals (and therefore, chemical bonds) can extend over more than two nuclei. It is this aspect in particular that sets molecular orbital theory apart from the valence bond theory, in which only two different atoms are permitted to share a pair of electrons. It also permits the molecular orbital theory to avoid the concept of resonance.

One of the most important examples of this is the benzene molecule, C_6H_6, which consists of six carbon atoms arranged in a ring as shown in the skeletal structure below.

Valence bond theory treats this molecule as a resonance hybrid of the two structures shown below.

These are usually drawn as

It is understood that at each apex of the hexagon, there is a carbon that is also bonded to a hydrogen.

   If we examine any one carbon atom in the benzene ring, we see that it is bonded to three atoms: two carbons and a hydrogen. This means that it uses sp^2 hybrid orbitals, and it also means that at each carbon atom in the ring there is an unhybridized p orbital perpendicular to the plane of the ring. This is illustrated in Figure B.5a. According to molecular orbital theory, each of these p orbitals overlaps with *two* other p orbitals — one on each of the two neighboring carbon atoms. This gives one large circular π-type bond that extends around the six-member ring, with part of the bond below the plane of the ring and the other part above the plane of the ring as illustrated in Figure B.5b. In effect, the ring is sandwiched between two donut-shaped electron clouds.

   A molecular orbital that extends over three or more nuclei, such as the large

The benzene molecule is often represented by the symbol

The circle stands for the delocalized π-electron system in the molecule.

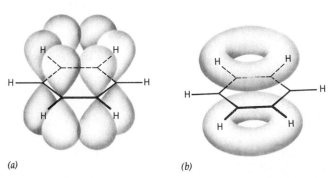

(a)   (b)

Figure B.5. (a) *At each carbon in the benzene ring there is a* p *orbital that is perpendicular to the plane of the ring.* (b) *Overlap of the unhybridized* p *orbitals produces a donut-shaped molecular orbital. The ring is sandwiched between the top and bottom halves of the molecular orbital.*

π-electron cloud in the benzene molecule, is said to be a **delocalized** molecular orbital, meaning that the electron pair is not *localized* between two atoms. By permitting delocalized bonds, molecular orbital theory is able to provide a single electronic picture of the bonding, even in molecules such as benzene or the NO_3^- ion. As a result, there is no need in molecular orbital theory to resort to the somewhat clumsy concept of resonance.

REVIEW QUESTIONS

(Questions whose numbers are in blue have their answers in Appendix D.)

B.1 From the point of view of molecular orbital theory, in what way is the electronic structure of an atom similar to the electronic structure of a molecule? In what way do they differ?

B.2 Sketch the shapes of the σ_{1s} and σ_{1s}^* molecular orbitals. How do their energies compare? Why is the σ_{1s}^* molecular orbital said to be antibonding?

B.3 Sketch the shapes of the σ_{2p_z} and $\sigma_{2p_z}^*$ molecular orbitals. Sketch the shapes of the π_{2p_x} and π_{2p_y} molecular orbitals.

B.4 Describe the bonding in the N_2 molecule according to molecular orbital theory.

B.5 Predict the relative stabilities of the species N_2^+, N_2, and N_2^-. What is the net bond order in each of them? From our discussion in Chapter 8, how would you expect the bond lengths in these species to compare?

B.6 Predict the relative stabilities of the species O_2^+, O_2, and O_2^-. On the basis of the relationship between bond order and bond properties, which we discussed in Chapter 8, predict the variations of bond lengths and bond vibrational frequencies for these species.

B.7 How would you expect the bond order, bond energy, bond length, and bond vibrational frequency of the peroxide ion, O_2^{2-}, to compare to those of O_2?

B.8 Use Figure B.4a to draw molecular orbital energy-level diagrams for Li_2, Be_2, B_2, and C_2. Which of these molecules should not exist, which should be paramagnetic?

B.9 What can you predict about the stabilities of the species in Question B.8 when one electron is (a) removed from each, (b) added to each?

B.10 How does molecular orbital theory avoid the concept of resonance?

B.11 The species H_2^+ and He_2^+ have been observed. Use molecular orbital theory to account for their existence.

B.12 Give the valence bond and molecular orbital descriptions for the following species, which can be drawn as two or more resonance structures: (a) SO_2, (b) NO_3^-, and (c)

$$H-C \underset{O^-}{\overset{O}{\diagup\hspace{-0.3em}\diagdown}}$$

TABLES OF
SELECTED DATA

The following pages contain tables of selected data, compiled from a variety of sources, that you may find helpful in this and other chemistry courses that you may take. Another useful source of chemical and physical data is the *Handbook of Chemistry and Physics* edited by Robert C. Weast and published by CRC Press. A copy of a recent edition is no doubt available in your school library.

Table C.1. Thermodynamic data for selected elements, compounds, and ions (25 °C)

Substance	ΔH_f° (kJ/mol)	S° (J/mol K)	ΔG_f° (kJ/mol)
Aluminum			
$Al(s)$	0	28.3	0
$Al^{3+}(aq)$	-524.7	—	-481.2
$AlCl_3(s)$	-704	110.7	-629
$Al_2O_3(s)$	-1676	51.0	-1577
$Al_2(SO_4)_3(s)$	-3441	239	-3100
Arsenic			
$As(s)$	0	35.1	0
$AsH_3(g)$	$+66.4$	223	$+68.9$
$As_4O_6(s)$	-1314	214	-1153
$As_2O_5(s)$	-925	105	-782
$H_3AsO_3(aq)$	-742.2	—	—
$H_3AsO_4(aq)$	-902.5	—	—
Barium			
$Ba(s)$	0	66.9	0
$Ba^{2+}(aq)$	-537.6	9.6	-560.8
$BaCO_3(s)$	-1219	112	-1139
$BaCrO_4(s)$	-1428.0	—	—
$BaCl_2(s)$	-860.2	125	-810.8

Table C.1. Continued

Substance	ΔH_f° (kJ/mol)	S° (J/mol K)	ΔG_f° (kJ/mol)
BaO(s)	−553.5	70.4	−525.1
Ba(OH)$_2$(s)	−998.22	95	−875.3
Ba(NO$_3$)$_2$(s)	−992	214	−795
BaSO$_4$(s)	−1465	132	−1353
Beryllium			
Be(s)	0	9.50	0
BeCl$_2$(s)	−468.6	89.9	−426.3
BeO(s)	−611	14	−582
Bismuth			
Bi(s)	0	56.9	0
BiCl$_3$(s)	−379	177	−315
Bi$_2$O$_3$(s)	−576	151	−497
Boron			
B(s)	0	5.87	0
BCl$_3$(g)	−404	290	−389
B$_2$H$_6$(g)	+36	232	+87
B$_2$O$_3$(s)	−1273	53.8	−1194
B(OH)$_3$(s)	−1094	88.8	−969
Bromine			
Br$_2$(l)	0	152.2	0
Br$_2$(g)	+30.9	245.4	+3.11
HBr(g)	−36	198.5	−53.1
Br$^-$(aq)	−121.55	82.4	−103.96
Cadmium			
Cd(s)	0	51.8	0
Cd^{2+}(aq)	−75.90	−73.2	−77.61
CdCl$_2$(s)	−392	115	−344
CdO(s)	−258.2	54.8	−228.4
CdS(s)	−162	64.9	−156
CdSO$_4$(s)	−933.5	123	−822.6
Calcium			
Ca(s)	0	41.4	0
Ca^{2+}(aq)	−542.83	−53.1	−553.58
CaCO$_3$(s)	−1207	92.9	−1128.8
CaF$_2$(s)	−741	80.3	−1166
CaCl$_2$(s)	−795.8	104.6	−748.1
CaBr$_2$(s)	−682.8	130	−663.6
CaI$_2$(s)	−535.9	143	—
CaO(s)	−635.5	39.8	−604.2
Ca(OH)$_2$(s)	−986.6	76.1	−896.6
Ca$_3$(PO$_4$)$_2$(s)	−4119	241	−3852
CaSO$_3$(s)	−1156	—	—
CaSO$_4$(s)	−1433	107	−1320.3
CaSO$_4 \cdot \frac{1}{2}$H$_2$O(s)	−1573	131	−1435.2
CaSO$_4 \cdot 2$H$_2$O(s)	−2020	194.0	−1795.7
Carbon			
C(s, graphite)	0	5.69	0
C(s, diamond)	+1.88	2.4	+2.9
CCl$_4$(l)	−134	214.4	−65.3
CO(g)	−110	197.9	−137.3
CO$_2$(g)	−394	213.6	−395
CO$_2$(aq)	−413.8	117.5	−386.02

Table C.1. *Continued*

Substance	ΔH_f° (kJ/mol)	S° (J/mol K)	ΔG_f° (kJ/mol)
$H_2CO_3(aq)$	-699.65	187.4	-623.16
$HCO_3^-(aq)$	-691.99	91.2	-586.77
$CO_3^{2-}(aq)$	-677.14	-56.9	-527.81
$CS_2(l)$	$+89.5$	151.3	$+65.3$
$CS_2(g)$	$+117$	237.7	$+67.2$
$HCN(g)$	$+135.1$	201.8	$+124.7$
$CN^-(aq)$	$+150.6$	94.1	$+172.4$
$CH_4(g)$	-74.9	186.2	-50.6
$C_2H_2(g)$	$+227$	200.8	$+209$
$C_2H_4(g)$	$+51.9$	219.8	$+68.2$
$C_2H_6(g)$	-84.5	229.5	-32.9
$C_3H_8(g)$	-104	269.9	-23
$C_4H_{10}(g)$	-126	310.2	-17.0
$C_6H_6(l)$	$+49.0$	173.3	$+124.3$
$CH_3OH(l)$	-238	126.8	-166.2
$C_2H_5OH(l)$	-278	160.7	-174.8
$HCHO_2(g)$ (formic acid)	-363	251	$+335$
$HC_2H_3O_2(l)$ (acetic acid)	-487.0	159.8	-392.5
$CH_2O(g)$ (formaldehyde)	-108.6	218.8	-102.5
$CH_3CHO(g)$ (acetaldehyde)	-167	250	-129
$(CH_3)_2CO(l)$ (acetone)	-248.1	200.4	-155.4
$C_6H_5CO_2H(s)$ (benzoic acid)	-385.1	167.6	-245.3
$CO(NH_2)_2(s)$ (urea)	-333.5	104.6	-197.2
$CO(NH_2)_2(aq)$	-319.2	173.8	-203.8
$CH_2(NH_2)CO_2H(s)$ (glycine)	-532.9	103.5	-373.4
Chlorine			
$Cl_2(g)$	0	223.0	0
$Cl^-(aq)$	-167.2	56.5	-131.2
$HCl(g)$	-92.5	186.7	-95.4
$HCl(aq)$	-167.2	56.5	-131.2
$HClO(aq)$	-131.3	106.8	-80.21
Chromium			
$Cr(s)$	0	23.8	0
$Cr^{3+}(aq)$	-232	—	—
$CrCl_2(s)$	-326	115	-282
$CrCl_3(s)$	-563.2	126	-493.7
$Cr_2O_3(s)$	-1141	81.2	-1059
$CrO_3(s)$	-585.8	72.0	-506.2
$(NH_4)_2Cr_2O_7(s)$	-1807	—	—
$K_2Cr_2O_7(s)$	-2033.01	292.5	-1864
Cobalt			
$Co(s)$	0	30.0	0
$Co^{2+}(aq)$	-59.4	-110	-53.6
$CoCl_2(s)$	-325.5	106	-282.4
$Co(NO_3)_2(s)$	-422.2	192	-230.5

Table C.1. *Continued*

Substance	ΔH_f° (kJ/mol)	S° (J/mol K)	ΔG_f° (kJ/mol)
CoO(s)	−237.9	53.0	−214.2
CoS(s)	−80.8	67.4	−82.8
Copper			
Cu(s)	0	33.15	0
$Cu^{2+}(aq)$	+64.77	−99.6	+65.49
CuCl(s)	−137.2	86.2	−119.87
$CuCl_2(s)$	−172	119	−131
$Cu_2O(s)$	−168.6	93.1	−146.0
CuO(s)	−155	42.6	−127
$Cu_2S(s)$	−79.5	121	−86.2
CuS(s)	−53.1	66.5	−53.6
$CuSO_4(s)$	−771.4	109	−661.8
$CuSO_4 \cdot 5H_2O(s)$	−2279.7	300.4	−1879.7
Fluorine			
$F_2(g)$	0	202.7	0
$F^-(aq)$	−332.6	−13.8	−278.8
HF(g)	−271	173.5	−273
Gold			
Au(s)	0	47.7	0
$Au_2O_3(s)$	+80.8	125	+163
$AuCl_3(s)$	−118	148	−48.5
Hydrogen			
$H_2(g)$	0	130.6	0
$H_2O(l)$	−286	69.96	−237.2
$H_2O(g)$	−242	188.7	−228
$H_2O_2(l)$	−187.8	109.6	−120.3
$H_2Se(g)$	+76	219	+62.3
$H_2Te(g)$	+154	234	+138
Iodine			
$I_2(s)$	0	116.1	0
$I_2(g)$	+62.4	260.7	+19.3
HI(g)	+26	206	+1.30
Iron			
Fe(s)	0	27.3	0
$Fe^{2+}(aq)$	−89.1	−137.7	−78.9
$Fe^{3+}(aq)$	−48.5	−315.9	−4.7
$Fe_2O_3(s)$	−822.2	87.4	−741.0
$Fe_3O_4(s)$	−1118.4	146.4	−1015.4
FeS(s)	−100.0	60.3	−100.4
$FeS_2(s)$	−178.2	52.9	−166.9
Lead			
Pb(s)	0	64.8	0
$Pb^{2+}(aq)$	−1.7	10.5	−24.4
$PbCl_2(s)$	−359.4	136	−314.1
PbO(s)	−217.3	68.7	−187.9
$PbO_2(s)$	−277	68.6	−219
$Pb(OH)_2(s)$	−515.9	88	−420.9
PbS(s)	−100	91.2	−98.7
$PbSO_4(s)$	−920.1	149	−811.3
Lithium			
Li(s)	0	29.1	0
$Li^+(aq)$	−278.6	10.3	—

Table C.1. *Continued*

Substance	ΔH_f° (kJ/mol)	S° (J/mol K)	ΔG_f° (kJ/mol)
LiF(s)	−611.7	35.7	−583.3
LiCl(s)	−408.8	59.33	−384.4
LiBr(s)	−350.3	66.9	−338.87
Li$_2$O(s)	−596.5	37.9	560.5
Li$_3$N(s)	−199	37.7	−155.4
Magnesium			
Mg(s)	0	32.5	0
Mg^{2+}(aq)	−466.9	−138.1	−454.8
MgCO$_3$(s)	−1113	65.7	−1029
MgF$_2$(s)	−1113	79.9	−1056
MgCl$_2$(s)	−641.8	89.5	−592.5
MgCl$_2 \cdot$2H$_2$O(s)	−1280	180	−1118
Mg$_3$N$_2$(s)	−463.2	87.9	−411
MgO(s)	−601.7	26.9	−569.4
Mg(OH)$_2$(s)	−924.7	63.1	−833.9
Manganese			
Mn(s)	0	32.0	0
Mn^{2+}(aq)	−223	−74.9	−228
MnO$_4^-$(aq)	−542.7	191	−449.4
KMnO$_4$(s)	−813.4	171.71	−737.6
MnO(s)	−385	60.2	−363
Mn$_2$O$_3$(s)	−959.8	110	−882.0
MnO$_2$(s)	−520.9	53.1	−466.1
Mn$_3$O$_4$(s)	−1387	149	−1280
MnSO$_4$(s)	−1064	112	−956
Mercury			
Hg(l)	0	76.1	0
Hg(g)	+61.32	175	+31.8
Hg$_2$Cl$_2$(s)	−265.2	192.5	−210.8
HgCl$_2$(s)	−224.3	146.0	−178.6
HgO(s)	−90.83	70.3	−58.54
HgS(s,red)	−58.2	82.4	−50.6
Nickel			
Ni(s)	0	30	0
NiCl$_2$(s)	−305	97.5	−259
NiO(s)	−244	38	−216
NiO$_2$(s)	—	—	−199
NiSO$_4$(s)	−891.2	77.8	−773.6
NiCO$_3$(s)	−664.0	91.6	−615.0
Ni(CO)$_4$(g)	−220	399	−567.4
Nitrogen			
N$_2$(g)	0	191.5	0
NH$_3$(g)	−46.0	192.5	−16.7
NH$_4^+$(aq)	−132.5	113	−79.37
N$_2$H$_4$(l)	+50.6	121.2	+149.4
NH$_4$Cl(s)	−314.4	94.6	−202.9
NO(g)	+90.4	210.6	+86.8
NO$_2$(g)	+34	240.5	+51.9
N$_2$O(g)	+81.5	220.0	+103.6
N$_2$O$_4$(g)	+9.16	304.3	+97.9
N$_2$O$_5$(g)	+11	356	+115
HNO$_3$(l)	−174.1	155.6	−79.9
NO$_3^-$(aq)	−205.0	146.4	−108.74

Table C.1. *Continued*

Substance	ΔH_f° (kJ/mol)	S° (J/mol K)	ΔG_f° (kJ/mol)
Oxygen			
$O_2(g)$	0	205.0	0
$O_3(g)$	+143	238.8	+163
$OH^-(aq)$	−230.0	−10.75	−157.24
Phosphorus			
P(s,white)	0	41.09	0
$P_4(g)$	+314.6	163.2	+278.3
$PCl_3(g)$	−287.0	311.8	−267.8
$PCl_5(g)$	−374.9	364.6	−305.0
$PH_3(g)$	+5.4	210.2	+12.9
$P_4O_6(s)$	−1640	—	—
$P_4O_{10}(s)$	−2984	228.9	−2698
$H_3PO_4(s)$	−1279	110.5	−1119
Potassium			
K(s)	0	64.18	0
$K^+(aq)$	−252.4	102.5	−283.3
KF(s)	−567.3	66.6	−537.8
KCl(s)	−436.8	82.59	−409.1
KBr(s)	−393.8	95.9	−380.7
KI(s)	−327.9	106.3	−324.9
KOH(s)	−424.8	78.9	−379.1
$K_2O(s)$	−361	98.3	−322
$K_2SO_4(s)$	−1433.7	176	−1316.4
Silicon			
Si(s)	0	19	0
$SiH_4(g)$	+33	205	+52.3
$SiO_2(s,alpha)$	−910.0	41.8	−856
Silver			
Ag(s)	0	42.55	0
$Ag^+(aq)$	+105.58	72.68	+77.11
AgCl(s)	−127.1	96.2	−109.8
AgBr(s)	−100.4	107.1	−96.9
$AgNO_3(s)$	−124	141	−32
$Ag_2O(s)$	−31.1	121.3	−11.2
Sodium			
Na(s)	0	51.0	0
$Na^+(aq)$	−240.12	59.0	−261.91
NaF(s)	−571	51.5	−545
NaCl(s)	−413	72.8	−384.0
NaBr(s)	−360	83.7	−349
NaI(s)	−288	91.2	−286
$NaHCO_3(s)$	−947.7	155	−851.9
$Na_2CO_3(s)$	−1131	136	−1048
$Na_2O_2(s)$	−504.6	94.6	−447.7
$Na_2O(s)$	−510	72.8	−376
NaOH(s)	−426.8	64.5	−379.5
$Na_2SO_4(s)$	−1384.49	149.6	−1270.2
Sulfur			
S(s,rhombic)	0	31.8	0
$SO_2(g)$	−297	248	−300
$SO_3(g)$	−396	256	−370
$H_2S(g)$	−20.6	206	−33.6

Table C.1. *Continued*

Substance	ΔH_f° (kJ/mol)	S° (J/mol K)	ΔG_f° (kJ/mol)
$H_2SO_4(l)$	-813.8	157	-689.9
$H_2SO_4(aq)$	-909.3	20.1	-744.5
$SF_6(g)$	-1209	292	-1105
Tin			
$Sn(s,white)$	0	51.6	0
$Sn^{2+}(aq)$	-8.8	-17	-27.2
$SnCl_4(l)$	-511.3	258.6	-440.2
$SnO(s)$	-285.8	56.5	-256.9
$SnO_2(s)$	-580.7	52.3	-519.6
Zinc			
$Zn(s)$	0	41.8	0
$Zn^{2+}(aq)$	-153.9	-112.1	-147.06
$ZnCl_2(s)$	-415.1	111	-369.4
$ZnO(s)$	-348.3	43.5	-318.3
$ZnS(s)$	-205.6	57.7	-201.3
$ZnSO_4(s)$	-982.8	120	-874.5

Table C.2. Heats of formation of gaseous atoms from elements in their standard states

Element	ΔH_f° (kJ mol^{-1})^a	Element	ΔH_f° (kJ mol^{-1})
Group IA		Group IVA	
H	217.89	C	715
Li	161.5	Si	454
Na	108.2	Group VA	
K	89.62	N	472.68
Rb	82.0	P	332.2
Cs	78.2	Group VIA	
Group IIA		O	249.17
Be	327	S	276.98
Mg	146.4	Group VIIA	
Ca	178.2	F	79.14
Sr	163	Cl	121.47
Ba	177	Br	112.38
Group IIIA		I	107.48
B	555		
Al	329.7		

^{a} All values in this table are positive because formation of the gaseous atoms from the elements is endothermic. It involves bond breaking.

Table C.3. Average bond energies

Bond	Bond Energy (kJ mol^{-1})	Bond	Bond Energy (kJ mol$^{-1)}$
C—C	348	C—Br	276
C=C	607	C—I	238
C≡C	833	H—H	436

Table C.3. *Continued*

Bond	Bond Energy (kJ mol^{-1})	Bond	Bond Energy (kJ mol^{-1})
C—H	415	H—F	563
C—N	292	H—Cl	432
C=N	619	H—Br	366
C≡N	879	H—I	299
C—O	356	H—N	391
C=O	724	H—O	463
C—F	484	H—S	338
C—Cl	338	H—Si	376

Table C.4. Vapor pressure of water as a function of temperature

Temp. (°C)	Vapor Pressure (torr)	Temp. (°C)	Vapor Pressure (torr)	Temp. (°C)	Vapor Pressure (torr)	Temp. (°C)	Vapor Pressure (torr)
0	4.58	26	25.2	51	97.2	76	301.4
1	4.93	27	26.7	52	102.1	77	314.1
2	5.29	28	28.3	53	107.2	78	327.3
3	5.68	29	30.0	54	112.5	79	341.0
4	6.10	30	31.8	55	118.0	80	355.1
5	6.54	31	33.7	56	123.8	81	369.7
6	7.01	32	35.7	57	129.8	82	384.9
7	7.51	33	37.7	58	136.1	83	400.6
8	8.04	34	39.9	59	142.6	84	416.8
9	8.61	35	41.2	60	149.4	85	433.6
10	9.21	36	44.6	61	156.4	86	450.9
11	9.84	37	47.1	62	163.8	87	468.7
12	10.5	38	49.7	63	171.4	88	487.1
13	11.2	39	52.4	64	179.3	89	506.1
14	12.0	40	55.3	65	187.5	90	525.8
15	12.8	41	58.3	66	196.1	91	546.0
16	13.6	42	61.5	67	205.0	92	567.0
17	14.5	43	64.8	68	214.2	93	588.6
18	15.5	44	68.3	69	223.7	94	610.9
19	16.5	45	71.9	70	233.7	95	633.9
20	17.5	46	75.6	71	243.9	96	657.6
21	18.6	47	79.6	72	254.6	97	682.1
22	19.8	48	83.7	73	265.7	98	707.3
23	21.1	49	88.0	74	277.2	99	733.2
24	22.4	50	92.5	75	289.1	100	760.0
25	23.8					101	787.6

Table C.5. Solubility product constants

Salt	Solubility Equilibrium	K_{sp}
Fluorides		
MgF_2	$MgF_2(s) \rightleftharpoons Mg^{2+}(aq) + 2F^-(aq)$	7.3×10^{-9}
CaF_2	$CaF_2(s) \rightleftharpoons Ca^{2+}(aq) + 2F^-(aq)$	1.7×10^{-10}
SrF_2	$SrF_2(s) \rightleftharpoons Sr^{2+}(aq) + 2F^-(aq)$	2.9×10^{-9}
BaF_2	$BaF_2(s) \rightleftharpoons Ba^{2+}(aq) + 2F^-(aq)$	1.7×10^{-6}
LiF	$LiF(s) \rightleftharpoons Li^+(aq) + F^-(aq)$	1.7×10^{-3}
PbF_2	$PbF_2(s) \rightleftharpoons Pb^{2+}(aq) + 2F^-(aq)$	3.2×10^{-8}
Chlorides		
$CuCl$	$CuCl(s) \rightleftharpoons Cu^+(aq) + Cl^-(aq)$	1.9×10^{-7}
$AgCl$	$AgCl(s) \rightleftharpoons Ag^+(aq) + Cl^-(aq)$	1.7×10^{-10}
Hg_2Cl_2	$Hg_2Cl_2(s) \rightleftharpoons Hg_2^{2+}(aq) + 2Cl^-(aq)$	2×10^{-18}
$TlCl$	$TlCl(s) \rightleftharpoons Tl^+(aq) + Cl^-(aq)$	1.8×10^{-4}
$PbCl_2$	$PbCl_2(s) \rightleftharpoons Pb^{2+}(aq) + 2Cl^-(aq)$	1.6×10^{-5}
$AuCl_3$	$AuCl_3(s) \rightleftharpoons Au^{3+}(aq) + 3Cl^-(aq)$	3.2×10^{-25}
Bromides		
$CuBr$	$CuBr(s) \rightleftharpoons Cu^+(aq) + Br^-(aq)$	5×10^{-9}
$AgBr$	$AgBr(s) \rightleftharpoons Ag^+(aq) + Br^-(aq)$	5.0×10^{-15}
Hg_2Br_2	$Hg_2Br_2(s) \rightleftharpoons Hg_2^{2+}(aq) + 2Br^-(aq)$	5.6×10^{-23}
$HgBr_2$	$HgBr_2(s) \rightleftharpoons Hg^{2+}(aq) + 2Br^-(aq)$	1.3×10^{-19}
$PbBr_2$	$PbBr_2(s) \rightleftharpoons Pb^{2+}(aq) + 2Br^-(aq)$	2.1×10^{-6}
Iodides		
CuI	$CuI(s) \rightleftharpoons Cu^+(aq) + I^-(aq)$	1×10^{-12}
AgI	$AgI(s) \rightleftharpoons Ag^+(aq) + I^-(aq)$	8.5×10^{-17}
Hg_2I_2	$Hg_2I_2(s) \rightleftharpoons Hg_2^{2+}(aq) + 2I^-(aq)$	4.7×10^{-29}
HgI_2	$HgI_2(s) \rightleftharpoons Hg^{2+}(aq) + 2I^-(aq)$	1.1×10^{-28}
PbI_2	$PbI_2(s) \rightleftharpoons Pb^{2+}(aq) + 2I^-(aq)$	1.4×10^{-8}
Hydroxides		
$Mg(OH)_2$	$Mg(OH)_2(s) \rightleftharpoons Mg^{2+}(aq) + 2OH^-(aq)$	7.1×10^{-12}
$Ca(OH)_2$	$Ca(OH)_2(s) \rightleftharpoons Ca^{2+}(aq) + 2OH^-(aq)$	6.5×10^{-6}
$Mn(OH)_2$	$Mn(OH)_2(s) \rightleftharpoons Mn^{2+}(aq) + 2OH^-(aq)$	1.2×10^{-11}
$Fe(OH)_2$	$Fe(OH)_2(s) \rightleftharpoons Fe^{2+}(aq) + 2OH^-(aq)$	2×10^{-15}
$Fe(OH)_3$	$Fe(OH)_3(s) \rightleftharpoons Fe^{3+}(aq) + 3OH^-(aq)$	1.1×10^{-36}
$Co(OH)_2$	$Co(OH)_2(s) \rightleftharpoons Co^{2+}(aq) + 2OH^-(aq)$	1×10^{-15}
$Co(OH)_3$	$Co(OH)_3(s) \rightleftharpoons Co^{3+}(aq) + 3OH^-(aq)$	3×10^{-45}
$Ni(OH)_2$	$Ni(OH)_2(s) \rightleftharpoons Ni^{2+}(aq) + 2OH^-(aq)$	1.6×10^{-14}
$Cu(OH)_2$	$Cu(OH)_2(s) \rightleftharpoons Cu^{2+}(aq) + 2OH^-(aq)$	4.8×10^{-20}
$V(OH)_3$	$V(OH)_3(s) \rightleftharpoons V^{3+}(aq) + 3OH^-(aq)$	4×10^{-35}
$Cr(OH)_3$	$Cr(OH)_3(s) \rightleftharpoons Cr^{3+}(aq) + 3OH^-(aq)$	2×10^{-30}
Ag_2O	$Ag_2O(s) + H_2O \rightleftharpoons 2Ag^+(aq) + 2OH^-(aq)$	1.9×10^{-8}
$Zn(OH)_2$	$Zn(OH)_2(s) \rightleftharpoons Zn^{2+}(aq) + 2OH^-(aq)$	4.5×10^{-17}
$Cd(OH)_2$	$Cd(OH)_2(s) \rightleftharpoons Cd^{2+}(aq) + 2OH^-(aq)$	5.0×10^{-15}
$Al(OH)_3$	$Al(OH)_3(s) \rightleftharpoons Al^{3+}(aq) + 3OH^-(aq)$	2×10^{-33}
Cyanides		
$AgCN$	$AgCN(s) \rightleftharpoons Ag^+(aq) + CN^-(aq)$	1.6×10^{-14}
$Zn(CN)_2$	$Zn(CN)_2(s) \rightleftharpoons Zn^{2+}(aq) + 2CN^-(aq)$	3×10^{-16}
Sulfites		
$CaSO_3$	$CaSO_3(s) \rightleftharpoons Ca^{2+}(aq) + SO_3^{2-}(aq)$	3×10^{-7}
Ag_2SO_3	$Ag_2SO_3(s) \rightleftharpoons 2Ag^+(aq) + SO_3^{2-}(aq)$	1.5×10^{-14}
$BaSO_3$	$BaSO_3(s) \rightleftharpoons Ba^{2+}(aq) + SO_3^{2-}(aq)$	8×10^{-7}

Table C.5. *Continued*

Salt	Solubility Equilibrium	K_{sp}
Sulfates		
$CaSO_4$	$CaSO_4(s) \rightleftharpoons Ca^{2+}(aq) + SO_4^{2-}(aq)$	2×10^{-4}
$SrSO_4$	$SrSO_4(s) \rightleftharpoons Sr^{2+}(aq) + SO_4^{2-}(aq)$	3.2×10^{-7}
$BaSO_4$	$BaSO_4(s) \rightleftharpoons Ba^{2+}(aq) + SO_4^{2-}(aq)$	1.5×10^{-9}
$RaSO_4$	$RaSO_4(s) \rightleftharpoons Ra^{2+}(aq) + SO_4^{2-}(aq)$	4.3×10^{-11}
Ag_2SO_4	$Ag_2SO_4(s) \rightleftharpoons 2Ag^+(aq) + SO_4^{2-}(aq)$	1.5×10^{-5}
Hg_2SO_4	$Hg_2SO_4(s) \rightleftharpoons Hg_2^{2+}(aq) + SO_4^{2-}(aq)$	7.4×10^{-7}
$PbSO_4$	$PbSO_4(s) \rightleftharpoons Pb^{2+}(aq) + SO_4^{2-}(aq)$	6.3×10^{-7}
Chromates		
$BaCrO_4$	$BaCrO_4(s) \rightleftharpoons Ba^{2+}(aq) + CrO_4^{2-}(aq)$	2.4×10^{-10}
$CuCrO_4$	$CuCrO_4(s) \rightleftharpoons Cu^{2+}(aq) + CrO_4^{2-}(aq)$	3.6×10^{-6}
Ag_2CrO_4	$Ag_2CrO_4(s) \rightleftharpoons 2Ag^+(aq) + CrO_4^{2-}(aq)$	1.9×10^{-12}
Hg_2CrO_4	$Hg_2CrO_4(s) \rightleftharpoons Hg_2^{2+}(aq) + CrO_4^{2-}(aq)$	2.0×10^{-9}
$CaCrO_4$	$CaCrO_4(s) \rightleftharpoons Ca^{2+}(aq) + CrO_4^{2-}(aq)$	1.0×10^{-4}
$PbCrO_4$	$PbCrO_4(s) \rightleftharpoons Pb^{2+}(aq) + CrO_4^{2-}(aq)$	1.8×10^{-14}
Carbonates		
$MgCO_3$	$MgCO_3(s) \rightleftharpoons Mg^{2+}(aq) + CO_3^{2-}(aq)$	3.5×10^{-8}
$CaCO_3$	$CaCO_3(s) \rightleftharpoons Ca^{2+}(aq) + CO_3^{2-}(aq)$	9×10^{-9}
$SrCO_3$	$SrCO_3(s) \rightleftharpoons Sr^{2+}(aq) + CO_3^{2-}(aq)$	9.3×10^{-10}
$BaCO_3$	$BaCO_3(s) \rightleftharpoons Ba^{2+}(aq) + CO_3^{2-}(aq)$	8.9×10^{-9}
$MnCO_3$	$MnCO_3(s) \rightleftharpoons Mn^{2+}(aq) + CO_3^{2-}(aq)$	5.0×10^{-10}
$FeCO_3$	$FeCO_3(s) \rightleftharpoons Fe^{2+}(aq) + CO_3^{2-}(aq)$	2.1×10^{-11}
$CoCO_3$	$CoCO_3(s) \rightleftharpoons Co^{2+}(aq) + CO_3^{2-}(aq)$	1.0×10^{-10}
$NiCO_3$	$NiCO_3(s) \rightleftharpoons Ni^{2+}(aq) + CO_3^{2-}(aq)$	1.3×10^{-7}
$CuCO_3$	$CuCO_3(s) \rightleftharpoons Cu^{2+}(aq) + CO_3^{2-}(aq)$	2.3×10^{-10}
Ag_2CO_3	$Ag_2CO_3(s) \rightleftharpoons 2Ag^+(aq) + CO_3^{2-}(aq)$	8.2×10^{-12}
Hg_2CO_3	$Hg_2CO_3(s) \rightleftharpoons Hg_2^{2+}(aq) + CO_3^{2-}(aq)$	8.9×10^{-17}
$ZnCO_3$	$ZnCO_3(s) \rightleftharpoons Zn^{2+}(aq) + CO_3^{2-}(aq)$	1.0×10^{-10}
$CdCO_3$	$CdCO_3(s) \rightleftharpoons Cd^{2+}(aq) + CO_3^{2-}(aq)$	1.8×10^{-14}
$PbCO_3$	$PbCO_3(s) \rightleftharpoons Pb^{2+}(aq) + CO_3^{2-}(aq)$	7.4×10^{-14}
Phosphates		
$Mg_3(PO_4)_2$	$Mg_3(PO_4)_2(s) \rightleftharpoons 3Mg^{2+}(aq) + 2PO_4^{3-}(aq)$	6.3×10^{-26}
$SrHPO_4$	$SrHPO_4(s) \rightleftharpoons Sr^{2+}(aq) + HPO_4^{2-}(aq)$	1.2×10^{-7}
$BaHPO_4$	$BaHPO_4(s) \rightleftharpoons Ba^{2+}(aq) + HPO_4^{2-}(aq)$	4.0×10^{-8}
$LaPO_4$	$LaPO_4(s) \rightleftharpoons La^{3+}(aq) + PO_4^{3-}(aq)$	3.7×10^{-23}
$Fe_3(PO_4)_2$	$Fe_3(PO_4)_2(s) \rightleftharpoons 3Fe^{2+}(aq) + 2PO_4^{3-}(aq)$	1×10^{-36}
Ag_3PO_4	$Ag_3PO_4(s) \rightleftharpoons 3Ag^+(aq) + PO_4^{3-}(aq)$	2.8×10^{-18}
$FePO_4$	$FePO_4(s) \rightleftharpoons Fe^{3+}(aq) + PO_4^{3-}(aq)$	4.0×10^{-27}
$Zn_3(PO_4)_2$	$Zn_3(PO_4)_2(s) \rightleftharpoons 3Zn^{2+}(aq) + 2PO_4^{3-}(aq)$	5×10^{-36}
$Pb_3(PO_4)_2$	$Pb_3(PO_4)_2(s) \rightleftharpoons 3Pb^{2+}(aq) + 2PO_4^{3-}(aq)$	3.0×10^{-44}
$Ba_3(PO_4)_2$	$Ba_3(PO_4)_2(s) \rightleftharpoons 3Ba^{2+}(aq) + 2PO_4^{3-}(aq)$	5.8×10^{-38}
Ferrocyanides		
$Zn_2[Fe(CN)_6]$	$Zn_2[Fe(CN)_6](s) \rightleftharpoons 2Zn^{2+}(aq) + Fe(CN)_6^{4-}(aq)$	2.1×10^{-16}
$Cd_2[Fe(CN)_6]$	$Cd_2[Fe(CN)_6](s) \rightleftharpoons 2Cd^{2+}(aq) + Fe(CN)_6^{4-}(aq)$	4.2×10^{-18}
$Pb_2[Fe(CN)_6]$	$Pb_2[Fe(CN)_6](s) \rightleftharpoons 2Pb^{2+}(aq) + Fe(CN)_6^{4-}(aq)$	9.5×10^{-19}

Table C.6. Formation constants of complexes (25 °C)

Complex Ion Equilibrium	K_{form}
Halide complexes	
$Al^{3+} + 6F^- \rightleftharpoons [AlF_6]^{3-}$	6.7×10^{19}
$Al^{3+} + 4F^- \rightleftharpoons [AlF_4]^-$	2.0×10^8
$Be^{2+} + 4F^- \rightleftharpoons [BeF_4]^{2-}$	1.3×10^{13}
$Sn^{4+} + 6F^- \rightleftharpoons [SnF_6]^{2-}$	1×10^{25}
$Cu^+ + 2Cl^- \rightleftharpoons [CuCl_2]^-$	3×10^5
$Ag^+ + 2Cl^- \rightleftharpoons [AgCl_2]^-$	1.8×10^5
$Pb^{2+} + 4Cl^- \rightleftharpoons [PbCl_4]^{2-}$	2.5×10^{15}
$Zn^{2+} + 4Cl^- \rightleftharpoons [ZnCl_4]^{2-}$	1.6
$Hg^{2+} + 4Cl^- \rightleftharpoons [HgCl_4]^{2-}$	5.0×10^{15}
$Cu^+ + 2Br^- \rightleftharpoons [CuBr_2]^-$	8×10^5
$Ag^+ + 2Br^- \rightleftharpoons [AgBr_2]^-$	1.7×10^7
$Hg^{2+} + 4Br \rightleftharpoons [HgBr_4]^{2-}$	1×10^{21}
$Cu^+ + 2I^- \rightleftharpoons [CuI_2]^-$	8×10^8
$Ag^+ + 2I^- \rightleftharpoons [AgI_2]^-$	1×10^{11}
$Pb^{2+} + 4I^- \rightleftharpoons [PbI_4]^{2-}$	3×10^4
$Hg^{2+} + 4I^- \rightleftharpoons [HgI_4]^{2-}$	1.9×10^{30}
Ammonia complexes	
$Ag^+ + 2NH_3 \rightleftharpoons [Ag(NH_3)_2]^+$	1.7×10^7
$Zn^{2+} + 4NH_3 \rightleftharpoons [Zn(NH_3)_4]^{2+}$	7.8×10^8
$Cu^{2+} + 4NH_3 \rightleftharpoons [Cu(NH_3)_4]^{2+}$	4.8×10^{12}
$Hg^{2+} + 4NH_3 \rightleftharpoons [Hg(NH_3)_4]^{2+}$	1.8×10^{19}
$Co^{2+} + 6NH_3 \rightleftharpoons [Co(NH_3)_6]^{2+}$	7.7×10^4
$Co^{3+} + 6NH_3 \rightleftharpoons [Co(NH_3)_6]^{3+}$	5.0×10^{33}
$Cd^{2+} + 6NH_3 \rightleftharpoons [Cd(NH_3)_6]^{2+}$	2.6×10^5
$Ni^{2+} + 6NH_3 \rightleftharpoons [Ni(NH_3)_6]^{2+}$	5.0×10^8
Cyanide complexes	
$Fe^{2+} + 6CN^- \rightleftharpoons [Fe(CN)_6]^{4-}$	1.0×10^{35}
$Fe^{3+} + 6CN^- \rightleftharpoons [Fe(CN)_6]^{3-}$	9.1×10^{41}
$Ag^+ + 2CN^- \rightleftharpoons [Ag(CN)_2]^-$	5.3×10^{18}
$Cu^+ + 2CN^- \rightleftharpoons [Cu(CN)_2]^-$	1.0×10^{16}
$Cd^{2+} + 4CN^- \rightleftharpoons [Cd(CN)_4]^{2-}$	7.7×10^{16}
$Au^+ + 2CN^- \rightleftharpoons [Au(CN)_2]^-$	2×10^{38}
Complexes with other monodentate ligands	
$Ag^+ + 2CH_3NH_2 \rightleftharpoons [Ag(CH_3NH_2)_2]^+$ methylamine	7.8×10^6
$Cd^{2+} + 4SCN^- \rightleftharpoons [Cd(SCN)_4]^{2-}$ thiocyanate ion	1×10^3
$Cu^{2+} + 2SCN^- \rightleftharpoons [Cu(SCN)_2]$	5.6×10^3
$Fe^{3+} + 3SCN^- \rightleftharpoons [Fe(SCN)_3]$	2×10^6
$Hg^{2+} + 4SCN^- \rightleftharpoons [Hg(SCN)_4]^{2-}$	5.0×10^{21}
$Cu^{2+} + 4OH^- \rightleftharpoons [Cu(OH)_4]^{2-}$	1.3×10^{16}
$Zn^{2+} + 4OH^- \rightleftharpoons [Zn(OH)_4]^{2-}$	2.8×10^{15}
Complexes with bidentate ligands[a]	
$Mn^{2+} + 3\,en \rightleftharpoons [Mn(en)_3]^{2+}$	6.5×10^5
$Fe^{2+} + 3\,en \rightleftharpoons [Fe(en)_3]^{2+}$	5.2×10^9
$Co^{2+} + 3\,en \rightleftharpoons [Co(en)_3]^{2+}$	1.3×10^{14}
$Co^{3+} + 3en \rightleftharpoons [Co(en)_3]^{3+}$	4.8×10^{48}
$Ni^{2+} + 3\,en \rightleftharpoons [Ni(en)_3]^{2+}$	4.1×10^{17}
$Cu^{2+} + 2\,en \rightleftharpoons [Cu(en)_2]^{2+}$	3.5×10^{19}
$Mn^{2+} + 3\,bipy \rightleftharpoons [Mn(bipy)_3]^{2+}$	1×10^6
$Fe^{2+} + 3\,bipy \rightleftharpoons [Fe(bipy)_3]^{2+}$	1.6×10^{17}

Table C.6. *Continued*

Complex Ion Equilibrium	K_{form}
Complexes with bidentate ligands[a]	
$Ni^{2+} + 3\ bipy \rightleftharpoons [Ni(bipy)_3]^{2+}$	3.0×10^{20}
$Co^{2+} + 3\ bipy \rightleftharpoons [Co(bipy)_3]^{2+}$	8×10^{15}
$Mn^{2+} + 3\ phen \rightleftharpoons [Mn(phen)_3]^{2+}$	2×10^{10}
$Fe^{2+} + 3\ phen \rightleftharpoons [Fe(phen)_3]^{2+}$	1×10^{21}
$Co^{2+} + 3\ phen \rightleftharpoons [Co(phen)_3]^{2+}$	6×10^{19}
$Ni^{2+} + 3\ phen \rightleftharpoons [Ni(phen)_3]^{2+}$	2×10^{24}
$Co^{2+} + 3C_2O_4^{2-} \rightleftharpoons [Co(C_2O_4)_3]^{4-}$	4.5×10^{6}
$Fe^{3+} + 3C_2O_4^{2-} \rightleftharpoons [Fe(C_2O_4)_3]^{3-}$	3.3×10^{20}
Complexes of other ligands	
$Zn^{2+} + EDTA^{4-} \rightleftharpoons [Zn(EDTA)]^{2-}$	3.8×10^{16}
$Mg^{2+} + 2NTA^{3-} \rightleftharpoons [Mg(NTA)_2]^{4-}$	1.6×10^{10}
$Ca^{2+} + 2NTA^{3-} \rightleftharpoons [Ca(NTA)_2]^{4-}$	3.2×10^{11}

[a] en = ethylenediamine

bipy = 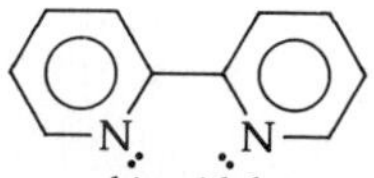
bipyridyl

phen = 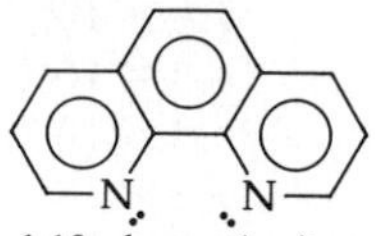
1,10-phenanthroline

$EDTA^{4-}$ = ethylenediaminetetraacetate ion
NTA^{3-} = nitrilotriacetate ion, $N(CH_2CO_2)_3^{3-}$

Table C.7. Ionization constants of weak acids and bases

Monoprotic Acids		K_a
$HC_2O_2Cl_3$	Trichloroacetic acid (Cl_3CCO_2H)	2.2×10^{-1}
HIO_3	Iodic acid	1.69×10^{-1}
$HC_2HO_2Cl_2$	Dichloroacetic acid (Cl_2CHCO_2H)	5.0×10^{-2}
$HC_2H_2O_2Cl$	Chloroacetic acid (ClH_2CCO_2H)	1.36×10^{-3}
HF	Hydrofluoric acid	6.5×10^{-4}
HNO_2	Nitrous acid	4.5×10^{-4}
$HOCN$	Cyanic acid	3.5×10^{-4}
$HCHO_2$	Formic acid (HCO_2H)	1.8×10^{-4}
$HC_3H_5O_3$	Lactic acid ($CH_3CH(OH)CO_2H$)	1.38×10^{-4}
$HC_7H_5O_2$	Benzoic acid ($C_6H_5CO_2H$)	6.5×10^{-5}

Table C.7. Continued

Monoprotic Acids		K_a
$HC_4H_7O_2$	Butanoic acid ($CH_3CH_2CH_2CO_2H$)	1.52×10^{-5}
HN_3	Hydrazoic acid	1.8×10^{-5}
$HC_2H_3O_2$	Acetic acid (CH_3CO_2H)	1.8×10^{-5}
$HC_3H_5O_2$	Propanoic acid ($CH_3CH_2CO_2H$)	1.34×10^{-5}
$HC_4H_3N_2O_3$	Barbituric acid	1.0×10^{-5}
$HOCl$	Hypochlorous acid	3.1×10^{-8}
$HOBr$	Hypobromous acid	2.1×10^{-9}
HCN	Hydrocyanic acid	4.9×10^{-10}
HC_6H_5O	Phenol	1.3×10^{-10}
HOI	Hypoiodous acid	2.3×10^{-11}
H_2O_2	Hydrogen peroxide	1.8×10^{-12}

Polyprotic Acids		K_{a_1}	K_{a_2}	K_{a_3}
H_2SO_4	Sulfuric acid	Large	1.2×10^{-2}	
H_2CrO_4	Chromic acid	5.0	1.5×10^{-6}	
$H_2C_2O_4$	Oxalic acid	5.6×10^{-2}	5.4×10^{-5}	
H_3PO_3	Phosphorous acid	3×10^{-2}	1.6×10^{-7}	
H_2SO_3	Sulfurous acid	1.5×10^{-2}	1.0×10^{-7}	
H_2SeO_3	Selenous acid	4.5×10^{-3}	1.1×10^{-8}	
H_2TeO_3	Tellurous acid	3.3×10^{-3}	2.0×10^{-8}	
$H_2C_3H_2O_4$	Malonic acid ($HO_2CCH_2CO_2H$)	1.4×10^{-3}	2.0×10^{-6}	
$H_2C_8H_4O_4$	Phthalic acid	1.1×10^{-3}	3.9×10^{-6}	
$H_2C_4H_4O_6$	Tartaric acid	9.2×10^{-4}	4.3×10^{-5}	
$H_2C_6H_6O_6$	Ascorbic acid	7.9×10^{-5}	1.6×10^{-12}	
H_2CO_3	Carbonic acid	4.3×10^{-7}	5.6×10^{-11}	
H_3PO_4	Phosphoric acid	7.5×10^{-3}	6.2×10^{-8}	2.2×10^{-12}
H_3AsO_4	Arsenic acid	5.6×10^{-3}	1.7×10^{-7}	4.0×10^{-12}
$H_3C_6H_5O_7$	Citric acid	7.1×10^{-4}	1.7×10^{-5}	6.3×10^{-6}

Weak Bases		K_b
$(CH_3)_2NH$	Dimethylamine	9.6×10^{-4}
CH_3NH_2	Methylamine	3.7×10^{-4}
$CH_3CH_2NH_2$	Ethylamine	4.3×10^{-4}
$(CH_3)_3N$	Trimethylamine	7.4×10^{-5}
NH_3	Ammonia	1.8×10^{-5}
N_2H_4	Hydrazine	1.7×10^{-6}
NH_2OH	Hydroxylamine	1.1×10^{-8}
C_5H_5N	Pyridine	1.7×10^{-9}
$C_6H_5NH_2$	Aniline	3.8×10^{-10}
PH_3	Phosphine	10^{-28}

Table C.8. Standard reduction potentials (25 °C)

$E°$ (volts)	Half-Cell Reaction
+2.87	$F_2(g) + 2e^- \rightleftharpoons 2F^-(aq)$
+2.08	$O_3(g) + 2H^+(aq) + 2e^- \rightleftharpoons O_2(g) + H_2O$
+2.00	$S_2O_8^{2-}(aq) + 2e^- \rightleftharpoons 2SO_4^{2-}(aq)$
+1.82	$Co^{3+}(aq) + e^- \rightleftharpoons Co^{2+}(aq)$
+1.78	$H_2O_2(aq) + 2H^+(aq) + 2e^- \rightleftharpoons 2H_2O$
+1.695	$MnO_4^-(aq) + 4H^+(aq) + 3e^- \rightleftharpoons MnO_2(s) + 2H_2O$
+1.69	$PbO_2(s) + SO_4^{2-}(aq) + 4H^+(aq) + 2e^- \rightleftharpoons PbSO_4(s) + 2H_2O$
+1.63	$2HOCl(aq) + 2H^+(aq) + 2e^- \rightleftharpoons Cl_2(g) + 2H_2O$
+1.51	$Mn^{3+}(aq) + e^- \rightleftharpoons Mn^{2+}(aq)$
+1.49	$MnO_4^-(aq) + 8H^+(aq) + 5e^- \rightleftharpoons Mn^{2+}(aq) + 4H_2O$
+1.47	$2ClO_3^-(aq) + 12H^+(aq) + 10e^- \rightleftharpoons Cl_2(g) + 6H_2O$
+1.46	$PbO_2(s) + 4H^+(aq) + 2e^- \rightleftharpoons Pb^{2+}(aq) + 2H_2O$
+1.44	$BrO_3^-(aq) + 6H^+(aq) + 6e^- \rightleftharpoons Br^-(aq) + 3H_2O$
+1.42	$Au^{3+}(aq) + 3e^- \rightleftharpoons Au(s)$
+1.36	$Cl_2(g) + 2e^- \rightleftharpoons 2Cl^-(aq)$
+1.33	$Cr_2O_7^{2-}(aq) + 14H^+(aq) + 6e^- \rightleftharpoons 2Cr^{3+}(aq) + 7H_2O$
+1.28	$MnO_2(s) + 4H^+(aq) + 2e^- \rightleftharpoons Mn^{2+}(aq) + 2H_2O$
+1.24	$O_3(g) + H_2O + 2e^- \rightleftharpoons O_2(g) + 2OH^-(aq)$
+1.23	$O_2(g) + 4H^+(aq) + 4e^- \rightleftharpoons 2H_2O$
+1.20	$Pt^{2+}(aq) + 2e^- \rightleftharpoons Pt(s)$
+1.09	$Br_2(aq) + 2e^- \rightleftharpoons 2Br^-(aq)$
+0.96	$NO_3^-(aq) + 4H^+(aq) + 3e^- \rightleftharpoons NO(g) + 2H_2O$
+0.94	$NO_3^-(aq) + 3H^+(aq) + 2e^- \rightleftharpoons HNO_2(aq) + H_2O$
+0.91	$2Hg^{2+}(aq) + 2e^- \rightleftharpoons Hg_2^{2+}(aq)$
+0.87	$HO_2^-(aq) + H_2O + 2e^- \rightleftharpoons 3OH^-(aq)$
+0.80	$NO_3^-(aq) + 4H^+(aq) + 2e^- \rightleftharpoons 2NO_2(g) + 2H_2O$
+0.80	$Ag^+(aq) + e^- \rightleftharpoons Ag(s)$
+0.77	$Fe^{3+}(aq) + e^- \rightleftharpoons Fe^{2+}(aq)$
+0.69	$O_2(g) + 2H^+(aq) + 2e^- \rightleftharpoons H_2O_2(aq)$
+0.54	$I_2(s) + 2e^- \rightleftharpoons 2I^-(aq)$
+0.52	$Cu^+(aq) + e^- \rightleftharpoons Cu(s)$
+0.49	$NiO_2(s) + 2H_2O + 2e^- \rightleftharpoons Ni(OH)_2(s) + 2OH^-(aq)$
+0.45	$SO_2(aq) + 4H^+(aq) + 4e^- \rightleftharpoons S(s) + 2H_2O$
+0.401	$O_2(g) + 2H_2O + 4e^- \rightleftharpoons 4OH^-(aq)$
+0.34	$Cu^{2+}(aq) + 2e^- \rightleftharpoons Cu(s)$
+0.27	$Hg_2Cl_2(s) + 2e^- \rightleftharpoons 2Hg(l) + 2Cl^-(aq)$
+0.25	$PbO_2(s) + H_2O + 2e^- \rightleftharpoons PbO(s) + 2OH^-(aq)$
+0.2223	$AgCl(s) + e^- \rightleftharpoons Ag(s) + Cl^-(aq)$
+0.172	$SO_4^{2-}(aq) + 4H^+(aq) + 2e^- \rightleftharpoons H_2SO_3(aq) + H_2O$
+0.169	$S_4O_6^{2-}(aq) + 2e^- \rightleftharpoons 2S_2O_3^{2-}(aq)$
+0.16	$Cu^{2+}(aq) + e^- \rightleftharpoons Cu^+(aq)$
+0.15	$Sn^{4+}(aq) + 2e^- \rightleftharpoons Sn^{2+}(aq)$
+0.14	$S(s) + 2H^+(aq) + 2e^- \rightleftharpoons H_2S(g)$
+0.07	$AgBr(s) + e^- \rightleftharpoons Ag(s) + Br^-(aq)$
0.00	$2H^+(aq) + 2e^- \rightleftharpoons H_2(g)$
−0.04	$Fe^{3+}(aq) + 3e^- \rightleftharpoons Fe(s)$

Table C.8. *Continued*

$E°$ (volts)	Half-Cell Reaction
−0.13	$Pb^{2+}(aq) + 2e^- \rightleftharpoons Pb(s)$
−0.14	$Sn^{2+}(aq) + 2e^- \rightleftharpoons Sn(s)$
−0.15	$AgI(s) + e^- \rightleftharpoons Ag(s) + I^-(aq)$
−0.25	$Ni^{2+}(aq) + 2e^- \rightleftharpoons Ni(s)$
−0.28	$Co^{2+}(aq) + 2e^- \rightleftharpoons Co(s)$
−0.34	$In^{3+}(aq) + 3e^- \rightleftharpoons In(s)$
−0.34	$Tl^+(aq) + e^- \rightleftharpoons Tl(s)$
−0.36	$PbSO_4(s) + 2e^- \rightleftharpoons Pb(s) + SO_4^{2-}(aq)$
−0.40	$Cd^{2+}(aq) + 2e^- \rightleftharpoons Cd(s)$
−0.44	$Fe^{2+}(aq) + 2e^- \rightleftharpoons Fe(s)$
−0.56	$Ga^{3+}(aq) + 3e^- \rightleftharpoons Ga(s)$
−0.58	$PbO(s) + H_2O + 2e^- \rightleftharpoons Pb(s) + 2OH^-(aq)$
−0.74	$Cr^{3+}(aq) + 3e^- \rightleftharpoons Cr(s)$
−0.76	$Zn^{2+}(aq) + 2e^- \rightleftharpoons Zn(s)$
−0.81	$Cd(OH)_2(s) + 2e^- \rightleftharpoons Cd(s) + 2OH^-(aq)$
−0.83	$2H_2O + 2e^- \rightleftharpoons H_2(g) + 2OH^-(aq)$
−0.88	$Fe(OH)_2(s) + 2e^- \rightleftharpoons Fe(s) + 2OH^-(aq)$
−0.91	$Cr^{2+}(aq) + e^- \rightleftharpoons Cr(s)$
−1.03	$Mn^{2+}(aq) + 2e^- \rightleftharpoons Mn(s)$
−1.16	$N_2(g) + 4H_2O + 4e^- \rightleftharpoons N_2O_4(aq) + 4OH^-(aq)$
−1.18	$V^{2+}(aq) + 2e^- \rightleftharpoons V(s)$
−1.216	$ZnO_2^-(aq) + 2H_2O + 2e^- \rightleftharpoons Zn(s) + 4OH^-(aq)$
−1.63	$Ti^{2+}(aq) + 2e^- \rightleftharpoons Ti(s)$
−1.67	$Al^{3+}(aq) + 3e^- \rightleftharpoons Al(s)$
−1.79	$U^{3+}(aq) + 3e^- \rightleftharpoons U(s)$
−2.02	$Sc^{3+}(aq) + 3e^- \rightleftharpoons Sc(s)$
−2.36	$La^{3+}(aq) + 3e^- \rightleftharpoons La(s)$
−2.37	$Y^{3+}(aq) + 3e^- \rightleftharpoons Y(s)$
−2.38	$Mg^{2+}(aq) + 2e^- \rightleftharpoons Mg(s)$
−2.71	$Na^+(aq) + e^- \rightleftharpoons Na(s)$
−2.76	$Ca^{2+}(aq) + 2e^- \rightleftharpoons Ca(s)$
−2.89	$Sr^{2+}(aq) + 2e^- \rightleftharpoons Sr(s)$
−2.90	$Ba^{2+}(aq) + 2e^- \rightleftharpoons Ba(s)$
−2.92	$Cs^+(aq) + e^- \rightleftharpoons Cs(s)$
−2.92	$K^+(aq) + e^- \rightleftharpoons K(s)$
−2.93	$Rb^+(aq) + e^- \rightleftharpoons Rb(s)$
−3.05	$Li^+(aq) + e^- \rightleftharpoons Li(s)$

D ANSWERS TO SELECTED QUESTIONS AND PROBLEMS

CHAPTER 1

1.11 $1 \text{ N} = 1 \text{ kg m/s}^2$

1.14 (a) 3.4×10^3 g (c) 4.0×10^6 mol (e) 7.2×10^{-2} m
(b) 5.7×10^{-9} s (d) 6.4×10^{-3} g

1.18 (a) 50.8 kg (c) 4.17×10^9 m³ (e) 13 m²
(b) 1.0×10^2 mi/hr (d) 2.1×10^3 cm/s

1.19 For the Pferdburper, fuel consumption is 24 mi/gal. Therefore, the Pferdburper is better than the Smokebelcher.

1.21 6.9 km

1.26 Mp = 86 °F, bp = 3601 °F

1.28 -78 °C $= -108$ °F $= 195$ K

1.30 °N $= (°C - 80) \times (100 °N/138 °C)$. Therefore, fp of water $= -58$ °N and bp of water $= 14$ °N

1.33 (a) 5, (b) 3, (c) 3, (d) 4, (e) 5

1.35 (a) 1.25×10^3 g (d) 2.1457×10^5 mg
(b) 1.3×10^7 m (e) 3.147×10^1 g
(c) 6.023×10^{22} atoms

1.37 (a) 30,000,000,000 m (d) 0.00000034 g
(b) 0.0000254 m (e) 32,500 cm
(c) 1.22 g

1.40 (a) 7.7 cm² (c) 0.781 g/cm³ (e) 81.4 g
(b) 73.3 m² (d) 3.478 g

1.42 (a) 2.14×10^3 cm² (d) 5.9 m²
(b) 1.21×10^4 cm³ (e) 2.61×10^2 mm²
(c) 4.1×10^1 cm² (f) 0.690 m²

1.44 (a) 0.63 km (d) 7.905×10^{-5} km/s
(b) 3.66×10^4 km (e) 5.06 g/cm³
(c) 3.04×10^{-2} m

1.46 1.434×10^4 in.

1.53 4.54 g/mL

1.55 4.55 g/cm³

1.57 (a) 6.702 mL (b) 14.92 g

1.60 (a) 0.787 (b) 0.787 g/mL

1.62 Calculated densities: For 1, $d = 1.00$ g/mL; for 2, $d = 1.59$ g/mL; for 3, $d = 1.13$ g/mL. Choose 1 because its density is the same as water.

1.66 The ratio of the mass of Si to the mass of O is the same for both samples: 3.44 g Si/3.91 g O $=$ 6.42 g Si/7.30 g O $= 0.880$.

1.73 (a) 2K, 1S (d) 3N, 12H, 1P, 4O
(b) 2Na, 1C, 3O (e) 3Na, 1Ag, 4S, 6O
(c) 4K, 1Fe, 6C, 6N

1.75 1Al, 24H, 20O, 1K, 2S

1.77 6.0×10^5 m

1.79 1.33×10^5 ft³

CHAPTER 2

2.6 Carbon, 0.63222 u; hydrogen, 0.053053 u

2.8 35.5 u

2.11 1.44×10^2 g C

2.13 Data support the law. For both samples, the C-to-F mass ratio is 0.158, the C-to-Cl mass ratio is 0.0857, and the F-to-Cl mass ratio is 0.542.

2.15 For samples with the same mass of oxygen, the ratio of the masses of phosphorus is 5-to-3.

2.16 4.58 g O

2.19 (a) 2-to-3 (c) 3 mol O
(b) 2-to-3 (d) 0.3 mol O

2.21 (a) 1-to-1   (c) 1-to-3   (e) 1-to-3
(b) 1-to-1   (d) 1-to-1   (f) 1-to-3

2.22 For parts a, b, e, and f:

(a) $\dfrac{1 \text{ mol Na}}{1 \text{ mol H}}$ or $\dfrac{1 \text{ mol H}}{1 \text{ mol Na}}$

(b) $\dfrac{1 \text{ mol Na}}{1 \text{ mol C}}$ or $\dfrac{1 \text{ mol C}}{1 \text{ mol Na}}$

(e) $\dfrac{1 \text{ mol H}}{3 \text{ mol O}}$ or $\dfrac{3 \text{ mol O}}{1 \text{ mol H}}$

(f) $\dfrac{1 \text{ mol C}}{3 \text{ mol O}}$ or $\dfrac{3 \text{ mol O}}{1 \text{ mol C}}$

2.24 (a) 2-to-5   (c) 2.50 mol H   (e) 0.750 mol H_2
(b) 12.0 mol C (d) 0.240 mol C

2.26 3.00 mol S

2.28 1.00 mol CO_2

2.30 3.75 mol $BaSO_4$

2.33 (a) 2.17 mol Na   (c) 0.962 mol Cr (e) 1.28 mol K
(b) 0.668 mol As (d) 1.85 mol Al   (f) 0.463 mol Ag

2.35 (a) 40.3   (c) 208.2   (e) 163.9
(b) 111.0   (d) 135.0

2.37 For parts a and b:

(a) $\dfrac{1 \text{ mol MgO}}{40.3 \text{ g MgO}}$ or $\dfrac{40.3 \text{ g MgO}}{1 \text{ mol MgO}}$

(b) $\dfrac{1 \text{ mol CaCl}_2}{111.0 \text{ g CaCl}_2}$ or $\dfrac{111.0 \text{ g CaCl}_2}{1 \text{ mol CaCl}_2}$

2.38 (a) 60.1   (c) 246.5   (e) 176.1
(b) 58.3   (d) 812.4   (f) 342.3

2.40 779 g

2.43 2.88 mol $NaHCO_3$

2.45 0.870 mol H_2SO_4

2.47 1.68 mol K

2.49 7.81 mol FeS_2

2.50 9.274×10^{-23} g per atom of Fe

2.52 5.684×10^{-22} g/molecule

2.53 1.64×10^{15} atoms C

2.55 0.844 g Cu

2.56 (a) 34.43% Fe, 65.75% Cl
(b) 42.07% Na, 18.89% P, 39.04% O
(c) 28.71% K, 0.74% H, 23.55% S, 47.00% O
(d) 21.21% N, 6.87% H, 23.46% P, 48.46% O
(e) 84.98% Hg, 15.02% Cl

2.59 (a) 10.5 g Fe   (c) 6.94 g Na
(b) 3.94 g Al   (d) 12.3 g Mg

2.60 5.60 g N

2.63 (a) 2.55 g C   (c) 60.6% C, 13.4% H, 26.6% O
(b) 1.13 g O

2.65 (a) NH_4SO_4 (d) CH   (f) CH_2O
(b) Fe_2O_3 (e) $C_3H_8O_3$ (g) Hg_2SO_4
(c) $AlCl_3$

2.66 $C_2H_6O_2$ and CH_3O

2.69 SO_3

2.71 P_4S_3

2.73 CCl_2

2.75 SCl_2

2.77 $C_8H_8O_3$

2.79 $CrCl_3$

2.81 (a) 25.1% C, 4.23% H, 9.79% N, 49.7% Cl, 11.2% O
(b) $C_3H_6NCl_2O$

2.83 0.1875 g C, 0.02097 g H, 0.2915 g O   The empirical formula is $C_6H_8O_7$, and the molecular formula is also $C_6H_8O_7$.

2.85 179.972

2.87 37.5 g O

2.90 3.3×10^3 atoms thick

2.91 Ag_2S

2.93 (a) 2.00 mol S   (c) 6.00 mol S
(b) 0.720 mol Fe (d) 3.00 mol FeS_2

CHAPTER 3

3.2 Coefficients are (a) 1, 2, 1, 1   (d) 1, 1, 2
(b) 2, 1, 1, 1   (e) 2, 1, 2, 1
(c) 8, 3, 4, 9

3.4 Coefficients are (a) 2, 13, 8, 10   (d) 4, 11, 2, 8
(b) 2, 15, 14, 6   (e) 4, 5, 4, 6
(c) 1, 6, 4

3.7 (a) 2.50 mol C_2H_2 (c) 6.40 mol H_2O
(b) 13.0 g C_2H_2 (d) 79.8 g $Ca(OH)_2$

3.9 (a) $P_4 + 5O_2 \rightarrow P_4O_{10}$ (c) 21.8 g P_4
(b) 0.100 mol P_4O_{10} (d) 19.4 g P_4

3.11 (a) 52.5 mol CO   (c) 45.5 g Fe_2O_3 (e) 24.7 g Fe
(b) 1.50 mol Fe_2O_3 (d) 0.911 mol CO

3.13 1.574×10^3 kg DDT

3.15 (a) $(CH_3)_2NNH_2 + 2N_2O_4 \rightarrow 4H_2O + 2CO_2 + 3N_2$
(b) 153 kg N_2O_4

3.17 (a) 12.5 g white lead   (b) 12.8 g CO_2

3.19 96.2 tons Ca

3.21 3.66×10^3 lb NH_3

3.23 (a) HCl   (b) 0.38 mol H_2 (c) 0.02 mol Fe

3.24 4.43 g $HClO_3$

3.26 (a) C_2H_2 (b) 83.8 g C_2H_3Cl (c) 2 g HCl

3.28 1.02 g Ag_2S

3.30 20 g $C_2H_2Br_4$ and 24.9 g $C_2H_4Br_2$

3.32 (a) 14.8 g CCl_2F_2 (b) 8.62 g CCl_2F_2 (c) 58.2%

3.34 (a) 30.1 g C_6H_5Br (c) 28.5 g C_6H_5Br
(b) 0.828 g C_6H_6 (d) 94.7%

3.36 0.590 g, 508 g

3.42 (a) 0.625 M NaCl   (c) 2.27 M H_2SO_4
(b) 4.20 M sucrose   (d) 7.13 M KOH

3.44 (a) 0.250 mol/L   (c) M = mol/L = mmol/mL
(b) 0.250 M

3.48 0.7583 g KNO_3

3.49 18.5 g $MgSO_4 \cdot 7H_2O$

3.52 38.5 mL

3.54 66.7 mL

3.58 (a) 973 mL   (b) 8.00×10^{-2} mol H_2

3.60 (a) 25.0 mL   (c) 0.00525 mol Na_2SO_4
(b) 270 mL

3.62 0.1000 M $AgNO_3$

3.64 (a) $2Al + 6HCl \rightarrow 3H_2 + 2AlCl_3$ (d) 0.200 mol Al
(b) 0.450 mol H_2 (e) 2.25 g Al
(c) 0.133 mol $AlCl_3$

3.66 (a) $2Al + 6HCl \rightarrow 3H_2 + 2AlCl_3$ (c) $0.0500\ M\ AlCl_3$
 (b) 0.405 g Al
3.68 3.87 kg white lead

CHAPTER 4

4.5 $2Na + 2H_2O \rightarrow 2NaOH + H_2$
4.25 Al^{3+}
4.27 3.19×10^5 g/cm³
4.29 0.0951 mi
4.34 ^{132}Cs has 55 protons, 77 neutrons, and 55 electrons.
 $^{115}Cd^{2+}$ has 48 protons, 67 neutrons, and 46 electrons.
 ^{194}Tl has 81 protons, 113 neutrons, and 81 electrons.
 $^{105}Ag^+$ has 47 protons, 58 neutrons, and 46 electrons.
 $^{78}Se^{2-}$ has 34 protons, 44 neutrons, and 36 electrons.
4.36 (a) $^{55}_{26}Fe$ (c) $^{204}_{81}Tl$ (e) $^{169}_{70}Yb$
 (b) $^{86}_{37}Rb$ (d) $^{170}_{71}Lu$
4.38 151.9 g/mol
4.40 207 g/mol
4.42 48.17%
4.57 $RaCl_2$
4.64 See Table 4.3. (a) $CrCl_2$, $CrCl_3$, CrS, and Cr_2S_3
 (b) $MnCl_2$, $MnCl_3$, MnS, and Mn_2S_3
 (c) $FeCl_2$, $FeCl_3$, FeS, and Fe_2S_3
 (d) $CoCl_2$, $CoCl_3$, CoS, and Co_2S_3
 (e) $NiCl_2$ and NiS
 (f) $CuCl$, $CuCl_2$, Cu_2S, and CuS
 (g) $AgCl$ and Ag_2S
4.68 (a) Na_2CO_3 (c) SrS (e) $Ti(ClO_4)_4$
 (b) $Ca(ClO_3)_2$ (d) $CrCl_3$
4.70 (a) $Fe_2(HPO_4)_3$ (c) $Ni(NO_3)_2$ (e) $BaSO_3$
 (b) K_3N (d) $Cu(C_2H_3O_2)_2$
4.72 (a) PH_3 (c) HBr (e) SbH_3
 (b) H_2S (d) SiH_4
4.74 PCl_3 and PCl_5
4.78 AlF_3 and CaF_2
4.85 (a) K, $+1$; Cl, $+3$; O, -2
 (b) Ba, $+2$; Mn, $+6$; O, -2
 (c) Fe, $+8/3$; O, -2
 (d) O, $+1$; F, -1
 (e) I, $+5$; F, -1
 (f) H, $+1$; O, -2; Cl, $+1$
 (g) Ca, $+2$; S, $+6$; O, -2
 (h) Cr, $+3$, S, $+6$; O, -2
 (i) O, 0
 (j) Hg, $+1$; Cl, -1
4.87 (a) oxidation (c) oxidation (e) reduction
 (b) reduction (d) oxidation

4.89

	Oxidized	Reduced	Oxidizing Agent	Reducing Agent
(a)	NaI	$NaIO_3$	$NaIO_3$	NaI
(b)	Cu	HNO_3	HNO_3	Cu
(c)	Cu	HNO_3	HNO_3	Cu
(d)	Cu	H_2SO_4	H_2SO_4	Cu
(e)	SO_2	HNO_3	HNO_3	SO_2

4.90 (a) NaBr sodium bromide
 (b) CaO calcium oxide
 (c) $FeCl_3$ ferric chloride; iron(III) chloride
 (d) $CuCO_3$ cupric carbonate; copper(II) carbonate
 (e) CBr_4 carbon tetrabromide
 (f) P_4O_6 tetraphosphorus hexoxide
 (g) $AsCl_5$ arsenic pentachloride
 (h) $Mn(HCO_3)_2$ manganous hydrogen carbonate;
 manganous bicarbonate; manganese(II) hydrogen carbonate; manganese(II) bicarbonate
 (i) $NaMnO_4$ sodium permanganate
 (j) O_2F_2 dioxygen difluoride
4.91 (a) $Al(NO_3)_3$ (e) PCl_3 (h) $Mg(OH)_2$
 (b) $FeSO_4$ (f) N_2O_4 (i) H_2Se
 (c) $NH_4H_2PO_4$ (g) $KMnO_4$ (j) NaH
 (d) IF_5
4.95 (a) strontium chloride (g) bromic acid
 (b) calcium nitrate (h) mercury(II) bromide
 (c) copper(II) sulfide (i) cobalt(II) sulfate
 (d) tin(II) phosphate (j) potassium dihydrogen
 (e) nickel(II) chlorate arsenate
 (f) zinc acetate
4.97 (a) $Pb(C_2H_3O_2)_4$ (e) $HI(aq)$ (h) $Ag_2C_2O_4$
 (b) Na_2Se (f) PBr_3 (i) H_2CrO_4
 (c) $Ba_3(PO_4)_2$ (g) $Ca(OCl)_2$ (j) SiF_4
 (d) HI [or $Ca(ClO)_2$]
4.99 (a) S_3O_9 (e) PbS (h) $Hg(NO_3)_2$
 (b) ICl_5 (f) HIO_4 (i) $Au_2(SO_4)_3$
 (c) $Cr_2(SO_4)_3$ (g) $LiOI$ [or $LiIO$] (j) Bi_2O_5
 (d) $Fe_2(SO_4)_3$
4.101 (a) $Fe_2(SO_4)_3$ (d) $CuCl$ (g) $AuCl_3$
 (b) $FeCl_2$ (e) $SnCl_4$ (h) $Cr(C_2H_3O_2)_3$
 (c) $Hg_2(NO_3)_2$ (f) $Co(OH)_2$

CHAPTER 5

5.11 $KCl(aq) \rightarrow K^+(aq) + Cl^-(aq)$
 $(NH_4)_2SO_4(aq) \rightarrow 2NH_4^+(aq) + SO_4^{2-}(aq)$
 $Na_3PO_4(aq) \rightarrow 3Na^+(aq) + PO_4^{3-}(aq)$
 $NaOH(aq) \rightarrow Na^+(aq) + OH^-(aq)$
 $HCl(aq) \rightarrow H^+(aq) + Cl^-(aq)$
5.14 $CdSO_4(aq) \rightleftharpoons Cd^{2+}(aq) + SO_4^{2-}(aq)$
5.25 Ionic equation:
 $Ca^{2+}(aq) + 2Cl^-(aq) + 2K^+(aq) + CO_3^{2-}(aq) \rightarrow$
 $CaCO_3(s) + 2K^+(aq) + 2Cl^-(aq)$
 Net ionic equation:
 $Ca^{2+}(aq) + CO_3^{2-}(aq) \rightarrow CaCO_3(s)$
5.27 (a) $Cu^{2+} + 2NO_3^- + 2Na^+ + 2OH^- \rightarrow$
 $Cu(OH)_2(s) + 2Na^+ + 2NO_3^-$
 $Cu^{2+}(aq) + 2OH^-(aq) \rightarrow Cu(OH)_2(s)$ (net ionic)
 (b) $3Ba^{2+} + 6Cl^- + 2Al^{3+} + 3SO_4^{2-} \rightarrow$
 $3BaSO_4(s) + 2Al^{3+} + 6Cl^-$
 $Ba^{2+}(aq) + SO_4^{2-}(aq) \rightarrow BaSO_4(s)$ (net ionic)
 (c) $Hg_2^{2+} + 2NO_3^- + 2H^+ + 2Cl^- \rightarrow$
 $Hg_2Cl_2(s) + 2H^+ + 2NO_3^-$
 $Hg_2^{2+}(aq) + 2Cl^-(aq) \rightarrow Hg_2Cl_2(s)$ (net ionic)

(d) $2Bi^{3+} + 6NO_3^- + 6Na^+ + 3S^{2-} \rightarrow$
$$Bi_2S_3(s) + 6Na^+ + 6NO_3^-$$
$2Bi^{3+}(aq) + 3S^{2-}(aq) \rightarrow Bi_2S_3(s)$ (net ionic)

(e) $Ca^{2+} + 2Cl^- + 2Na^+ + SO_4^{2-} \rightarrow$
$$CaSO_4(s) + 2Na^+ + 2Cl^-$$
$Ca^{2+}(aq) + SO_4^{2-}(aq) \rightarrow CaSO_4(s)$ (net ionic)

5.36 (a) acidic   (c) acidic   (e) basic
(b) basic   (d) acidic

5.37 H_3PO_4

5.44 $N_2H_4(aq) + H_2O \rightleftharpoons N_2H_5^+(aq) + OH^-(aq)$

5.46 (a) $KOH + HCl \rightarrow KCl + H_2O$
(b) $NaOH + HC_2H_3O_2 \rightarrow NaC_2H_3O_2 + H_2O$
(c) $NH_3(aq) + HCl \rightarrow NH_4Cl(aq)$
(d) $CuO + 2HBr \rightarrow CuBr_2 + H_2O$
(e) $Fe_2O_3 + 3H_2SO_4 \rightarrow Fe_2(SO_4)_3 + 3H_2O$

5.49

Soluble	Insoluble
KCl	$PbSO_4$
$(NH_4)_2SO_4$	$Mn(OH)_2$
$AgNO_3$	$FePO_4$
$Zn(ClO_4)_2$	$CaCO_3$
$Ba(C_2H_3O_2)_2$	NiO

5.51 (a) Ionic: $Al(OH)_3(s) + 3H^+(aq) + 3Cl^-(aq) \rightarrow$
$$Al^{3+}(aq) + 3Cl^-(aq) + 3H_2O$$
Net ionic:
$Al(OH)_3(s) + 3H^+(aq) \rightarrow Al^{3+}(aq) + 3H_2O$
(b) Ionic: $CuCO_3(s) + 2H^+(aq) + SO_4^{2-}(aq) \rightarrow$
$$Cu^{2+}(aq) + SO_4^{2-}(aq) + H_2O + CO_2(g)$$
Net ionic: $CuCO_3(s) + 2H^+(aq) \rightarrow$
$$Cu^{2+}(aq) + H_2O + CO_2(g)$$
(c) Ionic: $Cr_2(CO_3)_3(s) + 6H^+(aq) + 6NO_3^-(aq) \rightarrow$
$$2Cr^{3+}(aq) + 6NO_3^-(aq) + 3H_2O + 3CO_2(g)$$
Net ionic: $Cr_2(CO_3)_3(s) + 6H^+(aq) \rightarrow$
$$2Cr^{3+}(aq) + 3H_2O + 3CO_2(g)$$

5.53 (a) $CoS(s) + 2H^+(aq) \rightarrow H_2S(g) + Co^{2+}(aq)$
(b) $PbCO_3(s) + 2H^+(aq) \rightarrow$
$$H_2O + CO_2(g) + Pb^{2+}(aq)$$
(c) $PbCO_3(s) + 2H^+(aq) + SO_4^{2-}(aq) \rightarrow$
$$PbSO_4(s) + H_2O + CO_2(g)$$
(d) $Sn^{2+}(aq) + 2OH^-(aq) \rightarrow Sn(OH)_2(s)$
(e) $Ag_2O(s) + 2H^+(aq) + 2Cl^-(aq) \rightarrow$
$$2AgCl(s) + H_2O$$
(f) This reaction does not have a driving force.

5.55 (a) No reaction between reactants.
(b) No reaction between reactants.
(c) $K_2S(aq) + Ni(C_2H_3O_2)_2(aq) \rightarrow$
$$2KC_2H_3O_2(aq) + NiS(s)$$
$2K^+(aq) + S^{2-}(aq) + Ni^{2+}(aq) + 2C_2H_3O_2^-(aq) \rightarrow$
$$2K^+(aq) + 2C_2H_3O_2^-(aq) + NiS(s)$$
$Ni^{2+}(aq) + S^{2-}(aq) \rightarrow NiS(s)$
(d) $MgSO_4(aq) + 2LiOH(aq) \rightarrow$
$$Li_2SO_4(aq) + Mg(OH)_2(s)$$
$Mg^{2+}(aq) + SO_4^{2-}(aq) + 2Li^+(aq) + 2OH^-(aq) \rightarrow$
$$2Li^+(aq) + SO_4^{2-}(aq) + Mg(OH)_2(s)$$
$Mg^{2+}(aq) + 2OH^-(aq) \rightarrow Mg(OH)_2(s)$

(e) $AgC_2H_3O_2(aq) + KCl(aq) \rightarrow$
$$AgCl(s) + KC_2H_3O_2(aq)$$
$Ag^+(aq) + C_2H_3O_2^-(aq) + K^+(aq) + Cl^-(aq) \rightarrow$
$$AgCl(s) + K^+(aq) + C_2H_3O_2^-(aq)$$
$Ag^+(aq) + Cl^-(aq) \rightarrow AgCl(s)$

5.58 (a) $8H^+ + 2NO_3^- + 3Cu \rightarrow 2NO + 3Cu^{2+} + 4H_2O$
(b) $10H^+ + NO_3^- + 4Zn \rightarrow NH_4^+ + 4Zn^{2+} + 3H_2O$
(c) $2Cr + 6H^+ \rightarrow 2Cr^{3+} + 3H_2$
(d) $8H^+ + Cr_2O_7^{2-} + 3H_3AsO_3 \rightarrow$
$$2Cr^{3+} + 4H_2O + 3H_3AsO_4$$
(e) $10H^+ + SO_4^{2-} + 8I^- \rightarrow 4I_2 + H_2S + 4H_2O$
(f) $4H_2O + 8Ag^+ + AsH_3 \rightarrow H_3AsO_4 + 8Ag + 8H^+$
(g) $H_2O + S_2O_8^{2-} + HNO_2 \rightarrow$
$$NO_3^- + 2SO_4^{2-} + 3H^+$$
(h) $4H^+ + MnO_2 + 2Br^- \rightarrow Mn^{2+} + Br_2 + 2H_2O$
(i) $2S_2O_3^{2-} + I_2 \rightarrow 2I^- + S_4O_6^{2-}$
(j) $IO_3^- + 3HSO_3^- \rightarrow I^- + 3SO_4^{2-} + 3H^+$

5.60 (a) $H_2O + CN^- + AsO_4^{3-} \rightarrow$
$$AsO_2^- + CNO^- + 2OH^-$$
(b) $2CrO_2^- + 3HO_2^- \rightarrow 2CrO_4^{2-} + H_2O + OH^-$
(c) $7OH^- + 4Zn + NO_3^- + 6H_2O \rightarrow$
$$4Zn(OH)_4^{2-} + NH_3$$
(d) $4OH^- + Cu(NH_3)_4^{2+} + S_2O_4^{2-} \rightarrow$
$$2SO_3^{2-} + Cu + 4NH_3 + 2H_2O$$
(e) $N_2H_4 + 2Mn(OH)_3 \rightarrow 2Mn(OH)_2 + 2NH_2OH$
(f) $4OH^- + 2MnO_4^- + 3C_2O_4^{2-} \rightarrow$
$$2MnO_2 + 6CO_3^{2-} + 2H_2O$$
(g) $6OH^- + 7ClO_3^- + 3N_2H_4 \rightarrow$
$$6NO_3^- + 7Cl^- + 9H_2O$$

5.64 (a) $3HSO_3^-(aq) + 5H^+(aq) + Cr_2O_7^{2-}(aq) \rightarrow$
$$2Cr^{3+}(s) + 4H_2O + 3SO_4^{2-}(aq)$$
(b) $2S_2O_3^{2-}(aq) + I_2(aq) \rightarrow 2I^-(aq) + S_4O_6^{2-}(aq)$
(c) $S_2O_3^{2-}(aq) + 4Cl_2(aq) + 5H_2O \rightarrow$
$$8Cl^-(aq) + 2SO_4^{2-}(aq) + 10H^+(aq)$$

5.66 $3SO_3^{2-}(aq) + 2MnO_4^-(aq) + H_2O \rightarrow$
$$3SO_4^{2-}(aq) + 2MnO_2(s) + 2OH^-(aq)$$

5.71 (a) $1 \times 10^{-4}\%$ F^- (by mass)
(b) 1 ppm F^-
(c) 1×10^3 ppb F^-

5.73 (a) 0.750 M   (c) 0.556 M   (e) 0.962 M
(b) 0.992 M   (d) 1.36 M

5.75 4.77 g Na_2CO_3

5.77 24.01 M

5.79 (a) 0.100 M Li^+ and 0.100 M Cl^-
(b) 0.250 M Ca^{2+} and 0.500 M Cl^-
(c) 2.40 M NH_4^+ and 1.20 M SO_4^{2-}
(d) 0.600 M Na^+ and 0.600 M HSO_4^-
(e) 0.800 M Fe^{3+} and 1.20 M SO_4^{2-}

5.81 0.100 M Na_2SO_4

5.83 0.0700 M

5.84 (a) 0.0100 mol Na^+ and 0.0100 mol Cl^-
(b) 0.00480 mol Ca^{2+} and 0.00960 mol Cl^-
(c) 0.0351 mol Na^+ and 0.0176 mol SO_4^{2-}
(d) 0.221 mol NH_4^+ and 0.111 mol SO_4^{2-}
(e) 0.0375 mol Al^{3+} and 0.0562 mol SO_4^{2-}

5.86 100 mL

5.88 (a) $2H^+(aq) + CuCO_3(s) \rightarrow$
$$CO_2(g) + H_2O + Cu^{2+}(aq)$$
(b) 29.6 mL $HClO_4$
(c) 2.47 g $CuCO_3$

5.90 50.0 mL $BaCl_2$ soln.

5.92 122 g/mol

5.94 0.18 g AgCl

5.96 0.059 M H^+

5.97 (a) $3Ba^{2+}(aq) + 6OH^-(aq) + 2Al^{3+}(aq) +$
$$3SO_4^{2-}(aq) \rightarrow 3BaSO_4(s) + 2Al(OH)_3(s)$$
(b) 3.08 g
(c) $\sim 0\ M\ Ba^{2+}, \sim 0\ M\ OH^-, 0.215\ M\ SO_4^{2-},$
$0.143\ M\ Al^{3+}$

5.99 $TiCl_3$

5.101 34.2%

5.104 0.1065 M H_2SO_4

5.106 0.064 g lactic acid and 0.036 g caproic acid

5.110 0.400 eq $Ba(OH)_2$

5.112 0.280 eq H_3AsO_4

5.115 1.55 g $MnSO_4 \cdot 6H_2O$

5.117 (a) 49.00 g H_3PO_4/eq   (d) 35.58 g $NaIO_3$/eq
(b) 100.5 g $HClO_4$/eq   (e) 26.00 g $Al(OH)_3$/eq
(c) 32.98 g $NaIO_3$/eq

5.119 44.8 g/eq

5.121 10.5 N HCl or 10.5 M HCl

5.123 46 mL

5.126 50.0% NaCl, 20% NaBr, 30% NaI

5.128 19.1 mL HNO_3 soln.

5.130 56.88% $CaCO_3$

5.132 (a) 1 eq   (b) 5 eq

5.134 (a) 0.009820 mol CaC_2O_4 (b) 45.70% $CaCl_2$

CHAPTER 6

6.2 By a factor of four.

6.6 Since the charges on the electron and the proton of the hydrogen atom have opposite signs, when they are moved apart the potential energy increases, as would be the case in the separation of any two objects that have attractive forces.

6.10 For the temperature change from 100 °C to 200 °C (373 K to 473 K), the average K.E. increases by a factor of 1.27. For the change from 27 °C (300 K) to 372 °C (600 K), the average K.E. increases by a factor of 2.00.

6.13 (a) exothermic   (c) The kinetic energy increases.
(b) $2Mg + O_2$

6.17 (a) 82.5 cal   (c) 55.9 kcal
(b) 2.28×10^3 J   (d) 5.259 kJ

6.19 1.54×10^3 J

6.21 5.32 kg

6.23 2.06×10^4 J

6.26 7.82×10^3 J/°C

6.28 (a) 0.0638 °C increase   (b) 0.233 °C decrease

6.30 1.57×10^4 J

6.32 Specific heat = 2.56 J g^{-1} $°C^{-1}$; molar heat capacity = 81.9 J mol^{-1} $°C^{-1}$

6.35 1187 J/°C

6.37 (a) -2.22×10^5 J   (b) -222 kJ mol^{-1}

6.39 (a) 6.367×10^4 J liberated
(b) 3.911×10^6 J mol^{-1} liberated

6.45 (a) 1.5×10^3 J mol^{-1}
(b) 4.1×10^3 J mol^{-1}
(c) -2.6×10^3 J mol^{-1}
The sum of values for parts b and c equals the value for part a.

6.51 The ΔH for $SO_2 + \frac{1}{2}O_2 \rightarrow SO_3$ is not labeled ΔH_f because in this reaction SO_3 is not formed from its elements.

6.53 125 kJ

6.55 -53.0 kJ

6.57 -1268 kJ

6.59 -188 kJ

6.61 29 kJ

6.63 24 kJ

6.65 (a) -854 kJ   (c) -402 kJ   (e) -136.4 kJ
(b) -1427 kJ   (d) -87 kJ

6.67 48.8 kJ/mol H^+

6.69 630 g hr^{-1}

6.71 -267 kJ/mol

CHAPTER 7

7.7 (a) 38 nm   (b) 1.50×10^{15} Hz

7.9 (a) 2.97 m   (b) 341 m

7.14 5.49×10^{14} Hz

7.16 $\lambda = 589$ nm; $v = 5.09 \times 10^{14}$ Hz

7.21 (a) 2×10^{-18} J   (b) 1×10^{-5} m

7.23 (a) 1.0×10^5 J   (b) 2.19×10^5 J

7.27 (a) 486.273 nm
(b) 1.09411×10^{-4} cm = 1094.11 nm

7.29 (a) -2.42×10^{-19} J   (d) $v = 4.57 \times 10^{14}$ Hz
(b) -5.45×10^{-19} J   (e) $\lambda = 656$ nm
(c) $\Delta E = 3.03 \times 10^{-19}$ J   (f) red-orange

7.32 2.4×10^{-17} J

7.42 $s, p, d, f,$ and g

7.44 25 orbitals

7.46 (a) $s = 2, p = 6, d = 10, f = 14, g = 18,$ and $h = 22$
(b) $n = 6$
(c) $-5, -4, -3, -2, -1, 0, 1, 2, 3, 4, 5$

7.53 18

7.59 K $4s$, Al $3s^23p^1$, F $2s^22p^5$, S $3s^23p^4$, Tl $6s^26p^1$, Bi $6s^26p^3$

7.62 (a) 15   (b) 8   (c) 18

7.64 (a) Sn $\underset{5s}{\uparrow\downarrow}$ $\underset{5p}{\uparrow\ \uparrow\ \_\_}$ (c) Ba $\underset{6s}{\uparrow\downarrow}$

(b) Br $\underset{4s}{\uparrow\downarrow}$ $\underset{4p}{\uparrow\downarrow\ \uparrow\downarrow\ \uparrow}$

7.76 (a) 2+   (c) 4+   (e) 7+
(b) 3+   (d) 6+

7.77 (a) 2+   (c) 2+   (e) 2+
(b) 2+   (d) 2+

7.80 (a) Cr^{3+} [Ar] $3d^3$ [Ar] $\underset{3d}{\uparrow\ \uparrow\ \uparrow\ \_\_\ \_\_}$

(b) Mn^{2+} [Ar] $3d^5$ [Ar] $\underline{\uparrow}\ \underline{\uparrow}\ \underline{\uparrow}\ \underline{\uparrow}\ \underline{\uparrow}$
$\qquad\qquad\qquad\qquad\qquad\qquad\ \ 3d$

(c) Mn^{3+} [Ar] $3d^4$ [Ar] $\underline{\uparrow}\ \underline{\uparrow}\ \underline{\uparrow}\ \underline{\uparrow}\ \underline{\ \ }$
$\qquad\qquad\qquad\qquad\qquad\qquad\ \ 3d$

(d) Co^{2+} [Ar] $3d^7$ [Ar] $\underline{\uparrow\downarrow}\ \underline{\uparrow\downarrow}\ \underline{\uparrow}\ \underline{\uparrow}\ \underline{\uparrow}$
$\qquad\qquad\qquad\qquad\qquad\qquad\ \ 3d$

(e) Co^{3+} [Ar] $3d^6$ [Ar] $\underline{\uparrow\downarrow}\ \underline{\uparrow}\ \underline{\uparrow}\ \underline{\uparrow}\ \underline{\uparrow}$
$\qquad\qquad\qquad\qquad\qquad\qquad\ \ 3d$

(f) Ni^{2+} [Ar] $3d^8$ [Ar] $\underline{\uparrow\downarrow}\ \underline{\uparrow\downarrow}\ \underline{\uparrow\downarrow}\ \underline{\uparrow}\ \underline{\uparrow}$
$\qquad\qquad\qquad\qquad\qquad\qquad\ \ 3d$

7.82 (a) Se   (c) Fe^{2+} (e) S^{2-}
      (b) C   (d) O^-

7.89 (a) Cl   (c) P
      (b) S   (d) S

7.92 $E = 2.04 \times 10^{-18}$ J, $v = 3.08 \times 10^{15}$ Hz, $\lambda = 97.4$ nm

7.94 Seven

7.96 [Rn] $5f^{14}6d^47s^2$

7.98 [Ar] $4s^1$

7.100 1.3×10^4 g

CHAPTER 8

8.2 (a) Ba^{2+} $1s^22s^22p^63s^23p^63d^{10}4s^24p^64d^{10}5s^25p^6$ or [Xe]
     (b) Se^{2-} $1s^22s^22p^63s^23p^63d^{10}4s^24p^6$ or [Kr]
     (c) Al^{3+} $1s^22s^22p^6$ or [Ne]
     (d) Na^+ $1s^22s^22p^6$ or [Ne]
     (e) Br^- $1s^22s^22p^63s^23p^63d^{10}4s^24p^6$ or [Kr]

8.4 Na_2O and Mg_3N_2

8.7 (a) $+148$ kJ   (b) $+4365$ kJ   The lattice energy for $NaCl_2$ would have to be more than 29.5 times the lattice energy for NaCl before $NaCl_2$ would become more stable than NaCl.

8.12 Pb^{2+} [Xe] $4f^{14}5d^{10}6s^2$
      Pb^{4+} [Xe] $4f^{14}5d^{10}$
      Mn^{2+} [Ar] $3d^5$
      Mn^{3+} [Ar] $3d^4$
      Sb^{3+} [Kr] $4d^{10}5s^2$
      Sc^{3+} [Ar]
      Ti^{2+} [Ar] $3d^2$

8.15 (a) K^+ and Cl^- (e) Mg^{2+} and C^{4-}
      (d) Sr^{2+} and Br^-

8.19 (a) :S̈e·   (c) ·A̋l·   (e) ·G̈e·

      (b) :B̈r·   (d) ·Ba·   (f) ·P̈·

8.22 (a) $Ba^{2+}\left[:\ddot{O}:\right]^{2-}$ (c) $K^+\left[:\ddot{F}:\right]^-$ (e) $3Li^+, \left[:\ddot{N}:\right]^{3-}$

      (b) $2Na^+, \left[:\ddot{O}:\right]^{2-}$ (d) $Ca^{2+}\left[:\ddot{S}:\right]^{2-}$

8.26 (a) 4   (c) 1   (e) 3
      (b) 2   (d) 4

8.29 None. (Nature prefers paired electrons whenever possible.)

8.33 (a) 4   (c) 9
      (b) 4   (d) 9

8.36 (a) 5   (c) 4   (e) 6
      (b) 7   (d) 4

8.38 Hydrogen normally forms only one bond, so it cannot be bonded to more than one atom.

8.40

Boron does not have an octet in the last structure. To form an octet one of the chlorines would need to provide two more electrons for covalent bonding. Then that chlorine would have a double bond, which chlorine does not normally form.

8.42

8.44

8.47 (a) Each C—Cl bond has a bond order of 1.
      (b) In HCN the H—C bond order is 1 and the C≡N bond order is 3.
      (c) In CO_2 each C=O has a bond order of 2.
      (d) In NO^+ the bond order is 3.
      (e) In CH_3NCO the bond order between each hydrogen and the carbon is 1, between one of the carbons and nitrogen it is 1, between nitrogen and the other carbon it is 2, and between carbon and the oxygen it is 2.

8.52 Bond order and bond energy increase in the order

154 pm < 146 pm < 140 pm < 137 pm

8.54

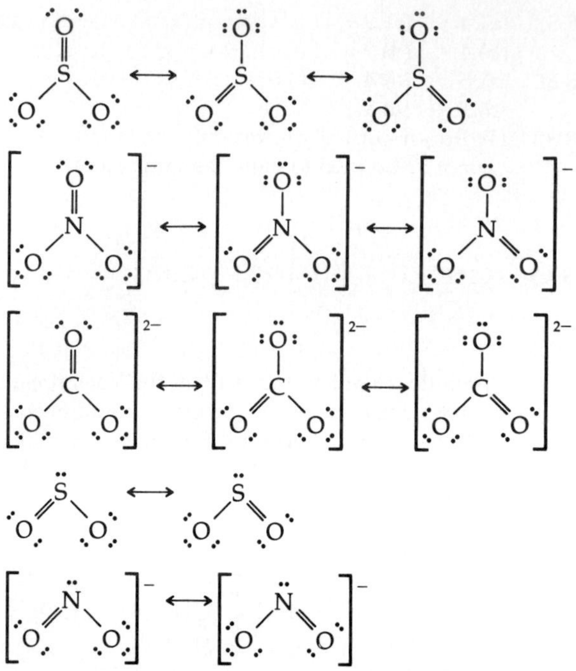

8.57 The N—O bond order in NO_2^- is an average of 1.5, and in NO_3^- it is an average of 1.3. Therefore, the bond length should be a little shorter and the bond energy a little higher in NO_2^-.

8.59

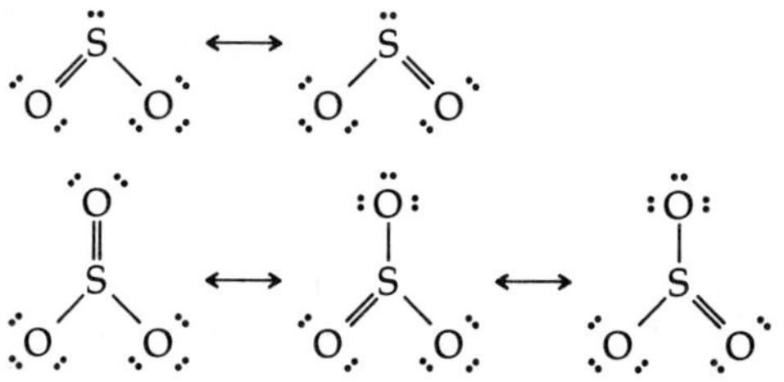

Molecule	Average Bond Order	Bond Length	Bond Energy	Vib. Frequency
SO_2	1.5	shorter	more	higher
SO_3	1.3	longer	less	lower

8.61 (See Question 8.59)

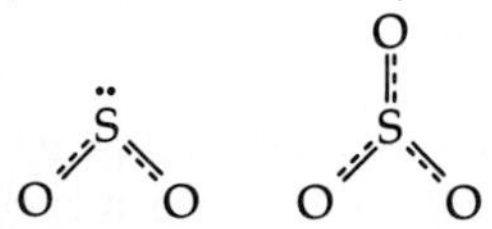

8.66 (a) 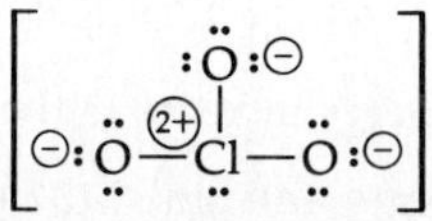 (b)

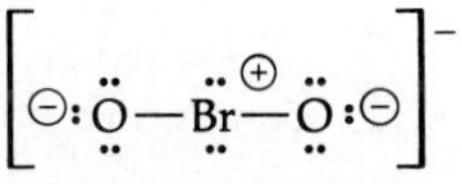

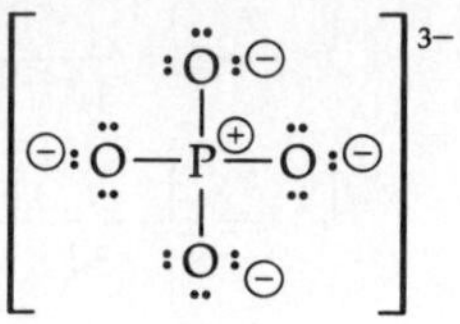

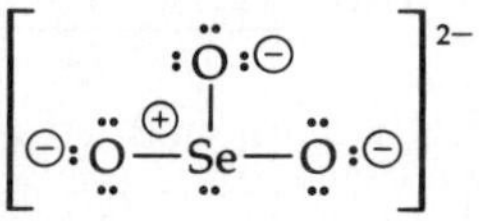

8.68 In each of the ions in Question 8.66 a better Lewis structure can be obtained if the octet rule is not obeyed.

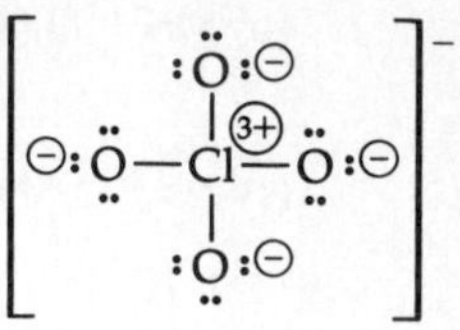

(a)

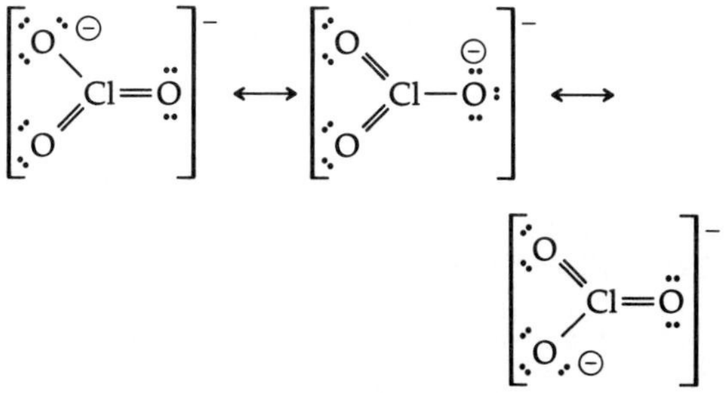

This has each atom, except the indicated oxygen, with a formal charge of zero. Remember: "Those (structures) with the smallest formal charges are the most stable and are preferred."

(b)

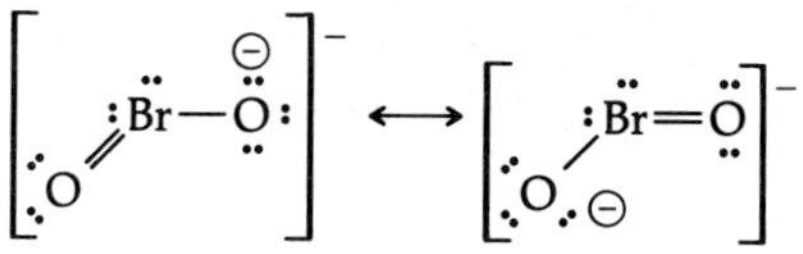

(c)

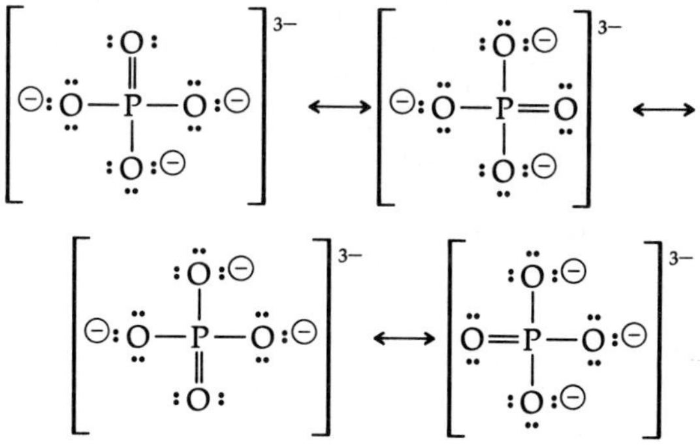

(d)

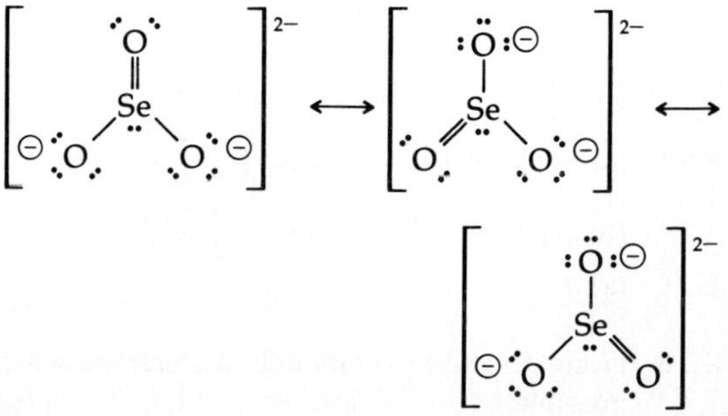

(e)
$$\left[\begin{matrix} :\!O\!: \\ \| \\ \ominus:\!\overset{..}{O}\!-\!\overset{..}{Cl}\!=\!O\!: \\ | \\ :\!O\!: \end{matrix}\right]^{-} \longleftrightarrow \left[\begin{matrix} :\!\overset{..}{O}\!:\!\ominus \\ \| \\ \overset{..}{O}\!=\!\overset{..}{Cl}\!=\!O\!: \\ | \\ :\!O\!: \end{matrix}\right]^{-} \longleftrightarrow$$

$$\left[\begin{matrix} :\!O\!: \\ \| \\ \overset{..}{O}\!=\!\overset{..}{Cl}\!-\!\overset{..}{O}\!:\!\ominus \\ | \\ :\!O\!: \end{matrix}\right]^{-} \longleftrightarrow \left[\begin{matrix} :\!O\!: \\ \| \\ \overset{..}{O}\!=\!\overset{..}{Cl}\!=\!\overset{..}{O}\!: \\ | \\ :\!\overset{..}{O}\!:\!\ominus \end{matrix}\right]^{-}$$

8.70 Nitrogen (a period 2 element) cannot have more than an octet in its valence shell.

8.72 The structure $:\!\overset{..}{O}\!=\!C\!=\!\overset{..}{O}\!:$ has formal charges of zero on all its atoms.

8.74 $H\!-\!\overset{..}{\underset{..}{O}}\!-\!\overset{..}{\overset{\oplus}{F}}\!-\!\overset{..}{O}\!:\!\ominus$ Fluorine is more electronegative than oxygen, yet in this structure it is forced to carry the positive formal charge.

8.77 $:\!\overset{..}{Cl}\!-\!\overset{\overset{\displaystyle :O:}{\|}}{\underset{\underset{\displaystyle :O:}{\|}}{S}}\!-\!\overset{..}{Cl}\!:$ (All formal charges are zero.)

8.79 $:\!\overset{..}{Cl}\!-\!\overset{\overset{\displaystyle :\overset{..}{Cl}:}{|}}{\underset{\underset{\displaystyle :\overset{..}{Cl}:}{|}}{Al}} + \left[:\!\overset{..}{\underset{..}{Cl}}\!:\right]^{-} \longrightarrow \left[:\!\overset{..}{Cl}\!-\!\overset{\overset{\displaystyle :\overset{..}{Cl}:}{|}}{\underset{\underset{\displaystyle :\overset{..}{Cl}:}{|}}{Al}}\!\leftarrow\!\overset{..}{Cl}\!:\right]^{-}$

8.80 It has no lone electron pairs, and its atoms cannot accept additional electrons.

8.85 (a) P—F   (b) Al—Cl   (c) Se—Cl
8.87 NH_3, BCl_3, $MgCl_2$, BeI_2, and NaH
8.89 Rb

8.91

The bonds are polar but, since the molecule is symmetrical, it is a nonpolar molecule.

8.92 $\Delta H = -424$ kJ. The lattice energy.
8.95 The absolute difference between a calculated bond energy and an experimental bond energy is proportional to the electronegativity difference between the bonded atoms.

Compounds	HF	HCl	HBr	HI
Difference between calc. and exp. bond energies (kJ/mol)	270	94	50	10

Since the electronegativity of hydrogen is constant and less than the electronegativity of the element to which it is bonded, the differences must be due to the presence of the other elements. The change in bond energy indicates a decrease in electronegativity from F to I.

CHAPTER 9

9.5 (a) 4   (b) 6   (c) 5
9.7 (a) MX_3E; trigonal pyramidal
(b) MX_4; tetrahedral
(c) MX_3; planar triangular
(d) MX_2E; nonlinear or V-shaped
(e) MX_4E; unsymmetrical tetrahedral
(f) MX_2E_3; linear
(g) MX_5E; square pyramidal
(h) MX_4; tetrahedral
(i) MX_6; octahedral
(j) MX_5; trigonal bipyramidal
(k) MX_2E; nonlinear, bent, angular, or V-shaped

9.10

	(1)	(2)
(a)	tetrahedral	bent
(b)	tetrahedral	bent
(c)	octahedral	octahedral
(d)	tetrahedral	tetrahedral
(e)	octahedral	square planar
(f)	tetrahedral	tetrahedral
(g)	planar triangular	planar triangular
(h)	trigonal bipyramidal	T-shaped
(i)	planar triangular	planar triangular
(j)	tetrahedral	trigonal pyramidal
(k)	tetrahedral	tetrahedral

9.12 (a) In Figure 9.2, the X—M—X bond angles in compounds such as SO_2 will be less than 120°.
(b) In Figure 9.3, the X—M—X bond angles will be less than 109.5° in compounds such as NH_3 and H_2O
(c) In Figure 9.4, the axial bonds in the distorted tetrahedron will be bent in the direction of the two bonds in the equatorial plane. In the T-shaped molecule, the axial bonds will be bent toward the equatorial bond.
(d) In Figure 9.5, the four bonds in the square plane of the square pyramidal structure will be pushed toward the atom above the plane.

9.16 (a), (c), (d), (g), (h), and (j)
9.20 The p orbital with one electron on one Cl atom overlaps with the p orbital with one electron on the other Cl atom similar to the overlap of p_z orbitals. For the drawing imagine two atoms like the fluorine in Figure 9.9 arranged so that the half-filled p orbitals overlap. The spins of the electrons must be paired if bonding is to occur.

9.22 See Figure 8.2.
9.30 The overlap of the unpaired electrons in the p orbitals of As with the unpaired electron in the s orbital of the hydrogens will yield bonding with a bond angle of close to 90°.

9.32   The bond angle of 104° is much closer to the 109.5° of sp^3 orbital bonding than it is to the 90° of p orbital bonding.

(unhybridized) P $\underset{3s}{\uparrow\downarrow}$ $\underset{3p}{\uparrow\ \uparrow\ \uparrow}$

(hybridized) P $\underset{sp^3}{\uparrow\downarrow\ \uparrow\ \uparrow\ \uparrow}$

F $\underset{2s}{\uparrow\downarrow}$ $\underset{2p}{\uparrow\downarrow\ \uparrow\downarrow\ \uparrow}$ F $\underset{2s}{\uparrow\downarrow}$ $\underset{2p}{\uparrow\downarrow\ \uparrow\downarrow\ \uparrow}$

F $\underset{2s}{\uparrow\downarrow}$ $\underset{2p}{\uparrow\downarrow\ \uparrow\downarrow\ \uparrow}$

The hybridized sp^3 orbitals of P have three unpaired electrons. Each of the sp^3 orbitals of P that contains only a single electron can bond with an unpaired electron in a $2p$ orbital of one of the F's. The result will be three bonds each formed by the overlap of an sp^3 and a p orbital.

9.35   (a) planar triangular, sp^2 bonding
        (b) tetrahedral, sp^3 bonding
        (c) trigonal bipyramidal, sp^3d bonding
        (d) octahedral, sp^3d^2 bonding
        (e) linear, sp bonding
        (f) octahedral, sp^3d^2 bonding
        (g) trigonal pyramidal, sp^3 bonding
        (h) unsymmetrical tetrahedral, sp^3d bonding
        (i) tetrahedral, sp^3 bonding

9.36   (a) sp^3 (d) sp^2 (g) sp^3d^2 (j) sp^3d
        (b) sp^3 (e) sp^3d (h) sp^3 (k) sp^2
        (c) sp^2 (f) sp^3d (i) sp^3d^2

9.38   Sb $\underset{5s}{\uparrow\downarrow}$ $\underset{5p}{\uparrow\ \uparrow\ \uparrow}$ $\underset{5d}{\rule{0.3cm}{0.4pt}\ \rule{0.3cm}{0.4pt}\ \rule{0.3cm}{0.4pt}\ \rule{0.3cm}{0.4pt}\ \rule{0.3cm}{0.4pt}}$

SbCl$_5$ $\underset{sp^3d}{\uparrow x\ \uparrow x\ \uparrow x\ \uparrow x\ \uparrow x}$ $\rule{0.3cm}{0.4pt}\ \rule{0.3cm}{0.4pt}\ \rule{0.3cm}{0.4pt}\ \rule{0.3cm}{0.4pt}$

x = Cl electrons   unhybridized
                          5d orbitals

In SbCl$_5$, sp^3d hybrid orbitals are involved in bonding.

9.40   Boron in BCl$_3$ is sp^2-hybridized. Nitrogen in NH$_3$ is sp^3-hybridized. In order for boron to accept two electrons from the nitrogen to form Cl$_3$BNH$_3$, it must first become sp^3-hybridized. The geometry of B changes from planar triangular to tetrahedral. The N does not change geometry or hybridization.

9.42   Sn is sp^3-hybridized. The Sn—Cl bond would be formed by sp^3–p overlap.

9.45   The Lewis structure is [$\cdot$C≡N$\cdot$]$^-$. The bonding consists of one sigma bond and two pi bonds.

CHAPTER 10

10.4   Mn(s) + 2H$^+$(aq) → H$_2$(g) + Mn^{2+}(aq)
10.9   4Zn(s) + NO$_3^-$(aq) + 10H$^+$(aq) →
                   4Zn^{2+}(aq) + NH$_4^+$(aq) + 3H$_2$O

10.12   (a) 2Cr(s) + 6HCl(aq) → 2CrCl$_3$(aq) + 3H$_2$(g)
        (b) Ni(s) + H$_2$SO$_4$(aq) → NiSO$_4$(aq) + H$_2$(g)
10.13   For 10.11: (a) Mg(s) + 2H$^+$(aq) → Mg^{2+}(aq) + H$_2$(g)
        (b) 2Al(s) + 6H$^+$(aq) →
                        2Al^{3+}(aq) + 3H$_2$(g)
        For 10.12: (a) 2Cr(s) + 6H$^+$(aq) →
                        2Cr^{3+}(aq) + 3H$_2$(g)
        (b) Ni(s) + 2H$^+$(aq) → Ni^{2+}(aq) + H$_2$(g)
        (The oxidizing agent is H$^+$ in each of these reactions.)
10.15   The order of increasing ease of oxidation is
        Ag < Cu < Sn < Cd < Mg.
10.16   (a) 2Al(s) + 3Zn^{2+}(aq) → 2Al^{3+}(aq) + 3Zn(s)
        (b) Sn(s) + Cu^{2+}(aq) → Sn^{2+}(aq) + Cu(s)
        (c) Ag(s) + Co^{2+}(aq) → no reaction
        (d) Mn(s) + Pb^{2+}(aq) → Mn^{2+}(aq) + Pb(s)
        (e) Cu(s) + Mg^{2+}(aq) → no reaction
        (f) Hg(l) + H$^+$(aq) → no reaction
        (g) Ni(s) + 2H$^+$(aq) → Ni^{2+}(aq) + H$_2$(g)
        (h) Cd(s) + H$_2$O → no reaction
        (i) Ba(s) + 2H$_2$O → Ba^{2+}(aq) + 2OH$^-$(aq) + H$_2$(g)
        (j) H$_2$(g) + Pt^{2+}(aq) → Pt(s) + 2H$^+$(aq)
10.18   (a) Rb   (c) Na
        (b) Rb   (d) Ca
10.23   (a) 2Na(s) + 2H$_2$O → H$_2$(g) + 2Na$^+$(aq) + 2OH$^-$(aq)
        (b) 2Rb(s) + 2H$_2$O → H$_2$(g) + 2Rb$^+$(aq) + 2OH$^-$(aq)
        (c) Sr(s) + 2H$_2$O → H$_2$(g) + Sr^{2+}(aq) + 2OH$^-$(aq)
10.25   (a) C < N < O < F
        (b) I < Br < Cl < F
10.26   (a) F$_2$ + 2Cl$^-$ → 2F$^-$ + Cl$_2$ (c) I$_2$ + Cl$^-$ → N.R.
        (b) Br$_2$ + Cl$^-$ → N.R.   (d) Br$_2$ + 2I$^-$ → 2Br$^-$ + I$_2$
10.32   (a) 4Fe(s) + 3O$_2$(g) → 2Fe$_2$O$_3$(s)
        (b) 4Li(s) + O$_2$(g) → 2Li$_2$O(s)
        (c) 2Ca(s) + O$_2$(g) → 2CaO(s)
        (d) 2Mg(s) + O$_2$(g) → 2MgO(s)
        (e) 4Al(s) + 3O$_2$(g) → 2Al$_2$O$_3$(s)
10.35   (a) C$_9$H$_{20}$ + 14O$_2$ → 9CO$_2$ + 10H$_2$O
        (b) 2C$_2$H$_4$(OH)$_2$ + 5O$_2$ → 4CO$_2$ + 6H$_2$O
        (c) 2(CH$_3$)$_2$S + 9O$_2$ → 4CO$_2$ + 6H$_2$O + 2SO$_2$
10.37   2C$_{20}$H$_{42}$ + 21O$_2$ → 40C + 42H$_2$O
        2C$_{20}$H$_{42}$ + 41O$_2$ → 40CO + 42H$_2$O
        2C$_{20}$H$_{42}$ + 61O$_2$ → 40CO$_2$ + 42H$_2$O
10.40   (a) NH$_2^-$ (c) C$_2$H$_3$O$_2^-$ (e) NO$_3^-$
        (b) NH$_3$ (d) H$_2$PO$_4^-$
10.41   (a) H$_2$SO$_4$ (c) H$_3$O$^+$ (e) HCHO$_2$ (or HCOOH)
        (b) HSO$_4^-$ (d) HCl
10.43   Acid–base conjugate pairs (acid written first in each pair).
        (a) HC$_2$H$_3$O$_2$, C$_2$H$_3$O$_2^-$ and H$_2$O, OH$^-$
        (b) HF, F$^-$ and NH$_4^+$, NH$_3$
        (c) HSO$_4^-$, SO$_4^{2-}$ and H$_2$PO$_4^-$, HPO$_4^{2-}$
        (d) Al(H$_2$O)$_6^{3+}$, Al(H$_2$O)$_5$OH^{2+} and H$_2$O, OH$^-$
        (e) N$_2$H$_5^+$, N$_2$H$_4$ and H$_2$O, OH$^-$
        (f) NH$_3$OH$^+$, NH$_2$OH and HCl, Cl$^-$
        (g) OH$^-$, O^{2-} and H$_2$O, OH$^-$
        (h) H$_2$, H$^-$ and H$_2$O, OH$^-$
        (i) NH$_3$, NH$_2^-$ and N$_2$H$_4$, N$_2$H$_3^-$
        (j) HNO$_3$, NO$_3^-$ and H$_3$SO$_4^+$, H$_2$SO$_4$

10.47 $HCO_3^- + H_2O \leftrightarrows CO_3^{2-} + H_3O^+$ (as an acid)
$HCO_3^- + H_2O \leftrightarrows H_2CO_3 + OH^-$ (as a base)

10.48 The highly charged Cr^{3+} ion polarizes the O—H bonds of the water molecules attached to it, thereby making it easier for the H^+ to be transferred to a neighboring H_2O molecule.

$$Cr(H_2O)_6^{3+} + H_2O \Longleftrightarrow Cr(H_2O)_5(OH)^{2+} + H_3O^+$$

10.49 (a) $HClO_3$ (d) H_2SO_4 (f) $HBrO_3$ (h) HBr
(b) HNO_3 (e) $HClO_3$ (g) H_2Se (i) PH_3
(c) H_3PO_4

10.51 CH_3SH. This compound and CH_3OH act like weak binary acids. The CH_3SH has a much weaker S—H bond than is the O—H bond in CH_3OH.

10.56 $HClO_4$ is a stronger acid than $HClO_3$ because the presence of another electronegative oxygen in $HClO_4$ makes the O—H bond more polar. The anion from $HClO_4$ will also have the negative charge spread over a greater number of atoms.

10.60

10.62 $H_3N \rightarrow Ag^+ \leftarrow NH_3 \longrightarrow [H_3N—Ag—NH_3]^+$

CHAPTER 11

11.3 (a) 1.14×10^3 torr (e) 2.20×10^4 Pa
(b) 1.03 atm (f) 3.38 atm
(c) 3.50×10^5 Pa (g) 86.3 torr
(d) 350 kPa

11.8 158 torr
11.10 239 torr
11.12 882 torr
11.22 288 mL
11.24 802 torr
11.26 102 °C
11.27 34 lb/in.^{2}
11.28 1.88 L
11.30 2.49 L
11.32 375 °C
11.34 61.4 cm
11.36 513 torr
11.38 350 °C
11.40 26.2 mL
11.43 8.31 Pa m^3 mol^{-1} K^{-1}
11.44 (a) 4.48 L (at STP) (c) 3.36 L (total at STP)
(b) 3.92 L (at STP)
11.45 0.701 g SO_2
11.47 43.9 g/mol
11.51 403 mL
11.53 (a) 66.0% C, 15.2% H, 19.2% N
(b) $C_4H_{11}N$
11.54 0.195 g N_2
11.56 200 mL N_2, 600 mL H_2
11.58 (a) 0.144 g O_2 (b) 0.368 g $KClO_3$

11.60 975 torr
11.62 (a) 2.10×10^{-6} mol I_2 (d) 4.70×10^{-2} mL
(b) 2.10×10^{-6} mol I_2 (e) 0.000235 ppm
(c) 2.10×10^{-6} mol O_3
11.64 (a) 1.36×10^{-3} mol O_2 (c) 44.4%
(b) 0.111 g $KClO_3$
11.66 850 torr
11.68 680 torr
11.70 81.5 mL
11.72 $X_{N_2} = 0.749$, $X_{O_2} = 0.153$, $X_{CO_2} = 0.037$,
$X_{H_2O} = 0.062$
11.74 19.6 mL
11.76 1560 torr (rounded)
11.78 255 mL
11.81 Helium effuses 2.25 times faster than neon.
11.83 1.99 g/mol
11.96 1.002 atm
11.98 22.385 L/mol at STP
11.99 8.98 g/eq; aluminum
11.101 871 torr

CHAPTER 12

12.6

	Dipole–Dipole	Hydrogen Bonding	London Forces
(a) HCl	x		x
(b) Ar			x
(c) CH_4			x
(d) HF	x	x	x
(e) NO	x		x
(f) CO_2			x
(g) H_2S	x		x
(h) SO_2	x		x

12.9 London forces should increase from helium to neon to argon as the atoms become larger.
12.22 They sublime at the low pressures found at high altitudes.
12.26 Methyl alcohol has the weaker intermolecular attractions at this temperature.
12.29 Substance X should have the higher boiling point. Substance Y would be less likely to hydrogen bond.
12.31 ΔH_{vap} should increase $PH_3 < AsH_3 < SbH_3$ because of the increased size.
12.34 The gasoline evaporates very rapidly, thereby removing heat *faster* than the evaporation of water.
12.38 46.1 kJ
12.40 59.3 kJ/mol, 14.2 kcal/mol
12.42 2.50 kJ of energy absorbed
12.44 40.2 °C
12.47 Ethyl alcohol
12.51 Vapor pressure is the equilibrium pressure of the gas in the equilibrium, liquid $\leftrightarrows$ gas. The value of the equilibrium gas pressure is independent of the volume of liquid or gas. Increased surface area of

the liquid will increase the rate at which equilibrium can be attained but not the value at equilibrium. A larger surface area will allow more molecules to enter the gaseous phase, but at the same time, if equilibrium exists, more will be condensing to the liquid phase.

12.57   Below -267.8 °C (its critical temperature).

12.60   Intermolecular attractions are weaker in I_2 than in NaCl.

12.65   The equilibrium will shift to the left.

12.68   Approximately 73 °C; between 73 and 75 °C.

12.82   (a) 6.57°   (b) 34.9°

12.84   20.6° and 44.8°

12.86 $r = 144.20$ pm

12.88   176 pm

12.90   (a) 7.50 g/cm³   (c) 10.6 g/cm³; Therefore, (b) 9.74 g/cm³   face-centered cubic

12.92 $r = 81$ pm   This is the maximum radius of the Li^+. If it is smaller, it simply will not fill all the available space between the bromide ions.

12.94   0.43 nm

12.96   101 pm

12.98   (a) molecular   (d) ionic   (f) ionic
(b) molecular   (e) molecular   (g) molecular
(c) metallic

12.100 $SnCl_4$, molecular solid; $SnCl_2$, ionic solid

12.102   Molecular solid

12.104   Ionic solid

12.110   Prepare a diagram like that in Figure 12.41. Position the triple point at 150 torr and 20 °C. Have the solid–liquid line pass through 760 torr and 25 °C and the liquid–gas line pass through 760 torr and 95 °C, with both intersecting at the triple point.

12.111

	At 22 °C		At 10 °C	
State	Pressure (torr)	State	Pressure (torr)	
vapor	up to 160	vapor	up to 75	
vapor–liquid	160	solid–vapor	75	
liquid	160 to 250	solid	75 to 1,000	
solid–liquid	250			
solid	250 to 1,000			

12.113   The density of the solid is greater than that of the liquid because the solid–liquid line leans to the right (connecting the triple point, 20 °C–150 torr, and the melting point, 25 °C–760 torr).

12.117   For (b) boiling point, (c) heat of vaporization, (d) surface tension, (f) heat of sublimation, and (g) critical temperature to be arranged from highest value to lowest value, the compounds would follow the order propylene glycol, isopropyl alcohol, acetone, methyl ethyl ether, and butane. For (a) vapor pressure and (e) rate of evaporation, the order would be reversed with butane having the highest value.

12.118 5.56×10^4 J

CHAPTER 13

13.7

	Dispersing Phase	Dispersed Phase	Kind of Colloid
(a) Styrofoam	solid	gas	solid foam
(b) cream	liquid	liquid	emulsion
(c) lard	solid	liquid	solid emulsion
(d) jelly	liquid	solid	gel, sol
(e) liq. rubber cement	liquid	solid	sol, gel

13.11   First AgCl forms and begins to form dispersed particles. The excess Cl^- absorbs on the AgCl particles and acts to prevent them from growing too large to remain suspended. As more Ag^+ is added, the electrical charge is neutralized by the formation of more AgCl. The particles grow in size and will settle out of solution.

13.17   An interstitial solid solution would be more likely to exist under the conditions described in this question.

13.20   For glycerin, $X = 0.0810$, $w = 0.310$, weight percent $= 31.0\%$, molality $= 4.89$ m.

13.22   (a) weight percent $= 11.00\%$
(b) molality $= 0.653$ m
(c) mole fraction $= 0.0116$
(d) molarity $= 0.6431$ M

13.24   (a) $1.24 \times 10^{-3}\%$ (b) 1.04×10^{-4} M

13.26 $X_{NaHCO_3} = 0.019$, molality $= 1.1$ m

13.28 $X_{Na_2CO_3} = 0.0269$, molality $= 1.53$ m

13.30   Molality $= 2.4 \times 10^2$ m H_2SO_4, $X_{H_2SO_4} = 0.82$, $X_{H_2O} = 0.18$

13.32   (a) 76.0 mol % $CHCl_3$ (d) 17.1% C_6H_6 and
(b) 2.64 m C_6H_6 82.9% $CHCl_3$
(c) 40.5 m $CHCl_3$

13.34   23.9% $C_2H_4(OH)_2$, $X_{C_2H_4(OH)_2} = 0.0837$, 5.07 m $C_2H_4(OH)_2$

13.36   13.5 g $Na_2CO_3 \cdot 10H_2O$

13.41   Ammonia can form hydrogen bonds to water.

13.44   The ions of the salt increase the effective polarity of the solvent and, therefore, decrease the solubility of solutes that are not as polar.

13.52   Water forms hydrogen bonds to acetone, but acetone molecules can't form hydrogen bonds to each other.

13.57 A large exothermic hydration energy for Al^{3+}.
13.59 24.1 kJ liberated
13.66 572 torr
13.68 0.216 g/L
13.73 92.0 torr
13.75 549 torr
13.77 144 g/mol
13.81 Approximately four or five times.
13.86 $-4.44\ °C$
13.88 $C_8H_8O_4$
13.90 $100.68\ °C$
13.92 $-33.3\ °C$
13.97 2000 g/mol
13.98 $-0.56\ °C$
13.102 29.1 atm A pressure greater than 29.1 atm.
13.103 $-0.23\ °C$
13.105 $Al_2(SO_4)_3$
13.108 89.5 torr
13.109 736 torr

CHAPTER 14

14.16 First step: $q = -w = 75.0$ L atm,
$$\Delta E_{sys} = \Delta E_{surroundings} = 0$$
Second step: $q = -w = 130$ L atm,
$$\Delta E_{sys} = \Delta E_{surroundings} = 0$$
14.18 $q = 12.6$ kJ, $w = 0.0760$ kJ, $\Delta E_{sys} = 12.7$ kJ,
$\Delta E_{surroundings} = -12.7$ kJ
14.19 (a) $q = 35$ J, $w = 40$ J, $\Delta E = 75$ J
(b) $q = -35$ J, $w = -40$ J, $\Delta E = -75$ J
14.21 Only for reaction (a).
14.24 203 kJ/mol
14.26 8 kJ/mol
14.33 (a), (b), and (d)
14.35 (a) positive   (c) positive   (e) negative
(b) negative   (d) positive
14.45 (a) -273.3 J mol^{-1} K^{-1} (d) -274 J mol^{-1} K^{-1}
(b) -305.0 J mol^{-1} K^{-1} (e) -99.2 J mol^{-1} K^{-1}
(c) -358 J mol^{-1} K^{-1}
14.47 (a) -38.4 J/K   (c) -189 J/K   (e) -121 J/K
(b) -196 J/K   (d) 48.6 J/K
14.49 (a) -836 kJ   (c) -346 kJ   (e) -101 kJ
(b) -1364 kJ   (d) -101 kJ
14.51 -2880 kJ (rounded)
14.53 -122 kJ
14.56 -2074 kJ   A real process does not follow a reversible path.
14.59 $\Delta S_{vap} = 109$ J mol^{-1} K^{-1}, $\Delta S_{fus} = 22.1$ J mol^{-1} K^{-1}
14.66 (a) $\Delta G° = +51.9$ kJ (No)
(b) $\Delta G° = -57$ kJ (Yes)
(c) $\Delta G° = -511$ kJ (Yes)
(d) $\Delta G° = +92$ kJ (No)
(e) $\Delta G° = -34.9$ kJ (Yes)
(f) $\Delta G° = +1637$ kJ (No)

CHAPTER 15

15.4 (a) $K_P = \dfrac{p_{CH_3OH}}{p_{CO}p_{H_2}^2}$, $K_c = \dfrac{[CH_3OH]}{[CO][H_2]^2}$

(b) $K_P = \dfrac{p_{CO_2}p_{H_2}}{p_{CO}p_{H_2O}}$, $K_c = \dfrac{[CO_2][H_2]}{[CO][H_2O]}$

(c) $K_P = \dfrac{p_{PCl_5}}{p_{PCl_3}p_{Cl_2}}$, $K_c = \dfrac{[PCl_5]}{[PCl_3][Cl_2]}$

(d) $K_P = \dfrac{p_{N_2}p_{H_2O}^4}{p_{NO_2}^2 p_{H_2}^4}$, $K_c = \dfrac{[N_2][H_2O]^4}{[NO_2]^2[H_2]^4}$

(e) $K_P = \dfrac{p_{H_2O}^2 p_{SO_2}^2}{p_{H_2S}^2 p_{O_2}^3}$, $K_c = \dfrac{[H_2O]^2[SO_2]^2}{[H_2S]^2[O_2]^3}$

15.6 Mass action expression $= 5.5 \pm 0.1$. $K_c = 5.5$
15.8 (a) 7.1×10^{-8} (b) 2.0×10^{14}
15.10 1.0×10^{38}
15.12 The tendency to proceed toward completion increases in the order (b) < (c) < (d) < (a).
15.14 reaction (a)
15.18 2×10^{62}
15.20 (a) $+80$ kJ   (b) 1.3×10^{-14}
15.22 $+17.1$ kJ
15.24 3×10^{-14}
15.26 The system is not at equilibrium, and the reaction will proceed spontaneously to the left.
15.28 4.05
15.30 2.0 L mol^{-1}
15.33 (a) $K_c = [CO_2(g)]$

(b) $K_c = \dfrac{[Ni(CO)_4(g)]}{[CO(g)]^4}$

(c) $K_c = \dfrac{[I_2(g)][CO_2(g)]^5}{[CO(g)]^5}$

(d) $K_c = \dfrac{[CO_2(g)]}{[Ca(HCO_3)_2(aq)]}$

(e) $K_c = [Ag^+(aq)][Cl^-(aq)]$
15.34 (a) shifts to the right   (d) shifts to the right
(b) shifts to the left   (e) no effect
(c) shifts to the right
15.35 None of them affects the value of K_c.
15.38 (a) increased   (c) increased
(b) decreased   (d) no change
15.41 (a) no change   (c) decreased
(b) increased   (d) increased
15.44 The system is not at equilibrium. The reaction will proceed to the right.
15.46 (a) 1.40×10^{-1} atm   (c) 4.87 kJ/mol
(b) 5.72×10^{-3} mol L^{-1}
15.48 0.30 atm^{-1}
15.52 $[SO_2] = 0.0005\ M$, $[NO_2] = 0.0105\ M$,
$[NO] = 0.0195\ M$, $[SO_3] = 0.0245\ M$
15.54 $[CO] = [Cl_2] = 2.1 \times 10^{-6}\ M$, $[COCl_2] = 0.020\ M$
15.56 $[HI] = 0.113\ M$, $[H_2] = [I_2] = 0.0153\ M$
15.58 $[CO] = 2.2 \times 10^{-10}\ M$, $[Cl_2] = 0.15\ M$,
$[COCl_2] = 0.15\ M$

15.60 $[BrF_3] = 2.4 \times 10^{-19}\ M$, $[F_2] = 2.4 \times 10^{-19}\ M$, $[BrF_5] = 0.0500\ M$

CHAPTER 16

16.5 (a) $[H^+] = 1.0 \times 10^{-3}\ M$ $[OH^-] = 1.0 \times 10^{-11}\ M$
   pH = 3.00
   (b) $[H^+] = 1.25 \times 10^{-1}\ M$ $[OH^-] = 8.00 \times 10^{-14}\ M$
   pH = 0.903
   (c) $[H^+] = 3.2 \times 10^{-12}\ M$ $[OH^-] = 3.1 \times 10^{-3}\ M$
   pH = 11.49
   (d) $[H^+] = 4.2 \times 10^{-13}\ M$ $[OH^-] = 2.4 \times 10^{-2}\ M$
   pH = 12.38
   (e) $[H^+] = 2.1 \times 10^{-4}\ M$ $[OH^-] = 4.8 \times 10^{-11}\ M$
   pH = 3.68
   (f) $[H^+] = 1.3 \times 10^{-5}\ M$ $[OH^-] = 7.7 \times 10^{-10}\ M$
   pH = 4.89
   (g) $[H^+] = 1.2 \times 10^{-12}\ M$ $[OH^-] = 8.4 \times 10^{-3}\ M$
   pH = 11.92
   (h) $[H^+] = 2.1 \times 10^{-13}\ M$ $[OH^-] = 4.8 \times 10^{-2}\ M$
   pH = 12.68

16.8 (a) acidic   (c) neutral   (e) basic
   (b) basic   (d) acidic

16.11 (a) $[H^+] = 0.050\ M$ $[OH^-] = 2.0 \times 10^{-13}\ M$
   (b) $[H^+] = 1.9 \times 10^{-6}\ M$ $[OH^-] = 5.3 \times 10^{-9}\ M$
   (c) $[H^+] = 1.0 \times 10^{-4}\ M$ $[OH^-] = 1.0 \times 10^{-10}\ M$
   (d) $[H^+] = 1.6 \times 10^{-8}\ M$ $[OH^-] = 6.25 \times 10^{-7}\ M$
   (e) $[H^+] = 1.1 \times 10^{-11}\ M$ $[OH^-] = 9.1 \times 10^{-4}\ M$
   (f) $[H^+] = 2.5 \times 10^{-13}\ M$ $[OH^-] = 4.0 \times 10^{-2}\ M$

16.14 (a) $HOBr + H_2O \rightleftharpoons H_3O^+ + OBr^-$
   (b) $HCN + H_2O \rightleftharpoons H_3O^+ + CN^-$
   (c) $(CH_3)_3NH^+ + H_2O \rightleftharpoons H_3O^+ + (CH_3)_3N$
   (d) $C_5H_5NH^+ + H_2O \rightleftharpoons H_3O^+ + C_5H_5N$
   (e) $HCO_3^- + H_2O \rightleftharpoons H_3O^+ + CO_3^{2-}$

16.15 (a) $C_5H_5N + H_2O \rightleftharpoons C_5H_5NH^+ + OH^-$
   (b) $CO_3^{2-} + H_2O \rightleftharpoons HCO_3^- + OH^-$
   (c) $H_2PO_4^- + H_2O \rightleftharpoons H_3PO_4 + OH^-$
   (d) $NO_2^- + H_2O \rightleftharpoons HNO_2 + OH^-$
   (e) $C_6H_5NH_2 + H_2O \rightleftharpoons C_6H_5NH_3^+ + OH^-$

16.17 (d) < (c) < (a) < (b)
16.19 PH_2^- is a stronger base than HS^-.
16.21 Ammonia
16.24 1.4×10^{-4}
16.26 (a) $1.2 \times 10^{-2}\ M$ (d) $1.2 \times 10^{-3}\ M$
   (b) $2.5 \times 10^{-2}\ M$ (e) $7.1 \times 10^{-4}\ M$
   (c) $3.5 \times 10^{-6}\ M$
16.28 (a) $8.3 \times 10^{-13}\ M$ (d) $8.3 \times 10^{-12}\ M$
   (b) $4.0 \times 10^{-13}\ M$ (e) $1.4 \times 10^{-11}\ M$
   (c) $2.9 \times 10^{-9}\ M$
16.30 9.6×10^{-3}
16.32 3.1×10^{-10}
16.33 (a) 1.3%   (c) 0.014%   (e) 0.024%
   (b) 3.7%   (d) 0.63%   (f) 100%
16.36 9.1 g HCl
16.37 Essentially 100%
16.39 $0.57\ M$
16.41 7.8×10^{-8}

16.43 $[H^+] = 1.3 \times 10^{-3}\ M$, $[CHO_2^-] = 1.3 \times 10^{-3}\ M$,
   $[HCHO_2] = 0.009\ M$, $[OH^-] = 7.7 \times 10^{-12}\ M$
16.47 (a) $3.0 \times 10^{-5}\ M$ (d) $3.3 \times 10^{-10}\ M$
   (b) $1.8 \times 10^{-4}\ M$ (e) $1.0 \times 10^{-8}\ M$
   (c) $3.4 \times 10^{-4}\ M$
16.48 0.41
16.50 210 g
16.53 (a) +0.18   (c) +0.59   (e) +7.30
   (b) +0.37   (d) +4.03
16.57 (a) neutral   (c) basic
   (b) acidic   (d) acidic
16.59 (a) 7.88   (c) 8.38   (e) 3.37
   (b) 5.08   (d) 11.15
16.61 2.1×10^{-6}
16.63 8.86
16.65 8.84
16.67 10.11
16.70 pH = 1.4, $[H_2SeO_3] = 0.46\ M$, $[HSeO_3^-] = 0.04\ M$,
   $[SeO_3^{2-}] = 5 \times 10^{-8}\ M$
16.72 $[H^+] = [HC_6H_6O_6^-] = 2.0 \times 10^{-3}\ M$,
   $[OH^-] = 5.0 \times 10^{-12}\ M$, $[H_2C_6H_6O_6] = 0.048\ M$,
   $[C_6H_6O_6^{2-}] = 1.6 \times 10^{-12}\ M$, pH = 2.70
16.74 $[HCO_3^-] = 4.3 \times 10^{-5}\ mol/L$,
   $[CO_3^{2-}] = 2.4 \times 10^{-12}\ M$
16.81 (a) 7.00   (c) 2.10
   (b) 30.0 mL   (d) 11.00
16.83 (a) 2.17   (b) 3.19   (c) 8.00
16.91 Green
16.94 1.3 mL
16.96 Acidic, pH = 6.98
16.98 0.57

CHAPTER 17

17.3 (a) $K_{sp} = [Pb^{2+}][F^-]^2$ (d) $K_{sp} = [Li^+]^2[CO_3^{2-}]$
   (b) $K_{sp} = [Ag^+]^3[PO_4^{3-}]$ (e) $K_{sp} = [Ca^{2+}][IO_3^-]^2$
   (c) $K_{sp} = [Fe^{2+}]^3[PO_4^{3-}]^2$ (f) $K_{sp} = [Ag^+]^2[Cr_2O_7^{2-}]$
17.5 3.2×10^{-14}
17.7 1.0×10^{-4}
17.9 3.20×10^{-8}
17.11 1.6×10^{-5}
17.13 8.7×10^{-5}
17.14 (a) $3.3 \times 10^{-7}\ M$ (d) $8 \times 10^{-7}\ M$
   (b) $8 \times 10^{-6}\ M$ (e) $3.5 \times 10^{-4}\ M$
   (c) $3.9 \times 10^{-5}\ M$ (f) $9.3 \times 10^{-3}\ M$
17.16 0.8 g $CaSO_4$
17.18 $2 \times 10^{-12}\ mol/L$
17.20 (a) no precipitate   (c) $CaSO_4$ will precipitate.
   (b) BaF_2 will precipitate.
17.22 $[Ag^+] = 0.050\ M$, $[NO_3^-] = 0.10\ M$, $[H^+] = 0.050\ M$,
   $[Cl^-] = 3.4 \times 10^{-9}\ M$, $[OH^-] = 2.0 \times 10^{-13}\ M$
17.25 $2 \times 10^{-8}\ mol/L$
17.27 $3.6 \times 10^{-3}\ mol/L$
17.29 $2.2 \times 10^{-6}\ mol/L$
17.31 2.1 g NaF
17.33 (a) $4.0 \times 10^{-3}\ mol/L$ (b) $6.5 \times 10^{-4}\ mol/L$
17.35 (a) 2.2 g $Fe(OH)_2$ precipitated
   (b) $[Fe^{2+}] = 5 \times 10^{-12}\ M$

17.37 $[Mn^{2+}] = 0.100\ M$, $[Fe^{2+}] \approx 2 \times 10^{-5}\ M$, pH $= 9.04$

17.39 pH of less than 1.66

17.41 pH of less than 5.62

17.43 (a) $Fe(CN)_6^{4-} \rightleftharpoons Fe^{2+} + 6CN^-$ $K_{inst} = \dfrac{[Fe^{2+}][CN^-]^6}{[Fe(CN)_6^{4-}]}$

(b) $K_{inst} = \dfrac{[Cu^{2+}][Cl^-]^4}{[CuCl_4^{2-}]}$

(c) $K_{inst} = \dfrac{[Ni^{2+}][NH_3]^6}{[Ni(NH_3)_6^{2+}]}$

17.46 4.9×10^{-3} mol/L

17.48 9.7×10^{-3} mol/L

17.51 A precipitate will form.

17.53 More than 1.8 g of $NaC_2H_3O_2$.

17.55 4.2×10^{-6} mol/L

17.57 $9.1 \times 10^{-15}\ M$

CHAPTER 18

18.9 anode: $2I^- \longrightarrow I_2 + 2e^-$
cathode: $Ni^{2+} + 2e^- \longrightarrow Ni$
cell: $Ni^{2+} + 2I^- \longrightarrow Ni + I_2$
or $NiI_2(aq) \longrightarrow Ni(s) + I_2(aq)$

18.22 (a) 2 (c) 5 (e) 8
(b) 1 (d) 2

18.24 (a) 1 (c) 10 (e) 8
(b) 4 (d) 2

18.26 54 min

18.28 (a) 1.79×10^{-3} mol e^- (c) 0.217 mol e^-
(b) 1.11 mol e^- (d) 0.412 mol e^-

18.30 0.15 g O_2 and 0.019 g H_2; 0.10 L O_2 and 0.21 L H_2 (at STP)

18.32 (a) 272 g Ag (b) 1.02×10^4 cm^2

18.34 9.10 hr

18.36 1.23 A

18.38 0.0798 A

18.39 434 min

18.41 0.0833 mol Cr

18.48 (a) $Fe(s)|Fe^{2+}\ (1\ M)\|Ni^{2+}\ (1\ M)|Ni(s)$
(b) $Cr(s)|Cr^{3+}\ (1\ M)\|Sn^{2+}\ (1\ M)|Sn(s)$
(c) $Cu(s)|Cu^{2+}\ (1\ M)\|Ag^+\ (1\ M)|Ag$

18.49 (a) $Zn(s) + Pb^{2+}\ (1\ M) \rightarrow Pb(s) + Zn^{2+}\ (1\ M)$
Zn is the anode; Pb is the cathode.
(b) $Al(s) + 3Ag^+\ (1\ M) \rightarrow Al^{3+}\ (1\ M) + 3Ag(s)$
Al is the anode; Ag is the cathode.
(c) $3Mn(s) + 2Fe^{3+}\ (1\ M) \rightarrow 3Mn^{2+}\ (1\ M) + 2Fe(s)$
Mn is the anode; Fe is the cathode.

18.51 (a) anode: $Zn(s) \rightarrow Zn^{2+}(aq) + 2e^-$
cathode: $Ga^{3+}(aq) + 3e^- \rightarrow Ga(s)$
(b) $3Zn(s) + 2Ga^{3+}(aq) \rightarrow 3Zn^{2+}(aq) + 2Ga(s)$
(c) $E_{Ga^{3+}} = -0.53$ V
(d) $E_{cell} = 0.87$ V

18.53 (a) ClO_3^- (c) MnO_4^-
(b) $Cr_2O_7^{2-}$ (d) PbO_2

18.55 (a) Na (c) Cu (e) H_2
(b) Cl_2 (d) Sn

18.57 (a) 0.63 V (c) 0.99 V
(b) 2.47 V

18.58 (a) $2Al(s) + 3NiSO_4(aq) \rightarrow 3Ni(s) + Al_2(SO_4)_3(aq)$
(b) $3PbO_2(s) + H_2O + Cr_2(SO_4)_3(aq) + K_2SO_4(aq) \rightarrow K_2Cr_2O_7(aq) + 3PbSO_4(s) + H_2SO_4(aq)$
(c) $2AgNO_3(aq) + Pb(s) \rightarrow 2Ag(s) + Pb(NO_3)_2(aq)$
(d) $Cl_2(g) + MnCl_2(aq) + 2H_2O \rightarrow 4HCl(aq) + MnO_2(s)$
(e) $2HCl(aq) + Mn(s) \rightarrow MnCl_2(aq) + H_2(g)$

18.60 (a) -0.38 V, nonspontaneous
(b) -1.49 V, nonspontaneous
(c) $+0.64$ V, spontaneous
(d) -0.16 V, nonspontaneous
(e) -0.13 V, nonspontaneous

18.62 (a) 1.42 V (c) 0.93 V (e) 1.03 V
(b) 0.36 V (d) 0.08 V

18.63 (a) 0.91 V (c) -0.71 V (e) 1.42 V
(b) -0.34 V (d) 0.27 V

18.64 (a) $K_c = 5 \times 10^3$ (c) $K_c = 3$
(b) $K_c = 1 \times 10^9$

18.66 (a) 1×10^{-13} (c) 1×10^{22} (e) 2×10^{-9}
(b) 5×10^{-51} (d) 1×10^{-81}

18.70 (a) $E° = 1.42$ V, $E = 1.45$ V, $\Delta G = -840$ kJ
(b) $E° = 0.11$ V, $E = 0.17$ V, $\Delta G = -33$ kJ
(c) $E° = 1.56$ V, $E = 1.54$ V, $\Delta G = -297$ kJ

18.72 0.23 V

18.75 5.43 M

18.76 (a) 3.15 V (b) 3.13 V (c) 3.14 V

18.77 0.39 hr

18.79 33.0 kJ

18.81 $1.54 \times 10^{-7}\ M$

18.92 0.249 A

18.94 3×10^3 A (one significant figure)

18.96 1×10^{-14}

18.98 $K_c = 0.38$

CHAPTER 19

19.6 (a) Mg (d) iron nail in 1.00 M HCl
(b) Zn in 1.00 M HCl at 40 °C
(c) powdered zinc

19.9 For CO_2, rate $= 0.16$ mol L^{-1} s^{-1}. For H_2O, rate $= 0.32$ mol L^{-1} s^{-1}.

19.10 (a) 2.0 mol L^{-1} s^{-1} (c) -1.0 mol L^{-1} s^{-1}
(b) -1.4 mol L^{-1} s^{-1}

19.15 Zero order

19.17 (a) L mol^{-1} s^{-1} (c) L^2 mol^{-2} s^{-1}
(b) L mol^{-1} s^{-1}

19.19 -1

19.21 (a) 2.35×10^{-6} mol L^{-1} s^{-1}
(b) 1.91×10^{-7} mol L^{-1} s^{-1}

19.23 (a) Rate $= k[NO_2][O_3]$
(b) $k = 4.4 \times 10^7$ L mol^{-1} s^{-1}

19.24 (a) Rate of formation of $C = k[A][B]^2$
(b) $k = 1.2 \times 10^3$ L^2 mol^{-2} s^{-1}
(c) 1.7×10^{-1} mol D L^{-1} s^{-1}

19.30 (a) 2.4×10^{-4} mol/L (c) $5.62 \times 10^{-4}\ M$
(b) 4.75 s

19.31   21 min
19.33   (a) 22 s
         (b) 0.068 M
19.38   (a) Rate = $k[NO][Br_2]$
         (b) Rate = $k[NO]^2[Br_2]$
         (c) Step 2 is rate-determining.
         (d) It is termolecular; we prefer bimolecular processes.
         (e) No, because the mechanism is just a theory. More information is needed.
19.40 $(CH_3)_3CBr \rightarrow (CH_3)_3C^+ + Br^-$ slow
         $(CH_3)_3C^+ + OH^- \rightarrow (CH_3)_3COH$ fast
         for which Rate = $k[(CH_3)_3CBr]$
19.42 $NO_2 + O_3 \rightarrow NO_3 + O_2$ slow
         $NO_3 + NO_2 \rightarrow N_2O_5$ fast
19.46 $NO + O_2 \rightleftharpoons NO_3$ fast
         $NO_3 + NO \rightarrow 2NO_2$ slow
19.52 $E_a = 163$ kJ/mol, $A = 1.41 \times 10^{16}$ L mol^{-1} s^{-1}
19.54 1.2×10^{-4} L mol^{-1} s^{-1}
19.56   18.6 kJ/mol
19.58 1.6×10^2 kJ/mol
19.66 8.72×10^{14} Hz, 344 nm
19.69   (a) initiation step 1   (c) termination step 2
         (b) propagation steps 3, 5, 6
19.71   (a) 16 chirps at 20 °C, 21 chirps at 25 °C, 26 chirps at 30 °C, 31 chirps at 35 °C.
         (b) 32 kJ/mol
         (c) Probably none; it would be dead!
19.73 $2O_3 + h\nu \rightarrow 3O_2$ The ClO consumed in the first step is regenerated in the last two steps.

CHAPTER 20

20.11   188 °C (461 K)
20.13   3180 °C (3450 K)
20.15   At 100 °C, $K_P = 6.65 \times 10^{-93}$; at 500 °C, $K_P = 3.11 \times 10^{-38}$; at 1000 °C, $K_P = 1.36 \times 10^{-4}$.
20.25   (a) Li   (c) Cs   (e) Ga
         (b) Al   (d) Sn
20.27 Al_2O_3
20.31   (a) $GeCl_4$ (c) PbS   (e) MgS
         (b) Bi_2O_5 (d) Li_2S
20.34   (a) Ag_2S (c) SnS_2
         (b) CuBr   (d) Al_2S_3
20.39 -842.1 kJ/mol
20.41   103 J/K

CHAPTER 21

21.14 CrO_4^{2-}
21.16 Cr^{2+}
21.22   Calculated, 19.5 g/cm^3; reported, 19.3 g/cm^3.
21.29   Oxides of metals in high oxidation states tend to be acidic anhydrides since when they hydrate they tend to have one or more nonhydrated oxygens that, along with an increased charge of the central atom, strengthen the X—O bond and weakens the O—H bond.

21.31   The energy needed to remove four electrons is so high that Ti^{4+} ion does not have a real existence.
21.40 $2HCl + Mn \rightarrow MnCl_2 + H_2$
21.46 $2H^+(aq) + Fe(s) \rightarrow Fe^{2+}(aq) + H_2(g)$
21.77   NTA can coordinate to four sites in an octahedral complex in much the same manner as EDTA.

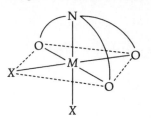

The remaining two sites, X, can be occupied by other ligands or by H_2O molecules.
        The NTA would increase the solubility of metal salts by shifting equilibria such as

$$MX_n(s) + NTA \rightleftharpoons M(NTA)(aq) + X(aq)$$

to the right.
21.79   Isomers of $[Co(NH_3)_2Cl_4]$

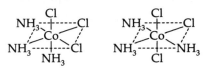

Isomers of $[Co(NH_3)_3Cl_3]$

21.83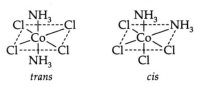

21.85   (a) four (high spin)
         (b) two (low spin)
21.93   (a) 2   (c) 5   (e) 3
         (b) 8   (d) 6

21.94

21.95 $[Co(NO_2)_6]^{4-}$ should be easier to oxidize than $[Co(NO_2)_6]^{3-}$ because the chromium in $[Co(NO_2)_6]^{4-}$

has an electron in the e_g level, but in $[Co(NO_2)_6]^{3-}$ it does not.

CHAPTER 22

22.16 Hydrides are binary compounds of hydrogen. NaH is sodium hydride.

$$2Na(s) + H_2(g) \rightarrow 2NaH(s)$$

$$NaH(s) + H_2O(l) \rightarrow NaOH(aq) + H_2(g)$$

22.18 (a) nonlinear (c) tetrahedral
(b) trigonal pyramidal

22.32 From elemental carbon: $C + O_2 \rightarrow CO_2$
From lower oxides: $2CO + O_2 \rightarrow 2CO_2$
From hydride: $CH_4 + 2O_2 \rightarrow CO_2 + 2H_2O$

22.58 The formula of potassium amide is KNH_2. In water, the NH_2^- hydrolyzes immediately to give NH_3 because NH_2^- is a very strong base.

22.66 The ΔG_f° of N_2O (+104 kJ/mol), NO (+86.8 kJ/mol), NO_2 (+51.9 kJ/mol) are all positive. Their decompositions, therefore, have negative ΔG°, so their decomposition reaction should proceed far toward completion. The oxides are stable because their rate of decomposition is slow.

22.70

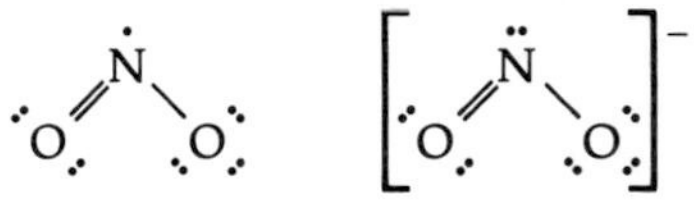

NO_2 should have the larger O—N—O bond angle.

22.76 (a) −5075 kJ (b) −7112 kJ

22.97 Boiler scale can be removed by washing with dilute acid to dissolve the $CaCO_3$.

22.99 $Al_4C_3(s) + 12H_2O \rightarrow 4Al(OH)_3 + 3CH_4(g)$

22.107 (a) $C_{30}H_{62}$ (b) $C_{27}H_{54}$ (c) $C_{33}H_{64}$

22.109

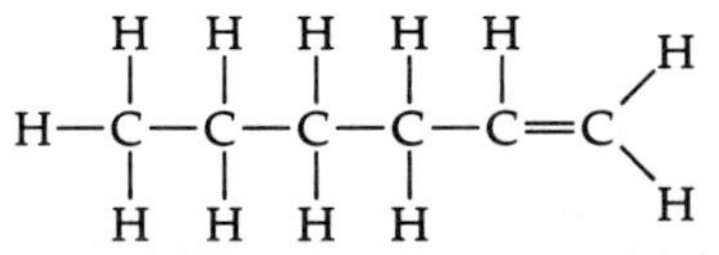

1-hexene

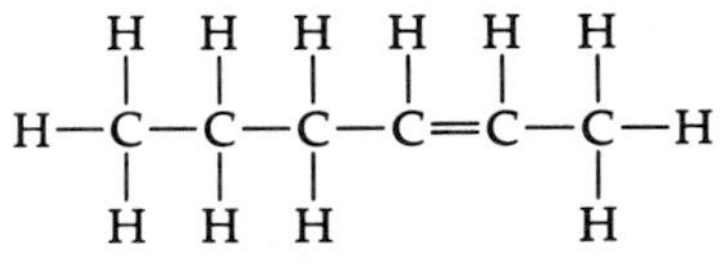

2-hexene; gives geometrical isomers

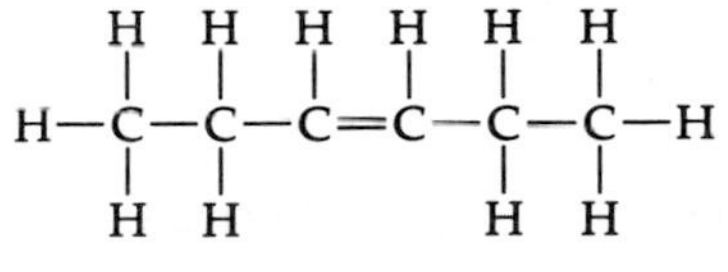

3-hexene; gives geometrical isomers

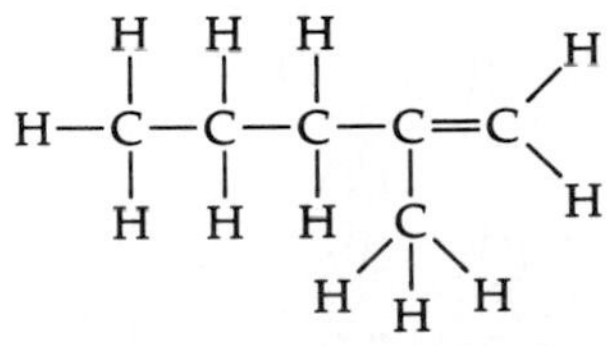

2-methyl-1-pentene

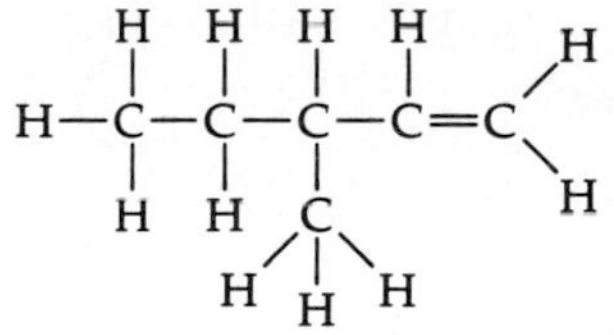

3-methyl-1-pentene; gives geometrical isomers

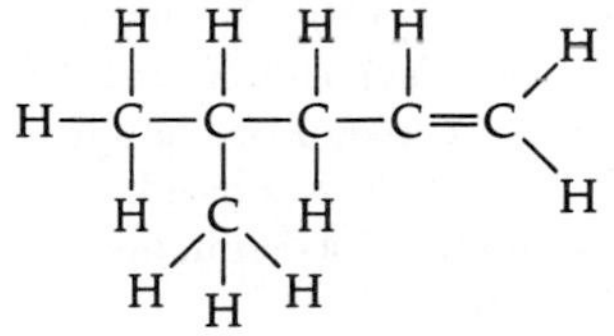

4-methyl-1-pentene

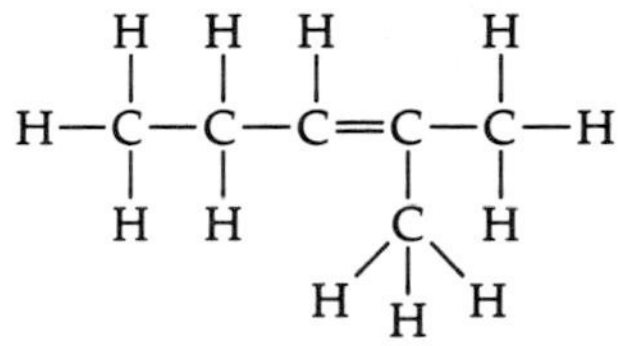

2-methyl-2-pentene

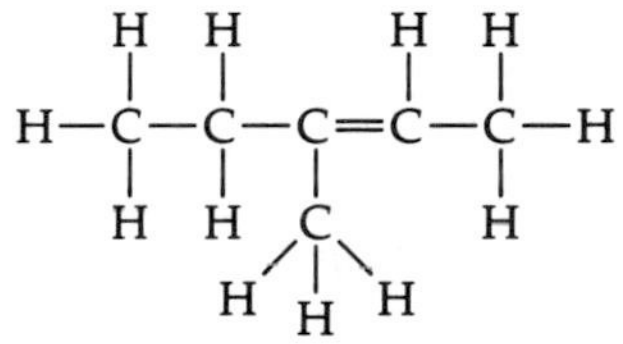

3-methyl-2-pentene; gives geometrical isomers

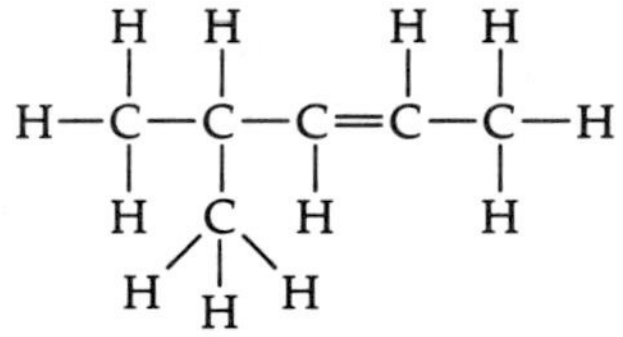

4-methyl-2-pentene; gives geometrical isomers

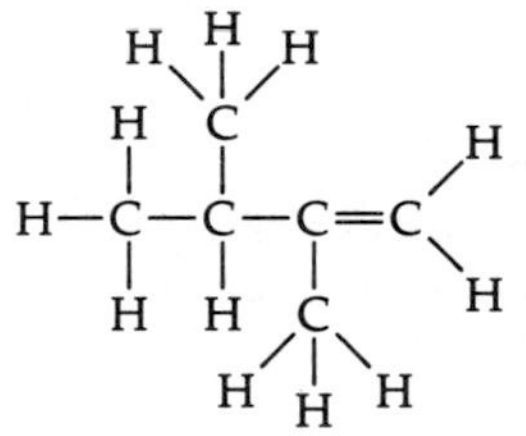

2,3-dimethyl-1-butene

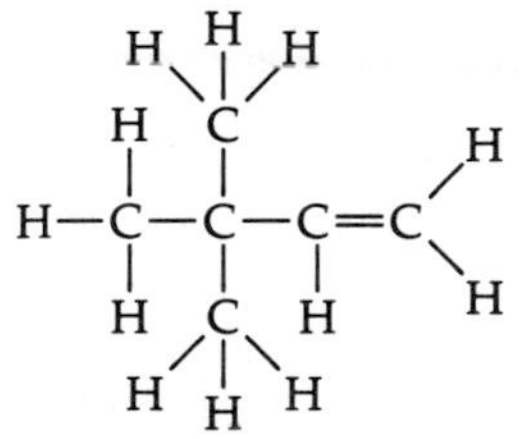

3,3-dimethyl-1-butene

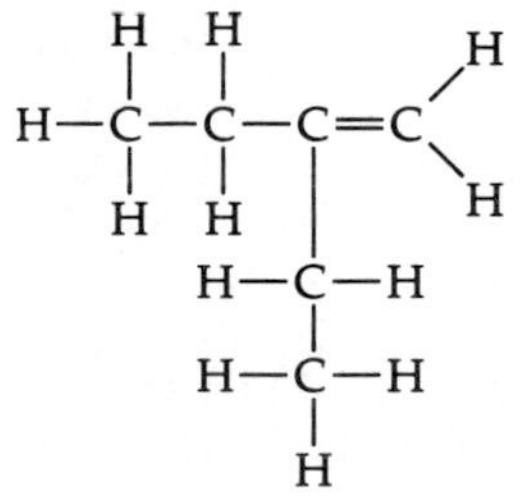

2-ethyl-1-butene

22.111

cis   trans

22.117 C_2Cl_4 is planar because of the double bond between the carbons. Each carbon uses sp^2 hybrid orbitals, which gives a planar configuration. There is no rotation about the double bond.

22.120 (a) 4-ethyl-3,5-dimethyl-2,4-heptadiene
(b) 5-methyl-3-heptyne
(c) 2,3,3,4,4-pentamethylhexane
(d) 4-methyl-2-pentyne
(e) 3,4-dimethyl-3,5-octadiene

22.124 carbonyl, hydroxyl, ether

CHAPTER 23

23.16 $Mg(H_2PO_4)_2$ magnesium dihydrogen phosphate
$MgHPO_4$ magnesium hydrogen phosphate
$Mg_3(PO_4)_2$ magnesium phosphate

23.23 (a) sodium hydrogen phosphate
(b) sodium dihydrogen phosphate

23.24 $12Na + P_4 \rightarrow 4Na_3P$

$$Na_3P + 3H_2O \longrightarrow 3NaOH + PH_3$$

23.29 3 mol NaH_2PO_4 to 2 mol Na_2HPO_4 (The HPO_4^{2-} units terminate the ends of the chains.)

23.31 Phosphorous acid. $Mg(H_2PO_3)_2$ and $MgHPO_3$. (H_3PO_3 is a diprotic acid.)

23.33 See Figure 23.4. PCl_5 exists as $PCl_4^+PCl_6^-$ in the solid. In PCl_3, phosphorous uses sp^3 hybrids; in PCl_5, it uses sp^3d hybrids.

23.52 (a) sodium sulfite
(b) sodium hydrogen sulfate

23.54 $\left[:N \equiv C - \ddot{\underset{\cdot\cdot}{O}} : \right]^-$

23.68 (a) $Cl_2 + 2KI \rightarrow I_2 + 2KCl$
(b) $F_2 + 2KBr \rightarrow Br_2 + 2KF$
(c) $I_2 + NaCl \rightarrow$ N.R.
(d) $Br_2 + 2NaI \rightarrow I_2 + 2NaBr$

23.79 (a) hypobromous acid   (e) periodic acid
(b) sodium hypochlorite   (f) bromic acid
(c) potassium bromate   (g) sodium iodate
(d) magnesium perchlorate   (h) potassium chlorite

23.82 $CaCl(OCl) + 2H^+ \rightarrow Ca^{2+} + H_2O + Cl_2$

23.86 (a) nonlinear   (c) T-shaped
(b) trigonal bipyramidal   (f) distorted tetrahedral
(c) octahedral   (g) square pyramidal
(d) trigonal pyramidal   (h) tetrahedral

23.98 Because Si does not form stable π bonds to oxygen or any other atoms.

23.110

23.112

23.118

CHAPTER 24

24.3 (a) $^{81}_{36}Kr + ^{0}_{-1}e \longrightarrow ^{81}_{35}Br$
(b) $^{104}_{47}Ag \longrightarrow ^{0}_{1}e + ^{104}_{46}Pd$
(c) $^{73}_{31}Ga \longrightarrow ^{0}_{-1}e + ^{73}_{32}Ge$
(d) $^{104}_{48}Cd \longrightarrow ^{104}_{47}Ag + ^{0}_{1}e$
(e) $^{54}_{25}Mn + ^{0}_{-1}e \longrightarrow ^{54}_{24}Cr$

24.5 (a) $^{135}_{53}I \longrightarrow ^{135}_{54}Xe + ^{0}_{-1}e$
(b) $^{245}_{97}Bk \longrightarrow ^{4}_{2}He + ^{241}_{95}Am$
(c) $^{238}_{92}U + ^{12}_{6}C \longrightarrow ^{246}_{98}Cf + 4\,^{1}_{0}n$
(d) $^{96}_{42}Mo + ^{2}_{1}H \longrightarrow ^{1}_{0}n + ^{97}_{43}Tc$
(e) $^{20}_{8}O \longrightarrow ^{20}_{9}F + ^{0}_{-1}e$

24.7 (a) $^{11}_{5}B \longrightarrow ^{4}_{2}He + ^{7}_{3}Li$
(b) $^{90}_{38}Sr \longrightarrow ^{0}_{-1}e + ^{90}_{39}Y$
(c) $^{107}_{47}Ag + ^{1}_{0}n \longrightarrow ^{108}_{47}Ag$
(d) $^{88}_{35}Br \longrightarrow ^{1}_{0}n + ^{87}_{35}Br$
(e) $^{116}_{51}Sb + ^{0}_{-1}e \longrightarrow ^{116}_{50}Sn$
(f) $^{70}_{33}As \longrightarrow ^{0}_{1}e + ^{70}_{32}Ge$
(g) $^{41}_{19}K \longrightarrow ^{1}_{1}H + ^{40}_{18}Ar$

24.10 41

24.12 (a) 0.500 g   (d) 0.0313 g
(b) 0.125 g

24.14 164 days

24.16 $2.89 \times 10^{-7} \text{ s}^{-1}$

24.18 (a) $6.37 \times 10^{-3} \text{ hr}^{-1}$
(b) $t_{1/2} = 109 \text{ hr}$

24.22 (a) $4.6 \times 10^{10} \text{ Bq}$
(b) $3.1 \times 10^{8} \text{ Bq/g}$

24.25 $8.3 \times 10^{8} \text{ yr}$

24.27 ON + O*NO   (bond breaking at 1)

ONO + *NO   (bond breaking at 2)

24.30 180 mL

24.35 $^{4}_{2}He > ^{58}_{28}Ni > ^{39}_{20}Ca > ^{71}_{32}Ge > ^{10}_{5}B$

24.42 (a) $^{27}_{13}Al + ^{4}_{2}He \longrightarrow ^{1}_{0}n + ^{30}_{15}P$
(b) $^{209}_{83}Bi + ^{2}_{1}H \longrightarrow ^{1}_{0}n + ^{210}_{84}Po$
(c) $^{15}_{7}N + ^{1}_{1}H \longrightarrow ^{4}_{2}He + ^{12}_{6}C$
(d) $^{12}_{6}C + ^{1}_{1}H \longrightarrow ^{13}_{7}N + \gamma$
(e) $^{14}_{7}N + ^{4}_{2}He \longrightarrow ^{1}_{1}H + ^{17}_{8}O$

24.45 (a) Na_2X (c) XO_2
      (b) H_2X (d) metallic
24.48 (a) ununpentium
      (b) unbiseptium
24.52 2.3009×10^9 kJ mol^{-1}
24.54 ^{7}Li 3.7870×10^9 kJ mol^{-1}, 39.250 MeV
      ^{19}F 1.4261×10^{10} kJ mol^{-1}, 147.73 MeV
      ^{14}N 1.0098×10^{10} kJ mol^{-1}, 104.61 MeV
24.66 (a) 2.3009×10^9 kJ/mol
      (b) 1.345×10^9 kJ/mol
      Reaction (a) produces more energy per mole of
      product. On the basis of energy per gram of reactant,
      reaction (a) wins by a margin of more than 10.1.

APPENDIX B

B.5 N_2^+ and N_2^- are both less stable than N_2. Bond
     orders: N_2, 3; N_2^+ and N_2^-, 2.5. Both N_2^+ and N_2^-
     would have longer bonds than N_2.
B.7 Bond energy of O_2 would be greater, its bond length
     less, and its vibrational frequency greater than those
     of O_2^{2-}.
B.9 (a) Li_2^+, B_2^+, and C_2^+ are less stable than Li_2, B_2, and
         C_2; Be^{2+} is more stable than Be_2
     (b) Li_2^- is less stable than Li_2; Be_2^-, B_2^- and C_2^- are
         more stable than Be_2, B_2, and C_2.
B.11 H_2^+ $(\sigma_{1s})^1$ has bond order $= \frac{1}{2}$
     He_2^+ $(\sigma_{1s})^2(\sigma_{1s}{}^*)^1$ has bond order $= \frac{1}{2}$

PHOTO CREDITS

CHAPTER 1

Figure 1.1*a*: Courtesy of James Brady and Kathy Bendo. Figure 1.1*b*: OPC, Inc. Figure 1.2: OPC, Inc. Page 4 (bottom): Helmut Schwesinger/The Image Bank. Figure 1.3: National Bureau of Standards. Figure 1.5: Peter Lerman. Figure 1.10*a*: Dan McCoy/Rainbow. Figure 1.10*b*: OPC, Inc. Figure 1.11: Dick Luria/FPG. Figure 1.12: OPC, Inc. Figure 1.14*a–c*: Peter Lerman. Figure 1.16*a,b*: Peter Lerman. Figure 1.17: Peter Lerman.

CHAPTER 2

Figure 2.2: Peter Lerman. Figure 2.3: Peter Lerman.

CHAPTER 3

Page 68: Ken Karp. Figure 3.2: Ken Karp. Figure 3.3: Peter Lerman. Figure 3.5: Peter Lerman. Page 59: Peter Lerman.

CHAPTER 4

Figure 4.1: The Rockefeller Group. Figure 4.2: General Cable Corporation. Figure 4.3*a*: Kathy Bendo. Figure 4.3*b*: Russ Kinne/ Photo Researchers. Figure 4.4: James Brady and Kathy Bendo. Figure 4.5: Peter Lerman. Figure 4.6: Courtesy of Eaton Corporation. Page 89 (bottom): Edgar Fahs Smith Collection, University of Pennsylvania. Figure 4.10: Peter Lerman. Page 101: Murray Alcosser/The Image Bank. Page 103: OPC, Inc. Page 110: Peter Lerman. Page 112: Robert J. Capece.

CHAPTER 5

Figure 5.1: OPC, Inc. Figure 5.4: Ken Karp. Figure 5.5: Robert J. Capece. Figure 5.6: Ken Karp. Figure 5.7: Mark E. Gibson. Figure 5.8: Peter Lerman. Page 137: USDA. Page 141 (left): Peter Lerman. Page 141 (right): Ken Karp. Figure 5.9: Peter Lerman.

CHAPTER 6

Page 169: Courtesy of Mason & Sullivan. Page 172: David Hundley/The Stock Market.

CHAPTER 7

Figure 7.3: From *The Gift of Color*, Eastman Kodak Company.

CHAPTER 9

Figure 9.23: William Rivelli/The Image Bank. Figure 9.29: Larry Burrows, *LIFE* Magazine. © 1966, Time, Inc.

CHAPTER 10

Page 299: Ken Karp. Figure 10.1: Peter Lerman. Figure 10.2: Peter Lerman. Figure 10.3: Peter Lerman. Page 305 (top): © Ben Asen. Figure 10.6: OPC, Inc. Figure 10.7: Ken Karp/Omni-Photo Communications, Inc. Figure 10.8: Robert J. Capece. Figure 10.9: Courtesy of Bethlehem Steel. Figure 10.10: Peter Lerman. Page 310: Ken Karp. Figure 10.11: Ken Karp. Figure 10.15: OPC, Inc. Figure 10.16: Peter Lerman.

CHAPTER 11

Figure 11.2: OPC, Inc. Figure 11.7: C. J. Collins/Photo Researchers.

CHAPTER 12

Page 371: Courtesy of Koehring Cranes & Excavators, Inc. Page 373: Adrian Davies/Bruce Coleman. Page 374 (all): OPC, Inc. Page 375: D. S. Henderson/The Image Bank. Page 384: Courtesy of Oregon Freeze Dry. Figure 12.20: Courtesy of Corning, Inc. Page 387: Robert J. Capece. Page 388: National Center for Atmospheric Research/National Science Foundation. Page 393: Robert J. Capece.

CHAPTER 13

Figure 13.2: OPC, Inc. Page 411 (margin): © 1982 E. Alan McGee/FPG. Page 413: John Colwell/Grant Heilman. Page 414: Courtesy of Teledyne Firth Sterling. Page 422: Ken Karp. Page 428: Courtesy of Johnson and Johnson/Ken Karp. Page 430: Ken Karp. Page 438: Brett Froomer/The Image Bank. Page 443: Dan McCoy/ Black Star. Page 446: Courtesy of Recovery Engineering, Inc.

CHAPTER 14

Page 467: David Stoecklein/The Stock Market. Figure 14.10: Ira Wyman/Sygma. Page 475: Jeff Zaruba/The Stock Market.

CHAPTER 15

Figure 15.3: Ken Karp/Omni-Photo Communications, Inc.

CHAPTER 16

Figure 16.2: Peter Lerman. Figure 16.3: Fisher Scientific. Page 539: Chris Stocker. Page 552: Peter Lerman. Page 553: OPC, Inc.

CHAPTER 17

Page 559: Ken Karp.

CHAPTER 18

Figure 18.4: Peter Lerman. Figure 18.5: OPC, Inc. Figure 18.6: James Brady and Kathy Bendo. Page 586: Kaiser Aluminum. Page 588: Kennecott. Figure 18.19: Fisher Scientific. Page 611: OPC, Inc.

CHAPTER 19

Figure 19.1: USDA. Figure 19.16b: Courtesy of AC Spark Plug.

CHAPTER 20

Page 653 (top): Smithsonian Institution. Page 653 (bottom): Courtesy NOAA. Page 654: David Brownell/The Image Bank. Figure 20.2: Courtesy of Bethlehem Steel. Figure 20.5: Arthur d'Arazien/The Image Bank. Page 663: Robert J. Capece. Figure 20.9: Peter Lerman. Figure 20.10a: International Salt Company, Clarks Summit, PA. Figure 20.10b: Dan McCoy/Rainbow. Figure 20.11: Courtesy of S. Wuelpern/Leslie Salt Company. Figure 20.12: Yoav/Phototake. Figure 20.15: Ken Karp/Omni-Photo Communications, Inc. Page 677: Courtesy of Dow Chemical. Figure 20.17: Yoav/Phototake. Page 679: OPC, Inc. Page 681 (left): Thomas Braise/The Stock Market. Page 681 (center/right): Ken Karp. Figure 20.19: Lester V. Bergman and Associates. Page 682 (margin): OPC, Inc. Page 685: Mark M. Lawrence/The Stock Market. Page 686: OPC, Inc. Figure 20.20: Peter Lerman. Page 688 (all): OPC, Inc.

CHAPTER 21

Page 693: Dahlgren/The Stock Market. Figure 21.6: OPC, Inc. Figure 21.7: OPC, Inc. Figure 21.8: OPC, Inc. Figure 21.9: OPC, Inc. Figure 21.10: OPC, Inc. Figure 21.11: OPC, Inc. Page 702 (bottom): Grant Heilman/Grant Heilman Photography. Page 703: Par/NYC. Figure 21.13: OPC, Inc. Figure 21.14: Peter Lerman. Page 705: OPC, Inc. Figure 21.15: Argonne National Laboratory Page 707 (top): OPC, Inc. Page 707 (bottom): Steve McCutcheon.

CHAPTER 22

Figure 22.1: United Press International. Page 727: Ken Karp. Figure 22.3: Ken Karp/Omni-Photo Communications, Inc. Page 735: Courtesy of Interdynamics, Inc./Ken Karp. Figure 22.5: OPC, Inc. Page 740: Robert J. Capece. Figure 22.7: OPC, Inc. Page 743 (top): OPC, Inc. Figure 22.8: Peter Lerman. Page 746: Ken Karp/Omni-Photo Communications, Inc. Page 747: Peter Arnold. Figure 22.11: OPC, Inc. Page 751: Sumitomo Electric Industries, Ltd. Page 752: Calgon Carbon Corporation. Page 753: OPC, Inc. Page 755: James Brady. Figure 22.12: Courtesy of Betz Company. Page 758: Robert J. Capece. Page 760: Robert J. Capece. Page 763 (left): USDA. Page 763 (right): Robert J. Capece. Page 764: Robert J. Capece.

CHAPTER 23

Figure 23.1: Peter Lerman. Page 779: Ken Karp/Omni-Photo Communications, Inc. Figure 23.6: © Leo Touchet. Figure 23.7: OPC, Inc. Figure 23.8: Peter Lerman. Figure 23.9: Peter Lerman. Page 786: Kollar/TVA. Figure 23.10: Schmidt–Thomsen, Landesdenkmalamt, Westfalen–Lippe; Münster, Germany. Figure 23.11a: © Erich Hartman/Magnum. Figure 23.11b: Ted Spiegel/Black Star. Figure 23.12: OPC, Inc. Page 792: OPC, Inc. Page 793: Kathy Bendo and James Brady. Page 797: OPC, Inc. Page 802 (left): A. W. Ambler/Photo Researchers. Page 802 (right): Andrew McClenaghan/Photo Researchers. Page 804 (all): H. Steen/Photo Researchers. Page 805: Asbestos Institute. Page 806: Runk/Schoenberger/Grant Heilman. Figure 23.20: Murray Alcosser/The Image Bank. Page 808: Buick Motor Division. Page 809: Courtesy of Regal Ware. Page 810: Ken Karp/Omni-Photo Communications, Inc. Figure 23.21: Ken Karp. Figure 23.22: Safra Nimrod. Figure 23.23: Courtesy of H. R. Allcock, Pennsylvania State University.

CHAPTER 24

Figure 24.5: Ludlum Measurements, Inc. Page 840: Brookhaven National Laboratory. Page 844: U.S. Council for Energy Awareness. Page 847: University of Rochester. Figure 24.14: U.S. Department of Energy.

INDEX